Metals

Semimetals

Nonmetals

10 VIII	11 IB	12 IIB	13 IIIA	14 IVA	15 VA	16 VIA	17 VIIA	18 VIIIA
								2 He 4.00 helium
			5 B 10.81 boron	6 C 12.01 carbon	7 N 14.01 nitrogen	8 O 16.00 oxygen	9 F 19.00 fluorine	10 Ne 20.18 neon
			13 Al 26.98 aluminum	14 Si 28.09 silicon	15 P 30.97 phosphorus	16 S 32.07 sulfur	17 Cl 35.45 chlorine	18 Ar 39.95 argon
28 Ni 58.69 nickel	29 Cu 63.55 copper	30 Zn 65.39 zinc	31 Ga 69.72 gallium	32 Ge 72.61 germanium	33 As 74.92 arsenic	34 Se 78.96 selenium	35 Br 79.90 bromine	36 Kr 83.80 krypton
46 Pd 106.42 palladium	47 Ag 107.87 silver	48 Cd 112.41 cadmium	49 In 114.82 indium	50 Sn 118.71 tin	51 Sb 121.75 antimony	52 Te 127.60 tellurium	53 I 126.90 iodine	54 Xe 131.29 xenon
78 Pt 195.08 platinum	79 Au 196.97 gold	80 Hg 200.59 mercury	81 Tl 204.38 thallium	82 Pb 207.2 lead	83 Bi 208.98 bismuth	84 Po (209) polonium	85 At (210) astatine	86 Rn (222) radon
110 Ds (281) darmstadtium	111 Rg (280) roentgenium	112* — (285)	113 — (284)	114 — (289)	115 — (288)	116 — (292)		118 — (294)

64 Gd 157.25 gadolinium	65 Tb 158.93 terbium	66 Dy 162.50 dysprosium	67 Ho 164.93 holmium	68 Er 167.26 erbium	69 Tm 168.93 thulium	70 Yb 173.04 ytterbium	71 Lu 174.97 lutetium
96 Cm (247) curium	97 Bk (247) berkelium	98 Cf (251) californium	99 Es (252) einsteinium	100 Fm (257) fermium	101 Md (258) mendelevium	102 No (259) nobelium	103 Lr (260) lawrencium

*Element 112 has a proposed name of Copernicium which is, at the time of this publication, under review by IUPAC.

SIXTH EDITION

Introductory Chemistry

Concepts and Critical Thinking

Charles H. Corwin

American River College

Prentice Hall

Boston Columbus Indianapolis New York San Francisco Upper Saddle River

Amsterdam Cape Town Dubai London Madrid Milan Munich Paris Montréal Toronto

Delhi Mexico City São Paulo Sydney Hong Kong Seoul Singapore Taipei Tokyo

Editor in Chief, Chemistry and Geosciences: Nicole Folchetti
Acquisitions Editor: Terry Haugen
Marketing Manager: Erin Gardner
Assistant Editor: Laurie Hoffman
Editorial Assistant: Kristen Wallerius
Marketing Assistant: Nicola Houston
Managing Editor, Chemistry and Geosciences: Gina M. Cheselka
Project Manager, Science: Maureen Pancza
Art Project Manager: Connie Long
Illustrations: Argosy Publishing
Art Director: Maureen Eide
Interior and Cover Design: Tamara Newnam
Senior Operations Supervisor: Nick Sklitsis
Operations Supervisor: Maura Zaldivar
Senior Media Producer: Angela Bernhardt
Associate Media Producer: Kristin Mayo
Senior Media Production Supervisor: Liz Winer
Media Production Coordinator: Shannon Kong
Photo Research Manager: Elaine Soares
Photo Researcher: Eric Schrader
Composition/Full Service: GEX Publishing Services

Cover: Pyramids of white salt, Salar de Uyuni in southwest Bolivia; Kazuyoshi Nomachi /Corbis

Library of Congress Cataloging-in-Publication Data
Corwin, Charles H.
Introductory chemistry : concepts and critical thinking / Charles H. Corwin. — 6th ed.
p. cm.
Includes bibliographical references and index.
ISBN-13: 978-0-321-66305-4 (alk. paper)
ISBN-10: 0-321-66305-5 (alk. paper)
1. Chemistry—Textbooks. I. Title.
QD33.2.C67 2011
540—dc22

2009036529

Prentice Hall
is an imprint of

www.pearsonhighered.com

ISBN-13: 978-0-321-66305-4
ISBN-10: 0-321-66305-5

10 9 8 7 6 5 4 3 2 1

Brief Contents

About the Author

Charles H. Corwin has taught over 12,000 students in a formal lecture setting as well as a traditional laboratory. He acknowledges the diversity of basic chemistry students and employs a variety of classroom pedagogies based on Bloom's taxonomy, Myers-Briggs inventory, research on scientific misconceptions, and algorithmic versus conceptual heuristics. His teaching strategies include frequent testing, learning by objective, collaborative learning, chemical demonstrations, Internet assignments, and multimedia presentations.

The author was awarded his degrees from San Jose State University where he was a member of Tau Delta Phi honor society. He did graduate research on dialysis membranes at Stanford Research Institute, and attended the University of Akron, Institute of Polymer Science. He spent 1997–99 at Purdue University with Professor George Bodner and the chemical education group. Previously, he received a National Science Foundation grant for a sabbatical year at Grand Valley State University where he evaluated self-paced, mastery learning, and is currently interested in developing a hybrid online chemistry course with a laboratory component.

Professor Corwin was recognized as teacher of the year at American River College in 1994, and received a teaching award from Purdue University in 1998. In addition, he has been department chair, faculty mentor, served on the ACS General Chemistry Examinations Committee, judged science projects at California state fairs, and for two decades was the examiner for the Sacramento Area Chemistry Contest.

TO THE STUDENT — A PERSONAL NOTE

I have been with *you* in lecture and answered your questions. I have been with *you* in laboratory and supervised your experiments while giving you encouragement. I have listened to *you* during office hours and discovered that more often than not, you have come not to talk about chemistry, but the numerous issues in your life that are hindering you from being as successful as you would like.

Perhaps, I have not spoken to *you* personally, but I have had conversations with thousands of chemistry students much like you who are trying to juggle college life, financial aid, employment hours, parental expectations, relationship issues, test anxiety, and stress in general. If you are a first time student, or a reentry student, my advice is to maintain regular student and instructor contacts, and find a mentor on campus who understands the complexity of college life and is willing to help guide you through the maze of responsibilities as you settle into your new environment.

I consider teaching chemistry a high calling, as the completion of an introductory chemistry course begins to open doors to a rewarding career. A rewarding career is a source of personal satisfaction that spreads to all aspects of your life and helps you avoid making poor life choices. The first day of class I ask students to attend every class, as each time I will give them an assignment. I have had students who are academically gifted achieve less than their potential, and others who are less adept achieve beyond their expectations. One of the keys to success is a disciplined, organized weekly schedule that allows time for recreation as well as classes and other activities.

Chemistry is a subject that builds systematically on fundamental topics and culminates in a knowledge base for general chemistry, the life sciences, and the health sciences. Therefore, it is essential to set aside time each day to study chemistry, and try not to cram for examinations. On days when you lack motivation, open the textbook to some attractive art that illustrates the topic you are covering in class. If you like going online, visit the textbook website and watch an interesting video, or view one of the presentations in Mastering Chemistry®.

I know *you* can be successful in your chemistry class. I have seen others like you with different expressions on their faces at the end of the semester when they turn in their final exams. It is a look of satisfaction and confidence after accepting the challenge, doing their best, and knowing they are better for their experience.

Contents

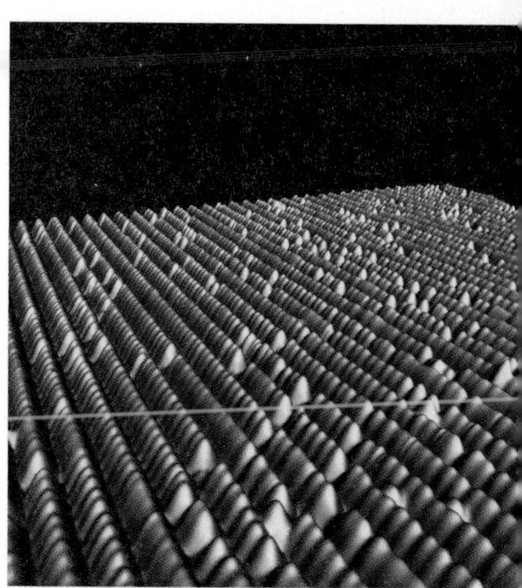

17 Oxidation and Reduction 480

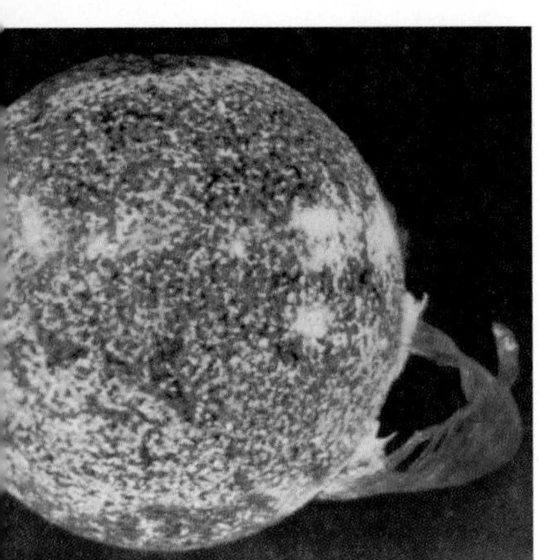

18 Nuclear Chemistry 512

19 Organic Chemistry 536

20 Biochemistry 568

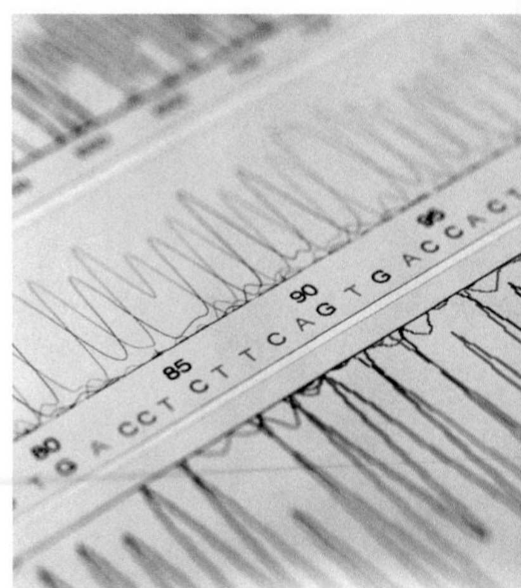

Appendices

Preface

Learning chemistry has a positive effect on our lives, so the study of chemistry should be as accessible and compelling as possible. The goals of this book are to help develop skillful problem solving, learn the language of chemistry, gain an understanding of chemical concepts, and to exercise critical judgment of scientific information. Hopefully, students will find chemistry an interesting subject so that they leave their first course with a positive impression, a new set of skills, and a desire to know more.

The sixth edition of *Introductory Chemistry: Concepts and Critical Thinking* retains all the strengths of the previous editions, while adding emphasis on conceptual understanding and critical thinking. Chemical educators agree that it is necessary to learn underlying concepts to acquire long-term, meaningful understanding. And in today's high-tech society, we are bombarded with information that makes it necessary to exercise cautious, critical judgment.

The sixth edition offers a traditional arrangement of topics for introductory chemistry. However, there is an unusual degree of flexibility owing to the original development of the text. Rather than being created in a void, the author developed the textbook surrounded by a support system of colleagues who teach diverse sequences of topics in their basic chemistry courses. Through progressive refinement, flexibility has been achieved without sacrificing continuity. Thus, topics including the periodic table, chemical bonding, and stoichiometry of gases can be covered earlier or later at the discretion of the instructor.

The goal of this text is twofold. First, we offer numerous strategies to make the subject understandable, and use illustrations and photographs abundantly to ensure that students have a positive initial impression of chemistry. Second, we offer tightly integrated ancillaries and media resources so as to give both instructors and students unparalleled support.

NEW TO THIS EDITION

- **New Concept Exercises** to accompany each example exercise and practice exercise. The concept exercise serves to present the topic at a higher level, whereas the example exercises and practice exercises are specific in nature.
- **An extensively revised design and layout** includes refined colors and style of rendering to achieve a new level of sophistication in illustrations and photographs.
- **The molecular art program has been enhanced significantly** by providing molecular structures for chemical formulas and chemical reactions.
- **Photos are enhanced with macro/micro molecular art.** This visual presentation not only enhances interest in the topics, but adds an effective pedagogical tool for understanding concepts that students find difficult to grasp.
- **New critical-thinking vignettes** on topics ranging from "Nanotechnology" to "Left- and Right-Handed Molecules." In addition, the self-test at the end of each chapter has questions requiring critical thinking.
- **A more flexible chapter construction** enables a seamless presentation of alternate sequences. For example, chemical bonding (Chapter 11) can immediately follow atomic theory and periodicity (Chapters 5 and 6).

AN INTEGRATED LEARNING SYSTEM

More than a textbook, *Introductory Chemistry: Concepts and Critical Thinking, Sixth Edition* is a comprehensive learning system that offers print and media resources as well as an extensive website. Unlike other introductory chemistry texts, all the materials are coherently integrated including the textbook, student solutions manual, laboratory manual, instructor's manual, and test item file. Moreover, the genesis of the package is based on actual classroom experience, student feedback, instructor feedback, and multiple rounds of reviewer feedback from dozens of college institutions across the country.

Students are presented with the same topics, in the same sequence, using the same vocabulary, as they refer to the textbook, solutions manual, or laboratory manual. Instructors are presented with a tightly integrated package including instructor's solution and resource manual, a 3000-question test item file. The media resources include Instructor's Resource Center on CD-ROM, Companion Website, course management in WebCT and Blackboard; and new to this edition is the Prentice Hall Mastering Chemistry, an online homework and tutorial system that provides thousands of problems and tutorials with automated grading.

PROBLEM SOLVING

An important objective of this text is to help students become effective problem solvers. This is accomplished by a walk-through discussion of each new quantitative topic, an example exercise, a problem solving organizer, and consistent use of the unit analysis method of problem solving.

The fifth edition received positive reviews for its systematic problem solving. In as much as basic chemistry students often have weak math skills, algebra has been scrupulously avoided in the early chapters in favor of the unit analysis method of problems solving. However, since many instructors prefer an algebraic approach to the gas laws, Chapter 11 offers both an algebraic and unit analysis approach to solving gas law problems.

There are numerous example exercises in each chapter each of which is preceded by a walk-through exercise that illustrates a chemical principle. Introductory chemistry students who have difficulty reading a chemistry text, can quickly gain easy access to the principles presented. New to the sixth edition, each example exercise and practice exercise is accompanied by a concept exercise. The concept exercise serves to present the topic at a higher level, whereas the example exercises and practice exercises are specific in nature.

LANGUAGE OF CHEMISTRY

Another important objective of this text is to help students learn the language of chemistry. To this end, each chapter has a unique matching exercise for all key terms that allows students to verify their definition in Appendix H. The student solutions manual has computer-generated crossword puzzles with clues for the key terms in each chapter.

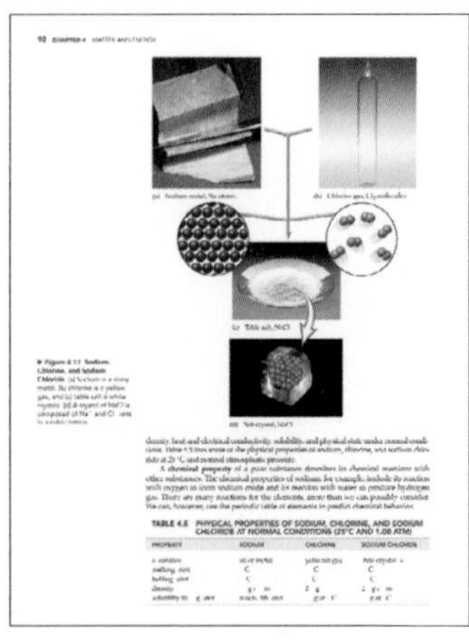

CONCEPTUAL LEARNING

An introductory chemistry student often requires motivation to learn this subject. To this end, in the sixth edition we have refined the palette of colors and style of rendering to achieve a new level of sophistication in illustrations and photographs. The molecular art program has been enhanced significantly by providing molecular structures for chemical formulas and chemical reactions. Moreover, photographs are enhanced by providing macro/micro molecular art. This visual presentation not only enhances interest in the topics, but adds an effective pedagogical tool for understanding concepts that students find difficult to grasp.

CRITICAL THINKING

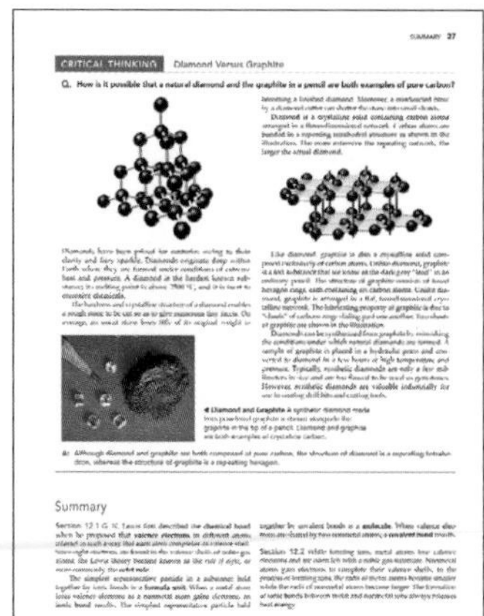

Cognitive scientists define "*critical thinking*" as mental activity associated with three types of thinking: applying reason, making decisions, and problem solving. In this text, critical thinking is introduced within the context of chemical principles. Critical thinking is undertaken specifically in the chapter vignettes and the end-of-chapter self-tests, and generally in unit analysis problem solving.

In the new sixth edition, we have introduced critical thinking vignettes on topics ranging from "Iron versus Steel" to "Diamond versus Graphite." In addition, the self-test at the end of each chapter has questions requiring critical thinking.

FLEXIBLE CHAPTER SEQUENCE

A survey of reviewers using previous editions indicated that many introductory chemistry classes have a unique course format. Moreover, for a variety of reasons, most instructors present topics in a novel sequence. Accordingly, the chapter sequence in *Introductory Chemistry: Concepts and Critical Thinking, Sixth Edition,* is constructed in such a way that there is a seamless presentation of alternate sequences. For example, chemical bonding (Chapter 12) can immediately follow atomic theory and periodicity (Chapters 5 and 6).

PRINT RESOURCES

For the Instructor

Instructor Manual and Printed Test Bank, Sixth Edition (0321677781)

Written by the author, Charles H. Corwin, the *Instructor Manual* provides complete solutions to all even-numbered exercises, lists learning objectives for course planning, recommends media resources, and suggests chemical demonstrations. It also includes the *Printed Test Bank* with over 3000 class-tested questions that have undergone item analysis and address each topic in the textbook.

For the Student

Study Guide and Selected Solutions Manual, Sixth Edition (0321675142)

Written by the author, Charles H. Corwin, this study aid includes diagnostic test questions for each topic covered in the text, crossword puzzles using key terms, and complete solutions to all odd-numbered exercises. Flashcards and study tips are all contained in this essential supplement.

MEDIA RESOURCES

For the Instructor

MasteringChemistry® (www.masteringchemistry.com)

MasteringChemistry® is an online homework and tutorial system. Instructors can create online assignments for their students by choosing from a wide range of items, including end-of-chapter problems and enhanced tutorials. Assignments are automatically graded with up-to-date diagnostic information, helping instructors recognize areas in which students are having difficulty either individually or as a class.

Instructor Resource CD (032167779X)

This lecture resource contains selected art from the textbook, three pre-built PowerPoint® presentations, movies, animations, interactive activities, the *Instructor Manual* lecture outlines in Word® format, the Test Bank in Word® format, Transparency Acetates in electronic format, and the TestGen computerized testing software.

For the Student

MasteringChemistry® (www.masteringchemistry.com)

MasteringChemistry® provides students with an extensive self-study diagnostic with an interactive eBook. In addition, MasteringChemistry® offers a chemistry homework and tutorial system if the instructor chooses to give online homework assignments.

ACKNOWLEDGMENTS

First, I would like to thank the many students, instructors, and reviewers who have helped make this learning package so successful. I teach at a large community college (~35,000 students) with one of the highest introductory chemistry enrollments in the country. I am fortunate to have colleagues who are committed to our mission of empowering students and preparing them for the health sciences and general chemistry. From the outset, this textbook has been shaped and honed by fellow faculty who class-tested the material in module form. And each semester, new adjunct faculty provide "fresh eyes" and suggest material that can be improved.

I would be remiss not to specifically acknowledge the following colleagues: Norman Allen, Kristin Casale, Darren Gottke, Ronald Grider, Tamilyn Hong, Greg Jorgensen, Michael Maddox, Dianne Meador, Chris Meadows, Ed Niedzinski, Luther Nolen, Dustin Nouri, Karen Pesis, Deboleena Roy, Daniel Stewart, Brian Weissbart, Veronica Wheaton, and Linda Zarzana.

The reviewers of this and previous editions of *Introductory Chemistry* continue to define that vague line between the simplifications that students require and the explanations that accuracy and detailed breadth of coverage demand. I thank them for working with me to make this a better book.

REVIEWERS OF THE SIXTH EDITION

David Ball
Cleveland State University
Sean Birke
Jefferson College
Angela Carraway
Meridian Community College
Taiya Fabre
Houston Baptist University
Gerald Ittenbach
Fayetteville Technical Community College
Graham Stoner
Houston Community College
Terri Beam
Mount San Antonio College
Simon Bott
University of Houston
Bernadette R. Corbett
Metropolitan Community College
Mary Hadley
Minnesota State University, Mankato
Kasem K. Kasem
Indiana University–Kokomo
Harold Trimm
SUNY Broome Community College

REVIEWERS OF PREVIOUS EDITIONS

David Ball
Cleveland State University
Karen Bender
Grossmont College
Sean Birke
Jefferson College
Henry Brenner
New York University
Nicole K. Carlson
Indiana University
Kaiguo Chang
University of Arkansas, Fort Smith
Bernadette R. Corbett
Metropolitan Community College
Terri Beam
Mount San Antonio College
Ralph Benedetto
Wayne Community College
Simon Bott
University of Houston
John Brundage
Cuesta College
Angela Carraway
Meridian Community College

Anne Chevalier
Louisiana State University, Alexandria
Dwaine Davis
Forsyth Technical Community College
John Dolhun
Norwalk Community College
Robert Fremland
Mesa College
Wesley Fritz
College of Dupage
David R. Gano
Minot State University
Rebecca K. Hanckel
Charleston Southern University
Leslie Hersh
Delta College
P. D. Hooker
Colby Community College
Ray Kennedy
Massachusetts Bay Community College
Anthony Lazzaro
CA University of Pennsylvania
Anne Lenhert
Kansas State University
Anne Loeb
College of Lake County
Sharyl Majorski-Briggs
Central Michigan University
Jerome Gerard May
Southeastern Louisiana University
Deborah Miller
CNMCC
Alice Monroe
St. Petersburg College
Lily Ng
Cleveland State University
Gita Perkins
Estrella Mountain Community College
Jeffrey Rahn
Eastern Washington University
Scott M. Savage
Northern State University
Julianne M. Smist
Springfield College
Charles E. Sundin
University of Wisconsin–Platteville
Harold Trimm
SUNY Broome Community College
Mine Ucak-Astarlioglu
The Pennsylvania State University
Marie Villarba
Glendale Community College
Brian Weissbart
American River College
Taiya Fabre
Houston Baptist University
Donna Friedman
St. Louis Community College, Florissant Valley
Ana Gaillat
Greenfield Community College
Mary Hadley
Minnesota State University, Mankato
Helen Haur
Delaware Technical & Community College
Mary Hickey
Henry Ford Community College
Kasem K. Kasem
Indiana University–Kokomo
Richard Laidlaw
Laramie Community College
Jeff Lehman
Grossmont College
David Lippmann
Southwest Texas State University
Richard Lomneth
University of Nebraska, Omaha
Greg J. Maloney
Northland Pioneer College
Johanna Mazlo
University of North Carolina, Greensboro
Richard S. Mitchell
Arkansas State University
Robert N. Nelson
Georgia Southern University
Gayle Nicoll
Metropolitan Community College
Sacha Place
Gonzaga Prep
Hussein Samha
Southern Utah University
Mary Setzer
University of Alabama, Huntsville
Clarissa Sorensen-Unruh
CNMCC
Jerry Swanson
Daytona Beach Community College
Paul E. Tyner
Schenectady County Community College
Mary Underdown
Montgomery College
William H. Voige
James Madison University

Charles H. Corwin
Department of Chemistry
American River College
Sacramento, CA 95841
corwinc@arc.losrios.edu

Inspiring Students to Succeed in Chemistry

A working knowledge of chemistry can be an essential skill for today's students, empowering them to make informed decisions throughout their lives. In this text, Corwin's primary goals are to make problems easier to solve—and to help students think critically about chemical concepts. Chemistry is presented in a clear and interesting way, encouraging students to leave their first chemistry course with a positive impression, a set of new skills, and the desire to learn more.

Problem-Solving Study Aids

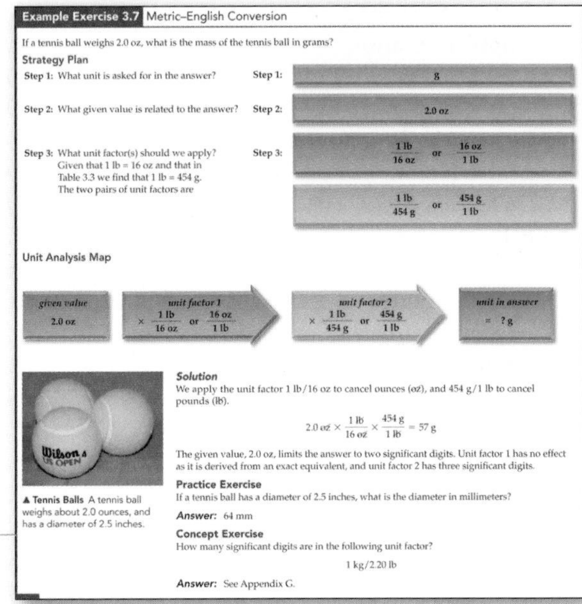

Example Exercise 3.7 Metric–English Conversion

If a tennis ball weighs 2.0 oz, what is the mass of the tennis ball in grams?

Strategy Plan

Step 1: What unit is asked for in the answer? — **Step 1:** g

Step 2: What given value is related to the answer? — **Step 2:** 2.0 oz

Step 3: What unit factor(s) should we apply? Given that 1 lb = 16 oz and that in Table 3.3 we find that 1 lb = 454 g. The two pairs of unit factors are — **Step 3:** $\frac{1\ lb}{16\ oz}$ or $\frac{16\ oz}{1\ lb}$; $\frac{1\ lb}{454\ g}$ or $\frac{454\ g}{1\ lb}$

Unit Analysis Map

given value 2.0 oz → unit factor 1 $\times \frac{1\ lb}{16\ oz}$ or $\frac{16\ oz}{1\ lb}$ → unit factor 2 $\times \frac{1\ lb}{454\ g}$ or $\frac{454\ g}{1\ lb}$ → unit in answer = ? g

▲ **Tennis Balls** A tennis ball weighs about 2.0 ounces, and has a diameter of 2.5 inches.

Solution

We apply the unit factor 1 lb/16 oz to cancel ounces (oz), and 454 g/1 lb to cancel pounds (lb).

$$2.0\ oz \times \frac{1\ lb}{16\ oz} \times \frac{454\ g}{1\ lb} = 57\ g$$

The given value, 2.0 oz, limits the answer to two significant digits. Unit factor 1 has no effect as it is derived from an exact equivalent, and unit factor 2 has three significant digits.

Practice Exercise

If a tennis ball has a diameter of 2.5 inches, what is the diameter in millimeters?

Answer: 64 mm

Concept Exercise

How many significant digits are in the following unit factor?

1 kg/2.20 lb

Answer: See Appendix G.

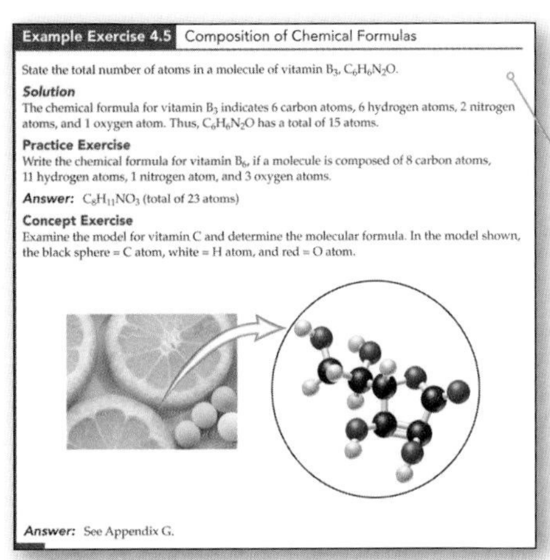

Example Exercise 4.5 Composition of Chemical Formulas

State the total number of atoms in a molecule of vitamin B_3, $C_6H_6N_2O$.

Solution

The chemical formula for vitamin B_3 indicates 6 carbon atoms, 6 hydrogen atoms, 2 nitrogen atoms, and 1 oxygen atom. Thus, $C_6H_6N_2O$ has a total of 15 atoms.

Practice Exercise

Write the chemical formula for vitamin B_6, if a molecule is composed of 8 carbon atoms, 11 hydrogen atoms, 1 nitrogen atom, and 3 oxygen atoms.

Answer: $C_8H_{11}NO_3$ (total of 23 atoms)

Concept Exercise

Examine the model for vitamin C and determine the molecular formula. In the model shown, the black sphere = C atom, white = H atom, and red = O atom.

Answer: See Appendix G.

TWO-COLUMN EXAMPLE EXERCISES

Using the unit analysis method, these exercises show students a logical way to apply the three-step problem-solving process in the worked-out example problems.

NEW! CONCEPT EXERCISES

References to the updated **Concept Exercises** at the end of the example above emphasize the concept underlying the problem, helping students build a conceptual bridge between quantitative operations and qualitative reasoning.

PROBLEM-SOLVING ORGANIZERS

These excellent review tools are found at the end of the chapters dealing with calculations. In the first column is a synopsis of the types of calculations in the chapter. In the second column is the correct procedure for solving each type of problem covered in the chapter. In the third column is an illustrated example of the problem type.

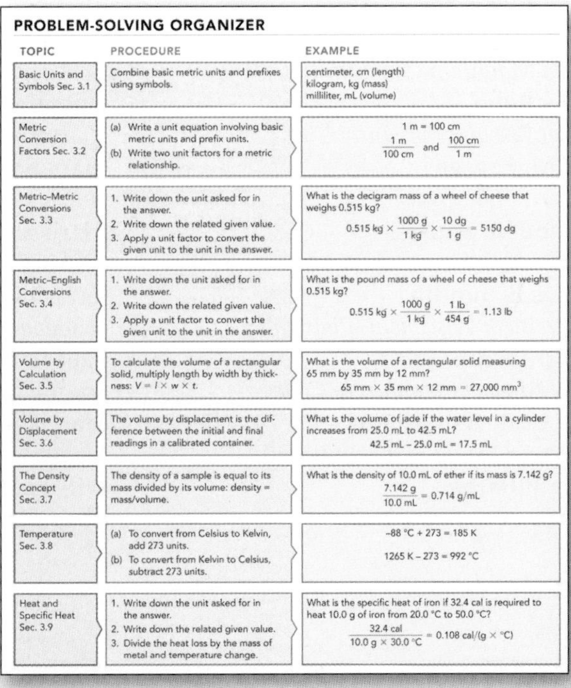

PROBLEM-SOLVING ORGANIZER

TOPIC	PROCEDURE	EXAMPLE
Basic Units and Symbols Sec. 3.1	Combine basic metric units and prefixes using symbols.	centimeter, cm (length) kilogram, kg (mass) milliliter, mL (volume)
Metric Conversion Factors Sec. 3.2	(a) Write a unit equation involving basic metric units and prefix units. (b) Write two unit factors for a metric relationship.	1 m = 100 cm $\frac{1\ m}{100\ cm}$ and $\frac{100\ cm}{1\ m}$
Metric–Metric Conversions Sec. 3.3	1. Write down the unit asked for in the answer. 2. Write down the related given value. 3. Apply a unit factor to convert the given unit to the unit in the answer.	What is the decigram mass of a wheel of cheese that weighs 0.515 kg? $0.515\ kg \times \frac{1000\ g}{1\ kg} \times \frac{10\ dg}{1\ g} = 5150\ dg$
Metric–English Conversions Sec. 3.4	1. Write down the unit asked for in the answer. 2. Write down the related given value. 3. Apply a unit factor to convert the given unit to the unit in the answer.	What is the pound mass of a wheel of cheese that weighs 0.515 kg? $0.515\ kg \times \frac{1000\ g}{1\ kg} \times \frac{1\ lb}{454\ g} = 1.13\ lb$
Volume by Calculation Sec. 3.5	To calculate the volume of a rectangular solid, multiply length by width by thickness: $V = l \times w \times t$.	What is the volume of a rectangular solid measuring 65 mm by 35 mm by 12 mm? 65 mm × 35 mm × 12 mm = 27,000 mm^3
Volume by Displacement Sec. 3.6	The volume by displacement is the difference between the initial and final readings in a calibrated container.	What is the volume of jade if the water level in a cylinder increases from 25.0 mL to 42.5 mL? 42.5 mL − 25.0 mL = 17.5 mL
The Density Concept Sec. 3.7	The density of a sample is equal to its mass divided by its volume: density = mass/volume.	What is the density of 10.0 mL of ether if its mass is 7.142 g? $\frac{7.142\ g}{10.0\ mL} = 0.714\ g/mL$
Temperature Sec. 3.8	(a) To convert from Celsius to Kelvin, add 273 units. (b) To convert from Kelvin to Celsius, subtract 273 units.	−88 °C + 273 = 185 K 1265 K − 273 = 992 °C
Heat and Specific Heat Sec. 3.9	1. Write down the unit asked for in the answer. 2. Write down the related given value. 3. Divide the heat loss by the mass of metal and temperature change.	What is the specific heat of iron if 32.4 cal is required to heat 10.0 g of iron from 20.0 °C to 50.0 °C? $\frac{32.4\ cal}{10.0\ g \times 30.0\ °C} = 0.108\ cal/(g \times °C)$

Conceptual Understanding and Critical Thinking

CRITICAL THINKING **Neon Lights**

Q: Do all "neon" advertising lights contain neon gas?

At the end of the nineteenth century, J. J. Thomson discovered the electron using a gas discharge tube. Thomson constructed his gas discharge tube out of thin glass, placed a metal electrode in each end, and introduced a small amount of gas. When the gas in the tube was electrically excited, he noticed that the tube glowed. He identified the glowing rays as a stream of small negatively charged particles. These particles acquired the name electrons, and Thomson is given credit for their discovery. The glass tubes used by Thomson were forerunners of cathode-ray tubes, which have been used in television sets and computer screens.

In 1913, Niels Bohr explained that exciting gases with electricity caused electrons to be temporarily promoted to higher energy states within the atom. However, the excited electrons quickly lose energy by emitting light and return to their original state. The light emitted by different gases varies because the energy levels within atoms vary for each element. For example, a gas discharge tube containing hydrogen gas produces a reddish-purple glow, whereas mercury vapor gives a blue glow.

In 1898, the Scottish chemist William Ramsay discovered the gas neon. Unlike argon, which comprises about 1% of air, neon is much more rare. When neon gas is placed in a narrow glass tube and electrically excited, it produces a reddish-orange light that is very eye-catching. The fact that gas discharge tubes produce an attractive array of colors led naturally to their use as advertising lights. Light from excited neon gas is very intense, and the term "neon light" has become a generic term for all lighted advertising displays.

Obviously, not all lights used in advertising are the same color. That is, "neon lights" can be red, green, blue, and so on. To produce a given color, a gas discharge tube must be filled with a specific gas. For a pink light, helium gas is used, and for a blue light, argon gas can be used. Only when we wish to produce a reddish-orange light is an advertising display actually filled with neon gas.

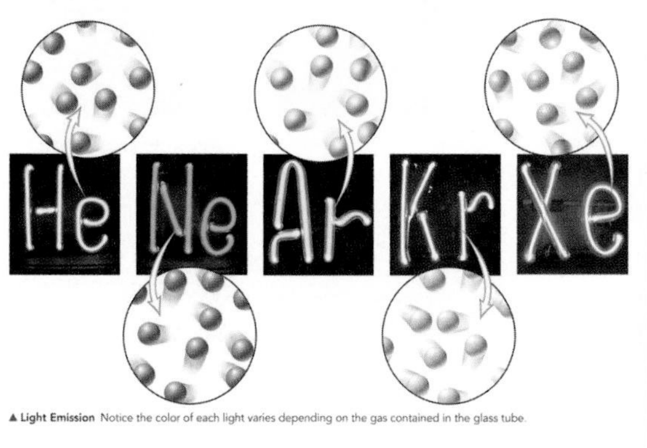

▲ **Light Emission** Notice the color of each light varies depending on the gas contained in the glass tube.

A: Not all "neon lights" contain neon gas; only "neon lights" that have a reddish-orange glow contain neon gas.

For students to gain a true understanding of chemistry, they need to focus on the concepts that underlie the reactions being studied. Conceptualizing chemistry takes students beyond memorization and helps make the information "theirs."

NEW! CRITICAL-THINKING VIGNETTE

New **Critical-Thinking Vignettes** in the Sixth Edition focus on topics ranging from "Iron versus Steel" to "Diamond versus Graphite." In addition, Key Concept and Critical Thinking exercises are included in each Chapter Self-Test.

Key Concepts

11. Identify the physical state (solid, liquid, or gas) that corresponds to each of the following illustrations:

(a)

(b)

(c)

12. Classify each of the following as an element, compound, homogeneous mixture, or heterogeneous mixture:
   (a) iron ore
   (b) iron oxide
   (c) iron metal
   (d) iron alloy
13. Classify the mineral (photo) and molecule (model) of sulfur as an element, compound, homogeneous mixture, or heterogeneous mixture:

KEY CONCEPTS QUESTIONS

Key Concepts questions help students identify the core concepts of chemistry and guide their conceptual understanding of these topics.

2 Chapter Self-Test

Answers to Self-Test are in Appendix J.

1. Which of the following balances has the most uncertainty? (Sec. 2.1)
   (a) platform balance, ± 0.1 g
   (b) triple-beam balance, ± 0.01 g
   (c) electronic balance, ± 0.001 g
   (d) analytical balance, ± 0.0001 g
   (e) all balances have the same uncertainty
2. State the number of significant digits in a mass of 1.050 g. (Sec. 2.2)
   (a) 2 (b) 3
   (c) 4 (d) 5
   (e) none of the above
3. Round off the following measurement to three significant digits: 7265 mL. (Sec. 2.3)
   (a) 720 mL (b) 726 mL
   (c) 727 mL (d) 7260 mL
   (e) 7270 mL
4. Add 55.5 cm to 0.50 cm and round off the answer. (Sec. 2.4)
   (a) 56 cm (b) 56.0 cm
   (c) 56.05 cm (d) 56.1 cm
   (e) 57 cm
5. Multiply 7.2 cm by 3.75 cm by 1.555 cm and round off the answer. (Sec. 2.5)
   (a) 41.9 cm^3 (b) 41.99 cm^3
   (c) 42 cm^3 (d) 42.0 cm^3
   (e) 42.00 cm^3
6. What is 5.55×10^{-2} expressed as an ordinary number? (Sec. 2.6)
   (a) 0.055 5 (b) 0.005 55
   (c) 0.000 555 (d) 5,550
   (e) none of the above
7. There are 26,800,000,000,000,000,000,000 molecules in one liter of oxygen gas at room temperature and pressure. Express the number of gas molecules in scientific notation. (Sec. 2.7)
   (a) 2.68×10^{-22} (b) 2.68×10^{-23}
   (c) 2.68×10^{22} (d) 2.68×10^{23}
   (e) none of the above
8. Which of the following unit factors is derived from: 1.00 yard = 91.4 centimeters? (Sec. 2.8)
   (a) 1.00 yard /1.00 centimeter
   (b) 91.4 centimeters/1.00 yard
   (c) 91.4 yards /1.00 centimeter
   (d) 91.4 centimeters/91.4 yards
   (e) none of the above
9. If a patient is injected with 0.500 liter of IV saline, what is the volume in quarts (given that 1 quart = 0.946 liters)? (Sec. 2.9)
   (a) 0.473 quarts (b) 0.500 quarts
   (c) 0.529 quarts (d) 0.946 quarts
   (e) 1.89 quarts
10. A sample of 18-karat gold contains the following by mass: 18.0 grams gold, 3.0 grams silver, and 3.0 grams copper. What is the percent gold? (Sec. 2.10)
   (a) 6.0% (b) 18%
   (c) 25% (d) 33%
   (e) 75%

Key Concepts

11. What is the best estimate for the diameter of a 1¢ coin: 1 cm, 2 cm, or 3 cm?
12. What is the best estimate for the thickness of a 10¢ coin: 0.1 cm, 0.5 cm, or 1 cm?
13. What is the best estimate for the mass of a 25¢ coin: 1 g, 5 g, or 10 g?
14. What is the first step in the unit analysis method of problem solving?
15. What is the second step in the unit analysis method?
16. In the third step in the unit analysis method, how do you decide which of the two reciprocal unit factors to select?
17. While using a scientific calculator to perform multistep multiplication or division, is it better to round off after each step, or round off the final answer in the display?

◄ **Nickel Coins** Since a dollar is equivalent to 20 nickels, we can write the unit equation 1 dollar = 20 nickels. The two associated unit factors are 1 dollar/20 nickels, and 20 nickels/1 dollar.

Critical Thinking

18. The mass of an alpha particle is 6.64×10^{-24} g, and the mass of a beta particle is 9.11×10^{-28} g. Which is heavier: an alpha particle or a beta particle?
19. An ounce of copper weighs 28.4 g, and an ounce of silver weighs 31.1 g. Which is heavier: an ounce of copper or an ounce of silver?
20. A popular "gold" jewelry contains 27% silver, 63% copper, and 10% zinc. What is the percent of gold in the jewelry?

39

CHAPTER SELF-TESTS

The final problem set at the end of each chapter is a **Chapter Self-Test,** which includes sample test questions (multiple-choice), Key Concepts questions (short answer), and Critical-Thinking questions (short answer).

Enhanced Visualization and Relevant Applications

The Sixth Edition offers an extensively revised design and layout to help today's visual learners grasp key concepts. The molecular art program has been augmented with molecular structures for chemical formulas and chemical reactions. Photos are enhanced with macro/micro molecular art. This visual presentation not only increases student interest, but acts as an effective pedagogical tool for conveying concepts that students may find to be difficult.

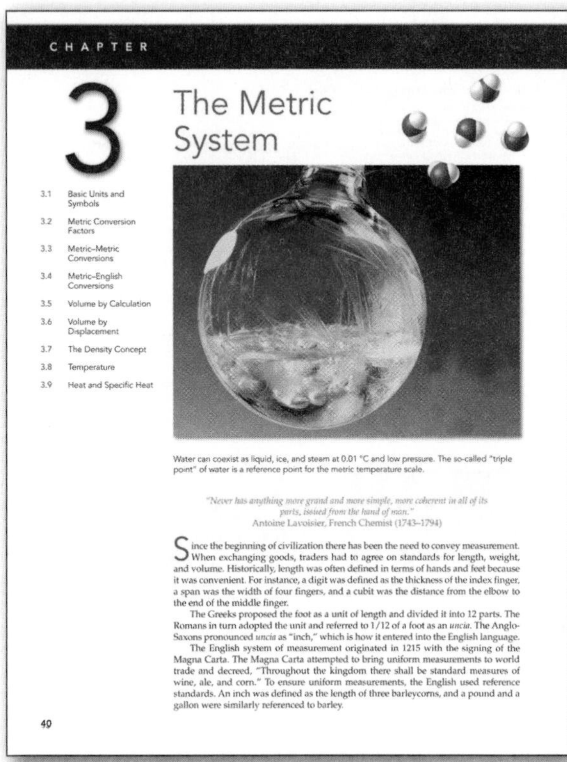

CHAPTER

3 The Metric System

3.1 Basic Units and Symbols
3.2 Metric Conversion Factors
3.3 Metric–Metric Conversions
3.4 Metric–English Conversions
3.5 Volume by Calculation
3.6 Volume by Displacement
3.7 The Density Concept
3.8 Temperature
3.9 Heat and Specific Heat

Water can coexist as liquid, ice, and steam at 0.01 °C and low pressure. The so-called "triple point" of water is a reference point for the metric temperature scale.

"Never has anything more grand and more simple, more coherent in all of its parts, issued from the hand of man."
Antoine Lavoisier, French Chemist (1743–1794)

Since the beginning of civilization there has been the need to convey measurement. When exchanging goods, traders had to agree on standards for length, weight, and volume. Historically, length was often defined in terms of hands and feet because it was convenient. For instance, a digit was defined as the thickness of the index finger, a span was the width of four fingers, and a cubit was the distance from the elbow to the end of the middle finger.

The Greeks proposed the foot as a unit of length and divided it into 12 parts. The Romans in turn adopted the unit and referred to 1/12 of a foot as an *uncia*. The Anglo-Saxons pronounced *uncia* as "inch," which is how it entered into the English language.

The English system of measurement originated in 1215 with the signing of the Magna Carta. The Magna Carta attempted to bring uniform measurements to world trade and decreed, "Throughout the kingdom there shall be standard measures of wine, ale, and corn." To ensure uniform measurements, the English used reference standards. An inch was defined as the length of three barleycorns, and a pound and a gallon were similarly referenced to barley.

40

CHAPTER OPENERS

Fully redesigned, the Sixth Edition offers a striking new layout for today's visual learners.

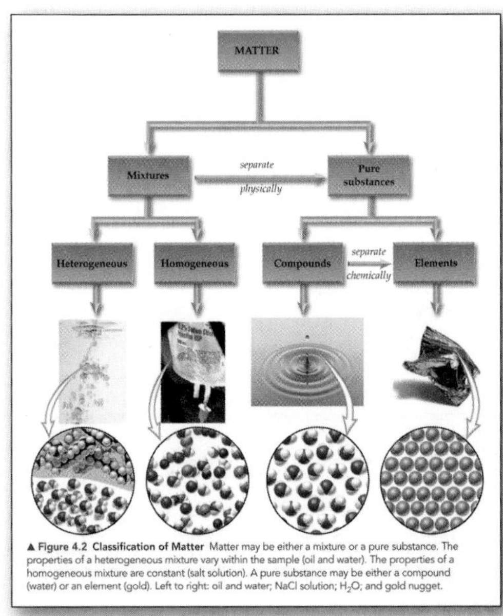

▲ **Figure 4.2 Classification of Matter** Matter may be either a mixture or a pure substance. The properties of a heterogeneous mixture vary within the sample (oil and water). The properties of a homogeneous mixture are constant (salt solution). A pure substance may be either a compound (water) or an element (gold). Left to right: oil and water; NaCl solution; H_2O; and gold nugget.

MACRO/MICRO ART

Molecular art is found throughout the text to illustrate chemical compounds on the atomic level. **Macro/Micro Art** helps students connect the things they encounter in everyday life (the macro view) with depictions of molecules that chemists visualize every day (the micro view). In Corwin's text, the two representations are presented together.

UPDATED CHEMISTRY CONNECTION BOXES

Chemistry Connection boxes demonstrate how chemistry is part of students' day-to-day existence. Each begins with a question that focuses on the subject being investigated. Approximately 25% of the Chemistry Connection boxes in the Sixth Edition are new or updated.

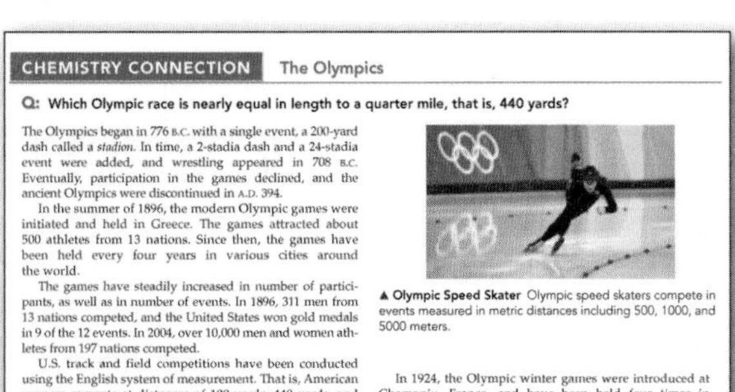

CHEMISTRY CONNECTION The Olympics

Q: Which Olympic race is nearly equal in length to a quarter mile, that is, 440 yards?

The Olympics began in 776 B.C. with a single event, a 200-yard dash called a *stadion*. In time, a 2-stadia dash and a 24-stadia event were added, and wrestling appeared in 708 B.C. Eventually, participation in the games declined, and the ancient Olympics were discontinued in A.D. 394.

In the summer of 1896, the modern Olympic games were initiated and held in Greece. The games attracted about 500 athletes from 13 nations. Since then, the games have been held every four years in various cities around the world.

The games have steadily increased in number of participants, as well as in number of events. In 1896, 311 men from 13 nations competed, and the United States won gold medals in 9 of the 12 events. In 2004, over 10,000 men and women athletes from 197 nations competed.

U.S. track and field competitions have been conducted using the English system of measurement. That is, American runners compete at distances of 100 yards, 440 yards, and 1 mile. In Olympic competitions, events are conducted using the metric system owing to its international acceptance. Olympic runners compete at comparable distances such as 100, 400, and 1500 meters. The marathon, however, is an exception. It corresponds to the distance run by the Greek messenger who carried news of the Athenian victory on the plains of Marathon in 490 B.C. That legendary distance is 26 miles, 385 yards.

▲ **Olympic Speed Skater** Olympic speed skaters compete in events measured in metric distances including 500, 1000, and 5000 meters.

In 1924, the Olympic winter games were introduced at Chamonix, France, and have been held four times in the United States, twice at Lake Placid, once at Squaw Valley, and in 2002 at Salt Lake City. The winter games feature skating, skiing, bobsledding, luge, tobogganing, and ice hockey. In Nordic cross-country skiing, the races are 10, 20, 30, and 50 kilometers; in ski jumping, the ramps are 70 and 90 meters. Thus, Olympic competitions—winter and summer—employ the metric system of measurement.

A: The 400-meter Olympic race is nearly identical to 440 yards, and the world records for the two races are within tenths of a second.

Extend Learning Beyond the Classroom

Make Learning Part of the Grade®

www.masteringchemistry.com

MasteringChemistry emulates the instructor's office-hour environment, coaching students on problem-solving techniques by asking questions that reveal gaps in understanding, and giving them the power to answer questions on their own. It tutors individually—with feedback specific to students' errors, offering optional simpler steps.

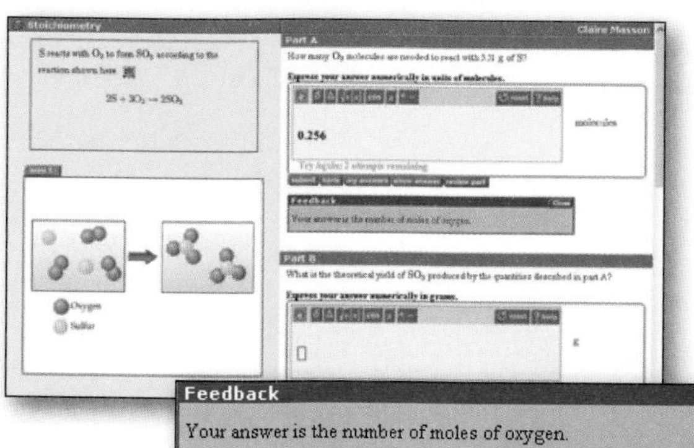

STUDENT TUTORIAL

MasteringChemistry is the only system to provide instantaneous feedback specific to the most-common wrong answers. Students can submit an answer and receive immediate, error-specific feedback. Simpler sub-problems—"hints"—are provided upon request.

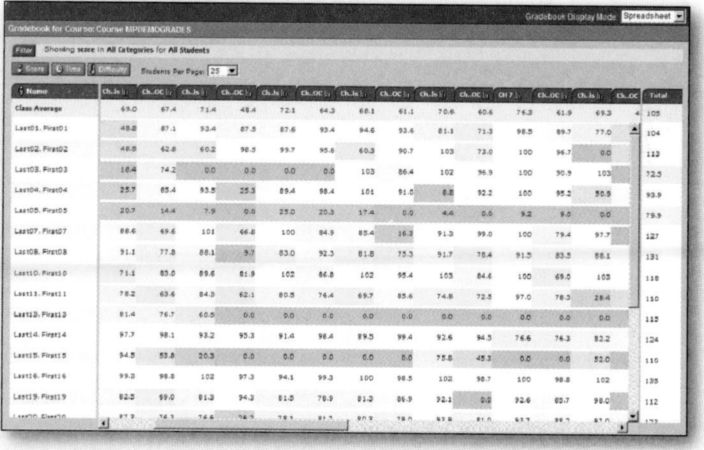

UNMATCHED GRADEBOOK CAPABILITY

MasteringChemistry is the only system to capture the step-by-step work of each student in your class, including wrong answers submitted, hints requested, and time taken on every step. This data powers an unprecedented gradebook.

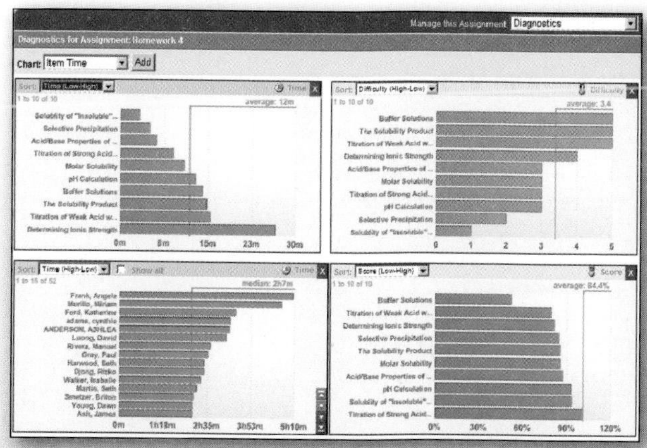

AT-A-GLANCE DIAGNOSTICS

Instructors can identify students who are having difficulty at a glance with the color-coded gradebook, identify the most difficult problem (and step within that problem) in your last assignment, or critique the detailed work of anyone who needs more help. They can even compare results on any problem and any step with a previous class, or with the national average.

SIXTH EDITION

Introductory Chemistry

Concepts and Critical Thinking

CHAPTER

1

Introduction to Chemistry

1.1 Evolution of Chemistry

1.2 Modern Chemistry

1.3 Learning Chemistry

Luca Galuzzi /www.galuzzi.it

Thousands of years ago, a giant salt lake dried up leaving a salt desert in what is now Bolivia. These piles of salt account for 25,000 tons of salt annually from a deposit of 10,000,000,000 tons.

> *"There are in fact two things, science and opinion; the former begets knowledge, the latter ignorance."*
> Hippocrates, Greek Physician (ca. 460–377 B.C.)

In the United States, Canada, and other developed countries, we enjoy a standard of living that could not have been imagined a century ago. Owing to the evolution of science and technology, we have abundant harvests, live in comfortable, climate-controlled buildings, and travel the world via automobiles and airplanes. We also have extended life spans free of many diseases that previously ravaged humanity.

The development of technology has provided machinery and equipment to perform tedious tasks, which gives us time for more interesting activities. The arrival of the computer chip has given us electronic appliances that afford ready convenience and dazzling entertainment. We can select from a multitude of audio and video resources that offer remarkable sound and brilliant color. We can access these audio and video resources from the internet or a compact disc (Figure 1.1).

Our present standard of living requires scientists and technicians with educational training in chemistry. The health sciences as well as the life sciences, physical sciences,

◀ **Figure 1.1 Compact Disc** A laser beam reflects off the surface of the compact disc to read the information on the CD; for example, the information can be printed text, musical sounds, or visual images.

and earth sciences demand an understanding of chemical principles. In fact, chemistry is sometimes referred to as the central science because it stands at the crossroads of biology, physics, geology, and medicine. Just as personal computers are becoming indispensable in our everyday activities, chemistry is assuming an essential role in our daily lives.

1.1 EVOLUTION OF CHEMISTRY

OBJECTIVES

- To describe the early practice of chemistry.
- To identify the three steps in the scientific method.

The earliest concept of science began with the ancient Chinese, Egyptian, and Greek civilizations. The Chinese believed that the universe was created from the interaction of two forces. Yin, the feminine force, was manifested in darkness, cold, and wetness. Yang, the masculine force, was manifested in light, heat, and dryness. When the yin and yang forces interacted, they brought the earthly world into existence and were responsible for everything in nature.

As early as 600 B.C., the Greeks began to speculate that the universe was composed of a single element. Thales, the founder of Greek science, mathematics, and philosophy, suggested that water was the single element. He claimed that Earth was a dense, flat disc floating in a universe of water. He also believed that air and space were less dense forms of water.

A few years later, another Greek philosopher proposed that air was the basic element. This theory was followed by the proposals that fire, and later earth, was the basic element. About 450 B.C., the Greek philosopher Empedocles observed that when wood burned, smoke was released (air), followed by a flame (fire). He also noticed that a cool surface held over a fire collected moisture (water) and that the only remains were ashes (earth). Empedocles interpreted his observations as evidence for air, fire, water, and earth as basic elements. He further speculated that other substances were examples of these four elements combined in varying proportions as illustrated in Figure 1.2.

In about 350 B.C., Aristotle adopted the idea that air, earth, fire, and water were basic elements. In addition, he added a fifth element, ether, that he believed filled all space. Aristotle's influence was so great that his opinions dominated other Greek philosophers and shaped our understanding of nature for nearly 2000 years.

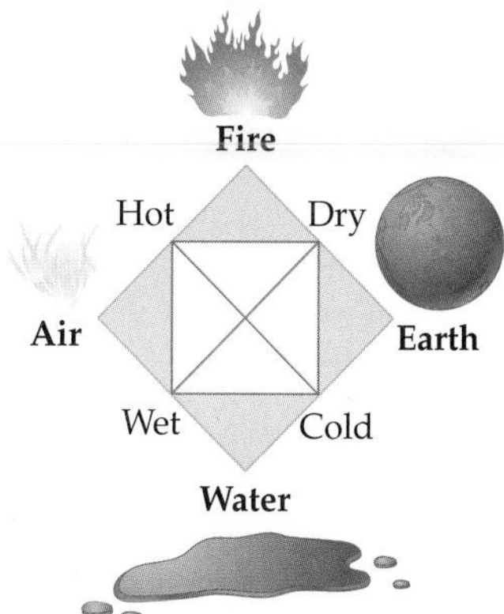

▲ **Figure 1.2 The Four Greek Elements** The four elements proposed by the Greeks: air, earth, fire, and water. Notice the properties hot, cold, wet, and dry associated with each element.

The Scientific Method

In 1661, the English scientist Robert Boyle (1627–1691) published *The Sceptical Chymist*. In his classic book, Boyle stated that theoretical speculation was worthless unless it was supported by experimental evidence. This principle led to development of the scientific method, which marked a turning point in scientific inquiry and the beginning of modern science.

Science can be defined as the methodical exploration of nature followed by a logical explanation of the observations. The practice of science entails planning an investigation, carefully recording observations, gathering data, and analyzing the results. In

▲ **Robert Boyle** This stamp honors Boyle for his invention of the vacuum pump in 1659. Boyle's classic textbook, *The Sceptical Chymist*, laid the foundation for the scientific method.

an **experiment**, scientists explore nature according to a planned strategy and make observations under controlled conditions.

The **scientific method** is a systematic investigation of nature and requires proposing an explanation for the results of an experiment in the form of a general principle. The initial, tentative proposal of a scientific principle is called a **hypothesis**.

After further experimentation, the initial hypothesis may be rejected, modified, or elevated to the status of a scientific principle. However, for a hypothesis to become a scientific principle, many additional experiments must support and verify the original proposal. Only after there is sufficient evidence does a hypothesis rise to the level of a scientific **theory**. We can summarize the three steps in the scientific method as follows:

Applying the Scientific Method

Step 1: Perform a planned experiment, make observations, and record data.
Step 2: Analyze the data and propose a tentative hypothesis to explain the experimental observations.
Step 3: Conduct additional experiments to test the hypothesis. If the evidence supports the initial proposal, the hypothesis may become a theory.

We should note that scientists exercise caution before accepting a theory. Experience has shown that nature reveals its secrets slowly and only after considerable probing. A scientific theory is not accepted until rigorous testing has established that the hypothesis is a valid interpretation of the evidence. For example, in 1803, John Dalton (1766–1844) proposed that all matter was composed of small, indivisible particles called atoms. However, it took nearly 100 years of gathering additional evidence before his proposal was universally accepted and elevated to the status of the atomic theory.

Although the terms "theory" and "law" are related, there is a distinction between the two terms. A theory is a model that explains the behavior of nature. A **natural law** does not explain behavior, but rather states a measurable relationship. To illustrate, it is a *law* that heat flows from a hotter object to a cooler one because we can measure experimentally the change in temperature if we drop an ice cube into water. It is a *theory* that the transfer of heat is due to changes in the motion of molecules in the ice and water.

We can distinguish between a theory and a law by simply asking the question, "Is the proposal measurable?" If the answer is yes, the statement is a law; otherwise, the statement is a theory. Figure 1.3 summarizes the relationship of a hypothesis, a scientific theory, and a natural law.

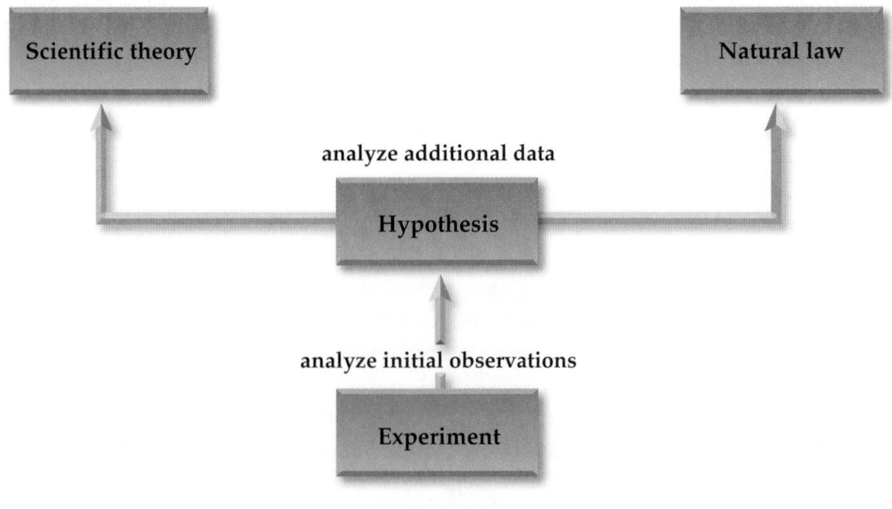

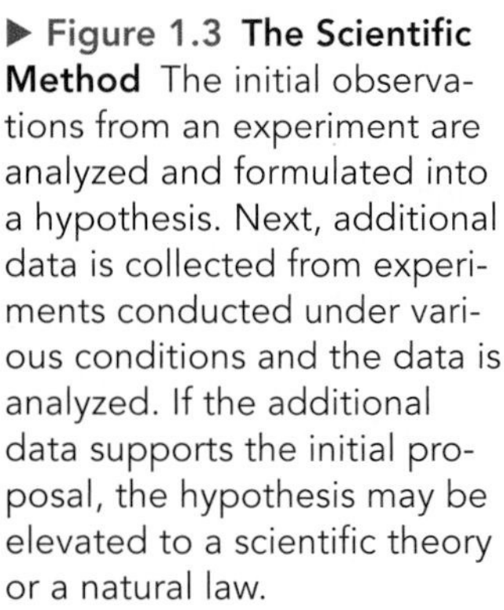
▶ **Figure 1.3 The Scientific Method** The initial observations from an experiment are analyzed and formulated into a hypothesis. Next, additional data is collected from experiments conducted under various conditions and the data is analyzed. If the additional data supports the initial proposal, the hypothesis may be elevated to a scientific theory or a natural law.

CRITICAL THINKING Rise and Fall of a Theory

Q: What is the difference between wood burning, iron rusting, and gasoline exploding?

In the late seventeenth century, the *phlogiston theory* was proposed to explain how matter underwent combustion. The theory stated that for a material to burn it must contain phlogiston (from the Greek *phlogistos*, meaning flammable), which was believed to be a colorless, odorless essence. Through the process of burning, a material was reduced to a substance called calx. According to the theory, calx was a basic element that had lost phlogiston. For example, wood burns to an ash (calx) after losing phlogiston. Similarly, iron is reduced to rust (calx) after releasing phlogiston.

Oxygen was first discovered by the Swedish chemist Carl Scheele (1742–1786), although he did not publish his work. In 1774, the English chemist Joseph Priestley (1733–1804) rediscovered the element, published his finding, and is usually given credit for the discovery of oxygen. However, it was the French chemist Antoine Lavoisier who gave us the name oxygen for the element.

As time passed, skeptics began to doubt the phlogiston theory. They pointed out that metals, such as magnesium, *gain* mass when they burn even though they *lose* phlogiston. Some defenders of the theory explained that phlogiston was lighter than air. Other die-hard proponents suggested that phlogiston had negative mass.

The phlogiston theory remained credible until Lavoisier showed that combustion requires oxygen gas from the air in order to burn. In the late eighteenth century, Lavoisier conducted experiments in which he burned yellow sulfur in a closed container and simultaneously burned yellow sulfur in an open container. He found that the sulfur in the closed container stopped burning, while the sulfur exposed to air continued to burn. He also observed that the product weighed more than the original sulfur, thus offering evidence that sulfur gained mass when it combined with oxygen from the air.

The study of combustion is a good example of how scientists apply critical thinking and the scientific method in order to explore nature and unravel the truth of its most closely guarded secrets.

▲ **Combustion** We observe burning when wood combines with oxygen from the air, not when wood loses phlogiston.

A: Wood burning, iron rusting, and gasoline exploding are all examples of combustion. That is, wood, iron, and gasoline are combining with oxygen gas from the air. Burning is rapid, rusting is a slow process, and an explosion is instantaneous.

1.2 MODERN CHEMISTRY

OBJECTIVE

▸ To describe the modern practice of chemistry.

In the A.D. eighth century, the Arabs introduced the pseudoscience of **alchemy**. Alchemists conducted simple experiments and believed in the existence of a magic potion that had miraculous healing powers and could transmute lead into gold. Although alchemy did not withstand the test of time, it preceded the planned, systematic, scientific experiments that are the cornerstone of modern chemical research.

In the late eighteenth century, the French chemist Antoine Lavoisier (1743–1794) organized chemistry and wrote two important textbooks. Lavoisier also built a magnificent laboratory and invited scientists from around the world to view it; his many visitors included Benjamin Franklin and Thomas Jefferson. Lavoisier was a prolific experimenter and published his work in several languages. For his numerous contributions, he is considered the founder of modern chemistry.

Today, we can define **chemistry** as the science that studies the composition of matter and its properties. Chemists have accumulated so much information during the past two centuries that we now divide the subject into several branches or specialties. The branch of chemistry that studies substances containing the element carbon is called **organic chemistry**. The study of all other substances, those that do not contain the element carbon, is called **inorganic chemistry**.

The branch of chemistry that studies substances derived from plants and animals is called **biochemistry**. Another branch, analytical chemistry, includes qualitative analysis (what substances are in a sample) and quantitative analysis (how much of each substance is present). Physical chemistry is a specialty that proposes theoretical and mathematical explanations for chemical behavior. Recently, environmental chemistry has become an important specialty that focuses on the safe disposal of chemical waste.

Chemistry plays a meaningful role in medicine, especially in the dispensing of pharmaceutical prescriptions. Chemists help ensure agricultural harvests by formulating fertilizers and pesticides. Chemistry is indispensable to many industries including the manufacture of automobiles, electronic components, aluminum, steel, paper, and plastics. One of the largest industries is the petrochemical industry. Petrochemicals are chemicals derived from petroleum and natural gas. They can be used to manufacture a wide assortment of consumer products including paints, plastics, rubber, textiles, dyes, detergents, and explosives.

In each chapter of this textbook, you will find example exercises that put learning into action. Each example exercise poses a question and shows the solution. There is

CHEMISTRY CONNECTION "Worth Your Salt?"

Q: What are the sources of ordinary table salt?

The uses for salt predate modern history. In the ancient world, towns and settlements were near salt reservoirs, as salt is a dietary necessity and a food preservative. Hippocrates, the Greek founder of medicine, urged physicians to soak their patients in salt water as treatment for various ailments. Since most natural salt is not suitable for consumption, pure salt was a rare and valuable commodity. So called "salt roads" were used by caravans of camels to transport salt long distances in trade for gold and textiles.

The familiar phrases "salt of the earth" and "worth your salt" refer to people who are deserving of respect. The origin of "worth your salt" goes back to Roman times when soldiers were given rations of salt and other necessities. These rations were referred to as *sal* (Latin, salt), and when soldiers were paid money, the stipend was called a salarium. Our modern term salary is derived from the phrase meaning "salt money."

Salt is a necessity in the diet of humans and animals, but toxic to most plants. Table salt comes from three sources: (1) salt mining, (2) solution mining, and (3) solar evaporation of salt water. The United States and Canada have extensive deposits of salt, and the Great Salt Lake in Utah is so concentrated and dense that humans easily float.

(1) In salt mining, salt is obtained by drilling shafts deep into the earth. The salt is excavated using a "room and pillar" system of mining that offers support while the salt is removed. After crushing, the salt is hauled to the surface on conveyor belts. (2) In solution mining, wells are placed over salt beds and water is injected to dissolve the salt. The resulting salt solution is pumped to a nearby plant for evaporation. The brine is then evaporated to dryness and refined. (3) Salt can also be obtained by the solar evaporation of seawater and salt lakes. The wind and sun evaporate the water in shallow pools, leaving solid salt. The salt is collected when the crust reaches a certain thickness; the salt is then washed and allowed to recrystallize.

Table salt (99% sodium chloride) is necessary in the human diet; however, too much sodium has been linked to high blood pressure that can lead to diabetes and heart problems. A teaspoon of salt contains approximately 2400 mg of sodium. Surprisingly, most salt in the human diet does not come from table salt, but from processed foods, especially ketchup, pickles, snack foods, and soy sauce. Table salt contains iodine in the form of potassium iodide. Humans require iodine in small quantities for proper function of the thyroid gland. The hormone thyroxine, which contains iodine, is largely responsible for maintaining our metabolic rate. The amount of iodine in one teaspoon of iodized table salt contains about 0.3 mg, which is twice the minimum recommended daily allowance (RDA).

◀ The Great Salt Lake was created in prehistoric times and contains far more salt than seawater. Although its habitat are brine shrimp and aquatic birds, it is called "America's Dead Sea."

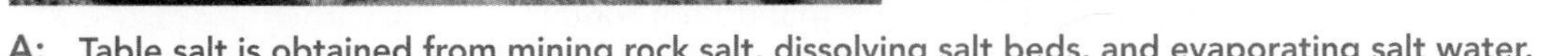

A: Table salt is obtained from mining rock salt, dissolving salt beds, and evaporating salt water.

also a practice exercise and a concept exercise to further your understanding. Example Exercise 1.1 illustrates a topic question, practice exercise, and concept exercise.

▲ **Antoine Lavoisier** This stamp honors Lavoisier for his numerous achievements, including the establishment of a magnificent laboratory that attracted scientists from around the world.

Example Exercise 1.1 Introduction to Chemistry

What is the difference between ancient chemistry and modern chemistry?

Solution

The principal difference is that modern chemistry is founded on the scientific method. Ancient chemistry was based on speculation, while modern chemistry is based on planned experiments and the analysis of data.

Practice Exercise

What question can we ask to distinguish between a scientific theory and a natural law?

Answer: We can distinguish a theory from a law by asking the question, "Is the proposed statement measurable?" If we take measurements and verify a relationship by a mathematical equation, the statement is a law; if not, it is a theory.

Concept Exercise

Which of the line segments in the image below appears to be longer, AB or BC?

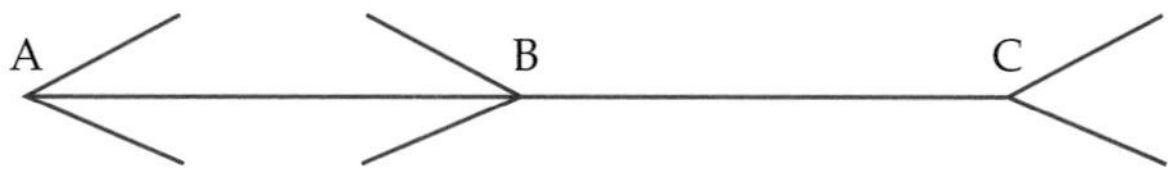

Answer: See Appendix G

1.3 LEARNING CHEMISTRY

OBJECTIVE

▸ To appreciate that chemistry is an interesting and relevant subject.

In a survey published by the American Chemical Society, entering college students were asked to express their attitudes about science courses. The students rated chemistry as the most relevant science course, and as highly relevant to their daily lives. Unfortunately, many of the students thought chemistry is a difficult subject. In view of the results of the student survey, perhaps we should take a moment to consider perceptions in general, and attitudes about chemistry in particular.

You are probably familiar with the expression that some people see a glass of water as half full, while others see the same glass as half empty. This expression implies that different people can respond to the same experience with optimism or pessimism. Moreover, experimental psychologists have found that they can use abstract visual images to discover underlying attitudes regarding a particular perception. A practical lesson involving two perceptions obtained from the same image is revealed by the following picture.

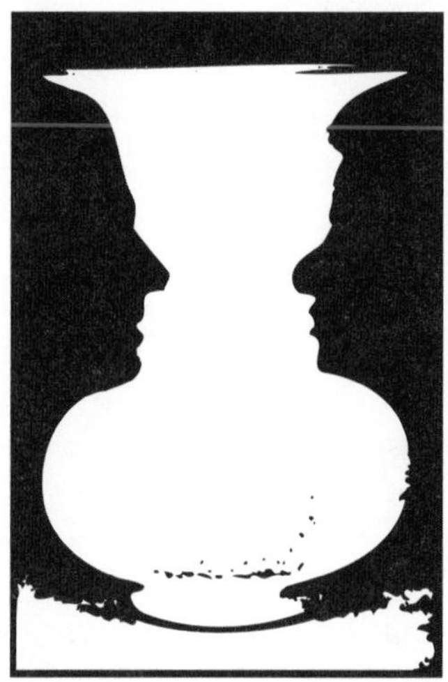

What do you see? Some students see a white vase on a dark background; others see two dark profiles facing each other. After a short period of time, one image switches to the other. If you concentrate, can you view only one of the images? Can you choose to switch the images back and forth? This exercise is an example of our brains registering dual perceptions from the same image.

Your experience of learning chemistry may be somewhat like the preceding exercise that tests your perspective. Sometimes your perception may be that chemistry is challenging, while a short time later your attitude may be that chemistry is easy and fun.

Perception is often affected by unconscious assumptions. Let's consider a type of problem that is slightly different from the vase perception. In the following problem try to connect each of the nine dots using only *four* straight, continuous lines.

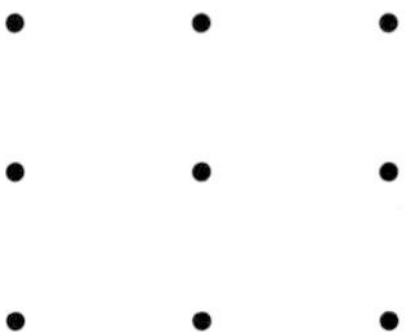

We can begin to solve the problem by experimenting. For example, let's start with the upper left dot and draw a line to the upper right dot. We can continue to draw straight lines as follows:

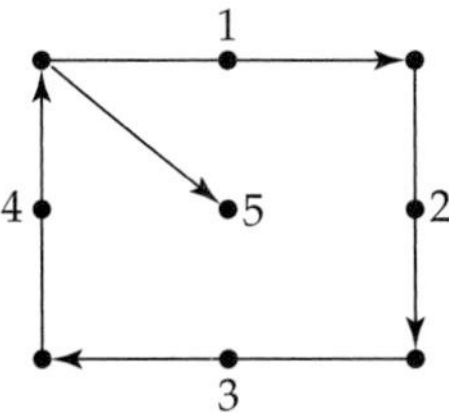

Notice that we connected the nine dots, but that it was necessary to use *five* straight, continuous lines. If we start with a different dot, we find that *five* lines are required no matter where we start. Perhaps we are bringing an underlying assumption to the nine-dot problem. That is, we may be unconsciously framing the nine dots, thus limiting the length of the four straight lines.

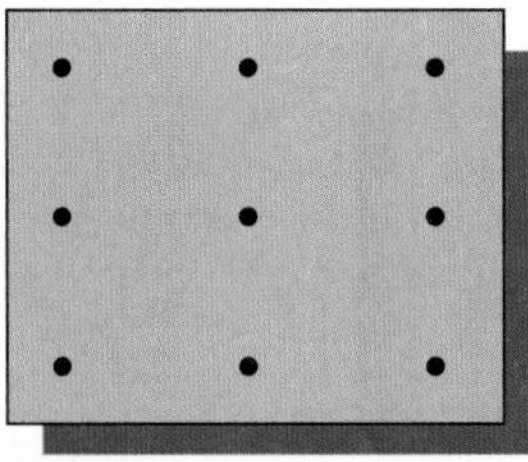

What will happen if we start with the upper left dot and draw a line through the upper right dot? If we continue, we can complete the problem with *four* straight, continuous lines as follows:

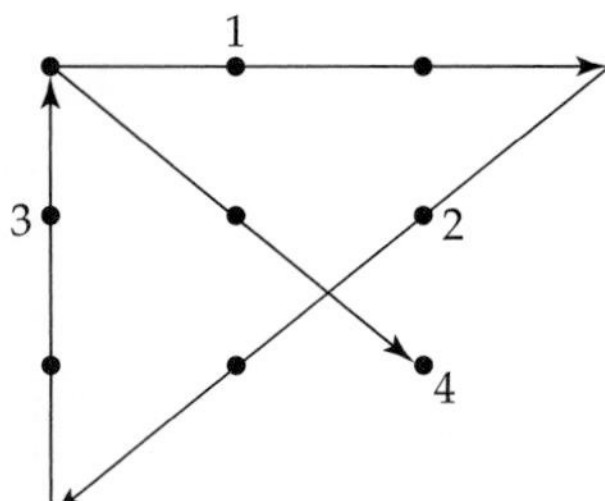

The "secret" to solving this nine-dot problem is to recognize that we may be unconsciously confining our thinking and making it impossible to solve. Similarly, we should not confine our concept of chemistry to a preconceived attitude that learning chemistry will be difficult. Or better yet, we should choose positive associations for our concept of chemistry.

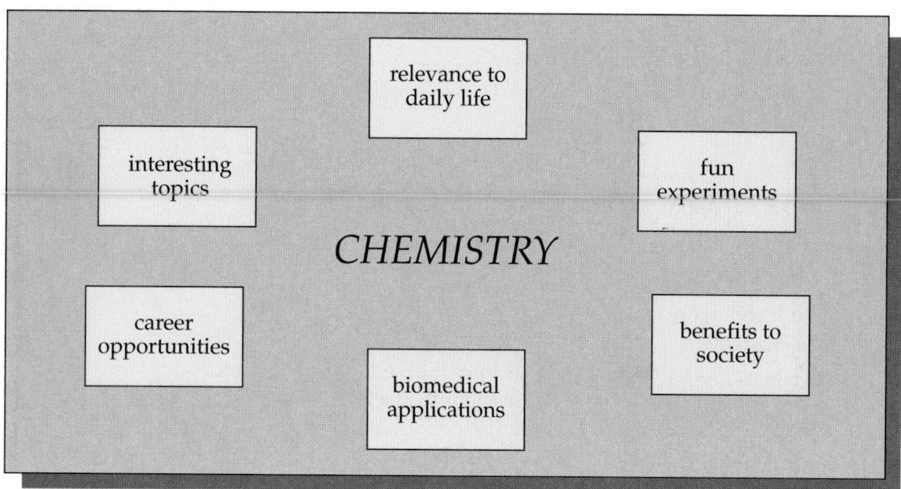

Summary

Section 1.1 This chapter traces the development of chemistry from a historical point of view. Beginning in the period 600–350 B.C., the early Greeks used reason and thoughtful mental exercises to understand the laws of nature. Although they often arrived at conclusions based on speculation, they did unveil some of nature's secrets and had a profound influence on Western civilization that lasted for 20 centuries.

The term **science** implies a rigorous, systematic investigation of nature. Moreover, a scientist must accumulate significant evidence before attempting to explain the results. In the seventeenth century, Robert Boyle founded the **scientific method**, and laboratory experimentation became essential to an investigation. After an **experiment**, scientists use their observations to formulate an initial proposal, which is called a **hypothesis**. However, a hypothesis must be tested repeatedly before it is accepted as valid. After a hypothesis has withstood extensive testing, it becomes either a scientific **theory** or a **natural law**. A scientific theory is an accepted explanation for the behavior of nature, whereas a natural law states a relationship under different experimental conditions and is often expressed as a mathematical equation.

Section 1.2 The pseudoscience of **alchemy** introduced the practice of laboratory experimentation and was the forerunner of modern **chemistry**. Today, chemistry is quite diverse and has several branches, including **inorganic chemistry**, **organic chemistry**, and **biochemistry**. The impact of chemistry is felt in medicine and agriculture, as well as in the electronics, pharmaceutical, petrochemical, and other industries.

Section 1.3 In this section we examined some dual perceptions and pointed out that our brains have the ability to respond to the same image in two ways. Before beginning to learn chemistry, most students have already made associations with the subject. It is hoped that you will be able to focus on chemistry as being an interesting and relevant subject and put aside any preconceived limiting attitudes.

Key Terms *Answers to Key Terms are in Appendix H.*

Select the key term that corresponds to each of the following definitions.

_____ **1.** the methodical exploration of nature and the logical explanation of the observations

_____ **2.** a scientific procedure for gathering data and recording observations under controlled conditions

_____ **3.** a systematic investigation that entails performing an experiment, proposing a hypothesis, testing the hypothesis, and stating a theory or law

_____ **4.** a tentative proposal of a scientific principle that attempts to explain the meaning of a set of data collected in an experiment

_____ **5.** an extensively tested proposal of a scientific principle that explains the behavior of nature

(a) alchemy *(Sec. 1.2)*
(b) biochemistry *(Sec. 1.2)*
(c) chemistry *(Sec. 1.2)*
(d) experiment *(Sec. 1.1)*
(e) hypothesis *(Sec. 1.1)*
(f) inorganic chemistry *(Sec. 1.2)*
(g) natural law *(Sec. 1.1)*

_____ 6. an extensively tested proposal of a scientific principle that states a measurable relationship under different experimental conditions

_____ 7. a pseudoscience that attempted to convert a base metal, such as lead, to gold; a medieval science that sought to discover a universal cure for disease and a magic potion for immortality

_____ 8. the branch of science that studies the composition and properties of matter

_____ 9. the study of chemical substances that do not contain the element carbon

_____ 10. the study of chemical substances that contain the element carbon

_____ 11. the study of chemical substances derived from plants and animals

(h) organic chemistry *(Sec. 1.2)*

(i) science *(Sec. 1.1)*

(j) scientific method *(Sec. 1.1)*

(k) theory *(Sec. 1.1)*

Exercises *Answers to odd-numbered Exercises are in Appendix I.*

Evolution of Chemistry (Sec. 1.1)

1. According to ancient Chinese beliefs, what two forces were responsible for bringing the earthly world into existence?
2. According to the Greek philosopher Thales in 600 B.C., what single element composed Earth, air, and space?
3. According to the Greek philosopher Empedocles in 450 B.C., what four basic elements composed everything in nature?
4. According to the Greek philosopher Aristotle in 350 B.C., what five basic elements composed everything in nature?
5. What is the first step in the scientific method?
6. What is the second step in the scientific method?
7. What is the third step in the scientific method?
8. Who is considered the founder of the scientific method?
9. What is the difference between a scientific theory and a natural law?
10. What is the difference between a hypothesis and a theory?
11. Which of the following statements is a scientific theory?
    (a) The total mass of reacting substances remains constant after reaction.
    (b) The total energy of two gas molecules remains constant after colliding.
    (c) The volumes of two gases combine in the ratio of small whole numbers.
    (d) The nucleus of an atom contains protons and neutrons.
12. Which of the following statements is a natural law?
    (a) The energy emitted from an atomic nucleus is found by $E = mc^2$.
    (b) There is the same number of molecules in equal volumes of gases.
    (c) If the temperature of a gas doubles, the pressure doubles.
    (d) The region surrounding the nucleus of an atom contains electrons.

Modern Chemistry (Sec. 1.2)

13. Why is chemistry considered the central science?
14. Who is considered the founder of modern chemistry?
15. Name at least five professions that require a training in chemistry.
16. Name at least five industries in which chemistry plays an important role.

Learning Chemistry (Sec. 1.3)

17. It is possible to solve the nine-dot problem with only *three* straight, continuous lines. Solve the problem and identify the unconscious assumption.

• • •
• • •
• • •

18. It is possible to solve the nine-dot problem with only *one* straight, continuous line. Solve the problem and identify the unconscious assumption.

• • •
• • •
• • •

19. Stare at the image below and attempt to "flip" the stack of blocks upside down. (*Hint:* If you cannot see the second perception, stare at the point where the three blocks come together and mentally pull the point toward you.)

20. Explain how the following image creates a contradiction in perception.

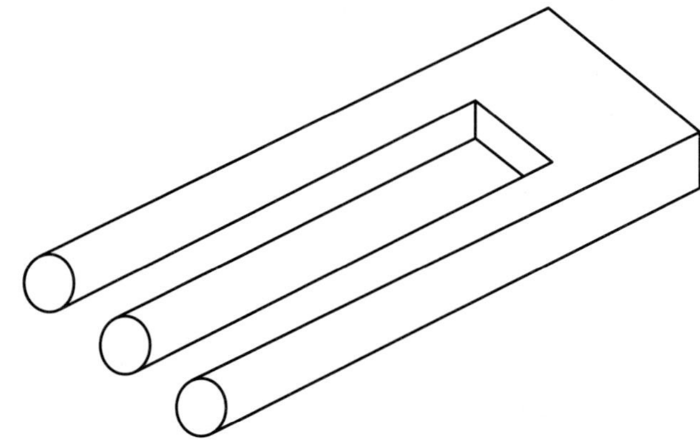

1 Chapter Self-Test

Answers to Self-Test are in Appendix J.

1. Which of the following is a basic step in the scientific method? (Sec. 1.1)
   (a) perform an experiment and collect data
   (b) analyze experimental data and propose a hypothesis
   (c) test a hypothesis and state a theory or law
   (d) all of the above
   (e) none of the above
2. Which of the following is an important branch of chemistry? (Sec. 1.2)
   (a) biochemistry
   (b) environmental chemistry
   (c) inorganic chemistry
   (d) organic chemistry
   (e) all of the above
3. Which of the following is a positive reason for learning chemistry? (Sec. 1.3)
   (a) chemistry is relevant to daily life
   (b) chemistry offers career opportunities
   (c) chemistry studies interesting topics
   (d) chemistry benefits society
   (e) all of the above

Key Concepts

4. Stare at the image below and attempt to change the view of the box. (*Hint:* If you cannot see the second perception, stare at the red dot and mentally "push and pull" the dot.)

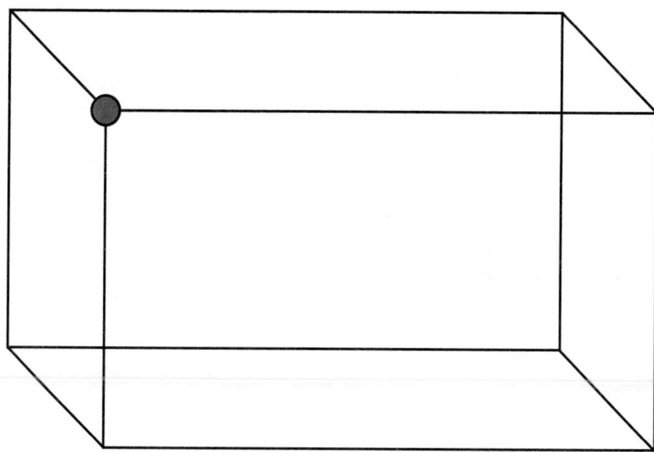

Critical Thinking

5. Biologists note Darwin's principle of natural selection ("survival of the fittest") and propose biological evolution of species. Is biological evolution an example of a scientific theory or a natural law?
6. Chemists note the periodic table is arranged by the number of protons in an atom (that is, hydrogen (1), helium (2), lithium (3), etc.) and propose chemical evolution of elements. Is chemical evolution an example of a scientific theory or a natural law?

CHAPTER

2 Scientific Measurements

For very precise scientific measurements, an instrument using an argon laser beam can measure distance to one-billionth of a meter.

"Make everything as simple as possible, but not simpler."
Albert Einstein, German/Swiss/American Physicist (1879–1955)

In this chapter we will establish an important foundation for the chemical concepts and calculations discussed in later chapters. You are aware, of course, that we live in an electronic age where calculators and computers are part of our daily lives. In addition, hundreds of instruments are available that use state-of-the-art technology. In the laboratory, scientists use instruments that provide very sensitive measurements. For instance, chemists routinely use electronic balances that are so sensitive you can weigh your fingerprints!

Our discussion begins with the instruments commonly found in an introductory chemistry laboratory. Later, we will learn to add, subtract, multiply, and divide measurements obtained from these instruments. Last, and perhaps most importantly, we will learn a powerful method for solving problems in three simple steps.

2.1 UNCERTAINTY IN MEASUREMENTS

OBJECTIVES

- To identify typical instruments in a chemistry laboratory.
- To explain why instrumental measurements are never exact.

We can define a **measurement** as a number with an attached unit. For example, a 5¢ coin may have a mass of 5.005 grams. We measure the mass of the nickel with an **instrument** called a balance. The exactness of the measurement depends on the balance. For instance, electronic balances are common that measure the mass of a sample to 1/1000 of a gram.

Lab technicians routinely inject liquid samples with hypodermic syringes that measure volume to one-millionth of a liter, and electronic stopwatches are available that measure time to a nanosecond, that is, one-billionth of a second. Nevertheless, it is not possible to make exact measurements. An exact measurement is impossible because no instrument measures exactly. That is, an instrument may give a very sensitive reading, but every measurement has **uncertainty** or a degree of inexactness.

A measurement must always include a unit, such as inches, attached to a numerical value. For example, a length measurement may be 12 inches. In general, we will avoid English units such as inch (in.), pound (lb), and quart (qt), although they will be used on occasion in reference to similar metric units. In chemistry measurements, we use metric units such as **centimeter** (symbol cm), **gram** (symbol **g**), and **milliliter** (symbol **mL**).

In the metric system, a centimeter is a unit of length, a gram is a unit of mass, and a milliliter is a unit of volume. For reference, it is interesting to note that a 5¢ coin has a diameter of about 2 cm and a mass of about 5 g. Twenty drops from an eyedropper is approximately 1 mL. Figure 2.1 offers some common references for the estimation of length, mass, and volume.

Length Measurements

To help you understand uncertainty, suppose we measure a candy cane. We have two metric rulers available that differ as shown in Figure 2.2. Both rulers are satisfactory for the task. However, Ruler B provides a more exact measurement than Ruler A.

Notice that Ruler A has ten 1 cm divisions. Since the divisions are large, we can imagine ten subdivisions. Thus, we can estimate to one-tenth of a division, that is, ±0.1 cm. On Ruler A, we see that the candy cane measures about 4.2 cm. Since the uncertainty is ±0.1 cm, a reading of 4.1 cm or 4.3 cm is also acceptable.

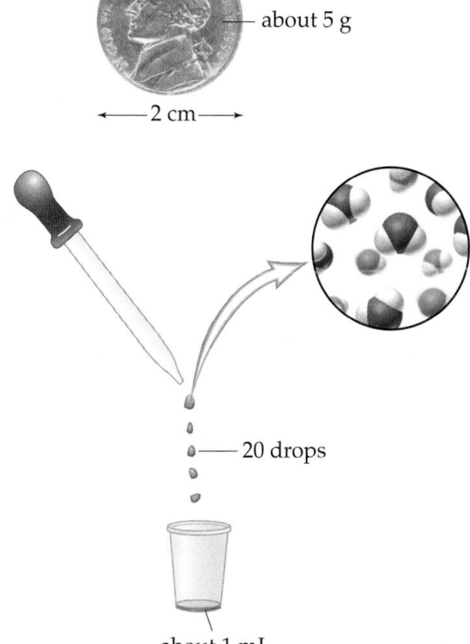

◀ **Figure 2.1 Estimation of Length, Mass, and Volume** The diameter of a 5¢ coin is about 2 cm, and its mass is about 5 g. The volume of 20 drops from an eyedropper is about 1 mL.

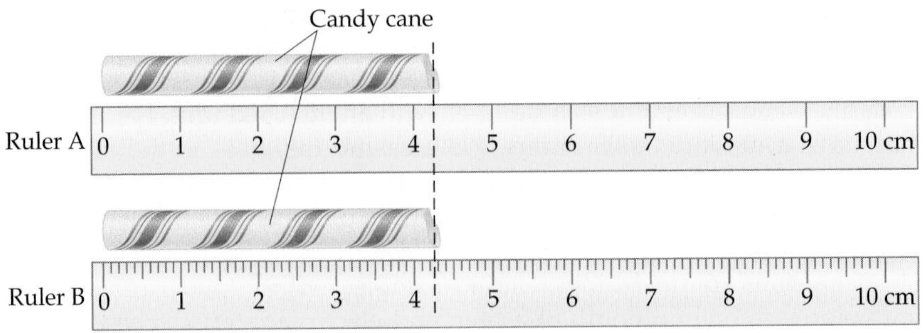

▶ **Figure 2.2 Metric Rulers for Measuring Length** On Ruler A, each division is 1 cm. On Ruler B, each division is 1 cm and each subdivision is 0.1 cm.

Notice that Ruler B has ten 1 cm divisions and ten 0.1 cm subdivisions. On Ruler B, the subdivisions are smaller. So, with Ruler B we can estimate to one-half of a subdivision, that is, we can estimate to ±0.05 cm. On Ruler B, we see that the candy cane measures about 4.25 cm. Since the uncertainty is ±0.05 cm, a reading of 4.20 cm or 4.30 cm is also acceptable.

We can compare the length of the candy cane measured with Rulers A and B as follows:

Ruler A: 4.2 ± 0.1 cm **Ruler B:** 4.25 ± 0.05 cm

In summary, Ruler A has more uncertainty and gives less precise measurements. Conversely, Ruler B has less uncertainty and gives more precise measurements. Example Exercise 2.1 further illustrates the uncertainty in recorded measurements.

Example Exercise 2.1 Uncertainty in Measurement

Which measurements are consistent with the metric rulers shown in Figure 2.2?

(a) Ruler A: 2 cm, 2.0 cm, 2.05 cm, 2.5 cm, 2.50 cm
(b) Ruler B: 3.0 cm, 3.3 cm, 3.33 cm, 3.35 cm, 3.50 cm

Solution

Ruler A has an uncertainty of ±0.1 cm, and Ruler B has an uncertainty of ±0.05 cm. Thus,

(a) Ruler A can give the measurements 2.0 cm and 2.5 cm.
(b) Ruler B can give the measurements 3.35 cm and 3.50 cm.

Practice Exercise

Which measurements are consistent with the metric rulers shown in Figure 2.2?

(a) Ruler A: 1.5 cm, 1.50 cm, 1.55 cm, 1.6 cm, 2.00 cm
(b) Ruler B: 0.5 cm, 0.50 cm, 0.055 cm, 0.75 cm, 0.100 cm

Answers: (a) 1.5 cm, 1.6 cm; (b) 0.50 cm, 0.75 cm

Concept Exercise

What high-tech instrument is capable of making an exact measurement?

Answer: See Appendix G.

Mass Measurements

The **mass** of an object is a measure of the amount of matter it possesses. Mass is measured with a balance and is not affected by gravity. You can think of a balance as a teeter-totter, or seesaw, with two pans. After an object is placed on one pan, weights are added onto the other pan until the balance is level.

The measurement of mass has uncertainty and varies with the balance. A typical mechanical balance in a laboratory may weigh a sample to 1/100 of a gram. Thus, its mass measurements have an uncertainty of ±0.01 g. Many laboratories have electronic balances with digital displays. These balances may have uncertainties ranging from ±0.1 g to ±0.0001 g. Figure 2.3 shows three common types of balances.

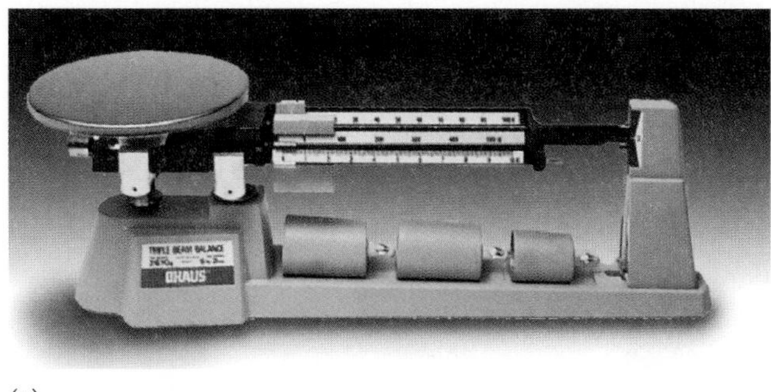
(a)

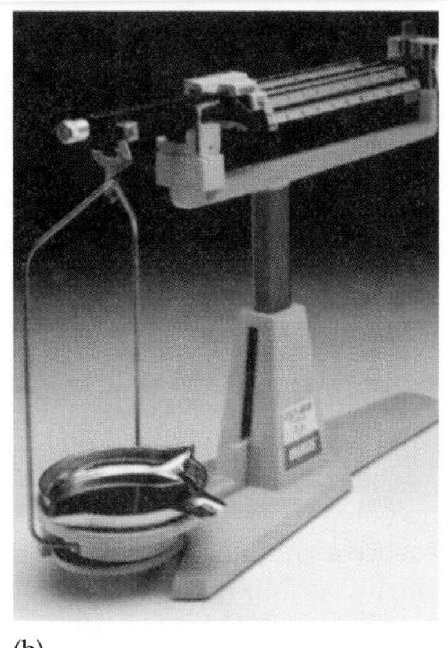
(b)

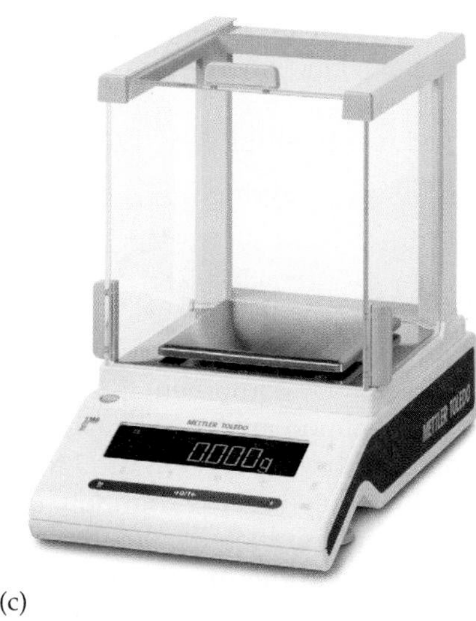
(c)

◀ **Figure 2.3 Balances for Measuring Mass** (a) A platform balance having an uncertainty of ±0.1 g. (b) A beam balance having an uncertainty of ±0.01 g. (c) A digital electronic balance having an uncertainty of ±0.001 g.

Although the term "weight" is often used to mean "mass," strictly speaking, the two terms are not interchangeable. **Weight** is the force exerted by gravity on an object. Since Earth is heavier than the Moon, Earth's gravity is greater and objects weigh more. Similarly, the same object weighs more on the huge planet Jupiter than on Earth. On the other hand, the mass of an object obtained using a balance is constant. This is because gravity operates equally on both pans of the balance, thereby canceling its effect. The mass of an object is constant whether it is measured on Earth, on the Moon, or on any other planet. Figure 2.4 illustrates the distinction between mass and weight.

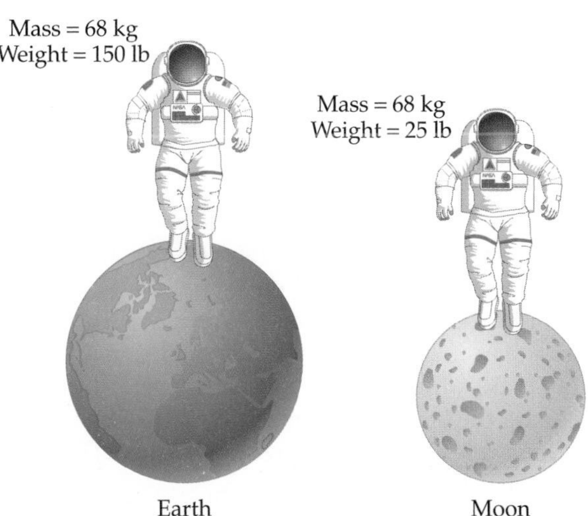

(a) (b)

(c)

◀ **Figure 2.4 Mass versus Weight** Weight is affected by gravity, whereas mass is not. (a) On Earth, the astronaut has a mass of 68 kilograms (kg) and a weight of 150 pounds (lb). (b) On the Moon, the mass remains 68 kg, but the weight is only 25 lb. (c) In space, the mass is still 68 kg although the astronaut is weightless.

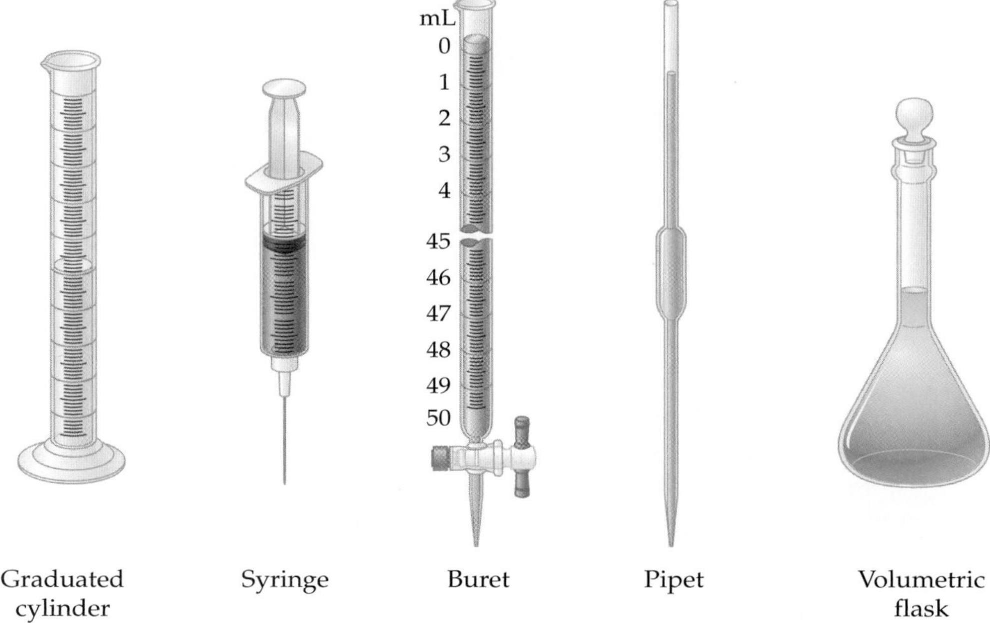

▶ **Figure 2.5 Instruments for Measuring Volume** A graduated cylinder, a syringe, and a buret are calibrated to measure a variable quantity of liquid, whereas a volumetric pipet and a volumetric flask measure only fixed quantities, for example, 10 mL and 250 mL.

Volume Measurements

The amount of space occupied by a solid, gas, or liquid is its volume. There are many pieces of laboratory equipment available for measuring the volume of a liquid. Three of the most common are a graduated cylinder, a pipet, and a buret. Figure 2.5 shows common laboratory equipment used for measuring volume.

A graduated cylinder is routinely used to measure a volume of liquid. The most common sizes of graduated cylinders are 10 mL, 50 mL, and 100 mL. The uncertainty of a graduated cylinder measurement varies, but usually ranges from 1/10 to 1/2 of a milliliter (±0.1 mL to ±0.5 mL).

There are many types of pipets. The volumetric pipet shown in Figure 2.5 is used to deliver a fixed volume of liquid. The liquid is drawn up until it reaches a calibration line etched on the pipet. The tip of the pipet is then placed in a container, and the liquid is allowed to drain from the pipet. The volume delivered varies, but the uncertainty usually ranges from 1/10 to 1/100 of a milliliter. For instance, a 10 mL pipet can deliver 10.0 mL (±0.1 mL) or 10.00 mL (±0.01 mL), depending on the uncertainty of the instrument.

A buret is a long, narrow piece of calibrated glass tubing with a valve called a "stopcock" at the bottom end. The flow of liquid is regulated by opening and closing the stopcock, and the initial and final liquid levels in the buret are observed and recorded. The volume delivered is found by subtracting the initial buret reading from the final buret reading. Burets usually have uncertainties ranging from 1/10 to 1/100 of a milliliter. For instance, the liquid level in a buret can read 22.5 mL (±0.1 mL) or 22.55 mL (±0.01 mL), depending on the uncertainty of the instrument.

2.2 SIGNIFICANT DIGITS

OBJECTIVE

▸ To identify the number of significant digits in a given measurement.

In a properly recorded measurement, each number is a **significant digit**, also referred to as a "significant figure." For instance, suppose we find the mass of a 5¢ coin on a platform balance, a beam balance, and an electronic balance. We may find that the mass of the coin on the different balances is 5.0 g, 5.00 g, and 5.000 g, respectively. Although the uncertainty of the mass varies for the three balances, every digit is significant in each measurement. Removing the last digit from any weighing changes the uncertainty of the measurement. In this example, the measurements of mass have two, three, and four significant digits, respectively.

In every measurement, the significant digits express the uncertainty of the instrument. By way of example, let's examine the chemical reaction shown in Figure 2.6.

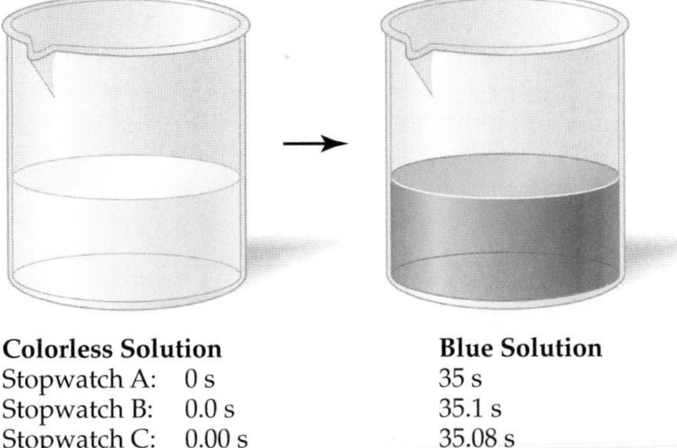

◀ **Figure 2.6 Significant Digits and a Timed Reaction** The data from the timed reaction demonstrates the uncertainty of three different stopwatches. Although each of the measurements is correct, Stopwatch C has the least uncertainty.

This is called a "clock reaction," and we note that the solution changes from colorless to blue after about 35 seconds (s). We can use three different stopwatches to time the reaction. Since a stopwatch can be calibrated in seconds (±1 s), tenths of a second (±0.1 s), or hundredths of a second (±0.01 s), we can use stopwatches having different uncertainties to time the reaction.

Stopwatch A displays 35 s, Stopwatch B displays 35.1 s, and Stopwatch C displays 35.08 s. Therefore, Stopwatch A has more uncertainty than B, and Stopwatch B has more uncertainty than C.

To determine the number of significant digits in a measurement, we simply count the number of digits from left to right, starting with the first nonzero digit. Therefore, 35 s has two significant digits, 35.1 s has three significant digits, and 35.08 s has four significant digits. Example Exercise 2.2 further illustrates how to determine the number of significant digits in a measurement.

Example Exercise 2.2 Significant Digits

State the number of significant digits in the following measurements:

(a) 12,345 cm (b) 0.123 g
(c) 0.5 mL (d) 102.0 s

Solution

In each example, we simply count the number of digits. Thus,

(a) 5 (b) 3
(c) 1 (d) 4

Notice that the leading zero in (b) and (c) is not part of the measurement but is inserted to call attention to the decimal point that follows.

Practice Exercise

State the number of significant digits in the following measurements:

(a) 2005 cm (b) 25.000 g
(c) 25.0 mL (d) 0.25 s

Answers: (a) 4; (b) 5; (c) 3; (d) 2

Concept Exercise

What type of measurement is exact?

Answer: See Appendix G.

Significant Digits and Placeholder Zeros

A measurement may contain placeholder zeros to properly locate the decimal point, for example, 500 cm and 0.005 cm. If the number is less than 1, a placeholder zero is never significant. Thus, 0.5 cm, 0.05 cm, and 0.005 cm each contain only one significant digit.

If the number is greater than 1, a placeholder zero is usually not significant. To avoid confusion, we will assume that placeholder zeros are never significant. Thus, 50 cm, 500 cm, and 5000 cm each contain only one significant digit. Example Exercise 2.3 further illustrates how to determine the number of significant digits.

Example Exercise 2.3 Significant Digits

State the number of significant digits in the following measurements:

(a) 0.025 cm (b) 0.2050 g
(c) 25.0 mL (d) 2500 s

Solution

In each example, we count the number of significant digits and disregard placeholder zeros. Thus,

(a) 2 (b) 4
(c) 3 (d) 2

Practice Exercise

State the number of significant digits in the following measurements:

(a) 0.050 cm (b) 0.0250 g
(c) 50.00 mL (d) 1000 s

Answers: (a) 2; (b) 3; (c) 4; (d) 1

Concept Exercise

What type of measurement has infinite significant digits?

Answer: See Appendix G.

Note If a placeholder zero is significant, we can express the number using a power of 10. For example, if one zero is significant in 100 cm, we can express the measurement as 1.0×10^2 cm. If both zeros are significant, we can write 1.00×10^2 cm. If neither zero is significant, we can write 1×10^2 cm. The power of 10 does not effect the number of significant digits; thus, 1.1×10^5 cm has two significant digits, and 1.11×10^{-5} cm has three significant digits. (Refer to Sections 2.6 and 2.7 for a discussion of exponents and scientific notation.)

▶ **Exact Numbers** We count seven coins in the photo, which is an exact number. This is not a measurement; thus, the number of significant digits is infinite.

Significant Digits and Exact Numbers

Since all measurements have uncertainty, a measurement never represents an exact number. However, we can obtain exact numbers when counting items. For instance, a chemistry laboratory may have 30 rulers, 3 balances, and 24 pipets. Since we have simply counted items, 30, 3, and 24 are exact numbers. As we soon learn, the rules of significant digits do not apply to exact numbers; they apply only to measurements.

We can summarize the directions for determining the number of significant digits with two simple rules.

Determining Significant Digits

Rule 1: Count the number of digits in a measurement from left to right.
(a) Start with the first nonzero digit.
(b) Do not count placeholder zeros (0.011, 0.00011, and 11,000 each have two significant digits).

Rule 2: The rules for significant digits apply only to measurements and not to exact numbers.

2.3 ROUNDING OFF NONSIGNIFICANT DIGITS

OBJECTIVE

▶ To round off a given value to a stated number of significant digits.

All digits in a correctly recorded measurement, except placeholder zeros, are significant. However, we often generate **nonsignificant digits** when using a calculator. These nonsignificant digits should not be reported, but they frequently appear in the calculator display. Since nonsignificant digits are not justified, we must eliminate them. We get rid of nonsignificant digits through a process of **rounding off**. We round off nonsignificant digits by following three simple rules.

Rounding Off Nonsignificant Digits

Rule 1: If the first nonsignificant digit is less than 5, drop all nonsignificant digits.

Rule 2: If the first nonsignificant digit is greater than 5, or equal to 5, increase the last significant digit by 1 and drop all nonsignificant digits.*

Rule 3: If a calculation has several multiplication or division operations, retain nonsignificant digits in your calculator display until the last operation. Not only is it more convenient, it is also more accurate.

* If the nonsignificant digit is 5, or 5 followed by zeros, an odd–even rule can be applied. That is, if the last significant digit is odd, round up; if it is even, drop the nonsignificant digits.

If a calculator displays 12.846239 and three significant digits are justified, we must round off. Since the first nonsignificant digit is 4 in 12.846239, we follow Rule 1, drop the nonsignificant digits, and round to 12.8. If a calculator displays 12.856239 and three significant digits are justified, we follow Rule 2. In this case, since the first nonsignificant digit is 5 in 12.856239, we round to 12.9.

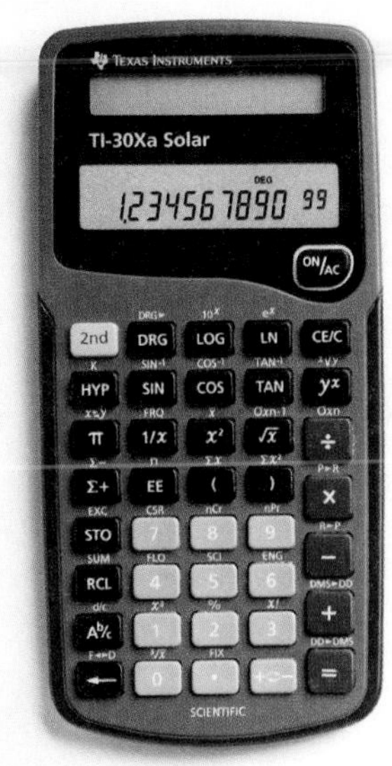

▲ **Scientific Calculator** A calculator display often shows nonsignificant digits, which must be rounded off.

Rounding Off and Placeholder Zeros

On occasion, rounding off can create a problem. For example, if we round off 151 to two significant digits, we obtain 15. Since 15 is only a fraction of the original value, we must insert a placeholder zero; thus, rounding off 151 to two significant digits gives 150. Similarly, rounding off 1514 to two significant digits gives 1500 or 1.5×10^3. Example Exercise 2.4 further illustrates how to round off numbers.

Example Exercise 2.4 Rounding Off

Round off the following numbers to three significant digits:

(a) 22.250 (b) 0.34548
(c) 0.072038 (d) 12,267

Solution

To locate the first nonsignificant digit, count three digits from left to right. If the first nonsignificant digit is less than 5, drop all nonsignificant digits. If the first nonsignificant digit is 5 or greater, add 1 to the last significant digit.

(a) 22.3 (Rule 2) (b) 0.345 (Rule 1)
(c) 0.0720 (Rule 1) (d) 12,300 (Rule 2)

In (d), notice that two placeholder zeros must be added to 123 to obtain the correct decimal place.

Practice Exercise

Round off the following numbers to three significant digits:

(a) 12.514748 (b) 0.6015261
(c) 192.49032 (d) 14652.832

Answers: (a) 12.5 (Rule 1); (b) 0.602 (Rule 2); (c) 192 (Rule 1); (d) 14,700 (Rule 2)

Concept Exercise

How many significant digits are in the exact number 155?

Answer: See Appendix G.

2.4 ADDING AND SUBTRACTING MEASUREMENTS

OBJECTIVE

▸ To add and subtract measurements and round off the answer to the proper significant digits.

When adding or subtracting measurements, *the answer is limited by the value with the most uncertainty;* that is, the answer is limited by the decimal place. Note the decimal place in the following examples:

$$\begin{array}{r} 5\ \text{g} \\ 5.0\ \text{g} \\ +\ 5.00\ \text{g} \\ \hline 15.00\ \text{g} \end{array}$$

The mass of 5 g has the most uncertainty because it measures only ±1 g. Thus, the sum should be limited to the nearest gram. If we round off the answer to the proper significant digit, the correct answer is **15 g**. In addition and subtraction, the unit (cm, g, mL) in the answer is the same as the unit in each piece of data. Example Exercise 2.5 illustrates the addition and subtraction of measurements.

Example Exercise 2.5 Addition/Subtraction and Rounding Off

Add or subtract the following measurements and round off your answer:

(a) 106.7 g + 0.25 g + 0.195 g (b) 35.45 mL − 30.5 mL

Solution

In addition or subtraction operations, the answer is limited by the measurement with the most uncertainty.

(a) Let's align the decimal places and perform the addition.

$$\begin{array}{r} 106.7\ \ \ \text{g} \\ 0.25\ \ \text{g} \\ +\ \ 0.195\ \text{g} \\ \hline 107.145\ \text{g} \end{array}$$

Since 106.7 g has the most uncertainty (±0.1 g), the answer rounds off to one decimal place. The correct answer is **107.1 g** and is read "**one hundred and seven point one grams.**"

(b) Let's align the decimal places and perform the subtraction.

$$\begin{array}{r} 35.45\ \text{mL} \\ -\ 30.5\ \ \text{mL} \\ \hline 4.95\ \text{mL} \end{array}$$

Since 30.5 mL has the most uncertainty (±0.1 mL), we round off to one decimal place. The answer is **5.0 mL** and is read "**five point zero milliliters.**"

Practice Exercise

Add or subtract the following measurements and round off your answer:

(a) 8.6 cm + 50.05 cm (b) 34.1 s − 0.55 s

Answers: (a) 58.7 cm; (b) 33.6 s

Concept Exercise

When adding or subtracting measurements, which measurement in a set of data limits the answer?

Answer: See Appendix G.

2.5 MULTIPLYING AND DIVIDING MEASUREMENTS

OBJECTIVE

- To multiply and divide measurements and round off the answer to the proper significant digits.

Significant digits are treated differently in multiplication and division than in addition and subtraction. In multiplication and division, *the answer is limited by the measurement with the least number of significant digits.* Let's multiply the following length measurements:

$$5.15\ \text{cm} \times 2.3\ \text{cm} = 11.845\ \text{cm}^2$$

The measurement of 5.15 cm has three significant digits, and 2.3 cm has two. Thus, the product should be limited to two digits. When we round off to the proper number of significant digits, the correct answer is **12 cm²**. Notice that the units must also be multiplied together, which we have indicated by the superscript 2. Example Exercise 2.6 illustrates the multiplication and division of measurements.

Example Exercise 2.6 Multiplication/Division and Rounding Off

Multiply or divide the following measurements and round off your answer:

(a) 50.5 cm × 12 cm (b) 103.37 g/20.5 mL

Solution

In multiplication and division operations, the answer is limited by the measurement with the least number of significant digits.

(a) In this example, 50.5 cm has three significant digits and 12 cm has two.

$$(50.5 \text{ cm})\ (12 \text{ cm}) = 606 \text{ cm}^2$$

The answer is limited to two significant digits and rounds off to **610 cm^2** after inserting a placeholder zero. The answer is read "**six hundred and ten square centimeters**."

(b) In this example, 103.37 g has five significant digits and 20.5 mL has three.

$$\frac{103.37 \text{ g}}{20.5 \text{ mL}} = 5.0424 \text{ g/mL}$$

The answer is limited to three significant digits and rounds off to **5.04 g/mL**. Notice that the unit is a ratio; the answer is read as "**five point zero four grams per milliliter**."

Practice Exercise

Multiply or divide the following measurements and round off your answer.

(a) (359 cm) (0.20 cm) (b) 73.950 g/25.5 mL

Answers: (a) 72 cm^2; (b) 2.90 g/mL

Concept Exercise

When multiplying or dividing measurements, which measurement in a set of data limits the answer?

Answer: See Appendix G.

2.6 EXPONENTIAL NUMBERS

OBJECTIVES

- To explain the concept of exponents and specifically powers of 10.
- To express a value as a power of 10 and as an ordinary number.

When a value is multiplied times itself, the process is indicated by a number written as a superscript. The superscript indicates the number of times the process is repeated. For example, if the number 2 is multiplied two times, the product is expressed as 2^2. Thus, (2) (2) = 2^2. If the number 2 is multiplied three times, the product is expressed as 2^3. Thus, (2) (2) (2) = 2^3.

A superscript number indicating that a value is multiplied times itself is called an **exponent**. If 2 has the exponent 2, the value 2^2 is read as "2 to the second power" or "2 squared." The value 2^3 is read as "2 to the third power" or "2 cubed."

Powers of 10

A **power of 10** is a number that results when 10 is raised to an exponential power. You know that an exponent raises any number to a higher power, but we are most interested in the base number 10. A power of 10 has the general form

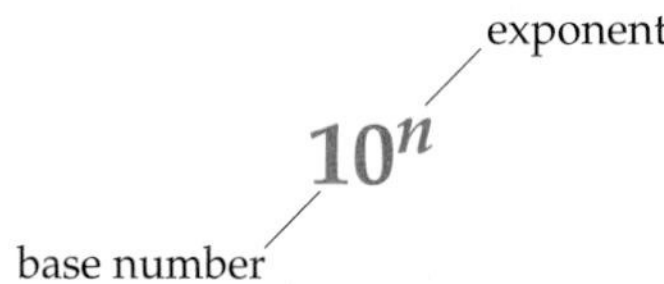

The number 10 raised to the n power is equal to 10 multiplied times itself n times. For instance, 10 to the second power (10^2) is equal to 10 times 10. When we write 10^2 as an ordinary number, we have 100. Notice that the exponent 2 corresponds to the number of zeros in 100. Similarly, 10^3 has three zeros (1000) and 10^6 has six zeros (1,000,000).

The exponent is positive for all numbers greater than 1. Conversely, the exponent is negative for numbers less than 1. For example, 10 to the negative first power (10^{-1}) is equal to 0.1, 10 to the negative second power (10^{-2}) is equal to 0.01, and 10 to the negative third power (10^{-3}) is equal to 0.001. Table 2.1 lists some powers of 10 along with the equivalent ordinary number.

Although you can easily carry out operations with exponents using an inexpensive scientific calculator, you will have greater confidence if you understand exponents. Example Exercises 2.7 and 2.8 further illustrate the relationship between ordinary numbers and exponential numbers.

TABLE 2.1 POWERS OF 10

EXPONENTIAL NUMBER	ORDINARY NUMBER
$1 \times 10^6 = 10 \times 10 \times 10 \times 10 \times 10 \times 10$	1,000,000
$1 \times 10^3 = 10 \times 10 \times 10$	1000
$1 \times 10^2 = 10 \times 10$	100
$1 \times 10^1 = 10$	10
$1 \times 10^0 = 1$	1
$1 \times 10^{-1} = \frac{1}{10}$	0.1
$1 \times 10^{-2} = \frac{1}{10} \times \frac{1}{10}$	0.01
$1 \times 10^{-3} = \frac{1}{10} \times \frac{1}{10} \times \frac{1}{10}$	0.001
$1 \times 10^{-6} = \frac{1}{10} \times \frac{1}{10} \times \frac{1}{10} \times \frac{1}{10} \times \frac{1}{10} \times \frac{1}{10}$	0.000 001

Example Exercise 2.7 Converting to Powers of 10

Express each of the following ordinary numbers as a power of 10:

(a) 100,000 (b) 0.000 000 01

Solution

The power of 10 indicates the number of places the decimal point has been moved.

(a) We must move the decimal five places to the left; thus, 1×10^5.

(b) We must move the decimal eight places to the right; thus, 1×10^{-8}.

Practice Exercise

Express each of the following ordinary numbers as a power of 10:

(a) 10,000,000 (b) 0.000 000 000 001

Answers: (a) 1×10^7; (b) 1×10^{-12}

Concept Exercise

Which of the following lengths is less: 1×10^3 cm or 1×10^{-3} cm?

Answer: See Appendix G.

▲ **Earth** A positive power of 10 can be used to express very large numbers; for example, the diameter of Earth is about 1×10^7 meters.

Example Exercise 2.8 Converting to Ordinary Numbers

Express each of the following powers of 10 as an ordinary number:

(a) 1×10^4 (b) 1×10^{-9}s

Solution

The power of 10 indicates the number of places the decimal point has been moved.

(a) The exponent in 1×10^4 is positive 4, and so we must move the decimal point four places to the right of 1, thus, 10,000.

(b) The exponent in 1×10^{-9} is negative 9, and so we must move the decimal point nine places to the left of 1, thus, 0.000 000 001.

Practice Exercise

Express each of the following powers of 10 as an ordinary number:

(a) 1×10^{10} (b) 1×10^{-5}

Answers: (a) 10,000,000,000; (b) 0.000 01

Concept Exercise

Which of the following masses is less: 0.000 001 g or 0.000 01 g?

Answer: See Appendix G.

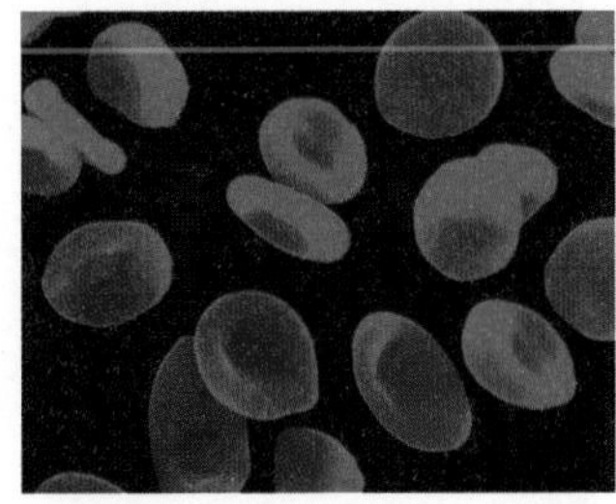

▲ **Red Blood Cells** A negative power of 10 can be used to express very small numbers; for example, the diameter of a red blood cell is about 1×10^{-6} meter.

OBJECTIVE

- To use scientific notation as a way of expressing any number.

2.7 SCIENTIFIC NOTATION

Science often deals with numbers that are very large or very small. These numbers may be awkward and incomprehensible when written with many zeros. The mass of an iron atom, for example, is 0.000 000 000 000 000 000 000 093 g. To overcome this problem, a standard notation has been devised that places the decimal after the first significant digit and sets the size of the number using a power of 10; this method is called **scientific notation**. The scientific notation format is

$$D.DD \times 10^n$$

(significant digits: $D.DD$; power of 10: 10^n)

To use scientific notation, write down the significant digits (D) in the number. Then, move the decimal point to follow the first nonzero digit. Indicate the number of places the decimal is moved using power-of-10 notation. For instance, you can write 555,000 in scientific notation by first moving the decimal five places to the left to give 5.55 and then adding the appropriate power of 10, which is 10^5. Thus, you can express 555,000 in scientific notation as 5.55×10^5.

CHEMISTRY CONNECTION A Student Success Story

Q: Which common inexpensive metal was more valuable than gold in the 19th century?

In 1885, Charles Martin Hall (1863–1914) was a 22-year-old student at Oberlin College in Ohio. One day his chemistry teacher told the class that anyone who could discover an inexpensive way to produce aluminum metal would become rich and benefit humanity. At the time, aluminum was a rare and expensive metal. In fact, Napoleon III, a nephew of Napoleon Bonaparte, entertained his most honored guests with utensils made from aluminum while other guests dined with utensils of silver and gold. Although aluminum is the most abundant metal in Earth's crust, it is not found free in nature; it is usually found combined with oxygen in minerals such as bauxite.

After graduation, Charles Hall set up a laboratory in a woodshed behind his father's church in Oberlin, Ohio. Using homemade batteries, he devised a simple method for producing aluminum by passing electricity through a molten mixture of minerals. After only 8 months of experimenting, he invented a successful method for reducing an aluminum mineral to aluminum metal. In February 1886, Charles Hall walked into his former teacher's office with a handful of metallic aluminum globules.

Just as his chemistry teacher had predicted, within a short period of time, Hall became rich and famous. In 1911, he received the Perkin Medal for achievement in chemistry, and in his will, he donated $5 million to Oberlin College. He also helped to establish the Aluminum Company of America (ALCOA), and the process for making aluminum metal gave rise to a huge industry. Aluminum is now second only to steel as a construction metal.

It is an interesting coincidence that the French chemist Paul Héroult, without knowledge of Hall's work, made a similar discovery at the same time. Thus, the industrial method for obtaining aluminum metal is referred to as the Hall–Héroult process. In 1886, owing to the discovery of this process, the price of aluminum plummeted from over $100,000 a pound. Today, the price of aluminum is less than $1 a pound.

◀ **Aluminum Globules** The notebook of Charles Hall along with globules of aluminum.

A: Before 1885, aluminum was an extremely rare and expensive metal.

You can also express numbers smaller than 1 in scientific notation. For example, you can write 0.000 888 in scientific notation by first moving the decimal four places to the right to give 8.88 and then adding the appropriate power of 10, which is 10^{-4}. Thus, 0.000 888 is expressed in scientific notation as 8.88×10^{-4}.

Regardless of the size of the number, in scientific notation the decimal point is always placed after the first nonzero digit. The size of the number is indicated by a power of 10. A positive exponent indicates a very large number, whereas a negative exponent indicates a very small number. To express a number in scientific notation, follow these two steps.

Applying Scientific Notation

Step 1: Place the decimal point after the first nonzero digit in the number, followed by the remaining significant digits.

Step 2: Indicate how many places the decimal is moved by the power of 10. When the decimal is moved to the left, the power of 10 is positive. When the decimal is moved to the right, the power of 10 is negative.

Example Exercise 2.9 illustrates the conversion between ordinary numbers and exponential numbers.

Example Exercise 2.9 Scientific Notation

Express each of the following values in scientific notation:

(a) There are 26,800,000,000,000,000,000,000 helium atoms in a one liter balloon filled with helium gas.

(b) The mass of one helium atom is 0.000 000 000 000 000 000 000 006 65 g.

Solution

We can write each value in scientific notation as follows:

(a) Place the decimal after the 2, followed by the other significant digits (2.68). Next, count the number of places the decimal has moved. The decimal is moved to the left 22 places, so the exponent is +22. Finally, we have the number of helium atoms in 1.00 L of gas: 2.68×10^{22} atoms.

(b) Place the decimal after the 6, followed by the other significant digits (6.65). Next, count the number of places the decimal has shifted. The decimal has shifted 24 places to the right, so the exponent is –24. Finally, we have the mass of a helium atom: 6.65×10^{-24} g.

Practice Exercise

Express each of the following values as ordinary numbers:

(a) The mass of one mercury atom is 3.33×10^{-22} g.

(b) The number of atoms in 1 mL of liquid mercury is 4.08×10^{22}.

Answers: (a) 0.000 000 000 000 000 000 000 333 g; (b) 40,800,000,000,000,000,000,000 atoms

Concept Exercise

Which of the following masses is greater: 1×10^{-6} g or 0.000 01 g?

Answer: See Appendix G.

▲ **Scientific Calculator** A scientific calculator has an exponent key (EXP) for expressing powers of ten.

Calculators and Scientific Notation

A scientific calculator is very helpful in performing calculations that we often encounter in chemistry. Before purchasing a calculator, check with your instructor for a recommended calculator that can display numbers in scientific notation.

OBJECTIVES

- To write a unit equation for an equivalent relationship.
- To write two unit conversion factors for a unit equation.

2.8 UNIT EQUATIONS AND UNIT FACTORS

In the next section we will tackle our single most important task: problem solving. We will learn a simple but powerful method called unit analysis, or dimensional analysis. To use the unit analysis method of problem solving, we must understand equivalent relationships. An equivalent relationship provides the unit conversion factors that will enable us to solve problems.

Unit Equations

What is an equivalent relationship? An equivalent relationship is a relationship between two quantities that are equal. For instance, 1 dollar equals 100 pennies. Written as an equation, we have

$$1 \text{ dollar} = 100 \text{ pennies}$$

This equivalent relationship is called a unit equation. A **unit equation** is a simple statement of two equivalent quantities.

A penny is 1/100 of a dollar, and we can write the unit equation 1 penny = 0.01 dollar. However, this unit equation is not recommended because it is easy to make an error using decimal fractions. Instead, we will write a whole-number relationship, which is less likely to cause an error. In this example, we will write the unit equation 1 dollar = 100 pennies.

▲ **One Hundred Pennies** Since a dollar is equivalent to 100 pennies, we can write the unit equation 1 dollar = 100 pennies. The two unit factors are 1 dollar/100 pennies, and 100 pennies/1 dollar.

Unit Factors

A unit conversion factor, or **unit factor**, is a ratio of two equivalent quantities. For example, 1 dollar/100 pennies is a unit factor. Since the numerator and denominator are equal, the **reciprocal**, 100 pennies/1 dollar, is equivalent. For the unit equation 1 dollar = 100 pennies, we can write a unit factor and its reciprocal as follows:

$$\frac{1 \text{ dollar}}{100 \text{ pennies}} \quad \text{and} \quad \frac{100 \text{ pennies}}{1 \text{ dollar}}$$

Let's write unit factors for the relationship between inches and yards. We know that there are 36 inches in a yard, so we can write the unit equation 1 yard = 36 inches. The unit factor and its reciprocal are

$$\frac{1 \text{ yard}}{36 \text{ inches}} \quad \text{and} \quad \frac{36 \text{ inches}}{1 \text{ yard}}$$

Example Exercise 2.10 further illustrates writing unit equations and unit factors.

Example Exercise 2.10 Unit Conversion Factors

Write the unit equation and the two corresponding unit factors for each of the following:

(a) pounds and ounces (b) quarts and gallons

Solution

We first write the unit equation and then the corresponding unit factors.

(a) There are 16 ounces in a pound, and so the unit equation is 1 pound = 16 ounces. The two associated unit factors are

$$\frac{1 \text{ pound}}{16 \text{ ounces}} \quad \text{and} \quad \frac{16 \text{ ounces}}{1 \text{ pound}}$$

(b) There are 4 quarts in a gallon, and so the unit equation is 1 gallon = 4 quarts. The two unit factors are

$$\frac{1 \text{ gallon}}{4 \text{ quarts}} \quad \text{and} \quad \frac{4 \text{ quarts}}{1 \text{ gallon}}$$

Practice Exercise

Write the unit equation and the two corresponding unit factors for each of the following:

(a) hours and days (b) hours and minutes

Answers: (a) 1 day = 24 hours; 1 day/24 hours and 24 hours/1 day; (b) 1 hour = 60 minutes; 1 hour/60 minutes and 60 minutes/1 hour

Concept Exercise

How many significant digits are in the following unit equation?

1 hour = 3600 seconds

Answer: See Appendix G.

Exactly Equivalent Relationships

A yard is an English unit of length equal to 36 inches. By definition, 1 yard is exactly 36 inches. Thus, we can write 1 yard = 36 inches, which is an exactly equal relationship, and is referred to as an **exact equivalent**. The rules for significant digits do not apply to exact equivalents. Therefore, the unit factor 1 yard/36 inches has an *infinite number* of significant digits.

Other exact equivalents include 1 foot = 12 inches, 1 pound = 16 ounces, and 1 gallon = 4 quarts. Examples of exact metric equivalents include 1 meter = 100 centimeters, 1 kilogram = 1000 grams, and 1 liter = 1000 milliliters. However, an equivalent that relates English and metric units is never exact because each measuring system is derived from different reference standards. Thus, 1 kilogram = 2.20 pounds is an equivalent, but not an exact equivalent.

2.9 UNIT ANALYSIS PROBLEM SOLVING

OBJECTIVES

- To describe the three steps in the unit analysis method.
- To apply the unit analysis method of problem solving.

Science has the reputation for tackling difficult problems. Fortunately, you are about to learn a powerful technique for solving problems. It is called the **unit analysis method**. This approach to problem solving is also referred to as *dimensional analysis,* or the *factor-label method*. Unit analysis cannot solve every problem, but it is very effective for the problems we encounter in introductory chemistry.

Solving problems by unit analysis is a simple three-step process. Step 1: Read the problem and determine the unit required in the answer. Step 2: Analyze the problem and determine the given value that is related to the answer. Step 3: Write one or more unit factors to convert the unit in the given value to the unit in the answer. You should follow these three steps consistently and systematically when using the unit analysis method of problem solving.

Applying the Unit Analysis Method

Step 1: Write down the unit asked for in the answer.

Step 2: Write down the given value related to the answer.

Step 3: Apply a unit factor to convert the unit in the given value to the unit in the answer. To illustrate,

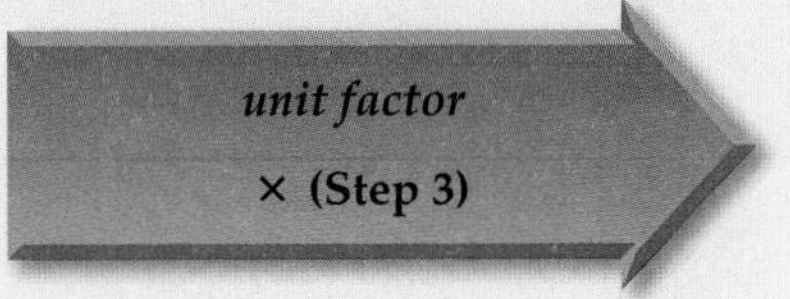

Suppose we wish to find the number of days in 2.5 years. *First*, we read the problem and determine the unit asked for (days). *Second*, we analyze the given information and select the value related to days (2.5 years). *Third*, we apply a unit factor using the following format:

$$\underset{\text{(Step 2)}}{2.5\,\text{years}} \times \underset{\text{(Step 3)}}{\frac{\text{unit}}{\text{factor}}} = \underset{\text{(Step 1)}}{\text{days}}$$

The unit equation that relates Steps (1) and (2) is 1 year = 365 days. Therefore, the two associated unit factors are

$$\frac{1\ \text{year}}{365\ \text{days}} \quad \text{and} \quad \frac{365\ \text{days}}{1\ \text{year}}$$

We choose the unit factor whose denominator has the same unit as the unit of the given value (years). After canceling (~~years~~), we have the unit asked for in this problem (days). Thus, we choose the unit factor 365 days/1 year.

$$2.5\ \cancel{\text{years}} \times \frac{365\ \text{days}}{1\ \cancel{\text{year}}} = 910\ \text{days}$$

After rounding off to two significant digits, we find that the answer is 910 days. Example Exercises 2.11 and 2.12 provide practical examples of the unit analysis method of problem solving.

Example Exercise 2.11 Unit Analysis Problem Solving

A can of soda contains 12 fluid ounces (fl oz). What is the volume in quarts (given that 1 qt = 32 fl oz)?

▲ **Dr. Pepper** A 12 fl oz can of soda contains 355 mL.

Strategy Plan

Step 1: What unit is asked for in the answer?

Step 1: qt

Step 2: What given value is related to the answer?

Step 2: 12 fl oz

Step 3: What unit factor should we apply? Since the unit equation is 1 qt = 32 fl oz, the two unit factors are 1 qt/32 fl oz, and its reciprocal 32 fl oz/1 qt.

Step 3: $\frac{1\text{ qt}}{32\text{ fl oz}}$ or $\frac{32\text{ fl oz}}{1\text{ qt}}$

Unit Analysis Map

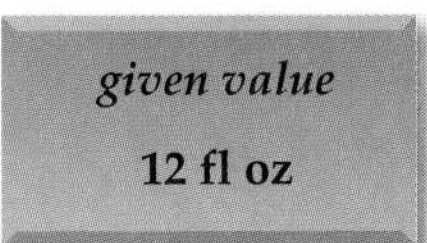

unit factor × $\frac{1\text{ qt}}{32\text{ fl oz}}$ or $\frac{32\text{ fl oz}}{1\text{ qt}}$

unit in answer = ? qt

Solution

We should apply the unit factor 1 qt/32 fl oz to cancel fluid ounces (fl oz), which appears in the denominator.

$$12\ \cancel{\text{fl oz}} \times \frac{1\text{ qt}}{32\ \cancel{\text{fl oz}}} = 0.38\text{ qt}$$

The given value, 12 fl oz, limits the answer to two significant digits. Since the unit factor 1 qt/32 fl oz is derived from an exact equivalent, 1 qt = 32 fl oz, it does not affect the significant digits in the answer.

Practice Exercise

A can of soda contains 355 mL. What is the volume in liters (given that 1 L = 1000 mL)?

Answer: 0.355 L

Concept Exercise

How many significant digits are in the following unit equation?

$$1\text{ L} = 1000\text{ mL}$$

Answer: See Appendix G.

Example Exercise 2.12 Unit Analysis Problem Solving

A marathon covers a distance of 26.2 miles (mi). If 1 mile is exactly equal to 1760 yards, what is the distance of the race in yards?

Strategy Plan

Step 1: What unit is asked for in the answer?

Step 1: yd

Step 2: What given value is related to the answer?

Step 2: 26.2 mi

Step 3: What unit factor should we apply? Since the unit equation is 1 mi = 1760 yd, the two unit factors are 1 mi/1760 yd, and its reciprocal 1760 yd/1 mi.

Step 3: $\frac{1\text{ mi}}{1760\text{ yd}}$ or $\frac{1760\text{ yd}}{1\text{ mi}}$

▲ **Boston Marathon** Marathon athletes run a distance of 26.2 miles.

Unit Analysis Map

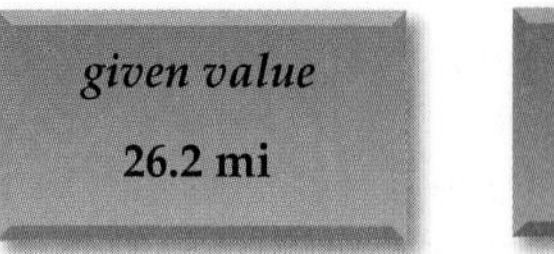

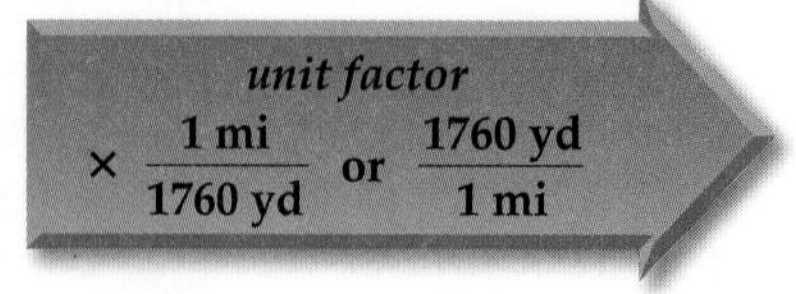

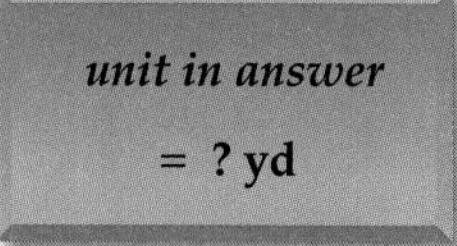

Solution

We should apply the unit factor 1760 yd/1 mi to cancel miles (mi), which appears in the denominator.

$$26.2\ \cancel{mi} \times \frac{1760\ yd}{1\ \cancel{mi}} = 46{,}100\ yd$$

The given value, 26.2 mi, limits the answer to three significant digits. Since the unit factor 1760 yd/1 mi is derived from an exact equivalent, 1 mi = 1760 yd, it does not affect the significant digits in the answer.

Practice Exercise

Given that a marathon is 26.2 miles, what is the distance in kilometers (given that 1 km = 0.62 mi)?

Answer: 42 km

Concept Exercise

How many significant digits are in the following unit equation?

$$1\ km = 0.62\ mi$$

Answer: See Appendix G.

CRITICAL THINKING World Trade Center

Q: What is the difference between a building that has an area footprint of 150 square feet, versus 150 feet square?

▲ **Proposed World Trade Center** The architect's sketch of the proposed World Trade Center in NYC.

When the architect's proposal was accepted for the new World Trade Center building in New York City, a national daily newspaper stated that the building "footprint" was to be 150 square feet. Given that Freedom Tower will be the highest structure in the world, does this statement reflect critical thinking?

If the building footprint is 150 square feet, the structure would measure only about 12 feet per side! Given that the building will be 1776 feet high, it is unreasonable that the base would be only 12 feet per side. It makes more sense that the footprint would be 150 feet square; that is, 150 feet per side.

Another example reflecting a lack of critical thinking occurred when NASA engineers mixed up English and metric units in the navigation program for the Mars Climate Orbiter. The Orbiter was to sample the Martian atmosphere and a navigation error caused a $100 million loss of the Orbiter as it approached the planet's atmosphere too closely.

The minimal safe approach distance to the planet was estimated to be 53 miles. When NASA engineers mixed up English and metric units, the Orbiter approached within *60 kilometers* (*37 miles*) and burned up in the Martian atmosphere.

A: An area footprint of 150 square feet is only 12 feet per side; whereas, 150 feet square, is 150 feet per side.

2.10 THE PERCENT CONCEPT

OBJECTIVES

- To explain the concept of percent.
- To apply percent as a unit factor.

A **percent** (symbol %) expresses the amount a single quantity compared to an entire sample. For example, a dime is 10% of a dollar, and a quarter is 25% of a dollar. The percent symbol (%) is derived from the ratio sign (/) and the two zeros in 100.

A percent expresses a ratio and can be referred to as parts per 100 parts. To calculate a percent, divide a single quantity by the total sample, and multiply that ratio by 100%.

$$\frac{\text{one quantity}}{\text{total sample}} \times 100\% = \%$$

As an example, consider that 5¢ coins minted after 1971 are composed of the two metals copper and nickel. If a 5¢ coin contains 3.80 g copper and 1.27 g nickel, what is the percent of copper in the coin? We can calculate the percent of copper as follows:

$$\frac{3.80\text{ g}}{(3.80 + 1.27)\text{ g}} \times 100\% = 75.0\%$$

Example Exercise 2.13 The Percent Concept

Sterling silver contains silver and copper metals. If a sterling silver chain contains 18.5 g of silver and 1.5 g of copper, what is the percent of silver?

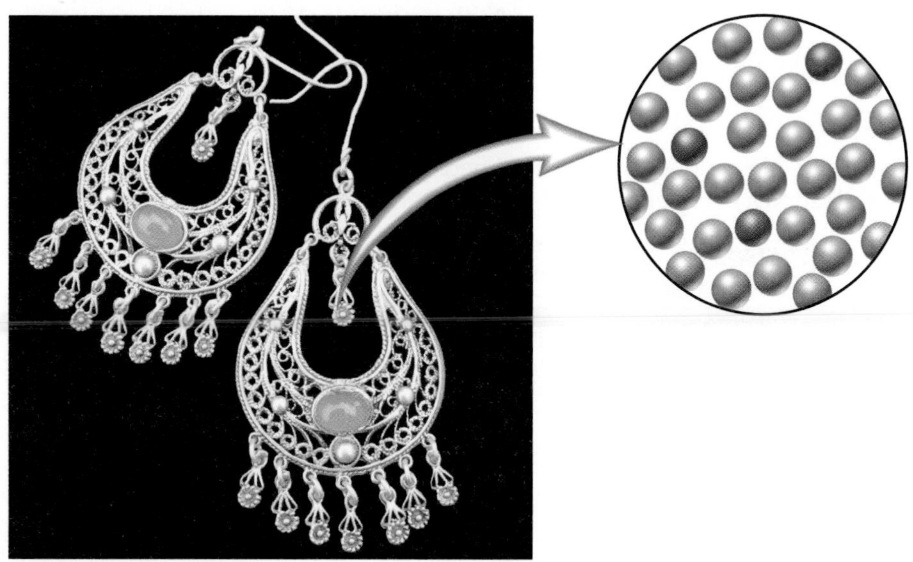

▲ **Sterling Silver** Sterling silver has a high luster and is found in fine utensils and jewelry.

Strategy Plan

Step 1: What is asked for in the answer?	**Step 1:**	%
Step 2: What given value is related to the answer?	**Step 2:**	**18.5 g and 1.5 g**
Step 3: What unit factor should we apply? No unit factor is required.	**Step 3:**	**none**

Solution

To find percent, we compare the mass of silver metal to the total mass of the silver and copper in the chain, and multiply by 100%.

$$\frac{18.5 \text{ g}}{(18.5 + 1.5) \text{ g}} \times 100\% = 92.5\%$$

Genuine sterling silver is cast from 92.5% silver and 7.5% copper. If you carefully examine a piece of sterling silver, you may see the jeweler's notation .925, which indicates the item is genuine sterling silver.

Practice Exercise

A 14-karat gold ring contains 7.45 g of gold, 2.66 g of silver, and 2.66 g of copper. Calculate the percent of gold in the 14-karat ring.

Answer: 58.3%

Concept Exercise

If a gold alloy contains 20% silver and 5% copper, what is the percent of gold in the alloy?

Answer: See Appendix G.

Percent Unit Factors

▲ **Moon** The Moon and Earth have a similar composition.

A percent can be expressed as parts per 100 parts. For instance, 10% and 25% correspond to 10 and 25 parts per 100 parts, respectively. Thus, we can write a percent as a ratio fraction; that is, we can write 10% as 10/100, and 25% as 25/100.

Let's try a type of problem in which the percent is given. *Apollo 11* traveled 394,000 km from Earth to land the first astronaut on the Moon. How many kilometers had the astronauts traveled when they had gone 15.5% of the distance? By definition, 15.5% means 15.5 parts per 100 parts. In this example, 15.5% is 15.5 km per 100 km. Thus, we can write the ratio as a unit factor and apply the unit analysis method.

$$394{,}000 \cancel{\text{km}} \times \frac{15.5 \text{ km}}{100 \cancel{\text{km}}} = 61{,}100 \text{ km}$$

In this problem, notice that we used percent as a unit factor. Example Exercise 2.14 further illustrates the application of percent as a unit factor.

Example Exercise 2.14 Percent as a Unit Factor

The Moon and Earth have a similar composition and each contains 4.70% iron, which supports the theory that the Moon and Earth were originally a single planet. What is the mass of iron in a lunar sample that weighs 235 g?

Strategy Plan

Step 1: What unit is asked for in the answer? — **Step 1:** **g iron**

Step 2: What given value is related to the answer? — **Step 2:** **235 g sample**

Step 3: What unit factor should we apply? From the definition of percent, 4.70 g iron = 100 g sample; the two unit factors are 4.70 g iron/100 g sample, and its reciprocal 100 g sample/4.70 g iron. — **Step 3:** $\frac{\textbf{4.70 g iron}}{\textbf{100 g sample}}$ **or** $\frac{\textbf{100 g sample}}{\textbf{4.70 g iron}}$

Unit Analysis Map

given value	*unit factor*	*unit in answer*
235 g sample	$\times \frac{4.70 \text{ g iron}}{100 \text{ g sample}}$ or $\frac{100 \text{ g sample}}{4.70 \text{ g iron}}$	= g iron

Solution

We should apply the unit factor 4.70 g iron/100 g sample to cancel grams sample (~~g sample~~), which appears in the denominator.

$$235\ \cancel{\text{g sample}} \times \frac{4.70\ \text{g iron}}{100\ \cancel{\text{g sample}}} = 11.0\ \text{g iron}$$

The given value and unit factor each limits the answer to three significant digits.

Practice Exercise

A Moon sample is found to contain 7.50% aluminum. What is the mass of the lunar sample if the amount of aluminum is 5.25 g?

Answer: 70.0 g sample

Concept Exercise

Water is 11.2% hydrogen by mass. What two unit factors express the percent hydrogen in water?

Answer: See Appendix G.

CHEMISTRY CONNECTION The Coinage Metals

Q: What common metal makes up most of the mass of a penny minted before 1982? after 1982?

Earth's core is thought to be composed of only two metals, iron and nickel. On the other hand, Earth's crust contains dozens of metals including aluminum, iron, calcium, sodium, potassium, magnesium, and titanium. These metals, however, do not occur naturally in the free state. They are found combined with other elements.

Copper, silver, and gold are found free in nature. Since ancient times, these metals have been used to make jewelry and precious objects. Not only do they have an attractive appearance, but they are relatively soft and can be easily altered into various shapes. Thousands of years ago in India and Egypt, silver and gold tokens were used for trade. Later, the Romans used silver and gold to make coins for monetary exchange. Today, copper, silver, and gold are often referred to as *coinage metals.*

The term "coinage metals" is somewhat misleading. In 1934, the United States took gold coins out of circulation. In 1971, the United States stopped using silver except for 50¢ coins and special collector issues. A 5¢ coin is minted from 75% copper and 25% nickel. Both 10¢ and 25¢ coins are minted as a "sandwich" composed of an outer layer of copper-nickel and an inner core of pure copper. When you examine a 10¢ or a 25¢ coin closely, the reddish-brown color of copper is clearly visible along the serrated edge.

The first 1¢ coin was minted in 1793 from pure copper and was the size of our current half dollar. The copper penny continued to be minted until 1856 when it was replaced by a penny made from a mixture of 88% copper and 12% nickel. The addition of nickel metal gave a white appearance to the coin. Five years later, the penny was changed to a mixture of copper, tin, and zinc metals. From 1962 to 1982, the penny was cast from an alloy mixture of 95% copper and 5% zinc.

In 1982, the high price of copper forced the U.S. Mint to alter the composition once again. In lieu of copper, zinc is the main component in the new 1¢ coin. These pennies are minted from a zinc core plated with an ultrathin coating of copper metal. To date, the U.S. Mint has produced over 200,000,000,000 of these coins, but most pennies do not stay in active circulation.

◀ **Coinage Metals** Copper, silver, and gold are found in selected coins minted in the United States.

A: Pre-1982 pennies are mostly copper; post-1982 pennies are almost entirely zinc.

Summary

Section 2.1 This chapter introduced the basic laboratory equipment for obtaining a **measurement**. We often record measurements in units of **centimeter (cm)**, **gram (g)**, **milliliter (mL)**, or second (s). No measurement is exact because every **instrument** has **uncertainty**. We use a balance to measure the **mass** of an object. Although we commonly say that we weigh an object, strictly speaking, we determine the mass of an object. **Weight** is affected by Earth's gravity, but mass is not.

Section 2.2 The **significant digits** in a measurement are those known with certainty from the instrument, plus one digit that is estimated.

Section 2.3 **Nonsignificant digits** exceed the certainty of the instrument and result from calculator computations. The process of eliminating nonsignificant digits is called **rounding off**.

Section 2.4 When adding or subtracting measurements, the answer is limited by the measurement having the most uncertainty. If the uncertainties of three mass measurements are ±0.1 g, ±0.01 g, and ±0.001 g, respectively, the measurement with the most uncertainty, ±0.1 g, limits the answer.

Section 2.5 When multiplying or dividing measurements, the answer is limited by the least number of significant digits in the data. Every calculated answer must include the correct number of significant digits, as well as the proper units.

Section 2.6 A number written as a superscript indicating that a value is multiplied by itself is called an **exponent**. If the value is 10, the exponent is a **power of 10**. A power of 10 can be positive or negative. If the exponent is positive (10^3), the number is greater than 1; if the exponent is negative (10^{-3}), it is less than 1.

Section 2.7 In science we use **scientific notation** to express numbers that are very large or very small. Scientific notation uses the following format to express a number with three significant digits: D.DD × 10^n. The three significant digits (D.DD) are multiplied by a power of 10 (10^n) to set the decimal point.

Section 2.8 We introduced our single most important task: namely, problem solving. We write a **unit equation** to generate a **unit factor**. For example, the unit equation 1 yard = 3 feet gives two unit factors: 1 yard/3 feet and its **reciprocal**, 3 feet/1 yard. An **exact equivalent** is an exactly equal relationship such as 1 yard = 36 inches.

Section 2.9 We systematically used the **unit analysis method** to solve problems. Step 1: Write down the unit asked for in the answer. Step 2: Write down the given value related to the unit in the answer. Step 3: Apply one or more unit factors to cancel the units in the given value and obtain the unit in the answer.

Section 2.10 A **percent** (%) expresses the amount of a single quantity compared to an entire sample. Since percent is a ratio, we can apply percent unit factors to perform calculations using the unit analysis method.

PROBLEM–SOLVING ORGANIZER

TOPIC	PROCEDURE	EXAMPLE
Uncertainty in Measurements Sec. 2.1	State the measurement and indicate the uncertainty.	11.50 cm ± 0.05 cm 0.545 g ± 0.001 g 10.0 mL ± 0.5 mL
Significant Digits Sec. 2.2	Count the significant digits in a measurement from left to right: (a) start with the first nonzero digit, (b) do not count placeholder zeros.	0.05 g (1 significant digit) 50.0 g (3 significant digits) 505 g (3 significant digits) 50,000 g (1 significant digit)
Rounding Off Nonsignificant Digits Sec. 2.3	1. If the first nonsignificant digit is less than 5, drop all nonsignificant digits. 2. If the first nonsignificant digit is 5 or greater, add 1 to the last significant digit. 3. If a calculation involves several steps, retain nonsignificant digits until the final step.	505.05 g rounds to 505 g (3 significant digits) 505.50 g rounds to 506 g (3 significant digits)
Adding and Subtracting Measurements Sec. 2.4	The answer is limited by the least certain measurement in a set of data.	5.05 g + 5.005 g = 10.06 g 5.05 mL − 2.1 mL = 3.0 mL
Multiplying and Dividing Measurements Sec. 2.5	The answer is limited by the least number of significant digits in the given data.	5.05 cm × 5.005 cm = 25.3 cm^2 5.0 g /10.0 mL = 0.50 g/mL

PROBLEM–SOLVING ORGANIZER *(Continued)*

TOPIC	PROCEDURE	EXAMPLE
Exponential Numbers Sec. 2.6	(a) Numbers greater than 1 have a positive exponent. (b) Numbers less than 1 have a negative exponent.	$1{,}000{,}000 = 1 \times 10^{6}$ $0.000\,001 = 1 \times 10^{-6}$
Scientific Notation Sec. 2.7	Place the decimal point after the first nonzero digit. Using an exponent, indicate how many places the decimal is moved.	$55{,}500 \text{ g} = 5.55 \times 10^{4} \text{ g}$ $0.000\,555 \text{ g} = 5.55 \times 10^{-4} \text{ g}$
Unit Equations and Unit Factors Sec. 2.8	(a) Write a unit equation that shows an equivalent relationship. (b) Write two unit factors corresponding to the unit equation.	1 quart = 2 pints $\frac{1 \text{ quart}}{2 \text{ pints}}$ and $\frac{2 \text{ pints}}{1 \text{ quart}}$
Unit Analysis Problem Solving Sec. 2.9	1. Write down the unit asked for in the answer. 2. Write down the related given value. 3. Apply a unit factor to convert the given unit to the unit in the answer.	What is the number of feet in 1.75 yd? Given: 1 yard = 3 feet $1.75 \text{ yd} \times \frac{3 \text{ ft}}{1 \text{ yd}} = 5.25 \text{ ft}$
The Percent Sec. 2.10	Compare the amount of a single multiply by 100%.	If a 14-karat ring contains 20.0 g gold and 14.3 g silver, what is the percent of gold? $\frac{20.0 \text{ g}}{(20.0 + 14.3) \text{ g}} \times 100\% = 58.3\%$

Key Terms *Answers to Key Terms are in Appendix H.*

Select the key term that corresponds to each of the following definitions.

_____ **1.** a numerical value and unit that expresses a physical quantity such as length, mass, or volume
_____ **2.** a common metric unit of length
_____ **3.** a common metric unit of mass
_____ **4.** a common metric unit of volume
_____ **5.** a device for recording a measurement such as length, mass, or volume
_____ **6.** the degree of inexactness in an instrumental measurement
_____ **7.** the quantity of matter in an object
_____ **8.** the force exerted by gravity on an object
_____ **9.** the certain digits in a measurement plus one estimated digit
_____ **10.** the digits in a measurement that exceed the certainty of the instrument
_____ **11.** the process of eliminating digits that are not significant
_____ **12.** a number written as a superscript that indicates a value is multiplied by itself, for example, $10^4 = 10 \times 10 \times 10 \times 10$, and $\text{cm}^3 = \text{cm} \times \text{cm} \times \text{cm}$
_____ **13.** a positive or negative exponent of 10
_____ **14.** a method for expressing numbers by moving the decimal place after the first significant digit and indicating the number of decimal moves by a power of 10
_____ **15.** a statement of two equivalent values; for example, 1 foot = 12 inches
_____ **16.** a ratio of two quantities that are equivalent and can be applied to convert from one unit to another, for example, 1 foot/12 inches

(a) centimeter (cm) *(Sec. 2.1)*
(b) exact equivalent *(Sec. 2.8)*
(c) exponent *(Sec. 2.6)*
(d) gram (g) *(Sec. 2.1)*
(e) instrument *(Sec. 2.1)*
(f) mass *(Sec. 2.1)*
(g) measurement *(Sec. 2.1)*
(h) milliliter (mL) *(Sec. 2.1)*
(i) nonsignificant digits *(Sec. 2.3)*
(j) percent (%) *(Sec. 2.10)*
(k) power of 10 *(Sec. 2.6)*
(l) reciprocal *(Sec. 2.8)*
(m) rounding off *(Sec. 2.3)*
(n) scientific notation *(Sec. 2.7)*
(o) significant digits *(Sec. 2.2)*
(p) uncertainty *(Sec. 2.1)*

_____	17. the relationship of a fraction and its inverse; for example, 1 yard/3 feet and 3 feet/1 yard	(q) unit analysis method *(Sec. 2.9)*
_____	18. a statement of two exactly equal values, for example, 1 yard = 36 inches	(r) unit equation *(Sec. 2.8)*
_____	19. a systematic procedure for solving problems that proceeds from a given value to a related answer by the conversion of units	(s) unit factor *(Sec. 2.8)*
_____	20. an expression for the amount of a single quantity compared to an entire sample	(t) weight *(Sec. 2.1)*

Exercises *Answers to odd-numbered Exercises are in Appendix I.*

Uncertainty in Measurements (Sec. 2.1)

1. What quantity (length, mass, volume, time) is expressed by the following units?
   (a) centimeter (b) gram
   (c) milliliter (d) second
2. What quantity (length, mass, volume, time) is expressed by the following units?
   (a) inch (b) pound
   (c) quart (d) hour
3. Which of the following measurements correspond to using metric Ruler A shown in Figure 2.2?
   (a) 10.0 cm (b) 10.00 cm
   (c) 10.05 cm (d) 10.5 cm
4. Which of the following measurements correspond to using metric Ruler B shown in Figure 2.2?
   (a) 50.0 cm (b) 50.00 cm
   (c) 50.05 cm (d) 50.5 cm
5. Which of the following measurements corresponds to using the beam balance shown in Figure 2.3?
   (a) 75 g (b) 75.0 g
   (c) 75.22 g (d) 75.518 g
6. Which of the following measurements corresponds to using the electronic balance shown in Figure 2.3?
   (a) 75 g (b) 75.0 g
   (c) 75.22 g (d) 75.518 g
7. Which of the following measurements correspond to using the graduated cylinder shown in Figure 2.5 having an uncertainty of ± 0.5 mL?
   (a) 25 mL (b) 25.0 mL
   (c) 25.5 mL (d) 25.50 mL
8. Which of the following measurements correspond to using the buret shown in Figure 2.5 having an uncertainty of ± 0.05 mL?
   (a) 25 mL (b) 25.0 mL
   (c) 25.5 mL (d) 25.50 mL

Significant Digits (Sec. 2.2)

9. State the number of significant digits in each of the following:
   (a) 0.05 cm (b) 0.50 cm
   (c) 25.0 cm (d) 20.50 cm
10. State the number of significant digits in each of the following:
    (a) 5.0 g (b) 5.05 g
    (c) 0.5 g (d) 1055 g
11. State the number of significant digits in each of the following:
    (a) 0.5 mL (b) 0.50 mL
    (c) 5.00 mL (d) 500 mL
12. State the number of significant digits in each of the following:
    (a) 100 s (b) 10.0 s
    (c) 10.00 s (d) 1.0 s
13. State the number of significant digits in each of the following:
    (a) 1.050×10^2 cm (b) 2×10^3 cm
    (c) 3.00×10^{-4} cm (d) 5.0×10^{-5} cm
14. State the number of significant digits in each of the following:
    (a) 1.2×10^0 g (b) 4.50×10^1 g
    (c) 5.02×10^{-1} g (d) 100×10^{-2} g

Rounding Off Nonsignificant Digits (Sec. 2.3)

15. Round off the following to three significant digits:
    (a) 10.25 (b) 10.20
    (c) 0.01029 (d) 10,248
16. Round off the following to three significant digits:
    (a) 20.155 (b) 0.204 500
    (c) 2055 (d) 0.2065
17. Round off the following to three significant digits:
    (a) 1.454×10^1 (b) 1.455×10^2
    (c) 1.508×10^{-3} (d) 1.503×10^{-4}
18. Round off the following to three significant digits:
    (a) 9.123×10^5 (b) 9.456×10^{10}
    (c) 9.000×10^{-7} (d) 9.075×10^{-12}

Adding and Subtracting Measurements (Sec. 2.4)

19. Add the following measurements and round off the answer:
    (a) 1.55 cm + 36.15 cm + 17.3 cm
    (b) 5.0 cm + 16.3 cm + 0.95 cm
20. Add the following measurements and round off the answer:
    (a) 0.4 g + 0.44 g + 0.444 g
    (b) 15.5 g + 7.50 g + 0.050 g
21. Subtract the following measurements and round off the answer:
    (a) 242.167 g − 175 g
    (b) 27.55 g − 14.545 g
22. Subtract the following measurements and round off the answer:
    (a) 22.10 cm − 10.5 cm
    (b) 10.0 cm − 0.15 cm

Multiplying and Dividing Measurements (Sec. 2.5)

23. Multiply the following measurements and round off the answer:
    (a) 3.65 cm × 2.10 cm (b) 8.75 cm × 1.15 cm
    (c) 16.5 cm × 1.7 cm (d) 21.1 cm × 20 cm
24. Multiply the following measurements and round off the answer:
    (a) 5.1 cm × 1.25 cm × 0.5 cm
    (b) 5.15 cm × 2.55 cm × 1.1 cm
    (c) 12.0 cm^2 × 1.00 cm
    (d) 22.1 cm^2 × 0.75 cm

25. Divide the following measurements and round off the answer:
(a) 26.0 cm^2/10.1 cm (b) 9.95 cm^3/0.15 cm^2
(c) 131.78 cm^3/19.25 cm (d) 131.78 cm^3/19.2 cm

26. Divide the following measurements and round off the answer:
(a) 66.3 g/7.5 mL (b) 12.5 g/4.1 mL
(c) 42.620 g/10.0 mL (d) 91.235 g/10.00 mL

Exponential Numbers (Sec. 2.6)

27. Express each of the following products as a power of 10:
(a) $10 \times 10 \times 10 \times 10$ (b) $1/10 \times 1/10 \times 1/10$

28. Express each of the following products as a power of 10:
(a) $10 \times 10 \times 10 \times 10 \times 10 \times 10$
(b) $1/10 \times 1/10 \times 1/10 \times 1/10$

29. Express each of the following products in exponential form:
(a) $2 \times 2 \times 2$ (b) $1/2 \times 1/2 \times 1/2$

30. Express each of the following products in exponential form:
(a) $3 \times 3 \times 3 \times 3$ (b) $1/3 \times 1/3 \times 1/3 \times 1/3$

31. Express each of the following powers of 10 as an ordinary number:
(a) 1×10^3 (b) 1×10^{-7}

32. Express each of the following powers of 10 as an ordinary number:
(a) 1×10^{12} (b) 1×10^{-22}

33. Express each of the following ordinary numbers as a power of 10:
(a) 1,000,000,000 (b) 0.000 000 01

34. Express each of the following ordinary numbers as a power of 10:
(a) 100,000,000,000,000,000 (b) 0.000 000 000 000 001

35. Express each of the following powers of 10 as an ordinary number:
(a) 1×10^1 (b) 1×10^{-1}

36. Express each of the following powers of 10 as an ordinary number:
(a) 1×10^0 (b) 1×10^{-10}

Scientific Notation (Sec. 2.7)

37. Express the following ordinary numbers in scientific notation:
(a) 1,010,000,000,000,000 (b) 0.000 000 000 000 456
(c) 94,500,000,000,000,000 (d) 0.000 000 000 000 000 019 50

38. Express the following ordinary numbers in scientific notation:
(a) 80,916,000 (b) 0.000 000 015
(c) 335,600,000,000,000 (d) 0.000 000 000 000 927

39. There are 26,900,000,000,000,000,000,000 atoms in one liter of helium gas at standard temperature and pressure. Express this number in scientific notation.

40. There are 269,000,000,000,000,000,000,000 molecules in ten liters of hydrogen gas at standard temperature and pressure. Express this number in scientific notation.

41. The mass of a neon atom is 0.000 000 000 000 000 000 000 03 35 g. Express the mass in scientific notation.

42. The mass of a chlorine molecule is 0.000 000 000 000 000 000 000 118 g. Express the mass in scientific notation.

Unit Equations and Unit Factors (Sec. 2.8)

43. Write a unit equation for each of the following:
(a) miles and feet (b) tons and pounds

44. Write a unit equation for each of the following:
(a) quarts and pints (b) centuries and years

45. Write the two unit factors associated with each of the following:
(a) miles and feet (b) tons and pounds

46. Write the two unit factors associated with each of the following:
(a) quarts and pints (b) centuries and years

47. Which of the following is an exact equivalent?
(a) 1 mile = 5280 feet (b) 1 mile = 1.61 kilometers

48. Which of the following is an exact equivalent?
(a) 1 kilogram = 1000 grams (b) 1 kilogram = 2.20 pounds

49. Which of the following is an exact equivalent?
(a) 1 liter = 1000 milliliters (b) 1 quart = 946 milliliters

50. Which of the following is an exact equivalent?
(a) 1 month = 4 weeks (b) 1 hour = 60 minutes

51. What is the number of significant digits in each of the following relationships?
(a) 1 mile = 5280 feet (b) 1 mile = 1.61 kilometers

52. What is the number of significant digits in each of the following relationships?
(a) 1 kilogram = 1000 grams (b) 1 kilogram = 2.20 pounds

53. What is the number of significant digits in each of the following relationships?
(a) 1 liter = 1000 milliliters (b) 1 quart = 946 milliliters

54. What is the number of significant digits in each of the following relationships?
(a) 1 month = 4 weeks (b) 1 hour = 60 minutes

Unit Analysis Problem Solving (Sec. 2.9)

55. Using the unit analysis method, find the number of centimeters in 36.0 inches (given that 1 inch = 2.54 centimeters).

56. Using the unit analysis method, find the number of kilometers in 20.5 miles (given that 1 mile = 1.61 kilometers).

57. Using the unit analysis method, find the mass in grams of a diamond that is 3/4 carat (0.750 carat) (given that 1 carat = 0.200 gram).

58. Using the unit analysis method, find the mass in grains of a 325-milligram aspirin tablet (given that 1000 milligrams = 15.4 grains).

59. Using the unit analysis method, find the number of cups of cream in 2.5 pints (given that 1 pint = 2 cups).

60. Using the unit analysis method, find the number of tablespoons of salad oil in 15 teaspoons (given that 1 tablespoon = 3 teaspoons).

61. Using the unit analysis method, calculate the number of months in 5.5 years.

62. Using the unit analysis method, calculate the number of days in 15 weeks.

63. If the United States has 275,000,000,000 troy ounces of gold on reserve, what is the mass of gold in troy pounds (given that 1 troy pound = 12 troy ounces)?

◀ **Gold** Bars of pure gold and gold coins.

64. If a bar of silver weighs 50.0 troy ounces, what is the mass of silver in grams (given that 1 troy ounce = 31.1 grams)?

◀ **Silver** A bar of 0.999 fine silver and silver coins.

65. How many meters does light travel in 1.00 hour (given that the velocity of light is 3.00×10^8 m/s)?

66. How many centimeters does light travel in 1.00 year (given that the velocity of light is 3.00×10^{10} cm/s)?

The Percent Concept (Sec. 2.10)

67. A sterling silver spoon has a mass of 65.5 g and contains 7.50% copper. Find the mass of copper in the spoon.

68. A stainless steel spoon has a mass of 50.0 g and contains 10.5% chromium. Find the mass of chromium in the spoon.

69. If a solution contains 255 mL of ethanol and 375 mL of water, what is the percent of ethanol in the solution?

70. If 12.5 gallons (gal) of gasohol contains 1.50 gal of ethyl alcohol, what is the percent of alcohol in the gasohol?

71. Water is composed of 11.2% hydrogen and 88.8% oxygen. What mass of water contains 15.0 g of oxygen?

72. Ordinary table salt, sodium chloride, is composed of 39.3% sodium and 60.7% chlorine. Calculate the mass of sodium in 0.375 g of salt.

73. Earth's crust has a mass of 2.37×10^{25} g and contains 49.2% oxygen, 25.7% silicon, and 7.50% aluminum. Calculate the mass of each element in the crust.

74. Uranus has five large moons with a combined mass of 3.87×10^{21} kg. What is the mass of the largest moon, Titania, if it is 54.3% of the total mass?

75. Before 1982 the U.S. Mint cast penny coins from a copper and zinc mixture. If a 1980 penny weighs 3.051 g and contains 0.153 g zinc, what is the percent of copper in the coin?

76. In 1982 the U.S. Mint stopped making copper pennies and began phasing in pennies made of zinc plated with a thin layer of copper. If a 1990 penny weighs 2.554 g and contains 2.490 g zinc, what is the percent of copper in the coin?

General Exercises

77. A Class A 10 mL pipet has an uncertainty of ±0.01 mL. Express the volume of the pipet as an ordinary number.

78. A student 10 mL pipet has an uncertainty of ±0.1 mL. Express the volume of the pipet as an ordinary number.

79. The velocity of light is 186,282.397 miles per second. Round off this value to three significant digits.

80. The velocity of light is 299,792,458 meters per second. Round off this value to three significant digits.

81. Find the total mass of two brass cylinders that weigh 126.457 g and 131.6 g.

82. Find the length of silk ribbon that remains after two 25.0 cm strips are cut from 255 cm of the ribbon.

83. Convert the following exponential numbers to scientific notation.
(a) 352×10^4 (b) 0.191×10^{-5}

84. Convert the following exponential numbers to scientific notation.
(a) 0.170×10^2 (b) 0.00350×10^{-1}

85. The mass of a neutron is 1.6749×10^{-24} g, and the mass of a proton is 1.6726×10^{-24} g. What is the total mass of a neutron and a proton?

86. The mass of a neutron is 1.6749×10^{-24} g, and the mass of a proton is 1.6726×10^{-24} g. What is the mass difference of a neutron and a proton?

87. A metric ton is defined as 1000 kg, or 2.200×10^3 lb. An English ton is defined as 2000 lb, or 2.000×10^3 lb. What is the difference in mass between a metric ton and an English ton expressed in pounds?

88. The distance from Earth to the Moon is 2.39×10^5 mi, while the distance from the Moon to Mars is 4.84×10^7 mi. What is the total distance from Earth to the Moon to Mars?

▲ **Earthrise** This photo, taken on the Moon, shows Earth rising over the lunar horizon.

89. The oldest rock found on Earth was discovered in South Africa and is estimated to be 4.0 billion years old. Express the age of the rock in hours using scientific notation.

90. *Apollo 15* brought back a lunar rock estimated to be 4.1 billion years old. Express the age of the rock in minutes using scientific notation.

91. Blood bank records show blood type and Rh factor for 55,368 patients as follows: 21,594 O+, 18,825 A+, 4706 B+, 1938 AB+, 3876 O–, 3322 A–, 831 B–, 276 AB–. What percent of the blood types is A positive?

92. Blood bank records show blood type and Rh factor for 55,368 patients as follows: 21,594 O+, 18,825 A+, 4706 B+, 1938 AB+, 3876 O–, 3322 A–, 831 B–, 276 AB–. What percent of the blood types is B negative?

93. If an ounce equals 28.4 g, what is the mass in grams of a pound of feathers (given that 1 pound = 16 ounces)?

94. If a troy ounce equals 31.1 g, what is the mass in grams of a troy pound of gold (given that 1 troy pound = 12 troy ounces)?

95. Refer to Exercises 93 and 94 and answer the timeless question: *"Which weighs more, a pound of feathers, or a pound of gold?"*

2 Chapter Self-Test

Answers to Self-Test are in Appendix J.

1. Which of the following balances has the most uncertainty? (Sec. 2.1)
   (a) platform balance, ± 0.1 g
   (b) triple-beam balance, ± 0.01 g
   (c) electronic balance, ± 0.001 g
   (d) analytical balance, ± 0.0001 g
   (e) all balances have the same uncertainty

2. State the number of significant digits in a mass of 1.050 g. (Sec. 2.2)
   (a) 2 **(b)** 3
   (c) 4 **(d)** 5
   (e) none of the above

3. Round off the following measurement to three significant digits: 7265 mL. (Sec. 2.3)
   (a) 720 mL **(b)** 726 mL
   (c) 727 mL **(d)** 7260 mL
   (e) 7270 mL

4. Add 55.5 cm to 0.50 cm and round off the answer. (Sec. 2.4)
   (a) 56 cm **(b)** 56.0 cm
   (c) 56.05 cm **(d)** 56.1 cm
   (e) 57 cm

5. Multiply 7.2 cm by 3.75 cm by 1.555 cm and round off the answer. (Sec. 2.5)
   (a) 41.9 cm^3 **(b)** 41.99 cm^3
   (c) 42 cm^3 **(d)** 42.0 cm^3
   (e) 42.00 cm^3

6. What is 5.55×10^{-2} expressed as an ordinary number? (Sec. 2.6)
   (a) 0.055 5 **(b)** 0.005 55
   (c) 0.000 555 **(d)** 5,550
   (e) none of the above

7. There are 26,800,000,000,000,000,000,000 molecules in one liter of oxygen gas at room temperature and pressure. Express the number of gas molecules in scientific notation. (Sec. 2.7)
   (a) 2.68×10^{-22} **(b)** 2.68×10^{-23}
   (c) 2.68×10^{22} **(d)** 2.68×10^{23}
   (e) none of the above

8. Which of the following unit factors is derived from: 1.00 yard = 91.4 centimeters? (Sec. 2.8)
   (a) 1.00 yard /1.00 centimeter
   (b) 91.4 centimeters/1.00 yard
   (c) 91.4 yards /1.00 centimeter
   (d) 91.4 centimeters/91.4 yards
   (e) none of the above

9. If a patient is injected with 0.500 liter of IV saline, what is the volume in quarts (given that 1 quart = 0.946 liters)? (Sec. 2.9)
   (a) 0.473 quarts **(b)** 0.500 quarts
   (c) 0.529 quarts **(d)** 0.946 quarts
   (e) 1.89 quarts

10. A sample of 18-karat gold contains the following by mass: 18.0 grams gold, 3.0 grams silver, and 3.0 grams copper. What is the percent gold? (Sec. 2.10)
   (a) 6.0% **(b)** 18%
   (c) 25% **(d)** 33%
   (e) 75%

Key Concepts

11. What is the best estimate for the diameter of a 1¢ coin: 1 cm, 2 cm, or 3 cm?
12. What is the best estimate for the thickness of a 10¢ coin: 0.1 cm, 0.5 cm, or 1 cm?
13. What is the best estimate for the mass of a 25¢ coin: 1 g, 5 g, or 10 g?
14. What is the first step in the unit analysis method of problem solving?
15. What is the second step in the unit analysis method?
16. In the third step in the unit analysis method, how do you decide which of the two reciprocal unit factors to select?
17. While using a scientific calculator to perform multistep multiplication or division, is it better to round off after each step, or round off the final answer in the display?

◀ **Nickel Coins** Since a dollar is equivalent to 20 nickels, we can write the unit equation 1 dollar = 20 nickels. The two associated unit factors are 1 dollar/20 nickels, and 20 nickels/1 dollar.

Critical Thinking

18. The mass of an alpha particle is 6.64×10^{-24} g, and the mass of a beta particle is 9.11×10^{-28} g. Which is heavier: an alpha particle or a beta particle?
19. An ounce of copper weighs 28.4 g, and an ounce of silver weighs 31.1 g. Which is heavier: an ounce of copper or an ounce of silver?
20. A popular "gold" jewelry contains 27% silver, 63% copper, and 10% zinc. What is the percent of gold in the jewelry?

CHAPTER

3 The Metric System

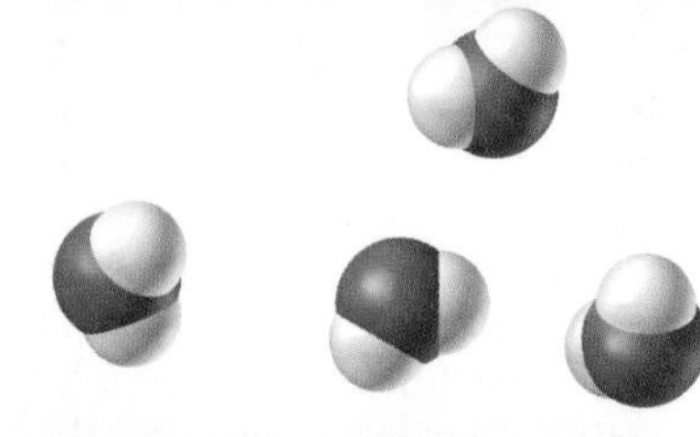

Water can coexist as liquid, ice, and steam at 0.01 °C and low pressure. The so-called "triple point" of water is a reference point for the metric temperature scale.

"Never has anything more grand and more simple, more coherent in all of its parts, issued from the hand of man."
Antoine Lavoisier, French Chemist (1743–1794)

Since the beginning of civilization there has been the need to convey measurement. When exchanging goods, traders had to agree on standards for length, weight, and volume. Historically, length was often defined in terms of hands and feet because it was convenient. For instance, a digit was defined as the thickness of the index finger, a span was the width of four fingers, and a cubit was the distance from the elbow to the end of the middle finger.

The Greeks proposed the foot as a unit of length and divided it into 12 parts. The Romans in turn adopted the unit and referred to 1/12 of a foot as an *uncia*. The Anglo-Saxons pronounced *uncia* as "inch," which is how it entered into the English language.

The English system of measurement originated in 1215 with the signing of the Magna Carta. The Magna Carta attempted to bring uniform measurements to world trade and decreed, "Throughout the kingdom there shall be standard measures of wine, ale, and corn." To ensure uniform measurements, the English used reference standards. An inch was defined as the length of three barleycorns, and a pound and a gallon were similarly referenced to barley.

3.1 BASIC UNITS AND SYMBOLS

OBJECTIVES

- To list the basic units and symbols of the metric system.
- To list the prefixes for multiples and fractions of basic units.

Discontent over the lack of worldwide standards in measurement reached a peak in the late 1700s. Although the **English system** was common, it was used primarily within the British empire. In 1790, the French government appointed a committee of scientists to investigate the possibility of a universal measuring system.

The committee spent nearly 10 years devising and agreeing on a new system. The resulting **metric system** offers simplicity and basic units. The basic unit of length is the **meter** (symbol **m**), the basic unit of mass is the **gram** (symbol **g**), the basic unit of volume is the **liter** (symbol **L**), and the basic unit of time is the **second** (symbol **s**). The basic units and symbols of the metric system are shown in Table 3.1.

The metric system is simple and coherent for two reasons: First, it uses a single basic unit for each quantity measured, that is, meter, gram, and liter. Second, it is a decimal system that uses prefixes to enlarge or reduce a basic unit. For instance, a kilometer is 1000 meters and a centimeter is 1/100 meter.

The metric committee defined a meter, kilogram, and liter as follows. A meter was defined as one ten-millionth of the distance from the North Pole to the equator. A kilogram was defined as the mass of a cube of water one-tenth of a meter on a side. A liter was equal to the volume occupied by a kilogram of water (at 4 °C).

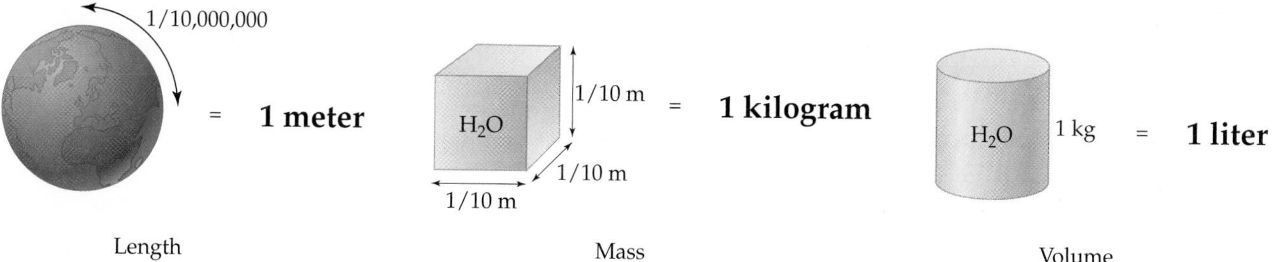

▲ **The Original Metric References** A meter was originally defined as one ten-millionth of the distance from the North Pole to the equator. A kilogram was equal to the mass of a cube of water one-tenth of a meter on each side. A liter was equal to the volume of one kilogram of water at 4 °C.

In addition, the metric committee members made reference standards for the meter, kilogram, and liter. They cast a platinum bar (alloyed with 10% iridium metal) as the reference standard for one meter. They cast a solid platinum cylinder as the reference for one kilogram and a platinum container as the standard for one liter. These platinum castings became the reference standards for world measurement. In 1875 all the major nations of the world, including the United States, signed the Treaty of the Meter. Duplicates of the original castings were made and distributed to all member nations. The original casting of the standard reference meter is shown in Figure 3.1.

TABLE 3.1 THE METRIC SYSTEM

PHYSICAL QUANTITY	BASIC UNIT	SYMBOL
length	meter*	m
mass	gram	g
volume	liter*	L
time	second	s

*The U.S. Metric Association recommends the spellings "meter" and "liter." All other English-speaking nations use the spellings "metre" and "litre."

▶ **Figure 3.1 The Official Meter** An official duplicate meter is an X-shaped bar cast from platinum and iridium metals (shown in foreground). The original meter was cast in 1790 and remains at the International Bureau of Weights and Measures outside Paris, France.

Metric Prefixes

The United States uses the dollar ($1.00) as the basic unit of currency. In our monetary system we move the decimal to express a multiple or fraction of a dollar ($10.00 and $0.10). Similarly, the metric system is a decimal system, but it uses a prefix to express a multiple or a fraction of a basic unit. A metric prefix increases or decreases a basic unit by a factor of 10. For instance, the prefix *kilo-* increases a basic unit by 1000, whereas the prefix *centi-* decreases a basic unit by 100. Thus, a kilometer is 1000 m, and a centimeter is 0.01 m. Table 3.2 lists some common prefixes for basic metric units.

The names of metric units can be abbreviated using symbols. We can illustrate this by writing symbols for the following metric units: kilometer (km), milligram (mg), microliter (μL), and nanosecond (ns). Example Exercise 3.1 further illustrates how to write names of metric units and symbols.

Example Exercise 3.1 Metric Basic Units and Prefixes

Give the symbol for each of the following metric units and state the quantity measured by each unit:

(a) gigameter
(b) kilogram
(c) centiliter
(d) microsecond

Solution

We compose the symbol for each unit by combining the prefix symbol and the basic unit symbol. If we refer to Tables 3.1 and 3.2, we have the following:

(a) Gm, length
(b) kg, mass
(c) cL, volume
(d) μs, time

Practice Exercise

Give the symbol for each of the following metric units and state the quantity measured by each unit:

(a) nanosecond
(b) milliliter
(c) decigram
(d) megameter

Answers: (a) ns, time; (b) mL, volume; (c) dg, mass; (d) Mm, length

Concept Exercise

What is the basic unit of length, mass, and volume in the metric system?

Answer: See Appendix G.

Note The metric system is sometimes referred to as the International System of measurement. Although the **International System** (symbol **SI**) is similar to the metric system, SI is more comprehensive and sophisticated. In this text, we will use the metric system and utilize SI units only in special instances.

TABLE 3.2 METRIC PREFIXES

PREFIX	SYMBOL	MULTIPLE/FRACTION
tera-	T	$1{,}000{,}000{,}000{,}000 = 1 \times 10^{12}$
giga-	G	$1{,}000{,}000{,}000 = 1 \times 10^{9}$
mega-	M	$1{,}000{,}000 = 1 \times 10^{6}$
kilo-	k	$1000 = 1 \times 10^{3}$
Basic unit: meter, gram, liter, second		
deci-	d	$0.1 = 1 \times 10^{-1}$
centi-	c	$0.01 = 1 \times 10^{-2}$
milli-	m	$0.001 = 1 \times 10^{-3}$
micro-	m*	$0.000\ 001 = 1 \times 10^{-6}$
nano-	n	$0.000\ 000\ 001 = 1 \times 10^{-9}$
pico-	p	$0.000\ 000\ 000\ 001 = 1 \times 10^{-12}$

*The Greek letter mu (μ) is the symbol for micro.

CRITICAL THINKING Nanotechnology

Q: What is nanotechnology?

The transistor is considered one of the greatest inventions of the twentieth century. After the transistor, integrated circuit boards soon followed, which revolutionized the electronics industry. No longer did radio and television use large vacuum tubes, but rather microscale chips. The term microelectronics is named for the manufacturing process of these devices, which uses technology at the level of one micrometer (10^{-6} m).

Recently, a second electronics revolution began that uses nanoscale technology at the level of one nanometer (10^{-9} m). The term nanotechnology refers to devices and processes operating in the range of 1–100 nm. This level of precision is much less than the wavelength of visible light (400–700 nm). A human hair is approximately 100,000 times thicker than a nanometer!

There are many products on the market that use nanotechnology. For example, nanoparticles of titanium dioxide are added to sunscreens to make them colorless; nanoparticles of silver are added to sports clothing to act as an anti-bacterial agent; other nanoparticles are being developed as pharmaceutical agents to fight cancer and other diseases.

Recently, scientists discovered that nanoscale hollow cylinders are capable of allowing fluid flow thousands of times faster than predicted. Despite its small inner diameter, a so-called "nanotube" allows for relatively fast flow of water and gas. One reason that small molecules speed through this tiny enclosure is that the inner wall of the nanotube has a slippery surface. In the accompanying figure, we see methane molecules, CH_4, moving through a carbon nanotube.

◀ **Nanotube** A carbon nanotube has a slippery inner surface that allows the flow of water molecules and gases such as methane shown here.

A DNA molecule is an example of a nanoscale substance with the diameter of its double helix measuring about a nanometer. Scientists hypothesize that understanding how molecules behave as they move through a nanotube will further our understanding of biology and how molecules move through cell membranes.

Scientists and engineers are hopeful that nanotechnology will lead to faster microchips for computers; medical applications, such as tiny devices for clearing restricted arteries; and practical applications, such as removal of minerals from salt water to produce drinking water. The present methods for producing drinking water in locations that suffer from a lack of potable water are laborious and expensive. A nanotube membrane is capable of achieving the same result more efficiently, and at a lower cost. A similar benefit applies to kidney patients who require artificial dialysis. That is, a nanotube membrane could reduce dialysis time and purify blood by separating toxins from proteins and other biomolecules.

Another potential application for nanoscale membranes is the separation of gases. Currently, much of our electrical energy is provided by coal-burning power plants. Unfortunately, burning coal releases carbon dioxide and other gases, which pollute the environment and contribute to global warming. The use of carbon nanotube membranes could more effectively separate greenhouse gases and cut the cost of operating a coal-burning power plant.

As with most technologies, there are concerns that compromise the benefits. Nanoscale materials constitute a health hazard as small particles have been shown to cause eye irritation, skin sensitivity, and impaired breathing. A chief concern is that when you make a substance very small, you may change its chemical properties. For example, gold is ordinarily a quite inert substance that does not oxidize or corrode. However, a nanoparticle of gold has been shown to be reactive. This is due to its small physical size, and the fact that the surface area of many small particles is much greater than the surface area of a large particle.

To view a video on nanotechnology, log on to the public television website: http://www.kqed.org/quest/television/view/189.

A: A working definition of *nanotechnology* is the science of materials, tools, and devices having dimensions of 1–100 nm.

OBJECTIVES

- To write the unit equation for a basic metric unit and a prefix unit.
- To write the two unit factors derived from a metric unit equation.

3.2 METRIC CONVERSION FACTORS

Recall from Section 2.8 that a **unit equation** relates two quantities that are equal. As an example, let's find a unit equation for kilometers and meters. Since the metric prefix *kilo-* means 1000 basic units, 1 kilometer equals 1000 meters. We can write the **exact equivalent** as 1 kilometer = 1000 meters, or simply 1 km = 1000 m.

A centimeter is 1/100 of a meter, and we can write the relationship as 1 cm = 0.01 m. However, working with fractions is more confusing than working with multiples of basic units. Fortunately, we can avoid decimal fractions by simply restating the relationship. Since there are 100 cm in a meter, we can write the exact equivalent as 1 m = 100 cm. Example Exercise 3.2 further illustrates how to write unit equations for metric relationships.

Example Exercise 3.2 Metric Unit Equations

Complete the unit equation for each of the following exact metric equivalents:

(a) 1 Mm = ? m (b) 1 kg = ? g
(c) 1 L = ? dL (d) 1 s = ? ns

Solution

We can refer to Table 3.2 as necessary.

(a) The prefix *mega-* means 1,000,000 basic units; thus, 1 Mm = 1,000,000 m.
(b) The prefix *kilo-* means 1000 basic units; thus, 1 kg = 1000 g.
(c) The prefix *deci-* means 0.1 of a basic unit, thus. 1 L = 10 dL.
(d) The prefix *nano-* means 0.000 000 001 of a basic unit, thus; 1 s = 1,000,000,000 ns.

Alternatively, we can express the unit equation using exponential numbers:

(a) 1 Mm $= 1 \times 10^6$ m (b) 1 kg $= 1 \times 10^3$ g
(c) 1 L $= 1 \times 10^1$ dL (d) 1 s $= 1 \times 10^9$ ns

Practice Exercise

Complete the unit equation for each of the following exact metric equivalents:

(a) 1 Gm = ? m (b) 1 g = ? cg
(c) 1 L = ? μL (d) 1 s = ? ms

Answers:

(a) 1 Gm $= 1 \times 10^9$ m (b) 1 g $= 1 \times 10^2$ cg
(c) 1 L $= 1 \times 10^6$ μL (d) 1 s $= 1 \times 10^3$ ms

Concept Exercise

How many significant digits are in the following unit equation?

1 m = 1000 mm

Answer: See Appendix G.

Writing Metric Unit Factors

Once we are able to write unit equations, we can easily generate two unit factors. A **unit factor** is the ratio of the two equivalent quantities. That is, the quantity in the numerator is equal to the quantity in the denominator.

Since the numerator and the denominator are equivalent, we can generate a second unit factor by inverting the ratio and writing the **reciprocal**. As an example, consider the unit equation 1 m = 100 cm. The two associated unit factors are

$$\frac{1\ \text{m}}{100\ \text{cm}} \quad \text{and} \quad \frac{100\ \text{cm}}{1\ \text{m}}$$

Example Exercise 3.3 further illustrates how to write unit factors corresponding to a given metric relationship:

It is helpful to consistently write the larger metric unit first and the smaller metric unit second in a unit equation; for example, 1 m = 100 cm, 1 kg = 1000 g, and 1 L = 1000 mL.

Example Exercise 3.3 Metric Unit Factors

Write two unit factors for each of the following metric relationships:

(a) kilometers and meters (b) grams and decigrams

Solution

We start by writing the unit equation to generate the two unit factors.

(a) The prefix *kilo-* means 1000 basic units; thus, 1 km = 1000 m. The two unit factors are

$$\frac{1\text{ km}}{1000\text{ m}} \quad \text{and} \quad \frac{1000\text{ m}}{1\text{ km}}$$

(b) The prefix *deci-* means 0.1 basic unit; thus, 1 g = 10 dg. The two unit factors are

$$\frac{1\text{ g}}{10\text{ dg}} \quad \text{and} \quad \frac{10\text{ dg}}{1\text{ g}}$$

Practice Exercise

Write two unit factors for each of the following metric relationships:

(a) liters and milliliters (b) megaseconds and seconds

Answers:

(a) 1 L/1000 mL and 1000 mL/1 L (b) 1 Ms/1,000,000 s and 1,000,000 s/1 Ms

Concept Exercise

How many significant digits are in the following unit factor?

1 L/1000 mL

Answer: See Appendix G.

3.3 METRIC–METRIC CONVERSIONS

OBJECTIVE

- To express a given metric measurement with a different metric prefix.

In Section 2.9, we learned a powerful method for solving problems. It is called the **unit analysis method**. In Chapter 2, we solved problems requiring only one unit factor. In this chapter we solve problems requiring two or more unit factors. We will be successful if we consistently apply the following three steps.

Applying the Unit Analysis Method

Step 1: Write down the unit asked for in the answer.

Step 2: Write down the given value related to the answer.

Step 3: Apply unit factor(s) to convert the unit in the given value to the unit in the answer.

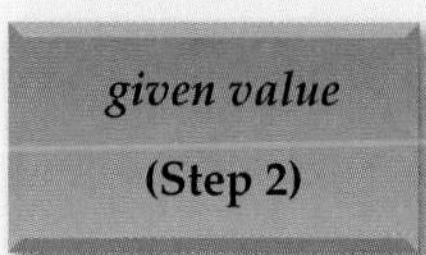

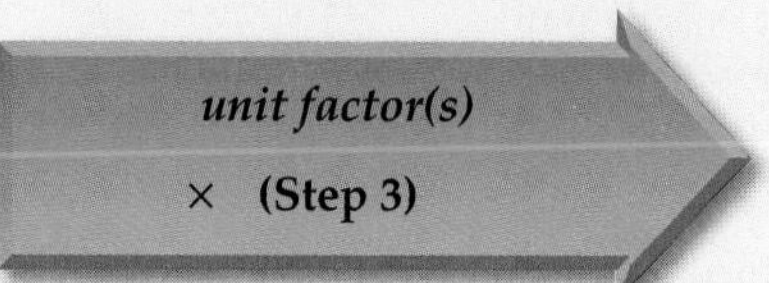

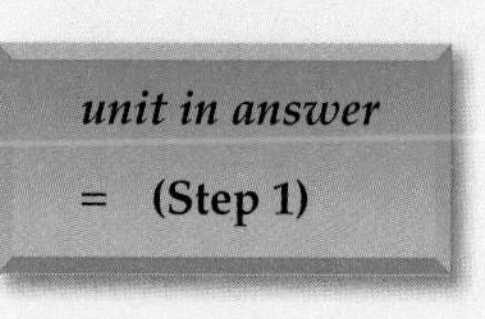

Let's begin with a simple problem and find the mass in grams of a 325 mg aspirin tablet. *First,* we write down the unit asked for in the answer (g). *Second,* we write down the related given value (325 mg). *Third,* we apply a unit factor using the following format:

$$\underset{\text{(Step 2)}}{325\text{ mg}} \times \underset{\text{(Step 3)}}{\frac{\text{unit}}{\text{factor}}} = \underset{\text{(Step 1)}}{\text{g}}$$

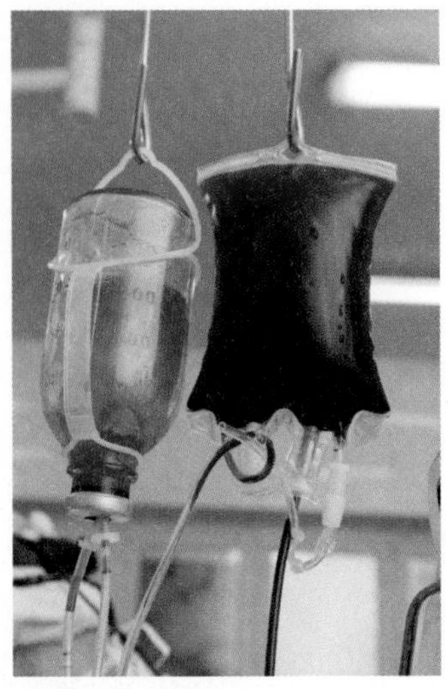

▲ **Blood Plasma** A blood plasma bag has a volume of 1 deciliter, that is, 0.1 liter.

The unit equation that relates Steps 1 and 2 is 1 g = 1000 mg. Therefore, the two associated unit factors are

$$\frac{1\ \text{g}}{1000\ \text{mg}} \quad \text{and} \quad \frac{1000\ \text{mg}}{1\ \text{g}}$$

To cancel the unit in the given value (325 mg), we should select the unit factor 1 g/1000 mg. After substituting, we have

$$325\ \cancel{\text{mg}} \times \frac{1\ \text{g}}{1000\ \cancel{\text{mg}}} = 0.325\ \text{g}$$

The given value has three significant digits, and the unit factor is an exact equivalent. Thus, the answer has three significant digits.

Example Exercises 3.4 and 3.5 further illustrate the unit analysis method of problem solving for metric–English conversions.

Example Exercise 3.4 Metric–Metric Conversion

A hospital has 125 deciliter bags of blood plasma. What is the volume of plasma expressed in milliliters?

Strategy Plan

Step 1: What unit is asked for in the answer? — **Step 1:** mL

Step 2: What given value is related to the answer? — **Step 2:** 125 dL

Step 3: What unit factor(s) should we apply? Given that 1 L = 10 dL, and 1 L = 1000 mL, the two pairs of unit factors are — **Step 3:**

$$\frac{1\ \text{L}}{10\ \text{dL}} \quad \text{or} \quad \frac{10\ \text{dL}}{1\ \text{L}}$$

$$\frac{1\ \text{L}}{1000\ \text{mL}} \quad \text{or} \quad \frac{1000\ \text{mL}}{1\ \text{L}}$$

Unit Analysis Map

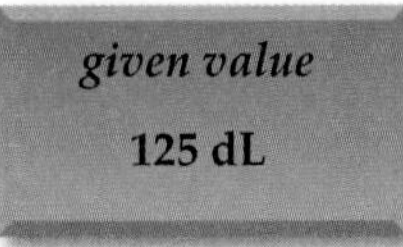

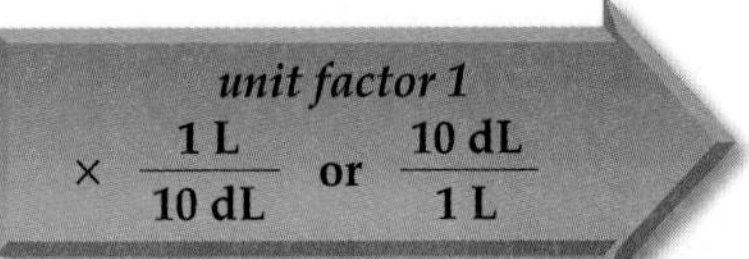

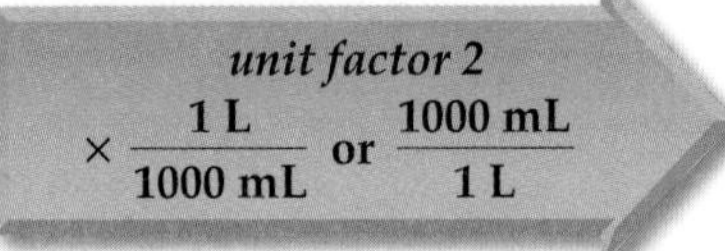

unit in answer
= ? mL

Solution

We apply the unit factor 1 L/10 dL to cancel deciliters ($\cancel{\text{dL}}$), and 1000 mL/1 L to cancel liters ($\cancel{\text{L}}$).

$$125\ \cancel{\text{dL}} \times \frac{1\ \cancel{\text{L}}}{10\ \cancel{\text{dL}}} \times \frac{1000\ \text{mL}}{1\ \cancel{\text{L}}} = 1.25 \times 10^4\ \text{mL}$$

The given value, 125 dL, limits the answer to three significant digits. Since each unit factor is derived from an exact equivalent, neither affects the number of significant digits in the answer.

Practice Exercise

A dermatology patient is treated with ultraviolet light having a wavelength of 375 nm. What is the wavelength expressed in centimeters?

Answer: 0.0000375 cm (3.75×10^{-5} cm)

Concept Exercise
Express the volume of a cube 1 cm on a side in milliliters.

Answer: See Appendix G.

Example Exercise 3.5 Metric–Metric Conversion

The mass of Earth is 5.98×10^{24} kg. What is the mass expressed in megagrams?

Strategy Plan

Step 1: What unit is asked for in the answer?

Step 1: Mg

Step 2: What given value is related to the answer?

Step 2: 5.98×10^{24} kg

Step 3: What unit factor(s) should we apply? Given that 1 kg = 1000 g, and 1 Mg = 1,000,000 g, the two pairs of unit factors are

Step 3: $\frac{1 \text{ kg}}{1000 \text{ g}}$ or $\frac{1000 \text{ g}}{1 \text{ kg}}$

$\frac{1 \text{ Mg}}{1{,}000{,}000 \text{ g}}$ or $\frac{1{,}000{,}000 \text{ g}}{1 \text{ Mg}}$

Earth

◀ **Earth** The mass of Earth is about 6,000,000,000,000,000,000,000,000 kilograms.

Unit Analysis Map

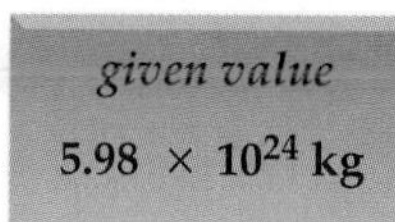

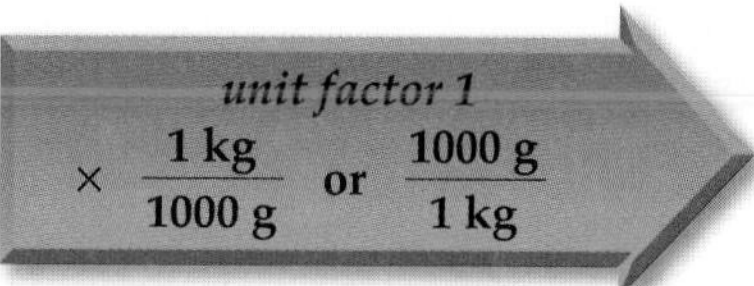

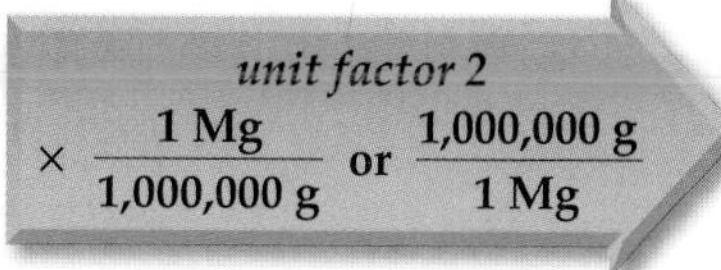

unit in answer

= ? Mg

Solution
We apply the unit factor 1000 g/1 kg to cancel kilograms (kg), and 1 Mg/1,000,000 g to cancel grams (g).

$$5.98 \times 10^{24} \text{ kg} \times \frac{1000 \text{ g}}{1 \text{ kg}} \times \frac{1 \text{ Mg}}{1{,}000{,}000 \text{ g}} = 5.98 \times 10^{21} \text{ Mg}$$

The given value limits the answer to three significant digits. Since each unit factor is derived from an exact equivalent, neither affects the number of significant digits in the answer.

Practice Exercise
Light travels through the universe at a velocity of 3.00×10^{10} cm/s. How many gigameters does light travel in one second?

Answer: 0.300 Gm (3.00×10^{-1} Gm)

Concept Exercise
How many significant digits are in the following unit factor?

1 g/1000 mg

Answer: See Appendix G.

OBJECTIVES

- To express a given measurement in metric units or English units.
- To gain a practical awareness of metric sizes.

3.4 METRIC–ENGLISH CONVERSIONS

Despite the superiority of the metric system, its adoption has been resisted from its inception. When the French asked the Americans and the British for their cooperation in creating a worldwide system of measurement, both countries declined. The initial objection, interestingly, was political rather than scientific. The meter was defined by a longitude passing through Dunkirk, France, and the original metric standards were kept in a special vault outside of Paris.

In 1975, President Gerald Ford signed an official metric conversion act. Thus, the United States became the last major nation in the world to formally adopt the metric system. Although the metric system is now taught in public schools, the United States is making slow progress in achieving full compliance.

In science, we regularly record measurements in metric units. In everyday activities, you may be more familiar with inches, pounds, and quarts. Therefore, let's compare the two systems to gain a practical awareness of metric sizes. Table 3.3 contrasts some common English and metric units.

Which race do you think is run faster: 100 meters or 100 yards? Obviously, it is the shorter race distance that is run faster. Given that 1 yd = 0.914 m, we can calculate the distance in yards for a 100.0-meter race.

First, we write down the unit asked for in the answer (yd). *Second*, we write down the related given value (100.0 m). *Third*, we apply a unit factor using the following format:

$$\underset{\text{(Step 2)}}{100.0\text{ m}} \times \underset{\text{(Step 3)}}{\frac{\text{unit}}{\text{factor}}} = \underset{\text{(Step 1)}}{\text{yd}}$$

The unit equation that relates Steps 1 and 2 is 1 yd = 0.914 m. Therefore, the two unit factors are

$$\frac{1\text{ yd}}{0.914\text{ m}} \quad \text{and} \quad \frac{0.914\text{ m}}{1\text{ yd}}$$

To cancel the unit in the given value (100.0 m), we select the unit factor (1 yd/0.914 m). After substituting, we have

$$100.0\text{ }\cancel{\text{m}} \times \frac{1\text{ yd}}{0.914\text{ }\cancel{\text{m}}} = 109\text{ yd}$$

Although the given value (100.0 m) has four significant digits, the unit factor has only three because 1 yd = 0.914 m is not an exact equivalent. Thus, the answer is rounded to three significant digits. Since 100 meters is a longer distance than 100 yards, the 100-yard race is run faster. In fact, the world record for 100 meters is more than 9 seconds, while the world record for 100 yards is less than 9 seconds!

TABLE 3.3 ENGLISH–METRIC EQUIVALENTS

PHYSICAL QUANTITY	ENGLISH UNIT	METRIC EQUIVALENT
length	1 inch (in.)	1 in. = 2.54 cm
mass	1 pound (lb)	1 lb = 454 g
volume	1 quart (qt)	1 qt = 946 mL
time	1 second (sec)	1 sec = 1.00 s

When applying these metric equivalents, assume three significant digits. Since the metric and English systems have different reference standards, these are not exact equivalents; however, the U.S. Bureau of Weights and Measures has redefined 1 in. as exactly equal to 2.54 cm.

Example Exercises 3.6 and 3.7 further illustrate the unit analysis method of problem solving for metric–English conversions.

Example Exercise 3.6 Metric–English Conversion

A half-gallon carton contains 64.0 fl oz of milk. How many milliliters of milk are in a carton (given that 1 qt = 32 fl oz)?

Strategy Plan

Step 1: What unit is asked for in the answer? — **Step 1:** **mL**

Step 2: What given value is related to the answer? — **Step 2:** **64.0 fl oz**

Step 3: What unit factor(s) should we apply? Given that 1 qt = 32 fl oz and that in Table 3.3 we find that 1 qt = 946 mL. The two pairs of unit factors are — **Step 3:**

$$\frac{1\text{ qt}}{32\text{ fl oz}} \text{ or } \frac{32\text{ fl oz}}{1\text{ qt}}$$

$$\frac{1\text{ qt}}{946\text{ mL}} \text{ or } \frac{946\text{ mL}}{1\text{ qt}}$$

Unit Analysis Map

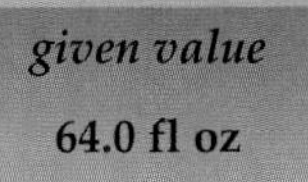

unit factor 1

$$\times \frac{1\text{ qt}}{32\text{ fl oz}} \text{ or } \frac{32\text{ fl oz}}{1\text{ qt}}$$

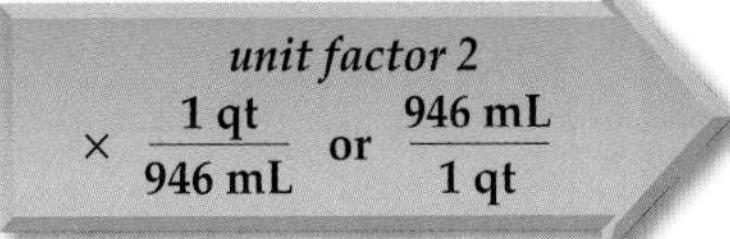

unit in answer

= ? mL

Solution

We apply the unit factor 1 qt/32 fl oz to cancel fluid ounces ($\cancel{\text{fl oz}}$), and 946 mL/1 qt to cancel quarts ($\cancel{\text{qt}}$).

$$64.0\ \cancel{\text{fl oz}} \times \frac{1\text{ qt}}{32\ \cancel{\text{fl oz}}} \times \frac{946\text{ mL}}{1\ \cancel{\text{qt}}} = 1{,}890\text{ mL}$$

The given value, and unit factor 2, each limits the answer to three significant digits. Since unit factor 1 is derived from an exact equivalent, it does not affect the number of significant digits in the answer.

Practice Exercise

A plastic bottle contains 5.00 gallons of distilled water. How many liters of distilled water are in the bottle (given that 1 gal = 4 qt)?

Answer: 18.9 L

Concept Exercise

How many significant digits are in the following unit factor?

1 qt/946 mL

Answer: See Appendix G.

▲ **Milk Carton** A half-gallon carton of milk has a volume of 64.0 fluid ounces.

Example Exercise 3.7 Metric–English Conversion

If a tennis ball weighs 2.0 oz, what is the mass of the tennis ball in grams?

Strategy Plan

Step 1: What unit is asked for in the answer? **Step 1:** g

Step 2: What given value is related to the answer? **Step 2:** 2.0 oz

Step 3: What unit factor(s) should we apply? Given that 1 lb = 16 oz and that in Table 3.3 we find that 1 lb = 454 g. The two pairs of unit factors are **Step 3:**

$$\frac{1\text{ lb}}{16\text{ oz}} \text{ or } \frac{16\text{ oz}}{1\text{ lb}}$$

$$\frac{1\text{ lb}}{454\text{ g}} \text{ or } \frac{454\text{ g}}{1\text{ lb}}$$

Unit Analysis Map

given value	*unit factor 1*	*unit factor 2*	*unit in answer*
2.0 oz	$\times \frac{1\text{ lb}}{16\text{ oz}}$ or $\frac{16\text{ oz}}{1\text{ lb}}$	$\times \frac{1\text{ lb}}{454\text{ g}}$ or $\frac{454\text{ g}}{1\text{ lb}}$	= ? g

▲ **Tennis Balls** A tennis ball weighs about 2.0 ounces, and has a diameter of 2.5 inches.

Solution

We apply the unit factor 1 lb/16 oz to cancel ounces (oz), and 454 g/1 lb to cancel pounds (lb).

$$2.0\ \cancel{\text{oz}} \times \frac{1\ \cancel{\text{lb}}}{16\ \cancel{\text{oz}}} \times \frac{454\text{ g}}{1\ \cancel{\text{lb}}} = 57\text{ g}$$

The given value, 2.0 oz, limits the answer to two significant digits. Unit factor 1 has no effect as it is derived from an exact equivalent, and unit factor 2 has three significant digits.

Practice Exercise

If a tennis ball has a diameter of 2.5 inches, what is the diameter in millimeters?

Answer: 64 mm

Concept Exercise

How many significant digits are in the following unit factor?

1 kg/2.20 lb

Answer: See Appendix G.

Measurements with Compound Units

Now, let's try a problem in which the answer involves a ratio of two units. For example, if a highway speed limit is posted at 55 miles per hour (mi/h), do you know the metric speed limit in kilometers per hour (given that 1 mi = 1.61 km)?

First, we write down the ratio of units asked for in the answer (km/h). *Second,* we write down the related given value (55 mi/h). *Third,* we apply a unit factor using the standard format.

$$\underset{\text{(Step 2)}}{\frac{55\text{ mi}}{1\text{ h}}} \times \underset{\text{(Step 3)}}{\frac{\text{unit}}{\text{factor}}} = \underset{\text{(Step 1)}}{\frac{\text{km}}{\text{h}}}$$

We can use the given unit equation 1 mi = 1.61 km to write the following unit factors: 1 mi/1.61 km and 1.61 km/1 mi. To cancel miles (~~mi~~) in the given value, we apply the second unit factor. Thus, we have

$$\frac{55\ \cancel{\text{mi}}}{1\ \text{h}} \times \frac{1.61\ \text{km}}{1\ \cancel{\text{mi}}} = 89\ \text{km/h}$$

Where the metric speed limit is posted, the value is usually 90 km/h. Example Exercise 3.8 illustrates problems requiring compound units in the answer.

Example Exercise 3.8 Conversion of a Unit Ratio

If a Mazda Miata is traveling at 95 km/h, what is the speed in meters per second (given that 1 km = 1000 m, and 1 h = 3600 s)?

Strategy Plan

Step 1: What unit is asked for in the answer?

Step 1: $\frac{\text{m}}{\text{s}}$

Step 2: What given value is related to the answer?

Step 2: $\frac{95\ \text{km}}{1\ \text{h}}$

Step 3: What unit factor(s) should we apply? Given that 1 km = 1000 m, and 1 h = 3600 s, the two pairs of unit factors are

Step 3: $\frac{1\ \text{km}}{1000\ \text{m}}$ or $\frac{1000\ \text{m}}{1\ \text{km}}$

$\frac{1\ \text{h}}{3600\ \text{s}}$ or $\frac{3600\ \text{s}}{1\ \text{h}}$

Unit Analysis Map

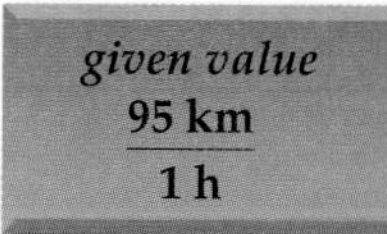

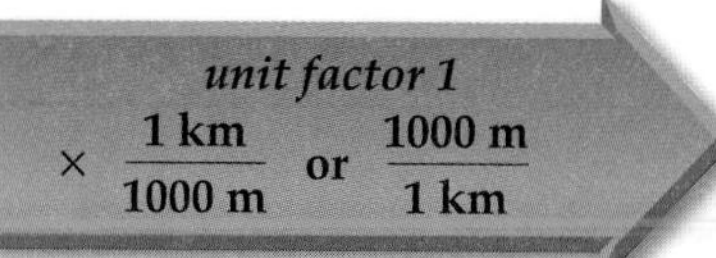

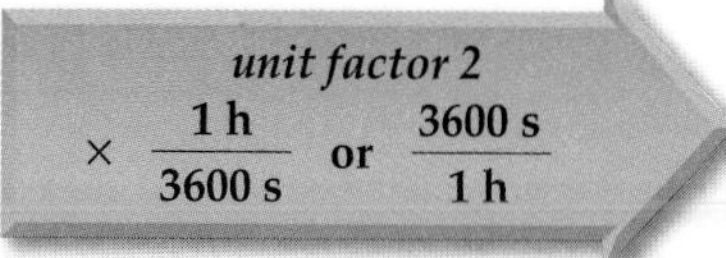

unit in answer $= ?\ \frac{\text{m}}{\text{s}}$

Solution

We apply the unit factor 1000 m/1 km to cancel kilometers (~~km~~), and 1 h/3600 s to cancel hours (~~h~~).

$$95\ \frac{\cancel{\text{km}}}{\cancel{\text{h}}} \times \frac{1000\ \text{m}}{1\ \cancel{\text{km}}} \times \frac{1\ \cancel{\text{h}}}{3600\ \text{s}} = 26\ \text{m/s}$$

The given value has two significant digits, so the answer is limited to two digits. Since each unit factor is derived from an exact equivalent, neither affects the number of significant digits in the answer.

▲ **Mazda Miata** A Mazda Miata.

Practice Exercise

If a runner completes a 10K race in 32.50 minutes (min), what is the 10.0 km pace in miles per hour (given that 1 mi = 1.61 km)?

Answer: 11.5 mi/h

Concept Exercise

Which speed is faster: 65 mi/h or 65 km/h?

Answer: See Appendix G.

CHEMISTRY CONNECTION The Olympics

Q: Which Olympic race is nearly equal in length to a quarter mile, that is, 440 yards?

The Olympics began in 776 B.C. with a single event, a 200-yard dash called a *stadion*. In time, a 2-stadia dash and a 24-stadia event were added, and wrestling appeared in 708 B.C. Eventually, participation in the games declined, and the ancient Olympics were discontinued in A.D. 394.

In the summer of 1896, the modern Olympic games were initiated and held in Greece. The games attracted about 500 athletes from 13 nations. Since then, the games have been held every four years in various cities around the world.

The games have steadily increased in number of participants, as well as in number of events. In 1896, 311 men from 13 nations competed, and the United States won gold medals in 9 of the 12 events. In 2004, over 10,000 men and women athletes from 197 nations competed.

U.S. track and field competitions have been conducted using the English system of measurement. That is, American runners compete at distances of 100 yards, 440 yards, and 1 mile. In Olympic competitions, events are conducted using the metric system owing to its international acceptance. Olympic runners compete at comparable distances such as 100, 400, and 1500 meters. The marathon, however, is an exception. It corresponds to the distance run by the Greek messenger who carried news of the Athenian victory on the plains of Marathon in 490 B.C. That legendary distance is 26 miles, 385 yards.

▲ **Olympic Speed Skater** Olympic speed skaters compete in events measured in metric distances including 500, 1000, and 5000 meters.

In 1924, the Olympic winter games were introduced at Chamonix, France, and have been held four times in the United States, twice at Lake Placid, once at Squaw Valley, and in 2002 at Salt Lake City. The winter games feature skating, skiing, bobsledding, luge, tobogganing, and ice hockey. In Nordic cross-country skiing, the races are 10, 20, 30, and 50 kilometers; in ski jumping, the ramps are 70 and 90 meters. Thus, Olympic competitions—winter and summer—employ the metric system of measurement.

A: The 400-meter Olympic race is nearly identical to 440 yards, and the world records for the two races are within tenths of a second.

OBJECTIVES

- To state the relationship of length, width, and thickness to the volume of a rectangular solid.
- To express a given volume in units of milliliters, cubic centimeters, or cubic inches.

3.5 VOLUME BY CALCULATION

Have you ever wondered how a building contractor arrives at the costs for a new home? For example, how does one estimate the cost of pouring a foundation? The cost is based on the volume of concrete that is poured. If the foundation is rectangular, the volume of concrete is found by multiplying length (l) times width (w) times thickness (t). We can express this relationship by the following equation:

$$l \times w \times t = \text{volume}$$

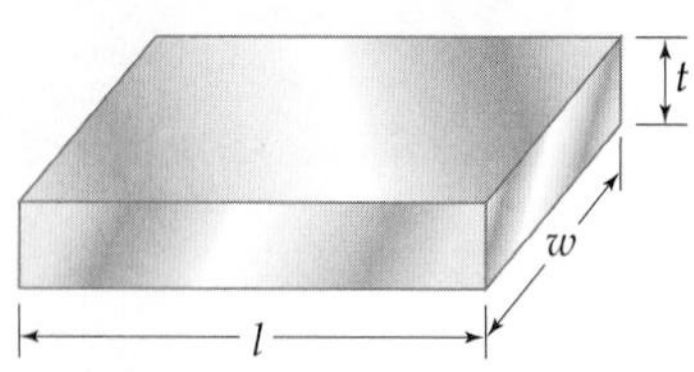

▲ **Volume by Calculation** The volume of a rectangular solid is equal to its length (l) times width (w) times thickness (t) or height.

Before calculating a volume, we must express each of the measurements in the same units. If we express length, width, and thickness in centimeters, the volume of the solid has the units of **cubic centimeters** (symbol cm^3). For example, if a rectangular solid measures 3 cm by 2 cm by 1 cm, we find the volume is 6 cm^3. Example Exercise 3.9 further illustrates how to determine the volume of a rectangular solid by calculation.

Example Exercise 3.9 Volume Calculation for a Rectangular Solid

If a stainless steel rectangular solid measures 5.55 cm long, 3.75 cm wide, and 2.25 cm thick, what is the volume in cubic centimeters?

Strategy Plan

Step 1: What unit is asked for in the answer? **Step 1:** cm^3

Step 2: What given value is related to the answer? **Step 2:** 5.55 cm, 3.75 cm, and 2.25 cm

Step 3: What unit factor(s) should we apply? No unit factor is required. **Step 3:** none

Solution

We can calculate the volume of the rectangular solid by multiplying length times width times thickness: $l \times w \times t$.

$$5.55 \text{ cm} \times 3.75 \text{ cm} \times 2.25 \text{ cm} = 46.8 \text{ cm}^3$$

The answer is rounded off to three significant digits because each given value has three significant digits.

Practice Exercise

If a rectangular brass solid measures 52.0 mm by 25.0 mm by 15.0 mm, what is the volume in cubic millimeters?

Answer: 19,500 mm^3 (1.95×10^4 mm^3)

Concept Exercise

Express the volume of a cube 10 cm on a side in liters.

Answer: See Appendix G.

Let's now consider a rectangular solid with a length of 4 cm, a width of 3 cm, and a volume of 24 cm^3. Given the length, width, and volume, we can calculate the thickness of the rectangular solid by dividing the volume by its length and its width; that is, we divide 24 cm^3 by 4 cm and 3 cm. In this example, the thickness of the solid is 2 cm. Example Exercise 3.10 illustrates finding the thickness of rectangular solid.

Example Exercise 3.10 Thickness Calculation for a Rectangular Solid

A sheet of aluminum foil measures 25.0 mm by 10.0 mm, and the volume is 3.75 mm^3. What is the thickness of the foil in millimeters?

Strategy Plan

Step 1: What unit is asked for in the answer? **Step 1:** mm

Step 2: What given value is related to the answer? **Step 2:** 3.75 mm^3, 25.0 mm, and 10.0 mm

Step 3: What unit factor(s) should we apply? No unit factor is required. **Step 3:** none

Solution

We can calculate the thickness of the foil by dividing the volume by length and width. Since the unit of volume is mm^3, we obtain the thickness in mm by unit cancellation.

$$3.75 \text{ mm}^3 \times \frac{1}{25.0 \text{ mm}} \times \frac{1}{10.0 \text{ mm}} = \frac{3.75 \text{ mm} \times \cancel{\text{mm}} \times \cancel{\text{mm}}}{25.0 \cancel{\text{mm}} \times 10.0 \cancel{\text{mm}}} = 0.0150 \text{ mm}$$

The answer is rounded off to three significant digits because each given value has three significant digits.

▲ **Aluminum foil** A thin sheet of aluminum foil.

Practice Exercise
A sheet of aluminum foil measures 35.0 cm by 25.0 cm, and the volume is 1.36 cm^3. What is the thickness of the foil in centimeters?

Answer: 0.00155 cm (1.55×10^{-3} cm)

Concept Exercise
Which of the following is the greatest thickness?

1 mm, 0.1 cm, or 0.001 m

Answer: See Appendix G.

Volumes of Solids, Liquids, and Gases

In the metric system, the basic unit of liquid volume is the liter. A liter is equivalent to the volume occupied by a cube exactly 10 cm on a side.

A cube is a regular solid with its length, width, and thickness being equal. We calculate the volume of the cube by multiplying length by width by thickness.

$$(10 \text{ cm})\,(10 \text{ cm})\,(10 \text{ cm}) = 1000 \text{ cm}^3$$

The calculated volume of 1 L is 1000 cm^3. Moreover, we recall the exact equivalent: 1 L = 1000 mL. We can combine the two equivalents as follows:

$$1000 \text{ cm}^3 = 1 \text{ L} = 1000 \text{ mL}$$

Simplifying, we have

$$1 \text{ cm}^3 = 1 \text{ mL}$$

We see that 1 cm^3 is exactly equivalent to 1 mL. Example Exercise 3.11 further illustrates the calculation of volume.

▲ **1-Liter Bottle** A 1-liter bottle contains 1000 mL of liquid.

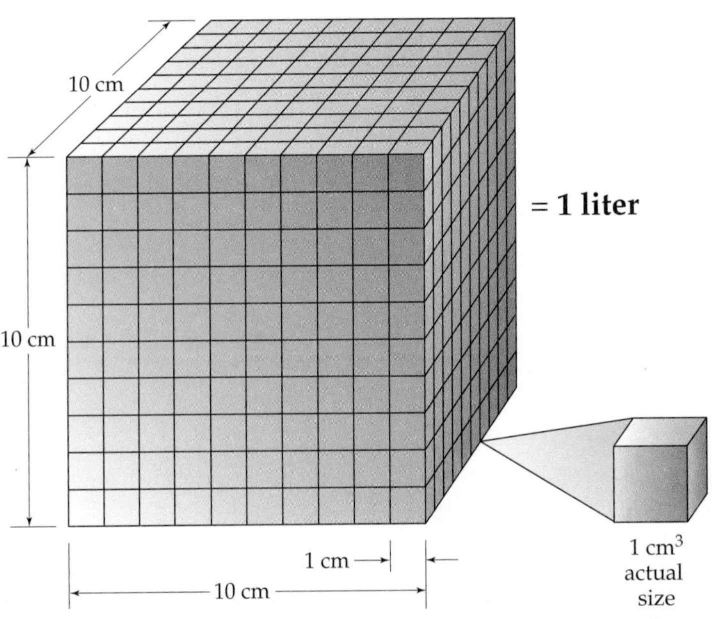

▲ **One Liter** A liter equals the volume of a cube 10 cm on a side, that is, 1000 cm^3.

Example Exercise 3.11 Metric–English Volume Conversion

Given that an automobile engine has a volume displacement of 498 cm^3 in each cylinder, express the volume in cubic inches.

Strategy Plan

Step 1: What unit is asked for in the answer? — **Step 1:** $in.^3$

Step 2: What given value is related to the answer? — **Step 2:** 498 cm^3

Step 3: What unit factor(s) should we apply? — **Step 3:** $\frac{1\text{ in.}}{2.54\text{ cm}}$ or $\frac{2.54\text{ cm}}{1\text{ in.}}$

Unit Analysis Map

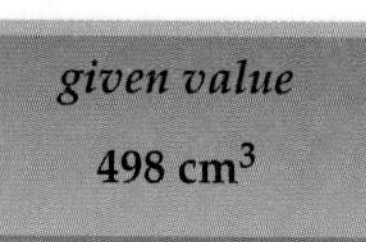

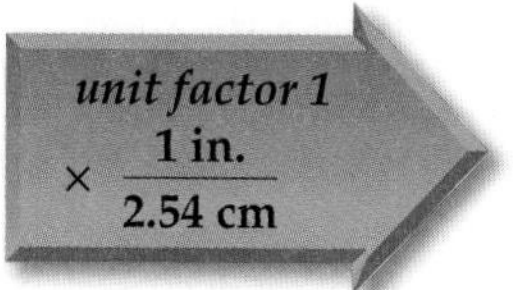

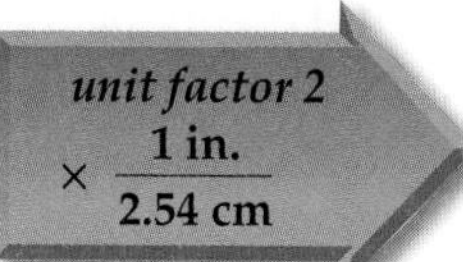

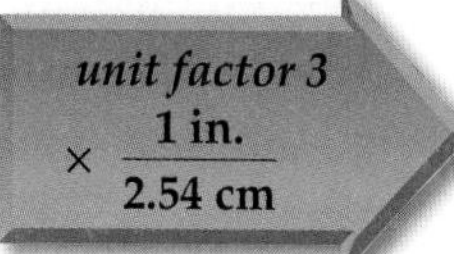

unit in answer
= ? $in.^3$

Solution

We can calculate the volume in cubic inches by converting the volume given in cubic centimeters. Notice that cm units do not cancel cm^3. To obtain cubic inches, we must apply the unit factor, 1 in./2.54 cm, three times. Thus,

$$498\text{ cm}^3 \times \frac{1\text{ in.}}{2.54\text{ cm}} \times \frac{1\text{ in.}}{2.54\text{ cm}} \times \frac{1\text{ in.}}{2.54\text{ cm}} = 30.4\text{ in.}^3$$

The given value, and unit factors, each limits the answer to three significant digits.

Practice Exercise

Given that an SUV has a 244 $in.^3$ engine, express the engine volume in liters.

Answer: 4.00 L

Concept Exercise

Which of the following is the greater volume?

500 mL or 500 cm^3

Answer: See Appendix G.

Jonathan Player/The New York Times

▲ **Mini Cooper** The Mini Cooper is built in England and marketed in the United States by BMW.

Note A cubic centimeter is sometimes abbreviated "cc," as in a 1 cc medical injection. Since it is not obvious that cc stands for cubic centimeters, use of this abbreviation is discouraged in science, even though it is common in medicine.

3.6 VOLUME BY DISPLACEMENT

OBJECTIVE

▸ To describe the technique of determining a volume by displacement.

The volumes of liquids and solids can be determined by several methods. The volumes of liquids can be determined directly with calibrated glassware, such as graduated cylinders, pipets, and burets. The volumes of regular solids (cubes, cylinders, spheres) can be calculated from measurements.

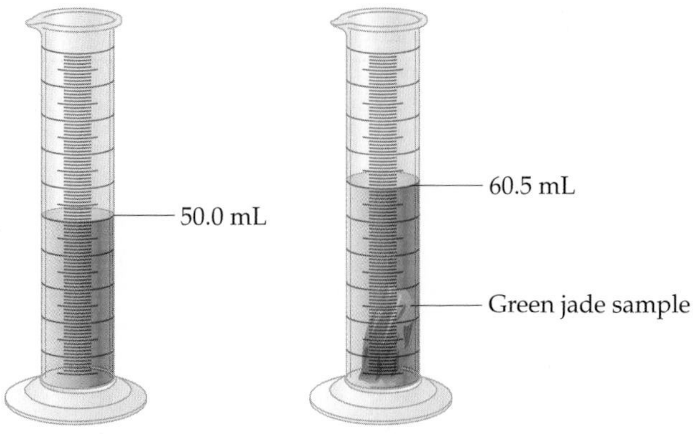

▶ **Figure 3.2 Volume of a Solid by Displacement** The difference between the final water level minus the initial water level in the graduated cylinder (10.5 mL) is equal to the volume of the solid piece of jade.

The volume of an irregular solid cannot be determined directly by multiplying its measurements. However, the volume can be determined *indirectly* by measuring the amount of water the solid displaces. This technique is called determination of **volume by displacement**.

Suppose we wish to determine the volume of a piece of jade. Since it is an irregular solid object, we cannot measure its volume directly. We have to use the technique of determining volume by displacement. We first fill a graduated cylinder halfway with water (Figure 3.2) and record the water level. Next, we carefully slip the piece of jade into the graduated cylinder and record the new water level. The difference between the final minus the initial water levels represents the volume of the piece of jade.

We can also use the volume by displacement method to determine the volume of a gas. When a gas is produced from a chemical reaction, its volume is equal to the amount of water it displaces. This technique provides good results for gases that do not dissolve, or only slightly dissolve, in water. For instance, heating a solid sample may release oxygen gas, which displaces water from a closed container. Since oxygen dissolves only slightly in water, the volume of water displaced is equal to the volume of gas released. Determining the volume of a gas by displacement is referred to as *collecting a gas over water* and is illustrated in Figure 3.3.

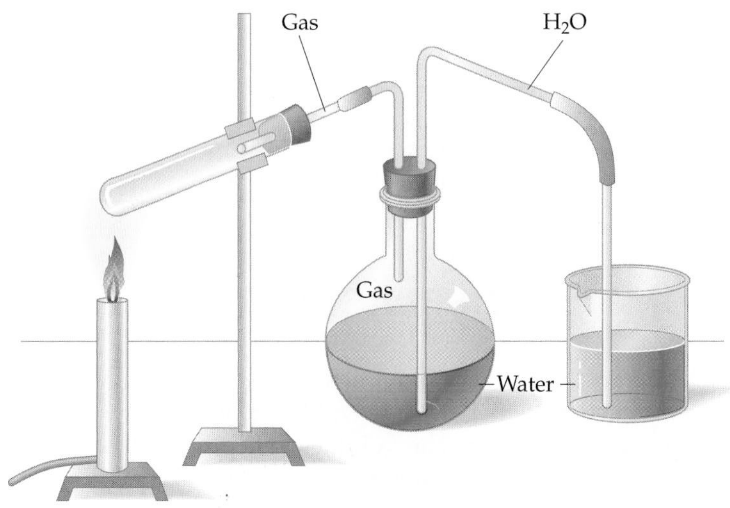

▶ **Figure 3.3 Volume of a Gas by Displacement** A flask is filled with water, and a sample is placed in the test tube. After heating, the sample decomposes and produces oxygen gas, which in turn displaces water from the flask. The volume of water displaced into the beaker equals the volume of oxygen gas produced by the sample.

Example Exercise 3.12 Volume by Displacement

A quartz stone weighing 30.475 g is dropped into a graduated cylinder. If the water level increases from 25.0 mL to 36.5 mL, what is the volume of the quartz stone?

Strategy Plan

Step 1: What unit is asked for in the answer? **Step 1:** mL

Step 2: What given value is related to the answer? **Step 2:** 25.0 mL and 36.5 mL

Step 3: What unit factor(s) should we apply? No unit factor is required. **Step 3:** none

Solution

We can calculate the displaced volume in milliliters by subtracting the initial volume from the final volume.

$$36.5\ \text{mL} - 25.0\ \text{mL} = 11.5\ \text{mL}$$

Practice Exercise

Hydrogen peroxide decomposes to give oxygen gas, which displaces a volume of water into a beaker. If the water level in the beaker increases from 50.0 mL to 105.5 mL, what is the volume of oxygen gas?

Answer: 55.5 mL

Concept Exercise

Which of the following has the greater volume?

$$1\ \text{mL or}\ 1\ \text{cm}^3$$

Answer: See Appendix G.

3.7 THE DENSITY CONCEPT

OBJECTIVES

- To explain the concept of density.
- To state the value for the density of water: 1.00 g/mL.
- To perform calculations that relate density to mass and volume.

If we hold a piece of wood in one hand and a piece of metal in the other (each having the same mass), the piece of metal may feel heavier. This contradiction is explained by the term **density** (symbol ***d***), which expresses the concentration of mass. The more concentrated the mass is in a sample, the greater its density. We sometimes say an object is heavy when we are trying to convey density. A 1 kg block of lead and a 1 kg block of ice have the same mass, but the lead is much more dense.

We define density as the amount of mass per unit volume. The term *per* indicates a ratio, as in "miles per hour." Thus, mass divided by volume equals density. We can write this relationship mathematically as

$$\frac{\text{mass}}{\text{volume}} = \text{density}$$

Density is expressed in different units. For solids and liquids, it is usually expressed in grams per cubic centimeter (g/cm^3) or grams per milliliter (g/mL). For gases, density is usually expressed in grams per liter (g/L). Figure 3.4 illustrates the concept of density.

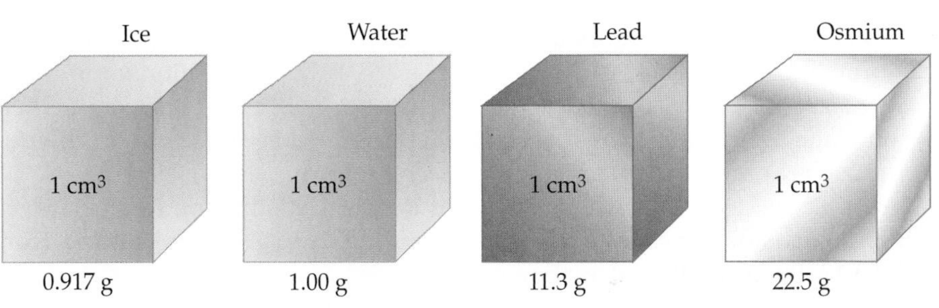

◀ **Figure 3.4 An Illustration of Density** Each cube represents a volume of 1 cm^3. Notice the mass of each cube increases as its density becomes greater. Osmium has the greatest density of any metal, 22.5 g/cm^3.

TABLE 3.4 DENSITY OF SELECTED SOLIDS, LIQUIDS, AND GASES

EXAMPLE	DENSITY (*d*)
Solids	
ice	0.917 g/cm^3 or g/mL
rubber	1.19
magnesium	1.74
aluminum	2.70
iron	7.87
lead	11.3
gold	19.3
Liquids	
ethyl ether	0.714 g/cm^3 or g/mL
ethyl alcohol	0.789
water	1.00
chloroform	1.48
mercury	13.6
Gases*	
hydrogen	0.090 g/L
helium	0.179
ammonia	0.760
air	1.29
oxygen	1.43

*The density value for each gas is given at 0 °C and a pressure equal to normal atmospheric pressure.

Recall that the original metric standards for mass and volume were based on water. A mass of 1 kg was set equal to a 1 L volume of water. Therefore, the density of water is 1 kg/L. A density of 1 kg/L is exactly equivalent to 1 g/mL.

Although the density of solids and liquids varies slightly with temperature, we will ignore this small effect. For our purposes, we will consider the density of water to be 1.00 g/mL. The density of gases, however, is an exception. The density of a gas varies greatly with changes in temperature and pressure. Table 3.4 lists the density for selected solids, liquids, and gases.

We can drop a solid object into water and estimate its density. An object that sinks in water has a density greater than 1.00 g/mL. An object that floats on water has a density less than 1.00 g/mL.

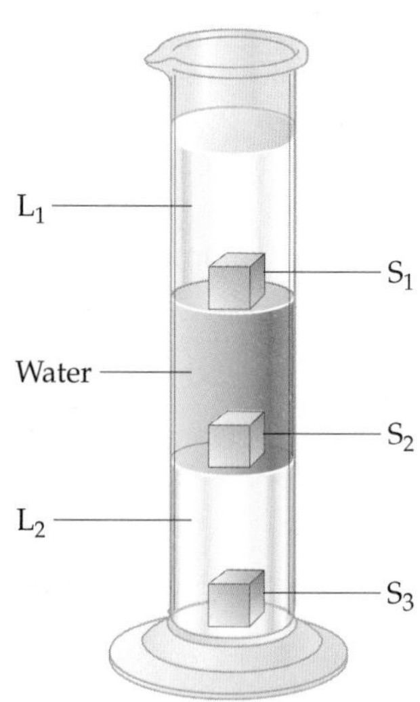

▲ **Figure 3.5 The Concept of Density** In a tall glass cylinder, liquid L_1 is shown floating on water, which is floating on liquid l_2. Solid S_1 is floating on water, solid S_2 is floating on liquid L_2, and solid S_3 has sunk to the bottom of the glass cylinder.

Consider the following experiment. Water, chloroform, and ethyl ether are poured into a tall glass cylinder. The liquids form three layers as shown in Figure 3.5. Identify liquids L_1 and L_2 by referring to the density of the liquids listed in Table 3.4. Next, ice, rubber, and aluminum cubes are dropped into the cylinder. Identify solids S_1, S_2, and S_3 by referring to the density of the solids listed in Table 3.4.

We know that liquid L_1 must be less dense than water ($d = 1.00$ g/mL) because it floats; liquid L_2 must be more dense than water because it sinks. Thus, liquid L_1 is ethyl ether ($d = 0.714$ g/mL), which is less dense than water, and liquid L_2 is chloroform ($d = 1.48$ g/mL), which is more dense than water.

We see in Figure 3.5 that S_1 floats on water, S_2 floats on chloroform, and S_3 sinks in chloroform. Table 3.4 indicates the densities of ice, rubber, and aluminum are 0.917, 1.19, and 2.70 g/cm^3, respectively. Thus, S_1 must be ice because it sinks in ether and floats on water, S_2 must be rubber because it sinks in water and floats on chloroform, and S_3 must be aluminum because it sinks in chloroform.

Now that we understand the concept of density, let's try some calculations. We can find the density of a sample by simply dividing its mass by its volume. Example Exercise 3.13 illustrates the calculation of density.

Example Exercise 3.13 Density Calculation

If a platinum nugget has a mass of 214.50 g and a volume 10.0 cm^3, what is the density of the metal?

Strategy Plan

Step 1: What unit is asked for in the answer? **Step 1:** g/cm^3

Step 2: What given value is related to the answer? **Step 2:** 214.50 g and 10.0 cm^3

Step 3: What unit factor(s) should we apply? No unit factor is required. **Step 3:** none

Solution

We can calculate the density of the platinum nugget by comparing the mass of metal, 214.50 g, to its volume, 10.0 cm^3.

$$\frac{214.50\ g}{10.0\ cm^3} = 21.5\ g/cm^3$$

The given volume has three significant digits, so the answer is rounded off to three digits. It is interesting to note that platinum metal is more dense than lead (d = 11.3 g/cm^3), and more valuable than gold.

Practice Exercise

Carbon tetrachloride is a solvent used for degreasing electronic parts. If 25.0 mL of carbon tetrachloride has a mass of 39.75 g, what is the density of the liquid?

Answer: 1.59 g/mL

Concept Exercise

Which of the following has the greater density: ice or water?

Answer: See Appendix G.

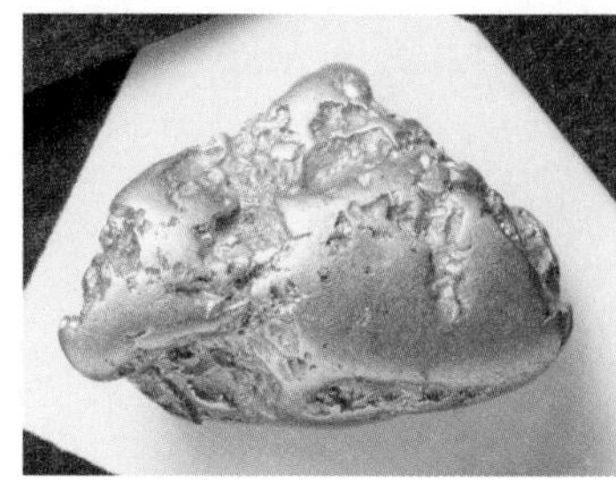

▲ **Platinum Nugget** A small nugget of precious platinum metal.

Applying Density as a Unit Factor

We can also solve problems that ask for the mass or volume of a sample by applying the unit analysis method of problem solving. The secret to solving these problems is to realize that density can be used as a unit factor because it relates mass and volume. For example, the density of mercury is 13.6 g/mL, and so we can write

$$\frac{13.6\ g}{1\ mL} \quad \text{and} \quad \frac{1\ mL}{13.6\ g}$$

Let's calculate the milliliter volume of 75.5 g of liquid mercury. *First*, we write down the unit asked for in the answer (mL). *Second*, we write down the related given value (75.5 g). *Third*, we apply a unit factor using the following format:

$$\underset{\text{(Step 2)}}{75.5\ g} \times \underset{\text{(Step 3)}}{\frac{\text{unit}}{\text{factor}}} = \underset{\text{(Step 1)}}{mL}$$

To cancel the units in the given value (75.5 g), we select the unit factor 1 mL/13.6 g. After substituting, we have

$$75.5\ \cancel{g} \times \frac{1\ mL}{13.6\ \cancel{g}} = 5.55\ mL$$

The given value and the unit factor each have three significant digits; thus, the answer is rounded to three digits. Example Exercises 3.14 and 3.15 further illustrate the use of density as a unit factor.

Example Exercise 3.14 Density as a Unit Factor

An automobile battery contains 1275 mL of sulfuric acid. If the density of battery acid is 1.84 g/mL, how many grams of acid are in the battery?

Strategy Plan

Step 1: What unit is asked for in the answer? **Step 1:** g

Step 2: What given value is related to the answer? **Step 2:** 1275 mL

Step 3: What unit factor(s) should we apply? From the definition of density, 1.84 g = 1 mL; thus, the two unit factors are 1.84 g/1 mL, and its reciprocal 1 mL /1.84 g. **Step 3:** $\frac{1.84\ \text{g}}{1\ \text{mL}}$ or $\frac{1\ \text{mL}}{1.84\ \text{g}}$

Unit Analysis Map

given value: 1275 mL

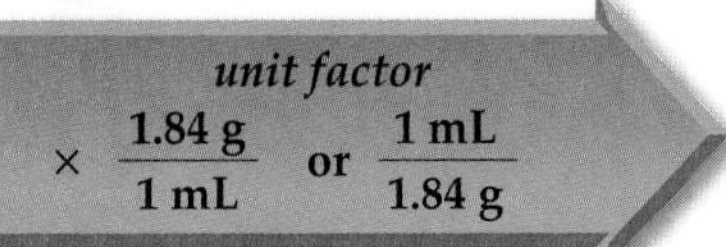

unit in answer: = g

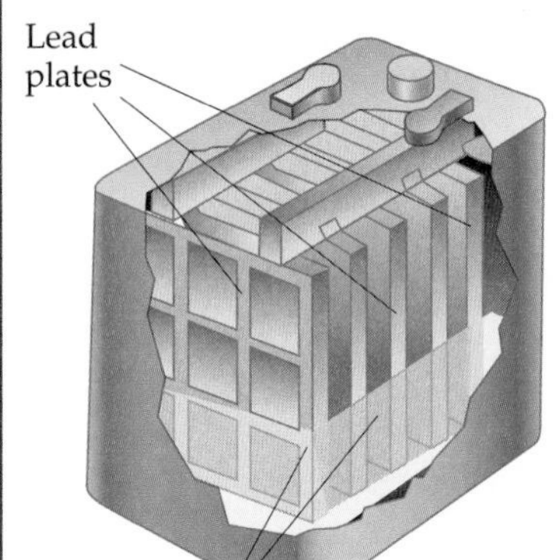

▲ **Automobile Battery** An automobile battery contains lead plates and is filled with sulfuric acid.

Solution

We apply the unit factor 1.84 g/1 mL to cancel milliliters (~~mL~~), which appears in the denominator.

$$1275\ \cancel{\text{mL}} \times \frac{1.84\ \text{g}}{1\ \cancel{\text{mL}}} = 2350\ \text{g}$$

The given value, 1275 mL, has four significant digits, but the unit factor has only three digits. Therefore, the answer is rounded off to three digits. Battery acid is the common name for sulfuric acid, which annually ranks as the most important industrial chemical.

Practice Exercise

The most abundant gases in our atmosphere are nitrogen, oxygen, and argon. What is the volume of 1.00 kg of air? (Assume the density of air is 1.29 g/L.)

Answer: 775 L

Concept Exercise

Which of the following is the greater density?

1 g/mL or 1 kg/L

Answer: See Appendix G.

Example Exercise 3.15 Density as a Unit Factor

A 1.00-in. cube of copper measures 2.54 cm on a side. What is the mass of the copper cube (given that *d* of copper = 8.96 g/cm^3)?

Strategy Plan

Step 1: What unit is asked for in the answer? **Step 1:** g

Step 2: What given value is related to the answer? **Step 2:** 2.54 cm cube

Step 3: What unit factor(s) should we apply? From the definition of density, 8.96 g = 1 cm^3; thus, the two unit factors are8.96 g/1 cm^3, and its reciprocal 1 cm^3/8.96 g. **Step 3:** $\frac{8.96\ \text{g}}{1\ \text{cm}^3}$ or $\frac{1\ \text{cm}^3}{8.96\ \text{g}}$

Unit Analysis Map

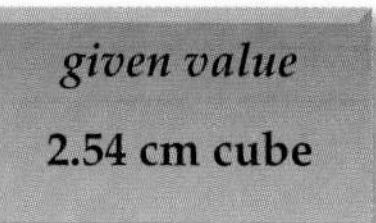

unit factor

$\times \frac{8.96\ g}{1\ cm^3}$ or $\frac{1\ cm^3}{8.96\ g}$

unit in answer

= g

Solution

First, we find the volume of the copper cube. We obtain the volume of the cube, 16.4 cm^3, by multiplying (2.54 cm) (2.54 cm) (2.54 cm). We use the given density, 8.96 g /1 cm^3, as a unit factor to cancel cubic centimeters ($\cancel{cm^3}$), which appears in the denominator.

$$16.4\ \cancel{cm^3} \times \frac{8.96\ g}{1\ \cancel{cm^3}} = 147\ g$$

The given value and unit factor each has three significant digits, so the answer is rounded off to three significant digits.

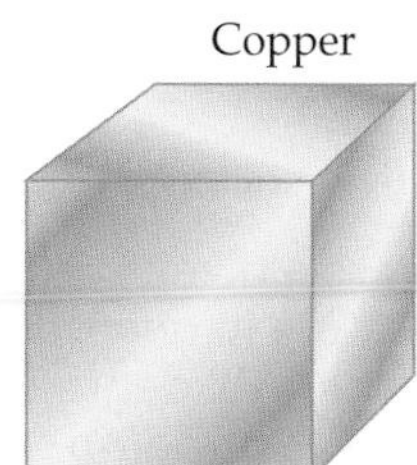

▲ **Copper Metal** A 1.00-inch cube of copper metal.

Practice Exercise

A cube of silver is 5.00 cm on a side and has a mass of 1312.5 g. What is the density of silver?

Answer: 10.5 g/cm^3

Concept Exercise

If some humans float in water and other sink, what is the approximate density of the human body?

Answer: See Appendix G.

Specific Gravity

The ratio of the density of a liquid to the density of water at 4 °C is called **specific gravity** (symbol **sp gr**). Since specific gravity is a ratio of two densities, the units cancel and specific gravity is a unitless quantity. The density of water is 1.00 g/mL, and the specific gravity of water is 1.00.

CRITICAL THINKING Lower Gasoline Bills

Q: In terms of expense, is it better to fill a gas tank in the cool morning, or in the warm afternoon?

Interestingly, you can reduce your automobile fuel bills by applying the density concept discussed in this chapter. The density of most liquids decreases as the temperature warms, and increases as the temperature cools. Thus, the density of gasoline decreases in the warm afternoon, and increases in the cool morning.

Gasoline is sold at the pump by the volume delivered (gallons), and not by the mass, which is greater when the density increases. Although $50.00 of fuel contains the *same volume* at different temperatures, $50.00 of fuel contains a *greater mass* of gasoline at 40 °F than at 70 °F because the density increases. Thus, filling up with gas in the cool morning saves money.

There are only two liquids that violate the principle of greater density at a lower temperature. The liquids are water and ammonia. At temperatures near their freezing points (0 °C and –77 °C, respectively), each of these liquids expands to form a crystalline solid. Consequently, the density of water decreases and solid ice demonstrates the unusual property of floating in water. Similarly, solid ammonia floats in liquid ammonia. All other solid substances are more dense and sink in their corresponding liquids.

▲ **Lower Gasoline Bills**

A: Although it is more economical to fill a gas tank in the cool morning, the density change is small and the mass is only about 1% greater at 40 °F than at 70 °F.

Diagnostic medical testing often includes the specific gravity of body fluids. For instance, the specific gravity of urine may be 1.02, and the specific gravity of blood may be 1.06. Both these values are considered to be in the normal range.

3.8 TEMPERATURE

OBJECTIVES

- To state the values for the freezing point and boiling point of water on the Fahrenheit, Celsius, and Kelvin scales.
- To express a given temperature in degrees Fahrenheit (°F), degrees Celsius (°C), or Kelvin units (K).

The "hotness" or "coolness" of the atmosphere is determined by how fast the tiny air molecules are moving. If the temperature is warmer, molecules move faster and have more energy. If the temperature is cooler, molecules move slower and have less energy. **Temperature** is the average energy of individual particles in motion; for example, the average energy of air molecules in the atmosphere. We measure temperature using a thermometer.

In 1724 Daniel Gabriel Fahrenheit (1686–1736), a German physicist, invented the mercury thermometer. In attempting to produce as cold a temperature as possible, Fahrenheit prepared an ice bath to which he added salt to lower the temperature further. He then assigned a value of zero to that temperature and marked his Fahrenheit scale accordingly.

Fahrenheit obtained a second reference point by recording his underarm body temperature; he assigned that temperature a value of 96. He divided the distance between the two reference points into 96 equal units, and each division was termed a **Fahrenheit degree** (symbol **°F**). Later, the freezing and boiling points of water were selected as the standard reference points. The freezing point of water was assigned a value of 32 °F, and the boiling point a value of 212 °F. The Fahrenheit degree eventually became a basic unit in the English system of measurement.

In 1742 Anders Celsius (1701–1744), a Swedish astronomer, proposed a scale similar in principle to the Fahrenheit scale. On the Celsius scale, the freezing point of water was assigned a value of 0 °C and the boiling point of water a value of 100 °C. The scale was then divided into 100 equal divisions, each division representing a **Celsius degree** (symbol **°C**). The Celsius degree is also referred to as a centigrade degree and is a basic unit in the metric system.

In 1848 William Thomson (1824–1907), an English physicist also known as Lord Kelvin, proposed a scale based on the lowest possible temperature. The unit of temperature was the **Kelvin unit** (symbol **K**), which is a basic unit in the SI system. On the Kelvin scale, the coldest temperature was assigned a value of 0 K, and each division on the scale is equal to one Celsius degree. The lowest temperature is referred to as absolute zero and corresponds to –273.15 °C. Although there is no highest temperature on the Kelvin scale, the interior of the Sun reaches about 10,000,000 K.

Fahrenheit and Celsius Temperature Conversions

One division on the Kelvin scale is equivalent to 1 ° on the Celsius scale. Since 0 K is equivalent to –273 °C, the freezing point of water is 273 K, and the boiling point of water is 373 K. Figure 3.6 illustrates how the three temperature scales are related.

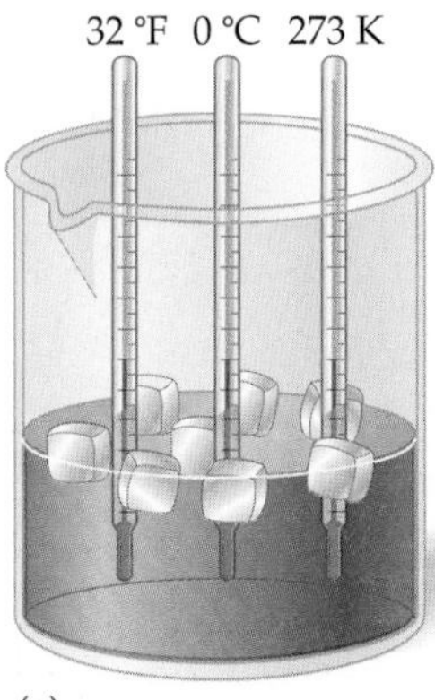

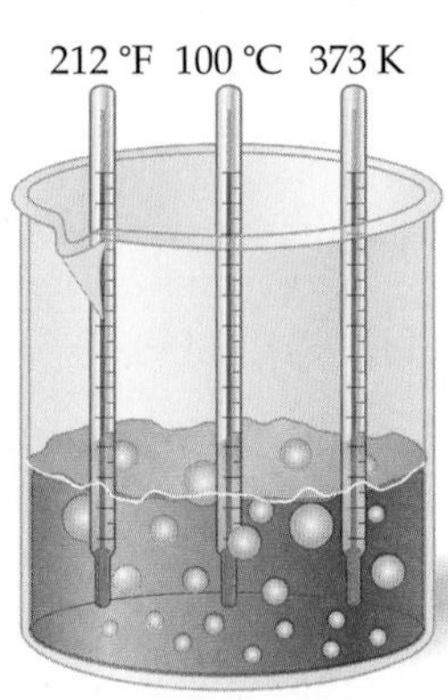

▶ **Figure 3.6 Temperature Scales** A Fahrenheit, a Celsius, and a Kelvin thermometer are placed in (a) ice water and (b) boiling water. Notice the freezing point and boiling point on each scale. The number of divisions is 180 units on the Fahrenheit scale, 100 units on the Celsius scale, and 100 units on the Kelvin scale.

Notice that 180 Fahrenheit units are equivalent to 100 Celsius units. Therefore, to convert from °F to °C, we first subtract 32 (the difference between the freezing point of water on the two scales) and then multiply by 100 °C/180 °F. That is,

$$(°\text{F} - 32\ °\text{F}) \times \frac{100\ °\text{C}}{180\ °\text{F}} = °\text{C}$$

To convert from °C to °F, we reverse the procedure of converting °F to °C. We first multiply the Celsius temperature by the ratio 180 °F/100 °C, and then add 32 °F. That is,

$$(°\text{C} \times \frac{180\ °\text{F}}{100\ °\text{C}}) + 32\ °\text{F} = °\text{F}$$

Example Exercise 3.16 °F and °C Temperature Conversions

Normal body temperature is 98.6 °F. What is normal body temperature in degrees Celsius?

Strategy Plan

Step 1: What unit is asked for in the answer? **Step 1:** °C

Step 2: What given value is related to the answer? **Step 2:** 98.6 °F

Step 3: What unit factor(s) should we apply? No unit factor is required. **Step 3:** none

Solution

To calculate °C, we refer to Figure 3.6 and compare the Celsius and Fahrenheit temperature scales. The conversion from °F to °C is as follows.

$$(98.6\ °\text{F} - 32\ °\text{F}) \times \frac{100\ °\text{C}}{180\ °\text{F}} = °\text{C}$$

Simplifying and canceling units gives

$$66.6\ \cancel{°\text{F}} \times \frac{100\ °\text{C}}{180\ \cancel{°\text{F}}} = 37.0\ °\text{C}$$

The given value, 98.6 °F, has three significant digits, so the answer is rounded off to three digits. Since 32 °F and 100 °C/180 °F are exact numbers, neither affects the significant digits in the answer.

Practice Exercise

The average surface temperature of Mars is –55 °C. What is the average temperature in degrees Fahrenheit?

Answer: –67 °F

Concept Exercise

What is the relationship between the Celsius and centigrade temperature scales?

Answer: See Appendix G.

▲ **Australian Stamp** The cartoon illustrates that 38 °C is approximately equal to 100 °F.

Celsius and Kelvin Temperature Conversions

Let's reexamine the temperature scales in Figure 3.6. Notice that the Kelvin scale is 273 units above the Celsius scale. Therefore, to convert from °C to K, we must add 273 units to the Celsius temperature.

$$°\text{C} + 273 = \text{K}$$

Conversely, to convert from K to °C, we subtract 273 units from the Kelvin temperature. It is helpful to remember that negative Kelvin temperatures are nonexistent. By definition, the lowest possible temperature is assigned a value of 0 K. Example Exercise 3.17 illustrates the conversion of Celsius and Kelvin temperatures.

Example Exercise 3.17 °C and K Temperature Conversions

Dermatologists use liquid nitrogen to freeze skin tissue. If the Celsius temperature of liquid nitrogen is –196 °C, what is the Kelvin temperature?

Strategy Plan

Step 1: What unit is asked for in the answer? **Step 1:** K

Step 2: What given value is related to the answer? **Step 2:** – 196 °C

Step 3: What unit factor(s) should we apply? No unit factor is required. **Step 3:**

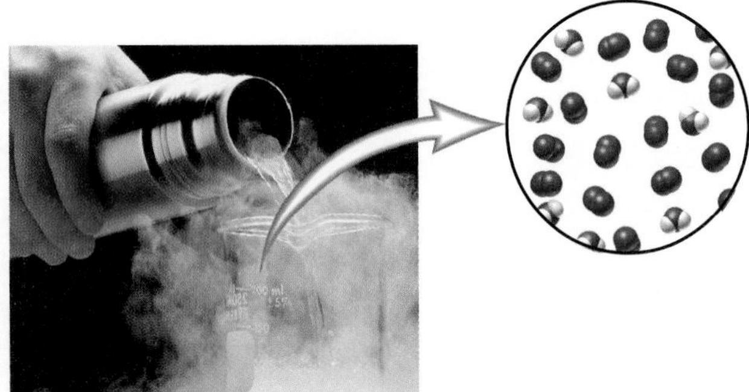

▲ **Liquid Nitrogen** Although nitrogen is normally a gas, it liquefies at –196 °C. When liquid nitrogen is poured from a Thermos, it is cold enough to freeze the moisture in air and form a white mist.

Solution

Given the Celsius temperature, we add 273 units to find the corresponding Kelvin temperature.

$$-196\ °C + 273 = 77\ k$$

Practice Exercise

The secret to "fire-walking" is to first walk barefoot through damp grass and then step lively on the red-hot coals. If the bed of coals is 1475 K, what is the Celsius temperature?

Answer: 1202 °C

Concept Exercise

Which of the following temperatures does not exist?

–100 °F, –100 °C, –100 K

Answer: See Appendix G.

Note We can convert Fahrenheit and Celsius temperatures by understanding the relationship of the two temperature scales (see Figure 3.6). In practice, nurses and other professionals may have a reference chart showing equivalent Fahrenheit and Celsius temperatures. Moreover, students can convert Fahrenheit and Celsius temperatures with a single keystroke using an inexpensive scientific calculator.

OBJECTIVES

- To state the value for the specific heat of water: 1.00 cal/(g × °C).
- To relate heat to the mass, specific heat, and temperature change of a substance by calculation.

3.9 HEAT AND SPECIFIC HEAT

The flow of heat is always from an object at a higher temperature to an object at a lower temperature. For example, if you hold an ice cube, the heat from your hand (higher temperature) flows into the ice cube (lower temperature) and your hand feels cold.

Heat and temperature are both a measure of energy. However, **heat** measures the *total energy*, whereas temperature measures the *average energy*. To distinguish heat from temperature, consider a cup of hot tea and a teaspoon of hot tea, each of which is at 100 °C. What would be more harmful: quickly drinking a cup of hot tea or a teaspoon of hot tea? Obviously, the cup of hot tea would more likely cause a burn than the teaspoon. The reason is that the cup of hot tea has more total heat than the teaspoon, even though each is at the same temperature. Figure 3.7 further illustrates the concepts of heat and temperature.

Heat energy is often expressed in units of calories or kilocalories. A **calorie** (symbol **cal**) is the amount of heat necessary to raise 1 gram of water 1 degree on the Celsius scale. A kilocalorie (kcal) is the amount of heat necessary to raise 1000 grams of water 1 degree on the Celsius scale. A nutritional Calorie (Cal) is spelled with a capital

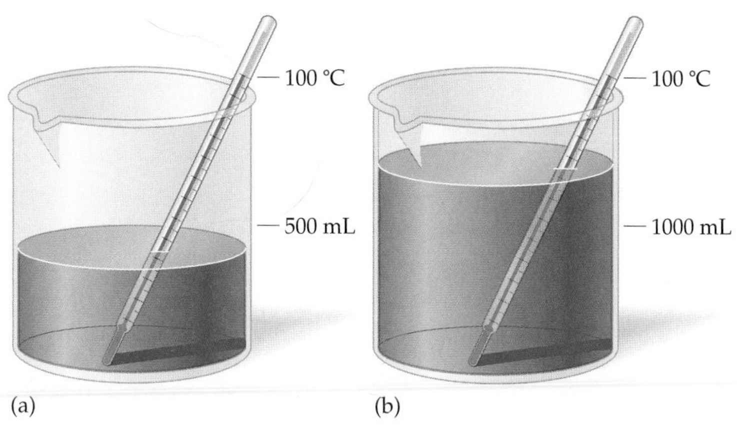

◀ **Figure 3.7 Heat versus Temperature** In (a) 500 mL of water is heated to 100 °C, and in (b) 1000 mL is heated to 100 °C. Although the temperatures are the same, the second beaker has twice the amount of heat.

letter to distinguish it from a metric calorie. One nutritional Calorie is equal to 1 kilocalorie, that is, 1000 calories.

A SI unit of energy is the **joule** (symbol **J**), where 1 cal = 4.184 J. The heat produced by chemical reactions is often expressed in kilocalories, as well as in kilojoules (kJ), where 1 kcal = 4.184 kJ. Example Exercise 3.18 illustrates the conversion of calories, kilocalories, and joules.

Example Exercise 3.18 Energy Conversion

Burning one liter of natural gas produces 9.46 kcal of heat energy. Express the energy in kilojoules (given that 1 kcal = 4.184 kJ).

Strategy Plan

Step 1: What unit is asked for in the answer? **Step 1:** **kJ**

Step 2: What given value is related to the answer? **Step 2:** **9.46 kcal**

Step 3: What unit factor(s) should we apply? Since the unit equation is 1 kcal = 4.184 kJ, the two unit factors are 1 kcal/4.184 kJ, and its reciprocal 4.184 kJ /1 kcal. **Step 3:** $\frac{1\ \text{kcal}}{4.184\ \text{kJ}}$ or $\frac{4.184\ \text{kJ}}{1\ \text{kcal}}$

Unit Analysis Map

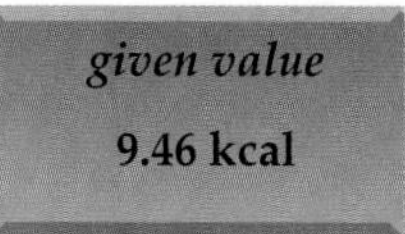

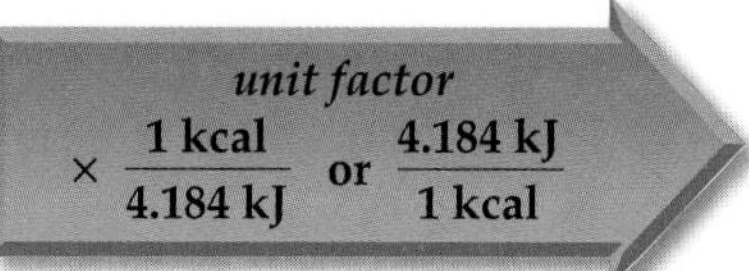

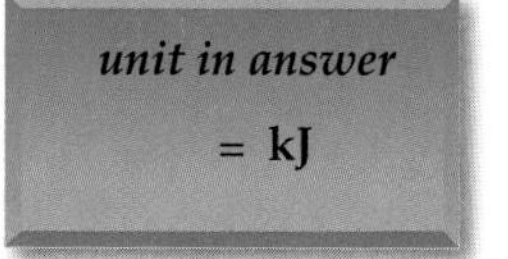

Solution

We apply the unit factor 4.184 kJ /1 kcal to cancel kilocalories (kcal), which appears in the denominator.

$$9.46\ \cancel{\text{kcal}} \times \frac{4.184\ \text{kJ}}{1\ \cancel{\text{kcal}}} = 39.6\ \text{kJ}$$

The given value has three significant digits, and the unit factor has four digits. Thus, we round off the answer to three significant digits.

Practice Exercise

Burning one gram of gasoline produces 47.9 kJ of energy. Express the heat energy in kilocalories (given that 1 kcal = 4.184 kJ).

Answer: 11.4 kcal

▲ **Bunsen Burner** A laboratory burner that uses natural gas for fuel.

Concept Exercise

If an aerosol can feels cold after releasing the spray, is heat flowing from the can or from your hand?

Answer: See Appendix G.

Specific Heat

The specific heat of a substance is the amount of heat required to bring about a given change in temperature. It is observed that the amount of heat necessary is unique for each substance. The specific heat of water is relatively high, and the change in temperature is minimal as water gains or loses heat. The surface of Earth is covered with water, and fortunately its high specific heat helps to regulate climates and maintain moderate temperatures. Figure 3.8 illustrates the increase in temperature for four substances each receiving 1 cal of heat.

We can define **specific heat** as the amount of heat required to raise the temperature of one gram of substance one degree Celsius; the units of specific heat are often given in calories per gram per degree Celsius. For reference, the specific heat of water has a value of 1.00 cal/(g × °C). The specific heats of ice and steam are much lower. Table 3.5 lists the specific heat for selected solids, liquids, and gases.

The units of specific heat can be expressed as calories per gram per degree Celsius. Thus, a gain or loss of heat divided by mass and change of temperature (Δt) equals specific heat. We can write this relationship mathematically as

$$\frac{\text{heat (cal)}}{\text{mass (g)} \times \text{temperature change } (\Delta t)} = \text{specific heat } \left(\frac{\text{cal}}{\text{g} \times {}^\circ\text{C}}\right)$$

Let's calculate the specific heat for zinc given that a 65.5-g sample of the metal loses 480 cal of heat when cooled from 100.0 °C to 21.0 °C. We divide the heat loss of the zinc metal by its mass and temperature change. Thus,

$$\frac{480 \text{ cal}}{65.5 \text{ g} \times (100.0 - 21.0)\ {}^\circ\text{C}} = \frac{\text{cal}}{\text{g} \times {}^\circ\text{C}}$$

We must first find the temperature change and then calculate the specific heat of the metal. The temperature change is 100.0 °C minus 21.0 °C, that is, 79.0 °C. Thus,

$$\frac{480 \text{ cal}}{65.5 \text{ g} \times 79.0\ {}^\circ\text{C}} = 0.093 \text{ cal/(g} \times {}^\circ\text{C)}$$

Example Exercise 3.19 illustrates a calculation using specific heat as a unit factor.

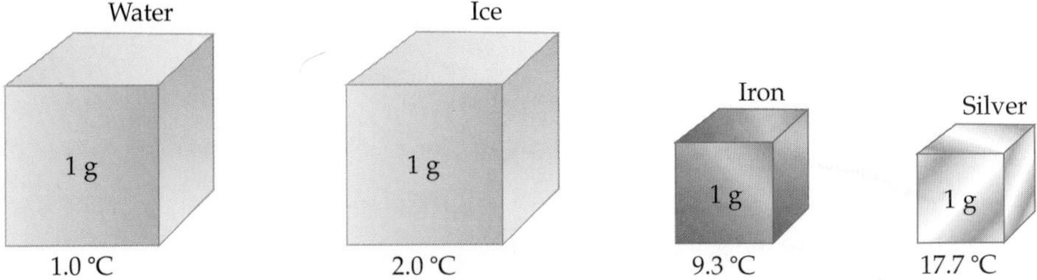

▲ **Figure 3.8 An Illustration of Specific Heat** Each cube represents a volume of 1 gram of substance receiving 1 calorie of heat. Notice the temperature change for each substance increases the number of degrees shown. Silver has a low specific heat and is a good conductor of heat.

TABLE 3.5 SPECIFIC HEAT FOR SELECTED SOLIDS, LIQUIDS, AND GASES

EXAMPLE	SPECIFIC HEAT
Solids	
ice	0.492 cal/(g × °C)
aluminum	0.215
carbon (graphite)	0.170
carbon (diamond)	0.124
iron	0.108
copper	0.0920
silver	0.0566
gold	0.0305
Liquids	
water	1.00 cal/(g × °C)
ethyl alcohol (ethanol)	0.587
methyl alcohol (methanol)	0.424
freon (CFC refrigerant)	0.232
mercury	0.0331
Gases	
steam	0.485 cal/(g × °C)
nitrogen	0.249
oxygen	0.219
argon	0.124
radon	0.0224

Example Exercise 3.19 Specific Heat

An energy-efficient home may have solar panels for heating water. If 350,000 cal heat water from 20.0 °C to 35.0 °C, what is the mass of water (specific heat = 1.00 cal/g × °C)?

Strategy Plan

Step 1: What unit is asked for in the answer? **Step 1:** g

Step 2: What given value is related to the answer? **Step 2:** 350,000 cal, 20.0 °C to 35.0 °C

Step 3: What unit factor(s) should we apply? **Step 3:** $\frac{1.00 \text{ cal}}{1 \text{ g} \times 1\ ^\circ\text{C}}$ or $\frac{1 \text{ g} \times 1\ ^\circ\text{C}}{1.00 \text{ cal}}$

Unit Analysis Map

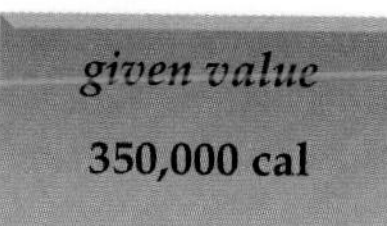

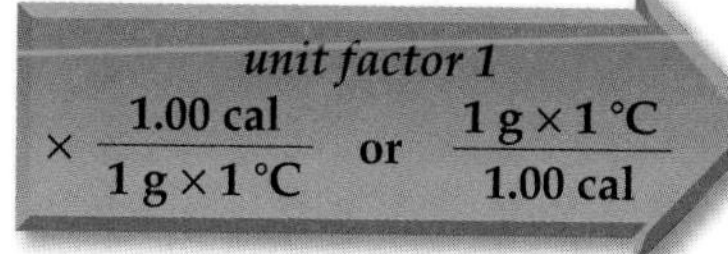

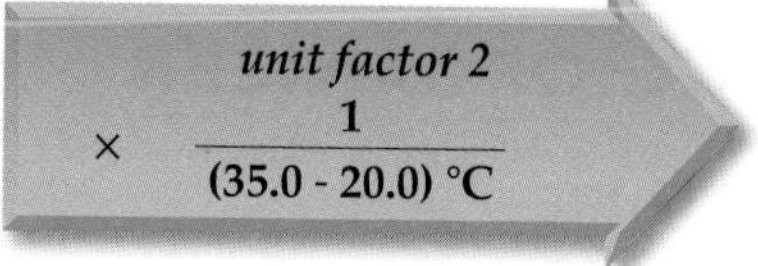

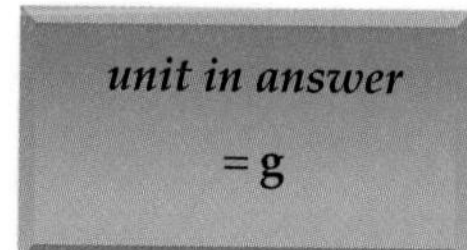

Solution

We apply the unit factor 1 g × °C/1.00 cal to cancel calories (cal), and (35.0 – 20.0) °C to cancel degrees Celsius (°C).

$$350{,}000 \cancel{\text{cal}} \times \frac{1 \text{ g} \times 1\ \cancel{^\circ\text{C}}}{1.00 \cancel{\text{cal}}} \times \frac{1}{(35.0 - 20.0)\ \cancel{^\circ\text{C}}} = 23{,}000 \text{ g (23 kg)}$$

▲ **Solar Panels on Rooftop**
Solar panels collect light energy from the Sun, which is converted to heat energy for the home.

Practice Exercise

A 725 g steel horseshoe is heated to 425 °C and dropped into a bucket of cold water. If the horseshoe cools to 20 °C and the specific heat of steel is 0.11 cal/g × °C, how much heat is released?

Answer: 32,000 cal (32 kcal)

Concept Exercise

If the crust of an apple pie cooks faster than the filling, which has the higher specific heat: the crust or apple filling?

Answer: See Appendix G.

Summary

Section 3.1 The **English system** of measurement has many unrelated units. On the other hand, the **metric system** of measurement is a decimal system with basic units: **meter (m)**, **gram (g)**, **liter (L)**, and **second (s)**. Metric prefixes provide multiples and fractions of basic units. The common prefixes include *giga-* (G), *mega-* (M), *kilo-* (k), *deci-* (d), *centi-* (c), *milli-* (m), *micro-* (μ), and *nano-* (n). The **International System (SI)** is based on the metric system but is more comprehensive.

Section 3.2 Metric conversion problems are solved by systematically writing a **unit equation** or an **exact equivalent** (1 m = 100 cm). An equal or equivalent relationship generates a **unit factor** and its **reciprocal** (1 m/100 cm and 100 cm/1 m).

Section 3.3 Metric problems are solved systematically by applying the **unit analysis method**: Step 1: Write down the units of the unknown. Step 2: Write down a given value that is related to the unknown. Step 3: Apply one or more unit factors to convert the units of the given value to the units asked for in the answer.

Section 3.4 Every nation in the world uses the metric system of measurement; however, the English system is still common in the United States. Therefore, we will relate the two systems and memorize the following: 2.54 cm = 1 in., 454 g = 1 lb, and 946 mL = 1 qt. These equivalents will be used for metric–English conversions.

Section 3.5 The volume of a rectangular solid is equal to length times width times thickness. The calculated volume is reported in cubic units, such as **cubic centimeters (cm^3)**. You should memorize the exact equivalent 1 mL = 1 cm^3.

Section 3.6 For irregularly shaped objects, as well as gases, we must determine volume indirectly. We can find the volume of an object or a gas from the amount of water it displaces; this technique involves **volume by displacement**.

Section 3.7 The **density** of water is 1.00 g/mL. A liquid or solid that floats on water is less dense than water; if it sinks, it is more dense. Density can be used as a unit factor, for example, 1.00 g/1 mL or 1 mL/1.00 g. **Specific gravity (sp gr)** is unitless and expresses the ratio of the density of a liquid to the density of water.

Section 3.8 **Temperature** is the average energy of molecules in motion. The **Fahrenheit**, **Celsius**, and **Kelvin** scales have related reference points. The freezing point of water is 32 °F, 0 °C, or 273 K; and the boiling point is 212 °F, 100 °C, or 373 K.

Section 3.9 **Heat** is the total energy of molecules in motion. Heat changes are expressed in **calories (cal)** or kilocalories (kcal). Heat changes can also be expressed in the SI unit **joule (J)**, where 1 cal = 4.184 J. **Specific heat** is the amount of heat necessary to raise 1 g of substance 1 °C; for water, the value is 1.00 cal/(g × °C).

PROBLEM-SOLVING ORGANIZER

TOPIC	PROCEDURE	EXAMPLE
Basic Units and Symbols Sec. 3.1	Combine basic metric units and prefixes using symbols.	centimeter, cm (length) kilogram, kg (mass) milliliter, mL (volume)
Metric Conversion Factors Sec. 3.2	(a) Write a unit equation involving basic metric units and prefix units. (b) Write two unit factors for a metric relationship.	1 m = 100 cm $\frac{1\text{ m}}{100\text{ cm}}$ and $\frac{100\text{ cm}}{1\text{ m}}$

PROBLEM–SOLVING ORGANIZER *(Continued)*

TOPIC	PROCEDURE	EXAMPLE
Metric–Metric Conversions Sec. 3.3	1. Write down the unit asked for in the answer. 2. Write down the related given value. 3. Apply a unit factor to convert the given unit to the unit in the answer.	What is the decigram mass of a wheel of cheese that weighs 0.515 kg? $0.515 \text{ kg} \times \frac{1000 \text{ g}}{1 \text{ kg}} \times \frac{10 \text{ dg}}{1 \text{ g}} = 5150 \text{ dg}$
Metric–English Conversions Sec. 3.4	1. Write down the unit asked for in the answer. 2. Write down the related given value. 3. Apply a unit factor to convert the given unit to the unit in the answer.	What is the pound mass of a wheel of cheese that weighs 0.515 kg? $0.515 \text{ kg} \times \frac{1000 \text{ g}}{1 \text{ kg}} \times \frac{1 \text{ lb}}{454 \text{ g}} = 1.13 \text{ lb}$
Volume by Calculation Sec. 3.5	To calculate the volume of a rectangular solid, multiply length by width by thickness: $V = l \times w \times t$.	What is the volume of a rectangular solid measuring 65 mm by 35 mm by 12 mm? $65 \text{ mm} \times 35 \text{ mm} \times 12 \text{ mm} = 27{,}000 \text{ mm}^3$
Volume by Displacement Sec. 3.6	The volume by displacement is the difference between the initial and final readings in a calibrated container.	What is the volume of jade if the water level in a cylinder increases from 25.0 mL to 42.5 mL? 42.5 mL – 25.0 mL = 17.5 mL
The Density Concept Sec. 3.7	The density of a sample is equal to its mass divided by its volume: density = mass/volume.	What is the density of 10.0 mL of ether if its mass is 7.142 g? $\frac{7.142 \text{ g}}{10.0 \text{ mL}} = 0.714 \text{ g/mL}$
Temperature Sec. 3.8	(a) To convert from Celsius to Kelvin, add 273 units. (b) To convert from Kelvin to Celsius, subtract 273 units.	–88 °C + 273 = 185 K 1265 K – 273 = 992 °C
Heat and Specific Heat Sec. 3.9	1. Write down the unit asked for in the answer. 2. Write down the related given value. 3. Divide the heat loss by the mass of metal and temperature change.	What is the specific heat of iron if 32.4 cal is required to heat 10.0 g of iron from 20.0 °C to 50.0 °C? $\frac{32.4 \text{ cal}}{10.0 \text{ g} \times 30.0 \text{ °C}} = 0.108 \text{ cal/(g} \times \text{°C)}$

Key Terms *Answers to Key Terms are in Appendix H.*

Select the key term that corresponds to each of the following definitions.

_____ **1.** a nondecimal system of measurement without basic units
_____ **2.** a decimal system of measurement with basic units
_____ **3.** the basic unit of length in the metric system
_____ **4.** the basic unit of mass in the metric system
_____ **5.** the basic unit of volume in the metric system
_____ **6.** the basic unit of time in the metric system
_____ **7.** a comprehensive system of measurement with seven base units
_____ **8.** a statement of two equivalent values, for example, 1 m = 39.4 in.
_____ **9.** a statement of two exactly equal values, for example, 1 m = 100 cm
_____ **10.** a ratio of two quantities that are equivalent, for example, 1 lb/454 g
_____ **11.** the relationship between a fraction and its inverse, for example, 1 qt/946 mL and 946 mL/1 qt

(a) calorie (cal) *(Sec. 3.9)*
(b) Celsius degree (°C) *(Sec. 3.8)*
(c) cubic centimeter (cm^3) *(Sec. 3.5)*
(d) density (d) *(Sec. 3.7)*
(e) English system *(Sec. 3.1)*
(f) exact equivalent *(Sec. 3.2)*
(g) Fahrenheit degree (°F) *(Sec. 3.8)*
(h) gram (g) *(Sec. 3.1)*
(i) heat *(Sec. 3.9)*
(j) International System (SI) *(Sec. 3.1)*
(k) joule (J) *(Sec. 3.9)*
(l) Kelvin unit (K) *(Sec. 3.8)*

_____ 12. a procedure for solving problems that proceeds from a given value to a related answer by the conversion of units
_____ 13. the volume occupied by a cube 1 cm on a side
_____ 14. a technique for determining volume from the amount of water displaced
_____ 15. the amount of mass in one unit of volume
_____ 16. the ratio of the density of a liquid compared to the density of water at 4 °C
_____ 17. the flow of energy from an object at a higher temperature to an object at a lower temperature
_____ 18. the average energy of molecules in motion
_____ 19. the basic unit of temperature in the English system
_____ 20. the basic unit of temperature in the metric system
_____ 21. the basic unit of temperature in the SI system
_____ 22. the amount of heat required to raise 1 g of substance 1 °C
_____ 23. the amount of heat required to raise 1 g of water 1 °C
_____ 24. a unit of energy in the SI system

(m) liter (L) *(Sec. 3.1)*
(n) meter (m) *(Sec. 3.1)*
(o) metric system *(Sec. 3.1)*
(p) reciprocal *(Sec. 3.2)*
(q) second (s) *(Sec. 3.1)*
(r) specific gravity (sp gr) *(Sec. 3.7)*
(s) specific heat *(Sec. 3.9)*
(t) temperature *(Sec. 3.8)*
(u) unit analysis method *(Sec. 3.3)*
(v) unit equation *(Sec. 3.2)*
(w) unit factor *(Sec. 3.2)*
(x) volume by displacement *(Sec. 3.6)*

Exercises

Answers to odd-numbered Exercises are in Appendix I.

Basic Units and Symbols (Sec. 3.1)

1. Which of the following are basic units in the metric system?
(a) kilometer **(b)** gram
(c) liter **(d)** millisecond

2. Which of the following are basic units in the English system?
(a) foot **(b)** ounce
(c) gallon **(d)** hour

3. Write the symbol for the following metric units:
(a) kilometer **(b)** gigagram
(c) deciliter **(d)** picosecond

4. Write the symbol for the following metric units:
(a) megameter **(b)** teragram
(c) centiliter **(d)** nanosecond

5. State the physical quantity expressed by the following metric units:
(a) kilometer **(b)** gigagram
(c) deciliter **(d)** picosecond

6. State the physical quantity expressed by the following metric units:
(a) megameter **(b)** teragram
(c) centiliter **(d)** nanosecond

7. State the physical quantity that is measured using the following instruments:
(a) ruler **(b)** platform balance
(c) pipet **(d)** stopwatch

8. State the physical quantity that is measured using the following instruments:
(a) micrometer **(b)** electronic balance
(c) syringe **(d)** atomic clock

9. State the name and symbol for the metric prefix that represents the following:
(a) 1×10^{12} **(b)** 1×10^{9}
(c) 1×10^{-3} **(d)** 1×10^{-6}

10. State the name and symbol for the metric prefix that represents the following:
(a) 1×10^{6} **(b)** 1×10^{3}
(c) 1×10^{-9} **(d)** 1×10^{-12}

11. Write the name of the metric unit indicated by the following symbols:
(a) Tm **(b)** Gg
(c) mL **(d)** μs

12. Write the name of the metric unit indicated by the following symbols:
(a) Mm **(b)** kg
(c) nL **(d)** ps

Metric Conversion Factors (Sec. 3.2)

13. Write a unit equation for each of the following metric equivalents:
(a) m and Tm **(b)** g and Gg
(c) L and mL **(d)** s and μs

14. Write a unit equation for each of the following metric equivalents:
(a) m and Mm **(b)** g and kg
(c) L and nL **(d)** s and ps

15. Write two unit factors for each of the following metric relationships:
(a) m and Tm **(b)** g and Gg
(c) L and mL **(d)** s and μs

16. Write two unit factors for each of the following metric relationships:
(a) m and Mm **(b)** g and kg
(c) L and nL **(d)** s and ps

Metric–Metric Conversions (Sec. 3.3)

17. Perform the following metric–metric conversions:
(a) 1.00 m to km **(b)** 100 g to cg
(c) 0.100 L to dL **(d)** 0.000 010 s to ns

18. Perform the following metric–metric conversions:
(a) 5.00 Mm to m **(b)** 50.0 μg to g
(c) 0.500 mL to L **(d)** 0.050 ds to s

19. Perform the following metric–metric conversions:
(a) 6.50 Tm to Mm **(b)** 650 Gg to kg
(c) 0.650 cL to dL **(d)** 0.000 650 ns to ps

20. Perform the following metric–metric conversions:
(a) 7.50 km to Gm **(b)** 750 Mg to Tg
(c) 0.750 pL to μL **(d)** 0.000 750 ms to ns

Metric–English Conversions (Sec. 3.4)

21. State the following metric–English equivalents:
(a) ? cm = 1 in. **(b)** ? g = 1 lb
(c) ? mL = 1 qt **(d)** ? s = 1 sec

22. State the following metric–English equivalents:
(a) 1 m = ? yd (b) 1 kg = ? lb
(c) 1 L = ? qt (d) 1 s = ? sec

23. Perform the following metric–English conversions:
(a) 86 cm to in. (b) 36 g to lb
(c) 0.500 qt to mL (d) 8.00×10^2 sec to s

24. Perform the following metric–English conversions:
(a) 66 in. to cm (b) 1.01 lb to g
(c) 750 mL to qt (d) 5.52×10^{-3} s to sec

25. Perform the following metric–English conversions:
(a) 72 in. to m (b) 175 lb to kg
(c) 0.500 qt to L (d) 1.05×10^{-4} min to ks

26. Perform the following metric–English conversions:
(a) 800.0 m to yards (b) 0.375 kg to lb
(c) 1250 mL to gallons (d) 1.52×10^3 ds to min

27. A Honda Civic gets 13 kilometers per liter in city driving. What is the mileage in miles per gallon (given that: 1 mi = 1.61 km and 1 gal = 3.784 L)?

28. A Honda Civic gets 16 kilometers per liter in highway driving. What is the mileage in miles per gallon (given that 1 mi = 1.61 km and 1 gal = 3.784 L)?

Volume by Calculation (Sec. 3.5)

29. A piece of quartz was cut into a rectangular solid measuring 5.00 cm by 5.00 cm by 25.0 mm. What is the volume in cubic millimeters?

30. A piece of amber was cut into a rectangular solid measuring 5.08 cm by 10.2 cm by 3.05 m. What is the volume in cubic centimeters?

31. A rectangular solid piece of brass measures 4.95 cm by 2.45 cm and has a volume of 15.3 cm^3. What is the thickness of the brass?

32. A sheet of aluminum foil measures 30.5 cm by 75.0 cm and has a mass of 9.94 g. What is the thickness of the foil if the volume is 3.68 cm^3?

33. Complete the following volume equivalents:
(a) 1 L = ? mL (b) 1 L = ? cm^3

34. Complete the following volume equivalents:
(a) 1 mL = ? cm^3 (b) 1 in.^{3} = ? cm^3

35. A Dodge Challenger has a 6.10 L engine. Express the volume in cubic inches.

36. A Chevrolet Camaro has a 378 in.^{3} engine. Express the volume in liters.

Volume by Displacement (Sec. 3.6)

37. The initial water level in a 10-mL graduated cylinder reads 4.5 mL. After a piece of jade is dropped into the cylinder, the water level reads 7.5 mL. What is the volume of the jade?

38. The initial water level in a 10-mL graduated cylinder is 4.5 mL. After a large ruby is added into the cylinder, the water level is 6.0 mL. What is the volume of the ruby gemstone?

39. Magnesium metal reacts with hydrochloric acid to produce hydrogen gas. The gas displaces water into a cylinder, and the level increases from 125 mL to 255 mL. What volume of hydrogen is produced by the reaction?

40. Calcium metal reacts with water to produce hydrogen gas. If the gas displaces 212 mL of water into a beaker, how many cubic centimeters of hydrogen are produced by the reaction?

The Density Concept (Sec. 3.7)

41. State whether the following will sink or float when dropped into water:
(a) ebony (d = 1.2 g/mL) (b) bamboo (d = 0.40 g/mL)

42. State whether the following will sink or float when dropped into water:
(a) paraffin wax (d = 0.90 g/cm^3)
(b) limestone (d = 2.8 g/cm^3)

43. Will a balloon filled with the given gas rise in the air or drop to the ground? (Assume the mass of the balloon is negligible and the density of air is 1.29 g/L.)
(a) helium (d = 0.178 g/L) (b) laughing gas (d = 1.96 g/L)

44. Will a balloon filled with the given gas rise in the air or drop to the ground? (Assume the mass of the balloon is negligible and the density of air is 1.29 g/L.)
(a) argon (d = 1.78 g/L) (b) ammonia (d = 0.759 g/L)

45. Calculate the volume in milliliters for each of the following liquids:
(a) 25.0 g of ether (d = 0.714 g/mL)
(b) 15.0 g of acetone (d = 0.792 g/mL)

46. Calculate the mass in grams for each of the following liquids:
(a) 250 mL of gasoline (d = 0.69 g/mL)
(b) 150 mL of methanol (d = 0.79 g/mL)

47. Calculate the mass in grams for each of the following solids:
(a) 7.50 cm^3 of rock salt (d = 2.18 g/cm^3)
(b) 455 cm^3 of borax (d = 1.715 g/cm^3)

48. Calculate the volume in milliliters for each of the following solids:
(a) 1.00 kg of iron (d = 7.87 g/cm^3)
(b) 1.00 kg of titanium (d = 4.51 g/cm^3)

49. Calculate the density in grams per milliliter for each of the following:
(a) 25.0 mL of ethyl alcohol having a mass of 19.7 g
(b) 11.6 g marble whose volume is found by displacement to be 4.1 mL

50. Calculate the density in grams per milliliter for each of the following:
(a) 10.0 g of ether having a volume of 14.0 mL
(b) 131.5 g bronze rectangular solid measuring 3.55 cm × 2.50 cm × 1.75 cm

Temperature (Sec. 3.8)

51. State the freezing point of water on the following temperature scales:
(a) Fahrenheit (b) Kelvin
(c) Celsius

52. State the boiling point of water on the following temperature scales:
(a) Fahrenheit (b) Kelvin
(c) Celsius

53. Express the following Fahrenheit temperatures in degrees Celsius:
(a) 100 °F (b) –215 °F

54. Express the following Celsius temperatures in degrees Fahrenheit:
(a) 19 °C (b) –175 °C

55. Express the following Celsius temperatures in Kelvin units:
(a) 495 °C (b) –185 °C

56. Express the following Kelvin temperatures in degrees Celsius:
(a) 273 K (b) 100 K

Heat and Specific Heat (Sec. 3.9)

57. Distinguish between the terms "temperature" and "heat."
58. State the metric value for the specific heat of water.
59. A beaker containing 250 g of water is heated from 23 °C to 100 °C. What is the heat gain in calories?
60. A pool containing 2.5×10^8 g of water cools from 25.0 °C to 17.0 °C. What is the heat loss in kilocalories?
61. Calculate the heat required to raise 25.0 g of iron [0.108 cal/(g × °C)] from 25.0 °C to 50.0 °C.
62. Calculate the heat released as 35.5 g of copper [0.0920 cal/(g × °C)] cools from 50.0 °C to 45.0 °C.
63. Find the specific heat of platinum if 35.7 cal is lost as 75.0 g of the metal cools from 43.9 °C to 28.9 °C.
64. Find the specific heat of gold if 25.0 cal is required to heat 30.0 g of gold from 27.7 °C to 54.9 °C.

General Exercises

65. If a computer hard disk has a capacity of 1.5 terabytes (TB), what is the storage capacity in megabytes (MB)?

◀ **Computer Hard Disk** Some computer hard disks have a storage capacity greater than 1000 gigabytes (GB). That is, some new hard disks can store terabytes (TB) of information.

66. If a CD can store 650 megabytes (MB) of data, and a computer hard disk has a capacity of 350 gigabytes (GB), how many CDs of information can be loaded onto the hard disk?

◀ **Computer Compact Disk** Compact disks (CDs) have generally replaced floppy disks for data backup and can store about 500 times the amount of information.

67. How many significant digits are in the following unit factors?
    (a) 1 m/10 dm **(b)** 1 lb/454 g
    (c) 1 L/1000 mL **(d)** 1 qt/946 mL
68. Which of the following are exact equivalent relationships?
    (a) 1 m = 10 dm **(b)** 1 lb = 454 g
    (c) 1 L = 1000 mL **(d)** 1 qt = 946 mL
69. An automobile airbag inflates to 50.0 L in 35 ms. What is the time of inflation in microseconds?
70. An automobile antilock brake system (ABS) operates the brakes at 30 pulses per second. How many times do the brakes pulse in 1.00 ds?
71. A light year is the distance light travels in 1.00 year. Given the velocity of light, 1.86×10^5 mi/s, how many miles does light travel in a light year?
72. A parsec is the distance light travels in 3.26 years. Given the velocity of light, 3.00×10^8 m/s, how many kilometers does light travel in a parsec?
73. Olympic athletes compete in a 1500-meter event but not in a mile event. Which race is shorter: 1500 meters or 1 mile?
74. An oxygen molecule travels 975 mi/h at room temperature. What is the velocity in meters per second?
75. How many 325 mg tablets can be produced from 2.50 kg of powdered aspirin?
76. How many molecules of water are in one drop if there are 3.34×10^{22} molecules in 1.00 g (given that 1 mL = 20 drops)?
77. A basketball court measures 94.0 feet by 50.0 feet. Calculate the playing area in square meters (given that 1 yd = 0.914 m)?
78. A football field measures 100.0 yards by 160.0 feet. Calculate the playing area in square yards.
79. What is the density of a metal sample if a 37.51 g sample added into a graduated cylinder increased the liquid level from 50.0 mL to 57.5 mL?
80. What is the mass of 275 L of seawater if the density is 1.025 g/cm^3?
81. What is the specific gravity of gasohol if the density is 0.801 g/mL?
82. What is the density of jet fuel if the specific gravity is 0.775?
83. Express the density of water in the English units of pounds per cubic foot.
84. Express the density of water in the English units of pounds per gallon.
85. The space shuttle uses liquefied hydrogen at a temperature of –422 °F. What is the equivalent Kelvin temperature?
86. The temperature in the Mojave Desert in California has reached 324 K. What is the equivalent Fahrenheit temperature?
87. A red-hot fireplace poker is plunged into 3500 g of water at 22 °C. If the water temperature rises to 51 °C, how many kilocalories of heat are lost by the poker?
88. A 10.0 g sample of metal at 19.0 °C is dropped into 75.0 g of water at 80.0 °C. If the resulting temperature is 78.3 °C, what is the metal? (*Hint:* Refer to Table 3.5.)
89. Calculate the mass of titanium [0.125 cal/(g × °C)] that requires 75.6 cal to heat the metal from 20.7 °C to 31.4 °C.
90. Calculate the mass of lead [0.0308 cal/(g × °C)] that releases 52.5 cal as the metal cools from 35.7 °C to 25.1 °C.
91. What is the temperature change when 10.5 g of silver [0.0566 cal/(g × °C)] absorbs 35.2 cal of heat?
92. What is the temperature change when 8.92 g of sodium [0.293 cal/(g × °C)] loses 750 cal of heat energy?
93. A 10.0 g sample of copper [0.0920 cal/(g × °C)] gains 27.8 cal of heat. If the initial temperature is 22.7 °C, what is the final temperature?
94. A 15.5 g sample of titanium [0.125 cal/(g × °C)] loses 32.9 cal of heat. If the initial temperature is 28.9 °C, what is the final temperature?
95. The density of mercury is 13.6 g/mL. Express the density in SI units (kg/m^3).
96. The specific heat of mercury is 0.0331 cal/(g × °C). Express the specific heat in SI units J/(kg × K).
97. The radius (r) of the international prototype kilogram cylinder is 1.95 cm. Assuming the density of the kilogram is 21.50 g/cm^3, calculate its height (h). The volume of a cylinder equals $\pi r^2 h$, where π is the constant 3.14.

3 Chapter Self-Test

Answers to Self-Test are in Appendix J.

1. Which of the following is a basic unit (symbol) in the metric system? (Sec. 3.1)
   (a) meter (m)
   (b) gram (gm)
   (c) milliliter (mL)
   (d) all of the above
   (e) none of the above
2. What are the two unit factors that relate meters and kilometers? (Sec. 3.2)
   (a) 1000 km/1 m and 1000 m/1 km
   (b) 1000 km/1 m and 1 m/1000 km
   (c) 1 km/1000 m and 1 m/1000 km
   (d) 1 km/1000 m and 1000 m/1 km
   (e) none of the above
3. If a computer chip switches off-on-off in 0.015 μs, what is the switching time in nanoseconds? (Sec. 3.3)
   (a) 0.000 000 015 ns **(b)** 15 ns
   (c) 15,000 ns **(d)** 15,000,000 ns
   (e) none of the above
4. If a bottle of mineral water contains 2.00 L, what is the volume in quarts? (Sec. 3.4)
   (a) 1.06 qt **(b)** 1.89 qt
   (c) 2.11 qt **(d)** 2110 qt
   (e) 1,890,000 qt
5. If a rectangular brass block measures 3.80 cm by 2.55 cm by 1.25 cm, what is the volume of the brass solid? (Sec. 3.5)
   (a) 0.0826 cm^3 **(b)** 1.19 cm^3
   (c) 1.86 cm^3 **(d)** 7.75 cm^3
   (e) 12.1 cm^3
6. What is the volume of water displaced by a sample of fool's gold that raises the water level from 25.5 mL to 37.0 mL in a graduated cylinder? (Sec. 3.6)
   (a) 11.5 mL **(b)** 25.5 mL
   (c) 37.0 mL **(d)** 62.5 mL
   (e) none of the above

◀ **Fool's Gold** Iron pyrite, FeS_2, is commonly referred to as fool's gold because of its yellow metallic luster.

7. If a 100.0 mL of gasohol has a mass of 74.5 g, what is the density of the gasohol in grams per cubic centimeter? (Sec. 3.7)
   (a) 0.0745 g/cm^3 **(b)** 0.745 g/cm^3
   (c) 7.45 g/cm^3 **(d)** 10.0 g/cm^3
   (e) 74.5 g/cm^3
8. The space shuttle uses liquid hydrogen at a temperature of 21 K. What is the temperature on the Celsius scale? (Sec. 3.8)
   (a) 294 °C **(b)** 252 °C
   (c) 21 °C **(d)** –252 °C
   (e) –294 °C
9. A beaker contains 1.00 kg of water at 50.0 °C. What is the heat loss in kilocalories when the water cools to 20.0 °C? (Sec. 3.9)
   (a) 20.0 kcal **(b)** 30.0 kcal
   (c) 50.0 kcal **(d)** 10,000 kcal
   (e) 30,000 kcal

Key Concepts

10. A tall glass cylinder contains water (d = 1.00 g/mL) floating on liquid mercury (d = 13.6 g/mL). If a piece of fool's gold (d = 5.00 g/mL) and a gold nugget (d = 18.3 g/mL) are dropped into the cylinder, where does each come to rest?
11. Which of the following temperatures is the coldest?

    0 °F, 0 °C, or 0 K

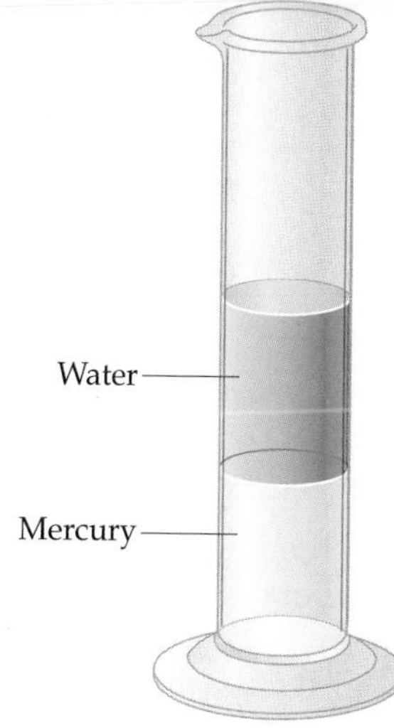

◀ **The Density Concept** A glass cylinder is shown containing water and liquid mercury.

Critical Thinking

12. Given two organic liquids, hexane (d = 0.70 g/mL) and chloroform (d = 1.5 g/mL), how could you identify each liquid given only a beaker of water?
13. A 10¢ silver dime is smaller and thinner than a 1¢ zinc penny. Why is the mass of the 10¢ coin heavier than the larger 1¢ coin?

CHAPTER

4 Matter and Energy

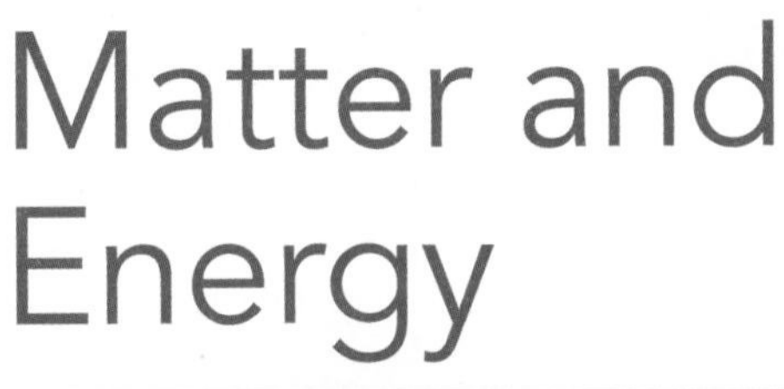

The snow pack at the top of the mountain represents potential energy. As the snow slides, the moving mass represents kinetic energy. The total of the potential and kinetic energy is constant.

"Matter is discontinuous and cannot be infinitely divided."
Democritus, Greek Philosopher (450–370 B.C.)

"Matter is continuous and can be infinitely divided."
Aristotle, Greek Philosopher (384–322 B.C.)

In the seventeenth century, the British scientist Robert Boyle (1627–1691) established the importance of experiments in the study of science. Boyle realized the value of laboratory research and described his work thoroughly so that other scientists could repeat his procedures and confirm his observations. In 1661, he published *The Sceptical Chymist,* which marked a turning point in science by suggesting that theories were no better than the experimental methods that supported them.

Boyle rejected the Greek notion that air, earth, fire, and water were basic elements. Instead, he proposed that an element was a substance that could not be broken down further. Thus, an element must gain weight when it undergoes a chemical change and combines with another element. This was a practical proposal that could be tested in the laboratory.

Following the lead of Boyle, others established scientific principles by performing experiments that could be verified by fellow scientists. Although scientists continued to propose theories about the behavior of matter, their theories were now based on laboratory evidence.

4.1 PHYSICAL STATES OF MATTER

OBJECTIVE

▶ To describe the motion of particles in the solid, liquid, and gaseous states of matter.

Matter is defined as a substance that has mass and occupies volume. It exists in one of three **physical states**: solid, liquid, or gas. In the solid state, matter has a fixed shape and a fixed volume. The particles in a solid are tightly packed, held in rigid positions, and cannot be compressed together. In the liquid state, matter has a variable shape, but a fixed volume. The particles in a liquid are loosely packed, free to move around, and compress only slightly.

In the gaseous state, tiny particles of matter are widely spaced apart and uniformly distributed throughout the container. If the volume increases, the gas expands and the particles move farther away from each other. If the volume decreases, the gas compresses and the particles move closer to one another. Table 4.1 summarizes the properties of the three states of matter.

TABLE 4.1 PHYSICAL STATES OF MATTER

PROPERTY	SOLID	LIQUID	GAS
shape	fixed	variable	variable
volume	fixed	fixed	variable
compressibility	negligible	negligible	significant

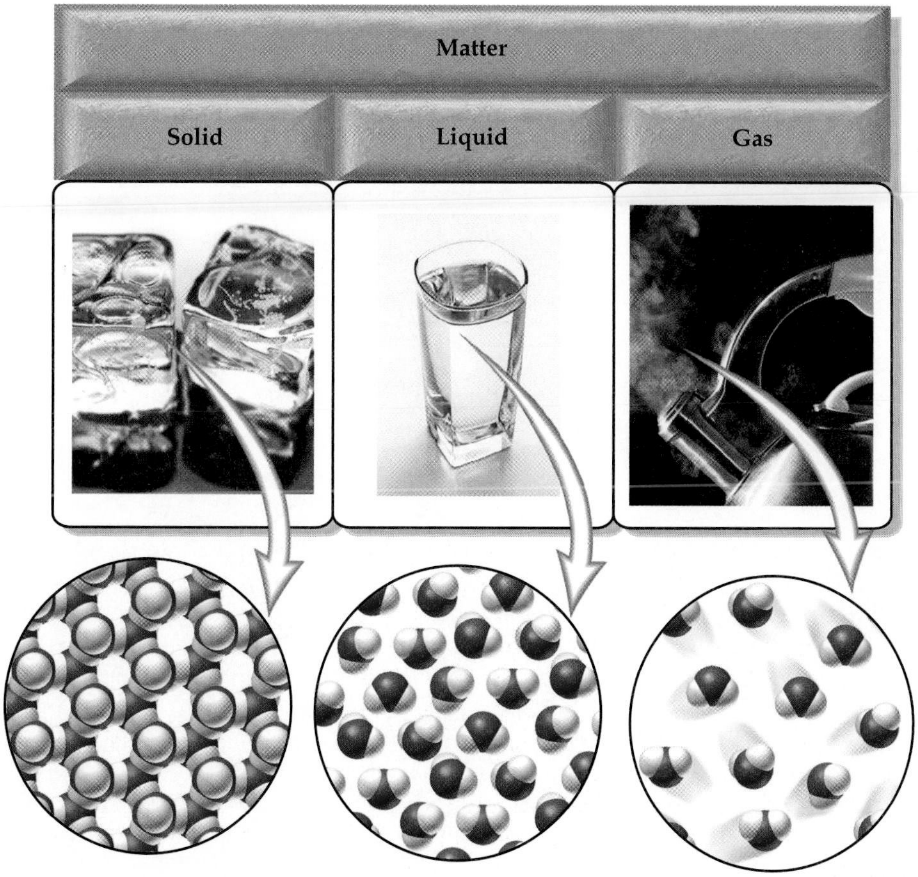

Matter

Solid | Liquid | Gas

◀ **Physical States of Matter**
Ice, water, and steam represent the three states of matter: solid, liquid, and gas.

▶ **Sublimation and Deposition** Iodine crystals in a heated beaker undergo sublimation to a purple gas. As the gas cools, it undergoes deposition to a solid on the upper rim of the beaker.

Even though we may describe a substance as being a solid, every substance can exist as a solid, liquid, or gas. We can change the physical state of a substance by changing the temperature. Water, for example, can be changed from a liquid to solid ice at 0 °C, and from a liquid to gaseous steam at 100 °C. Other substances behave similarly at different temperatures. Iron, for example, can be changed to a molten liquid at 1535 °C and to a gas at 2750 °C.

We can describe changes of physical state as follows. As temperature increases, a solid *melts* to a liquid, and then the liquid *vaporizes* to a gas. A direct change of state from a solid to a gas is called **sublimation**. Mothballs, for example, undergo sublimation and disappear as they change directly from a solid to a gas.

Conversely, as temperature decreases, a gas *condenses* to a liquid, and the liquid then *freezes* to a solid. A direct change of state from a gas to a solid is called **deposition**. The freezer compartment in a refrigerator may demonstrate deposition by collecting ice. Opening the refrigerator door allows moist air to enter, which deposits frost without a trace of liquid water. Figure 4.1 shows the relationship of temperature and physical state.

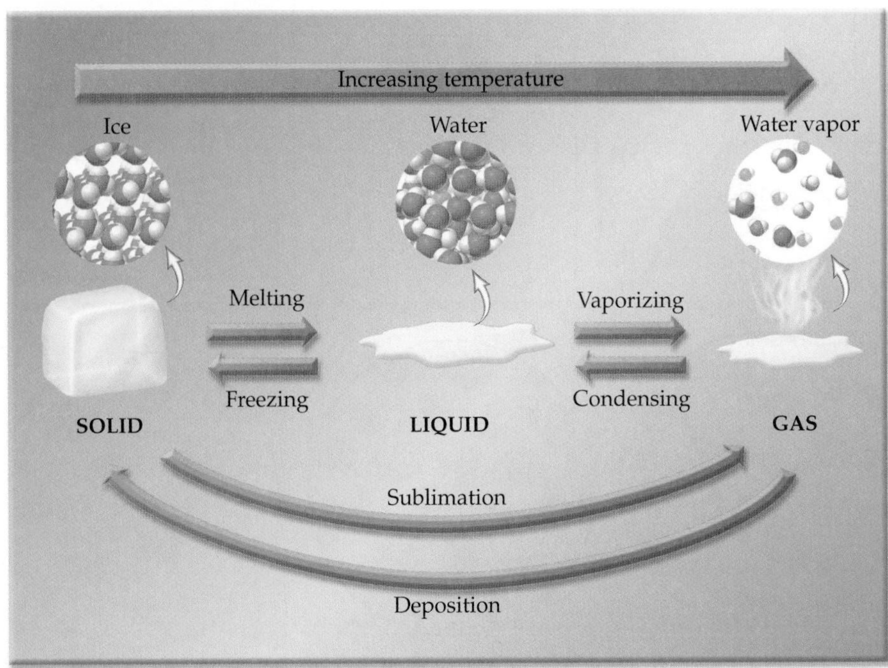

▶ **Figure 4.1 Changes in Physical State** As temperature increases, a solid melts to a liquid and then vaporizes into a gas. As temperature decreases, a gas condenses to a liquid and then freezes to a solid.

Example Exercise 4.1 Change of Physical State

State the term that applies to each of the following changes of physical state:

(a) Snow changes from a solid to a liquid.
(b) Gasoline changes from a liquid to a gas.
(c) Dry ice changes from a solid to a gas.

Solution

Refer to Figure 4.1 for the changes of physical state.

(a) The change from solid to liquid is called *melting*.
(b) The change from liquid to gas is called *vaporizing*.
(c) The change from solid to gas is called *sublimation*.

Practice Exercise

State the term that applies to each of the following changes of physical state:

(a) A refrigerant changes from a gas to a liquid.
(b) Water changes from a liquid to a solid.
(c) Iodine vapor changes from a gas to a solid.

Answers: (a) condensing; (b) freezing; (c) deposition

Concept Exercise

Identify the physical state (solid, liquid, gas) that corresponds to each of the following pictorial representations:

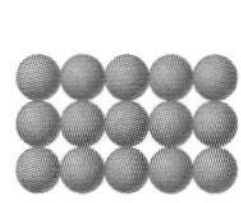

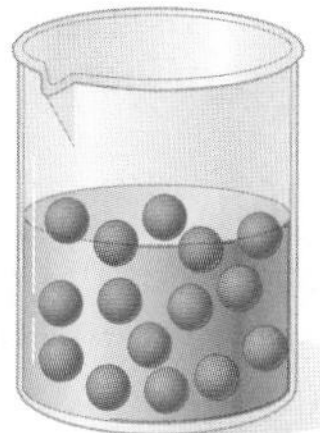

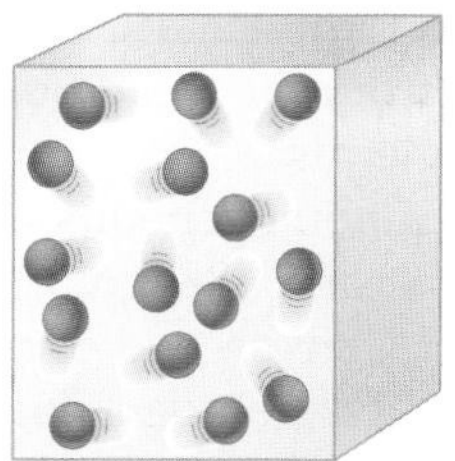

Answer: See Appendix G.

4.2 ELEMENTS, COMPOUNDS, AND MIXTURES

OBJECTIVE

▸ To classify a sample of matter as an element, compound, or mixture.

A sample of matter may have properties that are consistent throughout or that vary. One way to tell if the properties of matter are consistent is to melt the sample. A sample of pure gold always melts at 1064 °C and does not vary. It makes no difference whether it is a large gold nugget or a small gold flake; the yellow metal consistently melts at 1064 °C.

Now consider a sample of quartz rock that contains a vein of pure gold. Although the quartz sample is observed to melt over a broad range from 1000 °C to 1600 °C, pure gold always melts precisely at 1064 °C. Scientists classify the quartz rock as heterogeneous and gold as homogeneous. A sample of matter is said to be *heterogeneous* if its properties are indefinite and vary. A sample of pure gold is said to be *homogeneous* because its properties are definite and constant.

A **heterogeneous mixture** can be separated into pure substances by physical methods. For example, panning for gold uses the property of density to separate pure gold from sand and rock. Gold is quite dense, 19.3 g/cm^3, and remains at the bottom of the pan, while the less dense sand and rock are swirled away with water.

Air is not a heterogeneous mixture, but rather a homogeneous mixture of nitrogen, oxygen, and other gases. Similarly, salt water is a homogeneous mixture of salt and water. Unlike those of a heterogeneous mixture, the properties of a **homogeneous mixture** are constant for a given sample. However, a homogeneous mixture may have properties that vary from sample to sample. For example, samples

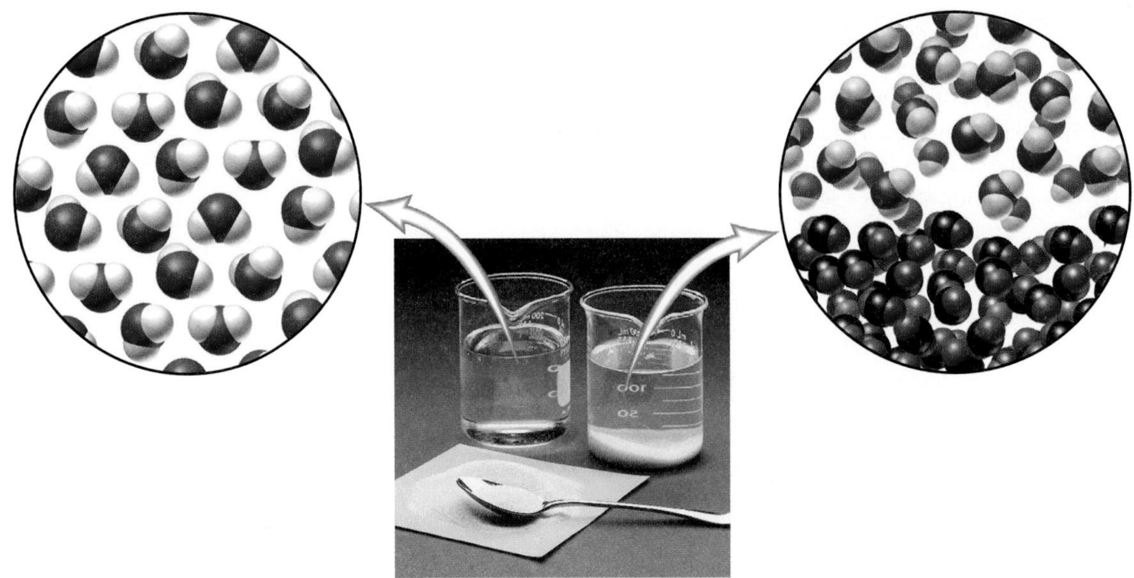

▲ **Heterogeneous Mixture** Sand and water are each pure substances. However, they form a heterogeneous mixture when added together, and the properties vary throughout the sample.

of salt water from the Pacific Ocean and from the Dead Sea have different properties. Salt water from the Dead Sea has a higher density and a greater concentration of dissolved minerals.

An **alloy** is a homogeneous mixture of two or more metals. Examples of alloys include 10K, 14K, and 18K gold jewelry. Although 10K, 14K, and 18K jewelry may contain only gold, silver, and copper metals, the amount of gold varies from 42% to 75%. Gold alloy is a homogeneous mixture; therefore the properties from different samples can vary. For instance, 10K gold is a harder alloy and is more scratch resistant than 18K gold.

A **substance** is matter that has definite composition and constant properties. A substance is either a compound or an element. A **compound** has predictable properties, but can be broken down into elements by an ordinary chemical reaction. Table sugar is an example of a compound that can be broken down into carbon, hydrogen, and oxygen.

An **element** is a substance that cannot be broken down further by a chemical reaction. For example, table salt is a compound composed of the elements sodium and chlorine, which cannot be broken down any further. Figure 4.2 shows the overall classification of matter.

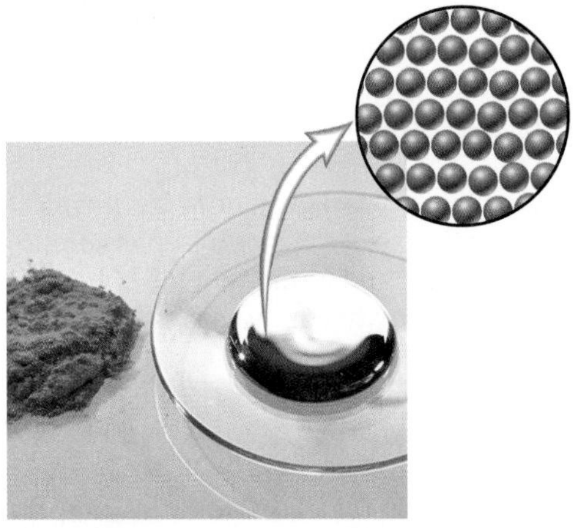

◀ **Compound, Element, Mixture** The orange compound mercury oxide decomposes to give the element mercury, which is a silver liquid. Adding the orange powder and the metallic liquid together produces a heterogeneous mixture.

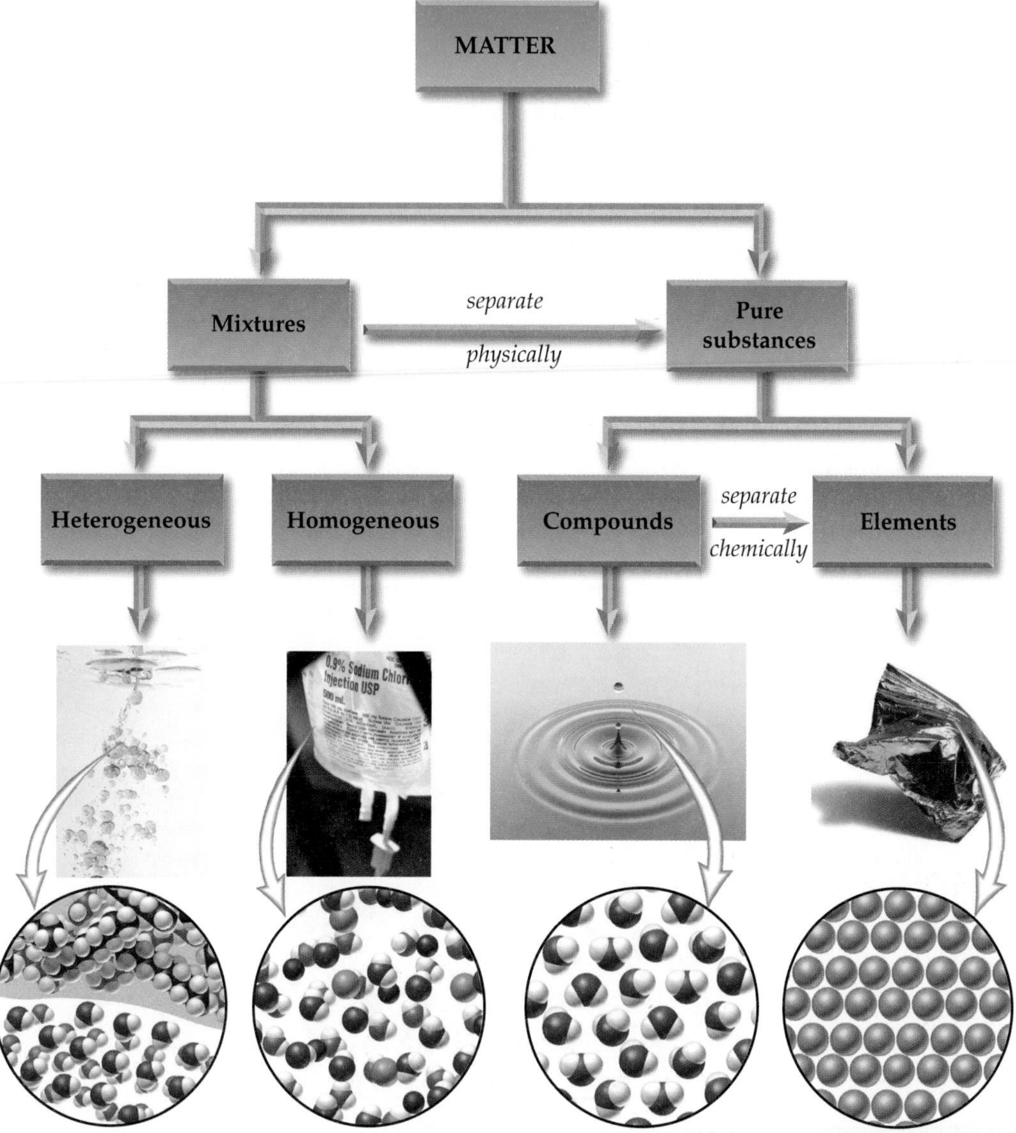

▲ **Figure 4.2 Classification of Matter** Matter may be either a mixture or a pure substance. The properties of a heterogeneous mixture vary within the sample (oil and water). The properties of a homogeneous mixture are constant (salt solution). A pure substance may be either a compound (water) or an element (gold). Left to right: oil and water; NaCl solution; H_2O; and gold nugget.

Example Exercise 4.2 Element, Compound, or Mixture

Consider the following properties of the element copper:

(a) Copper metal cannot be broken down by a chemical change.
(b) Copper reacts with oxygen in air to give copper oxide.
(c) Copper, in the form of malachite ore, is found worldwide.
(d) Copper and tin compose bronze alloy.

Classify each of the following copper samples as an element, a compound, a homogeneous mixture, or a heterogeneous mixture:

(a) copper wire (b) copper oxide
(c) malachite ore (d) bronze alloy

Solution

Refer to Figure 4.2 to classify each sample.

(a) Copper wire is a metallic *element*.
(b) Copper oxide is a *compound* of the elements copper and oxygen.

(c) Malachite ore is a *heterogeneous mixture* of copper and other substances.
(d) Bronze alloy is a *homogeneous mixture* of copper and tin.

Practice Exercise

Consider the following properties of the element mercury:

(a) Mercury liquid cannot be broken down by a chemical change.
(b) Mercury oxide can be heated to give mercury and oxygen gas.
(c) Mercury, in the form of cinnabar ore, is found in Spain and Italy.
(d) Mercury and silver compose the alloy used for dental fillings.

Classify each of the following mercury samples as an element, a compound, a homogeneous mixture, or a heterogeneous mixture:

(a) mercury liquid (b) mercury oxide
(c) cinnabar ore (d) dental alloy

Answers: (a) element; (b) compound; (c) heterogeneous mixture; (d) homogeneous mixture

Concept Exercise

Classify each of the following as an element, a compound, or a mixture as shown in the illustration:

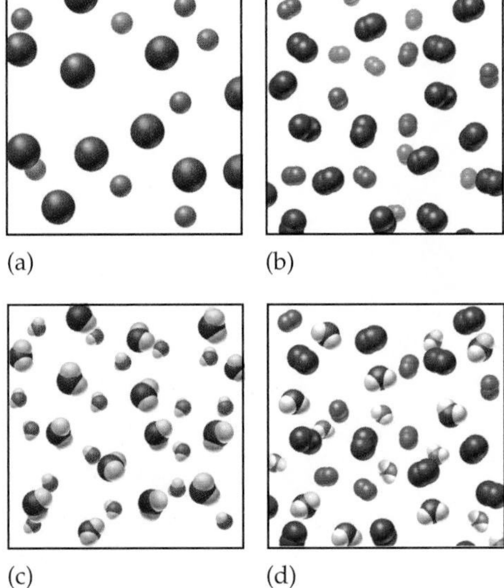

Answer: See Appendix G.

OBJECTIVE

▸ To relate the names and symbols of 48 common elements.

4.3 NAMES AND SYMBOLS OF THE ELEMENTS

There are 81 stable elements that occur in nature. In addition, there are a few other naturally occurring elements, such as uranium, which are unstable. In total, there are over 100 elements, but only 10 account for 95% of the mass of Earth's crust, water, and atmosphere.

Oxygen is the most abundant element on Earth. It is found combined with hydrogen in water and with silicon in sand and rocks. Oxygen is found in air and constitutes about 21% of our atmosphere. The mass of oxygen in Earth's crust, water, and atmosphere is about equal to the total mass of all other elements. Table 4.2 lists the ten most abundant elements.

Oxygen, silicon, and aluminum are the three most abundant elements on Earth. The elements oxygen, carbon, hydrogen, nitrogen, calcium, and phosphorus account for over 99% of the mass of the human body. The remaining 1% consists of trace elements, many of which are essential to human life. For example, a trace of iron is necessary to bind oxygen to hemoglobin in blood. Figure 4.3 compares the distribution of elements in Earth's crust and in the human body.

TABLE 4.2 ELEMENTS IN EARTH'S CRUST, WATER, AND ATMOSPHERE

ELEMENT	MASS PERCENT	ELEMENT MASS	PERCENT
oxygen	49.5%	sodium	2.6%
silicon	25.7%	potassium	2.4%
aluminum	7.5%	magnesium	1.9%
iron	4.7%	hydrogen	0.9%
calcium	3.4%	titanium	0.6%
		all other elements	0.5%

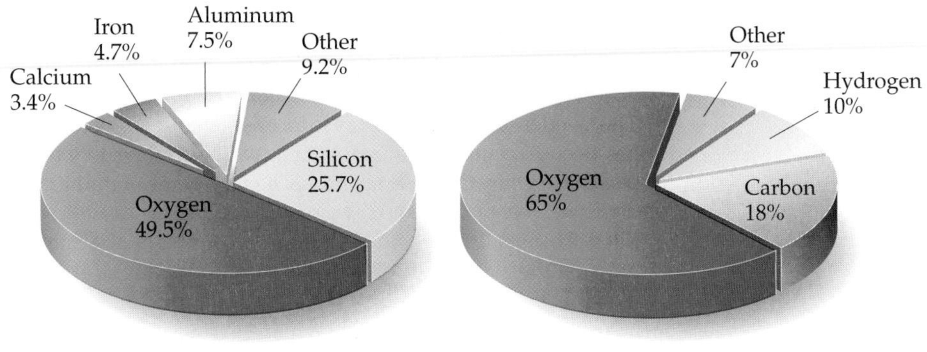

▲ **Figure 4.3 Abundance of Elements** Notice that oxygen is the most abundant element in both Earth's crust and the human body. Although aluminum is abundant in Earth, there is evidence that this element is toxic to the human body.

The names of elements are derived from various sources. For example, hydrogen is derived from the Greek word meaning "water former." Carbon is derived from the Latin word *carbo*, meaning "coal." Calcium is derived from the Latin word *calcis*, which translates as "lime," a mineral source of calcium. Some elements are named for their region of discovery. For example, germanium comes from Germany, and scandium from Scandinavia. Several elements are named for a famous scientist, such as curium (Marie Curie) and nobelium (Alfred Nobel).

The name of each element is abbreviated using a **chemical symbol**. In 1803, the English chemist John Dalton (1766–1844) proposed that elements are composed of indivisible, spherical particles. Dalton referred to each of these individual particles as an **atom**, from the Greek word *atomos* meaning "indivisible." He suggested the use of circles with enclosed markings as symbols for the elements. Figure 4.4 shows selected symbols chosen by Dalton to represent elements.

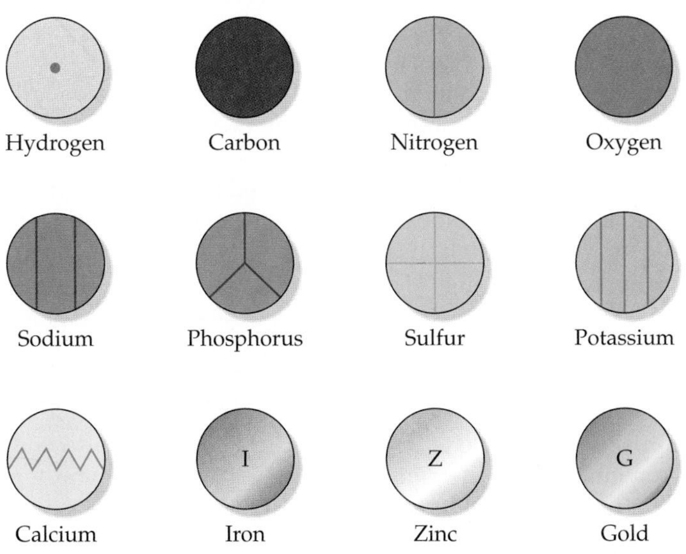

◀ **Figure 4.4 Dalton's Symbols for Selected Elements** Dalton symbolized each element using a circle with an inscribed marking. Dalton's system was not practical, but it did convey the idea that elements were composed of atomic spheres.

CRITICAL THINKING Aluminum or Aluminium?

Q: What suffix is commonly attached to the name of an element to indicate the element is a metal?

At the turn of the nineteenth century, many new elements were isolated and added to a growing list. For example, sodium, potassium, magnesium, calcium, strontium, and barium were all discovered in the period 1807–1808 by the English chemist Humphry Davy (1778–1829). Metallic elements were systematically named by adding an *–ium* suffix to its root name. The nonmetal selenium was discovered a few years later in 1817, and named for the Greek word *selene,* which means moon. The semimetal tellurium had been previously discovered in 1782, and named for the Latin word *tellus,* which means earth.

◀ **Aluminum Cap on the Washington Monument** The Washington Monument, in Washington, D.C., is capped by a shiny metal pyramid. The thin metal pyramid weighing 2.85 kilograms is composed of aluminum, which was an expensive metal at the time of its construction in 1884.

In 1808, Davy proposed the name *alumium* for an element that he was attempting to isolate from the mineral alumina. A couple of years later he changed the name to *aluminum* to correspond to its Latin root (e.g., ferrum for iron). Later, Davy decided to use the *–ium* suffix to be consistent with other newly discovered elements. Thus, Davy proposed the name *aluminium* for the element that was not isolated in a large quantity until Charles Martin Hall accomplished the feat in 1885 (see Chemistry Connection in Chapter 2).

Today, the official IUPAC name for the element is *aluminium;* however, both Americans and Canadians refer to it as aluminum. This condensed ending, which is inconsistent with other metals, has been explained in a couple of ways. According to Oberlin College in Ohio, Charles Hall misspelled the name of the element on a notice that publicized his discovery of the process that separated the metal from its ore. Another account is that a spelling error on the stationery of the Aluminum Company of America (ALCOA), a company that Charles Hall helped to found, was instrumental in dictating the name. In either case, it appears that a spelling error was responsible for the irregular ending for the name of the element that is extremely important today, and second only to steel as a construction metal.

For most of the nineteenth century, Americans used the *–ium* suffix and *aluminium* appeared in Webster's dictionary of 1828. However, the misspelling of the metal was common, and in 1926 the American Chemical Society decided officially to call the metal aluminum. Meantime, in Europe the *–ium* ending was the accepted spelling. In 1990, IUPAC adopted *aluminium* as the official international name for the metal, which is used by French, German, Danish, Dutch, Norwegian, and Swedish scientists. In 1993, IUPAC officially approved the alternate spelling aluminum, but stated a preference for the name *aluminium.*

A: An *–ium* suffix is the most common ending for metallic elements although there are a few exceptions such as platinum (Pt), molybdenum (Mo), and tantalum (Ta). In addition, the Latin names of elements known since ancient times such as cuprum (Cu), argentum (Ag), and aurum (Au) are exceptions.

In 1813, the Swedish chemist J. J. Berzelius (1779–1848) proposed our current system of symbols for the elements. He suggested that a symbol correspond to the first letter of the name of the element, for example, H for hydrogen, O for oxygen, and C for carbon. Furthermore, when elements start with the same letter, he recommended using two letters in the name, for example, Ca for calcium, Cd for cadmium, and Cl for chlorine.

In some instances, the chemical symbol is derived from the original Latin name of the element. For example, the symbol Pb for lead is derived from the Latin *plumbum.* Interestingly, the Romans used lead pipe to transport water, which explains the derivation of our word "plumbing." The symbol Hg for mercury is derived from the Greek *hydrargyro,* meaning "liquid silver." Similarly, the symbols for gold (Au), copper (Cu), iron (Fe), mercury (Hg), potassium (K), sodium (Na), antimony (Sb), and tin (Sn), are derived from the original Latin and Greek names.

When writing symbols for the names of elements, it is important to follow convention to avoid confusion. The first letter is always capitalized, and the second is lowercase. For example, the symbol for the metallic element cobalt is Co. In contrast, notice that the formula for the deadly gas carbon monoxide is CO. Table 4.3 lists the names and symbols of selected elements.

TABLE 4.3 NAMES AND SYMBOLS OF SELECTED ELEMENTS

NAME OF ELEMENT	SYMBOL	NAME OF ELEMENT	SYMBOL	NAME OF ELEMENT	SYMBOL	NAME OF ELEMENT	SYMBOL
aluminum	Al	chlorine	Cl	lead	Pb	radium	Ra
antimony	Sb	chromium	Cr	lithium	Li	selenium	Se
argon	Ar	cobalt	Co	magnesium	Mg	silicon	Si
arsenic	As	copper	Cu	manganese	Mn	silver	Ag
barium	Ba	fluorine	F	mercury	Hg	sodium	Na
beryllium	Be	germanium	Ge	neon	Ne	strontium	Sr
bismuth	Bi	gold	Au	nickel	Ni	sulfur	S
boron	B	helium	He	nitrogen	N	tellurium	Te
bromine	Br	hydrogen	H	oxygen	O	tin	Sn
cadmium	Cd	iodine	I	phosphorus	P	titanium	Ti
calcium	Ca	iron	Fe	platinum	Pt	xenon	Xe
carbon	C	krypton	Kr	potassium	K	zinc	Zn

4.4 METALS, NONMETALS, AND SEMIMETALS

OBJECTIVES

- To distinguish between the properties of metals and nonmetals.
- To predict whether an element is a metal, nonmetal, or semimetal given its position in the periodic table.
- To predict whether an element is a solid, liquid, or gas at 25 °C and normal atmospheric pressure.

An element is a pure substance that cannot be broken down further and still maintain its unique properties. A **metal** is an element that typically is a solid, has a bright metallic luster, a high density, a high melting point, and is a good conductor of heat and electricity. A metal can usually be hammered into a thin sheet of foil and is said to be **malleable**. If it can be drawn into a fine wire, it is said to be **ductile**. Aluminum, copper, and silver are familiar examples of metals.

A **nonmetal** is an element that usually has a low density, a low melting point, and is a poor conductor of heat and electricity. Many nonmetals occur naturally in the solid state and have a dull appearance, for example, phosphorous and sulfur. These solid nonmetals are neither malleable nor ductile and crush to a powder if hammered. Eleven nonmetals occur naturally in the gaseous state; for example, hydrogen and oxygen are familiar examples that are colorless, gaseous nonmetals. Table 4.4 summarizes the properties of metals and nonmetals.

TABLE 4.4 GENERAL CHARACTERISTICS OF METALS AND NONMETALS*

PROPERTY	METALS	NONMETALS
physical state	solid	solid, gas
appearance	metallic luster	dull
pliability	malleable, ductile	brittle
conductivity	heat, electricity	nonconductor
density	usually high	usually low
melting point	usually high	usually low
chemical reactivity	react with nonmetals	react with metals and nonmetals

*There are numerous exceptions to these general characteristics. For example, mercury and bromine are liquids, magnesium metal has a low density, and gallium metal has a melting point below 30 °C.

A **semimetal** (or metalloid) is an element that typically has properties midway between those of metals and nonmetals. Silicon is a familiar semimetal used in the semiconductor industry for making transistors and integrated circuits. Example Exercise 4.3 further illustrates the identification of properties of metals and nonmetals.

Example Exercise 4.3 Properties of Metals

Which of the following properties is *not* characteristic of a metal?

(a) good conductor of heat (b) malleable
(c) high melting point (d) reacts with other metals

Solution

Refer to Table 4.4 to classify each of the following properties:

(a) Metals are good conductors of heat.
(b) Metals are malleable.
(c) Metals usually have high melting points.
(d) Metals do *not* react with other metals; they mix to form alloys.

Practice Exercise

Which of the following properties is *not* characteristic of a nonmetal?

(a) insulator of electricity
(b) ductile
(c) low density
(d) reacts with nonmetals

Answer: (b) Nonmetals crush to a powder and are *not* malleable or ductile.

Concept Exercise

Which of the following is a solid metal under normal conditions: calcium, phosphorus, mercury, or silicon? (Refer to Figure 4.7.)

Answer: See Appendix G.

Periodic Table of the Elements

In Chapter 6, we will devote our entire discussion to the chemical elements. For now, realize that each element is assigned a number. The number that identifies a particular element is called the **atomic number**. For example, the atomic number of hydrogen is 1, helium is 2, lithium is 3, and so on, and the atomic number of uranium is 92.

All the elements have been arranged by atomic number and placed in a special chart. This chart is called the periodic table of the elements, or simply the **periodic table**. Metals are placed on the left side of the table, and nonmetals are placed on the right side. Hydrogen is an exception; although it is a nonmetal, hydrogen has unusual properties and has been placed by itself at the top center of the periodic tables in this text.

Metals and nonmetals are separated by semimetals that include boron, silicon, germanium, arsenic, antimony, and tellurium. The unstable elements polonium and astatine are also considered semimetals. Figure 4.5 shows the overall arrangement of metals, nonmetals, and semimetals in the periodic table of the elements.

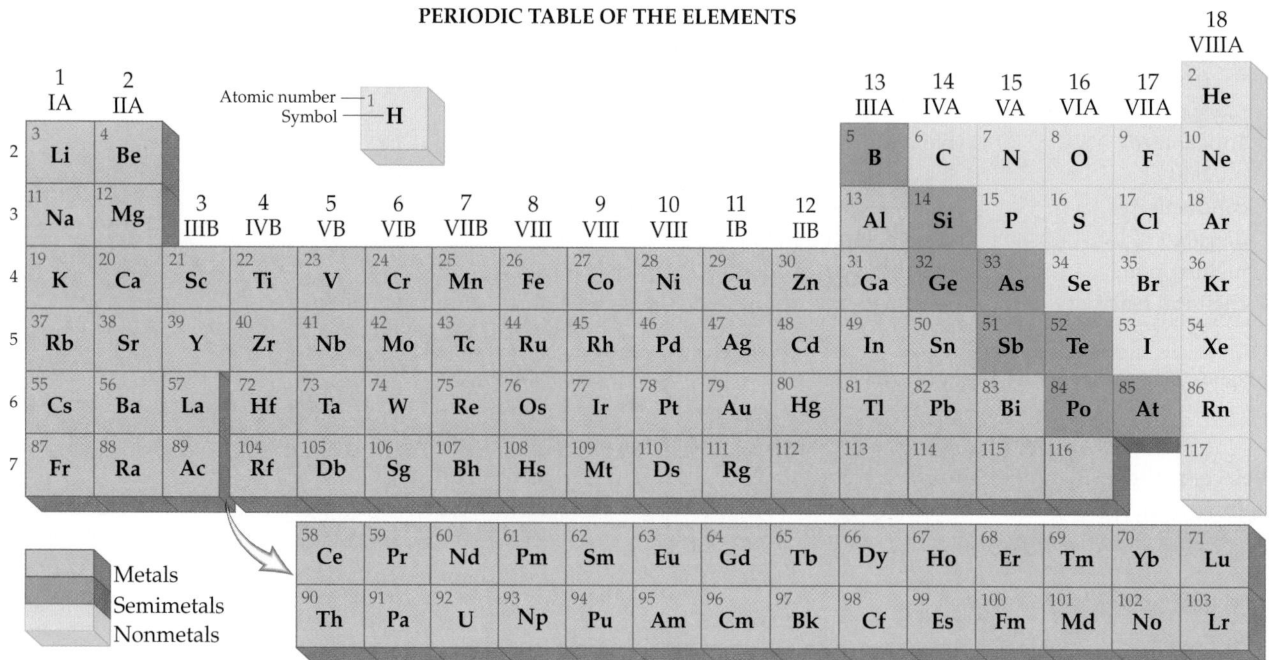

▲ **Figure 4.5 Metals, Nonmetals, and Semimetals** The symbols of elements having metallic properties are on the left side of the periodic table, nonmetallic are on the right side, and semimetallic are midway between. Notice the special placement of hydrogen, a nonmetallic element.

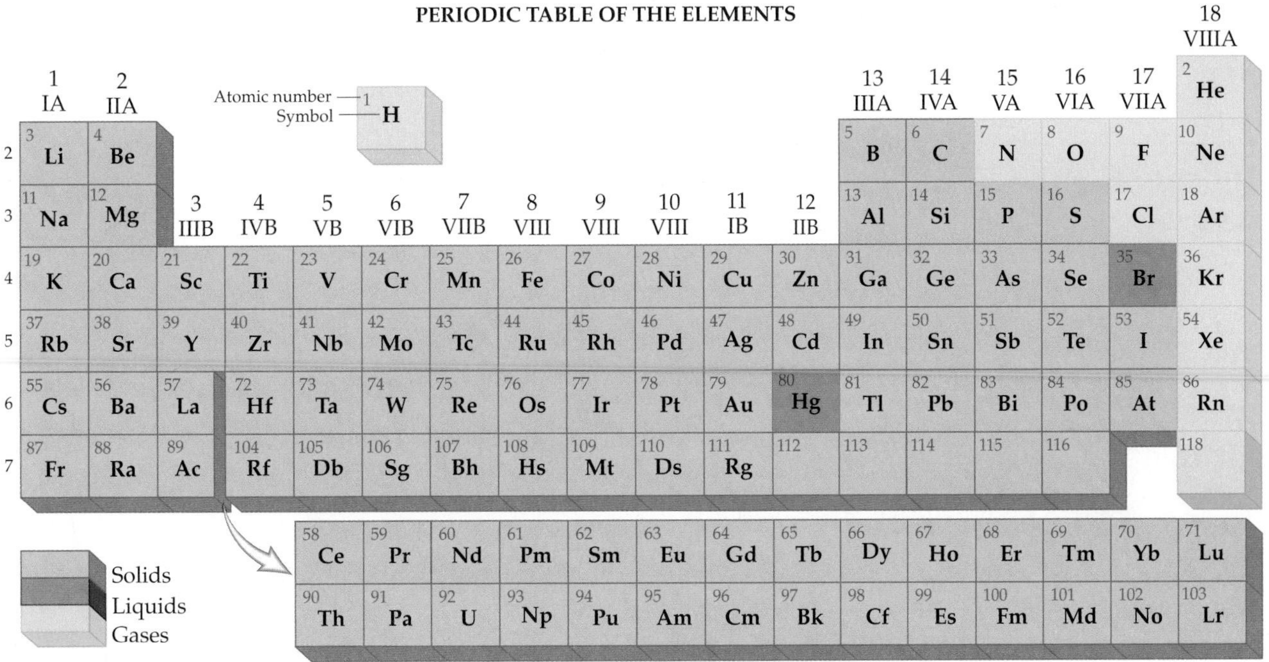

▲ **Figure 4.6 Physical States of the Elements** At 25 °C and normal atmospheric pressure, all metals are in the solid state except Hg. Most nonmetals are gases except C, P, S, Se, and I, which are solids. The only elements in the liquid state at normal conditions are Hg and Br.

Physical States of the Elements

The periodic table can help you master a large amount of information. With little effort, you can correctly predict the physical state of most elements. Excluding mercury, all the metals are in the solid state at normal conditions of 25 °C and normal atmospheric pressure. All semimetals are also in the solid state at normal temperature and pressure.

Nonmetals, on the other hand, show great diversity in physical state. At 25 °C and normal pressure, five nonmetals are solids, one is a liquid, and eleven are gases. The solids are carbon, phosphorus, sulfur, selenium, and iodine. The only nonmetal liquid is reddish-brown bromine. The eleven gases are hydrogen, nitrogen, oxygen, fluorine, chlorine, helium, neon, argon, krypton, xenon, and radon. All these gases are colorless except fluorine and chlorine, which are greenish-yellow. Figure 4.6 illustrates the normal physical state of each element.

Example Exercise 4.4 Physical States of the Elements

Indicate the physical state for each of the following elements at 25 °C and normal pressure; classify each element as a metal, nonmetal, or semimetal:

(a) barium (b) boron
(c) bismuth (d) bromine

Solution

Referring to Figures 4.5 and 4.6, we observe the following:

(a) Barium (Ba) is on the left side of the periodic table; it is a *solid metal* under normal conditions.
(b) Boron (B) is in the middle of the periodic table; it is a *solid semimetal.*
(c) Bismuth (Bi) is to the right, but below the semimetals in the periodic table, it is a *solid metal.*
(d) Bromine (Br) is on the right side of the periodic table; it is a *liquid nonmetal* at normal conditions.

Practice Exercise

Indicate the physical state for each of the following elements at 25 °C and normal pressure; classify each element as a metal, nonmetal, or semimetal:

(a) aluminum (b) hydrogen
(c) helium (d) radium

Now, let's show an overall representation for all of the elements—metals, semimetals, and nonmetals. Figure 4.7 presents a pictorial periodic table of the elements.

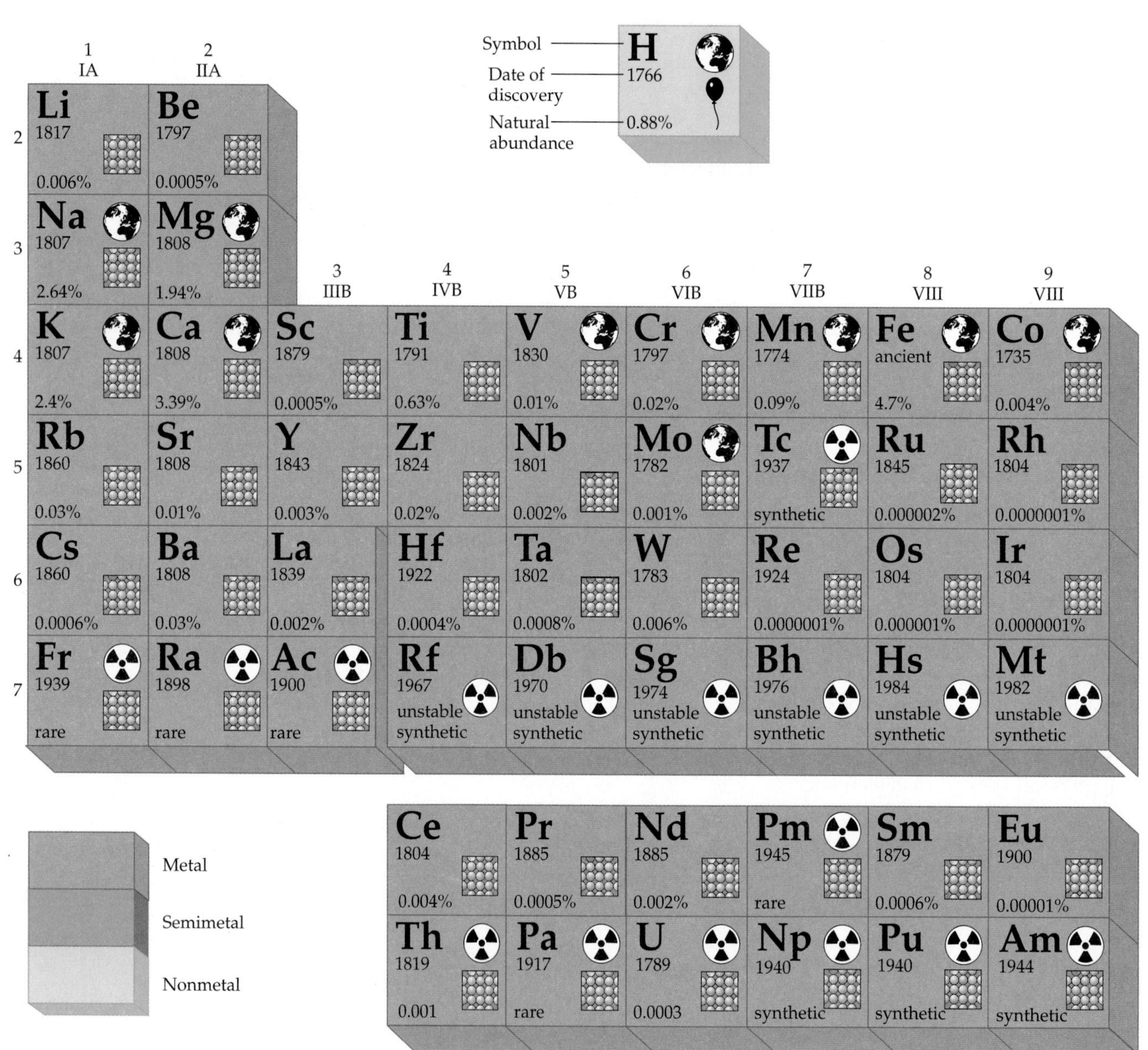

▲ **Figure 4.7 Pictorial Periodic Table of the Elements** The natural abundance is the percent by mass of an element in Earth's crust, oceans, and atmosphere. The natural abundance of an element listed as rare is less than 1 mg per metric ton (1000 kg). An element listed as synthetic is made artificially and does not occur naturally. An element listed as unstable often disintegrates in a fraction of a second.

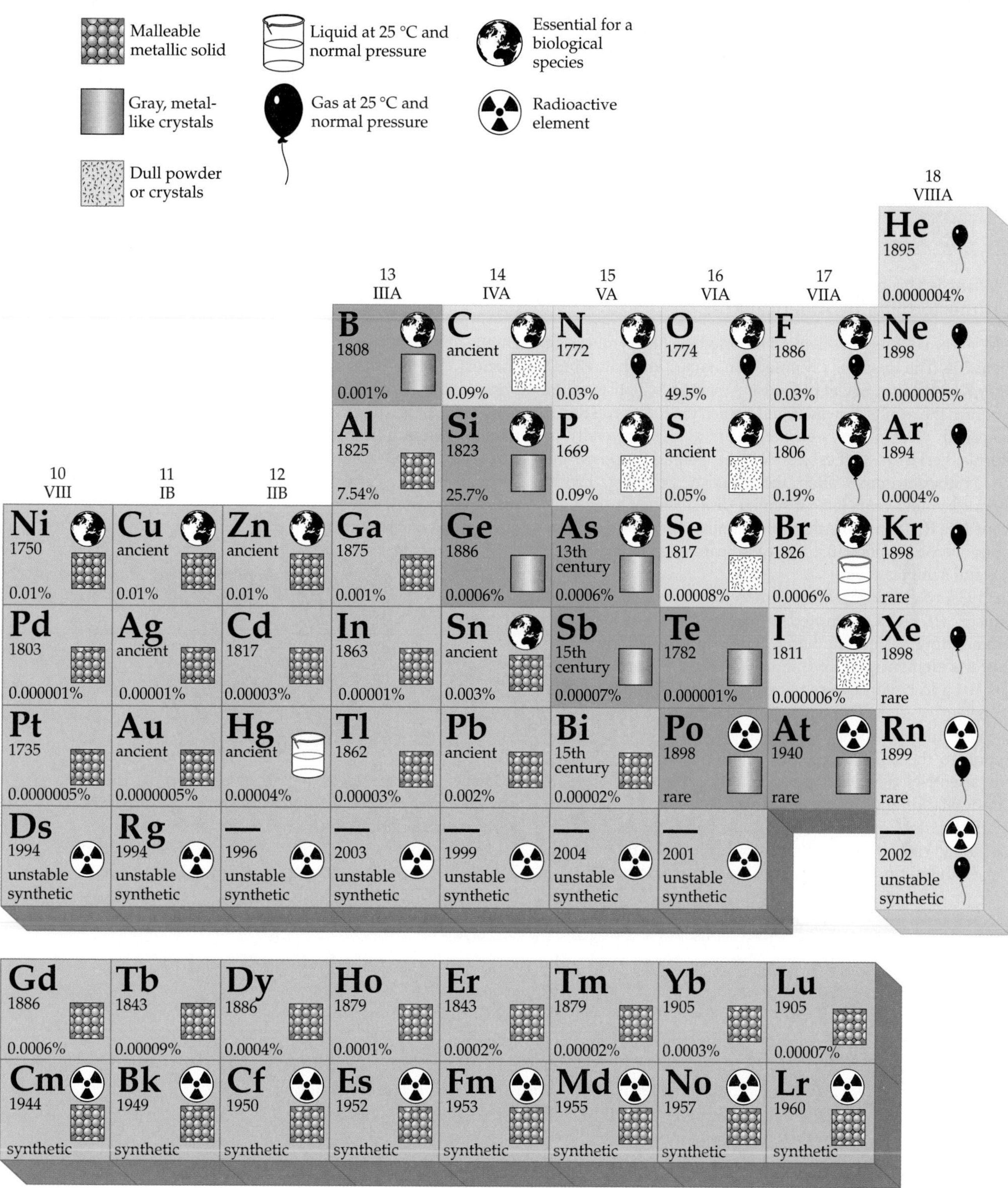
Malleable metallic solid
Gray, metal-like crystals
Dull powder or crystals
Liquid at 25 °C and normal pressure
Gas at 25 °C and normal pressure
Essential for a biological species
Radioactive element
13 IIIA
14 IVA
15 VA
16 VIA
17 VIIA
18 VIIIA
10 VIII
11 IB
12 IIB
He 1895 0.0000004%
B 1808 0.001%
C ancient 0.09%
N 1772 0.03%
O 1774 49.5%
F 1886 0.03%
Ne 1898 0.0000005%
Al 1825 7.54%
Si 1823 25.7%
P 1669 0.09%
S ancient 0.05%
Cl 1806 0.19%
Ar 1894 0.0004%
Ni 1750 0.01%
Cu ancient 0.01%
Zn ancient 0.01%
Ga 1875 0.001%
Ge 1886 0.0006%
As 13th century 0.0006%
Se 1817 0.00008%
Br 1826 0.0006%
Kr 1898 rare
Pd 1803 0.000001%
Ag ancient 0.00001%
Cd 1817 0.00003%
In 1863 0.00001%
Sn ancient 0.003%
Sb 15th century 0.00007%
Te 1782 0.000001%
I 1811 0.000006%
Xe 1898 rare
Pt 1735 0.0000005%
Au ancient 0.0000005%
Hg ancient 0.00004%
Tl 1862 0.00003%
Pb ancient 0.002%
Bi 15th century 0.00002%
Po 1898 rare
At 1940 rare
Rn 1899 rare
Ds 1994 unstable synthetic
Rg 1994 unstable synthetic
— 1996 unstable synthetic
— 2003 unstable synthetic
— 1999 unstable synthetic
— 2004 unstable synthetic
— 2001 unstable synthetic
— 2002 unstable synthetic
Gd 1886 0.0006%
Tb 1843 0.00009%
Dy 1886 0.0004%
Ho 1879 0.0001%
Er 1843 0.0002%
Tm 1879 0.00002%
Yb 1905 0.0003%
Lu 1905 0.00007%
Cm 1944 synthetic
Bk 1949 synthetic
Cf 1950 synthetic
Es 1952 synthetic
Fm 1953 synthetic
Md 1955 synthetic
No 1957 synthetic
Lr 1960 synthetic

Answers: (a) solid metal; (b) gaseous nonmetal; (c) gaseous nonmetal; (d) solid metal

Concept Exercise

Which of the following is a liquid nonmetal under normal conditions: sodium, mercury, bromine, or sulfur? (Refer back to Figure 4.7.)

Answer: See Appendix G.

CHEMISTRY CONNECTION Elements 104 and Beyond

Q: What is the atomic number of the element that was first named kurchatovium by Russian scientists and rutherfordium by Americans?

In 1964, a team of Russian scientists reported the first synthesis of element 104. The Russians proposed the name "kurchatovium" for Igor Kurchatov, a Soviet physicist. American scientists could not confirm the results, however, and denied the Russian claim of discovery. In 1969, a heavy-element team at the University of California, Berkeley, synthesized element 104 and proposed the name "rutherfordium" after the English physicist Ernest Rutherford. A few years later, a similar disagreement arose over element 105. Russian and American scientists each claimed they had discovered element 105, and each team proposed different names and symbols.

In 1985, the International Union of Pure and Applied Chemistry (IUPAC) attempted to resolve the controversy by recommending systematic names based on the atomic numbers of the elements. According to IUPAC, the names for elements 104 and beyond were to be formed from Latin prefixes plus the suffix *–ium*. The name for element 104, for example, was to be unnilquadium. Interestingly, there was a strong objection to the IUPAC recommendation because it did not honor scientists who had made significant contributions to our understanding of heavy elements.

In 1997, IUPAC issued a new set of official names. Element 104 is named rutherfordium for Ernest Rutherford. Element 105 is named dubnium for the Russian nuclear research facility at Dubna. Element 106 is named seaborgium for the American chemist, Glenn Seaborg. Element 107 is named bohrium in honor of the Danish physicist Niels Bohr. Element 108 is named hassium for the province of Hesse in Germany. Element 109 is named meitnerium in honor of the Austrian nuclear physicist Lise Meitner. Element 110 is named darmstadtium for the German nuclear research facility at Darmstadt. Recently, element 111 was named roentgenium for William Roentgen, the German physicist who discovered X rays.

▲ **Particle Accelerators** The "Doomsday Supercollider" is a powerful particle accelerator on the French-Swiss border. Particle accelerators have been used to create new elements and study the subatomic nature of matter by smashing atoms and particles together that are traveling near the speed of light.

ELEMENT	DISCOVERY	PROPOSED NAME	IUPAC NAME
104	Russia, 1964	kurchatovium, Ku	rutherfordium, Rf
	United States, 1969	rutherfordium, Rf	
105	Russia, 1967	nielsbohrium, Ns	dubnium, Db
	United States, 1970	hahnium, Ha	
106	United States, 1974	seaborgium, Sg	seaborgium, Sg
107	Germany, 1974	bohrium, Bh	bohrium, Bh
108	Germany, 1984	hassium, Hs	hassium, Hs
109	Germany, 1982	meitnerium, Mt	meitnerium, Mt
110	Germany, 1994	darmstadtium, Ds	darmstadtium, Ds
111	Germany, 1994	roentgenium, Rg	roentgenium, Rg
112	Germany, 1996	—	—
113	Russia, 2003	—	—
114	Russia, 1998	—	—
115	Russia, 2004	—	—
116	Russia, 2000	—	—
117	not yet confirmed	—	—
118	Russia, 2002	—	—

A: Element 104 was named kurchatovium for the Russian physicist Igor Kurchatov, and rutherfordium for the English scientist Ernest Rutherford.

4.5 COMPOUNDS AND CHEMICAL FORMULAS

OBJECTIVES

- To explain the law of definite composition for a compound.
- To state the number of atoms of each element in a compound given the chemical formula.

In the late 1700s, the French chemist Joseph Louis Proust (1754–1826) painstakingly analyzed the compound copper carbonate. No matter how he prepared the compound, he found that the elements copper, carbon, and oxygen were always present in the same proportion by mass. Proust studied many other compounds and obtained similar constant proportions for the elements. In 1799, Proust stated that "*Compounds always contain the same elements in a constant proportion by mass.*" This statement is now called the **law of definite composition** or the law of constant proportion.

Ordinary table salt is the compound sodium chloride, NaCl. According to the law of definite composition, salt contains sodium and chlorine in a constant proportion by mass. For NaCl, the ratio is 39.3% sodium and 60.7% chlorine. Moreover, the proportion of sodium and chlorine is the same whether we have a tiny crystal of salt, a block of salt, or a mountain of salt.

Similarly, the compound water contains 11.2% hydrogen and 88.8% oxygen by mass. Regardless of the amount, water always contains hydrogen and oxygen in a constant proportion by mass. Figure 4.8 illustrates the law of definite composition with water as an example.

Chemical Formulas

Most elements occur naturally as a collection of individual atoms. A few nonmetal elements, such as hydrogen and oxygen, occur naturally as particles composed of two or more atoms. A single particle composed of two or more nonmetal atoms is called a **molecule**. Hydrogen, H_2, and oxygen, O_2, occur naturally as molecules containing two atoms.

We have learned to use chemical symbols for the names of elements. In a similar fashion, we use chemical formulas for the names of compounds. A **chemical formula** expresses the number of atoms of each element in a compound. The number of atoms is indicated with a subscript, unless the number is 1 in which case the subscript is omitted. A molecule of water contains two atoms of hydrogen and one atom of oxygen. Therefore, its chemical formula is H_2O. A molecule of ammonia has one atom of nitrogen and three atoms of hydrogen; its chemical formula is NH_3. The chemical formula of sulfuric acid, H_2SO_4, is interpreted in Figure 4.9.

▲ **Figure 4.8 Law of Definite Composition** A drop of water, a glass of water, and a lake of water all contain hydrogen and oxygen in the same percent by mass, that is, 11.2% hydrogen and 88.8% oxygen.

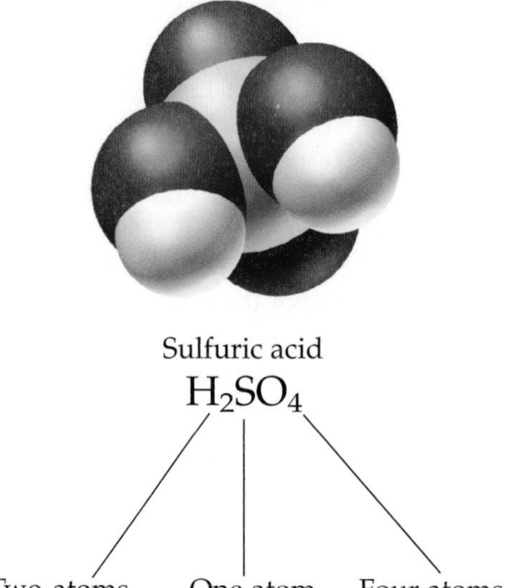

▶ Figure 4.9 Interpretation of a Chemical Formula The chemical formula of sulfuric acid, H_2SO_4, indicates 2 hydrogen atoms, 1 sulfur atom, and 4 oxygen atoms. Thus, sulfuric acid has a total of 7 atoms.

Example Exercise 4.5 Composition of Chemical Formulas

State the total number of atoms in a molecule of vitamin B_3, $C_6H_6N_2O$.

Solution

The chemical formula for vitamin B_3 indicates 6 carbon atoms, 6 hydrogen atoms, 2 nitrogen atoms, and 1 oxygen atom. Thus, $C_6H_6N_2O$ has a total of 15 atoms.

Practice Exercise

Write the chemical formula for vitamin B_6, if a molecule is composed of 8 carbon atoms, 11 hydrogen atoms, 1 nitrogen atom, and 3 oxygen atoms.

Answer: $C_8H_{11}NO_3$ (total of 23 atoms)

Concept Exercise

Examine the model for vitamin C and determine the molecular formula. In the model shown, the black sphere = C atom, white = H atom, and red = O atom.

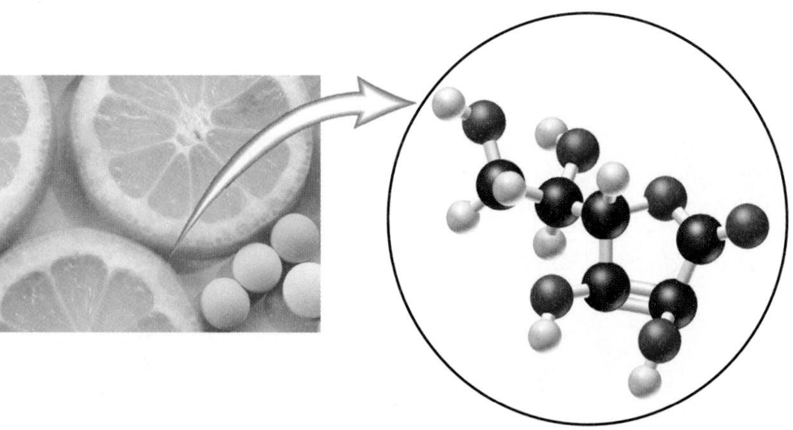

Answer: See Appendix G.

Some chemical formulas use parentheses to clarify the chemical composition. For example, antifreeze has the chemical formula $C_2H_4(OH)_2$. A molecule of ethylene glycol is composed of 2 atoms of carbon, 4 atoms of hydrogen, and 2 units of OH. The parentheses around (OH) in the chemical formula indicate that OH is a two-atom unit. The chemical formula of ethylene glycol, $C_2H_4(OH)_2$, is interpreted in Figure 4.10.

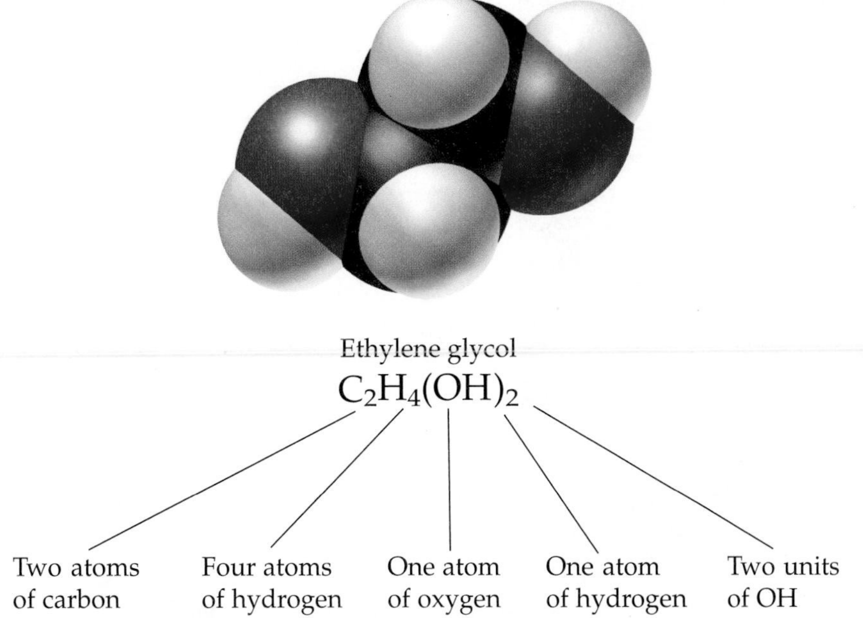

◀ **Figure 4.10 Interpretation of a Chemical Formula** The chemical formula of ethylene glycol, $C_2H_4(OH)_2$, indicates 2 carbon atoms, 4 hydrogen atoms, and 2 OH units. The oxygen atom and hydrogen atom are a single unit that appear twice in a single molecule. Thus, ethylene glycol has a total of 10 atoms.

Example Exercise 4.6 Composition of Chemical Formulas

State the total number of atoms in a molecule of glycerin, $C_3H_5(OH)_3$.

Solution

The chemical formula for glycerin indicates 3 carbon atoms, 5 hydrogen atoms, and 3 OH units. Thus, $C_3H_5(OH)_3$ has a total of 14 atoms.

Practice Exercise

Write the chemical formula for nitroglycerin if a molecule is composed of 3 carbon atoms, 5 hydrogen atoms, 3 oxygen atoms, and 3 NO_2 units.

Answer: $C_3H_5O_3(NO_2)_3$

Concept Exercise

State the total number of atoms in a molecule of nitroglycerin, $C_3H_5O_3(NO_2)_3$.

Answer: See Appendix G.

Note In Chapter 7 we will learn how to systematically name compounds. It is helpful to begin referring to compounds by pronouncing their chemical formulas. Even professional chemists pronounce chemical formulas of compounds; for instance, water is H_2O and is referred to as *"H two O."* Sulfuric acid is H_2SO_4 and is referred to as *"H two, S, O four."* We can pronounce the chemical formula for milk of magnesia, $Mg(OH)_2$, as *"Mg, OH taken twice."*

4.6 PHYSICAL AND CHEMICAL PROPERTIES

OBJECTIVE

▶ To classify a property of a substance as physical or chemical.

Previously, we classified matter as either a mixture or a pure substance. Recall that the physical and chemical properties of a pure substance are consistent throughout a sample. Moreover, no two substances have the same physical and chemical properties.

Furthermore, we said that a pure substance is either a compound or an element. However, the physical and chemical properties of a compound are never the same as the properties of its constituent elements. For example, sodium is a soft, silver metal, and chlorine is a poisonous, yellow gas. In a reaction, these two elements combine to form white crystals of table salt, NaCl (Figure 4.11).

A **physical property** refers to those characteristics of a pure substance that we can observe without changing the composition of the substance. The list of physical properties is extensive, but those considered important are appearance, melting and boiling points,

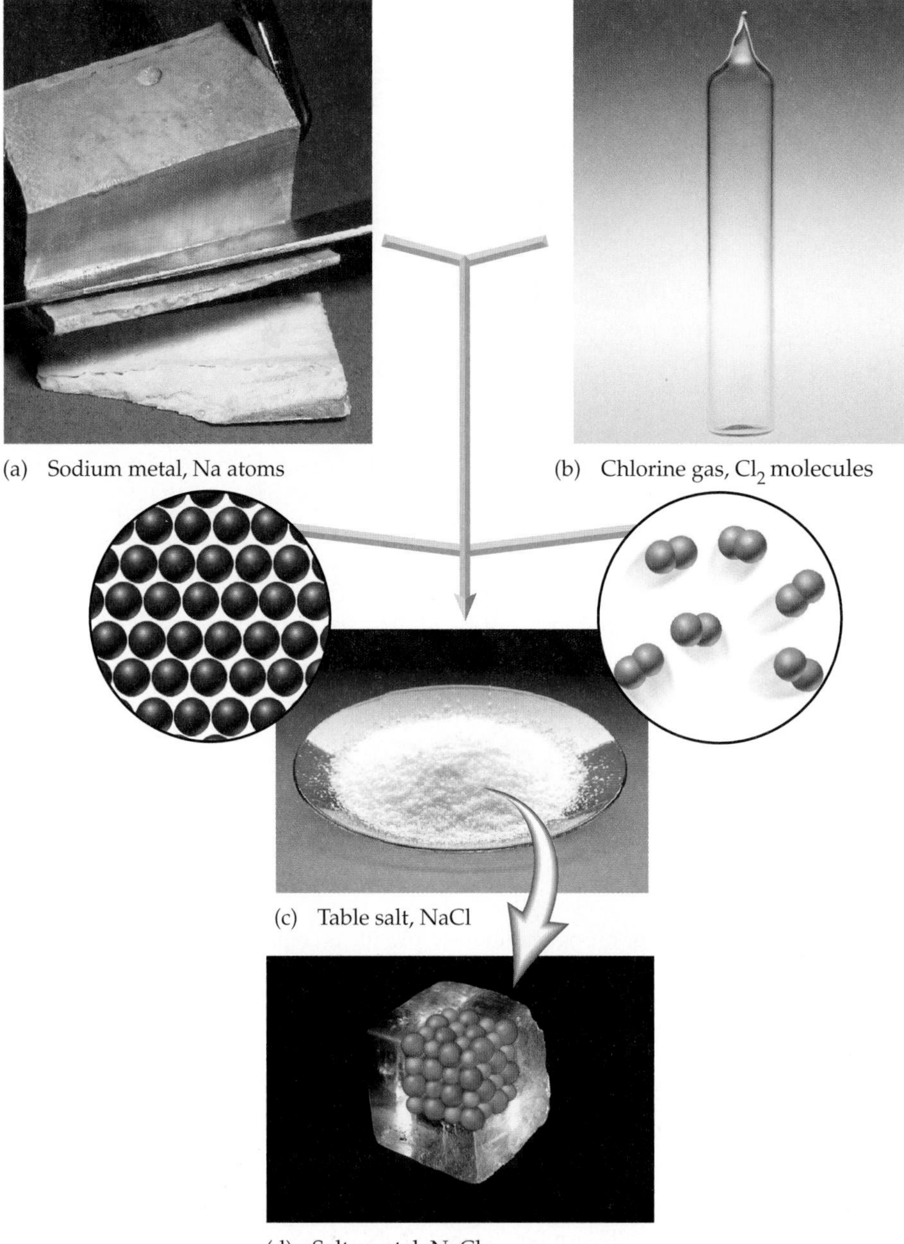

▶ **Figure 4.11 Sodium, Chlorine, and Sodium Chloride** (a) Sodium is a shiny metal, (b) chlorine is a yellow gas, and (c) table salt is white crystals. (d) A crystal of NaCl is composed of Na^+ and Cl^- ions in a cubic lattice.

density, heat and electrical conductivity, solubility, and physical state under normal conditions. Table 4.5 lists some of the physical properties of sodium, chlorine, and sodium chloride at 25 °C and normal atmospheric pressure.

A **chemical property** of a pure substance describes its chemical reactions with other substances. The chemical properties of sodium, for example, include its reaction with oxygen to form sodium oxide and its reaction with water to produce hydrogen gas. There are many reactions for the elements, more than we can possibly consider. We can, however, use the periodic table of elements to predict chemical behavior.

TABLE 4.5 PHYSICAL PROPERTIES OF SODIUM, CHLORINE, AND SODIUM CHLORIDE AT NORMAL CONDITIONS (25 °C AND 1.00 ATM)

PROPERTY	SODIUM	CHLORINE	SODIUM CHLORIDE
appearance	silver metal	yellowish gas	white crystals
melting point	98 °C	–101 °C	801 °C
boiling point	883 °C	–35 °C	1413 °C
density	0.97 g/cm^3	2.90 g/L	2.17 g/cm^3
solubility in 100 g water	reacts with water	0.51 g at 30 °C	35.7 g at 0 °C

The periodic table is arranged according to families of elements that in general give similar reactions. For instance, the metals in Group IA/1 on the far left side of the periodic table react with oxygen gas to give products with similar chemical formulas, namely, Li_2O, Na_2O, K_2O, Rb_2O, and Cs_2O. Notice that in each case, the number of metal to oxygen atoms is in the ratio of 2 to 1.

In Chapter 6 we will study families of elements. For now, review Table 4.6, which illustrates that elements in the same family have similar chemical properties and give products with similar chemical formulas.

TABLE 4.6 CHEMICAL PROPERTIES FOR FAMILIES OF ELEMENTS

FAMILY AND ELEMENT	REACTION WITH OXYGEN	REACTION WITH WATER	REACTION HYDROCHLORIC ACID
Group IA/1			
lithium	Li_2O	$H_2 + LiOH$	$H_2 + LiCl$
sodium	Na_2O	$H_2 + NaOH$	$H_2 + NaCl$
potassium	K_2O	$H_2 + KOH$	$H_2 + KCl$
Group IIA/2			
calcium	CaO	$H_2 + Ca(OH)_2$	$H_2 + CaCl_2$
strontium	SrO	$H_2 + Sr(OH)_2$	$H_2 + SrCl_2$
barium	BaO	$H_2 + Ba(OH)_2$	$H_2 + BaCl_2$
Group IVA/14			
carbon	CO_2	no reaction	no reaction
silicon	SiO_2	no reaction	no reaction
germanium	GeO_2	no reaction	no reaction
Group VIIIA/18			
helium	no reaction	no reaction	no reaction
neon	no reaction	no reaction	no reaction
argon	no reaction	no reaction	no reaction

Classifying Physical and Chemical Properties

We can distinguish between a physical and a chemical property as follows. Physical properties are observed without altering the composition of the substance, whereas chemical properties always involve a chemical change. When a chemical change occurs, the composition of a substance changes and another substance is formed. Example Exercise 4.7 illustrates the distinction between physical and chemical properties.

Example Exercise 4.7 Physical and Chemical Properties

Classify each of the following properties as physical or chemical:

(a) Water appears colorless and odorless at 20 °C.
(b) Water dissolves sugar crystals.
(c) Water produces a gas with calcium metal.
(d) Water exists as ice at –10 °C.

Solution

If a reaction occurs, there is a change in composition and the property is chemical. Otherwise, the property is physical.

(a) Color and odor are *physical* properties.
(b) Solubility is a *physical* property.
(c) A chemical reaction is a *chemical* property.
(d) A physical state is a *physical* property.

Practice Exercise

Classify each of the following properties as physical or chemical:

(a) Water appears hard and crystalline at 0 °C.
(b) Water is insoluble in gasoline.

(c) Water is a very weak conductor of electricity.
(d) Water produces a gas with sodium metal.

Answers: (a) physical; (b) physical; (c) physical; (d) chemical

Concept Exercise
Copper melts at 1083 °C, has a density of 8.92 g/mL, is a good conductor of electricity, and turns black when heated. Which of these is an example of a chemical property?

Answer: See Appendix G.

OBJECTIVE

- To classify a change in a substance as physical or chemical.

4.7 PHYSICAL AND CHEMICAL CHANGES

When a pure substance is altered, the alteration is classified as a physical or a chemical change. In a **physical change**, the chemical composition of the sample does not change. If we melt ice to water, heat alcohol to a vapor, or recycle aluminum cans into aluminum foil, we are performing a physical change. The chemical composition of the ice, alcohol, and aluminum is the same before and after the change. In other words, the formula of each substance (H_2O, C_2H_5OH, and Al) remains constant. Thus, altering the shape or physical state of a pure substance is an indication of a physical change.

In a **chemical change**, there is a chemical reaction. The composition of the sample changes, and we observe a new set of properties. If a banana ripens from green to yellow, an antacid tablet in water evolves gas bubbles, or fireworks display a shower of colorful lights, we are observing a chemical change. The composition of the banana, antacid tablet, and fireworks is different before and after the change.

Classifying Physical and Chemical Changes

We can distinguish between a physical and a chemical change as follows: a physical change can be observed without altering the composition of the substance, whereas a chemical change involves the formation of a new substance. When a new substance is formed, we usually observe one of the following: a permanent color change, an odor or bubbles from the release of a gas, or light or heat from the release of energy.

In the laboratory, we may observe that two solutions added together produce an insoluble substance. The formation of a solid substance in a solution is a practical observation that indicates a chemical change. Figure 4.12 illustrates three examples of chemical changes.

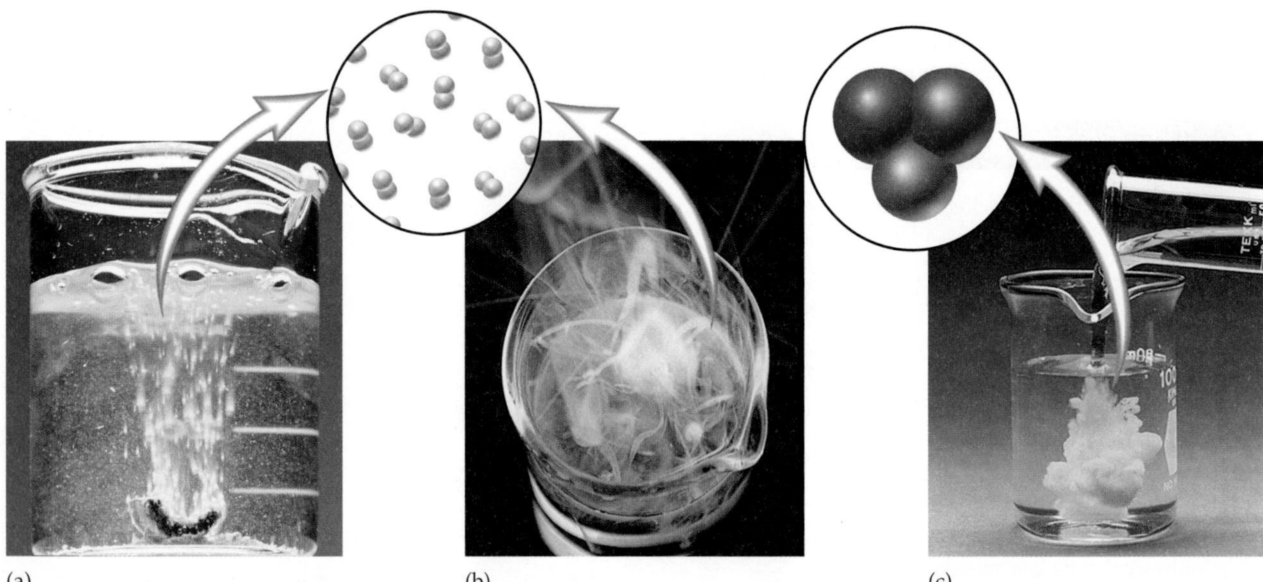

(a) (b) (c)

▲ **Figure 4.12 Evidence of Chemical Change** The evidence for a chemical change includes (a) gas bubbles from the reaction of calcium metal in water, (b) heat and light energy from the reaction of potassium metal in water, and (c) a colorful, insoluble substance from the reaction of two solutions.

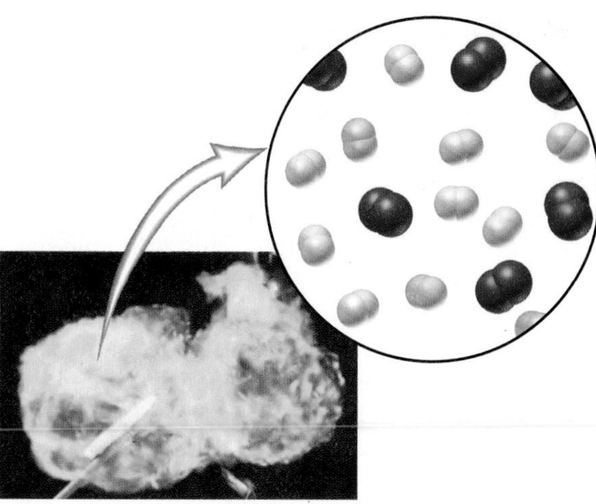

▲ **Hydrogen Explosion** Soap bubbles filled with hydrogen gas undergo an explosion (chemical change) after being ignited by a candle flame.

Example Exercise 4.8 further illustrates the distinction between physical and chemical changes.

Example Exercise 4.8 Physical and Chemical Changes

Classify each of the following observations as a physical or a chemical change:

(a) Touching a lit candle to hydrogen soap bubbles gives an explosion.
(b) Heating water in a flask produces moisture on the glass.
(c) Combining two colorless solutions gives a yellow solid.
(d) Pouring vinegar on baking soda produces gas bubbles.

Solution

An observation that indicates a physical change is a change of physical state. The observations that suggest a chemical change include burning, fizzing, changing color, or forming an insoluble substance in solution.

(a) Hydrogen explodes; thus, it is a *chemical* change.
(b) Water is boiled; thus, it is a *physical* change.
(c) Two solutions give an insoluble substance; thus, it is a *chemical* change.
(d) Baking soda fizzes; thus, it is a *chemical* change.

Practice Exercise

Classify each of the following observations as a physical or a chemical change:

(a) Freezing water in a refrigerator makes cubes of ice.
(b) Adding silver nitrate to tap water gives a cloudy solution.
(c) Burning sulfur gives a light blue flame.
(d) Grinding aspirin tablets produces a powder.

Answers: (a) physical; (b) chemical; (c) chemical; (d) physical

Concept Exercise

An Alka-Seltzer tablet dissolves in water and produces gas bubbles. Is this an example of a physical change or a chemical change?

Answer: See Appendix G.

▲ **Alka-Seltzer** An Alka-Seltzer tablet in water releases carbon dioxide gas bubbles.

OBJECTIVE

▸ To apply the conservation of mass law to chemical changes.

4.8 CONSERVATION OF MASS

In the late 1700s, the French chemist Antoine Lavoisier (1743–1794) performed exacting laboratory experiments and found that the mass of substances before a chemical change was always equal to the mass of substances after the change. He concluded that matter was neither created nor destroyed during a chemical reaction. In 1789, he announced the conservation of mass principle. This principle has become known as the **law of conservation of mass**.

As an example of conservation of mass, consider the reaction of hydrogen and oxygen. Hydrogen and oxygen always combine in the same ratio by mass to form water. That is, if 1 g of hydrogen combines with 8 g of oxygen, then 2 g of hydrogen combines with 16 g of oxygen, 3 g of hydrogen reacts with 24 g of oxygen, and so on. We can predict the mass of water produced by each of these reactions. From the conservation of mass law, we know that the mass of hydrogen plus the mass of oxygen must equal the mass of water produced. That is,

$$1.0 \text{ g hydrogen} + 8.0 \text{ g oxygen} = 9.0 \text{ g water}$$

Conversely, we can predict the masses of hydrogen and oxygen produced by the decomposition of water. Passing an electric current through water produces hydrogen gas and oxygen gas. This process is called *electrolysis*. If the electrolysis of 45.0 g of water produces 5.0 g of hydrogen, how many grams of oxygen are evolved?

The conservation of mass law states that the mass of the hydrogen and oxygen is equal to the mass of the water, 45.0 g. Therefore, the mass of oxygen must equal the mass of the water minus the mass of the hydrogen. That is,

$$45.0 \text{ g water} - 5.0 \text{ g hydrogen} = 40.0 \text{ g oxygen}$$

The conservation of mass law is one of the most important principles in chemistry. Historically, it was used to determine the formulas of compounds and to develop the first periodic table of elements. As we will discuss in Chapter 10, it is the cornerstone of all quantitative relationships in chemistry. Example Exercise 4.9 further illustrates the conservation of mass law.

Example Exercise 4.9 Conservation of Mass Law

In an experiment, 2.430 g of magnesium metal was ignited and burned with oxygen in the air. If 4.030 g of white magnesium oxide powder, MgO, was collected, what was the mass of oxygen gas that reacted?

Solution

Applying the conservation of mass law, we find that the mass of the magnesium metal plus the mass of the oxygen gas equals the mass of the magnesium oxide powder. That is,

$$2.430 \text{ g Mg} + \text{mass of oxygen} = 4.030 \text{ g MgO}$$

$$\text{mass of oxygen} = 4.030 \text{ g MgO} - 2.430 \text{ g Mg}$$

$$\text{mass of oxygen} = 1.600 \text{ g}$$

Practice Exercise

If 0.654 g of zinc metal reacts with 0.321 g of yellow powdered sulfur, what is the mass of the zinc sulfide produced?

Answer: 0.975 g

Concept Exercise

Heating 1.000 g copper metal with yellow sulfur produces 1.252 g of black copper sulfide. What is the mass of sulfur that reacted with the copper metal?

Answer: See Appendix G.

4.9 POTENTIAL AND KINETIC ENERGY

OBJECTIVES

- To distinguish between potential and kinetic energy.
- To relate kinetic energy, temperature, and physical state.

Energy makes things happen and is often defined as the ability to do work. There are two types of energy: potential and kinetic. **Potential energy (PE)** is stored energy that matter possesses as a result of its position or composition. **Kinetic energy (KE)** is the energy matter has as a result of its motion. A boulder perched at the top of a mountain has potential energy. If the boulder rolls down the mountain, it loses potential energy, but gains kinetic energy. In the process of rolling down the mountain, the potential energy of the boulder is converted to kinetic energy. Figure 4.13 illustrates the conversion from potential to kinetic energy.

In our study of chemistry, we are usually interested in the potential energy associated with the chemical composition of a pure substance. For example, gasoline has potential energy that is converted to kinetic energy when it undergoes combustion. As gasoline burns, the potential energy of the cool liquid is converted to the kinetic energy of the hot, expanding gas mixture.

Kinetic Energy, Temperature, and Physical State

We can view a sample of matter as a large collection of very small particles. The attraction between particles is one of the criteria that influences the physical state. In solids, this force of attraction holds particles in fixed positions and allows little or no movement.

When a substance in the *solid state* is heated, the temperature increases and particles begin to randomly vibrate in fixed positions. As a solid acquires more heat energy, particles vibrate so intensely that the force of attraction is overcome and the particles break free from each other. This is a description of the *liquid state.*

A liquid is therefore a state of matter in which particles possess sufficient energy to overcome their mutual attraction and have limited movement. When a liquid has enough energy to completely overcome the force of attraction, the particles are free to fly about and strike the walls of the container. At this point, the liquid has changed to the *gaseous state.* Table 4.7 summarizes the relationship between the properties of particles and the physical state.

We can review the relationship between kinetic energy and physical state as follows. If we heat a solid, it acquires enough energy to overcome the attraction between particles and melts to a liquid. As the temperature increases, particles gradually gain more energy and demonstrate increasing motion. If we continue to

TABLE 4.7 PROPERTIES AND PHYSICAL STATES

PROPERTY OF PARTICLES	SOLID	LIQUID	GAS
kinetic energy	very low	high	very high
movement	negligible	restricted	unrestricted

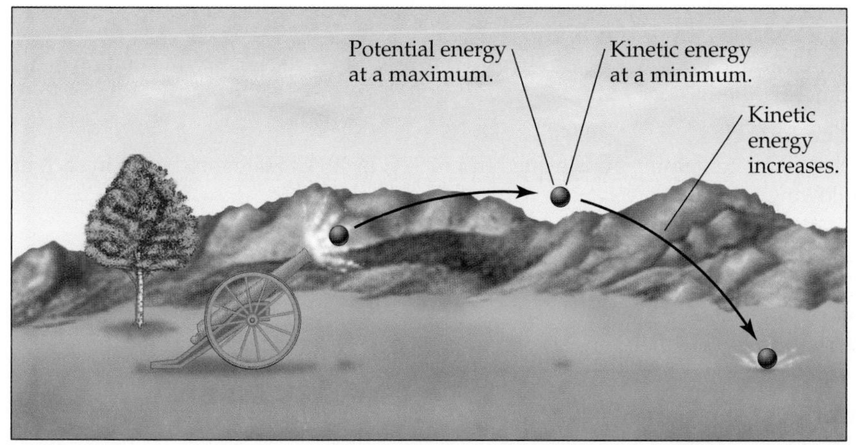

◀ **Figure 4.13 Potential Energy and Kinetic Energy** A cannonball shot into the air loses kinetic energy while gaining potential energy. At the top of its flight, the potential energy is at a maximum and the kinetic energy is at a minimum. As the cannonball falls to Earth, the potential energy decreases as the kinetic energy increases.

▶ **Figure 4.14 Energy and Physical State** The energy of individual H_2O molecules increases as solid ice is heated. As the attraction between molecules is overcome, solid ice changes to liquid water and then to gaseous steam.

heat the liquid, it acquires enough energy to vaporize to a gas. Figure 4.14 illustrates the relationship between increasing energy and physical state.

The kinetic energy of a gas is directly related to its temperature. If the temperature increases, the kinetic energy increases; if the temperature decreases, the kinetic energy decreases. Moreover, if the temperature increases, the motion of the particles increases; if the temperature decreases, the motion decreases.

Example Exercise 4.10 further illustrates how temperature, kinetic energy, and the motion of gas particles are related.

Example Exercise 4.10 Kinetic Energy and Molecular Motion

A balloon filled with helium gas is cooled from 25 °C to –25 °C. State the change in each of the following:

(a) kinetic energy of the gas (b) motion of helium atoms

Solution

Temperature, kinetic energy, and velocity are related as follows:

(a) As the temperature cools from 25 °C to –25 °C, the kinetic energy of helium atoms *decreases*.

(b) Since a drop in temperature produces a decrease in kinetic energy, the motion of helium atoms *decreases*.

Practice Exercise

A steel cylinder containing air is heated from 25 °C to 50 °C. State the change in each of the following:

(a) kinetic energy of the gas (b) motion of air molecules

Answers: (a) increases; (b) increases

Concept Exercise

What happens to the kinetic energy and velocity of air molecules when the temperature increases?

Answer: See Appendix G.

4.10 CONSERVATION OF ENERGY

OBJECTIVES

- To apply the conservation of energy law to physical and chemical changes.
- To identify the following forms of energy: chemical, electrical, mechanical, nuclear, heat, and light.

In the middle of the nineteenth century, several English and German scientists proposed that the total energy in the universe is constant. This proposal was worded in different ways, but it essentially stated that energy cannot be created nor destroyed. Energy can, however, be converted from one form to another. This principle is known as the **law of conservation of energy**. This principle applies experimentally, and we can make energy predictions for physical and chemical changes.

Energy and Physical Changes

Let's consider the conservation of energy law applied to a change of physical state. When water is converted to steam, a given amount of heat energy is required. Since energy cannot be destroyed, the same amount of heat must be released if the steam condenses to a liquid. In practice, it requires 540 cal of heat energy to convert 1 g of water at 100 °C to steam at 100 °C. It therefore follows that 540 cal of heat are released when 1 g of steam at 100 °C condenses to water. Figure 4.15 illustrates the conservation of energy for a physical change.

Energy and Chemical Changes

The conservation of energy law applied to a chemical change is analogous to the conservation of mass law. That is, the total energy before a reaction is the same as the total energy after the reaction. It is important to understand, however, that some chemical changes release energy, while others absorb energy. Yet, the total energy before and after a reaction must always be constant.

How can the energy be constant for a reaction that involves a change in heat? The explanation is that a reaction involves the two types of energy discussed previously, *potential* chemical energy and *kinetic* heat energy. In a reaction, the initial reactants may lose potential energy, but the resulting products gain kinetic energy. In reactions that produce heat, the initial reactants have more potential energy than the products. Conversely, in reactions that absorb heat, the initial reactants have less potential energy than the products.

If we consider the conservation of energy law, we can say that the energy released from a given change is equal to the energy required for the reverse change. For example, we know from experimentation that 3200 cal of heat is released when hydrogen

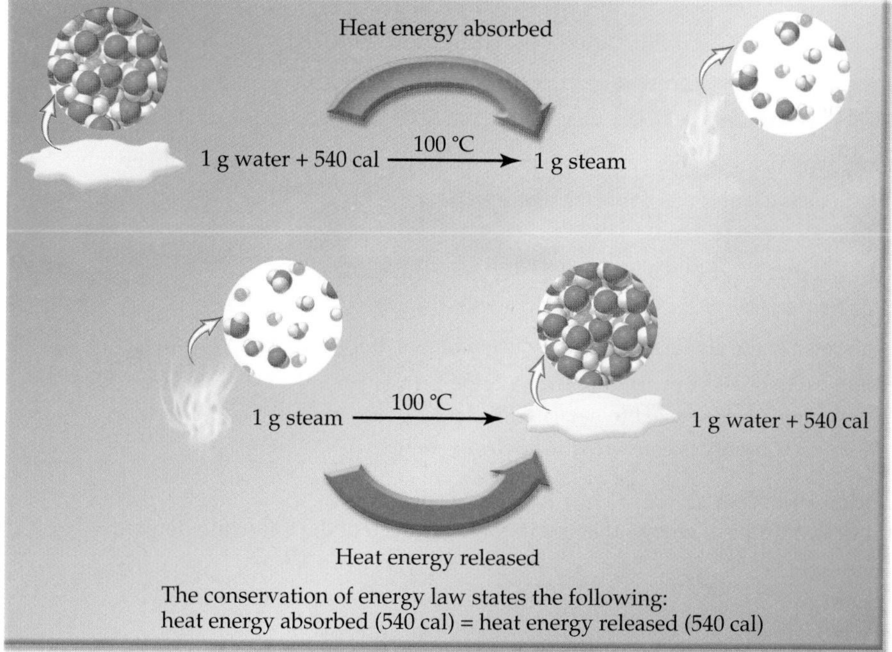

◀ **Figure 4.15 Conservation of Energy for a Physical Change** The physical change of state from water to steam, and the reverse process, involve the same amount of heat energy.

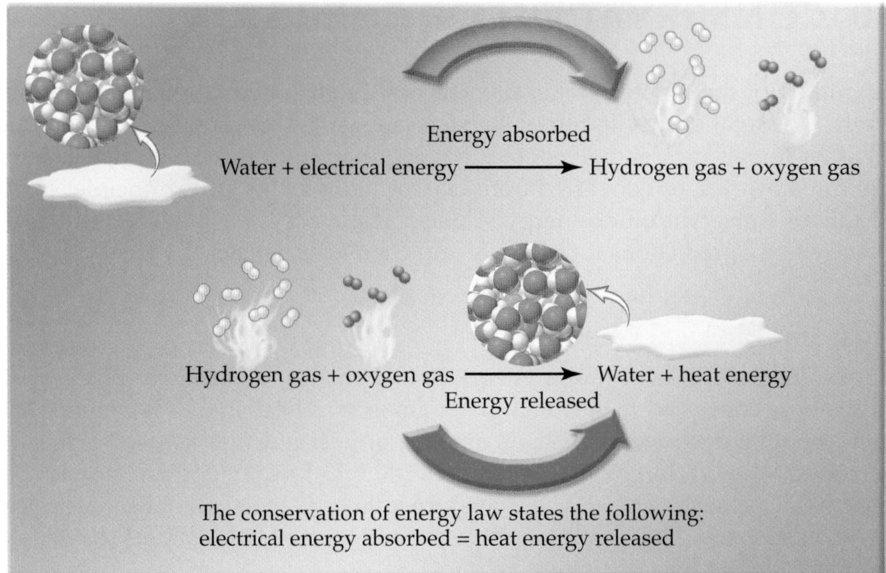

▶ **Figure 4.16 Conservation of Energy for a Chemical Change** The chemical change from water to hydrogen and oxygen gases, and the reverse process, involve the same amount of energy.

gas and oxygen gas react to produce 1 g of water. Thus, 3200 cal of heat is required for the reverse process of decomposing 1 g of water into hydrogen gas and oxygen gas. If we decompose water into the two gases using an electric current, the electrical energy required is equivalent to 3200 cal. Figure 4.16 illustrates the conservation of energy for a chemical change.

Forms of Energy

There are six forms of energy: heat, light, chemical, electrical, mechanical, and nuclear. Each of these can be converted from one form of energy to another. For example, we can use solar energy panels to convert light energy from the Sun to heat energy, which we can store as hot water.

Although we convert sunlight to electrical energy using solar energy panels, the current process is only 15% efficient. We can also convert mechanical energy to electrical energy, which is a more efficient process. For example, the mechanical energy of water spilling over a dam is used to spin a turbine that in turn drives an electrical generator. This hydroelectric process operates at nearly 90% efficiency. Example Exercise 4.11 illustrates the conversion of various forms of energy.

Example Exercise 4.11 Forms of Energy

Identify two forms of energy that are involved in each of the following conversions:

(a) Radioactive emissions vaporize water to steam.
(b) Steam drives a turbine.
(c) A turbine spins and drives an electrical generator.

Solution

We can refer to the six forms of energy listed above. It follows that

(a) *Nuclear* energy is converted to *heat* energy.
(b) *Heat* energy is converted to *mechanical* energy.
(c) *Mechanical* energy is converted to *electrical* energy.

Practice Exercise

Identify two forms of energy that are involved in each of the following devices:

(a) flashlight

(b) solar calculator
(c) lead–acid battery

Answers: (a) chemical and light; (b) light and electrical; (c) chemical and electrical

Concept Exercise

Which of the following is *not* a basic form of energy: chemical, electrical, heat, light, mechanical, nuclear, solar?

Answer: See Appendix G.

CHEMISTRY CONNECTION Recycling Aluminum

Q: How many aluminum cans must be recycled to achieve an energy savings equal to the energy produced by burning a gallon of gasoline?

Aluminum is a very common metal; in fact, it is the most abundant metal in Earth's crust. Why, then, is there so much concern about recycling aluminum? One reason is that it takes a great deal more energy to extract aluminum from its ore than to recycle it. To produce 1 ton of aluminum from its raw ore requires the combustion of about 8 tons of coal. However, to produce 1 ton of aluminum from recycled scrap requires 0.4 ton of coal. Thus, recycling aluminum requires only 5% of the energy required by processing aluminum ore. Recycling iron requires only 25% of the energy required for iron ore, and recycling paper requires only 30% of the energy required for wood pulp.

According to the Environmental Protection Agency (EPA), it is estimated that more than one million tons of aluminum cans are produced annually in the United States. Although we are using more aluminum each year, we are also recycling more. From 1970 to 2005, we increased our use of aluminum substantially, but during that same period of time we also increased the amount of recycled aluminum dramatically. The following bar graph shows the rising percentage of recycled aluminum.

It is estimated that we recycle more than 80% of the aluminum used for disposable containers in the United States. Thus, more than a million tons of aluminum metal will be recycled, rather than being processed from aluminum ore. Recycling 24 aluminum cans provides an energy savings approximately equivalent to the energy provided by a gallon of gasoline.

▲ **Aluminum Scrap Metal** In addition to aluminum cans, aluminum scrap metal is also recycled.

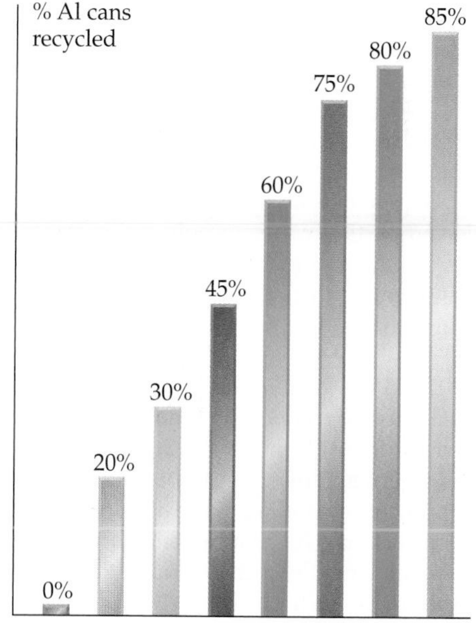

▲ **Aluminum Can Recycling** Notice the dramatic increase in the percentage of recycled aluminum cans.

A: **The energy saved by recycling 24 aluminum cans, rather than refining aluminum ore, is about equal to the energy produced by burning a gallon of gasoline.**

▲ **Albert Einstein** The relationship between mass and energy is expressed by Einstein's famous equation: $E = mc^2$.

Conservation of Mass and Energy

In 1905, Albert Einstein (1879–1955) presented a paper on the special theory of relativity that proposed that matter and energy are interrelated, stating this relationship in the form $E = mc^2$. In this famous equation, Einstein showed that energy (E) and mass (m) are related by the velocity of light (c) squared.

Since mass and energy are related, it naturally follows that the conservation of mass and the conservation of energy are related. Thus, a more accurate statement combines these two principles into the **law of conservation of mass and energy**. The combined law states that the total mass and energy in the universe is constant. When applied to a chemical reaction, the law indicates that the total mass and energy before and after a chemical change is constant.

By applying the combined law, we realize that any decrease in mass corresponds to an increase in energy. Even a slight decrease in mass, however, corresponds to an enormous amount of energy. For instance, a mass loss of 0.001 g produces enough heat energy to raise the temperature of a large swimming pool about 30 °C. Since the mass loss for ordinary chemical reactions is undetectable from a practical laboratory point of view, the original conservation of mass law is still valid.

Summary

Section 4.1 Matter exists in three **physical states**: solid, liquid, and gas. A *solid* is characterized by a fixed shape and fixed volume. A *liquid* has a variable shape, but a fixed volume. A *gas* has a variable shape and a variable volume.

As temperature increases, a solid *melts* to a liquid and then *vaporizes* to a gas. A direct change of state from a solid to a gas is called **sublimation**. Conversely, as temperature decreases, a gas *condenses* to a liquid and then *freezes* to a solid. A direct change of state from a gas to a solid is called **deposition**.

Section 4.2 We can classify matter according to its properties. A **heterogeneous mixture** has an indefinite composition and variable properties; it can be separated into two or more substances. A **homogeneous mixture** has a definite composition, but variable properties from sample to sample; examples include **alloys**, solutions, and mixtures of gases. A **substance** can be either an **element** or a **compound**. A substance has a definite composition and constant properties.

Section 4.3 In 1803, the English chemist John Dalton proposed that elements are composed of indivisible spherical particles called **atoms**. In 1813, the Swedish chemist J. J. Berzelius suggested that the name of an element be abbreviated with a **chemical symbol** corresponding to the first letter of the name of the element, for instance, N for nitrogen. If two or more names start with the same letter, he recommended using two letters in the name, for example, Ni for nickel.

Section 4.4 Most elements can be classified as **metals** or **nonmetals**. If an element can be hammered into a thin sheet, it is said to **malleable**. If it can be drawn into a fine wire, it is said to be **ductile**. A few elements have properties between those of the metals and nonmetals and are termed **semimetals**, or metalloids. All elements can be ordered by **atomic number** and arranged in a **periodic table**. The periodic table shows a general trend in properties, with metals on the left side and nonmetals on the right side.

Section 4.5 The **law of definite composition** states that a compound always contains the same elements in the same proportion by mass. The composition of a compound is given by its **chemical formula**. Vitamin C has the formula $C_6H_8O_6$; one **molecule** contains 6 carbon atoms, 8 hydrogen atoms, and 6 oxygen atoms. Thus, the total number of atoms in the chemical formula of vitamin C is 20.

Section 4.6 The properties of substances are classified as physical or chemical. A **physical property** refers to a characteristic such as color, density, melting point, conductivity of heat and electricity, solubility, and physical state. A **chemical property** describes a chemical reaction of a substance with another substance.

Section 4.7 The changes in substances are classified as physical or chemical. A **physical change** does not alter the chemical composition and involves a change in mass, volume, or physical state. A **chemical change** involves a change in chemical composition. Evidence for a chemical change includes burning, producing a gas, forming a solid substance in solution, and releasing heat or light energy.

Section 4.8 The **law of conservation of mass** states that matter is neither created nor destroyed. This fundamental principle has practical application in the chemistry laboratory; that is, the total mass of substances produced from a reaction can be calculated because the mass of the products must equal the mass of the reactants.

Section 4.9 The **potential energy (PE)** of a substance is related to its chemical composition, whereas the **kinetic energy (KE)** of a substance is related to the motion of its particles. Kinetic energy is directly related to temperature. That is, when we heat a solid substance and increase its temperature, we also increase its kinetic energy. As the temperature increases, the particles gain more energy and begin to overcome their attraction for each other. If we continue heating, we will eventually produce a change of state from a solid to a liquid, and from a liquid to a gas.

Section 4.10 The six forms of energy are heat, light, chemical, electrical, mechanical, and nuclear. The **law of conservation of energy** states that energy is neither created nor destroyed. Energy may be converted, however, from one form to another. In practice, chemical energy is used to provide heat and electrical energy. The laws of conservation of mass and conservation of energy can be treated as a single principle—the **law of conservation of mass and energy**. For ordinary chemical reactions, energy changes are sufficiently small that we can ignore this principle and apply conservation of mass and conservation of energy individually.

Key Terms *Answers to Key Terms are in Appendix H.*

Select the key term that corresponds to each of the following definitions.

_____ **1.** the condition of matter existing as a solid, liquid, or gas
_____ **2.** a direct change of state from a solid to a gas
_____ **3.** a direct change of state from a gas to a solid
_____ **4.** matter having an indefinite composition and variable properties
_____ **5.** matter having a definite composition, but variable properties from sample to sample; for example, alloys and solutions
_____ **6.** matter having a definite composition and constant properties; for example, an element or compound
_____ **7.** a homogeneous mixture of two or more metals
_____ **8.** a substance that can be broken down into simpler substances
_____ **9.** a substance that cannot be broken down by chemical reaction
_____ **10.** the smallest particle that represents an element
_____ **11.** an abbreviation for the name of a chemical element
_____ **12.** an element that is generally shiny in appearance, has a high density, high melting point, and is a good conductor of heat and electricity
_____ **13.** the property of a metal that allows it to be machined into a foil
_____ **14.** the property of a metal that allows it to be drawn into a wire
_____ **15.** an element that is generally dull in appearance, has a low density, low melting point, and is not a good conductor of heat and electricity
_____ **16.** an element that is generally metallic in appearance and has properties midway between those of a metal and a nonmetal
_____ **17.** a number that identifies a particular element
_____ **18.** a chart that arranges elements according to their properties
_____ **19.** the principle that a compound always contains the same mass ratio of elements
_____ **20.** a single particle composed of two or more nonmetal atoms
_____ **21.** an abbreviation for the name of a chemical compound that indicates the number of atoms of each element
_____ **22.** a property that can be observed without changing the chemical formula of a substance
_____ **23.** a property that cannot be observed without changing the chemical formula of a substance
_____ **24.** a modification of a substance that alters the chemical composition
_____ **25.** a modification of a substance that does not alter the chemical composition
_____ **26.** the stored energy that matter possesses owing to its position or composition
_____ **27.** the energy associated with the mass and velocity of a particle
_____ **28.** the law that states mass cannot be created or destroyed
_____ **29.** the law that states energy cannot be created or destroyed
_____ **30.** the law that states the total mass and energy in the universe is constant

(a) alloy *(Sec. 4.2)*
(b) atom *(Sec. 4.3)*
(c) atomic number *(Sec. 4.4)*
(d) chemical change *(Sec. 4.7)*
(e) chemical formula *(Sec. 4.5)*
(f) chemical property *(Sec. 4.6)*
(g) chemical symbol *(Sec. 4.3)*
(h) compound *(Sec. 4.2)*
(i) deposition *(Sec. 4.1)*
(j) ductile *(Sec. 4.4)*
(k) element *(Sec. 4.2)*
(l) heterogeneous mixture *(Sec. 4.2)*
(m) homogeneous mixture *(Sec. 4.2)*
(n) kinetic energy (KE) *(Sec. 4.9)*
(o) law of conservation of energy *(Sec. 4.10)*
(p) law of conservation of mass *(Sec. 4.8)*
(q) law of conservation of mass and energy *(Sec. 4.10)*
(r) law of definite composition *(Sec. 4.5)*
(s) malleable *(Sec. 4.4)*
(t) metal *(Sec. 4.4)*
(u) molecule *(Sec. 4.5)*
(v) nonmetal *(Sec. 4.4)*
(w) periodic table *(Sec. 4.4)*
(x) physical change *(Sec. 4.7)*
(y) physical property *(Sec. 4.6)*
(z) physical state *(Sec. 4.1)*
(aa) potential energy (PE) *(Sec. 4.9)*
(bb) semimetal *(Sec. 4.4)*
(cc) sublimation *(Sec. 4.1)*
(dd) substance *(Sec. 4.2)*

Exercises *Answers to odd-numbered Exercises are in Appendix I.*

Physical States of Matter (Sec. 4.1)

1. Which physical state is described as having a variable shape and fixed volume?
2. Which physical state is described as having a variable shape and variable volume?
3. Which physical state is described as having a variable shape and compresses negligibly?
4. Which physical state is described as having a variable shape and compresses significantly?
5. Which physical state is described as having particles closely packed and free to move around?
6. Which physical state is described as having particles closely packed and held rigidly in position?
7. Supply the term that describes each of the following changes of physical state:
   (a) gas to liquid **(b)** solid to liquid
8. Supply the term that describes each of the following changes of physical state:
   (a) liquid to gas **(b)** liquid to solid
9. Indicate whether the following changes of physical state require heating or cooling:
   (a) gas to liquid **(b)** solid to liquid
10. Indicate whether the following changes of physical state require heating or cooling:
    (a) liquid to gas **(b)** liquid to solid
11. State whether the process of sublimation requires heating or cooling.
12. State whether the process of deposition requires heating or cooling.

Elements, Compounds, and Mixtures (Sec. 4.2)

13. Classify each of the following as an element, compound, or mixture:
    (a) sodium metal **(b)** chlorine gas
    (c) sodium chloride **(d)** salt water
14. Classify each of the following as an element, compound, or mixture:
    (a) iron metal **(b)** oxygen gas
    (c) iron oxide **(d)** iron ore
15. Classify each of the following as an element, compound, homogeneous mixture, or heterogeneous mixture:
    (a) copper wire **(b)** copper ore
    (c) copper oxide **(d)** brass alloy

◀ **Copper Wire** Copper metal is ductile and can be drawn into a thin wire.

16. Classify each of the following as an element, compound, homogeneous mixture, or heterogeneous mixture:
    (a) silver ingot **(b)** silver ore
    (c) silver oxide **(d)** silver alloy

Names and Symbols of the Elements (Sec. 4.3)

17. Write the chemical symbol for each of the following elements:
    (a) lithium **(b)** argon
    (c) magnesium **(d)** manganese
    (e) beryllium **(f)** silicon
    (g) mercury **(h)** titanium
18. Write the chemical symbol for each of the following elements:
    (a) strontium **(b)** xenon
    (c) oxygen **(d)** zinc
    (e) fluorine **(f)** sodium
    (g) copper **(h)** nickel
19. Write the name of the element for each of the following chemical symbols:
    (a) Cl **(b)** Ne
    (c) Cd **(d)** Ge
    (e) Co **(f)** Ra
    (g) Cr **(h)** Te
20. Write the name of the element for each of the following chemical symbols:
    (a) P **(b)** S
    (c) Kr **(d)** Fe
    (e) Sb **(f)** N
    (g) Pt **(h)** Ag
21. Refer to the periodic table and find the atomic number for each of the following:
    (a) hydrogen **(b)** boron
    (c) aluminum **(d)** titanium
    (e) arsenic **(f)** strontium
    (g) tin **(h)** bismuth
22. Refer to the periodic table and find the atomic number for each of the following:
    (a) helium **(b)** carbon
    (c) potassium **(d)** copper
    (e) selenium **(f)** silver
    (g) lead **(h)** radon

Metals, Nonmetals, and Semimetals (Sec. 4.4)

23. State whether each of the following properties is more typical of a metal or a nonmetal:
    (a) dense solid **(b)** low boiling point
    (c) ductile **(d)** reacts with metals
24. State whether each of the following properties is more typical of a metal or a nonmetal:
    (a) shiny luster
    (b) low melting point
    (c) malleable
    (d) reacts with metals and nonmetals
25. State whether each of the following properties is more typical of a metal or a nonmetal:
    (a) shiny solid **(b)** brittle solid
    (c) low density **(d)** forms alloys
26. State whether each of the following properties is more typical of a metal or a nonmetal:
    (a) dull powder **(b)** colorless gas
    (c) high density **(d)** reacts with metals
27. Refer to the periodic table and classify each of the following elements as a metal, nonmetal, or semimetal:
    (a) Na **(b)** B
    (c) Hg **(d)** Kr

28. Refer to the periodic table and classify each of the following elements as a metal, nonmetal, or semimetal:
(a) Sb (b) S
(c) Ar (d) Te

29. Refer to the periodic table and classify each of the following elements as a metal, nonmetal, or semimetal:
(a) beryllium (b) germanium
(c) selenium (d) xenon

30. Refer to the periodic table and classify each of the following elements as a metal, nonmetal, or semimetal:
(a) radium (b) fluorine
(c) phosphorus (d) manganese

31. Refer to the periodic table and indicate the physical state for each of the following elements at 25 °C and normal pressure:
(a) Na (b) N
(c) Si (d) Kr

32. Refer to the periodic table and indicate the physical state for each of the following elements at 25 °C and normal pressure:
(a) Sb (b) S
(c) Ar (d) Ge

33. Refer to the periodic table and indicate the physical state for each of the following elements at 25 °C and normal pressure:
(a) hydrogen (b) titanium
(c) fluorine (d) bromine

34. Refer to the periodic table and indicate the physical state for each of the following elements at 25 °C and normal pressure:
(a) helium (b) manganese
(c) iodine (d) mercury

Compounds and Chemical Formulas (Sec. 4.5)

35. State the composition for a molecule of each of the following vitamins:
(a) vitamin B_1, $C_{12}H_{18}Cl_2N_8OS$
(b) vitamin B_2, $C_{17}H_{20}N_4O_6$

36. State the composition for a molecule of each of the following compounds:
(a) cholesterol, $C_{27}H_{45}OH$ (b) quinine, $C_{20}H_{24}N_2O_2$

37. Write the chemical formula for each of the following compounds:
(a) vitamin A: 20 carbon, 30 hydrogen, and 1 oxygen atom
(b) vitamin K: 31 carbon, 46 hydrogen, and 2 oxygen atoms

38. Write the chemical formula for each of the following compounds:
(a) lysine (an amino acid): 6 carbon, 14 hydrogen, 2 nitrogen, and 2 oxygen atoms
(b) methionine (an amino acid): 5 carbon, 11 hydrogen, 1 nitrogen, 2 oxygen, and 1 sulfur atom

39. What is the total number of atoms in a molecule of each of the following compounds?
(a) stearic acid (in animal fat), $CH_3(CH_2)_{16}COOH$
(b) triolein (in vegetable oil), $C_3H_5(C_{17}H_{33}COO)_3$

40. What is the total number of atoms in a molecule of each of the following compounds?
(a) citric acid (in citrus fruit), $C_3H_4OH(COOH)_3$
(b) lactic acid (in sore muscles), $C_2H_4(OH)COOH$

41. The mass ratio of H:O in a bottle of water is 1:8. Predict the mass ratio of water in 10 bottles of water.

42. The mass ratio of C:H:O in natural vitamin C found in rose hips is 9:1:12. Predict the mass ratio of synthetic vitamin C synthesized in the laboratory.

Physical and Chemical Properties (Sec. 4.6)

43. State whether each of the following is a physical or a chemical property of gold:
(a) yellow color (b) high density
(c) conducts electricity (d) unreactive in acid

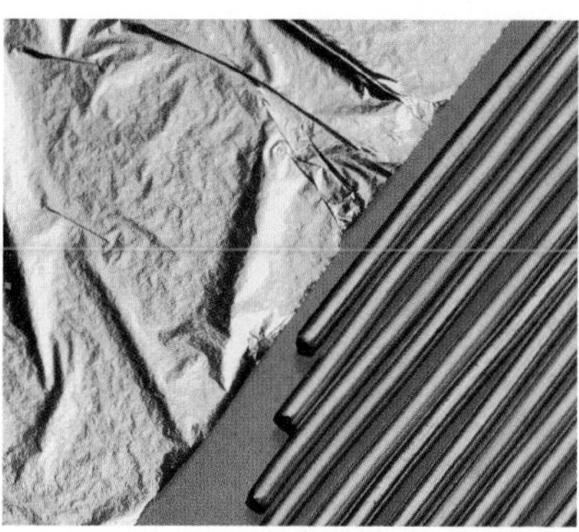

▲ **Gold Foil and Copper Rods** Gold and copper are both malleable and ductile metals; that is, each can be hammered into a foil and drawn into thin rods.

44. State whether each of the following is a physical or a chemical property of copper:
(a) shiny metallic luster (b) high melting point
(c) conducts heat (d) reacts with oxygen

45. State whether each of the following is a physical or a chemical property:
(a) Titanium pulls into a fine wire.
(b) Zinc generates an odorless gas in acid.
(c) Yellow phosphorus ignites in air.
(d) Chromium conducts heat.

46. State whether each of the following is a physical or a chemical property:
(a) Magnesium is a lightweight metal.
(b) Silver shows no reaction in acid.
(c) Lead turns dark slowly in air.
(d) Carbon crystallizes as diamond.

47. State whether each of the following is a physical or a chemical property:
(a) Ether is flammable.
(b) Bromine is a reddish-orange liquid.
(c) Ethanol vaporizes at 78 °C.
(d) Steam condenses to water at 100 °C.

48. State whether each of the following is a physical or a chemical property:
(a) Electricity decomposes water.
(b) Ice floats on water.
(c) Potassium metal fizzes in water.
(d) Acetone dissolves in water.

49. State whether each of the following is a physical or a chemical property:
(a) Sugar ferments to ethanol.
(b) Ethanol dissolves in gasoline.
(c) Ethylene glycol freezes at –12 °C.
(d) A liqueur has a fragrant bouquet.

50. State whether each of the following is a physical or a chemical property:
(a) Iron oxidizes to rust.
(b) Dry ice pellets sublimate.
(c) Lithium metal turns gray in air.
(d) Plutonium radioactively decays.

Physical and Chemical Changes (Sec. 4.7)

51. State whether each of the following is a physical or a chemical change:
(a) combustion **(b)** vaporization
(c) sublimation **(d)** oxidation

52. State whether each of the following is a physical or a chemical change:
(a) burning **(b)** condensing
(c) freezing **(d)** rusting

53. State whether each of the following is a physical or a chemical change:
(a) slicing an orange into wedges
(b) fermenting apples into apple cider
(c) grinding sugar crystals into powder
(d) digesting carbohydrates for energy

54. State whether each of the following is a physical or a chemical change:
(a) cutting aluminum foil **(b)** caramelizing sugar
(c) adding air to a tire **(d)** igniting a sparkler

55. State whether each of the following is a physical or a chemical change:
(a) steam condenses and freezes to ice
(b) hydrogen and oxygen gases explode
(c) bromine vaporizes to a brown gas
(d) sodium metal changes to a white powder

56. State whether each of the following is a physical or a chemical change:
(a) zinc metal produces a colorless gas in acid
(b) nickel alloys with copper in a 5¢ coin
(c) liquid helium boils at –269 °C
(d) sodium metal reacts with chlorine gas

57. State whether each of the following is a physical or a chemical change:
(a) baking soda fizzes in vinegar
(b) vinegar and oil separate into two layers
(c) a flaming wood log releases heat
(d) a helium balloon decreases in size

58. State whether each of the following is a physical or a chemical change:
(a) mercury cools to a silvery solid
(b) natural gas burns with a blue flame
(c) ammonia and copper solution turns deep blue
(d) sugar crystals dissolve in water

Conservation of Mass (Sec. 4.8)

59. If 5.00 g of gray iron powder reacts with 2.77 g of red phosphorus powder, what is the mass of the iron phosphide product?

60. If 1.00 g of gray zinc powder reacts with 0.49 g of yellow sulfur powder, what is the mass of the zinc sulfide product?

61. If 0.750 g of orange mercury oxide decomposes to 0.695 g of liquid mercury and oxygen gas, what is the mass of oxygen produced?

62. If 10.11 g of limestone decomposes by heat to give 8.51 g of solid calcium oxide and carbon dioxide gas, what is the mass of carbon dioxide produced?

Potential and Kinetic Energy (Sec. 4.9)

63. Which physical state has particles with the highest kinetic energy?

64. Which physical state has particles with the most restricted movement?

65. If the temperature increases, does the kinetic energy increase or decrease?

66. If the temperature increases, does the motion of particles increase or decrease?

67. If a steel cylinder of helium gas is heated from 50 °C to 100 °C, does the motion of helium atoms increase or decrease?

68. If a steel cylinder of hydrogen gas is cooled from 50 °C to 0 °C, does the motion of hydrogen molecules increase or decrease?

Conservation of Energy (Sec. 4.10)

69. Steam at 100 °C passing through a radiator condenses to water and releases heat. If 1500 kcal of heat are released, how much energy is required to vaporize the water, at 100 °C, to steam?

70. An ice cube at 0 °C melts to water by absorbing heat. If 10.5 kcal of heat are required to melt the ice, how much energy must be lost to freeze the water, at 0 °C, to ice?

71. A 10.0 g sample of mercury absorbs 110 cal as it is heated from 25 °C to its boiling point at 356 °C. It then requires an additional 697 cal to vaporize. How much energy is released as the mercury vapor cools from 356 °C to 25 °C?

72. A 25.0 g sample of molten iron releases 1230 cal as it cools from 2000 °C to its freezing point, 1535 °C. It then releases an additional 1590 cal as it solidifies. How much energy is required to heat the solid iron from 1535 °C to 2000 °C?

73. The reaction of hydrogen and oxygen gases produces 15.0 g of water and releases 48.0 kcal of heat. How much energy is required to decompose 15.0 g of water into hydrogen and oxygen gases?

74. The electrolysis of 25.0 g of water requires 335 kJ of energy to produce hydrogen and oxygen gases. How much energy is released when hydrogen and oxygen gases react to produce 25.0 g of water?

75. Identify two forms of energy that are involved in each of the following energy conversions relating to a fossil fuel power plant:
(a) Burning coal converts water to steam.
(b) Steam drives a turbine.
(c) A turbine turns a generator.
(d) A generator makes electricity.

76. Identify two forms of energy that are involved in each of the following energy conversions relating to a nuclear power plant:
(a) Uranium converts water to steam.
(b) Steam drives a turbine.
(c) A turbine turns a generator.
(d) A generator makes electricity.

77. Identify two forms of energy that are involved in each of the following energy conversions relating to an automobile:
(a) hot gases moving pistons
(b) pistons powering the crankshaft

(c) crankshaft turning the generator
(d) generator charging the battery

78. Identify two forms of energy that are involved in each of the following energy conversions relating to an automobile:
(a) lead–acid battery making electricity
(b) electricity turning a starter motor
(c) starter motor turning a flywheel
(d) gasoline exploding into hot gases

General Exercises

79. A pre-1982 penny was minted from an alloy of copper and zinc. Is a pre-1982 penny an example of a homogeneous or heterogeneous mixture?

80. A post-1982 penny is minted by plating copper onto a zinc disk. Is a post-1982 penny an example of a homogeneous or heterogeneous mixture?

81. Gasoline contains a complex mixture of hydrocarbons. Is gasoline an example of a heterogeneous or a homogeneous mixture?

82. A dry cleaning solvent contains chlorofluorocarbons (CFCs). Is this dry cleaning solvent an example of a homogeneous mixture or a pure substance?

83. The electronics industry manufactures semiconductor chips from silicon. Refer to the periodic table and predict an element that may substitute for silicon.

84. The electronics industry manufactures transistors using arsenic diffusion. Refer to the periodic table and predict an element that may substitute for arsenic.

85. Name the three most abundant elements in Earth's crust.

86. Name the three most abundant elements in the human body.

87. State the atomic number for each of the following elements:
(a) argentum (b) plumbum
(c) stannum (d) aurum

88. State the atomic number for each of the following elements:
(a) rutherfordium (b) dubnium
(c) seaborgium (d) bohrium

89. State whether the following describes a physical or a chemical change: changing physical state, but not chemical formula.

90. State whether the following describes a physical or a chemical change: changing chemical formula, but not physical state.

91. The reaction of hydrogen and nitrogen gases produces 5.00 g of ammonia and releases 3250 cal of heat. How much energy is required to decompose 5.00 g of ammonia into hydrogen and nitrogen gases?

92. The decomposition of 10.0 g of ammonia requires 27,200 J of energy to give hydrogen and nitrogen gases. How much energy is released when hydrogen and nitrogen gases react to produce 10.0 g of ammonia?

93. Iron and sulfur react to produce iron sulfide and heat energy. An experiment shows that the mass of iron and sulfur is equal to the mass of the iron sulfide. In *theory*, should the products weigh slightly more or slightly less than the reactants?

94. Hydrogen and iodine react to give hydrogen iodide while absorbing heat energy. An experiment shows that the total mass of the reactants is equal to the product. In *theory*, should the product weigh slightly more or slightly less than the reactants?

95. Using a playground swing as an example, when is the potential energy minimum and the kinetic energy maximum?

96. Using a roller coaster as an example, when is the potential energy maximum and the kinetic energy minimum?

97. Define each symbol in Einstein's equation: $E = mc^2$.

98. State the law of conservation of mass and energy as it applies to a chemical reaction.

4 Chapter Self-Test

Answers to Self-Test are in Appendix J.

1. Which of the following describes a substance in the solid physical state? (Sec. 4.1)
(a) The substance has a fixed shape.
(b) The substance has a fixed volume.
(c) The substance compresses negligibly.
(d) all of the above
(e) none of the above

2. Air contains nitrogen, oxygen, argon, and other gases. Which of the following describes air? (Sec. 4.2)
(a) compound
(b) element
(c) heterogeneous mixture
(d) homogeneous mixture
(e) none of the above

3. Which of the following elements is *not* one of the ten most abundant in Earth's crust, oceans, and atmosphere? (Sec. 4.3)
(a) aluminum (b) gold
(c) iron (d) oxygen
(e) silicon

4. Refer to the periodic table and predict which of the following elements is an example of a semimetal. (Sec. 4.4)
(a) Al (b) Ca
(c) Fe (d) Se
(e) Si

5. Nitroglycerin is an explosive made from glycerin and nitric acid. If the formula is $C_3H_5O_3(NO_2)_3$, what is the number of atoms in one molecule? (Sec. 4.5)
(a) 17 (b) 18
(c) 20 (d) 42
(e) none of the above

6. Which of the following properties of gasoline is a chemical property? (Sec. 4.6)
   (a) burns with air
   (b) insoluble in water
   (c) boiling point, 97 °C
   (d) density, 0.692 g/mL
   (e) colorless liquid
7. Which of the following is an example of a chemical change? (Sec. 4.7)
   (a) condensation
   (b) crystallization
   (c) oxidation
   (d) sublimation
   (e) none of the above
8. If 2.43 g of magnesium burns in oxygen to produce 4.03 g of magnesium oxide, what is the mass of reacting oxygen gas? (Sec. 4.8)
   (a) 1.60 g
   (b) 2.43 g
   (c) 4.03 g
   (d) 6.46 g
   (e) impossible to predict from the given information
9. What happens to the kinetic energy of oxygen molecules when the gas is heated? (Sec. 4.9)
   (a) kinetic energy decreases
   (b) kinetic energy increases
   (c) kinetic energy remains constant
   (d) depends on the gas
   (e) none of the above
10. A nuclear power plant uses radioactive uranium to convert water to steam. The steam drives a turbine that turns a generator to produce electricity. What two forms of energy are represented by the *uranium* and the *turbine*? (Sec. 4.10)
   (a) chemical energy and mechanical energy
   (b) chemical energy and heat energy
   (c) electrical energy and heat energy
   (d) nuclear energy and heat energy
   (e) nuclear energy and mechanical energy

Key Concepts

11. Identify the physical state (solid, liquid, or gas) that corresponds to each of the following illustrations:

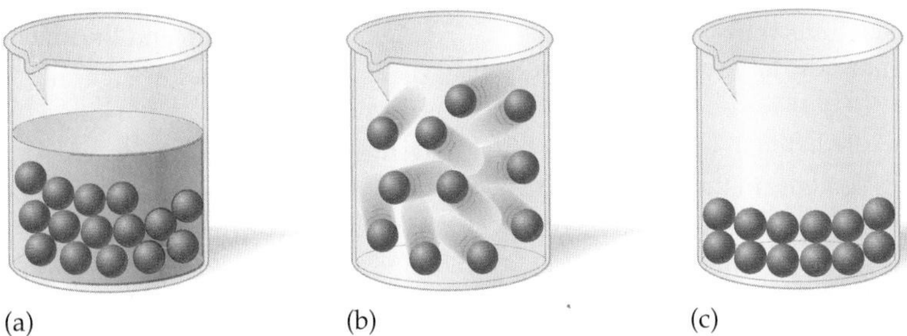

12. Classify each of the following as an element, compound, homogeneous mixture, or heterogeneous mixture:
   (a) iron ore
   (b) iron oxide
   (c) iron metal
   (d) iron alloy
13. Classify the mineral (photo) and molecule (model) of sulfur as an element, compound, homogeneous mixture, or heterogeneous mixture:

14. Examine the model for aspirin and determine the molecular formula. In the model shown, the black sphere = C atom, white = H atom, and red = O atom.

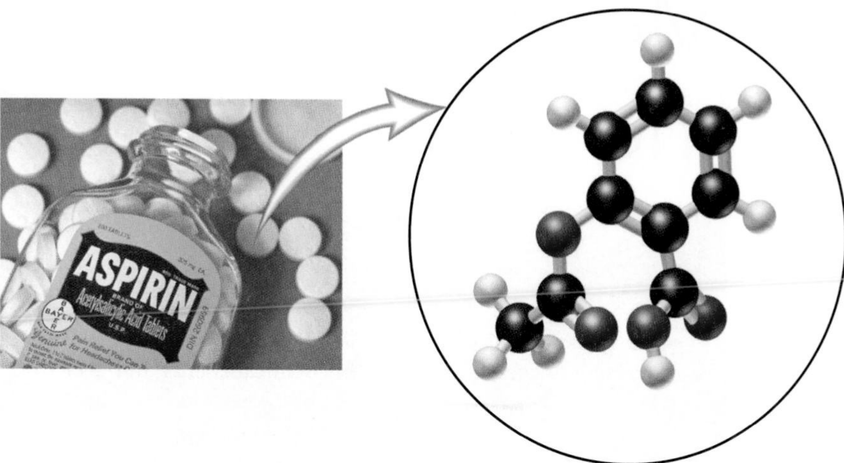

15. When a cannonball is shot from a cannon, at what point in its flight is the potential energy of the cannonball at a maximum?

16. Which has greater mass: a gallon of gasoline or the gaseous products from the combustion of a gallon of gasoline?

Critical Thinking

17. Is ice floating in water an example of a heterogeneous or a homogeneous mixture?

18. Which of the following compounds has physical and chemical properties similar to iron metal: FeO, Fe_2O_3, $Fe(NO_3)_3$?

CHAPTER

5

Models of the Atom

Is it possible to see tiny, tiny atoms? Yes, indirectly. We can obtain pictures of atoms using an electron microscope technique. This image shows chains of gold atoms.

"Anyone who says that they can contemplate quantum mechanics without becoming dizzy has not understood the concept in the least."
Niels Bohr, Danish Physicist (1885–1962)

The concept of an atom was born in Greece about 450 B.C. One of the great issues of the day was whether matter and motion were continuous or discontinuous. Democritus, a Greek philosopher, argued that *matter is discontinuous* and could not be infinitely divided. He believed that at some point a fundamental indivisible particle would emerge. He called this ultimate particle an atom, from the Greek word *atomos*, which means "indivisible."

About 350 B.C., Aristotle—the most influential philosopher of his time—argued that both *matter and motion are continuous*. He believed that matter could be divided an infinite number of times, and that even the smallest particle could be further divided. So powerful was Aristotle's influence that the existence of atoms was not believable, and his point of view prevailed for 21 centuries.

5.1 DALTON MODEL OF THE ATOM

OBJECTIVE

▶ To describe the Dalton model of the atom.

In 1803, John Dalton (1766–1844), a modest English schoolteacher, proposed that matter was discontinuous as Democritus had argued over 20 centuries before. But unlike the Greeks, Dalton offered experimental evidence for particles. Although he was not a distinguished scientist, Dalton presented convincing evidence based on the work of Boyle, Lavoisier, and Proust, three of the great scientists who had preceded him.

In the seventeenth century, Robert Boyle, the noted English physicist, had concluded from experiments that a gas was made up of tiny particles. Dalton expanded on Boyle's conclusion and proposed that *all* matter was composed of particles. In addition to Boyle's experiments, Dalton relied on two other scientific principles, the law of conservation of mass and the law of definite composition.

In 1789, Antoine Lavoisier established the law of conservation of mass. By carefully weighing substances before and after a chemical reaction, Lavoisier demonstrated that

CHEMISTRY CONNECTION John Dalton

Q: How was John Dalton able to accept an honorary degree from Oxford University dressed in a scarlet robe when wearing scarlet was forbidden by his Quaker faith?

John Dalton was born the son of a weaver into a devoutly religious family. At the age of 12, he began teaching at a Quaker school and developed an interest in science. He was most interested in meteorology and kept a lifelong daily journal of atmospheric conditions for his hometown of Manchester, England.

◀ **John Dalton (1766–1844)**

By all accounts, Dalton was not an inspiring lecturer, and he was hindered as a researcher by being color-blind. He had only a minimal education and limited finances, but he compensated for this with persistent and meticulous work habits. Dalton's daily study of the weather led him to conclude, like Robert Boyle and Isaac Newton before him, that the air was made up of gas particles. Over time, he began to construct his atomic theory as follows.

According to the law of definite composition, carbon and oxygen always react in the same mass ratio to produce carbon dioxide. Dalton proposed that one atom of carbon combines with two atoms of oxygen to produce a molecule of CO_2, as shown using Dalton's symbolism.

Carbon dioxide, CO_2

◀ **Carbon dioxide, CO_2**

Similarly, he proposed that two atoms of hydrogen combine with one atom of oxygen to give a molecule of H_2O.

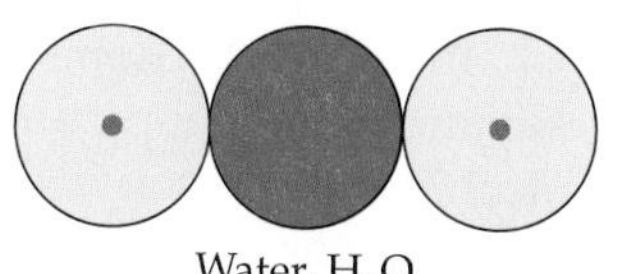

Water, H_2O

◀ **Water, H_2O**

Finally, Dalton reasoned that two atoms of hydrogen in water could substitute for each of the oxygen atoms in carbon dioxide. This would result in a molecule with one carbon atom and four hydrogen atoms, that is, CH_4. At the time, this compound was known as marsh gas, but we now refer to it as methane.

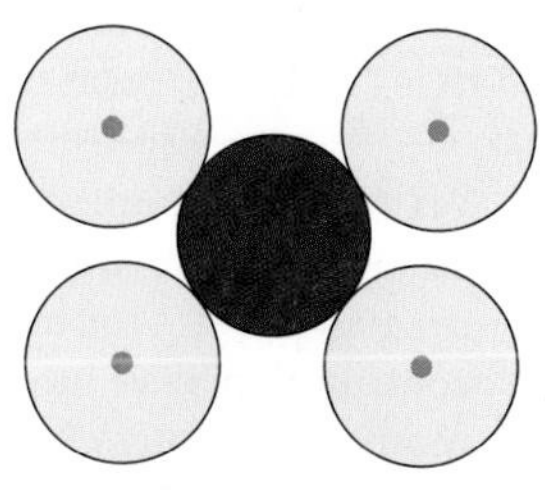

Methane, CH_4

◀ **Methane, CH_4**

Through experimentation, Dalton indeed found that the combining ratio of carbon to hydrogen in methane was 1 to 4; this agreed perfectly with his prediction. Thus, Dalton had laboratory evidence to support the atomic theory.

A: Dalton rationalized that since he was color-blind and saw the ceremonial robe as gray, he was not violating Quaker doctrine.

matter was neither created nor destroyed. Although atoms are much too small to weigh directly, this principle allows us to predict the masses of substances involved in a chemical change. In 1799, Joseph Proust established the law of definite composition, which demonstrated that a compound always contains the same elements in the same proportion by mass.

Dalton initially presented his evidence for the existence of atoms to the Literary and Philosophical Society of Manchester, England. He proposed that an element is composed of tiny, indivisible, indestructible particles. Furthermore, he argued that compounds are simply combinations of two or more atoms of different elements. In 1808, Dalton published the atomic theory in his classic textbook *A New System of Chemical Philosophy*.

In a surprisingly short period of time, the atomic theory was generally accepted by the scientific community. The theory can be summarized as follows:

1. An element is composed of tiny, indivisible, indestructible particles called atoms.
2. All atoms of an element are identical and have the same properties.
3. Atoms of different elements combine to form compounds.
4. Compounds contain atoms in small whole number ratios.
5. Atoms can combine in more than one ratio to form different compounds.

As we will learn, Dalton's first two proposals were incorrect, but the atomic theory was nonetheless an important step toward understanding the nature of matter.

OBJECTIVES

- To describe the Thomson plum pudding model of the atom.
- To state the relative charge on an electron and a proton.

5.2 THOMSON MODEL OF THE ATOM

Toward the end of the 1800s, there was disturbing evidence that the atom was divisible after all. Experiments involving the passage of electricity through gases led to the discovery of a negatively charged particle called an **electron (e^-)**, and a positively charged particle called a **proton (p^+)**. The electron had a relative charge of –1, and the proton had an opposite, but equal charge of +1.

In 1903, J. J. Thomson (1856–1940) proposed a subatomic model of the atom. The subatomic model pictured a positively charged atom containing negatively charged electrons. Thomson visualized electrons in homogeneous spheres of positive charge in a way that was analogous to raisins in English plum pudding (Figure 5.1). Thus, the Thomson proposal became popularly known as the *plum pudding model* or *raisin pudding model of the atom*.

Originally, Thomson was only able to determine the relative charge-to-mass ratio of the electron and the proton. However in 1911, after 5 years of tedious observations, the American physicist Robert Millikan (1868–1953) determined the actual charge on the electron. This allowed Thomson to calculate the actual mass of the electron and the proton. Thomson calculated that the mass of the electron is 9.11×10^{-28} g, and that the mass of the proton is 1.67×10^{-24} g.

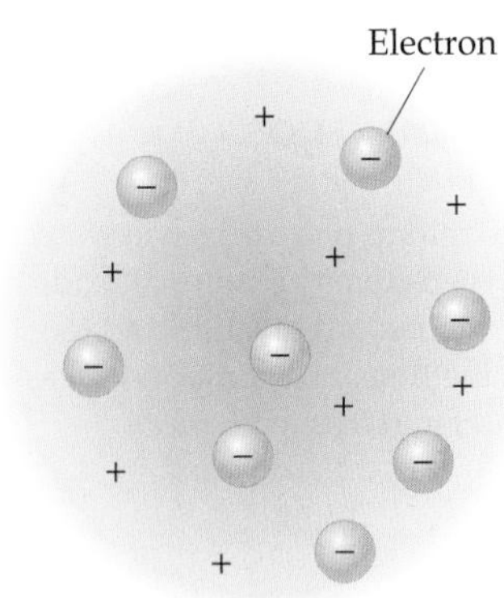

▲ **Figure 5.1 Thomson Model of the Atom** Atoms are pictured as spheres of positive charge. The small negative particles in the sphere represent electrons.

Note J. J. Thomson spent his academic career at Cambridge University and, at the age of 27, became director of the prestigious Cavendish Laboratory. In 1906 he won the Nobel prize in physics for his work on the electron, and 2 years later he was knighted. Thomson's contribution was far-reaching—seven of his former students and assistants went on to receive Nobel prizes.

OBJECTIVES

- To describe the Rutherford nuclear model of the atom.
- To state the relative charge and approximate mass of the electron, proton, and neutron.

5.3 RUTHERFORD MODEL OF THE ATOM

Ernest Rutherford (1871–1937) was digging potatoes on his father's farm in New Zealand when he received the news that he had won a scholarship to Cambridge University. He had actually come in second, but the winner had declined the scholarship to get married. The 24-year-old Rutherford postponed his own wedding plans and immediately set out for England.

At Cambridge, Ernest Rutherford studied subatomic particles and earned his doctorate working for J. J. Thomson. Upon graduation, Rutherford went to McGill

University in Canada and began to work in the field of radioactivity, which studies unstable atoms that radiate subatomic particles and energy. After a year he went back to New Zealand, married, and returned to Manchester University in England. There he continued to study radioactivity and coined the terms "alpha ray" and "beta ray" for two types of radiation. He discovered a third type of radiation that was not affected by a magnetic field and gave it the name "gamma ray." In 1908, Rutherford received the Nobel prize in chemistry for his work on radioactivity.

In 1906, Rutherford found that alpha rays contained particles identical to those of helium atoms stripped of electrons. He experimented with alpha rays by firing them at thin gold foils. As expected, the particles passed straight through the foil or, on occasion, deflected slightly (Figure 5.2). This observation seemed reasonable since the plum pudding model of the atom pictured homogeneous spheres.

A few years later the true picture of the atom was unveiled. The person who did the experiment was Hans Geiger, Rutherford's assistant and the inventor of the Geiger counter. Here is a description of the experiment in Rutherford's own words:

> *One day Geiger came to me and said, "Don't you think that young Marsden, whom I am training in radioactive methods, ought to begin a small research?" Now I had thought that too, so I said, "Why not let him see if any alpha particles can be scattered through a large angle?" I may tell you in confidence that I did not believe that they would be, since we knew that the alpha particle was a very fast massive particle, with a great deal of energy. Then I remember two or three days later Geiger coming to me in great excitement and saying, "We have been able to get some of the alpha particles coming backwards." It was quite the most incredible event that has ever happened to me in my life. It was almost as incredible as if you fired a 15-inch shell at a piece of tissue paper and it came back and hit you.*

Rutherford interpreted the alpha-scattering results as follows. He believed that most of the alpha particles passed directly through the foil because an atom is largely empty space with electrons moving about. But in the center of the atom is the **atomic nucleus** containing protons. Rutherford reasoned that, compared to the atom, the nucleus is tiny and has a very high density. The alpha particles that bounced backward recoiled after striking the dense nucleus. The scattering of alpha particles by the atomic nuclei in the gold foil is illustrated in Figure 5.3.

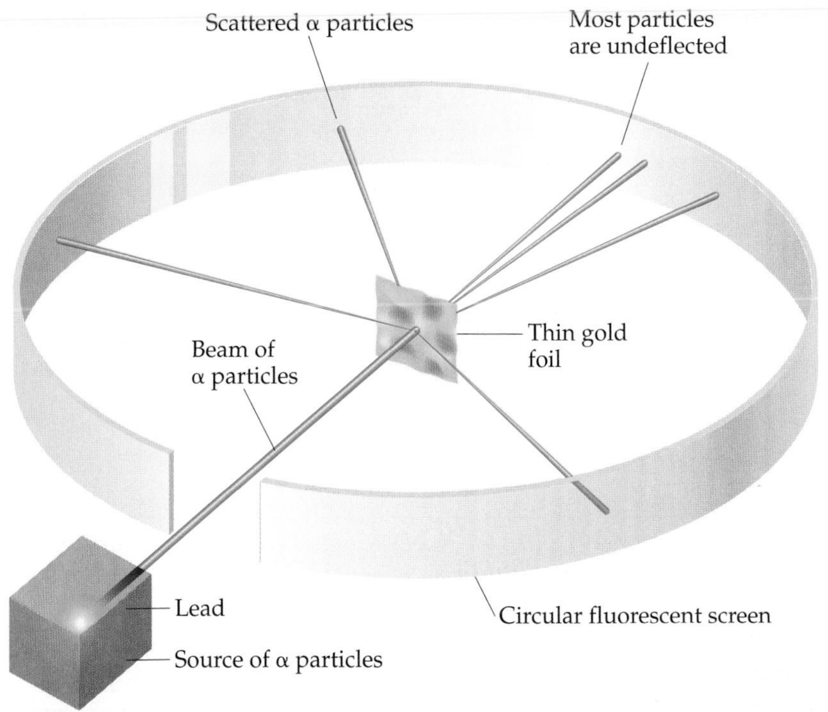

◀ **Figure 5.2 Alpha-Scattering Experiment** The diagram shows the deflection of alpha particles by a thin gold foil. Although the foil was only 0.5 μm thick, a few alpha particles actually rebounded backward.

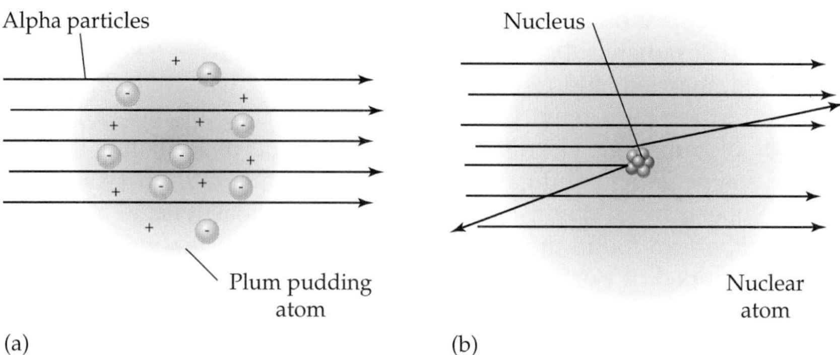

▲ **Figure 5.3 Explanation of Alpha-Scattering** (a) Powerful alpha particles should pass through a homogeneous plum pudding gold atom. (b) Since positively-charged, alpha particles are deflected, Rutherford reasoned correctly that a gold atom has a dense, positively charged nucleus.

In 1911, Rutherford proposed a new model of the atom. He suggested that negatively charged electrons were distributed about a positively charged nucleus. Rutherford was able to estimate the size of the atom and its nucleus. He calculated that an atom has a diameter of about 1×10^{-8} cm, and that the nucleus has a diameter of about 1×10^{-13} cm. The Rutherford model of the atom is illustrated in Figure 5.4.

Since the diameters of atoms and nuclei are extremely small and difficult to comprehend, consider the following analogy to visualize an atom. The size of an atom compared to the size of its nucleus is similar to a kilometer compared to a centimeter. Moreover, if the atom were as large as the Astrodome, the nucleus would be the size of a small marble. This analogy clarifies why so few alpha particles were scattered by gold nuclei in the thin foil.

Owing to the heaviness of the nucleus, Rutherford predicted that it contained neutral particles in addition to protons. Twenty years later, one of Rutherford's former students found the elusive neutral particle. In 1932, James Chadwick (1891–1974) discovered the **neutron (n^0)** and 3 years later was awarded the Nobel prize in physics. We can summarize the data for the electron, proton, and neutron as shown in Table 5.1.

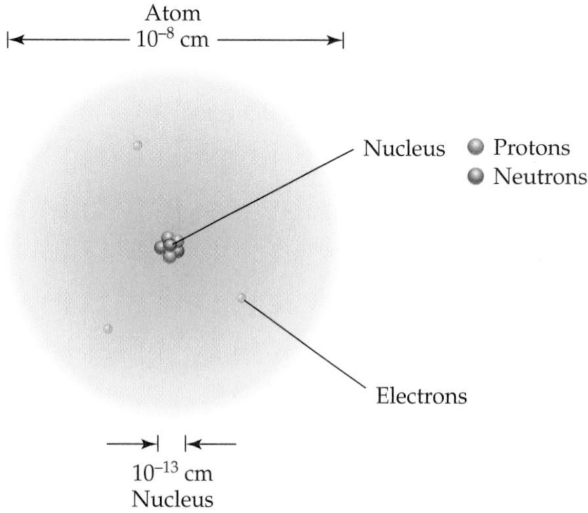

◀ **Figure 5.4 Rutherford Model of the Atom** Most of the mass of the atom is found in the nucleus. Note that the atoms and nuclei are not to scale, as the size of an atom is about 100,000 times larger than the size of the nucleus.

TABLE 5.1 SUBATOMIC PARTICLES

SUBATOMIC PARTICLE	SYMBOL	LOCATION	RELATIVE CHARGE	RELATIVE MASS
electron	e^-	outside nucleus	−1	1/1836
proton	p^+	inside nucleus	+1	1
neutron	n^0	inside nucleus	0	1

5.4 ATOMIC NOTATION

OBJECTIVES

- To draw a diagram of an atom given its atomic notation.
- To explain and illustrate the concept of isotopes.

Each element has a characteristic number of protons in its atomic nucleus. This value is called the **atomic number**. The total number of protons and neutrons in the nucleus of an atom is called the **mass number**. A shorthand method for keeping track of protons and neutrons in the nucleus of an atom is called **atomic notation**. By convention, the symbol of the element (Sy) is preceded by a subscript and superscript. The subscript, designated the *Z* value, represents the atomic number. The superscript, designated the *A* value, represents the mass number. Thus,

mass number (p^+ and n^0) → $^{A}_{Z}\mathrm{Sy}$ — symbol of the element

atomic number (p^+) →

As an example, an atom of sodium can be written as $^{23}_{11}\mathrm{Na}$. Here the atomic number (number of protons) is 11, and the mass number (number of protons + neutrons) is 23. From this information, we can determine the number of neutrons. The nucleus of this sodium atom contains 12 neutrons ($A - Z = 23 - 11 = 12$). Since atoms are neutral, the number of negative electrons must equal the number of positive protons. Thus, there are 11 electrons surrounding the nucleus. We can show a simple diagram of the sodium atom as follows:

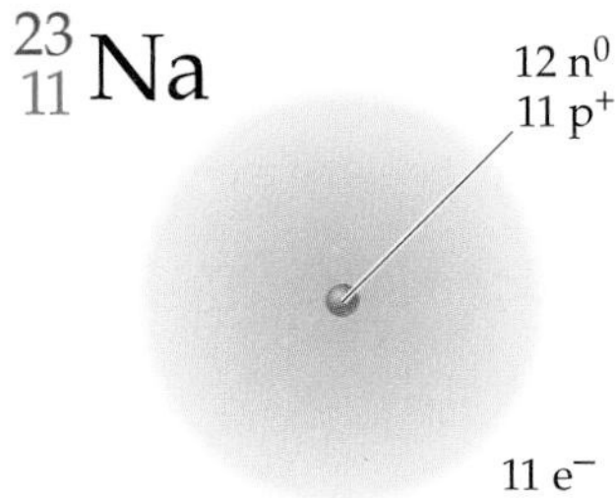

Example Exercise 5.1 further illustrates the interpretation of atomic notation to state the composition of an atom.

Example Exercise 5.1 Atomic Notation

Given the atomic notation for the following atoms, draw a diagram showing the arrangement of protons, neutrons, and electrons.

(a) $^{19}_{9}\mathrm{F}$ (b) $^{109}_{47}\mathrm{Ag}$

Solution

We can draw a diagram of an atom by showing protons and neutrons in the nucleus surrounded by electrons.

(a) Since the atomic number is 9 and the mass number is 19, the number of neutrons is 10 (19 – 9). If there are 9 protons, there must be 9 electrons.

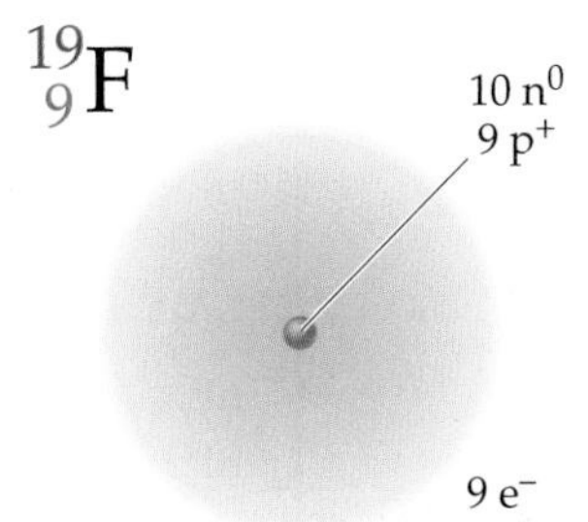

(b) Since the atomic number is 47 and the mass number is 109, the number of neutrons is 62 (109 – 47). If there are 47 protons, there must be 47 electrons.

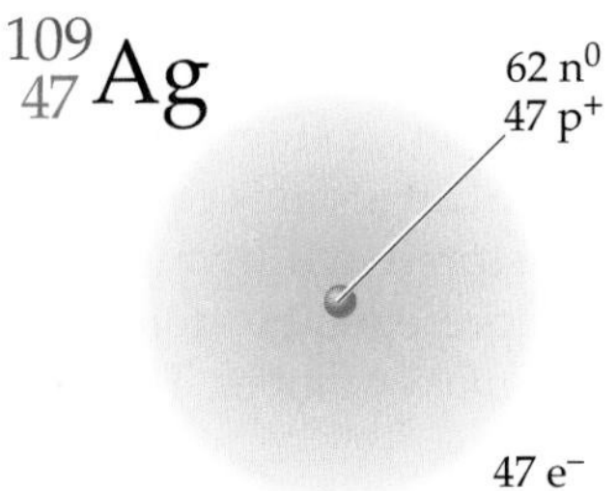

Practice Exercise

Given the following diagram, indicate the nucleus using atomic notation.

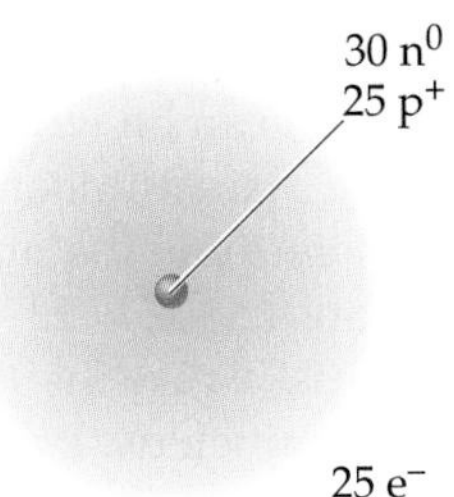

Answer: $^{55}_{25}Mn$

Concept Exercise

Can atoms of different elements have the same atomic number?

Answer: See Appendix G.

Isotopes

Approximately 20 elements occur naturally with a fixed number of neutrons. For most elements, the number of neutrons in the nucleus varies. Atoms of the same element that have a different number of neutrons in the nucleus are called **isotopes**.

For isotopes, as the number of neutrons varies, so does the mass number. For example, hydrogen occurs naturally as two stable isotopes, protium (^1_1H) and deuterium (^2_1H). Protium has only one proton in its nucleus, whereas deuterium has a proton and a neutron. A third isotope of hydrogen, tritium (^3_1H), is unstable and radioactive. Tritium has one proton and two neutrons in its nucleus. Table 5.2 lists the naturally occurring stable isotopes of the first 10 elements.

TABLE 5.2 NATURALLY OCCURRING STABLE ISOTOPES OF THE FIRST TEN ELEMENTS

ATOMIC NUMBER	MASS NUMBER	ATOMIC NOTATION	ATOMIC NUMBER	MASS NUMBER	ATOMIC NOTATION
1	1	^1_1H	6	13	$^{13}_6C$
1	2	^2_1H	7	14	$^{14}_7N$
2	3	^3_2He	7	15	$^{15}_7N$
2	4	^4_2He	8	16	$^{16}_8O$
3	6	^6_3Li	8	17	$^{17}_8O$
3	7	^7_3Li	8	18	$^{18}_8O$
4	9	^9_4Be	9	19	$^{19}_9F$
5	10	$^{10}_5B$	10	20	$^{20}_{10}Ne$
5	11	$^{11}_5B$	10	21	$^{21}_{10}Ne$
6	12	$^{12}_6C$	10	22	$^{22}_{10}Ne$

◀ **Neon Isotopes** The element neon consists of three naturally occurring isotopes: Ne–20, Ne–21, and Ne–22. We can deduce by the atomic mass of neon (20.18 amu) that Ne–20 is the most abundant.

We often refer to an isotope by stating the name of the element followed by its mass number, for example, carbon-12. To write the atomic notation, we must also know the atomic number. In this textbook, the atomic number is found above the symbol of the element in the periodic table. If we refer to the inside front cover of this textbook, we find that the atomic number of C is 6. Thus, we can write the atomic notation for carbon-12 as ${}^{12}_{6}C$.

Example Exercise 5.2 further illustrates the relationship between atomic notation and the composition of an atomic nucleus.

Example Exercise 5.2 Nuclear Composition of Isotopes

State the number of protons and the number of neutrons in an atom of each of the following isotopes.

(a) ${}^{37}_{17}Cl$ (b) mercury-202

Solution

The subscript value refers to the atomic number (p^+), and the superscript value refers to the mass number (p^+ and n^0).

(a) Thus, ${}^{37}_{17}Cl$ has 17 p^+ and 20 n^0 (37 – 17 = 20).

(b) In the periodic table, we find that the atomic number of mercury is 80. Thus, the atomic notation, ${}^{202}_{80}Hg$, indicates 80 p^+ and 122 n^0 (202 – 80 = 122).

Practice Exercise

State the number of protons and the number of neutrons in an atom of each of the following isotopes.

(a) ${}^{120}_{50}Sn$ (b) uranium-238

Answers: (a) 50 p^+ and 70 n^0; (b) 92 p^+ and 146 n^0

Concept Exercise

Can atoms of different elements have the same mass number?

Answer: See Appendix G.

Note In general, the properties of isotopes of an element are similar. Consider a carbon-12 atom, which has six protons and six neutrons in the nucleus. If the mass number increases by one neutron, we have a carbon-13 atom. Other than a slight change in mass, the properties of carbon-12 and carbon-13 are nearly identical.

On the other hand, when the atomic number of carbon-12 increases by one proton, we have a nitrogen-13 atom. The additional proton changes the properties dramatically. Carbon occurs naturally as coal, diamond, and graphite, whereas nitrogen is a colorless, odorless gas in the atmosphere.

5.5 ATOMIC MASS

OBJECTIVES

- To explain the concept of relative atomic mass.
- To calculate the atomic mass for an element given the mass and abundance of the naturally occurring isotopes.

We realize, of course, that atoms are much too small to weigh directly on a balance. A carbon atom, for instance, has a mass of only 1.99×10^{-23} g. Instead of weighing atoms, scientists determine the mass of atoms relative to each other. With a special magnetic-field instrument, the mass of an atom can be compared to the mass of a carbon-12 atom. The carbon-12 isotope has been chosen as a reference standard and is assigned a value of exactly 12 atomic mass units. Stated differently, an **atomic mass unit (amu)** is equal to 1/12 the mass of a carbon-12 atom.

Simple and Weighted Averages

Before proceeding, let's contrast two kinds of averages. Let's suppose a company manufactures 16 lb and 8 lb shot put balls for Olympic men and women athletes. Of all the shot put balls manufactured, 75% are 16 lb and 25% are 8 lb. What is the average mass of a shot put ball? If we add the two masses together and divide by 2, we obtain the simple average mass. That is,

$$\frac{16\text{ lb} + 8\text{ lb}}{2} = 12\text{ lb}$$

The simple average mass of a shot put ball is 12 lb, but the true average mass must be heavier, and weighted in favor of the higher percentage of 16 lb balls. Since the percentages are 75% and 25%, we will write the decimal fractions 0.75 and 0.25, and proceed as follows:

$$\begin{aligned} \text{16 lb balls: } 16\text{ lb} \times 0.75 &= 12\text{ lb} \\ \text{8 lb balls: } 8\text{ lb} \times 0.25 &= \underline{\ \ 2\text{ lb}} \\ & \quad\ 14\text{ lb} \end{aligned}$$

Although the weighted average mass is 14 lb, we should note that no shot put ball weighs 14 lb. An actual shot put ball weighs either 16 lb or 8 lb. A mass of 14 lb represents a theoretical average shot put ball.

▲ **Olympic Shot Put** In the women's Olympic shot put event, female athletes toss an 8 lb steel ball.

Atomic Mass of an Element

The **atomic mass** of an element is the weighted average mass of all naturally occurring isotopes. To calculate the atomic mass, we will use a method similar to the one used in the shot put example. That is, to calculate the weighted average mass of an atom, we must consider the mass as well as the percent abundance of each isotope.

Carbon has two naturally occurring stable isotopes: carbon-12 and carbon-13. We can calculate the atomic mass of carbon given the mass and natural abundance of each isotope.

ISOTOPE	MASS	ABUNDANCE
^{12}C	12.000 amu	98.89%
^{13}C	13.003 amu	1.11%

The natural abundance of each isotope, expressed as a decimal, is 0.9889 and 0.0111. To calculate the true weighted average mass of a carbon atom, we must consider the mass and percentage of each isotope. We proceed as follows:

$$
\begin{aligned}
{}^{12}\text{C}: 12.000 \text{ amu} \times 0.9889 &= 11.87 \text{ amu} \\
{}^{13}\text{C}: 13.003 \text{ amu} \times 0.0111 &= \underline{0.144 \text{ amu}} \\
&\ \ 12.01 \text{ amu}
\end{aligned}
$$

Even though the atomic mass of carbon is 12.01 amu, we should note that no carbon atom weighs 12.01 amu. An actual carbon atom weighs either 12.000 amu or 13.003 amu. A mass of 12.01 amu represents the mass of a hypothetical average carbon atom. Example Exercise 5.3 further illustrates the calculation of atomic mass from isotopic mass and abundance data.

Example Exercise 5.3 Calculation of Atomic Mass

Silicon is the second most abundant element in Earth's crust. Calculate the atomic mass of silicon given the following data for its three natural isotopes:

ISOTOPE	MASS	ABUNDANCE
^{28}Si	27.977 amu	92.21%
^{29}Si	28.976 amu	4.70%
^{30}Si	29.974 amu	3.09%

Solution

We can find the atomic mass of silicon as follows:

$$
\begin{aligned}
{}^{28}\text{Si}: 27.977 \text{ amu} \times 0.9221 &= 25.80 \text{ amu} \\
{}^{29}\text{Si}: 28.976 \text{ amu} \times 0.0470 &= 1.36 \text{ amu} \\
{}^{30}\text{Si}: 29.974 \text{ amu} \times 0.0309 &= \underline{0.926 \text{ amu}} \\
&\ 28.09 \text{ amu}
\end{aligned}
$$

The average mass of a silicon atom is 28.09 amu, although we should note that there are no silicon atoms with a mass of 28.09 amu.

Practice Exercise

Calculate the atomic mass of copper given the following data:

ISOTOPE	MASS	ABUNDANCE
^{63}Cu	62.930 amu	69.09%
^{65}Cu	64.928 amu	30.91%

Answer: 63.55 amu

Concept Exercise

A bag of marbles has 75 large marbles with a mass of 2.00 g each, and 25 small marbles with a mass of 1.00 g each. Calculate (a) the simple average mass, and (b) the weighted average mass of the marble collection.

Answer: See Appendix G.

The Periodic Table

We often need to refer to the atomic number and atomic mass of an element. For convenience, this information is listed in the periodic table. In this textbook the atomic number is indicated above the symbol, and the atomic mass is given below the symbol. Figure 5.5 illustrates carbon as it appears in the periodic table.

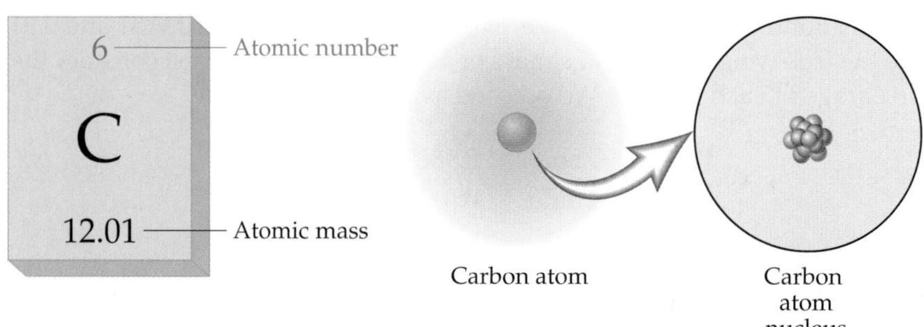

▲ **Figure 5.5 Carbon in the Periodic Table** The atomic number designates the number of protons (6). The atomic mass designates the weighted average mass of the naturally occurring stable isotopes of the element.

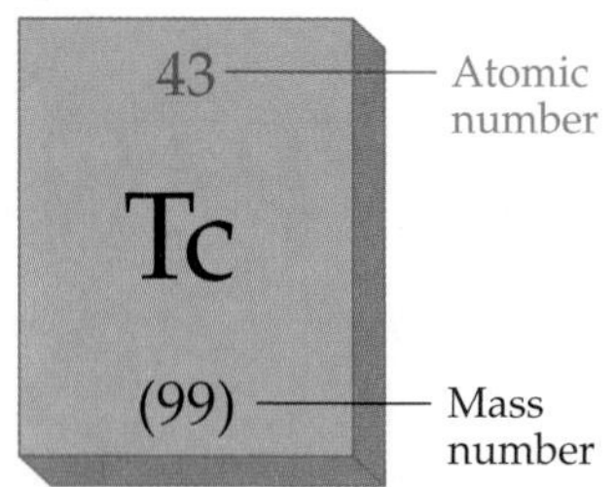

▲ **Figure 5.6 Technetium in the Periodic Table** The atomic number designates the number of protons (43). The mass number for this radioactive isotope is 99. Technetium-99 has a total of 43 protons and 56 neutrons in its nucleus.

When we refer to the periodic table, it is not possible to determine the number of naturally occurring isotopes for a given element. Of the first 83 elements, however, 81 elements have one or more stable isotopes. Only technetium (element 43) and promethium (element 61) are radioactive and unstable. In the periodic table on the inside front cover of this textbook, the mass number of unstable elements is given in parentheses as shown in Figure 5.6.

A value in parentheses below the symbol of an element indicates the mass number of the most stable or best known radioactive isotope. The mass number refers to the total number of protons and neutrons in the unstable nucleus. From the periodic table you can easily distinguish the elements that are stable from those that are unstable and radioactive. Example Exercise 5.4 further illustrates how to use the periodic table as a reference.

Example Exercise 5.4 Nuclear Composition from the Periodic Table

Refer to the periodic table on the inside cover of this textbook and determine the atomic number and atomic mass for iron.

Solution

In the periodic table we observe

The atomic number of iron is 26, and the atomic mass is 55.85 amu. From the periodic table information, we should note that it is not possible to determine the number of isotopes for iron or their mass numbers.

Practice Exercise

Refer to the periodic table on the inside cover of this text and determine the atomic number and mass number for the given radioactive isotope of radon gas.

Answers: 86 and (222)

Concept Exercise

Which of the following is *never* a whole number value: atomic number, atomic mass, or mass number?

Answer: See Appendix G.

CHEMISTRY CONNECTION Heavy Water

Q: Do all water molecules have the same mass?

Hydrogen atoms usually have one proton as their nucleus. However, about 1 in 6000 hydrogen atoms has both a proton and a neutron as its nucleus. This isotope of hydrogen is called deuterium (D), and is listed in Table 5.2. The mass of deuterium is about twice that of hydrogen. When deuterium atoms replace the hydrogen atoms in water, the resulting compound is referred to as *heavy water*. The systematic name of heavy water is deuterium oxide, and its formula is D_2O.

Heavy water was discovered by the American chemist Harold Urey, who won the Nobel prize in 1934 for his discovery. Heavy water is colorless, odorless, and tasteless, but it is toxic to humans and animals. The physical and chemical properties of heavy water are similar to those of ordinary light water. The following table lists a few properties of light water versus heavy water.

Heavy water has two important uses in nuclear reactors. When an atomic nucleus splits (i.e., fissions), it releases neutrons. However, these neutrons are moving at a high velocity and must be slowed in order to be captured and cause further nuclear fissions of a radioactive substance such as uranium–235. Heavy water is ideal for slowing down fast neutrons released during the fission process. In addition, heavy water has a higher specific heat than ordinary water, and is used to cool the nuclear reactor and prevent meltdown.

PROPERTIES OF WATER VERSUS HEAVY WATER

PROPERTY	LIGHT WATER, H_2O	HEAVY WATER, D_2O
appearance	colorless, odorless, tasteless liquid	colorless, odorless, tasteless liquid
molecular mass	18.02 amu	20.03 amu
density	1.000 g/mL at 4 °C	1.105 g/mL at 4 °C
freezing point	0.00 °C	3.82 °C
boiling point	100.00 °C	101.42 °C

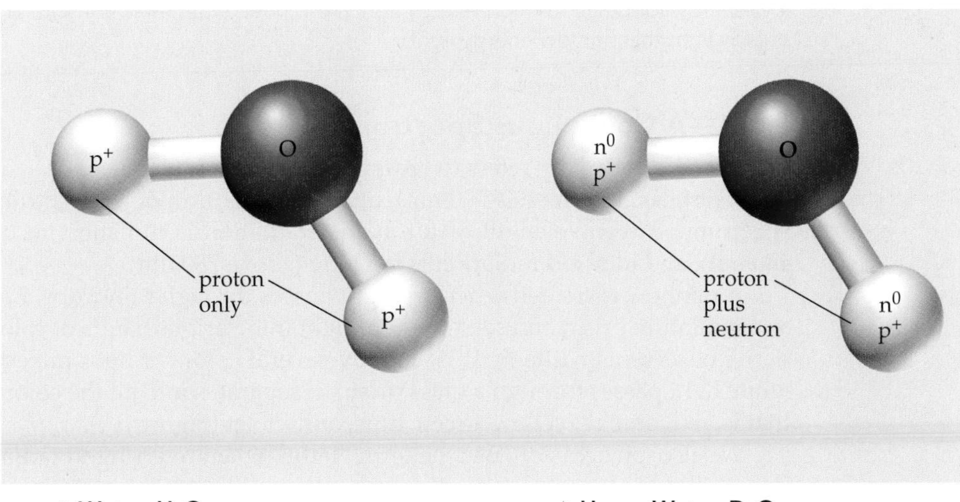

▲ Water, H_2O ▲ Heavy Water, D_2O

A: Light water, H_2O, has a molecular mass of 18.02 amu, while heavy water, D_2O, is 20.03 amu. Heavy water molecules have a greater mass because deuterium atoms have one more neutron than hydrogen atoms.

5.6 THE WAVE NATURE OF LIGHT

OBJECTIVES

- To explain the wave nature of light.
- To state the relationship of wavelength, frequency, and energy of light.

We can use our imagination to visualize light traveling through space in a fashion similar to an ocean wave. **Wavelength** refers to the distance the light wave travels to complete one cycle. **Frequency** refers to the number of wave cycles completed in each second. The velocity of light is constant, as all wavelengths and frequencies travel at 3.00×10^8 m/s. Figure 5.7 illustrates the wave nature of light.

As the wavelength of light decreases, the frequency of a light wave increases. Conversely, as the wavelength increases, the frequency decreases. As an example, the wavelength decreases from red to violet as the frequency increases. We can visualize a high-frequency light wave moving rapidly up and down while completing many cycles per second. Consequently, high-frequency light is more energetic than low-frequency light, and short-wavelength light is more energetic than long-wavelength light.

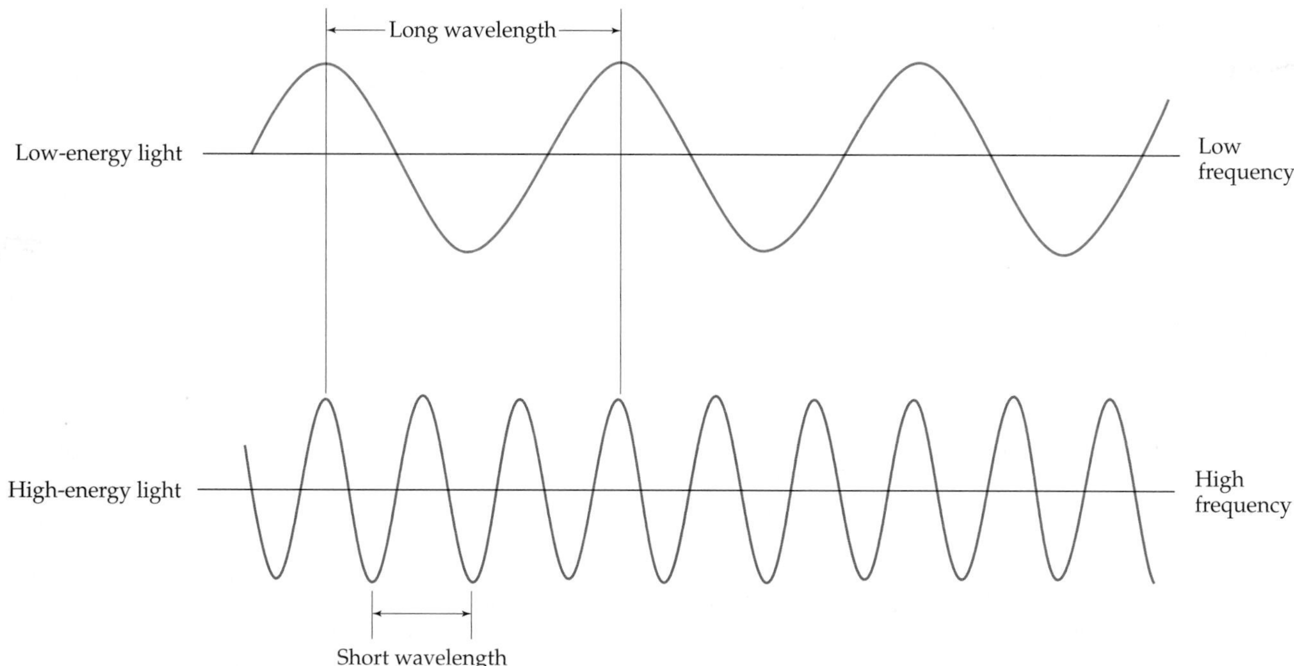

▲ **Figure 5.7 Wavelength and Frequency of Light** Notice that the wavelength is longer for low-energy light than for high-energy light. Also notice that the frequency is greater for high-energy light than for low energy light.

Light—A Continuous Spectrum

The term **light** usually refers to radiant energy that is visible. Our eyes can see light in the **visible spectrum** (400–700 nm), but not in the ultraviolet spectrum or the infrared spectrum. The wavelength of ultraviolet radiation is too short to be visible (below 400 nm), and infrared radiation is too long (above 700 nm).

A rainbow is created when sunlight passes through raindrops. Each raindrop acts as a miniature prism and separates sunlight into various bands of color. What we ordinarily observe as white light is actually several colors of light mixed together. When white light passes through a glass prism, it separates into all the colors of the rainbow, that is, red, orange, yellow, green, blue, and violet (Figure 5.8).

We also, on occasion, use the term "light" when referring to radiant energy that is not visible. The complete **radiant energy spectrum** is an uninterrupted band, or **continuous spectrum**, of visible and invisible light that ranges from short to long wavelengths. The radiant energy spectrum includes invisible gamma rays, X rays, and microwaves, as well as visible light (Figure 5.9).

Example Exercise 5.5 illustrates how the wavelength, frequency, and energy of light are related.

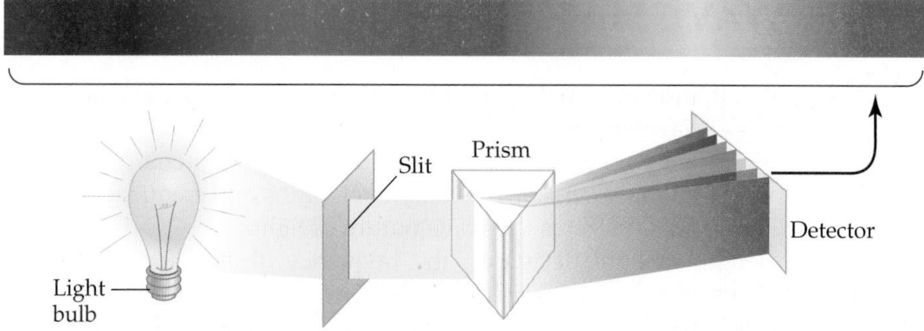

▲ **Figure 5.8 White Light Passing through a Prism** Notice that white light separates into a rainbow of colors when it passes through a glass prism. Similarly, sunlight produces a rainbow when it passes through raindrops.

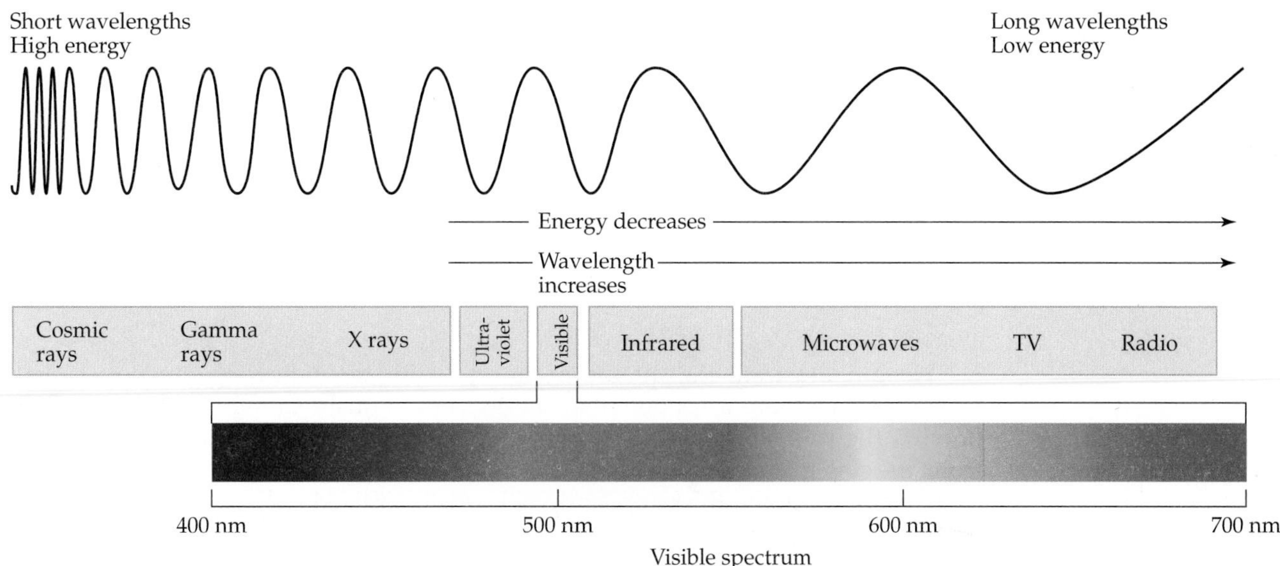

▲ **Figure 5.9 The Radiant Energy Spectrum** The complete radiant energy spectrum includes short-wavelength gamma rays through long-wavelength microwaves. Notice that the visible spectrum is only a narrow window in a broad band of radiant energy.

Example Exercise 5.5 Properties of Light

Considering blue light and yellow light, which has the

(a) longer wavelength? (b) higher frequency?
(c) higher energy?

Solution

Referring to Figure 5.9, we notice that the wavelength of yellow light is about 600 nm and that of blue light is about 500 nm. Thus,

(a) *yellow light* has a longer wavelength than blue light.
(b) *blue light* has a higher frequency because it has a shorter wavelength.
(c) *blue light* has a higher energy because it has a higher frequency.

Practice Exercise

Considering infrared light and ultraviolet light, which has the

(a) longer wavelength? (b) higher frequency?
(c) higher energy?

Answers: (a) infrared; (b) ultraviolet; (c) ultraviolet

Concept Exercise

The energy of light (increases/decreases) as the wavelength increases. The energy of light (increases/decreases) as the frequency increases.

Answer: See Appendix G.

5.7 THE QUANTUM CONCEPT

OBJECTIVE

▸ To explain the quantum concept applied to matter and energy.

In 1900, Max Planck (1858–1947), a German physicist, introduced a revolutionary idea. Planck proposed that the energy radiated by a heated object is not continuous, but rather that the radiation is emitted in small bundles. The idea that energy is composed of discrete units is referred to as the quantum concept.

Although radiant energy has a wave nature, it also has a particle nature, and an individual unit of light energy is referred to as a **photon**. According to the quantum theory, an ordinary light bulb radiates light which has both a wave and particle nature.

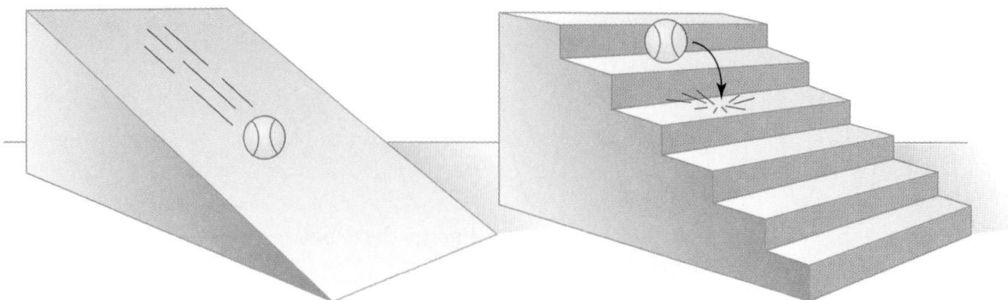

▶ **Figure 5.10 Stair Analogy for the Quantum Principle** A ball rolling down a ramp loses potential energy continuously. Conversely, a ball rolling down a flight of stairs loses potential energy in quantized amounts each time it drops from one step to another one.

We can illustrate the quantum concept with the following analogy. A ball rolling down a ramp loses potential energy continuously. Conversely, a ball rolling down a flight of stairs loses potential energy in discrete units each time it drops from one step to another. In this example, the loss of potential energy is continuous as the ball rolls down the ramp, and the loss is quantized as the ball rolls down the stairs. Figure 5.10 contrasts a continuous change in energy versus a quantized energy change.

The following Example Exercise 5.6 further illustrates practical applications of the quantum theory.

Example Exercise 5.6 Quantum Concept

State whether each of the following scientific instruments gives a continuous or a quantized measurement of mass:

(a) triple-beam balance (b) digital electronic balance

Solution

Refer to Figure 2.3 if you have not used these balances in the laboratory.

(a) On a triple-beam balance a small metal rider is moved along a beam. Since the metal rider can be moved to any position on the beam, a triple-beam balance gives a *continuous* mass measurement.

(b) On a digital electronic balance the display indicates the mass of an object to a particular decimal place, for example, 5.015 g. Since the last digit in the display must be a whole number, a digital balance gives a *quantized* mass measurement.

Practice Exercise

State whether each of the following musical instruments produces continuous or quantized musical notes:

(a) acoustic guitar (b) electronic keyboard

Answers: (a) continuous; (b) quantized

Concept Exercise

Complete the following quantum analogy: a water wave is to a drop of water, as a light wave is to a ______.

Answer: See Appendix G.

OBJECTIVES

- To describe the Bohr model of the atom.
- To explain the relationship between energy levels in an atom and lines in an emission spectrum.

5.8 BOHR MODEL OF THE ATOM

In 1911, a brilliant young Danish physicist, Niels Bohr (1885–1962), completed his doctorate and left for England to begin postdoctoral work under J. J. Thomson. After a few months, he left Cambridge to join Rutherford at Manchester where the atomic nucleus had been discovered. It took only 2 years for the young Dane to raise our understanding of the atom to yet another level.

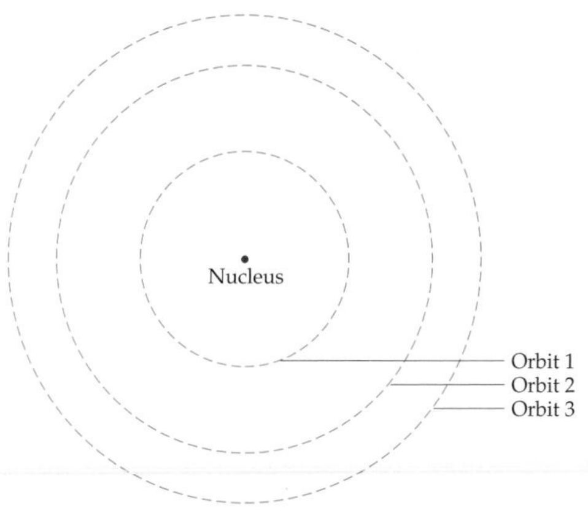

◀ **Figure 5.11 Bohr Model of the Hydrogen Atom** The electron is a specific distance from the nucleus and occupies an orbit of discrete energy. According to the Bohr model, the electron is found only in a given energy level.

In 1913, Bohr speculated that electrons orbit around the atomic nucleus just as planets orbit around the Sun. He further suggested that electron orbits were at a fixed distance from the nucleus and had a definite energy. The electron was said to travel in a fixed-energy orbit referred to as an **energy level**. Moreover, electrons could be found only in specific energy levels and nowhere else. Figure 5.11 illustrates the model of the hydrogen atom proposed by Niels Bohr. This model is referred to as the Bohr model of the atom, or simply, the **Bohr atom**.

▲ **Niels Bohr Stamp** The planetary model of the atom was proposed by Niels Bohr.

Evidence for Energy Levels

The Bohr model was a beautiful mental picture of electrons in atoms. However, no one knew whether the model was right or wrong because there was no experimental evidence to support the theory. Coincidentally, Bohr received a paper on the emission of light from hydrogen gas. The paper showed that excited hydrogen gas emits separate lines of light, rather than a continuous band of color.

Passing electricity through a sealed glass tube containing a gas produces an emission spectrum. For example, passing an electrical current through a glass tube containing hydrogen produces light, which separates into a series of narrow lines when passed through a prism. This collection of narrow bands of light energy is referred to as an **emission line spectrum**, and the individual bands of light are called spectral lines. The emission spectrum of hydrogen gas is shown in Figure 5.12.

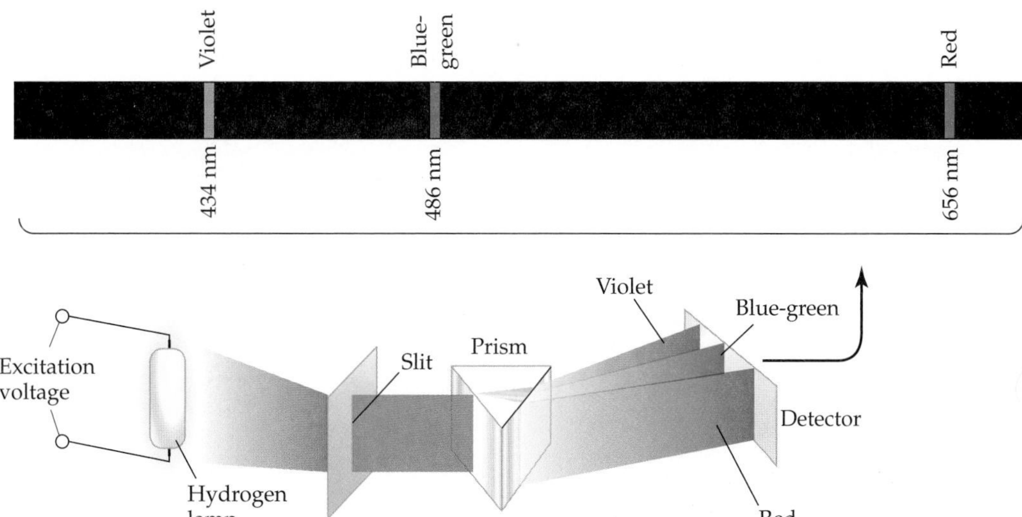

▲ **Figure 5.12 Hydrogen Emission Spectrum** An emission line spectrum is produced when hydrogen gas is excited by an electrical voltage. After the emitted light is passed through a prism, three discrete vivid lines are observed.

After examining the emission spectrum of hydrogen, Bohr realized that he had experimental evidence to support his model of the atom. The concept of electron energy levels was supported by the line spectrum of hydrogen. In a gas discharge tube, excited atoms of hydrogen have electrons in a high energy orbit; for example, the electron may temporarily occupy the second orbit. Since this state is unstable the electron quickly drops from the higher level back to a lower level, that is, from level 2 to level 1. In the process, the electron loses a discrete amount of energy. This discrete energy loss corresponds to a photon of light energy. The energy of the photon of light equals the amount of energy lost by the electron as it drops from the higher to the lower energy level.

In the hydrogen spectrum, there are three bright lines: red, blue-green, and violet. The red line has the longest wavelength of the three and the lowest energy. Bohr found that the red line corresponds to an excited electron dropping from energy level 3 to 2. The blue-green line is more energetic than the red line and corresponds to an excited electron dropping from energy level 4 to 2. The violet line is the most energetic of the three and is produced when an electron drops from energy level 5 to 2. Figure 5.13 illustrates the correlation between Bohr's electron energy levels and the observed lines in the hydrogen spectrum.

Since a single photon is emitted each time an electron drops to a lower level, it follows that several photons are emitted when several electrons change levels. For example, if the electron drops from energy level 5 to 2 in ten hydrogen atoms, ten photons would be emitted. Each of the photons would have the same energy, and collectively they would be observed as a violet line in the emission spectrum. Example Exercise 5.7 further illustrates the relationship between energy levels and emission lines.

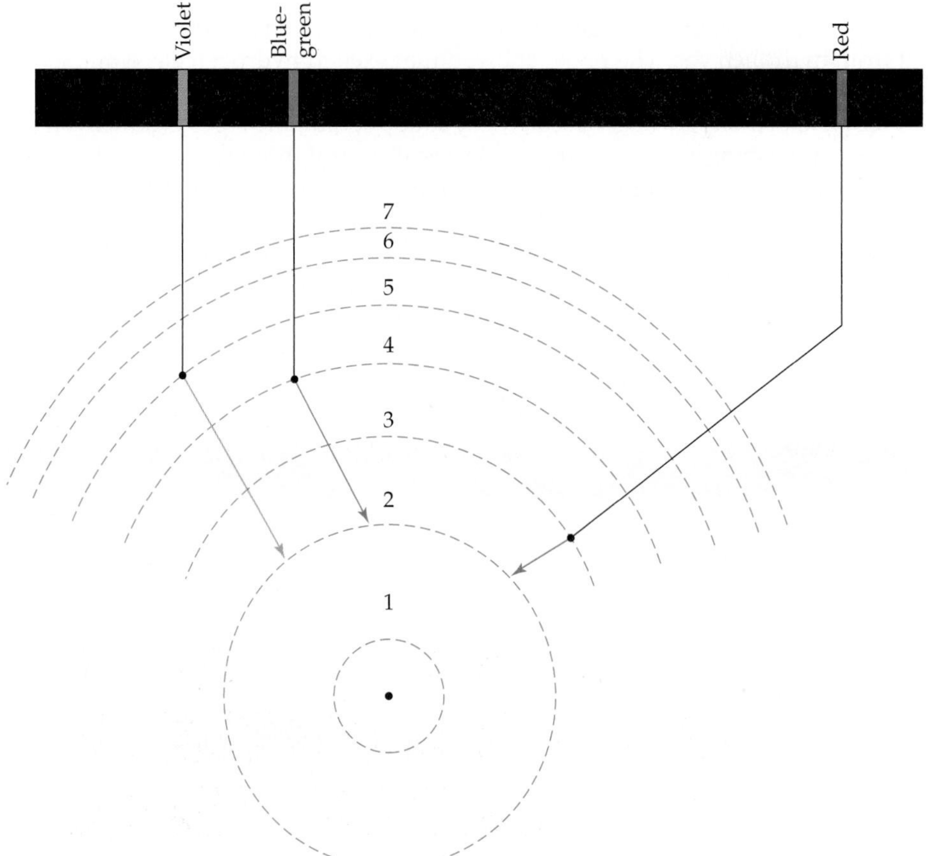

▲ **Figure 5.13 Spectral Lines and Energy Levels in Hydrogen** When an electron drops from energy level **5** to **2**, a violet photon is released. When an electron drops from level **4** to **2**, a blue-green photon is released; from level **3** to **2**, a red photon is released. A visible spectral line corresponds to a large number of identical photons.

Example Exercise 5.7 Emission Spectra and Energy Levels

Explain the relationship between an observed emission line in a spectrum and electron energy levels.

Solution

When an electron drops from a higher to a lower energy level, light is emitted. For each electron that drops, a single photon of light energy is emitted. The energy lost by the electron that drops equals the energy of the photon that is emitted. Several photons of light having the same energy are observed as a spectral line.

Practice Exercise

Indicate the number and color of the photons emitted for each of the following electron transitions in hydrogen atoms:

(a) 1 e^- dropping from energy level 3 to 2
(b) 10 e^- dropping from energy level 3 to 2
(c) 100 e^- dropping from energy level 4 to 2
(d) 500 e^- dropping from energy level 5 to 2

Answers: (a) 1 red photon; (b) 10 red photons; (c) 100 blue-green photons; (d) 500 violet photons

Concept Exercise

Which of the following statements are true according to the Bohr model of the atom?

(a) Electrons are attracted to the atomic nucleus.
(b) Electrons have fixed energy as they circle the nucleus.
(c) Electrons lose energy as they drop to an orbit closer to the nucleus.

Answer: See Appendix G.

Emission Spectra as "Atomic Fingerprints"

Further study of emission spectra revealed that each element produced a unique set of spectral lines. This observation indicated that the energy levels must be unique for atoms of each element. Therefore, a line spectrum is sometimes referred to as an "atomic fingerprint."

Atomic fingerprints are useful in the identification of elements. For instance, in 1868, the atomic fingerprint of a new element was observed in the spectrum from the Sun. This element was named helium, after *helios*, the Greek word for "sun." In 1895, an element was discovered in uranium ore with an atomic fingerprint identical to that observed for helium in the Sun's spectrum. Thus, helium was discovered on Earth 27 years after it had first been observed in the solar spectrum. Figure 5.14 compares a continuous spectrum to the emission line spectra of four elements.

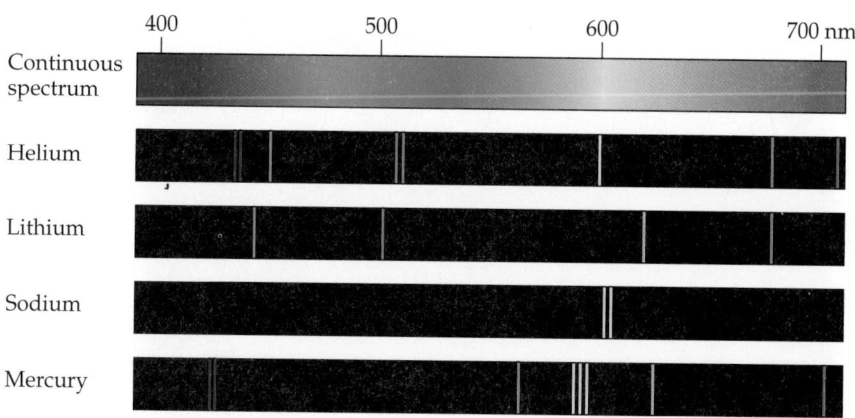

▲ **Figure 5.14 A Continuous Spectrum versus Line Spectra** A continuous spectrum is produced from an ordinary light bulb. The emission line spectra are produced by excited atoms of elements in the gaseous state.

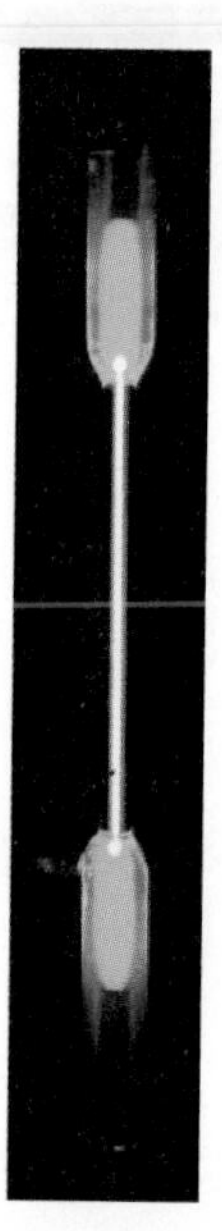

▲ **Mercury Gas Discharge Tube** The emission lines from mercury vapor illuminates the glass tube with a blue glow.

CRITICAL THINKING Neon Lights

Q: Do all "neon" advertising lights contain neon gas?

At the end of the nineteenth century, J. J. Thomson discovered the electron using a gas discharge tube. Thomson constructed his gas discharge tube out of thin glass, placed a metal electrode in each end, and introduced a small amount of gas. When the gas in the tube was electrically excited, he noticed that the tube glowed. He identified the glowing rays as a stream of small negatively charged particles. These particles acquired the name electrons, and Thomson is given credit for their discovery. The glass tubes used by Thomson were forerunners of cathode-ray tubes, which have been used in television sets and computer screens.

In 1913, Niels Bohr explained that exciting gases with electricity caused electrons to be temporarily promoted to higher energy states within the atom. However, the excited electrons quickly lose energy by emitting light and return to their original state. The light emitted by different gases varies because the energy levels within atoms vary for each element. For example, a gas discharge tube containing hydrogen gas produces a reddish-purple glow, whereas mercury vapor gives a blue glow.

In 1898, the Scottish chemist William Ramsay discovered the gas neon. Unlike argon, which comprises about 1% of air, neon is much more rare. When neon gas is placed in a narrow glass tube and electrically excited, it produces a reddish-orange light that is very eye-catching. The fact that gas discharge tubes produce an attractive array of colors led naturally to their use as advertising lights. Light from excited neon gas is very intense, and the term "neon light" has become a generic term for all lighted advertising displays.

Obviously, not all lights used in advertising are the same color. That is, "neon lights" can be red, green, blue, and so on. To produce a given color, a gas discharge tube must be filled with a specific gas. For a pink light, helium gas is used, and for a blue light, argon gas can be used. Only when we wish to produce a reddish-orange light is an advertising display actually filled with neon gas.

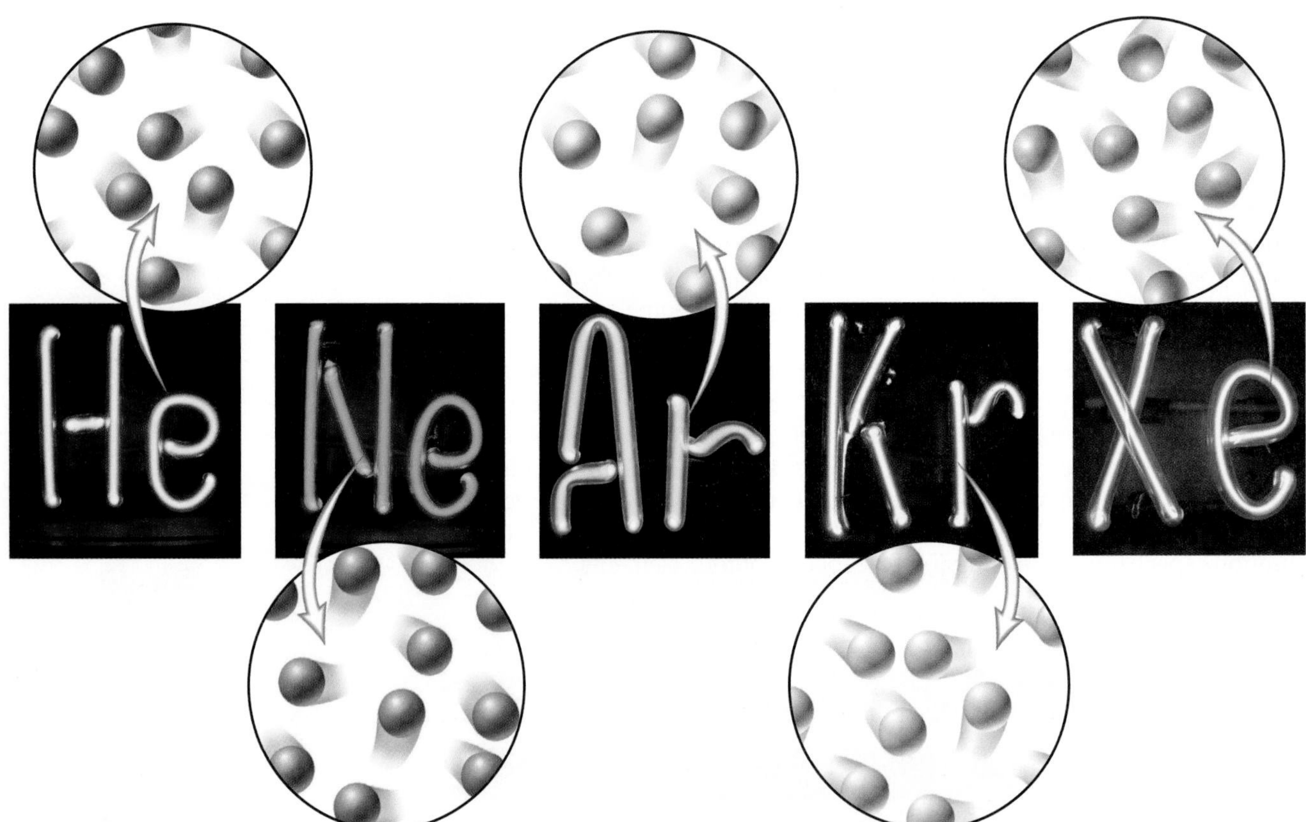

▲ **Light Emission** Notice the color of each light varies depending on the gas contained in the glass tube.

A: Not all "neon lights" contain neon gas; only "neon lights" that have a reddish-orange glow contain neon gas.

5.9 ENERGY LEVELS AND SUBLEVELS

OBJECTIVES

- To state each energy sublevel within a given energy level.
- To sum the maximum number of electrons that can occupy any given energy level.

As we learned in the previous section, in 1913 Niels Bohr proposed a model for the atom that pictured electrons circling around the nucleus in fixed-energy levels. His proposal was supported experimentally by the lines in the emission spectrum of hydrogen. The emission spectra of other elements, however, had far too many lines to interpret. Although Bohr could not explain the spectra of other elements, he did suggest the idea of sublevels within a main energy level. The model that eventually emerged had electrons occupying an **energy sublevel** within a main energy level. These energy sublevels were designated *s*, *p*, *d*, and *f*—in reference to the *sharp*, *principal*, *diffuse*, and *fine* lines, respectively, in the emission spectra of the elements.

The number of sublevels in each level corresponds to the number of the main energy level. That is, the first energy level (1) has one sublevel designated 1*s*. The second energy level (2) has two sublevels designated 2*s* and 2*p*. The third energy level (3) has three sublevels designated 3*s*, 3*p*, and 3*d*. The fourth energy level (4) is composed of 4*s*, 4*p*, 4*d*, and 4*f* sublevels (Figure 5.15).

The maximum number of electrons in each of the energy sublevels depends on the type of sublevel. That is, an *s* sublevel can hold a maximum of 2 electrons. A *p* sublevel can have a maximum of 6 electrons. A *d* sublevel can have 10 electrons, and an *f* sublevel can hold a maximum of 14 electrons.

To find the maximum number of electrons in a main energy level, we add up the electrons in each sublevel. The first energy level has one *s* sublevel; it can contain only 2 e^-. The second major energy level has two sublevels, 2*s* and 2*p*. The 2*s* can hold 2 e^- and the 2*p* can hold 6 e^-. Thus, the second energy level can hold a maximum of 8 e^-. Example Exercise 5.8 further illustrates how energy levels, sublevels, and number of electrons are related.

Example Exercise 5.8 Energy Levels, Sublevels, and Electrons

What is the maximum number of electrons that can occupy the third energy level?

Solution

The third energy level is split into three sublevels: 3*s*, 3*p*, and 3*d*. The maximum number of electrons that can occupy each sublevel is as follows:

$$\begin{aligned} s \text{ sublevel} &= 2\ e^- \\ p \text{ sublevel} &= 6\ e^- \\ d \text{ sublevel} &= 10\ e^- \end{aligned}$$

The maximum number of electrons in the third energy level is found by adding the three sublevels together:

$$3s + 3p + 3d = \text{total electrons}$$
$$2\ e^- + 6\ e^- + 10\ e^- = 18\ e^-$$

The third energy level can hold a maximum of 18 electrons. Of course, in elements where the third energy level of an atom is not filled, there are fewer than 18 electrons.

Practice Exercise

What is the maximum number of electrons that can occupy the fourth energy level?

Answers: 4*s*, 4*p*, 4*d*, 4*f*; 32 e^- (2 e^- + 6 e^- + 10 e^- + 14 e^-)

Concept Exercise

What is the theoretical number of sublevels in the tenth energy level?

Answer: See Appendix G.

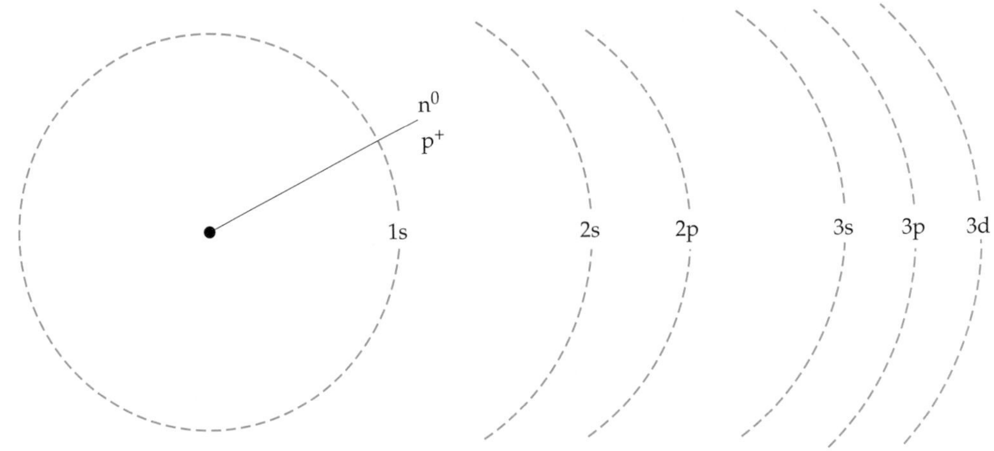

▶ **Figure 5.15 A Cross Section of an Atom** The first energy level has one sublevel (1*s*). The second energy level has two sublevels (2*s* and 2*p*). The third energy level has three sublevels (3*s*, 3*p*, and 3*d*). Although the diagram suggests electrons travel in circular orbits, this is a simplification and is not actually the case.

TABLE 5.3 DISTRIBUTION OF ELECTRONS BY ENERGY LEVEL

ENERGY LEVEL	ENERGY SUBLEVEL	MAXIMUM e^- SUBLEVEL	MAXIMUM e^- IN ENERGY LEVEL
1	1*s*	2 e^-	2 e^-
2	2*s*	2 e^-	
	2*p*	6 e^-	8 e^-
3	3*s*	2 e^-	
	3*p*	6 e^-	
	3*d*	10 e^-	18 e^-
4	4*s*	2 e^-	
	4*p*	6 e^-	
	4*d*	10 e^-	
	4*f*	14 e^-	32 e^-

In summary, electrons are arranged around the nucleus in sublevels that have a specific fixed energy. Electrons farther from the nucleus occupy higher energy levels than those closer to the nucleus. Table 5.3 summarizes how main energy levels, sublevels, and number of electrons are related.

5.10 ELECTRON CONFIGURATION

OBJECTIVES

- ▶ To list the order of sublevels according to increasing energy.
- ▶ To write the predicted electron configurations for selected elements.

Electrons are arranged about the nucleus in a systematic order. The first electrons fill the energy sublevel closest to the nucleus. Additional electrons fill energy sublevels further and further from the nucleus. In other words, each energy level is filled sublevel by sublevel. The *s* sublevel is filled before a *p* sublevel, a *p* sublevel is filled before a *d* sublevel, and a *d* sublevel is filled before an *f* sublevel.

In general, sublevels are higher in energy as the energy level increases. Therefore, we would expect the order of sublevel filling to be 1*s*, 2*s*, 2*p*, 3*s*, 3*p*, 3*d*, 4*s*, and so on. This is not quite accurate because there are exceptions. For instance, the 4*s* sublevel is lower in energy than the 3*d* sublevel, and the 5*s* is lower than the 4*d* sublevel. A partial list of sublevels in order of increasing energy is $1s < 2s < 2p < 3s < 3p < 4s < 3d < 4p < 5s < 4d < 5p < 6s$.

In Chapter 6 we will learn how to predict the order in which sublevels are filled from the position of the element in the periodic table. In fact, the unusual shape of the periodic table reflects the order of sublevels according to increasing energy. For now, you should memorize the order of sublevel filling or refer to Figure 5.16.

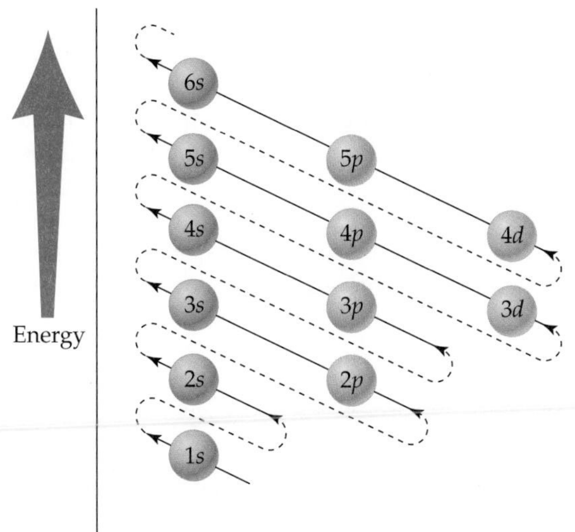

◀ **Figure 5.16 Filling Diagram for Energy Sublevels** The order of sublevel filling is arranged according to increasing energy. Electrons first fill the 1*s* sublevel followed by the 2*s*, 2*p*, 3*s*, 3*p*, 4*s*, 3*d*, 4*p*, 5*s*, 4*d*, 5*p*, and 6*s* sublevels.

Example Exercise 5.9 Order of Sublevels

According to increasing energy, what is the next energy sublevel after each of the following sublevels?

(a) 3*p* (b) 4*d*

Solution

If you do not know the order of sublevels, refer to the filling diagram in Figure 5.16.

(a) Although the third energy level has 3*s*, 3*p*, and 3*d* sublevels, the 3*d* sublevel does not immediately follow the 3*p*. Instead, the 4*s* sublevel follows the 3*p* and precedes the 3*d*. Thus,

3*s*, 3*p*, **4*s***

(b) Although the fourth energy level has 4*s*, 4*p*, 4*d*, and 4*f* sublevels, the 4*f* sublevel does not immediately follow the 4*d*. Instead, the 5*p* sublevel begins accepting electrons after the 4*d* is filled. Thus,

4*p*, 5*s*, 4*d*, **5*p***

Practice Exercise

Which sublevel gains electrons after each of the following sublevels is filled?

(a) 2*s* (b) 5*p*

Answers: (a) 2*p*; (b) 6*s*

Concept Exercise

The energy difference between sublevels (increases/decreases) moving away from the nucleus.

Answer: See Appendix G.

Writing Electron Configurations

The **electron configuration** of an atom is a shorthand statement describing the location of electrons by sublevel. First, the sublevel is written, followed by a superscript indicating the number of electrons. For example, if the 2*p* sublevel contains two electrons, the standard notation is $2p^2$. Thus,

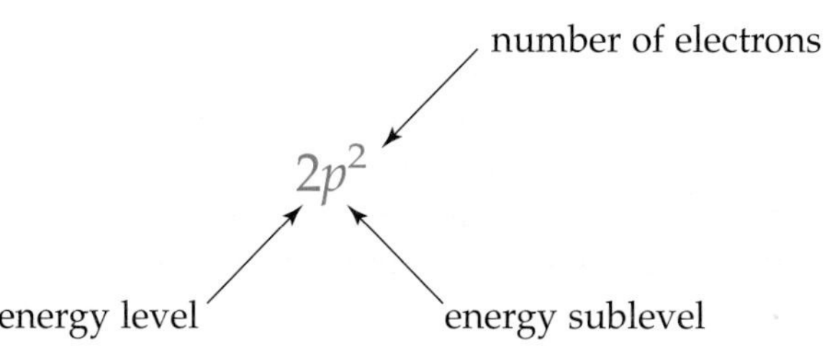

Writing the electron configuration for an atom is a straightforward procedure. First, find the atomic number for the element in the periodic table. The atomic number corresponds to the number of protons in the nucleus, which equals the number of electrons in a neutral atom. Next, write the sublevels according to increasing energy. Each sublevel is filled with electrons in sequence until the total of all the superscripts equals the atomic number of the element.

As an example, let's write the electron configuration for iron. If we refer to the periodic table, we find that the atomic number of iron is 26. Given the atomic number, we know that an iron nucleus must have 26 protons and is surrounded by 26 electrons. The order in which the electron sublevels are arranged according to increasing energy is as follows:

$$1s\ 2s\ 2p\ 3s\ 3p\ 4s\ 3d\ \ldots$$

The energy sublevels in iron are filled beginning with the 1*s* sublevel and ending when there is a total of 26 electrons. The electron configuration for iron is as follows:

$$\text{Fe:}\quad 1s^2\,2s^2\,2p^6\,3s^2\,3p^6\,4s^2\,3d^6$$

Notice that the sum of the superscripts equals the atomic number of iron (26). Example Exercise 5.10 illustrates the electron configuration for other elements.

Example Exercise 5.10 Electron Configuration

Write the predicted electron configuration for each of the following elements:

(a) F (b) Sr

Solution

Refer to the periodic table to find the atomic number of an element.

(a) The atomic number of fluorine is 9; therefore, the number of electrons is 9. We can fill sublevels with 9 electrons as follows.

$$\text{F:}\quad 1s^2\,2s^2\,2p^5$$

(b) The atomic number of strontium is 38; therefore, the number of electrons is 38. We can fill sublevels with 38 electrons as follows:

$$\text{Sr:}\quad 1s^2\,2s^2\,2p^6\,3s^2\,3p^6\,4s^2\,3d^{10}\,4p^6\,5s^2$$

To check your answer, find the total number of electrons by adding up the superscripts. The total is 38 e^-; this agrees with the atomic number for Sr.

Practice Exercise

Write the predicted electron configuration for each of the following elements:

(a) argon (b) cadmium

Answers: (a) $1s^2\,2s^2\,2p^6\,3s^2\,3p^6$; (b) $1s^2\,2s^2\,2p^6\,3s^2\,3p^6\,4s^2\,3d^{10}\,4p^6\,5s^2\,4d^{10}$

Concept Exercise

Refer to the periodic table and state whether Cr or Mn has more electrons in the outermost *d* sublevel.

Answer: See Appendix G.

5.11 QUANTUM MECHANICAL MODEL OF THE ATOM

OBJECTIVES

- To describe the quantum mechanical model of the atom.
- To compare the relative sizes and shapes of *s* and *p* orbitals.

In the mid-1920s, a new model of the atom began to emerge. A more powerful theory evolved because the behavior of electrons could not be fully explained by the Bohr model of the atom. The German physicist Werner Heisenberg (1901–1976) concluded that it was not possible to accurately determine both the location and energy of an electron. In his **uncertainty principle**, Heisenberg stated that it is impossible to precisely measure both the location and momentum (or energy) of a small particle simultaneously. In fact, the more accurately the location of an electron in an atom is known, the less precisely its energy can be determined.

In 1932, Heisenberg won the Nobel prize in physics for his uncertainty principle. Not everyone, however, subscribed to the principle of uncertainty. Some physicists found it unsettling to consider that they might live in a universe ruled by chance. Albert Einstein (1879–1955) was sufficiently troubled by the uncertainty principle that he offered the famous quote: "*It seems hard to look into God's cards, but I cannot for a moment believe He plays dice as the current quantum theory alleges He does.*" Although the uncertainty principle was initially controversial, it was a crucial concept for explaining the new model of the atom.

Gradually, the deeper nature of the atom came into focus. The new model retained the idea of quantized energy levels, but incorporated the concept of uncertainty. The new model that emerged became known as the **quantum mechanical atom**. Recall that in the Bohr model the energy of an electron is defined in terms of a fixed-energy orbit about the nucleus. In the quantum mechanical model the energy of an electron can be described in terms of the probability of it being within a spatial volume surrounding the nucleus. This region of high probability (~95%) for finding an electron of given energy is called an **orbital**.

▲ **Albert Einstein and Niels Bohr** The two famous scientists disagreed on quantum theory, but each provided important insights to further our understanding of the atom.

Sizes and Shapes of *s* and *p* Orbitals

In the quantum mechanical atom, orbitals are arranged about the nucleus according to their size and shape. In general, electrons having higher energy are found in larger orbitals. Similar to the energy levels in the Bohr atom, the energy of orbitals is quantized and assigned a whole number value such as 1, 2, 3, 4, ... As the number increases, the energy and size of an orbital also increases.

We can describe the shapes of orbitals by the letters *s*, *p*, *d*, and *f*. For example, the shape of an *s* orbital is that of a sphere, and the shape of a *p* orbital is that of a dumbbell. We can designate the size and shape of an orbital by combining the number that indicates its energy, and the letter that indicates its shape. For example, the designations 1*s*, 2*p*, and 3*d* indicates three orbitals that differ in size, energy, and shape. All *s* orbitals are spherical, but they are not all the same size. A 3*s* orbital is a larger sphere than a 2*s*, and a 2*s* orbital is larger than a 1*s*. That is, the size and energy of the orbital increase as the energy level increases. Figure 5.17 illustrates the relationship between *s* orbitals about the nucleus.

All *p* orbitals have the shape of a dumbbell, but they are not all equal in size or energy. A 3*p* orbital is larger than a 2*p* orbital and is at a higher energy level. A *p* orbital is said to resemble a dumbbell because it has two lobes. Electrons in a *p* orbital can occupy either of the lobes.

There are three different 2*p* orbitals. Although these three orbitals are identical in size and shape, they differ in their orientation to each other. That is, the three 2*p* orbitals intersect at the nucleus, but they are oriented at right angles to each other. Figure 5.18 illustrates the relationships between the $2p_x$, $2p_y$, and $2p_z$ orbitals. The p_x orbital is oriented along the x-axis of a three-dimensional system, and the p_y and p_z orbitals are oriented along the y-axis and z-axis, respectively.

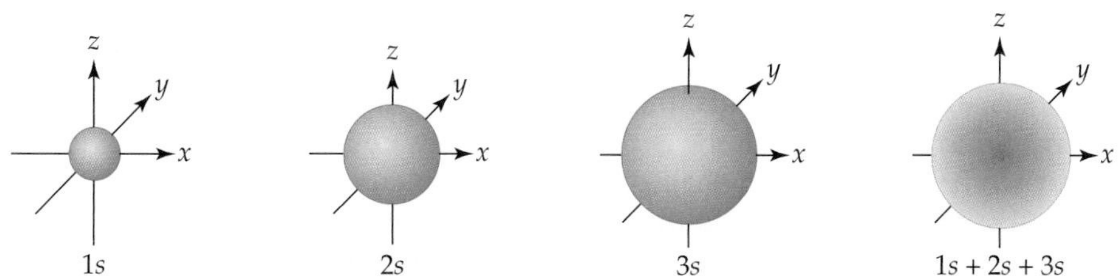

▲ **Figure 5.17 Relative Sizes of *s* Orbitals** The relative sizes of 1*s*, 2*s*, and 3*s* orbitals are shown. As the main energy level increases, the size and energy of the orbital increases. The nucleus of the atom is located in the center where the three axes intersect. The sketch on the far right illustrates the relationship of the 1*s*, 2*s*, and 3*s* orbitals.

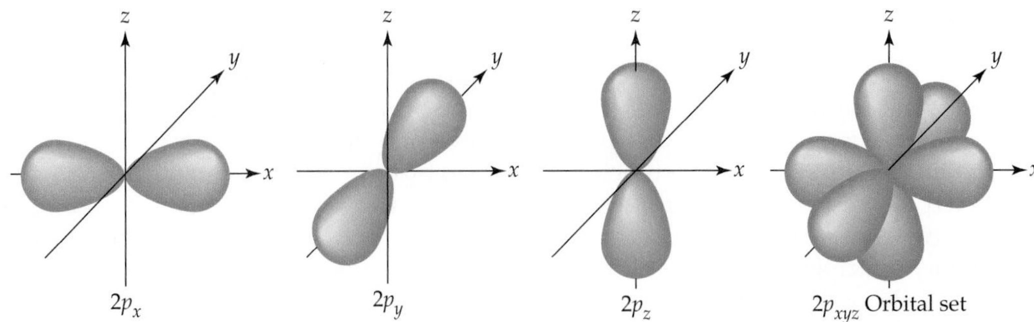

▲ **Figure 5.18 Orientation of 2*p* Orbitals** The size and shape of the three 2*p* orbitals are identical. The 2*p* orbitals do not have a fixed orientation, but rather are perpendicular to each other. An electron in a $2p_x$ orbital has the same probability of occupying the $2p_y$ or $2p_z$ orbital.

Example Exercise 5.11 provides further practice in describing the relative sizes and shapes of orbitals.

Example Exercise 5.11 Atomic Orbitals

Describe the relative size, energy, and shape for each of the following orbitals:

(a) 4*s* versus 3*s* and 5*s* (b) 4*p* versus 3*p* and 5*p*

Solution

The size and energy of an orbital is indicated by the number; the shape of the orbital is designated by the letter.

(a) Size and energy are greater for a 4*s* orbital than for a 3*s* orbital, but less than for a 5*s* orbital. The shape of a 4*s* orbital—and all *s* orbitals—is similar to the shape of a sphere.

(b) Size and energy are greater for a 4*p* orbital than for a 3*p* orbital, but less than for a 5*p* orbital. The shape of a 4*p* orbital—and all *p* orbitals—is similar to the shape of a dumbbell.

Practice Exercise

Select the orbital in each of the following pairs that fits the description:

(a) the higher energy orbital: 3*p* or 4*p* (b) the larger size orbital: 4*d* or 5*d*

Answers: (a) 4*p*; (b) 5*d*

Concept Exercise

Which of the following statements are true according to the quantum mechanical model of the atom?

(a) Orbitals represent quantum energy levels for electrons.

(b Orbitals represent probability boundaries for electrons.

(c) Orbitals can have different shapes.

Answer: See Appendix G.

By analogy, try to visualize a flying insect trapped inside two bottles with the open ends held together. The insect is free to fly about the entire volume of the two bottles. In this analogy, the insect represents an electron, and the two bottles represent the two lobes of a *p* orbital. Thus, there is a high probability of finding the electron anywhere within the volume of the entire *p* orbital. Figure 5.19 illustrates this analogy.

So far we have shown the shapes of *s* and *p* orbitals. However, there are many other orbitals; for example, in the third energy level there are 3*s*, 3*p*, and 3*d* orbitals, and in the fourth energy level there are 4*s*, 4*p*, 4*d*, and 4*f* orbitals.

The 3*s* sublevel contains a single spherical 3*s* orbital. The 3*p* sublevel contains three dumbbell-shaped orbitals: $3p_x$, $3p_y$, and $3p_z$. The 3*p* orbitals are similar in shape, but are larger than 2*p* orbitals. The 3*d* sublevel contains five 3*d* orbitals. Although it is not necessary to know the shapes of *d* orbitals, the five different 3*d* orbitals are shown in Figure 5.20.

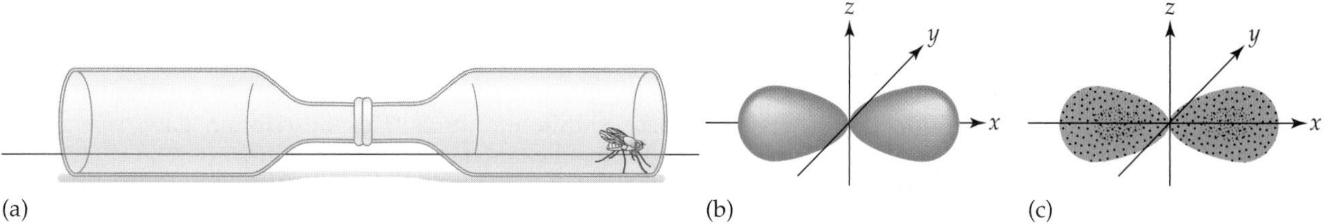

▲ **Figure 5.19 Analogy for a *p* Orbital** (a) Notice the insect trapped within the bottles held end to end. The insect can be in the left bottle, or the right bottle. (b) Similarly, an electron has a probability of being found anywhere within the two lobes of the *p* orbital. (c) The concentration of black dots indicates the most probable location of the electrons.

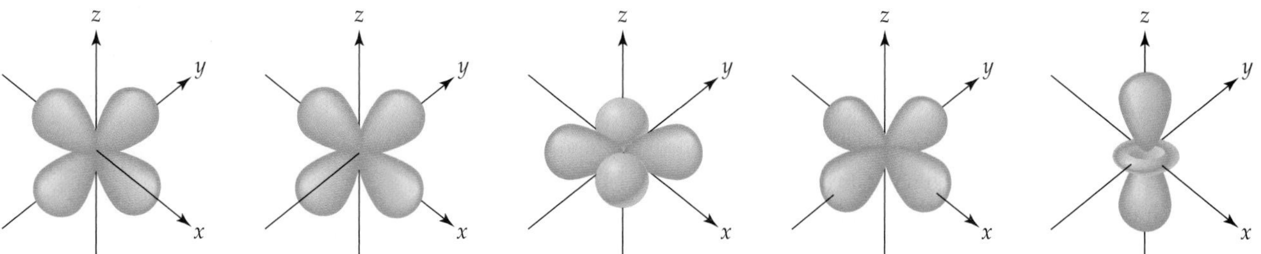

▲ **Figure 5.20 Shapes of 3*d* Orbitals** The shapes of the five 3*d* orbitals are not the same. Although four of the 3*d* orbitals are similar in shape, they have a different orientation to each other.

Summary

Section 5.1 In 1803, John Dalton proposed that matter consisted of atoms and supported the atomic theory with experiments on the behavior of gases and the laws of definite composition and conservation of mass.

Section 5.2 Toward the end of the 1800s, there was evidence that the atom was divisible. When electricity was applied to a gas in a sealed glass tube, negative and positive rays were observed in the tube. Scientists discovered the rays from hydrogen gas were composed of tiny, charged particles they named **electrons** and **protons**.

Section 5.3 In 1911, Ernest Rutherford supervised a classic experiment in which alpha particles were fired at a thin sheet of gold foil. Much to his astonishment, some of the alpha particles bounced backward. He interpreted the results as evidence for a tiny, dense **atomic nucleus** at the center of the atom. The nucleus contains positively charged protons surrounded by negatively charged electrons. Twenty years later, the nucleus was also found to contain a neutral particle called a **neutron**.

Section 5.4 Chemists use a symbolic shorthand called **atomic notation** to designate the composition of a nucleus. Atoms of an element always contain the same number of protons, but the number of neutrons in the nucleus can vary. The number of protons is called the atomic number (Z), and the sum of the protons and neutrons is called the mass number (A). Atoms with the same **atomic number**, but a different **mass number** are called **isotopes**.

Section 5.5 Although the mass of an atom is much too small to measure directly, we can determine its relative mass. Carbon-12 is used as the reference isotope and is assigned a mass of exactly 12 **atomic mass units (amu)**. The masses of all other atoms are related to the mass of carbon-12. To find a representative value for the atomic mass of an element, we average the mass of each isotope. The weighted average mass of all isotopes is termed the **atomic mass** of the element.

Section 5.6 **Light** travels through space as a wave of radiant energy. The crest-to-crest distance between waves is the **wavelength**, and the number of cycles completed in a second is the **frequency**. As the wavelength decreases, the frequency and energy of light increase. The **visible spectrum** extends from 400–700 nm, but the entire **radiant energy spectrum** also includes gamma rays, X rays, and microwaves. Thus, radiant energy is a **continuous spectrum** of visible and invisible light.

Section 5.7 In 1900, Max Planck introduced the quantum concept. Planck stated that the energy radiated by a object is not continuous, but rather that the radiation is emitted in small bundles. When an object radiates light, it releases a unit of radiant energy called a **photon**.

Section 5.8 In 1913, Niels Bohr suggested that electrons travel in circular orbits about the nucleus. The electron possesses a specific energy and is said to occupy an **energy level**. If an electron changes orbits in the **Bohr atom**, there is a quantum energy change. Bohr argued that an **emission line spectrum** results from electrons dropping from higher energy levels to lower levels. When an electron drops, a photon of light is released having an energy that corresponds to the difference in energy between the two levels.

Section 5.9 A closer examination of emission line spectra from gases revealed **energy sublevels** within main levels. The number of sublevels corresponds to the number of the energy level. For example, the fourth energy level has four sublevels (*s*, *p*, *d*, *f*). An *s* sublevel can hold 2 electrons, a *p* sublevel 6 electrons, a *d* sublevel 10 electrons, and an *f* sublevel a maximum of 14 electrons.

Section 5.10 Electrons fill sublevels in order of increasing energy as follows: $1s < 2s < 2p < 3s < 3p < 4s < 3d < 4p < 5s < 4d < 5p < 6s$. Notice that the 4*s* sublevel fills before the 3*d* and the 5*s* sublevel before the 4*d*. A description of sublevel filling for an element is given by the **electron configuration**. A superscript following each sublevel indicates the number of electrons in a sublevel; for instance, $1s^2\ 2s^2\ 2p^6\ 3s^1$ is the electron configuration for sodium (atomic number 11).

Section 5.11 In the 1920s our understanding of electrons in atoms became very sophisticated. In 1925 Werner Heisenberg suggested the **uncertainty principle**; that is, it is impossible to simultaneously know both the precise location and the energy of an electron. Instead, the energy of an electron can be known only in terms of its probability of being located somewhere within the atom. This description gave rise to the **quantum mechanical atom**. A location within the atom where there is a high probability of finding an electron having a certain energy is called an **orbital**. An orbital is a region about the nucleus having a given energy, size, and shape. The shape of an *s* orbital is spherical, and a *p* orbital resembles the shape of a dumbbell.

Key Terms *Answers to Key Terms are in Appendix H.*

Select the key term that corresponds to each of the following definitions.

_____ **1.** a negatively charged subatomic particle having a tiny mass
_____ **2.** a positively charged subatomic particle having an approximate mass of 1 amu
_____ **3.** a neutral subatomic particle having an approximate mass of 1 amu
_____ **4.** a region in the center of an atom containing protons and neutrons
_____ **5.** a value indicating the number of protons in the nucleus of an atom
_____ **6.** a value indicating the number of protons and neutrons in the nucleus of an atom
_____ **7.** a symbolic method for expressing the composition of an atomic nucleus
_____ **8.** atoms of the same element that have a different number of neutrons
_____ **9.** a unit of mass exactly equal to 1/12 the mass of a C-12 atom
_____ **10.** the average mass of all the naturally occurring isotopes of an element
_____ **11.** the distance a light wave travels to complete one cycle
_____ **12.** the number of times a light wave completes a cycle in one second
_____ **13.** a general term that can refer to either visible or invisible radiant energy
_____ **14.** a range of light energy extending from violet to red, that is, approximately 400–700 nm
_____ **15.** a range of light energy extending from gamma rays to microwaves
_____ **16.** a band of light energy that is uninterrupted
_____ **17.** a particle of radiant energy
_____ **18.** a model of the atom that describes electrons circling the nucleus in orbits
_____ **19.** an orbit of specific energy that electrons occupy as they circle the nucleus
_____ **20.** a collection of narrow bands of light produced by atoms of a given element releasing energy
_____ **21.** an electron energy level that results from splitting a main energy level
_____ **22.** a shorthand description of the arrangement of electrons by sublevels according to increasing energy
_____ **23.** the statement that it is impossible to precisely measure the location and energy of a particle at the same time
_____ **24.** a sophisticated model of the atom that describes the energy of an electron in terms of its probability of being found in a particular location about the nucleus
_____ **25.** a region about the nucleus in which there is a high probability of finding an electron with a given energy

(a) atomic mass *(Sec. 5.5)*
(b) atomic mass unit (amu) *(Sec. 5.5)*
(c) atomic notation *(Sec. 5.4)*
(d) atomic nucleus *(Sec. 5.3)*
(e) atomic number (Z) *(Sec. 5.4)*
(f) Bohr atom *(Sec. 5.8)*
(g) continuous spectrum *(Sec. 5.6)*
(h) electron (e^-) *(Sec. 5.2)*
(i) electron configuration *(Sec. 5.10)*
(j) emission line spectrum *(Sec. 5.8)*
(k) energy level *(Sec. 5.8)*
(l) energy sublevel *(Sec. 5.9)*
(m) frequency *(Sec. 5.6)*
(n) isotopes *(Sec. 5.4)*
(o) light *(Sec. 5.6)*
(p) mass number (*A*) *(Sec. 5.4)*
(q) neutron (n^o) *(Sec. 5.3)*
(r) orbital *(Sec. 5.11)*
(s) photon *(Sec. 5.7)*
(t) proton (p^+) *(Sec. 5.2)*
(u) quantum mechanical atom *(Sec. 5.11)*
(v) radiant energy spectrum *(Sec. 5.6)*
(w) uncertainty principle *(Sec. 5.11)*
(x) visible spectrum *(Sec. 5.6)*
(y) wavelength *(Sec. 5.6)*

Exercises *Answers to odd-numbered Exercises are in Appendix I.*

Dalton Model of the Atom (Sec. 5.1)

1. Which of the following of Dalton's proposals proved to be invalid?
   (a) Atoms of different elements combine to form compounds.
   (b) Atoms can combine in small whole number ratios.
   (c) Atoms are indivisible.

2. Which of the following of Dalton's proposals proved to be invalid?
   (a) Atoms are indestructible.
   (b) Atoms of the same element are identical.
   (c) Atoms can combine in more than one whole number ratio.

3. State three scientists whose work Dalton used to support the atomic theory.

4. State two experimental laws Dalton used to support the atomic theory.

Thomson Model of the Atom (Sec. 5.2)

5. What is the simplest negative particle in an atom?

6. What is the simplest positive particle in an atom?

7. What is the relative charge on an electron?

8. What is the relative charge on a proton?

9. What do the raisins represent in the raisin pudding model of the atom?

10. What does the pudding represent in the plum pudding model of the atom?

Rutherford Model of the Atom (Sec. 5.3)

11. What is the approximate size (in cm) of a nucleus?

12. What is the approximate size (in cm) of an atom?

13. What is the relative charge of an electron and proton?

14. What is the relative mass of an electron and proton?

15. Where is the location of electrons in an atom?

16. Where is the location of protons and neutrons in an atom?

Atomic Notation (Sec. 5.4)

17. State the number of neutrons in an atom of each of the following isotopes:
(a) $^{23}_{11}\text{Na}$ (b) $^{65}_{30}\text{Zn}$
(c) $^{27}_{13}\text{Al}$ (d) $^{107}_{47}\text{Ag}$

18. State the number of neutrons in an atom of each of the following isotopes:
(a) $^{20}_{10}\text{Ne}$ (b) $^{40}_{18}\text{Ar}$
(c) $^{80}_{36}\text{Kr}$ (d) $^{131}_{54}\text{Xe}$

19. State the number of neutrons in an atom of each of the following isotopes:
(a) hydrogen-2 (b) carbon-14
(c) cobalt-60 (d) iodine-131

20. State the number of neutrons in an atom of each of the following isotopes:
(a) hydrogen-1 (b) carbon-13
(c) cobalt-59 (d) iodine-127

21. Complete the following table and provide the missing information:

Atomic Notation	Atomic Number	Mass Number	Number of Protons	Number of Neutrons	Number of Electrons
$^{11}_{5}\text{B}$					
$^{15}_{7}\text{N}$					
$^{40}_{20}\text{Ca}$					
$^{200}_{80}\text{Hg}$					

22. Complete the following table and provide the missing information:

Atomic Notation	Atomic Number	Mass Number	Number of Protons	Number of Neutrons	Number of Electrons
	9			10	
		31	15		
	28			59	
		210	82		

23. Draw a diagram of the arrangement of protons, neutrons, and electrons in an atom of each of the following isotopes:
(a) $^{7}_{3}\text{Li}$ (b) $^{13}_{6}\text{C}$
(c) $^{16}_{8}\text{O}$ (d) $^{20}_{10}\text{Ne}$

24. Draw a diagram of the arrangement of protons, neutrons, and electrons in an atom of each of the following isotopes:
(a) ^{31}P (b) ^{35}Cl
(c) ^{40}Ar (d) ^{131}I

Atomic Mass (Sec. 5.5)

25. Distinguish between isotopic mass and atomic mass.

26. Distinguish between atomic number and mass number.

27. What is the current atomic mass scale reference isotope?

28. What is the assigned mass for the current reference isotope?

29. Given that the only naturally occurring isotope of sodium is ^{23}Na, determine its mass from the periodic table.

30. Given that the only naturally occurring isotope of aluminum is ^{27}Al, determine its mass from the periodic table.

31. Given that the only naturally occurring isotope of fluorine is ^{19}F, determine its mass from the periodic table.

32. Given that the only naturally occurring isotope of phosphorus is ^{31}P, determine its mass from the periodic table.

33. Calculate the atomic mass for lithium given the following data for its natural isotopes:

^{6}Li	6.015 amu	7.42%
^{7}Li	7.016 amu	92.58%

34. Calculate the atomic mass for magnesium given the following data for its natural isotopes:

^{24}Mg	23.985 amu	78.70%
^{25}Mg	24.986 amu	10.13%
^{26}Mg	25.983 amu	11.17%

35. Calculate the atomic mass for iron given the following data for its natural isotopes:

^{54}Fe	53.940 amu	5.82%
^{56}Fe	55.935 amu	91.66%
^{57}Fe	56.935 amu	2.19%
^{58}Fe	57.933 amu	0.33%

36. Calculate the atomic mass for zinc given the following data for its natural isotopes:

^{64}Zn	63.929 amu	48.89%
^{66}Zn	65.926 amu	27.81%
^{67}Zn	66.927 amu	4.11%
^{68}Zn	67.925 amu	18.57%
^{70}Zn	69.925 amu	0.62%

37. Bromine has only two natural isotopes, and they occur in about equal abundance. If ^{79}Br is one of the isotopes, what is the other isotope given the atomic mass of bromine is approximately 80 amu?

38. Silver has only two natural isotopes, and they occur in about equal abundance. If ^{109}Ag is one of the isotopes, what is the other isotope given the atomic mass of silver is approximately 108 amu?

The Wave Nature of Light (Sec. 5.6)

39. Which has the shorter wavelength: blue light or violet light?
40. Which has the longer wavelength: red light or yellow light?
41. Which has the higher frequency: blue light or violet light?
42. Which has the lower frequency: red light or yellow light?
43. Which has the higher energy: blue light or violet light?
44. Which has the lower energy: red light or yellow light?
45. Which wavelength has the higher frequency: 550 nm or 650 nm?
46. Which wavelength has the lower frequency: 280 nm or 750 nm?

The Quantum Concept (Sec. 5.7)

47. What particle represents the quantum nature of light energy?
48. What particle represents the quantum nature of electrical energy?
49. State whether each of the following is continuous or quantized:
    (a) a rainbow
    (b) a line spectrum
50. State whether each of the following is continuous or quantized:
    (a) a spiral staircase
    (b) an elevated ramp
51. State whether each of the following instruments gives a continuous or a quantized measurement of length:
    (a) a metric ruler
    (b) a digital laser
52. State whether each of the following instruments gives a continuous or a quantized measurement of volume:
    (a) 10 mL volumetric pipet
    (b) 10 mL graduated cylinder

Bohr Model of the Atom (Sec. 5.8)

53. Draw the Bohr model of the atom.
54. What is the experimental evidence for electron energy levels in an atom?
55. Which of the following energy level changes for an electron is most energetic: $5 \rightarrow 2$, $4 \rightarrow 2$, or $3 \rightarrow 2$?
56. Which of the following energy level changes for an electron is least energetic: $4 \rightarrow 1$, $3 \rightarrow 1$, or $2 \rightarrow 1$?

◀ **Hydrogen Gas Discharge Tube** The emission lines from hydrogen gas illuminate the glass tube with a reddish-purple glow.

57. In a hydrogen atom, what color is the emission line observed when electrons drop from the fourth to the second energy level?
58. In a hydrogen atom, what color is the emission line observed when electrons drop from the fifth to the second energy level?
59. In the emission spectrum from hydrogen gas, what type of energy is released when electrons drop from the fifth to the first energy level?
60. In the emission spectrum from hydrogen gas, what type of energy is released when electrons drop from the fifth to the fourth energy level?
61. Which of the following lines in the emission spectrum of hydrogen has the least energy: red, blue-green, or violet?
62. Which of the following lines in the emission spectrum of hydrogen has the shortest wavelength: red, blue-green, or violet?
63. How many photons of light are emitted for each of the following?
    (a) 1 e^- drops from energy level 3 to 1
    (b) 1 e^- drops from energy level 3 to 2
64. How many photons of light are emitted for each of the following?
    (a) 100 e^- drop from energy level 3 to 2
    (b) 100 e^- drop from energy level 4 to 2
65. What is the color of the spectral line emitted for each of the following electron energy changes in excited hydrogen gas?
    (a) Electrons drop from energy level 2 to 1
    (b) Electrons drop from energy level 3 to 2
    (c) Electrons drop from energy level 4 to 3
66. What is the color of the spectral line emitted for each of the following electron energy changes in excited hydrogen gas?
    (a) Electrons drop from energy level 5 to 1
    (b) Electrons drop from energy level 5 to 2
    (c) Electrons drop from energy level 5 to 4

Energy Levels and Sublevels (Sec. 5.9)

67. What experimental evidence suggests the concept of electrons in energy levels?
68. What experimental evidence suggests main energy levels split into sublevels?
69. State the number of sublevels in each of the following main energy levels:
    (a) 1st **(b)** 2nd
    (c) 3rd **(d)** 4th
70. Designate all the sublevels within each of the following main energy levels:
    (a) 1st **(b)** 2nd
    (c) 3rd **(d)** 4th
71. What is the maximum number of electrons in each of the following sublevels?
    (a) 2*s* **(b)** 4*p*
    (c) 3*d* **(d)** 5*f*
72. What is the maximum number of electrons in each of the following?
    (a) an *s* sublevel **(b)** a *p* sublevel
    (c) a *d* sublevel **(d)** an *f* sublevel

73. What is the maximum number of electrons in the second energy level?

74. What is the maximum number of electrons in the fourth energy level?

Electron Configuration (Sec. 5.10)

75. Draw a filling diagram and predict the sublevel that follows 4*s*.

76. Draw a filling diagram and predict the sublevel that follows 5*s*.

77. Write the predicted electron configuration for each of the following elements:
(a) He (b) Be
(c) Co (d) Cd

78. Write the predicted electron configuration for each of the following elements:
(a) boron (b) argon
(c) manganese (d) nickel

79. Which element corresponds to each of the following electron configurations?
(a) $1s^2\,2s^1$
(b) $1s^2\,2s^2\,2p^6\,3s^2\,3p^2$
(c) $1s^2\,2s^2\,2p^6\,3s^2\,3p^6\,4s^2\,3d^2$
(d) $1s^2\,2s^2\,2p^6\,3s^2\,3p^6\,4s^2\,3d^{10}\,4p^6\,5s^2$

80. Which element corresponds to each of the following electron configurations?
(a) $1s^2\,2s^2\,2p^5$
(b) $1s^2\,2s^2\,2p^6\,3s^2\,3p^6$
(c) $1s^2\,2s^2\,2p^6\,3s^2\,3p^6\,4s^2\,3d^{10}\,4p^6\,5s^2\,4d^5$
(d) $1s^2\,2s^2\,2p^6\,3s^2\,3p^6\,4s^2\,3d^{10}\,4p^6\,5s^2\,4d^{10}\,5p^5$

Quantum Mechanical Model of the Atom (Sec. 5.11)

81. What is the distinction between an orbit and an orbital?

82. What are two significant differences between the Bohr model of the atom and the quantum mechanical model?

83. Sketch a three-dimensional representation for each of the following orbitals. Label the *x*–, *y*–, and *z*–axes.
(a) $1s$ (b) $2p_x$
(c) $3p_y$ (d) $4p_z$

84. Sketch a three-dimensional representation for each of the following orbital sets. Label the *x*–, *y*–, and *z*–axes.
(a) $1s$, $2s$, $2p_x$ (b) $3p_x$, $3p_y$, $3p_z$

85. Which orbital in each of the following pairs has the higher energy?
(a) $2s$ or $3s$ (b) $2p_x$ or $3p_x$
(c) $2p_x$ or $2p_y$ (d) $4p_y$ or $4p_z$

86. Which orbital in each of the following pairs has the larger size?
(a) $2s$ or $3s$ (b) $2p_x$ or $3p_x$
(c) $2p_x$ or $2p_y$ (d) $4p_y$ or $4p_z$

87. Designate the orbital that fits each of the following descriptions:
(a) a spherical orbital in the fifth energy level
(b) a dumbbell-shaped orbital in the fourth energy level

88. Designate the orbital that fits each of the following descriptions:
(a) a spherical orbital in the sixth energy level
(b) a dumbbell-shaped orbital in the third energy level

89. State the maximum number of electrons that can occupy each of the following orbitals:
(a) $1s$ (b) $2p$
(c) $3d$ (d) $4f$

90. State the maximum number of electrons that can occupy each of the following sublevels:
(a) $1s$ (b) $2p$
(c) $3d$ (d) $4f$

General Exercises

91. If the electron charge-to-mass ratio is 1.76×10^8 coulomb/g and the absolute charge is 1.60×10^{-19} coulomb, what is the mass of an electron in grams?

92. If the proton charge-to-mass ratio is 9.57×10^4 coulomb/g and the absolute charge is 1.60×10^{-19} coulomb, what is the mass of a proton in grams?

93. Silver occurs naturally as ^{107}Ag and ^{109}Ag. Given the mass and abundance of silver-109 (108.91 amu and 48.16%), what is the mass of silver-107?

94. Gallium occurs naturally as ^{69}Ga and ^{71}Ga. Given the mass and abundance of gallium-69 (68.92 amu and 60.10%), what is the mass of gallium-71?

95. Element 61 was named for the mythological Greek Prometheus who stole fire from the gods. Refer to the periodic table and state whether Pm has any stable isotopes.

96. Element 84 was discovered by Marie Curie and is named polonium for her native Poland. Refer to the periodic table and state whether Po has any stable isotopes.

97. Indicate the region of the spectrum (infrared, visible, or ultraviolet) for each of the following wavelengths of light:
(a) 200 nm
(b) 500 nm
(c) 1200 nm

98. Which of the following light frequencies has the higher energy, 5×10^{10} cycles/s or 5×10^{11} cycles/s?

99. Explain why the electron configuration for copper is $1s^2\,2s^2\,2p^6\,3s^2\,3p^6\,4s^1\,3d^{10}$ rather than the predicted $1s^2\,2s^2\,2p^6\,3s^2\,3p^6\,4s^2\,3d^9$.

100. Explain why the electron configuration for silver is $1s^2\,2s^2\,2p^6\,3s^2\,3p^6\,4s^2\,3d^{10}\,4p^6\,5s^1\,4d^{10}$ rather than the predicted $1s^2\,2s^2\,2p^6\,3s^2\,3p^6\,4s^2\,3d^{10}\,4p^6\,5s^2\,4d^9$.

5 Chapter Self-Test

Answers to Self-Test are in Appendix J.

1. Which of the following of Dalton's proposals is still valid? (Sec. 5.1)
   (a) Atoms of different elements can combine to form compounds.
   (b) Atoms can combine in small whole number ratios.
   (c) Atoms can combine in more than one whole number ratio.
   (d) all of the above
   (e) none of the above

2. Where is the mass of an atom according to the plum pudding model? (Sec. 5.2)
   (a) The mass is distributed homogeneously.
   (b) The mass is found in the electrons.
   (c) The mass is found in the protons.
   (d) The mass is found in the nucleus.
   (e) none of the above

3. Which subatomic particle has a charge of +1 and a mass of ~0 amu? (Sec. 5.3)
   (a) electron **(b)** neutron
   (c) proton **(d)** all of the above
   (e) none of the above

4. How many neutrons are in the nucleus of an atom of gold-197? (Sec. 5.4)
   (a) 79 **(b)** 118
   (c) 197 **(d)** 276
   (e) none of the above

5. Element X has two natural isotopes: X-6 (6.015 amu) and X-7 (7.016 amu). Calculate the atomic mass of element X given the abundance of X-7 is 92.5%. (Sec. 5.5)
   (a) 6.09 amu **(b)** 6.50 amu
   (c) 6.52 amu **(d)** 6.94 amu
   (e) 12.5 amu

6. Which of the following colors of light has the lowest energy? (Sec. 5.6)
   (a) violet **(b)** blue
   (c) green **(d)** yellow
   (e) red

7. Which of the following is an example of a quantized change in energy? (Sec. 5.7)
   (a) a ball rolling down a ramp
   (b) a ball rolling down a flight of stairs
   (c) a ball rolling across a playground
   (d) all of the above
   (e) none of the above

8. How many photons of light are emitted when the electron in a hydrogen atom drops from energy level 5 to 1? (Sec. 5.8)
   (a) 1 **(b)** 2
   (c) 4 **(d)** 5
   (e) none of the above

9. What is the maximum number of electrons that can occupy the 3rd energy level? (Sec. 5.9)
   (a) 2 **(b)** 8
   (c) 18 **(d)** 32
   (e) none of the above

10. What is the electron configuration for an atom of zinc? (Sec. 5.10)
    (a) $1s^2\ 2s^2\ 2p^6\ 3s^2\ 3p^6\ 3d^{10}$
    (b) $1s^2\ 2s^2\ 2p^6\ 3s^2\ 3p^6\ 4s^2\ 3d^{10}$
    (c) $1s^2\ 2s^2\ 2p^6\ 3s^2\ 3p^6\ 4s^2\ 4p^{10}$
    (d) $1s^2\ 2s^2\ 2p^6\ 3s^2\ 3p^6\ 4s^2\ 4d^{10}$
    (e) none of the above

11. Which of the following is true for both an orbit and an orbital? (Sec. 5.11)
    (a) An orbit and an orbital must contain two electrons.
    (b) An orbit and an orbital can be described as elliptical.
    (c) An orbit and an orbital represent a fixed energy state.
    (d) An orbit and an orbital represent an electron boundary.
    (e) all of the above

Key Concepts

12. An atomic nucleus has been described by the analogy "like a marble in the Astrodome." If a marble represents the atomic nucleus, what does the Astrodome represent?

▲ **Astrodome** By analogy, imagine a small marble on second base (representing the nucleus) compared to the entire Astrodome (representing an atom).

13. The scattering of alpha particles by a thin metal foil has been described by the analogy "like a missile shot through our solar system." If the missile represents an alpha particle, what do the planets represent?

14. Complete the following quantum analogy: an *electron* is to electricity, as a ________ is to light.

15. What is the experimental evidence for electrons occupying fixed energy states surrounding the nucleus?

◀ **Neon Gas Discharge Tube** When the reddish-orange light from neon gas passes through a prism, a line spectrum is observed, which is the "atomic fingerprint" for neon.

16. Distinguish between the Bohr model and the quantum mechanical model of the atom.

Critical Thinking

17. Copper has two natural isotopes, ^{63}Cu and ^{65}Cu. Which isotope is more abundant? (*Hint*: Refer to the periodic table.)

18. Can atoms of different elements have the same number of protons? Can atoms of different elements have the same number of neutrons?

CHAPTER

6 The Periodic Table

A spiral form of the periodic table conveys the concept of chemical evolution. Starting with H, all the elements evolved step by step. That is, two H fused to form He, H and He formed Li, H and Li formed Be, etc. Mendeleev's likeness is commemorated in the center of the spiral of elements.

"It is the function of science to discover the existence of a general reign of order in nature and to find the causes governing this order."
Dmitri Mendeleev, Russian Chemist (1834–1907)

Copper, silver, and gold have been used since prehistoric times. However, not until the 1700s were these metals defined as elements by Robert Boyle (1627–1691). In 1800, 30 elements had been isolated and identified, and by 1870, there were more than 60 known elements.

In 1829, the German chemist J. W. Döbereiner (1780–1849) observed that several elements could be classified into groups of three, or *triads.* All three elements in a triad showed remarkable similarity in their chemical properties. Each triad also revealed an orderly trend in physical properties such as density, melting point, and especially atomic mass.

In 1865, the English chemist J. A. R. Newlands (1837–1898) suggested that the 62 known elements be arranged into groups of seven according to increasing atomic mass. He proposed that an eighth element would then repeat the properties of the first element in the previous group. Interestingly, his theory, called the *law of octaves,* was received with ridicule and was not accepted for publication. Although Newlands's insight into the periodic relationships of the elements was essentially correct, it took 20 years for him to receive professional recognition. In 1887, Newlands graciously accepted the Davy Medal from the Royal Society of Great Britain.

6.1 CLASSIFICATION OF ELEMENTS

OBJECTIVE

- To state the original periodic law proposed by Mendeleev.

In the 1860s, the Russian chemist Dmitri Mendeleev proposed that properties of different elements repeat at regular intervals when they are arranged in order of *increasing atomic mass.* Today, Mendeleev is regarded as the architect of the modern periodic table, which depicts a systematic arrangement of the elements.

Mendeleev's achievement was the result of his patient and systematic study of the physical and chemical properties of the elements. Furthermore, he had the insight and courage to predict the existence and properties of three elements before their discovery. In 1869, Mendeleev predicted the existence of an element he called ekasilicon. In 1886, ekasilicon was discovered in Germany and given the name germanium. The properties of ekasilicon predicted by Mendeleev and the observed properties of germanium are compared in Table 6.1.

TABLE 6.1 MENDELEEV'S PREDICTIONS OF THE PROPERTIES FOR EKASILICON (EK)

PROPERTY	EKASILICON PREDICTED (1869)	GERMANIUM DISCOVERED (1886)
color	gray	gray
atomic mass	72 amu	72.6 amu
density	5.5 g/mL	5.32 g/mL
melting point	very high	937 °C
formula of oxide	EkO_2	GeO_2
density of oxide	4.7 g/mL	4.70 g/mL
formula of chloride	$EkCl_4$	$GeCl_4$
boiling point of chloride	100 °C	86 °C

In 1869, Mendeleev published his original periodic table with the elements arranged in vertical columns. Two years later, he created an updated version showing the elements in horizontal rows. He arranged the elements by increasing atomic mass and began a new row with an element that repeated the properties of a previous element. Figure 6.1 illustrates a portion of Mendeleev's 1871 periodic table of elements.

The Noble Gases

The arrangement of the elements expanded significantly with the discovery of the group of similar elements that appear on the far right side of the periodic table. Argon was isolated from air in 1894, and helium, neon, krypton, xenon, and radon were discovered in the ensuing 5 years. Originally, this collection of gaseous elements was referred to as the *inert gases* because they showed no chemical reactivity.

In 1962 a compound containing xenon gas was synthesized at the University of British Columbia. After the first compound was made, $PtXeF_6$, several other compounds containing inert gases quickly followed. Although compounds of xenon and krypton have been prepared, the elements argon, neon, and helium have yet to be combined. More recently, the term "noble gas" has been substituted for "inert gas" to convey the unreactive nature of the gases in this group. Similarly, copper, silver, and gold are referred to as noble metals because of their resistance to chemical reaction.

Group	I	II	III	IV	V	VI	VII	VIII
Formula of Oxide	R_2O	RO	R_2O_3	RO_2	R_2O_5	RO_3	R_2O_7	RO_4
	H							
	Li	Be	B	C	N	O	F	
	Na	Mg	Al	Si	P	S	Cl	
	K	Ca	**eka-**	Ti	V	Cr	Mn	Fe, Co
	Cu	Zn	**eka-**	**eka-**	As	Se	Br	& Ni
	Rb	Sr	Y	Zr	Nb	Mo	—	Ru, Rh
	Ag	Cd	In	Sn	Sb	Te	I	& Pd
	Cs	Ba	Di	Ce	—	—	—	
	—	—	—	—	—	—	—	
	—	—	Er	La	Ta	W	—	Os, Ir
	Au	Hg	Tl	Pb	Bi	—	—	& Pt
	—	—	—	Th	—	U	—	

▶ **Figure 6.1 Mendeleev's Classification of the Elements (1871)** The formula for the oxide of an element is indicated by the notation R_2O, RO, and so on. For example, Group I oxides include Li_2O, Na_2O, and K_2O; and Group II oxides include BeO, MgO, and CaO. In Groups III and IV, Mendeleev indicated the existence of three undiscovered elements, which he named ekaboron, ekaaluminum, and ekasilicon.

CHEMISTRY CONNECTION Dmitri Mendeleev

Q: Which element in the periodic table is named in honor of Dmitri Mendeleev?

Dmitri Ivanovich Mendeleev (1834–1907) was born in Siberia, the youngest of 14 to 17 children (records vary). Mendeleev's father was a high school principal, but blindness ended his career when he was quite young. Mendeleev's mother, a woman of remarkable energy and determination, subsequently started a glass factory to support the family. About the time Mendeleev graduated from high school, his father died and his mother's factory burned down. Although destitute, his mother used her influence to get him into college just a few months before her own death.

Mendeleev enrolled at the University of St. Petersburg where he graduated first in his class. Following graduate work in Europe, he returned to St. Petersburg as a professor of chemistry. Mendeleev's consuming interest was finding a common thread that linked the rapidly growing number of elements. In 1869, he published a table that related elements according to increasing atomic mass. He even predicted the existence and properties of three undiscovered elements. In 1875, the predicted element *gallium* was discovered in France (Gallia); in 1879, *scandium* was discovered in Scandinavia; and in 1886, *germanium* was discovered in Germany.

Gradually, scientists recognized the importance of the periodic table, and Mendeleev became the most famous chemist in the world. His professional reputation thrived, and he was invited to major universities throughout Europe and gave lectures in the United States and Canada.

His personal reputation suffered when he divorced his wife and married a young art student. In the eyes of the Russian Orthodox Church he was a bigamist because he did not wait the required 7 years before he remarried. When Czar Alexander II was questioned about Mendeleev's bigamy, he replied, "Yes, Mendeleev has two wives, but I have only one Mendeleev."

In the late nineteenth century, Russia was engulfed in political turmoil. Mendeleev was a liberal and voiced his concern about the rights of students in general and women in particular. His outspoken speech led to his forced resignation from the university in 1890. Fortunately, Mendeleev still had influential friends, and he was named director of the Bureau of Weights and Measures.

In 1906, he missed winning the Nobel prize by a single vote. Historians speculate that Mendeleev was more deserving of the recognition, but owing to his controversial personality, the prize was awarded instead to Henri Moissan, the French chemist who discovered the element fluorine.

◀ **Dmitri Mendeleev (1834–1907)**

A: Element 101 is named mendelevium, symbol Md, in honor of Mendeleev.

6.2 THE PERIODIC LAW CONCEPT

OBJECTIVE

▸ To explain the modern periodic law concept proposed by Moseley.

In 1869, Mendeleev proposed that elements showed recurring properties according to increasing atomic mass. In 1913, H. G. J. Moseley (1887–1915), a postdoctoral student, bombarded atomic nuclei with high-energy radiation. By studying the X rays that were subsequently emitted, Moseley discovered that the nuclear charge increased by 1 for each element in the periodic table. Thus,

H = 1							He = 2
Li = 3	Be = 4	B = 5	C = 6	N = 7	O = 8	F = 9	Ne = 10
Na = 11	Mg = 12	Al = 13	Si = 14	P = 15	S = 16	Cl = 17	Ar = 18

Moseley correctly concluded that arranging elements according to increasing nuclear charge, rather than atomic mass, more clearly explained their repeating properties. That is, arranging elements according to atomic number better explained the trends found in the periodic table. Elements, therefore, should be arranged according to the number of protons in their nucleus and not their atomic mass. The **periodic law** states that the properties of elements recur in a repeating pattern when arranged according to *increasing atomic number.* As it so happens, with only a few exceptions, these trends are identical.

In 1913, Niels Bohr introduced the concept of electron energy levels and Mendeleev's periodic table soon took on a new shape. The new shape resembled the arrangement used today, as shown in Figure 6.2.

If you closely examine the fourth row in the periodic table, you will notice that the sequence of atomic masses does not increase from Co to Ni, even though Co precedes Ni in the periodic table. But notice that the atomic number of Co (27) is less than the atomic number of Ni (28). It is also true that the properties of cobalt resemble those of the elements in Group 9, and that the properties of Ni are similar to those of the elements in Group 10.

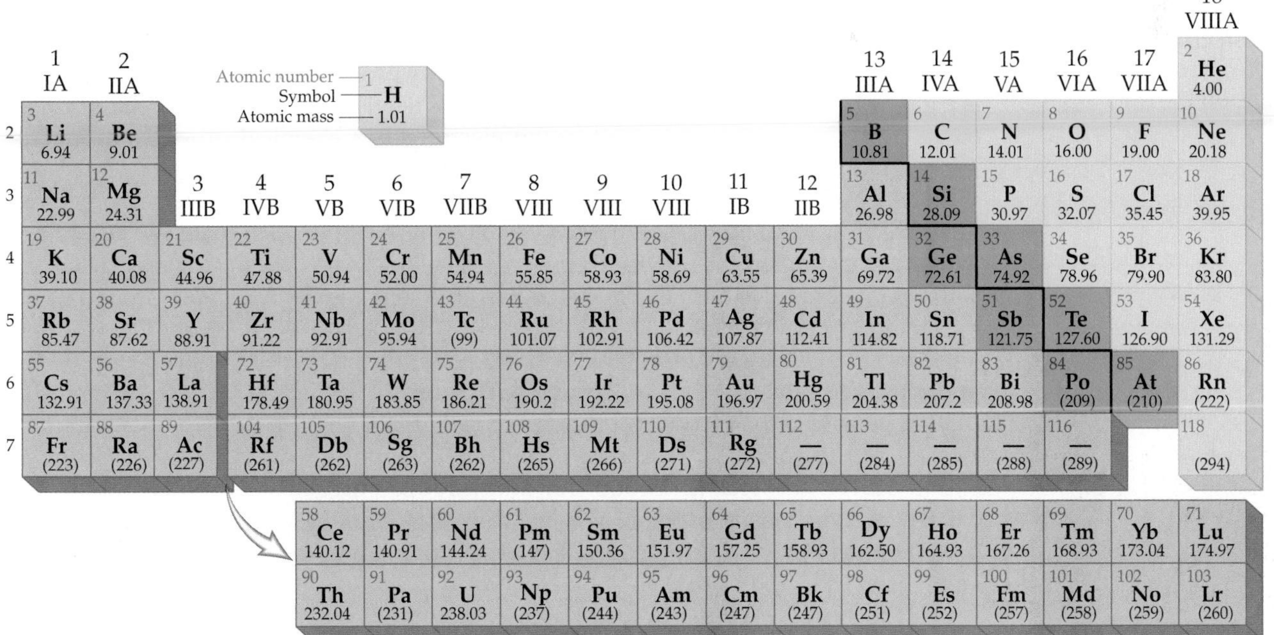

▲ **Figure 6.2 The Modern Periodic Table of Elements** The atomic numbers increase stepwise throughout the periodic table. Note that atomic masses, with few exceptions, also increase. However, the nuclear charge (that is, the atomic number), more clearly explains the recurring properties of elements.

Example Exercise 6.1 Periodic Law

Find the two elements in the fifth row of the periodic table that violate the original periodic law proposed by Mendeleev.

Solution

Mendeleev proposed that elements be arranged according to increasing atomic mass. Beginning with Rb, each of the elements in the fifth row increases in atomic mass until iodine. Although the atomic numbers of Te (52) and I (53) increase, the atomic masses of Te (127.60) and I (126.90) do not. Experimentally, it is I and not Te, with properties similar to those of F, Cl, and Br.

Practice Exercise

Find a pair of elements in the periodic table with atomic numbers less than 20 that are an exception to the original periodic law.

Answer: Ar and K

Concept Exercise

The modern periodic law states that elements are arranged according to increasing (atomic mass/atomic number). (Circle the correct choice.)

Answer: See Appendix G.

6.3 GROUPS AND PERIODS OF ELEMENTS

OBJECTIVES

- To classify the elements according to their groups (families) and periods (series) in the periodic table.
- To designate a group of elements in the periodic table using both the American convention (IA–VIIIA) and the IUPAC convention (1–18).

A vertical column in the periodic table is called a **group** or family of elements, and a horizontal row is called a **period** or series. When we examine the periodic table on the inside front cover of this textbook, we notice there are seven horizontal rows of elements. These periods of elements are numbered 1–7. The first period has only two elements, H and He. The second and third periods each have eight elements, Li through Ne, and Na through Ar. The fourth and fifth periods have 18 elements, K through Kr and Rb through Xe.

When Mendeleev constructed his original periodic table, he did not know where to place hydrogen as its properties were unlike any group of elements. Since hydrogen is a gas, it has physical properties similar to nonmetals. On the other hand, hydrogen reacts by losing an electron—and losing electrons is a chemical property of metals. Today, H appears at the top of the far left column in most periodic tables.

There are 18 vertical columns in the periodic table. In the past, American chemists have used a Roman numeral followed by the letter A or B to designate a group of elements, for example IA, IIA, IIB. In the periodic table, you will notice that Group IA contains the elements Li to Fr, Group IIA has the elements Be to Ra, and Group IIB has the elements Zn, Cd, Hg.

Since the 1920s, a controversy has existed over the numbering of the groups in the periodic table. Recently, the International Union of Pure and Applied Chemistry (IUPAC) resolved the dispute by proposing to number the groups using the numerals 1 to 18. For example, IUPAC recommends that Group IA be designated Group 1, Group IIB be designated Group 12, and Group VIIIA be designated Group 18. Currently, chemists are in the process of adopting the IUPAC convention. Hence, the periodic tables in this textbook will list both conventions, for example, IA/1, IIB/12, and VIIIA/18.

In the periodic table, groups of elements can also be referred to by their family names. The family name for Group IA/1 is **alkali metals**. The elements in Group IIA/2 are called **alkaline earth metals**. The elements in Group VIIA/17 are called **halogens**. Group VIIIA/18 elements are all gases that are usually unreactive under normal conditions; they are called **noble gases**.

Representative Elements and Transition Elements

Groups of elements are classified as either representative elements or transition elements. **Representative elements** (also called main-group elements) are found in the A groups, on the left and right sides of the periodic table. As a rule, the chemical behavior of representative elements is predictable. For example, magnesium always reacts with oxygen to produce MgO.

Transition elements are found in the B groups in the middle of the periodic table. The chemical behavior of transition elements is not as predictable as that of representative elements. For instance, in the presence of limited oxygen, iron reacts to form FeO; if excess oxygen gas is available, the product is Fe_2O_3.

Inner transition elements are found beneath the main portion of the periodic table. These two series are placed below the main portion to avoid an unduly wide periodic table. The first series of elements, Ce through Lu, lie between La and Hf in the periodic table and are considered part of Period 6. This series is called the **lanthanide series** because it follows lanthanum. The elements Ce through Lu have similar properties, and they occur together in nature along with Sc, Y, and La. The natural abundance of these elements in Earth's crust is less than 0.005%, and they are collectively referred to as **rare earth elements**.

The second series of inner transition elements, Th through Lr, is called the **actinide series** because it follows element 89, actinium. The actinide series is considered part of Period 7. All the elements in this series are radioactive; in fact, except for trace amounts, none of the elements past uranium is naturally occurring. Element 93 and beyond are created in particle accelerators called "atom smashers." The isotopes of these elements have very short lifetimes, often less than a millisecond. The elements following uranium, Np through element 118, are called **transuranium elements**. The periodic table shown in Figure 6.3 summarizes the names of the important groups and periods.

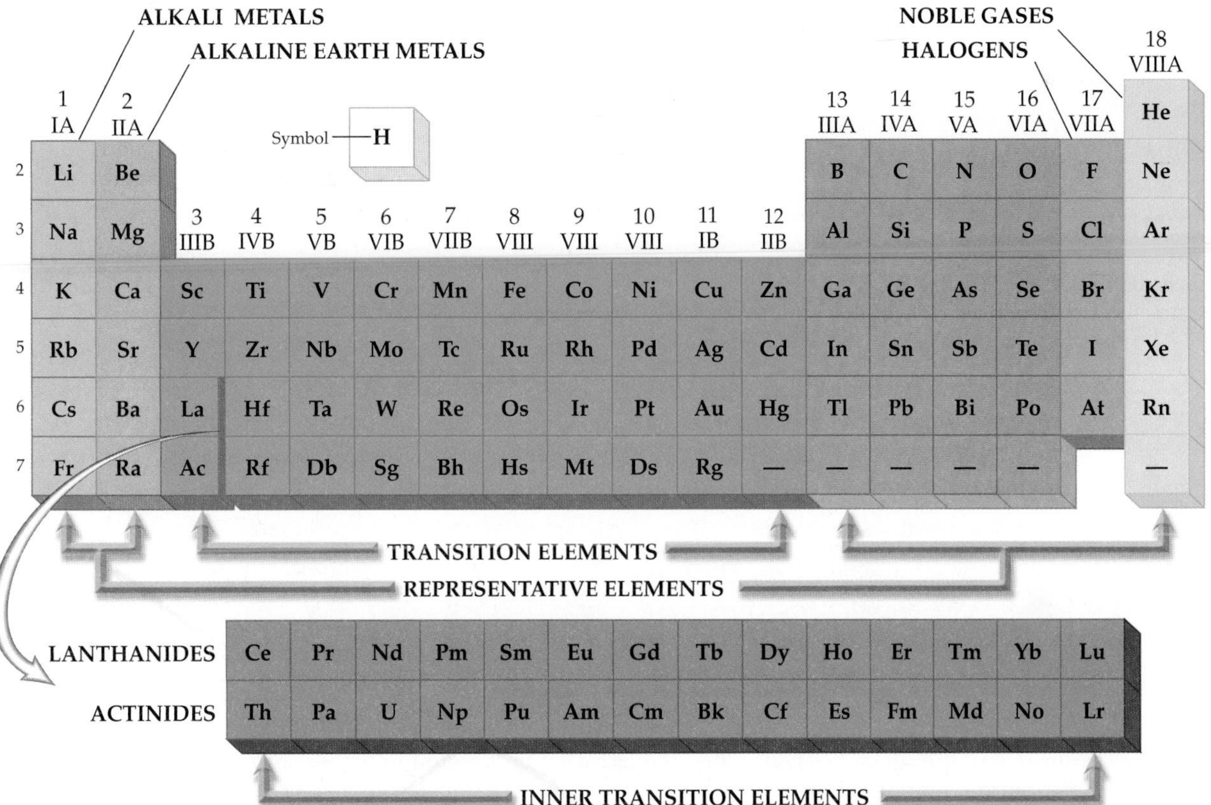

▲ **Figure 6.3 Names of Groups and Periods** The common names of groups and periods are shown for selected families and series.

CRITICAL THINKING IUPAC Group Numbers

Q: Which two groups of transition elements in the periodic table have the same designation according to both the American convention and European convention?

Recently, the designation of groups of elements in the periodic table has been a topic of interest. In the 1970s a lingering controversy surfaced that was addressed by IUPAC. Previously, in the United States, elements on the left side and right side of the periodic table had been designated A groups. The elements in the middle of the periodic table were labeled B groups. This so-called American convention is shown next.

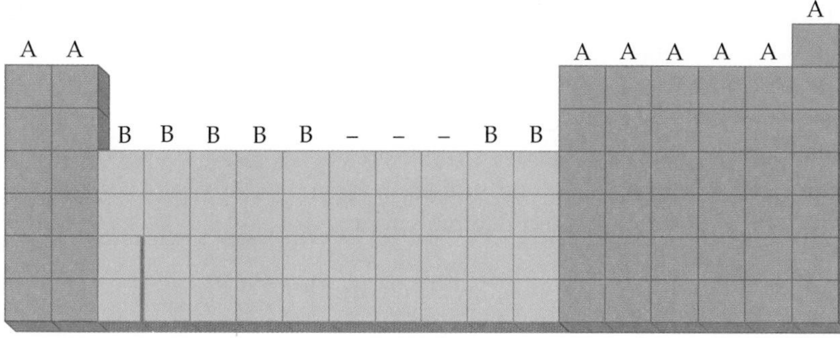

◀ **Periodic Table of Elements—American Convention** Representative elements are placed in A groups and transition elements are in B groups.

In Europe, elements on the left side of the periodic table had been previously designated A groups, and those on the right side had been labeled B groups. The so-called European convention is shown in the following illustration.

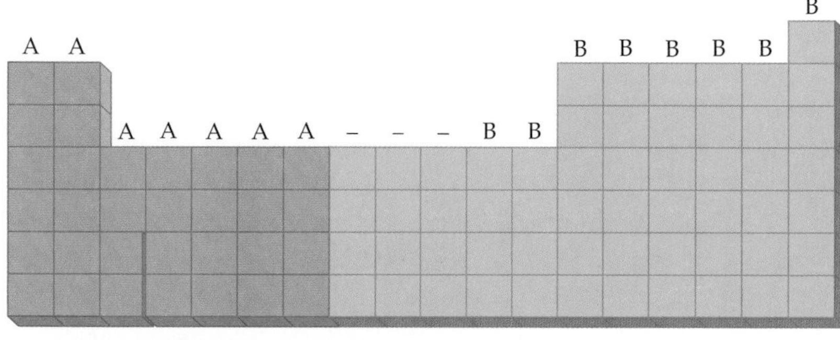

◀ **Periodic Table of Elements—European Convention** The elements on the left are A groups, and the elements on the right are B groups.

After much discussion in 1985 IUPAC proposed that groups of elements be designated by the numerals 1–18. Currently, IUPAC recommends that groups of elements be numbered as shown in the following table.

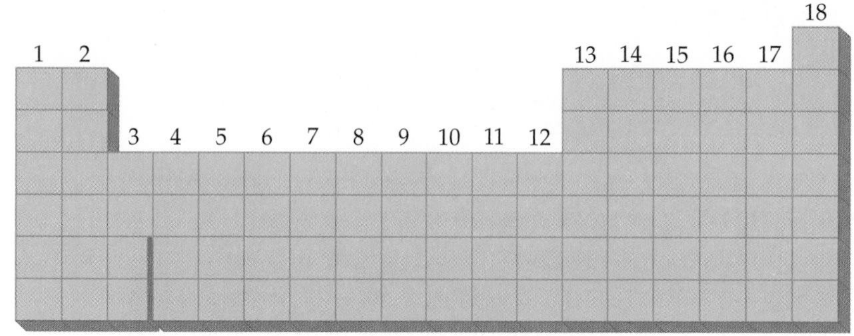

◀ **Periodic Table of Elements—IUPAC Convention** Groups of elements are indicated by the sequential numerals 1–18.

One of the reasons for IUPAC's recommendation is that the last digit in the group number corresponds to the Roman numeral in both the American and European conventions. For example, in group 13, the number 3 corresponds to the Roman numeral III in both the American (IIIA) and European (IIIB) versions.

A: Groups IB and IIB have the same designation in both the American and European versions of the periodic table. These two groups are labeled 11 and 12 using IUPAC convention.

Example Exercise 6.2 Groups and Periods of Elements

Select the symbol of the element that fits each of the following descriptions:
(a) the alkali metal in the fourth period
(b) the halogen in the third period
(c) the rare earth with the lowest atomic mass
(d) the metal in Group VIIB/7 and Period 4

Solution
Referring to the periodic table in Figure 6.3, we have

(a) K (b) Cl
(c) Sc (d) Mn

Practice Exercise
Select the symbol of the element that fits each of the following descriptions:
(a) the alkaline earth metal in the sixth period
(b) the noble gas in the third period
(c) the actinide with the highest atomic mass
(d) the semimetal in Group IIIA/13

Answers: (a) Ba; (b) Ar; (c) Lr; (d) B

Concept Exercise
What is the group number for each of the following families of elements?

(a) alkali metals (b) alkaline earth metals
(c) halogens (d) noble gases

Answer: See Appendix G.

6.4 PERIODIC TRENDS

OBJECTIVE

- To describe the trend in atomic size within a group or period of elements.
- To describe the trend in metallic character within a group or period.

We can visualize atoms as spheres and express their size in terms of atomic radius. The atomic radius is the distance from the nucleus to the outermost electrons. As this distance is very small, we will express the atomic radius in units of nanometers.

In the periodic table there are two general trends in atomic size. First, as we move through a group of elements from bottom to top, the radii of the atoms decrease. We can explain the decrease in radius because there are fewer energy levels of electrons surrounding the nucleus. As the number of energy levels decreases, the distance from the nucleus to the outermost electrons decreases. Therefore, the trend in *atomic radius decreases up a group.*

Second, as we move left to right within a period of elements, the radii of the atoms decrease. The explanation for this decrease is as follows: The atomic number increases from left to right, and therefore, the number of protons increases. As the number of protons in the nucleus increases across a period, the nuclear charge of the elements increases. This has the effect of pulling the electrons closer to the nucleus and reducing the size of the atom. Thus, the *atomic radius decreases across a period* from left to right.

Figure 6.4 illustrates trends in the size of atomic radii for a portion of the periodic table. Transition elements and inner transition elements have been omitted as there are numerous exceptions to the general trends. Moreover, the atomic radii of transition elements in a given sublevel are reasonably similar in size.

The periodic table shows a horizontal trend in metallic character of the elements. First, recall that metals are on the left side of the periodic table and nonmetals are on the right side. The most reactive metals are on the far left side, and the most reactive nonmetals are on the far right side of the table. Thus, as we move from left to right, *metallic character decreases across a period.*

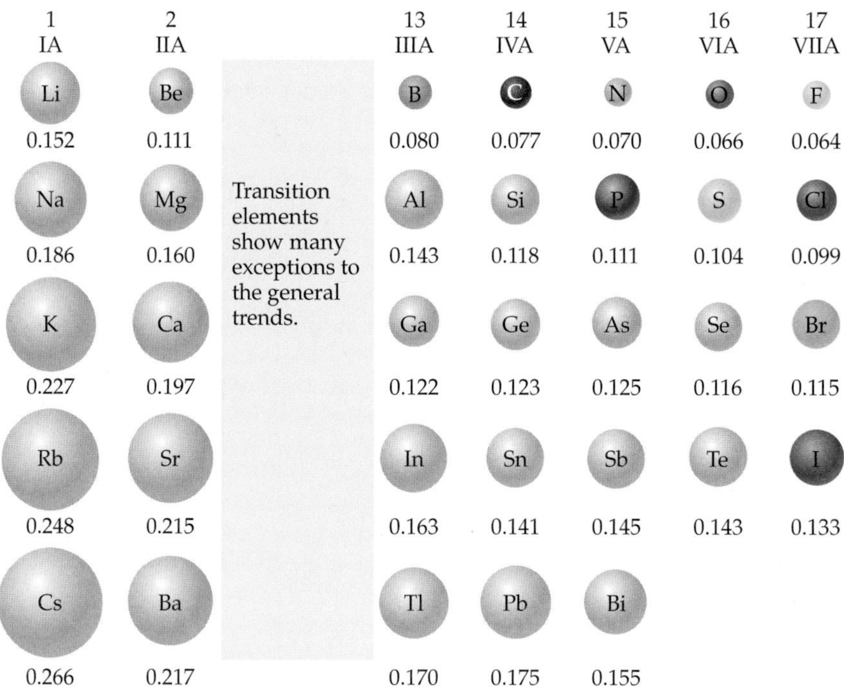

▶ **Figure 6.4 Atomic Radii of Selected Representative Elements** The values for atomic radii are given in nanometers (nm). Some references list atomic radii in picometers (pm); for example, the atomic radius of Li is 152 pm.

The periodic table also shows a vertical trend in metallic character of the elements. A metal reacts by losing its outermost valence electrons. As we move up a group, the atomic radius is smaller and the outermost electrons are closer to the nucleus. When the negatively charged electrons and the positively charged nucleus are closer together, it is more difficult for an atom to lose an electron. Therefore, *metallic character decreases up a group.* Figure 6.5 illustrates the general trends in the periodic table for atomic radius and metallic character.

Example Exercise 6.3 Periodic Table Predictions

Based on the general trends in the periodic table, predict which element in each of the following pairs has the smaller atomic radius:

(a) Na or K (b) P or N
(c) Ca or Ni (d) Si or S

Solution

The general trend in atomic radii decreases up a group and across a period from left to right. Referring to the periodic table,

(a) Na is above K in Group IA/1; the atomic radius of Na is smaller.
(b) N is above P in Group VA/15; the atomic radius of N is smaller.
(c) Ni is to the right of Ca in Period 4; the atomic radius of Ni is smaller.
(d) S is to the right of Si in Period 3; the atomic radius of S is smaller.

Practice Exercise

Based on the general trends in the periodic table, predict which element in each of the following pairs has the most metallic character:

(a) Sn or Pb (b) Ag or Sr
(c) Al or B (d) Br or As

Answers: (a) Pb; (b) Sr; (c) Al; (d) As

Concept Exercise

Which period represents the smallest atomic radius? Which group represents the most metallic character?

Answer: See Appendix G.

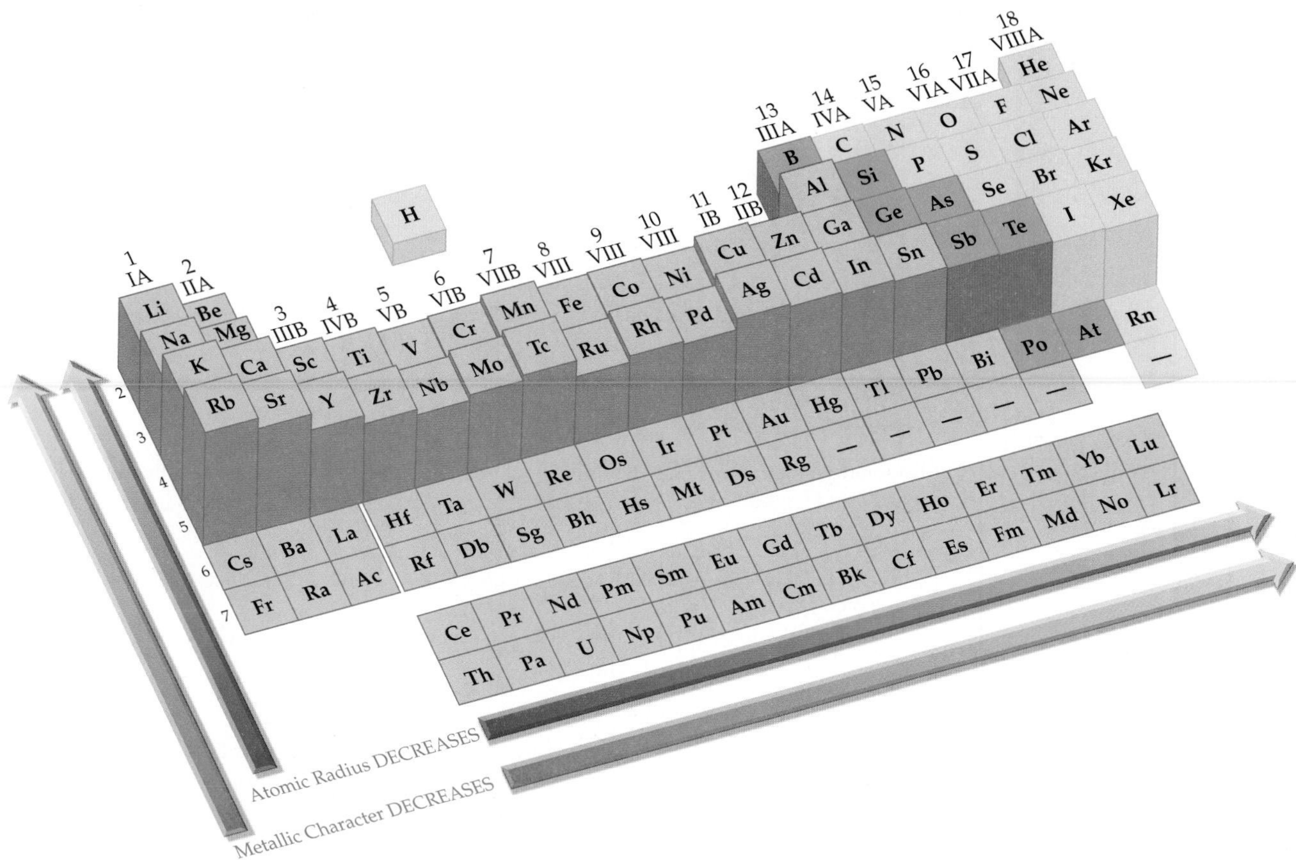

▲ **Figure 6.5 General Trends in Atomic Radii and Metallic Character** The general trend for the atomic radius is to decrease both up a group and across a period from left to right. The metallic character trend in the periodic table is similar.

6.5 PROPERTIES OF ELEMENTS

OBJECTIVES

- To predict a physical property for an element given the values of other elements in the same group.
- To predict a chemical formula for a compound given the formulas of other compounds containing an element in the same group.

Mendeleev designed the periodic table so that elements with similar properties were placed in the same group. He was then able to predict yet undiscovered elements from the gaps in his periodic table. He successfully predicted the properties of ekasilicon (i.e., germanium), for example, after noticing a trend in properties such as color, density, melting point, and atomic mass.

Predicting Physical Properties

Today, chemists have access to information regarding properties of elements cataloged in reference books such as the *Handbook of Chemistry and Physics*. However, we can use the same procedure that Mendeleev used to estimate the numerical value for a physical property. For example, if we know the density of two elements in the same group, we can make a reasonable estimate of the density for another element in the same group. Table 6.2 lists selected physical properties of alkali metals, which appear in Group IA/1 in the periodic table.

Notice that the properties of the radioactive element francium, Fr, are not included with those of the other alkali metals in Table 6.2. We can, however, make some reasonable predictions about its physical properties based on the trends shown by the other elements. Since Fr is below Cs in Group IA/1, we can predict that its atomic radius is greater than 0.266 nm, its density is greater than 1.87 g/mL, its melting point is less than 28.4 °C, and its atomic mass is greater than 132.91 amu.

TABLE 6.2 PHYSICAL PROPERTIES OF ALKALI METALS

ELEMENT	ATOMIC RADIUS (nm)	DENSITY (g/mL at 20 °C)	MELTING POINT (°C)	ATOMIC MASS (amu)
Li	0.152	0.53	180.5	6.94
Na	0.186	0.97	97.8	22.99
K	0.227	0.86	63.3	39.10
Rb	0.248	1.53	38.9	85.47
Cs	0.266	1.87	28.4	132.91

*Note the density of K is slightly less than that of Na. Small irregularities in group trends are not unusual.

Example Exercise 6.4 Predicting Physical Properties

Predict the missing value (?) for each physical property listed below. The (a) atomic radius, (b) density, and (c) melting point are given for two of three alkaline earth metals in Group IIA/2.

ELEMENT	ATOMIC RADIUS	DENSITY AT 20 °C	MELTING POINT
Ca	0.197 nm	1.54 g/mL	(?) °C
Sr	0.215 nm	(?) g/mL	769 °C
Ba	(?) nm	3.65 g/mL	725 °C

Solution

We can estimate a value for the physical property of an element by observing the trend in values for other elements within the same group.

(a) To determine the atomic radius value for Ba, we first find the increase from Ca to Sr; that is, 0.215 nm – 0.197 nm = 0.018 nm. We then add the difference (0.018 nm) to the atomic radius of Sr and obtain 0.215 nm + 0.018 nm = 0.233 nm. We assumed that the atomic radius increased the same amount from Sr to Ba as it did from Ca to Ba. (The *Handbook of Chemistry and Physics* value is *0.217 nm*.)

(b) Notice that Sr lies between Ca and Ba in Group IIA/2. Thus, we can estimate that the density of Sr lies midway between Ca and Ba. To find the density of Sr, we calculate the average value for Ca and Ba; that is, (1.54 g/mL + 3.65 g/mL)/2 = 2.60 g/mL. (The literature value is *2.63 g/mL*.)

(c) From the general trend, we can predict that the melting point of Ca is greater than that of Sr. To determine the value, let's find the increase in melting point from Ba to Sr. It is 769 °C – 725 °C = 44 °C. Now we add 44 °C to the value of Sr: 769 °C + 44 °C = 813 °C. Therefore we predict the melting point of Ca as 813 °C. (The literature value is *839 °C*.)

Practice Exercise

Predict the missing value (?) for each physical property listed below. The (a) atomic radius, (b) density, and (c) melting point are given for two of the metals in Group VIII/10.

ELEMENT	ATOMIC RADIUS	DENSITY AT 20 °C	MELTING POINT
Ni	0.125 nm	8.91 g/cm^3	(?) °C
Pd	0.138 nm	(?) g/cm^3	1554 °C
Pt	(?) nm	21.5 g/cm^3	1772 °C

Answers: (a) 0.151 nm; (b) 15.2 g/cm^3; (c) 1336 °C

Concept Exercise

Given the atomic radius (0.186 nm), density (0.97 g/mL), and melting point (98 °C) for sodium, estimate the corresponding values of lithium.

Answer: See Appendix G.

Predicting Chemical Properties

In Chapter 8 we will study chemical reactions systematically. However, it is possible to predict the products of chemical reactions by understanding the organization of elements in the periodic table. For example, if we know that magnesium and oxygen react to give magnesium oxide (MgO), we can predict that the other Group IIA/2 elements react in a similar fashion. That is, calcium, strontium, and barium should react with oxygen to give similar oxides (CaO, SrO, and BaO), and they do. There are exceptions to the general rule, but the principle of using the periodic table is helpful to our understanding of chemical properties.

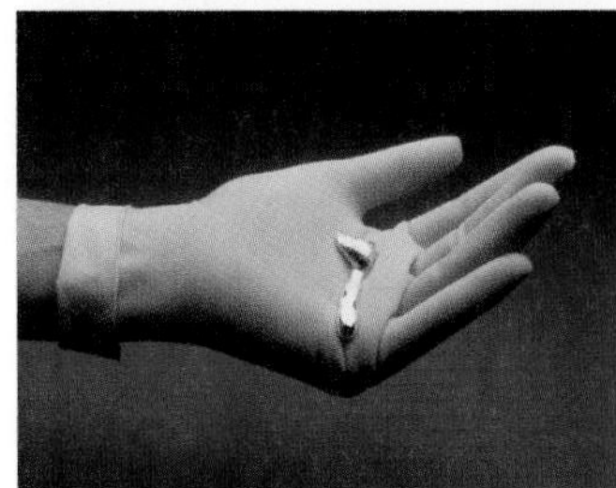

▲ **Gallium Metal** Gallium has chemical properties similar to aluminum. However, gallium has a much lower melting point (30 °C), and will melt in the hand.

Example Exercise 6.5 Predicting Chemical Properties

Metallic sodium reacts with chlorine gas to give sodium chloride, NaCl. Predict the products formed when (a) lithium and (b) potassium react with chlorine gas.

Solution

Since Li and K are in the same group as Na (Group IA/1), we can predict that the products are similar to NaCl. Thus,

(a) lithium metal should react with chlorine gas to give LiCl.

(b) potassium metal should react with chlorine gas to give KCl.

Practice Exercise

The chemical formulas for the oxides of potassium, calcium, gallium, and germanium are, respectively, K_2O, CaO, Ga_2O_3, and GeO_2. Refer to the periodic table and predict the chemical formula for each of the following compounds:

(a) rubidium oxide (b) strontium oxide

(c) indium oxide (d) lead oxide

Answers: (a) Rb_2O; (b) SrO; (c) In_2O_3; (d) PbO_2

Concept Exercise

Predict which of the following metals has chemical properties most similar to zinc: Fe, Cu, or Cd.

Answer: See Appendix G.

6.6 BLOCKS OF ELEMENTS

OBJECTIVES

- To predict the highest energy sublevel for an element given its position in the periodic table.
- To predict the electron configuration for an element given its position in the periodic table.

In Section 5.9 we learned that the order of electron energy sublevels according to increasing energy is $1s < 2s < 2p < 3s < 3p < 4s < 3d < 4p < 5s < 4d < 5p < 6s$. In this section, we will see that the unusual shape of the periodic table is the result of the ordering of energy sublevels. That is, the order of energy sublevels follows the systematic arrangement of elements by groups.

Elements in Groups IA/1 and IIA/2 fill *s* sublevels. The *s* sublevels are to the left in the periodic table and are collectively called the *s block* of elements. Elements in Groups IIIA/13 through VIIIA/18 fill *p* sublevels, and are referred to as the *p block* of elements. Transition elements in Groups IIIB/3 through IIB/12 fill *d* sublevels, and likewise make up the *d block* of elements. Finally, the inner transition elements (Ce–Lu and Th–Lr) fill *f* sublevels, and comprise the *f block* of elements. Figure 6.6 illustrates the *s*, *p*, *d*, and *f* block elements in the periodic table.

Previously, we had to memorize that the 3*d* energy sublevel comes after the 4*s* energy sublevel. Now, we can simply refer to the periodic table and see that *s* block elements precede each *d* sublevel. Also, recall that the number of sublevels corresponds to the number of the main energy level. For example, in the third main level, there are three sublevels (3*s*, 3*p*, and 3*d*). In the fourth main level, there are four sublevels (4*s*, 4*p*, 4*d*, and 4*f*).

We can see in Figure 6.6 that members of the lanthanide and actinide series are *f* block elements. The lanthanides are filling the 4*f* energy sublevel, and the actinides are filling the 5*f* sublevel.

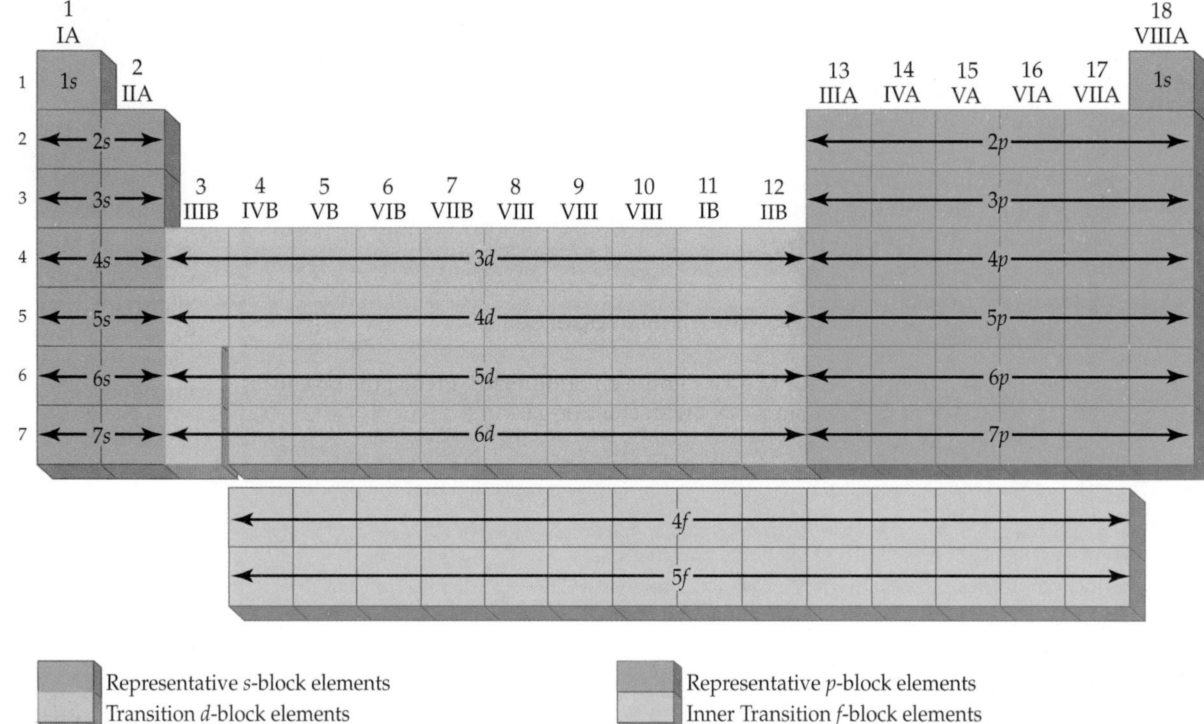

▲ **Figure 6.6 Blocks of Elements in the Periodic Table** The relationship between energy sublevels and the *s*, *p*, *d*, and *f* blocks of elements in the periodic table are shown.

Example Exercise 6.6 Energy Sublevels and the Periodic Table

State the highest energy sublevel in each of the following elements:

(a) H
(b) S
(c) Ni
(d) U

Solution

Refer to the periodic table and determine the energy sublevel based on the period and block of the elements.

(a) Hydrogen has only one electron; thus, H is filling a 1*s* sublevel.
(b) Sulfur is in the third period and is a *p* block element; S is filling a 3*p* sublevel.
(c) Nickel is in the first series of *d* block elements; Ni is filling a 3*d* sublevel.
(d) Uranium is in the second series of *f* block elements; U is filling a 5*f* sublevel.

Practice Exercise

State the energy sublevel being filled in each of the following series of elements:

(a) Cs – Ba
(b) Y – Cd
(c) In – Xe
(d) Ce – Lu

Answers: (a) 6*s*; (b) 4*d*; (c) 5*p*; (d) 4*f*

Concept Exercise

Refer to the periodic table and state the highest energy sublevel in a silver atom.

Answer: See Appendix G.

Electron Configuration of Elements

In Section 5.10, we learned to write the **electron configuration** for an element after memorizing the order of sublevel filling. Now that we understand the relationship of sublevels in the periodic table, we can write electron configurations by referring to blocks of elements and counting the number of electrons in each block.

The electron configuration for Na (atomic number 11) is $1s^2\,2s^2\,2p^6\,3s^1$. As a shorthand method, we can abbreviate the electron configuration by indicating the innermost electrons with the symbol of the preceding noble gas. The preceding noble gas with an atomic number less than 11 is neon (atomic number 10). The symbol for neon is placed in brackets, [Ne], followed by the outermost electrons. That is, the electron configuration for Na can be written as [Ne] $3s^1$.

Example Exercise 6.7 Electron Configuration and the Periodic Table

Refer to a periodic table and write the predicted electron configuration for each of the following elements by counting the number of electrons in each block:

(a) P (b) Co

Solution

Now that you understand blocks of elements in the periodic table, you can predict the order of sublevels according to increasing energy.

(a) Phosphorus is the third element in the $3p$ sublevel. The electron configuration for P is $1s^2\,2s^2\,2p^6\,3s^2\,3p^3$, or [Ne] $3s^2\,3p^3$.

(b) Cobalt is the seventh element in the $3d$ sublevel. The electron configuration for Co is $1s^2\,2s^2\,2p^6\,3s^2\,3p^6\,4s^2\,3d^7$ or [Ar] $4s^2\,3d^7$.

Practice Exercise

Refer to a periodic table and write the predicted electron configuration for each of the following elements.

(a) Zn (b) Se

Answers:

(a) $1s^2\,2s^2\,2p^6\,3s^2\,3p^6\,4s^2\,3d^{10}$ or [Ar] $4s^2\,3d^{10}$

(b) $1s^2\,2s^2\,2p^6\,3s^2\,3p^6\,4s^2\,3d^{10}\,4p^4$ or [Ar] $4s^2\,3d^{10}\,4p^4$

Concept Exercise

Refer to the periodic table and predict the number of $5d$ electrons in a Pt atom.

Answer: See Appendix G.

Note There are a few exceptions in the predicted electron configurations for some of the elements. That is, some elements have actual electron configurations that vary from their predicted arrangements. On occasion, electrons in a lower energy sublevel jump to an unfilled sublevel of slightly higher energy. Some of these exceptions are easy to explain, while others are not. For our purposes, we will disregard these exceptions and write electron configurations based on the general predictions for the order of sublevel filling.

6.7 VALENCE ELECTRONS

OBJECTIVE

- To predict the number of valence electrons for any representative element.

When an atom undergoes a chemical reaction, only the outermost electrons are involved. These electrons are farthest from the nucleus, and more available to react with another atom. These outermost electrons are called **valence electrons**. Valence electrons form chemical bonds between atoms and are responsible for the chemical behavior of the element. For our purposes, we will ignore transition elements and consider only the representative elements. In these cases, the number of valence electrons is equal to the total number of electrons in the outermost *s* and *p* sublevels.

We can predict the number of valence electrons by noting the group number in the periodic table. Using the Roman numeral convention, we see that the group number is identical to the total number of valence electrons. That is, elements in Group IA have one valence electron, and elements in Group IIA have two valence electrons. Group IIIA elements have two electrons from an *s* sublevel plus one electron from a *p* sublevel, for a total of three valence electrons. Group IVA elements have four valence electrons, Group VA elements have five valence electrons, and so on.

With the IUPAC designations for group numbers, the last digit (for example, the "8" in 18) indicates the number of valence electrons. Group 13 elements have three valence electrons, Group 14 elements have four valence electrons, Group 15 elements have five valence electrons, and so on. Note that the *s* and *p* sublevels can have a maximum of two and six electrons, respectively. Thus, the maximum number of valence electrons is equal to 8 (2 + 6).

Lithium is in Group IA/1, and so a lithium atom has only one valence electron. Oxygen is in Group VIA/16 and has six valence electrons. The following example further illustrates the relationship between group number and number of valence electrons:

Example Exercise 6.8 Valence Electrons and the Periodic Table

Refer to the periodic table and predict the number of valence electrons for an atom of each of the following representative elements:

(a) Na
(b) Al
(c) S
(d) Xe

Solution

Find the element in the periodic table, note the group number, and indicate the number of valence electrons.

(a) Since sodium is in Group IA/1, Na has 1 valence electron.
(b) Aluminum is in Group IIIA/13, and so Al has 3 valence electrons.
(c) Sulfur is in Group VIA/16, and so S has 6 valence electrons.
(d) Xenon is in Group VIIIA/18, and so Xe has 8 valence electrons.

Practice Exercise

Refer to the periodic table and state the number of valence electrons for any element in each of the following groups:

(a) Group IIA
(b) Group VA
(c) Group 14
(d) Group 17

Answers: (a) 2; (b) 5; (c) 4; (d) 7

Concept Exercise

Why are predictions for the number of valence electrons only for representative elements?

Answer: See Appendix G.

Note We can quickly predict the number of valence electrons for any main-group element by referring to its group number in the periodic table. However, the transition elements are filling *d* sublevels, not an *s* or *p* sublevel, which complicates the discussion. Therefore, we will not predict the number of valence electrons for a transition element.

6.8 ELECTRON DOT FORMULAS

OBJECTIVE

▶ To draw the electron dot formula for any representative element.

We said previously that only valence electrons are involved when an element undergoes a chemical reaction. That's because valence electrons are farthest from the nucleus and are the most accessible. To keep track of valence electrons, chemists have devised a notation called the **electron dot formula**. Electron dot formulas are also referred to as *Lewis structures* in honor of G. N. Lewis (1875–1946), a famous American chemist who conceived the notion of valence electrons.

An electron dot formula shows the symbol of an element surrounded by its valence electrons. The symbol of the element represents the **core** (or kernel) of the atom. That is, it represents the nucleus and the inner electrons. Dots are placed around the symbol to represent the valence electrons. Figure 6.7 illustrates the general symbol for an electron dot formula.

In practice, we will use the following guidelines for drawing electron dot formulas:

Core (nucleus + inner electrons)

:Sy:

Valence electrons

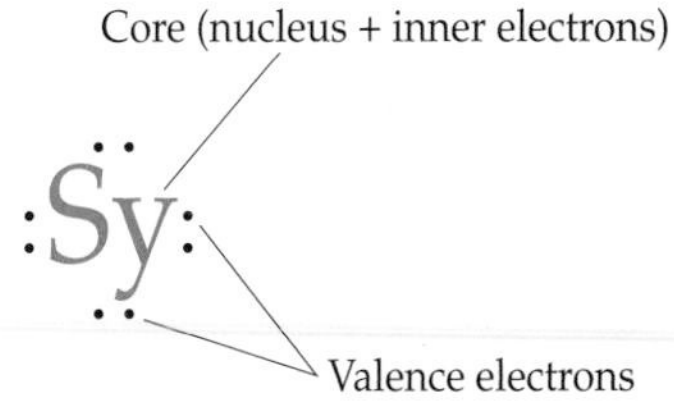

▲ **Figure 6.7 Representation of Electron Dot Formulas** The electron dot formula for an atom of an element uses the symbol for the element to represent the core of the atom. Each valence electron is represented by a dot arranged around the symbol.

Guidelines for Drawing Electron Dot Formulas of Atoms

1. Write the symbol of the element to represent the core of the atom.
2. By convention, assume that we can draw a maximum of two dots on each side of the symbol. A dot represents 1 valence electron, and the maximum number is 8 electrons (*s* sublevel + *p* sublevel = $2\ e^- + 6\ e^- = 8\ e^-$).
3. Determine the number of valence electrons from the group number of the element in the periodic table. Draw one dot around the symbol for each valence electron.

Although there is no absolute rule for placing dots, the first dot is usually placed to the right of the symbol, the second dot beneath the symbol, the third dot to the left of the symbol, and the fourth dot above the symbol. The fifth dot pairs with the first dot. The sixth, seventh, and eighth dots pair with the second, third, and fourth dots, respectively, moving clockwise around the symbol.

As an example, consider the element phosphorus. Phosphorus is in Group VA/15 and, therefore, has five valence electrons. We can draw the electron dot formula for phosphorus as follows. First, write the symbol for phosphorus, then add five dots, one at a time, as follows:

P	>	P•	>	P• (dot below)	>	•P• (dot below)	>	•P• (dots above and below)	>	•P: (dots above and below)
core	+	1 e⁻	+	2 e⁻	+	3 e⁻	+	4 e⁻	+	5 e⁻

Keep in mind that the placement of dots is not intended to show the actual positions of the electrons about the core of the atom. Actually, electrons are in constant motion, moving about the nucleus. Electron dot formulas are intended only to help us keep track of valence electrons. The following example exercise further illustrates electron dot formulas.

Example Exercise 6.9 Electron Dot Formulas

Draw the electron dot formula for each of the following elements:

(a) Si (b) Xe

Solution

Let's find the group number of the element in the periodic table and note the number of valence electrons. We write the symbol of the element and place the same number of dots around the symbol as there are valence electrons. In these examples, Si has 4 valence electrons and Xe has 8. The electron dot formulas are as follows:

(a) Si (b) Xe

$\cdot\dot{\underset{\cdot}{\text{Si}}}\cdot$ $:\ddot{\underset{\cdot\cdot}{\text{Xe}}}:$

Practice Exercise

Draw the electron dot formula for each of the following elements:

(a) K (b) I

Answers:

(a) K (b) I

$\text{K}\cdot$ $:\dot{\underset{\cdot\cdot}{\text{I}}}:$

Concept Exercise

Propose two ways of drawing the electron dot formula for Mg.

Answer: See Appendix G.

6.9 IONIZATION ENERGY

OBJECTIVE

- To state the general trends of ionization energy in the periodic table.
- To state the group with the highest and the lowest ionization energy.

Although electrons can be removed from all elements, metals lose electrons more easily than nonmetals. In fact, metals undergo chemical reactions by losing one or more valence electrons. Since electrons are negatively charged, metals become positively charged after losing an electron. Any atom bearing a charge is called an **ion**.

Energy is always required to remove an electron from an atom. By definition, the amount of energy necessary to remove an electron from a neutral atom in the gaseous state is called the **ionization energy**, or ionization potential. We can illustrate ionization of a sodium atom as follows:

$$\text{Na} \xrightarrow{\text{ionization energy}} \text{Na}^+ + e^-$$

Figure 6.8 shows the relative energy required to remove a single electron from an atom of elements 1–86.

Notice in Figure 6.8 that the elements having the highest ionization energy belong to the same group—the noble gases. One reason this group shows little tendency to undergo reaction is the difficulty with which an electron is removed. We can also reason that noble gases have high ionization energies because their valence shells are completely filled, and so they do not need to gain or lose electrons to become stable.

On the other hand, notice that the elements having the lowest ionization energy are alkali metals. Since they are in the same group, all alkali metals have a similar electron configuration. That is, they all have one electron in an *s* sublevel. If we remove an electron from an alkali metal, the resulting ion will have the same number of electrons as the preceding noble gas. For example, a lithium ion has 2 electrons (the same as He), a sodium ion has 10 electrons (the same as Ne), and a potassium ion has 18 electrons (the same as Ar). When an alkali metal loses an electron, it achieves a noble gas electron configuration, which is exceedingly stable.

There are general trends in the periodic table regarding ionization energy. Within a group of elements, the energy required to remove an electron increases as the atomic radius decreases. As we proceed up a group, the atomic radius decreases since the valence electrons are closer to the nucleus and are more tightly held. Thus, the ionization energy *increases up a group of elements.* As we move from left to right in the periodic table, the nuclear charge becomes greater. The energy required to remove an electron, therefore, increases. In general, the ionization energy *increases from left to right across a period of elements.*

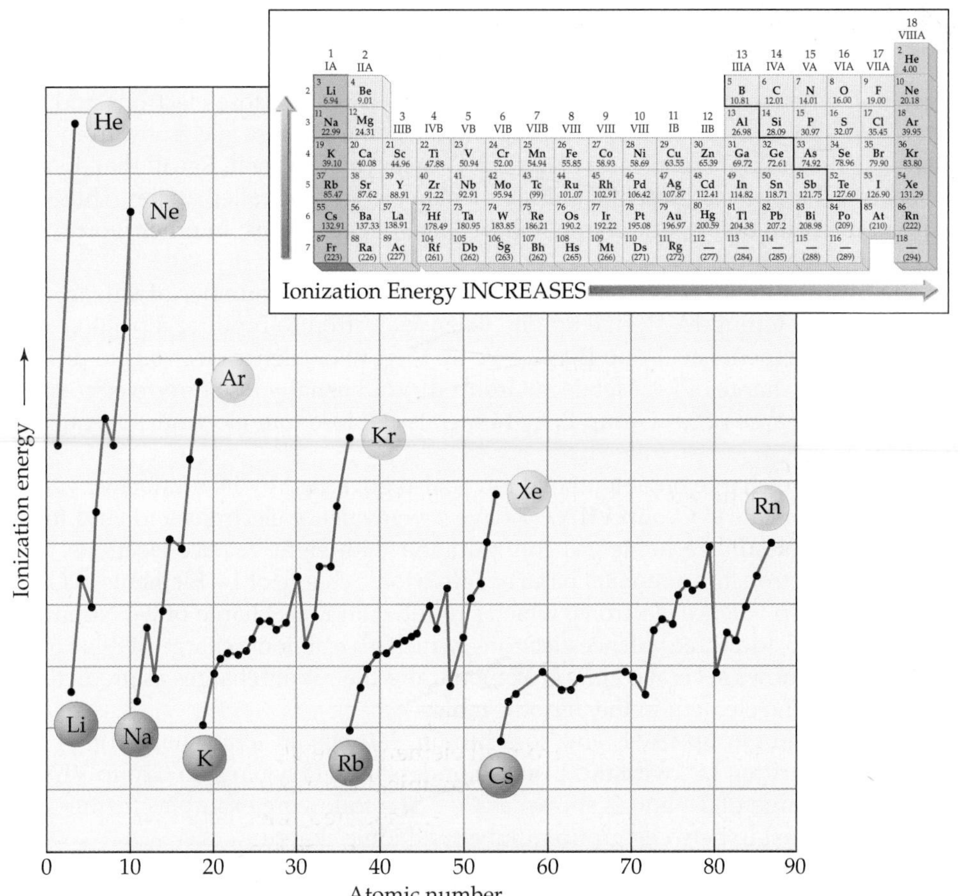

◀ **Figure 6.8 Ionization Energy** The diagram shows the energy required to remove a single electron from a neutral atom of an element in the gaseous state. Note the alkali metals have the lowest ionization energy, and the noble gases have the highest ionization energy.

Example Exercise 6.10 Ionization Energy and the Periodic Table

Based on the general trends in the periodic table, predict which element in each of the following pairs has the higher ionization energy:

(a) Li or Na (b) O or F

Solution

Let's refer to a periodic table and apply the general trends in ionization energy, which increases up a group and across a period.

Li is above Na in Group IA/1, and so Li has the higher ionization energy.

F is right of O in Period 2, and so F has the higher ionization energy.

Practice Exercise

Based on the general trends in the periodic table, predict which element in each of the following pairs has the higher ionization energy:

(a) Na or Mg (b) O or S

Answers: (a) Mg; (b) O

Concept Exercise

Which group in the periodic table has the highest ionization energy?

Answer: See Appendix G.

Note The amount of energy involved when a neutral atom gains an electron is called *electron affinity*. Metals show a weak tendency to gain an electron and have a low electron affinity. Nonmetals have a strong tendency to gain electrons and have a high electron affinity. The nonmetals closest to the noble gases have the highest electron affinity, so we can predict the electron affinity for F is greater than for O; and Cl is greater than S.

OBJECTIVES

- To predict the ionic charge for any representative element.
- To write the predicted electron configuration for selected ions.

6.10 IONIC CHARGES

In general, when metals and nonmetals react, metals tend to lose electrons and nonmetals tend to gain electrons. More specifically, metals lose electrons from their valence shell, and nonmetals add electrons to their valence shell. Recall that an atom that bears a charge as a result of gaining or losing electrons is called an ion. Thus, metals become positive ions, and nonmetals become negative ions; in other words, they acquire an **ionic charge**.

The positive ionic charge on a metal ion is related to its number of valence electrons. Metals in Group IA/1 give up one valence electron to produce a positive ionic charge of 1+. Elements in Group IIA/2 give up their two valence electrons to produce a positive ionic charge of 2+. Metals in Group IIIA/13 usually lose three electrons and have an ionic charge of 3+. Group IVA/14 metals can lose four electrons, producing a charge of 4+.

The negative charge on a nonmetal ion is also governed by its number of valence electrons. Nonmetals in Group VIIA/17 have seven valence electrons and tend to add one electron to assume a noble gas configuration with eight valence electrons. After gaining one electron, the nonmetal has a negative ionic charge of 1–. Elements in Group VIA/16 gain two valence electrons, which produces an ionic charge of 2–. Nonmetals in Group VA/15 add three valence electrons, which give an ionic charge of 3–.

Figure 6.9 shows several common elements and their ionic charges as predicted by the position of the element in the periodic table.

Lithium is in Group IA/1, and so a lithium ion has an ionic charge of 1+. The lithium ion is written Li^+, with the 1 (in 1+) understood. Oxygen is in Group VIA/16, has an ionic charge of 2–, and is shown as O^{2-}. The following examples further illustrate the relationship between group number and ionic charge.

Example Exercise 6.11 Ionic Charge and the Periodic Table

Predict the ionic charge for each of the following ions based on the group number of the element in the periodic table:

(a) Al ion (b) S ion

Solution

We refer to the periodic table to find the group number of the element.

(a) Aluminum is in Group IIIA/13. The metal atom loses 3 e^-, and so the ionic charge is 3+, that is, Al^{3+}.

(b) Sulfur is in Group VIA/16. The nonmetal atom gains 2 e^-, and so the ionic charge is 2–, that is, S^{2-}.

Practice Exercise

Predict the ionic charge for each of the following ions based on the group number of the element in the periodic table:

(a) Mg ion (b) Br ion

Answers: (a) Mg^{2+}; (b) Br^-

Concept Exercise

Predict the common ionic charge for Group IA/1 elements, for Group IIA/2 elements and for Group VIIA/17 elements.

Answer: See Appendix G.

In Section 6.9 we learned that Group IA/1 elements form ions by losing one electron. We showed that each Group IA/1 ion has an electron configuration identical to that of the previous noble gas element. Similarly, Group VIIA/17 elements gain one electron to assume an electron configuration identical to that of the next noble gas element.

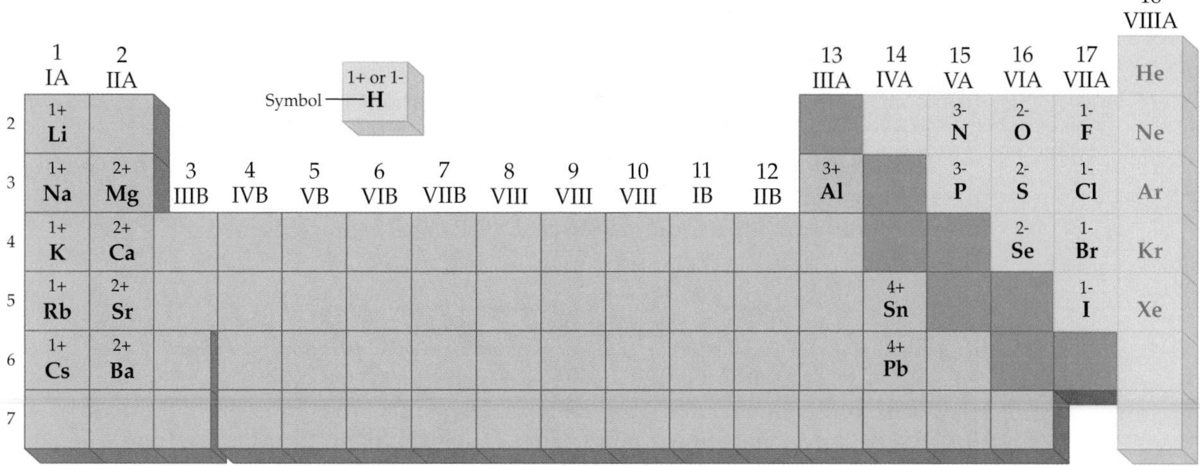

▲ **Figure 6.9 Periodic Table of Selected Ions** Notice that metals have a positive ionic charge equal to their group number. Nonmetals have a negative ionic charge equal to 8 minus their A group number in the periodic table. Both metals and nonmetals attempt to assume a stable noble gas electron configuration.

For example, a sodium ion (Na^+) and a fluoride ion (F^-) each have 10 electrons, as does a neutral neon atom. Although their properties are not related, their electron configurations are identical. By definition, two or more ions having the same number of electrons are said to be **isoelectronic**. The sodium ion, the fluoride ion, and the neon atom are said to be members of an isoelectronic series.

Example Exercise 6.12 Predicting Isoelectronic Ions

Refer to the periodic table and predict which of the following ions are isoelectronic with the noble gas argon:

(a) K^+
(b) Br^-
(c) Ca^{2+}
(d) O^{2-}

Solution

We refer to the periodic table to find the atomic number of the element.

(a) K^+ has 18 electrons (19–1); it is isoelectronic with argon (18 e^-).
(b) Br^- has 36 electrons (35+1); it is isoelectronic with krypton (36 e^-).
(c) Ca^{2+} has 18 electrons (20–2); it is isoelectronic with argon (18 e^-).
(d) O^{2-} has 10 electrons (8+2); it is isoelectronic with neon (10 e^-).

Practice Exercise

Refer to the periodic table and predict which of the following ions are isoelectronic with the noble gas xenon:

(a) Cs^+
(b) Cl^-
(c) La^{3+}
(d) Se^{2-}

Answers: (a) Cs^+ is isoelectronic with xenon; (b) Cl^- is isoelectronic with argon; (c) La^{3+} are isoelectronic with xenon; (d) Se^{2-} is isoelectronic with krypton

Concept Exercise

Predict the next ion in the isoelectronic series: Se^{2-}, Br^-, Kr, Rb^+, Sr^{2+}.

Answer: See Appendix G.

Electron Configuration of Ions

In Section 6.6 we learned how to predict the electron configuration of an element from the blocks of elements in the periodic table. Now we will learn how to write the electron configuration of an ion. First we find the element in the periodic table, and then we write out the electron configuration as we did in Section 6.6.

When writing the electron configuration for positive ions, we remove the number of electrons that corresponds to the positive charge on the ion. For example, the electron configuration for a sodium atom is $1s^2\,2s^2\,2p^6\,3s^1$. Since the sodium ion is Na^+, it loses one electron. That is,

$$\underset{1s^2\,2s^2\,2p^6\,3s^1}{\text{Na atom}} \xrightarrow{\text{loses 1 e}-} \underset{1s^2\,2s^2\,2p^6}{Na^+}$$

We should note that transition metals lose two electrons from the highest *s* sublevel first before losing electrons from their outer *d* sublevel. In the fourth-period elements, for example, 4*s* electrons are lost before 3*d* electrons. The electron configuration for a change manganese to manganese(II) ion is $1s^2\,2s^2\,2p^6\,3s^2\,3p^6\,3d^5$. Note that manganese forms a Mn^{2+} ion by losing two electrons from the 4*s* sublevel rather than the 3*d* sublevel. That is,

$$\underset{1s^2\,2s^2\,2p^6\,3s^2\,3p^6\,4s^2\,3d^5}{\text{Mn atom}} \xrightarrow{\text{loses 2 e}-} \underset{1s^2\,2s^2\,2p^6\,3s^2\,3p^6\,3d^5}{Mn^{2+}}$$

CHEMISTRY CONNECTION Evolution of Chemical Elements

Q: All the chemical elements may be derived from which single element?

While biologists are keenly interested in the "origin of species," chemists are equally curious about the origin of elements. Although the scientific community is not in total agreement, in general, scientists believe that the chemical elements evolved from the simplest element hydrogen. This theory that the chemical elements evolved from hydrogen was first hypothesized in the nineteenth century, and coincided with Darwin's "On the Origin of Species" published in 1859, which set forth the principle of biological evolution.

The universe is calculated to be about 14 billion years old beginning with an event referred to as the "big bang." The "big bang" signaled the creation of our universe and was characterized by inconceivably high temperatures. Subsequently, the universe cooled rapidly and subatomic particles including protons, neutrons, and electrons began to form. There is evidence for both hydrogen and helium being formed in the big bang, but we also know that helium is currently being formed by our Sun through the process of nuclear fusion. That is, two hydrogen nuclei ($_1H$) fuse together to form one helium nucleus ($_2He$). It is reasonable to speculate that heavier nuclei can form through further nuclear fusion by a process termed *nucleosynthesis*.

There are a number of mechanisms that have been put forth to explain the creation of elements heavier than helium. For example, the fusion of hydrogen and helium can produce lithium ($_3Li$), which in turn can yield beryllium ($_4Be$), and eventually carbon ($_6C$). There is also evidence for the direct fusion of three helium nuclei ($_2He$) to give carbon ($_6C$). Although the study of chemical elements is clearly within the scope of chemists, the ultimate answer(s) to the evolution of the elements may come from astrophysics, the branch of astronomy that studies the physical nature of stars and other planetary bodies. Astrophysicists apply the laws of physics and chemistry to interpret astronomical observations.

◀ **Milky Way Galaxy** Gravity acting on debris from the Big Bang produced regions of high density including galaxies having numerous stars. Some of these stars simply burned out, while others reached high core temperatures capable of nucleosynthesis of heavier elements. Eventually, these stars collapsed and exploded creating supernovas and sending chemical elements throughout the universe.

A: Hydrogen was first formed from the Big Bang and, directly or indirectly, gave rise to the evolution of heavier chemical elements.

As shown in Section 6.6, electron configurations can be simplified by using a noble gas symbol to represent the nucleus and inner electrons. This method of showing the core of an atom is called **core notation**. In the preceding example with a Mn atom, we can write the electron configuration as [Ar] $4s^2\ 3d^5$ using core notation. For Mn^{2+}, we can write the electron configuration as [Ar] $3d^5$.

When writing the electron configuration for negative nonmetal ions, we must add the number of electrons that corresponds to the negative charge on the ion. For example, the electron configuration for chlorine is [Ne] $3s^2\ 3p^5$. Since chloride ion is Cl^-, it gains one electron. That is,

$$\begin{array}{ccc} \text{Cl atom} & \xrightarrow{\text{gains 1 e}-} & Cl^- \\ [\text{Ne}]\ 3s^2\ 3p^5 & & [\text{Ne}]\ 3s^2\ 3p^6 \end{array}$$

We can write the electron configuration of Cl^- as [Ar]. The following example exercise further illustrates how to write the electron configuration of ions.

Example Exercise 6.13 Electron Configuration of Ions

Refer to the periodic table and write the predicted electron configuration for each of the following ions using core notation:

(a) Fe^{3+} (b) Se^{2-}

Solution

We refer to the periodic table to recall the blocks of elements so that we can write the electron configuration for the element.

(a) The electron configuration for an Fe atom is [Ar] $4s^2\ 3d^6$. The core electron configuration for an Fe^{3+} ion is [Ar] $3d^5$.

(b) The electron configuration for a Se atom is [Ar] $4s^2\ 3d^{10}\ 4p^4$. The electron configuration for a Se^{2-} ion is [Ar] $4s^2\ 3d^{10}\ 4p^6$ or simply [Kr].

Practice Exercise

Refer to the periodic table and write the predicted electron configuration for each of the following ions using core notation:

(a) Cd^{2+} (b) P^{3-}

Answers: (a) [Kr] $4d^{10}$; (b) [Ne] $3s^2\ 3p^6$, or [Ar].

Concept Exercise

What is the electron configuration for radon gas using core notation?

Answer: See Appendix G.

Summary

Section 6.1 In 1869, Dmitri Mendeleev explained the recurring properties of more than 60 elements by arranging them according to increasing atomic mass. In 1871, Mendeleev published a revised periodic table of elements, which even included undiscovered elements and their predicted properties.

Section 6.2 In 1913, H. G. J. Moseley showed that the positive charge on the nucleus of each element in the periodic table increases progressively. As a result, the **periodic law** was rewritten, and it is now stated that physical and chemical properties repeat periodically when the elements are arranged in order of increasing atomic number.

Section 6.3 The periodic table is organized by vertical columns called **groups** and horizontal rows called **periods**. The main-group elements in Groups 1, 2, and 13–18 are referred to as **representative elements**, and the elements in Groups 3–12 are called **transition elements**. The elements in Groups 1, 2, 17, and 18 are referred to as the **alkali metals**, **alkaline earth metals**, **halogens**, and **noble gases**, respectively. The elements following lanthanum (Ce–Lu) are referred to as **lanthanides** and the elements following actinium (Th–Lr) are called **actinides**. Collectively, lanthanides and actinides are known as **inner transition elements**. The elements Sc, Y, La, and Ce–Lu are referred to as **rare earth elements**. The elements beyond uranium are called **transuranium elements**.

Section 6.4 The periodic table is divided into metals on the left side and nonmetals on the right. Semimetals have intermediate properties and are found between the metals and the nonmetals. The atomic radius decreases up a group and from left to right. The metallic character also decreases up a group and from left to right.

Section 6.5 The trends in the periodic table enable us to predict physical and chemical properties. If we are given the atomic radius and oxide formula of two elements in a group, we can make a reasonable prediction for the atomic radius and oxide formula of another element in the same group.

Section 6.6 The elements in the periodic table are arranged by sublevels according to increasing energy. We can easily predict the energy sublevel containing the outermost electrons by referring to the periodic table. We can write the **electron configuration** by noting the sublevels in the *s*, *p*, *d*, and *f* blocks of elements.

Section 6.7 We can use the periodic table to quickly predict the number of electrons in the outermost *s* and *p* sublevels of an element, that is, the number of **valence electrons**. The number of valence electrons corresponds to the group number in the periodic table. For example, a Group IA/1 element has 1 valence electron, and a Group VIIA/17 element has 7 valence electrons.

Section 6.8 We can diagram an atom by letting the symbol of the element represent the **core** of the atom and drawing dots to represent the valence electrons. This diagram is referred to as the **electron dot formula** of the element.

Section 6.9 When an element loses or gains electrons, the resulting positively or negatively charged atoms are called **ions**. The amount of energy required to remove an electron from an atom is called the **ionization energy**. Since each noble gas has its valence shell filled, its ionization energy is extremely high. Conversely, each alkali metal has a low ionization energy because it has to lose only one valence electron to acquire a noble gas electron structure. Ionization energy increases up a group and from left to right in the periodic table.

Section 6.10 We can predict the **ionic charge** for a representative element from its group number. Metals lose valence electrons and nonmetals gain electrons to obtain a noble gas structure. For example, potassium is in Group IA/1 and loses 1 valence electron to form K^+. Sulfur is in Group VIA/16, has 6 valence electrons, and gains 2 electrons to form S^{2-}. The resulting K^+ and S^{2-} ions each have 18 electrons and are said to be **isoelectronic**. To write the electron configuration for an ion, we add or subtract electrons corresponding to the ionic charge. For the magnesium ion, Mg^{2+}, we subtract 2 electrons, and for the fluoride ion, F^-, we add 1 electron. Since Mg^{2+} and F^- each have 10 electrons, they are isoelectronic, and we can write their electron configurations as $1s^2\,2s^2\,2p^6$ or simply in **core notation** as [Ne].

Key Terms *Answers to Key Terms are in Appendix H.*

Select the key term that corresponds to each of the following definitions.

_____ **1.** the properties of elements recur in a repeating pattern when arranged by increasing atomic number
_____ **2.** a vertical column of elements in the periodic table
_____ **3.** a horizontal row of elements in the periodic table
_____ **4.** the Group IA/1 elements, excluding hydrogen
_____ **5.** the Group IIA/2 elements
_____ **6.** the Group VIIA/17 elements
_____ **7.** the relatively unreactive Group VIIIA/18 elements
_____ **8.** the Group A (1, 2, and 13–18) elements in the periodic table
_____ **9.** the Group B (3–12) elements in the periodic table
_____ **10.** the elements with atomic numbers 58–71
_____ **11.** the elements with atomic numbers 90–103
_____ **12.** the elements in the lanthanide and actinide series
_____ **13.** the elements with atomic numbers 21, 39, 57, and 58–71
_____ **14.** the elements beyond atomic number 92
_____ **15.** a shorthand description of the arrangement of electrons by sublevels according to increasing energy
_____ **16.** the portion of the atom that includes the nucleus and inner electrons
_____ **17.** the electrons that occupy the outermost *s* and *p* sublevels of an atom
_____ **18.** a symbolic diagram for an element and its valence electrons; the chemical symbol is surrounded by a dot for each valence electron
_____ **19.** an atom that bears a charge as a result of gaining or losing valence electrons
_____ **20.** the amount of energy necessary to remove an electron from an atom

(a) actinide series (Sec. 6.3)
(b) alkali metals *(Sec. 6.3)*
(c) alkaline earth metals *(Sec. 6.3)*
(d) core (kernel) *(Sec. 6.8)*
(e) core notation *(Sec. 6.10)*
(f) electron configuration *(Sec. 6.6)*
(g) electron dot formula *(Sec. 6.8)*
(h) group *(Sec. 6.3)*
(i) halogens *(Sec. 6.3)*
(j) inner transition elements *(Sec. 6.3)*
(k) ion *(Sec. 6.9)*
(l) ionic charge *(Sec. 6.10)*
(m) ionization energy *(Sec. 6.9)*
(n) isoelectronic *(Sec. 6.10)*
(o) lanthanide series *(Sec. 6.3)*
(p) noble gases *(Sec. 6.3)*
(q) period *(Sec. 6.3)*
(r) periodic law *(Sec. 6.2)*
(s) rare earth elements *(Sec. 6.3)*

_____ **21.** refers to the positive charge on a metal atom that has lost electrons or to the negative charge on a nonmetal atom that has gained electrons

_____ **22.** refers to ions having the same electron configuration; for example, Mg^{2+} and O^{2-} each have ten electrons

_____ **23.** a method of writing an electron configuration in which core electrons are represented by a noble gas symbol in brackets followed by the valence electrons, for example, [Ne] $3s^2$

(t) representative elements *(Sec. 6.3)*

(u) transition elements *(Sec. 6.3)*

(v) transuranium elements *(Sec. 6.3)*

(w) valence electrons *(Sec. 6.7)*

Exercises *Answers to odd-numbered Exercises are in Appendix I.*

Classification of Elements (Sec. 6.1)

1. Why did Mendeleev not include the noble gases in his periodic table of 1871?
2. Why did Mendeleev not include plutonium in his periodic table of 1871?
3. Refer to Figure 6.1 and predict the group number for germanium, given that the chemical formula for germanium oxide is GeO_2.
4. Refer to Figure 6.1 and predict the group number for gallium, given that the chemical formula for gallium oxide is Ga_2O_3.

The Periodic Law Concept (Sec. 6.2)

5. After Moseley's discovery in 1913, the periodic law stated that physical and chemical properties tend to repeat periodically when elements are arranged according to what trend?
6. Before Moseley's discovery in 1913, the periodic law stated that physical and chemical properties tend to repeat periodically when elements are arranged according to what trend?
7. Which two elements in the fourth period violate the original periodic law as stated by Mendeleev?
8. Which two elements in the fifth period violate the original periodic law as stated by Mendeleev?

Groups and Periods of Elements (Sec. 6.3)

9. Horizontal rows in the periodic table are referred to by what two terms?
10. Vertical columns in the periodic table are referred to by what two terms?
11. What is the term for the elements that belong to Groups IA–VIIIA (Groups 1, 2, and 13–18)?
12. What is the term for the elements that belong to Groups IIIB–IIB (Groups 3–12)?
13. What is the term for the elements in the series that follows element 57?
14. What is the term for the elements in the series that follows element 89?
15. What is the term for the two series of elements that include Ce–Lu and Th–Lr?
16. Which three elements are rare earth elements, but not lanthanides?
17. What is the descriptive term for the elements on the left side of the periodic table?
18. What is the descriptive term for the elements on the right side of the periodic table?
19. Which six nonradioactive elements are referred to as semimetals because their appearance resembles that of the metals?
20. Which two radioactive elements are referred to as semimetals because their appearance resembles that of the metals?
21. Identify the group number corresponding to each of the following families of elements:
    (a) alkali metals **(b)** alkaline earth metals
    (c) halogens **(d)** noble gases
22. Identify the group number corresponding to each of the following families of elements:
    (a) boron group **(b)** oxygen group
    (c) nickel group **(d)** copper group
23. According to IUPAC, what is the designation for each of the following groups indicated by the American convention?
    (a) Group IA **(b)** Group IIIA
    (c) Group VA **(d)** Group VIIA
    (e) Group IB **(f)** Group IIIB
    (g) Group VB **(h)** Group VIIB
24. According to the American convention using Roman numerals, what is the group number designation for each of the following?
    (a) Group 2 **(b)** Group 4
    (c) Group 6 **(d)** Group 8
    (e) Group 11 **(f)** Group 12
    (g) Group 16 **(h)** Group 18
25. Refer to the periodic table and select the symbol of the element that fits each of the following descriptions:
    (a) the Group IVA/14 semimetal in the fourth period
    (b) the third-period alkali metal
    (c) the halogen that is normally a solid and not radioactive
    (d) the lanthanide that is not naturally occurring
26. Refer to the periodic table and select the symbol of the element that fits each of the following descriptions:
    (a) the Group VA/15 semimetal in the fifth period
    (b) the sixth-period representative element with properties similar to Sr
    (c) the fifth-period transition element with properties similar to Ti
    (d) the lighter-than-air noble gas used in blimps and party balloons

27. Refer to the periodic table and select the symbol of the element that fits each of the following descriptions:
   (a) the third-period alkaline earth metal
   (b) the halogen that exists as a reddish-brown liquid at normal conditions
   (c) the rare earth element whose atomic number is greatest
   (d) the actinide with properties similar to Ce

28. Refer to the periodic table and select the symbol of the element that fits each of the following descriptions:
   (a) the fourth-period representative element with properties similar to O
   (b) the sixth-period transition element with properties similar to Ni
   (c) the rare earth element whose atomic mass is lowest
   (d) the radioactive noble gas

Periodic Trends (Sec. 6.4)

29. According to the general trend, the atomic radius for a period of elements (increases/decreases) proceeding from left to right in the periodic table.

30. According to the general trend, metallic character for a period of elements (increases/decreases) proceeding from left to right in the periodic table.

31. According to the general trend, the atomic radius (increases/decreases) proceeding up a group of elements in the periodic table.

32. According to the general trend, metallic character (increases/decreases) proceeding up a group of elements in the periodic table.

33. According to general trends in the periodic table, predict which element in each of the following pairs has the larger atomic radius:
   (a) Li or Na **(b)** N or P
   (c) Mg or Ca **(d)** Ar or Kr

34. According to general trends in the periodic table, predict which element in each of the following pairs has the larger atomic radius:
   (a) Rb or Sr **(b)** As or Se
   (c) Pb or Bi **(d)** I or Xe

35. According to general trends in the periodic table, predict which element in each of the following pairs has greater metallic character:
   (a) B or Al **(b)** Na or K
   (c) Mg or Ba **(d)** H or Fe

36. According to general trends in the periodic table, predict which element in each of the following pairs has greater metallic character:
   (a) K or Ca **(b)** Mg or Al
   (c) Fe or Cu **(d)** S or Ar

Properties of Elements (Sec. 6.5)

37. Predict the missing value (?) for each property listed below. The atomic radius, density, and melting point are given for elements in Group IA/1.

Element	Atomic Radius	Density at 20 °C	Melting Point
K	(?) nm	0.86 g/mL	63.3 °C
Rb	0.248 nm	(?) g/mL	38.9 °C
Cs	0.266 nm	1.90 g/mL	(?) °C

38. Predict the missing value (?) for each property listed below. The atomic radius, density, and melting point are given for elements in Group VIB/6.

Element	Atomic Radius	Density at 20 °C	Melting Point
Cr	0.125 nm	7.14 g/mL	(?) °C
Mo	(?) nm	10.28 g/mL	2617 °C
W	0.137 nm	(?) g/mL	3410 °C

39. Predict the missing value (?) for each property listed below. The atomic radius, density, and boiling point are given for elements in Group VIIA/17.

Element	Atomic Radius	Density at 20 °C	Boiling Point
Cl	(?) nm	1.56 g/mL	–34.6 °C
Br	0.115 nm	(?) g/mL	58.8 °C
I	0.133 nm	4.97 g/mL	(?) °C

40. Predict the missing value (?) for each property listed below. The atomic radius, density, and boiling point are given for elements in Group VIIIA/18.

Element	Atomic Radius	Density at STP	Boiling Point
Ar	0.180 nm	1.78 g/L	(?) °C
Kr	(?) nm	3.74 g/L	–152 °C
Xe	0.210 nm	(?) g/L	–107 °C

41. The formulas for the chlorides of potassium, calcium, boron, and germanium are, respectively, KCl, $CaCl_2$, BCl_3, and $GeCl_4$. Using the periodic table, predict the chemical formulas for each of the following similar compounds:
   (a) potassium fluoride **(b)** calcium fluoride
   (c) boron bromide **(d)** germanium iodide

42. The formulas for the oxides of sodium, magnesium, aluminum, and silicon are, respectively, Na_2O, MgO, Al_2O_3, and SiO_2. Using the periodic table, predict the chemical formulas for each of the following similar compounds:
   (a) lithium oxide **(b)** barium oxide
   (c) gallium oxide **(d)** tin oxide

43. The chemical formula for barium chloride is $BaCl_2$. Predict the formulas for each of the following similar compounds:
   (a) calcium fluoride **(b)** calcium chloride
   (c) calcium bromide **(d)** calcium iodide

44. The chemical formula for zinc oxide is ZnO. Predict the formulas for each of the following similar compounds:
   (a) cadmium oxide **(b)** cadmium sulfide
   (c) cadmium selenide **(d)** cadmium telluride

45. Depending upon reaction conditions, nitrogen and oxygen can produce either N_2O_3 and N_2O_5. Predict two formulas for each of the following compounds:
   (a) arsenic oxide
   (b) antimony oxide

46. Depending upon reaction conditions, phosphorus and sulfur can produce either P_2S_3 and P_2S_5. Predict two formulas for each of the following compounds:
   (a) arsenic sulfide
   (b) antimony sulfide

Blocks of Elements (Sec. 6.6)

47. What type of energy sublevel is being filled by the elements in Groups IA/1 and IIA/2?

48. What type of energy sublevel is being filled by the elements in Groups IIIA/13 through VIIIA/18?

49. What type of energy sublevel is being filled by the transition elements?

50. What type of energy sublevel is being filled by the inner transition elements?

51. Which energy sublevel is being filled by the lanthanide series?

52. Which energy sublevel is being filled by the actinide series?

53. Refer to the periodic table and state the highest energy sublevel for each of the following elements:
(a) H (b) Na
(c) Sm (d) Br
(e) Sr (f) C
(g) Sn (h) Cs

54. Refer to the periodic table and state the highest energy sublevel for each of the following elements:
(a) He (b) K
(c) U (d) Pd
(e) Be (f) Co
(g) Si (h) Pt

55. Refer to the periodic table and write the predicted electron configuration for each of the following elements:
(a) Li (b) F
(c) Mg (d) P
(e) Ca (f) Mn
(g) Ga (h) Rb

56. Refer to the periodic table and write the predicted electron configuration for each of the following elements:
(a) B (b) Ti
(c) Na (d) O
(e) Ge (f) Ba
(g) Pd (h) Kr

Valence Electrons (Sec. 6.7)

57. State the number of valence electrons in each of the following groups as predicted from the periodic table:
(a) Group IA/1
(b) Group IIIA/13
(c) Group VA/15
(d) Group VIIA/17

58. State the number of valence electrons in each of the following groups as predicted from the periodic table:
(a) Group IIA/2
(b) Group IVA/14
(c) Group VIA/16
(d) Group VIIIA/18

59. State the number of valence electrons for each of the following elements:
(a) H (b) B
(c) N (d) F
(e) Ca (f) Si
(g) O (h) Ar

60. State the number of valence electrons for each of the following elements:
(a) He (b) Pb
(c) Se (d) Ne
(e) Cs (f) Ga
(g) Sb (h) Br

Electron Dot Formulas (Sec. 6.8)

61. Draw the electron dot formula for each of the following elements:
(a) H (b) B
(c) N (d) F
(e) Ca (f) Si
(g) O (h) Ar

62. Draw the electron dot formula for each of the following elements:
(a) He (b) Pb
(c) Se (d) Ne
(e) Cs (f) Ga
(g) Sb (h) Br

Ionization Energy (Sec. 6.9)

63. According to the general trend, the ionization energy for a period of elements (increases/decreases) proceeding from left to right in the periodic table.

64. According to the general trend, the ionization energy for a group of elements (increases/decreases) proceeding up a group in the periodic table.

65. Which group of elements has the highest ionization energy?

66. Which group of elements has the lowest ionization energy?

67. Refer to the periodic table and predict which element in each of the following pairs has the higher ionization energy:
(a) Mg or Ca (b) S or Se
(c) Sn or Pb (d) N or P

68. Refer to the periodic table and predict which element in each of the following pairs has the higher ionization energy:
(a) Ga or Ge (b) Si or P
(c) Br or Cl (d) As or Sb

69. Refer to the periodic table and predict which element in each of the following pairs has the lower ionization energy:
(a) Rb or Cs (b) He or Ar
(c) B or Al (d) F or I

70. Refer to the periodic table and predict which element in each of the following pairs has the lower ionization energy:
(a) Mg or Si (b) Pb or Bi
(c) Ca or Ga (d) P or Cl

Ionic Charges (Sec. 6.10)

71. State the predicted ionic charge of metal ions in each of the following groups of elements:
(a) Group IA/1 (b) Group IIA/2
(c) Group IIIA/13 (d) Group IVA/14

72. State the predicted ionic charge of nonmetal ions in each of the following groups of elements:
(a) Group IVA/14 (b) Group VA/15
(c) Group VIA/16 (d) Group VIIA/17

73. Write the ionic charge for each of the following ions as predicted from the group number in the periodic table:
(a) Cs ion (b) Ga ion
(c) O ion (d) I ion

74. Write the ionic charge for each of the following ions as predicted from the group number in the periodic table:
(a) Be ion (b) Sn ion
(c) P ion (d) S ion

75. Refer to the periodic table and predict which of the following ions are isoelectronic with the noble gas argon:
(a) Al^{3+} (b) Ca^{2+}
(c) S^{2-} (d) N^{3-}

76. Refer to the periodic table and predict which of the following ions are isoelectronic with the noble gas krypton:
(a) K^+ **(b)** Sr^{2+}
(c) Cl^- **(d)** Se^{2-}

77. Refer to the periodic table and write the predicted electron configuration for each of the following positive ions using core notation:
(a) Mg^{2+} **(b)** K^+
(c) Fe^{2+} **(d)** Zr^{2+}

78. Refer to the periodic table and write the predicted electron configuration for each of the following positive ions using core notation:
(a) Ti^{2+} **(b)** Zn^{2+}
(c) Y^{3+} **(d)** Cs^+

79. Refer to the periodic table and write the predicted electron configuration for each of the following negative ions using core notation:
(a) F^- (b) S^{2-}
(c) N^{3-} **(d)** I^-

80. Refer to the periodic table and write the predicted electron configuration for each of the following negative ions using core notation:
(a) Br^- **(b)** Te^{2-}
(c) As^{3-} **(d)** O^{2-}

General Exercises

81. Examine Figure 6.1 and refer to a modern periodic table to determine the name of the element that Mendeleev predicted in 1871 and called ekaboron.

82. Examine Figure 6.1 and refer to a modern periodic table to determine the name of the element that Mendeleev predicted in 1871 and called ekaaluminum.

83. Use the American convention to designate the group number corresponding to each of the following groups listed by the European convention:
(a) Group IA **(b)** Group IB
(c) Group IIIA **(d)** Group IIIB

84. Use the IUPAC convention to designate the group number corresponding to each of the following groups listed by the European convention:
(a) Group IIA **(b)** Group IIB
(c) Group IVA **(d)** Group IVB

85. Predict the atomic radius, density, and melting point for radioactive francium.

Element	Atomic Radius	Density at 20 °C	Melting Point
Rb	0.248 nm	1.53 g/mL	38.9 °C
Cs	0.266 nm	1.87 g/mL	28.4 °C
Fr	(?) nm	(?) g/mL	(?) °C

86. Predict the atomic radius, density, and melting point for radioactive radium.

Element	Atomic Radius	Density at 20 °C	Melting Point
Sr	0.215 nm	2.63 g/mL	769 °C
Ba	0.217 nm	3.65 g/mL	725 °C
Ra	(?) nm	(?) g/mL	(?) °C

87. Refer to the periodic table and write the predicted electron configuration for each of the following elements using core notation:
(a) Sr
(b) Ru
(c) Sb
(d) Cs

88. Refer to the periodic table and write the predicted electron configuration for each of the following elements using core notation:
(a) W
(b) Bi
(c) Ra
(d) Ac

89. Explain why the ionization energy for the alkaline earth metals is higher than the ionization energy for the alkali metals.

90. Explain why the ionization energy for aluminum, contrary to the general trend, is less than the ionization energy for magnesium.

91. Explain why the ionization energy for hydrogen is much higher than that of other Group IA/1 elements.

92. Predict two ionic charges for hydrogen. Write the formulas of the two ions and explain the ionic charges.

▲ In 2007 Spain issued a periodic table postage stamp commemorating the 100th anniversary of Mendeleev's death. The tiny white squares indicate the position of elements predicted by Mendeleev before their actual discovery.

6 Chapter Self-Test

Answers to Self-Test are in Appendix J.

1. Why was neon absent from Mendeleev's original periodic table of 1871? (Sec. 6.1)
(a) Neon is unstable.
(b) Neon is radioactive.
(c) Neon had not been discovered.
(d) Neon occurs only in compounds.
(e) none of the above

2. The modern periodic law states that the properties of the elements repeat when the periodic table is arranged according to which of the following? (Sec. 6.2)
(a) increasing atomic number
(b) increasing mass number
(c) increasing atomic mass
(d) increasing isotopic mass
(e) none of the above

3. Which of the following is a fourth period alkali metal? (Sec. 6.3)
(a) Ca **(b)** K
(c) Mg **(d)** Na
(e) Sr

4. Which of the following elements has the most metallic character? (Sec. 6.4)
(a) N **(b)** P
(c) As **(d)** Sb
(e) Bi

5. Predict the melting point for strontium, given the melting points of calcium, (839 °C) and barium, (725 °C). (Sec. 6.5)
(a) 114 °C **(b)** 725 °C
(c) 782 °C **(d)** 953 °C
(e) 1564 °C

6. Which energy sublevel is being filled by the lanthanides? (Sec. 6.6)
(a) 4*f* **(b)** 5*d*
(c) 5*f* **(d)** 6*d*
(e) 6*f*

7. Predict the number of valence electrons for a Group VIIIA/18 element. (Sec. 6.7)
(a) 1 **(b)** 2
(c) 6 **(d)** 8
(e) 18

8. What is the electron dot formula for an atom of oxygen? (Sec. 6.8)
(a) $\mathrm{O}\cdot$ **(b)** $\underset{\cdot}{\mathrm{O}}\cdot$
(c) $\cdot\underset{\cdot}{\dot{\mathrm{O}}}\cdot$ **(d)** $\cdot\underset{\cdot\cdot}{\dot{\mathrm{O}}}:$
(e) $:\underset{\cdot\cdot}{\dot{\mathrm{O}}}:$

9. Which group of elements has the lowest ionization energy? (Sec. 6.9)
(a) Group IA/1 **(b)** Group IIA/2
(c) Group IIB/12 **(d)** Group VIIA/17
(e) Group VIIIA/18

10. What is the predicted ionic charge for a Cl ion? (Sec. 6.10)
(a) 1+ **(b)** 1–
(c) 7+ **(d)** 7–
(e) none of the above

Key Concepts

11. State the trends in the periodic table for decreasing atomic radii.

12. State the trends in the periodic table for decreasing metallic character.

13. State the trends in the periodic table for increasing ionization energy.

14. Given the chemical formulas for Al_2O_3, SiO_2, and P_2O_5, predict the formula for arsenic oxide, $As_?O_?$.

Critical Thinking

15. Can ions of different elements have the same number of protons? Can ions of different elements have the same number of electrons?

16. The following photograph shows a watchglass containing antimony, sulfur, and antimony sulfide. Based on typical characteristics of semimetals and nonmetals, identify the (a) yellow, (b) gray, and (c) orange substances.

CHAPTER

7 Language of Chemistry

In regions having low rainfall, crystals of silver iodide, AgI, are dispersed in clouds to induce precipitation. Crystals of AgI attract sufficient moisture to form raindrops.

"Science is organized knowledge. Wisdom is organized life."
Immanuel Kant, German Philosopher (1724–1804)

During the Middle Ages, alchemists identified many substances that they named in a haphazard fashion. Since there were only a few dozen substances, this did not pose a great problem. Toward the end of the 1700s, however, they had identified more than 10,000 substances, and the number of compounds was growing rapidly. Chemists then faced the staggering task of providing names for all these substances. The problem of assigning names was eventually solved by using a set of systematic rules.

The French chemist Antoine Lavoisier developed the first systematic method of naming substances. He proposed that chemical names refer to the composition of the compound and be derived from Latin or Greek. In 1787, Lavoisier, with the aid of others, published *Methods of Chemical Nomenclature*. The naming system he proposed was so clear and logical that it was universally accepted in a short time. In fact, it was so simple and insightful that it became the basis for our current system of naming.

7.1 CLASSIFICATION OF COMPOUNDS

OBJECTIVES

- To classify a compound as a binary ionic, a ternary ionic, or a binary molecular compound.
- To classify an acid as a binary acid or a ternary oxyacid.
- To classify an ion as a monoatomic cation, a monoatomic anion, a polyatomic cation, or a polyatomic anion.

In 1921, the International Union of Pure and Applied Chemistry (IUPAC) formed the Commission on the Nomenclature of Inorganic Chemistry. In 1938, the IUPAC Committee on the Reform of Inorganic Nomenclature met in Berlin. Two years later it released a comprehensive set of rules. Although the rules have been expanded and revised, the 1940 rules remain as the official international system for naming chemical compounds. These rules are referred to as **IUPAC nomenclature**.

Classifying Inorganic Compounds

An **inorganic compound** does not contain the element carbon. However, carbon dioxide (CO_2), and compounds containing carbonate (CO_3^{2-}) or hydrogen carbonate (HCO_3^-) are exceptions and are considered inorganic. According to the 1940 IUPAC rules, inorganic compounds can be placed in one of several categories. Five common classes are binary ionic, ternary ionic, binary molecular, binary acid, and ternary oxyacid.

A **binary ionic compound** contains two elements, a metal and a nonmetal. Examples of binary ionic compounds are KCl and $AlCl_3$. A **ternary ionic compound** contains three elements with at least one metal and one nonmetal. Examples of ternary ionic compounds are KNO_3 and $Al(NO_3)_3$. A **binary molecular compound** contains two elements that are both nonmetals. Water, H_2O, is a common example, as is ammonia, NH_3.

An **aqueous solution** is produced when a compound dissolves in water. Aqueous solutions are indicated by the symbol (*aq*). A **binary acid** is an aqueous solution of a compound containing hydrogen and one other nonmetal. Formulas of acids begin with H, and examples of binary acids are HCl(*aq*) and H_2S(*aq*).

A **ternary oxyacid** is an aqueous solution of a compound containing hydrogen, a nonmetal, and oxygen. Examples of ternary oxyacids are HNO_3(*aq*) and H_2SO_4(*aq*). Figure 7.1 illustrates the relationships among the five types of compounds we will classify according to IUPAC rules.

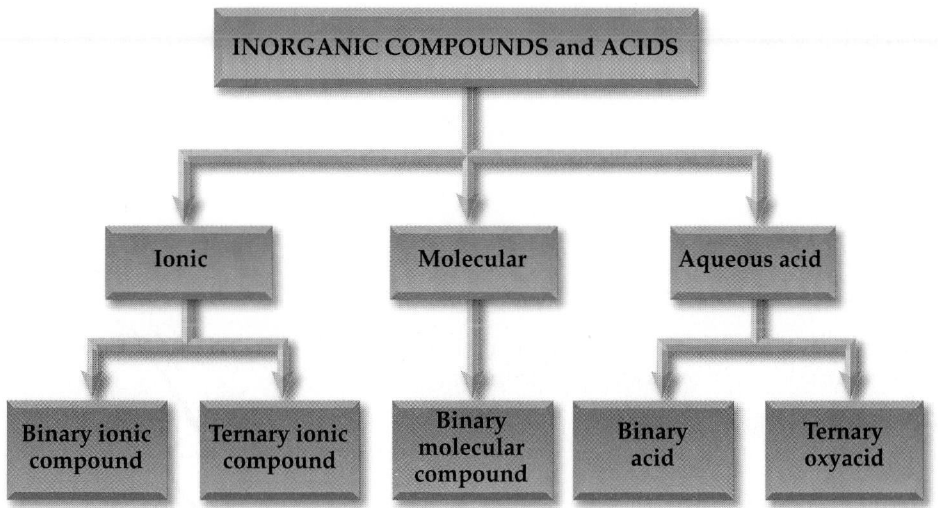

▲ **Figure 7.1 Classification of Inorganic Compounds and Acids** According to IUPAC nomenclature, inorganic compounds and acids are divided into five categories: binary ionic, ternary ionic, binary molecular, binary acid, and ternary oxyacid.

▶ Table salt, sodium chloride, NaCl, is an example of a binary ionic compound composed of metal and nonmetal ions.

The following example exercise illustrates the classification of different types of compounds and aqueous acids:

Example Exercise 7.1 Classifying Compounds and Acids

Classify each of the following as a binary ionic compound, ternary ionic compound, binary molecular compound, binary acid, or ternary oxyacid:

(a) calcium oxide, CaO
(b) sulfur dioxide, SO_2
(c) silver chromate, Ag_2CrO_4
(d) hydrofluoric acid, $HF(aq)$
(e) carbonic acid, $H_2CO_3(aq)$

Solution

We can refer to the periodic table and classify each compound or solution as follows:

(a) CaO contains two elements, a metal and nonmetal. Thus, CaO is a *binary ionic compound.*
(b) SO_2 contains two elements, both nonmetals. Thus, SO_2 is a *binary molecular compound.*
(c) Ag_2CrO_4 contains three elements, two metals and a nonmetal. Thus, Ag_2CrO_4 is a *ternary ionic compound.*
(d) $HF(aq)$ is a compound of hydrogen and a nonmetal dissolved in water. Thus, $HF(aq)$ is a *binary acid.*
(e) $H_2CO_3(aq)$ is a compound containing three elements, including hydrogen and oxygen, dissolved in water. Thus, $H_2CO_3(aq)$ is a *ternary oxyacid.*

Practice Exercise

Classify each of the following as a binary ionic compound, ternary ionic compound, binary molecular compound, binary acid, or ternary oxyacid:

(a) carbon disulfide, CS_2
(b) lithium dichromate, $Li_2Cr_2O_7$
(c) magnesium iodide, MgI_2
(d) nitric acid, $HNO_3(aq)$
(e) hydrochloric acid, $HCl(aq)$

Answers: (a) binary molecular compound; (b) ternary ionic compound; (c) binary ionic compound; (d) ternary oxyacid; (e) binary acid

Concept Exercise

How do you distinguish a binary molecular compound from a binary acid?

Answer: See Appendix G.

Classifying Ions

According to IUPAC nomenclature, ions are named systematically, depending on the category in which they are placed. A positive ion is referred to as a **cation**, and a negative ion is referred to as an **anion**. A single atom of a metal with a positive

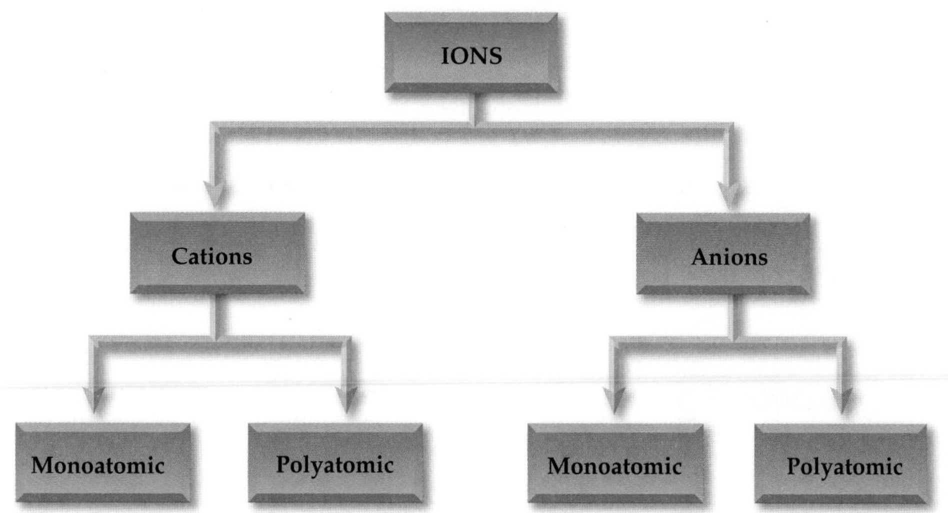

◀ **Figure 7.2 Classification of Ions** According to the IUPAC nomenclature, ions are divided into four categories: monoatomic cations, polyatomic cations, monoatomic anions, and polyatomic anions.

charge, or a single atom of a nonmetal with a negative charge, is called a **monoatomic ion**. A group of atoms bound together and having an overall positive or negative charge is called a **polyatomic ion**. Figure 7.2 illustrates the classification of and relationship of these ions.

The following example exercise illustrates the classification of different types of ions:

Example Exercise 7.2 Classifying Cations and Anions

Classify each of the following ions as a monoatomic cation, monoatomic anion, polyatomic cation, or polyatomic anion:

(a) barium ion, Ba^{2+}
(b) chloride ion, Cl^-
(c) nickel(II) ion, Ni^{2+}
(d) chlorate ion, ClO_3^-

Solution

We can classify each ion as follows:

(a) Ba^{2+} is a single atom with a positive charge. Thus, Ba^{2+} is a *monoatomic cation*.
(b) Cl^- is a single atom with a negative charge. Thus, Cl^- is a *monoatomic anion*.
(c) Ni^{2+} is a single atom with a positive charge. Thus, Ni^{2+} is a *monoatomic cation*.
(d) ClO_3^- has four atoms and a negative charge. Thus, ClO_3^- is a *polyatomic anion*.

Practice Exercise

Classify each of the following ions as a monoatomic cation, monoatomic anion, polyatomic cation, or polyatomic anion:

(a) ammonium ion, NH_4^+
(b) sulfide ion, S^{2-}
(c) permanganate ion, MnO_4^-
(d) stannous ion, Sn^{2+}

Answers: (a) polyatomic cation; (b) monoatomic anion; (c) polyatomic anion; (d) monoatomic cation

Concept Exercise

What is the distinction between Hg^{2+} and Hg_2^{2+}?

Answer: See Appendix G.

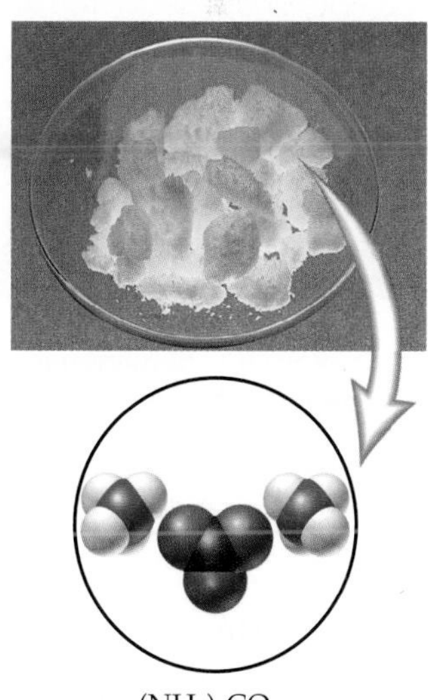

▲ **Ammonium Carbonate, $(NH_4)_2CO_3$** Ammonium carbonate decomposes in air to give ammonia, NH_3, and carbon dioxide, CO_2; it is used in smelling salts and is found in baking powder.

CHEMISTRY CONNECTION Antoine and Marie Lavoisier

Q: Who provided invaluable laboratory assistance to Lavoisier, translated the experimental results into other languages, and helped spread the enormous body of work throughout Europe and America?

Antoine Lavoisier (1743–1794) was born into an affluent French family and received an excellent education. He initially became a lawyer like his father but later turned to science. Lavoisier was a gifted experimenter and is generally considered to be the founder of modern chemistry. He established a magnificent laboratory that attracted scientists from all over the world.

◀ **Antoine Lavoisier** Antoine Lavoisier is shown with Marie in their laboratory.

In 1787, Lavoisier published a book entitled *Methods of Chemical Nomenclature*. The book designated systematic principles for naming chemical substances and was so logical that it became the basis for our present rules of nomenclature. In 1789, he published the landmark textbook *Elementary Treatise on Chemistry*, which offered the first modern view of chemistry.

In 1771, Lavoisier married Marie-Anne Pierrette Paulze (1758–1836). Marie was only thirteen years old and she had been raised and educated in a convent after her mother died. Marie's father arranged the marriage in order to thwart a proposal by an influential nobleman who was more than three times her age. Despite her youth, she proved invaluable to Lavoisier's work and joined him in the laboratory. Marie recorded the experimental work through her drawings and notes, and translated the results from French into other languages. Owing to their close working relationship, it is impossible to assign individual credit and separate their joint contribution to the advancement of chemistry.

Unfortunately, Marie's father was an important official in a firm that collected taxes. Although Lavoisier did not collect taxes, he received royalties from the same firm. When the French Revolution broke out, revolutionaries threw tax collectors into jail. Lavoisier was first barred from his laboratory and then arrested. After a farce trial, he was sentenced to the guillotine, along with his father-in-law and other tax collectors.

On May 8, 1794, Lavoisier was executed and buried in an unmarked grave. The French court declared that France had no need for scientists, but the famous mathematician Lagrange lamented, "Only a moment was required to strike off his head, and a hundred years may not be enough to produce another like it again." Within two years, the French people regretted their hasty decision and began unveiling statues of Lavoisier.

A: Lavoisier's multilingual wife, Marie-Anne, is recognized for keeping laboratory records and translating scientific papers into other languages.

OBJECTIVES

- To write systematic names and formulas for common monoatomic ions.
- To predict the ionic charge for ions of representative elements.

7.2 MONOATOMIC IONS

Metal atoms can lose valence electrons and become positively charged cations; examples include Na^+, Ca^{2+}, and Al^{3+}. Conversely, nonmetal atoms can gain valence electrons and become negatively charged anions; examples include, Cl^-, S^{2-}, and P^{3-}.

Naming Metal Ions

According to IUPAC nomenclature rules, cations are named for the parent metal followed by the word "ion." For example, Na^+ is named sodium ion, Mg^{2+} is named magnesium ion, and Al^{3+} is named aluminum ion.

Main-group metals usually form one cation; tin and lead are exceptions. Transition metals, however, often form more than one cation. Iron, for example, can form Fe^{2+}

and Fe^{3+}. To name a metal cation having more than one ionic charge, it is necessary to specify the charge. IUPAC recommends that the cation be named for the parent metal followed by its charge specified by Roman numerals in parentheses. Thus, Fe^{2+} is named iron(II) ion, and Fe^{3+} is named iron(III) ion. Similarly, Cu^{+} is named copper(I) ion and Cu^{2+} is named copper(II) ion. This method of naming transition metal cations is called the **Stock system**. We should note that the names of Ag^{+}, Zn^{2+}, and Cd^{2+} are transition metal exceptions and do not require Roman numerals.

In addition to the Stock system, IUPAC allows another method for naming metal cations having two common ionic charges. It is called the **Latin system** or the suffix system. This system takes the Latin name of the metal and then adds the suffix *-ous* or *-ic*. The lower of the two ionic charges receives the *-ous* suffix, and the higher charge receives the *-ic* suffix. For example, the Latin name for iron is *ferrum*. To name an iron ion, we take the stem *ferr-* and add the suffix *-ous* or *-ic*. Thus, Fe^{2+} is named ferrous ion, and Fe^{3+} is named ferric ion. The Latin name for copper is *cuprum*. Similarly, we add the suffix *-ous* or *-ic* to the stem *cupr-*. Thus, we have Cu^{+} for cuprous ion, and Cu^{2+} for cupric ion.

The formula for the mercury(I) ion is an exception. When mercury loses an electron, the resulting ion, Hg^{+}, becomes more stable by combining with another Hg^{+} ion. Thus, the mercury(I) ion is written $Hg_2{}^{2+}$. The Latin name for mercury, *hydrargyrum*, is difficult to pronounce, and so IUPAC recommends adding the suffix to the English name. Thus, $Hg_2{}^{2+}$ is named mercurous ion, and Hg^{2+} is named mercuric ion. Table 7.1 lists the systematic names for common cations.

TABLE 7.1 COMMON MONOATOMIC CATIONS

CATION	STOCK SYSTEM	LATIN SYSTEM
Al^{3+}	aluminum ion	
Ba^{2+}	barium ion	
Cd^{2+}	cadmium ion	
Ca^{2+}	calcium ion	
Co^{2+}	cobalt(II) ion	cobaltous ion
Co^{3+}	cobalt(III) ion	cobaltic ion
Cu^{+}	copper(I) ion	cuprous ion
Cu^{2+}	copper(II) ion	cupric ion
Cr^{3+}	chromium(III) ion	
H^{+}	hydrogen ion	
Fe^{2+}	iron(II) ion	ferrous ion
Fe^{3+}	iron(III) ion	ferric ion
Pb^{2+}	lead(II) ion	plumbous ion
Pb^{4+}	lead(IV) ion	plumbic ion
Li^{+}	lithium ion	
Mg^{2+}	magnesium ion	
Mn^{2+}	manganese(II) ion	
$Hg_2{}^{2+}$	mercury(I) ion*	mercurous ion
Hg^{2+}	mercury(II) ion	mercuric ion
Ni^{2+}	nickel(II) ion	
K^{+}	potassium ion	
Ag^{+}	silver ion	
Na^{+}	sodium ion	
Sr^{2+}	strontium ion	
Sn^{2+}	tin(II) ion	stannous ion
Sn^{4+}	tin(IV) ion	stannic ion
Zn^{2+}	zinc ion	

*Note that the mercury(I) ion is diatomic and is written $Hg_2{}^{2+}$.

Naming Nonmetal Ions

As we now know, nonmetal atoms can gain valence electrons and become negatively charged ions called anions. According to IUPAC rules, nonmetal ions are named by using the *nonmetal stem* plus the suffix *-ide*. Examples of nonmetal ions are Cl^-, chloride ion; S^{2-}, sulfide ion; and P^{3-}, phosphide ion. Table 7.2 lists the systematic names for common anions.

TABLE 7.2 COMMON MONOATOMIC ANIONS

ANION	IUPAC NAME	ANION	IUPAC NAME
Br^-	bromide ion	N^{3-}	nitride ion
Cl^-	chloride ion	O^{2-}	oxide ion
F^-	fluoride ion	P^{3-}	phosphide ion
I^-	iodide ion	S^{2-}	sulfide ion

Predicting Formulas of Monoatomic Cations

We can use the periodic table to help learn the names and formulas of the ions and predict their charge based on the group number of the element. Recall that Group IA/1 metals always form 1+ ions. Thus, we have Li^+, Na^+, and K^+. Also recall that the Group IIA/2 elements always form 2+ ions. Thus, we have Mg^{2+}, Ca^{2+}, Sr^{2+}, and Ba^{2+}. Similarly, we can predict that aluminum in Group IIIA/13 forms a 3+ ion, that is, Al^{3+}.

Not all metal ions are predictable from the periodic table. Tin and lead in Group IVA/14 can each form two ions: Sn^{2+} or Sn^{4+} and Pb^{2+} or Pb^{4+}. Therefore, you will have to memorize the formulas for these ions.

With few exceptions all transition elements have two *s* electrons as well as a variable number of *d* electrons. In general, a transition metal loses its two *s* electrons to form an ion with a 2+ charge, for example, Ni^{2+}. Many transition metals form additional ions having charges that are unpredictable. You will have to memorize the names of these ions. Figure 7.3 shows the relationship of ionic charge to the position of the element in the periodic table.

Although we should memorize the names and formulas of ions, the periodic table is a valuable resource for verifying ionic charges. The following example exercise illustrates the names and formulas of monoatomic cations:

Example Exercise 7.3 Names and Formulas of Monoatomic Cations

Provide the formula for the following monoatomic cations:

(a) barium ion (b) cobalt(II) ion

Solution

We can use the periodic table to predict the charge on a cation.

(a) Barium is found in Group IIA/2 and can lose two valence electrons. We predict the formula of the ion is Ba^{2+}.

(b) Cobalt is a transition metal and can lose two or more valence electrons.The Roman number (II) indicates the loss of two electrons, and so the formula of the ion is Co^{2+}.

Practice Exercise

Supply a systematic name for the following monoatomic cations:

(a) Al^{3+} (b) Co^{3+}

Answers: (a) aluminum ion; (b) cobalt(III) ion or cobaltic ion

Concept Exercise

How does the Stock system specify a variable charge on a cation?

Answer: See Appendix G.

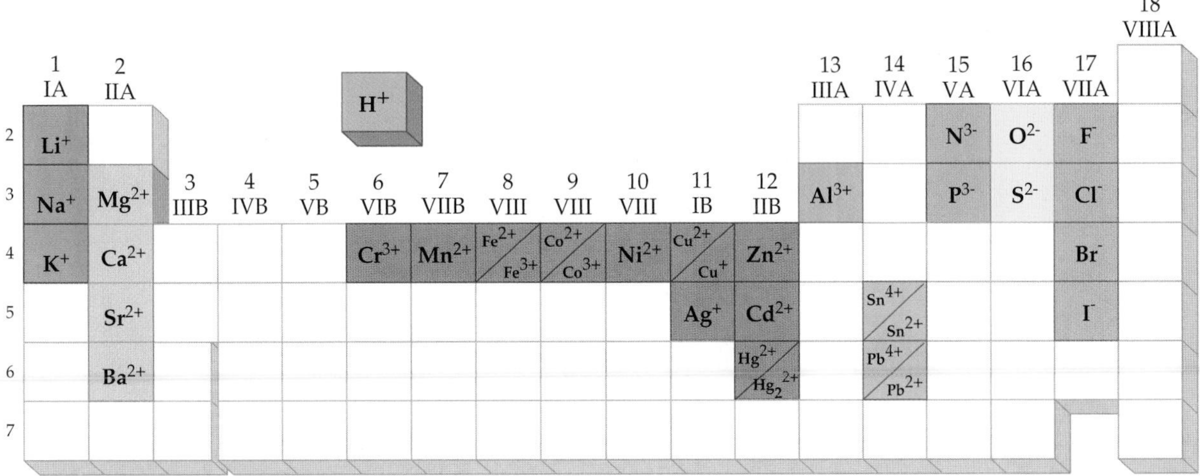

▲ **Figure 7.3 Periodic Table of Selected Ions** Note the correlation of ionic charge and group number. Although transition elements exhibit more than one ionic charge, nearly all these metals exhibit a charge of 2+.

Predicting Formulas of Monoatomic Anions

Nonmetals gain electrons to form negative ions in a predictable fashion. The halogens in Group VIIA/17 need only one electron to become isoelectronic with a noble gas. Thus, the halogens have a 1– charge and the formulas F^-, Cl^-, Br^-, and I^-. The elements in Group VIA/16 gain two electrons to become isoelectronic with a noble gas. These elements have a 2– charge and the formulas O^{2-}, S^{2-}, and Se^{2-}. Nitrogen and phosphorus are in Group VA/15 and gain three electrons. These elements have a 3– charge and the formulas N^{3-}, and P^{3-}.

Example Exercise 7.4 Names and Formulas of Monoatomic Anions

Provide the formula for each of the following monoatomic anions:

(a) fluoride ion (b) oxide ion

Solution

Recall that nonmetals gain electrons so as to acquire a noble gas electron configuration.

(a) Fluorine is found in Group VIIA/17 and can gain one electron to become isoelectronic with neon. We predict the formula for the fluoride ion is F^-.

(b) Oxygen is found in Group VIA/16 and can gain two electrons to become isoelectronic with neon. We predict the formula for the oxide ion is O^{2-}.

Practice Exercise

Supply a systematic name for each of the following monoatomic anions:

(a) Br^- (b) N^{3-}

Answers: (a) bromide ion; (b) nitride ion

Concept Exercise

Which one of the following is a monoatomic anion: iodide, iodate, or iodite?

Answer: See Appendix G.

7.3 POLYATOMIC IONS

OBJECTIVE

▶ To write systematic names and formulas for common polyatomic ions.

Polyatomic anions generally contain one or more elements combined with oxygen. Such an anion is referred to as an **oxyanion**, a term that reflects its composition. Most oxyanions have names ending in the suffix *-ate*. Examples include the nitrate ion, NO_3^-, and the sulfate ion, SO_4^{2-}.

A few oxyanions have names ending in the suffix *-ite*. Examples include the nitrite ion, NO_2^-, and the sulfite ion, SO_3^{2-}. Notice that in each case the formula for ions with the suffix *-ite* has one less oxygen than that for ions with the suffix *-ate*.

This suffix pattern can provide clues when naming some polyatomic ions. Given the formula for the chlorate ion, ClO_3^-, we can predict the formula for the chlorite ion. Since the suffix *-ate* has changed to the suffix *-ite*, the formula has one less oxygen. Therefore, the formula for the chlorite ion is ClO_2^-. The general relationship between the suffixes *-ate* and *-ite* allows us to simplify our task of memorizing formulas of ions.

There are two important polyatomic ions that have the suffix *-ide*. The cyanide ion, CN^-, and the hydroxide ion, OH^-, are exceptions, and their names should be memorized. In addition to these two polyatomic anions, there is the ammonium cation, NH_4^+, whose formula is derived from that for ammonia gas, NH_3. Table 7.3 lists the common polyatomic ions whose formulas should be memorized.

TABLE 7.3 COMMON POLYATOMIC IONS

CATION		IUPAC NAME	
NH_4^+		ammonium ion	
ANION	**IUPAC NAME**	**ANION**	**IUPAC NAME**
$C_2H_3O_2^-$	acetate ion	OH^-	hydroxide ion*
CO_3^{2-}	carbonate ion	ClO^-	hypochlorite ion
ClO_3^-	chlorate ion	NO_3^-	nitrate ion
ClO_2^-	chlorite ion	NO_2^-	nitrite ion
CrO_4^{2-}	chromate ion	ClO_4^-	perchlorate ion
CN^-	cyanide ion*	MnO_4^-	permanganate ion
$Cr_2O_7^{2-}$	dichromate ion	PO_4^{3-}	phosphate ion
HCO_3^-	hydrogen carbonate ion	SO_4^{2-}	sulfate ion
HSO_4^-	hydrogen sulfate ion	SO_3^{2-}	sulfite ion

*Note that the suffix *-ide* is an exception to the general *-ate* and *-ite* rule.

The following example exercise illustrates how general principles can be used to assist you in memorizing the names and formulas of polyatomic ions:

Example Exercise 7.5 Names and Formulas of Polyatomic Ions

Provide a systematic name for each of the following polyatomic oxyanions:

(a) CO_3^{2-}
(b) CrO_4^{2-}
(c) ClO_2^-
(d) HSO_4^-

Solution

We can make reasonable predictions for the names of many polyatomic ions. This makes the task of memorization much easier.

(a) CO_3^{2-} contains carbon, and we predict the name has the suffix *-ate*. Thus, we predict CO_3^{2-} is named the *carbonate ion*.

(b) CrO_4^{2-} contains chromium, and we predict the name has the suffix *-ate*. Thus, we predict CrO_4^{2-} is named the *chromate ion*.

(c) ClO_2^- is related to ClO_3^-, which is named the chlorate ion. Since ClO_2^- has one less oxygen atom, the suffix changes to *-ite*. Thus, we predict ClO_2^- is named the *chlorite ion*.

(d) HSO_4^- is related to the sulfate ion, SO_4^{2-}. With the addition of hydrogen, the name becomes the *hydrogen sulfate ion*.

Practice Exercise

Provide the formula for each of the following polyatomic oxyanions:

(a) acetate ion (b) dichromate ion

(c) perchlorate ion (d) hydrogen carbonate ion

Answers: (a) $C_2H_3O_2^-$; (b) $Cr_2O_7^{2-}$; (c) ClO_4^-; (d) HCO_3^-

Concept Exercise

How does systematic naming distinguish between a monoatomic anion and a polyatomic anion?

Answer: See Appendix G.

Note You may find it useful to make flashcards to help you memorize ions. Write the name of the ion on one side of a card and the formula of the ion on the other side. Your task will be easier if you recall the following:

1. There is one common polyatomic cation, NH_4^+.
2. Most oxyanions have the suffix *-ate*.
3. The suffix *-ate* changes to the suffix *-ite* for an oxyanion with one less oxygen atom; for example, nitrate, NO_3^-, changes to nitrite, NO_2^-.
4. There are two common polyatomic anions that are exceptions: OH^- and CN^- each have the suffix *-ide*.

7.4 WRITING CHEMICAL FORMULAS

OBJECTIVE

▸ To put together formula units for compounds composed of monoatomic ions and polyatomic ions.

An ionic compound is composed of positive and negative ions. A **formula unit** is the simplest representative particle in an ionic compound. Since a formula unit is neutral, the total positive charge must equal the total negative charge in both the formula unit and the ionic compound. That is, the total positive charge from the metal ions must be the same as the total negative charge from the nonmetal ions.

A formula unit of ordinary salt contains Na^+ and Cl^-. Since the positive and negative ions have equal but opposite charges, the formula for the compound is NaCl. In a formula unit containing Ca^{2+} and Cl^- the charges are not the same. It is necessary to have two Cl^- ions for each Ca^{2+} ion to balance the charges and have a neutral compound; therefore, the formula is $CaCl_2$. In a formula unit containing Al^{3+} and Cl^-, it is necessary to have three Cl^- ions for each Al^{3+} ion, and the formula for the compound is $AlCl_3$.

The following example exercise provides additional illustrations of how to write formulas for ionic compounds:

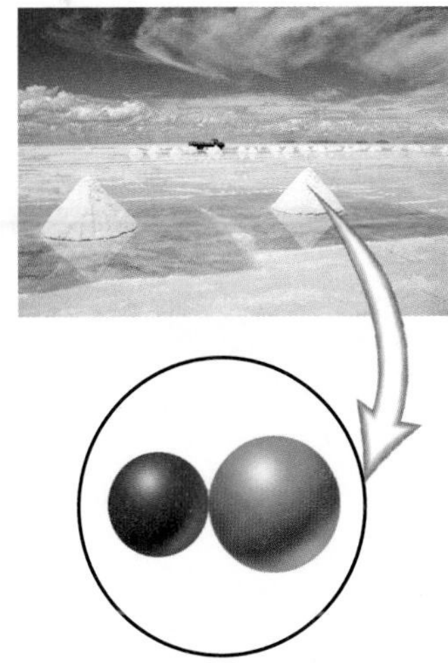

▲ **Salt, NaCl** Table salt is an example of a binary ionic compound composed of sodium ions and chloride ions.

Example Exercise 7.6 Writing Formulas of Binary Ionic Compounds

Write the chemical formula for the following binary compounds given their constituent ions:

(a) copper(I) oxide, Cu^+ and O^{2-}

(b) cadmium oxide, Cd^{2+} and O^{2-}

(c) cobalt(III) oxide, Co^{3+} and O^{2-}

Solution

(a) The copper(I) ion has a charge of 1+, and the oxide ion has a charge of 2–. Thus, two positive ions are required for each negative ion in a neutral formula unit. The formula of copper(I) oxide is written Cu_2O.

(b) Since the cadmium ion and oxide ion each have a charge of 2, the ratio is 1:1, that is, Cd_1O_1. It is not necessary to write the subscript 1, and so the formula of cadmium oxide is simply CdO.

(c) This example is more difficult. The cobalt(III) ion has a charge of 3+ and the oxide ion has a charge of 2–. Since the lowest common multiple is 6, two 3+ ions are required to cancel the charge of three 2– ions. The ratio is 2:3, and the formula of cobalt(III) oxide is written Co_2O_3.

Practice Exercise

Write the chemical formula for the following binary compounds given their constituent ions:

(a) iron(II) sulfide, Fe^{2+} and S^{2-}
(b) mercury(I) fluoride, $Hg_2{}^{2+}$ and F^-
(c) lead(IV) oxide, Pb^{4+} and O^{2-}

Answers: (a) FeS; (b) Hg_2F_2; (c) PbO_2

Concept Exercise

What is the formula for a metal oxide if the charge on the metal (M) is +3?

Answer: See Appendix G.

Note You can quickly verify that the chemical formula is written correctly by crossing over the charge on each ion. Consider aluminum oxide, which contains Al^{3+} and O^{2-}. The 3+ charge on the aluminum ion becomes the subscript 3 for the oxygen, and the 2– charge on the oxide ion becomes the subscript 2 for the aluminum ion. That is,

$$Al^{3+}\ O^{2-} = Al_2O_3$$

Formula Units Containing Polyatomic Ions

As mentioned earlier, formula units are the simplest particles representing an ionic compound, and the total positive charge is equal to the total negative charge. Previously, we learned to write neutral formula units for binary ionic compounds. Similarly, in a formula unit containing K^+ and $SO_4{}^{2-}$, two K^+ ions are required to balance the charge for each $SO_4{}^{2-}$. Thus, a neutral formula unit is written K_2SO_4.

Magnesium sulfate contains Mg^{2+} and $SO_4{}^{2-}$. Since the magnitude of the charge is the same on each ion, the ratio of positive ion to negative ion is 1:1. The formula unit is written $MgSO_4$.

Ammonium sulfate is a nitrogen-supplying component in fertilizer and contains $NH_4{}^+$ and $SO_4{}^{2-}$ ions. Since the negative charge is greater for $SO_4{}^{2-}$, two $NH_4{}^+$ ions are necessary to give a neutral formula unit. To avoid misunderstanding, parentheses are placed around the $NH_4{}^+$ ion. The correct formula is written $(NH_4)_2SO_4$. The following example exercise provides additional illustrations of how to write formulas for ionic compounds:

▲ **Chalk, $CaCO_3$** Chalk is an example of a ternary ionic compound composed of calcium ions and carbonate ions.

Example Exercise 7.7 Writing Formulas of Ternary Ionic Compounds

Write the chemical formula for each of the following ternary compounds given their constituent ions:

(a) calcium carbonate, Ca^{2+} and $CO_3{}^{2-}$
(b) calcium hydroxide, Ca^{2+} and OH^-
(c) calcium phosphate, Ca^{2+} and $PO_4{}^{3-}$

Solution

(a) Since the positive and negative ions each have a charge of 2, one positive ion and one negative ion are required to produce a neutral formula unit, and the formula is $CaCO_3$. Calcium carbonate occurs naturally as ordinary chalk.

(b) The positive ion has a charge of 2+, and the negative ion has a charge of 1–. Therefore, one positive ion and two negative ions are required to produce a neutral formula unit. Since OH^- is a polyatomic ion, parentheses are required, and the formula is written $Ca(OH)_2$. Calcium hydroxide is known as "slaked lime" and is sometimes used to mark the boundaries of an athletic field.

(c) The positive ion has a charge of 2+, and the negative ion has a charge of 3–. The lowest common multiple of the charges is 6. Three positive ions are required for every two negative ions to produce a neutral formula unit. A calcium phosphate formula unit is written $Ca_3(PO_4)_2$. Calcium phosphate is found in tooth enamel.

Practice Exercise

Write the chemical formula for each of the following ternary compounds given their constituent ions:

(a) copper(II) permanganate, Cu^{2+} and MnO_4^-

(b) iron(III) carbonate, Fe^{3+} and CO_3^{2-}

(c) potassium dichromate, K^+ and $Cr_2O_7^{2-}$

Answers: (a) $Cu(MnO_4)_2$; (b) $Fe_2(CO_3)_3$; (c) $K_2Cr_2O_7$

Concept Exercise

What is the formula for a metal carbonate if the charge on the metal (M) is +3?

Answer: See Appendix G.

Note As before, we can verify that the formula is correct by simply crossing over the charge on each ion. Consider calcium phosphate, which contains Ca^{2+} and PO_4^{3-}. The 2+ charge on the calcium ion becomes the subscript 2 for the phosphate ion. Since phosphate is a polyatomic ion, we must use parentheses, that is, (PO_4). Conversely, the 3– charge on the phosphate ion becomes the subscript 3 for the calcium ion.

$$Ca^{2+}\ PO_4^{3-} = Ca_3(PO_4)_2$$

CRITICAL THINKING Nutritional Potassium

Q: How can potassium be an essential mineral in the human diet since it reacts explosively in water?

The human body requires many minerals such as calcium, sodium, and potassium, as well as trace amounts of iron, zinc, and selenium. Minerals perform numerous functions in the body, such as building strong bones, transmitting nerve impulses, making hormones, and maintaining a regular heartbeat.

▲ **Potassium in Water** Potassium metal reacts explosively in water as shown; potassium chloride, KCl, simply dissolves in water.

One role of potassium is to maintain normal amounts of water in body tissues and avoid excessive water retention. Potassium is also important for building muscle and metabolizing proteins and carbohydrates. Although there is no recommended daily allowance (RDA) for potassium, the National Library of Medicine suggests about 2 grams of potassium a day.

The best dietary sources of potassium include fish (especially cod and sardines), meat, peas, tomatoes, potatoes, leafy green vegetables, citrus fruits, bananas, melons, dried fruit, nuts, and chocolate. People with high blood pressure are often put on a diet low in sodium, and a "salt substitute" high in potassium.

Pregnant women and people with chronic diseases are at risk to develop potassium deficiency. Depleted potassium levels may also be caused by vomiting, diarrhea, or use of diuretic or cortisone medications. Symptoms of potassium deficiency include fatigue, slow reflexes, muscle weakness, and dry skin. If potassium levels are depleted quickly, a person can experience heart problems. If potassium levels are not replenished quickly, a person can suffer heart failure and death.

Conversely, ingesting extreme amounts of potassium or other minerals can be unhealthy. In some cases, an excess mineral can make it difficult for the body to absorb other minerals and vitamins. The best remedy to avoid vitamin and mineral deficiencies is a balanced diet that includes a wide variety of fruits, green vegetables, protein sources, whole grains, and dairy products daily.

A: Nutritionists state the need for potassium in the diet. To be more precise, humans require potassium salt (not potassium metal) in their diet. In fact, individuals on low salt diets may use a "salt substitute" containing KCl, in lieu of table salt, NaCl.

OBJECTIVES

- To determine the ionic charge on a cation in a binary ionic compound.
- To write systematic names and formulas for binary ionic compounds.

7.5 BINARY IONIC COMPOUNDS

Positively charged metal ions and negatively charged nonmetal ions compose neutral binary ionic compounds. The oppositely charged cation and anion are attracted to each other and create compounds in which the overall charge is zero. That is, an ionic compound has an overall charge of zero even though it contains charged ions.

Fe_2O_3

▲ **Hematite, Fe_2O_3** Hematite is a mineral found in nature that contains iron.

Binary Compounds Containing a Transition Metal

Some binary ionic compounds contain transition metal ions, which have a variable ionic charge, for instance, Fe^{2+} and Fe^{3+}. To name such a compound correctly, we first have to determine the ionic charge of the metal cation. The following example exercise illustrates the determination of ionic charge:

Example Exercise 7.8 Determining Ionic Charge in a Compound

Determine the ionic charge for iron in the mineral hematite, Fe_2O_3.

Solution

The charge on an oxide ion is 2–, and there are three oxide ions. The total negative charge must be equal to six negative:

$$O^{2-} + O^{2-} + O^{2-} = 6 \text{ negative}$$

Since all compounds are electrically neutral, the total positive charge must equal the total negative charge: 6 negative = 6 positive. Thus, the two iron ions have a charge of six positive:

$$Fe^{x+} + Fe^{x+} = 6 \text{ positive}$$

$$Fe^{x+} = 3 \text{ positive}$$

The iron ion is therefore Fe^{3+}. The name of Fe_2O_3 is iron(III) oxide according to the Stock system. It is named ferric oxide according to the Latin system.

Practice Exercise

Determine the ionic charge for each transition metal in the following compounds:

(a) Cu_3P

(b) CoN

Answers: (a) Cu^+; (b) Co^{3+}

Concept Exercise

What is the ionic charge on a metal (M), given the formula of the oxide, MO_2?

Answer: See Appendix G.

Naming Binary Ionic Compounds

When writing a chemical formula, the positive cation is always written before the negative anion. Therefore, in naming binary ionic compounds, the metal cation is named first followed by the nonmetal anion. We learned in Section 7.2 that monoatomic anions end with the suffix *-ide*. For example, MgO is composed of the magnesium ion and the oxide ion and is named magnesium oxide.

The mineral cinnabar, HgS, is composed of the mercury ion and the sulfide ion. Mercury has two possible ions, $Hg_2{}^{2+}$ and Hg^{2+}. Since the ionic charge on a sulfide ion is 2–, the mercury ion is Hg^{2+}. Thus, we can name HgS mercury(II) sulfide according to the Stock system. Alternatively, we can name HgS mercuric sulfide according to the Latin system.

Example Exercise 7.9 Names of Binary Ionic Compounds

Supply a systematic name for each of the following binary ionic compounds:

(a) ZnO (b) SnF_2

Solution

We can name an ionic compound by designating the two ions.

(a) ZnO contains the zinc ion and the oxide ion; ZnO is named zinc oxide.

(b) SnF_2 contains the tin(II) ion and the fluoride ion. Thus, SnF_2 is named tin(II) fluoride. The Latin system name is stannous fluoride, which is an active ingredient in a popular toothpaste.

Practice Exercise

Supply a systematic name for each of the following binary ionic compounds:

(a) Mn_3P_2 (b) Fe_2S_3

Answers: (a) manganese(II) phosphide; (b) iron(III) sulfide or ferric sulfide

Concept Exercise

Which of the following is a binary ionic compound: sodium chloride, sodium chlorate, or sodium chlorite?

Answer: See Appendix G.

▲ **Cinnabar, HgS**
Cinnabar is a mineral found in nature that contains mercury.

Example Exercise 7.10 Formulas of Binary Ionic Compounds

Provide the formula for each of the following binary ionic compounds:

(a) lithium fluoride (b) lead(II) sulfide

Solution

We can write the formula by combining the cation and the anion into a neutral formula unit. Refer to Section 7.4 to review the writing of formula units.

(a) Lithium fluoride is composed of Li^+ and F^-; thus, the formula of the compound is written LiF.

(b) Lead(II) sulfide is composed of Pb^{2+} and S^{2-}; thus, the formula of the compound is written PbS.

Practice Exercise

Provide the formula for the following binary ionic compounds:

(a) copper(II) iodide (b) mercury(II) oxide

Answers: (a) CuI_2; (b) HgO

Concept Exercise

Which of the following is a binary ionic compound: LiCl, $LiClO_3$, or $LiClO_2$?

Answer: See Appendix G.

▲ **Galena, PbS**
Galena is a mineral found in nature that contains lead.

Predicting Formulas of Binary Ionic Compounds

We have learned the value of the periodic table in mastering the formulas of ions. Now we are going to use the periodic table to predict the chemical formulas of binary ionic compounds.

We can predict the formulas of compounds based on the formula of a similar compound. Let's begin with alkali metal chlorides. Sodium is in Group IA/1. The formula of sodium chloride is NaCl. All alkali metal chlorides, in fact, have a similar chemical formula. For the alkali metal family, the chloride compounds have the following formulas: LiCl, NaCl, and KCl.

We can predict the formulas for alkaline earth metal chlorides in a similar manner. Barium is in Group IIA/2, and the formula for barium chloride is $BaCl_2$. Therefore, the formulas for the other alkaline earth metal chlorides are $MgCl_2$, $CaCl_2$, and $SrCl_2$.

Example Exercise 7.11 Predicting Formulas of Binary Ionic Compounds

Predict the chemical formula for each of the following binary compounds given the formula of aluminum oxide, Al_2O_3:

(a) gallium oxide (b) aluminum sulfide

Solution

To predict the chemical formula, we compare the elements that are different in the similar compounds.

(a) The elements Ga and Al are both in Group IIIA/3, and so the formula is Ga_2O_3.
(b) The elements S and O are both in Group VIA/16, and so the formula is Al_2S_3.

Practice Exercise

Predict the chemical formula for each of the following binary compounds given the formula of magnesium chloride, $MgCl_2$:

(a) radium chloride (b) magnesium fluoride

Answers: (a) $RaCl_2$; (b) MgF_2

Concept Exercise

Predict the formula for potassium iodide, given the formula of sodium fluoride, NaF.

Answer: See Appendix G.

7.6 TERNARY IONIC COMPOUNDS

OBJECTIVES

- To determine the ionic charge on a cation in a ternary ionic compound.
- To write systematic names and formulas for ternary ionic compounds.

Compounds containing a metal and two other elements are classified as ternary ionic compounds. Typically, ternary ionic compounds have a monoatomic metal cation and a polyatomic oxyanion, for example, $NaNO_3$.

Like all ionic compounds, the cation is written first in the formula. The names of ternary ionic compounds usually have the suffix *-ate* or *-ite*. For example, $KClO_3$ is potassium chlorate, and $KClO_2$ is potassium chlorite.

Ternary Compounds Containing a Transition Metal

Some ternary compounds contain transition metal ions that have a variable ionic charge, for instance, Fe^{2+} and Fe^{3+}. To name such a compound correctly, we first have to determine the ionic charge of the transition metal cation.

Example Exercise 7.12 Determining Ionic Charge in a Compound

Determine the ionic charge for iron in $Fe_3(PO_4)_2$.

Solution

The charge on a phosphate ion is 3– and there are two phosphate ions. Therefore, the total negative charge must be equal to six negative:

$$PO_4^{\ 3-} + PO_4^{\ 3-} = 6 \text{ negative}$$

Since all compounds are electrically neutral, the total positive charge must equal the total negative charge: 6 negative = 6 positive. Thus, the three iron ions have a charge of six positive.

$$Fe^{x+} + Fe^{x+} + Fe^{x+} = 6 \text{ positive}$$
$$Fe^{x+} = 2 \text{ positive}$$

The iron ion is therefore Fe^{2+}. The name of $Fe_3(PO_4)_2$ is iron(II) phosphate according to the Stock system. It is named ferrous phosphate according to the Latin system.

Practice Exercise

Determine the ionic charge for the metal cation in each of the following compounds:

(a) $Hg(OH)_2$ (b) $Co(ClO_3)_3$

Answers: (a) Hg^{2+}; (b) Co^{3+}

Concept Exercise
What is the ionic charge on a metal (M), given the formula of the carbonate, MCO_3?

Answer: See Appendix G.

Naming Ternary Ionic Compounds

Like all ionic compounds, ternary ionic compounds are named by designating the cation followed by the anion. The mineral marble, $CaCO_3$, is composed of the calcium ion and the carbonate ion and is named calcium carbonate. $Ca(NO_3)_2$ is composed of the calcium ion and two nitrate ions and is named calcium nitrate.

The compound $CuSO_4$ is composed of a copper ion and a sulfate ion. Copper, however, has two possible ions, Cu^+ and Cu^{2+}. Since the sulfate charge is 2–, the charge on copper must be 2+ because ionic compounds are neutral. Since the copper ion is Cu^{2+}, the name of the compound is copper(II) sulfate, or cupric sulfate.

$CaCO_3$

▲ **Marble, $CaCO_3$** Marble, $CaCO_3$, is a mineral found in nature that was used to construct the Taj Mahal in India.

Example Exercise 7.13 Names of Ternary Ionic Compounds

Supply a systematic name for the following ternary ionic compounds:

(a) $KMnO_4$ (b) $Hg(NO_3)_2$

Solution
We can name an ionic compound by designating the two ions.

(a) $KMnO_4$ is composed of the potassium ion and the permanganate ion. Thus, the compound is named potassium permanganate.
(b) $Hg(NO_3)_2$ contains the mercury(II) ion and the nitrate ion. Therefore, it is named mercury(II) nitrate, or mercuric nitrate.

Practice Exercise
Supply a systematic name for each of the following ternary ionic compounds.

(a) $BaCrO_4$ (b) $Cu(NO_2)_2$

Answers: (a) barium chromate; (b) copper(II) nitrite or cupric nitrite

Concept Exercise
Which of the following is a ternary ionic compound: potassium nitride, potassium nitrate, or potassium nitrite?

Example Exercise 7.14 Formulas of Ternary Ionic Compounds

Provide the formula for each of the following ternary ionic compounds:

(a) nickel(II) acetate (b) iron(III) sulfate

Solution
We can write the formula by combining the cation and polyatomic anion into a neutral formula unit.

(a) Nickel(II) acetate is composed of Ni^{2+} and $C_2H_3O_2^-$. The formula of the compound is written $Ni(C_2H_3O_2)_2$.
(b) Iron(III) sulfate contains Fe^{3+} and SO_4^{2-}; the formula is written $Fe_2(SO_4)_3$.

Practice Exercise
Provide the formula for each of the following ternary ionic compounds.

(a) mercury(I) nitrite (b) tin(IV) permanganate

Answers: (a) $Hg_2(NO_2)_2$; (b) $Sn(MnO_4)_4$

Concept Exercise
Which of the following is a ternary ionic compound: LiCl, LiClO, or $LiClO_2$?

Answer: See Appendix G.

Predicting Formulas of Ternary Ionic Compounds

The periodic table can be used to predict the formulas of ternary ionic compounds. For example, if we know that the formula of potassium nitrate is KNO_3, we can predict the formulas of sodium nitrate and lithium nitrate. Since Li, Na, and K are all in Group IA/1, we can predict that the compounds have similar chemical formulas, that is, $LiNO_3$, $NaNO_3$, and KNO_3. The following example exercise illustrates how to predict formulas of ternary ionic compounds.

Example Exercise 7.15 Predicting Formulas of Ternary Ionic Compounds

Predict the chemical formula for each of the following ternary ionic compounds given the formula of calcium carbonate, $CaCO_3$:

(a) radium carbonate (b) calcium silicate

Solution

To predict the formula, we compare the elements that are different in the similar compounds.

(a) The elements Ra and Ca are both in Group IIA/2. Thus, the formula for radium carbonate is $RaCO_3$.
(b) The elements Si and C are both in Group IVA/14. Therefore, the formula for calcium silicate is $CaSiO_3$.

Practice Exercise

Predict the chemical formula for each of the following ternary compounds given the formula of potassium chlorate, $KClO_3$:

(a) lithium chlorate (b) potassium bromate

Answers: (a) $LiClO_3$; (b) $KBrO_3$

Concept Exercise

Predict the formula for strontium sulfate, given the formula of calcium sulfate, $CaSO_4$.

Answer: See Appendix G.

7.7 BINARY MOLECULAR COMPOUNDS

OBJECTIVE

▶ To write systematic names and formulas for binary molecular compounds.

Recall that a binary molecular compound is composed of two nonmetal elements. The simplest representative particle in a binary molecular compound is a **molecule**. In general, the more nonmetallic element is written second in the chemical formula. IUPAC prescribes the following order for writing the elements in a compound: C, P, N, H, S, I, Br, Cl, O, F. Notice that hydrogen is in the middle of the series. Thus, the binary compounds of hydrogen are written as follows: CH_4, PH_3, NH_3, H_2S, HI, HBr, HCl, H_2O, and HF.

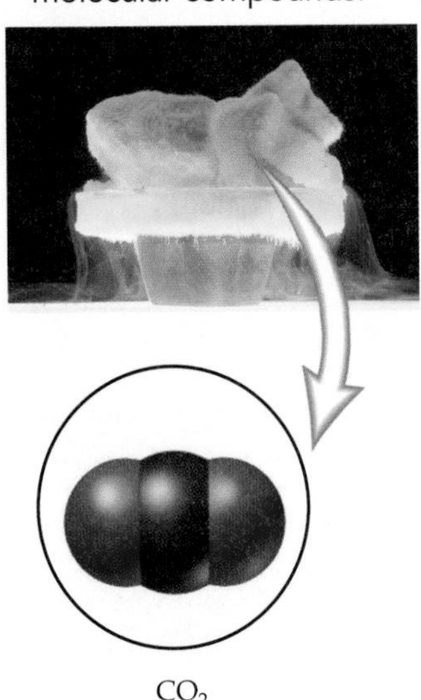

▲ **Carbon Dioxide, CO_2**
Dry ice is solid CO_2.

Naming Binary Molecular Compounds

In the naming of binary molecular compounds, IUPAC specifies that the first element in the compound be named followed by the second element bearing the suffix *-ide*. Thus, binary ionic and binary molecular compounds both end in the suffix *-ide*. It also specifies that the number of atoms of each element be indicated by Greek prefixes. The Greek prefixes in Table 7.4 indicate the number of atoms of each element that correspond to the subscripts in the chemical formula.

According to the official IUPAC recommendations, the prefix *mono-* is always omitted unless its presence is necessary to avoid confusion. The only common exceptions are CO and NO. The name of CO includes the prefix *mono-* in front of oxygen and is written carbon monoxide. Similarly, NO is named nitrogen monoxide.

TABLE 7.4 GREEK PREFIXES FOR BINARY MOLECULAR COMPOUNDS

ATOMS	PREFIX	ATOMS	PREFIX
1	*mono-*	6	*hexa-*
2	*di-*	7	*hepta-*
3	*tri-*	8	*octa-*
4	*tetra-*	9	*nona-**
5	*penta-*	10	*deca-*

*Although the Latin prefix *nona-* is commonly used, IUPAC prefers the Greek prefix *ennea-*.

Let's consider the binary molecular compound P_4S_3. This compound is found on match tips and ignites in air when struck on a rough surface (Figure 7.4). Since the ratio is four phosphorus atoms to three sulfur atoms, the Greek prefixes are *tetra-* and *tri-*, respectively. Thus, the name of P_4S_3 is tetraphosphorus trisulfide. A different compound, P_4S_7, has the same elements but is named tetraphosphorus heptasulfide.

Example Exercise 7.16 Names of Binary Molecular Compounds

Give the IUPAC systematic name for each of the following binary molecular compounds:

(a) IF_6 (b) Br_3O_8

Solution

We name binary molecular compounds by attaching the suffix *-ide* to the second nonmetal and indicate the atomic ratios by Greek prefixes.

(a) IF_6 is first named iodine fluoride. After supplying the Greek prefixes for the atomic ratios, we have *iodine hexafluoride*.

(b) Br_3O_8 is first named bromine oxide. After supplying the Greek prefixes for the atomic ratios, we have *tribromine octaoxide*.

Practice Exercise

Give the IUPAC systematic name for each of the following binary molecular compounds:

(a) Cl_2O_5 (b) P_4S_{10}

Answers: (a) dichlorine pentaoxide; (b) tetraphosphorus decasulfide

Concept Exercise

Which of the following is named using Greek prefixes to specify the number of atoms of each element: Fe_2O_3 or P_2O_3?

Answer: See Appendix G.

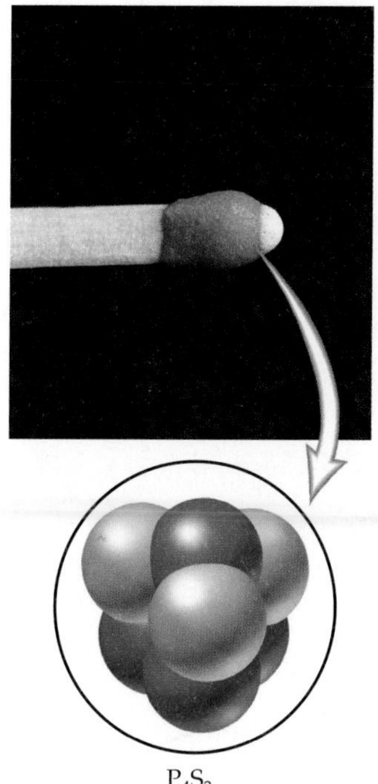

▲ **Figure 7.4 Chemistry of Matches** The substances P_4S_3 and $KClO_3$ are present on the tip of a match. When the match is struck on a rough surface, the two chemicals ignite and produce a flame. The products from the reaction are P_2O_5, KCl, and SO_2, the last of which is responsible for the characteristic sulfur smell.

Example Exercise 7.17 Formulas of Binary Molecular Compounds

Provide the formula for each of the following binary molecular compounds:

(a) diphosphorus pentasulfide (b) tetraiodine nonaoxide

Solution

To write the formula, we give the symbol for each element followed by a subscript indicating the number of atoms.

(a) Diphosphorus pentasulfide is composed of two phosphorus atoms and five sulfur atoms. The formula of the compound is written P_2S_5.

(b) Tetraiodine nonaoxide is composed of four iodine atoms and nine oxygen atoms. The formula of the compound is written I_4O_9.

Practice Exercise

Provide the formula for each of the following binary molecular compounds:

(a) diphosphorus tetraiodide
(b) sulfur hexafluoride

Answers: (a) P_2I_4; (b) SF_6

Concept Exercise

Which of the following is a binary molecular compound: magnesium oxide or nitrogen oxide?

Answer: See Appendix G.

Note For ease of pronunciation, double vowels are sometimes avoided in naming compounds with Greek prefixes. For example, if the Greek prefix ends in "a" or "o" and the nonmetal is oxygen, two pronunciations are common. That is, the name may be pronounced tetraoxide or tetroxide, pentaoxide or pentoxide, etc. However, according to the official IUPAC recommendations, vowels are not to be dropped. The one exception is monoxide, not monooxide.

7.8 BINARY ACIDS

OBJECTIVE

▸ To write systematic names and formulas for binary acids.

A binary acid is an aqueous solution of a compound containing hydrogen and a nonmetal. The chemical formulas of acids always begin with H, for example, HCl(*aq*). A binary acid is produced by dissolving a binary molecular compound, such as HCl, in water. The resulting aqueous solution, HCl(*aq*), is a binary acid.

Binary acids are systematically named by using the prefix *hydro-* before the nonmetal stem and adding the suffix *-ic acid*. As an example, consider the acid used to treat swimming pools and sold in supermarkets under the common name muriatic acid. Muriatic acid is aqueous hydrogen chloride, that is, HCl(*aq*). The IUPAC name for muriatic acid is formed as follows: hydro + chlor + ic acid. The systematic name for HCl(*aq*) is hydrochloric acid.

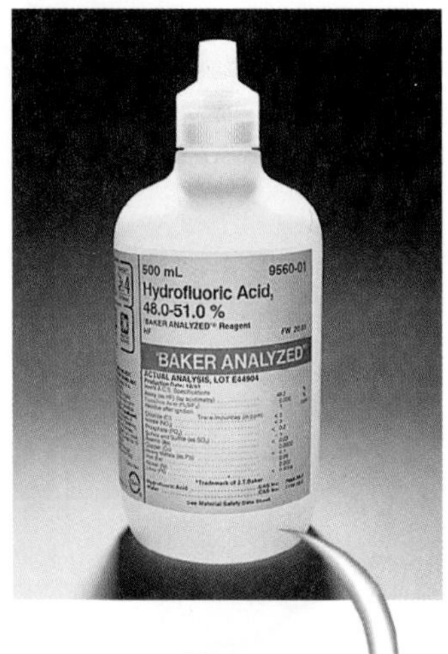

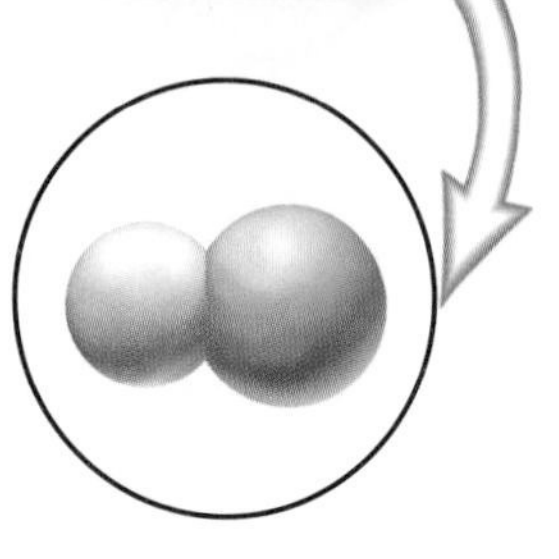

▲ **Hydrofluoric Acid, HF**
Aqueous hydrofluoric acid, HF, is used to etch silicon oxide during the manufacture of computer chips.

Example Exercise 7.18 Names of Binary Acids

Give the IUPAC systematic name for HF(*aq*), a binary acid.

Solution

Binary acids are named as *hydro-* plus nonmetal stem plus *-ic acid*. Since HF(*aq*) contains the nonmetal fluorine, we construct the systematic name as follows: hydro + fluor + ic acid gives *hydrofluoric acid*.

Practice Exercise

Give the IUPAC systematic name for H_2S(*aq*).

Answer: hydrosulfuric acid

Concept Exercise

Which of the following acids is named using a *hydro-* prefix: HBr(*aq*), $HBrO_2$(*aq*), $HBrO_3$(*aq*)?

Answer: See Appendix G.

Note Be careful not to confuse the names of a binary acid and a binary molecular compound. A binary acid is a compound of hydrogen and a nonmetal dissolved in water. For example, HCl(*aq*) is a binary acid and is named hydrochloric acid. On the other hand, HCl(*g*) is a gaseous binary molecular compound and is named hydrogen chloride.

7.9 TERNARY OXYACIDS

OBJECTIVE

- To write systematic names and formulas for ternary oxyacids.

A ternary oxyacid is an aqueous solution of a compound containing hydrogen and an oxyanion. Most ternary oxyacids are named by attaching the suffix *-ic acid* to the *nonmetal stem* of the oxyanion. For example, the name of $HNO_3(aq)$ is systematically formed as follows: *nonmetal stem* plus *ic acid*. The name of $HNO_3(aq)$ is nitr + ic acid, or *nitric acid.*

Some ternary oxyacids are named by attaching the suffix *-ous acid* to the *nonmetal stem* of the oxyanion. For example, the name of $HNO_2(aq)$ is systematically formed as follows: *nonmetal stem + ous acid*. The name of $HNO_2(aq)$ is nitr + ous acid, or *nitrous acid*.

CHEMISTRY CONNECTION IUPAC Nomenclature

Q: What is the official systematic name for stomach acid?

Alchemists, the forerunners of modern chemists, are usually associated with attempts to convert lead to gold. They did, however, produce a rich vocabulary describing chemical substances. Unfortunately, their names for substances did not offer a clue to the chemical composition.

◀ The Official IUPAC "Red Book"

In the eighteenth century, Antoine Lavoisier, the founder of modern chemistry, provided the first published naming system that indicated the chemical composition of substances. In the nineteenth century, the great Swedish chemist J. J. Berzelius adopted Lavoisier's ideas and extended the system into the Germanic languages, which in turn were translated into English.

In 1892, a conference held in Geneva, Switzerland, laid the foundation for an internationally accepted system of nomenclature for organic compounds. Subsequently, the International Union of Pure and Applied Chemistry (IUPAC) was formed.

In 1921, IUPAC addressed the issue of systematic nomenclature for *inorganic compounds*. Almost 20 years elapsed before IUPAC issued a systematic set of rules. Interestingly, these so-called 1940 IUPAC rules were consistent with the methods proposed in Lavoisier's text 150 years earlier. The official recommendations suggest that there is more than one acceptable system of nomenclature. For example, $CuSO_4$ can be named copper(II) sulfate according to the preferred Stock system of nomenclature or cupric sulfate according to the Latin system.

Although common names persist, IUPAC disapproves the use of common names and discourages the use of names that do not indicate the chemical composition. Examples of common names include "baking soda" or "sodium bicarbonate" for sodium hydrogen carbonate, $NaHCO_3$. Hydrochloric acid, $HCl(aq)$, is commonly referred to as "stomach acid" or "swimming pool acid," and is sold in the supermarket under the common name "muriatic acid."

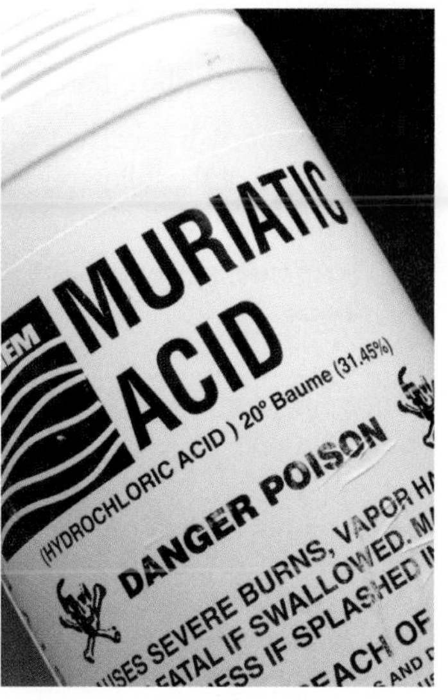

◀ Muriatic Acid

In truth, the chemistry community is not in complete agreement on subtle aspects of nomenclature. The most authoritative guide to systematic naming is the IUPAC recommendations contained in *Nomenclature of Inorganic Chemistry*. This official publication is referred to as the "Red Book."

A: Stomach acid is aqueous HCl, and its systematic name is hydrochloric acid.

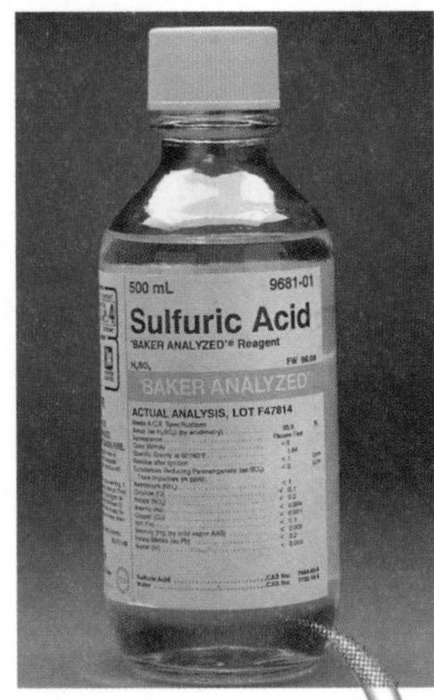

H_2SO_4

▲ **Sulfuric Acid, H_2SO_4** Aqueous sulfuric acid, $H_2SO_4(aq)$, is found in automobile batteries; hence, it is commonly termed "battery acid."

A ternary oxyacid with the suffix *-ic acid* contains an oxyanion ending in *-ate*. Thus, the name of a compound containing the oxyanion will bear the suffix *-ate*. Consider sulfuric acid, $H_2SO_4(aq)$, which contains the sulfate oxyanion, SO_4^{2-}. The name of the compound Na_2SO_4 is sodium sulfate.

Similarly, a ternary oxyacid with the suffix *-ous acid* contains an oxyanion ending in *-ite*. Consider sulfurous acid, $H_2SO_3(aq)$, which contains the sulfite oxyanion, SO_3^{2-}. The name of the compound Na_2SO_3 is sodium sulfite. We can illustrate the ternary oxyacid relationship to its compounds as shown in Table 7.5.

TABLE 7.5 TERNARY OXYACIDS AND RELATED COMPOUNDS

TERNARY OXYACID		TERNARY COMPOUND	
$HClO$	**hypo**chlor**ous** acid	$NaClO$	sodium **hypo**chlor**ite**
$HClO_2$	chlor**ous** acid	$NaClO_2$	sodium chlor**ite**
$HClO_3$	chlor**ic** acid	$NaClO_3$	sodium chlor**ate**
$HClO_4$	**per**chlor**ic** acid	$NaClO_4$	sodium **per**chlor**ate**

Now that we have provided an overview of ternary oxyacids, let's try the following example exercise that illustrates systematic naming:

Example Exercise 7.19 Names of Ternary Oxyacids

Give the IUPAC systematic name for $H_3PO_4(aq)$, a ternary oxyacid.

Solution

Ternary oxyacids are named as *-ic acids* or *-ous acids*. Since $H_3PO_4(aq)$ contains the phosphate oxyanion, it is an *-ic acid*. We construct the systematic name as follows: phosphor + ic acid gives *phosphoric acid*.

Practice Exercise

Give the IUPAC systematic name for $H_3PO_3(aq)$, a ternary oxyacid.

Answer: phosphorous acid

Concept Exercise

Which of the following acids is named *nonmetal stem* plus *-ic acid:* $HBr(aq)$, $HBrO_2(aq)$, $HBrO_3(aq)$?

Answer: See Appendix G.

Summary

Section 7.1 **Inorganic compounds** do not contain carbon and are named according to a systematic set of rules referred to as **IUPAC nomenclature**. The IUPAC rules classify substances as belonging to one of five categories: **binary ionic**, **ternary ionic**, **binary molecular**, **binary acid**, or **ternary oxyacid**. A substance dissolved in water is referred to as an **aqueous solution**. Ionic compounds are composed of positive **cations** and negative **anions**. A **monoatomic ion** is a single atom having a charge; whereas a **polyatomic ion** is a group of two or more atoms bearing a charge.

Section 7.2 The names of monoatomic cations are derived from a metal, for example, the sodium ion, Na^+. Cations having two possible ionic charges require further identification. The ionic charge may be indicated using either the Stock system or the Latin system. The **Stock system** indicates the charge on the metal with Roman numerals in parentheses. The **Latin system** adds the suffix *-ous* or *-ic* to the Latin name of the element. Thus, Cu^{2+} is named copper(II) ion according to the Stock system, and cupric ion using the Latin system.

The names of monoatomic anions are derived from a nonmetal. A monoatomic anion is named using the nonmetal stem and adding the suffix *-ide*. Examples of nonmetal ions include the chloride ion, Cl^-, and the oxide ion, O^{2-}.

Section 7.3 A polyatomic anion that contains one or more elements combined with oxygen is called an **oxyanion**. The names of most polyatomic oxyanions have the suffix *-ate*, for example, the nitrate ion, NO_3^-. A few oxyanions have the suffix *-ite*, for example, the nitrite ion, NO_2^-. The nitrite ion has one less oxygen atom than the nitrate ion. Oxyanions having one less oxygen atom end with the suffix *-ite*, such as the sulfite ion, SO_3^{2-}, with one less oxygen than the sulfate ion, SO_4^{2-}.

Section 7.4 The simplest representative particle in an ionic compound is called a **formula unit**. The net charge for a formula unit is zero. That is, the total positive charge for the cations must be the same as the total negative charge for the anions.

Section 7.5 Binary ionic compounds have names ending with the suffix *-ide*. For example, NaCl is named sodium chloride, $CaBr_2$ is named calcium bromide, and AlF_3 is named aluminum fluoride.

Section 7.6 Ternary ionic compounds have names ending with the suffix *-ate* or *-ite*. For example, $CaSO_4$ is named calcium sulfate, and $CaSO_3$ is named calcium sulfite. A ternary compound containing CN^- or OH^- is an exception and each ends with the suffix *-ide*. For example, NaOH is named sodium hydroxide, and KCN is named potassium cyanide.

Section 7.7 Binary molecular compounds contain two nonmetals, and the simplest representative particle is called a **molecule**. Binary molecular compounds are named using a Greek prefix and the suffix *-ide;* for example, SO_2 and SO_3 are named sulfur dioxide and sulfur trioxide, respectively.

Section 7.8 A binary acid is an aqueous solution of a hydrogen containing compound. Binary acids are named *hydro + nonmetal stem + -ic acid*. For example, hydrogen chloride dissolved in water, HCl(*aq*), is named hydrochloric acid.

Section 7.9 Ternary oxyacids are usually named *nonmetal stem + -ic acid*. For example, an aqueous solution of $HNO_3(aq)$ is named nitric acid. A few ternary oxyacids are named *nonmetal stem + -ous acid*. An aqueous solution of $HNO_2(aq)$ is named nitrous acid.

NOMENCLATURE ORGANIZER

TOPIC	PROCEDURE	EXAMPLE
Classification of Compounds Sec. 7.1	When a formula has two elements, a metal and a nonmetal, it is a **binary ionic compound.** When a formula has three elements and at least one metal, it is a **ternary ionic compound.** When a formula has two nonmetal elements, it is a **binary molecular compound.** When a formula has H and a nonmetal in an aqueous solution, it is a **binary acid.** When a formula has H, a nonmetal, and oxygen in an aqueous solution, it is a **ternary oxyacid.**	binary ionic compound, KCl ternary ionic compound, $KClO_3$ binary molecular compound, HCl binary acid, HCl(*aq*) ternary oxyacid, $HClO_3(aq)$
Monoatomic Ions Sec. 7.2	Representative metal ions are named for the element. Transition metal ions are named using (1) the **Stock system**, or (2) the **Latin system.** Nonmetals are named for the element plus the **suffix *-ide*.**	K^+, potassium ion Fe^{2+}, iron(II) or ferrous ion Cl^-, chloride ion
Polyatomic Ions Sec. 7.3	Polyatomic oxyanions are named with (1) the **suffix *-ate*,** or (2) the **suffix *-ite*** when there is one less oxygen atom in the oxyanion. There are two common polyatomic anion exceptions that have the **suffix *-ide*.**	ClO_3^-, chlorate ion ClO_2^-, chlorite ion OH^-, hydroxide ion CN^-, cyanide ion
Writing Chemical Formulas Sec. 7.4	In writing formulas of ionic compounds, the ratio of cations and anions must provide a total positive charge equal to the total negative charge so that the compound is electrically neutral.	Fe^{2+} and O^{2-} = FeO Fe^{3+} and O^{2-} = Fe_2O_3 Fe^{2+} and ClO_3^- = $Fe(ClO_3)_2$ Fe^{3+} and ClO_3^- = $Fe(ClO_3)_3$
Binary Ionic Compounds Sec. 7.5	Binary ionic compounds end in the **suffix *-ide*.** The metal cation is named first, followed by the name of the anion.	KCl, potassium chloride $FeCl_2$, iron(II) chloride, or ferrous chloride

NOMENCLATURE ORGANIZER *(Continued)*

TOPIC	PROCEDURE	EXAMPLE
Ternary Ionic Compounds Sec. 7.6	Ternary ionic compounds end in the **suffix *-ate*** or the **suffix *-ite***. The metal cation is named first followed by the name of the oxyanion.	$KClO_3$, potassium chlorate $KClO_2$, potassium chlorite
Binary Molecular Compounds Sec. 7.7	Binary molecular compounds end in the **suffix *-ide***. The number of nonmetal atoms is indicated by a Greek prefix.	CO, carbon monoxide CO_2, carbon dioxide
Binary Acids Sec. 7.8	Binary acids are named ***hydro-*** plus ***nonmetal stem*** plus ***-ic acid***.	HCl(*aq*), hydrochloric acid
Ternary Oxyacids Sec. 7.9	Ternary oxyacids are named with (1) a **nonmetal stem** plus ***-ic acid***, or (2) a **nonmetal stem** plus ***-ous acid***.	$HClO_3(aq)$, chloric acid $HClO_2(aq)$, chlorous acid

Key Terms *Answers to Key Terms are in Appendix H.*

Select the key term that corresponds to each of the following definitions.

_____ **1.** the international system of rules for naming chemical compounds
_____ **2.** a compound that does not contain the element carbon
_____ **3.** a compound that contains one metal and one nonmetal
_____ **4.** a compound that contains three elements, including at least one metal
_____ **5.** a compound that contains two nonmetals
_____ **6.** a substance dissolved in water
_____ **7.** an aqueous compound that contains hydrogen and a nonmetal
_____ **8.** an aqueous compound that contains hydrogen, a nonmetal, and oxygen
_____ **9.** any positively charged ion
_____ **10.** any negatively charged ion
_____ **11.** a single atom that has a positive or negative charge
_____ **12.** a group of atoms bound together and having an overall positive or negative charge
_____ **13.** a system that designates the charge on a cation with Roman numerals
_____ **14.** a system that designates the charge on a cation with the suffix *-ic* or *-ous*
_____ **15.** a polyatomic anion that contains one or more elements combined with oxygen, for example, CO_3^{2-}
_____ **16.** the simplest representative particle in a compound composed of ions
_____ **17.** the simplest representative particle in a compound composed of nonmetal atoms

(a) anion *(Sec. 7.1)*
(b) aqueous solution *(Sec. 7.1)*
(c) binary acid *(Sec. 7.1)*
(d) binary ionic compound *(Sec. 7.1)*
(e) binary molecular compound *(Sec. 7.1)*
(f) cation *(Sec. 7.1)*
(g) formula unit *(Sec. 7.4)*
(h) inorganic compound *(Sec. 7.1)*
(i) IUPAC nomenclature *(Sec. 7.1)*
(j) Latin system *(Sec. 7.2)*
(k) molecule *(Sec. 7.7)*
(l) monoatomic ion *(Sec. 7.1)*
(m) oxyanion *(Sec. 7.3)*
(n) polyatomic ion *(Sec. 7.1)*
(o) Stock system *(Sec. 7.2)*
(p) ternary ionic compound *(Sec. 7.1)*
(q) ternary oxyacid *(Sec. 7.1)*

Exercises *Answers to odd-numbered Exercises are in Appendix I.*

Classification of Compounds (Sec. 7.1)

1. Classify each of the following as a binary ionic compound, ternary ionic compound, binary molecular compound, binary acid, or ternary oxyacid:
(a) ammonia, NH_3 **(b)** lithium nitrate, $LiNO_3$
(c) iron(III) nitride, FeN **(d)** nitric acid, $HNO_3(aq)$

2. Classify each of the following as a binary ionic compound, ternary ionic compound, binary molecular compound, binary acid, or ternary oxyacid:
(a) hydrazine, N_2H_4 **(b)** sodium nitrite, $NaNO_2$
(c) cobalt(II) nitride, Co_3N_2 **(d)** nitrous acid, $HNO_2(aq)$

3. Classify each of the following as a monoatomic cation, monoatomic anion, polyatomic cation, or polyatomic anion:
(a) ammonium ion, NH_4^+ (b) nickel(II) ion, Ni^{2+}
(c) bromide ion, Br^- (d) carbonate ion, CO_3^{2-}

4. Classify each of the following as a monoatomic cation, monoatomic anion, polyatomic cation, or polyatomic anion:
(a) hydronium ion, H_3O^+ (b) cobalt(III) ion, Co^{3+}
(c) phosphide ion, P^{3-} (d) acetate ion, $C_2H_3O_2^-$

Monoatomic Ions (Sec. 7.2)

5. Supply the Stock system name for each of the following monoatomic cations:
(a) K^+ (b) Ba^{2+}
(c) Fe^{2+} (d) Co^{3+}

6. Supply the Stock system name for each of the following monoatomic cations:
(a) Ag^+ (b) Cd^{2+}
(c) Hg^{2+} (d) Cu^{2+}

7. Provide the formula for each of the following monoatomic cations:
(a) lithium ion (b) aluminum ion
(c) lead(II) ion (d) nickel(II) ion

8. Provide the formula for each of the following monoatomic cations:
(a) zinc ion (b) strontium ion
(c) tin(IV) ion (d) manganese(II) ion

9. Supply the Latin system name for each of the following monoatomic cations:
(a) Cu^+ (b) Fe^{3+}
(c) Sn^{4+} (d) Pb^{2+}

10. Supply the Latin system name for each of the following monoatomic cations:
(a) Cu^{2+} (b) Fe^{2+}
(c) Sn^{2+} (d) Pb^{4+}

11. Provide the formula for each of the following monoatomic cations:
(a) cobaltous ion (b) cobaltic ion

12. Provide the formula for each of the following monoatomic cations:
(a) mercurous ion (b) mercuric ion

13. Supply a systematic name for each of the following monoatomic anions:
(a) F^- (b) Cl^-
(c) Br^- (d) I^-

14. Supply a systematic name for each of the following monoatomic anions:
(a) O^{2-} (b) S^{2-}
(c) N^{3-} (d) P^{3-}

Polyatomic Ions (Sec. 7.3)

15. Supply a systematic name for each of the following oxyanions:
(a) ClO_3^- (b) NO_2^-
(c) CO_3^{2-} (d) HSO_4^-

16. Supply a systematic name for each of the following oxyanions:
(a) ClO_2^- (b) NO_3^-
(c) $C_2H_3O_2^-$ (d) HCO_3^-

17. Provide the formula for each of the following polyatomic ions:
(a) dichromate ion (b) phosphate ion
(c) hydroxide ion (d) hypochlorite ion

◀ **Sodium Hydroxide, NaOH** Sodium hydroxide, NaOH, is used in oven cleaner and drain cleaner. Since it is strongly caustic and contains sodium, it is commonly referred to as "caustic soda."

18. Provide the formula for each of the following polyatomic ions:
(a) chromate ion (b) permanganate ion
(c) cyanide ion (d) perchlorate ion

Writing Chemical Formulas (Sec. 7.4)

19. Write the chemical formula for the following binary compounds given their constituent ions:
(a) lithium chloride, Li^+ and Cl^-
(b) silver oxide, Ag^+ and O^{2-}
(c) chromium(III) iodide, Cr^{3+} and I^-
(d) tin(II) nitride, Sn^{2+} and N^{3-}

20. Write the chemical formula for the following binary compounds given their constituent ions:
(a) aluminum bromide, Al^{3+} and Br^-
(b) cadmium sulfide, Cd^{2+} and S^{2-}
(c) magnesium phosphide, Mg^{2+} and P^{3-}
(d) lead(IV) fluoride, Pb^{4+} and F^-

21. Write the chemical formula for the following ternary compounds given their constituent ions:
(a) potassium nitrate, K^+ and NO_3^-
(b) ammonium dichromate, NH_4^+ and $Cr_2O_7^{2-}$
(c) aluminum sulfite, Al^{3+} and SO_3^{2-}
(d) chromium(III) hypochlorite, Cr^{3+} and ClO^-

22. Write the chemical formula for the following ternary compounds given their constituent ions:
(a) iron(III) hydrogen carbonate, Fe^{3+} and HCO_3^-
(b) cuprous carbonate, Cu^+ and CO_3^{2-}
(c) mercury(II) cyanide, Hg^{2+} and CN^-
(d) stannic acetate, Sn^{4+} and $C_2H_3O_2^-$

23. Write the chemical formula for the following ternary compounds given their constituent ions:
(a) strontium nitrite, Sr^{2+} and NO_2^-
(b) zinc permanganate, Zn^{2+} and MnO_4^-
(c) calcium chromate, Ca^{2+} and CrO_4^{2-}
(d) chromium(III) perchlorate, Cr^{3+} and ClO_4^-

24. Write the chemical formula for the following ternary compounds given their constituent ions:
(a) lead(IV) sulfate, Pb^{4+} and $SO_4{}^{2-}$
(b) stannous chlorite, Sn^{2+} and $ClO_2{}^{-}$
(c) cobalt(II) hydroxide, Co^{2+} and OH^{-}
(d) mercurous phosphate, $Hg_2{}^{2+}$ and $PO_4{}^{3-}$

Binary Ionic Compounds (Sec. 7.5)

25. Supply a systematic name for each of the following binary ionic compounds:
(a) MgO (b) ZnO
(c) CdO (d) BaO

26. Supply a Stock system name for each of the following binary ionic compounds:
(a) CuO (b) FeO
(c) HgO (d) SnO

27. Supply a systematic name for each of the following binary ionic compounds:
(a) LiBr (b) SrI_2
(c) Na_3N (d) AlF_3

28. Supply a Stock system name for each of the following binary ionic compounds:
(a) $MnCl_2$ (b) NiF_2
(c) $SnBr_2$ (d) CrP

29. Provide the formula for each of the following binary ionic compounds:
(a) copper(I) oxide (b) iron(III) nitride
(c) mercury(II) chloride (d) lead(IV) sulfide

◀ **Cuprite** Cuprite, copper(I) oxide, is a mineral found in nature that contains copper.

30. Provide the formula for each of the following binary ionic compounds:
(a) cuprous sulfide (b) ferrous phosphide
(c) mercuric iodide (d) plumbic oxide

31. Predict the chemical formula for each of the following binary ionic compounds given the formula of sodium chloride, NaCl:
(a) rubidium chloride (b) sodium bromide

32. Predict the chemical formula for each of the following binary ionic compounds given the formula of calcium oxide, CaO:
(a) beryllium oxide (b) calcium selenide

33. Predict the chemical formula for each of the following binary ionic compounds given the formula of aluminum nitride, AlN:
(a) gallium nitride (b) aluminum arsenide

34. Predict the chemical formula for each of the following binary ionic compounds given the formula of titanium oxide, TiO_2:
(a) zirconium oxide (b) titanium sulfide

Ternary Ionic Compounds (Sec. 7.6)

35. Supply a systematic name for each of the following ternary ionic compounds:
(a) $LiMnO_4$ (b) $Sr(ClO_2)_2$
(c) $BaCO_3$ (d) $(NH_4)_2Cr_2O_7$

36. Supply a systematic name for each of the following ternary ionic compounds:
(a) $KClO_3$ (b) $Mg(NO_3)_2$
(c) Ag_2SO_4 (d) $Al_2(CrO_4)_3$

37. Supply a Stock system name for each of the following ternary ionic compounds:
(a) $CuSO_4$ (b) $FeCrO_4$
(c) $Hg(NO_2)_2$ (d) $Pb(C_2H_3O_2)_2$

38. Supply a Latin system name for each of the following ternary ionic compounds:
(a) Cu_2SO_4 (b) $Fe_2(CrO_4)_3$
(c) $Hg_2(NO_2)_2$ (d) $Pb(C_2H_3O_2)_4$

39. Provide the formula for each of the following ternary ionic compounds:
(a) manganese(II) acetate (b) copper(II) chlorite
(c) tin(II) phosphate (d) iron(III) hypochlorite

40. Provide the formula for each of the following ternary ionic compounds:
(a) cuprous chlorite (b) plumbic sulfite
(c) mercuric chlorate (d) ferrous chromate

41. Predict the chemical formula for each of the following ternary ionic compounds given the formula of sodium sulfate, Na_2SO_4:
(a) francium sulfate (b) sodium sulfite

42. Predict the chemical formula for each of the following ternary ionic compounds given the formula of scandium nitrate, $Sc(NO_3)_3$:
(a) lanthanum nitrate (b) scandium nitrite

43. Predict the chemical formula for each of the following ternary ionic compounds given the formula of barium chlorate, $Ba(ClO_3)_2$:
(a) radium chlorate (b) barium bromate

44. Predict the chemical formula for each of the following ternary ionic compounds given the formula of iron(III) sulfate, $Fe_2(SO_4)_3$:
(a) cobalt(III) sulfate (b) iron(III) selenate

Binary Molecular Compounds (Sec. 7.7)

45. Give a systematic name for each of the following binary molecular compounds:
(a) BrF (b) CF_4
(c) I_2O_4 (d) Cl_2O_3

46. Give a systematic name for each of the following binary molecular compounds:
(a) BrCl (b) SF_6
(c) I_4O_9 (d) Cl_2O_7

47. Provide the formula for each of the following binary molecular compounds:
(a) dinitrogen pentaoxide (b) carbon tetrachloride
(c) iodine monobromide (d) dihydrogen sulfide

48. Provide the formula for each of the following binary molecular compounds:
(a) chlorine dioxide (b) sulfur tetrafluoride
(c) iodine monochloride (d) nitrogen monoxide

Binary Acids (Sec. 7.8)

49. Give the IUPAC systematic name for each of the following binary acids:
(a) $HBr(aq)$ (b) $HI(aq)$

50. Provide the formula for each of the following binary acids:
(a) hydrosulfuric acid (b) hydroselenic acid

Ternary Oxyacids (Sec. 7.9)

51. Give the IUPAC systematic name for each of the following ternary oxyacids:
(a) $HClO_2(aq)$ (b) $H_3PO_4(aq)$

52. Give the IUPAC systematic name for each of the following ternary oxyacids:
(a) $HClO_4(aq)$ (b) $H_2SO_3(aq)$

53. Provide the formula for each of the following ternary oxyacids:
(a) acetic acid (b) nitric acid

54. Provide the formula for each of the following ternary oxyacids:
(a) carbonic acid (b) nitrous acid

55. Predict the chemical formula for each of the following ternary oxyacids given the formula of hypochlorous acid, $HClO(aq)$:
(a) hypoiodous acid (b) hypobromous acid

56. Predict the chemical formula for each of the following ternary oxyacids given the formula of phosphoric acid, $H_3PO_4(aq)$:
(a) arsenic acid (b) phosphorous acid

General Exercises

57. State the ionic charge for each of the following substances:
(a) iron metal atoms
(b) ferrous ions
(c) iron(III) ions
(d) iron compounds

58. State the ionic charge for each of the following substances:
(a) chlorine gas molecules
(b) chloride ions
(c) hypochlorite ion
(d) chlorine compounds

59. Predict which of the following polyatomic anions has an ionic charge of 2–. (*Hint:* The total number of valence electrons must be an even number.)
(a) periodate ion, $IO_4^{?-}$
(b) silicate ion, $SiO_3^{?-}$

60. Predict which of the following polyatomic anions has an ionic charge of 1–. (*Hint:* The total number of valence electrons must be an even number.)
(a) thiocyanate ion, $CNS^{?-}$
(b) thiosulfate ion, $S_2O_3^{?-}$

61. Complete the following table by combining cations and anions into chemical formulas. Give the Stock system name for each of the compounds.

IONS	fluoride ion	oxide ion	nitride ion
copper(II) ion	CuF_2 copper(II) fluoride		
cobalt(II) ion			
lead(II) ion			

62. Complete the following table by combining cations and anions into chemical formulas. Give the Stock system name for each of the compounds.

IONS	chloride ion	sulfide ion	phosphide ion
copper(I) ion	$CuCl$ copper(I) chloride		
cobalt(III) ion			
lead(IV) ion			

63. Complete the following table by combining cations and oxyanions into chemical formulas. Give the Latin name for each of the compounds.

IONS	hydroxide ion	sulfate ion	phosphate ion
mercurous ion	$Hg_2(OH)_2$ mercurous hydroxide		
ferrous ion			
stannous ion			

64. Complete the following table by combining cations and oxyanions into chemical formulas. Give the Latin system name for each of the compounds.

IONS	cyanide ion	sulfite ion	permanganate ion
mercuric ion	$Hg(CN)_2$ mercuric cyanide		
ferric ion			
stannic ion			

65. State the suffix in the name for each of the following:
(a) Na_2S (b) $H_2S(aq)$

66. State the suffix in the name for each of the following:
(a) NaI (b) $HI(aq)$

67. State the suffix in the name for each of the following:
(a) Na_2SO_3 (b) $H_2SO_3(aq)$

68. State the suffix in the name for each of the following:
(a) $NaIO_2$ (b) $HIO_2(aq)$

69. State the suffix in the name for each of the following:
(a) Na_2SO_4 (b) $H_2SO_4(aq)$

70. State the suffix in the name for each of the following:
(a) $NaIO_3$ (b) $HIO_3(aq)$

71. Write the chemical formula for each of the following household chemicals:
(a) dihydrogen oxide (common liquid)
(b) sodium hypochlorite (bleach)
(c) sodium hydroxide (caustic soda)
(d) sodium bicarbonate (baking soda)

72. Write the chemical formula for each of the following household chemicals:
(a) acetic acid (vinegar solution)
(b) aqueous nitrogen trihydride (ammonia solution)
(c) aqueous magnesium hydroxide (milk of magnesia)
(d) aqueous sodium hydrogen sulfate (porcelain cleaner)

73. Give the name for each of the following compounds containing a semimetal. (*Hint:* Name the formula as a binary molecular compound.)
(a) BF_3 (b) $SiCl_4$
(c) As_2O_5 (d) Sb_2O_3

74. Write the formula for each of the following compounds containing a semimetal:
(a) boron tribromide (b) trisilicon tetranitride
(c) diarsenic trioxide (d) diantimony pentaoxide

75. The compound titanium oxide occurs in the mineral rutile. Use the periodic table to predict the formula for titanium oxide, given the formula of zirconium oxide, ZrO_2.

◀ **Rutile** Rutile is a mineral found in nature that contains titanium.

76. The compound zirconium silicate occurs in cubic zirconia. Use the periodic table to predict the formula for zirconium silicate, given the formula of titanium silicate, $TiSiO_4$.

77. The element actinium is unstable and radioactive. Use the periodic table to predict the formula for actinium chloride, given the formula of lanthanum chloride, $LaCl_3$.

78. The element lawrencium is unstable and radioactive. Use the periodic table to predict the chemical formula for lawrencium chloride, given the formula of lutetium chloride, $LuCl_3$.

7 Chapter Self-Test

Answers to Self-Test are in Appendix J.

1. Gaseous HCl is classified as which of the following? (Sec. 7.1)
(a) binary ionic (b) ternary ionic
(c) binary molecular (d) binary acid
(e) ternary oxyacid

2. What is the Stock system name for Hg^{2+}? (Sec. 7.2)
(a) mercury(I) ion (b) mercury(II) ion
(c) mercuric ion (d) mercurous ion
(e) none of the above

3. What is the chemical formula for the hypochlorite ion? (Sec. 7.3)
(a) ClO^- (b) ClO_2^-
(c) ClO_3^- (d) ClO_4^-
(e) none of the above

4. What is the chemical formula for the ternary compound composed of Cr^{3+} and OH^- ions? (Sec. 7.4)
(a) CrOH (b) $CrOH_3$
(c) $Cr(OH)_3$ (d) Cr_3OH
(e) none of the above

5. What is the Latin system name for $CuCl_2$? (Sec. 7.5)
(a) cuprous chloride (b) cupric chloride
(c) copper(I) chloride (d) copper(II) chloride
(e) none of the above

6. What is the Stock system name for $CuSO_4$? (Sec. 7.6)
(a) cuprous sulfate (b) cupric sulfite
(c) copper(I) sulfate (d) copper(II) sulfate
(e) none of the above

7. What is the systematic name for gaseous H_2S? (Sec. 7.7)
(a) hydrogen sulfide (b) hydrosulfuric acid
(c) sulfurous acid (d) sulfuric acid
(e) none of the above

8. What is the systematic name for aqueous H_2S? (Sec. 7.8)
(a) hydrogen sulfide (b) hydrosulfuric acid
(c) sulfurous acid (d) sulfuric acid
(e) none of the above

9. What is the systematic name for aqueous H_2SO_4? (Sec. 7.9)
(a) hydrogen sulfide (b) hydrosulfuric acid
(c) sulfurous acid (d) sulfuric acid
(e) none of the above

Key Concepts

10. Classify each of the following as binary ionic, ternary ionic, binary molecular, binary acid, or ternary oxyacid: NaCl, HCl, HCl(*aq*), $NaClO_3$, $HClO_3$(*aq*).

11. Which of the following compounds is named using the suffix *-ide*: NaCl, $NaClO_2$, $NaClO_3$?

12. Which of the following compounds is named using the suffix *-ate*: NaCl, $NaClO_2$, $NaClO_3$?

13. Which of the following compounds is named using the suffix *-ite*: NaCl, $NaClO_2$, $NaClO_3$?

14. Which of the following acids is named *hydro + nonmetal stem + -ic acid*: HCl(*aq*), $HClO_3$(*aq*), $HClO_2$(*aq*)?

15. Which of the following acids is named *nonmetal stem + -ic acid*: HCl(*aq*), $HClO_3$(*aq*), $HClO_2$(*aq*)?

16. Which of the following acids is named *nonmetal stem + -ous acid*: HCl(*aq*), $HClO_3$(*aq*), $HClO_2$(*aq*)?

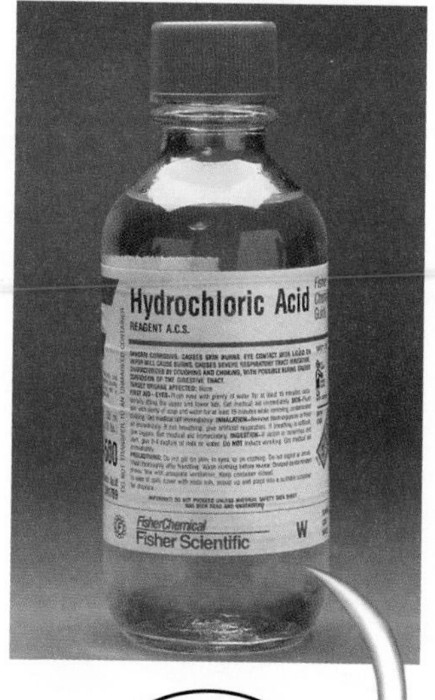

◀ **Hydrochloric Acid, HCl** Aqueous hydrochloric acid, HCl, is found in swimming pools and is produced in the stomach to aid digestion; hence, it is commonly termed "stomach acid."

Critical Thinking

17. Examine the periodic table and propose an explanation why copper ions can have a charge of either 1+ or 2+.

18. Examine the periodic table and propose an explanation why iron ions can have a charge of either 2+ or 3+.

CHAPTER

8 Chemical Reactions

After igniting magnesium metal in a fireworks display, the magnesium continues to burn in air and produces a white smoke of magnesium oxide, MgO. This reaction is exothermic, as evidenced by the release of light.

"Science is built of facts the way a house is built of bricks; but an accumulation of facts is no more science than a pile of bricks is a house."
Henri Poincare, French Physicist (1854–1912)

Every day substances are constantly being altered in some way. These alterations can be classified as a physical or a chemical change. In a *physical change,* the chemical composition of the substance remains constant. For instance, when we melt ice or boil water, we cause a physical change. The chemical composition of the ice and the water, H_2O, is the same before and after the change.

In a *chemical change,* the chemical composition of the substance does not remain constant. For example, when we drop an antacid tablet into a glass of water, we produce a chemical change. The chemical composition of the antacid tablet is not the same before and after the change. The evidence for the formation of a new substance is the gas bubbles that are released in the glass of water.

We may observe a *chemical reaction* when a substance undergoes a chemical change and forms a new substance. When we form a new substance, we usually observe one of the following: a change in color, an odor or bubbles indicating the liberation of a gas, or light or heat from the release of energy. For instance, we can mix flour, yeast, and water and put the mixture into an oven, and in a short time we can smell the odor of bread baking. A chemical reaction has occurred in which the flour, yeast, and water have combined to produce a new substance, evidenced by a change in color and smell.

In the laboratory, we may observe that two colorless solutions added together produce a solid substance that is not soluble. The formation of an insoluble solid in a solution is a practical observation that indicates a chemical reaction has occurred.

CHEMISTRY CONNECTION Fireworks

Q: What element is responsible for the red-orange color in the fireworks display?

Have you ever watched a Fourth of July fireworks display and wondered how the colors were produced? The color of fireworks is produced by the chemicals in a rocket shell, which is fitted with a fuse and fired into the air. When the fuse burns, gunpowder ignites and sets off an explosion that shoots chemicals throughout the sky. If the rocket shell is packed with a sodium compound, an orange-yellow color is observed. If the shell is packed with a barium compound, a yellow-green color is observed.

◀ **Fourth of July Fireworks**

However, not all chemicals produce a colored display. You may recall having seen fireworks that simply produce a shower of white sparks. Powdered aluminum or magnesium metals are used to produce this effect. The following table lists the chemicals used to produce various colors in fireworks:

CHEMICAL	FIREWORKS COLOR
Na compounds	orange-yellow
Ba compounds	yellow-green
Ca compounds	red-orange
Sr compounds	bright red
Li compounds	scarlet-red
Cu compounds	blue-green
Al or Mg metals	white sparks

Interestingly, the colors of fireworks and the colors of neon signs are based on the same principle. In a fireworks display an element is energized by heat, whereas in a "neon light" an element is energized by electricity. In either case, electrons are first excited and then immediately lose energy by emitting light. The observed color corresponds to the wavelength of the light emitted. For example, if the emitted light has a wavelength near 650 nm, it is observed as red. If the light has a wavelength near 450 nm, it appears blue.

Chemists can identify an element by the characteristic color of its flame test. In the laboratory, a flame test is performed by placing a small amount of chemical on the tip of a wire and holding the wire in a hot flame. For example, sodium gives an orange-yellow flame test, and barium a yellow-green flame test. Flame-test colors are identical to fireworks colors containing the same element. The red-orange flame-test color of calcium is illustrated.

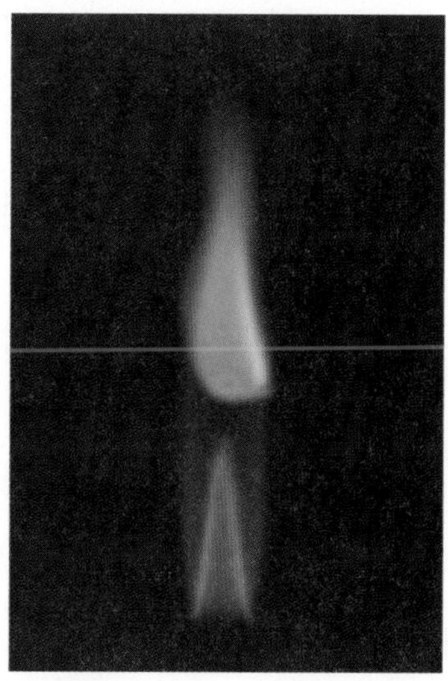

▲ **Calcium Flame Test** The flame test for calcium is red-orange.

A: The red-orange color suggests that the element is calcium.

8.1 EVIDENCE FOR CHEMICAL REACTIONS

OBJECTIVE

▶ To state four observations that are evidence for a chemical reaction.

If we combine substances, the resulting change can be physical or chemical. We can distinguish between a physical and a chemical change by observing the change. Each of the following indicates a chemical change; that is, a **chemical reaction** has taken place.

1. **A gas is produced.** We can observe a gas in a number of ways; for example, we may see bubbles in a solution. When we add an Alka-Seltzer tablet to water, it begins to fizz. The fizzing is evidence for a chemical reaction and the release of carbon dioxide gas, which is produced by the reaction of citric acid and baking soda in the tablet.
2. **An insoluble solid is produced in a solution.** A substance dissolves in water to give an **aqueous solution**. If we add two aqueous solutions together, we may observe solid particles in the solution. This insoluble solid is called a **precipitate**. The formation of a precipitate is evidence for a reaction.
3. **A permanent color change is observed.** Many chemical reactions involve a permanent change in color. For example, adding aqueous ammonia to a copper(II) sulfate solution changes the color from light blue to deep blue. To observe an acid reacting with a base, we use an indicator, a substance that changes color. The indicator enables us to indirectly follow a reaction that would otherwise not be visible.
4. **An energy change is observed.** In chemical reactions there is always a change in energy. A reaction that releases heat is said to be an **exothermic reaction**; conversely, a reaction that absorbs heat is said to be an **endothermic reaction**. For example, when an emergency flare is ignited, heat and light are observed. Heat and light are two forms of energy indicating that an exothermic chemical reaction has taken place.

In summary, each of the above four criteria—the production of a gas, the formation of a precipitate, a change in color, or a change in energy—is a signal that a chemical reaction has occurred. This is shown in Figure 8.1. A chemical reaction, however, may have occurred even though no change is obvious. In some reactions, for example, acid–base reactions, the energy change may be too subtle to notice and an external indicator may be necessary.

(a)

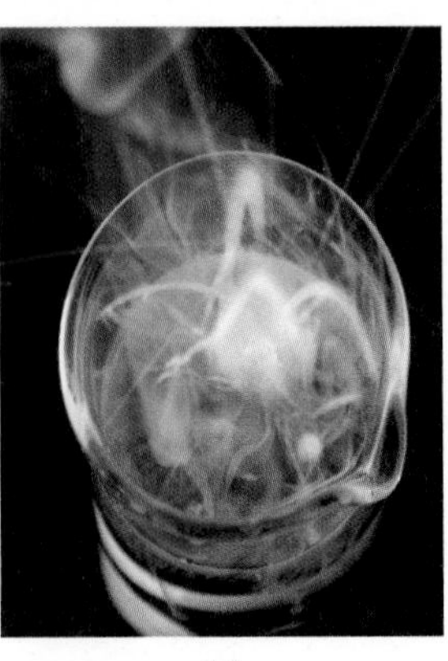

(b)

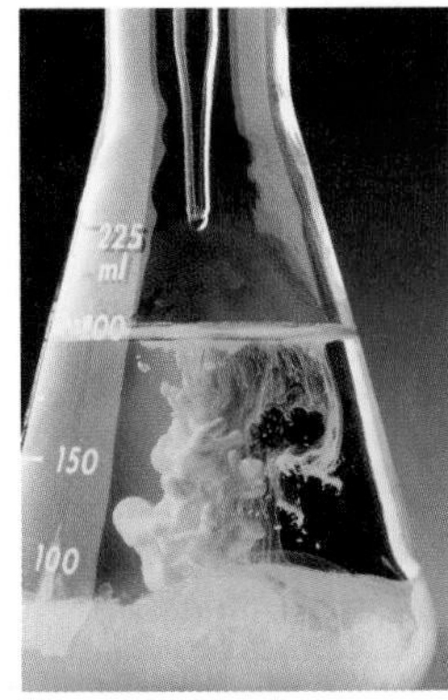

(c)

▶ **Figure 8.1 Evidence for a Chemical Reaction** Each of these images is evidence for a chemical reaction: (a) the gas bubbles released from magnesium metal in acid; (b) the heat and light released from potassium metal in water; and (c) the color change and the precipitate produced by two aqueous solutions.

Example Exercise 8.1 Evidence for a Reaction

Which of the following is experimental evidence for a chemical reaction?

(a) Pouring vinegar on baking soda gives foamy bubbles.
(b) Mixing two solutions produces insoluble particles.
(c) Mixing two colorless solutions gives a yellow solution.
(d) Mixing two solutions produces a temperature increase.

Solution

We can analyze each of these observations based on the criteria for a chemical reaction.

(a) The bubbles produced indicate a chemical reaction is occurring.
(b) The insoluble particles formed indicate a chemical reaction.
(c) The yellow color produced indicates a chemical reaction.
(d) The temperature increase indicates heat energy is being released and, thus, an exothermic chemical reaction.

Practice Exercise

What are four observations that a chemical reaction has occurred?

Answers: (a) a gas is released; (b) a precipitate is produced; (c) a permanent color change is observed; (d) an energy change is noted

Concept Exercise

What four observations are evidence for a chemical reaction?

Answer: See Appendix G.

8.2 WRITING CHEMICAL EQUATIONS

OBJECTIVES

- To identify seven elements that occur naturally as diatomic molecules: H_2, N_2, O_2, F_2, Cl_2, Br_2, I_2.
- To write a chemical equation from the description of a chemical reaction.

In this chapter we will use formulas and symbols to describe a chemical reaction; that is, we will write a **chemical equation**. Consider the following description of a chemical reaction: substance A and substance B undergo a chemical change that produces substances C and D. We can state this description in symbols as follows:

$$A + B \rightarrow C + D$$

In this general chemical equation, A and B are called the **reactants**, and C and D are called the **products**.

We can provide more information about the reaction if we specify the physical state of each substance, that is, solid, liquid, or gas. The physical state is indicated by the abbreviation (*s*), (*l*), or (*g*). By convention, we indicate an aqueous solution by the symbol (*aq*) and a precipitate by the symbol (*s*). For example, an aqueous solution of A reacts with a gaseous substance B to yield a precipitate C and an aqueous solution D. This statement can be written as follows:

$$A(aq) + B(g) \rightarrow C(s) + D(aq)$$

A **catalyst** is a substance that speeds up a reaction without being consumed or permanently altered. For example, the catalytic converter in an automobile contains a metallic catalyst that speeds up the conversion of unburned fuel to carbon dioxide and water. The presence of a catalyst is indicated by placing its formula above the arrow. Table 8.1 lists the symbols that are used to describe chemical reactions.

Let's interpret the symbols in the following chemical equation for the reaction of acetic acid and baking soda.

$$HC_2H_3O_2(aq) + NaHCO_3(s) \rightarrow NaC_2H_3O_2(aq) + H_2O(l) + CO_2(g)$$

The formulas and symbols are read as follows: aqueous acetic acid is added to solid sodium hydrogen carbonate, which produces aqueous sodium acetate, water, and carbon dioxide gas.

TABLE 8.1 CHEMICAL EQUATION SYMBOLS

SYMBOL	INTERPRETATION OF CHEMICAL EQUATION SYMBOL
$\rightarrow$	produces, yields, gives (separates reactants from products)
+	reacts with, added to, plus (separates two or more reactants or products)
$\xrightarrow{\Delta}$	heat is a catalyst for the reaction
$\xrightarrow{Fe}$	iron is a catalyst for the reaction
NR	no reaction
(*s*)	solid substance or precipitate
(*l*)	liquid substance
(*g*)	gaseous substance
(*aq*)	aqueous solution

Naturally Occurring Diatomic Molecules

H_2, N_2, O_2, F_2, Cl_2, Br_2, and I_2

Chemical reactions often involve hydrogen, nitrogen, or oxygen gases. These nonmetals occur naturally as **diatomic molecules**, that is, H_2, N_2, and O_2. Moreover, the halogens also occur naturally as diatomic molecules and are therefore written as F_2, Cl_2, Br_2, and I_2. If one of these elements appears in a chemical equation, it is written as a diatomic molecule.

The following example exercise further illustrates how to write a chemical equation from the description of a chemical reaction.

Example Exercise 8.2 Writing Chemical Equations

Write a chemical equation for each of the following chemical reactions:

(a) Mercury liquid and fluorine gas react to give solid mercury(II) fluoride.

(b) Zinc metal reacts with sulfuric acid to give aqueous zinc sulfate and hydrogen gas.

Solution

To write the chemical equation, we must provide formulas and symbols for each substance. We can describe each of the preceding chemical reactions as follows:

(a) $Hg(l) + F_2(g) \rightarrow HgF_2(s)$

(b) $Zn(s) + H_2SO_4(aq) \rightarrow ZnSO_4(aq) + H_2(g)$

Practice Exercise

Write a chemical equation for each of the following chemical reactions:

(a) Aqueous solutions of sodium iodide and silver nitrate yield silver iodide precipitate and aqueous sodium nitrate.

(b) Acetic acid reacts with aqueous potassium hydroxide to give aqueous potassium acetate plus water.

Answers:

(a) $NaI(aq) + AgNO_3(aq) \rightarrow AgI(s) + NaNO_3(aq)$

(b) $HC_2H_3O_2(aq) + KOH(aq) \rightarrow KC_2H_3O_2(aq) + H_2O(l)$

Concept Exercise

Which seven nonmetals occur naturally as diatomic molecules?

Answer: See Appendix G.

OBJECTIVE

▸ To write balanced chemical equations.

8.3 BALANCING CHEMICAL EQUATIONS

In the previous section we translated the description of a chemical reaction into a chemical equation. In each case, the number of atoms of each element was equal in the reactants and products. Because the number of atoms of each element is the same on each side of the arrow, the chemical reaction is said to be a balanced chemical equation.

More often, the number of atoms of each element in the reactants and products is not the same. Therefore, it is necessary to balance the number of atoms of each element on each side of the chemical reaction. We balance a chemical reaction by placing a whole number **coefficient** in front of each substance. It is important to note that a coefficient multiplies all subscripts in the chemical formula that follows. That is, 3 H_2O indicates six hydrogen atoms and three oxygen atoms.

Let's consider the reaction of hydrogen gas with chlorine gas to give hydrogen chloride, a gas with a sharp odor. We can write the chemical reaction as follows:

$$H_2(g) + Cl_2(g) \longrightarrow HCl(g)$$

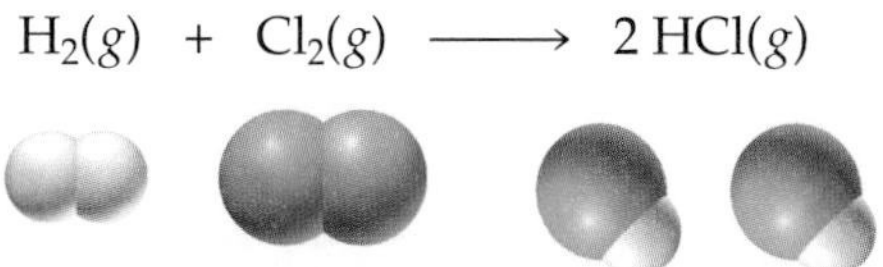

Notice that the **subscript** for H_2 and Cl_2 is 2, but that only one H atom and one Cl atom appear in the product. Thus, the equation is not balanced. To balance the number of H atoms on both sides of the equation, we place the coefficient 2 in front of the HCl. This gives us

$$H_2(g) + Cl_2(g) \longrightarrow 2\,HCl(g)$$

After adding the reaction coefficient 2, we have two molecules of HCl. That is, we have two H atoms and two Cl atoms on each side of the equation. The equation is now balanced.

Let's try a more difficult example. Aluminum metal is heated with oxygen gas to give solid aluminum oxide. The formula for aluminum oxide is Al_2O_3. The chemical equation is

$$Al(s) + O_2(g) \longrightarrow Al_2O_3(s)$$

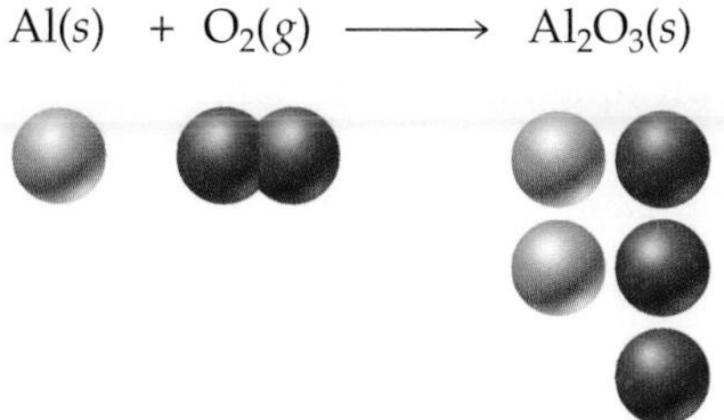

Notice that two O atoms appear in the reactants but that three O atoms appear in the product. This reaction is not balanced. To balance the numbers of O atoms, we will use the lowest common multiple of 2 and 3; the number is 6. We place the coefficient 3 in front of the O_2 to give six oxygen atoms in the reactants and the coefficient 2 in front of Al_2O_3 to give six oxygen atoms in the products. That gives us

$$Al(s) + 3\,O_2(g) \longrightarrow 2\,Al_2O_3(s)$$

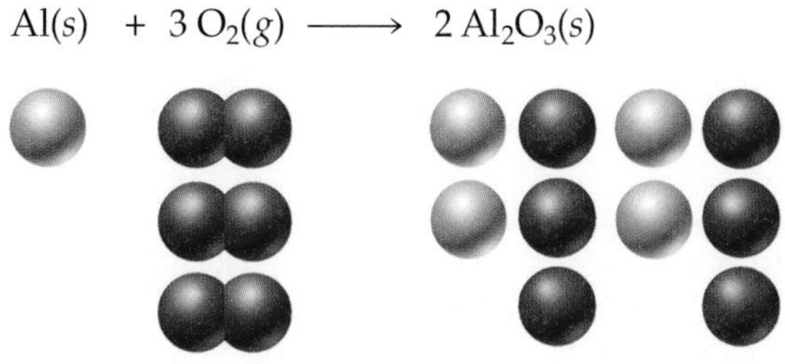

The numbers of Al atoms, however, are not balanced. On the reactant side we have one Al atom. On the product side, we have two units of Al_2O_3, for a total of four Al atoms. We can place the coefficient 4 in front of the reacting Al metal. This gives a balanced chemical equation:

$$4\,Al(s) + 3\,O_2(g) \longrightarrow 2\,Al_2O_3(s)$$

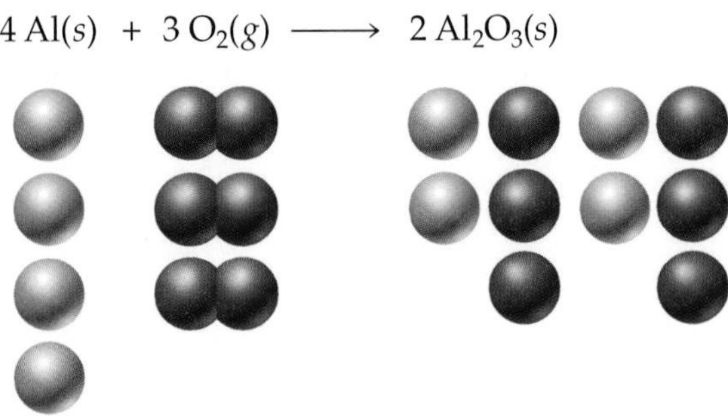

This method of placing coefficients by systematically analyzing each side of an equation is called *balancing by inspection*. Although there is no formal prescription for balancing an equation by inspection, we can list some general guidelines.

General Guidelines for Balancing a Chemical Equation

1. Before placing a coefficient in an equation, verify the formula of each substance for correct subscripts. Note that subscripts of formulas are never changed to balance an equation. In fact, if an equation is difficult to balance, it is usually because of an incorrect subscript in the formula.
2. Balance each element in the equation by placing a coefficient in front of each substance. Coefficients of 1 are assumed and do not appear in the balanced chemical equation.
   (a) Begin balancing the equation starting with the most complex formula.
   (b) Balance polyatomic ions as a single unit. However, if a polyatomic ion breaks down, you must balance each atom separately.
   (c) The coefficients in an equation must be whole numbers. Occasionally, it is helpful to use a fractional coefficient to balance an element in a diatomic molecule; for example,

$$H_2(g) + \tfrac{1}{2}O_2(g) \longrightarrow H_2O(l)$$

   If a fraction is used, such as 1/2, we must then multiply the equation by 2 to obtain whole number coefficients. Thus, the balanced equation is

$$2\,H_2(g) + O_2(g) \longrightarrow 2\,H_2O(l)$$

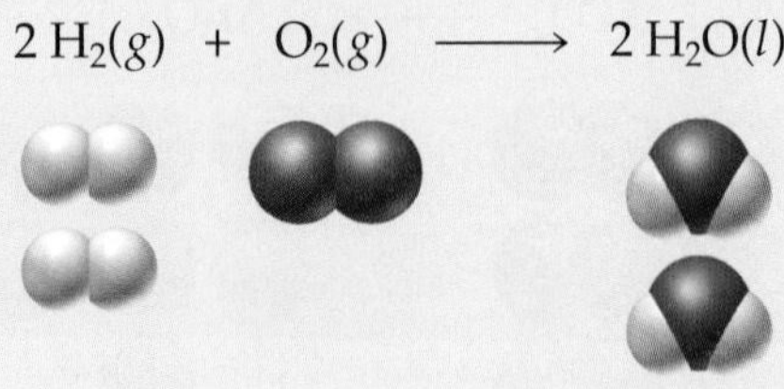

3. After balancing the equation, check (√) each symbol of every element (or polyatomic ion) to verify that the coefficients are correct. Proceed back and forth between reactants and products. The procedure for verification is to multiply the coefficient times the subscript of each element; the totals should be the same on both sides of the equation.

   Finally, check the coefficients to make sure they represent the smallest whole number ratio. It may be possible to obtain a smaller set of coefficients. Although the following equation is balanced,

$$2\,N_2(g) + 6\,H_2(g) \rightarrow 4\,NH_3(g)$$

   you must reduce the coefficients by a factor of 2 to obtain the correct balanced equation:

$$N_2(g) + 3\,H_2(g) \rightarrow 2\,NH_3(g)$$

We can write an equation for the formation of the yellow precipitate shown in Figure 8.1c. In the reaction, aqueous solutions of lead(II) nitrate and potassium iodide produce a yellow precipitate of lead(II) iodide and an aqueous solution of potassium nitrate. We can write a chemical equation for the reaction as follows:

$$Pb(NO_3)_2(aq) + KI(aq) \rightarrow PbI_2(s) + KNO_3(aq)$$

Let's begin balancing with $Pb(NO_3)_2$ because it is the most complex formula. Notice that one Pb atom appears on each side of the equation. However, there are two NO_3^- ions on the reactant side and only one NO_3^- ion on the product side. We can balance the NO_3^- polyatomic ion by placing a 2 in front of KNO_3. Thus,

$$Pb(NO_3)_2(aq) + KI(aq) \rightarrow PbI_2(s) + 2\,KNO_3(aq)$$

We now have two K on the right side of the equation and only one K on the left. Thus, we need the coefficient 2 in front of KI:

$$Pb(NO_3)_2(aq) + 2KI(aq) \rightarrow PbI_2(s) + 2\,KNO_3(aq)$$

Notice the coefficient 2 in front of KI gives two I, which is equal to the two I in PbI_2. As a final inspection, let's check off each element and polyatomic ion to verify that we indeed have a balanced equation:

$$\overset{\surd}{Pb}(\overset{\surd}{NO_3})_2(aq) + 2\overset{\surd\surd}{KI}(aq) \rightarrow \overset{\surd\surd}{PbI_2}(s) + 2\,\overset{\surd}{K}\overset{\surd}{NO_3}(aq)$$

The following example exercises further illustrate the general guidelines for balancing a chemical equation:

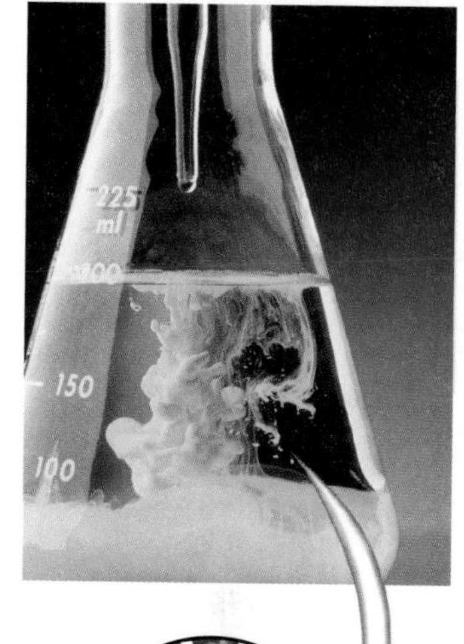

▲ **Insoluble Lead(II) Iodide** Lead(II) iodide, PbI_2, forms a yellow precipitate in an aqueous solution.

Example Exercise 8.3 Balancing Chemical Equations

Aqueous solutions of calcium nitrate and sodium carbonate react to give a white precipitate of calcium carbonate and aqueous sodium nitrate. Write a balanced chemical equation given

$$Ca(NO_3)_2(aq) + Na_2CO_3(aq) \rightarrow CaCO_3(s) + NaNO_3(aq)$$

Solution

We see that $Ca(NO_3)_2$ has two NO_3^- ions in the formula. There is only one NO_3^- ion on the right side, so we insert the coefficient 2 as follows:

$$Ca(NO_3)_2(aq) + Na_2CO_3(aq) \rightarrow CaCO_3(s) + 2NaNO_3(aq)$$

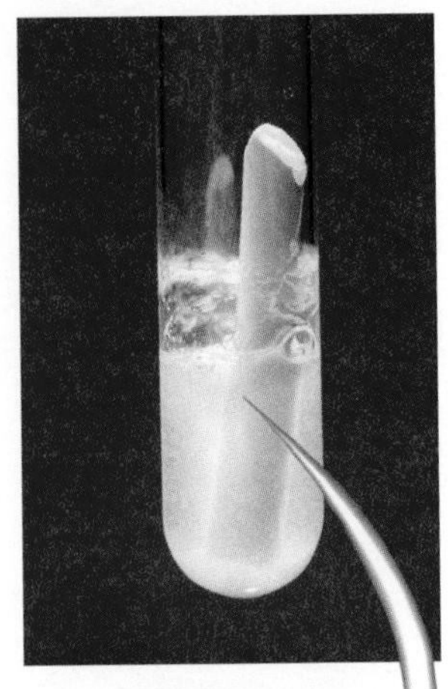

▲ **Insoluble Calcium Carbonate** Chalk, calcium carbonate, $CaCO_3$, is insoluble in an aqueous solution.

The coefficient 2 in front of $NaNO_3$ generates two Na on the right side, and we have two Na on the left side. Notice we have one CO_3 on each side, and one Ca on each side. Finally, we can check off each element and polyatomic ion to verify that the chemical equation is balanced.

$$\overset{\surd}{Ca}(\overset{\surd}{NO_3})_2(aq) + \overset{\surd}{Na_2}\overset{\surd}{CO_3}(aq) \rightarrow \overset{\surd}{Ca}\overset{\surd}{CO_3}(s) + 2\,\overset{\surd}{Na}\overset{\surd}{NO_3}(aq)$$

Practice Exercise

Aqueous solutions of aluminum sulfate and barium nitrate react to yield a white precipitate of barium sulfate and aqueous aluminum nitrate. Write a balanced chemical equation given

$$Al_2(SO_4)_3(aq) + Ba(NO_3)_2(aq) \rightarrow BaSO_4(s) + Al(NO_3)_3(aq)$$

Answer: $\overset{\surd}{Al_2}(\overset{\surd}{SO_4})_3(aq) + 3\,\overset{\surd}{Ba}(\overset{\surd}{NO_3})_2(aq) \rightarrow 3\,\overset{\surd}{Ba}\overset{\surd}{SO_4}(s) + 2\,\overset{\surd}{Al}(\overset{\surd}{NO_3})_3(aq)$

Concept Exercise

Do you change a subscript in a chemical formula to balance an equation?

Answer: See Appendix G.

Example Exercise 8.4 Balancing Chemical Equations

Sulfuric acid reacts with aqueous sodium hydroxide to give aqueous sodium sulfate and water. Write a balanced chemical equation given

$$H_2SO_4(aq) + NaOH(aq) \rightarrow Na_2SO_4(aq) + H_2O(l)$$

Solution

Let's start with Na_2SO_4, which contains the same number of atoms as H_2SO_4. There is one SO_4 on each side of the equation, and so SO_4 is balanced. However, there are two Na on the right side of the equation and one Na on the left. We must place the coefficient 2 in front of NaOH.

$$H_2SO_4(aq) + 2\,NaOH(aq) \rightarrow Na_2SO_4(aq) + HOH(l)$$

To balance this reaction more easily, we can write the formula for water as HOH. Notice that we now have two OH on the left side of the equation, and one OH on the right side. By placing a 2 in front of the HOH, we can balance the OH. The two H on the right side are balanced by the two H in H_2SO_4. Finally, we check off each atom and polyatomic ion to verify that we have a balanced chemical equation.

$$\overset{\surd}{H_2}\overset{\surd}{SO_4}(aq) + 2\,\overset{\surd}{Na}\overset{\surd}{OH}(aq) \rightarrow \overset{\surd}{Na_2}\overset{\surd}{SO_4}(aq) + 2\,\overset{\surd}{H}\overset{\surd}{OH}(l)$$

Practice Exercise

Nitric acid reacts with aqueous barium hydroxide to give aqueous barium nitrate and water. Write a balanced chemical equation given

$$HNO_3(aq) + Ba(OH)_2(aq) \rightarrow Ba(NO_3)_2(aq) + HOH(l)$$

Answer: $2\,\overset{\surd}{H}\overset{\surd}{NO_3}(aq) + \overset{\surd}{Ba}(\overset{\surd}{OH})_2(aq) \rightarrow \overset{\surd}{Ba}(\overset{\surd}{NO_3})_2(aq) + 2\,\overset{\surd}{H}\overset{\surd}{OH}(l)$

Concept Exercise

Why write water as HOH for the product of a neutralization reaction?

Answer: See Appendix G.

Note Balancing an equation is a straightforward task. Start with the most complex formula and balance systematically, back and forth, between reactants and products. If you encounter difficulty in the final step, it is usually because the original equation had an incorrect subscript in a chemical formula.

8.4 CLASSIFYING CHEMICAL REACTIONS

OBJECTIVES

- To distinguish five types of chemical reactions.
- To classify a chemical reaction as one of the following types: combination, decomposition, single-replacement, double-replacement, or neutralization.

Since there are millions of chemical reactions, how can we attempt to master such a large topic? The answer lies in the same way that we learned previously to master chemical formulas. We can *classify chemical reactions* and put them in categories.

In this chapter, we will begin with five simple types of reactions: combination, decomposition, single-replacement, double-replacement, and neutralization. We can briefly describe these five types as follows:

1. A **combination reaction** involves simpler substances being combined into a more complex compound. It is also referred to as a *synthesis reaction*. In the general case, substance A combines with substance Z to produce compound AZ. The chemical equation is

$$A + Z \rightarrow AZ$$

2. In a **decomposition reaction** a single compound is broken down into two or more simpler substances. In this case, heat is usually applied to decompose the compound. In the general case, compound AZ decomposes into substances A and Z. The chemical equation is

$$AZ \xrightarrow{\Delta} A + Z$$

3. In a **single-replacement reaction** a more reactive element displaces another element from a compound or aqueous solution. The substance that is displaced shows less tendency to undergo reaction; that is, it is less active. In one case, metal A replaces metal B in BZ to give AZ and B. The chemical equation is

$$A + BZ \rightarrow AZ + B$$

4. In a **double-replacement reaction** two compounds exchange anions. Compound AX reacts with compound BZ to yield the products AZ and BX. The general form of the reaction is

$$AX + BZ \rightarrow AZ + BX$$

5. An acid and a base react with each other in a **neutralization reaction**. Acid HX reacts with base BOH to give the compound BX and water. We can write the general form of the reaction as

$$HX + BOH \rightarrow BX + HOH$$

When we examine a neutralization reaction, we see that it is actually a special type of double-replacement reaction. An acid and a base simply switch anions. Notice that we wrote water as HOH rather than H_2O. Using the formula HOH shows the double-replacement nature of the neutralization reaction and makes it easier to balance the equation.

Example Exercise 8.5 Classifying Chemical Reactions

Classify each of the following reactions as combination, decomposition, single-replacement, double-replacement, or neutralization.

(a) Copper metal heated with oxygen gas produces solid copper(II) oxide.

$$2\,Cu(s) + O_2(g) \xrightarrow{\Delta} 2\,CuO(s)$$

(b) Heating powdered iron(III) carbonate produces solid iron(III) oxide and carbon dioxide gas.

$$Fe_2(CO_3)_3(s) \xrightarrow{\Delta} Fe_2O_3(s) + 3\,CO_2(g)$$

(c) Aluminum metal reacts with aqueous manganese(II) sulfate to give aqueous aluminum sulfate and manganese metal.

$$2\,Al(s) + 3\,MnSO_4(aq) \rightarrow Al_2(SO_4)_3(aq) + 3\,Mn(s)$$

(d) Aqueous sodium chromate reacts with aqueous barium chloride to give insoluble barium chromate and aqueous sodium chloride.

$$Na_2CrO_4(aq) + BaCl_2(aq) \rightarrow BaCrO_4(s) + 2\,NaCl(aq)$$

(e) Nitric acid reacts with aqueous potassium hydroxide to give aqueous potassium nitrate and water.

$$HNO_3(aq) + KOH(aq) \rightarrow KNO_3(aq) + H_2O(l)$$

Solution

We can classify the type of each reaction as follows:

(a) The two elements Cu and O_2 synthesize a single compound; this is an example of a *combination* reaction.

(b) The compound $Fe_2(CO_3)_3$ is heated and breaks down into a simpler compound and a gas; this is an example of a *decomposition* reaction.

(c) The metal Al displaces the metal Mn from aqueous $MnSO_4$; this is an example of a *single-replacement* reaction.

(d) The two compounds Na_2CrO_4 and $BaCl_2$ exchange anions; this is an example of a *double-replacement* reaction.

(e) The acid HNO_3 reacts with the base KOH to form KNO_3 and water; this is an example of a *neutralization* reaction.

Practice Exercise

Classify the following types of reactions as combination, decomposition, single-replacement, double-replacement, or neutralization:

(a) $Zn(s) + CuSO_4(aq) \rightarrow ZnSO_4(aq) + Cu(s)$

(b) $2\,Sr(s) + O_2(g) \rightarrow 2\,SrO(s)$

(c) $Cd(HCO_3)_2(s) \rightarrow CdCO_3(s) + H_2O(g) + CO_2(g)$

(d) $HC_2H_3O_2(aq) + NaOH(aq) \rightarrow NaC_2H_3O_2(aq) + H_2O(l)$

(e) $AgNO_3(aq) + KCl(aq) \rightarrow AgCl(s) + KNO_3(aq)$

Answers: (a) single-replacement; (b) combination; (c) decomposition; (d) neutralization; (e) double-replacement

Concept Exercise

How can a decomposition reaction be easily distinguished from other types of chemical reactions?

Answer: See Appendix G.

8.5 COMBINATION REACTIONS

OBJECTIVES

- To write a balanced chemical equation for the reaction of a metal and oxygen gas.
- To write a balanced chemical equation for the reaction of a nonmetal and oxygen gas.
- To write a balanced chemical equation for the reaction of a metal and a nonmetal.

There are many examples of combination reactions. For instance, heating a metal in air can cause the metal to combine with oxygen and form a compound. In this section, we will study three important kinds of combination reactions.

Metal and Oxygen Gas

In one kind of combination reaction, a metal and oxygen gas react to give a metal oxide. In the following example, magnesium metal and oxygen gas combine to give magnesium oxide:

metal + oxygen gas → metal oxide

$$2\,Mg(s) + O_2(g) \xrightarrow{\Delta} 2\,MgO(s)$$

A metal and oxygen react to give a metal oxide compound. We can usually predict the formulas of metal oxides containing a main-group metal. On the other hand, we cannot predict the formulas of most metal oxides containing a transition metal. For transition metal oxides, the ionic charge of the metal must be given to write the chemical formula of the metal oxide.

The following example exercise further illustrates how to write balanced equations for a combination reaction of a metal and oxygen gas:

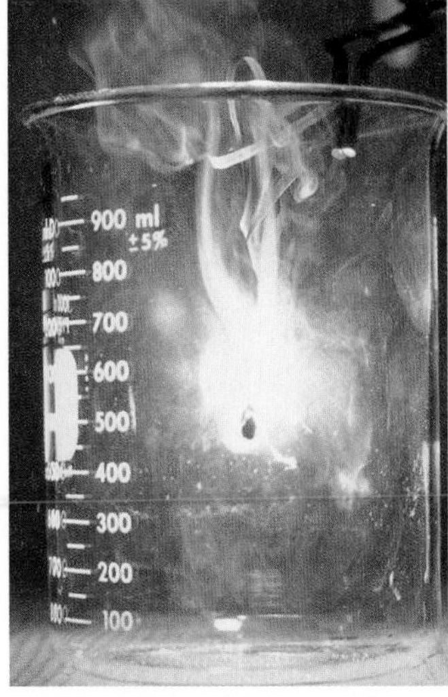

▲ Igniting Magnesium Metal A strip of magnesium metal reacts with oxygen in air to give white smoke, MgO, and a bright white light.

Example Exercise 8.6 Combination Reaction of a Metal and Oxygen Gas

Write a balanced chemical equation for each of the following combination reactions:

(a) Zinc metal is heated with oxygen gas in air to yield solid zinc oxide.

(b) Chromium metal is heated with oxygen gas to give chromium(III) oxide.

Solution

A metal and oxygen react to produce a metal oxide.

(a) Zinc is a metal with a predictable charge, that is, Zn^{2+}. The formula of zinc oxide is ZnO. The balanced equation for the reaction is

$$2\ Zn(s) + O_2(g) \xrightarrow{\Delta} 2\ ZnO(s)$$

(b) Chromium is a metal with a variable charge. From the name chromium(III) oxide, we know the ion is Cr^{3+}. The formula of the oxide is, therefore, Cr_2O_3, and the balanced equation for the reaction is

$$4\ Cr(s) + 3\ O_2(g) \xrightarrow{\Delta} 2\ Cr_2O_3(s)$$

Practice Exercise

Write a balanced chemical equation for each of the following combination reactions:

(a) Lead metal is heated with oxygen in air to yield solid lead(IV) oxide.

(b) Cobalt metal is heated with oxygen gas to give solid cobalt(III) oxide.

Answers:

(a) $Pb(s) + O_2(g) \xrightarrow{\Delta} PbO_2(s)$

(b) $4\ Co(s) + 3\ O_2(g) \xrightarrow{\Delta} 2\ Co_2O_3(s)$

Concept Exercise

What is the product from heating a metal and oxygen gas?

Answer: See Appendix G.

Nonmetal and Oxygen Gas

In this kind of combination reaction, a nonmetal and oxygen gas react to give a nonmetal oxide. In the following example, yellow sulfur powder and oxygen gas combine to give sulfur dioxide gas.

nonmetal + oxygen gas → nonmetal oxide

$$S(s) + O_2(g) \xrightarrow{\Delta} SO_2(g)$$

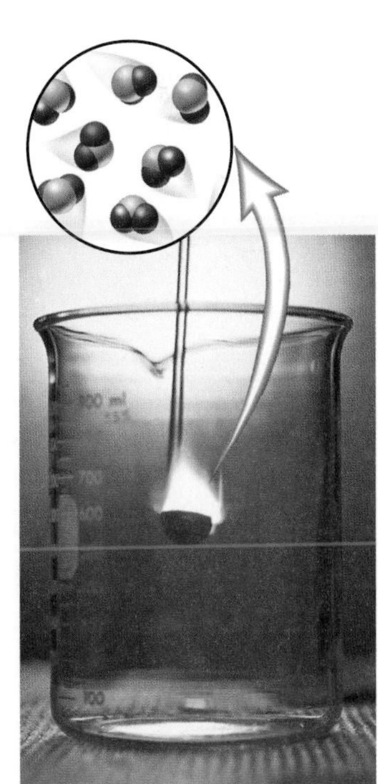

▲ Igniting Sulfur Powder A small portion of yellow sulfur burns with the oxygen in air to give an intense blue flame and sulfur dioxide gas, SO_2, which has a sharp odor.

Nonmetal oxides demonstrate multiple combining capacities, for example, CO and CO_2. In general, the formula of a nonmetal oxide product is unpredictable and varies with temperature and pressure. For example, nitrogen and oxygen gases can combine to give all of the following: NO, NO_2, N_2O, N_2O_3, N_2O_4, and N_2O_5. Therefore, to complete and balance an equation of a nonmetal and oxygen gas, we must be given the name or formula of the nonmetal oxide.

The following example exercise further illustrates how to write balanced equations for a combination reaction of a nonmetal and oxygen gas:

Example Exercise 8.7 Combination Reaction of a Nonmetal and Oxygen Gas

Write a balanced chemical equation for each of the following combination reactions:

(a) Carbon is heated with oxygen gas to produce carbon dioxide gas.

(b) Phosphorus and oxygen gas react to give solid diphosphorus pentaoxide.

Solution

A nonmetal and oxygen combine to produce a nonmetal oxide.

(a) The formula of the nonmetal oxide is unpredictable. We are given that the product is carbon dioxide and not carbon monoxide. The balanced equation for the reaction is

$$C(s) + O_2(g) \xrightarrow{\Delta} CO_2(g)$$

(b) The formula for the oxide of phosphorus is not predictable, but we have the name of the nonmetal oxide product. The formula for diphosphorus pentaoxide is P_2O_5. The balanced equation is

$$4\,P(s) + 5\,O_2(g) \rightarrow 2\,P_2O_5(s)$$

Practice Exercise

Write a balanced chemical equation for each of the following combination reactions:

(a) Nitrogen gas is heated with oxygen to give dinitrogen trioxide gas.

(b) Chlorine gas is heated with oxygen to give dichlorine monoxide gas.

Answers:

(a) $2\,N_2(g) + 3\,O_2(g) \xrightarrow{\Delta} 2\,N_2O_3(g)$

(b) $2\,Cl_2(g) + O_2(g) \xrightarrow{\Delta} 2\,Cl_2O(g)$

Concept Exercise

What is the product from heating a nonmetal and oxygen gas?

Answer: See Appendix G.

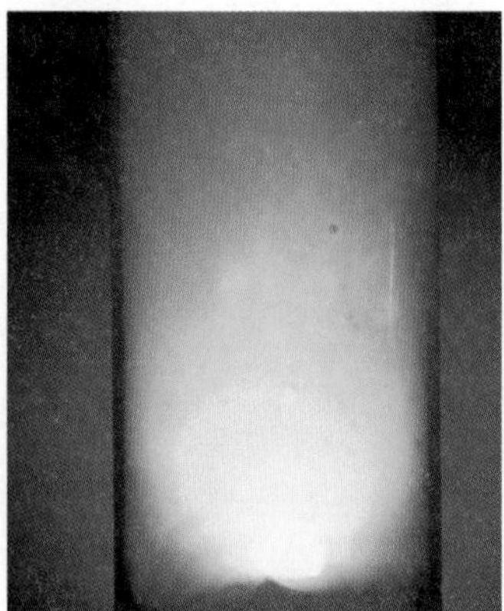

▲ **Igniting Sodium and Chlorine** Sodium metal reacts strongly with chlorine gas to give a bright yellow light and sodium chloride, NaCl, powder.

Metal and Nonmetal

In this kind of combination reaction, a metal and a nonmetal react to give an ionic compound. In the following example, silvery sodium metal and yellow chlorine gas combine to give solid sodium chloride:

metal + nonmetal → binary ionic compound

$$2\,Na(s) + Cl_2(g) \xrightarrow{\Delta} 2\,NaCl(s)$$

When a metal and a nonmetal react, they produce a binary ionic compound. If the compound contains a main-group metal, the formula is usually predictable. If the compound contains a transition metal, the formula is not predictable.

The following example exercise further illustrates how to write a balanced equation for a combination reaction of a metal and a nonmetal:

Example Exercise 8.8 Combination Reaction of a Metal and a Nonmetal

Write a balanced chemical equation for each of the following combination reactions:

(a) Aluminum metal is heated with sulfur and gives a solid product.

(b) Chromium metal is heated with iodine and produces powdered chromium(III) iodide.

Solution

A metal and a nonmetal react to produce a binary ionic compound.

(a) The formula of the product is predictable. Aluminum combines with sulfur to give aluminum sulfide, Al_2S_3. The balanced equation for the reaction is

$$2\,Al(s) + 3\,S(s) \xrightarrow{\Delta} Al_2S_3(s)$$

(b) Chromium is a transition metal, and so we cannot predict the formula for the product. We are given the name of the compound, chromium(III) iodide, and so the formula is CrI_3. The equation is

$$2\,Cr(s) + 3\,I_2(s) \xrightarrow{\Delta} 2\,CrI_3(s)$$

Practice Exercise

Write a balanced chemical equation for each of the following combination reactions:

(a) Calcium metal is heated with fluorine gas to yield solid calcium fluoride.

(b) Manganese metal reacts with bromine vapor to give crystalline manganese(IV) bromide.

Answers:

(a) $Ca(s) + F_2(g) \xrightarrow{\Delta} CaF_2(s)$

(b) $Mn(s) + 2\,Br_2(g) \rightarrow MnBr_4(s)$

Concept Exercise

What is the product from heating a metal and nonmetal?

Answer: See Appendix G.

8.6 DECOMPOSITION REACTIONS

OBJECTIVES

- To write a balanced equation for the decomposition of a metal hydrogen carbonate.
- To write a balanced equation for the decomposition of a metal carbonate.
- To write a balanced equation for the decomposition of a compound that releases oxygen gas.

There are many examples of decomposition reactions. For instance, heating a compound may cause it to decompose into two or more simpler substances. In this section, we will study three important kinds of decomposition reactions.

Metal Hydrogen Carbonates

A metal hydrogen carbonate undergoes a decomposition reaction when it is heated. For example, you may be aware that baking soda is a natural fire extinguisher. That is, heat from a fire decomposes baking soda (sodium hydrogen carbonate) and releases carbon dioxide gas. Since carbon dioxide is more dense than air, it can smother a fire by excluding oxygen from the flame.

The decomposition of a metal hydrogen carbonate yields a metal carbonate, water, and carbon dioxide. In the following example, sodium hydrogen carbonate is decomposed by heating to give sodium carbonate, water, and carbon dioxide:

$$\textbf{metal hydrogen carbonate} \xrightarrow{\Delta} \textbf{metal carbonate} + \textbf{water} + \textbf{carbon}$$

$$2\,NaHCO_3(s) \xrightarrow{\Delta} Na_2CO_3(s) + H_2O(g) + CO_2(g)$$

During the decomposition reaction of a metal hydrogen carbonate, the ionic charge of the metal does not change. Therefore, the formula of the metal carbonate is predictable. Even transition metal hydrogen carbonates are predictable. If nickel(II) hydrogen carbonate decomposes, one of the products is nickel(II) carbonate. That is,

$$Ni(HCO_3)_2(s) \xrightarrow{\Delta} NiCO_3(s) + H_2O(g) + CO_2(g)$$

The following example exercise further illustrates how to write balanced equations for a decomposition reaction of a metal hydrogen carbonate:

Example Exercise 8.9 Decomposition Reaction of a Hydrogen Carbonate

Write a balanced chemical equation for each of the following decomposition reactions:

(a) Lithium hydrogen carbonate decomposes on heating.

(b) Lead(II) hydrogen carbonate decomposes on heating.

Solution

A metal hydrogen carbonate decomposes with heat to give a metal carbonate, water, and carbon dioxide gas.

(a) All the formulas are predictable, including that of the product, Li_2CO_3. The balanced equation for the reaction is

$$2LiHCO_3(s) \xrightarrow{\Delta} Li_2CO_3(s) + H_2O(g) + CO_2(g)$$

(b) Although the ionic charge of lead is variable, we are given that lead(II) hydrogen carbonate is the reactant. Therefore, the product is lead(II) carbonate, $PbCO_3$. The balanced equation for the reaction is

$$Pb(HCO_3)_2(s) \xrightarrow{\Delta} PbCO_3(s) + H_2O(g) + CO_2(g)$$

Practice Exercise

Write a balanced chemical equation for each of the following decomposition reactions:

(a) Barium hydrogen carbonate is decomposed with heat.

(b) Copper(I) hydrogen carbonate is decomposed with heat.

Answers:

(a) $Ba(HCO_3)_2(s) \xrightarrow{\Delta} BaCO_3(s) + H_2O(g) + CO_2(g)$

(b) $2\ CuHCO_3(s) \xrightarrow{\Delta} Cu_2CO_3(s) + H_2O(g) + CO_2(g)$

Concept Exercise

What is the product from heating a metal hydrogen carbonate?

Answer: See Appendix G.

Metal Carbonates

After a metal hydrogen carbonate decomposes into a metal carbonate, the metal carbonate can further decompose with prolonged heating. The carbonate can decompose into a metal oxide while releasing carbon dioxide gas.

In the following example, calcium carbonate is decomposed by heating to give calcium oxide and carbon dioxide:

$$\textbf{metal carbonate} \xrightarrow{\Delta} \textbf{metal oxide} + \textbf{carbon dioxide}$$

$$CaCO_3(s) \xrightarrow{\Delta} CaO(s) + CO_2(g)$$

The ionic charge of the metal does not change during the decomposition reaction of a metal carbonate. Therefore, we can predict the formula for the metal oxide product from the decomposition of a metal carbonate. If nickel(II) carbonate is decomposed, one of the products is nickel(II) oxide. That is,

$$NiCO_3 \xrightarrow{\Delta} NiO(s) + CO_2(g)$$

The following example exercise further illustrates how to write balanced equations for a decomposition reaction of a metal carbonate:

Example Exercise 8.10 Decomposition Reaction of a Carbonate

Write a balanced chemical equation for each of the following decomposition reactions:
(a) Magnesium carbonate decomposes on heating.
(b) Copper(I) carbonate decomposes on heating.

Solution

A metal carbonate decomposes with heat to a metal oxide and carbon dioxide gas.

(a) All the formulas are predictable including that of the metal oxide, MgO. The balanced equation for the reaction is

$$MgCO_3(s) \xrightarrow{\Delta} MgO(s) + CO_2(g)$$

(b) The ionic charge of copper can be either 1+ or 2+. Since the reactant is copper(I) carbonate, the product is copper(I) oxide, that is, Cu_2O. Thus, the balanced equation for the reaction is

$$Cu_2CO_3(s) \xrightarrow{\Delta} Cu_2O(s) + CO_2(g)$$

Practice Exercise

Write a balanced chemical equation for each of the following decomposition reactions:
(a) Aluminum carbonate is decomposed on heating.
(b) Iron(II) carbonate is decomposed on heating.

Answers:

(a) $Al_2(CO_3)_3(s) \xrightarrow{\Delta} Al_2O_3(s) + 3\,CO_2(g)$

(b) $FeCO_3(s) \xrightarrow{\Delta} FeO(s) + CO_2(g)$

Concept Exercise

What is the product from heating a metal carbonate?

Answer: See Appendix G.

▲ **Decomposing Mercury(II) Oxide** A small portion of orange HgO powder decomposes with heat to give oxygen gas, O_2, and beads of silvery, metallic mercury.

Miscellaneous Oxygen-Containing Compounds

We can often decompose an oxygen-containing ionic compound, which releases oxygen gas. However, we cannot always predict the products. In this kind of decomposition reaction, we must be given the formulas of the products.

In the following example, orange mercury(II) oxide powder is decomposed by heating to give droplets of shiny liquid mercury and oxygen gas:

$$\textbf{oxygen-containing compound} \xrightarrow{\Delta} \textbf{substance} + \textbf{oxygen gas}$$

$$2\,HgO(s) \xrightarrow{\Delta} 2\,Hg(l) + O_2(g)$$

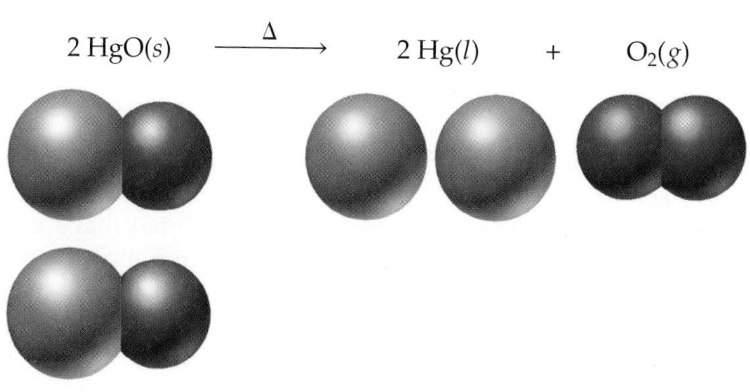

The following example exercise illustrates how to write balanced equations for miscellaneous decomposition reactions:

Example Exercise 8.11 Decomposition Reaction Producing Oxygen Gas

Write a balanced chemical equation for each of the following decomposition reactions:

(a) Solid potassium chlorate decomposes upon heating to give solid potassium chloride and oxygen gas.

(b) Solid sodium nitrate decomposes upon heating to give solid sodium nitrite and oxygen gas.

Solution

There is no general format for these reactions; however, we are given the names of the reactants and products.

(a) We can write the reaction for the decomposition of potassium chlorate and balance the chemical equation as follows:

$$2\ KClO_3(s) \xrightarrow{\Delta} 2\ KCl(s) + 3\ O_2(g)$$

(b) We can write the reaction for the decomposition of sodium nitrate and balance the chemical equation as follows:

$$2\ NaNO_3(s) \xrightarrow{\Delta} 2\ NaNO_2(s) + O_2(g)$$

Practice Exercise

Write a balanced chemical equation for each of the following decomposition reactions:

(a) Solid lead(IV) oxide decomposes upon heating to give solid lead(II) oxide and oxygen gas.

(b) Aqueous hydrogen peroxide, H_2O_2, decomposes upon heating to give water and oxygen gas.

Answers:

(a) $2\ PbO_2(s) \xrightarrow{\Delta} 2\ PbO(s) + O_2(g)$

(b) $2\ H_2O_2(aq) \xrightarrow{\Delta} 2\ H_2O(l) + O_2(g)$

Concept Exercise

What is the usual product from heating an oxygen-containing compound?

Answer: See Appendix G.

8.7 THE ACTIVITY SERIES CONCEPT

OBJECTIVES

- To explain the concept of an activity series for metals.
- To predict whether a single-replacement reaction occurs by referring to the activity series.

When a metal undergoes a single replacement reaction, it displaces another metal from a compound or aqueous solution. The metal that displaces the other metal does so because it has a greater tendency to undergo a reaction; in other words, it is more active. If metal A is more active than metal B, metal A will displace metal B from compound BZ. We can show the reaction as

$$A + BZ \rightarrow AZ + B$$

The activity of a metal is a measure of its ability to compete in a single replacement reaction. In an **activity series** (or electromotive series), a sequence of metals is arranged according to their ability to undergo reaction. Metals that are most reactive appear first in the activity series. Metals that are less reactive appear last in the series. The relative activity of several metals is listed in the margin.

Activity Series for Metals

Li > K > Ba > Sr > Ca > Na > Mg > Al > Mn > Zn > Fe > Cd > Co > Ni > Sn > Pb > (H) > Cu > Ag > Hg > Au

◀ Iron in a Copper Solution An iron nail reacts in an aqueous blue $CuSO_4$ solution to give a reddish-brown deposit of Cu metal on the nail.

We can see that lithium precedes potassium in the series, which in turn precedes barium, and so on. Although hydrogen (H) is not a metal, it is included in the series as a reference. Metals that precede (H) in the series react with an aqueous acid; metals that follow (H) in the series do not react with acids. Notice that the metals Cu, Ag, Hg, and Au do not react with acids.

Consider the reaction of iron metal in an aqueous solution of copper(II) sulfate:

$$Fe(s) + CuSO_4(aq) \rightarrow FeSO_4(aq) + Cu(s)$$

Iron precedes copper in the series, and so Fe displaces Cu from the aqueous solution. If we perform the reaction in the laboratory, a reddish-brown copper deposit appears on the iron metal. It is this type of experiment that establishes the order of metals in the activity series; that is, Fe > > > Cu.

Conversely, there is no reaction if we place copper metal in an aqueous solution of iron(II) sulfate. That is, Cu cannot displace Fe from the solution because copper follows iron in the activity series. Thus, if we put copper wire into a solution of iron(II) sulfate, we observe

$$Cu(s) + FeSO_4(aq) \rightarrow NR$$

Now let's consider the reaction of iron wire with sulfuric acid. The reaction releases gas bubbles in the solution, and the equation is

$$Fe(s) + H_2SO_4(aq) \rightarrow FeSO_4(aq) + H_2(g)$$

Iron precedes (H) in the activity series, and so Fe displaces H_2 gas from a sulfuric acid solution. When we observe the reaction, we see tiny bubbles of hydrogen gas on the iron wire. Now consider the reaction of copper metal and sulfuric acid:

$$Cu(s) + H_2SO_4(aq) \rightarrow NR$$

Copper is after (H) in the activity series, so Cu cannot displace H_2 gas from sulfuric acid. When we observe the copper wire in sulfuric acid, there is no evidence for reaction.

Active Metals

List of Active Metals
Li > K > Ba > Sr > Ca > Na

There are a few metals that are so reactive they react directly with water at room temperature. These metals are called the **active metals** and include most of the metals in Groups IA/1 and IIA/2. Specifically, Li, Na, K, Rb, Cs, Ca, Sr, and Ba react with water. The most common active metals are listed in the margin.

Consider an experiment in which sodium metal is dropped into water. Sodium is an active metal and reacts with water according to the following equation:

$$2\,Na(s) + 2\,H_2O(l) \rightarrow 2\,NaOH(aq) + H_2(g)$$

The following example exercise further illustrates how to predict the reactivity of metals with aqueous solutions, acids, and water:

Example Exercise 8.12 Predictions Based on the Activity Series

Predict whether or not a reaction occurs for each of the following:

(a) Aluminum foil is added to an iron(II) sulfate solution.
(b) Iron wire is added to an aluminum sulfate solution.
(c) Manganese metal is added to acetic acid.
(d) Magnesium metal is added to water.

Solution

We refer to the activity series for the reactivity of each of the metals.

(a) Aluminum precedes iron in the activity series: Al > Fe. Thus, a reaction occurs, and Al displaces Fe from the solution.
(b) Conversely, iron follows aluminum in the activity series: Al > Fe. Thus, there is *no reaction*.
(c) Manganese precedes hydrogen in the activity series: Mn > (H). Thus, a reaction occurs, and Mn produces H_2 gas that bubbles from the solution.
(d) Magnesium is not an active metal. Therefore, magnesium does not react with water, and there is *no reaction*.

Practice Exercise

Predict whether or not a reaction occurs for each of the following:

(a) A gold ring is dropped into sulfuric acid.
(b) A zinc granule is dropped into hydrochloric acid.
(c) A cadmium foil is put into a lead(II) nitrate solution.
(d) A chromium strip is put into water.

Answers:

(a) There is *no reaction* because Au follows (H) in the activity series.
(b) There is a reaction because Zn precedes (H) in the activity series.
(c) There is a reaction because Cd precedes Pb in the activity series.
(d) There is *no reaction* because Cr is not an active metal.

Concept Exercise

Which of the following metals reacts with aqueous silver nitrate: Zn, Cu, Au?

Answer: See Appendix G.

▲ **Gold in Acid** A gold ring gives no reaction in an aqueous sulfuric acid, H_2SO_4, solution.

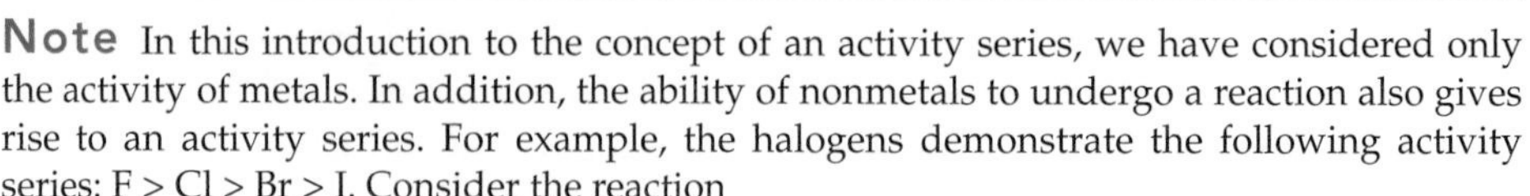

Note In this introduction to the concept of an activity series, we have considered only the activity of metals. In addition, the ability of nonmetals to undergo a reaction also gives rise to an activity series. For example, the halogens demonstrate the following activity series: F > Cl > Br > I. Consider the reaction

$$Cl_2(g) + 2\,NaBr(aq) \rightarrow 2\,NaCl(aq) + Br_2(l)$$

A reaction occurs because Cl precedes Br in the activity series; thus, the more active Cl displaces the less active Br from the solution. Conversely, Cl follows F in the series. Therefore, Cl is less reactive than F and cannot displace it from the solution. Thus,

$$Cl_2(g) + NaF(aq) \rightarrow NR$$

8.8 SINGLE-REPLACEMENT REACTIONS

OBJECTIVES

- To write a balanced chemical equation for the reaction of a metal in an aqueous solution.
- To write a balanced chemical equation for the reaction of a metal in an acid.
- To write a balanced chemical equation for the reaction of an active metal in water.

There are many examples of single-replacement reactions in which a more active metal displaces a less active metal from a solution. In general, the activity of a metal follows the metallic character trend in the periodic table. To determine the relative reactivity of a specific metal, we refer to the activity series.

Metal and Aqueous Solution

One kind of single-replacement reaction involves a metal reacting with an aqueous solution of an ionic compound. The general equation for the reaction is

$$\textbf{metal}_1 + \textbf{aqueous solution}_1 \rightarrow \textbf{metal}_2 + \textbf{aqueous solution}_2$$

Consider the reaction of metallic copper wire in an aqueous solution of silver nitrate. The balanced equation for the reaction is

$$Cu(s) + 2\,AgNO_3(aq) \rightarrow 2\,Ag(s) + Cu(NO_3)_2(aq)$$

If we refer to the activity series, we note that copper is the more active metal. That is, copper lies above silver in the series. Therefore, Cu metal can displace Ag metal from an aqueous solution.

Now let's consider the reaction of metallic silver wire in an aqueous solution of copper nitrate. Since Ag metal is less active than Cu metal, we can predict that there is no reaction. That is,

$$Ag(s) + Cu(NO_3)_2(aq) \rightarrow \text{NR}$$

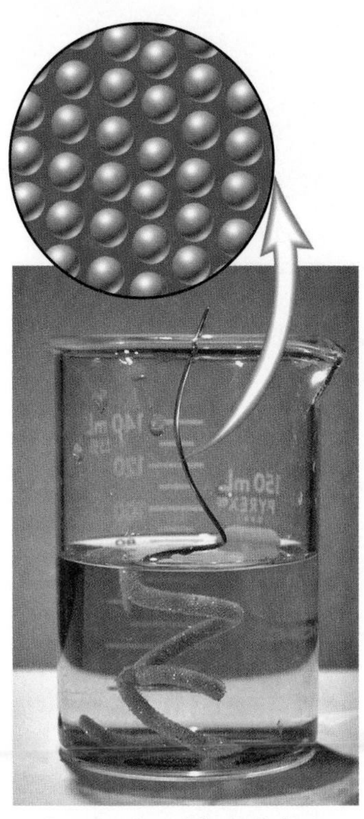

▲ **Copper in a Silver Solution**
A copper wire reacts in a colorless $AgNO_3$ solution to give white crystals of Ag and a light blue aqueous solution.

Metal and Aqueous Acid Solution

Another kind of single-replacement reaction is that of a metal with an aqueous acid. A metal and an aqueous acid react to produce hydrogen gas and an aqueous solution of an ionic compound. The general equation for the reaction is

$$\textbf{metal} + \textbf{aqueous acid} \rightarrow \textbf{aqueous solution} + \textbf{hydrogen gas}$$

Consider the reaction of iron metal and sulfuric acid. First, refer to the activity series to check the position of iron. Iron is above hydrogen in the series. Therefore, iron displaces hydrogen gas from the acid. If we assume that the reaction produces iron(II) sulfate, we can write the balanced chemical equation.

$$Fe(s) + H_2SO_4(aq) \longrightarrow FeSO_4(aq) + H_2(g)$$

Active Metal and Water

In this type of single-replacement reaction, an active metal and water react to give a metal hydroxide and hydrogen gas. The general equation for the reaction is

$$\textbf{metal} + \textbf{water} \rightarrow \textbf{metal hydroxide} + \textbf{hydrogen gas}$$

▲ **Calcium in Water** A piece of calcium metal in water reacts to give bubbles of hydrogen gas.

Consider the reaction of calcium metal in water to give calcium hydroxide and hydrogen gas. The balanced equation for the reaction is

$$Ca(s) + 2\,H_2O(l) \longrightarrow Ca(OH)_2(aq) + H_2(g)$$

The following example exercise further illustrates how to write balanced chemical equations for the three types of single-replacement reactions:

Example Exercise 8.13 Single-Replacement Reactions

Write a balanced chemical equation for each of the following single-replacement reactions.

(a) Nickel metal is placed in a tin(II) sulfate solution.
(b) Cobalt metal is put in a cadmium nitrate solution.
(c) Manganese chips are added to sulfuric acid.
(d) A chunk of potassium is dropped into water.

Solution

First, we will refer to the activity series for the relative positions of each of the metals. Then we will write the equations.

(a) Nickel is above tin in the activity series: Ni > Sn. Therefore, a reaction occurs and Ni displaces Sn from the solution.

$$Ni(s) + SnSO_4(aq) \rightarrow NiSO_4(aq) + Sn(s)$$

(b) Cobalt is below cadmium in the series: Cd > Co. Therefore, there is no reaction.

$$Co(s) + Cd(NO_3)_2(aq) \rightarrow NR$$

(c) Manganese is above hydrogen in the series: Mn > (H). Therefore, a reaction occurs, and Mn releases H_2 gas bubbles from the solution.

$$Mn(s) + H_2SO_4(aq) \rightarrow MnSO_4(aq) + H_2(g)$$

(d) Potassium is a Group IA/1 active metal. Therefore, a reaction occurs, and K releases H_2 gas from the solution.

$$2\,K(s) + 2\,H_2O(l) \rightarrow 2\,KOH(aq) + H_2(g)$$

Practice Exercise

Write a balanced chemical equation for each of the following single-replacement reactions:

(a) Nickel metal is placed in a silver nitrate solution.
(b) Gold metal is placed in a silver nitrate solution.
(c) A chunk of cadmium metal is dropped into hydrochloric acid.
(d) A small piece of strontium metal is dropped into water.

Answers:

(a) $Ni(s) + 2\,AgNO_3(aq) \rightarrow 2\,Ag(s) + Ni(NO_3)_2(aq)$
(b) $Au(s) + AgNO_3(aq) \rightarrow NR$
(c) $Cd(s) + 2\,HCl(aq) \rightarrow CdCl_2(aq) + H_2(g)$
(d) $Sr(s) + 2\,H_2O(l) \rightarrow Sr(OH)_2(aq) + H_2(g)$

Concept Exercise

Which of the following metals reacts with an aqueous acid: Zn, Cu, Au?

Answer: See Appendix G.

Note In a single-replacement reaction, it may not be possible to predict the ionic charge of the resulting metal cation. For example, when copper metal reacts with silver nitrate solution, the resulting compound may contain either Cu^{+} or Cu^{2+}.

8.9 SOLUBILITY RULES

OBJECTIVE

▸ To predict whether an ionic compound dissolves in water given the general rules for solubility.

In a double-replacement reaction, which we will discuss in Section 8.10, two aqueous solutions may react to form a compound that is not soluble, that is, a precipitate. When we observe that a precipitate is produced in a solution, we have evidence for a chemical reaction.

It is important to be able to identify an insoluble compound in an aqueous solution. In Table 8.2, there are general rules for predicting the solubility of ionic compounds in water. However, there are exceptions to the general rules, and a few of these exceptions are indicated; for example, Na_2CO_3 is soluble in water (Rule 6).

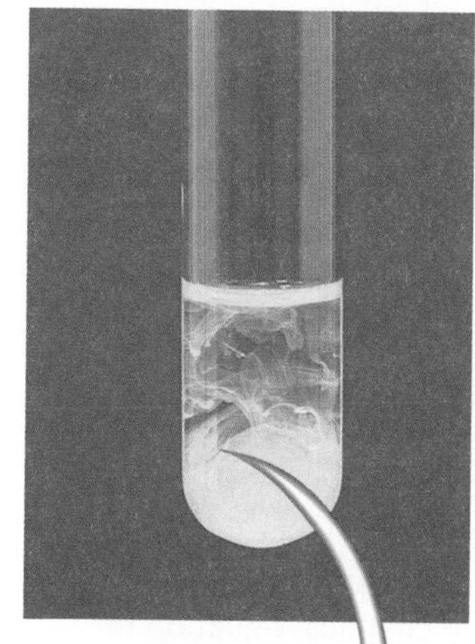

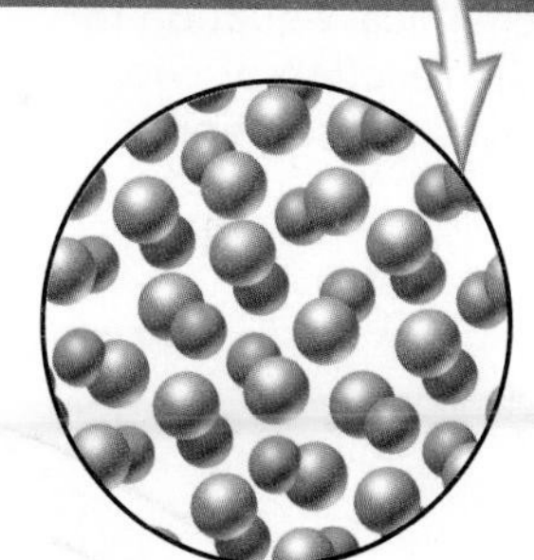

▲ **Insoluble Silver Chloride** Aqueous solutions of silver nitrate and sodium chloride react to give white, insoluble AgCl.

TABLE 8.2 SOLUBILITY RULES FOR IONIC COMPOUNDS

Compounds containing the following ions are generally *soluble* in water:

1. alkali metal ions and the ammonium ion, Li^{+}, Na^{+}, K^{+}, NH_4^{+}
2. acetate ion, $C_2H_3O_2^{-}$
3. nitrate ion, NO_3^{-}
4. halide ions (X = Cl^{-}, Br^{-}, I^{-}) (AgX, Hg_2X_2, and PbX_2 are exceptions and *insoluble*)
5. sulfate ion, an SO_4^{2-} ($SrSO_4$, $BaSO_4$, and $PbSO_4$ are exceptions and insoluble)

Compounds containing the following ions are generally *insoluble* in water:*

6. carbonate ion, CO_3^{2-} (see Rule 1 exceptions, which are *soluble*)
7. chromate ion, CrO_4^{2-} (see Rule 1 exceptions, which are *soluble*)
8. phosphate ion, PO_4^{3-} (see Rule 1 exceptions, which are *soluble*)
9. sulfide ion, S^{2-} (CaS, SrS, BaS, and Rule 1 exceptions are *soluble*)
10. hydroxide ion, OH^{-} [$Ca(OH)_2$, $Sr(OH)_2$, $Ba(OH)_2$, and Rule 1 exceptions are *soluble*]

*These compounds are actually slightly soluble, or very slightly soluble, in water.

Now let's apply the solubility rules to selected compounds and determine whether or not the compound is soluble in water.

Example Exercise 8.14 Applying Solubility Rules

State whether each of the following compounds is soluble or insoluble in water:

(a) sodium sulfate, Na_2SO_4
(b) aluminum nitrate, $Al(NO_3)_3$
(c) barium sulfate, $BaSO_4$
(d) lead(II) chromate, $PbCrO_4$
(e) ammonium sulfide, $(NH_4)_2S$

Solution

We refer to the solubility rules in Table 8.2.

(a) Sodium sulfate contains the alkali metal ion Na^{+}. According to Rule 1, Na_2SO_4 is *soluble.*
(b) Aluminum nitrate contains the nitrate ion NO_3^{-}. According to Rule 3, $Al(NO_3)_3$ is *soluble.*
(c) Barium sulfate contains the sulfate ion SO_4^{2-}. According to the Rule 5 exception, $BaSO_4$ is *insoluble.*

(d) Lead(II) chromate contains the chromate ion CrO_4^{2-}. According to Rule 7, $PbCrO_4$ is *insoluble.*
(e) Ammonium sulfide contains the ammonium ion NH_4^+. According to Rule 1, $(NH_4)_2S$ is *soluble.*

Practice Exercise

State whether each of the following compounds is soluble or insoluble in water:

(a) lead(II) acetate, $Pb(C_2H_3O_2)_2$
(b) mercury(II) bromide, $HgBr_2$
(c) magnesium carbonate, $MgCO_3$
(d) zinc phosphate, $Zn_3(PO_4)_2$
(e) calcium hydroxide, $Ca(OH)_2$

Answers:

(a) soluble (Rule 2)
(b) soluble (Rule 4)
(c) insoluble (Rule 6)
(d) insoluble (Rule 8)
(e) soluble (Rule 10 exception)

Concept Exercise

Which of the following compounds is soluble in water: $Pb(NO_3)_2$, $PbCl_2$, $PbCO_3$?

Answer: See Appendix G.

OBJECTIVE

▸ To write a balanced chemical equation for the reaction of two aqueous solutions.

8.10 DOUBLE-REPLACEMENT REACTIONS

In a double-replacement precipitation reaction, two ionic compounds in an aqueous solution switch anions and produce an insoluble substance. The general form of the reaction is

$$AX + BZ \rightarrow AZ + BX$$

In the reaction of AX and BZ, we can assume that either AZ or BX is an insoluble precipitate. If both AZ and BX are soluble, there is no precipitate, and hence there is no chemical reaction. Consider the double-replacement reaction of two aqueous solutions:

$$\textbf{aqueous solution}_1 + \textbf{aqueous solution}_2 \rightarrow \textbf{precipitate} + \textbf{aqueous solution}_3$$

As an example, when we add an aqueous solution of silver nitrate to an aqueous solution of sodium carbonate, we observe a white precipitate. In this reaction, the two compounds switch anions to give insoluble silver carbonate and sodium nitrate. We can write the balanced chemical equation as follows:

$$2AgNO_3(aq) + Na_2CO_3(aq) \rightarrow Ag_2CO_3(s) + 2\,NaNO_3(aq)$$

▲ **Insoluble Silver Carbonate** Aqueous solutions of silver nitrate and sodium carbonate react to give an insoluble precipitate of Ag_2CO_3.

By referring to solubility Rule 6, we find that most compounds containing the carbonate ion, CO_3^{2-}, are insoluble. Since silver carbonate is insoluble, it is written as $Ag_2CO_3(s)$ in the above equation. The (*s*) indicates that Ag_2CO_3 is an insoluble precipitate. The following example exercise further illustrates how to write balanced equations for double-replacement reactions:

Example Exercise 8.15 Double-Replacement Reactions

Write a balanced chemical equation for each of the following double-replacement reactions.

(a) Aqueous barium chloride is added to a potassium chromate solution.

(b) Aqueous strontium acetate is added to a lithium hydroxide solution.

Solution

For double-replacement reactions, we switch anions for the two compounds and check the solubility rules for an insoluble compound.

(a) Barium chloride and potassium chromate give barium chromate and potassium chloride. According to solubility Rule 7, we find that barium chromate is insoluble. The balanced equation is

$$BaCl_2(aq) + K_2CrO_2(aq) \rightarrow BaCrO_4(s) + 2\,KCl(aq)$$

(b) Strontium acetate and lithium hydroxide react as follows:

$$Sr(C_2H_3O_2)_2(aq) + 2\,LiOH(aq) \rightarrow Sr(OH)_2(aq) + 2\,LiC_2H_3O_2(aq)$$

However, the solubility rules indicate that $Sr(OH)_2$ and $LiC_2H_3O_2$ are soluble. Therefore, the equation is written

$$Sr(C_2H_3O_2)_2(aq) + LiOH(aq) \rightarrow NR$$

Practice Exercise

Write a balanced chemical equation for each of the following double-replacement reactions:

(a) Aqueous zinc sulfate is added to a sodium carbonate solution.

(b) Aqueous manganese(II) nitrate is added to a potassium hydroxide solution.

Answers:

(a) $ZnSO_4(aq) + Na_2CO_3(aq) \rightarrow ZnCO_3(s) + Na_2SO_4(aq)$

(b) $Mn(NO_3)_2(aq) + 2\,KOH(aq) \rightarrow Mn(OH)_2(s) + 2\,KNO_3(aq)$

Concept Exercise

Identify the insoluble product from the reaction of silver nitrate and sodium chloride.

Answer: See Appendix G.

Note In a double-replacement reaction, the ionic charge of a metal cation does not change. For example, if a reactant contains Fe^{2+}, a product will contain Fe^{2+}. Although transition metals have cations with variable charge, the ionic charge does not change during a double-replacement reaction.

8.11 NEUTRALIZATION REACTIONS

OBJECTIVE

▸ To write a balanced chemical equation for the reaction of an acid and a base.

In a neutralization reaction, an acid and a base react to produce an aqueous ionic compound and water. An **acid** is a substance that releases hydrogen ions, H^+, in water; a **base** is a substance that releases hydroxide ions, OH^-, in water. The resulting ionic compound is called a **salt** and is composed of the cation from the base and the anion from the acid. The general form of the reaction is

$$HX + B\mathbf{OH} \rightarrow BX + \mathbf{HOH}$$

▲ **Phenolphthalein Acid–Base Indicator** In an acid–base neutralization reaction, the phenolphthalein indicator signals a change. If we have acid in a flask and add base, the indicator changes from colorless to pink.

A neutralization reaction is a special case of a double-replacement reaction. Notice that we have written water as HOH to emphasize the switching of ions. You will also discover that writing water as HOH is helpful in balancing equations for neutralization reactions. In a neutralization reaction, an acid and a base react to produce a salt and water:

aqueous acid + aqueous base → aqueous salt + water

As an example, when we add hydrochloric acid to sodium hydroxide solution, we obtain sodium chloride and water. We can write the balanced chemical equation as follows:

$$HCl(aq) + NaOH(aq) \rightarrow NaCl(aq) + HOH(l)$$

Other than a slight warming, there is no evidence of reaction. To provide evidence for this neutralization reaction, we can add a drop of the acid–base indicator phenolphthalein to the HCl. Phenolphthalein is colorless in acid and pink in base. Initially, the HCl acid solution is colorless; however, as we add NaOH base, we see flashes of color and eventually the solution turns pink. With an acid–base indicator we can observe a color change, which is evidence for a chemical reaction.

The following example exercise provides additional practice in writing balanced equations for neutralization reactions:

Example Exercise 8.16 Neutralization Reactions

Write a balanced chemical equation for each of the following neutralization reactions:

(a) Nitric acid neutralizes an ammonium hydroxide solution.
(b) Sulfuric acid neutralizes a potassium hydroxide solution.

Solution

A neutralization reaction produces a salt and water.

(a) Nitric acid and ammonium hydroxide produce ammonium nitrate and water. The balanced equation is

$$HNO_3(aq) + NH_4OH(aq) \rightarrow NH_4NO_3(aq) + HOH(l)$$

(b) Sulfuric acid and potassium hydroxide produce potassium sulfate and water. The balanced equation is

$$H_2SO_4(aq) + 2\ KOH(aq) \rightarrow K_2SO_4(aq) + 2\ HOH(l)$$

Practice Exercise

Write a balanced chemical equation for each of the following neutralization reactions:

(a) Chloric acid neutralizes a strontium hydroxide solution.
(b) Phosphoric acid neutralizes a sodium hydroxide solution.

Answers:

(a) $2\ HClO_3(aq) + Sr(OH)_2(aq) \rightarrow Sr(ClO_3)_2(aq) + 2\ HOH(l)$
(b) $H_3PO_4(aq) + 3\ NaOH \rightarrow Na_3PO_4(aq) + 3\ HOH(l)$

Concept Exercise

Identify the salt product from the reaction of hydrochloric acid and sodium hydroxide.

Answer: See Appendix G.

CRITICAL THINKING Household Chemicals

Q: Is it possible that mixing two common household chemicals can produce a deadly gas?

Salt and sugar are two familiar chemicals on your kitchen table. Vinegar, another kitchen chemical, is a solution of acetic acid. Citrus fruit contains citric acid and is responsible for the sour taste of lemons and limes. Aspirin contains acetylsalicylic acid, which can irritate the lining of the stomach if taken in excess.

Hydrochloric acid, used to acidify swimming pools and clean concrete, is sold in the supermarket as muriatic acid. Sulfuric acid, found in automobile batteries, is a dangerous chemical that should be handled with great caution.

Perhaps the most dangerous chemical in the home is caustic soda, NaOH, which is sold as drain cleaner and as oven cleaner. If it contacts your skin, it gives a slippery sensation and causes redness and irritation. If taken internally, caustic soda could be fatal, especially to an infant. Household ammonia is also potentially dangerous and should not be used with bleach. Together, these two chemicals produce a poisonous gas.

▶ Common Household Chemicals

COMMON HOUSEHOLD CHEMICALS

NAME	FORMULA	PRODUCT/USE	SAFETY
Acids			
acetic acid	$HC_2H_3O_2$	vinegar	
carbonic acid	H_2CO_3	carbonated drinks	
hydrochloric acid	HCl	swimming pools	avoid contact*
sulfuric acid	H_2SO_4	battery acid	avoid contact*†
Bases			
ammonia	NH_3	window cleaner	
sodium hydrogen carbonate	$NaHCO_3$	baking soda, antacid	
magnesium hydroxide	$Mg(OH)_2$	milk of magnesia	
sodium hydroxide	$NaOH$	drain and oven cleaner	avoid contact*†
Miscellaneous			
aluminum hydroxide	$Al(OH)_3$	antacid tablets	
dry ice	solid CO_2	keep items cold	avoid contact
Epsom salts	$MgSO_4 \cdot 7H_2O$	salt bath, laxative	
sodium hypochlorite	$NaClO$	bleach	avoid contact*
Organic			
ethylene glycol	$C_2H_4(OH)_2$	antifreeze	avoid ingestion†
isopropyl alcohol	$CH_3CHOHCH_3$	rubbing alcohol	avoid ingestion†
naphthalene	$C_{10}H_8$	mothballs	avoid ingestion†
trichloroethane	$C_2H_3Cl_3$	spot remover	avoid ingestion†

*In the event of contact, flush with water.
†In the event of ingestion, seek medical attention immediately.

A: Caution must be exercised when using any chemicals; for example, household ammonia and ordinary bleach, when mixed, release a poisonous gas.

Summary

Section 8.1 In this chapter we learned that the evidence for a **chemical reaction** is any of the following: a gas is detected, a **precipitate** is formed in an **aqueous solution**, or a color change or energy change is observed. An **exothermic reaction** releases heat energy, and an **endothermic reaction** absorbs heat energy.

Section 8.2 A **reactant** substance undergoes a chemical change to become a **product**. A **catalyst** is a substance that speeds up a reaction without being permanently altered. A **chemical equation** describes a reaction using formulas and symbols. Seven nonmetal elements occur naturally as **diatomic molecules**, and their formulas are written as H_2, N_2, O_2, F_2, Cl_2, Br_2, and I_2.

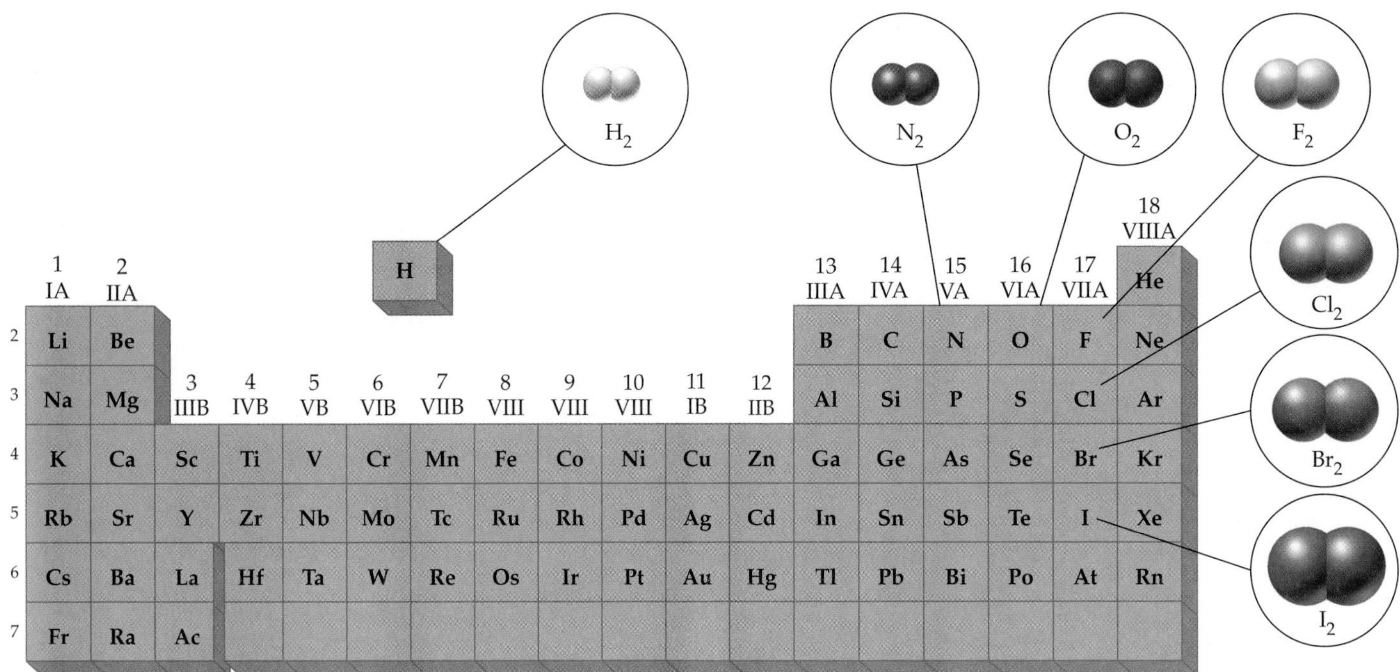

▲ **Diatomic Molecules** Seven elements occur naturally as diatomic molecules: H_2, N_2, O_2, F_2, and Cl_2 are gases; bromine, Br_2, occurs as a liquid; and iodine, I_2, as solid crystals.

Section 8.3 To balance a chemical equation, **coefficients** are systematically placed in front of each substance until there is the same number of atoms or ions of each element on both sides of the equation. If the reaction coefficient is 1, it is omitted. Similarly, if a formula **subscript** is 1, such as oxygen in H_2O, it is omitted.

Section 8.4 Although there are thousands of chemical reactions, most can be classified as one of the following: combination, decomposition, single-replacement, double-replacement, or neutralization. Table 8.3 lists these five basic types.

TABLE 8.3 BASIC TYPES OF CHEMICAL REACTIONS

REACTION TYPE	GENERAL FORMAT
Combination reaction	A + Z → AZ
Decomposition reaction	AZ → A + Z
Single-replacement reaction	A + BZ → AZ + B
Double-replacement reaction	AX + BZ → AZ + BX
Neutralization reaction	HX + BOH → BX + HOH

Section 8.5 We have examined three kinds of combination reactions: a metal and oxygen, a nonmetal and oxygen, and a metal and a nonmetal. Often, we can predict the products from the reaction. Nonmetals, however, combine with oxygen to give a variety of oxides, and so the product is not predictable. In summary,

1. **metal + oxygen gas → metal oxide**
2. **nonmetal + oxygen gas → nonmetal oxide**
3. **metal + nonmetal → binary ionic compound**

Section 8.6 We have examined three kinds of decomposition reactions. In the first type, a metal hydrogen carbonate is decomposed by heating to give a metal carbonate product. In the second, a metal carbonate is decomposed to a metal oxide. Most oxygen-containing polyatomic ions, such as chlorate and nitrate, decompose to give oxygen gas. In summary,

1. **metal hydrogen carbonate → metal carbonate + water + carbon dioxide**
2. **metal carbonate → metal oxide + carbon dioxide**
3. **oxygen-containing compound → oxygen gas**

Section 8.7 The **activity series** lists the relative ability of a metal to undergo reaction. A metal higher in the series always displaces a metal lower in the series. Metals that are more active than (H) displace hydrogen gas from an acid solution. **Active metals** include Li, Na, K, Ba, Sr, and Ca; each of these metals displaces hydrogen gas from water.

Section 8.8 We have examined three kinds of single-replacement reactions: a metal and aqueous solution, a metal and a dilute acid, and an active metal and water. In summary,

1. **metal$_1$ + aqueous solution$_1$ → metal$_2$ + aqueous solution$_2$**
2. **metal + aqueous acid → aqueous solution + hydrogen gas**
3. **metal + water → metal hydroxide + hydrogen gas**

Section 8.9 An insoluble substance does not dissolve completely in water although it may dissolve slightly. To determine whether a substance is soluble or insoluble, we will simply refer to the solubility rules in Table 8.2.

Section 8.10 In a double-replacement precipitation reaction, two ionic compounds react in an aqueous solution. The two compounds react by switching anions and forming an insoluble substance. In general,

aqueous solution$_1$ + aqueous solution$_2$ → precipitate + aqueous solution$_3$

Section 8.11 In a neutralization reaction an **acid** and a **base** react to yield a **salt** and water.k A neutralization reaction is a special type of double-replacement reaction. It is helpful to write the formula for water as HOH in balancing neutralization reactions:

aqueous acid + aqueous base → aqueous salt + water

Key Terms *Answers to Key Terms are in Appendix H.*

Select the key term that corresponds to each of the following definitions.

_____ **1.** the process of undergoing a chemical change
_____ **2.** a chemical reaction that releases heat energy
_____ **3.** a chemical reaction that absorbs heat energy
_____ **4.** a substance dissolved in water
_____ **5.** an insoluble solid substance produced from a reaction in aqueous solution
_____ **6.** a representation using formulas and symbols to describe a chemical change
_____ **7.** a substance undergoing a chemical reaction
_____ **8.** a substance resulting from a chemical reaction
_____ **9.** a substance that speeds up a reaction without being permanently changed
_____ **10.** a particle composed of two nonmetal atoms
_____ **11.** a digit in front of a chemical formula that helps to balance a chemical equation
_____ **12.** a digit in a chemical formula that represents the number of atoms or ions appearing in the substance
_____ **13.** a type of reaction in which two substances produce a single compound
_____ **14.** a type of reaction in which a single compound produces two or more substances
_____ **15.** a type of reaction in which a more active metal displaces a less active metal from a solution
_____ **16.** a type of reaction in which two cations in different compounds exchange anions
_____ **17.** a type of reaction in which an acid and a base produce a salt and water
_____ **18.** a relative order of metals arranged by their ability to undergo a reaction
_____ **19.** a metal that reacts with water at room temperature
_____ **20.** a substance that releases hydrogen ions in water
_____ **21.** a substance that releases hydroxide ions in water
_____ **22.** an ionic compound produced by an acid–base reaction

(a) acid *(Sec. 8.11)*
(b) active metal *(Sec. 8.7)*
(c) activity series *(Sec. 8.7)*
(d) aqueous solution (*aq*) *(Sec. 8.1)*
(e) base *(Sec. 8.11)*
(f) catalyst *(Sec. 8.2)*
(g) chemical equation *(Sec. 8.2)*
(h) chemical reaction *(Sec. 8.1)*
(i) coefficient *(Sec. 8.3)*
(j) combination reaction *(Sec. 8.4)*
(k) decomposition reaction *(Sec. 8.4)*
(l) diatomic molecule *(Sec. 8.2)*
(m) double-replacement reaction *(Sec. 8.4)*
(n) endothermic reaction *(Sec. 8.1)*
(o) exothermic reaction *(Sec. 8.1)*
(p) neutralization reaction *(Sec. 8.4)*
(q) precipitate *(Sec. 8.1)*
(r) product *(Sec. 8.2)*
(s) reactant *(Sec. 8.2)*
(t) salt *(Sec. 8.11)*
(u) single-replacement reaction *(Sec. 8.4)*
(v) subscript *(Sec. 8.3)*

Exercises *Answers to odd-numbered Exercises are in Appendix I.*

Evidence for Chemical Reactions (Sec. 8.1)

1. Which of the following is evidence for a chemical reaction?
   (a) Heating ammonium carbonate gives ammonia gas.
   (b) Heating baking soda gives a flame-extinguishing gas.
2. Which of the following is evidence for a chemical reaction?
   (a) Heating iodine crystals produces a purple gas.
   (b) Heating a solid compound produces a liquid.
3. Which of the following is evidence for a chemical reaction?
   (a) Mixing two aqueous solutions produces a precipitate.
   (b) Mixing two solutions gives an increased volume.
4. Which of the following is evidence for a chemical reaction?
   (a) Mixing two aqueous solutions produces gas bubbles.
   (b) Mixing two colorless solutions gives a blue solution.
5. Which of the following is evidence for a chemical reaction?
   (a) Dropping potassium metal into water produces fizzing.
   (b) Dropping sodium metal into water gives a yellow flame.
6. Which of the following is evidence for a chemical reaction?
   (a) Heating gold in air produces a black solid.
   (b) Heating gold in a vacuum jar gives a yellow film.

Writing Chemical Equations (Sec. 8.2)

7. Write a chemical equation for magnesium metal combining with bromine liquid to give magnesium bromide solid.
8. Write a chemical equation for iron metal combining with chlorine gas to give yellow ferrous chloride solid.

◀ **Iron in Chlorine Gas** Iron wool reacts in chlorine gas to give sparks and a yellow-brown $FeCl_3$ residue.

9. Write a chemical equation for solid zinc hydrogen carbonate decomposing to yield solid zinc carbonate, water, and carbon dioxide gas.
10. Write a chemical equation for solid cadmium hydrogen carbonate decomposing to yield solid cadmium carbonate, water, and carbon dioxide gas.
11. Write a chemical equation for the reaction of cadmium metal with aqueous cobalt(II) nitrate to produce aqueous cadmium nitrate and cobalt metal.
12. Write a chemical equation for the reaction of manganese metal with aqueous nickel(II) nitrate to produce aqueous manganese(II) nitrate and nickel metal.
13. Write a chemical equation for the reaction of aqueous solutions of lithium iodide and silver nitrate to give silver iodide precipitate and aqueous lithium nitrate.
14. Write a chemical equation for the reaction of aqueous solutions of potassium chromate and calcium sulfate to give calcium chromate precipitate and aqueous potassium sulfate.
15. Write a chemical equation for the neutralization of nitric acid with aqueous sodium hydroxide to give aqueous sodium nitrate plus water.
16. Write a chemical equation for the neutralization of acetic acid with aqueous ammonium hydroxide to give aqueous ammonium acetate plus water.

Balancing Chemical Equations (Sec. 8.3)

17. Balance each of the following chemical equations by inspection:
   (a) $H_2(g) + N_2(g) \rightarrow NH_3(g)$
   (b) $Al_2(CO_3)_3(s) \rightarrow Al_2O_3(s) + CO_2(g)$
   (c) $Sr(s) + H_2O(l) \rightarrow Sr(OH)_2(aq) + H_2(g)$
   (d) $K_2SO_4(aq) + Ba(OH)_2(aq) \rightarrow BaSO_4(s) + KOH(aq)$
   (e) $H_3PO_4(aq) + Mn(OH)_2(s) \rightarrow Mn_3(PO_4)_2(s) + H_2O(l)$
18. Balance each of the following chemical equations by inspection:
   (a) $Co(s) + O_2(g) \rightarrow Co_2O_3(s)$
   (b) $LiClO_3(s) \rightarrow LiCl(s) + O_2(g)$
   (c) $Cu(s) + AgC_2H_3O_2(aq) \rightarrow Cu(C_2H_3O_2)_2(aq) + Ag(s)$
   (d) $Pb(NO_3)_2(aq) + LiCl(aq) \rightarrow PbCl_2(s) + LiNO_3(aq)$
   (e) $H_2SO_4(aq) + Al(OH)_3(aq) \rightarrow Al_2(SO_4)_3(aq) + H_2O(l)$
19. Balance each of the following chemical equations by inspection:
   (a) $H_2CO_3(aq) + NH_4OH(aq) \rightarrow (NH_4)_2CO_3(aq) + H_2O(l)$
   (b) $Hg_2(NO_3)_2(aq) + NaBr(aq) \rightarrow Hg_2Br_2(s) + NaNO_3(aq)$
   (c) $Mg(s) + HC_2H_3O_2(aq) \rightarrow Mg(C_2H_3O_2)_2(aq) + H_2(g)$
   (d) $LiNO_3(s) \rightarrow LiNO_2(s) + O_2(g)$
   (e) $Pb(s) + O_2(g) \rightarrow PbO(s)$
20. Balance each of the following chemical equations by inspection:
   (a) $HClO_4(aq) + Ba(OH)_2(s) \rightarrow Ba(ClO_4)_2(s) + H_2O(l)$
   (b) $Co(NO_3)_2(aq) + H_2S(g) \rightarrow CoS(s) + HNO_3(aq)$
   (c) $Fe(s) + Cd(NO_3)_2(aq) \rightarrow Fe(NO_3)_3(aq) + Cd(s)$
   (d) $Fe_2(CO_3)_3(s) \rightarrow Fe_2O_3(s) + CO_2(g)$
   (e) $Sn(s) + P(s) \rightarrow Sn_3P_2(s)$

Classifying Chemical Reactions (Sec. 8.4)

21. Classify each reaction in Exercise 17 as one of the following: combination, decomposition, single-replacement, double-replacement, or neutralization.
22. Classify each reaction in Exercise 18 as one of the following: combination, decomposition, single-replacement, double-replacement, or neutralization.
23. Classify each reaction in Exercise 19 as one of the following: combination, decomposition, single-replacement, double-replacement, or neutralization.

24. Classify each reaction in Exercise 20 as one of the following: combination, decomposition, single-replacement, double-replacement, or neutralization.

Combination Reactions (Sec. 8.5)

25. Write a balanced equation for each of the following combination reactions:
   (a) Nickel metal is heated with oxygen gas to produce nickel(II) oxide.
   (b) Iron wool is heated with oxygen gas to produce gray iron(III) oxide.

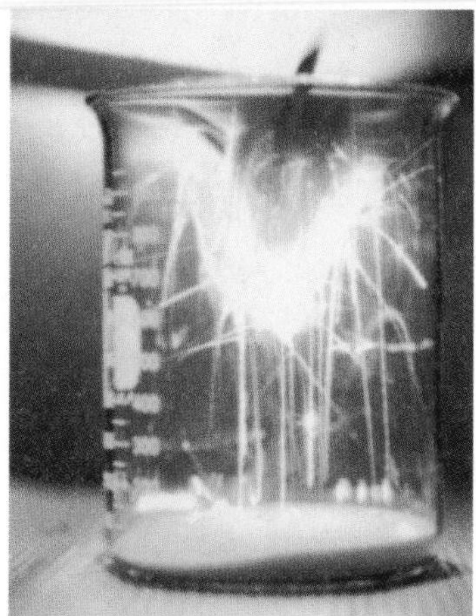

◀ **Iron in Oxygen Gas** Iron wool reacts in oxygen gas to give bright white light and a gray Fe_2O_3 residue.

26. Write a balanced equation for each of the following combination reactions:
   (a) Tin metal is heated with oxygen gas to yield tin(IV) oxide.
   (b) Lead metal is heated with oxygen gas to yield lead(II) oxide.

27. Write a balanced equation for each of the following combination reactions:
   (a) Carbon is heated with oxygen to give carbon monoxide gas.
   (b) Phosphorus is heated with oxygen to give diphosphorus pentaoxide.

28. Write a balanced equation for each of the following combination reactions:
   (a) Nitrogen is heated with oxygen to form dinitrogen pentaoxide gas.
   (b) Chlorine is heated with oxygen to form dichlorine trioxide gas.

29. Write a balanced equation for each of the following combination reactions:
   (a) Copper is heated with chlorine gas to produce solid copper(I) chloride.
   (b) Cobalt is heated with sulfur powder to produce cobalt(II) sulfide.

30. Write a balanced equation for each of the following combination reactions:
   (a) Iron is heated with fluorine gas to yield iron(III) fluoride.
   (b) Lead is heated with phosphorus powder to yield lead(IV) phosphide.

31. Write a balanced equation for each of the following combination reactions:
   (a) Chromium is heated with oxygen to give chromium(III) oxide.
   (b) Chromium is heated with nitrogen to give chromium(III) nitride.

32. Write a balanced equation for each of the following combination reactions:
   (a) Sulfur is heated with oxygen to form sulfur dioxide gas.
   (b) Sulfur is heated with oxygen and Pt catalyst to form sulfur trioxide gas.

33. Complete and balance each of the following combination reactions:
   (a) $Li + O_2 \rightarrow$ **(b)** $Ca + O_2 \rightarrow$

34. Complete and balance each of the following combination reactions:
   (a) $Sr + O_2 \rightarrow$ **(b)** $Al + O_2 \rightarrow$

35. Complete and balance each of the following combination reactions:
   (a) $Na + I_2 \rightarrow$ **(b)** $Ba + N_2 \rightarrow$

36. Complete and balance each of the following combination reactions:
   (a) $Zn + P \rightarrow$ **(b)** $Al + S \rightarrow$

Decomposition Reactions (Sec. 8.6)

37. Write a balanced equation for each of the following decomposition reactions:
   (a) Potassium hydrogen carbonate decomposes by heating to give solid potassium carbonate, water, and carbon dioxide gas.
   (b) Strontium hydrogen carbonate decomposes by heating to give solid strontium carbonate, water, and carbon dioxide gas.

38. Write a balanced equation for each of the following decomposition reactions:
   (a) Silver hydrogen carbonate decomposes by heating to give solid silver carbonate, water, and carbon dioxide gas.
   (b) Nickel(II) hydrogen carbonate decomposes by heating to give nickel(II) carbonate, water, and carbon dioxide gas.

39. Write a balanced equation for each of the following decomposition reactions:
   (a) Cobalt(III) carbonate decomposes with heat to give solid cobalt(III) oxide, and carbon dioxide gas.
   (b) Tin(IV) carbonate decomposes with heat to give solid tin(IV) oxide, and carbon dioxide gas.

40. Write a balanced equation for each of the following decomposition reactions:
   (a) Chromium(III) carbonate decomposes with heat to give chromium(III) oxide and carbon dioxide gas.
   (b) Lead(IV) carbonate decomposes with heat to give lead(IV) oxide and carbon dioxide gas.

41. Write a balanced equation for each of the following decomposition reactions:
   (a) Calcium nitrate decomposes on heating to give calcium nitrite and oxygen gas.
   (b) Silver sulfate decomposes on heating to give silver sulfite and oxygen gas.

42. Write a balanced equation for each of each of the following decomposition reactions:
   (a) Stannous chlorate decomposes on heating to give stannous chloride and oxygen gas.
   (b) Plumbic oxide decomposes on heating to give plumbous oxide and oxygen gas.

43. Complete and balance each of the following decomposition reactions:
(a) $AgHCO_3(s) \rightarrow$ **(b)** $Zn(HCO_3)_2(s) \rightarrow$
44. Complete and balance each of the following decomposition reactions:
(a) $Li_2CO_3(s) \rightarrow$ **(b)** $CdCO_3(s) \rightarrow$
45. Balance each of the following decomposition reactions:
(a) $NaClO_3(s) \rightarrow NaCl(s) + O_2(g)$
(b) $Ca(NO_3)_2(s) \rightarrow Ca(NO_2)_2(s) + O_2(g)$
46. Balance each of the following decomposition reactions:
(a) $AlPO_4(s) \rightarrow AlPO_3(s) + O_2(g)$
(b) $SnSO_4(s) \rightarrow SnSO_3(s) + O_2(g)$

The Activity Series Concept (Sec. 8.7)

(Refer to the activity series in Section 8.7, or refer to Appendix C, to answer the following questions.)

47. Predict which of the following metals reacts with aqueous nickel(II) nitrate:
(a) Ag **(b)** Sn
(c) Co **(d)** Mn
48. Predict which of the following metals reacts with aqueous iron(II) nitrate:
(a) Hg **(b)** Zn
(c) Cd **(d)** Mg
49. Predict which of the following metals reacts with dilute hydrochloric acid:
(a) Ni **(b)** Zn
(c) Cu **(d)** Al
50. Predict which of the following metals reacts with dilute sulfuric acid:
(a) Fe **(b)** Mn
(c) Cd **(d)** Au
51. Predict which of the following metals reacts with water at room temperature:
(a) Li **(b)** Mg
(c) Ca **(d)** Al
52. Predict which of the following metals reacts with water at room temperature:
(a) Ba **(b)** Mn
(c) Sn **(d)** K

Single-Replacement Reactions (Sec. 8.8)

53. Write a balanced equation for each of the following single-replacement reactions:
(a) Copper wire is placed in an aluminum nitrate solution.
(b) Aluminum wire is placed in a copper(II) nitrate solution.
54. Write a balanced equation for each of the following single-replacement reactions:
(a) Silver wire is put into a cadmium nitrate solution.
(b) Cadmium wire is put into a silver nitrate solution.
55. Write a balanced equation for each of the following single-replacement reactions:
(a) Nickel metal is put into a lead(II) acetate solution.
(b) Lead metal is put into a nickel(II) acetate solution.
56. Write a balanced equation for each of the following single-replacement reactions:
(a) Iron filings are added to a mercury(II) sulfate solution.
(b) Drops of liquid mercury are added to a ferrous sulfate solution.
57. Write a balanced equation for each of the following single-replacement reactions:
(a) Magnesium metal is added to nitric acid.
(b) Manganese metal is added to hydrochloric acid.
58. Write a balanced equation for each of the following single-replacement reactions:
(a) Zinc granules are placed in carbonic acid.
(b) Cadmium metal is added to acetic acid.
59. Write a balanced equation for each of the following single-replacement reactions:
(a) A soft, gray piece of lithium is added to water.
(b) A small, shiny piece of barium is added to water.
60. Write a balanced equation for each of the following single-replacement reactions:
(a) A gray chunk of cesium is added to water.
(b) A piece of radioactive radium is added to water.
61. Complete and balance each of the following single-replacement reactions:
(a) $Zn(s) + Pb(NO_3)_2(aq) \rightarrow$
(b) $Cd(s) + Fe(NO_3)_2(aq) \rightarrow$

▶ **Zinc in a Lead Solution**
Zinc metal reacts in an aqueous $Pb(NO_3)_2$ solution to give a gray deposit of spongy Pb metal.

62. Complete and balance each of the following single-replacement reactions:
(a) $Mg(s) + NiSO_4(aq) \rightarrow$
(b) $Al(s) + SnSO_4(aq) \rightarrow$
63. Complete and balance each of the following single-replacement reactions:
(a) $Zn(s) + HNO_3(aq) \rightarrow$
(b) $Cd(s) + HNO_3(aq) \rightarrow$

64. Complete and balance each of the following single-replacement reactions:
(a) $Mg(s) + H_2SO_4(aq) \rightarrow$
(b) $Al(s) + H_2SO_4(aq) \rightarrow$

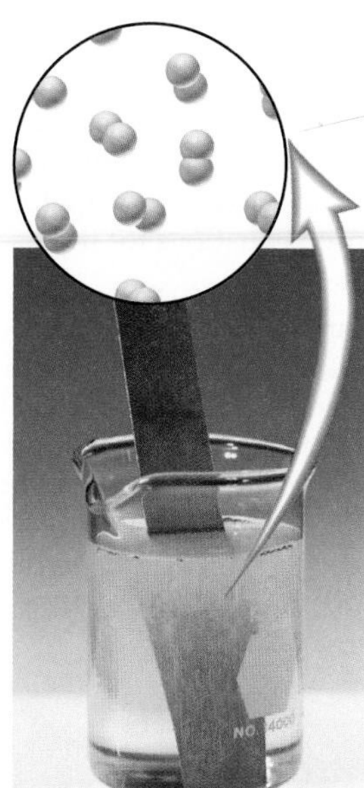

◀ **Iron in Acid** A strip of iron metal reacts in an aqueous sulfuric acid, H_2SO_4, solution to give bubbles of hydrogen gas.

65. Complete and balance each of the following single-replacement reactions:
(a) $K(s) + H_2O(l) \rightarrow$
(b) $Ba(s) + H_2O(l) \rightarrow$

66. Complete and balance each of the following single-replacement reactions:
(a) $Mg(s) + H_2O(l) \rightarrow$
(b) $Ca(s) + H_2O(l) \rightarrow$

Solubility Rules (Sec. 8.9)

(Refer to the solubility rules in Table 8.2, or Appendix D, to answer the following questions.)

67. Predict which of the following compounds are soluble in water:
(a) aluminum nitrate, $Al(NO_3)_3$
(b) sodium sulfate, Na_2SO_4
(c) cobalt(II) hydroxide, $Co(OH)_2$
(d) iron(III) phosphate, $FePO_4$

68. Predict which of the following compounds are soluble in water:
(a) ammonium sulfide, $(NH_4)_2S$
(b) zinc carbonate, $ZnCO_3$
(c) tin(II) chromate, $SnCrO_4$
(d) lead(II) acetate, $Pb(C_2H_3O_2)_2$

69. Predict which of the following compounds are insoluble in water:
(a) mercury(I) chloride, Hg_2Cl_2
(b) mercury(II) chloride, $HgCl_2$
(c) silver bromide, AgBr
(d) lead(II) iodide, PbI_2

70. Predict which of the following compounds are insoluble in water:
(a) strontium carbonate, $SrCO_3$
(b) calcium hydroxide, $Ca(OH)_2$
(c) nickel(II) sulfide, NiS
(d) mercury(II) bromide, $HgBr_2$

Double-Replacement Reactions (Sec. 8.10)

71. Write a balanced equation for each of the following double-replacement reactions:
(a) Aqueous solutions of chromium(III) iodide and sodium hydroxide react to give aqueous sodium iodide and a chromium(III) hydroxide precipitate.
(b) Aqueous solutions of nickel(II) sulfate and mercury(I) nitrate react to give aqueous nickel(II) nitrate and a mercury(I) sulfate precipitate.

72. Write a balanced equation for each of the following double-replacement reactions:
(a) Aqueous solutions of tin(II) chloride and sodium sulfide react to give aqueous sodium chloride and a tin(II) sulfide precipitate.
(b) Aqueous solutions of cobalt(II) nitrate and potassium chromate react to give aqueous potassium nitrate and a cobalt(II) chromate precipitate.

73. Complete and balance each of the following double-replacement reactions:
(a) $AgC_2H_3O_2(aq) + SrI_2(aq) \rightarrow$
(b) $FeSO_4(aq) + Ca(OH)_2(aq) \rightarrow$

74. Complete and balance each of the following double-replacement reactions:
(a) $MgSO_4(aq) + BaCl_2(aq) \rightarrow$
(b) $AlBr_3(aq) + Na_2CO_3(aq) \rightarrow$

Neutralization Reactions (Sec. 8.11)

75. Write a balanced equation for each of the following neutralization reactions:
(a) Lithium hydroxide solution is added to carbonic acid.
(b) Calcium hydroxide solution is added to nitric acid.

76. Write a balanced equation for each of the following neutralization reactions:
(a) Potassium hydroxide solution is added to phosphoric acid.
(b) Strontium hydroxide solution is added to acetic acid.

77. Complete and balance each of the following neutralization reactions:
(a) $HNO_3(aq) + Ba(OH)_2(aq) \rightarrow$
(b) $H_2SO_4(aq) + NaOH(aq) \rightarrow$

78. Complete and balance each of the following neutralization reactions:
(a) $HNO_2(aq) + Ba(OH)_2(aq) \rightarrow$
(b) $H_2SO_3(aq) + NaOH(aq) \rightarrow$

General Exercises

79. If a chemical equation is impossible to balance, what is most likely the problem?

80. If the sum of the coefficients of reactants is equal to 5, is the sum of the coefficents of products equal to 5?

81. Balance each of the following chemical equations by inspection:
(a) $Fe(s) + H_2O(g) \rightarrow Fe_3O_4(s) + H_2(g)$
(b) $FeS(s) + O_2(g) \rightarrow Fe_2O_3(s) + SO_2(g)$

82. Balance each of the following chemical equations by inspection:
(a) $FeO(l) + Al(l) \rightarrow Al_2O_3(l) + Fe(l)$
(b) $MnO_2(l) + Al(l) \rightarrow Al_2O_3(l) + Mn(l)$

83. Balance each of the following chemical equations by inspection:
(a) $F_2(g) + NaBr(aq) \rightarrow Br_2(l) + NaF(aq)$
(b) $Sb_2S_3(s) + HCl(aq) \rightarrow SbCl_3(aq) + H_2S(g)$

84. Balance each of the following chemical equations by inspection:
(a) $PCl_5(s) + H_2O(l) \rightarrow H_3PO_4(aq) + HCl(aq)$
(b) $TiCl_4(s) + H_2O(g) \rightarrow TiO_2(s) + HCl(g)$

85. Balance each of the following combustion reactions by inspection:
(a) $CH_4(g) + O_2(g) \rightarrow CO_2(g) + H_2O(g)$
(b) $C_3H_8(g) + O_2(g) \rightarrow CO_2(g) + H_2O(g)$

86. Balance each of the following combustion reactions by inspection:
(a) $CH_4O(l) + O_2(g) \rightarrow CO_2(g) + H_2O(g)$
(b) $C_3H_8O(l) + O_2(g) \rightarrow CO_2(g) + H_2O(g)$

87. Iron is prepared industrially by passing carbon monoxide gas through molten iron ore, Fe_2O_3, in a blast furnace at 1500°C. Assuming the only products are molten iron and carbon dioxide gas, write a balanced chemical equation for the manufacture of iron.

88. Chlorine is prepared industrially by heating hydrogen chloride gas with oxygen gas. Assuming the only products are water and chlorine gas, write a balanced chemical equation for the manufacture of chlorine.

89. The Ostwald process is an important industrial method for making nitric acid. In the first step, ammonia and oxygen are heated to give nitrogen monoxide gas and water. In the second step, nitrogen monoxide and oxygen react to produce nitrogen dioxide. In the third step, nitrogen dioxide gas is passed through water to yield nitric acid and nitrogen monoxide gas. Write three balanced equations for the Ostwald process.

90. The Contact process is the industrial method for manufacturing sulfuric acid. In the first step, sulfur is burned with oxygen gas to give sulfur dioxide. In the second step, sulfur dioxide and oxygen gas react in the presence of a platinum catalyst to produce sulfur trioxide. In the third step, sulfur trioxide gas is passed through water to yield sulfuric acid. Write three balanced equations for the Contact process.

8 Chapter Self-Test

Answers to Self-Test are in Appendix J.

1. Which of the following is evidence for a chemical reaction? (Sec. 8.1)
(a) a gas is released
(b) a precipitate is formed
(c) a permanent color change
(d) all of the above
(e) none of the above

2. Which of the following represents an element in its natural state? (Sec. 8.2)
(a) F_2 (b) Cl_2
(c) Br_2 (d) I_2
(e) all of the above

3. What is the coefficient of hydrogen gas after balancing the following equation? (Sec. 8.3)

$$\_N_2(g) + \_H_2(g) \rightarrow \_NH_3(g)$$

(a) 1 (b) 2
(c) 3 (d) 4
(e) none of the above

4. Which type of chemical reaction is illustrated in question 3? (Sec. 8.4)
(a) combination reaction
(b) decomposition reaction
(c) single-replacement reaction
(d) double-replacement reaction
(e) neutralization reaction

5. Predict the product from the following reaction: (Sec. 8.5)

$$Mg(s) + O_2 \xrightarrow{\Delta}$$

(a) MgO (b) Mg_2O
(c) MgO_2 (d) Mg_2O_3
(e) Mg_3O_2

6. Predict the products from the following reaction: (Sec. 8.6)

$$LiHCO_3(s) \xrightarrow{\Delta}$$

(a) Li, H_2, and CO_2 (b) Li, H_2O, and CO_2
(c) Li_2CO_3, H_2, and CO_2 (d) Li_2CO_3, H_2O, and CO_2
(e) Li_2CO_3 and H_2O

7. Which of the following metals reacts with sulfuric acid? (Sec. 8.7)
(a) Ag (b) Cu
(c) Fe (d) all of the above
(e) none of the above

8. What are the products from the following reaction? (Sec. 8.8)

$$Mg(s) + H_2SO_4(aq) \rightarrow$$

(a) MgO and H_2SO_3 (b) MgO and H_2S
(c) $MgSO_4$ and H_2 (d) $MgSO_4$ and H_2O
(e) no reaction

9. Which of the following compounds is *insoluble* in water? (Sec. 8.9)
   (a) $(NH_4)_2CO_3$ **(b)** $BaSO_4$
   (c) K_2CrO_4 **(d)** Na_2S
   (e) $Sr(OH)_2$

10. What are the products from the following reaction? (Sec. 8.10)

$$Fe(NO_3)_3(aq) + Na_3PO_4(aq) \rightarrow$$

   (a) FeP and $NaNO_3$ **(b)** $FePO_3$ and $NaNO_2$
   (c) $FePO_3$ and $NaNO_3$ **(d)** $FePO_4$ and $NaNO_2$
   (e) $FePO_4$ and $NaNO_3$

11. What are the predicted products from the following neutralization reaction? (Sec. 8.11)

$$H_2SO_4(aq) + NH_4OH(aq) \rightarrow$$

   (a) NH_4SO_3 and H_2O **(b)** NH_4SO_4 and O_2
   (c) $(NH_4)_2SO_4$ and O_2 **(d)** $(NH_4)_2SO_4$ and H_2O
   (e) no reaction

Key Concepts

12. Write a balanced equation for the combination of iron metal with oxygen gas to give solid iron(III) oxide.
13. Write a balanced equation for the decomposition of solid calcium hydrogen carbonate with heat to give calcium carbonate powder, steam, and carbon dioxide gas.
14. Write a balanced equation for the reaction of zinc metal with aqueous nickel(II) nitrate to produce aqueous zinc nitrate and nickel metal.
15. Write a balanced equation for the neutralization of nitric acid with aqueous ammonium hydroxide to give aqueous ammonium nitrate and water.
16. Write a balanced equation for the combustion of ethanol, C_2H_5OH, with oxygen to give carbon dioxide and water.

Critical Thinking

17. Write a balanced equation for the reaction of hydrochloric acid dripping into a test tube containing baking soda as shown in the illustration.

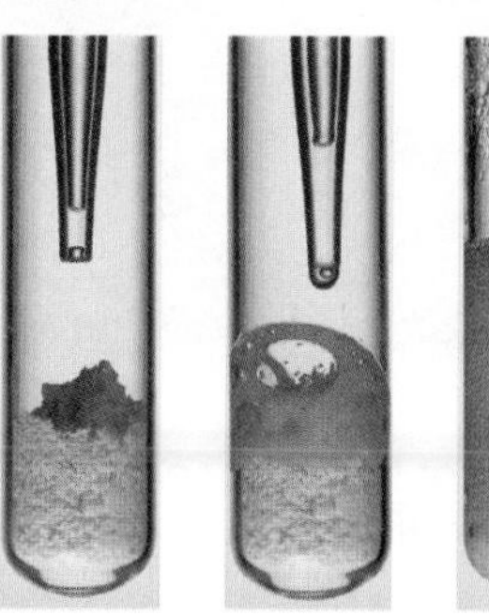

18. Write a balanced equation for the reaction of potassium iodide solution pouring into a beaker containing lead(II) nitrate solution as shown in the illustration.

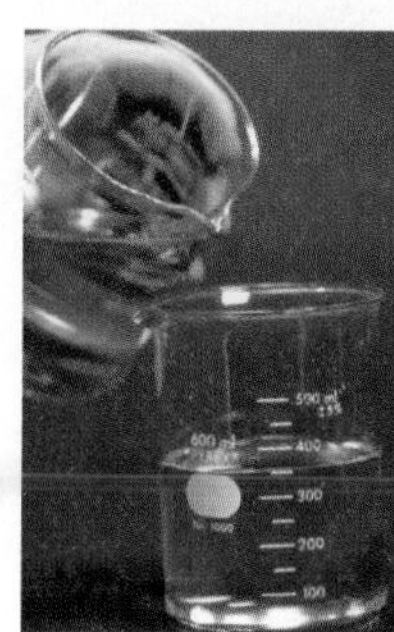

CHAPTER

9 The Mole Concept

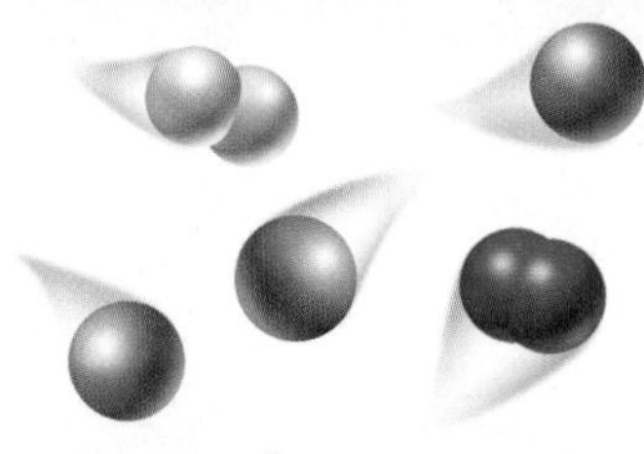

The Moon's atmosphere has only about 100 atoms of helium, neon, argon, and hydrogen per cubic centimeter. The Earth's atmosphere has 30,000,000,000,000,000,000,000,000 molecules per cubic centimeter.

"Chance favors the prepared mind."
Louis Pasteur, French Chemist (1822–1895)

Is it possible to keep track of atoms by counting them? The answer is yes, although we cannot count them directly. Atoms are much too small to count individually, so we count them in groups. We count atoms in the same way we count eggs by the dozen (12), pencils by the gross (144), and sheets of paper by the ream (500). Since atoms are so tiny, we must consider a very large group of atoms.

When considering a very large number, it may be difficult to comprehend its magnitude. In this chapter, we will encounter an extremely large number—602,000,000,000,000,000,000,000. This number refers to a group of atoms and is called Avogadro's number. To appreciate the enormity of this number, consider the following analogy. If you had Avogadro's number of marbles, the volume of the group of marbles would be the size of the Moon.

9.1 AVOGADRO'S NUMBER

OBJECTIVES

- To state the value of Avogadro's number: 6.02×10^{23}.
- To state the mass of Avogadro's number of atoms for any element by referring to the periodic table.

To count the number of atoms in a sample, consider the following experiment. Let's weigh enough atoms so that the mass in grams has the same numerical value as the atomic mass expressed in atomic mass units. For example, let's weigh enough carbon atoms to equal a mass of 12.01 g (atomic mass = 12.01 amu).

But how many atoms are in 12.01 g of carbon? Since atoms are so tiny, the number of carbon atoms in 12.01 g must be extremely large. Experiments have shown that there are 6.02×10^{23} carbon atoms in 12.01 g. This very large number is referred to as **Avogadro's number** (symbol *N*) in honor of the Italian scientist Amedeo Avogadro (1776–1856). Avogadro's number is sometimes referred to as the "chemist's dozen."

To count Avogadro's number of atoms we only have to know the atomic mass of the element. Since one carbon atom has a mass of 12.01 amu, then 6.02×10^{23} carbon atoms have a mass of 12.01 g. The following example exercise further illustrates how to count Avogadro's number of atoms.

CHEMISTRY CONNECTION Analogies for Avogadro's Number

Q: If the Moon were composed of a collection of small marbles, how many marbles correspond to the size of the Moon?

Length Analogy: If 6.02×10^{23} hydrogen atoms were laid side by side, the total length would be long enough to encircle Earth about a million times.

Mass Analogy: The mass of 6.02×10^{23} Olympic shotput balls would be about equal to the mass of Earth.

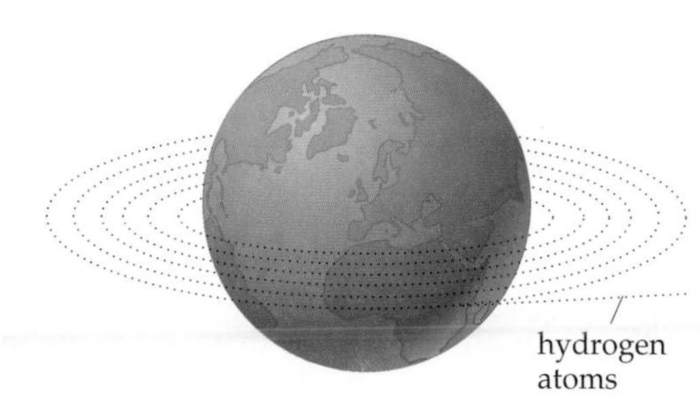

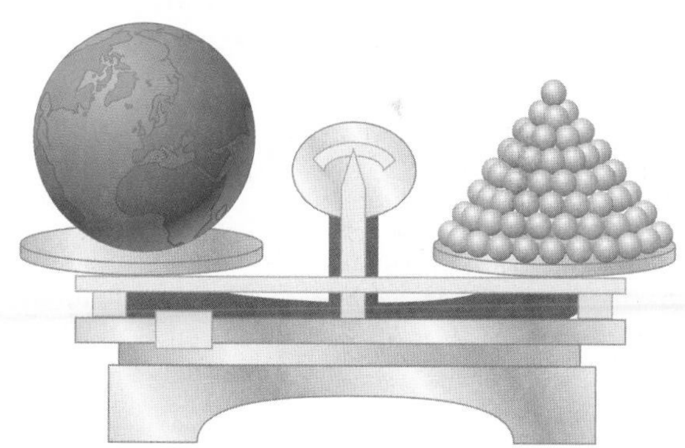

Volume Analogy: The volume occupied by 6.02×10^{23} softballs would be about the size of Earth.

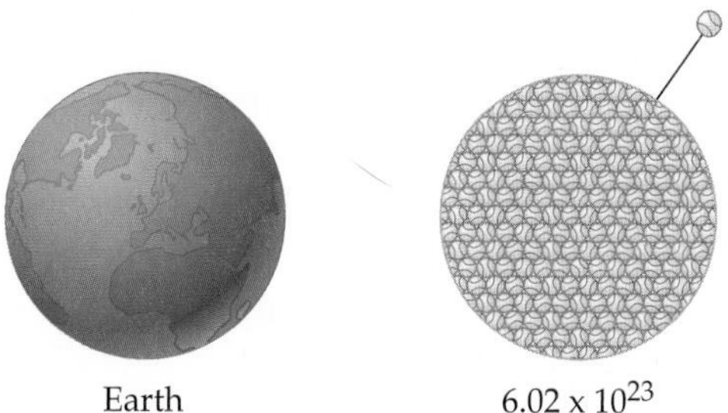

A: It would require 6.02×10^{23} ordinary small marbles to equal the size of the Moon.

Example Exercise 9.1 Atomic Mass and Avogadro's Number

Refer to the atomic masses in the periodic table inside the front cover of this textbook. State the mass of Avogadro's number of atoms for each of the following elements:

(a) copper
(b) mercury
(c) sulfur
(d) helium

Solution

The atomic mass of each element is listed below the symbol of the element in the periodic table: Cu = 63.55 amu, Hg = 200.59 amu, S = 32.07 amu, and He = 4.00 amu. The mass of Avogadro's number of atoms is the atomic mass expressed in grams. Therefore, 6.02×10^{23} atoms of

(a) Cu = 63.55 g
(b) Hg = 200.59 g
(c) S = 32.07 g
(d) He = 4.00 g

Practice Exercise

Refer to the periodic table and state the mass for each of the following:

(a) 1 atom of Au
(b) 6.02×10^{23} atoms of Au

Answers: (a) 196.97 amu; (b) 196.97 g

Concept Exercise

What is the mass of an average platinum atom? What is the mass of Avogadro's number of Pt atoms?

Answer: See Appendix G.

9.2 MOLE CALCULATIONS I

OBJECTIVE

▸ To relate the moles of a substance to the number of particles.

The **mole** (symbol **mol**) is a unit of measure for an amount of a chemical substance. We define a mole as the amount of substance that contains Avogadro's number of particles, that is, 6.02×10^{23} particles. The individual particles may be atoms, molecules, formula units, or any other particles. Thus,

$$1\ mol = \text{Avogadro's number } (N) = 6.02 \times 10^{23} \text{ atoms, molecules, or formula units}$$

The mole relationship allows us to convert between the number of particles and the mass of substance. Now we are ready to try some mole calculations. Remember to apply the following three steps in the unit analysis method of problem solving:

Applying the Unit Analysis Method

Step 1: Write down the unit asked for in the answer.
Step 2: Write down the given value related to the answer.
Step 3: Apply unit factor(s) to convert the unit in the given value to the unit in the answer. To illustrate,

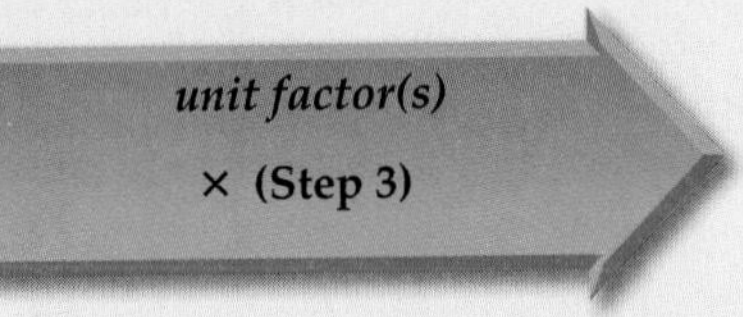

We can perform calculations that relate number of moles and number of particles. For instance, we can determine how many molecules of chlorine are in 0.250 mol of the yellow gas. By applying the unit analysis method of problem solving, we have

$$0.250\ \cancel{\text{mol Cl}_2} \times \frac{6.02 \times 10^{23}\ \text{molecules Cl}_2}{1\ \cancel{\text{mol Cl}_2}} = 1.51 \times 10^{23}\ \text{molecules Cl}_2$$

The following example exercises will help to reinforce the unit analysis method of problem solving for calculations involving the mole concept:

Example Exercise 9.2 Mole Calculations I

Calculate the number of sodium atoms in 0.120 mol Na.

Strategy Plan

Step 1: What unit is asked for in the answer?

Step 1: atoms Na

Step 2: What given value is related to the answer?

Step 2: 0.120 mol Na

Step 3: What unit factor(s) should we apply?

Step 3:

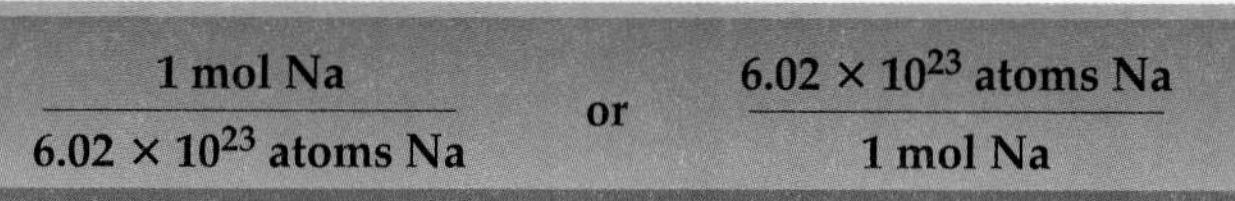

Since 1 mol Na = 6.02×10^{23} atoms Na, the two unit factors are 1 mol Na/6.02×10^{23} atoms Na, and its reciprocal 6.02×10^{23} atoms Na/1 mol Na.

Unit Analysis Map

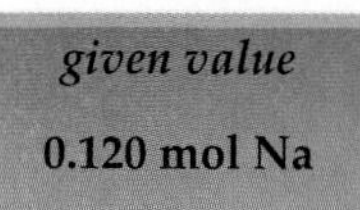

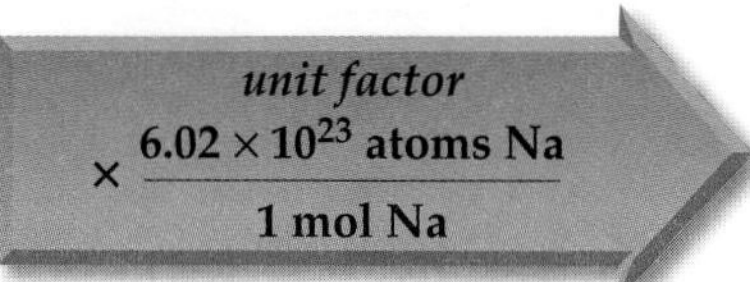

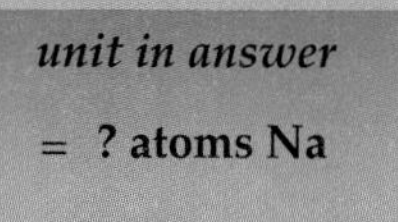

Solution

We apply the unit factor 6.02×10^{23} atoms Na/1 mol Na to cancel moles Na($\cancel{\text{mol Na}}$), which appears in the denominator.

$$0.120\ \cancel{\text{mol Na}} \times \frac{6.02 \times 10^{23}\ \text{atoms Na}}{1\ \cancel{\text{mol Na}}} = 7.22 \times 10^{22}\ \text{atoms Na}$$

The answer is rounded to three digits because the given value and unit factor each have three significant digits.

Practice Exercise

Calculate the number of formula units in 0.0763 mol of sodium chloride, NaCl.

Answers: 4.59×10^{22} formula units NaCl

Concept Exercise

What is the number of molecules in 1.00 mol of any gas?

Answer: See Appendix G.

We can also perform the reverse procedure and determine the number of moles of substance given the number of particles. For instance, we can calculate the number of moles of violet iodine crystals corresponding to 2.50×10^{23} molecules of I_2. Applying the unit analysis method of problem solving, we have

$$2.50 \times 10^{23}\ \cancel{\text{molecules I}_2} \times \frac{1\ \text{mol I}_2}{6.02 \times 10^{23}\ \cancel{\text{molecules I}_2}} = 0.415\ \text{mol I}_2$$

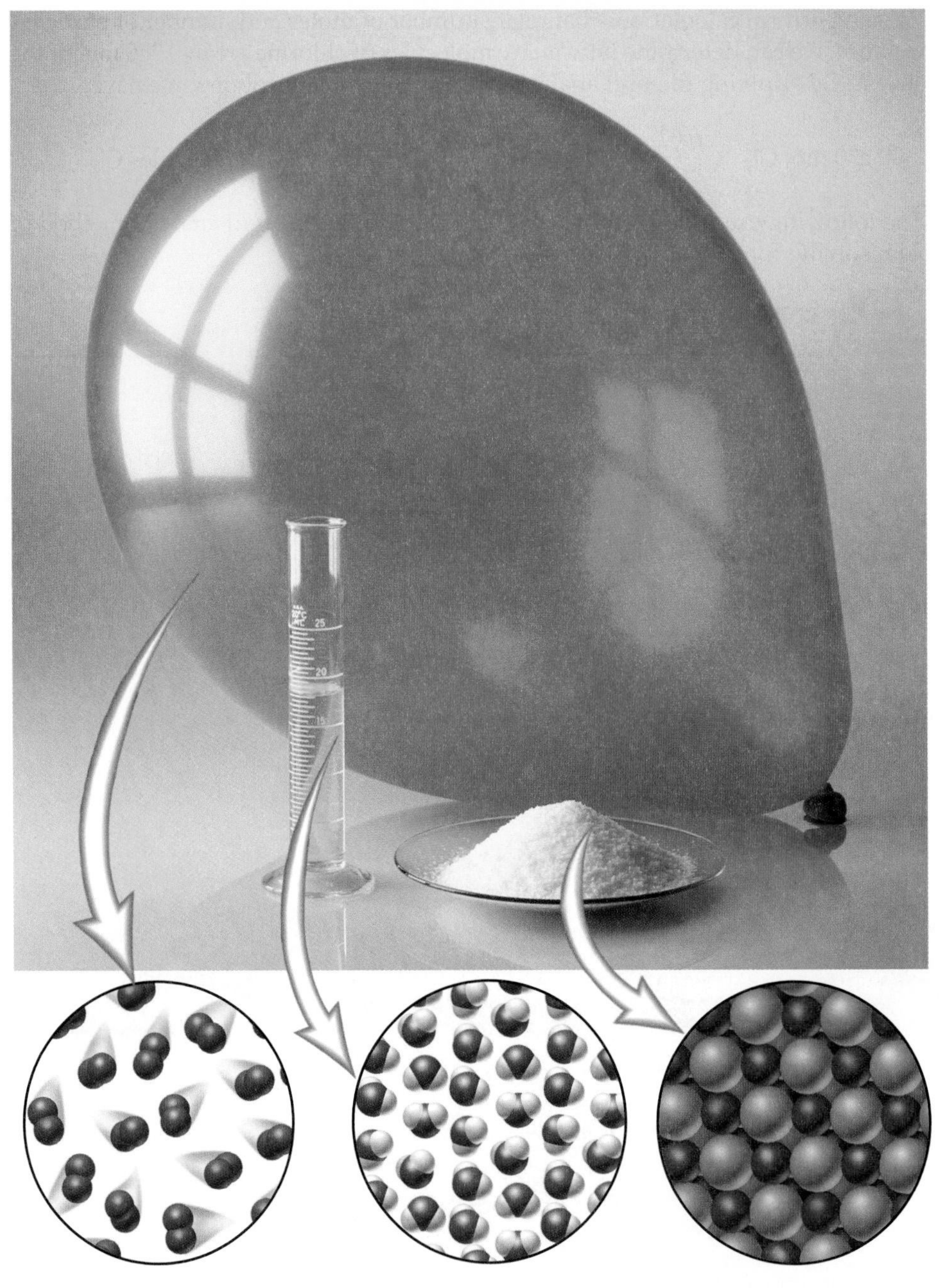

▶ **The Mole Concept — Avogadro's Number** The photograph shows a mole of oxygen, a mole of water, and a mole of table salt. The balloon contains 6.02×10^{23} molecules of O_2 gas, the graduated cylinder holds 6.02×10^{23} molecules of H_2O, and the watchglass has 6.02×10^{23} formula units of NaCl.

Example Exercise 9.3 Mole Calculations I

Calculate the number of moles of potassium in 1.25×10^{21} atoms K.

Strategy Plan

Step 1: What unit is asked for in the answer?

Step 2: What given value is related to the answer?

Step 3: What unit factor(s) should we apply?

Since 1 mol K = 6.02×10^{23} atoms K, the two unit factors are 1 mol K/6.02×10^{23} atoms K, and its reciprocal 6.02×10^{23} atoms K/1 mol K.

Step 1: **mol K**

Step 2: **1.25×10^{21} atoms K**

Step 3: $\frac{1 \text{ mol K}}{6.02 \times 10^{23} \text{ atoms K}}$ or $\frac{6.02 \times 10^{23} \text{ atoms K}}{1 \text{ mol K}}$

Unit Analysis Map

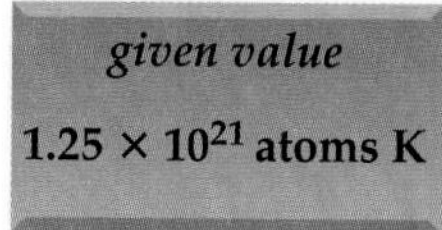

× *unit factor* $\frac{1 \text{ mol K}}{6.02 \times 10^{23} \text{ atoms K}}$ *unit in answer* = ? mol K

Solution

We apply the unit factor 1 mol K/6.02×10^{23} atoms K to cancel atoms K (~~atoms K~~), which appears in the denominator.

$$1.25 \times 10^{21} \text{ \sout{atoms K}} \times \frac{1 \text{ mol K}}{6.02 \times 10^{23} \text{ \sout{atoms K}}} = 2.08 \times 10^{-3} \text{ mol K}$$

The answer is rounded to three digits because the given value and unit factor each have three significant digits.

Practice Exercise

Calculate the number of moles of potassium iodide in 5.34×10^{25} formula units of KI.

Answer: 88.7 mol KI

Concept Exercise

What is the number of molecules in 1.00 mol of iodine crystals?

Answer: See Appendix G.

9.3 MOLAR MASS

OBJECTIVE

- To calculate the molar mass of a substance given its chemical formula.

A mole is an amount of substance that indicates the number of particles in a sample. That is, 1 mol of an element contains 6.02×10^{23} atoms. Since the atomic mass of carbon is 12.01 amu, we know that 1 mol of carbon has a mass of 12.01 g. In fact, the atomic mass of any substance expressed in grams corresponds to 1 mol of the substance. In our discussion, we will refer to the atomic mass of a substance expressed in grams as the **molar mass** (symbol **MM**).

The molar mass of an element is equal to its atomic mass expressed in grams. By referring to the periodic table, we find that the atomic mass of iron is 55.85 amu. Thus, the molar mass of iron is 55.85 g/mol. Since naturally occurring oxygen is O_2, the molar mass of oxygen gas is equal to two times 16.00 g, or 32.00 g/mol.

We can calculate the molar mass of a compound by adding the molar masses of each element. For example, we can find the molar mass of iron(III) oxide, Fe_2O_3, by summing the masses as follows:

$$Fe_2O_3\text{: } 2(55.85 \text{ g/mol Fe}) + 3(16.00 \text{ g/mol O}) = 159.70 \text{ g/mol}$$

$$Fe_2O_3\text{: } 111.70 \text{ g/mol Fe} \quad + \quad 48.00 \text{ g/mol O} = 159.70 \text{ g/mol}$$

The calculated molar mass of Fe_2O_3 is 159.70 g/mol. Notice that we used 16.00 g as the molar mass of oxygen. Even though the element occurs naturally as molecules of oxygen, it is atoms of oxygen that are combined in compounds. The following example exercise further illustrates the calculation of molar mass:

Example Exercise 9.4 Molar Mass Calculations

Calculate the molar mass for each of the following substances:

(a) silver metal, Ag
(b) ammonia gas, NH_3
(c) magnesium nitrate, $Mg(NO_3)_2$

Solution

We begin by finding the atomic mass of each element in the periodic table. The molar mass equals the sum of the atomic masses expressed in g/mol.

(a) The atomic mass of Ag is 107.87 amu, and the molar mass of silver equals 107.87 g/mol.

(b) The sum of the atomic masses for NH_3 is 14.01 amu + 3(1.01)amu = 17.04 amu. The molar mass of ammonia equals 17.04 g/mol.

(c) The sum of the atomic masses for $Mg(NO_3)_2$ is
24.31 amu + 2(14.01 + 16.00 + 16.00 + 16.00)amu = 148.33 amu.
The molar mass of magnesium nitrate equals 148.33 g/mol.

Practice Exercise

Calculate the molar mass for each of the following substances:

(a) manganese metal, Mn

(b) sulfur hexafluoride, SF_6

(c) strontium acetate, $Sr(C_2H_3O_2)_2$

Answers: (a) 54.94 g/mol; (b) 146.07 g/mol; (c) 205.72 g/mol

Concept Exercise

The molecular mass of water is 18.02 amu. What is the mass of Avogadro's number of water molecules?

Answer: See Appendix G.

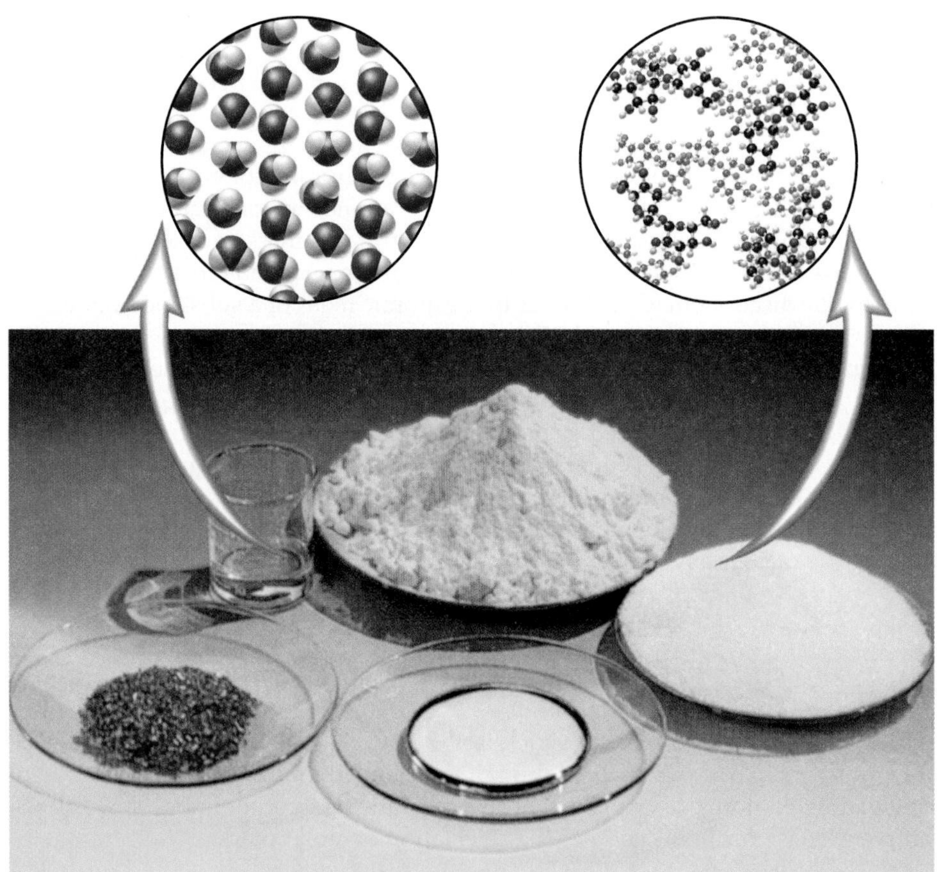

▶ **The Mole Concept—Molar Mass** The photograph shows a mole of water, sulfur, table sugar, mercury, and copper (clockwise). The molar masses are as follows: 18.02 g/mol H_2O, 32.07 g/mol S, 342.34 g/mol $C_{11}H_{22}O_{11}$, 200.59 g/mol Hg, and 63.55 g/mol Cu.

OBJECTIVE

▶ To relate the mass of a substance to the number of particles.

9.4 MOLE CALCULATIONS II

The mole is the central unit in chemistry. It is an amount of a substance and relates the number of particles to the mass of the substance, that is,

$$6.02 \times 10^{23} \text{ particles} = 1 \text{ mol} = \text{molar mass of substance}$$

We can perform calculations that relate mass to the number of particles. For instance, we can find the mass of 2.55×10^{23} atoms of lead. By applying the unit analysis method of problem solving, we first find the number of moles:

$$2.55 \times 10^{23} \text{ atoms Pb} \times \frac{1 \text{ mol Pb}}{6.02 \times 10^{23} \text{ atoms Pb}} = 0.424 \text{ mol Pb}$$

To calculate the mass of lead, we multiply the number of moles of lead by its molar mass. From the periodic table we find that the molar mass of Pb is 207.2 g/mol:

$$0.424 \text{ mol Pb} \times \frac{207.2 \text{ g Pb}}{1 \text{ mol Pb}} = 87.8 \text{ g Pb}$$

The following example exercises reinforce mole calculations involving Avogadro's number and molar mass:

Example Exercise 9.5 Mole Calculations II

What is the mass in grams of 2.01×10^{22} atoms of sulfur?

Strategy Plan

Step 1: What unit is asked for in the answer?

Step 1: g S

Step 2: What given value is related to the answer?

Step 2: 2.01×10^{22} atoms S

Step 3: What unit factor(s) should we apply?

Step 3: $\frac{1 \text{ mol S}}{6.02 \times 10^{23} \text{ atoms S}}$ or $\frac{6.02 \times 10^{23} \text{ atoms S}}{1 \text{ mol S}}$

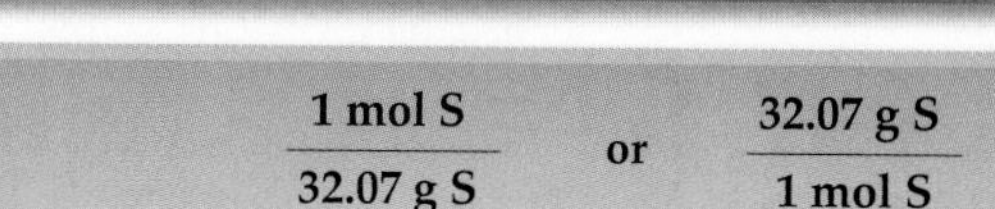

By definition, 1 mol S = 6.02×10^{23} atoms S, and the molar mass from the periodic table is 32.07 g S = 1 mol S; the two pairs of unit factors are shown in Step 3.

Unit Analysis Map

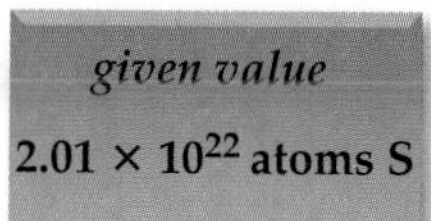

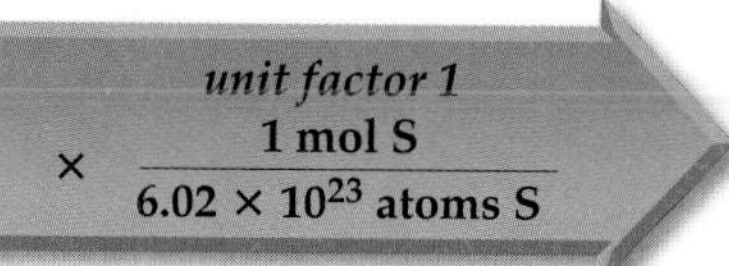

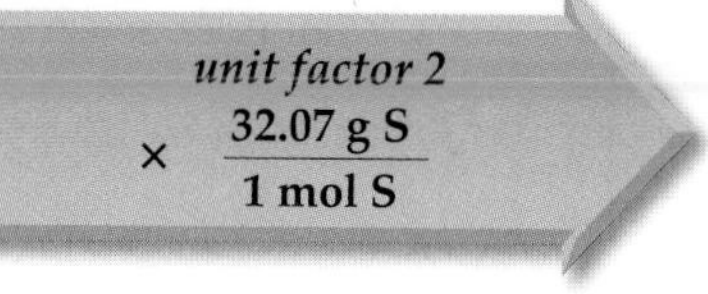

unit in answer

= ? g S

Solution

We apply the unit factor 1 mol S/6.02×10^{23} atoms S to cancel atoms S (~~atoms S~~), and 32.07 g S/1 mol S to cancel mol S (~~mol S~~):

$$2.01 \times 10^{22} \text{ atoms S} \times \frac{1 \text{ mol S}}{6.02 \times 10^{23} \text{ atoms S}} \times \frac{32.07 \text{ g S}}{1 \text{ mol S}} = 1.07 \text{ g S}$$

Practice Exercise

What is the mass of 7.75×10^{22} formula units of lead(II) sulfide, PbS?

Answer: 30.8 g PbS

Concept Exercise

Sulfur occurs naturally as S_8 molecules. What is the mass of Avogadro's number of S_8 molecules?

Answer: See Appendix G.

Example Exercise 9.6 Mole Calculations II

How many O_2 molecules are present in 0.470 g of oxygen gas?

Strategy Plan

Step 1: What unit is asked for in the answer? — **Step 1:** molecules O_2

Step 2: What given value is related to the answer? — **Step 2:** 0.470 g O_2

Step 3: What unit factor(s) should we apply? — **Step 3:** $\frac{1 \text{ mol } O_2}{32.00 \text{ g } O_2}$ or $\frac{32.00 \text{ g } O_2}{1 \text{ mol } O_2}$

The molar mass from the periodic table is 32.00 g O_2 = 1 mol O_2, and by definition, 1 mol O_2 = 6.02×10^{23} molecules O_2. Thus, the two pairs of unit factors are shown in Step 3.

$\frac{1 \text{ mol } O_2}{6.02 \times 10^{23} \text{ molecules } O_2}$ or $\frac{6.02 \times 10^{23} \text{ molecules } O_2}{1 \text{ mol } O_2}$

Unit Analysis Map

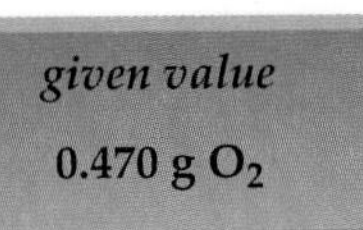

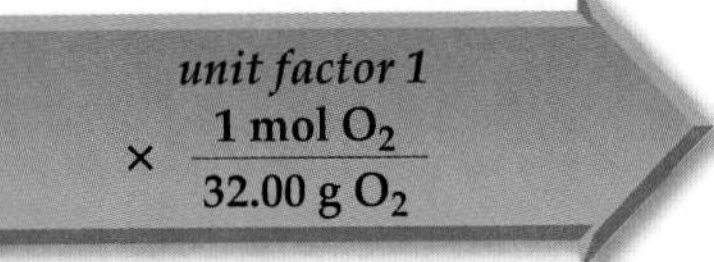

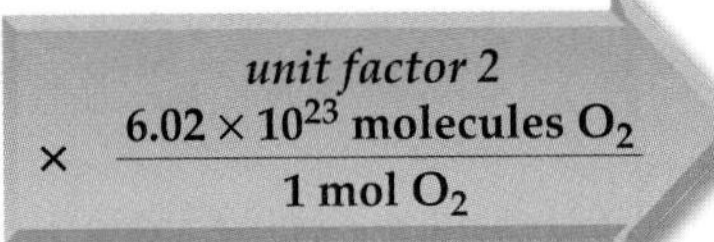

unit in answer
= ? molecules O_2

Solution

We apply the unit factor 1 mol O_2/32.00 g O_2 to cancel g O_2 (g O_2 struck through), and 6.02×10^{23} molecules O_2/1 mol O_2 to cancel moles O_2 (mol O_2 struck through). Thus,

$$0.470 \text{ g } O_2 \times \frac{1 \text{ mol } O_2}{32.00 \text{ g } O_2} \times \frac{6.02 \times 10^{23} \text{ molecules } O_2}{1 \text{ mol } O_2} = 8.84 \times 10^{21} \text{ molecules } O_2$$

Practice Exercise
How many formula units of lithium fluoride are found in 0.175 g LiF?

Answer: 4.06×10^{21} formula units LiF

Concept Exercise
Ozone occurs naturally as O_3 molecules. What is the mass of Avogadro's number of ozone molecules?

Answer: See Appendix G.

Mass of an Atom or Molecule

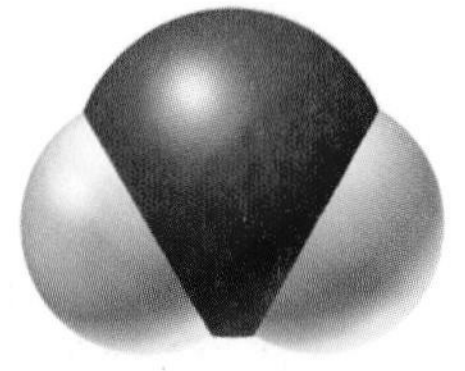

▲ Water molecule—H_2O

Now, let's try a different type of problem. What is the mass of a single molecule of water, given that the molar mass of water is 18.02 g/mol? This problem requires a ratio of units in the answer, that is, g/molecule:

$$\text{given value} \times \frac{\text{unit}}{\text{factor}} = \frac{\text{g}}{\text{molecule}}$$

To arrive at an answer having compound units, we start with a ratio of two units. Since the molar mass of water is 18.02 g/mol, we can write

$$\frac{18.02 \text{ g}}{1 \text{ mol } H_2O} \times \frac{\text{unit}}{\text{factor}} = \frac{\text{g}}{\text{molecule}}$$

Avogadro's number provides the unit factor 1 mol H_2O/6.02 × 10^{23} molecules. We can apply the unit factor and cancel units as follows:

$$\frac{18.02 \text{ g}}{1 \text{ mol } H_2O} \times \frac{1 \text{ mol } H_2O}{6.02 \times 10^{23} \text{ molecules}} = 2.99 \times 10^{-23} \text{ g/molecule}$$

The following example exercise further illustrates how to apply the mole concept to determine the mass of an individual atom or molecule:

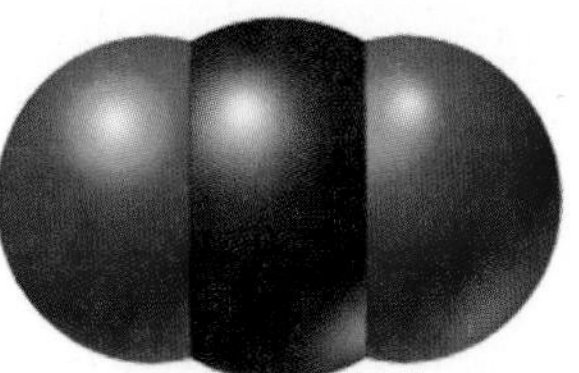

▲ Carbon Dioxide molecule—CO_2

Example Exercise 9.7 Mass of a Molecule

Calculate the mass in grams for a single molecule of carbon dioxide, CO_2 (given that MM = 44.01 g/mol).

Strategy Plan

Step 1: What unit is asked for in the answer? **Step 1:** g CO_2/molecule CO_2

Step 2: What given value is related to the answer? **Step 2:** 44.01 g CO_2/1 mol CO_2

Step 3: What unit factor(s) should we apply? **Step 3:** $\frac{1 \text{ mol } CO_2}{6.02 \times 10^{23} \text{ molecules } CO_2}$ or $\frac{6.02 \times 10^{23} \text{ molecules } CO_2}{1 \text{ mol } CO_2}$

By definition, 1 mol CO_2 = 6.02 × 10^{23} molecules. Thus, the unit factors are shown in Step 3.

Unit Analysis Map

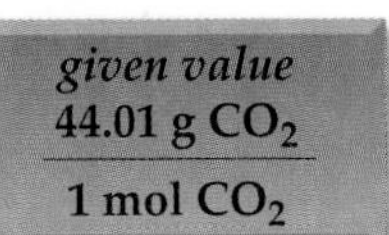

×

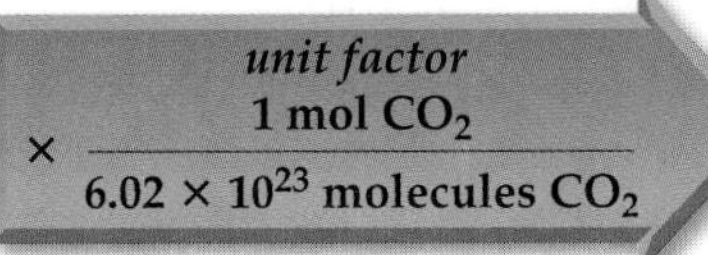

= $\frac{\text{g } CO_2}{\text{molecule } CO_2}$

Solution

We apply the unit factor 1 mol CO_2/6.02 × 10^{23} molecules CO_2 to cancel mol CO_2 (~~mol CO_2~~).

$$\frac{44.01 \text{ g}}{1 \text{ mol } CO_2} \times \frac{1 \text{ mol } CO_2}{6.02 \times 10^{23} \text{ molecules}} = 7.31 \times 10^{-23} \text{ g /molecule}$$

Practice Exercise

Calculate the mass in grams for a single molecule of carbon monoxide, CO.

Answer: 4.65 × 10^{-23} g /molecule

Concept Exercise

What is the mass of a carbon monoxide molecule expressed in amu?

Answer: See Appendix G.

9.5 MOLAR VOLUME

OBJECTIVES

- To state the value for the molar volume of any gas at STP: 22.4 L/mol.
- To relate the density of a gas at STP to its molar mass and volume.

Recall that 1 mol of any gas contains 6.02 × 10^{23} molecules. The gas can be hydrogen, oxygen, or any other gas. In 1811, Avogadro proposed that two gases containing equal numbers of molecules occupy equal volumes under similar conditions. This statement is known as **Avogadro's theory**, and is also referred to as Avogadro's law or Avogadro's hypothesis. It naturally follows, therefore, that 1 mol of hydrogen gas and 1 mol of oxygen gas occupy the same volume. Stated differently, 6.02 × 10^{23} molecules of H_2

▲ **Figure 9.1 Molar Volume of a Gas** A balloon containing 1 mol of gas at STP has a diameter of about 35 cm. A mole of any gas contains Avogadro's number of molecules and occupies a volume of 22.4 L at STP.

TABLE 9.1 MOLE RELATIONSHIPS FOR SELECTED GASES

GAS	NO. OF MOLES	NO. OF MOLECULES	MOLAR MASS	MOLAR VOLUME AT STP
hydrogen, H_2	1.00	6.02×10^{23}	2.02 g/mol	22.4 L/mol
oxygen, O_2	1.00	6.02×10^{23}	32.00 g/mol	22.4 L/mol
carbon dioxide, CO_2	1.00	6.02×10^{23}	44.01 g/mol	22.4 L/mol
ammonia, NH_3	1.00	6.02×10^{23}	17.04 g/mol	22.4 L/mol
argon, Ar*	1.00	6.02×10^{23}	39.95 g/mol	22.4 L/mol

*Argon gas is composed of atoms rather than molecules.

occupy the same volume as 6.02×10^{23} molecules of O_2. In fact, 6.02×10^{23} molecules of any gas occupy the same volume as 6.02×10^{23} molecules of hydrogen or oxygen.

What volume of gas contains Avogadro's number of molecules? At **standard temperature and pressure** (symbol STP) the volume is 22.4 L. Standard temperature is 0 °C. Standard pressure is 1 atm, which is the atmospheric pressure exerted by air at sea level. The volume occupied by 1 mol of any gas at STP is called the **molar volume** (Figure 9.1).

Table 9.1 compares 1 mol of five different gases. Notice that the number of molecules and the volume are constant, whereas the molar mass varies.

Gas Density

The density of a gas is much less than the density of a liquid or a solid. In Section 3.7 we found that the density of water at 4 °C is 1.00 g/mL. However, the density of air at 4 °C is only 0.00129 g/mL (1.29 g/L). Therefore, the density of water is almost a 1000 times greater than that of air.

We also learned in Section 3.7 that the definition of density is mass divided by volume. The density of a gas, such as hydrogen or oxygen, is its mass divided by its volume. We can easily calculate the density of a gas at STP because 1 mol of any gas has a mass equal to its molar mass and a volume equal to its molar volume. The formula for **gas density** is

$$\frac{\text{molar mass in grams}}{\text{molar volume in liters}} = \text{density, g/L (at STP)}$$

The following example exercises illustrate how density, molar mass, and molar volume of a gas are related:

▶ **The Mole Concept—Molar Volume** The photograph depicts a molar volume of gas. For reference, a basketball contains about nine liters of air at STP.

Example Exercise 9.8 Density of a Gas at STP

Calculate the density of ammonia gas, NH_3, at STP. (MM = 17.04 g/mol)

Strategy Plan

Step 1: What unit is asked for in the answer? **Step 1:** g/L

Step 2: What given value is related to the answer? **Step 2:** 17.04 g/mol

Step 3: What unit factor(s) should we apply? **Step 3:** $\frac{1 \text{ mol}}{22.4 \text{ L}}$ or $\frac{22.4 \text{ L}}{1 \text{ mol}}$

From the molar volume of a gas at STP, 1 mol = 22.4 L; thus, the unit factors are 1 mol /22.4 L, and 22.4 L/1 mol.

Unit Analysis Map

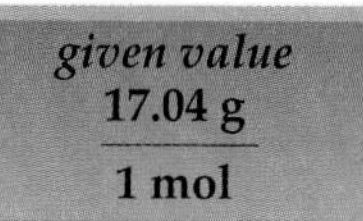

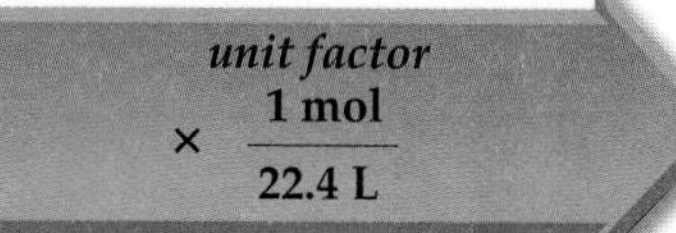

unit in answer = $\frac{\text{g}}{\text{L}}$

Solution

We apply the unit factor 1 mol /22.4 L to cancel moles (~~mol~~), which appears in the denominator.

$$\frac{17.04 \text{ g}}{1 \cancel{\text{mol}}} \times \frac{1 \cancel{\text{mol}}}{22.4 \text{ L}} = 0.761 \text{ g/L}$$

If we filled a balloon with ammonia gas (0.761 g/L), the balloon would float in air (1.29 g/L). Similarly, a less dense liquid floats on a more dense liquid; for example, gasoline (0.7 g/mL) floats on water (1.0 g/mL).

Practice Exercise

Calculate the density of hydrogen sulfide gas, H_2S, at STP.

Answer: 1.52 g/L

Concept Exercise

Why does the balloon filled with helium gas rise, while the balloon filled with nitrogen gas does not?

Answer: See Appendix G.

▲ Gas Balloons

Example Exercise 9.9 Molar Mass of a Gas

A fire extinguisher releases 1.96 g of an unknown gas that occupies 1.00 L at STP. What is the molar mass (g/mol) of the unknown gas?

Strategy Plan

Step 1: What unit is asked for in the answer? **Step 1:** g/mol

Step 2: What given value is related to the answer? **Step 2:** 1.96 g/1.00 L

Step 3: What unit factor(s) should we apply? **Step 3:** $\frac{1 \text{ mol}}{22.4 \text{ L}}$ or $\frac{22.4 \text{ L}}{1 \text{ mol}}$

From the molar volume of a gas at STP, 1 mol = 22.4 L; thus, the unit factors are 1 mol/22.4 L, and 22.4 L/1 mol.

Unit Analysis Map

given value		*unit factor*		*unit in answer*
$\frac{1.96\text{ g}}{1.00\text{ L}}$	×	$\frac{22.4\text{ L}}{1\text{ mol}}$	=	$\frac{\text{g}}{\text{mol}}$

Solution

We apply the unit factor 1 mol/22.4 L to cancel liters (L), which appears in the denominator.

$$\frac{1.96\text{ g}}{1.00\text{ L}} \times \frac{22.4\text{ L}}{1\text{ mol}} = 43.9\text{ g/mol}$$

Because the unknown gas is from a fire extinguisher, we suspect that it is carbon dioxide. By summing the molar mass for CO_2(44.01 g/mol), we help confirm that the unknown gas is carbon dioxide.

Practice Exercise

Boron trifluoride gas is used in the manufacture of computer chips. Given that 1.505 g of the gas occupies 497 mL at STP, what is the molar mass of boron fluoride gas?

Answer: 67.8 g/mol

Concept Exercise

What is the volume of one mole of any gas at STP?

Answer: See Appendix G.

OBJECTIVE

- To relate the volume of a gas at STP to its mass and number of particles.

9.6 MOLE CALCULATIONS III

As mentioned previously, the mole is a central unit in chemical calculations. A mole has three interpretations:

1. A mole is Avogadro's number of particles.
2. A mole of a substance has a mass equal to its atomic mass in grams.
3. A mole of any gas at STP occupies a volume of 22.4 L.

We can state the three interpretations in the form of equations as follows:

$$1\ mol = 6.02 \times 10^{23}\text{ particles}$$

$$1\ mol = \text{molar mass}$$

$$1\ mol = 22.4\text{ L at STP}$$

From the preceding mole equations, we can relate the number of particles to the mass or volume of a gaseous substance at STP. The process of interchanging these three quantities involves mole calculations.

The following example exercises illustrate calculations involving the number of moles of a substance. In these calculations, we will use the concepts of Avogadro's number, molar mass, and molar volume.

Example Exercise 9.10 Mole Calculations III

What is the mass of 3.36 L of ozone gas, O_3, at STP?

Strategy Plan

Step 1: What unit is asked for in the answer?

Step 1: g O_3

Step 2: What given value is related to the answer?

Step 2: 3.36 L O_3

Step 3: What unit factor(s) should we apply?

Step 3: $\frac{1 \text{ mol } O_3}{22.4 \text{ L } O_3}$ or $\frac{22.4 \text{ L } O_3}{1 \text{ mol } O_3}$

From the molar volume of a gas at STP, 1 mol O_3 = 22.4 L O_3. The molar mass from the periodic table is 48.00 g O_3 = 1 mol O_3. Thus, the two pairs of unit factors are shown in Step 3.

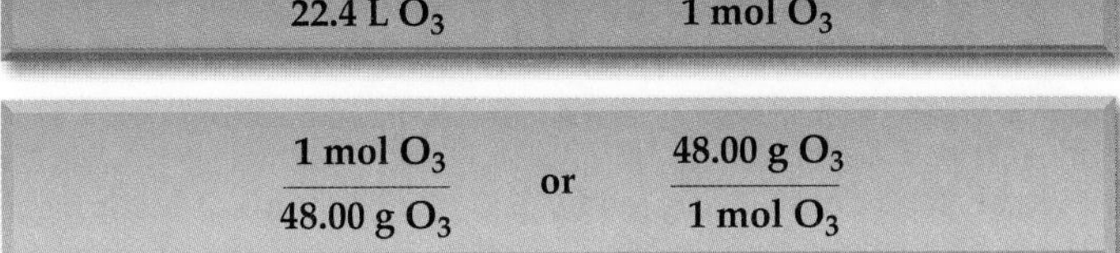

Unit Analysis Map

given value 3.36 L O_3 → *unit factor 1* $\times \frac{1 \text{ mol } O_3}{22.4 \text{ L } O_3}$ →

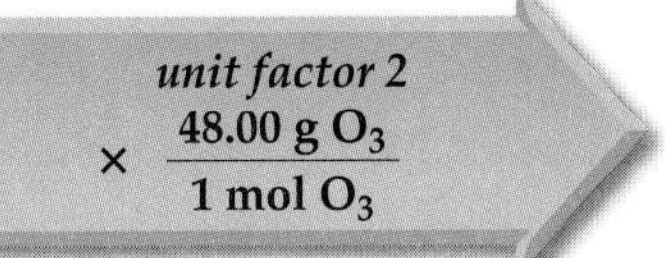

→ *unit in answer* = ? g O_3

Solution

We apply the unit factor 1 mol O_3/22.4 L O_3 to cancel L O_3 ($\cancel{\text{L } O_3}$), and 48.00 g O_3/1 mol O_3 to cancel moles O_3 ($\cancel{\text{mol } O_3}$),

$$3.36 \cancel{\text{L } O_3} \times \frac{1 \cancel{\text{mol } O_3}}{22.4 \cancel{\text{L } O_3}} \times \frac{48.00 \text{ g } O_3}{1 \cancel{\text{mol } O_3}} = 7.20 \text{ g } O_3$$

Practice Exercise

What volume is occupied by 0.125 g of methane gas, CH_4, at STP?

Answer: 0.174 L CH_4 (174 mL)

Concept Exercise

How many molecules of ozone are in one mole of gas?

Answer: See Appendix G.

Example Exercise 9.11 Mole Calculations III

How many molecules of hydrogen gas, H_2, occupy 0.500 L at STP?

Strategy Plan

Step 1: What unit is asked for in the answer?

Step 1: molecules H_2

Step 2: What given value is related to the answer?

Step 2: 0.500 L H_2

Step 3: What unit factor(s) should we apply?

Step 3: $\frac{1 \text{ mol } H_2}{22.4 \text{ L } H_2}$ or $\frac{22.4 \text{ L } H_2}{1 \text{ mol } H_2}$

From the molar volume of a gas at STP, 1 mol H_2 = 22.4 L H_2 and by definition, 1 mol H_2 = 6.02×10^{23} molecules H_2. Thus, the two pairs of unit factors are shown in Step 3.

$\frac{1 \text{ mol } H_2}{6.02 \times 10^{23} \text{ molecules } H_2}$ or $\frac{6.02 \times 10^{23} \text{ molecules } H_2}{1 \text{ mol } H_2}$

Unit Analysis Map

given value		*unit factor 1*		*unit factor 2*	*unit in answer*
0.500 L H_2	×	$\frac{1 \text{ mol } H_2}{22.4 \text{ L } H_2}$	×	$\frac{6.02 \times 10^{23} \text{ molecules } H_2}{1 \text{ mol } H_2}$	= ? molecules H_2

Solution

We apply the unit factor 1 mol H_2/22.4 L H_2 to cancel L H_2 (~~L H_2~~), and 6.02×10^{23} molecules H_2/1 mol H_2 to cancel moles 77H_2 (~~mol H_2~~). Thus,

$$0.500 \cancel{\text{L H}_2} \times \frac{1 \cancel{\text{mol H}_2}}{22.4 \cancel{\text{L H}_2}} \times \frac{6.02 \times 10^{23} \text{ molecules H}_2}{1 \cancel{\text{mol H}_2}} = 1.34 \times 10^{22} \text{ molecules H}_2$$

Practice Exercise

What is the volume occupied by 3.33×10^{21} atoms of helium gas, He, at STP?

Answer: 0.124 L He (124 mL)

Concept Exercise

What is the mass of one mole of hydrogen gas?

Answer: See Appendix G.

OBJECTIVE

- To calculate the percent composition of a compound given its chemical formula.

9.7 PERCENT COMPOSITION

Previously, in Section 2.10 we defined percent as parts per 100 parts. Another way to define percent is to say that it is the amount of a given quantity compared to an entire sample. For example, a dime is 10% of a dollar. Now, let's apply the percent concept to the composition of a compound.

The **percent composition** of a compound lists the mass percent of each element. The percent composition of water, H_2O, for example, is 11% hydrogen and 89% oxygen. According to the law of definite composition discussed in Section 4.5, the elements in a compound are always present in the same proportion by mass. Therefore, water always contains 11% hydrogen and 89% oxygen regardless of the amount. A drop of water, a milliliter of water, and a pool of water all contain 11% hydrogen and 89% oxygen.

The percent of each element in water is given, but how are the values obtained? We can calculate the percent composition of water as follows. Let's begin by assuming

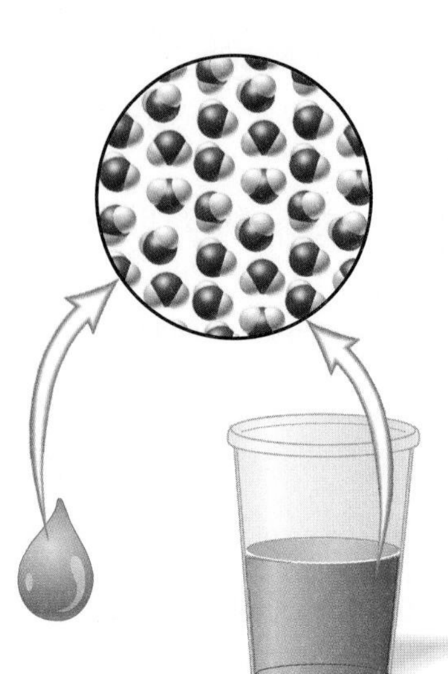

▶ **Percent Composition** The percent composition of H_2O is the same for a drop of water and a glass of water, that is, 11% hydrogen and 89% oxygen.

that we have 1 mol of H_2O. A mole of H_2O contains 2 mol of hydrogen and 1 mol of oxygen. Thus,

$$2\text{ (mol H)} + 1\text{ mol O} = 1\text{ mol H}_2\text{O}$$
$$2\text{ (molar mass H)} + 1\text{ molar mass O} = 1\text{ molar mass H}_2\text{O}$$
$$2\text{ (1.01 g H)} + 16.00\text{ g O} = \text{g H}_2\text{O}$$
$$2.02\text{ g H} + 16.00\text{ g O} = 18.02\text{ g H}_2\text{O}$$

Next, we find the percent composition of water by comparing the molar masses of hydrogen and oxygen to the molar mass of the whole compound:

$$\frac{2.02\text{ g H}}{18.02\text{ g H}_2\text{O}} \times 100\% = 11.2\%\text{ H}$$

$$\frac{16.00\text{ g O}}{18.02\text{ g H}_2\text{O}} \times 100\% = 88.79\%\text{ O}$$

The following example exercise further illustrates the calculation of the percent composition of a compound:

Example Exercise 9.12 Percent Composition of a Substance

TNT (trinitrotoluene) is a white crystalline substance that explodes at 240 °C. Calculate the percent composition of TNT, $C_7H_5(NO_2)_3$.

Solution

Let's calculate the percent composition assuming there is 1 mol of TNT. For compounds with parentheses, it is necessary to count the number of atoms of each element carefully. That is, 1 mol of $C_7H_5(NO_2)_3$ contains 7 mol of C atoms, 5 mol of H atoms, 3 mol of NO_2 (that is, 3 mol N atoms and 6 mol of O atoms).

We begin the calculation by finding the molar mass of $C_7H_5(NO_2)_3$ as follows:

$$7\text{ (12.01 g C)} + 5\text{ (1.01 g H)} + 3\text{ (14.01 g N} + 32.00\text{ g O)} = \text{g C}_7\text{H}_5(\text{NO}_2)_3$$
$$84.07\text{ g C} + 5.05\text{ g H} + 42.03\text{ g N} + 96.00\text{ g O} = 227.15\text{ g C}_7\text{H}_5(\text{NO}_2)_3$$

Now, let's compare the mass of each element to the total molar mass of the compound, that is, 227.15 g.

$$\frac{84.07\text{ g C}}{227.15\text{ g C}_7\text{H}_5(\text{NO}_2)_3} \times 100\% = 37.01\%\text{ C}$$

$$\frac{5.05\text{ g H}}{227.15\text{ g C}_7\text{H}_5(\text{NO}_2)_3} \times 100\% = 2.22\%\text{ H}$$

$$\frac{42.03\text{ g N}}{227.15\text{ g C}_7\text{H}_5(\text{NO}_2)_3} \times 100\% = 18.50\%\text{ N}$$

$$\frac{96.00\text{ g O}}{227.15\text{ g C}_7\text{H}_5(\text{NO}_2)_3} \times 100\% = 42.26\%\text{ O}$$

The percent composition of TNT reveals that the explosive is mostly oxygen by mass. When we sum the individual percentages $(37.01 + 2.22 + 18.50 + 42.26 = 99.99\%)$, we verify our calculation and, find that the total is approximately 100%.

Practice Exercise

EDTA (ethylenediaminetetraacetic acid) is used as a food preservative and in the treatment of metallic lead poisoning. Calculate the percent composition of EDTA, $C_{10}H_{16}N_2O_8$.

Answers: 41.09% C, 5.53% H, 9.587% N, and 43.79% O

Concept Exercise

If an analysis of sugar, $C_xH_yO_z$, gave 40.0% C and 6.7% H, what is the percent oxygen?

Answers: See Appendix G.

OBJECTIVES

- To calculate the empirical formula of a compound given its mass composition.
- To calculate the empirical formula of a compound given its percent composition.

9.8 EMPIRICAL FORMULA

During the late 1700s, chemists experimented to see how elements reacted to form compounds. In particular, they were interested in the reactions of elements with oxygen to form oxides. By measuring the mass of an element and the mass of the oxide after reaction, chemists could determine the formula of the compound.

The **empirical formula** of a compound corresponds to the simplest whole-number ratio of ions in a formula unit, or to the simplest ratio of atoms of each element in a molecule. The following example illustrates the determination of an empirical formula. A 1.640 g sample of radioactive radium was heated with oxygen to produce 1.755 g of radium oxide. By subtracting the mass of radium from the mass of the oxide (1.755 g – 1.640 g), we find that the mass gain was 0.115 g. Thus, the radium reacted with 0.115 g of oxygen gas.

Now, let's find the empirical formula for radium oxide, $Ra_?O_?$. The empirical formula is the simplest whole-number ratio of radium and oxide ions in a formula unit of the compound. We can determine this ratio from the moles of each reactant. If the molar mass of radium is given as 226.03 g/mol, we can proceed as follows:

$$1.640 \cancel{\text{g Ra}} \times \frac{1 \text{ mol Ra}}{226.03 \cancel{\text{g Ra}}} = 0.00726 \text{ mol Ra}$$

We can calculate the moles of oxygen (16.00 g/mol) as follows:

$$0.115 \cancel{\text{g O}} \times \frac{1 \text{ mol O}}{16.00 \cancel{\text{g O}}} = 0.00719 \text{ mol O}$$

The mole ratio of the elements in radium oxide is $Ra_{0.00726}O_{0.00719}$. We can simplify the mole ratio by dividing by the smaller number:

$$Ra_{\frac{0.00726}{0.00719}}O_{\frac{0.00719}{0.00719}} = Ra_{1.01}O_{1.00}$$

Since the empirical formula must be a small whole-number ratio, we will round the calculated chemical formula of radium oxide from $Ra_{1.01}O_{1.00}$ to RaO.

Notice that the empirical formula for radium oxide is similar to other Group IIA/2 elements. Magnesium, calcium, and barium are in Group IIA/2 because their compounds have similar empirical formulas. That is, the empirical formulas of the metal oxides all have a metal-to-oxygen ratio of 1:1, that is, MgO, CaO, and BaO. Accordingly, the empirical formula of radium oxide, which has a Group IIA/2 metal, is RaO.

Example Exercise 9.13 Empirical Formula from Mass Composition

In a laboratory experiment, 0.500 g of scandium was heated and allowed to react with oxygen from the air. The resulting product oxide had a mass of 0.767 g. Now, let's find the empirical formula for scandium oxide, $Sc_?O_?$.

Solution

The empirical formula is the simplest whole-number ratio of scandium and oxygen in the compound scandium oxide. This ratio is experimentally determined from the moles of each reactant. The moles of scandium are calculated as follows:

$$0.500 \cancel{\text{g Sc}} \times \frac{1 \text{ mol Sc}}{44.96 \cancel{\text{g Sc}}} = 0.0111 \text{ mol Sc}$$

The moles of oxygen are calculated after first subtracting the mass of Sc from the product, $Sc_?O_?$:

$$0.767 \text{ g } Sc_xO_y - 0.500 \text{ g Sc} = 0.267 \text{ g O}$$

The moles of oxygen are calculated from the mass of oxygen that reacted:

$$0.267 \text{ g O} \times \frac{1 \text{ mol O}}{16.00 \text{ g O}} = 0.0167 \text{ mol O}$$

The mole ratio in scandium oxide is $Sc_{0.0111}O_{0.0167}$. To simplify the ratio and obtain small whole numbers, we divide by the smaller number:

$$Sc\frac{0.0111}{0.0111}O\frac{0.0167}{0.0111} = Sc_{1.00}O_{1.50}$$

Since the calculated ratio is not close to whole numbers, we cannot round off the experimental ratio $Sc_{1.00}O_{1.50}$. However, we can double the ratio to obtain whole numbers, that is, $Sc_{2.00}O_{3.00}$. Thus, the empirical formula is Sc_2O_3.

Practice Exercise
Iron can react with chlorine gas to give two different compounds, $FeCl_2$ and $FeCl_3$. If 0.558 g of metallic iron reacts with chlorine gas to yield 1.621 g of iron chloride, which iron compound is produced in the experiment?

Answer: $FeCl_3$

Concept Exercise
If 0.500 mol of yellow powder sulfur reacts with 0.500 mol of oxygen gas, what is the empirical formula of the sulfur oxide?

Answer: See Appendix G.

Empirical Formulas from Percent Composition

Benzene was a common liquid solvent until the Environmental Protection Agency (EPA) discovered that it was a carcinogen. Let's calculate the formula for benzene if its percent composition is 92.2% carbon and 7.83% hydrogen.

The empirical formula expresses the simplest whole-number ratio of carbon to hydrogen atoms in a molecule of benzene. To calculate the moles of each element, let's assume we have 100 g of sample. In 100 g of benzene there are 92.2 g of carbon and 7.83 g of hydrogen. That is, the percentage of each element corresponds to its mass in 100 g of the compound. We find the moles of C and H as we did in the previous examples:

$$92.2 \text{ g C} \times \frac{1 \text{ mol C}}{12.01 \text{ g C}} = 7.68 \text{ mol C}$$

$$7.83 \text{ g H} \times \frac{1 \text{ mol H}}{1.01 \text{ g H}} = 7.75 \text{ mol H}$$

The mole ratio of the elements in benzene is $C_{7.68}H_{7.75}$. We can simplify the mole ratio by dividing both values by the smaller number, 7.68:

$$C\frac{7.68}{7.68}H\frac{7.75}{7.68} = C_{1.00}H_{1.01}$$

Since the ratio $C_{1.00}H_{1.01}$ is close to whole numbers, we can round off to CH to obtain the empirical formula for benzene. The following example exercise further illustrates the calculation of an empirical formula from percent composition data:

Example Exercise 9.14 Empirical Formula from Percent Composition

Glycine is an amino acid found in protein. An analysis of glycine gave the following data: 32.0% carbon, 6.7% hydrogen, 18.7% nitrogen, and 42.6% oxygen. Calculate the empirical formula of the amino acid.

Solution

If we assume a 100 g sample, then the percentage of each element equals its mass in 100 g of glycine, that is, 32.0 g C, 6.7 g H, 18.7 g N, and 42.6 g O. We can determine the empirical formula as follows:

$$32.0\ \text{g C} \times \frac{1\ \text{mol C}}{12.01\ \text{g C}} = 2.66\ \text{mol C}$$

$$6.7\ \text{g H} \times \frac{1\ \text{mol H}}{1.01\ \text{g H}} = 6.6\ \text{mol H}$$

$$18.7\ \text{g H} \times \frac{1\ \text{mol N}}{14.01\ \text{g N}} = 1.33\ \text{mol N}$$

$$42.6\ \text{g O} \times \frac{1\ \text{mol O}}{16.00\ \text{g O}} = 2.66\ \text{mol O}$$

The mole ratio of the elements in the amino acid is $C_{2.66}H_{6.6}N_{1.33}O_{2.66}$. We can find a small whole-number ratio by dividing by the smallest number:

$$C_{\frac{2.66}{1.33}}H_{\frac{6.6}{1.33}}N_{\frac{1.33}{1.33}}O_{\frac{2.66}{1.33}} = C_{2.00}H_{5.0}N_{1.00}O_{2.00}$$

Simplifying, we find that the empirical formula for the amino acid glycine is $C_2H_5NO_2$.

Practice Exercise

Calculate the empirical formula for caffeine given the following percent composition: 49.5% C, 5.15% H, 28.9% N, and 16.5% O.

Answer: $C_4H_5N_2O$

Concept Exercise

If 0.500 mol of yellow powder sulfur reacts with 0.750 mol of oxygen gas, what is the empirical formula of the sulfur oxide?

Answer: See Appendix G.

9.9 MOLECULAR FORMULA

OBJECTIVE

- To calculate the molecular formula for a compound given its empirical formula and molar mass.

Molecular compounds are represented by individual molecules composed of nonmetal atoms. To see what this means in terms of empirical formulas, recall that in the previous section we found the empirical formula for benzene is CH. The empirical formula for benzene—1 C atom and 1 H atom—does not represent a benzene molecule. Instead, it illustrates the relative number of carbon and hydrogen atoms in a molecule. The actual **molecular formula** for benzene is some multiple of the empirical formula. We can represent the formula as $(CH)_n$, where n is some multiple of CH, such as 2, 3, 4, 5, or 6.

Now consider acetylene, used in oxyacetylene welding, which also has the empirical formula CH. Even though acetylene and benzene have unrelated properties, each shares the same empirical formula, CH. In addition, the compound styrene, used in manufacturing styrofoam cups, also shares the same empirical formula, CH.

Since benzene, acetylene, and styrene are different compounds, their molecular formulas must be different. Experiments have provided the molar mass for each of these compounds. For benzene, the molar mass is 78 g/mol; for acetylene, it is 26 g/mol; and for styrene, it is 104 g/mol. We can indicate the number of multiples of the empirical formula for each compound as follows:

$$\text{Benzene: } (CH)n = 78 \text{ g/mol}$$
$$\text{Acetylene: } (CH)n = 26 \text{ g/mol}$$
$$\text{Styrene: } (CH)n = 104 \text{ g/mol}$$

The mass of the empirical formula CH is found by adding the molar masses of C and H, that is, 12 g + 1 g = 13 g/mol. Now we can determine how many multiples of the empirical formula are in benzene. We have

$$\text{Benzene: } \frac{(CH)_n}{CH} = \frac{78 \text{ g/mol}}{13 \text{ g/mol}}$$

$$n = 6$$

Therefore, the molecular formula of benzene is $(CH)_6$, which we can write as C_6H_6. Similarly, we can find the molecular formula for acetylene:

$$\text{Acetylene: } \frac{(CH)_n}{CH} = \frac{26 \text{ g/mol}}{13 \text{ g/mol}}$$

$$n = 2$$

The molecular formula of acetylene is $(CH)_2$ or C_2H_2. In a similar fashion, we can find the number of multiples of the empirical formula in styrene:

$$\text{Styrene: } \frac{(CH)_n}{CH} = \frac{104 \text{ g/mol}}{13 \text{ g/mol}}$$

$$n = 8$$

Thus, the molecular formula of styrene is $(CH)_8$ or C_8H_8.

Even though benzene, acetylene, and styrene have the same empirical formula, their molecular formulas are different. Thus, benzene, acetylene, and styrene are different compounds. The following example exercise further illustrates how to determine a molecular formula from an empirical formula:

Example Exercise 9.15 Molecular Formula from Empirical Formula

The empirical formula for fructose, or fruit sugar, is CH_2O. If the molar mass of fructose is 180 g/mol, find the actual molecular formula for the sugar.

Solution

We can indicate the molecular formula of fructose as $(CH_2O)n$. The molar mass of the empirical formula CH_2O is 12 g C + 2(1 g H) + 16 g O = 30 g/mol. Thus, the number of multiples of the empirical formula is

$$\text{Fructose: } \frac{(CH_2O)_n}{CH_2O} = \frac{180 \text{ g/mol}}{30 \text{ g/mol}}$$

$$n = 6$$

Thus, the molecular formula of fructose is $(CH_2O)_6$ or $C_6H_{12}O_6$.

Practice Exercise

Ethylene dibromide was used as a grain pesticide until it was banned. Calculate (a) the empirical formula and (b) the molecular formula for ethylene dibromide given its approximate molar mass of 190 g/mol and its percent composition: 12.7% C, 2.1% H, and 85.1% Br.

Answers: (a) CH_2Br; (b) $C_2H_4Br_2$

Concept Exercise

If the molecular formula of hydrogen peroxide is H_2O_2, what is the empirical formula?

Answer: See Appendix G.

Note For molecular compounds, the empirical and molecular formulas are usually different. For ionic compounds, the actual formulas are almost always identical to the empirical formulas. For example, the actual formula for sodium chloride is NaCl, not Na_2Cl_2 or some other multiple. The common exception, however, is the mercury(I) compounds. Notice that the actual formula of mercury(I) chloride, Hg_2Cl_2, is twice the empirical formula.

CRITICAL THINKING Avogadro's Number

Q: What mass of pure silicon contains Avogadro's number of atoms?

In 1911, Ernest Rutherford determined a value for Avogadro's number based on the principle of radioactivity. Rutherford was studying the emission of alpha particles from radioactive radium and was able to count these particles using a Geiger counter, which had been invented by his assistant Hans Geiger. After the alpha particles decayed into helium, he measured the volume of the helium gas. From the data, Rutherford calculated a value of 6.11×10^{23} for Avogadro's number.

This experimental value was found with an X ray method. X rays were used to determine the spacing of atoms in an ultrapure crystal of silicon (see figure). After striking the nuclei of silicon atoms, the X rays were diffracted; that is, the path of the X ray beam was altered. The volume occupied by a single atom can be calculated from the spacing between two adjacent X rays. The value for Avogadro's number is obtained by dividing the volume of an individual silicon atom into the total volume of a crystal with a mass equal to the atomic mass of silicon.

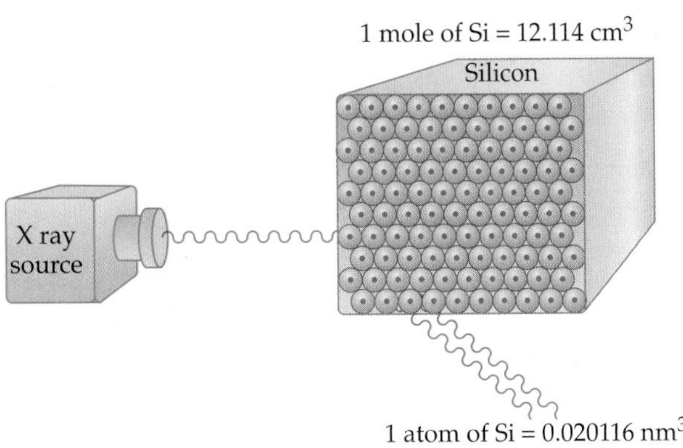

We can illustrate how to calculate a value for Avogadro's number as follows. In the figure notice that the volume occupied by each silicon atom is 0.020116 nm^3 and that the volume of the crystal is 12.114 cm^3. To find Avogadro's number we simply divide the volume of one atom of silicon into the volume of the silicon crystal. However, we must use the same units for the volumes of the atom and the crystal. For example, we can convert the volume of the crystal from cm^3 to nm^3:

$$12.114\ \text{cm}^3 \times \left(\frac{1\ \text{m}}{10^2\ \text{cm}}\right)^3 \times \left(\frac{10^9\ \text{nm}}{1\ \text{m}}\right)^3 = \text{nm}^3$$

$$12.114\ \cancel{\text{cm}^3} \times \frac{1\ \cancel{\text{m}^3}}{10^6\ \cancel{\text{cm}^3}} \times \frac{10^{27}\ \text{nm}^3}{1\ \cancel{\text{m}^3}} = 1.2114 \times 10^{22}\ \text{nm}^3$$

When we divide the volume of the silicon atom into the volume of the crystal, we find the number of silicon atoms:

$$1.2114 \times 10^{22}\ \cancel{\text{nm}^3} \times \frac{1\ \text{atom Si}}{0.020116\cancel{\text{nm}^3}} = 6.0221 \times 10^{23}\ \text{atoms Si}$$

In this example, we obtained an experimental value of 6.0221×10^{23} atoms for Avogadro's number. Currently, the most accurate value for Avogadro's number is 6.0221367×10^{23}.

◀ **X Ray Diffraction of a Silicon Crystal** We can calculate the volume of a Si atom after using an X ray beam to find the spacing of atoms in a silicon crystal.

A: The mass of silicon containing Avogadro's number of atoms corresponds to the molar mass, that is, 28.09 g.

Summary

Section 9.1 **Avogadro's number** (*N*) is the value that corresponds to the number of atoms in 12.01 g of carbon. Moreover, the number of atoms in a gram atomic mass of any element is 6.02×10^{23}.

Section 9.2 A **mole (mol)** is the amount of substance that contains Avogadro's number of particles, that is, 6.02×10^{23} particles. A mole is the central unit in performing chemical formula calculations. These computations, based on moles of substance, are known as mole calculations.

Section 9.3 The mass of 1 mol of any substance is called the **molar mass (MM)**. The molar mass of an element corresponds to the atomic mass of the element. For instance, the molar mass of carbon is 12.01 g/mol, and the molar mass of oxygen is 16.00 g/mol. The molar mass of a compound corresponds to the sum of the gram atomic masses for all the atoms of each element in the compound. For example, the molar mass of carbon monoxide, CO, equals 28.01 g/mol (12.01 g + 16.00 g).

Section 9.4 The mole is perhaps the most important unit in chemistry. A mole relates the number of particles (atoms, molecules, or formula units) to the amount of a substance. Given the moles of a substance, we can calculate the mass of the substance as well as the number of particles.

Section 9.5 Avogadro's theory states that equal volumes of gases, at the same conditions, contain equal numbers of molecules. The volume of 1 mol of a gaseous substance at standard conditions is called the **molar volume**. The molar volume of any gas at **standard temperature and pressure (STP)** is 22.4 L/mol. Standard conditions are 0 °C and 1 atm pressure. **Gas density** is found by dividing the molar mass of a gas by the molar volume; that is, we divide the molar mass by 22.4 L/mol.

Section 9.6 The mole concept relates the number of particles, molar mass, and molar volume of a gas. Given the number of liters of a gaseous substance, we can apply the mole concept to find the mass of the substance as well as the number of molecules.

Section 9.7 The **percent composition** of a compound is found by comparing the mass contributed by each element to the molar mass of the substance. A list of the resulting percentages of each element is the percent composition of the compound.

Section 9.8 The **empirical formula** is the simplest whole-number ratio of the atoms of the elements in a compound. The empirical formula can be calculated from (1) experimental synthesis data, or (2) the percent composition of the compound. After calculating the moles of each element in a compound, the mole ratio is simplified to small whole numbers.

Section 9.9 The **molecular formula** is the actual ratio of the atoms in a molecular compound. For example, lactic acid and glucose each have the empirical formula CH_2O. The molecular formula of lactic acid, $C_3H_6O_3$, is three times the empirical formula. The actual formula of glucose, $C_6H_{12}O_6$, is six times the empirical formula.

▲ **The Mole Concept** The photograph shows a mole of sugar, water, mercury, sulfur, salt, copper, lead (clockwise), and orange potassium dichromate (center). Each sample contains 6.02×10^{23} particles; however, the molar masses differ as follows: 342.34 g/mol $C_{12}H_{22}O_{11}$, 18.02 g/mol H_2O, 200.59 g/mol Hg, 32.07 g/mol S, 58.44 g/mol NaCl, 63.55 g/mol Cu, 207.2 g/mol Pb, and 294.20 g/mol $K_2Cr_2O_7$.

PROBLEM–SOLVING ORGANIZER

TOPIC	PROCEDURE	EXAMPLE
Avogadro's Number Sec. 9.1	Express the atomic mass of an element in grams to find the mass of Avogadro's number of atoms.	Avogadro's number of Fe atoms has a mass of 55.85 g.
Mole Calculations I Sec. 9.2	1. Write down the unit asked for in the answer. 2. Write down the related given value. 3. Apply a unit factor to convert the given unit to the unit in the answer.	How many iron atoms are in 0.500 mol of Fe? $0.500\ \text{mol} \times \frac{6.02 \times 10^{23}\ \text{atoms}}{1\ \text{mol}} = 3.01 \times 10^{23}$ atoms
Molar Mass Sec. 9.3	Sum the masses of each atom of each element in a compound to calculate the molar mass.	What is the molar mass of iron(III) oxide, Fe_2O_3 $2(55.85\ \text{g/mol}) + 3(16.00\ \text{g/mol}) = 159.70\ \text{g/mol}$
Mole Calculations II Sec. 9.4	1. Write down the unit asked for in the answer. 2. Write down the related given value. 3. Apply a unit factor to convert the given unit to the unit in the answer.	What is the mass of 5.00×10^{23} formula units of Fe_2O_3? $5.00 \times 10^{23}\ \text{formula units} \times \frac{1\ \text{mol}}{6.02 \times 10^{23}\ \text{formula units}} \times \frac{159.70\ \text{g}\ Fe_2O_2}{1\ \text{mol}} = 133\ \text{g}\ Fe_2O_3$
Molar Volume Sec. 9.5	The volume occupied by 1 mol of any gas at STP is 22.4 L.	The volume of 1.00 mol of O_2 at STP is 22.4 L.
Mole Calculations III Sec. 9.6	1. Write down the unit asked for in the answer. 2. Write down the related given value. 3. Apply a unit factor to convert the given unit to the unit in the answer.	What is the mass of 2.50 L of O_2 gas at STP? $2.50\ \text{L} \times \frac{1\ \text{mol}}{22.4\ \text{L}} \times \frac{32.00\ \text{g}\ O_2}{1\ \text{mol}} = 3.57\ \text{g}\ O_2$
Percent Composition	Find the mass of each element compared to the total molar mass of the compound, all multiplied by 100%.	What is the percent composition of hydrogen peroxide, H_2O_2 (MM = 34.02 g/mol)? $\frac{2(1.01)\text{g}}{34.02\ \text{g}} \times 100\% = 5.94\%\ \text{H}$ $\frac{2(16.00)\text{g}}{34.02\ \text{g}} \times 100\% = 94.06\%\ \text{O}$
Empirical Formula	Find the mol ratio of each element in the compound, and simplify to whole numbers.	What is the empirical formula of hydrogen peroxide, given 2.25 mol H and 2.22 mol O? $\text{H}_{\frac{2.25}{2.22}}\text{O}_{\frac{2.22}{2.22}} = H_{1.01}O_{1.00}$ thus, H_1O_1
Molecular Formula Sec. 9.9	Find the mass of each element compared to the total molar mass of the compound, all multiplied by 100%.	What is the molecular formula of hydrogen peroxide, given the molar mass is about 34 g/mol? $\frac{(HO)_n}{HO} = \frac{34\ \text{g/mol}}{17\ \text{g/mol}} = 2$ thus, H_2O_2

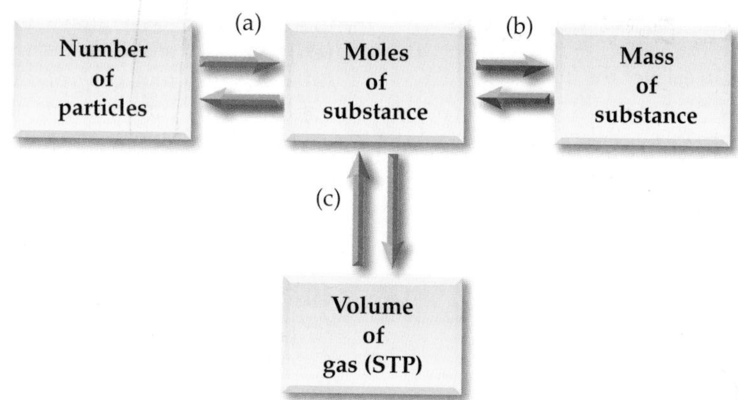

(a) Apply Avogadro's number as a unit factor: 1 mol = 6.02×10^{23} particles.
(b) Apply molar mass as a unit factor: 1 mol = ? g (refer to Periodic Table).
(c) Apply molar volume as a unit factor: 1 mol = 22.4 L at STP.

▲ **Concept Map — Summary of Mole Calculations** We use a strategy map to show the application of unit factors to solve problems by unit analysis. In this chapter, we have related the number of particles, the mass of substance, and the volume of a gas. Now, let's summarize mole calculations using a concept map that relates these concepts.

Key Terms *Answers to Key Terms are in Appendix H.*

Select the key term that corresponds to each of the following definitions.

_____ **1.** the value that corresponds to the number of atoms in 12.01 g of carbon, that is, 6.02×10^{23} individual particles
_____ **2.** the amount of substance that contains 6.02×10^{23} particles
_____ **3.** the mass of 1 mol of pure substance expressed in grams; the mass of Avogadro's number of atoms, molecules, or formula units
_____ **4.** the volume occupied by 1 mol of any gas at standard conditions
_____ **5.** a temperature of 0 °C and 1 atm pressure that have been chosen as standard conditions
_____ **6.** equal volumes of gases, under the same conditions of temperature and pressure, contain equal numbers of molecules
_____ **7.** the ratio of mass per unit volume for a gas, expressed in grams per liter
_____ **8.** a list of the mass percent of each element in a compound
_____ **9.** the chemical formula of a compound that expresses the simplest whole-number ratio of atoms of elements in a molecule, or ions in a formula unit
_____ **10.** the chemical formula of a compound that expresses the actual number of atoms of each element in a molecule

(a) Avogadro's number (N) (*Sec. 9.1*)
(b) Avogadro's theory (*Sec. 9.5*)
(c) empirical formula (*Sec. 9.8*)
(d) gas density (*Sec. 9.5*)
(e) molar mass (MM) (*Sec. 9.3*)
(f) molar volume (*Sec. 9.5*)
(g) mole (mol) (*Sec. 9.2*)
(h) molecular formula (*Sec. 9.9*)
(i) percent composition (*Sec. 9.7*)
(j) standard temperature and pressure (STP) (*Sec. 9.5*)

Exercises *Answers to odd-numbered Exercises are in Appendix I.*

Avogadro's Number (Sec. 9.1)

1. Refer to the periodic table and state the atomic mass (in amu) of one atom for each of the following elements:
(a) Be **(b)** Br
(c) Sc **(d)** Se

2. Refer to the periodic table and state the atomic mass (in amu) of one atom for each of the following elements:
(a) titanium **(b)** tungsten
(c) platinum **(d)** cerium

3. Refer to the periodic table and state the mass of 6.02×10^{23} atoms of each of the following elements:
(a) Be **(b)** Br
(c) Sc **(d)** Se

4. Refer to the periodic table and state the mass of 6.02×10^{23} atoms of each of the following elements:
(a) titanium **(b)** tungsten
(c) platinum **(d)** cerium

Mole Calculations I (Sec. 9.2)

5. State the number of particles in each of the following:
(a) 1 mol of lithium atoms, Li
(b) 1 mol of fluorine molecules, F_2

6. State the number of particles in each of the following:
(a) 1 mol of silicon atoms, Si
(b) 1 mol of silane molecules, SiH_4

7. State the number of moles represented by each of the following:
(a) 6.02×10^{23} formula units of sodium chloride, NaCl
(b) 6.02×10^{23} molecules of ammonia, NH_3

8. State the number of moles represented by each of the following:
(a) 6.02×10^{23} formula units of potassium iodide, KI
(b) 6.02×10^{23} molecules of hydrogen sulfide, H_2S

9. Calculate the number of particles in each of the following:
(a) 0.125 mol zinc atoms, Zn
(b) 0.250 mol chlorine molecules, Cl_2
(c) 0.675 mol zinc chloride formula units, $ZnCl_2$

10. Calculate the number of particles in each of the following:
(a) 0.150 mol cadmium atoms, Cd
(b) 0.325 mol fluorine molecules, F_2
(c) 0.750 mol cadmium fluoride formula units, CdF_2

11. Calculate the number of moles containing each of the following:
(a) 1.25×10^{22} atoms of manganese, Mn
(b) 2.50×10^{23} molecules of sulfur trioxide, SO_3
(c) 6.75×10^{24} formula units of manganese(II) sulfate, $MnSO_4$

12. Calculate the number of moles containing each of the following:
(a) 1.50×10^{22} atoms of silver, Ag
(b) 3.25×10^{23} molecules of carbon dioxide, CO_2
(c) 7.50×10^{24} formula units of silver carbonate, Ag_2CO_3

Molar Mass (Sec. 9.3)

13. State the molar mass for each of the following:
(a) aluminum, Al (b) silicon, Si
(c) oxygen, O_2 (d) sulfur, S_8

14. State the molar mass for each of the following:
(a) arsenic, As (b) germanium, Ge
(c) ozone, O_3 (d) phosphorus, P_4

15. Calculate the molar mass for each of the following:
(a) calcium sulfide, CaS
(b) calcium sulfate, $CaSO_4$
(c) iron(II) acetate, $Fe(C_2H_3O_2)_2$
(d) iron(II) phosphate, $Fe_3(PO_4)_2$

16. Calculate the molar mass for each of the following:
(a) bromine iodide, BrI
(b) tribromine octaoxide, Br_3O_8
(c) glycerin, $C_3H_5(OH)_3$
(d) nitroglycerin, $C_3H_5O_3(NO_2)_3$

Mole Calculations II (Sec. 9.4)

17. Calculate the mass in grams for each of the following:
(a) 2.95×10^{23} atoms of mercury, Hg
(b) 1.16×10^{22} molecules of nitrogen, N_2
(c) 5.05×10^{21} formula units of barium chloride, $BaCl_2$

18. Calculate the mass in grams for each of the following:
(a) 1.21×10^{24} atoms krypton, Kr
(b) 6.33×10^{22} molecules of dinitrogen oxide, N_2O
(c) 4.17×10^{21} formula units of magnesium perchlorate, $Mg(ClO_4)_2$

19. Calculate the number of particles in each of the following:
(a) 1.50 g potassium, K (b) 0.470 g oxygen, O_2
(c) 0.555 g silver chlorate, $AgClO_3$

20. Calculate the number of particles in each of the following:
(a) 7.57 g platinum, Pt (b) 3.88 g ethane, C_2H_6
(c) 0.152 g aluminum chloride, $AlCl_3$

21. Calculate the mass in grams for a single atom of the following elements:
(a) beryllium, Be (b) sodium, Na
(c) cobalt, Co (d) arsenic, As

22. Calculate the mass in grams for a single molecule of the following compounds:
(a) methane, CH_4 (b) ammonia, NH_3
(c) sulfur trioxide, SO_3 (d) nitrogen dioxide, NO_2

Molar Volume (Sec. 9.5)

23. State standard conditions for a gas in degrees Celsius and atmospheres.

24. State standard conditions for a gas in Kelvin units and atmospheres.

25. Calculate the density for each of the following gases at STP:
(a) argon, Ar (b) chlorine, Cl_2
(c) methane, CH_4 (d) ethane, C_2H_6

26. Calculate the density for each of the following gases at STP:
(a) xenon, Xe (b) fluorine, F_2
(c) propane, C_3H_8 (d) butane, C_4H_{10}

27. Calculate the molar mass for each of the following gases given the STP density:
(a) oxygen, 1.43 g/L (b) phosphine, 1.52 g/L
(c) nitrous oxide, 1.97 g/L (d) Freon-12, 5.40 g/L

28. Calculate the molar mass for each of the following gases given the STP density:
(a) ozone, 2.14 g/L (b) silane, 1.43 g/L
(c) nitric oxide, 1.34 g/L (d) Freon-22, 3.86 g/L

29. Given 1 mol of each gas listed, complete the following table:

Gas	Molecules	Mass	Volume at STP
chlorine, Cl_2			
hydrogen chloride, HCl			
sulfur tetrachloride, SCl_4			

30. Given 1 mol of each gas listed, complete the following table:

Gas	Molecules	Mass	Volume at STP
fluorine,			
hydrogen fluoride, HF			
sulfur tetrachloride,			

F_2

Mole Calculations III (Sec. 9.6)

31. Calculate the volume in liters for each of the following gases at STP:
(a) 0.250 g of helium, He (b) 5.05 g of nitrogen, N_2

32. Calculate the volume in liters for each of the following gases at STP:
(a) 2.22×10^{22} molecules of methane, CH_4
(b) 4.18×10^{24} molecules of ethane, C_2H_6

33. Calculate the mass in grams for each of the following gases at STP:
(a) 1.05 L of hydrogen sulfide, H_2S
(b) 5.33 L of dinitrogen trioxide, N_2O_3

34. Calculate the mass in grams for each of the following gases at STP:
(a) 5.42×10^{22} molecules of propane, C_3H_8
(b) 1.82×10^{23} molecules of butane, C_4H_{10}

35. Calculate the number of molecules in each of the following gases at STP:
(a) 100.0 mL of hydrogen, H_2
(b) 70.5 mL of ammonia, NH_3

36. Calculate the number of molecules in each of the following gases at STP:
(a) 0.150 g of carbon monoxide, CO
(b) 2.75 g of nitrogen monoxide, NO

37. Use the given quantity for each gas listed to complete the following table:

Gas	Molecules	Mass	Volume at STP
hydrogen, H_2	1.50×10^{23}		
ammonia, NH_3		4.25 g	
methane, CH_4			5.58 L

38. Use the given quantity for each gas listed to complete the following table:

Gas	Molecules	Mass	Volume at STP
nitrogen, N_2	2.50×10^{22}		
hydrazine, N_2H_4		1.33 g	
nitrogen dioxide, NO_2			0.930 L

Percent Composition (Sec. 9.7)

39. If the percent hydrogen in a drop of water is 11%, what is the percent hydrogen in a kilogram of water?

40. If the percent oxygen in a drop of water is 89%, what is the percent oxygen in a liter of water?

41. Lactose is a sugar found in milk. Calculate the percent composition of lactose, $C_{12}H_{22}O_{11}$.

42. Benzoyl peroxide is the active ingredient in an acne cream. Calculate the percent composition of benzoyl peroxide, $C_7H_6O_3$.

43. Methionine is an essential amino acid in protein. Calculate the percent composition of methionine, $C_5H_{11}NSO_2$.

44. Monosodium glutamate (MSG) is added to food to enhance the flavor. Calculate the percent composition of MSG, $NaC_5H_8NO_4$.

45. Mercurochrome dissolved in water is a reddish liquid used as an antiseptic. Calculate the percent composition of mercurochrome, $HgNa_2C_{20}H_8Br_2O_6$.

46. Chlorophyll is a dark green plant pigment. Calculate the percent composition of chlorophyll, $C_{55}H_{70}MgN_4O_6$.

Empirical Formula (Sec. 9.8)

47. In an experiment 0.500 g of tin reacted with nitric acid to give tin oxide. If the oxide had a mass of 0.635 g, what is the empirical formula of tin oxide?

48. In an experiment 0.500 g of nickel reacted with air to give 0.704 g of nickel oxide. What is the empirical formula of the oxide?

49. In an experiment 2.410 g of copper oxide produced 1.925 g of copper metal after heating with hydrogen gas. What is the empirical formula of copper oxide?

50. In an experiment 1.550 g of mercury oxide decomposed to give oxygen gas and 1.435 g of liquid mercury. What is the empirical formula of mercury oxide?

51. A 1.115 g sample of cobalt was heated with sulfur to give 2.025 g of cobalt sulfide. What is the empirical formula of cobalt sulfide?

52. A 0.715 g sample of titanium was heated with chlorine gas to give 2.836 g of titanium chloride. What is the empirical formula of titanium chloride?

53. The percent composition of copper chloride is 64.1% Cu and 35.9% Cl. Calculate the empirical formula.

54. The percent composition of vanadium oxide is 68.0% V and 32.0% O. Calculate the empirical formula.

55. The percent composition of tin bromide is 42.6% Sn and 57.4% Br. Calculate the empirical formula.

56. The percent composition of bismuth oxide is 89.7% Bi and 10.3% O. Calculate the empirical formula.

57. Trichloroethylene (TCE) is a common solvent used to degrease machine parts. Calculate the empirical formula for TCE if the percent composition is 18.25% C, 0.77% H, and 80.99% Cl.

58. Dimethyl sulfoxide (DMSO) is a liniment for horses and has been used in the treatment of arthritis for humans. Calculate the empirical formula for DMSO, given that the percent composition is 30.7% C, 7.74% H, 20.5% O, and 41.0% S.

Molecular Formula (Sec. 9.9)

59. Aspirin has a molar mass of 180 g/mol. If the empirical formula is $C_9H_8O_4$, what is the molecular formula of aspirin?

60. Quinine is used to treat malaria and has a molar mass of 325 g/mol. If the empirical formula is $C_{10}H_{12}NO$, what is the molecular formula of quinine?

61. Hexamethylene diamine is used to make nylon. If the molar mass is 115 g/mol and the empirical formula is C_3H_8N, what is the molecular formula of the compound?

62. Adipic acid is used to make nylon. If the molar mass is 147 g/mol and the empirical formula is $C_3H_5O_2$, what is the molecular formula of the compound?

63. Ethylene glycol is used as permanent antifreeze. If the molar mass is 62 g/mol and the percent composition is 38.7% C, 9.74% H, and 51.6% O, what is the molecular formula of ethylene glycol?

64. Dioxane is a common solvent for plastics. If the molar mass is 88 g/mol and the percent composition is 54.5% C, 9.15% H, and 36.3% O, what is the molecular formula of dioxane?

65. Lindane is an insecticide. If the molar mass is 290 g/mol and the percent composition is 24.8% C, 2.08% H, and 73.1% Cl, what is the molecular formula of lindane?

66. Mercurous chloride is a fungicide. If the molar mass is 470 g/mol and the percent composition is 85.0% Hg and 15.0% Cl, what is the actual formula of mercurous chloride?

67. Nicotine has a molar mass of 160 g/mol. If the percent composition is 74.0% C, 8.70% H, and 17.3% N, what is the molecular formula of nicotine?

68. Allose is a sugar with a molar mass of 180 g/mol. If the percent composition is 40.0% C, 6.72% H, and 53.3% O, what is the molecular formula of allose?

General Exercises

69. A mole of photons is referred to as an Einstein after the German-Swiss-American physicist Albert Einstein. How many photons are in an Einstein?

70. A mole of electrons is referred to as a Faraday after the English scientist Michael Faraday. How many electrons are in a Faraday?

71. A quadrillion, 1×10^{15}, is the approximate number of red blood cells in 50,000 people. Which is greater: a quadrillion red blood cells or the number of nickel atoms in a 5-g nickel coin?

72. By rubbing the Lincoln profile on a penny coin, a student removed just enough mass to be detected by an analytical balance, 0.0001 g. How many copper atoms were rubbed off the coin?

73. If the mass of Earth is 6×10^{24} kg and an average rodent mole has a mass of 100 g, which weighs more, Earth or a mole of furry moles?

74. If a light year, 9.5×10^{12} km, is the distance light travels in a year and an average rodent mole has a length of 17 cm, which is longer, a light year or a mole of furry moles?

75. In 1871 Mendeleev predicted the undiscovered element *ekaaluminum*. In 1875 the element was discovered in Gaul (France) and was given the name gallium. If 0.500 g of gallium reacts with oxygen gas to give 0.672 g of gallium oxide, what is its empirical formula?

76. In 1871 Mendeleev predicted the undiscovered element *ekasilicon*. In 1886 the element was discovered in Germany and was given the name germanium. If 0.500 g of germanium reacts with chlorine gas to give 1.456 g of germanium chloride, what is its empirical formula?

77. Calculate the volume in cubic centimeters occupied by one molecule of water in a beaker of water. Recall that the density of water is 1.00 g/cm^3.

78. Calculate the volume in milliliters occupied by one molecule of ethyl alcohol, C_2H_5OH, in a beaker of alcohol. The density of ethyl alcohol is 0.789 g/mL.

79. Calculate the number of carbon atoms in 1.00 g of table sugar, $C_{12}H_{22}O_{11}$.

80. Calculate the number of carbon atoms in 1.00 g of blood sugar, $C_6H_{12}O_6$.

81. Vitamin K has a molar mass of 173 g/mol and is 76.3% carbon by mass. How many carbon atoms are in one molecule of vitamin K?

82. What is the mass of rust, Fe_2O_3, that contains 10.0 g of iron?

83. The volume occupied by each copper atom in a 1-mole crystal is 0.0118 nm^3. If the density of the copper crystal is 8.92 g/cm^3, what is the experimental value of Avogadro's number?

84. Each atom in a crystal of aluminum metal occupies a theoretical cube that is 0.255 nm on a side. If the density of the aluminum crystal is 2.70 g/cm^3, what is the experimental value of Avogadro's number?

9 Chapter Self-Test

Answers to Self-Test are in Appendix J.

1. How many atoms of silver equal a mass of 107.87 amu? (Sec. 9.1)
   (a) 1 (b) 47
   (c) 107.87 (d) 108
   (e) 6.02×10^{23}

2. How many atoms of silver are in 2.50 mol of Ag metal? (Sec. 9.2)
   (a) 1.51×10^{22} atoms (b) 1.51×10^{23} atoms
   (c) 1.51×10^{24} atoms (d) 2.41×10^{23} atoms
   (e) 2.41×10^{24} atoms

3. What is the molar mass of ammonium dichromate, $(NH_4)_2Cr_2O_7$? (Sec. 9.3)
   (a) 238.09 g/mol (b) 252.10 g/mol
   (c) 260.18 g/mol (d) 372.18 g/mol
   (e) 386.19 g/mol

4. What is the mass of 3.30×10^{23} atoms of silver, Ag? (Sec. 9.4)
   (a) 0.549 g (b) 25.8 g
   (c) 59.1 g (d) 85.7 g
   (e) 197 g

5. How many ozone molecules are in 22.4 liters of O_3 gas at STP? (Sec. 9.5)
(a) 1.20×10^{24}
(b) 1.35×10^{25}
(c) 1.81×10^{24}
(d) 2.69×10^{22}
(e) 6.02×10^{23}

6. What is the volume of 1.20×10^{22} molecules of ozone gas, O_3, at STP? (Sec. 9.6)
(a) 0.0199 L
(b) 0.447 L
(c) 2.24 L
(d) 5.02 L
(e) 1120 L

7. An emerald has the chemical formula $Be_3Al_2Si_{16}O_{18}$. What is the percentage of aluminum in the gemstone? (Sec. 9.7)
(a) 3.30%
(b) 3.37%
(c) 6.59%
(d) 35.2%
(e) 54.9%

8. If 0.500 mol of silver combines with 0.250 mol of sulfur, what is the empirical formula of the silver sulfide product? (Sec. 9.8)
(a) AgS
(b) Ag_2S
(c) AgS_2
(d) Ag_5S_5
(e) none of the above

9. A common name for fructose is fruit sugar. What is the molecular formula of fructose if the empirical formula is CH_2O, and the approximate molar mass is 180 g/mol? (Sec. 9.9)
(a) CHO
(b) CH_2O
(c) CH_2O_6
(d) $C_6H_{12}O_6$
(e) $C_{12}H_{22}O_{11}$

Key Concepts

10. Given that F–19 is the only natural isotope of fluorine, what is the mass of one F atom and Avogadro's number of F atoms, respectively?

11. Given 1 mol of fluorine gas at STP, what is the mass, volume, and number of F_2 molecules?

12. Given Avogadro's number of steel balls, which of the following is the best estimate of the total mass: an automobile, the Empire State building, or Earth?

13. Given Avogadro's number of small marbles, which of the following is the best estimate of the total volume: a basketball, the Astrodome, or the Moon?

Critical Thinking

14. If an ultra-sensitive balance can weigh samples having a mass of 1 μg or greater, can it weigh 10 trillion copper atoms?

15. A 1-carat diamond is pure, crystalline carbon and has a mass of 0.200 g. Which has more atoms, a 1-carat diamond or a googol (1×10^{100}) of atoms?

CHAPTER

10 Chemical Equation Calculations

A match tip contains a sulfur compound that ignites when struck. Sulfur combines wth oxygen in air to produce sulfur dioxide, SO_2, and sulfur trioxide, SO_3.

"In the middle of difficulty lies opportunity."
Albert Einstein, German/Swiss/American Physicist (1879–1955)

The manufacturing of chemicals is one of the most important industries in the United States. Millions of tons of chemicals are produced each year for use both in the United States and in other countries. These chemicals are used to manufacture cars, computer chips, electronic instruments, prescription drugs, glass, paper, fertilizers, pesticides, plumbing supplies, synthetic fibers, and other products.

Chemists and chemical engineers routinely perform calculations based on balanced chemical equations to determine the cost of producing chemicals. These calculations are critical to the manufacture of products for the automobile, electronic, pharmaceutical, plastics, agricultural, textile, and other industries. The amounts of chemicals used industrially are published annually, and sulfuric acid, found in automobile batteries and every chemistry laboratory, consistently tops this list. The amount of sulfuric acid produced indicates the general activity of manufacturing and for that reason is referred to as the barometer of the chemical industry.

10.1 INTERPRETING A CHEMICAL EQUATION

OBJECTIVES

- To relate the coefficients in a balanced chemical equation to
  - (a) moles of reactants and products, and
  - (b) liters of gaseous reactants and products.

Let's examine the information we can obtain from a balanced chemical equation. Consider nitrogen monoxide gas, which is present in automobile emissions. In the atmosphere, ultraviolet light catalyzes the reaction of nitrogen monoxide and oxygen to produce reddish-brown nitrogen dioxide smog. The balanced chemical equation is

$$2\,NO(g) + O_2(g) \xrightarrow{UV} 2\,NO_2(g)$$

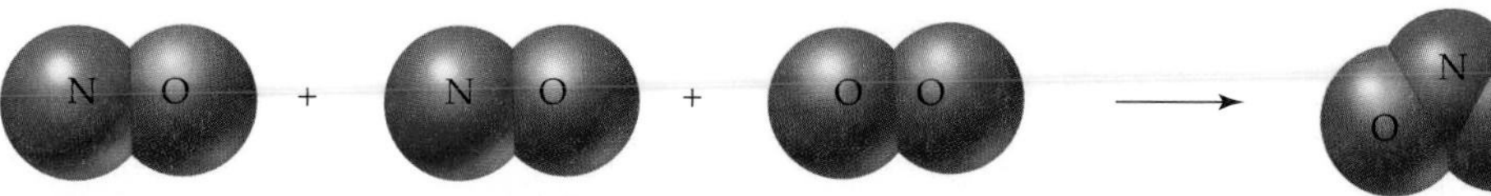

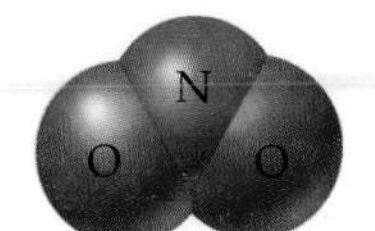

The Mole Interpretation of Equation Coefficients

The coefficients in the balanced equation indicate that 2 molecules of NO react with 1 molecule of O_2 to produce 2 molecules of NO_2. These coefficients indicate relative numbers of reactant and product molecules. It therefore follows that multiples of these coefficients are in the same ratios. For example, 2000 molecules of NO react with 1000 molecules of O_2 to give 2000 molecules of NO_2. Let's consider a number even larger than 2000; let's consider Avogadro's number (symbol *N*) of molecules, that is, 6.02×10^{23} molecules. When we substitute this very large number of molecules for the coefficients, the equation remains balanced. That is,

$$2\,N\,NO + 1\,N\,O_2 \longrightarrow 2\,N\,NO_2$$

We read the above equation as *two times Avogadro's number of nitrogen monoxide molecules reacts with Avogadro's number of oxygen molecules to produce two times Avogadro's number of nitrogen dioxide molecules.* Since Avogadro's number is the number of molecules in 1 mol, we can write the equation in terms of moles of substance:

$$2\text{ mol }NO + 1\text{ mol }O_2 \longrightarrow 2\text{ mol }NO_2$$

Similarly, this equation reads *two moles of nitrogen monoxide react with one mole of oxygen to give two moles of nitrogen dioxide.* Furthermore, the coefficients indicate the ratio of moles, or the **mole ratio**, of reactants and products in every balanced chemical equation.

▲ This photo shows a smoggy day in Los Angeles; however, the Capitol Records Tower is visible in the foreground.

The Volume Interpretation of Equation Coefficients

According to **Avogadro's theory**, there are equal numbers of molecules in equal volumes of gas at the same temperature and pressure. It therefore follows that twice the number of molecules occupies twice the volume of gas. In the preceding equation, 2 molecules of NO react with 1 molecule of O_2 to give 2 molecules of NO_2. Accordingly, 2 volumes of NO react with 1 volume of O_2 gas to give 2 volumes of NO_2 gas. If we choose the liter as our unit of volume, we can write the ratio of volumes of gas for the reaction.

$$2\text{ L }NO + 1\text{ L }O_2 \longrightarrow 2\text{ L }NO_2$$

Regardless of the units, the numbers in the ratio of volumes correspond to the coefficients of the balanced chemical equation, that is, 2:1:2. For example, 2 mL of NO reacts with 1 mL of O_2 to give 2 mL of NO_2. Similarly, 20 mL of NO reacts with 10 mL of O_2 to give 20 mL of NO_2. Although the ratio of gaseous volumes is 20:10:20, the smallest whole-number ratio is 2:1:2. Table 10.1 summarizes the information that can be obtained from the coefficients of a balanced chemical equation.

TABLE 10.1 INTERPRETATION OF CHEMICAL EQUATION COEFFICIENTS

FOR THE GENERAL EQUATION	2 A	+	3 B	→	C	+	2 D
The ratio of molecules	2	:	3	:	1	:	2
The ratio of moles	2	:	3	:	1	:	2
The ratio of volumes of gas	2	:	3	:	1	:	2

The following example exercise illustrates the information that can be obtained by interpreting the coefficients in a balanced chemical equation:

Example Exercise 10.1 Interpreting Chemical Equation Coefficients

Given the chemical equation for the combustion of methane, CH_4, balance the equation and interpret the coefficients in terms of (a) moles and (b) liters:

$$CH_4(g) + O_2(g) \xrightarrow{\text{spark}} CO_2(g) + H_2O(g)$$

Solution

The balanced chemical equation is

$$CH_4(g) + 2\,O_2(g) \xrightarrow{\text{spark}} CO_2(g) + 2\,H_2O(g)$$

(a) The coefficients in the equation (1:2:1:2) indicate the ratio of moles as well as molecules.

$$1 \text{ mol } CH_4 + 2 \text{ mol } O_2 \longrightarrow 1 \text{ mol } CO_2 + 2 \text{ mol } H_2O$$

(b) The coefficients in the equation (1:2:1:2) indicate the ratio of volumes of gases. If we express the volume in liters, we have

$$1 \text{ L } CH_4 + 2 \text{ L } O_2 \longrightarrow 1 \text{ L } CO_2 + 2 \text{ L } H_2O$$

Practice Exercise

Given the chemical equation for the combustion of propane, C_3H_8, balance the equation and interpret the coefficients in terms of (a) moles and (b) milliliters:

$$C_3H_8(g) + O_2(g) \xrightarrow{\text{spark}} CO_2(g) + H_2O(g)$$

Answers:

(a) $1 \text{ mol } C_3H_8 + 5 \text{ mol } O_2 \longrightarrow 3 \text{ mol } CO_2 + 4 \text{ mol } H_2O$

(b) $1 \text{ ml } C_3H_8 + 5 \text{ ml } O_2 \longrightarrow 3 \text{ ml } CO_2 + 4 \text{ mL } H_2O$

Concept Exercise

Which of the following is in the same ratio as the coefficients in a balanced equation: moles of gas, mass of gas, volume of gas?

Answer: See Appendix G.

Verifying the Conservation of Mass Law

In 1787 after conducting numerous experiments, Antoine Lavoisier formulated the **law of conservation of mass**. He stated that mass is neither created nor destroyed during a chemical reaction. As a result, the combined masses of the reactants must equal the combined masses of the products.

We can verify the conservation of mass law for any balanced equation. Let's consider the previous equation for the formation of nitrogen dioxide:

$$2\ NO(g) + O_2(g) \xrightarrow{UV} 2\ NO_2(g)$$

According to the coefficients in the balanced equation, the ratio of the substances is 2:1:2. Therefore, the mole ratio of each substance is 2:1:2.

$$2 \text{ mol NO} + 1 \text{ mol } O_2 \longrightarrow 2 \text{ mol } NO_2$$

By summing the gram atomic masses for each substance, we find that the **molar mass** for NO is 30.01 g/mol; for O_2, 32.00 g/mol; and for NO_2, 46.01 g/mol. Substituting the molar mass for a mole of each substance gives

$$2\ (30.01 \text{ g}) + 1\ (32.00 \text{ g}) \longrightarrow 2\ (46.01 \text{ g})$$

$$\text{Simplifying, } 60.02 \text{ g} + 32.00 \text{ g} \longrightarrow 92.02 \text{ g}$$

$$\text{and, } 92.02 \text{ g} \longrightarrow 92.02 \text{ g}$$

In this example, the total mass of the reactants is 92.02 g, which is equal to the total mass of the products. Thus from our calculation, we have verified the conservation of mass law.

10.2 MOLE–MOLE RELATIONSHIPS

OBJECTIVE

▶ To relate the number of moles of two substances in a balanced chemical equation.

As we demonstrated in Section 10.1, the coefficients in a chemical equation indicate the mole ratio of the reactants and products. Consider the combination reaction of nitrogen and oxygen to give nitrogen monoxide. The equation is

$$N_2(g) + O_2(g) \xrightarrow{\Delta} 2\ NO(g)$$

From the coefficients in the balanced equation, we see that 1 mol of nitrogen reacts with 1 mol of oxygen to produce 2 mol of nitrogen monoxide. We can write several mole ratios for this equation.

Since 1 mol of N_2 reacts with 1 mol of O_2, the corresponding mole ratios are

$$\frac{1 \text{ mol } N_2}{1 \text{ mol } O_2} \quad \text{and} \quad \frac{1 \text{ mol } O_2}{1 \text{ mol } N_2}$$

Since 1 mol of N_2 produces 2 mol of NO, the corresponding mole ratios are

$$\frac{1 \text{ mol } N_2}{2 \text{ mol NO}} \quad \text{and} \quad \frac{2 \text{ mol NO}}{1 \text{ mol } N_2}$$

Since 1 mol of O_2 produces 2 mol of NO, the corresponding mole ratios are

$$\frac{1 \text{ mol } O_2}{2 \text{ mol NO}} \quad \text{and} \quad \frac{2 \text{ mol NO}}{1 \text{ mol } O_2}$$

Suppose we want to find out how many moles of oxygen react with 2.25 mol of nitrogen. As usual, we apply the unit analysis method of problem solving:

$$2.25 \text{ mol } N_2 \times \frac{\text{unit}}{\text{factor}} = \text{mol } O_2$$

To cancel units, we select the mole ratio 1 mol O_2/1 mol N_2 as our unit factor:

$$2.25 \cancel{\text{mol } N_2} \times \frac{1 \text{ mol } O_2}{1 \cancel{\text{mol } N_2}} = 2.25 \text{ mol } O_2$$

We can also calculate the moles of nitrogen monoxide produced by the reaction. To cancel units, we select the mole ratio 2 mol NO/1 mol N_2 as our unit factor:

$$2.25\ \cancel{\text{mol N}_2} \times \frac{2\ \text{mol NO}}{1\ \cancel{\text{mol N}_2}} = 4.50\ \text{mol NO}$$

Whenever we have a balanced chemical equation, we can always convert from moles of one substance to moles of another substance using a mole ratio as a unit factor. Example Exercise 10.2 further illustrates the relationship between moles of reactants and moles of products.

Example Exercise 10.2 Mole–Mole Relationships

Carbon monoxide is produced in a blast furnace by passing oxygen gas over hot coal. How many moles of oxygen react with 2.50 mol of carbon according to the balanced equation?

$$2\ C(s) + O_2(g) \xrightarrow{\Delta} 2\ CO(g)$$

Strategy Plan

Step 1: What unit is asked for in the answer? **Step 1:** mol O_2

Step 2: What given value is related to the answer? **Step 2:** 2.50 mol C

Step 3: What unit factor(s) should we apply? **Step 3:** $\frac{2\ \text{mol C}}{1\ \text{mol O}_2}$ or $\frac{1\ \text{mol O}_2}{2\ \text{mol C}}$

Using the balanced equation, we see that 2 mol C = 1 mol O_2. Thus, the mole ratio and the two unit factors are

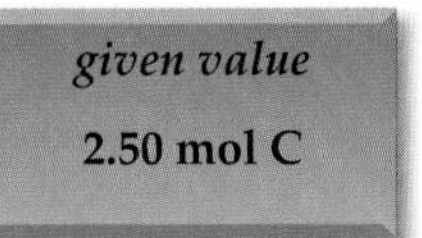

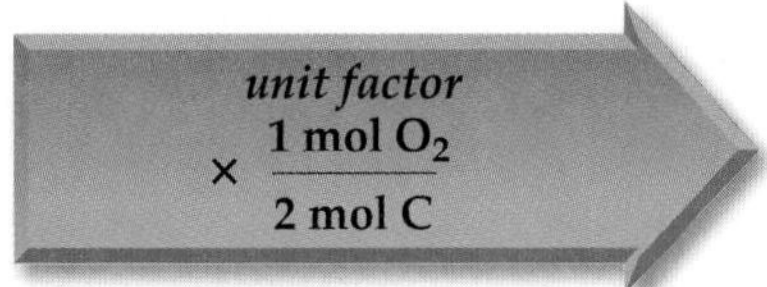

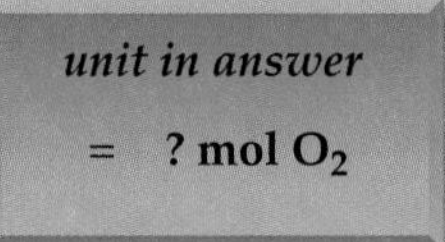

Unit Analysis Map

Solution

We select the unit factor that cancels the unit in the given value ($\cancel{\text{mol C}}$). Thus,

$$2.50\ \cancel{\text{mol C}} \times \frac{1\ \text{mol O}_2}{2\ \cancel{\text{mol C}}} = 1.25\ \text{mol O}_2$$

Practice Exercise

Iron is produced from iron ore in a blast furnace by passing carbon monoxide gas through molten iron(III) oxide. The balanced equation is

$$Fe_2O_3(l) + 3\ CO(g) \xrightarrow{\Delta} 2\ Fe(l) + 3\ CO_2(g)$$

(a) How many moles of carbon monoxide react with 2.50 mol of Fe_2O_3?
(b) How many moles of iron are produced from 2.50 mol of Fe_2O_3?

Answers: (a) 7.50 mol CO; (b) 5.00 mol Fe

Concept Exercise

How many unit factors are required to solve a mole–mole stoichiometry problem?

Answer: See Appendix G.

CRITICAL THINKING Iron versus Steel

Q: What is the difference between iron and steel?

The Iron Age began about 1000 B.C. in western Asia and Europe. The Iron Age followed the Bronze Age after it was discovered that iron could be made by heating iron ore with coal. Interestingly, iron is still obtained by a similar method.

Today, an industrial blast furnace converts iron ore to iron. In the process, iron ore, coke (a form of coal), and limestone, $CaCO_3$, are added into the top of a heated furnace while a blast of hot air is blown in at the bottom.

Many reactions take place in the process, but first, oxygen in the hot air reacts with carbon in coke to produce carbon monoxide:

$$2\,C(s) + O_2(g) \xrightarrow{\Delta} 2\,CO(g)$$

▲ **Blast Furnace** Molten pig iron reacts with a "blast" of oxygen gas that converts carbon and sulfur impurities into CO_2 and SO_2 gases.

Iron ore is an impure mixture of hematite, Fe_2O_3, and magnetite, Fe_3O_4. In the blast furnace hematite and magnetite are converted to iron by reaction with carbon monoxide gas. The reactions are as follows:

$$Fe_2O_3(l) + 3\,CO(g) \longrightarrow 2\,Fe(l) + 3\,CO_2(g)$$

$$Fe_3O_4(l) + 4\,CO(g) \longrightarrow 3\,Fe(l) + 4\,CO_2(g)$$

The iron ore also contains silicon impurities that must be removed. This is done by reacting the impurities with CaO, which is produced from limestone, that is, $CaCO_3$. Silicon compounds such as $CaSiO_3$ form and are referred to as "slag." This less dense slag is easily removed, as it floats on the molten iron at the bottom of the furnace.

The molten iron obtained from the blast furnace is still not pure and is referred to as "pig iron." Pig iron contains impurities such as carbon and sulfur. In a second stage of the process, these impurities are removed by blowing oxygen gas through the molten pig iron. The impurities are converted to oxides, and are released as gases such as CO_2 and SO_2.

The iron obtained after the oxygen process contains a trace amount of carbon. Manganese is then often added for strength and flexibility, and the resulting product is referred to as "steel." A large number of alloy steels can be produced by adding other metals. For example, stainless steel contains about 10% chromium.

A: Iron is an element, whereas steel is an alloy of iron with traces of carbon and metals, such as manganese, nickel, and chromium, which impart special properties.

10.3 TYPES OF STOICHIOMETRY PROBLEMS

OBJECTIVES

- To classify the three basic types of stoichiometry problems: mass–mass, mass–volume, and volume–volume.
- To state the procedure for solving a stoichiometry problem given the balanced equation.

In Chapter 9, we performed mole calculations based on chemical formulas. In this chapter, we will perform mole calculations based on chemical equations. That is, we will apply mole ratios to relate quantities of reactants and products. The term **stoichiometry** (stoy-kee-AHM-uh-tree) is used to refer to the relationship between quantities in a chemical reaction according to the balanced chemical equation. The term is derived from the Greek words *stoicheion* meaning "element" and *metron* meaning "measure."

Some stoichiometry calculations relate a given amount of reactant to an unknown amount of product. Conversely, other stoichiometry calculations relate a given amount of product to an unknown amount of reactant. Whether we relate a reactant to a product, or vice versa, the calculations use the same basic method of problem solving.

We can classify three types of stoichiometry problems based on the given quantity and the quantity to be calculated. Consider the general equation for any reaction, where *a*, *b*, *c*, and *d* are the respective coefficients for the substances A, B, C, and D:

$$a\,A + b\,B \longrightarrow c\,C + d\,D$$

If we are given grams of A and asked to find how many grams of C are produced, we classify this as a **mass–mass problem**. If we are given grams of A and asked to calculate the milliliters of a gas D, we classify this as a **mass–volume problem**. If we are given liters of a gas A and asked to find the liters of a gas D, we classify this as a **volume–volume problem.** In summary, stoichiometry problems can be classified as one of these basic types: mass–mass, mass–volume, or volume–volume.

After classifying a stoichiometry problem, we can apply the unit analysis method of problem solving to its solution. Example Exercise 10.3 offers practice in classifying types of stoichiometry problems.

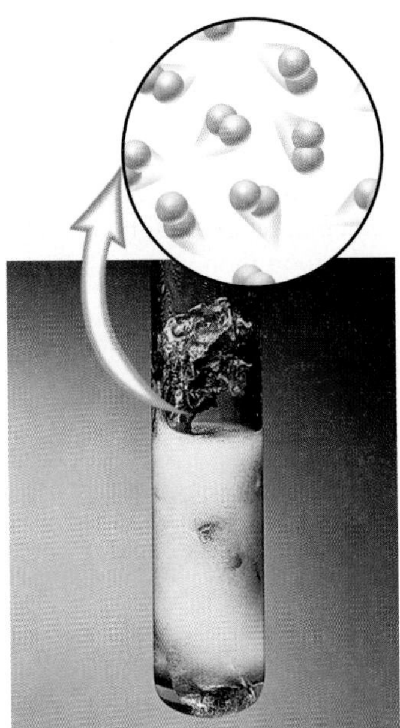

▲ **Reaction of Zn Metal** Zinc metal reacts with hydrochloric acid to give $ZnCl_2$ and H_2 bubbles.

Example Exercise 10.3 Classifying Stoichiometry Problems

Classify the type of stoichiometry problem for each of the following:

(a) How many grams of Zn metal react with hydrochloric acid to give 0.500 g of zinc chloride?

(b) How many liters of H_2 gas react with chlorine gas to yield 50.0 cm^3 of hydrogen chloride gas?

(c) How many kilograms of Fe react with sulfuric acid to produce 50.0 mL of hydrogen gas?

Solution

After analyzing a problem for the unknown quantity and the relevant given value, we classify the type of problem.

(a) The problem asks for grams of Zn (mass) that react to give 0.500 g of $ZnCl_2$ (mass). This is a *mass–mass* type of problem.

(b) The problem asks for liters of H_2 gas (volume) that react to yield 50.0 cm^3 of HCl gas (volume). This is a *volume–volume* type of problem.

(c) The problem asks for kilograms of Fe (mass) that react to produce 50.0 mL of H_2 gas (volume). This is a *mass–volume* type of problem.

Practice Exercise

Classify the type of stoichiometry problem for each of the following:

(a) How many grams of HgO decompose to give 0.500 L of oxygen gas at STP?

(b) How many grams of AgCl are produced from the reaction of 0.500 g of solid sodium chloride with silver nitrate solution?

(c) How many milliliters of H_2 gas react with nitrogen gas to yield 1.00 L of ammonia gas?

Answers: (a) mass–volume; (b) mass–mass; (c) volume–volume

Concept Exercise

What are the three types of stoichiometry problems?

Answer: See Appendix G.

10.4 MASS–MASS PROBLEMS

OBJECTIVE

▸ To perform mass–mass stoichiometry calculations.

A mass–mass stoichiometry problem is so named because an *unknown mass* of substance is calculated from a *given mass* of reactant or product in a chemical equation. After balancing the chemical equation, we proceed as follows:

(a) Convert the given mass of substance to moles using the molar mass of the substance as a unit factor.

(b) Convert the moles of the given to moles of the unknown using the coefficients in the balanced equation.

(c) Convert the moles of the unknown to grams using the molar mass of the substance as a unit factor.

To see how to solve a mass–mass problem, consider the high-temperature conversion of 14.4 g of iron(II) oxide to elemental iron with aluminum metal. The balanced equation for the reaction is

$$3\ FeO(l) + 2\ Al(l) \xrightarrow{\Delta} 3\ Fe(l) + Al_2O_3(l)$$

To calculate the mass of aluminum necessary for the reaction, we must first find the moles of iron(II) oxide. Given that the molar mass of FeO is 71.85 g/mol, the unit factor conversion is as follows:

(a) $$14.4\ \cancel{\text{g FeO}} \times \frac{1\ \text{mol FeO}}{71.85\ \cancel{\text{g FeO}}} = 0.200\ \text{mol FeO}$$

Next, we convert the moles of FeO to moles of Al by applying a mole ratio using the coefficients in the balanced equation. Since 3 mol FeO = 2 mol Al, the mole ratio unit factor is

(b) $$0.200\ \cancel{\text{mol FeO}} \times \frac{2\ \text{mol Al}}{3\ \cancel{\text{mol FeO}}} = 0.134\ \text{mol Al}$$

Last, we use the molar mass of aluminum (26.98 g/mol) as a unit factor to obtain the mass of Al reacting with 14.4 g of FeO.

(c) $$0.134\ \cancel{\text{mol Al}} \times \frac{26.98\ \text{g Al}}{1\ \cancel{\text{mol Al}}} = 3.60\ \text{g Al}$$

After you gain confidence in solving stoichiometry problems, it will be more convenient to perform one continuous calculation. For example, the solution to the preceding problem can be shown as

$$14.4\ \cancel{\text{g FeO}} \times \underset{(a)}{\frac{1\ \cancel{\text{mol FeO}}}{71.85\ \cancel{\text{g FeO}}}} \times \underset{(b)}{\frac{2\ \cancel{\text{mol Al}}}{3\ \cancel{\text{mol FeO}}}} \times \underset{(c)}{\frac{26.98\ \text{g Al}}{1\ \cancel{\text{mol Al}}}} = 3.60\ \text{g Al}$$

The following example exercises give additional illustrations of mass–mass stoichiometry calculations using unit factors:

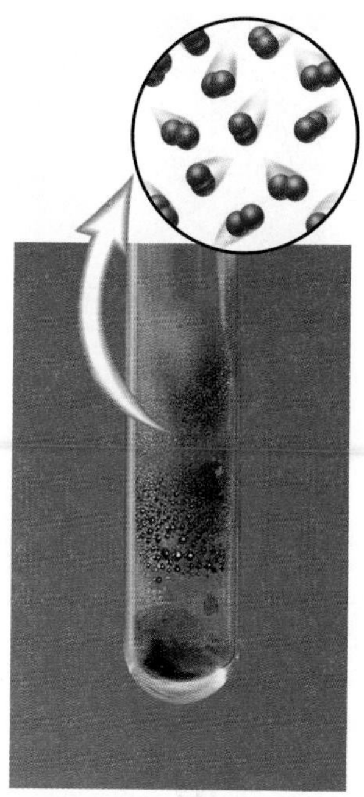

▲ **Decomposition of HgO** Orange mercury(II) oxide decomposes by heating to give droplets of silver Hg liquid and O_2 gas.

Example Exercise 10.4 Mass–Mass Stoichiometry

Calculate the mass of mercury produced from the decomposition of 1.25 g of orange mercury(II) oxide (MM = 216.59 g/mol):

$$2\ \text{HgO}(s) \xrightarrow{\Delta} 2\text{Hg}(l) + \text{O}_2(g)$$

Strategy Plan

Step 1: What unit is asked for in the answer? **Step 1:** g Hg

Step 2: What given value is related to the answer? **Step 2:** 1.25 g HgO

Step 3: What unit factor(s) should we apply? **Step 3:**

$$\frac{1\ \text{mol HgO}}{216.59\ \text{g HgO}} \text{ or } \frac{216.59\ \text{g HgO}}{1\ \text{mol HgO}}$$

$$\frac{2\ \text{mol HgO}}{2\ \text{mol Hg}} \text{ or } \frac{2\ \text{mol Hg}}{2\ \text{mol HgO}}$$

$$\frac{1\ \text{mol Hg}}{200.59\ \text{g Hg}} \text{ or } \frac{200.59\ \text{g Hg}}{1\ \text{mol Hg}}$$

We are given that 1 mol HgO = 216.59 g HgO; the mole ratio from the balanced equation is 2 mol HgO = 2 mol Hg. From the periodic table, we find that 1 mol Hg = 200.59 g Hg. Thus, the unit factors are

Unit Analysis Map

given value		*unit factor 1*		*unit factor 2*		*unit factor 3*	*unit in answer*
1.25 g HgO	×	$\frac{1\ \text{mol HgO}}{216.59\ \text{g HgO}}$	×	$\frac{2\ \text{mol Hg}}{2\ \text{mol HgO}}$	×	$\frac{200.59\ \text{g Hg}}{1\ \text{mol Hg}}$	= ? g Hg

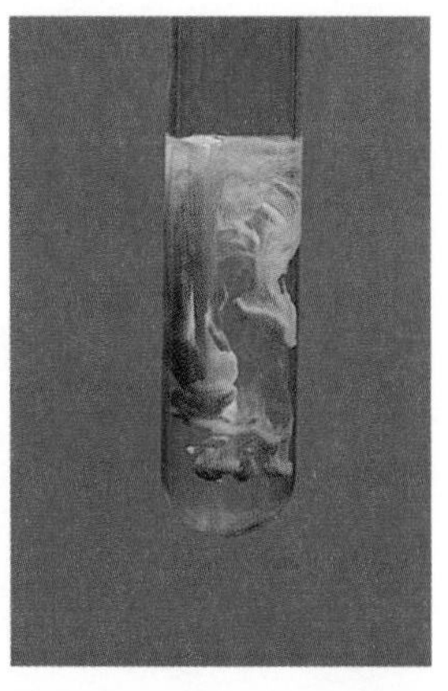

▲ **Precipitation of HgI_2** Potassium iodide and mercury(II) nitrate react to give an orange precipitate of HgI_2.

Solution

We select from each of the three pairs of ratios a unit factor that cancels the previous units. Thus,

$$1.25\ \cancel{g\ HgO} \times \frac{1\ \cancel{mol\ HgO}}{216.59\ \cancel{g\ HgO}} \times \frac{2\ \cancel{mol\ Hg}}{2\ \cancel{mol\ HgO}} \times \frac{200.59\ g\ Hg}{1\ \cancel{mol\ Hg}} = 1.16\ g\ Hg$$

Practice Exercise

Calculate the mass of carbon dioxide released from 10.0 g of cobalt(III) carbonate given the *unbalanced* equation for the reaction:

$$Co_2(CO_3)_3(s) \xrightarrow{\Delta} Co_2O_3(s) + CO_2(g)$$

Answer: 4.43 g CO_2

Concept Exercise

What are the three steps in the unit analysis method of problem solving?

Answer: See Appendix G.

Example Exercise 10.5 Mass–Mass Stoichiometry

Calculate the mass of potassium iodide (166.00 g/mol) required to yield 1.78 g of mercury(II) iodide precipitate (454.39 g/mol):

$$2\ KI(s) + Hg(NO_3)_2(aq) \longrightarrow HgI_2(s) + 2\ KNO_3(aq)$$

Strategy Plan

Step 1: What unit is asked for in the answer? **Step 1:** g KI

Step 2: What given value is related to the answer? **Step 2:** 1.78 g HgI_2

Step 3: What unit factor(s) should we apply? **Step 3:**

$$\frac{1\ mol\ HgI_2}{454.39\ g\ HgI_2} \text{ or } \frac{454.39\ g\ HgI_2}{1\ mol\ HgI_2}$$

$$\frac{2\ mol\ KI}{1\ mol\ HgI_2} \text{ or } \frac{1\ mol\ HgI_2}{2\ mol\ KI}$$

$$\frac{1\ mol\ KI}{166.00\ g\ KI} \text{ or } \frac{166.00\ g\ KI}{1\ mol\ KI}$$

From the given molar mass, we see that 1 mol HgI_2 = 454.39 g HgI_2. The mole ratio from the balanced equation is 2 mol KI = 1 mol HgI_2, and from the given molar mass, 1 mol KI = 166.00 g KI. Thus, the three pairs of unit factors are

Unit Analysis Map

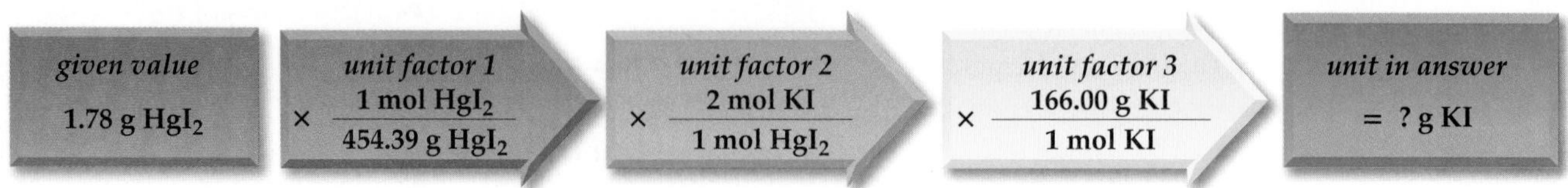

Solution

We select from each of the three pairs of ratios a unit factor that cancels the previous units. Thus,

$$1.78\ \cancel{g\ HgI_2} \times \frac{1\ \cancel{mol\ HgI_2}}{454.39\ \cancel{g\ HgI_2}} \times \frac{2\ \cancel{mol\ KI}}{1\ \cancel{mol\ HgI_2}} \times \frac{166.00\ g\ KI}{1\ \cancel{mol\ KI}} = 1.30\ g\ KI$$

Practice Exercise

Calculate the mass of iron filings required to produce 0.455 g of silver metal given the *unbalanced* equation for the reaction:

$$Fe(s) + AgNO_3(aq) \longrightarrow Fe(NO_3)_3(aq) + Ag(s)$$

Answer: 0.0785 g Fe

Concept Exercise

In general, how many unit factors are required to solve a mass–mass stoichiometry problem?

Answer: See Appendix G.

10.5 MASS–VOLUME PROBLEMS

OBJECTIVE

- To perform mass–volume stoichiometry calculations.

A mass–volume stoichiometry problem is so named because an *unknown volume* of substance is calculated from a *given mass* of reactant or product in a chemical equation. After balancing the chemical equation, we proceed as follows:

(a) Convert the given mass of substance to moles using the molar mass of the substance as a unit factor.

(b) Convert the moles of the given to moles of the unknown using the coefficients in the balanced equation.

(c) Convert the moles of the unknown to liters using the **molar volume** of a gas as a unit factor. At **standard temperature and pressure (STP)**, the molar volume of a gas is 22.4 L/mol.

Of course, it is also possible to perform the reverse calculation; that is, we can find the mass of an unknown substance from a given volume of gas. For simplicity, this reverse process is also referred to as a mass–volume stoichiometry problem.

To understand how to solve a mass–volume stoichiometry problem, consider the reaction of 0.165 g of aluminum metal with dilute hydrochloric acid. The balanced equation for the reaction is

$$2\ Al(s) + 6\ HCl(aq) \longrightarrow 2\ AlCl_3(aq) + 3\ H_2(g)$$

Let's calculate the volume of hydrogen gas produced by the reaction at STP. This is a mass–volume stoichiometry problem. First, we calculate the moles of aluminum whose molar mass is 26.98 g/mol:

(a)
$$0.165\ \cancel{g\ Al} \times \frac{1\ mol\ Al}{26.98\ \cancel{g\ Al}} = 0.00611\ mol\ Al$$

Next, we use the coefficients in the balanced equation to find the moles of hydrogen gas. According to the balanced equation, 2 mol Al = 3 mol H_2. Applying the mole ratio unit factor, we have

(b)
$$0.00611\ \cancel{mol\ Al} \times \frac{3\ mol\ H_2}{2\ \cancel{mol\ Al}} = 0.00917\ mol\ H_2$$

Last, we multiply by the molar volume, 22.4 L/mol, to obtain the volume of H_2 gas produced at STP.

(c)
$$0.00917\ \cancel{mol\ H_2} \times \frac{22.4\ L\ H_2}{1\ \cancel{mol\ H_2}} = 0.205\ L\ H_2$$

The solution to the preceding problem can also be shown as one continuous calculation:

$$0.165\ \cancel{g\ Al} \times \underset{(a)}{\frac{1\ \cancel{mol\ Al}}{26.98\ \cancel{g\ Al}}} \times \underset{(b)}{\frac{3\ \cancel{mol\ H_2}}{2\ \cancel{mol\ Al}}} \times \underset{(c)}{\frac{22.4\ L\ H_2}{1\ \cancel{mol\ H_2}}} = 0.205\ L\ H_2$$

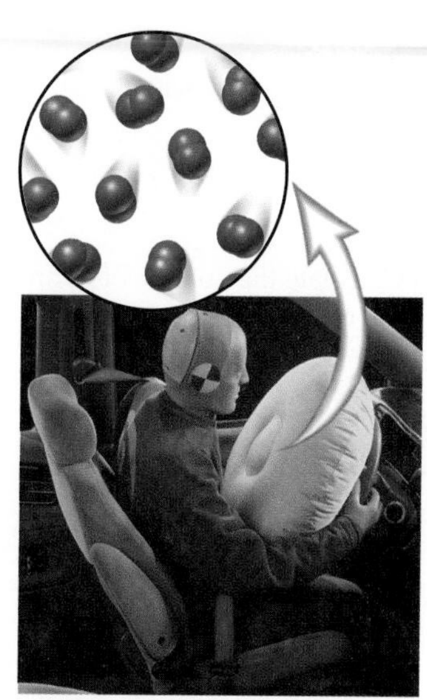

▲ **Automobile Air Bag** The decomposition of sodium azide releases N_2 gas, which inflates the air bag.

The following example exercises provide additional practice in solving mass–volume stoichiometry problems:

Example Exercise 10.6 Mass–Volume Stoichiometry

In an automobile collision, sodium azide, NaN_3, decomposes and fills an air bag with nitrogen gas. If an air bag contains 100.0 g of NaN_3 (65.02 g/mol), what is the volume of nitrogen gas produced at STP? The equation for the reaction is

$$2\,NaN_3(s) \xrightarrow{\text{spark}} 2\,Na(s) + 3\,N_2(g)$$

Strategy Plan

Step 1: What unit is asked for in the answer? **Step 1:** $L\ N_2$

Step 2: What given value is related to the answer? **Step 2:** $100.0\ g\ NaN_3$

Step 3: What unit factor(s) should we apply? **Step 3:**

Given the molar mass, 1 mol NaN_3 = 65.02 g NaN_3. The mole ratio from the balanced equation is 2 mol NaN_3 = 3 mol N_2, and the molar volume of a gas is 1 mol N_2 = 22.4 L N_2. Thus, the unit factors are

$$\frac{1\ \text{mol NaN}_3}{65.02\ \text{g NaN}_3} \quad \text{or} \quad \frac{65.02\ \text{g NaN}_3}{1\ \text{mol NaN}_3}$$

$$\frac{2\ \text{mol NaN}_3}{3\ \text{mol N}_2} \quad \text{or} \quad \frac{3\ \text{mol N}_2}{2\ \text{mol NaN}_3}$$

$$\frac{1\ \text{mol N}_2}{22.4\ \text{L N}_2} \quad \text{or} \quad \frac{22.4\ \text{L N}_2}{1\ \text{mol N}_2}$$

Unit Analysis Map

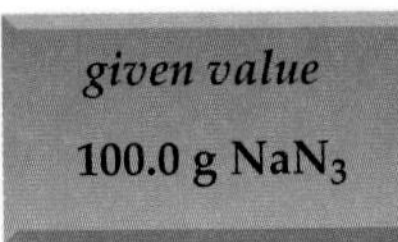

×

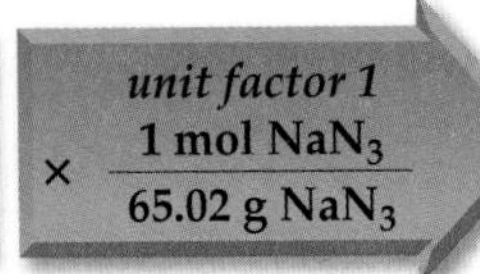

×

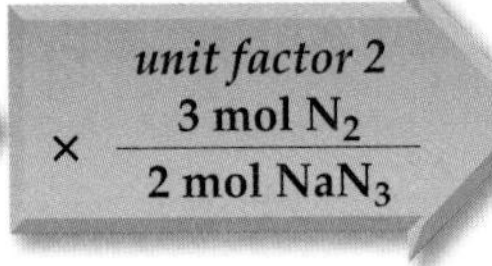

×

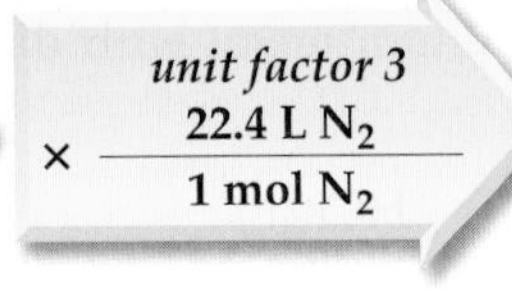

unit in answer = ? L N_2

Solution

We select from each of the three pairs of ratios a unit factor that cancels the previous units. Thus,

$$100.\ \cancel{\text{g NaN}_3} \times \frac{1\ \cancel{\text{mol NaN}_3}}{65.02\ \cancel{\text{g NaN}_3}} \times \frac{3\ \cancel{\text{mol N}_2}}{2\ \cancel{\text{mol NaN}_3}} \times \frac{22.4\ \text{L N}_2}{1\ \cancel{\text{mol N}_2}} = 51.7\ \text{L N}_2$$

Practice Exercise

Calculate the volume of hydrogen gas produced at STP from 1.55 g of sodium metal in water given the *unbalanced* equation for the reaction:

$$Na(s) + H_2O(l) \longrightarrow NaOH(aq) + H_2(g)$$

Answer: 0.755 L H_2

Concept Exercise

What are the three steps in the unit analysis method of problem solving?

Answer: See Appendix G.

Example Exercise 10.7 Mass–Volume Stoichiometry

Baking soda can be used as a fire extinguisher. When heated, it decomposes to carbon dioxide gas, which can smother a fire. If a sample of $NaHCO_3$ (84.01 g/mol) produces 0.500 L of carbon dioxide gas at STP, what is the mass of the sample?

$$2\,NaHCO_3(s) \xrightarrow{\Delta} Na_2CO_3(s) + H_2O(g) + CO_2(g)$$

Strategy Plan

Step 1: What unit is asked for in the answer? **Step 1:** g $NaHCO_3$

Step 2: What given value is related to the answer? **Step 2:** 0.500 L CO_2

Step 3: What unit factor(s) should we apply? **Step 3:**

From molar volume, we know that 1 mol CO_2 = 22.4 L CO_2. The mole ratio from the balanced equation is 2 mol $NaHCO_3$ = 1 mol CO_2, and from the given molar mass, 1 mol $NaHCO_3$ = 84.01 g $NaHCO_3$. Thus, the unit factors are

$$\frac{1 \text{ mol } CO_2}{22.4 \text{ L } CO_2} \text{ or } \frac{22.4 \text{ L } CO_2}{1 \text{ mol } CO_2}$$

$$\frac{2 \text{ mol } NaHCO_3}{1 \text{ mol } CO_2} \text{ or } \frac{1 \text{ mol } CO_2}{2 \text{ mol } NaHCO_3}$$

$$\frac{1 \text{ mol } NaHCO_3}{84.01 \text{ g } NaHCO_3} \text{ or } \frac{84.01 \text{ g } NaHCO_3}{1 \text{ mol } NaHCO_3}$$

Unit Analysis Map

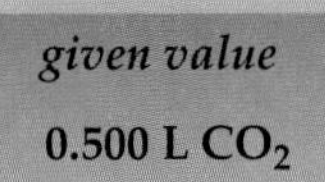

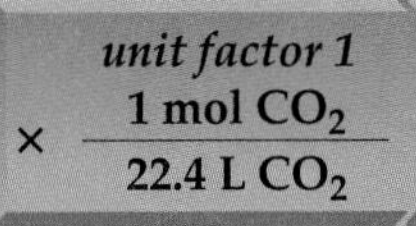

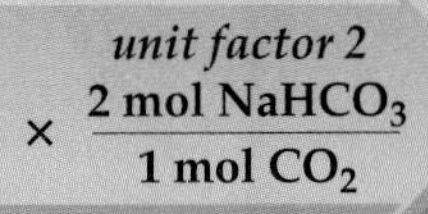

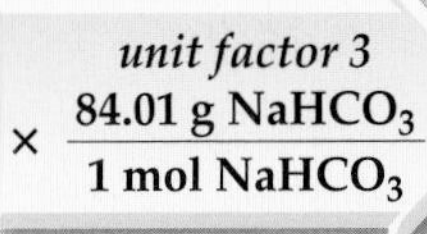

unit in answer

= ? g $NaHCO_3$

Solution

We select from each of the three pairs of ratios a unit factor that cancels the previous units. Thus,

$$0.500 \cancel{\text{L } CO_2} \times \frac{1 \cancel{\text{mol } CO_2}}{22.4 \cancel{\text{L } CO_2}} \times \frac{2 \cancel{\text{mol } NaHCO_3}}{1 \cancel{\text{mol } CO_2}} \times \frac{84.01 \text{ g } NaHCO_3}{1 \cancel{\text{mol } NaHCO_3}} = 3.75 \text{ g } NaHCO_3$$

Practice Exercise

Calculate the mass of aluminum metal required to release 2160 mL of hydrogen gas at STP from sulfuric acid given the *unbalanced* equation for the reaction:

$$Al(s) + H_2SO_4(aq) \longrightarrow Al_2(SO_4)_3(aq) + H_2(g)$$

Answer: 1.73 g Al

Concept Exercise

In general, how many unit factors are required to solve a mass–volume stoichiometry problem?

Answer: See Appendix G.

◀ **Baking Soda, $NaHCO_3$** Ordinary baking soda can be used to extinguish a fire. When heated, it decomposes and releases carbon dioxide gas, which is more dense than air and displaces oxygen from the flame.

OBJECTIVE

▸ To perform volume–volume stoichiometry calculations.

10.6 VOLUME–VOLUME PROBLEMS

In 1804, the French chemist Joseph Gay-Lussac ascended to 23,000 feet in a hydrogen-filled balloon. During his record-setting ascent, he took samples of the atmosphere and later studied their composition.

In 1808, Gay-Lussac proposed the **law of combining volumes**. This law states that volumes of gases, under similar conditions, combine in small whole-number ratios. For example, 10.0 mL of hydrogen gas combines with 10.0 mL of chlorine gas to give 20.0 mL of hydrogen chloride gas:

$$\underset{10.0\ mL}{H_2(g)} + \underset{10.0\ mL}{Cl_2(g)} \longrightarrow \underset{20.0\ mL}{2\ HCl(g)}$$

Notice that the volumes of gas (10.0 mL:10.0 mL:20.0 mL) are in the ratio of small whole numbers (1:1:2). Moreover, the ratio of the volumes is identical to the mole ratio in the balanced chemical equation (1:1:2).

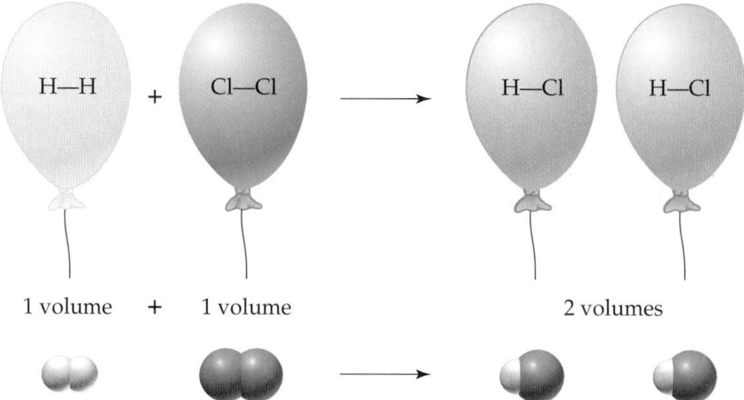

Similarly, 100.0 mL of hydrogen gas reacts with 50.0 mL of oxygen gas to produce 100.0 mL of water vapor:

$$\underset{100.0\ mL}{2\ H_2(g)} + \underset{50.0\ mL}{O_2(g)} \rightarrow \underset{100.0\ mL}{2\ H_2O(g)}$$

In this example, the volumes of gas (100.0 mL:50.0 mL:100.0 mL) are also in the ratio of small whole numbers (2:1:2). Once again, the ratio of the volumes is identical to the mole ratio in the balanced equation (2:1:2).

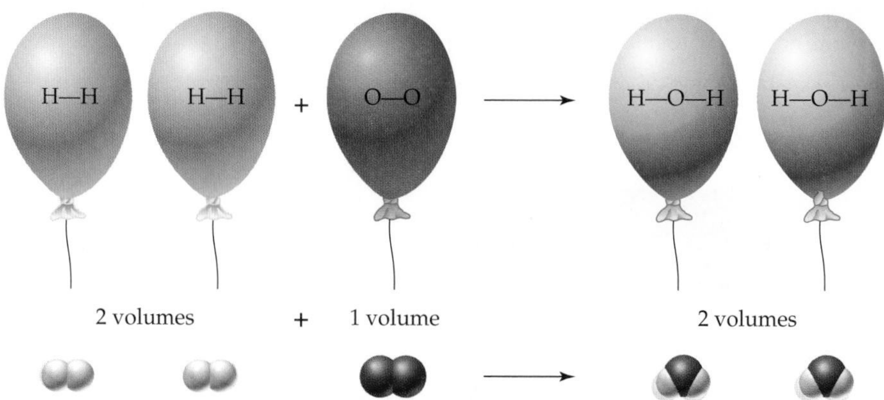

The *law of combining volumes* states that gases combine in small whole-number ratios. Since the volume ratio is identical to the coefficients in the balanced equation, we can convert from a given volume to an unknown volume of gas in a single conversion using the mole ratio. A volume–volume stoichiometry problem is so named because an *unknown volume* of gas is calculated from a *given volume* of gaseous reactant or product in a chemical equation.

We can illustrate a volume–volume problem using the industrial method for preparing sulfuric acid. As mentioned earlier, H_2SO_4 production is the barometer of the chemical industry. Sulfuric acid is manufactured by the Contact process, which begins with the conversion of sulfur dioxide to sulfur trioxide using heat and a catalyst. The sulfur trioxide gas is then passed through water to produce sulfuric acid. The balanced equation for the conversion of SO_2 to SO_3 is

$$2\,SO_2(g) + O_2(g) \xrightarrow{\Delta} 2\,SO_3(g)$$

Let's calculate the liters of oxygen gas that react with 37.5 L of sulfur dioxide. From the balanced equation, we see that 2 volumes of SO_2 = 1 volume of O_2. When we express the volume in liters, we have 2 L SO_2 = 1 L O_2. Using the volume ratio as a unit factor, we have

$$37.5\,\cancel{\text{L SO}_2} \times \frac{1\text{ L O}_2}{2\,\cancel{\text{L SO}_2}} = 18.8\text{ L O}_2$$

Now let's calculate the volume, in liters, of sulfur trioxide produced. From the balanced equation we see that 2 volumes of SO_2 = 2 volumes of SO_3. Therefore,

$$37.5\,\cancel{\text{L SO}_2} \times \frac{2\text{ L SO}_3}{2\,\cancel{\text{L SO}_2}} = 37.5\text{ L SO}_3$$

The following example exercise further illustrates the application of stoichiometry principles to a problem of the volume–volume type:

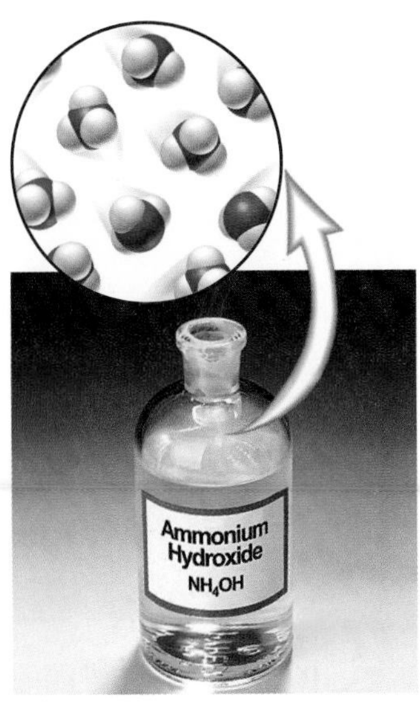

▲ **Ammonia, NH_3** Aqueous ammonia, NH_3, is found in laboratory reagent bottles labeled "ammonium hydroxide," NH_4OH.

Example Exercise 10.8 Volume–Volume Stoichiometry

In the Haber process, nitrogen and hydrogen gases combine to give ammonia gas. If 5.55 L of nitrogen gas is available, calculate the volume of ammonia that is produced. Assume that all volumes of gas are measured under constant conditions of 500 °C and 300 atm pressure:

$$N_2(g) + 3\,H_2(g) \xrightarrow{Fe/Al_2O_3} 2\,NH_3(g)$$

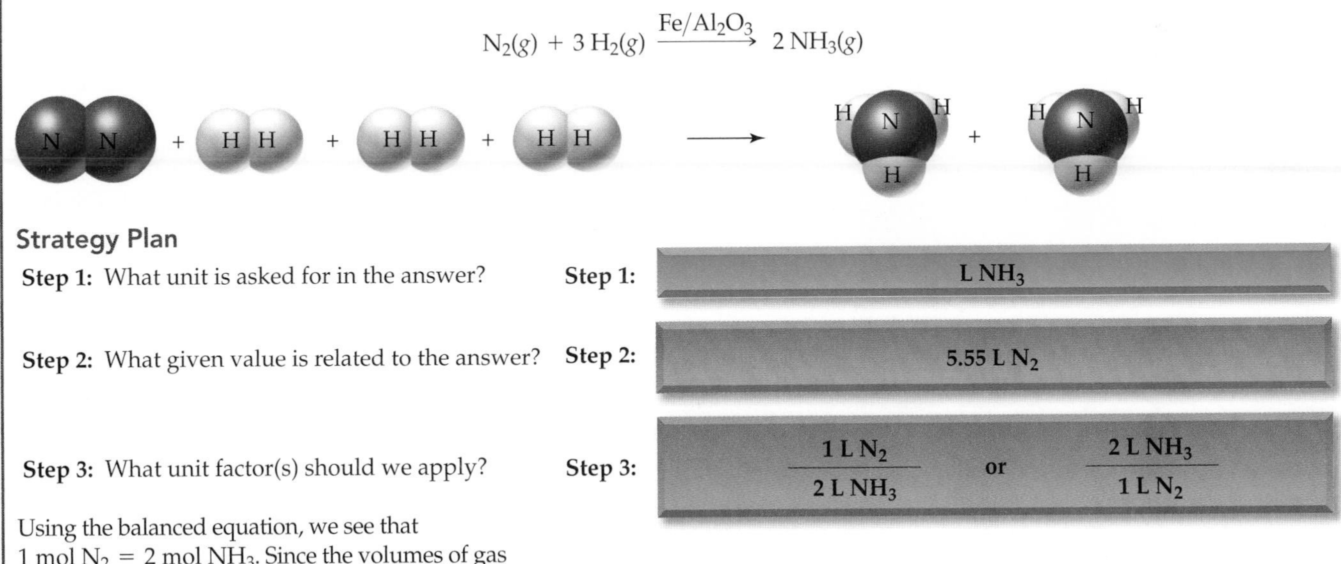

Strategy Plan

Step 1: What unit is asked for in the answer? **Step 1:** L NH_3

Step 2: What given value is related to the answer? **Step 2:** 5.55 L N_2

Step 3: What unit factor(s) should we apply? **Step 3:** $\frac{1\text{ L N}_2}{2\text{ L NH}_3}$ or $\frac{2\text{ L NH}_3}{1\text{ L N}_2}$

Using the balanced equation, we see that 1 mol N_2 = 2 mol NH_3. Since the volumes of gas have the same ratio as the mole ratio, the unit factors are

Unit Analysis Map

given value	*unit factor(s)*	*unit in answer*
5.55 L N_2	$\times \frac{2\text{ L NH}_3}{1\text{ L N}_2}$	= ? L NH_3

Solution

We select the unit factor that cancels the unit in the given value (L N_2). Thus,

$$5.55 \text{ L } N_2 \times \frac{2 \text{ L } NH_3}{1 \text{ L } N_2} = 11.1 \text{ L } NH_3$$

Practice Exercise

Calculate the volumes of (a) hydrogen chloride gas and (b) oxygen gas that react to yield 50.00 mL of chlorine gas given the following unbalanced equation. Assume that all gases are at the same temperature and pressure:

$$HCl(g) + O_2(g) \longrightarrow Cl_2(g) + H_2O(g)$$

Answers: (a) 100.0 mL HCl; (b) 25.00 mL O_2

Concept Exercise

In general, how many unit factors are required to solve a volume–volume stoichiometry problem?

Answer: See Appendix G.

CHEMISTRY CONNECTION Manufacturing Ammonia

Q: What are the main uses of ammonia, NH_3, which is one of the ten most important industrial chemicals?

Over 15 million tons of ammonia are produced annually in the United States. Ammonia, NH_3, is primarily used in agriculture to fertilize the soil and replenish nitrogen depleted by crops. Although our atmosphere is a vast potential source of nitrogen, the gaseous element is unreactive and does not readily combine with other elements. Nitrogen is essentially a colorless, odorless, inert gas.

In 1905, the German chemist Fritz Haber (1868–1934) successfully prepared ammonia in the laboratory for the first time. He discovered that nitrogen and hydrogen gases combine directly at high temperatures and pressures in the presence of metal oxide catalysts. The Haber process thus provides a method for converting unreactive atmospheric nitrogen into the versatile compound ammonia, NH_3. Ammonia is easily converted to important ammonium compounds and nitric acid, HNO_3. Today, the Haber process is the main source for manufacturing nitrogen compounds throughout the world.

▲ **Soil Nitrogen** Spraying anhydrous ammonia provides nitrogen to soil for optimum crop yields.

Until the beginning of the 1900s, the bulk of the world's nitrate was supplied by the rich saltpeter (KNO_3) deposits in South America. Although nitrates were used principally as fertilizer, they were also used to make explosives. Indirectly, the Haber process had a pronounced effect on World War I. Soon after the outbreak of the war, the British blockade halted the supply of Chilean saltpeter to Germany. Without a source of nitrate, Germany would have soon run out of ammunition. But by late 1914, Germany had built factories that applied the Haber process for the manufacture of ammonia and nitrate compounds.

During World War I, Haber helped to develop weapons used in chemical warfare, such as chlorine gas and even more lethal mustard gas. Haber's wife was so troubled by his crimes against humanity that she committed suicide at a dinner party honoring his role in the use of poison gas. She shot herself with his service revolver, but did not die until the following morning as Haber was leaving for the Eastern front to personally oversee the use of poison gas against the Russians. In 1918, the same year World War I ended, Haber was awarded the Nobel prize in chemistry. Following the announcement of the Nobel prize, he was denounced by American, English, and French scientists for his involvement in the war.

Ironically, the patriotic Haber was later condemned by his own country. When the Nazis rose to power in 1933, Haber had to resign as director of the Kaiser Wilhelm Institute because he was of Jewish descent. In distress, he fled to Cambridge University in England where he spent a few months before passing away while traveling in Switzerland.

A: The chief use of ammonia is as an agricultural fertilizer, but it is also used in explosives and household products such as window cleaner.

10.7 LIMITING REACTANT CONCEPT

OBJECTIVES

- To explain the concept of a limiting reactant.
- To identify the limiting reactant in a chemical reaction given the number of moles of each reactant.

In world-class cycling, competitors carry extra bicycle frames and tires. If a cyclist participating in the Tour de France has 5 frames and 20 tires, how many complete bicycles can be assembled? Since 5 complete frames require 10 tires, there are 10 spare tires. In this example, the frames limit the number of bicycles that can be assembled. Similarly, we will learn that one of the reactants in a chemical reaction can limit the amount of product.

We have learned how to perform calculations using information from a balanced chemical equation. For example, we can calculate the amount of iron(II) sulfide produced by the reaction of metallic iron and powdered sulfur:

$$Fe(s) + S(s) \xrightarrow{\Delta} FeS(s)$$

Previously, we assumed that there is sufficient iron and sulfur to react completely with each other. However, this assumption may not be true. What if there is not enough sulfur to react with the iron? If that were the case, the sulfur would *limit* the amount of FeS produced. In a stoichiometry problem, the reactant that controls the amount of product is called the **limiting reactant** and must be identified to correctly calculate the amount of product. Figure 10.1 is an example model to help you visualize the concept of a limiting reactant.

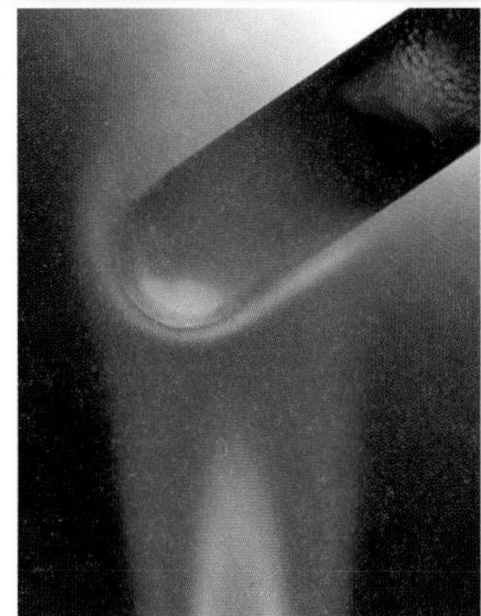

▲ **Producing FeS** Heating a mixture of iron powder and sulfur powder produces iron(II) sulfide; either the iron or the sulfur can be the limiting reactant.

Determining the Limiting Reactant

Now that we have a visual model for the limiting reactant concept, let's examine data from an experiment and solve a problem. Suppose we heat 2.50 mol of iron with 3.00 mol of sulfur. How many moles of FeS are formed? We will analyze the problem by considering the amount of each substance before and after the reaction.

Before the reaction begins, there are 2.50 mol of Fe, 3.00 mol of S, and 0.00 mol of FeS. According to the balanced equation, 1 mol of Fe reacts with 1 mol of S to give 1 mol of FeS. Therefore, 2.50 mol of Fe reacts with 2.50 mol of S to give 2.50 mol of FeS. At the outset we have 3.00 mol of S, so there is not enough iron to react with all the sulfur. Therefore, iron is the limiting reactant and sulfur is the excess reactant. After the reaction, the excess sulfur is 0.50 mol (3.00 mol – 2.50 mol). Table 10.2 summarizes the amounts of each substance before and after the reaction.

TABLE 10.2 SUMMARY OF EXPERIMENTAL SYNTHESIS OF FES

EXPERIMENT	mol Fe	mol S	mol FeS
before reaction:	2.50	3.00	0.00
after reaction:	0.00	0.50	2.50

◀ **Tour de France** The relationship between bicycle frames and tires illustrates the limiting reactant concept.

Fe S Fe^{2+} S^{2-}

(a)

(b)

▶ **Figure 10.1 Limiting Reactant Concept** (a) In this reaction there is excess S; thus, Fe is the limiting reactant. Notice that an unreacted S atom remains. (b) In another reaction, there is excess Fe; thus, S is the limiting reactant. Notice that an unreacted Fe atom remains.

The following example exercise further illustrates the concept of a limiting reactant:

Example Exercise 10.9 Limiting Reactant Concept

A 1.00 mol sample of iron(II) oxide is heated with 1.00 mol of aluminum metal and converted to molten iron. Identify the limiting reactant, and calculate the moles of iron produced:

$$3\ FeO(l) + 2\ Al(l) \xrightarrow{\Delta} 3\ Fe(l) + Al_2O_3(s)$$

Strategy Plan

Step 1: What unit is asked for in the answer? **Step 1:** **mol Fe**

Step 2: What given value is related to the answer? **Step 2:** **1.00 mol FeO and 1.00 mol Al**

Step 3: What unit factor(s) should we apply? **Step 3:** $\frac{3\ mol\ FeO}{3\ mol\ Fe}$ or $\frac{3\ mol\ Fe}{3\ mol\ FeO}$

From the balanced equation, we have the mole ratio 3 mol FeO/3 mol Fe, and 2 mol Al/3 mol Fe.

$\frac{2\ mol\ Al}{3\ mol\ Fe}$ or $\frac{3\ mol\ Fe}{2\ mol\ Al}$

Unit Analysis Map

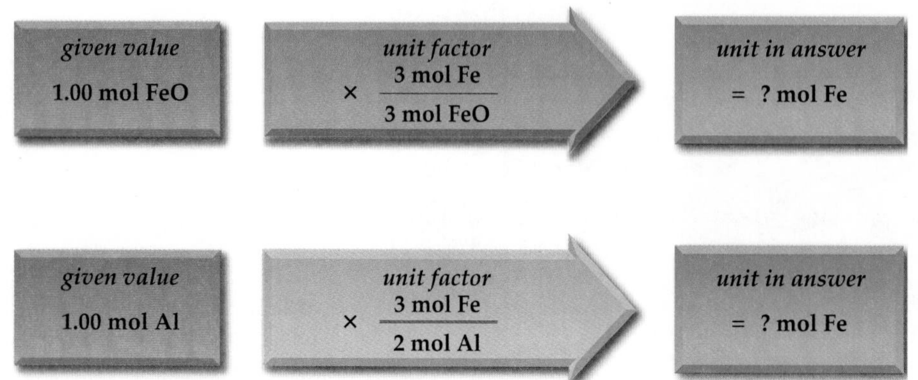

Solution

We select the unit factor that cancels the unit in the given value (~~mol FeO~~). Thus,

$$1.00 \text{ mol FeO} \times \frac{3 \text{ mol Fe}}{3 \text{ mol FeO}} = 1.00 \text{ mol Fe}$$

Second, we select the unit factor that cancels the unit in the given value (~~mol Al~~). Thus,

$$1.00 \text{ mol Al} \times \frac{3 \text{ mol Fe}}{2 \text{ mol Al}} = 1.50 \text{ mol Fe}$$

Notice that FeO produces 1.00 mol Fe, whereas Al produces 1.50 mol Fe. Thus, FeO is the *limiting reactant* and the amount of product is **1.00 mol Fe**.

Practice Exercise

A 5.00 mol sample of iron(III) oxide is heated with 5.00 mol aluminum metal and converted to molten iron. Identify the limiting reactant, and calculate the moles of iron produced given the *unbalanced* equation for the reaction:

$$Fe_2O_3(l) + Al(l) \xrightarrow{\Delta} Fe(l) + Al_2O_3(s)$$

Answer: The limiting reactant is Al, which produces 5.00 mol of Fe.

Concept Exercise

A Tour de France cyclist has 10 tires and 3 frames. How many complete bicycles can be assembled?

Answer: See Appendix G.

▲ **Thermite reaction** The thermite reaction of iron oxide and aluminum metal produces molten iron and a shower of sparks.

10.8 LIMITING REACTANT PROBLEMS

OBJECTIVES

- To perform mass–mass stoichiometry calculations involving a limiting reactant.
- To perform volume–volume stoichiometry calculations involving a gaseous limiting reactant.

Suppose we wish to find how much molten iron is produced from the reaction of 25.0 g of FeO with 25.0 g of Al. The balanced equation for the reaction is

$$3\ FeO(l) + 2\ Al(l) \xrightarrow{\Delta} 3\ Fe(l) + Al_2O_3(s)$$

Since the mass of each reactant is given, it is not obvious whether FeO or Al limits the amount of iron produced. In problems where we are given the amounts of two reactants, we can determine the limiting reactant in three steps.

I. *Calculate the mass of product that can be produced from the first reactant.*
  (a) Calculate the moles of reactant.
  (b) Calculate the moles of product.
  (c) Calculate the mass of product.

II. *Calculate the mass of product that can be produced from the second reactant.*
  (a) Calculate the moles of reactant.
  (b) Calculate the moles of product.
  (c) Calculate the mass of product.

III. *State the limiting reactant and the corresponding mass of product formed.* The limiting reactant gives the least amount of product. The actual mass of product obtained from the chemical reaction is the lesser of the two product masses calculated in Steps I and II.

Applying I: We can calculate the amount of iron produced from ferrous oxide in the preceding problem. Using the balanced equation for the reaction, we can find the mass of Fe produced from 25.0 g of FeO. Let's outline the solution to the problem:

$$\text{g FeO} \underset{\text{I(a)}}{\longrightarrow} \text{mol FeO} \underset{\text{I(b)}}{\longrightarrow} \text{mol Fe} \underset{\text{I(c)}}{\longrightarrow} \text{g Fe}$$

From the periodic table, we find that for Fe, the molar mass is 55.85 g/mol, and for FeO, the molar mass is 71.85 g/mol. The unit analysis solution to the problem is

$$25.0 \text{ g FeO} \times \frac{1 \text{ mol FeO}}{71.85 \text{ g FeO}} \times \frac{3 \text{ mol Fe}}{3 \text{ mol FeO}} \times \frac{55.85 \text{ g Fe}}{1 \text{ mol Fe}} = 19.4 \text{ g Fe}$$

Applying II: We calculate the amount of iron that can be produced by the aluminum. Using the balanced equation, we can find the mass of Fe produced from 25.0 g of Al. We can outline the solution to the problem as follows:

$$\text{g Al} \xrightarrow[\text{II(a)}]{} \text{mol Al} \xrightarrow[\text{II(b)}]{} \text{mol Fe} \xrightarrow[\text{II(c)}]{} \text{g Fe}$$

From the periodic table, we find that for Al, the molar mass is 26.98 g/mol. The unit analysis solution to the problem is

$$25.0 \text{ g Al} \times \frac{1 \text{ mol Al}}{26.98 \text{ g Al}} \times \frac{3 \text{ mol Fe}}{2 \text{ mol Al}} \times \frac{55.85 \text{ g Fe}}{1 \text{ mol Fe}} = 77.6 \text{ g Fe}$$

Applying III: We compare the grams of Fe produced from each of the reactants:

$$25.0 \text{ g FeO} \longrightarrow 19.4 \text{ g Fe}$$

$$25.0 \text{ g Al} \longrightarrow 77.6 \text{ g Fe}$$

We see that the FeO yields less mass of product; thus, FeO is the *limiting reactant*. Conversely, aluminum metal is the excess reactant, since not all the Al is used in the reaction. Therefore, the maximum yield of metallic iron from the reaction is limited to 19.4 g of Fe.

Example Exercise 10.10 provides additional practice in solving a limiting reactant problem.

Example Exercise 10.10 Mass–Mass Limiting Reactant

In a reaction, 50.0 g of manganese(IV) oxide reacts with 25.0 g of aluminum. Identify the limiting reactant and calculate the mass of manganese metal produced. The equation for the reaction is

$$3\,MnO_2(l) + 4\,Al(l) \xrightarrow{\Delta} 3\,Mn(l) + 2\,Al_2O_3(s)$$

Solution

Step 1: We calculate the mass of Mn produced from 50.0 g of MnO_2. We can outline the solution as follows:

$$\text{g } MnO_2 \longrightarrow \text{mol } MnO_2 \longrightarrow \text{mol Mn} \longrightarrow \text{g Mn}$$

From the periodic table, we find the molar mass of MnO_2 is 86.94 g/mol and Mn is 54.94 g/mol. The unit analysis solution to the problem is

$$50.0 \text{ g } MnO_2 \times \frac{1 \text{ mol } MnO_2}{86.94 \text{ g } MnO_2} \times \frac{3\text{mol Mn}}{3 \text{ mol } MnO_2} \times \frac{54.94 \text{ g Mn}}{1 \text{ mol Mn}} = 31.6 \text{ g Mn}$$

Step 2: We calculate the mass of Mn produced from 25.0 g of Al. We can outline the solution as follows:

$$\text{g Al} \longrightarrow \text{mol Al} \longrightarrow \text{mol Mn} \longrightarrow \text{g Mn}$$

From the periodic table, we find that the molar mass of Al is 26.98 g/mol. Starting with 25.0 g of Al, the unit analysis solution is

$$25.0 \text{ g Al} \times \frac{1 \text{ mol Al}}{26.98 \text{ g Al}} \times \frac{3 \text{ mol Mn}}{4 \text{ mol Al}} \times \frac{54.94 \text{ g Mn}}{1 \text{ mol Mn}} = 38.2 \text{ g Mn}$$

Step 3: We compare the mass of product obtained from each of the reactants:

$$50.0 \text{ g } MnO_2 \longrightarrow 31.6 \text{ g Mn}$$

$$25.0 \text{ g Al} \longrightarrow 38.2 \text{ g Mn}$$

In this example, MnO_2 is the *limiting reactant* because it yields less product. Al is the excess reactant and is not completely consumed in the reaction. Thus, the maximum amount of product from the reaction is 31.6 g of Mn.

Practice Exercise
In a reaction, 75.0 g of manganese(IV) oxide react with 30.0 g of aluminum. Identify the limiting reactant and calculate the mass of aluminum oxide produced:

$$3\ MnO_2(l) + 4\ Al(l) \xrightarrow{\Delta} 3\ Mn(l) + 2\ Al_2O_3(s)$$

Answer: The limiting reactant is Al, which gives 56.7 g of Al_2O_3.

Concept Exercise
Indicate the limiting reactant shown in the following illustration:

$$Zn(s) + S(s) \xrightarrow{\Delta} ZnS(s)$$

Answer: See Appendix G.

Limiting Reactants Involving Volumes of Gas

Now let's try a limiting reactant problem that involves reactants and products in the gaseous state. For example, consider the reaction of nitrogen monoxide and oxygen to give nitrogen dioxide. The balanced equation for the reaction is

$$2\ NO(g) + O_2(g) \xrightarrow{UV} 2\ NO_2(g)$$

Suppose 5.00 L of NO gas reacts with 5.00 L of O_2 gas. Since the volumes of both reactants are given, it is not obvious which gas limits the amount of product. We will assume that the temperature and pressure remain constant and use the following three steps to determine the limiting reactant:

I. *Calculate the volume of product that can be produced from the first reactant.*
Recall that the coefficients in the balanced equation are in the same mole ratio as the volumes of gas.

II. *Calculate the volume of product that can be produced from the second reactant.*
Again, the coefficients in the balanced equation are in the same mole ratio as the volumes of gas.

III. *State the limiting reactant and the corresponding volume of product.*
The limiting reactant is the gas that gives the least amount of product. The actual volume of product obtained from the chemical reaction is the lesser of the two volumes calculated in Steps I and II.

Applying I: We calculate the volume of nitrogen dioxide produced from 5.00 L of nitrogen monoxide gas. From the balanced equation, we see that 2 mol of NO produce 2 mol of NO_2. Accordingly, 2 L of NO produce 2 L of NO_2. Hence,

$$5.00\ \cancel{L\ NO} \times \frac{2\ L\ NO_2}{2\ \cancel{L\ NO}} = 5.00\ L\ NO_2$$

▲ **Oxyacetylene Welding** The reaction of oxygen and acetylene gases releases enough heat to melt iron.

Applying II: We calculate the volume of nitrogen dioxide produced from 5.00 L of oxygen gas. From the balanced equation, we see that 1 mol of O_2 produces 2 mol of NO_2. Accordingly, 1 L of O_2 produces 2 L of NO_2. Thus,

$$5.00\ \cancel{L\ O_2} \times \frac{2\ L\ NO_2}{1\ \cancel{L\ O_2}} = 10.0\ L\ NO_2$$

Applying III: We compare the liters of NO_2 gas from each of the reactants.

$$5.00\ L\ NO \longrightarrow 5.00\ L\ NO_2$$

$$5.00\ L\ O_2 \longrightarrow 10.0\ L\ NO_2$$

We see that the NO yields less volume of product; thus, NO is the *limiting reactant*. Conversely, O_2 is the excess reactant, and not all the O_2 gas is used in the reaction. Therefore, the maximum yield of product from the reaction is 5.00 L of NO_2. The following example exercise provides additional practice in solving a limiting reactant problem involving gases.

Example Exercise 10.11 Volume–Volume Limiting Reactant

In oxyacetylene welding, acetylene reacts with oxygen to give carbon dioxide and water. If 25.0 mL of C_2H_2 reacts with 75.0 mL of O_2, what is the limiting reactant? Assuming constant conditions, what is the volume of CO_2 produced?

$$2\ C_2H_2(g) + 5\ O_2(g) \xrightarrow{\text{spark}} 4\ CO_2(g) + 2\ H_2O(g)$$

Strategy Plan

Step 1: What unit is asked for in the answer? **Step 1:** mL CO_2

Step 2: What given value is related to the answer? **Step 2:** 25.0 mL C_2H_2 and 75.0 mL O_2

Step 3: What unit factor(s) should we apply? **Step 3:** $\frac{2\ mL\ C_2H_2}{4\ mL\ CO_2}$ or $\frac{4\ mL\ CO_2}{2\ mL\ C_2H_2}$; $\frac{5\ mL\ O_2}{4\ mL\ CO_2}$ or $\frac{4\ mL\ CO_2}{5\ mL\ O_2}$

From the balanced equation, we have the mole ratio so the volume ratio is 2 mL C_2H_2/4 mL CO_2, and 5 mL O_2/4 mL CO_2.

Unit Analysis Map

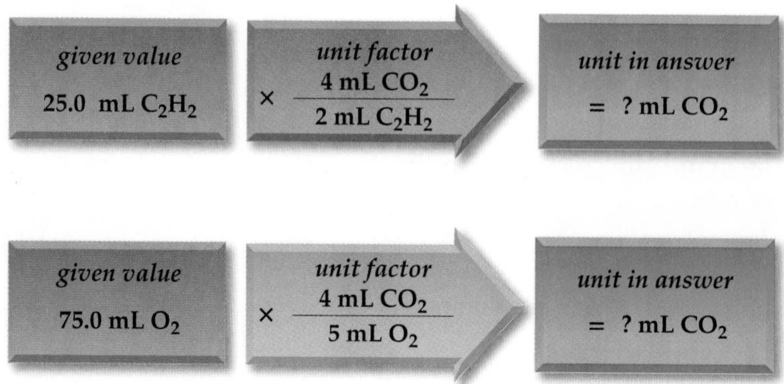

Solution

First, we find the amount of product that can be produced from the first reactant (mL C_2H_2). Thus,

$$25.0\ \cancel{mL\ C_2H_2} \times \frac{4\ mL\ CO_2}{2\ \cancel{mL\ C_2H_2}} = 50.0\ mL\ CO_2$$

Second, we find the amount of product that can be produced from the second reactant (mL O_2):

$$75.0\ \cancel{\text{mL } O_2} \times \frac{4 \text{ mL } CO_2}{5\ \cancel{\text{mL } O_2}} = 60.0 \text{ mL } CO_2$$

Third, we compare the amount of product from each reactant and see that C_2H_2 produces 50.0 mL CO_2, whereas O_2 produces 60.0 mL CO_2. Thus, C_2H_2 is the *limiting reactant* and the amount of product is limited to 50.0 mL CO_2.

Practice Exercise

Ethane undergoes combustion to give carbon dioxide and water. If 10.0 L of C_2H_6 reacts with 25.0 L of O_2, what is the limiting reactant? Assuming constant conditions, what is the volume of CO_2 produced? The equation is

$$2\ C_2H_6(g) + 7\ O_2(g) \xrightarrow{\text{spark}} 4\ CO_2(g) + 6\ H_2O(g)$$

Answer: The limiting reactant is O_2, which gives 14.3 L of CO_2.

Concept Exercise

Indicate the limiting reactant shown in the following illustration:

$$Zn(s) + S(s) \xrightarrow{\Delta} ZnS(s)$$

Answer: See Appendix G.

10.9 PERCENT YIELD

OBJECTIVE

- To calculate the percent yield for a reaction given the actual yield and theoretical yield.

To understand the concept of yield, consider the following laboratory experiment. First, a student weighs a 1.00 g sample of cadmium chloride and dissolves it in water. Then, the student adds aqueous sodium sulfide solution to obtain a precipitate of cadmium sulfide, CdS. The balanced equation for the reaction is

$$CdCl_2(aq) + Na_2S(aq) \longrightarrow CdS(s) + 2\ NaCl(aq)$$

The student collects the insoluble CdS precipitate in filter paper and determines its mass on a balance. The mass of the precipitate is called the **actual yield**.

The mass of cadmium sulfide obtained, starting with the 1.00 g sample of cadmium chloride, can also be calculated as a mass–mass problem. The calculated amount of precipitate is called the **theoretical yield**. In practice, there is always inherent experimental error. Thus, we compare the actual and theoretical yields and express the results in terms of percent yield. The **percent yield** is the amount of the actual yield compared to the theoretical yield expressed as a percent:

$$\frac{\text{actual yield}}{\text{theoretical yield}} \times 100\% = \text{percent yield}$$

▲ **Percent Yield of CdS** The aqueous solution reaction of $CdCl_2$ and Na_2S yields a yellow precipitate of CdS.

To find the percent yield for this laboratory experiment, we start with the 1.00 g sample of cadmium chloride. Applying stoichiometry to the balanced equation, we calculate that the amount of cadmium sulfide precipitate is 0.788 g. Suppose the student weighs the CdS precipitate and finds that it has an actual mass of 0.775 g. To find the percent yield, we compare the actual yield (0.775 g) to the theoretical yield (0.788 g) as follows:

$$\frac{0.775 \text{ g}}{0.788 \text{ g}} \times 100\% = 98.4\%$$

Suppose that the student performs a second trial for the same experiment, using a 1.00 g sample of cadmium chloride. In this second trial, the precipitate has a mass of

▲ **Percent Yield of $CuCO_3$**
The aqueous solution reaction of $Cu(NO_3)_2$ and Na_2CO_3 yields a precipitate of $CuCO_3$.

0.805 g. Since the student started with 1.00 g of cadmium chloride, the theoretical yield for the two trials is the same, 0.788 g. As before, to find the percent yield we compare the actual yield to the theoretical yield.

$$\frac{0.805 \text{ g}}{0.788 \text{ g}} \times 100\% = 102\%$$

Notice that the percent yield for this second trial is greater than 100%. Although surprising, this result is possible. In an experiment, some errors lead to high results and other errors lead to low results. In the two trials, the student made different types of experimental errors. For example, high results will be obtained when impurities are trapped, or when the precipitate is not completely dry.

The following example exercise provides additional practice in calculating percent yield:

Example Exercise 10.12 Percent Yield

A student dissolves 1.50 g of copper(II) nitrate in water. After adding aqueous sodium carbonate solution, the student obtains 0.875 g of $CuCO_3$ precipitate. If the theoretical yield is 0.988 g, what is the percent yield?

$$Cu(NO_3)_2(aq) + Na_2CO_3(aq) \longrightarrow CuCO_3(s) + 2\ NaNO_3(aq)$$

Solution

The percent yield is the ratio of the actual yield compared to the theoretical yield. In this experiment, the actual yield is 0.875 g and the theoretical yield is 0.988 g. The percent yield, therefore, is

$$\frac{\text{actual yield}}{\text{theoretical yield}} \times 100\% = \text{percent yield}$$

$$\frac{0.875 \text{ g}}{0.988 \text{ g}} \times 100\% = 88.6\%$$

The percent yield obtained by the student is 88.6%.

Practice Exercise

Ammonium nitrate is used in explosives and is produced by the reaction of ammonia, NH_3, and nitric acid. The equation for the reaction is

$$NH_3(g) + HNO_3(aq) \longrightarrow NH_4NO_3(s)$$

If 15.0 kg of ammonia gives an actual yield of 65.3 kg of ammonium nitrate, what is the percent yield? The calculated yield of ammonium nitrate for the experiment is 70.5 kg.

Answer: 92.6%

Concept Exercise

Is it possible to have a percent yield greater than 100%?

Answer: See Appendix G.

Summary

Section 10.1 We learned to interpret the coefficients in a balanced equation as a **mole ratio**. We also showed that by applying **Avogadro's theory** we can express the mole ratio in terms of volumes of gases. Furthermore, we verified the **law of conservation of mass** by substituting the **molar mass** for a mole of each reactant and product in the chemical equation.

Section 10.2 We learned to relate the moles of reactants to the moles of products by simply applying the coefficients in the balanced chemical reaction.

Section 10.3 **Stoichiometry** relates the quantities of substances involved in a chemical reaction. There are three basic types of stoichiometry problems. In a **mass–mass problem**, the mass of one substance is related to the mass of another substance according to a balanced chemical equation. In a **mass–volume problem**, the mass of a substance is related to the volume of a gas. In a **volume–volume problem**, the volumes of two gaseous substances are related.

Section 10.4 In a mass–mass problem, we first convert the given mass of a substance to the number of moles using the molar mass. If the molar mass is not shown in the problem, we refer to the periodic table and add up the gram atomic masses for each element in the substance. Next, we apply the mole ratio to convert the moles of the given substance to the moles of the unknown substance using the coefficients from the balanced equation. Finally, we determine the mass of the unknown by multiplying by the molar mass of the unknown.

Section 10.5 In a mass–volume problem, we first convert the given mass of substance to the number of moles using the molar mass. Next, we apply the mole ratio to convert the moles of the given substance to the moles of the unknown substance using the coefficients from the balanced equation. Finally, we determine the volume of the unknown gas at STP by multiplying by the **molar volume**. The volume occupied by 1 mol of gas is 22.4 L at STP, that is, 0 °C and 1 atm pressure.

Section 10.6 In a volume–volume problem, we can take advantage of Gay-Lussac's **law of combining volumes**. Since the volumes of gases combine in the same whole-number ratio as the coefficients in the balanced equation, we can solve this type of problem in one conversion step. We simply multiply the given volume by the ratio of the coefficients in the equation.

Section 10.7 In more difficult stoichiometry problems, we cannot assume that there is a sufficient quantity of each reactant available. The amount of one reactant may not be sufficient to completely consume another reactant that is in excess. The reactant that controls the reaction and limits the amount of product formed is referred to as the **limiting reactant**.

Section 10.8 If we are given the amounts of two reactants, we can determine the limiting reactant by finding the amount of product yield from each of the reactants. The reactant that produces the least amount of product controls the reaction and is the limiting reactant. A limiting reactant problem may involve the masses of two reacting substances as well as the volumes of two gases.

Section 10.9 The percent yield provides an evaluation of the laboratory method and the skill of the analyst. A **percent yield** from an experiment expresses the amount of the **actual yield** compared to the **theoretical yield**.

PROBLEM-SOLVING ORGANIZER

TOPIC	PROCEDURE	EXAMPLE
Interpreting a Chemical Equation Sec. 10.1	Express the mole ratio and volume ratio of gaseous substances in a balanced equation using the coefficients.	Given that $N_2(g) + O_2(g) \rightarrow 2\ NO(g)$, interpret the mole ratio and the volume ratio in liters. $1\ mol\ N_2 + 1\ mol\ O_2 \longrightarrow 2\ mol\ NO$ $1\ L\ N_2 + 1\ L\ O_2 \longrightarrow 2\ L\ NO$
Mole–Mole Relationships Sec. 10.2	Convert moles of a given substance to moles of an unknown using the mole ratio.	How many moles of NO are produced from 0.500 mol N_2? $0.500\ \cancel{mol\ N_2} \times \frac{2\ mol\ NO}{1\ \cancel{mol\ N_2}} = 1.00\ mol\ NO$
Types of Stoichiometry Problems Sec. 10.3	Classify a stoichiometry problem as a mass–mass, mass–volume, or volume–volume type of problem.	What type of problem is the following: How many grams of NO are produced from 11.2 L N_2? This problem is a mass-volume type.
Mass–Mass Problems Sec. 10.4	(a) Convert mass of given to moles of given. (b) Convert moles of given to moles of unknown. (c) Convert moles of unknown to mass of unknown.	How many grams of NO are produced from 10.0 g N_2? $10.0\ \cancel{g\ N_2} \times \frac{1\ \cancel{mol\ N_2}}{28.02\ \cancel{g\ N_2}} \times \frac{2\ \cancel{mol\ NO}}{1\ \cancel{mol\ N_2}} \times \frac{30.01\ g\ NO}{1\ \cancel{mol\ NO}}$ $= 21.4\ g\ NO$
Mass–Volume Problems Sec. 10.5	(a) Convert mass of given to moles of given. (b) Convert moles of given to moles unknown. (c) Convert moles unknown to volume unknown.	How many liters of NO at STP are produced from 10.0 g N_2? $10.0\ \cancel{g\ N_2} \times \frac{1\ \cancel{mol\ N_2}}{28.02\ \cancel{g\ N_2}} \times \frac{2\ \cancel{mol\ NO}}{1\ \cancel{mol\ N_2}} \times \frac{22.4\ L\ NO}{1\ \cancel{mol\ NO}}$ $= 16.0\ L\ NO$
Volume–Volume Problems Sec. 10.6	Convert volume of given to volume of unknown.	How many liters of NO at STP are produced from 10.0 L N_2? $10.0\ \cancel{L\ N_2} \times \frac{2\ L\ NO}{1\ \cancel{L\ N_2}} = 20.0\ L\ NO$

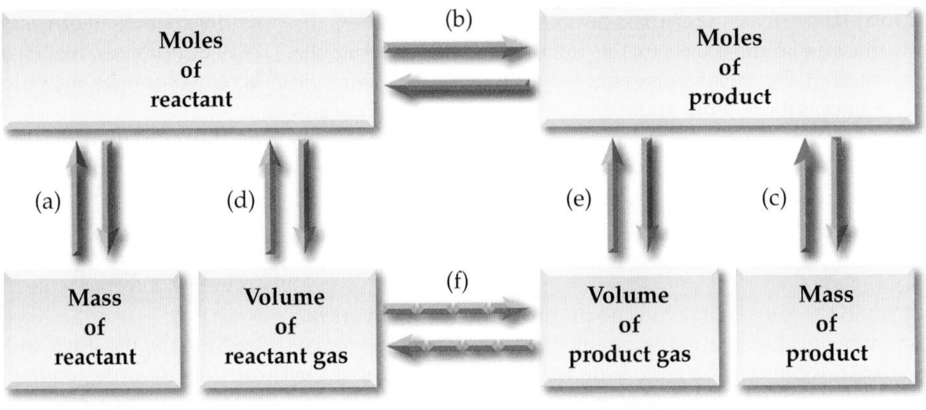

▶ **Concept Map—Summary of Stoichiometry** We use a strategy map to show the application of unit factors to solve problems by unit analysis. In this chapter, we have related the masses of substances and volumes of gases according to the coefficients of a balanced equation. Now, let's summarize stoichiometry calculations using a concept map that relates these concepts.

Key Terms *Answers to Key Terms are in Appendix H.*

Select the key term that corresponds to each of the following definitions.

_____ **1.** the ratio of moles of reactants and products according to the coefficients in the balanced chemical equation

_____ **2.** the principle that equal volumes of gases contain equal numbers of molecules, at the same temperature and pressure

_____ **3.** the principle that states that the total mass of reactants in a chemical reaction is equal to the total mass of products

_____ **4.** the mass in grams of 1 mol of substance

_____ **5.** the relationship between amounts of substance (mass of a substance or volume of a gas) in a chemical reaction according to the balanced equation

_____ **6.** a type of stoichiometry calculation that relates the masses of two substances

_____ **7.** a type of stoichiometry calculation that relates the mass of a substance to the volume of a gas

_____ **8.** a type of stoichiometry calculation that relates the volumes of two gases at the same temperature and pressure

_____ **9.** the volume of 1 mol of gas at STP, that is, 22.4 L/mol

_____ **10.** a temperature of 0 °C and a pressure of 1 atm

_____ **11.** the principle that states that the volumes of gases that react in a chemical reaction are in the ratio of small whole numbers

_____ **12.** the substance in a chemical reaction that controls the maximum amount of product formed

_____ **13.** the amount of product experimentally obtained from a given amount of reactant

_____ **14.** the amount of product calculated to be obtained from a given amount of reactant

_____ **15.** an expression for the actual yield compared to the theoretical yield

(a) actual yield *(Sec. 10.9)*
(b) Avogadro's theory *(Sec. 10.1)*
(c) law of combining volumes *(Sec. 10.6)*
(d) law of conservation of mass *(Sec. 10.1)*
(e) limiting reactant *(Sec. 10.7)*
(f) mass–mass problem *(Sec. 10.3)*
(g) mass–volume problem *(Sec. 10.3)*
(h) molar mass *(Sec. 10.1)*
(i) molar volume *(Sec. 10.5)*
(j) mole ratio *(Sec. 10.1)*
(k) percent yield *(Sec. 10.9)*
(l) standard temperature and pressure (STP) *(Sec. 10.5)*
(m) stoichiometry *(Sec. 10.3)*
(n) theoretical yield *(Sec. 10.9)*
(o) volume–volume problem *(Sec. 10.3)*

Exercises *Answers to odd-numbered Exercises are in Appendix I.*

Interpreting a Chemical Equation (Sec. 10.1)

1. Consider the general chemical equation
$A + 2\,B \longrightarrow 3\,C + 2\,D$.
(a) How many moles of C are produced from 1 mol of A?
(b) How many liters of gas B must react to give 2 L of gas D at the same temperature and pressure?

2. Consider the general chemical equation
$2\,A + 3\,B \longrightarrow 2\,C + D$.
(a) How many moles of C are produced from 2 mol of A?
(b) How many liters of gas B must react to give 1 L of gas D at the same temperature and pressure?

3. Consider the general chemical equation 2 A + 3 B $\longrightarrow$ C.
   (a) If 1.00 g of A reacts with 1.50 g of B, what is the mass of C?
   (b) If 1.00 g of A reacts to produce 2.75 g of C, what is the mass of B?
4. Consider the general chemical equation 3 A + B $\longrightarrow$ 2 C.
   (a) If 1.50 g of A reacts with 1.65 g of B, what is the mass of C?
   (b) If 1.50 g of A reacts to produce 3.75 g of C, what is the mass of B?
5. Verify the conservation of mass law using the molar masses of reactants and products for each substance in the following balanced equations:
   (a) $2\,N_2(g) + O_2(g) \longrightarrow 2\,N_2O(g)$
   (b) $N_2(g) + 2\,O_2(g) \longrightarrow N_2O_4(g)$
6. Verify the conservation of mass law using the molar masses of reactants and products for each substance in the following balanced equations:
   (a) $P_4(s) + 3\,O_2(g) \longrightarrow 2\,P_2O_3(s)$
   (b) $P_4(s) + 5\,O_2(g) \longrightarrow 2\,P_2O_5(s)$

Mole-Mole Relationships (Sec. 10.2)

7. Given the decomposition of hydrogen peroxide, calculate the moles of oxygen gas produced from 5.00 mol of H_2O_2.
   $$2\,H_2O_2(l) \longrightarrow 2\,H_2(g) + O_2(g)$$
8. Given the reaction of hydrogen and chlorine gases, calculate the moles of hydrogen that react to produce 5.00 mol of HCl.
   $$H_2(g) + Cl_2(g) \longrightarrow 2\,HCl(g)$$
9. How many moles of chlorine gas react with 0.100 mol of metallic iron? How many moles of iron(III) chloride are produced?
   $$Fe(s) + Cl_2(g) \longrightarrow FeCl_3(s)$$
10. How many moles of barium metal react to produce 0.100 mol of barium nitride? How many moles of nitrogen gas react?
    $$Ba(s) + N_2(g) \longrightarrow Ba_3N_2(s)$$
11. How many moles of propane gas, C_3H_8, react with 2.50 mol of oxygen gas? How many moles of carbon dioxide are produced?
    $$C_3H_8(g) + O_2(g) \longrightarrow CO_2(g) + H_2O(g)$$

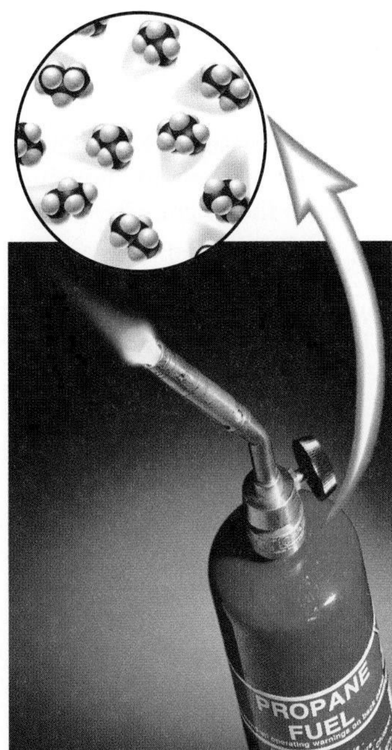

◀ Propane Torch The reaction of propane gas, C_3H_8, and oxygen in the air releases heat energy.

12. How many moles of butane gas, C_4H_{10}, react to produce 2.50 mol of water? How many moles of oxygen gas react?
    $$C_4H_{10}(g) + O_2(g) \longrightarrow CO_2(g) + H_2O(g)$$

Types of Stoichiometry Problems (Sec. 10.3)

13. Classify the type of stoichiometry problem: How many liters of carbon dioxide gas are produced from the decomposition of 25.0 g $NaHCO_3$?
14. Classify the type of stoichiometry problem: How many grams of baking soda must decompose to give 25.0 mL CO_2 gas?
15. Classify the type of stoichiometry problem: How many milliliters of hydrogen bromide gas are produced from 25.0 mL of H_2 gas and excess bromine liquid?
16. Classify the type of stoichiometry problem: How many milliliters of hydrogen gas react with bromine liquid to yield 25.0 cm^3 of HBr gas?
17. Classify the type of stoichiometry problem: How many grams of ferric oxide are produced from steam and 1.00 g of iron?
18. Classify the type of stoichiometry problem: How many grams of potassium chlorate must decompose to give 1.00 g of KCl?

Mass–Mass Problems (Sec. 10.4)

19. What is the mass of silver that can be prepared from 1.00 g of copper metal?
    $$Cu(s) + 2\,AgNO_3(aq) \longrightarrow Cu(NO_3)_2(aq) + 2\,Ag(s)$$
20. How many grams of silver nitrate must react to give 1.00 g of Ag according to the equation in Exercise 19?
21. Given the balanced equation, calculate the mass of product that can be prepared from 1.00 g of zinc metal.
    $$2\,Zn(s) + O_2(g) \longrightarrow 2\,ZnO(s)$$
22. How many grams of oxygen gas must react to give 1.00 g of ZnO according to the equation in Exercise 21?
23. Given the balanced equation, calculate the mass of product that can be prepared from 3.45 g of bismuth metal.
    $$2\,Bi(s) + 3\,Cl_2(g) \longrightarrow 2\,BiCl_3(s)$$
24. How many grams of chlorine gas must react to give 3.52 g of $BiCl_3$ according to the equation in Exercise 23?
25. What is the mass of mercury that can be prepared from 1.25 g of cobalt metal?
    $$Co(s) + HgCl_2(aq) \longrightarrow CoCl_3(aq) + Hg(l)$$
26. How many grams of mercuric chloride must react to give 5.11 g of Hg according to the equation in Exercise 25?
27. What is the mass of calcium phosphate that can be prepared from 1.78 g of Na_3PO_4?
    $$Na_3PO_4(aq) + Ca(OH)_2(aq) \longrightarrow Ca_3(PO_4)_2(s) + NaOH(aq)$$
28. How many grams of calcium hydroxide must react to give 2.39 g of $Ca_3(PO_4)_2$ according to the equation in Exercise 27?

Mass–Volume Problems (Sec. 10.5)

29. How many milliliters of carbon dioxide gas at STP are produced from the decomposition of 1.59 g of lithium hydrogen carbonate?
    $$2\,LiHCO_3(s) \longrightarrow Li_2CO_3(s) + H_2O(l) + CO_2(g)$$

30. How many milliliters of carbon dioxide gas at STP are produced from the decomposition of 5.00 g of ferric carbonate?

$$Fe_2(CO_3)_3(s) \longrightarrow Fe_2O_3(s) + 3\ CO_2(g)$$

31. How many milliliters of oxygen gas at STP are released from heating 5.00 g of calcium chlorate?

$$Ca(ClO_3)_2(s) \longrightarrow CaCl_2(s) + 3\ O_2(g)$$

32. How many milliliters of oxygen gas at STP are released from heating 2.50 g of mercuric oxide?

$$2\ HgO(s) \longrightarrow 2\ Hg(l) + O_2(g)$$

33. What mass of magnesium metal reacts with sulfuric acid to produce 225 mL of hydrogen gas at STP?

$$Mg(s) + H_2SO_4(aq) \longrightarrow MgSO_4(aq) + H_2(g)$$

34. What mass of manganese(II) chloride must react with sulfuric acid to release 49.5 mL of hydrogen chloride gas at STP?

$$MnCl_2(s) + H_2SO_4(aq) \longrightarrow MnSO_4(aq) + 2\ HCl(g)$$

35. How many grams of sodium metal react with water to give 75.0 mL of hydrogen gas at STP?

$$Na(s) + H_2O(l) \longrightarrow NaOH(aq) + H_2(g)$$

36. How many grams of hydrogen peroxide must decompose to give 55.0 mL of oxygen gas at STP?

$$H_2O_2(l) \longrightarrow H_2O(l) + O_2(g)$$

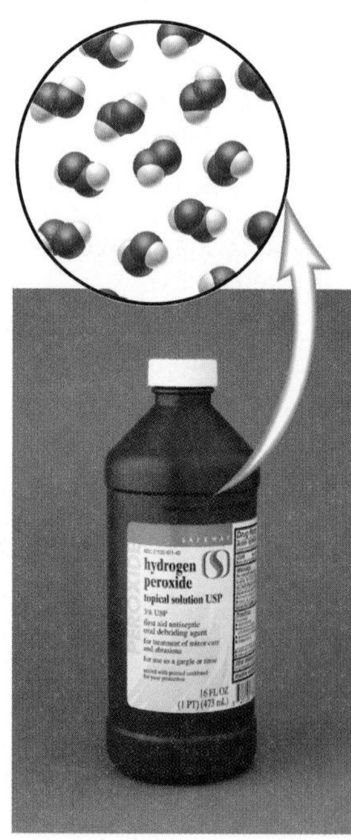

◀ **Hydrogen Peroxide** Hydrogen peroxide is unstable and breaks down to give water and oxygen gas. When H_2O_2 is poured on a wound, the fizzing bubbles are O_2 gas.

Volume–Volume Problems (Sec. 10.6)

37. Assuming all gases are at the same temperature and pressure, how many milliliters of iodine vapor react with 125 mL of hydrogen gas?

$$H_2(g) + I_2(g) \longrightarrow 2\ HI(g)$$

38. Assuming all gases are at the same temperature and pressure, how many milliliters of hydrogen iodide are produced from 125 mL of H_2 according to the equation in Exercise 37?

39. Assuming all gases are at the same temperature and pressure, how many milliliters of nitrogen gas react to give 45.0 mL of ammonia gas?

$$3\ H_2(g) + N_2(g) \longrightarrow 2\ NH_3(g)$$

40. Assuming all gases are at the same temperature and pressure, how many milliliters of hydrogen gas must react to give 45.0 mL of NH_3 according to the equation in Exercise 39?

41. Assuming all gases are at the same temperature and pressure, how many liters of oxygen gas react with 10.0 L of carbon monoxide?

$$2\ CO(g) + O_2(g) \longrightarrow 2\ CO_2(g)$$

42. Assuming all gases are at the same temperature and pressure, how many liters of carbon dioxide are produced from 10.0 L of CO according to the equation in Exercise 41?

43. Assuming all gases are at the same temperature and pressure, how many milliliters of chlorine gas react to yield 1.75 L of dichlorine trioxide?

$$Cl_2(g) + O_2(g) \longrightarrow Cl_2O_3(g)$$

44. Assuming all gases are at the same temperature and pressure, how many milliliters of oxygen gas must react to give 1.75 L of Cl_2O_3 according to the equation in Exercise 43?

45. What volume of sulfur trioxide gas is produced from 25.0 L of oxygen gas? The volume of each gas is measured at 875 °C and 1.00 atm pressure.

$$SO_2(g) + O_2(g) \longrightarrow SO_3(g)$$

46. What volume of sulfur dioxide gas reacts with 25.0 L of O_2 according to the equation in Exercise 45? The reaction conditions are 875 °C and 1.00 atm pressure.

47. What volume of nitrogen gas reacts to yield 500.0 cm^3 of dinitrogen pentaoxide? The volume of each gas is measured at 350 °C and 2.50 atm pressure.

$$N_2(g) + O_2(g) \longrightarrow N_2O_5(g)$$

48. What volume of oxygen gas reacts to yield 500.0 cm^3 of N_2O_5 according to the equation in Exercise 47? The reaction conditions are 350 °C and 2.50 atm pressure.

The Limiting Reactant Concept (Sec. 10.7)

49. If 1.00 mol of nitrogen monoxide gas and 1.00 mol of oxygen gas react, what is the limiting reactant and how many moles of NO_2 are produced from the reaction?

$$2\ NO(g) + O_2(g) \longrightarrow 2\ NO_2(g)$$

50. If 1.00 mol of nitrogen monoxide gas and 2.00 mol of oxygen gas react, what is the limiting reactant and how many moles of NO_2 are produced according to the equation in Exercise 49?

51. If 5.00 mol of nitrogen gas and 5.00 mol of oxygen gas react, what is the limiting reactant and how many moles of NO are produced from the reaction?

$$N_2(g) + O_2(g) \longrightarrow 2\ NO(g)$$

52. If 0.125 mol of nitrogen gas and 0.125 mol of oxygen gas react, what is the limiting reactant and how many moles of NO are produced according to the equation in Exercise 51?

53. If 5.00 mol of hydrogen gas and 5.00 mol of oxygen gas react, what is the limiting reactant and how many moles of water are produced from the reaction?

$$2\,H_2(g) + O_2(g) \longrightarrow 2\,H_2O(l)$$

54. If 5.00 mol of hydrogen gas and 1.50 mol of oxygen gas react, what is the limiting reactant and how many moles of water are produced according to the equation in Exercise 53?

55. If 1.00 mol of ethane gas and 5.00 mol of oxygen gas react, what is the limiting reactant and how many moles of water are produced from the reaction?

$$2\,C_2H_6(g) + 7\,O_2(g) \longrightarrow 4\,CO_2(g) + 6\,H_2O(g)$$

56. If 1.00 mol of ethane gas and 3.00 mol of oxygen gas react, what is the limiting reactant and how many moles of water are produced according to the equation in Exercise 55?

57. Complete the following table for the reaction of cobalt and sulfur and calculate the moles of Co, S, and CoS after reaction according to the balanced equation:

$$Co(s) + S(s) \longrightarrow CoS(s)$$

Experiment	mol Co	mol S	mol CoS
(a) before reaction:	1.00	1.00	0.00
after reaction:			
(b) before reaction:	2.00	3.00	0.00
after reaction:			

58. Complete the following table for the reaction of cobalt and sulfur and calculate the moles of Co, S, and Co_2S_3 after reaction according to the balanced equation:

$$2\,Co(s) + 3\,S(s) \longrightarrow Co_2S_3(s)$$

Experiment	mol Co	mol S	mol Co_2S_3
(a) before reaction:	1.00	1.00	0.00
after reaction:			
(b) before reaction:	2.00	3.00	0.00
after reaction:			

Limiting Reactant Problems (Sec. 10.8)

59. If 40.0 g of molten iron(II) oxide reacts with 10.0 g of magnesium, what is the mass of iron produced?

$$FeO(l) + Mg(l) \longrightarrow Fe(l) + MgO(s)$$

60. If 10.0 g of molten iron(II) oxide reacts with of 40.0 g of magnesium according to the equation in Exercise 59, what is the mass of iron produced?

61. If 175 g of molten iron(III) oxide reacts with 37.5 g of aluminum, what is the mass of iron produced?

$$Fe_2O_3(l) + Al(l) \longrightarrow Fe(l) + Al_2O_3(s)$$

62. If 37.5 g of molten iron(III) oxide reacts with 175 g of aluminum according to the equation in Exercise 61, what is the mass of iron produced?

63. If 1.00 g of magnesium hydroxide reacts with 0.605 g of sulfuric acid, what is the mass of magnesium sulfate produced?

$$Mg(OH)_2(s) + H_2SO_4(l) \longrightarrow MgSO_4(s) + H_2O(l)$$

64. If 0.605 g of magnesium hydroxide reacts with 1.00 g of sulfuric acid according to the equation in Exercise 63, what is the mass of magnesium sulfate produced?

65. If 1.00 g of aluminum hydroxide reacts with 3.00 g of sulfuric acid, what is the mass of water produced?

$$Al(OH)_3(s) + H_2SO_4(l) \longrightarrow Al_2(SO_4)_3(aq) + H_2O(l)$$

66. If 3.00 g of aluminum hydroxide reacts with 1.00 g of sulfuric acid according to the equation in Exercise 65, what is the mass of water produced?

67. If 45.0 mL of nitrogen gas reacts with 95.0 mL of oxygen gas, what is the volume of NO_2 produced? Assume all gases are at the same temperature and pressure.

$$N_2(g) + O_2(g) \longrightarrow NO_2(g)$$

68. If 95.0 mL of nitrogen gas reacts with 45.0 mL of oxygen gas according to the equation in Exercise 67, what is the volume of NO_2 produced? Assume all gases are at the same temperature and pressure.

69. If 70.0 mL of nitrogen gas reacts with 45.0 mL of oxygen gas, what is the volume of N_2O_3 produced? Assume all gases are at the same temperature and pressure.

$$N_2(g) + O_2(g) \longrightarrow N_2O_3(g)$$

70. If 45.0 mL of nitrogen gas reacts with 70.0 mL of oxygen gas according to the equation in Exercise 69, what is the volume of N_2O_3 produced? Assume all gases are at the same temperature and pressure.

71. If 3.00 L of sulfur dioxide gas reacts with 1.25 L of oxygen gas, what is the volume of sulfur trioxide gas produced? Assume all gases are at the same temperature and pressure.

$$SO_2(g) + O_2(g) \longrightarrow SO_3(g)$$

72. If 1.25 L of sulfur dioxide gas reacts with 3.00 L of oxygen gas according to the equation in Exercise 71, what is the volume of sulfur trioxide gas produced? Assume all gases are at the same temperature and pressure.

73. If 50.0 L of hydrogen chloride gas reacts with 10.0 L of oxygen gas, what is the volume of chlorine gas produced? Assume all gases are at the same temperature and pressure.

$$HCl(g) + O_2(g) \longrightarrow Cl_2(g) + H_2O(g)$$

74. If 10.0 L of hydrogen chloride gas reacts with 50.0 L of oxygen gas according to the equation in Exercise 73, what is the volume of chlorine gas produced? Assume all gases are at the same temperature and pressure.

Percent Yield (Sec. 10.9)

75. A chemistry student prepared lead(II) iodide from 10.0 g of lead(II) nitrate and aqueous potassium iodide. If the student collected 12.5 g of PbI_2 and the theoretical yield is 13.9 g, what is the percent yield?

76. A chemistry student prepared acetone by decomposing 10.0 g of calcium acetate. If the student collected 3.50 g of acetone and the theoretical yield is 3.67 g, what is the percent yield?

77. A 1.50 g sample of sodium nitrate is decomposed by heating. If the resulting sodium nitrite has a mass of 1.29 g and the calculated yield is 1.22 g, what is the percent yield?

78. A 1.00 g sample of potassium bicarbonate is decomposed by heating. If the resulting potassium carbonate weighs 0.725 g and the calculated yield is 0.690 g, what is the percent yield?

General Exercises

79. What units are associated with molar mass?
80. What units are associated with molar volume?
81. The "volcano reaction" produces chromium(III) oxide, water, and nitrogen gas from the decomposition of ammonium dichromate. Calculate the mass of chromium(III) oxide produced by the decomposition of 1.54 g of ammonium dichromate.
82. What STP volume of nitrogen gas is produced in the "volcano reaction" from the decomposition of 1.54 g of ammonium dichromate according to the reaction described in Exercise 81?
83. Manganese metal and aluminum oxide are produced by the reaction of manganese(IV) oxide and aluminum metal. Calculate the mass of aluminum that is necessary to yield 1.00 kg of manganese metal.
84. What mass of aluminum oxide must react to produce 1.00 kg of manganese metal according to the reaction described in Exercise 83?
85. The mineral stibnite, antimony(III) sulfide, is treated with hydrochloric acid to give antimony(III) chloride and hydrogen sulfide gas. What STP volume of H_2S is produced from a 3.00 g sample of stibnite?
86. What mass of $SbCl_3$ is produced from a 3.00 g sample of stibnite according to the reaction described in Exercise 85?
87. When an electric current is passed through water, hydrogen and oxygen gases are liberated. What STP volume of hydrogen gas is obtained from the electrolysis of 100.0 mL of water? (*Hint:* The density of water is 1.00 g/mL.)
88. What STP volume of oxygen gas is obtained from the electrolysis of 100.0 mL of water according to the reaction described in Exercise 87?
89. Recreational vehicles use propane for cooking and heating. The combustion of propane and oxygen produces carbon dioxide and water. What mass of water is produced from the combustion of 10.0 g of propane, C_3H_8?
90. What STP volume of carbon dioxide gas is produced from the combustion of 10.0 g of propane according to the reaction described in exercise 89?
91. A chemical reaction stops when one of the reactants is consumed. For example, the combustion reaction in an automobile stops when the gasoline "runs out." In an automobile, what is the limiting reactant? What is the excess reactant?

10 Chapter Self-Test

Answers to Self-Test are in Appendix J.

1. Hydrogen gas reacts with 5.00 g of oxygen gas to give 5.63 g of water. Use the conservation of mass law to predict the mass of reacting hydrogen. (Sec. 10.1)

$$2\,H_2(g) + O_2(g) \longrightarrow 2\,H_2O(l)$$

(a) 0.63 g
(b) 5.00 g
(c) 5.63 g
(d) 10.63 g
(e) none of the above

2. How many moles of water must decompose to yield 5.00 mol of oxygen gas? (Sec. 10.2)

$$2\,H_2O(l) \longrightarrow 2\,H_2(g) + O_2(g)$$

(a) 1.00 mol
(b) 2.00 mol
(c) 5.00 mol
(d) 10.0 mol
(e) none of the above

3. Classify the following type of stoichiometry problem: *How many liters of oxygen are produced from decomposing 5.00 g of water?* (Sec. 10.3)
(a) mass–mass problem
(b) mass–volume problem
(c) volume–volume problem
(d) mole–mole problem
(e) none of the above

4. What mass of oxygen gas is produced from decomposing 5.00 g of water (18.02 g/mol)? (Sec. 10.4)

$$2\,H_2O(l) \longrightarrow 2\,H_2(g) + O_2(g)$$

(a) 2.22 g
(b) 2.50 g
(c) 4.44 g
(d) 5.00 g
(e) 8.88 g

5. What STP volume of hydrogen gas is produced from decomposing 5.00 g of water (18.02 g/mol)? (Sec. 10.5)

$$2\,H_2O(l) \longrightarrow 2\,H_2(g) + O_2(g)$$

(a) 1.55 L
(b) 3.11 L
(c) 5.00 L
(d) 6.22 L
(e) 12.4 L

6. What volume of oxygen gas is released from decomposing water if 5.00 mL of hydrogen gas is produced (assume constant temperature and pressure)? (Sec. 10.6)

$$2\,H_2O(l) \longrightarrow 2\,H_2(g) + O_2(g)$$

(a) 2.50 mL
(b) 5.00 mL
(c) 7.50 mL
(d) 10.0 mL
(e) 20.0 mL

7. Consider the limiting reactant concept and predict how many moles of steam are released from the reaction of 5.00 mol of hydrogen gas and 5.00 mol of oxygen gas (assume constant temperature and pressure). (Sec. 10.7)

$$2\ H_2(g) + O_2(g) \longrightarrow 2\ H_2O(g)$$

(a) 2.50 mol
(b) 5.00 mol
(c) 10.0 mol
(d) 15.0 mol
(e) none of the above

8. Consider the limiting reactant concept and predict how many liters of steam are released from the reaction of 5.00 L of hydrogen gas and 5.00 L of oxygen gas (assume constant temperature and pressure). (Sec. 10.8)

$$2\ H_2(g) + O_2(g) \longrightarrow 2\ H_2O(g)$$

(a) 2.50 L
(b) 5.00 L
(c) 10.0 L
(d) 15.0 L
(e) none of the above

9. The decomposition of 5.00 g of water releases 4.00 g of oxygen gas. If the calculated mass of oxygen is 4.44 g, what is the percent yield? (Sec. 10.9)

$$2\ H_2O(l) \longrightarrow 2\ H_2(g) + O_2(g)$$

(a) 80.0%
(b) 88.8%
(c) 90.1
(d) 111%
(e) 125%

Key Concepts

10. What must be done first before applying unit analysis to solve a stoichiometry problem?
11. How many unit factors are typically required to solve a mass–mass problem?
12. How many unit factors are typically required to solve a mass–volume problem?
13. How many unit factors are typically required to solve a volume–volume problem?
14. Why must mass–volume problems specify the gas at STP?
15. To solve volume–volume problems, is it necessary for reactants and products to be at STP?

Critical Thinking

16. Silver metal heated with yellow sulfur powder gives Ag_2S.

$$2\ Ag(s) + S(s) \xrightarrow{\Delta} Ag_2S(s)$$

Complete the following illustrations by drawing the product and excess reactant. Indicate the element that is the limiting reactant in each example.

(a)

(b)

CHAPTER

11 The Gaseous State

Nitrogen, oxygen, and water molecules are responsible for wind velocities greater than 100 miles per hour in Hurricane Katrina shown in the Gulf of Mexico. In the "eye" of the hurricane, the pressure is less than half the normal atmospheric pressure.

"When we attempt to conceive the number of particles in an atmosphere, it is somewhat like attempting to conceive the number of stars in the universe."
John Dalton, English Chemist (1766–1844)

The English chemist Robert Boyle is best known for establishing the scientific method, and his work with gases. He studied the atmosphere and found that a volume of air could be compressed. He envisioned air as being composed of invisible particles that can be squeezed closer together. Furthermore, he found that other gases could be compressed and reasoned that all gases were composed of tiny, discrete particles. His experiments were remarkable because they offered the first evidence of the particle nature of matter. In 1803, John Dalton proposed the atomic theory, which in part was supported by Boyle's experiments with gases.

11.1 PROPERTIES OF GASES

OBJECTIVE

▸ To list five observed properties of a gas.

In the late 1700s, high-altitude ballooning became popular in France. Even French scientists filled balloons with hydrogen gas and ascended miles into the sky. Ballooning in turn helped to create a great deal of interest in the study of gases. After numerous experiments, scientists concluded that all gases have common characteristics. We can summarize these properties as follows.

1. **Gases have a variable shape and volume.** A gas has the same shape and volume as its container. If the volume of the container increases or decreases, the volume of the gas increases or decreases.
2. **Gases expand uniformly.** When a container increases its volume, a gas expands, and the gas molecules distribute uniformly throughout the container.
3. **Gases compress uniformly.** When a container decreases its volume, a gas compresses, and gas molecules are closer to each other. If the volume of the container is reduced sufficiently, the gas will eventually liquefy.
4. **Gases have a low density.** The density of air is about 1 g/L (0.001 g/mL). The density of water is 1 g/mL. Thus, water is about 1000 times more dense than air.
5. **Gases mix uniformly with other gases in the same container.** Air is a common example of a gaseous mixture. The burning of fossil fuels releases carbon dioxide into the atmosphere where it mixes with the gases in air. During photosynthesis, plants consume carbon dioxide from air and release oxygen gas into the atmosphere.

Although most gases are colorless and invisible, we can observe their behavior indirectly. For example, if we decrease the volume of a gas, we observe an increase in gas pressure. Thus, it is possible to gather information about a gas by studying its properties.

▲ **Hot Air Balloon** The hot air inside the balloon is less dense than the air surrounding the balloon. As the less dense air rises, the balloon ascends.

11.2 ATMOSPHERIC PRESSURE

OBJECTIVES

▸ To state standard atmospheric pressure in the following units: atm, mm Hg, torr, cm Hg, in. Hg, psi, and kPa.

▸ To convert a given gas pressure to a different unit of measurement.

Gas pressure is the result of constantly moving molecules striking the inside surface of their container. The pressure that a gas exerts depends on how often and how hard these molecules strike the walls of the container.

1. If the molecules collide more often, the gas pressure *increases*.
2. If the molecules collide with more energy, the pressure *increases*.

As the temperature increases, the molecules move faster and collide more frequently and with more energy. Thus, the gas pressure increases (Figure 11.1).

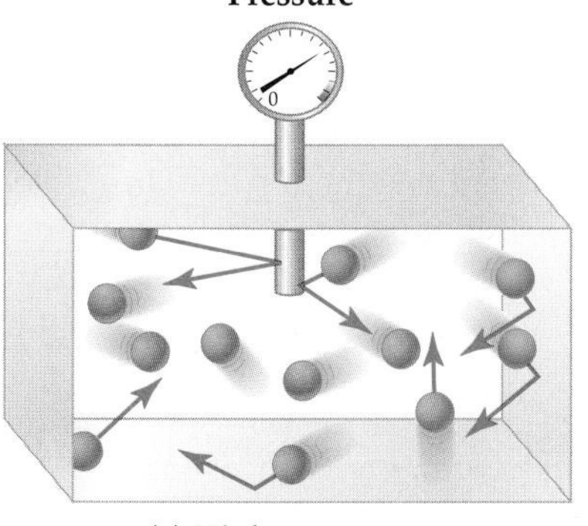

(a) Higher temperature
Higher pressure

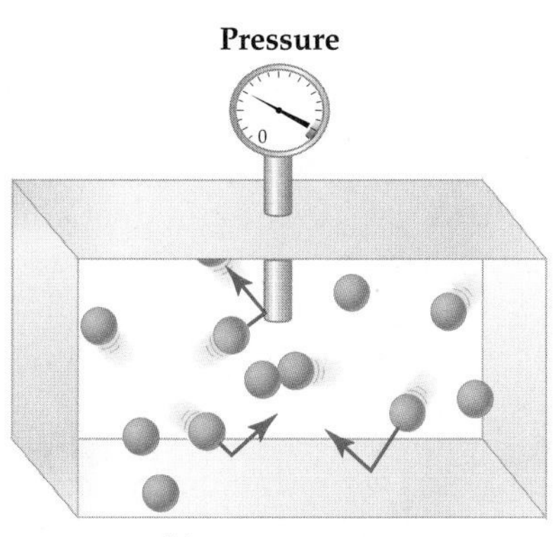

(b) Lower temperature
Lower pressure

◀ **Figure 11.1 Gas Pressure and Temperature** (a) The molecules are moving faster and colliding more frequently with the container.
(b) The molecules are moving slower and colliding less frequently with the container.

CHEMISTRY CONNECTION The Greenhouse Effect

Q: Why has planting millions of trees been suggested as a means of reducing global warming?

When we read about the greenhouse effect, we may be alarmed and in fear that it is life-threatening to our planet. However, the greenhouse effect is necessary for life on Earth as it helps to maintain a moderately mild climate. The current concern is that our atmosphere has an abundance of so-called greenhouse gases that will cause drastic global warming. Although carbon dioxide is singled out as the chief offender, other gases such as chlorofluorocarbons, CFCs, methane, CH_4, and the oxides of nitrogen, NO_x, also contribute to global warming.

Scientists explain the greenhouse effect as follows: Earth's atmosphere is transparent to short-wavelength radiation from the Sun. After striking Earth's surface, the radiation from the Sun is emitted back into space as long-wavelength heat energy. Long-wavelength heat energy can be absorbed by carbon dioxide and other trace gases in the atmosphere. Subsequently, these gas molecules release heat energy in all directions, some of which is radiated back toward Earth. The net effect is that carbon dioxide and other gases in the atmosphere form an invisible canopy that traps heat energy.

Our present concern is that additional gases in the atmosphere are increasing the temperature of our planet at an alarming rate. Global warming can result from high concentrations of carbon dioxide in the atmosphere. Carbon dioxide is produced by burning fossil fuels (coal, oil, and gas) in automobiles, homes, and power plants. The problem is further accelerated by massive deforestation, especially in South America. The loss of forests is especially harmful because trees convert carbon dioxide in the atmosphere to oxygen.

Some scientists think global warming is a severe threat and believe we must do something immediately. One futuristic suggestion is to add chemicals to the oceans that will stimulate the growth of plankton. Plankton remove carbon dioxide from the air through the process of photosynthesis. Currently, our efforts are directed at reforestation and reducing the emission of greenhouse gases into the atmosphere.

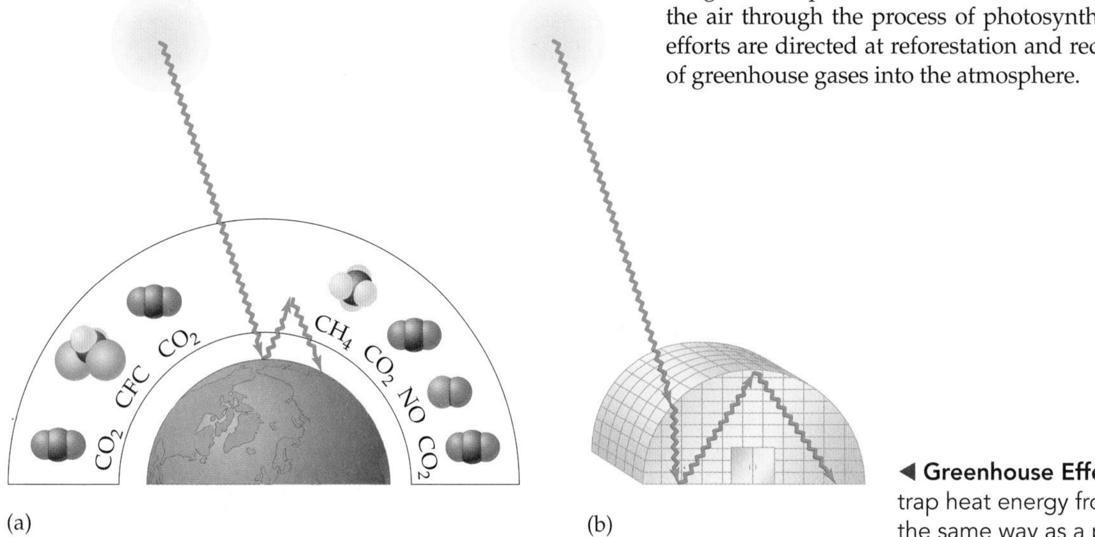

◀ **Greenhouse Effect** Greenhouse gases trap heat energy from the Sun in much the same way as a plant greenhouse.

A: The greenhouse gas that contributes most to global warming is carbon dioxide. The planting of millions of trees will allow the process of photosynthesis to reduce the amount of carbon dioxide and return more oxygen to the atmosphere.

Atmospheric Pressure

The Italian scientist Evangelista Torricelli (1608–1647) published the first scientific explanation of a vacuum. He proposed that a "sea of air" surrounds Earth and argued that this sea of air exerts a pressure on everything it contacts. We now understand that there are no gas molecules in a **vacuum** and that the gas pressure is therefore zero.

By the end of the 1600s, the concept of atmospheric pressure was well established. Today, we understand that **atmospheric pressure** is the result of air molecules in the environment. The pressure of the atmosphere is considerable, about 15 pounds on every square inch of surface. In fact, the atmosphere exerts a total weight of nearly 20 tons on an average-size human body! More effects of atmospheric pressure are illustrated in Figure 11.2.

In 1643, Torricelli invented the **barometer** to measure atmospheric pressure (Figure 11.3). He showed that at sea level, the height of mercury measures 29.9 inches, which we define as standard atmospheric pressure, that is, 1 atmosphere (symbol atm). Above sea level, the height of mercury is less than 29.9 inches, indicating that atmospheric pressure is less at higher elevations.

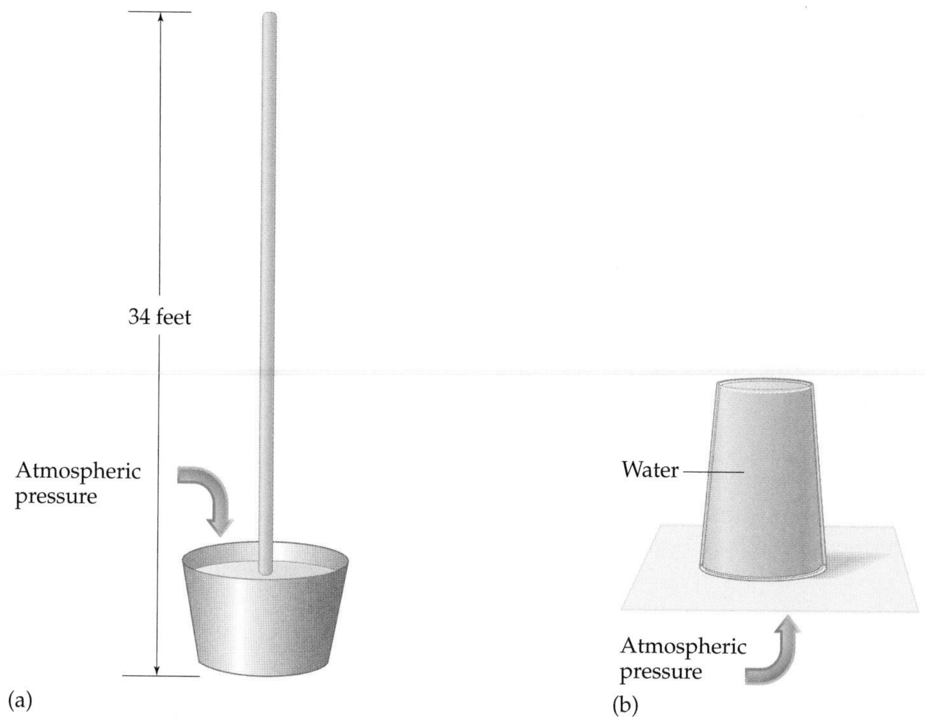

◀ **Figure 11.2 Atmospheric Pressure** (a) The pressure of air molecules in the atmosphere supports a 34-foot column of water.
(b) Since atmospheric pressure operates in all directions, air molecules strike the bottom of the card and support the water in the inverted glass.

We can express standard pressure in many units besides atmospheres and inches of mercury. For example, 29.9 in. of Hg is equivalent to 760 mm Hg. In honor of Torricelli, standard pressure is stated as 760 torr, where 1 **torr** equals 1 mm Hg. Table 11.1 lists units that are frequently used to express pressure.

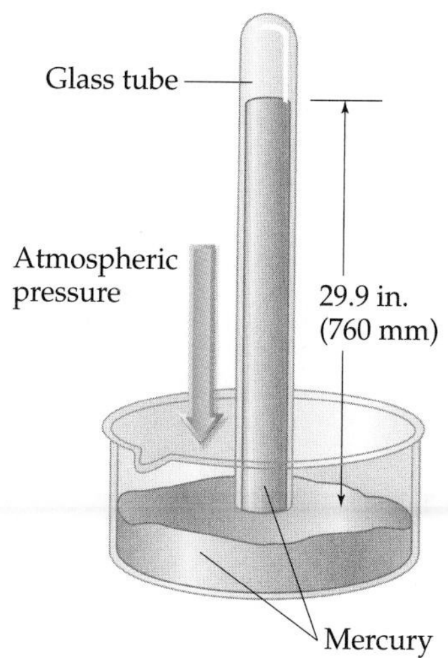

▲ **Figure 11.3 Torricelli's Mercury Barometer** A sealed glass tube is filled with liquid mercury, inverted, and placed into a dish of mercury. There are no gas molecules above the column of mercury inside the glass tube. The pressure of the atmosphere at sea level supports a column of mercury 29.9 inches (760 mm) high.

TABLE 11.1 UNITS OF GAS PRESSURE

UNIT	STANDARD PRESSURE
atmosphere	1 atm (exactly)
inches of mercury	29.9 in. Hg
centimeters of mercury	76 cm Hg (exactly)
millimeters of mercury	760 mm Hg (exactly)
torr[1]	760 torr (exactly)
pounds per square inch	14.7 psi
kilopascal[2]	101 kPa

[1]A torr is defined as exactly equal to 1 mm Hg pressure; thus, standard pressure is 760 torr.
[2]The kilopascal (kPa) is the official SI unit of pressure.

Since each of the values in Table 11.1 is an expression for the same standard pressure, we can relate one unit to another. For example, if the pressure of propane gas in a steel tank is 2550 torr, we can relate the pressure to atmospheres. We can perform the conversion by applying the unit analysis method. The unknown unit is atm, and the given value is 2550 torr. Thus,

$$2550\ \cancel{\text{torr}} \times \frac{\text{unit}}{\text{factor}} = \text{atm}$$

To convert from one unit of pressure to another, we derive a unit factor from the relationship among units of standard pressure. Since 1 atm = 760 torr, the two related unit factors are 1 atm/760 torr, and its reciprocal 760 torr/1 atm. We select the first unit factor to cancel units.

$$2550\ \cancel{\text{torr}} \times \frac{1\ \text{atm}}{760\ \cancel{\text{torr}}} = 3.36\ \text{atm}$$

▲ Adding air to an automobile tire increases the pressure because there are more molecules, which cause an increase in the number of collisions inside the tire.

The following example exercise further illustrates the conversion of gas pressures:

Example Exercise 11.1 Gas Pressure Conversion

Meteorologists state that a "falling" barometer indicates an approaching storm. Given a barometric pressure of 27.5 in. Hg, express the pressure in each of the following units of pressure:

(a) atm (b) mm Hg
(c) psi (d) kPa

Solution

For each conversion, we apply a unit conversion factor related to units of standard pressure.

(a) To express the pressure in atmospheres, we derive a unit factor related to the equivalent relationship 29.9 in. Hg = 1 atm.

$$27.5 \text{ \cancel{in. Hg}} \times \frac{1 \text{ atm}}{29.9 \text{ \cancel{in. Hg}}} = 0.920 \text{ atm}$$

(b) To convert to millimeters of mercury, we derive a unit factor related to the equivalent relationship 29.9 in. Hg = 760 mm Hg.

$$27.5 \text{ \cancel{in. Hg}} \times \frac{760 \text{ mm Hg}}{29.9 \text{ \cancel{in. Hg}}} = 699 \text{ mm Hg}$$

(c) To calculate the pressure in pounds per square inch, we derive a unit factor related to the equivalent relationship 29.9 in. Hg = 14.7 psi.

$$27.5 \text{ \cancel{in. Hg}} \times \frac{14.7 \text{ psi}}{29.9 \text{ \cancel{in. Hg}}} = 13.5 \text{ psi}$$

(d) To find the pressure in kilopascals, we derive a unit factor related to the equivalent relationship 29.9 in. Hg = 101 kPa.

$$27.5 \text{ \cancel{in. Hg}} \times \frac{101 \text{ kPa}}{29.9 \text{ \cancel{in. Hg}}} = 92.9 \text{ kPa}$$

Since 1 atm and 760 mm Hg are exact values, the answers have been rounded to three significant digits, which is consistent with the given value, 27.5 in. Hg.

Practice Exercise

Given that the air pressure inside an automobile tire is 34.0 psi, express the pressure in each of the following units:

(a) atm (b) cm Hg
(c) torr (d) kPa

Answers:

(a) 2.31 atm (b) 176 cm Hg
(c) 1760 torr (d) 234 kPa

Concept Exercise

Which of the following expresses the greatest gas pressure: 1 atm, 1 in. Hg, 1 torr, 1 cm Hg, 1 mm Hg, or 1 psi?

Answer: See Appendix G.

11.3 VARIABLES AFFECTING GAS PRESSURE

OBJECTIVES

- To identify the three variables that affect the pressure of a gas.
- To state whether gas pressure increases or decreases for a given change in volume, temperature, or the number of moles of gas.

In Section 11.2, we learned that gas pressure is related to the frequency and energy of gas molecule collisions. Experimentally, there are only three ways to change the number of collisions, and hence the pressure.

1. **Increase or decrease the volume of the container.** When we increase the volume, gas molecules are further apart and collide less frequently, and the pressure decreases. When we decrease the volume, gas molecules are closer together and collide more frequently, and the pressure increases. The pressure is said to be

inversely related to the volume. That is, when the volume increases, the pressure decreases. When the volume decreases, the pressure increases.

2. **Increase or decrease the temperature of the gas.** When we increase the temperature, gas molecules move faster and collide with a greater frequency and energy. When we decrease the temperature, the gas molecules move slower and collide less frequently. The pressure is said to be *directly* related to the temperature. When the temperature increases, the pressure increases. When the temperature decreases, the pressure decreases.
3. **Increase or decrease the number of molecules in the container. Avogadro's theory,** also called Avogadro's law, states that equal volumes of gases, at the same temperature and pressure, contain equal numbers of molecules. When we increase the number of gas molecules, there are more collisions and the pressure increases. When we decrease the number of gas molecules, there are fewer collisions and the pressure decreases. The pressure is said to be *directly* related to the number of moles of gas, or the number of gas molecules. When the number of gas molecules increases, the pressure increases. When the number of molecules decreases, the pressure decreases.The effects of these three variables (volume, temperature, and number of molecules) on pressure are illustrated in Figure 11.4.

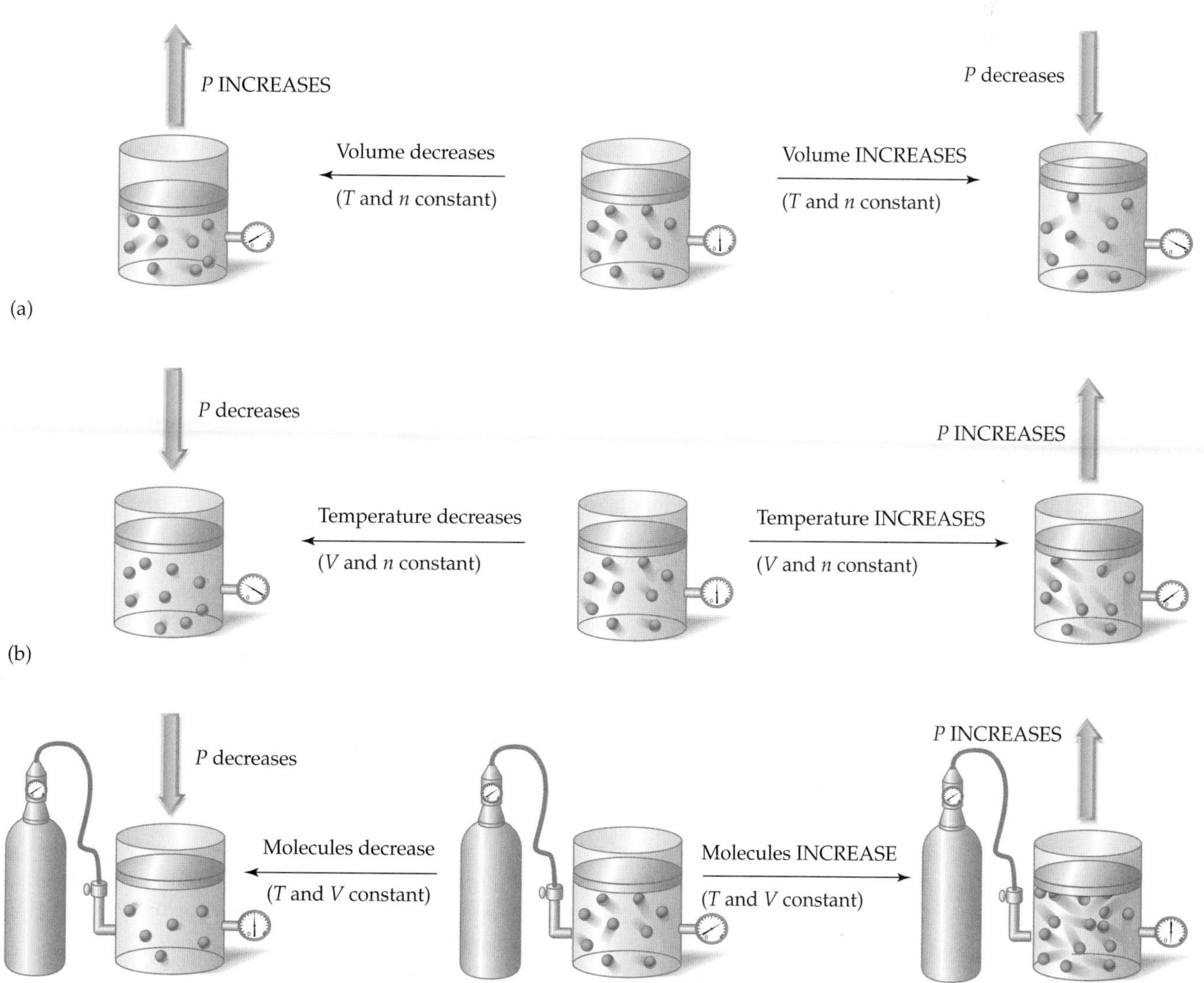

▲ **Figure 11.4 Variables Affecting Gas Pressure** A gas is contained in a cylinder with a moving piston. The gas pressure, *P*, is affected by (a) changing volume, *V*, (b) changing temperature, *T*, and (c) changing the number of gas molecules, *n*, in the cylinder.

Example Exercise 11.2 Gas Pressure Changes

State whether the pressure of a gas in a sealed container increases or decreases with each of the following changes:

(a) The volume changes from 250 mL to 500 mL.
(b) The temperature changes from 20 °C to −80 °C.
(c) The moles of gas change from 1.00 mol to 1.50 mol.

Solution

For each of the changes we must consider whether the number of molecular collisions increases or decreases.

(a) The volume increases, and so the number of collisions decreases; thus the *pressure decreases.*
(b) The temperature decreases, and so the number of collisions decreases; thus, the *pressure decreases.*
(c) When the moles of gas increase, the number of molecules increases. With more molecules, there are more collisions, and the *pressure increases.*

Practice Exercise

Indicate whether gas pressure increases or decreases with each of the following changes in a sealed container:

(a) increasing the temperature
(b) increasing the volume
(c) increasing the number of gas molecules

Answers: (a) pressure increases; (b) pressure decreases; (c) pressure increases

Concept Exercise

Which of the following decreases gas pressure: increasing volume, increasing temperature, or increasing the number of molecules?

Answer: See Appendix G.

11.4 BOYLE'S LAW: PRESSURE–VOLUME RELATIONSHIPS

OBJECTIVES

- To sketch a graph of the pressure–volume relationship for a gas.
- To calculate the pressure or volume of a gas after a change in conditions.

We have mentioned that Robert Boyle was the founder of the scientific method. In one of his experiments, Boyle trapped air in a J-shaped tube with liquid mercury. As he added mercury into the tube, he found that the volume of trapped air decreased (Figure 11.5). When he doubled the pressure, the volume was reduced to a half. When he tripled the pressure, the volume was reduced to a third. Thus, Boyle established the relationship between the pressure and the volume of a gas.

The results of Boyle's experiments are known as **Boyle's law**, which states that the volume of a gas is inversely proportional to the pressure when the temperature remains constant. **Inversely proportional** means that two variables have a reciprocal relationship. As one variable gets larger, the other variable gets smaller. We can express Boyle's law mathematically, where the symbol "$\propto$" is read "proportional to."

$$V \propto \frac{1}{P} \qquad (T \textit{ remains constant})$$

That is, the volume (V) is proportional to the reciprocal pressure ($1/P$) when the temperature (T) remains constant. Using experimental data, we can plot pressure versus volume and obtain the curve shown in Figure 11.6.

The relationship between pressure and volume can be written as an equation by using a proportionality constant (k):

$$V = k \times \frac{1}{P}$$

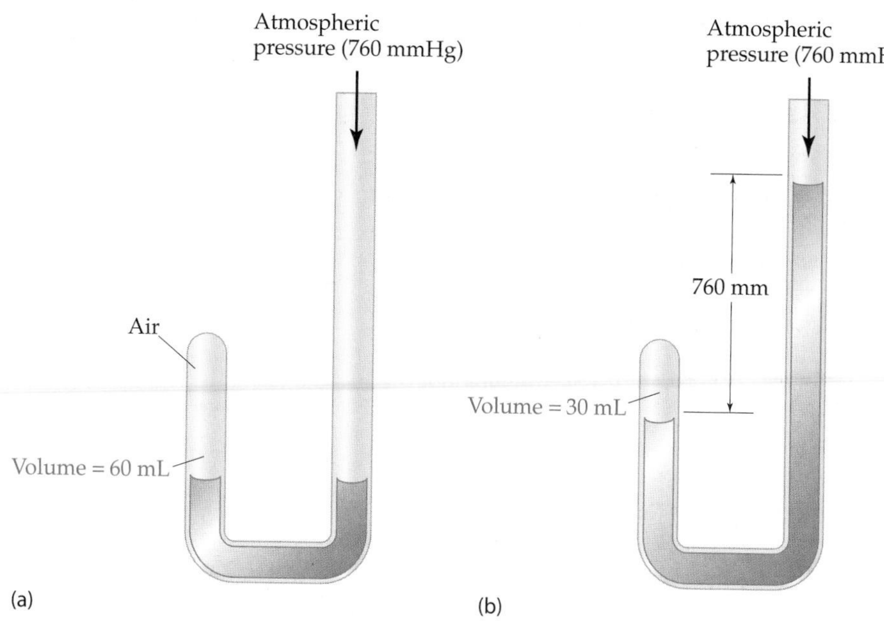

◀ **Figure 11.5 Boyle's Law** (a) The initial volume of gas at atmospheric pressure is 60 mL. (b) After adding mercury, the volume of gas is 30 mL. Note that the volume is halved when the pressure is doubled. Initially, the pressure is 760 mm Hg; after adding mercury, the pressure is 1520 mm Hg (760 mm Hg + 760 mm Hg).

Multiplying both sides of the equation by P, we see that the product of pressure and volume is equal to the constant (k). The equation becomes

$$PV = k$$

We can consider a sample of a gas under changing conditions but with temperature remaining the same. Let the initial conditions of pressure and volume be P_1 and V_1, respectively. After a change in conditions, we can indicate the final pressure and volume as P_2 and V_2. Since the product of pressure and volume is a constant, we write

$$P_1 V_1 = k = P_2 V_2$$

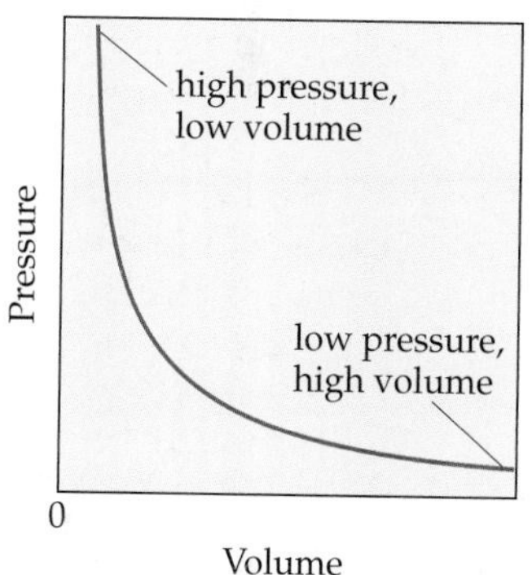

▲ **Figure 11.6 Gas Pressure versus Volume** As the volume of a gas increases, the pressure decreases. Conversely, as the volume decreases, the pressure increases. The volume and the pressure of a gas are inversely proportional.

Solving Boyle's Law Problems

In this textbook, we will present two methods for solving gas law problems. Firstly, we will solve problems using a conceptual approach and a reasoning method. Secondly, we will solve problems using an algebraic method.

Since the pressure and the volume of a gas are inversely related, we can apply a proportionality factor to find an unknown variable. To find the gas pressure after a change in volume, we apply a volume factor to the initial pressure (P_1).

$$\boxed{P_1 \times V_{factor} = P_2}$$

Similarly, to find the volume after a change in gas pressure, we apply a pressure factor to the initial volume (V_1).

$$\boxed{V_1 \times P_{factor} = V_2}$$

Let's consider the following Boyle's law gas problem. In an experiment, 5.00 L of propane gas is compressed and the pressure increases from 1.00 atm to 1.50 atm. To calculate the final volume (V_2), we write

$$5.00 \text{ L} \times P_{\text{factor}} = V_2$$

Since the pressure of the propane gas increases from 1.00 atm to 1.50 atm, the final volume must decrease. Therefore, the P_{factor} must be less than 1. That is, the smaller pressure value must appear in the numerator.

$$5.00 \text{ L} \times \frac{1.00 \cancel{\text{atm}}}{1.50 \cancel{\text{atm}}} = 3.33 \text{ L}$$

Notice that the cancellation of units takes place in the unit factor (P_{factor}). Although the numerator and the denominator are not equal, this method uses a format very much like that of the unit analysis method of problem solving. Now, let's reinforce our understanding of Boyle's law with an example exercise:

Example Exercise 11.3 Boyle's Law

A 1.50 L sample of methane gas exerts a pressure of 1650 mm Hg. Calculate the final pressure if the volume changes to 7.00 L. Assume the temperature remains constant.

Conceptual Solution

We can find the final pressure (P_2) by applying Boyle's law and using the relationship

$$P_1 \times V_{factor} = P_2$$

The volume increases from 1.50 L to 7.00 L. Thus, the pressure decreases. The V_{factor} must be less than 1. Hence,

$$1650 \text{ mm Hg} \times \frac{1.50 \not{L}}{7.00 \not{L}} = 354 \text{ mm Hg}$$

We can visually summarize the Boyle's law solution as follows:

Algebraic Solution

Alternatively, we can solve this problem using the equation

$$P_1 V_1 = P_2 V_2$$

Solving for P_2 gives

$$\frac{P_1 V_1}{V_2} = P_2$$

Substituting for each variable and simplifying, we obtain

$$\frac{1650 \text{ mm Hg} \times 1.50 \not{L}}{7.00 \not{L}} = 354 \text{ mm Hg}$$

Practice Exercise

A sample of ethane gas has a volume of 125 mL at 20 °C and 725 torr. What is the volume of the gas at 20 °C when the pressure decreases to 475 torr?

Answer: 191 mL

Concept Exercise

When air in a steel cylinder is compressed from 10 L to 5 L, and temperature remains constant, the gas pressure inside the cylinder (increases/decreases).

Answer: See Appendix G.

Note A conceptual solution for solving gas law problems is valuable because it emphasizes reason and a deeper understanding of gaseous behavior. A conceptual solution requires that we understand how changes in pressure, volume, and temperature of a gas affect each other. On the other hand, mathematics is a powerful tool for solving problems, and an algebraic solution is valuable where the concepts are initially difficult to grasp.

CHEMISTRY CONNECTION Robert Boyle

Q. What instrument did Robert Boyle invent that demonstrated a feather and a lump of lead are affected identically by gravity?

Robert Boyle (1627–1691) was a child prodigy who was speaking Latin and Greek by the age of 8. He was the son of wealthy English aristocrats and traveled throughout Europe with a private tutor to gain a broad education. When he was 18 years old, his father died and left him with a lifetime income. At 27, he enrolled in Oxford University to continue his studies. There he became interested in experimentation, although laboratory work was considered to be of minor importance at the time. Most influential scientists believed, like the Greeks, that reason was far superior to experimentation.

In 1657, Boyle designed a vacuum pump that was an improvement over the first one developed a few years earlier. In one experiment, he evacuated most of the air from a sealed chamber and showed that a ticking clock could not be heard in a vacuum. He then correctly concluded that sound does not exist in the absence of air. Boyle was fascinated by the behavior of gases and formulated the law stating that pressure and volume of a gas are inversely related.

In 1661, Boyle published *The Sceptical Chymist* and argued that theories were no better than the experiments on which they were based. Gradually, this point of view was accepted, and his text marked a turning point for recognition of the importance of experimentation. Because of his numerous contributions to chemistry and physics, Boyle is generally regarded as the founder of the modern scientific method.

Boyle felt strongly about keeping meticulous notes and reporting experimental results. He thought that everyone should be able to learn from published accounts of laboratory inquiry. Boyle felt that experiments should be held up to scientific scrutiny and that others should have the opportunity to confirm or disprove the results. This practice has become a cornerstone of science and, except for information involving military and industrial secrets, research experiments are published and made available to the scientific community and to the public at large.

◀ Robert Boyle (1627–1691)

A: Boyle invented a vacuum pump to remove air from a cylinder. When he released a feather and a lump of lead from the same height, he found that they landed simultaneously at the bottom of the cylinder.

11.5 CHARLES'S LAW: VOLUME–TEMPERATURE RELATIONSHIPS

OBJECTIVES

- To sketch a graph of the volume–temperature relationship for a gas.
- To calculate the volume or temperature of a gas after a change in conditions.

In 1783, the French scientist Jacques Charles (1746–1823) filled a balloon with hydrogen gas and ascended to 10,000 feet—nearly two miles! Although hot air ballooning had been pioneered by the Montgolfier brothers, Jacques Charles was the first to make an ascent in a hydrogen-filled balloon.

In 1787, Charles discovered the effect of temperature on the volume of a gas. From his experiments, he correctly deduced that the volume of a gas is **directly proportional** to the Kelvin temperature. That is, when we double the Kelvin temperature of a gas, we double its volume. When we halve the Kelvin temperature of a gas, we halve its volume. We can state **Charles's law** as follows: the volume of a gas is directly proportional to the Kelvin temperature if the pressure remains constant. Using experimental data, we can plot volume versus Kelvin temperature and obtain the straight-line graph shown in Figure 11.7.

We can express the direct relationship between volume and Kelvin temperature as

$$V \propto T \qquad (P \text{ remains constant})$$

The relationship of volume and Kelvin temperature can also be written as an equation using a proportionality constant (k).

$$V = k\,T$$

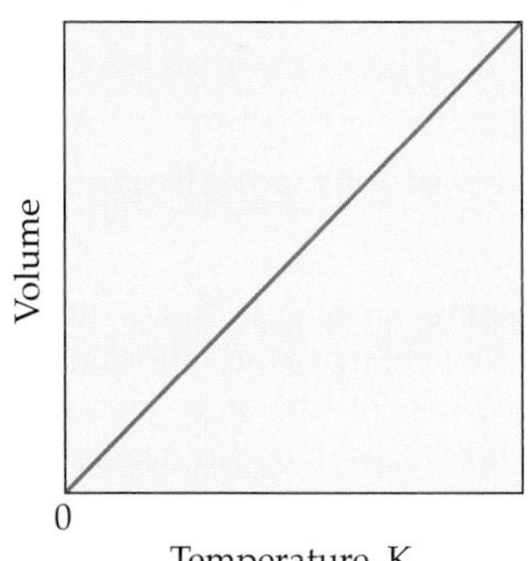

▲ **Figure 11.7 Gas Volume versus Temperature** As the Kelvin temperature decreases, the volume of a gas decreases. The Kelvin temperature and the volume of a gas are directly proportional.

Dividing both sides of the equation by T, we see that the ratio of volume to temperature equals the constant k. The equation becomes

$$\frac{V}{T} = k$$

Let's consider a sample of gas under changing conditions, but with pressure remaining the same. We indicate the initial conditions of volume and temperature as V_1 and T_1, respectively. After a change in conditions, we indicate the final volume and temperature as V_2 and T_2. Since the ratio of volume to temperature equals the constant k, we write

$$\frac{V_1}{T_1} = k = \frac{V_2}{T_2}$$

Solving Charles's Law Problems

Since the volume and temperature of a gas are directly related, we can apply a proportionality factor to find an unknown variable. To find the volume after a change in temperature, we apply a temperature factor to the initial volume (V_1).

$$\boxed{V_1 \times T_{\text{factor}} = V_2}$$

Similarly, to find the temperature after a change in volume, we apply a volume factor to the initial temperature (T_1).

$$\boxed{T_1 \times V_{\text{factor}} = T_2}$$

Let's consider the following Charles's law gas problem. In an experiment, a sample of argon gas at 225 K is heated and the volume increases from 3.50 L to 12.5 L. To calculate the final temperature (T_2), we write

$$225 \text{ K} \times V_{\text{factor}} = T_2$$

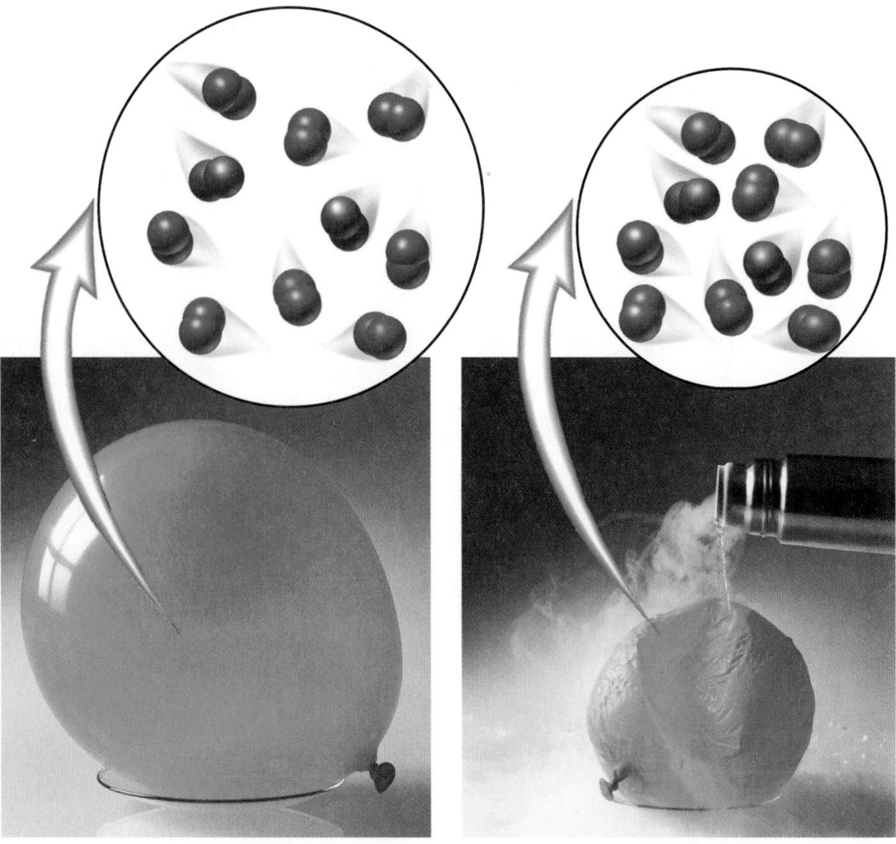

▶ **Illustration of Charles's Law** (a) A balloon contains air at room temperature. (b) As liquid nitrogen at −196 °C is poured on the balloon, the air inside the balloon cools, causing it to decrease in volume.

Since the volume of argon gas increases from 3.50 L to 12.5 L, the final temperature must increase. Therefore, the V_{factor} must be greater than 1. That is, the larger value must appear in the numerator.

$$225 \text{ K} \times \frac{12.5 \cancel{\text{L}}}{3.50 \cancel{\text{L}}} = 804 \text{ K}$$

The final temperature is 804 K, or 531 °C. Now, let's reinforce our understanding of Charles's law with an example exercise:

Example Exercise 11.4 Charles's Law

A 275 L helium balloon is heated from 20 °C to 40 °C. Calculate the final volume assuming the pressure remains constant.

Conceptual Solution

We first convert the Celsius temperatures to Kelvin by adding 273 units.

$$20 \text{ °C} + 273 = 293 \text{ K}$$

$$40 \text{ °C} + 273 = 313 \text{ K}$$

We can find the final volume, (V_2), by applying Charles's law and using the relationship

$$V_1 \times T_{factor} = V_2$$

The temperature increases from 293 K to 313 K. It follows that the volume increases and the T_{factor} must be greater than 1.

$$275 \text{ L} \times \frac{313 \cancel{\text{K}}}{293 \cancel{\text{K}}} = 294 \text{ L}$$

We can visually summarize the Charles's law solution as follows:

Algebraic Solution

Alternatively, we can solve this problem using the equation

$$\frac{V_1}{T_1} = \frac{V_2}{T_2}$$

Solving for V_2 gives

$$\frac{V_1 T_2}{T_1} = V_2$$

Substituting for each variable and simplifying, we obtain

$$\frac{275 \text{ L} \times 313 \cancel{\text{K}}}{293 \cancel{\text{K}}} = 294 \text{ L}$$

Practice Exercise

A krypton balloon has a volume of 555 mL at 21 °C. If the balloon is cooled and the volume decreases to 475 mL, what is the final temperature? Assume that the pressure remains constant.

Answer: –21 °C

Concept Exercise

When air in an elastic balloon cools from 25 °C to 20 °C, the volume of the balloon (increases/decreases).

Answer: See Appendix G.

Note In Charles's law problems we can state the volume in any units. For example, we can express the volume in liters or cubic centimeters. However, we must *always* express the temperature in Kelvin units.

11.6 GAY-LUSSAC'S LAW: PRESSURE–TEMPERATURE RELATIONSHIPS

OBJECTIVES

- To sketch a graph of the pressure–temperature relationship for a gas.
- To calculate the pressure or temperature of a gas after a change in conditions.

Like Jacques Charles, the French scientist Joseph Gay-Lussac (1778–1850) was also interested in ballooning and made a record ascent to 23,000 feet in a hydrogen-filled balloon. During his ascent, he collected samples of the atmosphere for study in his laboratory.

In 1802, Gay-Lussac established the relationship between pressure and temperature. He correctly interpreted his experiments to mean that the pressure of a gas is directly proportional to the Kelvin temperature. That is, when we double the Kelvin temperature of a gas, we double its pressure. When we halve the Kelvin temperature of a gas, we halve its pressure. We can state **Gay-Lussac's law** as follows: the pressure of a gas is directly proportional to the Kelvin temperature if the volume remains constant. Using experimental data, we can plot pressure versus Kelvin temperature and obtain the straight-line graph shown in Figure 11.8.

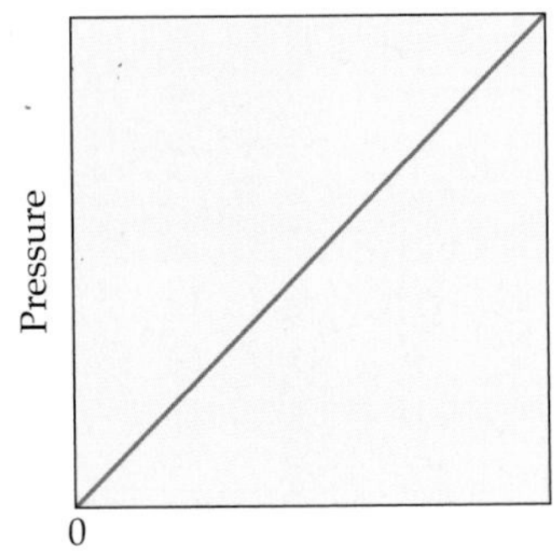

▲ **Figure 11.8 Gas Pressure versus Temperature** As the Kelvin temperature decreases, the pressure of a gas decreases. The Kelvin temperature and the pressure exerted by a gas are directly proportional.

We can express the relationship between pressure and Kelvin temperature as

$$P \propto T \qquad (V \text{ remains constant})$$

The relationship between pressure and temperature can also be written as an equation using the proportionality constant (k).

$$P = kT$$

Dividing both sides of the equation by T, we see that the ratio of pressure to temperature equals the constant k. The equation becomes

$$\frac{P}{T} = k$$

Let's consider a sample of a gas under changing conditions, but with volume remaining the same. We indicate the initial conditions of pressure and temperature as P_1 and T_1, respectively. After a change in conditions, we indicate the final pressure and temperature as P_2 and T_2. Since the ratio of pressure to temperature equals the constant (k), we can write

$$\frac{P_1}{T_1} = \text{k} = \frac{P_2}{T_2}$$

Solving Gay-Lussac's Law Problems

Since the pressure and temperature of a gas are directly related, we can apply a proportionality factor to find an unknown variable. To find the pressure after a change in temperature, we apply a temperature factor to the initial pressure (P_1).

$$\boxed{P_1 \times T_{\text{factor}} = P_2}$$

Similarly, to find the temperature after a change in pressure, we apply a pressure factor to the initial temperature (T_1).

$$\boxed{T_1 \times P_{\text{factor}} = T_2}$$

Let's consider the following problem involving Gay-Lussac's law. An automobile tire is inflated to 28.0 psi at a temperature of 20 °C. After it has traveled at high speed, the tire pressure is 36.0 psi. To calculate the final temperature (T_2), we first convert the initial temperature to Kelvin (20 °C + 273 = 293 K), and then write

$$293 \text{ K} \times P_{\text{factor}} = T_2$$

Since the volume of the tire remains constant and the pressure increases from 28.0 psi to 36.0 psi, the final temperature must increase. Therefore, the P_{factor} must be greater than 1. That is, the larger pressure value must appear in the numerator.

$$293\text{ K} \times \frac{36.0\text{ psi}}{28.0\text{ psi}} = 377\text{ K}$$

The final temperature is 377 K, or 104 °C. Now, let's reinforce our understanding of Gay-Lussac's law with an example exercise:

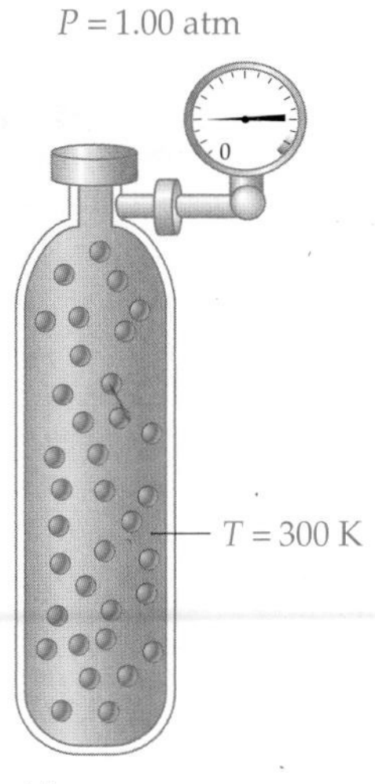

(a)

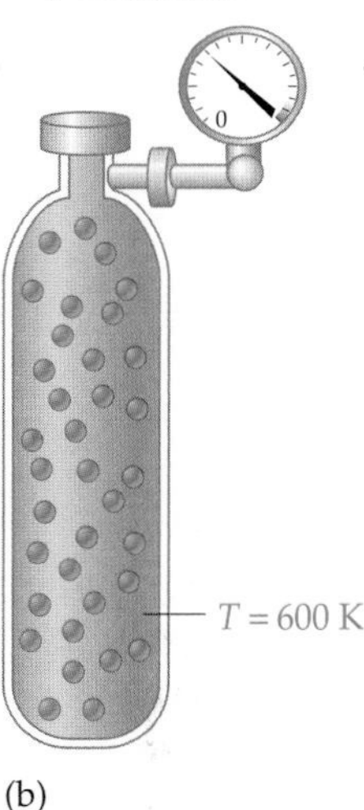

(b)

▲ **Illustration of Gay-Lussac's Law** (a) A cylinder contains helium gas at 300 K and a pressure of 1.00 atm. (b) After doubling the temperature of the helium gas to 600 K, the pressure doubles to 2.00 atm.

Example Exercise 11.5 Gay-Lussac's Law

A steel container filled with nitrous oxide at 15.0 atm is cooled from 2 °C to –40 °C. Calculate the final pressure assuming the volume remains constant.

Conceptual Solution

We must first convert the Celsius temperatures to Kelvin by adding 273.

$$25\ °\text{C} + 273 = 298\text{ K}$$

$$-40\ °\text{C} + 273 = 233\text{ K}$$

We can find the final pressure (P_2) by applying Gay-Lussac's law and using the relationship

$$P_1 \times T_{factor} = P_2$$

The volume of the container remains constant, but the temperature decreases from 298 K to 233 K. Therefore, the pressure decreases. The T_{factor} must be less than 1. Hence,

$$15.0\text{ atm} \times \frac{233\text{ K}}{298\text{ K}} = 11.7\text{ atm}$$

We can visually summarize the Gay-Lussac's law solution as follows:

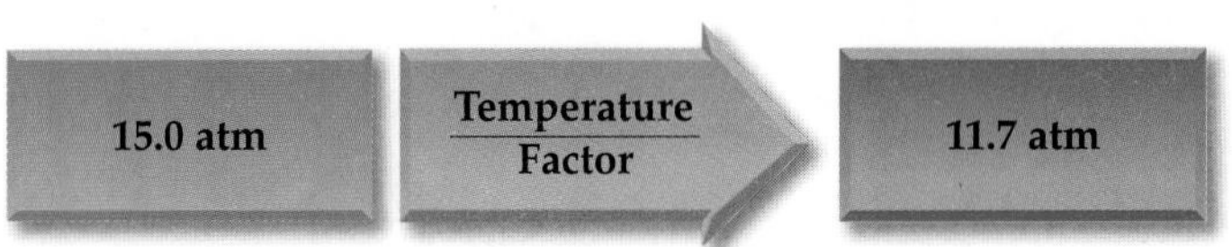

Algebraic Solution

Alternatively, we can solve this problem using the equation

$$\frac{P_1}{T_1} = \frac{P_2}{T_2}$$

Solving for P_2 gives

$$\frac{P_1 T_2}{T_1} = P_2$$

Substituting for each variable and simplifying, we have

$$\frac{15.0\text{ atm} \times 233\text{ K}}{298\text{ K}} = 11.7\text{ atm}$$

Practice Exercise

A copper container has a volume of 555 mL and is filled with air at 25 °C. The container is immersed in dry ice, and the pressure of the gas drops from 761 torr to 495 torr. What is the final temperature of the air in the copper container?

Answer: 194 K (–79 °C)

Concept Exercise

When air in a rigid steel tank cools from 25 °C to 20 C, the pressure inside the tank (increases/decreases).

Answer: See Appendix G.

OBJECTIVE

▸ To calculate the pressure, volume, or temperature for a gas after a change in conditions.

11.7 COMBINED GAS LAW

We began our discussion of gas laws with a simplifying assumption. In discussing Boyle's, Charles's, and Gay-Lussac's laws, we assumed that we could limit our treatment of gases to two variables. Experimentally, all three variables (pressure, volume, and temperature) usually change simultaneously. Now, we will bring all three variables together in a single expression. The resulting expression is called the **combined gas law**, and the equation is

$$\frac{P_1 V_1}{T_1} = \frac{P_2 V_2}{T_2}$$

Since the variables P, V, and T are proportional, we can solve combined gas law problems using the unit analysis approach. To calculate a final pressure (P_2), we apply a volume factor and a temperature factor to the initial pressure (P_1).

$$\boxed{P_1 \times V_{\text{factor}} \times T_{\text{factor}} = P_2}$$

To calculate a final volume (V_2), we apply a pressure factor and a temperature factor to the initial volume (V_1).

$$\boxed{V_1 \times P_{\text{factor}} \times T_{\text{factor}} = V_2}$$

To calculate a final temperature (T_2), we apply a pressure factor and a volume factor to the initial temperature (T_1).

$$\boxed{T_1 \times P_{\text{factor}} \times V_{\text{factor}} = T_2}$$

In a combined gas law problem, we have three variables at initial and final conditions. That gives us six pieces of data, and so it is helpful to use a table. We place the data in a table and organize the information as follows:

CONDITIONS	P	V	T
initial	P_1	V_1	T_1
final	P_2	V_2	T_2

Let's apply the combined gas law to 10.0 L of carbon dioxide gas at 300 K and 1.00 atm. When both the volume and the Kelvin temperature double, what is the final pressure in atmospheres? The initial conditions are 300 K and 1.00 atm. Doubling the temperature gives 600 K; doubling the volume gives 20.0 L. Now let's enter the data in a table:

CONDITIONS	P	V	T
initial	1.00 atm	10.0 L	300 K
final	P_2	20.0 L	600 K

We calculate the final pressure (P_2) by applying a V_{factor} and a T_{factor} to the initial pressure, 1.00 atm.

$$1.00 \text{ atm} \times V_{\text{factor}} \times T_{\text{factor}} = P_2$$

Since the volume of the gas increases from 10.0 L to 20.0 L, P_2 must decrease. The V_{factor} is less than 1 and the smaller value is placed in the numerator.

$$1.00 \text{ atm} \times \frac{10.0 \not{L}}{20.0 \not{L}} \times T_{\text{factor}} = P_2$$

The temperature of the gas increases from 300 K to 600 K; therefore, P_2 increases. The T_{factor} is greater than 1 and the larger value is placed in the numerator.

$$1.00 \text{ atm} \times \frac{10.0 \text{ L}}{20.0 \text{ L}} \times \frac{600 \text{ K}}{300 \text{ K}} = 1.00 \text{ atm}$$

Notice that the pressure did not change even though both the volume and the temperature each doubled. The pressure remained unchanged because pressure and volume are inversely proportional, while pressure and temperature are directly proportional. Thus, in effect the volume and temperature changes cancel and the pressure remains constant.

Standard Conditions of Temperature and Pressure

The **standard temperature and pressure** (symbol **STP**) for a gas are 0 °C and 1 atm. We could also express standard temperature as 273 K and standard pressure in other units such as 760 mm Hg, 760 torr, or 76 cm Hg. The following example exercise illustrates how to convert a set of conditions to STP conditions:

Example Exercise 11.6 Combined Gas Law

A nitrogen gas sample occupies 50.5 mL at –80 °C and 1250 torr. What is the volume at STP?

Conceptual Solution

Although the final conditions are not given, we know that STP conditions are 273 K and 760 torr. We can summarize the information as follows:

CONDITIONS	P	V	T
initial	1250 torr	50.5 mL	−80 °C + 273 = 193 K
final	760 torr	V_2	273 K

We can calculate the final volume by applying a P_{factor} and a T_{factor} to the initial volume.

$$V_1 \times P_{factor} \times T_{factor} = V_2$$

The pressure decreases, and so the volume increases; thus, the P_{factor} is greater than 1. The temperature increases, and so the volume increases; thus, the T_{factor} is also greater than 1.

$$50.5 \text{ mL} \times \frac{1250 \text{ torr}}{760 \text{ torr}} \times \frac{273 \text{ K}}{193 \text{ K}} = 117 \text{ mL}$$

We can visually summarize the combined gas law solution as follows:

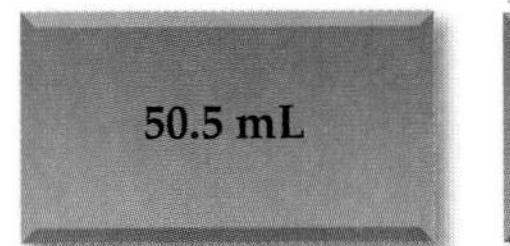

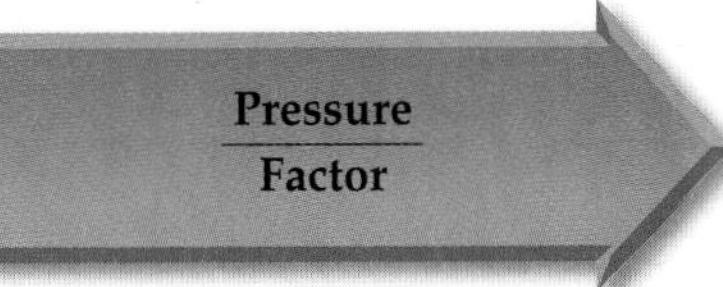

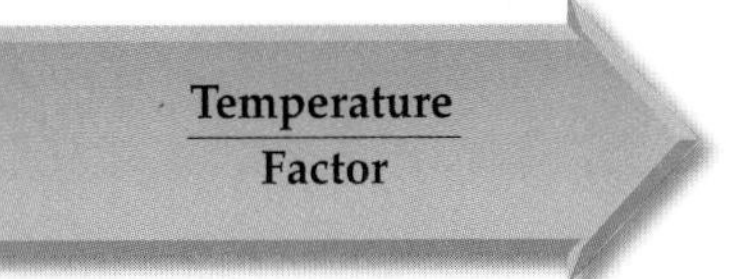

117 mL

Algebraic Solution

Alternatively, we can solve this problem using the equation

$$\frac{P_1 V_1}{T_1} = \frac{P_2 V_2}{T_2}$$

Rearranging variables and solving for V_2,

$$\frac{P_1 V_1 T_2}{T_1 P_2} = V_2$$

Substituting for each variable and simplifying, we obtain

$$\frac{1250 \text{ torr} \times 50.5 \text{ mL} \times 273 \text{ K}}{193 \text{ K} \times 760 \text{ torr}} = 117 \text{ mL}$$

Practice Exercise

An oxygen gas sample occupies 50.0 mL at 27 °C and 765 mm Hg. What is the final temperature if the gas is cooled to a volume of 35.5 mL and a pressure of 455 mm Hg?

Answer: 127 K (–146 °C)

Concept Exercise

A sample of air is at an initial pressure of 1.00 atm. If both the volume and Kelvin temperature each double, what is the final pressure?

Answer: See Appendix G.

OBJECTIVES

- To explain the concept of vapor pressure.
- To state the relationship between vapor pressure and temperature.

11.8 THE VAPOR PRESSURE CONCEPT

Everyone has observed that water in an open container evaporates. Vaporization occurs when molecules have enough energy to escape from the liquid. When the container is enclosed, the vapor molecules are trapped above the liquid. We should note that molecules continuously escape from the liquid into the vapor. Simultaneously, other molecules return from the vapor to the liquid. The **vapor pressure** is the pressure exerted by the gaseous vapor above a liquid when the rates of evaporation and condensation are equal (Figure 11.9).

The vapor pressure of a liquid can be determined with a mercury barometer (Figure 11.10). A drop of liquid is introduced at the bottom of the barometer. The less dense liquid droplet floats to the top of the mercury and partially vaporizes. The vaporized liquid exerts a gas pressure, thus driving the column of mercury down. *The decrease in the mercury level corresponds to the increase in vapor pressure.* If the mercury level drops from 760 mm to 740 mm, the vapor pressure of the liquid is recorded as 20 mm Hg.

We can observe that vapor pressure increases as the temperature increases. This is because molecules in the liquid evaporate faster and the vapor molecules have more kinetic energy. At 25 °C the vapor pressure of water is 23.8 mm Hg. At 50 °C the vapor pressure increases to 92.5 mm Hg. Table 11.2 lists selected values for the vapor pressure of water.

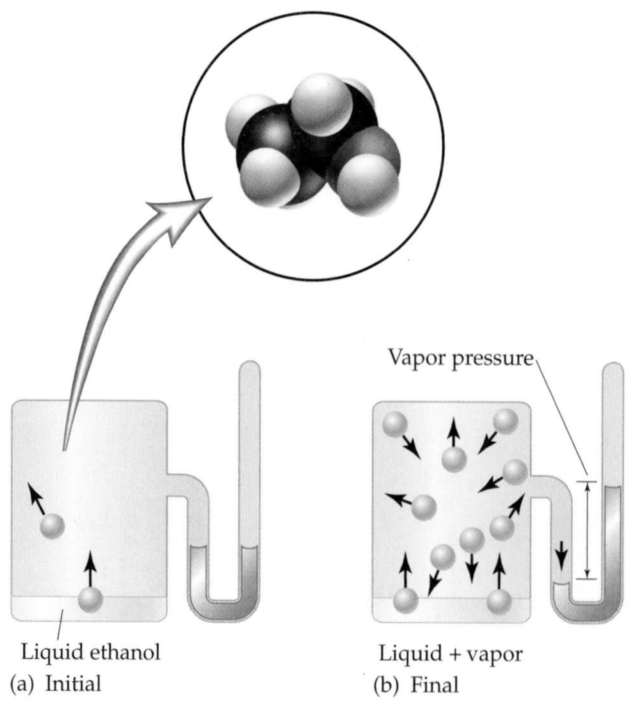

▶ **Figure 11.9 Vapor Pressure of a Liquid** (a) Initially, ethanol is placed in a sealed container and molecules begin to escape from the liquid. After awhile, some of the ethanol molecules in the vapor return to the liquid. (b) Eventually, the ethanol molecules are evaporating and condensing at the same rate. The vapor pressure is measured by the difference in the height of mercury in the side tube.

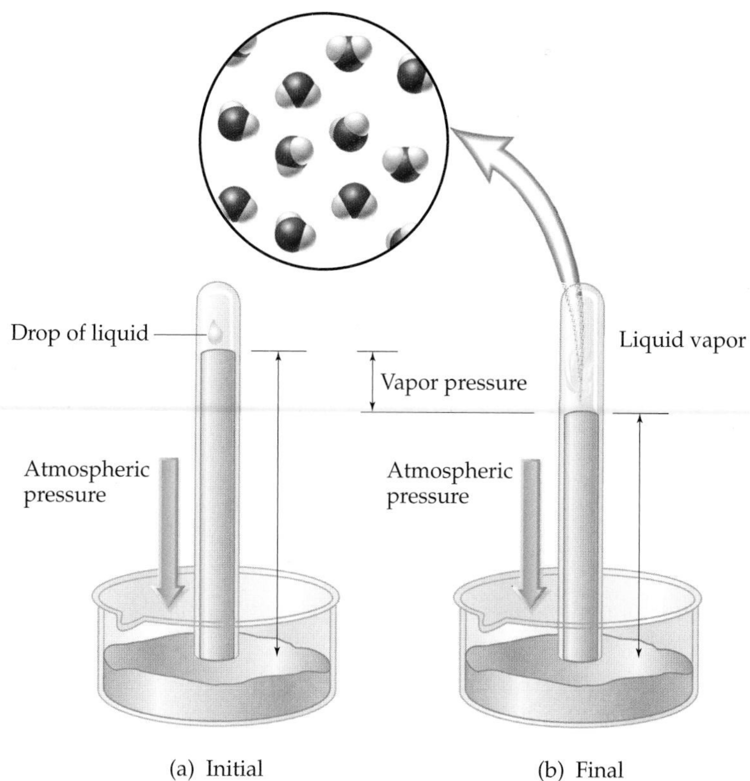

◀ **Figure 11.10 Measuring Vapor Pressure** (a) Initially, a single drop of water is introduced into the tube of mercury. As molecules of the liquid evaporate, the gaseous vapor molecules exert a pressure on the column of mercury. (b) Eventually, molecules are evaporating and condensing at the same rate. The difference in the height of the mercury before and after a drop of liquid is added is the vapor pressure of the liquid at the given temperature.

TABLE 11.2 VAPOR PRESSURE OF WATER

TEMPERATURE °C	PRESSURE *mm Hg*	TEMPERATURE °C	PRESSURE *mm Hg*
5	6.5	55	118.0
10	9.2	60	149.4
15	12.8	65	187.5
20	17.5	70	233.7
25	23.8	75	289.1
30	31.8	80	355.1
35	41.2	85	433.6
40	55.3	90	525.8
45	71.9	95	633.9
50	92.5	100	760.0

11.9 DALTON'S LAW OF PARTIAL PRESSURES

OBJECTIVE

▶ To apply Dalton's law of partial pressures to a mixture of gases.

When John Dalton proposed the atomic theory, his evidence was largely based on experiments with gases. His knowledge of gases was related to his interest in meteorology, and in fact he kept daily records of the weather throughout his life. In 1801, Dalton proposed that each gas in a mixture of gases contributes to the total pressure of the mixture. He then formulated a law, now called **Dalton's law of partial pressures**, that states that the total pressure of a gaseous mixture is equal to the sum of the individual pressures of each gas. We can express this law as follows:

$$P_1 + P_2 + P_3 + \ldots = P_{total}$$

The pressure exerted by each gas in a mixture of gases is called the **partial pressure**. In the equation, P_1, P_2, and P_3 represent the partial pressures of each of the gases in the mixture; P_{total} symbolizes the total of all the partial pressures.

In 1952, Stanley Miller conducted a historic experiment at the University of Chicago. In an attempt to prove the chemical evolution of life, Miller attempted to simulate Earth's primordial atmosphere. He placed hydrogen, ammonia, methane, and water vapor in a glass container and energized the mixture with ultraviolet light and electric sparks. After a week, Miller analyzed the contents of the glass container. In the "primordial soup," he found amino acids and nucleic acids—the basic building blocks of life! Biologists frequently cite this famous experiment as evidence for the chemical evolution of life. Miller won a Nobel prize for his work.

We can use Stanley Miller's gas mixture to explore Dalton's law in more detail. Using a mixture of gases, we can write an expression for the total pressure of the gaseous mixture.

$$P_{\text{hydrogen}} + P_{\text{ammonia}} + P_{\text{methane}} + P_{\text{water vapor}} = P_{\text{total}}$$

If the partial pressures of hydrogen, ammonia, and methane were 275 torr, 125 torr, and 340 torr, respectively, what was the partial pressure of the water vapor? Assuming the pressure in the container was standard pressure, 760 torr, we have

$$275 \text{ torr} + 125 \text{ torr} + 340 \text{ torr} + P_{\text{water vapor}} = 760 \text{ torr}$$

Subtracting the partial pressures of hydrogen, ammonia, and methane from each side of the equation, we have

$$P_{\text{water vapor}} = 760 \text{ torr} - (275 \text{ torr} + 125 \text{ torr} + 340 \text{ torr})$$

$$P_{\text{water vapor}} = 760 \text{ torr} - 740 \text{ torr} = 20 \text{ torr}$$

The following example exercise illustrates an application of Dalton's law:

▲ **Primordial Soup** Stanley Miller placed H_2, NH_3, CH_4, and H_2O vapor in his apparatus and energized the gaseous mixture with ultraviolet light and electric sparks.

Example Exercise 11.7 Dalton's Law of Partial Pressures

An atmospheric sample contains nitrogen, oxygen, argon, and traces of other gases. If the partial pressure of nitrogen is 587 mm Hg, oxygen is 158 mm Hg, and argon is 7 mm Hg, what is the observed pressure as read on the barometer?

Solution

The sum of the individual partial pressures equals the total atmospheric pressure; therefore,

$$P_{\text{nitrogen}} + P_{\text{oxygen}} + P_{\text{argon}} = P_{\text{total}}$$

Substituting the values for the partial gas pressures, we have

$$587 \text{ mm Hg} + 158 \text{ mm Hg} + 7 \text{ mm Hg} = 752 \text{ mm Hg}$$

Thus, the atmospheric pressure as read on the barometer is 752 mm Hg.

Practice Exercise

The regulator on a steel scuba tank containing compressed air indicates that the pressure is 2250 psi. If the partial pressure of nitrogen is 1755 psi and that of argon is 22 psi, what is the partial pressure of oxygen in the tank?

Answer: 473 psi

Concept Exercise

A rigid steel cylinder contains N_2, O_2, and NO at a total pressure of 2.50 atm. What is the partial pressure of NO gas if N_2 and O_2 are each 1.00 atm?

Answer: See Appendix G.

Collecting a Gas over Water

One way of measuring the volume of a gas is to measure the amount of water it displaces. This method is called determining **volume by displacement**. The volume of gas equals the volume of water displaced.

Let's consider a laboratory experiment that produces hydrogen gas from the reaction of zinc metal and sulfuric acid.

$$Zn(s) + H_2SO_4(aq) \rightarrow ZnSO_4(aq) + H_2(g)$$

To collect the hydrogen gas produced in a laboratory experiment, we can invert a graduated cylinder filled with water over the zinc metal. Figure 11.11 shows the experimental apparatus.

When the reaction is complete, the pressure exerted by the hydrogen gas inside the graduated cylinder is balanced by the atmospheric pressure outside. That is, the hydrogen gas pressure on the water inside the graduated cylinder is equal to the atmospheric pressure on the surface of the solution in the beaker.

However, there are actually two gases inside the graduated cylinder. Since the hydrogen gas is collected over an aqueous solution, water vapor is also present. We can use Dalton's law to determine the partial pressure of the hydrogen gas.

$$P_{\text{hydrogen}} + P_{\text{water vapor}} = P_{\text{atmosphere}}$$

If the reaction takes place at 25 °C and the barometer reads 767 mm Hg, what is the partial pressure of the hydrogen gas? We refer to Table 11.2 and find that the vapor pressure of water is 23.8 mm Hg at 25 °C. Then we calculate the pressure of the hydrogen gas.

$$P_{hydrogen} + 23.8 \text{ mm Hg} = 767 \text{ mm Hg}$$

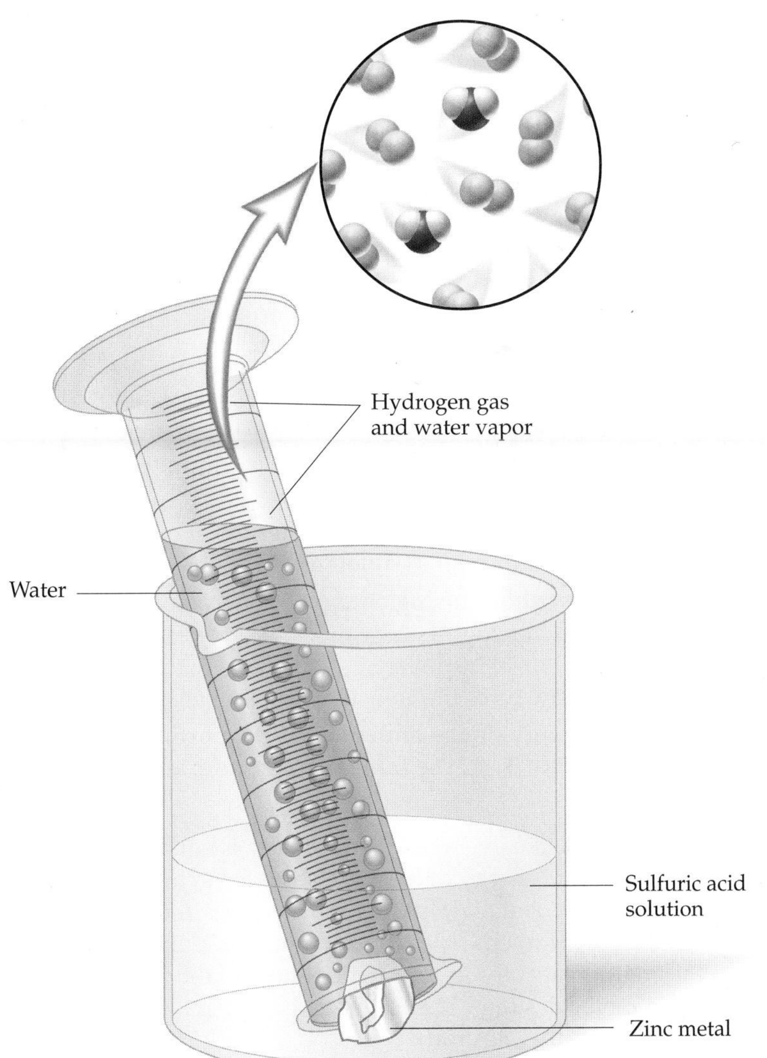

◀ **Figure 11.11 Collecting a Gas over Water** Zinc metal reacts with sulfuric acid to give bubbles of hydrogen gas. A graduated cylinder full of water is placed over the metal to collect the gas bubbles. The volume of water displaced from the graduated cylinder equals the volume of hydrogen gas liberated from the acid.

After rearranging, we have

$$P_{hydrogen} = 767 \text{ mm Hg} - 23.8 \text{ mm Hg} = 743 \text{ mm Hg}$$

Note When a gas is collected over water it is often called a "wet" gas. That is, the collected gas contains water vapor. A gas that does not contain water vapor is sometimes referred to as a "dry" gas.

11.10 IDEAL GAS BEHAVIOR

OBJECTIVES

- To list five characteristics of an ideal gas according to the kinetic theory of gases.
- To determine the value of absolute zero from a graph of volume or pressure versus temperature.

By the early 1800s, experiments had provided a lot of information about the gaseous state. There was, however, no clear way to interpret this information. By 1850, scientists had enough data to begin to construct a theory to explain the behavior of gases. The British physicist James Joule (1818–1889) suggested that the temperature of a gas is related to the energy of its molecules. Joule proposed that temperature and the motion of molecules in a gas are related. At higher temperatures, gas molecules move faster, and at lower temperatures, gas molecules move slower.

Kinetic Theory of Gases

In the period from 1850 to 1870, scientists attempted to construct a model that would explain the behavior of individual gas molecules. They reasoned that given a theoretical model, they would be able to explain the behavior of real gases. This model would also allow predictions of ideal gas behavior. An **ideal gas** is a gas that always behaves in a consistent and predictable manner. Experimentally, a **real gas**, such as hydrogen or oxygen, does not behave ideally under all conditions. Specifically, a real gas does not behave ideally at low temperatures and high pressures.

In due time, the behavior of an ideal gas was described by a model called the kinetic molecular theory, or simply the **kinetic theory** of gases. According to the kinetic theory, an ideal gas has the following characteristics:

1. **Gases are made up of tiny molecules.** The size of gas molecules is tiny, and the distance between molecules is quite large. Therefore, gases are mostly empty space. Gas molecules occupy zero volume in an ideal gas.
2. **Gas molecules demonstrate rapid motion, move in straight lines, and travel in random directions.**
3. **Gas molecules have no attraction for one another.** After colliding with each other, molecules simply bounce off in different directions.
4. **Gas molecules collide without losing energy.** Two gas molecules are said to have an **elastic collision**. If a high-energy molecule strikes a less energetic molecule, part of the energy can be transferred. However, the total kinetic energy of both molecules, before and after the collision, does not change.
5. **The average kinetic energy of gas molecules is proportional to the Kelvin temperature that is, KE $\propto$ T.** At the same temperature, all gas molecules have equal kinetic energy.

Even for different gases, the average kinetic energy is equal at the same temperature. For example, the kinetic energy of hydrogen and oxygen gases is equal at 25 °C. Although the kinetic energy is equal, hydrogen molecules move faster than oxygen molecules because they are lighter. Given the equation for kinetic energy, $KE = \frac{1}{2}mv^2$, we can explain this mathematically. If the kinetic energy of the two gases is equal, then the lighter hydrogen molecules ($m = 2$ amu) must have a greater velocity (v) than the heavier oxygen molecules ($m = 32$ amu).

At higher temperatures, molecules have more kinetic energy, move faster, and collide more frequently. At lower temperatures, molecules have less kinetic energy, move slower, and collide less frequently. The following example exercise further illustrates the characteristics of an ideal gas according to the kinetic theory:

Example Exercise 11.8 Ideal Gas Behavior

Suppose we have two 5.00 L samples of gas at 25 °C. One sample is ammonia, NH_3, and the other nitrogen dioxide, NO_2. Which gas has the greater kinetic energy? Which gas has the faster molecules?

Solution

Since the temperature of each gas is 25 °C, we know that the kinetic energy is the same for NH_3 and NO_2. At the same temperature, we know that lighter molecules move faster than heavier molecules. The molecular mass of NH_3 is 17 amu and NO_2 is 46 amu. Since NH_3 is lighter than NO_2, the ammonia molecules have a higher velocity than the nitrogen dioxide molecules.

Practice Exercise

Which of the following statements is *not* true according to the kinetic theory of gases?

(a) Molecules occupy a negligible volume.
(b) Molecules move in straight-line paths.
(c) Molecules are attracted to each other.
(d) Molecules undergo elastic collisions.
(e) Molecules of different gases at the same temperature have the same average kinetic energy.

Answer: All these statements are true except (c). Molecules of an ideal gas are not attracted to each other and behave as independent particles.

Concept Exercise

Which of the following is an example of an ideal gas: H_2, O_2, or He?

Answer: See Appendix G.

Absolute Zero

The temperature at which the pressure and volume of a gas theoretically reach zero is referred to as **absolute zero**. Absolute zero is the coldest possible temperature and corresponds to –273 °C, or 0 K. An ideal gas at absolute zero has no kinetic energy, and therefore no molecular motion.

How do we determine the value of absolute zero if –273 °C is impossible to attain experimentally? We determine absolute zero from the data for the volume of a gas over a range of temperatures above 0 K. When we plot the volume and temperature data points and extend the graph to zero volume, we have the temperature corresponding to absolute zero. Figure 11.12 illustrates an extrapolated value for absolute zero.

We can also determine absolute zero by obtaining data for the pressure and temperature of a gas. We can plot the pressure and temperature data and extrapolate the graph to absolute zero. Figure 11.13 illustrates the determination of absolute zero.

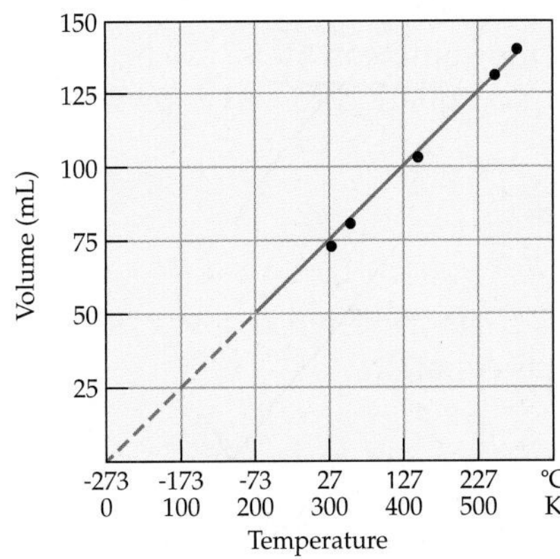

◀ **Figure 11.12 Absolute Zero at Zero Volume** As the temperature of the gas decreases, the volume decreases. As an ideal gas approaches zero volume, the temperature approaches absolute zero.

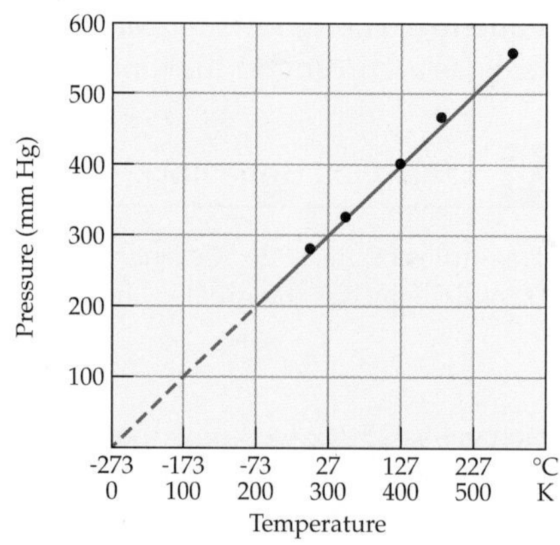

▲ **Figure 11.13 Absolute Zero at Zero Pressure** As the temperature of the gas decreases, the pressure decreases. As an ideal gas approaches zero pressure, the temperature approaches absolute zero.

OBJECTIVE

▶ To calculate the pressure, volume, temperature, or moles of gas from the ideal gas equation.

11.11 IDEAL GAS LAW

In Section 11.3 we learned that the pressure (P) of a gas is inversely proportional to the volume (V). We also learned that pressure is directly proportional to the number of molecules, or moles of gas (n), and directly proportional to the Kelvin temperature (T). We can therefore write a relationship that pressure is proportional to the moles of gas, multiplied by the temperature, divided by the volume.

$$P \propto \frac{nT}{V}$$

By introducing the proportionality constant (R), we can write this relationship as an equation.

$$P = \frac{RnT}{V}$$

After rearranging,

$$PV = nRT$$

This equation is called the **ideal gas law**. The constant R is the **ideal gas constant** and can be expressed as 0.0821 L • atm/mol • K. To use this value of R in a gas law calculation, the pressure must be expressed in atmospheres, the volume in liters, and the temperature in Kelvin. The ideal gas law is a powerful equation that summarizes the behavior of gases. That is, any change in the pressure, volume, temperature, or amount of a gas (moles of gas) is related through this single equation.

Let's try a calculation using the ideal gas law. If a sealed container holds 1.10 mol of nitrogen gas at 25 °C and 3.75 atm, what is the volume in liters? We need to find volume, so we begin by rearranging the ideal gas equation to solve for volume (V). If $PV = nRT$, then

$$V = \frac{nRT}{P}$$

To convert the temperature to Kelvin, we add 273 to 25 °C which gives 298 K. Now, we can substitute for each variable in the ideal gas equation.

$$V = \frac{1.10\ \cancel{\text{mol}} \times \dfrac{0.0821\ \text{L} \cdot \cancel{\text{atm}}}{1\ \cancel{\text{mol}} \cdot \cancel{\text{K}}} \times 298\ \cancel{\text{K}}}{375\ \cancel{\text{atm}}} = 7.18\ \text{L}$$

CRITICAL THINKING Conceptualizing Gases

Q: Can you describe the distribution of helium atoms in a steel tank containing helium gas at 200 °C, –200 °C, and –273 °C?

One of the first studies to offer evidence that most students can solve traditional gas law problems, but many students find it difficult to describe the behavior of gases, was *Concept Learning versus Problem Solving: Is There a Difference.*[1] Research carried out at the University of Wisconsin, Stout, and the University of Missouri, Kansas City, found that most chemistry students could correctly answer a traditional gas law problem, but only one-third gave a correct response to the corresponding conceptual question.

Three years later, in *Concept Learning versus Problem Solving: Revisited,*[2] a more detailed study at the University of California, San Diego, confirmed the research. At UCSD, a study of 285 freshman chemistry students found that nearly 90% answered a traditional gas law problem correctly, but only about 30% of the students gave a correct response to a related conceptual question.

Conceptual Questions

Now, let's see if you can answer three conceptual questions about a gas at different temperatures. The following diagram represents a steel tank filled with helium gas at 20 °C and 1 atm pressure. The dots represent helium atoms in the steel container.

Initial Conditions:

Diagram of a steel tank with helium gas at 20 °C and 1 atm pressure.

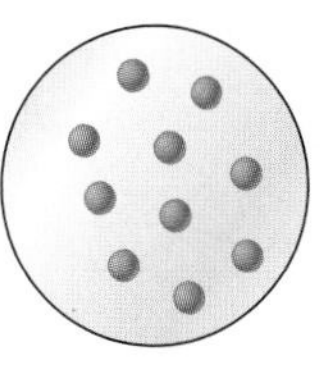

Which of the following diagrams represents the steel container filled with helium gas heated to 200 °C?

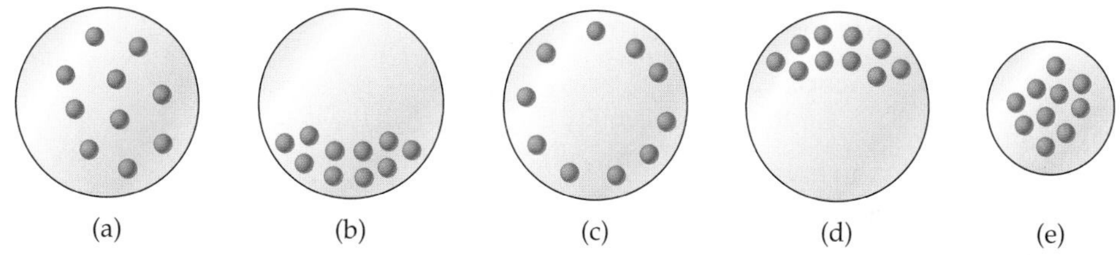

Which of the following diagrams represents the steel container filled with helium gas cooled to –200 °C? (The boiling point of helium is –269 °C.)

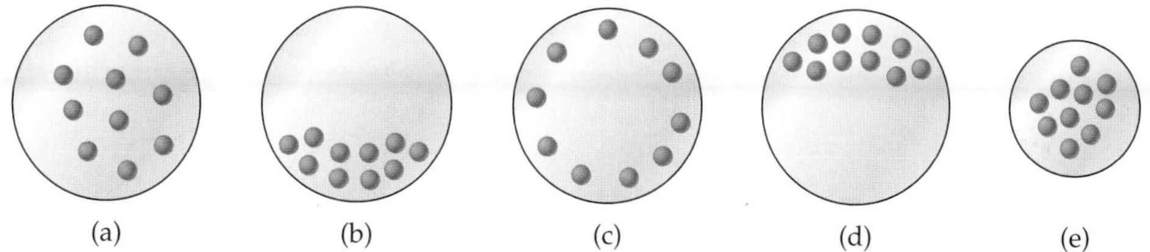

Which of the following diagrams represents the steel container filled with helium gas cooled to –273 °C? (Recall the concept of absolute zero.)

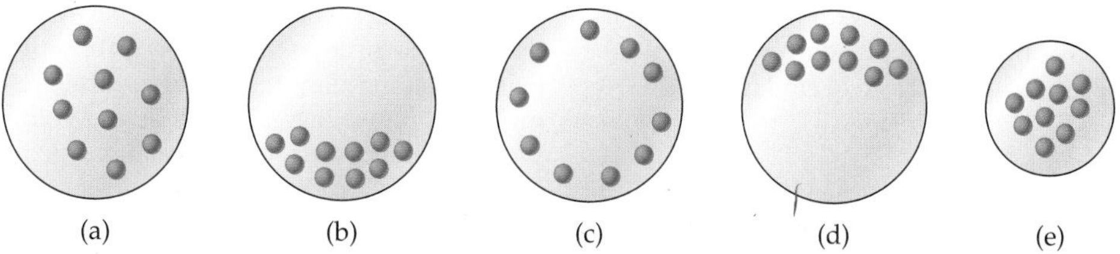

[1]*Journal of Chemical Education*, 64, 508–510.

[2]*Journal of Chemical Education*, 67, 253–254.

A: The helium atoms are randomly distributed in the steel container at 200 °C, and –200 °C as shown in (a). At absolute zero, –273 °C, the gas condenses to a liquid and freezes to a solid as shown in (b). Since the tank is steel, the volume of the gas cannot change as shown in (e).

The following example exercise further illustrates the ideal gas law:

Example Exercise 11.9 Ideal Gas Law

How many moles of hydrogen gas occupy a volume of 0.500 L at STP?

Solution

We begin by rearranging the ideal gas equation, $PV = nRT$, and solving for n.

$$n = \frac{PV}{RT}$$

The temperature at STP is 273 K, and the pressure is 1 atm. Then we substitute for each variable in the ideal gas equation.

$$n = \frac{1\ \cancel{\text{atm}} \times 0.500\ \cancel{\text{L}}}{\dfrac{0.0821\ \cancel{\text{L}} \cdot \cancel{\text{atm}}}{1\ \text{mol} \cdot \cancel{\text{K}}} \times 273\ \cancel{\text{K}}} = 0.0223\ \text{mol}$$

Alternatively, we can display the value for the ideal gas constant separately to more clearly show the cancellation of units.

$$n = \frac{1\ \cancel{\text{atm}} \times 0.500\ \cancel{\text{L}}}{273\ \cancel{\text{K}}} \times \frac{1\ \text{mol} \cdot \cancel{\text{K}}}{0.0821\ \cancel{\text{L}} \cdot \cancel{\text{atm}}} = 0.0223\ \text{mol}$$

Practice Exercise

What is the temperature of 0.250 mol of chlorine gas at 655 torr if the volume is 3.50 L?

Answer: 147 K (–126 °C)

Concept Exercise

Solve for the proportionality constant (R) in the ideal gas law.

Answer: See Appendix G.

Summary

Section 11.1 The shape and volume of a gas are variable and occupy the entire volume of a sealed container. A gas can expand and compress, and the gaseous state has a density much less than that of the liquid state. A gas mixes completely with other gases in the same container.

Section 11.2 **Gas pressure** results from gas molecules striking the inside surface of their container. The pressure exerted by a gas depends on how often and how hard the molecules strike the container. In a **vacuum**, there are no gas molecules and the pressure is zero. **Atmospheric pressure** results from molecules in the air and is measured with a **barometer**. Standard pressure can be expressed as 1 atm, 760 mm Hg, 760 **torr**, 76 cm Hg, 29.9 in. Hg, 14.7 psi, or 101 kPa.

Section 11.3 We can increase gas pressure by changing one of three variables. First, if we *decrease the volume*, molecules will collide more frequently with the inner surface of the container and the pressure will increase. Second, if we *increase the temperature*, molecules will collide more frequently with the walls of the container and the pressure will increase. Third, if we *increase the number of molecules*, there will be more collisions against the walls of the container and the pressure will increase.

Section 11.4 The relationship between the pressure and the volume of a gas is **inversely proportional**. In solving problems involving **Boyle's law**, the pressure increases when the volume decreases, while the temperature remains constant.

Section 11.5 The relationship between the volume and the temperature of a gas is **directly proportional**. In solving problems involving **Charles's law**, the volume increases when the temperature increases, while the pressure remains constant.

Section 11.6 The relationship between the pressure and the temperature of a gas is directly proportional. In solving problems involving **Gay-Lussac's law**, the pressure increases when the temperature increases, while the volume remains constant. We can summarize all the relationships for the gas laws as shown in Table 11.3.

Section 11.7 **Standard temperature and pressure** (**STP**) is 273 K and exactly 1 atm. In solving problems involving the **combined gas law**, simultaneous changes occur in pressure, volume, and temperature.

Section 11.8 **Vapor pressure** is the pressure exerted by gaseous vapor above a liquid in a sealed container, when the

TABLE 11.3 SUMMARY OF GAS LAW VARIABLES

GAS LAW	PRESSURE	VOLUME	TEMPERATURE
Boyle's	increases	decreases	*constant*
	decreases	increases	*constant*
Charles's	*constant*	increases	increases
	constant	decreases	decreases
Gay-Lussac's	increases	*constant*	increases
	decreases	*constant*	decreases

rates of evaporation and condensation are equal. A familiar example is the pressure of H_2O molecules in the vapor above water in a sealed bottle. When the temperature increases, the vapor pressure increases. Conversely, when the temperature decreases, the vapor pressure decreases.

Section 11.9 **Dalton's law of partial pressures** states that the total pressure of gases in a sealed container is equal to the sum of the **partial pressure** of each gas in the mixture. We can collect a gas over water to determine the **volume by displacement**. A "wet" gas collected over water contains water vapor. The partial pressure of the gas is found by subtracting the vapor pressure of water from the total pressure.

Section 11.10 According to the **kinetic theory**, a gas is made up of individual molecules behaving ideally. Molecules in an **ideal gas** move about rapidly and randomly in straight-line paths. Gas molecules show no attraction for one another and undergo **elastic collisions**. The kinetic energy is directly proportional to the Kelvin temperature. At **absolute zero** (0 K), the pressure of an ideal gas is zero. The velocity of gas molecules increases with the temperature and decreases with the molecular mass. The behavior of a **real gas** can deviate substantially at low temperatures and high pressures from that predicted by the kinetic theory.

Section 11.11 A more mathematical method for solving gas problems involves the **ideal gas law**. In the equation $PV = nRT$, the **ideal gas constant** (symbol $\mathbf{R}$) has a value of 0.0821 L • atm/mol • K.

PROBLEM–SOLVING ORGANIZER

TOPIC	PROCEDURE	EXAMPLE
Properties of Gases Sec. 11.1	Predict the shape and volume of a gas.	The shape and volume of argon gas are variable.
Atmospheric Pressure Sec. 11.2	Express a given gas pressure in units of atm, mm Hg, torr, cm Hg, in. Hg, psi, or kPa.	If argon gas is at 655 mm Hg, what is the pressure in atm? $655\ \cancel{\text{mm Hg}} \times \frac{1\ \text{atm}}{760\ \cancel{\text{mm Hg}}} = 0.862\ \text{atm}$
Variables Affecting Gas Pressure Sec. 11.3	Predict an increase or decrease in gas pressure for a change in volume, temperature, or number of molecules.	An increase in which variable causes a decrease in pressure? Answer: *volume*
Boyle's Law Sec. 11.4	Calculate a final pressure after a change in volume (temperature constant). Pressure and volume are inversely proportional.	If a gas is at 1.00 atm, what is the pressure when the volume decreases from 5.00 L to 2.00 L? $1.00\ \text{atm} \times \frac{5.00\ \cancel{\text{L}}}{2.00\ \cancel{\text{L}}} = 2.50\ \text{atm}$
Charles's Law Sec. 11.5	Calculate a final volume after a change in temperature (pressure constant). Volume and temperature are directly proportional.	If a gas occupies 50.0 mL, what is the volume when the gas is heated from 300 K to 450 K? $50.0\ \text{mL} \times \frac{450\ \cancel{\text{K}}}{300\ \cancel{\text{K}}} = 75.0\ \text{mL}$
Gay-Lussac's Law Sec. 11.6	Calculate a final pressure after a change in temperature (volume constant). Pressure and temperature are directly proportional.	If a gas is at 1.00 atm, what is the pressure when the gas is heated from 300 K to 450 K? $1.00\ \text{atm} \times \frac{450\ \cancel{\text{K}}}{300\ \cancel{\text{K}}} = 1.50\ \text{atm}$

PROBLEM–SOLVING ORGANIZER *(Continued)*

TOPIC	PROCEDURE	EXAMPLE
Combined Gas Law Sec. 11.7	Calculate a final volume after a change in pressure and temperature. Volume and pressure are inversely proportional; volume and temperature are directly proportional.	If 50.0 mL of gas is at 1.50 atm and 450 K, what is the volume at 1.00 atm and 273 K? $50.0 \text{ mL} \times \frac{1.50 \text{ atm}}{1.00 \text{ atm}} \times \frac{273 \text{ K}}{450 \text{ K}} = 45.5 \text{ mL}$
Dalton's Law Sec. 11.9	Calculate the total pressure of a gas mixture by summing the individual partial pressures.	If a gas sample contains He at 0.25 atm and Ar at 1.50 atm, what is the total pressure? $P_1 + P_2 + P_3 + \ldots = P_{total}$ 0.25 atm + 1.50 atm = 1.75 atm

Key Terms *Answers to Key Terms are in Appendix H.*

Select the key term that corresponds to each of the following definitions.

_____ **1.** a measure of the frequency and energy of molecules colliding against the walls of a container
_____ **2.** a volume of space that does not contain gas molecules
_____ **3.** the pressure exerted by the molecules in air
_____ **4.** an instrument for measuring atmospheric pressure
_____ **5.** a unit of pressure equal to 1 mm Hg
_____ **6.** the statement that equal volumes of gases, under the same conditions of temperature and pressure, contain equal numbers of molecules
_____ **7.** a relationship between two variables such that when one variable doubles, the other variable halves
_____ **8.** a relationship between two variables such that when one variable doubles, the other variable doubles
_____ **9.** the statement that the pressure and volume of a gas are inversely proportional at constant temperature
_____ **10.** the statement that the volume and the Kelvin temperature of a gas are directly proportional at constant pressure
_____ **11.** the statement that the pressure and the Kelvin temperature of a gas are directly proportional at constant volume
_____ **12.** the statement that the pressure exerted by a gas is inversely proportional to its volume and directly proportional to its Kelvin temperature
_____ **13.** the conditions of a gas at 273 K and 760 mm Hg
_____ **14.** the pressure exerted by gaseous molecules above a liquid in a sealed container when the rates of evaporation and condensation are equal
_____ **15.** the statement that the pressure exerted by a mixture of gases is equal to the sum of the individual pressures exerted by each gas
_____ **16.** the pressure exerted by an individual gas in a mixture of two or more gases
_____ **17.** a technique for determining the volume of gas from the volume of water it displaces
_____ **18.** a theoretical description of gas molecules demonstrating ideal behavior
_____ **19.** a theoretical gas that obeys the kinetic theory under all conditions
_____ **20.** an actual gas that deviates from ideal behavior under certain conditions
_____ **21.** an impact between gas molecules with no change in total energy
_____ **22.** the theoretical temperature at which the kinetic energy of a gas is zero
_____ **23.** the principle stated by the relationship $PV = nRT$
_____ **24.** the proportionality constant R in the equation $PV = nRT$

(a) absolute zero *(Sec. 11.10)*
(b) atmospheric pressure *(Sec. 11.2)*
(c) Avogadro's theory *(Sec. 11.3)*
(d) barometer *(Sec. 11.2)*
(e) Boyle's law *(Sec. 11.4)*
(f) Charles's law *(Sec. 11.5)*
(g) combined gas law *(Sec. 11.7)*
(h) Dalton's law of partial pressures *(Sec. 11.9)*
(i) directly proportional *(Sec. 11.5)*
(j) elastic collision *(Sec. 11.10)*
(k) gas pressure *(Sec. 11.2)*
(l) Gay-Lussac's law *(Sec. 11.6)*
(m) ideal gas *(Sec. 11.10)*
(n) ideal gas constant *(Sec. 11.11)*
(o) ideal gas law *(Sec. 11.11)*
(p) inversely proportional *(Sec. 11.4)*
(q) kinetic theory *(Sec. 11.10)*
(r) partial pressure *(Sec. 11.9)*
(s) real gas *(Sec. 11.10)*
(t) standard temperature and pressure *(Sec. 11.7)*
(u) torr *(Sec. 11.2)*
(v) vacuum *(Sec. 11.2)*
(w) vapor pressure *(Sec. 11.8)*
(x) volume by displacement *(Sec. 11.9)*

Exercises *Answers to odd-numbered Exercises are in Appendix I.*

Properties of Gases (Sec. 11.1)

1. Which of the following are observed properties of gases?
   (a) variable shape
   (b) expand uniformly
   (c) compress infinitely
   (d) mix with other gases
2. Which of the following are observed properties of gases?
   (a) variable volume
   (b) expand infinitely
   (c) compress uniformly
   (d) low density

Atmospheric Pressure (Sec. 11.2)

3. Give the value for standard atmospheric pressure in each of the following units:
   (a) millimeters of mercury
   (b) centimeters of mercury
4. Give the value for standard atmospheric pressure in each of the following units:
   (a) inches of mercury
   (b) pounds per square inch

◀ **The Atmosphere** The atmospheric pressure progressively decreases at higher altitudes. By way of example, the atmospheric pressure on Mt. Everest is about one-third the atmospheric pressure at sea level.

5. If oxygen gas in a steel container is at a pressure of 5.00 atm, what is the pressure expressed in each of the following units?
   (a) psi **(b)** cm Hg
6. If a piston in an automobile engine compresses a fuel–air mixture to a pressure of 9.50 atm, what is the pressure expressed in each of the following units?
   (a) in. Hg **(b)** mm Hg
7. An American newscast states that the barometer reads 30.8 in. Hg. Express the atmospheric pressure in each of the following units:
   (a) kPa **(b)** torr
8. A Canadian newscast states that the barometer reads 104 kPa. Express the atmospheric pressure in each of the following units:
   (a) psi **(b)** cm Hg

Variables Affecting Gas Pressure (Sec. 11.3)

9. Indicate what happens to the pressure of a gas with the following changes:
   (a) The volume increases.
   (b) The temperature increases.
   (c) The moles of gas increase.
10. Indicate what happens to the pressure of a gas with the following changes:
    (a) The volume decreases.
    (b) The temperature decreases.
    (c) The moles of gas decrease.
11. State whether the pressure of a gas in a sealed container increases or decreases with the following changes:
    (a) The volume changes from 2.50 L to 5.00 L.
    (b) The temperature changes from 20 °C to 100 °C.
    (c) The moles of gas change from 0.500 mol to 0.250 mol.
12. State whether the pressure of a gas in a sealed container increases or decreases with the following changes:
    (a) The volume changes from 75.0 mL to 50.0 mL.
    (b) The temperature changes from 0 °C to –195 °C.
    (c) The moles of gas change from 1.00 mol to 5.00 mol.
13. Explain how increasing the volume of a gas decreases its pressure.
14. Explain how increasing the temperature of a gas increases its pressure.

◀ **Tire Pressure** The pressure inside the bicycle tire increases as the number of air molecules increases.

Boyle's Law: Pressure–Volume Relationships (Sec. 11.4)

15. Sketch a graph of pressure versus volume, assuming temperature is constant. Label the vertical axis P and the horizontal axis V.
16. Sketch a graph of pressure versus inverse volume, assuming temperature is constant. Label the vertical axis P and the horizontal axis 1/V.
17. A sample of air at 5.00 atm expands from 1.75 L to 2.50 L. If the temperature remains constant, what is the final pressure in atm?

18. A sample of air at 15.0 psi compresses from 555 mL to 275 mL. If the temperature remains constant, what is the final pressure in psi?
19. Calculate the final volume when 5.00 L of argon gas undergoes a pressure change from 1.55 atm to 6.50 atm. Assume that the temperature remains constant.
20. Calculate the final volume when 125 mL of argon gas undergoes a pressure change from 705 mm Hg to 385 mm Hg. Assume that the temperature remains constant.

Charles's Law: Volume–Temperature Relationships (Sec. 11.5)

21. Sketch a graph of volume versus Kelvin temperature, assuming pressure is constant. Label the vertical axis V and the horizontal axis T, K.
22. Sketch a graph of volume versus Celsius temperature, assuming pressure is constant. Label the vertical axis V and the horizontal axis T, °C. Assume the Celsius temperature approaches zero at the origin.
23. A 555 mL sample of nitrous oxide at 25 °C is heated to 50 °C. If the pressure remains constant, what is the final volume?
24. A 2.50 L sample of nitric oxide gas at 100 °C is cooled to 20 °C. If the pressure remains constant, what is the final volume?
25. Calculate the final Celsius temperature when 125 mL of chlorine gas at 25 °C is heated to give a volume of 175 mL. Assume that the pressure remains constant.
26. Calculate the final Celsius temperature when 5.00 L of fluorine gas at 50 °C is cooled to give a volume of 3.75 L. Assume that the pressure remains constant.

Gay-Lussac's Law: Pressure–Temperature Relationships (Sec. 11.6)

27. Sketch a graph of pressure versus Kelvin temperature, assuming volume is constant. Label the vertical axis P and the horizontal axis T (K).
28. Sketch a graph of pressure versus Celsius temperature, assuming volume is constant. Label the vertical axis P and the horizontal axis T (°C). Assume the Celsius temperature approaches zero at the origin.
29. A 1.00 L sample of argon gas at 1.00 atm is heated from 20 °C to 110 °C. If the volume remains constant, what is the final pressure?
30. A 25.0 mL sample of neon gas at 455 mm Hg is cooled from 100 °C to 10 °C. If the volume remains constant, what is the final pressure?
31. Calculate the final Celsius temperature of sulfur dioxide gas if 50.0 mL of the gas at 20 °C and 0.450 atm is heated until the pressure is 0.750 atm. Assume that the volume remains constant.
32. Calculate the final Celsius temperature of nitrogen dioxide gas if 1.95 L of the gas at 0 °C and 375 mm Hg is cooled until the pressure is 225 mm Hg. Assume that the volume remains constant.

Combined Gas Law (Sec. 11.7)

33. A 5.00 L sample of air is collected at 500 °C and 5.00 atm. What is the volume of air at STP?
34. A 100.0 mL sample of air is collected at 21 °C and 772 mm Hg. What is the volume of air at STP?
35. If a sample of hydrogen gas occupies 2.00 L at –50 °C and 155 mm Hg, what is the volume at 75 °C and 365 mm Hg?
36. If a sample of oxygen gas occupies 25.0 mL at –25 °C and 650 mm Hg, what is the volume at 25 °C and 350 mm Hg?
37. A sample of hydrogen chloride gas occupies 0.750 L at STP. What is the pressure in atm if the volume is 0.100 L at 25 °C?
38. A sample of hydrogen fluoride gas occupies 1250 mL at STP. What is the pressure in mm Hg if the volume is 255 mL at 300 °C?
39. A sample of nitrogen gas has a volume of 1.00 L at STP. What is the temperature in °C if the volume is 10.0 L at 2.00 atm?
40. A sample of ammonia gas has a volume of 50.0 mL at STP. What is the Celsius temperature if the volume is 350.0 mL at 350 mm Hg?
41. A sample of argon gas occupies 0.500 L at –125 °C and 225 mm Hg. Calculate the pressure in mm Hg if the gas occupies 0.375 L at 100 °C.
42. A sample of xenon gas occupies 500.0 mL at –185 °C and 75.0 cm Hg. Calculate the Celsius temperature if the volume is 225.0 mL at 55.0 cm Hg.

The Vapor Pressure Concept (Sec. 11.8)

43. What is the relationship between the vapor pressure of a liquid and temperature?
44. Which of the following liquids has the higher vapor pressure at 25 °C: water or mercury?
45. Refer to Table 11.2 and state the vapor pressure for water in mm Hg at each of the following temperatures:
    (a) 25 °C **(b)** 50 °C
46. Refer to Table 11.2 and state the vapor pressure for water in atm at each of the following temperatures:
    (a) 75 °C **(b)** 100 °C

Dalton's Law of Partial Pressures (Sec. 11.9)

47. Air contains nitrogen, oxygen, argon, and trace gases. If the partial pressure of nitrogen is 592 mm Hg, oxygen is 160 mm Hg, argon is 7 mm Hg, and trace gas is 1 mm Hg, what is the atmospheric pressure?
48. Air contains nitrogen, oxygen, argon, and trace gases. If the partial pressure of nitrogen is 587 mm Hg, oxygen is 158 mm Hg, argon is 7 mm Hg, and trace gas pressure is negligible, what is the atmospheric pressure?
49. A steel cylinder with sulfur dioxide, sulfur trioxide, and oxygen gases is at 825 °C and 1.00 atm. If the partial pressure of sulfur dioxide is 150 mm Hg and sulfur trioxide is 475 mm Hg, what is the partial pressure of oxygen in mm Hg?
50. A steel cylinder with nitrogen, hydrogen, and ammonia gases is at 500 °C and 5.00 atm. If the partial pressure of nitrogen is 1850 mm Hg and hydrogen is 1150 mm Hg, what is the partial pressure of ammonia in mm Hg?
51. Describe the meaning of the expression "collecting a gas over water."

52. Distinguish between a "wet" gas and a "dry" gas.

53. If oxygen is collected over water at 20 °C and 766 mm Hg, what is the partial pressure of the oxygen? Refer to Table 11.2 for the vapor pressure of water.

54. If xenon is collected over water at 30 °C and 755 mm Hg, what is the partial pressure of the xenon? Refer to Table 11.2 for the vapor pressure of water.

Ideal Gas Behavior (Sec. 11.10)

55. State five characteristics of an ideal gas according to the kinetic theory.

56. Distinguish between a real gas and an ideal gas.

57. Under what conditions of temperature and pressure does a real gas behave most like an ideal gas?

58. At what Celsius temperature does a gas possess zero kinetic energy?

59. A sealed container holds the gases He, Ne, and Ar. Which of these gases corresponds to each of the following descriptions?
(a) highest kinetic energy
(b) lowest kinetic energy
(c) highest atomic velocity
(d) lowest atomic velocity

60. A sealed container holds the gases H_2, N_2, and O_2. Which of these gases corresponds to each of the following descriptions?
(a) highest kinetic energy
(b) lowest kinetic energy
(c) highest molecular velocity
(d) lowest molecular velocity

61. What pressure is exerted by an ideal gas at absolute zero?

62. What volume is occupied by an ideal gas at absolute zero?

Ideal Gas Law (Sec. 11.11)

63. If 0.500 mol of hydrogen gas occupies 5.00 L at 25 °C, what is the pressure in atmospheres?

64. If 1.25 mol of oxygen gas exerts a pressure of 1200 mm Hg at 25 °C, what is the volume in liters?

65. If 10.0 L of neon gas exerts a pressure of 125 psi at 373 K, what is the number of moles of gas?

66. If 0.100 mol of argon gas occupies 2.15 L at 725 mm Hg, what is the temperature in degrees Celsius?

General Exercises

67. If the surface area of a human body is 255 square inches, what is the total weight of the atmosphere on a human body expressed in pounds?

68. If the surface area of a human hand is 165 square centimeters, what is the total weight of the atmosphere on a human hand expressed in pounds?

69. What is standard atmospheric pressure (in cm) measured by a barometer filled with water (given that mercury is 13.6 times more dense than water)?

70. What is standard atmospheric pressure (in cm) measured by a barometer filled with alcohol (given that mercury is 19.2 times more dense than alcohol)?

71. A deep-sea scuba tank contains oxygen and helium gases. If the partial pressures of the gases are 35 atm and 125 atm, what is the total pressure (in psi)?

72. A bicycle tire contains nitrogen, oxygen, and argon gases. If the partial pressures are 51.0 psi, 13.5 psi, and 0.5 psi, what is the total pressure (in atm)?

73. The decomposition of baking soda produces carbon dioxide gas. If 42.5 mL of wet gas is collected over water at 22 °C and 764 mm Hg, what is the volume of dry gas at STP conditions? (The vapor pressure of water at 22 °C is 19.8 mm Hg.)

74. Calcium metal and water produce hydrogen gas. If the volume of wet gas collected over water is 95.0 mL at 20 °C and 758 mm Hg, what is the volume of dry gas at STP conditions? (The vapor pressure of water at 20 °C is 17.5 mm Hg.)

75. What pressure (in atm) is exerted by 1.51×10^{23} oxygen molecules at 25 °C in a 5.00 L container?

76. What volume (in liters) is occupied by 3.38×10^{22} nitrogen molecules at 100 °C and 255 mm Hg?

77. Natural gas contains methane, CH_4, ethane, C_2H_6, and hydrogen sulfide, H_2S. Which molecules have the fastest velocity?

78. Bottled gas for recreational vehicles contains propane, C_3H_8, and butane, C_4H_{10}. Which molecules have the faster velocity?

79. Which of the following gases has a density of 2.14 g/L at STP: H_2, O_2, O_3?

80. Which of the following gases has a density of 5.40 g/L at STP: H_2S, SO_3, CF_2Cl_2?

81. A sample of unknown gas weighs 1.95 g and occupies 3.00 L at 1.25 atm and 20 °C. What is the molar mass of the unknown gas?

82. A sample of unknown gas weighs 2.85 g and occupies 750 mL at 760 mm Hg and 100 °C. What is the molar mass of the unknown gas?

83. A sample of chlorine gas occupies 1550 mL at 0.945 atm and 50 °C. What is the mass of the sample?

84. A sample of fluorine gas occupies 855 mL at 710 mm Hg and 155 °C. What is the mass of the sample?

85. Express the ideal gas constant in the following units: L • torr/mol • K.

86. Express the ideal gas constant in the following SI units (given that 1 atm • L = 101.27 J): J/mol • K.

87. Deep-sea divers breathe a helium–oxygen mixture at depths below 200 feet. Explain why divers have high, squeaky voices while breathing the mixture. (Hint: As molecules move faster across the vocal cords, voice pitch rises.)

▲ **Deep-Sea Divers** These divers are breathing a mixture of helium and oxygen gases.

11 Chapter Self-Test

Answers to Self-Test are in Appendix J.

1. Which of the following is *not* an observed property of gases? (Sec. 11.1)
   (a) Gases vary in shape and volume.
   (b) Gases expand infinitely.
   (c) Gases compress infinitely.
   (d) Gases have low density.
   (e) Gases mix completely.
2. Which of the following expresses standard atmospheric pressure? (Sec. 11.2)
   (a) 29.9 in. Hg
   (b) 76.0 cm Hg
   (c) 760 mm Hg
   (d) 14.7 psi
   (e) all of the above
3. Which of the following changes decreases the pressure of a gas? (Sec. 11.3)
   (a) increasing the volume
   (b) increasing temperature
   (c) increasing the number of molecules
   (d) all of the above
   (e) none of the above
4. A volume of hydrogen gas at 1.00 atm increases from 0.250 L to 0.500 L. If the temperature remains constant, what is the final pressure? (Sec. 11.4)
   (a) 0.250 atm
   (b) 0.500 atm
   (c) 1.00 atm
   (d) 2.00 atm
   (e) none of the above
5. A 1.00 L volume of oxygen gas is cooled from 50.0 °C to 25.0 °C. If the pressure remains constant, what is the final volume? (Sec. 11.5)
   (a) 0.500 L
   (b) 0.923 L
   (c) 1.00 L
   (d) 1.08 L
   (e) 2.00 L
6. A sample of nitrogen gas at 1.00 atm is heated from 250 K to 500 K. If the volume remains constant, what is the final pressure? (Sec. 11.6)
   (a) 0.250 atm
   (b) 0.500 atm
   (c) 1.00 atm
   (d) 2.00 atm
   (e) none of the above
7. A sample of air occupies a volume of 10.0 L. If both the pressure and Kelvin temperature each double, what is the final volume? (Sec. 11.7)
   (a) 2.50 L
   (b) 5.00 L
   (c) 10.0 L
   (d) 20.0 L
   (e) 40.0 L
8. If the temperature of a liquid increases from 20.0 °C to 50.0 °C, what happens to the vapor pressure? (Sec. 11.8)
   (a) increases
   (b) decreases
   (c) remains constant
   (d) unpredictable
   (e) none of the above
9. What is the partial pressure of a gas collected over water at 25 °C and 775 mm Hg? The vapor pressure of water at 25 °C is 23.8 mm Hg. (Sec. 11.9)
   (a) 750 torr
   (b) 751 torr
   (c) 760 torr
   (d) 775 torr
   (e) 799 torr
10. Which of the following is true for an ideal gas? (Sec. 11.10)
    (a) All molecules have the same velocity.
    (b) All molecules have the same kinetic energy.
    (c) All molecular collisions have the same energy.
    (d) all of the above
    (e) none of the above
11. How many moles of helium occupy a volume of 5.00 L at 227 °C and 5.00 atm (given that $R = 0.0821$ atm • L/mol • K)? (Sec. 11.11)
    (a) 0.609 mol
    (b) 1.64 mol
    (c) 5.00 mol
    (d) 25.0 mol
    (e) 6090 mol

Key Concepts

12. A few drops of water in a gallon can are heated to steam, and the can is sealed with a cap. As the steam cools from 100 °C to 20 °C, the can is crushed as shown in the diagram. Explain the observation.

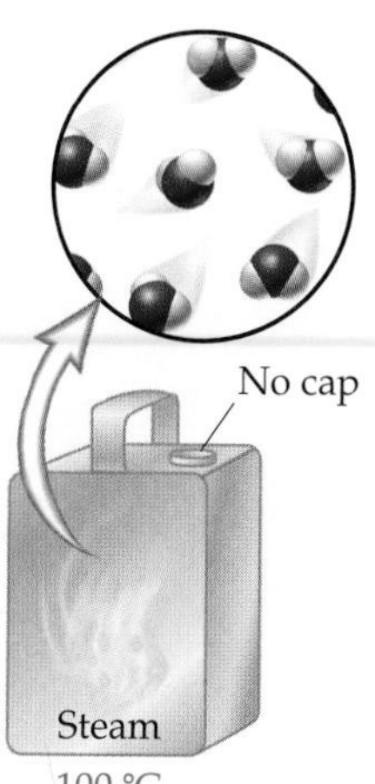

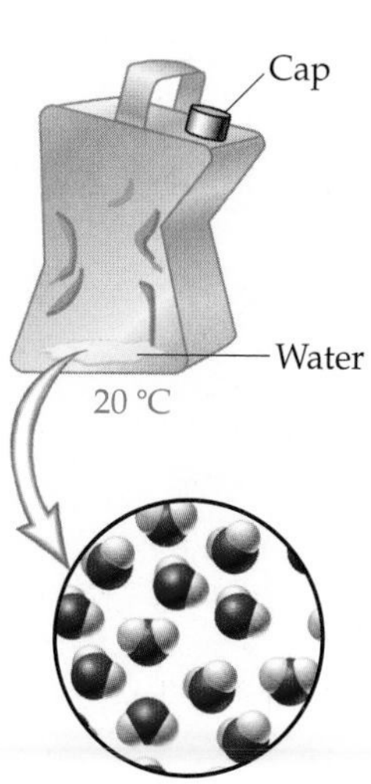

13. If you check automobile tire pressure on a cold morning, does the tire pressure increase or decrease after the automobile tire has rolled several miles?

14. Methyl acetate, $C_3H_6O_2$, smells fragrant, and hydrogen sulfide gas, H_2S, smells like rotten eggs. Would you be attracted or repulsed first, if both gases escaped from the same container at the same time?

Critical Thinking

15. When you heat the air inside an open hot air balloon, why does the balloon rise? (Assume that the air pressure inside the balloon equals that of the surrounding air.)

16. If you release a helium balloon into the air, why does the volume increase as the balloon ascends into the atmosphere? (Assume that the temperature stays constant.)

17. A plastic water bottle in a car on a hot day bulges after an hour. Explain the observation.

18. A plastic water bottle in a car on a cold night is crushed in the morning. Explain the observation.

CHAPTER

12 Chemical Bonding

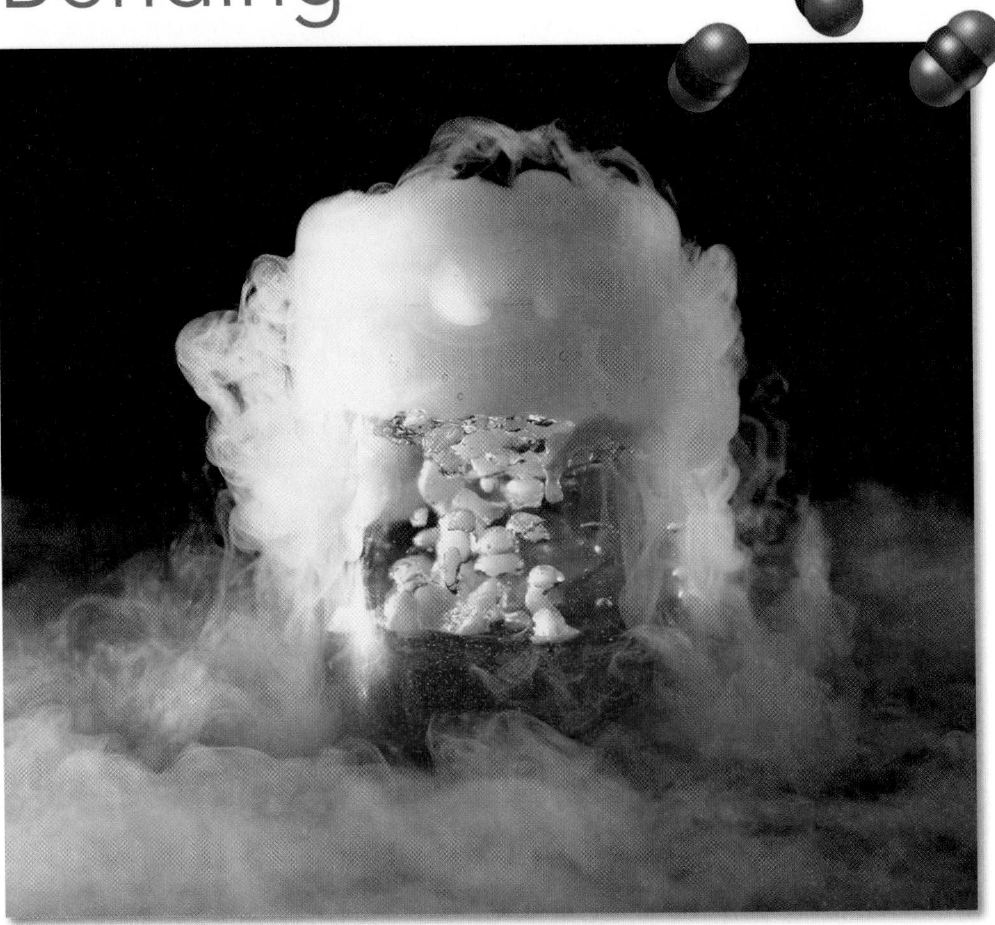

Dry ice is shown in a beaker of water containing blue dye. When dry ice changes to a gas, the bubbles freeze water vapor and produce the white fog shown above the beaker.

"The best way to have a good idea is to have lots of ideas."
Linus Pauling, American Chemist (1901–1994)

At the beginning of the twentieth century, scientists were rapidly unraveling the mysteries of the atom. In 1897, J. J. Thomson discovered the electron and, shortly thereafter, the proton. In 1911, Ernest Rutherford unveiled the atomic nucleus. And in 1913, Niels Bohr offered evidence that electrons occupied orbits and circled the nucleus. During this same period of time, scientists began to speculate that electrons were responsible for holding molecules together. They suggested that electrons provided the "glue" that bonded one atom to another.

In 1916, the American chemist G. N. Lewis (1875–1946) formulated one of the first theories of chemical bonding. He noted that the noble gases were unusually stable and that, with the exception of helium, all these gases had eight electrons in their outer shell. Lewis theorized that atoms bonded together to attain a stable noble gas electron structure. This principle, referred to as the *octet rule*, states that atoms bond in such a way that each atom attains eight electrons in its outer shell.

12.1 THE CHEMICAL BOND CONCEPT

OBJECTIVES

- To explain how valence electrons create a chemical bond.
- To predict whether a bond is ionic or covalent.

In Section 6.7, we learned that an atom has core electrons and valence electrons. Core electrons are found closer to the nucleus, whereas **valence electrons** are found in the most distant *s* and *p* energy subshells. Valence electrons are responsible for holding two or more atoms together in a **chemical bond**.

The **octet rule** states that an atom tends to bond in such a way that it acquires eight electrons in its outer shell. In the words of G. N. Lewis, the chemical bond concept can be described as follows:

> *Two atoms may conform to the rule of eight, or the octet rule, not only by the transfer of electrons from one atom to another, but also by sharing one or more pairs of electrons. These electrons which are held in common by two atoms may be considered to belong to the outer shells of both atoms.*
>
> *The simplest explanation of the predominant occurrence of an even number of electrons in the valence shells of molecules is that the electrons are definitely paired with one another. Two electrons thus coupled together, when lying between two atomic centers, and held jointly in the shells of the two atoms, I have considered to be the chemical bond.*

In an **ionic bond**, a metal cation is attracted to a nonmetal anion. The positively charged cation and negatively charged anion are held together by the attraction of opposite charges. A fundamental particle held together by an ionic bond is a **formula unit**. We can easily identify a formula unit by the fact that it contains metal and nonmetal ions. NaCl, MgO, and AlN all contain a metal cation and a nonmetal anion and are examples of formula units.

In a **covalent bond**, two nonmetal atoms share valence electrons. The formation of an ionic bond and a covalent bond is shown in Figure 12.1. A fundamental particle held together by covalent bonds is a **molecule**. We can easily identify a molecule by the fact that it contains nonmetal atoms. H_2O, NH_3, and CH_4 all contain nonmetal atoms and are examples of molecules.

A substance can be composed of either formula units or molecules. The distinction between the types of representative particles found in a substance is shown in Figure 12.2.

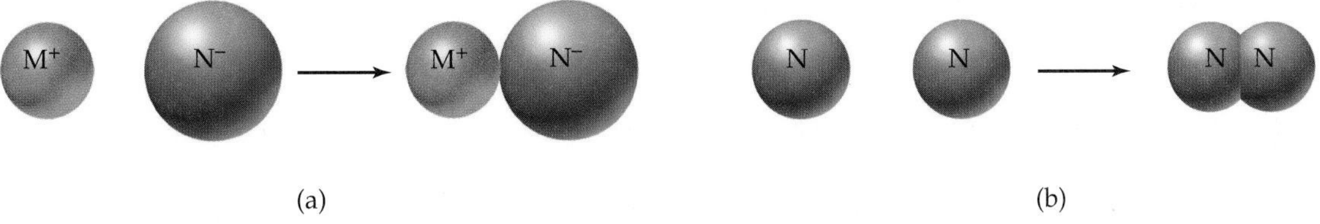

▲ **Figure 12.1 Chemical Bonds** (a) An ionic bond forms when a metal cation (M^+) is attracted to a nonmetal anion (N^-). (b) A covalent bond forms when valence electrons are shared by two nonmetal atoms (N).

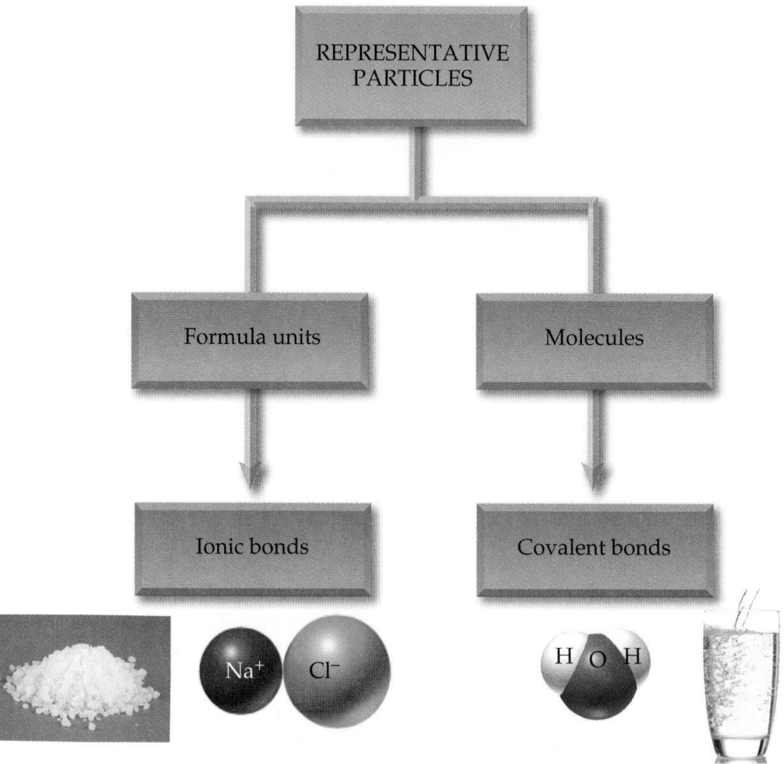

▶ **Figure 12.2 Classification of Representative Particles**
Particles can be formula units (for example, NaCl) held together by ionic bonds or they can be molecules (for example, H_2O) held together by covalent bonds.

The following example exercise further illustrates the relationship between bonding and the type of fundamental particles:

Example Exercise 12.1 Bond Predictions

Predict whether each of the following is held together by ionic or by covalent bonds:

(a) ammonia, NH_3 (b) magnesium nitride, Mg_3N_2

Solution

A metal ion and nonmetal ion are attracted in ionic bonds; two or more nonmetal atoms are attracted in covalent bonds.

(a) Ammonia contains the nonmetals nitrogen and hydrogen. It follows that NH_3 has *covalent* bonds.

(b) Magnesium nitride contains a metal (Mg) and a nonmetal (N). It follows that Mg_3N_2 has *ionic* bonds.

Practice Exercise

Predict whether each of the following is held together by ionic or by covalent bonds:

(a) aluminum oxide, Al_2O_3 (b) sulfur dioxide, SO_2

Answers: (a) ionic; (b) covalent

Concept Exercise

What type of chemical bond results from the attraction between a metal cation and a nonmetal anion? between two nonmetal atoms?

Answer: See Appendix G.

OBJECTIVE

▶ To describe the formation of an ionic bond between a metal atom and a nonmetal atom.

12.2 IONIC BONDS

An ionic bond results from the attraction between a positively charged cation and a negatively charged anion. In a crystal of table salt, NaCl, there is an attraction between Na^+ ions and Cl^- ions. This attraction between ions is similar to the attraction between the ends of two magnets. Figure 12.3 illustrates the attraction between ions, and the similar attraction between two magnets.

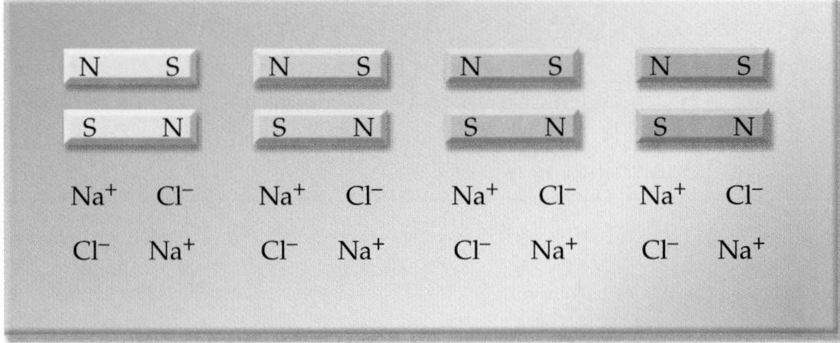

◀ **Figure 12.3 Electrostatic Attraction** Opposite ends of ordinary magnets are attracted to each other in much the same way that oppositely charged cations and anions are attracted. Notice that attraction can take place vertically as well as horizontally.

In ordinary table salt, NaCl, the ionic bonds between sodium ion and chloride ion are so strong that they can create a rigid crystalline structure (Figure 12.4). In fact, the high melting point of NaCl, 801 °C, is explained by the rigid crystal structure.

▲ **Figure 12.4 Crystalline Structure of Sodium Chloride** A crystal of sodium chloride is composed of sodium and chloride ions that repeat regularly. Notice that each sodium ion is attracted to several nearby chloride ions. The alternating pattern of cations and anions gives rise to a three-dimensional crystalline structure.

Formation of Cations

When a metal atom loses valence electrons to a nonmetal atom, the metal atom becomes positively charged. In Section 6.7, we learned that the number of valence electrons corresponds to the group number of the element in the periodic table. When a sodium atom (Group IA/1) loses its 1 valence electron, it becomes Na^+. Referring to the periodic table, we notice that Na^+ has 10 electrons ($11 - 1 = 10$ e−). This is the same number of electrons found in the preceding noble gas neon (10 e–).

Although there are exceptions, main group metals usually achieve a noble gas electron configuration after losing their valence electrons. Stated differently, main group metal ions are generally *isoelectronic* with a noble gas as explained in Section 6.10. When a magnesium atom (Group IIA/2) loses its 2 valence electrons, it becomes Mg^{2+}. A magnesium ion has 10 electrons ($12 - 2 = 10$ e−) and is also isoelectronic with neon.

In Section 6.8 we learned how to write electron dot formulas for atoms of elements. We can now use electron dot diagrams to clarify the formation of ions. Electron dot diagrams allow us to focus our attention on valence electrons involved in this process of forming ions. Figure 12.5 illustrates the formation of ions by third-period metals.

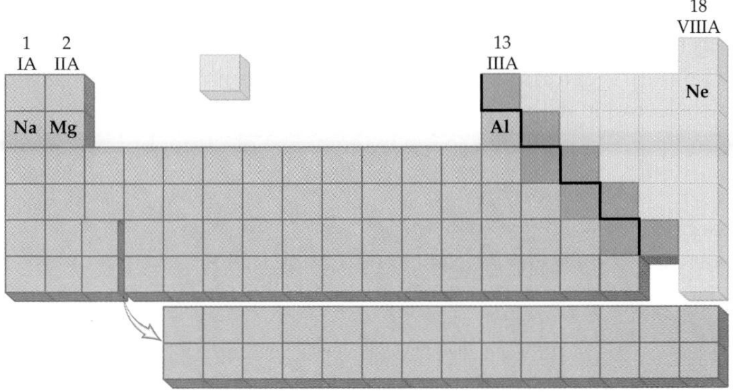

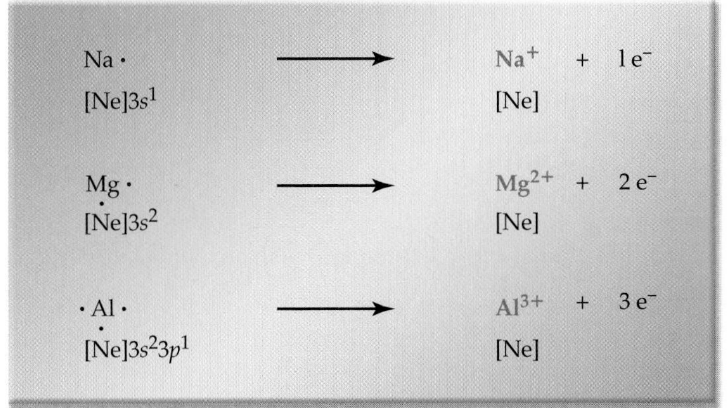

◀ **Figure 12.5 Formation of Metal Ions** The atoms of the metals in Period 3 form cations by losing 1, 2, and 3 electrons, respectively. In each case the metal atom becomes isoelectronic with the preceding noble gas neon, [Ne].

Note Notice that the electron dot diagrams of atoms in Figure 12.5 show electron dots, while the ions in Figure 12.5 do not show electron dots. Each of the ions has the electron configuration of neon. And each of these ions, like neon, has a stable electron configuration consisting of eight electrons. By convention, chemists do not draw the underlying octet of electrons because they are not valence electrons.

Formation of Anions

When a nonmetal atom gains valence electrons from a metal atom, the nonmetal atom becomes negatively charged. When a chlorine atom (Group VIIA/17) gains 1 valence electron, it becomes Cl^-. Referring to the periodic table, we notice that Cl^- has 18 electrons ($17 + 1 = 18$ e−). This is the same number of electrons found in the next noble gas argon (18 e–). When a sulfur atom (Group VIA/16) gains two valence electrons, it becomes S^{2-}. A sulfide ion has 18 electrons ($16 + 2 = 18$ e−) and is also isoelectronic with argon. Figure 12.6 illustrates the formation of ions by third-period nonmetals.

The following example exercise further illustrates the formation of cations and anions:

Example Exercise 12.2 Electron Configuration of Ions

Which noble gas has an electron configuration identical to that of each of the following ions?

(a) lithium ion
(b) oxide ion
(c) calcium ion
(d) bromide ion

Solution

Refer to the group number in the periodic table for the number of valence electrons.

(a) Lithium (Group IA/1) forms an ion by losing 1 valence electron; Li^+ has only 2 electrons remaining and is isoelectronic with *He*.
(b) Oxygen (Group VIA/16) forms an ion by gaining 2 electrons; thus, O^{2-} has 10 electrons and is isoelectronic with *Ne*.
(c) Calcium (Group IIA/2) forms an ion by losing 2 electrons; Ca^{2+} has 18 electrons remaining and is isoelectronic with *Ar*.
(d) Bromine (Group VIIA/17) forms an ion by gaining 1 electron; Br^- has 36 electrons and is isoelectronic with *Kr*.

Practice Exercise

Which noble gas element has an electron configuration identical to that of each of the following ions?

(a) potassium ion
(b) nitride ion
(c) strontium ion
(d) iodide ion

Answers:

(a) K^+ is isoelectronic with Ar.
(b) N^{3-} is isoelectronic with Ne.
(c) Sr^{2+} is isoelectronic with Kr.
(d) I^- is isoelectronic with Xe.

Concept Exercise

Which noble gas is isoelectronic with a barium ion? with an iodide ion?

Answer: See Appendix G.

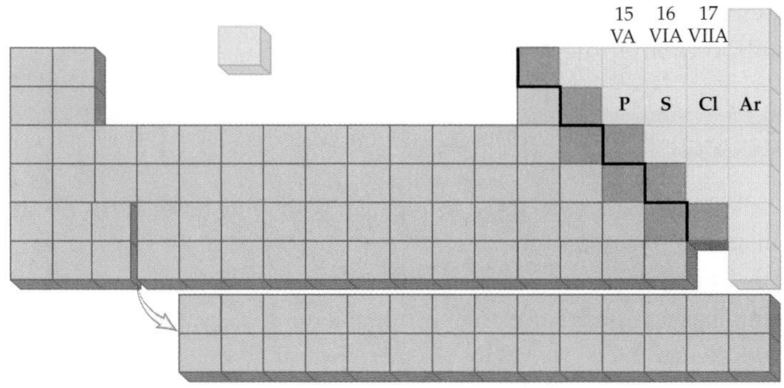

:Cl: $[Ne]3s^23p^5$	$+ 1 e^-$	⟶	$[:Cl:]^-$ $[Ne]3s^23p^6 = [Ar]$
·S: $[Ne]3s^23p^4$	$+ 2 e^-$	⟶	$[:S:]^{2-}$ $[Ne]3s^23p^6 = [Ar]$
·P: $[Ne]3s^23p^3$	$+ 3 e^-$	⟶	$[:P:]^{3-}$ $[Ne]3s^23p^6 = [Ar]$

◀ **Figure 12.6 Formation of Nonmetal Ions** The nonmetals in Period 3 gain 1, 2, and 3 electrons, respectively. In each case the nonmetal ion becomes isoelectronic with the next noble gas argon, [Ar].

Ionic Radii versus Atomic Radii

To visualize an ionic bond more clearly, we can picture a sodium atom becoming smaller after losing its valence electron. The radius of a sodium atom is 0.186 nm, whereas the radius of the sodium ion is 0.095 nm. The reason for this decrease in radius is that the sodium atom has lost its 3*s* energy sublevel. Moreover, in the sodium ion the positive nuclear charge (11+) is greater than the negative electron charge (10–) and pulls the electrons closer to the nucleus; thus, the radius decreases.

Conversely, we can picture a chlorine atom becoming larger after gaining a valence electron. The radius of a chlorine atom is 0.099 nm, whereas the radius of the chloride ion is 0.181 nm. The reason for this increase in radius is that the additional electron repels the electrons already present. In fact, the radius of the charged chloride ion is almost twice as large as that of the neutral chlorine atom.

The behavior of sodium and chlorine atoms show the same general trend observed for all metals and nonmetals with respect to atomic radius. The radius of a *cation is smaller* than that of the corresponding metal atom; the radius of an *anion is larger* than that of the corresponding nonmetal atom. The specific changes in the radius of a sodium atom and a chlorine atom are depicted in Figure 12.7.

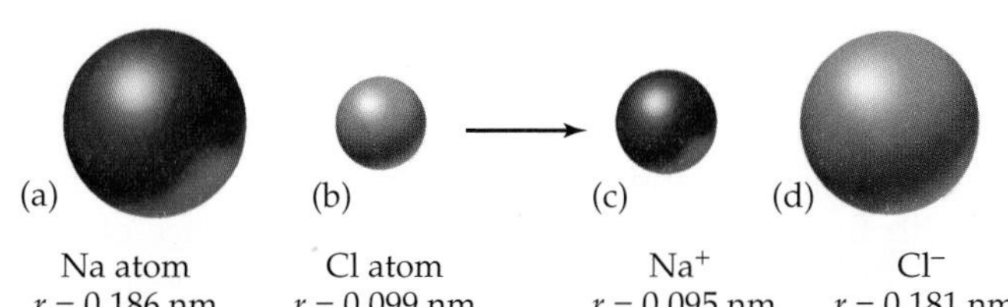

▲ **Figure 12.7 Formation of Sodium and Chloride Ions** The radius of the sodium atom decreases (a and c) as a result of losing an electron, whereas the radius of the chlorine atom increases (b and d) after gaining an electron.

The following example exercise further illustrates the characteristics associated with the formation of an ionic bond:

Example Exercise 12.3 Characteristics of Ionic Bonds

Which of the following statements are true regarding an ionic bond between an iron ion and a sulfide ion in an FeS formula unit?

(a) The iron atom loses electrons and the sulfur atom gains electrons.
(b) The iron atom has a larger radius than the iron ion.
(c) Iron and sulfide ions form a bond by electrostatic attraction.

Solution

All of these statements are true regarding an ionic bond.

(a) An iron atom loses electrons and a sulfur atom gains electrons.
(b) The radius of an iron atom is greater than its ionic radius.
(c) An ionic bond is from the attraction of positive and negative ions.

Practice Exercise

Which of the following statements are true regarding an ionic bond between a zinc ion and an oxide ion in a ZnO formula unit?

(a) Zinc and oxygen form a bond by sharing electrons.
(b) The oxygen atom is larger in radius than the oxide ion.
(c) Zinc and oxide ions form a bond by the repulsion of ions.

Answer:

(a) False, zinc and oxygen bond by transferring electrons.
(b) False, the radius of an oxygen atom is smaller than its ionic radius.
(c) False, zinc and oxygen form a bond by the attraction of ions.

Concept Exercise

Which of the following statements are true regarding the formation of an ionic bond between a metal and a nonmetal?

(a) The atomic radius of a metal atom is less than its ionic radius.
(b) The atomic radius of a nonmetal atom is greater than its ionic radius.
(c) Energy is released in the formation of an ionic bond.

Answer: See Appendix G.

OBJECTIVE

▸ To describe the formation of a covalent bond between two nonmetal atoms by sharing valence electrons.

12.3 COVALENT BONDS

A covalent bond results from the sharing of electrons by two nonmetal atoms. To create an accurate picture, we can visualize shared electrons roaming about both atoms. The electrons belong to both nonmetal atoms, and each atom uses these bonding electrons to complete an octet. That is, each nonmetal atom uses shared electrons to complete its valence shell. Since a filled valence shell is very stable, the resulting bond is also stable.

Now, let's see what happens during the formation of a covalent bond. As an example, consider the formation of hydrogen chloride, HCl, from a hydrogen atom and a chlorine atom. During bond formation, the hydrogen atom shares its one valence electron with the chlorine atom and its seven valence electrons. The additional electron gives the chlorine atom eight electrons in its valence shell, and thus fulfills the octet rule.

In the same process, the chlorine atom shares one of its valence electrons with the hydrogen atom. The additional electron gives the hydrogen atom two electrons in its valence shell. The hydrogen atom is thus similar to an atom of the noble gas helium, and is therefore stable. In this process, both the hydrogen atom and the chlorine atom become stable by sharing an electron pair and forming a covalent bond.

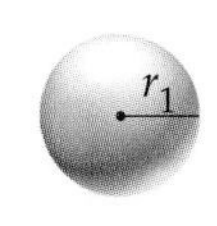

H atom
atomic radius
$r_1 = 0.037$ nm

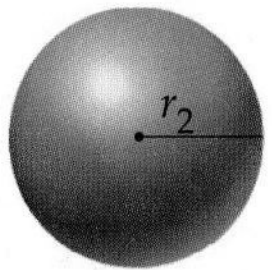

Cl atom
atomic radius
$r_2 = 0.099$ nm

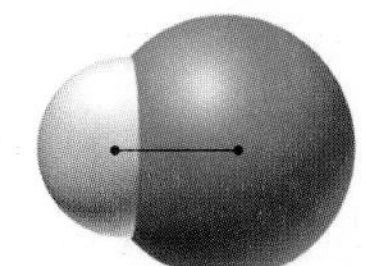

HCl molecule
bond length = 0.127 nm
$r_1 + r_2 = 0.136$ nm

◀ **Figure 12.8 Formation of a Covalent Bond** The sharing of an electron pair by a hydrogen atom and a chlorine atom produces a covalent bond. The two atoms are held together in a molecule of HCl. The bond length (0.127 nm) is less than the sum of the two atomic radii (0.037 + 0.099 = 0.136 nm).

Bond Length

To understand bond formation more clearly, try to visualize the valence shells of the two atoms overlapping each other. In a molecule of HCl, the valence shells of the two atoms overlap as the 1*s* energy sublevel of the hydrogen atom mixes with the 3*p* sublevel of the chlorine atom. This mixing of sublevels draws the two nuclei closer together. The radius of the hydrogen atom is 0.037 nm, and that of the chlorine atom is 0.099 nm. If the two atoms were simply next to each another, the distance from the hydrogen nucleus to the chlorine nucleus would be (0.037 + 0.099) nm or 0.136 nm. But experiments reveal that the distance between the two nuclei is actually 0.127 nm. Thus, the shells overlap as shown in Figure 12.8. The distance between the two nuclei is called the **bond length**.

Bond Energy

Energy is released when two ions are attracted to one another and form an ionic bond. For example, the attraction of Na^+ and Cl^- to form NaCl releases heat energy. Similarly, energy is released when two atoms are attracted and form a covalent bond. When hydrogen and chlorine atoms form an HCl molecule, heat energy is released.

$$H(g) + Cl(g) \longrightarrow HCl(g) + \textit{heat}$$

Conversely, energy is necessary to break an H—Cl bond. The amount of energy required to break a covalent bond between two atoms is referred to as the **bond energy**. The amount of energy necessary to break the H—Cl bond is identical to the amount of heat released when H and Cl form a bond in the creation of a hydrogen chloride molecule.

$$HCl(g) + \textit{heat} \longrightarrow H(g) + Cl(g)$$

The following example exercise summarizes the characteristics associated with the formation of a covalent bond:

Example Exercise 12.4 Characteristics of Covalent Bonds

Which of the following statements are true regarding a covalent bond between a hydrogen atom and an oxygen atom in an H_2O molecule?

(a) Valence electrons are shared by hydrogen and oxygen atoms.
(b) The H–O bond length is less than the sum of the two atomic radii.
(c) Breaking the bond between H–O requires energy.

Solution

All of the previous statements regarding a covalent bond in a molecule are true.

(a) Valence electrons are shared by the two nonmetal atoms.
(b) The bond length is less than the sum of the two atomic radii.
(c) The breaking of a covalent bond requires energy.

Practice Exercise

Which of the following statements are true regarding a covalent bond between a hydrogen atom and a sulfur atom in an H_2S molecule?

(a) Valence electrons are transferred from hydrogen to sulfur.
(b) The H–S bond length equals the sum of the two atomic radii.
(c) Forming a bond between H–S requires energy.

Answer:

(a) False, valence electrons are shared by hydrogen and sulfur.
(b) False, the bond length is less than the sum of the two atomic radii.
(c) False, the forming of a covalent bond releases energy.

Concept Exercise

Which of the following statements are true regarding the formation of a covalent bond between two nonmetal atoms?

(a) Bonding electrons are shared between two nonmetal atoms.
(b) The bond length is less than the sum of the two atomic radii.
(c) The bond energy is the energy require to break a covalent bond.

Answer: See Appendix G.

Note Energy is always released when two *atoms* form a covalent bond. However, energy is required to form an H—I bond from *molecules* of H_2 and I_2. Realize that energy is required to break H—H and I—I bonds, in order to form H—I bonds. The formation of HI is an *endothermic* reaction because more energy is required in breaking H—H and I—I bonds than is released in forming H—I bonds.

12.4 ELECTRON DOT FORMULAS OF MOLECULES

OBJECTIVES

- To draw the electron dot formula for a molecule.
- To draw the structural formula for a molecule.

In Section 6.8, we drew the electron dot formula for an atom of a representative element. We found the element in the periodic table and noted its group number to predict the number of valence electrons. For example, in the second period, we found that Li has 1 valence electron (Group IA/1), Be has 2 (Group IIA/2), B has 3 (Group IIIA/13), C has 4 (Group IVA/14), N has 5 (Group VA/15), O has 6 (Group VIA/16), F has 7 (Group VIIA/17), and the noble gas Ne has 8 (Group VIIIA/18). We represented each valence electron by a dot surrounding the symbol for the element.

Now, we will draw the **electron dot formula** (also called a *Lewis structure*) for a molecule. An electron dot formula shows the valence electrons that form bonds between atoms. These electron pairs shared by atoms are called **bonding electrons**. Other electron pairs that simply complete the octet around each atom are called **nonbonding electrons**, or lone pairs.

Guidelines for Drawing Electron Dot Formulas of Molecules†

1. Calculate the total number of valence electrons by adding all the valence electrons for each atom in the molecule. The total should be an even number in keeping with the octet rule. If the total is an odd number, check your calculations.
2. Divide the total number of valence electrons by 2 to find the number of electron pairs in the molecule.
3. Surround the central atom with four electron pairs. Use the remaining electron pairs to complete an octet around each of the other atoms. (Hydrogen is the sole exception as it requires only one electron pair.)
4. If there are not enough electron pairs to provide an octet for each atom, move a nonbonding electron pair between two atoms that already share an electron pair.

†When two or more atoms are bonded to a central atom, it is often difficult to recognize the central atom. Thus, this textbook will indicate the central atom in bold type to help you write electron dot formulas.

Electron Dot Formula for Water, H_2O

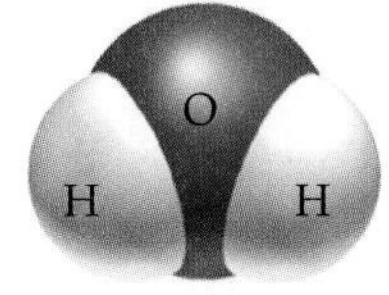

▲ Water, H_2O

To learn how to draw electron dot formulas for molecules, we will draw the electron dot formula for water, H_2**O**. The total number of valence electrons in the molecule is one for each hydrogen atom and six for the oxygen atom. Thus, the total number of electrons is 2(1 e−) + 6 e− = 8 e−. The number of electron pairs is 4 (8/2 = 4).

In the formula H_2**O**, oxygen is the central atom and is indicated in bold. We place the four electron pairs around the oxygen to provide the necessary octet.

We then place the two hydrogen atoms in any of the four electron pair positions.

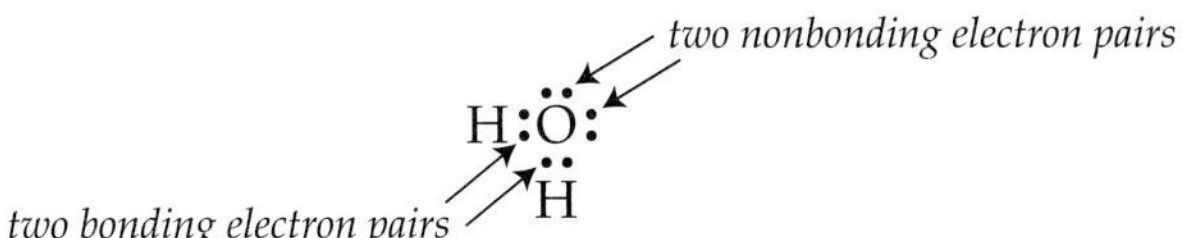

Notice that there are two bonding and two nonbonding electron pairs. All of the electrons, however, are identical, and the four electron pairs (eight valence electrons) move around the entire molecule. Thus, bonding and nonbonding valence electrons spread over the entire molecule and are said to be *delocalized*.

To simplify, we represent each pair of bonding electrons by a single dash, which is called a **single bond**. The resulting structure is referred to as the **structural formula** of the molecule. In a structural formula the nonbonding electrons are usually shown, but their inclusion is optional.

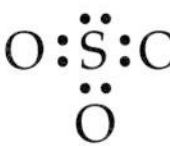

Electron Dot Formula for Sulfur Trioxide, SO_3

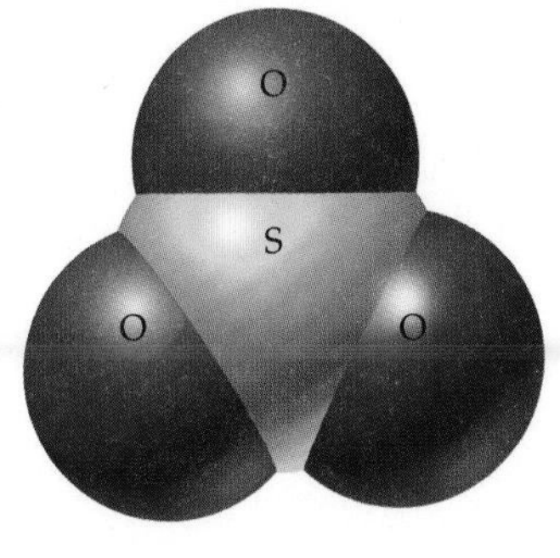

▲ Sulfur Trioxide, SO_3

Now we'll try a more difficult example. Let's draw the electron dot formula for sulfur trioxide, **S**O_3. Sulfur is the central atom in the molecule and is indicated in bold. The total number of valence electrons in the molecule is 6 e− + 3(6 e−) = 24 e−. The number of electron pairs is 12 (24/2 = 12).

Since sulfur is the central atom, we can begin by placing four electron pairs around the sulfur and then attaching the three oxygen atoms. That gives us

O:S̈:O
Ö

We started with 12 electron pairs, and so we have eight pairs remaining. We place the remaining pairs around the oxygen atoms to complete each octet.

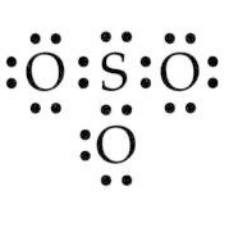

Notice that the bottom oxygen atom does not have an octet; it has only three electron pairs. According to the guidelines, we can move a nonbonding electron pair from the S atom so as to provide two bonding electron pairs between the sulfur and the oxygen atoms.

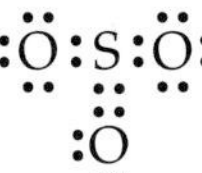

The two shared electron pairs constitute a **double bond**. All four electrons in the double bond are shared by the S and the O atoms. Therefore, the sulfur atom still has an octet and the bottom oxygen atom has gained two electrons to complete its octet. The structural formula for SO_3 has one double bond and two single bonds. We represent the double bond by a double dash and the single bond by a single dash.

```
 ..       ..
:O — S — O:
 ..  ||   ..
     O:
     ..
```

We can draw the structural formula for an SO_3 molecule with the double bond between any of the oxygen atoms. Experimentally, the double bond is found to be shared by each of the three oxygen atoms. This phenomenon is known as *resonance,* and we can draw three resonance structures for an SO_3 molecule.

```
O=S—O    O—S—O    O—S=O
  |        ||       |
  O        O        O
```

Although these structures may appear to be different, all three formulas are identical. If we were to construct the molecule using models, we could easily verify this statement. Each molecular model can be rotated and shown to be identical to the other two.

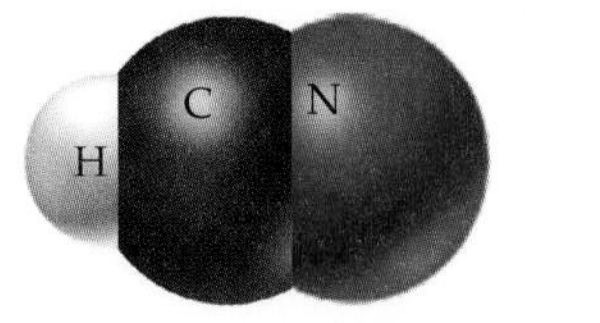

▲ Hydrogen Cyanide, HCN

Electron Dot Formula for Hydrogen Cyanide, HCN

Let's draw the electron dot formula for hydrogen cyanide, H**C**N. Carbon is the central atom in the molecule and is indicated in bold. The total number of valence electrons in the molecule is 1 e– + 4 e– + 5 e– = 10 e–. The number of electron pairs is 5 (10/2 = 5). Since carbon is the central atom, we begin by placing 4 electron pairs around the carbon and attaching the hydrogen and nitrogen atoms.

```
  ..
H:C:N
  ..
```

We started with five electron pairs, and there is only one pair remaining. Since the hydrogen atom has its two required electrons, we place the remaining pair next to the nitrogen atom.

```
  ..
H:C:N:
  ..
```

The nitrogen atom now has 4 electrons, which is 4 e– less than an octet. According to the guidelines, we can move nonbonding electron pairs between the C and the N atoms. For example,

```
H:C::N:
  ..
```

The nitrogen atom has only six electrons. If we move the other nonbonding electron pair from the carbon atom, we have

```
H:C:::N:
```

The three electron pairs produce a **triple bond**. All six electrons in the triple bond are shared by the C and the N atoms. Therefore, both the carbon atom and the nitrogen atom have obtained an octet. A triple bond is indicated by a triple dash. The structural formula for HCN is therefore written as

$$H—C\equiv N:$$

The following example exercises further illustrate how to draw electron dot and structural formulas of molecules:

Example Exercise 12.5 Electron Dot Formulas—Single Bonds

Draw the electron dot formula and the structural formula for a chloroform molecule, $CHCl_3$.

Solution

Carbon is the central atom in the chloroform molecule and is indicated in bold. The total number of valence electrons is 4 e− + 1 e− + 3(7 e−) = 26 e−. The number of electron pairs is 13 (26/2 = 13). We begin by placing 4 electron pairs around the carbon and adding the one H and three Cl atoms.

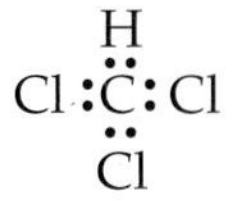

```
      H
      ••
  Cl :C: Cl
      ••
      Cl
```

We have 13 electron pairs minus the 4 pairs we used for the carbon. Since 9 electron pairs remain, let's place 3 pairs around each chlorine.

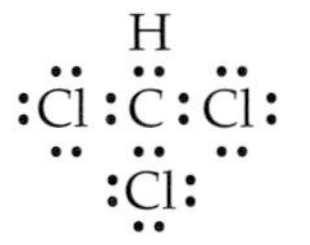

```
       H
  ••   ••   ••
 :Cl : C : Cl:
  ••   ••   ••
      :Cl:
       ••
```

Each atom is surrounded by an octet of electrons, except hydrogen, which has two. This is the correct electron dot formula. In the corresponding structural formula we replace each bonding electron pair by a single dash. The structural formula is

```
        H
        |
  ••         ••
 :Cl ── C ── Cl:
  ••    |    ••
       :Cl:
        ••
```

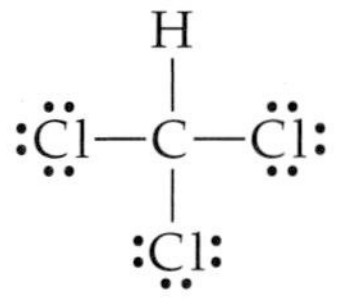

Note that the structural formula is free to rotate. We could have also written the hydrogen atom below or at the side of the carbon atom.

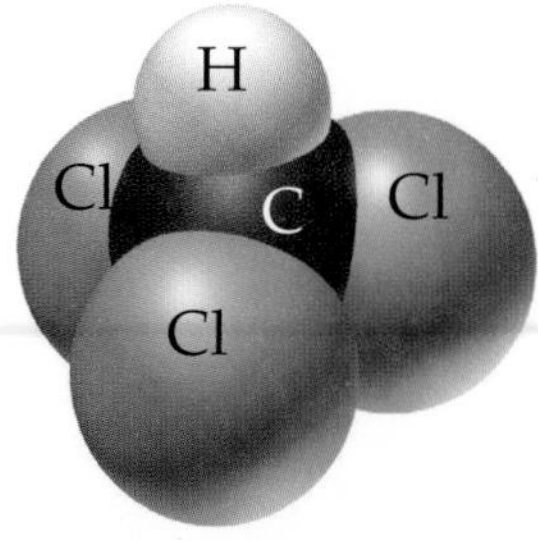

▲ Chloroform, $CHCl_3$

Practice Exercise

Draw the electron dot formula and the structural formula for a molecule of **Si**HClBrI.

Answers:

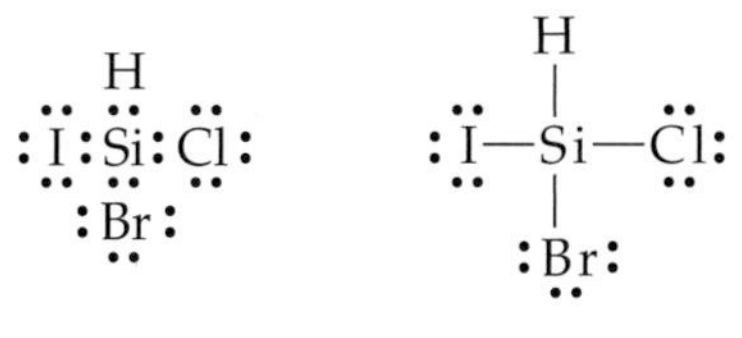

```
     H                         H
 ••  ••  ••                    |
: I :Si: Cl:           ••             ••
 ••  ••  ••           : I ── Si ── Cl:
    :Br:               ••      |     ••
     ••                      :Br:
                              ••
```

Concept Exercise

Draw the electron dot formula for H_2**S**. How many pairs of nonbonding electrons are in a hydrogen sulfide molecule?

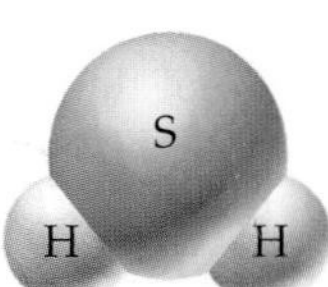

Answer: See Appendix G.

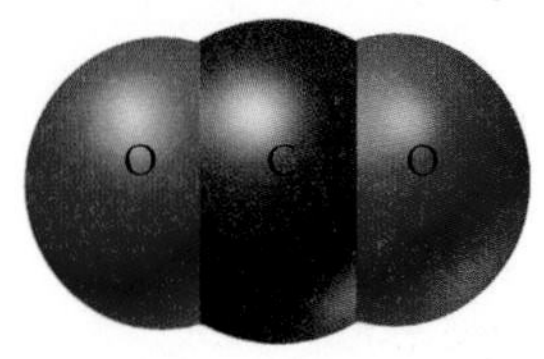

▲ Carbon Dioxide, CO_2

Example Exercise 12.6 Electron Dot Formulas—Double Bonds

Draw the electron dot formula and the structural formula for a carbon dioxide molecule, CO_2.

Solution

Carbon is the central atom in the carbon dioxide molecule as indicated in bold. The total number of valence electrons is 4 e− + 2(6 e−) = 16 e−. The number of electron pairs is 8 (16/2 = 8). We begin by placing 4 electron pairs around the carbon and adding the two O atoms.

$$\mathrm{O{:}\ddot{\underset{\cdot\cdot}{C}}{:}O}$$

The number of remaining electron pairs is 8 − 4 = 4 pairs. We can add 2 pairs to each oxygen.

$$\mathrm{\ddot{\underset{\cdot\cdot}{O}}{:}\ddot{\underset{\cdot\cdot}{C}}{:}\ddot{\underset{\cdot\cdot}{O}}}$$

Each oxygen atom shares six electrons, two less than an octet. We can use the nonbonding electron pairs around carbon to complete the octet around oxygen. We move one pair to the oxygen on the left, and the other pair to the oxygen on the right to complete the octet around each oxygen. Each carbon–oxygen bond shares two electron pairs.

$$\mathrm{\ddot{\underset{\cdot\cdot}{O}}{::}C{::}\ddot{\underset{\cdot\cdot}{O}}}$$

Now the octet rule is satisfied for each atom. This is the correct electron dot formula. There are two electron pairs between each oxygen and carbon. Thus, there are two double bonds in the carbon dioxide molecule. In the structural formula each double bond is indicated by a double dash.

$$\mathrm{{:}\ddot{O}{=}C{=}\ddot{O}{:}}$$

Practice Exercise

Draw the electron dot formula and the structural formula for a molecule of SiO_2.

Answers:

$$\mathrm{{:}\ddot{O}{::}Si{::}\ddot{O}{:}} \qquad \mathrm{{:}\ddot{O}{=}Si{=}\ddot{O}{:}}$$

Concept Exercise

Draw the electron dot formula for SO_2. How many pairs of nonbonding electrons are in a sulfur dioxide molecule?

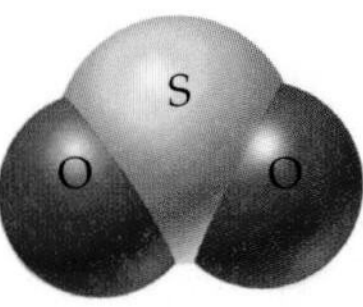

Answer: See Appendix G.

Note Writing electron dot formulas is a trial-and-error procedure that involves the use of valence electron pairs to satisfy the octet rule. However, there are common exceptions to the octet rule that are beyond the scope of this introduction to bonding. For example, in BeF_2 there are only 4 electrons around the **Be** atom, and in BF_3 there are only 6 electrons around the **B** atom. Moreover, there are exceptions that violate the octet rule by exceeding 8 electrons. For example, in PF_5 there are 10 electrons around the **P** atom; and in SF_6 there are 12 electrons around the **S** atom.

12.5 ELECTRON DOT FORMULAS OF POLYATOMIC IONS

OBJECTIVES

- To draw the electron dot formula for a polyatomic ion.
- To draw the structural formula for a polyatomic ion.

So far in our discussion of ions, we have considered only single atoms that have gained or lost electrons. This type of simple ion having a positive or negative charge is called a **monoatomic ion**. An ion can also contain two or more atoms held together and bearing a charge; this type of ion is called a **polyatomic ion**. Some common polyatomic ions are the ammonium ion, NH_4^+, found in fertilizer; the hydroxide ion, OH^-, found in household products like drain and oven cleaner; and the hydrogen carbonate ion, HCO_3^-, found in baking soda.

In Section 12.4 we learned to draw the electron dot formula for molecules. Now, we can draw the electron dot formula for polyatomic ions. The following directions for ions are similar to the previous guidelines for molecules:

Guidelines for Drawing Electron Dot Formulas of Polyatomic Ions†

1. Calculate the valence electrons by adding the valence electrons for each atom in the polyatomic ion. If the ion is negatively charged, add electrons—equal to the charge—to the number of valence electrons. If the ion is positively charged, subtract electrons—equal to the charge—from the number of valence electrons. In each case, the result will be the total number of valence electrons.
2. Divide the total number of valence electrons and charges by two to find the number of electron pairs in the polyatomic ion.
3. Surround the central atom with four electron pairs. Use the remaining electron pairs to complete an octet around each of the other atoms. (Hydrogen is the sole exception, as H requires only one electron pair.)
4. If there are not enough electron pairs to provide an octet for each atom, move a nonbonding electron pair between two atoms that already share an electron pair.

†When two or more atoms are bonded to a central atom, it is often difficult to recognize the central atom. Thus, this textbook will indicate the central atom in bold type to help you write electron dot formulas.

Electron Dot Formula for the Ammonium Ion, NH_4^+

We can draw the electron dot formula for the ammonium ion, $\mathbf{N}H_4^+$, as follows. The total number of valence electrons in this positive polyatomic ion is 5 e– + 4(1 e–) − 1 e– = 8 e–. (Note that we must subtract 1 e–, because the ion has a 1+ charge.) The number of electron pairs is 4 (8/2 = 4). In the formula $\mathbf{N}H_4^+$, nitrogen is the central atom. We place the 4 electron pairs around the nitrogen atom to provide the necessary octet.

```
[   H   ]+
[  ..   ]
[ H:N:H ]
[  ..   ]
[   H   ]
```

Notice that the electron dot formula is written correctly for the ammonium ion. The nitrogen atom has an octet, and each hydrogen atom shares two electrons.

Now, let's simplify the electron dot formula and draw the structural formula of this polyatomic ion. We represent each bonding pair of electrons by a single dash and draw the structural formula for the ammonium ion as follows:

```
[    H    ]+
[    |    ]
[ H—N—H ]
[    |    ]
[    H    ]
```

The usual way to draw structural formulas of polyatomic ions is to enclose them in brackets. The overall charge on the polyatomic ion is indicated outside the brackets. Placing the charge outside the brackets emphasizes that the positive charge is distributed over the entire ion.

Electron Dot Formula for the Chlorate Ion, ClO_3^-

Now, let's draw the electron dot formula for the chlorate ion, ClO_3^-. The total number of valence electrons for this negative ion is 7 e− + 3(6 e−) + 1 e− = 26 e−. (Note that we must add 1 e–, because the ion has a 1– charge.) The number of electron pairs is 13 (26/2 = 13). Since chlorine is the central atom, we begin by placing 4 electron pairs around the Cl atom and then add the 3 O atoms.

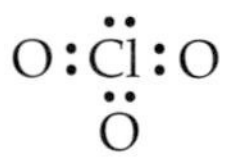

```
    ••
O : Cl : O
    ••
    O
```

We started with 13 electron pairs and placed 4 pairs around the central atom. We place the remaining 9 electron pairs around the oxygen atoms to complete each octet.

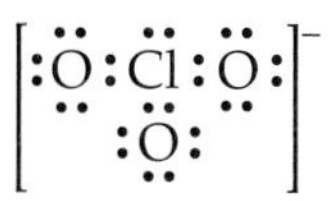

```
 ⎡ ••   ••   •• ⎤ −
 ⎢:O : Cl : O:  ⎥
 ⎢ ••   ••   •• ⎥
 ⎢     :O:      ⎥
 ⎣      ••      ⎦
```

In the structural formula, each bonding electron pair is represented by a single dash. Thus, we can show the structural formula for ClO_3^- as follows:

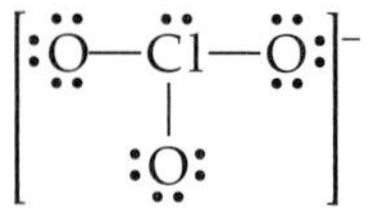

```
 ⎡ ••   ••   •• ⎤ −
 ⎢:O — Cl — O:  ⎥
 ⎢ ••   |    •• ⎥
 ⎢     :O:      ⎥
 ⎣      ••      ⎦
```

Electron Dot Formula for the Carbonate Ion, CO_3^{2-}

Let's try a more difficult example and draw the electron dot formula for the carbonate ion, CO_3^{2-}. The total number of valence electrons in this polyatomic ion is 4 e− + 3(6 e−) + 2 e− (because the ion has a 2– charge) = 24 e–. The total number of electron pairs is 12 (24/2 = 12). Since carbon is the central atom, we begin by placing 4 pairs of electrons around the C atom and then add 3 O atoms. We have

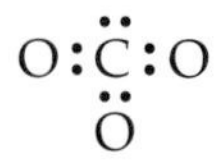

```
    ••
O : C : O
    ••
    O
```

We started with 12 electron pairs, and so we have 8 pairs remaining. We place the remaining pairs around the oxygen atoms to give octets as follows:

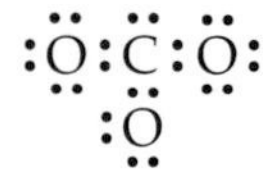

```
 ••   ••   ••
:O : C : O:
 ••   ••   ••
     :O
      ••
```

Notice that the bottom oxygen atom does not have an octet; it has only three electron pairs. According to our guidelines, we can move a nonbonding electron pair. Let's shift the nonbonding pair between the C and the O atoms.

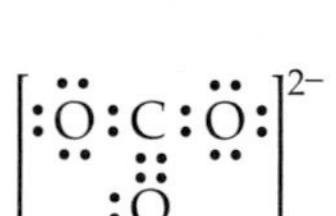

```
 ⎡ ••        •• ⎤ 2−
 ⎢:O : C : O:   ⎥
 ⎢ ••   ••   •• ⎥
 ⎢      ••      ⎥
 ⎢     :O       ⎥
 ⎣      ••      ⎦
```

The two electron pairs constitute a double bond. All four electrons in the double bond are shared by the C and the O atoms. Therefore, the central carbon atom has an octet and the oxygen atom has gained two electrons to complete its octet.

We use brackets to convey the idea that valence electrons are free to move over the entire polyatomic ion. Consequently, there appear to be three different structural formulas for the carbonate ion.

```
[O=C—O]2−     [O—C—O]2−     [O—C=O]2−
[   |  ]      [   ‖  ]      [   |  ]
[   O  ]      [   O  ]      [   O  ]
```

Although the structures for the carbonate ion appear to be different, they are all equivalent. No matter how we represent CO_3^{2-}, the structural formula has one double bond and two single bonds. Thus, we can draw three resonance structures for the carbonate ion. The following example exercise further illustrates how to draw electron dot and structural formulas for polyatomic ions:

Example Exercise 12.7 Electron Dot Formulas of Polyatomic Ions

Draw the electron dot formula and the structural formula for the sulfate ion, SO_4^{2-}.

Solution

The total number of valence electrons is the sum of each atom plus 2 for the negative charge: 6 e− + 4(6 e−) + 2 e− = 32 e−. The number of electron pairs is 16 (32/2 = 16). We begin by placing 4 electron pairs around the central sulfur atom and attaching the four oxygen atoms as follows:

```
    O
    ··
O : S : O
    ··
    O
```

We have 12 remaining electron pairs, and so we place 3 electron pairs around each oxygen atom.

```
[      ··      ]2−
[     :O:      ]
[  ··  ··  ··  ]
[ :O : S : O:  ]
[  ··  ··  ··  ]
[     :O:      ]
[      ··      ]
```

Notice that each atom is surrounded by an octet of electrons. This is the correct electron dot formula. In the structural formula each bonding electron pair is replaced by a single dash. The structural formula is

```
[      ··      ]2−
[     :O:      ]
[      |       ]
[  ··      ··  ]
[ :O — S — O:  ]
[  ··      ··  ]
[      |       ]
[     :O:      ]
[      ··      ]
```

Practice Exercise

Draw the electron dot formula and the structural formula for the nitrite ion, NO_2^{-}.

Answers:

```
[:Ö:N̈::Ö:]−      [:Ö—N̈=Ö:]−
```

Concept Exercise

Draw the electron dot formula for the polyatomic sulfite ion, SO_3^{2-}. How many pairs of nonbonding electrons are in a sulfite ion?

Answer: See Appendix G.

OBJECTIVES

- To describe and identify a polar covalent bond.
- To state the electronegativity trends in the periodic table.
- To apply delta notation (δ^+ and δ^-) to a polar bond.

12.6 POLAR COVALENT BONDS

Covalent bonds result from the sharing of valence electrons. To this point we have assumed that electrons are shared equally in a covalent bond. What about a covalent bond in which the two atoms do not share electrons equally? In many instances one of the two atoms holds the electron pair more tightly. When the electrons are drawn more closely to one of the atoms, the bond is said to be polarized. This type of bond is called a polar covalent bond, or simply a **polar bond**.

Electronegativity Trends

Each element has an innate ability to attract valence electrons. This ability is related to the nearness of the valence shell to the nucleus. It is also related to the magnitude of the positive charge in the nucleus. The ability of an atom to attract electrons in a chemical bond is referred to as its **electronegativity**. Atoms of elements that strongly attract bonding electrons are said to be highly electronegative.

Linus Pauling, an American chemist, devised a method for measuring the electronegativity values for each of the elements. He assigned carbon a value of 2.5 and then determined the ability of other elements to attract bonding electrons relative to carbon. He found that fluorine is the most electronegative element. It has a value of 4.0 compared to carbon. Other highly electronegative elements are oxygen, 3.5; nitrogen, 3.0; and chlorine, 3.0. The most electronegative elements are the nonmetals on the far right of the periodic table. Figure 12.9 shows selected elements in the periodic table and their Pauling electronegativity value.

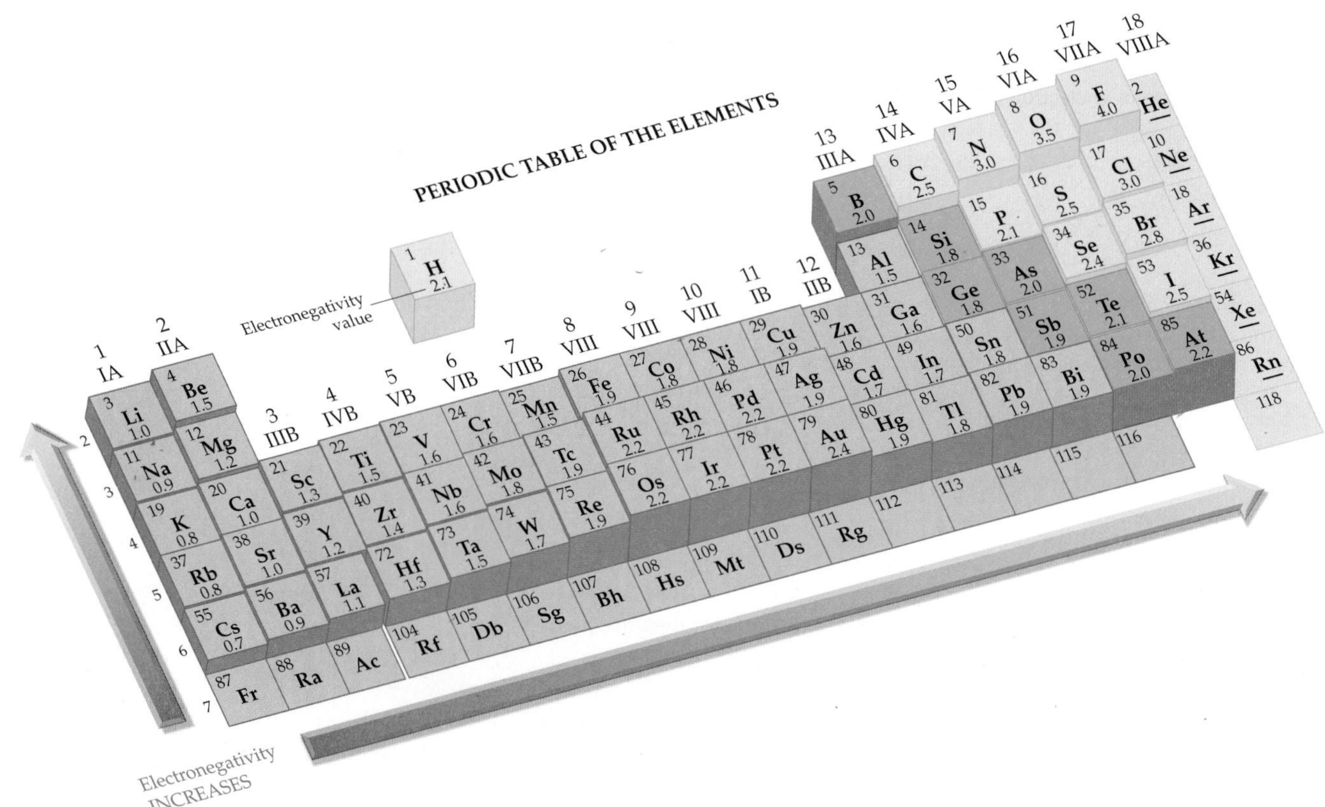

▲ **Figure 12.9 Electronegativity Values for the Elements** The Pauling electronegativity value for an element is shown below the symbol. In general, electronegativity increases across a period and up a group.

Note the electronegativity trends in Figure 12.9. First, the elements generally become more electronegative from *left to right* in the periodic table. Second, elements become more electronegative from *bottom to top* within a group.

We can see that the horizontal trend is consistent with the chemical properties of elements. That is, nonmetals react by gaining electrons, and metals react by losing electrons. The vertical trend follows the trend in nonmetallic character. As an example, examine Group VA/15. We see that Bi is a metal, Sb and As are semimetals, and P and N are nonmetals. Therefore, the trend in electronegativity increases corresponding to the trend in nonmetallic character. That is, nonmetals are more electronegative than semimetals, which in turn are more electronegative than metals.

The following example exercise illustrates the prediction of electronegativity values from the general trends in the periodic table:

Example Exercise 12.8 Electronegativity Predictions

Predict which element in each of the following pairs is more electronegative according to the general electronegativity trends in the periodic table:

(a) N or O (b) Br or Se
(c) F or Cl (d) Si or C

Solution

According to the trends in the periodic table, elements that lie to the right in a series or at the top of a group are more electronegative.

(a) O is more electronegative than N. (b) Br is more electronegative than Se.
(c) F is more electronegative than Cl. (d) C is more electronegative than Si.

Practice Exercise

Predict which element in each of the following pairs is more electronegative according to the general trends in the periodic table:

(a) H or Cl (b) Br or I
(c) P or S (d) As or Sb

Answers:

(a) Cl (b) Br
(c) S (d) As

Concept Exercise

State the general trends in the periodic table for increasing electronegativity.

Answer: See Appendix G.

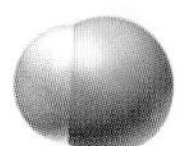

▲ **Hydrogen Halides** The two atoms in each hydrogen halide molecule are joined by a polar covalent bond owing to the different electronegativity value of each halogen.

Delta Notation for Polar Bonds

In Section 12.1, we introduced the covalent bond. In Figure 12.9, we see that the electronegativity value of H is 2.1 and of Cl is 3.0. Since there is a difference in electronegativity between the two elements (3.0 − 2.1 = 0.9), the bond in a H–Cl molecule is polar. Moreover, since Cl is the more electronegative element, the bonding electron pair is attracted toward the Cl atom and away from the H atom. The Cl atom thus becomes slightly negatively charged, whereas the H atom becomes slightly positively charged.

We can indicate a polar bond with a special symbol, the Greek letter delta (δ). We indicate the atom having a partially negative charge by the symbol δ^-. Similarly, we indicate the atom having a partially positive charge by the symbol δ^+. These symbols, δ^- and δ^+, are referred to as **delta notation**. We can apply delta notation to illustrate the polar bond in a HCl molecule as follows:

$$\delta^+ \text{H—Cl}\ \delta^-$$

CHEMISTRY CONNECTION Linus Pauling

Q. For which of his many accomplishments was Linus Pauling awarded the Nobel Prize in Chemistry?

Linus Pauling (1901–1994) is regarded by many as the most influential chemist of the twentieth century. Pauling was born in Portland, Oregon, to a family of modest means. His father died when Linus was quite young, and he had to take on a series of menial jobs to help support the family. At age 16, he enrolled at Oregon State University where he received a bachelor's degree in chemical engineering. There he met his wife and acknowledged inspiration, Ava Helen Miller, while teaching a chemistry class to home economics students.

Pauling decided on the California Institute of Technology for graduate school. In 1925, he received a Ph.D. and following graduation, was awarded a Guggenheim fellowship to study in Europe. In 1927, he returned to Caltech as a professor of chemistry and remained there for most of his career. Pauling received immediate recognition and was described by his department chairman in the following words:

> "Were all the rest of the chemistry department wiped away except Pauling, it would still be one of the most important departments of chemistry in the world."

◀ **Linus Pauling** (1901–1994)

Among his many accomplishments, Linus Pauling invented the concept of electronegativity and provided an ingenious method for calculating the electronegativity values of elements. One of his most brilliant achievements was an explanation for the unusual stability of the benzene molecule, C_6H_6, which was discussed in Section 9.9. His most celebrated work was investigating the shapes of proteins. He found that their structures resembled a helix and further established the relationship between abnormal molecular structure and genetic disease. His work in this area earned him the 1954 Nobel Prize in Chemistry.

In the 1960s, Pauling focused his energies on banning aboveground testing of nuclear weapons and pointed out the potential dangers of radiation. He spoke passionately against war and encouraged global disarmament. For his efforts, Pauling was awarded the 1962 Nobel Peace Prize. In the 1970s, he stirred controversy when he publicly advocated massive doses of vitamin C to combat the common cold. In addition to two Nobel prizes, Linus Pauling received numerous medals and awards, accepted honorary memberships in several scientific societies, published more than 400 research papers, and wrote the classic textbook *The Nature of the Chemical Bond.*

A: Linus Pauling won the 1954 Nobel Prize in Chemistry for his pioneering work on the structure of proteins.

Since the chlorine atom is more electronegative, it draws the electron pair away from the hydrogen atom. As a result, the hydrogen atom becomes slightly positively charged. The following example exercise illustrates the application of delta notation:

Example Exercise 12.9 Delta Notation for Polar Bonds

Calculate the electronegativity difference and apply delta notation to the bond between carbon and oxygen, C—O.

Solution

From Figure 12.9, we find that the electronegativity value of C is 2.5 and of O is 3.5. By convention, we always subtract the lesser value from the greater. The difference between the two elements is $3.5 - 2.5 = 1.0$. The result shows that the C—O bond is slightly polarized.

Since O is the more electronegative element, the bonding electron pair is drawn away from the C atom and toward the O atom. The O atom thus becomes slightly negatively charged, whereas the C atom becomes slightly positively charged. Applying the delta convention, we have

$$\delta^{+}\,\mathrm{C{-}O}\,\delta^{-}$$

Practice Exercise

From the trends in the periodic table, apply delta notation and label each atom in the following polar covalent bonds:

(a) N—O (b) Br—F

Answers: (a) δ^+ N—O δ^-; (b) δ^+ Br—F δ^-

Concept Exercise

Refer to the electronegativity values in Figure 12.9 and predict which of the following bonds is most polar: H–**N**, H–**O**, or H–**F**.

Answer: See Appendix G.

Note Delta convention indicates partially negative and partially positive atoms in a molecule. However, polar bonds in molecules differ from ionic bonds in formula units. Ionic bonds are formed between negative anions and positive cations. An electronegativity difference between two atoms that is greater than 1.8 indicates an ionic bond. For example, NaCl has an electronegativity difference of 2.1 (3.0 – 0.9) and it is clearly an example of an ionic bond.

12.7 NONPOLAR COVALENT BONDS

OBJECTIVES

- To describe and identify a nonpolar covalent bond.
- To identify seven elements that occur naturally as diatomic molecules: H_2, N_2, O_2, F_2, Cl_2, Br_2, I_2.

In Section 12.6, we learned that polar bonds result from the unequal sharing of bonding electrons. In a polar bond, one of the two atoms has a greater tendency to attract electrons. The atom having the greater electronegativity attracts the bonded electron pair more strongly and has a partial negative charge.

But what about a covalent bond in which each atom has the same or similar electronegativity? To answer this question, let's begin by reexamining the Pauling electronegativity values in Figure 12.9. We see that electronegativity increases across a period and up a group. Notice, however, that several elements have the same Pauling electronegativity value. For example, N and Cl have a value of 3.0, and C, S, and I all have a value of 2.5.

How then would we describe an S—I bond? Since the electronegativity of each atom is the same, the bond is not polarized. A covalent bond between two atoms with the same electronegativity is referred to as a nonpolar covalent bond, or simply a **nonpolar bond**. The following example exercise illustrates the classification of polar and nonpolar bonds:

Example Exercise 12.10 Polar versus Nonpolar Bonds

Classify each of the following as a polar or a nonpolar bond:

(a) Cl—Cl (b) Cl—N
(c) Cl—Br (d) Cl—H

Solution

From Figure 12.9 we find the following electronegativity values: Cl = 3.0, N = 3.0, Br = 2.8, and H = 2.1.

(a) The Cl—Cl bond (3.0 − 3.0 = 0) is *nonpolar*.
(b) The Cl—N bond (3.0 − 3.0 = 0) is *nonpolar*.
(c) The Cl—Br bond (3.0 − 2.8 = 0.2) is slightly *polar*.
(d) The Cl—H bond (3.0 − 2.1 = 0.9) is *polar*.

Practice Exercise

Classify each of the following as a polar or a nonpolar bond:

(a) C—C (b) C—O
(c) C—S (d) C—H

Answers: (a) nonpolar; (b) polar; (c) nonpolar; (d) slightly polar

Concept Exercise

Refer to the electronegativity values in Figure 12.9 and predict which of the following bonds is most nonpolar: C–**N**, C–**O**, or C–**I**.

Answer: See Appendix G.

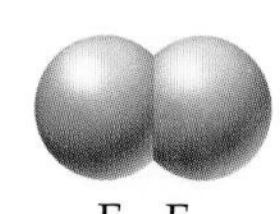

F—F

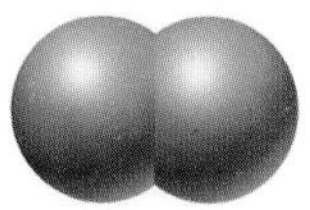

Cl—Cl

Br—Br

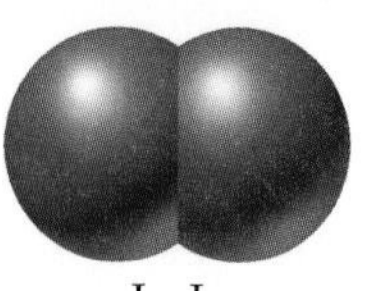

I—I

▲ **Halogens** The two atoms in each halogen molecule are joined by a nonpolar covalent bond.

Diatomic Nonpolar Molecules

A **diatomic molecule** consists of two nonmetal atoms joined by a covalent bond. The simplest example of a nonpolar bond is found between two identical atoms. For example, oxygen occurs naturally as a gas and contains O_2 molecules. Even though oxygen is quite electronegative, the bond in an O_2 molecule is nonpolar. This is because each oxygen atom attracts the electron pair with the same force.

In Section 6.3, we briefly studied the elements in Group VIIA/17. Recall that, collectively, these elements are called *halogens*. The halogens (F, Cl, Br, and I) are a family and have similar properties. All four elements occur naturally as diatomic molecules: F_2, Cl_2, Br_2, and I_2. Elemental fluorine is a yellowish-green gas, chlorine is a greenish-yellow gas, bromine is a reddish-brown liquid, and iodine is a dark-violet crystalline solid.

There are seven elements that exist naturally as diatomic molecules: H_2, N_2, O_2, F_2, Cl_2, Br_2, and I_2. Although each molecule is distinctly different, each exhibits a nonpolar covalent bond. For our discussion, we will classify a bond as nonpolar when there is no electronegativity difference between the two bonded atoms. In practice, a bond between two atoms having an electronegativity difference of 0.5 or less is usually considered a nonpolar bond.

OBJECTIVE

▸ To describe and identify a coordinate covalent bond.

12.8 COORDINATE COVALENT BONDS

We have described a covalent bond as an electron pair that is shared by two nonmetal atoms. Furthermore, each nonmetal atom is surrounded by nonbonding electron pairs to complete its octet.

Now, let's consider a special case of sharing electrons. An atom such as oxygen (six valence electrons) can attain an octet by sharing a nonbonding electron pair on another atom. A covalent bond resulting from one atom donating an electron pair to another atom is called a **coordinate covalent bond**. A good example of a coordinate covalent bond is found in a molecule of ozone, O_3. Here, an oxygen molecule donates a nonbonding electron pair to an oxygen atom to produce ozone.

$$:\ddot{O}::\ddot{O}: + \ddot{\underset{\cdot\cdot}{O}}: \longrightarrow :\ddot{O}::\ddot{O}:\ddot{\underset{\cdot\cdot}{O}}:$$

nonbonding electron pair   *coordinate covalent bond*

We can identify a coordinate covalent bond as follows. First, we draw the electron dot formula for the given molecule. If we can remove an atom and still have an octet around the remaining atoms, we know that the atom was joined by a coordinate covalent bond. The following example exercise further illustrates coordinate covalent bonds:

Example Exercise 12.11 Coordinate Covalent Bonds

Burning yellow sulfur powder produces sulfur dioxide, SO_2. Sulfur dioxide is a colorless gas with a suffocating odor. It is used to kill insect larvae and produces the odor released when a match is ignited. Draw the electron dot formula for $\mathbf{SO_2}$. Then show the formation of a coordinate covalent bond in $\mathbf{SO_3}$.

Solution

The total number of valence electrons in one molecule of SO_2 is 6 e− + 2(6 e−) = 18 e−. By trial and error, we find that the electron dot formula for SO_2 has a double bond. Since the sulfur atom has a nonbonding electron pair, it can bond to an additional oxygen atom. A diagram of the formation of the coordinate covalent bond is as follows:

$$:\ddot{O}:S: + \ddot{O}: \longrightarrow :\ddot{O}:S:\ddot{O}:$$

(with $:\ddot{O}$ bonded below S in each structure)

nonbonding electron pair *coordinate covalent bond*

Practice Exercise

A nitrogen molecule can form a coordinate covalent bond with an oxygen atom to give nitrous oxide, N_2O. Draw the electron dot formula for a nitrogen molecule and attach an oxygen atom.

Answers:

(a) :N:::N: (b) :N:::N:Ö:

Concept Exercise

Draw the electron dot formula for HClO and determine if the molecule contains a coordinate covalent bond.

Answer: See Appendix G.

12.9 HYDROGEN BONDS

OBJECTIVE

▶ To describe hydrogen bond attraction between two molecules.

When a molecule has a polar covalent bond, it has a region that is positively charged, and a region that is negatively charged. This usually produces a molecule that has a positive end and a negative end. We can visualize this type of molecule as being similar to a miniature magnet. Like two magnets that are attracted to each other, two polar molecules are attracted to each other.

This attraction between two molecules can be significant, but is much less than the attraction between two atoms in a covalent bond. The bond length between molecules is considerably longer than the bond length between atoms in a polar bond. The most common example of this type of attraction is found between water molecules.

Recall that the bonds between O–H in water are highly polar because of the large electronegativity difference between the two elements. Thus, in a water molecule the oxygen atom is partially negative, and each hydrogen atom is partially positive. As a result, a partially positive H atom in one molecule can be attracted to a partially negative oxygen atom in another molecule of water. This intermolecular attraction between H and O in different molecules is referred to as a **hydrogen bond**.

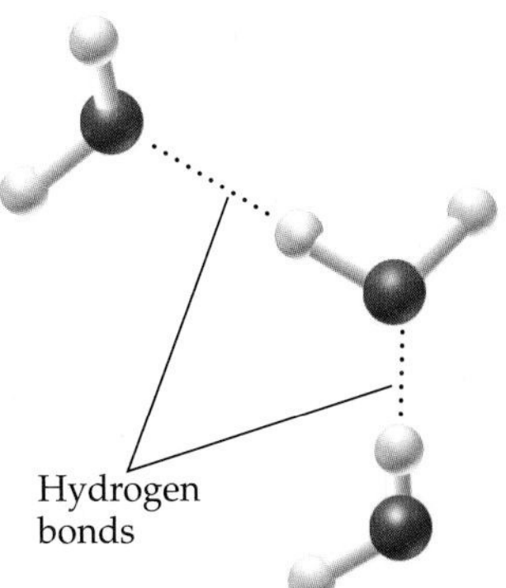

▲ **Hydrogen Bonding** This shows hydrogen bonds between water molecules.

Moreover, a hydrogen bond can result from the attraction of any two polar molecules having a hydrogen atom bonded to a highly electronegative atom (O, N, or F). Another example of hydrogen bonding occurs between ammonia molecules, NH_3. That is, the partially positive H atom in one ammonia molecule is attracted to the nonbonding electron pair of the electronegative nitrogen atom in a second molecule of ammonia.

Hydrogen bonds are especially important in living organisms where they are responsible for maintaining the structure of biochemical molecules (see Chapter 20). In large, complex biochemical molecules, hydrogen bonds are formed between H atoms and an electronegative atom (O or N) in the same molecule. In proteins, this can produce a single-stranded helix, while in DNA, hydrogen bonds are responsible for the double-stranded helix.

OBJECTIVE

- To determine the shape of a molecule by applying VSEPR theory.
- To explain how a molecule with polar bonds can be nonpolar.

12.10 SHAPES OF MOLECULES

In the 1950s, a simple theory was proposed to explain the shapes of molecules. This theory is an extension of electron dot formulas in which pairs of electrons surround a central atom. In essence, the theory states that the electron pairs surrounding an atom tend to repel each other and that the shape of the molecule is the result of this electron pair repulsion. This model is referred to as the **valence shell electron pair repulsion** (**VSEPR**, pronounced "vesper") theory. The VSEPR model portrays bonding and nonbonding electron pairs as occupying specific positions around the central atom in a molecule.

The VSEPR theory uses the term **electron pair geometry** to indicate the arrangement of bonding and nonbonding electron pairs around the central atom. VSEPR theory uses the term "molecular geometry," or **molecular shape**, to indicate the arrangement of atoms around the central atom as a result of electron pair repulsion. The angle formed by any two atoms bonded to the central atom is referred to as the **bond angle**.

Tetrahedral Molecules

Let's apply VSEPR theory to a methane molecule, CH_4, which has four pairs of bonding electrons. Since electron pairs repel each other, they try to get away from each other. The result is that in a molecule of CH_4, the central **C** atom is surrounded by four electron pairs that are repelled to the four corners of a three-dimensional figure called a tetrahedron. Each electron pair is bonded to a hydrogen atom, and so the shape of the molecule is said to be *tetrahedral*. Moreover, according to VSEPR theory, any molecule with four electron pairs around the central atom has a tetrahedral electron pair geometry.

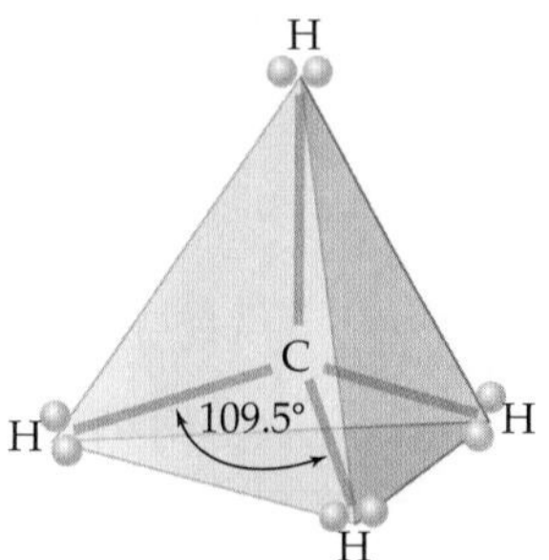

▲ The electron pair geometry in methane, CH_4, is *tetrahedral*.

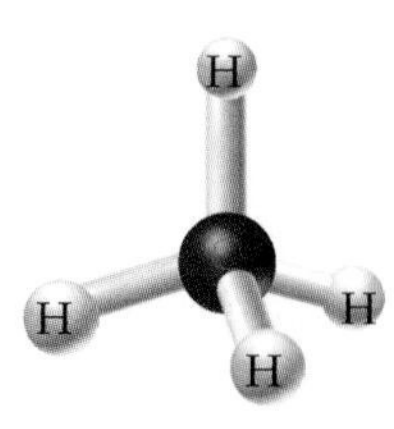

▲ The molecular shape of CH_4 is *tetrahedral*; the bond angle is 109.5°.

Trigonal Pyramidal Molecules

In a molecule of ammonia, NH_3, the central **N** atom is surrounded by three bonding electron pairs and one nonbonding electron pair. The electron pair geometry is once again tetrahedral, so ammonia should theoretically have bond angles of 109.5°. Experimentally, the bond angle in NH_3 is found to be 107°.

VSEPR theory explains the smaller bond angle by suggesting that the nonbonding electron pair exerts a stronger repelling force than the bonding pairs. Hence, the hydrogen atoms are pushed closer together, that is, from 109.5° to 107°. The molecular shape of NH_3 is said to be *trigonal pyramidal*.

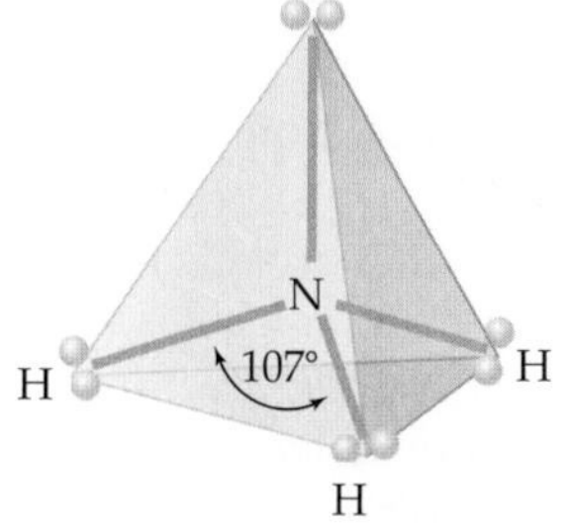

▲ The electron pair geometry in ammonia, NH_3, is *tetrahedral*.

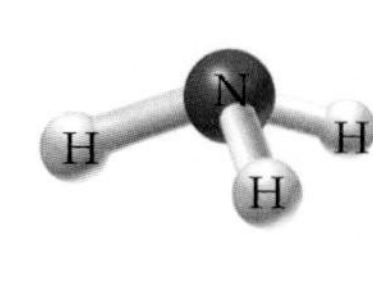

▲ The molecular shape of NH_3 is *trigonal pyramidal*; the bond angle is 107°.

Bent Molecules

In a molecule of water, $H_2\mathbf{O}$, the central **O** atom is surrounded by two bonding electron pairs and two nonbonding electron pairs. Since the electron pair geometry is tetrahedral, the predicted bond angle between the two hydrogen atoms is 109.5°. The experimentally measured bond angle in $H_2\mathbf{O}$ is only 104.5°. As in the preceding example, it is proposed that the nonbonding electrons in water exert a greater repelling force than the bonding pairs. The resulting bond angle is smaller, allowing more space for the unshared electron pairs. The molecular shape of $H_2\mathbf{O}$ is said to be *bent* or V-shaped.

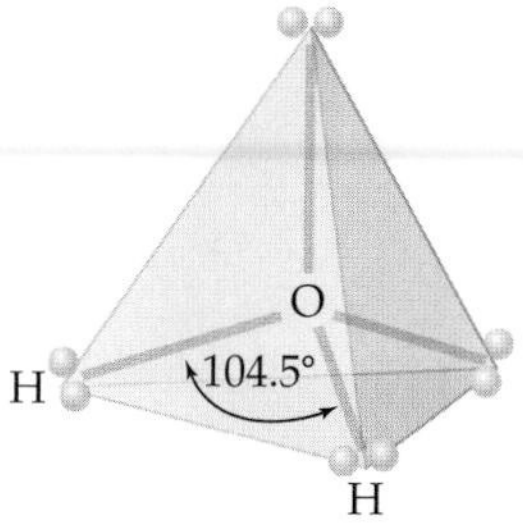

The electron pair geometry in water, $H_2\mathbf{O}$, is *tetrahedral.*

The molecular shape of $H_2\mathbf{O}$ is *bent*; the bond angle is 104.5°.

Linear Molecules

In a molecule of carbon dioxide, CO_2, the central **C** atom is bonded to each O atom by two bonding electron pairs; that is, there is a double bond to each O atom (O=**C**=O). Since the four pairs of electrons are on opposite sides of the **C** atom, the three atoms lie in a straight line. The electron pair geometry in CO_2 is said to be linear, and the predicted bond angle between the two oxygen atoms is 180°. Similarly, the molecular shape is said to be *linear*.

O::C::O

The electron pair geometry in carbon dioxide, CO_2, is *linear.*

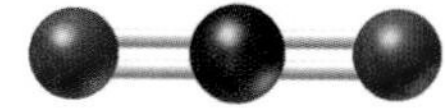

The molecular shape of is *linear*; the bond angle is 180°.

Trigonal Planar Molecules

In a molecule of formaldehyde, CH_2O, the central **C** atom is bonded to the O atom by a double bond (**C**=O), and to each hydrogen atom by a single bond. Since the bonding electron pairs repel one another, the three atoms are found at the corners of a triangle. The electron pair geometry in CH_2O is said to be trigonal planar, and the predicted bond angle between the atoms is 120°. Similarly, the molecular shape is said to be *trigonal planar*.

H:C::O
  ··
  H

The electron pair geometry in formaldehyde, CH_2O, is *trigonal planar.*

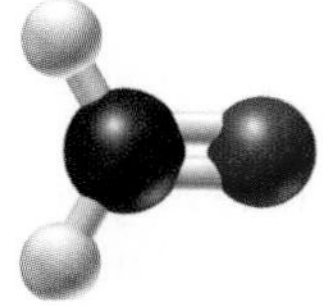

The molecular shape of CH_2O is *trigonal planar*; the bond angle is 120°.

We can summarize the information obtained from VSEPR theory about these types of molecules as shown in Table 12.1.

TABLE 12.1 SUMMARY OF VSEPR THEORY

BONDING/NONBONDING ELECTRON PAIRS	ELECTRON PAIR GEOMETRY	MOLECULAR SHAPE	BOND ANGLE	EXAMPLE MOLECULE
4 / 0	tetrahedral	tetrahedral	109.5°	CH_4
3 / 1	tetrahedral	trigonal pyramidal	107°	NH_3
2 / 2	tetrahedral	bent	104.5°	H_2O
4 / 0	linear	linear	180°	CO_2
4 / 0	trigonal planar	trigonal planar	120°	CH_2O

Nonpolar Molecules with Polar Bonds

Why is carbon tetrachloride, CCl_4, a nonpolar molecule even though it has polar bonds? Recall that a chlorine atom is more electronegative than a carbon atom; hence, the **C**–Cl bond is polar. Furthermore, there are four **C**–Cl polar bonds in a CCl_4 molecule.

To explain why a CCl_4 molecule is nonpolar, we can apply VSEPR theory. There are four bonding electron pairs in CCl_4; thus, the electron pair geometry and the molecular shape are both tetrahedral. We illustrate the CCl_4 molecule as having four chlorine atoms at the corners of a tetrahedron.

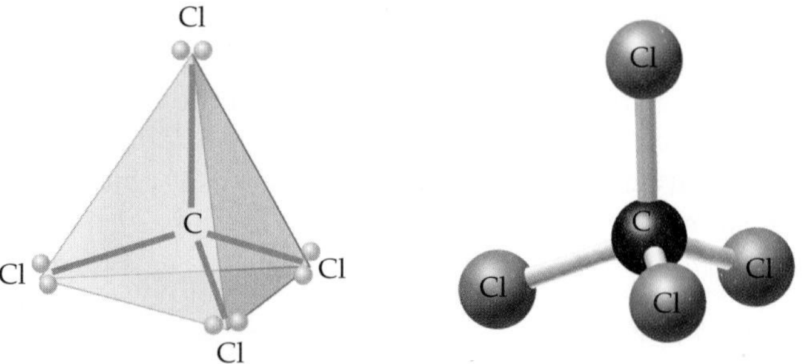

Note that each chlorine atom exerts equal attractive forces in four opposing directions. Thus, the attractions exerted by the four polar bonds cancel each other. The net effect of the polar bonds is zero, and overall, the molecule is considered nonpolar. We can illustrate that the CCl_4 molecule is nonpolar as follows:

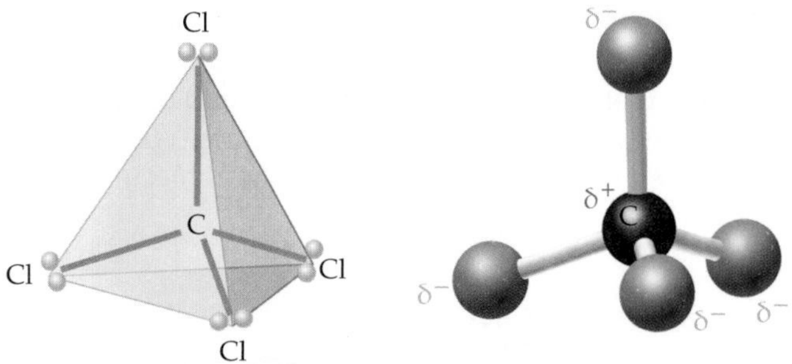

Consequently, a molecule may contain polar bonds and yet be nonpolar. In carbon dioxide, CO_2, there are two polar bonds. An O=**C**=O molecule is nonpolar because the more electronegative oxygen atoms pull equally in opposite directions.

On the other hand, formaldehyde, H_2**C**=O, is a polar molecule. The reason that formaldehyde is polar is that the H_2**C**=O molecule has only one oxygen atom and has a flat, triangular shape. The oxygen atom pulls electrons away from the central carbon atom and the two hydrogen atoms have little effect.

CRITICAL THINKING Diamond versus Graphite

Q: How is it possible that a natural diamond and the graphite in a pencil are both examples of pure carbon?

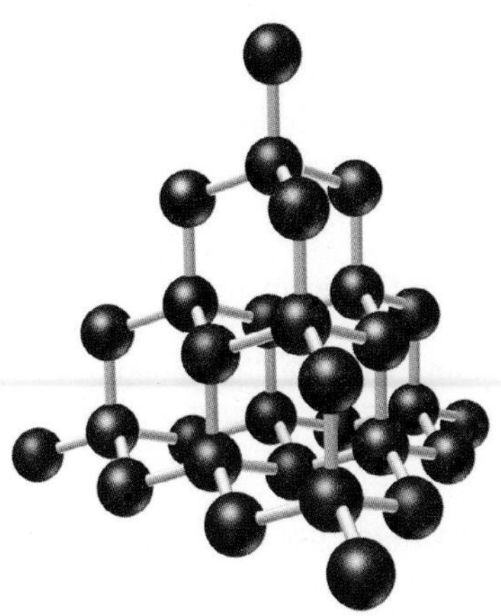
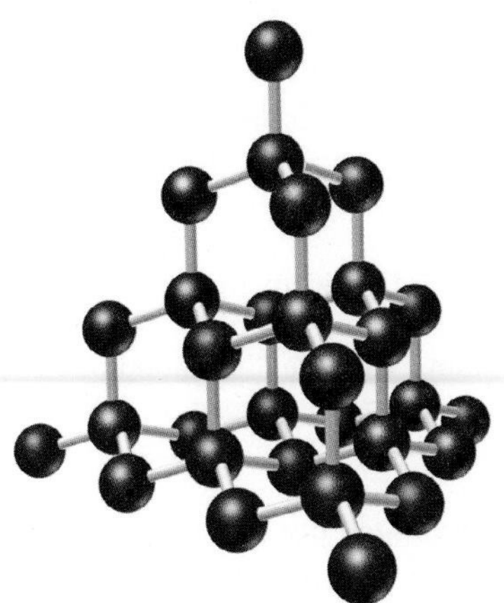

▲ **Diamond—tetrahedral structure**

Diamonds have been prized for centuries owing to their clarity and fiery sparkle. Diamonds originate deep within Earth where they are formed under conditions of extreme heat and pressure. A diamond is the hardest known substance; its melting point is above 3500 °C, and it is inert to corrosive chemicals.

The hardness and crystalline structure of a diamond enables a rough stone to be cut so as to give numerous tiny facets. On average, an uncut stone loses 50% of its original weight in becoming a finished diamond. Moreover, a misdirected blow by a diamond cutter can shatter the stone into small shards.

Diamond is a crystalline solid containing carbon atoms arranged in a three-dimensional network. Carbon atoms are bonded in a repeating tetrahedral structure as shown in the illustration. The more extensive the repeating network, the larger the actual diamond.

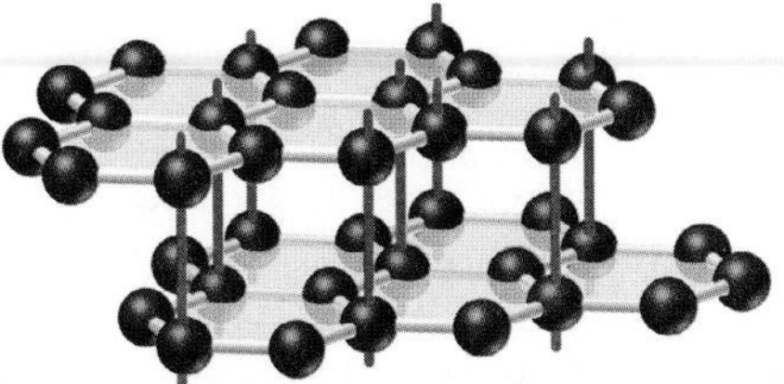

▲ **Graphite—planar structure**

Like diamond, graphite is also a crystalline solid composed exclusively of carbon atoms. Unlike diamond, graphite is a soft substance that we know as the dark gray "lead" in an ordinary pencil. The structure of graphite consists of fused hexagon rings, each containing six carbon atoms. Unlike diamond, graphite is arranged in a flat, two-dimensional crystalline network. The lubricating property of graphite is due to "sheets" of carbons rings sliding past one another.

Diamonds can be synthesized from graphite by mimicking the conditions under which natural diamonds are formed. A sample of graphite is placed in a hydraulic press and converted to diamond in a few hours at high temperature and pressure. Typically, synthetic diamonds are only a few millimeters in size and are too flawed to be used as gemstones. However, synthetic diamonds are valuable industrially for use in coating drill bits and cutting tools.

◀ **Diamond and Graphite** A synthetic diamond made from powdered graphite is shown alongside the graphite in the tip of a pencil. Diamond and graphite are both examples of crystalline carbon.

A: Although diamond and graphite are both composed of pure carbon, the structure of diamond is a repeating tetrahedron, whereas the structure of graphite is a repeating hexagon.

Summary

Section 12.1 G. N. Lewis first described the **chemical bond** when he proposed that **valence electrons** in different atoms interact in such a way that each atom completes its valence shell. Since eight electrons are found in the valence shells of noble gas atoms, the Lewis theory became known as the *rule of eight*, or more commonly, the **octet rule**.

The simplest representative particle in a substance held together by ionic bonds is a **formula unit**. When a metal atom loses valence electrons as a nonmetal atom gains electrons, an **ionic bond** results. The simplest representative particle held together by covalent bonds is a **molecule**. When valence electrons are shared by two nonmetal atoms, a **covalent bond** results.

Section 12.2 While forming ions, metal atoms lose valence electrons and are often left with a noble gas structure. Nonmetal atoms gain electrons to complete their valence shells. In the process of forming ions, the radii of metal atoms become smaller while the radii of nonmetal atoms become larger. The formation of ionic bonds between metal and nonmetal ions always releases heat energy.

Section 12.3 In the formation of covalent bonds, the outer valence shells of two nonmetal atoms overlap and share electrons. Each of the atoms in the bond shares the number of electrons necessary to complete their individual octets. The **bond length** between the two atoms is always less than the sum of their atomic radii. When atoms of different elements bond, energy is always released. The amount of energy released during the formation of a bond is exactly equal to the amount of energy required to break that bond. The amount of energy required to break a covalent bond between two atoms is called the **bond energy**.

Section 12.4 To write an **electron dot formula**, we first calculate the total number of valence electrons from all of the atoms in the molecule. We then divide the valence electron total by two to find the number of electron pairs. The electron pairs are placed around the central atom and then the remaining atoms in the molecule so as to provide octets. One pair of electrons shared by two atoms is a **single bond**. A molecule can also contain two or three electron pairs between two atoms. These bonds are referred to as a **double bond** and a **triple bond**, respectively.

Double and triple bonds result from an insufficient number of valence electrons around each atom in a molecule. To provide an octet, it may be necessary to move **nonbonding electrons** between two atoms; these electrons then become **bonding electrons**. In the **structural formula** of a molecule, a single bond is shown as a dash, a double bond as two dashes, and a triple bond as three dashes.

Section 12.5 An ionic bond can be formed by the attraction between a simple **monoatomic ion** and a complex **polyatomic ion**. Ammonium chloride, NH_4Cl, and cupric sulfate, $CuSO_4$, each contain a polyatomic ion. To draw the electron dot formula for a polyatomic ion, we must find the total number of valence electrons for each atom and then add or subtract the number of electrons equal to its ionic charge. For the ammonium cation, NH_4^+, the total number of valence electrons is 8 (9 e– minus 1 e–). For the sulfate ion, SO_4^{2-}, the total number of valence electrons is 32 e– (30 e– plus 2 e–).

Section 12.6 If one atom in a bond attracts electrons more strongly than the other, that atom is said to have a higher **electronegativity**. A **polar bond** results when one of the bonded atoms has a greater attraction for the electron pair. We can indicate a polar bond using **delta notation**: the more electronegative atom is labeled δ^- and the less electronegative atom is labeled δ^+.

Section 12.7 If each atom in a bond has an equal attraction for bonding electrons, the bond is said to be a **nonpolar bond**. A nonpolar bond always occurs in a **diatomic molecule** having two identical atoms: the bonds in H_2, N_2, O_2, F_2, Cl_2, Br_2, and I_2 are all examples of nonpolar bonds.

Section 12.8 A **coordinate covalent bond** is a special type of covalent bond and is formed when one atom donates an electron pair to another atom. A simple illustration of a coordinate covalent bond occurs when the chlorine atom in HCl donates an electron pair to an oxygen atom to give HClO.

Section 12.9 A **hydrogen bond** usually refers to the attraction between a hydrogen atom in one molecule, and an unshared electron pair on a highly electronegative atom (O or N) in a second molecule. For example, this attraction occurs between water molecules, when the partially positive H atom in one molecule attracts the unshared electron pair on the electronegative O atom in another molecule.

Section 12.10 After writing the electron dot formula for a molecule, we can draw its shape. That is, we can predict the shape of a molecule by applying valence shell electron pair repulsion (VSEPR) theory. According to **VSEPR theory**, bonding and nonbonding electron pairs surround the central atom. The arrangement of electron pairs around the central atom is termed **electron pair geometry**, and the arrangement of atoms is termed **molecular shape**. The angle formed by two atoms bonded to the central atom is referred to as the **bond angle**.

If there are four electron pairs around the central atom, they are located at the corners of a three-dimensional figure called a tetrahedron. Examples of molecules having four electron pairs around the central atom include methane, CH_4, ammonia, NH_3, and water, H_2O. In each of these examples, the electron pair geometry is said to be tetrahedral.

Key Terms *Answers to Key Terms are in Appendix H.*

Select the key term that corresponds to each of the following definitions.

_____ **1.** the general term for the attraction between two ions or two atoms
_____ **2.** the electrons in the outermost energy level that are available for bonding
_____ **3.** the statement that an atom tends to bond in such a way so as to acquire eight valence electrons
_____ **4.** a chemical bond characterized by the attraction between a cation and an anion
_____ **5.** a chemical bond characterized by the sharing of one or more pairs of valence electrons
_____ **6.** the simplest representative particle in a substance held together by ionic bonds
_____ **7.** the simplest representative particle in a substance held together by covalent bonds
_____ **8.** the distance between the nuclei of two atoms joined by a covalent bond
_____ **9.** the amount of energy required to break a covalent bond between two atoms
_____ **10.** the valence electrons in a molecule that are shared
_____ **11.** the valence electrons in a molecule that are not shared
_____ **12.** a diagram of a molecule that shows the chemical symbol of each atom and a dot representing each bonding and nonbonding electron

(a) bond angle *(Sec. 12.10)*
(b) bond energy *(Sec. 12.3)*
(c) bond length *(Sec. 12.3)*
(d) bonding electrons *(Sec. 12.4)*
(e) chemical bond *(Sec. 12.1)*
(f) coordinate covalent bond *(Sec. 12.8)*
(g) covalent bond *(Sec. 12.1)*
(h) delta (δ) notation *(Sec. 12.6)*
(i) diatomic molecule *(Sec. 12.7)*
(j) double bond *(Sec. 12.4)*
(k) electron dot formula *(Sec. 12.4)*

_____ 13. a diagram of a molecule that shows the chemical symbol of each atom and a dash representing each pair of bonding electrons
_____ 14. a bond composed of one electron pair shared by two atoms
_____ 15. a bond composed of two electron pairs shared by two atoms
_____ 16. a bond composed of three electron pairs shared by two atoms
_____ 17. a single atom that bears a negative or positive charge as the result of gaining or losing valence electrons
_____ 18. a group of atoms held together by covalent bonds and having an overall positive or negative charge
_____ 19. the ability of an atom to attract a shared pair of electrons
_____ 20. a method of indicating a partial positive and a partial negative charge in a chemical bond
_____ 21. a bond in which a pair of electrons is shared unequally
_____ 22. a bond in which a pair of electrons is shared equally
_____ 23. a molecule composed of two nonmetal atoms held together by a covalent bond
_____ 24. a bond in which an electron pair is shared but both electrons have been donated by a single atom
_____ 25. the attraction between a H atom bonded to O or N, and a nonbonding electron pair on an electronegative atom in another molecule
_____ 26. a model that explains the shapes of molecules as a result of electron pairs about the central atom repelling each other
_____ 27. the geometric shape formed by bonding and nonbonding electron pairs about the central atom in a molecule
_____ 28. the geometric shape formed by atoms bonded to the central atom in a molecule
_____ 29. the angle formed by two atoms attached to the central atom in a molecule

(l) electron pair geometry *(Sec. 12.10)*
(m) electronegativity *(Sec. 12.6)*
(n) formula unit *(Sec. 12.1)*
(o) hydrogen bond *(Sec. 12.9)*
(p) ionic bond *(Sec. 12.1)*
(q) molecular shape *(Sec. 12.10)*
(r) molecule *(Sec. 12.1)*
(s) monoatomic ion *(Sec. 12.5)*
(t) nonbonding electrons *(Sec. 12.4)*
(u) nonpolar bond *(Sec. 12.7)*
(v) octet rule *(Sec. 12.1)*
(w) polar bond *(Sec. 12.6)*
(x) polyatomic ion *(Sec. 12.5)*
(y) single bond *(Sec. 12.4)*
(z) structural formula *(Sec. 12.4)*
(aa) triple bond *(Sec. 12.4)*
(ab) valence electrons *(Sec. 12.1)*
(ac) VSEPR theory *(Sec. 12.10)*

Exercises *Answers to odd-numbered Exercises are in Appendix I.*

The Chemical Bond Concept (Sec. 12.1)

1. State the number of valence electrons for a potassium atom and an iodine atom.
2. State the number of valence electrons for a hydrogen atom and an iodine atom.
3. State the number of valence electrons for a potassium ion and an iodide ion in an ionic bond.
4. State the number of valence electrons for a hydrogen atom and an iodine atom in a covalent bond.
5. Predict whether each of the following is held together by ionic or covalent bonds:
(a) aluminum chloride, $AlCl_3$ (b) water, H_2O
(c) sulfur trioxide, SO_3 (d) iron(II) sulfate, $FeSO_4$
6. Predict whether each of the following is held together by ionic or covalent bonds:
(a) zinc bromide, $ZnBr_2$ (b) carbon dioxide, CO_2
(c) iodine heptafluoride, IF_7 (d) lead(II) sulfate, $PbSO_4$
7. State whether the representative particle in each of the following substances is a formula unit or a molecule:
(a) acetic acid, $HC_2H_3O_2$ (b) chlorine dioxide, ClO_2
(c) sodium chlorate, $NaClO_3$ (d) titanium(IV) oxide, TiO_2
8. State whether the representative particle in each of the following substances is a formula unit or a molecule:
(a) methanol, CH_3OH (b) cobaltous chloride, $CoCl_2$
(c) acetone, CH_3COCH_3 (d) stannous carbonate, $SnCO_3$
9. State whether the representative particle in each of the following substances is an atom, a formula unit, or a molecule:
(a) propane, C_3H_8 (b) platinum, Pt
(c) hematite, Fe_2O_3 (d) sulfur, S_8
10. State whether the representative particle in each of the following substances is an atom, a formula unit, or a molecule:
(a) xenon, Xe (b) phosphorus, P_4
(c) xenon trioxide, XeO_3 (d) magnetite, Fe_3O_4

Ionic Bonds (Sec. 12.2)

11. Use the periodic table to predict an ionic charge for each of the following metal ions:
(a) Li ion (b) Sr ion
(c) Al ion (d) Pb ion
12. Use the periodic table to predict an ionic charge for each of the following metal ions:
(a) Na ion (b) Ba ion
(c) Ga ion (d) Sn ion
13. Use the periodic table to predict the ionic charge for each of the following nonmetal ions:
(a) Cl ion (b) I ion
(c) S ion (d) P ion
14. Use the periodic table to predict the ionic charge for each of the following nonmetal ions:
(a) F ion (b) Br ion
(c) Se ion (d) N ion
15. Write out the electron configuration for each of the following metal ions:
(a) Li^+ (b) Al^{3+}
(c) Ca^{2+} (d) Mg^{2+}
16. Write out the electron configuration for each of the following metal ions:
(a) Sc^{3+} (b) K^+
(c) Ti^{4+} (d) Ba^{2+}

17. Write out the electron configuration for each of the following nonmetal ions:
(a) Cl^- (b) I^-
(c) S^{2-} (d) P^{3-}

18. Write out the electron configuration for each of the following nonmetal ions:
(a) Br^- (b) O^{2-}
(c) Se^{2-} (d) N^{3-}

19. Refer to the periodic table and state the noble gas with an electron configuration identical to each of the following ions:
(a) S^{2-} (b) Cl^-
(c) K^+ (d) Ca^{2+}

20. Refer to the periodic table and state the noble gas with an electron configuration identical to each of the following ions:
(a) Se^{2-} (b) Br^-
(c) Rb^+ (d) Sr^{2+}

21. Which noble gas is isoelectronic with each of the following metal ions?
(a) Li^+ (b) Al^{3+}
(c) Ca^{2+} (d) Mg^{2+}

22. Which noble gas is isoelectronic with each of the following metal ions?
(a) Sc^{3+} (b) K^+
(c) Ti^{4+} (d) Ba^{2+}

23. Which noble gas is isoelectronic with each of the following nonmetal ions?
(a) Cl^- (b) I^-
(c) S^{2-} (d) P^{3-}

24. Which noble gas is isoelectronic with each of the following nonmetal ions?
(a) Br^- (b) O^{2-}
(c) Se^{2-} (d) N^{3-}

25. In each of the following pairs, which has the larger radius?
(a) Li atom or Li ion (b) Mg atom or Mg ion
(c) F atom or F ion (d) O atom or O ion

26. In each of the following pairs, which has the larger radius?
(a) Al atom or Al ion (b) Pb atom or Pb ion
(c) Se atom or Se ion (d) N atom or N ion

27. Which of the following statements are false regarding an ionic bond between cobalt and sulfur in a CoS formula unit?
(a) Cobalt and sulfide ions bond by electrostatic attraction.
(b) Cobalt atoms gain electrons and sulfur atoms lose electrons.
(c) The ionic radius of a cobalt ion is greater than its atomic radius.
(d) Breaking an ionic bond between cobalt and sulfur requires energy.

28. Which of the following statements are false regarding an ionic bond between aluminum and nitrogen in an AlN formula unit?
(a) Aluminum and nitride ions bond by covalent attraction.
(b) Aluminum atoms lose electrons and nitrogen atoms gain electrons.
(c) The ionic radius of an aluminum ion is less than its atomic radius.
(d) Breaking an ionic bond between aluminum and nitrogen releases energy.

Covalent Bonds (Sec. 12.3)

29. Which of the following is less?
(a) the sum of the H and Cl atomic radii, or the bond length in H–Cl
(b) the sum of the O and O atomic radii, or the bond length in O=O

30. Which of the following is less?
(a) the sum of the H and Br atomic radii, or the bond length in H–Br
(b) the sum of the S and O atomic radii, or the bond length in S=O

31. Which of the following statements are false regarding a covalent bond between carbon and oxygen atoms in a carbon monoxide, CO, molecule?
(a) Valence electrons are transferred from carbon to oxygen atoms.
(b) Bonding electrons are distributed over the entire CO molecule.
(c) The bond length is less than the sum of the two atomic radii.
(d) Energy is released when a covalent bond is broken.

32. Which of the following statements are false regarding a covalent bond between nitrogen and oxygen atoms in a nitric oxide, NO, molecule?
(a) Valence electrons are shared between nitrogen and oxygen atoms.
(b) Bonding electrons are found only between the bonded atoms.
(c) The bond length is greater than the sum of the two atomic radii.
(d) Energy is released when a covalent bond is formed.

Electron Dot Formulas of Molecules (Sec. 12.4)

33. Write the electron dot formula and draw the structural formula for each of the following molecules. *(The central atom is indicated in* **bold**.*)*
(a) H_2 (b) F_2
(c) H**Br** (d) **N**H_3

34. Write the electron dot formula and draw the structural formula for each of the following molecules:
(a) Cl_2 (b) O_2
(c) HI (d) PH_3

35. Write the electron dot formula and draw the structural formula for each of the following molecules:
(a) HONO (b) SO_2
(c) C_2H_4 (d) C_2H_2

36. Write the electron dot formula and draw the structural formula for each of the following molecules:
(a) N_2 (b) PI_3
(c) CS_2 (d) HOCl

37. Write the electron dot formula and draw the structural formula for each of the following molecules:
(a) CH_4 (b) OF_2
(c) H_2O_2 (d) NF_3

38. Write the electron dot formula and draw the structural formula for each of the following molecules:
(a) CCl_4 (b) $HONO_2$
(c) CH_3OH (d) HOCN

Electron Dot Formulas of Polyatomic Ions (Sec. 12.5)

39. Write the electron dot formula and draw the structural formula for each of the following polyatomic ions. *(Central atoms are indicated in **bold**.)*
(a) **Br**O^- (b) **Br**O_2^-
(c) **Br**O_3^- (d) **Br**O_4^-

40. Write the electron dot formula and draw the structural formula for each of the following polyatomic ions:
(a) IO^- (b) IO_2^-
(c) IO_3^- (d) IO_4^-

41. Write the electron dot formula and draw the structural formula for each of the following polyatomic ions:
(a) SO_4^{2-} (b) HSO_4^-
(c) SO_3^{2-} (d) HSO_3^-

42. Write the electron dot formula and draw the structural formula for each of the following polyatomic ions:
(a) PO_4^{3-} (b) HPO_4^{2-}
(c) PO_3^{3-} (d) HPO_3^{2-}

43. Write the electron dot formula and draw the structural formula for each of the following polyatomic ions:
(a) H_3**O**$^+$ (b) OH^-
(c) HS^- (d) CN^-

44. Write the electron dot formula and draw the structural formula for each of the following polyatomic ions:
(a) PH_4^+ (b) **Se**O_3^{2-}
(c) CO_3^{2-} (d) **B**O_3^{3-}

Polar Covalent Bonds (Sec. 12.6)

45. What is the general trend in electronegativity within a group of elements in the periodic table?

46. What is the general trend in electronegativity within a period of elements in the periodic table?

47. Which elements are more electronegative: metals or nonmetals?

48. Which elements are more electronegative: semimetals or nonmetals?

49. Predict which element in each of the following pairs is more electronegative according to the general trends in the periodic table:
(a) Br or Cl (b) O or S
(c) Se or As (d) N or F

50. Predict which element in each of the following pairs is more electronegative according to the general trends in the periodic table:
(a) Se or Br (b) C or B
(c) Te or S (d) Ba or Be

51. Refer to the values in Figure 12.9 and calculate the electronegativity difference for each of the following bonds:
(a) Br—Cl (b) Br—F
(c) I—Cl (d) I—Br

52. Refer to the values in Figure 12.9 and calculate the electronegativity difference for each of the following bonds:
(a) H—Cl (b) H—Br
(c) N—O (d) C—O

53. Refer to Figure 12.9 and label each atom in the following polar covalent bonds using delta notation (δ^+ and δ^-):
(a) H—S (b) O—S
(c) N—F (d) S—Cl

54. Refer to Figure 12.9 and label each atom in the following polar covalent bonds using delta notation:
(a) C—H (b) Se—O
(c) P—I (d) H—Br

Nonpolar Covalent Bonds (Sec. 12.7)

55. Refer to Figure 12.9 and indicate which of the following are nonpolar covalent bonds:
(a) Cl—Cl (b) Cl—N
(c) N—H (d) H—P

56. Refer to Figure 12.9 and indicate which of the following are nonpolar covalent bonds:
(a) I—C (b) C—S
(c) S—H (d) H—Br

57. Which of the following elements occur naturally as diatomic molecules: H, N, Cl, Br, Ar?

58. Which of the following elements occur naturally as diatomic molecules: He, O, F, I, Ne?

Coordinate Covalent Bonds (Sec. 12.8)

59. An oxygen atom can bond to a hydrogen bromide molecule to give HBrO. Draw the electron dot formula for H**Br**O and label a coordinate covalent bond.

60. An oxygen atom can bond to a hydrogen iodide molecule to give HIO. Draw the electron dot formula for HIO and label a coordinate covalent bond.

61. An oxygen atom can bond to a HBrO molecule to give $HBrO_2$. Draw the electron dot formula for H**Br**O_2 and label a coordinate covalent bond.

62. An oxygen atom can bond to a HIO molecule to give HIO_2. Draw the electron dot formula for HIO_2 and label a coordinate covalent bond.

63. A hydrogen ion can bond to an ammonia molecule, NH_3, forming NH_4^+. Draw the electron dot formula for **N**H_4^+ and label a coordinate covalent bond.

64. A hydrogen ion can bond to a phosphine molecule, PH_3, forming PH_4^+. Draw the electron dot formula for **P**H_4^+ and label a coordinate covalent bond.

65. A nitrite ion, NO_2^-, can bond to an oxygen atom to form the nitrate ion. Draw the electron dot formula for **N**O_3^- and label a coordinate covalent bond.

66. A phosphite ion, PO_3^{3-}, can bond to an oxygen atom to form the phosphate ion. Draw the electron dot formula for **P**O_4^{3-} and label a coordinate covalent bond.

Hydrogen Bonds (Sec. 12.9)

67. Diagram a hydrogen bond between two molecules of ammonia, NH_3.

68. Diagram a hydrogen bond between two molecules of hydrogen fluoride, HF.

69. Which is greater, the energy to break a hydrogen bond or a polar covalent bond?

70. Which is longer, the length of a hydrogen bond or a polar covalent bond?

Shapes of Molecules (Sec. 12.10)

71. Predict the electron pair geometry, the molecular shape, and the bond angle for a silane molecule, SiH_4, using VSEPR theory.
72. Predict the electron pair geometry, the molecular shape, and the bond angle for a carbon tetrabromide molecule, CBr_4, using VSEPR theory.
73. Predict the electron pair geometry, the molecular shape, and the bond angle for a nitrogen triiodide molecule, NI_3, using VSEPR theory.
74. Predict the electron pair geometry, the molecular shape, and the bond angle for a phosphine molecule, PH_3, using VSEPR theory.
75. Predict the electron pair geometry, the molecular shape, and the bond angle for a hydrogen sulfide molecule, H_2S, using VSEPR theory.
76. Predict the electron pair geometry, the molecular shape, and the bond angle for a dichlorine monoxide molecule, Cl_2O, using VSEPR theory.
77. Apply VSEPR theory to explain why CF_4 is a nonpolar molecule even though it has four polar bonds.
78. Apply VSEPR theory to explain why CO_2 is a nonpolar molecule even though it has two polar bonds.

General Exercises

79. State whether the representative particle in each of the following substances is an atom, a molecule, or a formula unit:
(a) uranium, U
(b) fluorine, F_2
(c) uranium hexafluoride, UF_6
(d) hydrogen fluoride, HF
80. State whether the representative particle in each of the following substances is an atom, a molecule, or a formula unit:
(a) plutonium, Pu
(b) oxygen, O_2
(c) plutonium(III) oxide, Pu_2O_3
(d) hydrogen peroxide, H_2O_2
81. Write formula units by combining the cations and anions in each of the following pairs:
(a) Sr^{2+} and As^{3-} (b) Ra^{2+} and O^{2-}
(c) Al^{3+} and $CO_3^{\ 2-}$ (d) Cd^{2+} and OH^-
82. Write formula units by combining the cations and anions in each of the following pairs:
(a) Sc^{3+} and N^{3-} (b) Ti^{4+} and O^{2-}
(c) $NH_4^{\ +}$ and $CO_3^{\ 2-}$ (d) $Hg_2^{\ 2+}$ and $PO_4^{\ 3-}$
83. Explain why the radius of a sodium ion (0.095 nm) is about half that of a sodium atom (0.186 nm).
84. Explain why the radius of a chloride ion (0.181 nm) is about twice that of a chlorine atom (0.099 nm).
85. Refer to the values in Figure 12.9 and calculate the electronegativity difference in a B—Cl bond.
86. Refer to the values in Figure 12.9 and calculate the electronegativity difference in a Sb—Cl bond.
87. Refer to the values in Figure 12.9 and calculate the electronegativity difference in a H—P bond.
88. Refer to the values in Figure 12.9 and calculate the electronegativity difference in a S—I bond.
89. Label the polar Ge—Cl bond using delta notation (δ^+ and δ^-).
90. Label the polar As—Cl bond using delta notation (δ^+ and δ^-).
91. Write the electron dot formula and draw the structural formula for silane, SiH_4, whose central atom is a semimetal.
92. Write the electron dot formula and draw the structural formula for stibine, SbH_3, whose central atom is a semimetal.
93. Write the electron dot formula and draw the structural formula for the arsenate ion, $AsO_3^{\ 3-}$, whose central atom is a semimetal.
94. Write the electron dot formula and draw the structural formula for the silicate ion, $SiO_3^{\ 2-}$, whose central atom is a semimetal.
95. Predict the electron pair geometry, the molecular shape, and the bond angle for a carbon disulfide molecule, CS_2, using VSEPR theory.
96. Predict the electron pair geometry, the molecular shape, and the bond angle for a silicon dioxide molecule, SiO_2, using VSEPR theory.
97. Predict the electron pair geometry, the molecular shape, and the bond angle for a phosgene molecule, CCl_2O, using VSEPR theory.
98. Predict the electron pair geometry, the molecular shape, and the bond angle for a carbonate ion, $CO_3^{\ 2-}$, using VSEPR theory.
99. Xenon dioxide is a slightly stable molecule that contains a noble gas. Write the electron dot formula and draw the structural formula for XeO_2.
100. Xenon trioxide is a slightly stable molecule that contains a noble gas. Write the electron dot formula and draw the structural formula for XeO_3.

12 Chapter Self-Test

Answers to Self-Test are in Appendix J.

1. Predict which of the following has a covalent bond. (Sec. 12.1)
(a) CaO (b) CoO
(c) CO (d) all of the above
(e) none of the above
2. Which of the following statements is true regarding an ionic bond between calcium and oxygen in a CaO formula unit? (Sec. 12.2)
(a) Calcium atoms lose electrons and oxygen atoms gain electrons.

(b) The ionic radius of a calcium ion is less than its atomic radius.
(c) Forming an ionic bond between Ca^{2+} and O^{2-} releases energy.
(d) all of the above
(e) none of the above

3. Which of the following statements is true regarding a covalent bond between hydrogen and oxygen atoms in an H_2O molecule? (Sec. 12.3)
(a) Bonding electrons are stationary between the H and O atoms.
(b) The bond length is greater than the sum of the two atomic radii.
(c) Forming a bond between H and O atoms requires energy.
(d) all of the above
(e) none of the above

4. Draw the electron dot formula for hydrogen peroxide, H_2O_2. How many pairs of nonbonding electrons are in a hydrogen peroxide molecule? (Sec. 12.4)
(a) 1 **(b)** 2
(c) 3 **(d)** 4
(e) none of the above

5. Draw the electron dot formula for the carbonate ion, CO_3^{2-}. How many pairs of nonbonding electrons are in a carbonate ion? (Sec. 12.5)
(a) 4 **(b)** 8
(c) 12 **(d)** 24
(e) none of the above

6. Which of the following shows a general increasing trend for electronegativity in the periodic table? (Sec. 12.6)
(a) up a group of metals
(b) up a group of nonmetals
(c) across a series of metals from left to right
(d) across a series of nonmetals from left to right
(e) all of the above

7. Given the values for H (2.1) and F (4.0), calculate the electronegativity difference in a hydrogen fluoride bond, HF. (Sec. 12.7)
(a) –1.9 **(b)** –6.1
(c) 1.9 **(d)** 6.1
(e) none of the above

8. Chlorine has an electronegativity value of 3.0. Given the electronegativity of C, N, O, and P (2.5, 3.0, and 3.5, respectively), which of the following molecules has nonpolar bonds? (Sec. 12.8)
(a) CCl_4 **(b)** NCl_3
(c) Cl_2O **(d)** all of the above
(e) none of the above

9. What is the electron pair geometry for a molecule of H_2S? (Sec. 12.9)
(a) bent **(b)** linear
(c) tetrahedral **(d)** trigonal pyramidal
(e) none of the above

10. What is the molecular shape for a molecule of H_2S? (Sec. 12.10)
(a) bent **(b)** linear
(c) tetrahedral **(d)** trigonal pyramidal
(e) none of the above

Key Concepts

11. Classify each of the following diagrams as illustrating a substance composed of atoms, formula units, or molecules:

12. Based on electronegativity trends in the periodic table, which of the following molecules is most polar: HF, HCl, HBr, or HI?

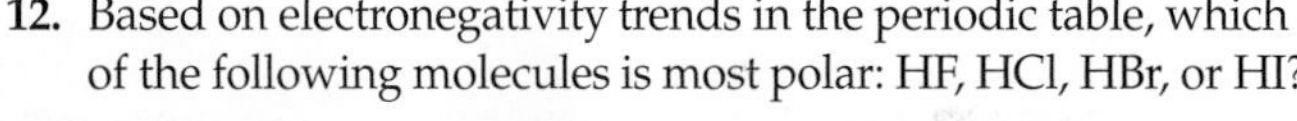

(a)

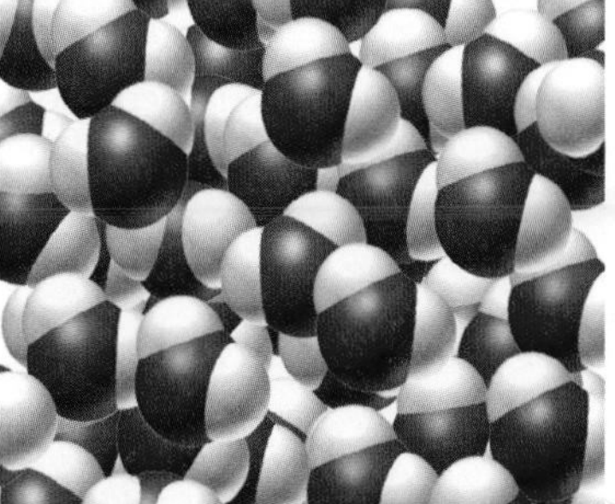

(b)

(c)

13. Given a H_2O molecule, how many additional H_2O molecules can attach directly to the given water molecule by hydrogen bonds?

14. Using VSEPR theory, predict the electron pair geometry and molecular shape for a chloroform molecule, $CHCl_3$.

Critical Thinking

15. "Acid rain" originates with the sulfur dioxide produced from coal-burning plants. Draw the structural formulas for two resonance structures of SO_2. Given that the bond energy of a double bond is greater than a single bond, predict the stronger bond in an SO_2 molecule.

16. Using VSEPR theory, contrast the molecular shape of a water molecule, H_2O, with that of a hydronium ion, H_3O^+.

17. Using VSEPR theory, contrast the molecular shape of an ammonia molecule, NH_3, with that of an ammonium ion, NH_4^+.

CHAPTER

13 Liquids and Solids

An iceberg floats in water because ice is less dense than water. When water freezes, it forms ice crystals with open holes. Thus, the ice has a greater volume and is less dense than water.

"If I have ever made any valuable discoveries it has been owing more to patient attention, than to any other talent."
Sir Isaac Newton, English Scientist (1643–1727)

In Chapter 11 we discussed the gaseous state of matter, where individual molecules are relatively distant from one another. In this chapter, we will discuss the liquid and solid states of matter in which particles contact one another. We will use water in much of our discussion to reinforce the concepts of specific heat, nomenclature, mole calculations, and bonding that we discussed in earlier chapters.

Water is the most important liquid on Earth and covers about three-fourths of the surface of our planet. It is necessary for all the chemical reactions that support plant and animal life, and about two-thirds of the mass of the human body is water. With the exception of oxygen, water is our most critical substance. We can survive quite some time without food, but only a few days without water.

The physical properties of water are unusual in several ways. One unusual property is its density. The density of a substance is generally greater in the solid state than in the liquid state. The two exceptions to this generalization are water and ammonia. Since the density of solid ice is less than that of liquid water, ice floats on water. If ice were more dense than water, marine life would not survive. Rivers and lakes would freeze into solid chunks of ice in wintry climates. Initially, ice would form on the surface and then sink, and eventually the entire body of water would change to solid ice. This, of course, would destroy aquatic life.

13.1 PROPERTIES OF LIQUIDS

OBJECTIVE

- To identify five observed properties of a liquid.

Unlike gases, liquids do not respond significantly to temperature and pressure changes. Also, the mathematical relationships that apply to gases, such as the combined gas law, do not apply to liquids. Rather, when we study the liquid state, we observe the following general properties:

1. **Liquids have a variable shape, but a fixed volume.** The shape of a liquid conforms to the shape of its container.
2. **Liquids usually flow readily.** Liquids flow at different rates; for example, petroleum flows more slowly than water. Glass is actually a liquid that flows very slowly. Panes of glass in 500-year-old chapels are now thicker at the bottom than at the top.
3. **Liquids do not compress or expand significantly.** The volume of a liquid varies very little with changes in temperature or pressure.
4. **Liquids have a high density compared to gases.** Gases and liquids are both fluids, that is, the shape is not fixed and individual particles are free to move throughout the container. However, the liquid state is about 1000 times more dense than the gaseous state because particles in a liquid are closer together. For example, the density of water is 1.00 g/mL, whereas the density of air at sea level is about 0.001 g/mL.
5. **Liquids that are soluble mix homogeneously.** Liquids diffuse more slowly than gases. However, liquids that are soluble eventually form a homogeneous mixture. For example, when food coloring is added to water, the liquids diffuse slowly and mix uniformly.

▲ **Scientific Misconception** It is a misconception that stained glass windows are thicker at the bottom due to liquid flow; the thickness varies owing to the glassmaking process.

13.2 THE INTERMOLECULAR BOND CONCEPT

OBJECTIVES

- To explain the concept of an intermolecular bond.
- To describe three types of attraction between molecules in a liquid.

We refer to highways as either interstates (between states) or intrastates (in state). Similarly, an intermolecular bond is an attraction between molecules, whereas an intramolecular bond is between atoms in a molecule.

Some properties of liquids, such as vapor pressure, are determined by the strength of attraction between molecules. This attraction is called an intermolecular bond, and it is much weaker than an ordinary covalent bond. The three types of intermolecular attraction are based on temporary dipoles, permanent dipoles, and hydrogen bonds.

In Section 12.9, we learned that a polar molecule results from one or more polar bonds. In a polar molecule, positive and negative charges are concentrated in different regions. These two charged regions are created by an uneven distribution of electrons in the molecule. A molecule with regions of positive and negative charge possesses a *dipole*.

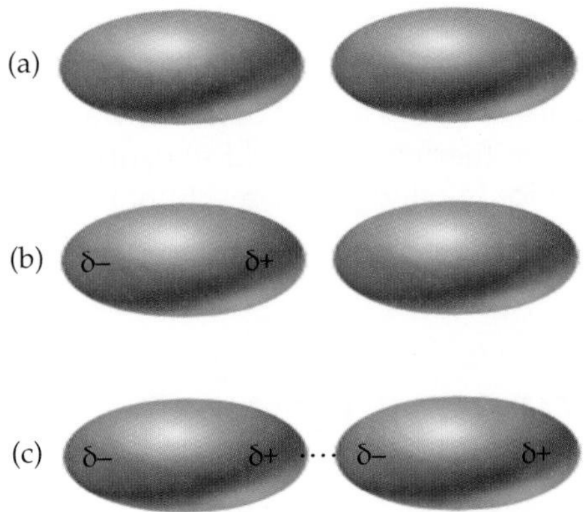

▶ **Figure 13.1 Intermolecular Dispersion Forces** (a)Two nonpolar molecules are shown. (b) The nonpolar molecule on the left forms a temporary dipole, which can induce a dipole in the molecule on the right. (c) The two nonpolar molecules are temporarily attracted to each other. In turn, these molecules can induce temporary dipole attractions in surrounding molecules in the liquid.

Let's first consider temporary dipole attraction. Even though atoms in a nonpolar molecule share electrons equally, the electrons are constantly shifting about. This shifting about of electrons produces regions in the molecule that are temporarily "electron rich" and slightly negative. Simultaneously, another region of the molecule is temporarily "electron poor" and slightly positive. A negative region in one molecule has a weak attraction for a positive region in another molecule. These temporary dipole attractions can occur between nonpolar molecules. This temporary attraction between molecules is referred to as a **dispersion force**.

Although dispersion forces last for only brief periods of time, they occur frequently and repeatedly between nonpolar molecules. At any one time, molecules in a liquid have a temporary dipole attraction. Figure 13.1 illustrates the nature of temporary dispersion forces.

Because of polar bonds, or nonbonding electron pairs, a polar molecule has regions of positive and negative charge. Similar to miniature magnets, these molecules are attracted to each other. In contrast to temporary dipole attraction, permanent dipole attraction operates continuously. A permanent **dipole force** operates between polar molecules as shown in Figure 13.2.

The dipole attraction between molecules is typically about 10% of the strength of a normal covalent bond. For instance, when a hydrogen atom is bonded to an oxygen atom or a nitrogen atom, a highly polar bond results. This highly polar bond is capable of producing a strong dipole force between molecules. This special type of dipole attraction is called a **hydrogen bond** (Figure 13.3). Hydrogen bonds are especially important in living organisms where they maintain the structure of protein molecules and DNA molecules.

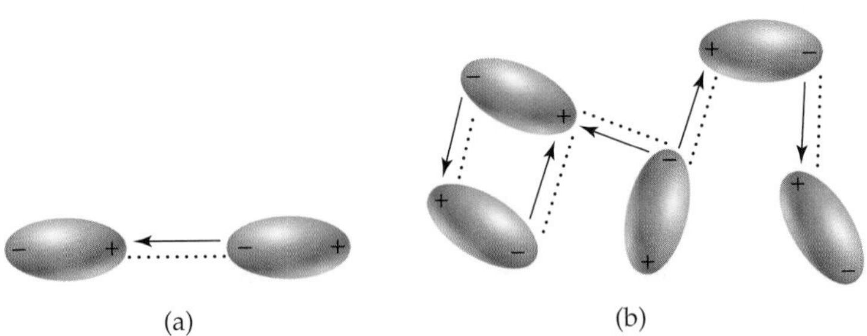

▲ **Figure 13.2 Intermolecular Dipole Forces** (a) A dipole force results from a dipole–dipole attraction between two molecules. (b) Dipole forces between several molecules in a liquid are shown.

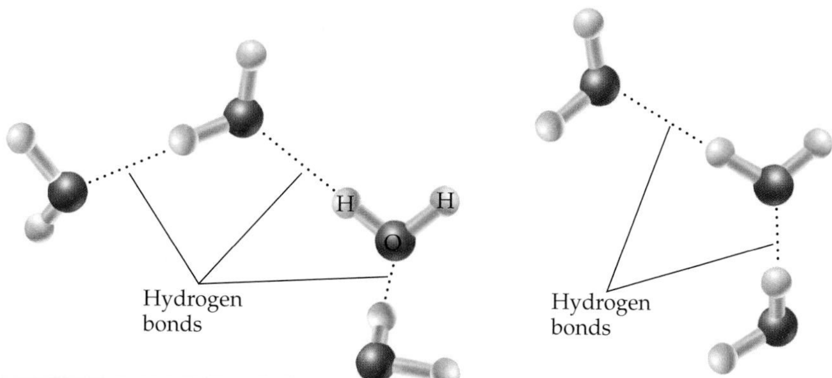

▲ **Figure 13.3 Intermolecular Hydrogen Bonds** Polar water molecules form intermolecular hydrogen bonds. Notice that the hydrogen atom is attracted to the nonbonding electrons on the highly electronegative oxygen atom.

The following example exercise illustrates the characteristics of intermolecular bonds:

Example Exercise 13.1 Intermolecular Attraction

In a liquid with polar molecules, what is the strongest type of intermolecular attraction?

Solution

In a liquid composed of polar molecules, the intermolecular attraction is the result of both permanent and temporary dipoles. Molecules that contain either H—O or H—N bonds have the strongest type of intermolecular attraction, that is, hydrogen bonds. Water, H_2O, and ammonia, NH_3, are examples of liquids with hydrogen bonds.

Practice Exercise

In a liquid with nonpolar molecules, what is the strongest type of intermolecular attraction?

Answer: The intermolecular attraction between nonpolar molecules is a dispersion force resulting from temporary dipoles.

Concept Exercise

Which of the following liquids has the stronger intermolecular attraction between molecules: $CH_3{-}CH_2{-}OH$ or $CH_3{-}O{-}CH_3$?

Answer: See Appendix G.

13.3 VAPOR PRESSURE, BOILING POINT, VISCOSITY, AND SURFACE TENSION

OBJECTIVES

- To understand the concepts of vapor pressure, boiling point, viscosity, and surface tension.
- To relate the vapor pressure, boiling point, viscosity, and surface tension of a liquid to the strength of attraction between molecules.

In Chapter 11, we introduced the kinetic theory of gases to explain the behavior of gases. Now, we will extend this theory to explain the properties of liquids.

According to the kinetic theory the attraction between gas molecules is negligible. This is not the case for liquids. Molecules in the liquid state are in contact with each other, and their attraction restricts their movement. Individual molecules do have enough energy to move about one another; that is, a liquid is free to flow. The liquid state and the gaseous state are analogous to honeybees swarming in a hive (liquid state), and individual bees flying off to gather pollen (gaseous state). Now let's consider some properties of liquids.

Vapor Pressure

The kinetic theory relates the average energy of molecules to temperature. The higher the temperature, the greater the kinetic molecular energy. However, not all molecules at the same temperature have identical energies. Some are more energetic than others. At the surface of a liquid, for example, some molecules have enough energy to completely escape the attraction of neighboring molecules. The molecules that escape enter the gaseous state, which is referred to as vapor. This process is called *vaporization.* In the reverse process, some molecules in the vapor return to the liquid. This process is called *condensation.*

When the rates of vaporization and condensation are equal, the pressure exerted by the gas molecules above a liquid is called the **vapor pressure**. However, the vapor pressure above a liquid depends on the attraction between molecules in the liquid. Water molecules have a strong attraction for each other, and ethyl ether molecules have a relatively weak attraction. Because of the lesser attraction between molecules, ether molecules escape from the liquid state more readily than water molecules. Therefore, the vapor pressure of ether is greater than that of water at the same temperature (Figure 13.4).

Table 13.1 compares the attraction between molecules, that is, *intermolecular attraction,* and the vapor pressure for a few selected liquids. In general, as the attraction between molecules increases, the vapor pressure decreases. In Table 13.1 we see that propionic acid and butyl alcohol both have a strong intermolecular attraction. Thus, their vapor pressures at 20 °C are quite low. Conversely, propyl chloride and ethyl ether each has a weak intermolecular attraction. Thus, their vapor pressures are considerably higher. When comparing vapor pressure values, it is necessary to choose two liquids that have approximately the same molar

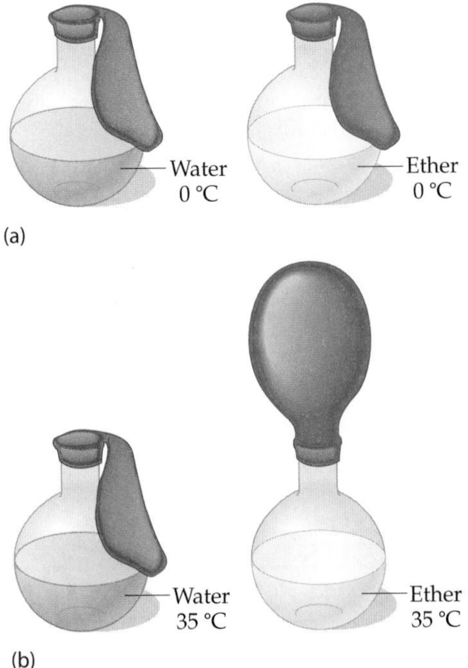

▶ **Figure 13.4 Vapor Pressure of Water and Ether** (a) At 0 °C neither water nor ether has sufficient vapor pressure to affect the balloons. (b) At 35 °C the vapor pressure of water is still low, however, the vapor pressure of ether is considerable and inflates the balloon.

TABLE 13.1 VAPOR PRESSURE OF SELECTED LIQUIDS

LIQUID	APPROXIMATE MOLAR MASS	INTERMOLECULAR ATTRACTION	VAPOR PRESSURE AT 20 °C
water	18 g/mol	strong	18 mm Hg
propionic acid, C_2H_5COOH	74 g/mol	strong	5 mm Hg
butyl alcohol, C_4H_9OH	74 g/mol	strong	6 mm Hg
propyl chloride, C_3H_7Cl	79 g/mol	weak	300 mm Hg
ethyl ether, $C_2H_5OC_2H_5$	74 g/mol	weak	450 mm Hg

mass. Note that in these comparisons, each liquid, except water, has an approximate molar mass of 74 g/mol.

Boiling Point

As the temperature of a liquid increases, its vapor pressure increases. At 30 °C the vapor pressure of water is about 30 mm Hg, and at 60 °C it is 150 mm Hg. Figure 13.5 shows the relationship between vapor pressure and temperature for ether, ethanol, and water.

A liquid begins to boil when the pressure of the vapor above the liquid equals the pressure of the atmosphere. Thus, we define the normal **boiling point** (symbol **Bp**) as the temperature at which the vapor pressure equals standard atmospheric pressure. Accordingly, liquids that have *high* boiling points must have *low* vapor pressures. The following example exercise illustrates the determination of boiling points from vapor pressure data:

Example Exercise 13.2 Boiling Point Predictions

Refer to Figure 13.5 and determine the approximate boiling point of ether.

Solution

The normal boiling point of a liquid is the temperature at which the vapor pressure equals standard atmospheric pressure, that is, 760 mm Hg. From the graph, we find that the vapor pressure of ether is 760 mm Hg at about 35 °C. Thus, the normal boiling point of ether is about 35 °C. The actual observed value for the boiling point of ether is 36 °C.

Practice Exercise

Refer to Figure 13.5 and determine the approximate boiling point of ethanol.

Answer: ~80 °C (The actual boiling point of ethanol is 78 °C.)

Concept Exercise

Which of the following liquids has the higher boiling point: CH_3-CH_2-OH or CH_3-O-CH_3?

Answer: See Appendix G.

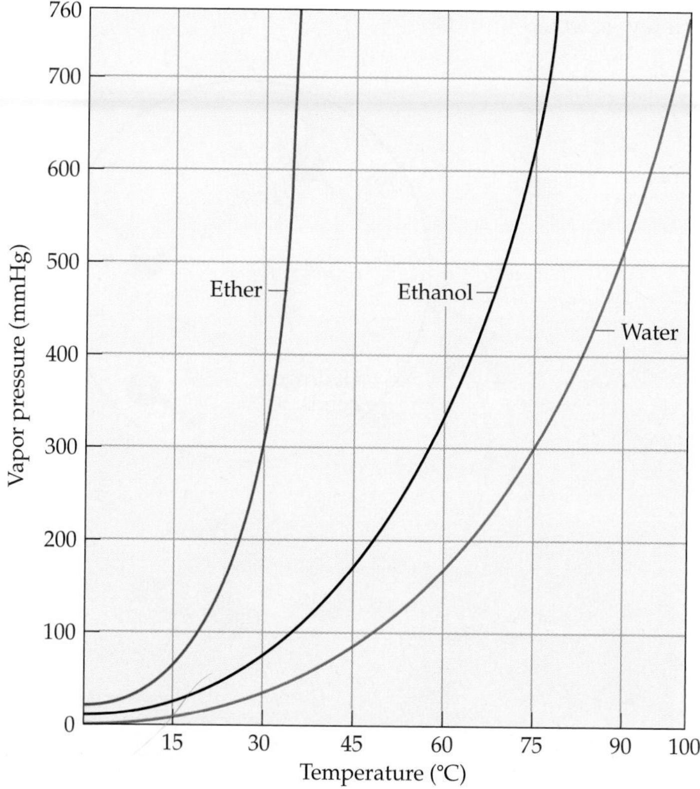

◀ **Figure 13.5 Vapor Pressure versus Temperature** The vapor pressure of a liquid increases as the temperature increases. The vapor pressure of water equals 760 mm Hg at 100 °C. Thus, at 100 °C the vapor pressure of water equals the standard atmospheric pressure (760 mm Hg), and water begins to boil.

▲ **Viscosity** The motor oil on the left is more viscous and flows more slowly than the motor oil on the right.

Viscosity

Some liquids are easier to pour than others. Water pours easily, whereas honey does not. The resistance of a liquid to flow is a property called **viscosity**. Viscosity is the result of an attraction between molecules. It is also affected by factors such as the size and shape of the molecules. In principle, the greater the attraction between molecules, the higher the viscosity. Table 13.2 compares the attraction between molecules and the viscosity for a few selected liquids.

TABLE 13.2 VISCOSITY OF SELECTED LIQUIDS

LIQUID	APPROXIMATE MOLAR MASS	INTERMOLECULAR ATTRACTION	VISCOSITY* AT 20 °C
water	18 g/mol	strong	1.00
propionic acid, C_2H_5COOH	74 g/mol	strong	1.10
butyl alcohol, C_4H_9OH	74 g/mol	strong	2.95
propyl chloride, C_3H_7Cl	79 g/mol	weak	0.35
ethyl ether, $C_2H_5OC_2H_5$	74 g/mol	weak	0.23

*Values are expressed in centipoise, a common unit of viscosity.

In Table 13.2 we see that propionic acid and butyl alcohol have a strong intermolecular attraction and that the viscosity of both is greater than that of water. We also see that propyl chloride and ethyl ether have a weak attraction between molecules and that the viscosity of both is much lower than that of water. Honey is an example of a viscous liquid. We can therefore predict that there is a strong attraction between molecules in honey.

Surface Tension

At some time you have probably noticed a small insect or some other object floating on water. For an insect or any object to sink in a liquid, it has to break through the surface. But the molecules on the surface of a liquid resist being pushed apart. The attraction between the surface molecules in a liquid is called **surface tension**. There are other factors to consider, but in general the greater the intermolecular attraction, the higher the surface tension. Table 13.3 compares the intermolecular attraction and surface tension for a few selected liquids.

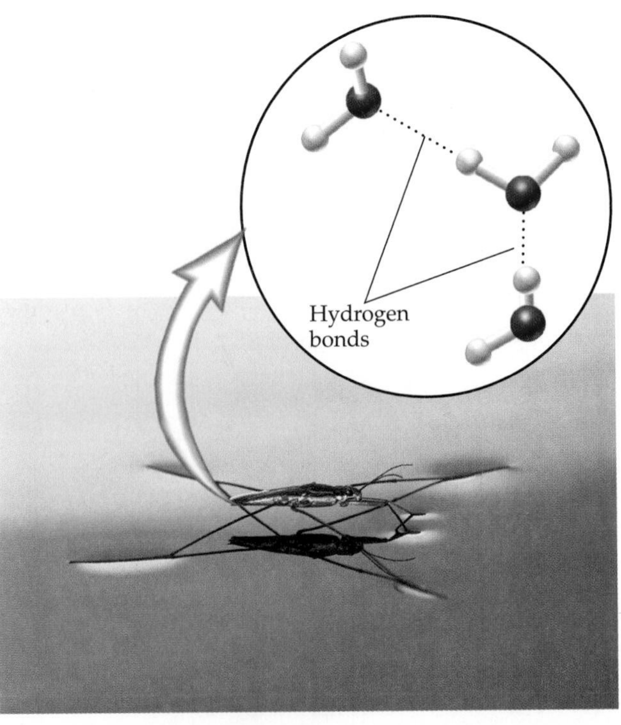

▶ **Surface Tension** Water has a high surface tension, and this allows the insect to "walk" on water.

TABLE 13.3 SURFACE TENSION OF SELECTED LIQUIDS

LIQUID	APPROXIMATE MOLAR MASS	INTERMOLECULAR ATTRACTION	SURFACE TENSION* AT 20 °C
water	18 g/mol	strong	70
propionic acid, C_2H_5COOH	74 g/mol	strong	27
butyl alcohol, C_4H_9OH	74 g/mol	strong	25
propyl chloride, C_3H_7Cl	79 g/mol	weak	18
ethyl ether, $C_2H_5OC_2H_5$	74 g/mol	weak	17

*Values are expressed in dynes per square centimeter, a common unit of surface tension.

We all know that rain forms drops as it falls. In fact, when we spray any liquid, it forms drops, and each drop has the shape of a small sphere. Drops of liquid are spherical because surface tension causes them to have the smallest possible surface area, and that area corresponds to a spherical droplet. Table 13.3 shows that water has an unusually high surface tension.

Example Exercise 13.3 illustrates the relationships between the properties of liquids and intermolecular attraction.

Example Exercise 13.3 Physical Property Predictions

Consider the following properties of liquids. State whether the value for each property is high or low for a liquid with a strong intermolecular attraction:

(a) vapor pressure
(b) boiling point
(c) viscosity
(d) surface tension

Solution

For a liquid having a strong attraction between molecules, properties (b), (c), and (d) are generally high; property (a) is low.

(a) Molecular attraction slows vaporization. Therefore, vapor pressure is *low* for liquids with a strong intermolecular attraction.
(b) Attraction between molecules inhibits boiling. Thus, the boiling point is *high* for liquids with a strong intermolecular attraction.
(c) Molecular attraction increases the resistance of a liquid to flow. Viscosity is *high* for liquids with a strong intermolecular attraction.
(d) Attraction between molecules causes a drop of liquid to form a sphere. Surface tension is *high* for liquids with a strong intermolecular attraction.

Practice Exercise

The intermolecular attraction is greater in isopropyl alcohol, C_3H_7OH, than in pentane, C_5H_{12}. Predict which liquid has the higher value for each of the following:

(a) vapor pressure
(b) boiling point
(c) viscosity
(d) surface tension

Answers:

(a) C_5H_{12}
(b) C_3H_7OH
(c) C_3H_7OH
(d) C_3H_7OH

Concept Exercise

Explain why the paper clip can float on water even though its density is much greater than the density of water.

Answer: See Appendix G.

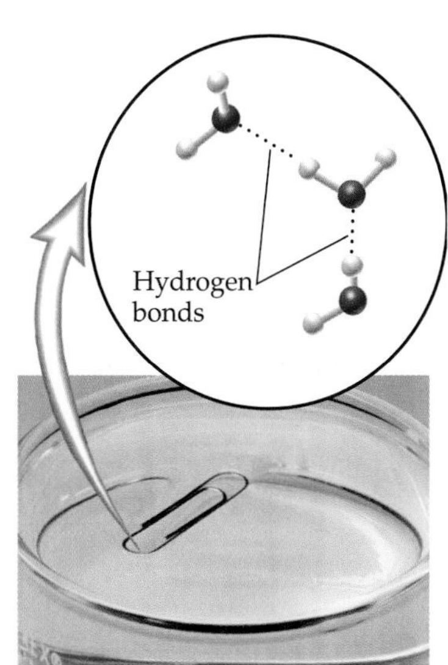

OBJECTIVE

▸ To identify five observed properties of a solid.

13.4 PROPERTIES OF SOLIDS

Unlike liquids, solids have a fixed shape. The reason for this is that the individual particles of the solid are not free to move. Unlike gases, the volume of a solid shows very little response to changes in temperature or pressure. The observed properties of the solid state are as follows:

1. **Solids have a fixed shape and a fixed volume.** Unlike liquids, solids are rigid and their shape is fixed.
2. **Solids are either crystalline or noncrystalline.** A **crystalline solid** contains particles arranged in a regular repeating pattern. Each particle occupies a fixed position in the crystal. The high degree of order of the molecules can produce a beautiful clear crystal, for example, a diamond or a ruby. An ordinary stone is a noncrystalline solid and is not transparent.
3. **Solids do not compress or expand to any degree.** Assuming no change in physical state, temperature and pressure have a negligible effect on the volume of a solid.
4. **Solids have a slightly higher density than their corresponding liquids.** For example, solid chunks of iron sink in a high temperature furnace containing molten iron. One important exception to this rule is water; ice is less dense than liquid water. As a result, ice floats on water.
5. **Solids do not mix by diffusion.** In a solid heterogeneous mixture, the particles are not free to diffuse and they cannot mix uniformly. In an alloy, which is a homogeneous mixture of metals, the particles mix uniformly in the molten liquid state before cooling to a solid.

OBJECTIVE

▸ To describe three types of crystalline solids: ionic, molecular, and metallic.

13.5 CRYSTALLINE SOLIDS

Chemists know that the particles of a crystalline solid are arranged in a regular geometric pattern. However, the particles can be of different types. They can be ionic, molecular, or metallic. In ionic solids the crystals are composed of regular patterns of ions. In molecular solids the molecules, not ions, form repeating patterns. In metallic solids, individual metal atoms are arranged geometrically.

Ionic Solids

A crystalline **ionic solid** is an ionic compound composed of positive and negative ions. Salt, for example, is a crystalline solid of NaCl. Here, sodium ions, Na^+, and chloride ions, Cl^-, are arranged in a regular three-dimensional structure referred to as a crystal lattice. Notice the clear crystal of table salt shown in Figure 13.6 has the shape of a cube. Other ionic crystalline compounds, such as NaF, CaF_2, and $CaCO_3$, have different geometric shapes.

Molecular Solids

A crystalline **molecular solid** has molecules arranged in a particular configuration. Crystalline sucrose (table sugar), for example, is composed of $C_{12}H_{22}O_{11}$ molecules. The sucrose molecules are arranged in a regular order that allows light to pass through the crystal. Therefore, a large crystal of sucrose appears transparent. Other molecular solids are sulfur and phosphorus. Sulfur crystals (Figure 13.7) are made from S_8 molecules held together by covalent bonds. Phosphorus powder is made from P_4 molecules.

Metallic Solids

A crystalline **metallic solid** has atoms of metals arranged in a definite pattern. That is, a metallic crystal is made up of positive metal ions surrounded by valence electrons. Metals are good conductors of electricity because the valence electrons are free to

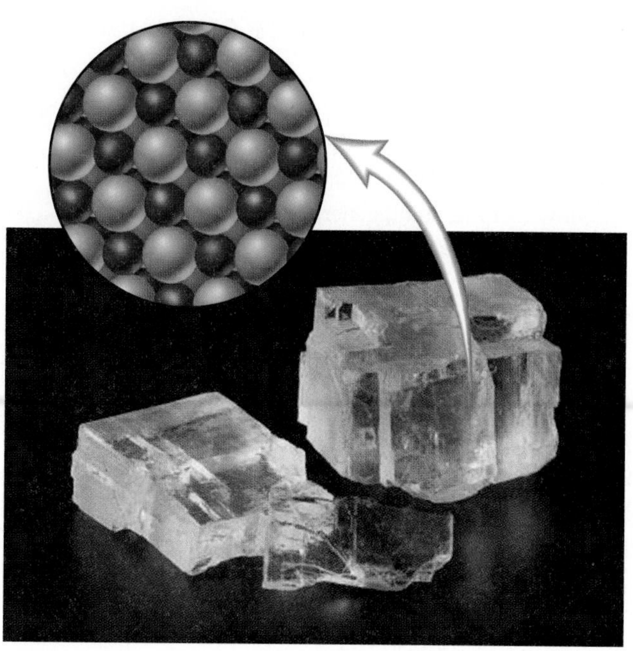

◀ **Figure 13.6 A Crystalline Ionic Solid** A sodium chloride crystal, NaCl, is an example of a crystalline ionic solid.

move about the crystal. This arrangement of atoms and freely moving electrons is referred to as the "electron sea" model. The flow of electricity is associated with the movement of electrons through a metal. Figure 13.8 shows a crystal of gold.

Classifying Crystalline Solids

It is helpful to classify crystalline solids to predict their properties. In general, the properties of ionic solids such as melting point, hardness, electrical conductivity, and water solubility are similar. The properties of molecular solids are usually similar as well. Metals show a range of physical properties, but all are malleable (form thin foils) and ductile (drawn into wire) and are good conductors of electricity. Table 13.4 lists the general properties for each of the three types of crystalline solids.

▲ **Figure 13.7 A Crystalline Molecular Solid** A sulfur crystal, S_8, is an example of a crystalline molecular solid.

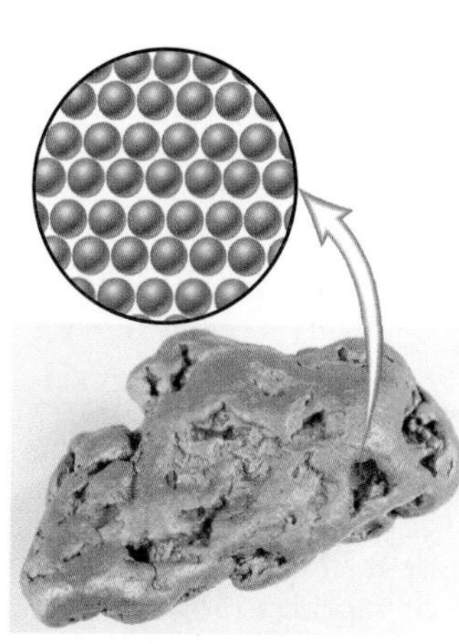

▲ **Figure 13.8 A Crystalline Metallic Solid** A gold nugget is an example of a crystalline metallic solid.

▲ **Figure 13.9 A Crystalline Network Solid** Diamond is a crystalline network solid of carbon atoms.

TABLE 13.4 GENERAL PROPERTIES OF CRYSTALLINE SOLIDS

TYPE OF SOLID	GENERAL PROPERTIES	EXAMPLES
ionic	high melting point, hard, brittle, at least slightly soluble in water, conductor of electricity when melted or in solution	NaCl, $CaCO_3$, $MgSO_4$
molecular	low melting point, generally insoluble in water, nonconductor of electricity	S_8, $C_{10}H_8$, $C_6H_{12}O_6$
metallic	low to high melting point, malleable, ductile, conductor of electricity, insoluble in most solvents	Cu, Ag, Au

Diamond, a form of carbon, is a special type of crystalline solid that has covalent bonds between a vast number of atoms. This type of crystalline solid is referred to as a *network solid.* Diamond is a hard, brittle, crystalline network solid with unusual properties, including a very high melting point (above 3500 °C).

The following example exercise further illustrates the classification of ionic, molecular, and metallic types of crystalline solids:

Example Exercise 13.4 Classifying Crystalline Solids

Classify each of the following crystalline solids as ionic, molecular, or metallic:

(a) nickel, Ni (b) nickel(II) oxide, NiO

Solution

The type of crystalline solid is dictated by the type of particle in the solid.

(a) Nickel is a metal composed of atoms; thus, Ni is a *metallic solid.*

(b) Nickel(II) oxide contains ions and is therefore an *ionic solid.*

Practice Exercise

Classify each of the following crystalline solids as ionic, molecular, or metallic:

(a) iodine, I_2 (b) silver iodide, AgI

Answers: (a) molecular; (b) ionic

Concept Exercise

Which type of crystalline solid has a high melting point and is a nonconductor of electricity in the solid state, but a good conductor of electricity when melted?

Answer: See Appendix G.

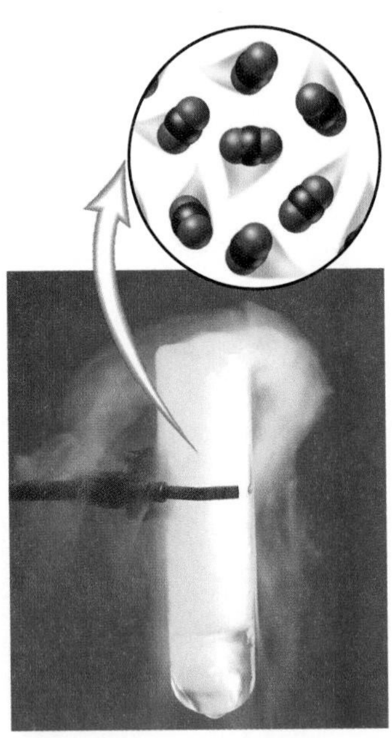

▲ **Dry Ice** Solid CO_2 is undergoing sublimation at the bottom of the test tube, which is a direct change of state from a solid to a gas.

Note Dry ice is a crystalline solid composed of CO_2 molecules having covalent bonds between carbon and oxygen atoms. In turn, each CO_2 molecule is attracted to other CO_2 molecules by dipole forces. Dry ice goes directly from a solid to a gas (sublimation) at –80 °C because dipole forces are relatively weak compared to covalent bonds.

13.6 CHANGES OF PHYSICAL STATE

OBJECTIVES

- To understand the concepts of specific heat, heat of fusion, and heat of vaporization.
- To calculate heat changes that involve the specific heat, heat of fusion, and heat of vaporization.

Heat is necessary to change the physical state of a substance. Specific heat is the amount of heat required to raise 1.00 g of a substance 1 °C. Every substance has a unique value for its specific heat. Water is considered a reference and its specific heat is 1.00 calorie per gram per degree Celsius, that is, 1.00 cal/(g × °C). The specific heats for ice and steam are approximately half that of liquid water.

Next, let's consider a substance changing state from a solid to a liquid at its melting point. The amount of heat required to melt 1.00 g of substance is called the **heat of fusion** (H_{fusion}). For water, the heat of fusion is 80.0 cal/g. Water releases the same amount of heat energy, 80 cal/g, when it changes from the liquid to the solid state. This heat change is called the *heat of solidification* (H_{solid}). For any substance, the heats of fusion and solidification are equal.

A substance rapidly changes state from a liquid to a vapor at its boiling point. The amount of heat required to vaporize 1.00 g of a substance is called the **heat of vaporization** (H_{vapor}). For water, it is 540 cal/g. Conversely, water releases the same amount of heat energy, 540 cal/g, when it condenses from a gas to a liquid. This heat change is called the *heat of condensation* (H_{cond}). For any substance, the heats of vaporization and condensation are equal. Table 13.5 lists the heat values for water, ice, and steam.

TABLE 13.5 HEAT VALUES FOR WATER

SUBSTANCE	SPECIFIC HEAT [cal/(g × °C)]	H_{FUSION} (cal/g)	H_{SOLID} (cal/g)	H_{VAPOR} (cal/g)	H_{COND} (cal/g)
ice, $H_2O(s)$	0.50	80.0			
water, $H_2O(l)$	1.00		80.0	540	
steam, $H_2O(g)$	0.48				540

To see the change in temperature with a constant application of heat, we draw a temperature–energy graph, sometimes called a heating curve. The heating curve for water is shown in Figure 13.10.

Let's combine the concept of a heating curve with the heat values in Table 13.5. For example, let's find the amount of heat energy necessary to convert 25.5 g of ice at –5.0 °C to steam at 100.0 °C. This problem requires four steps: (1) heat the ice from –5.0 °C to 0.0 °C, (2) convert the ice to liquid water at 0.0 °C, (3) heat the water from 0.0 °C to 100.0 °C, and (4) convert the water to steam at 100.0 °C.

1. To calculate the amount of energy required to heat the ice we use its mass (25.5 g), the temperature change (–5.0 °C to 0.0 °C), and the specific heat of ice (0.50 cal/(g × °C)). Using 0.50 cal/(g × °C) as a unit factor, we have

$$25.5\ \cancel{g} \times \frac{0.50\ \text{cal}}{1\ \cancel{g} \times \cancel{°C}} \times [0.0 - (-5.0)]\ \cancel{°C} = 64\ \text{cal}$$

2. The heat of fusion for ice, 80.0 cal/g, is found in Table 13.5. The energy required to melt 25.5 g of ice is

$$25.5\ \cancel{g} \times \frac{80.0\text{cal}}{1\ \cancel{g}} = 2040\ \text{cal}$$

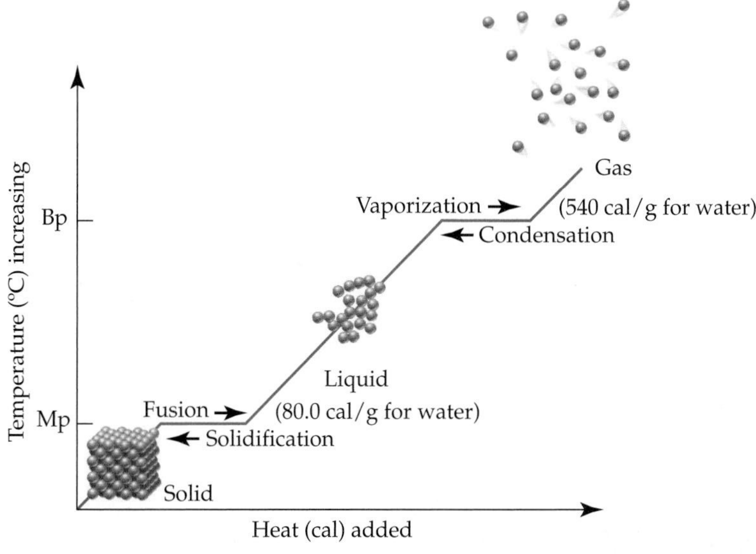

◀ **Figure 13.10 Temperature–Energy Graph** As heat is continuously added to a substance, the substance eventually changes its physical state. Notice the temperature remains constant during fusion (solid to liquid), vaporization (liquid to gas), condensation (gas to liquid), and solidification (liquid to solid).

3. To calculate the amount of energy required to heat the liquid water we must know its mass, the temperature change (0.0 °C to 100.0 °C), and the specific heat of water; therefore, we have

$$25.5\ \cancel{g} \times \frac{1.00\ \text{cal}}{1\ \cancel{g} \times \cancel{^\circ C}} \times (100.0 - 0.0)\cancel{^\circ C} = 2550\ \text{cal}$$

4. The heat of vaporization, 540 cal/g, is found in Table 13.5. The energy required to vaporize the water to steam is

$$25.5\ \cancel{g} \times \frac{540\ \text{cal}}{1\ \cancel{g}} = 13{,}800\ \text{cal}$$

The total heat energy required to heat and vaporize the ice is equal to the sum of the values obtained in Steps 1–4.

$$64\ \text{cal} + 2040\ \text{cal} + 2550\ \text{cal} + 13{,}800\ \text{cal} = 18{,}500\ \text{cal}$$

The heat required to raise the temperature of the ice at –5.0 °C to steam at 100.0 °C is 18,500 cal, or 18.5 kcal. The following example exercise illustrates the heat changes associated with the cooling of water and its solidification to ice:

Example Exercise 13.5 Change in Heat Problem

Calculate the amount of heat released when 15.5 g of liquid water at 22.5 °C cools to ice at –10.0 °C.

Solution

In this problem we have to consider (1) the specific heat of water, (2) the heat of solidification, and (3) the specific heat of ice.

1. To calculate the amount of heat released when cooling the water, consider the mass, the temperature change (22.5 °C to 0.0 °C), and the specific heat of water, 1.00 cal/(g × °C).

$$15.5\ \cancel{g} \times \frac{1.00\ \text{cal}}{1\ \cancel{g} \times \cancel{^\circ C} \times (22.5 - 0.0)\cancel{^\circ C}} = 349\ \text{cal}$$

We can summarize the solution visually as follows:

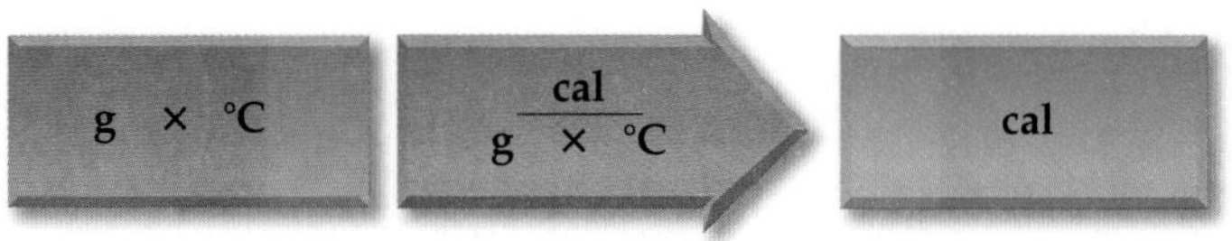

2. The heat of solidification, found in Table 13.5, is 80.0 cal/g. The heat released when water solidifies to ice is

$$15.5\ \cancel{g} \times \frac{80.0\ \text{cal}}{1\ \cancel{g}} = 1240\ \text{cal}$$

3. The specific heat of ice is 0.50 cal/(g × °C). The heat released as the ice cools to –10.0 °C is found as follows:

$$15.5\ \cancel{g} \times \frac{0.50\ \text{cal}}{1\ \cancel{g} \times \cancel{^\circ C}} \times [0.0 - (-10.0)]\cancel{^\circ C} = 78\ \text{cal}$$

The total heat energy released when the water cools to ice at –10.0 °C equals the sum of the values obtained in Steps 1–3.

$$349\ \text{cal} + 1240\ \text{cal} + 78\ \text{cal} = 1670\ \text{cal}$$

Thus, the heat released when the water cools is 1670 cal, or 1.67 kcal.

Practice Exercise

Calculate the amount of heat required to convert 50.0 g of steam at 100.0 °C to ice at 0.0 °C.

Answer: 3.60×10^4 cal (36.0 kcal)

Concept Exercise

Which of the following changes involves more energy: (a) heating 1 g of ice at 0 °C to water at 0 °C, or (b) heating 1 g of water at 100 °C to steam at 100 °C.

Answer: See Appendix G.

13.7 STRUCTURE OF WATER

OBJECTIVE

▶ To illustrate the bond angle and net dipole in a water molecule.

We learned in Section 13.3 that the boiling point and surface tension of water are unusually high. We also saw that water has a strong intermolecular attraction because of hydrogen bonding. To understand these properties more completely, let's review the covalent bonding in a water molecule. Specifically, let's consider the electron dot formula, the structural formula, the observed bond angle, polar covalent bonds, and the net dipole of water.

Electron Dot Formula

We begin by writing the electron dot formula for water, H_2O. The total number of valence electrons in one molecule is 2(1 e−) + 6 e− = 8 e−. Thus, there are four electron pairs. We can place the four pairs of electrons around the oxygen to provide the necessary octet.

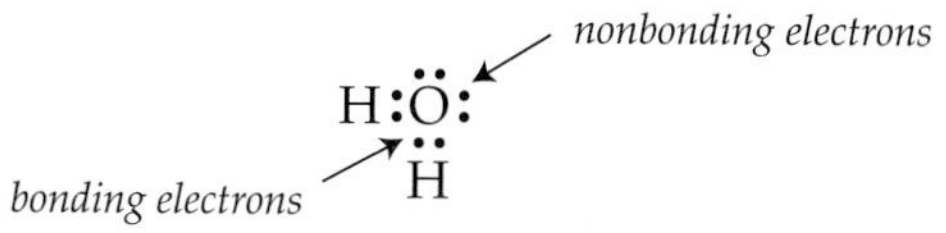

H:Ö:
H

nonbonding electrons

bonding electrons

Notice that there are two bonding and two nonbonding electron pairs. We learned in Section 12.9 that the molecular shape of a water molecule is bent, with a bond angle of 104.5°. Figure 13.11 shows the arrangement of bonding and nonbonding electron pairs.

Structural Formula and Bond Angle

The structural formula for a molecule uses dashes to represent covalent bonds. With this notation, we can show the water molecule with two hydrogen atoms at an angle to each other. Experimental evidence shows that the angle between the two hydrogen atoms is 104.5°. The angle formed by the central atom and two attached atoms is referred to as the **bond angle**.

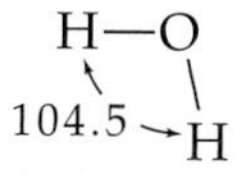

H—O
\
104.5 H

Polar Covalent Bonds

In a water molecule, each of the two covalent bonds is polar. That is, the more electronegative oxygen atom draws the bonding electrons closer to its nucleus. In turn, the hydrogen atoms become slightly positive. In Chapter 12 we used delta notation (δ^+ and δ^-) to indicate bond polarity. The covalent bond between H and O has a partially positive and a partially negative aspect, which creates a dipole.

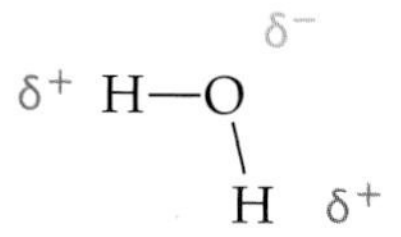

δ^-
δ^+ H—O
\
H δ^+

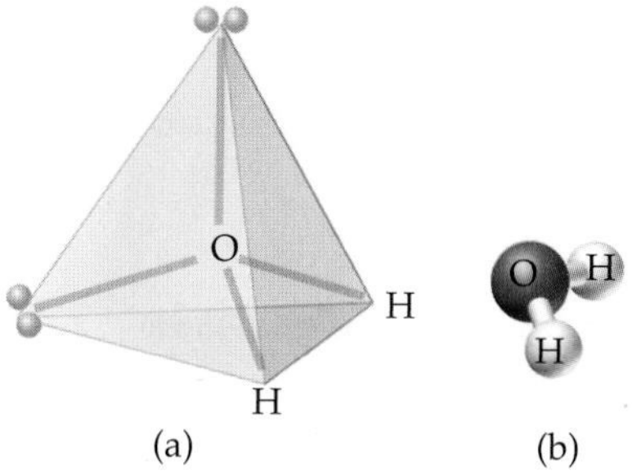

O
H
H
(a)

O H
H
(b)

◀ **Figure 13.11 Bonding in a Water Molecule** (a) The arrangement of the electron pairs in a water molecule forms the four corners of a tetrahedron. (b) A water molecule can be described as bent or V-shaped. The two hydrogen atoms are repelled slightly toward each other by the two nonbonding electron pairs.

Net Dipole

Notice that a water molecule has two dipoles, and the central atom is pulling each bonding electron pair toward itself. The effect of the two dipoles results in a single dipole that passes through the center of the molecule. The single, overall dipole for a molecule having two or more dipoles is called the **net dipole**.

The net dipole produces a negative end and a positive end in the water molecule. The negative end of the molecule is indicated by the arrow. The positive end is indicated by the plus sign on the opposite end of the net dipole.

CHEMISTRY CONNECTION Water Purification

Q: What is the process for changing hard water into soft water?

In some areas of the country, the minerals dissolved in water give it a high concentration of various ions that make the water suitable neither for drinking nor for agriculture. Such water is called **hard water**. Sometimes the mineral content in hard water is so great that it causes plumbing and corrosion problems. Hard water typically contains a high concentration of the following ions: Ca^{2+}, Mg^{2+}, Fe^{3+}, Cl^-, CO_3^{2-}, SO_4^{2-}, and PO_4^{3-}.

In homes having hard water, soap rings are sometimes found in bathtubs. A soap ring is formed by the cations in hard water reacting with soap to create a compound that deposits as an insoluble film. This problem can be eliminated by removing Ca^{2+}, Mg^{2+}, and Fe^{3+} from hard water by passing the water through a water softener. A water softener replaces Ca^{2+}, Mg^{2+}, and Fe^{3+} with Na^+; compounds containing Na^+ are generally soluble. Note, however, that although a water softener removes ions, **soft water** still has a high concentration of other ions. In addition to Na^+, soft water contains Cl^-, CO_3^{2-}, SO_4^{2-}, and PO_4^{3-}.

Soft water can interfere with the results of a chemical analysis. Therefore, chemists routinely use water that contains no ions. Chemists may use distilled water that is made by boiling water and collecting the condensed vapor. Alternatively, chemists use water with the minerals removed using an ion exchange system. Water purified by this method is called **deionized water**, or demineralized water. Deionized water is made by passing hard water through a system that exchanges both cations and anions. First, cations such as Na^+ in the water are exchanged for hydrogen ions on the resin. Second, anions such as Cl^- in the water are exchanged for hydroxide ions.

$$Na^+(aq) + H(resin) \longrightarrow Na(resin) + H^+(aq)$$

$$Cl^-(aq) + (resin)OH \longrightarrow (resin)Cl + OH^-(aq)$$

Notice that the ion exchange resin produces both hydrogen ions and hydroxide ions, which can readily combine to give water.

$$H^+(aq) + OH^-(aq) \longrightarrow H_2O(l)$$

The net result is that the resin removes *all ions* from water passing through the deionizing system.

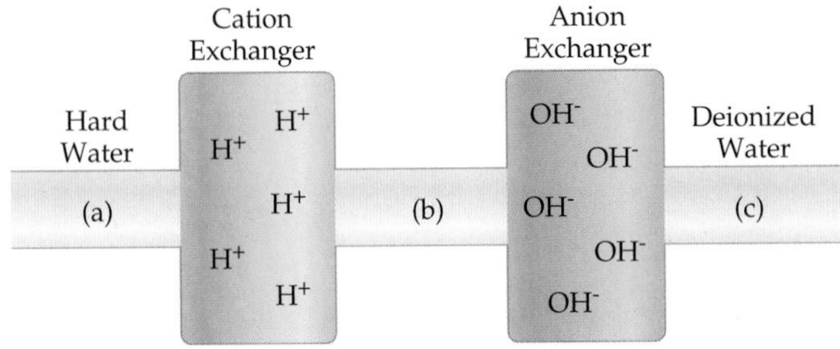

▲ **Water Softener** The cations in hard water are exchanged for H^+. The anions in hard water are exchanged for OH^-. The H^+ and OH^- combine to produce H_2O.

A: Hard water is changed to soft water using an ion-exchange resin, which exchanges Na^+ ions for Ca^{2+}, Mg^{2+}, and Fe^{3+}.

13.8 PHYSICAL PROPERTIES OF WATER

OBJECTIVE

- To explain the unusual physical properties of water.

Water is a colorless, odorless, tasteless liquid and a powerful solvent. At room temperature water has the highest specific heat, heat of fusion (except for ammonia), and heat of vaporization of any liquid.

Density

Generally, a substance in the solid state has a higher density than one in the liquid state. Therefore, we would predict that the density of ice is greater than that of water. But, we know that ice floats in water. That is, the solid form of water is *less dense* than the liquid form. The reason solid ice is less dense than liquid water relates to hydrogen bonding. Figure 13.12 illustrates three-dimensional hydrogen bonding in water.

When water freezes to ice, the hydrogen bonds produce a three–dimensional crystal. Figure 13.13 illustrates the structure of an ice crystal. Because of the arrangement of water molecules, however, the crystal has holes. These holes create a volume for ice that is greater than that for an equal mass of water. Thus, the density of solid ice is less than that of liquid water. At 0 °C, the density of ice is 0.917 g/mL, and the density of water is 1.00 g/mL.

Melting and Boiling Points

Water has an unusually high melting point and boiling point for a small molecule. To see how unusual these properties are, compare them with those of some hydrogen compounds of Group VIA/16 as outlined in Table 13.6.

TABLE 13.6 GROUP VIA /16 HYDROGEN COMPOUNDS

COMPOUND	MOLAR MASS (g/mol)	MP (°C)	BP (°C)	H_{FUSION} (cal/mol)	H_{VAPOR} (cal/mol)
H_2O	18	0.0	100.0	1440	9720
H_2S	34	−85.5	−60.7	568	4450
H_2Se	81	−60.4	−41.5	899	4620
H_2Te	130	−48.9	−2.2	1670	5570

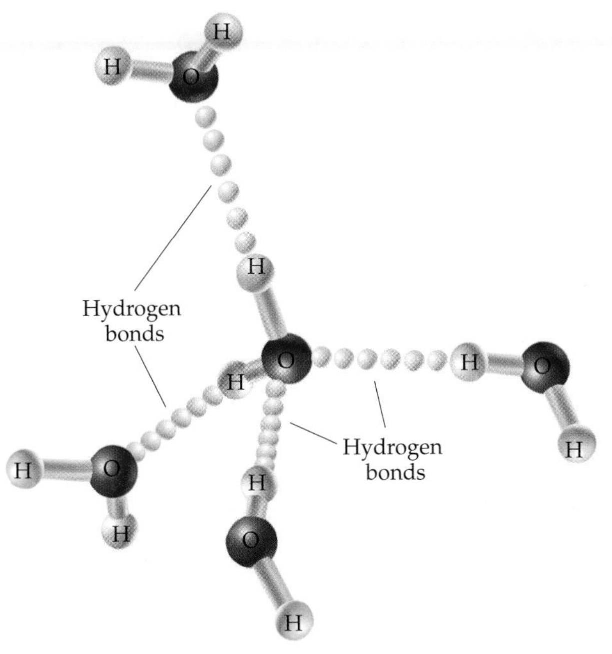

◀ **Figure 13.12 Hydrogen Bonding** Each water molecule is attracted to four other water molecules. These intermolecular hydrogen bonds are about 50% longer than an ordinary covalent bond. Since a hydrogen bond is longer, it is weaker and requires much less energy to break.

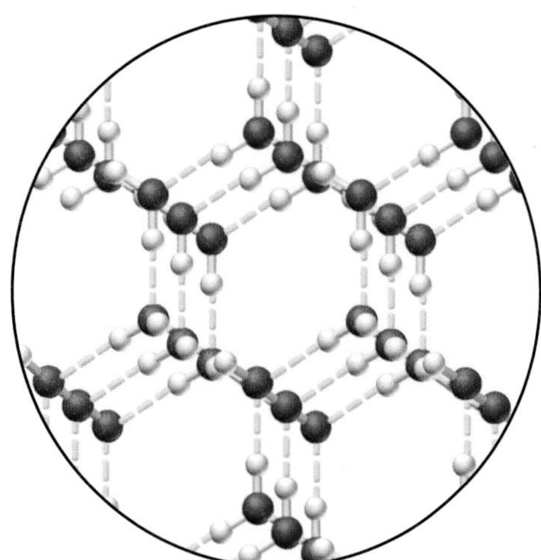

▲ **Figure 13.13 Structure of Ice Crystals** Water molecules hydrogen bond to form six-member rings. The rings in turn hydrogen bond to other rings, producing large, three-dimensional crystalline structures.

If we ignore water, we see a clear trend in the melting and boiling points of Group VIA/16 hydrogen compounds. First, notice the increase in molar mass for H_2S through H_2Te. Next, notice that the values for the melting point and the boiling point increase simultaneously. As with density, the properties of water are unusual because of hydrogen bonding. Hydrogen bonding produces a strong intermolecular attraction that resists the movement of molecules. Therefore, a higher temperature is needed to melt ice and to boil water. Figure 13.14 illustrates the melting and boiling points of Group VIA/16 hydrogen compounds.

Heats of Fusion and Vaporization

Water also has surprisingly high values for heat of fusion and heat of vaporization. This is illustrated in Table 13.6. Excluding water, we notice that as the molar mass of H_2S through H_2Te increases, the values for the heat of fusion and heat of vaporization increase. The explanation for this trend is that as the molecular size increases, the attractive forces increase slightly. Therefore, more energy is required to melt a solid or vaporize a liquid. The unusually high values for water are due to hydrogen bonding.

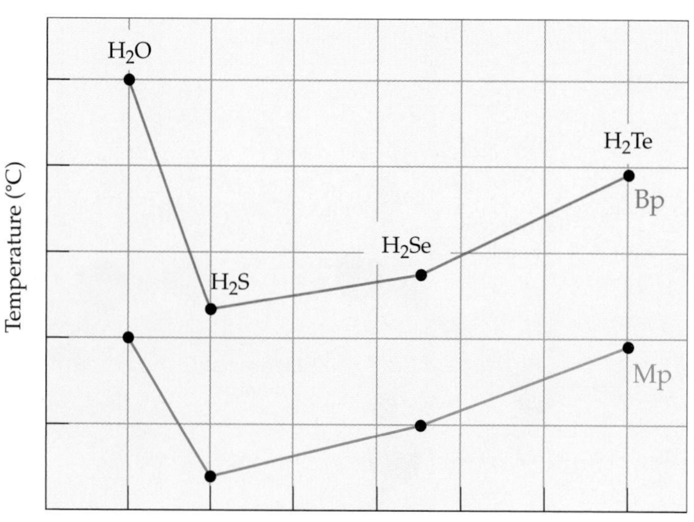

▶ **Figure 13.14 Mp and Bp of Group VIA/16 Hydrogen Compounds** Notice the systematic trend in melting point and boiling point as the molar mass increases. Water is a striking exception because of strong intermolecular hydrogen bonds.

13.9 CHEMICAL PROPERTIES OF WATER

OBJECTIVE

▶ To write chemical equations for the chemical reactions of water.

Recall that in Chapter 8 we studied five basic types of chemical reactions. Water is usually the solvent for these chemical reactions, although it does react under many conditions. One such condition is the passing of an electric current through water, which decomposes H_2O into hydrogen and oxygen gases. This process is called the **electrolysis** of water. From the balanced chemical equation, we notice that two volumes of hydrogen are produced for every volume of oxygen.

$$2\,H_2O(l) \xrightarrow{\text{electricity}} 2\,H_2(g) + O_2(g)$$

One of the five basic types of reactions is a replacement reaction. In this reaction an active metal (Li, Na, K, Ca, Sr, or Ba) reacts directly with water to give a metal hydroxide and hydrogen gas. These reactions occur rapidly at room temperature. For example, potassium metal reacts violently with water as follows:

$$2\,K(s) + 2\,H_2O(l) \longrightarrow 2\,KOH(aq) + H_2(g)$$

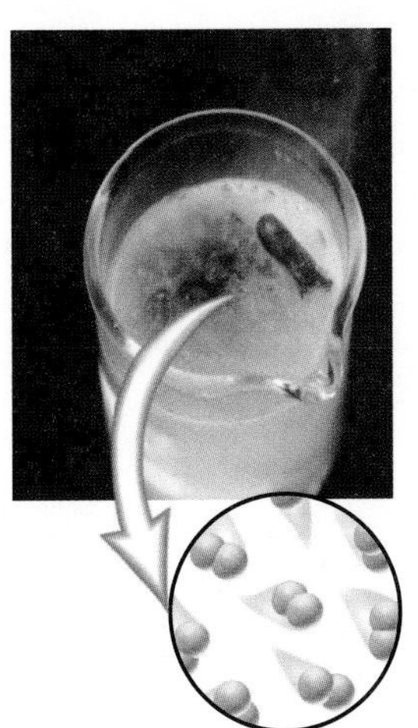

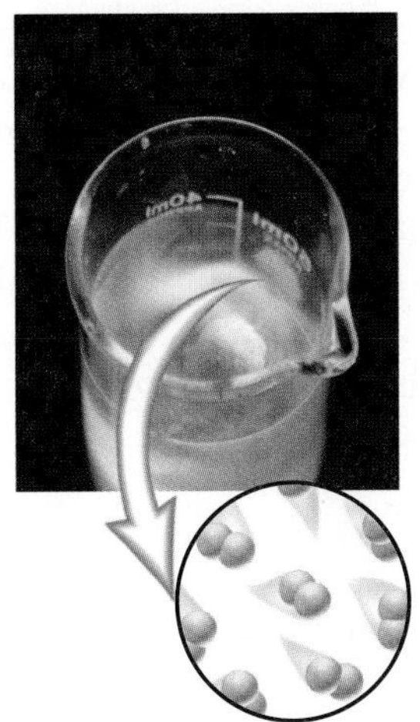

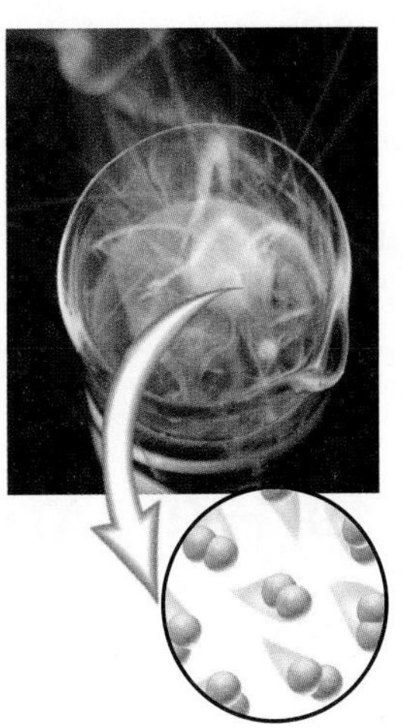

◀ **Alkali Metals in Water** The reaction of Li, Na, and K in water (left to right) produces hydrogen gas, which is flammable.

The oxides of many metals can react with water to yield a metal hydroxide. Hydroxide compounds are said to be basic, or alkaline. Since a **metal oxide** reacts with water to yield a basic solution, a metal oxide is referred to as a basic oxide. For example, calcium oxide reacts with water as follows:

$$CaO(s) + H_2O(l) \longrightarrow Ca(OH)_2(aq) \qquad \textit{(a base)}$$

The oxides of most nonmetals react with water to yield an acidic solution. Since a **nonmetal oxide** reacts with water to yield an acid, a nonmetal oxide is referred to as an acidic oxide. For example, carbon dioxide reacts with water as follows:

$$CO_2(g) + H_2O(l) \longrightarrow H_2CO_3(aq) \qquad \textit{(an acid)}$$

Reactions that Produce Water

Water is a produced by several types of reactions. The simplest reaction is the formation of water directly from hydrogen and oxygen. In this reaction, hydrogen and oxygen gases react to give H_2O. The reaction takes place very slowly at room temperature but explosively if exposed to a flame. From the balanced chemical equation, we note that two volumes of hydrogen react with one volume of oxygen.

$$2\,H_2(g) + O_2(g) \xrightarrow{\text{spark}} 2\,H_2O(l)$$

Another reaction that produces water is the combustion of hydrocarbons. Hydrocarbons are organic compounds containing hydrogen and carbon. They burn in oxygen to give carbon dioxide and water. For example, propane, C_3H_8, undergoes combustion as follows:

$$C_3H_8(g) + 5\,O_2(g) \xrightarrow{\text{spark}} 3\,CO_2(g) + 4\,H_2O(g)$$

Organic compounds containing hydrogen and oxygen also undergo combustion to give carbon dioxide and water. Ethanol, C_2H_5OH, for example, is currently blended with gasoline to produce gasohol. It undergoes combustion to give carbon dioxide and water as follows:

$$C_2H_5OH(g) + 3\,O_2(g) \xrightarrow{\text{spark}} 2\,CO_2(g) + 3\,H_2O(g)$$

Recall that neutralization reactions also produce water. An acid neutralizes a base to produce an aqueous salt and water. For example, battery acid, H_2SO_4, reacts with aqueous lye, NaOH, to produce sodium sulfate and water.

$$H_2SO_4(aq) + 2\,NaOH(aq) \longrightarrow Na_2SO_4(aq) + 2\,H_2O(l)$$

The decomposition of a hydrate compound also produces water. A **hydrate** is a crystalline compound that contains a specific number of water molecules attached to an ionic formula unit. Gypsum is a hydrate of calcium sulfate, $CaSO_4$. The formula $CaSO_4 \cdot 2H_2O$ indicates that two water molecules are attached to each formula unit. Heating a hydrate releases water from the compound. For example, heat decomposes gypsum to give $CaSO_4$ and two molecules of water. Thus,

$$CaSO_4 \cdot 2H_2O(s) \xrightarrow{\Delta} CaSO_4(s) + 2\,H_2O(g)$$

13.10 HYDRATES

OBJECTIVES

- To calculate the percentage of water in a hydrate.
- To determine the water of hydration for a hydrate.

A hydrate is a crystalline ionic compound containing water. Each formula unit in the hydrate has a specific number of water molecules attached to the compound. Common examples of hydrates include borax, $Na_2B_4O_7 \cdot 10H_2O$, and Epsom salts, $MgSO_4 \cdot 7H_2O$. The dot (•) in the hydrate formula indicates that water molecules are bonded directly to each unit of hydrate. In Epsom salts, $MgSO_4 \cdot 7H_2O$, for example, seven molecules of water are attached to each formula unit of $MgSO_4$.

Heating a hydrate drives off the water and produces an **anhydrous** compound (without water). For example, when we heat copper(II) sulfate pentahydrate, it decomposes to give anhydrous copper(II) sulfate and water. The equation for the reaction is

$$CuSO_4 \cdot 5H_2O(s) \xrightarrow{\Delta} CuSO_4(s) + 5\,H_2O(g)$$

The water molecules in the hydrate are referred to as the **water of hydration**, or *water of crystallization*. Thus, the water of hydration for copper(II) sulfate pentahydrate is 5. Figure 13.15 shows the hydrate and anhydrous forms of copper(II) sulfate.

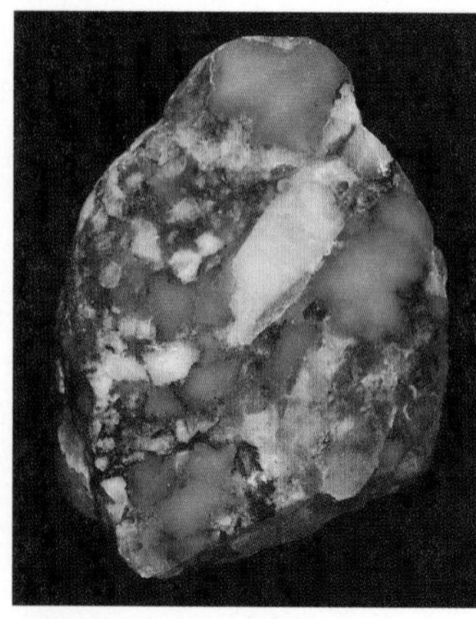

▲ **Mineral Turquoise** The mineral turquoise is an example of a tetrahydrate that contains copper, $CuAl_6(PO_4)_4(OH)_8 \cdot 4H_2O$.

IUPAC prescribes rules for the nomenclature of hydrate compounds. According to these rules, the name of the anhydrous compound appears first, followed by the number for the water of hydration (as indicated by a Greek prefix) and the word "hydrate." For example, gypsum, $CaSO_4 \cdot 2H_2O$, is systematically named calcium sulfate dihydrate; the *di-* indicates the water of hydration is 2. The following example exercise further illustrates how to name hydrate compounds:

Example Exercise 13.6 Naming Hydrates

Supply a systematic name for each of the following hydrate compounds:

(a) $Co(CN)_3 \cdot 3H_2O$ (b) $FeSO_4 \cdot H_2O$

Solution

First, name the anhydrous compound and then indicate the water of hydration by a Greek prefix. (Refer to Table 7.4 if necessary.)

(a) $Co(CN)_3$ is named cobalt(III) cyanide according to the Stock system, and cobaltic cyanide according to the Latin system. The Greek prefix for 3 is *tri-*, so the hydrate is named *cobalt(III) cyanide trihydrate*, or *cobaltic cyanide trihydrate.*

(b) $FeSO_4$ is named iron(II) sulfate according to the Stock system, and ferrous sulfate according to the Latin system. The Greek prefix for 1 is *mono-*, so the hydrate is named *iron(II) sulfate monohydrate*, or *ferrous sulfate monohydrate.*

Practice Exercise

Provide the chemical formula for each of the following hydrate compounds:

(a) zinc sulfate heptahydrate (b) sodium chromate tetrahydrate

Answers: (a) $ZnSO_4 \cdot 7H_2O$; (b) $Na_2CrO_4 \cdot 4H_2O$

Concept Exercise

What is the systematic name for the following hydrates: $CaCl_2 \cdot H_2O$, $CaCl_2 \cdot 2H_2O$, and $CaCl_2 \cdot 6H_2O$?

Answer: See Appendix G.

Percent Water in a Hydrate

In Section 9.7 we studied the percent composition of compounds. Here, we see that the percent water in a hydrate is the ratio of the mass of water in the hydrate to the mass of the hydrate, all multiplied by 100%.

$$\frac{\text{mass of water}}{\text{mass of hydrate}} \times 100\% = \%\ H_2O$$

▲ **Figure 13.15 Hydrate and Anhydrous Copper(II) Sulfate** Copper(II) sulfate pentahydrate, $CuSO_4 \cdot 5H_2O$, crystals are deep blue, whereas anhydrous copper(II) sulfate, $CuSO_4$, is a white powder. A few drops of water produce the blue color of the hydrate (right).

As an example, we'll find the percentage of water in gypsum, $CaSO_4 \cdot 2H_2O$. Using the periodic table, we find the molar mass of H_2O to be 18.02 g, and that of $CaSO_4$ to be 136.15 g. We find the percentage of water from the following ratio:

$$\frac{2\,(18.02)\,\cancel{g}}{136.15\,\cancel{g} + 2\,(18.02)\,\cancel{g}} \times 100\% = 20.93\%$$

Notice that in the numerator we multiplied the molar mass of water by 2 because the compound is a dihydrate. In the denominator, we added twice the molar mass of water to that of the anhydrous compound. The following example exercise provides additional practice in calculating the percent water in a hydrate:

Example Exercise 13.7 Percentage of Water in Hydrates

Calculate the percentage of water in each of the following hydrates:

(a) $CuSO_4 \cdot 5H_2O$ (b) $Na_2B_4O_7 \cdot 10H_2O$

Solution

In each example, first obtain the molar mass of the anhydrous compound from the periodic table. The molar mass of water is 18.02 g.

(a) Using the periodic table, we calculate that the molar mass of $CuSO_4$ is 159.62 g. Since the water of hydration for the hydrate is 5, we have

$$\frac{5\,(18.02)\,\cancel{g}}{159.62\,\cancel{g} + 5\,(18.02)\,\cancel{g}} \times 100\% = \%\ H_2O$$

$$= 36.08\%$$

(b) Using the periodic table, we calculate that the molar mass of $Na_2B_4O_7$ is 201.22 g. Since the water of hydration for the hydrate is 10, we have

$$\frac{10\,(18.02)\,\cancel{g}}{201.22\,\cancel{g} + 10\,(18.02)\,\cancel{g}} \times 100\% = \%\ H_2O$$

$$= 47.24\%$$

Practice Exercise

Calculate the percentage of water in each of the following hydrates:

(a) $NaC_2H_3O_2 \cdot 3H_2O$ (b) $Na_2S_2O_3 \cdot 5H_2O$

Answers: (a) 39.72%; (b) 36.30%

Concept Exercise

Which of the following hydrates has the lowest percentage of water: $CaCl_2 \cdot H_2O$, $CaCl_2 \cdot 2H_2O$, or $CaCl_2 \cdot 6H_2O$?

Answer: See Appendix G.

Determining the Chemical Formula of a Hydrate

In Section 9.8 we calculated the empirical formula for a compound from its percent composition. To determine the water of hydration for a hydrate we will proceed in a similar fashion.

The empirical formula of a compound is the simplest whole-number ratio of its elements. Similarly, the formula of a hydrate is the simplest whole-number ratio of water molecules to the anhydrous compound. In the chemical formula for washing soda, $Na_3PO_4 \cdot XH_2O$, X represents the water of hydration. To determine the value of X we must know the percent water. Let's assume that $Na_3PO_4 \cdot XH_2O$ is found by experiment to contain 52.3% water. The equation for the decomposition reaction is as follows:

$$Na_3PO_4 \cdot XH_2O(s) \xrightarrow{\Delta} Na_3PO_4(s) + XH_2O(g)$$

Since the hydrate contains 52.3% water, the percentage of Na_3PO_4 is 47.7% (100.0 − 52.3 = 47.7%). As in an empirical formula, let's assume we have a 100.0 g sample of hydrate. Therefore, our sample has 52.3 g of water and 47.7 g of anhydrous compound. The next step is to calculate the moles of water.

$$52.3 \text{ g } H_2O \times \frac{1 \text{ mol } H_2O}{18.02 \text{ g } H_2O} = 2.90 \text{ mol } H_2O$$

Using the periodic table, we calculate that the molar mass of Na_3PO_4 is 163.94 g. The moles of the anhydrous compound are

$$47.7 \text{ g } Na_3PO_4 \times \frac{1 \text{ mol } Na_3PO_4}{163.94 \text{ g } Na_3PO_4} = 0.291 \text{ mol } Na_3PO_4$$

We can write the mole ratio of the hydrate as $Na_3PO_4 \cdot (2.90/0.291)H_2O$, where the water of hydration is 2.90/0.291. This ratio simplifies to 10/1, meaning that the water of hydration is 10 and the formula of the hydrate is $Na_3PO_4 \cdot 10H_2O$.

Example Exercise 13.8 Water of Hydration for a Hydrate

Determine the water of hydration for a hydrate of magnesium iodide. In an experiment, $MgI_2 \cdot XH_2O$ was found to contain 34.0% water.

Solution

We begin by writing an equation for the decomposition.

$$MgI_2 \cdot XH_2O(s) \xrightarrow{\Delta} MgI_2(s) + XH_2O(g)$$

Since the hydrate contains 34.0% water, the percentage of MgI_2 is 66.0% (100.0 − 34.0 = 66.0%). Assume that we have a 100.0 g sample of hydrate. Therefore, we have 34.0 g of water and 66.0 g of anhydrous compound. The moles of water are

$$34.0 \text{ g } H_2O \times \frac{1 \text{ mol } H_2O}{18.02 \text{ g } H_2O} = 1.89 \text{ mol } H_2O$$

Using the periodic table, we calculate the molar mass of MgI_2 as 278.11 g. We find the moles of the anhydrous compound as follows:

$$66.0 \text{ g } MgI_2 \times \frac{1 \text{ mol } MgI_2}{278.11 \text{ g } MgI_2} = 0.237 \text{ mol } MgI_2$$

We can write the mole ratio of the hydrate as $MgI_2 \cdot (1.89/0.237)H_2O$. The ratio 1.89/0.237 reduces to 7.97 and rounds off to the whole number 8. The water of hydration is 8 and the chemical formula is $MgI_2 \cdot 8H_2O$. The name of the hydrate is *magnesium iodide octahydrate*.

Practice Exercise

Determine the water of hydration for the hydrate of copper(II) fluoride. In an experiment, $CuF_2 \cdot XH_2O$ was found to contain 26.2% water.

Answer: $CuF_2 \cdot 2H_2O$

Concept Exercise

Which of the following hydrates has the highest water of hydration: $CaCl_2 \cdot H_2O$, $CaCl_2 \cdot 2H_2O$, or $CaCl_2 \cdot 6H_2O$?

Answer: See Appendix G.

CRITICAL THINKING Bottled Water

Q: What is the difference between bottled water and tap water?

▲ **Periodic Table and Perrier, Dasani, and Aquafina** Bottles of purified water and natural spring water.

Recently, bottled water has become a huge industry and Americans consume nearly six billion gallons of bottled water each year. One survey found that more than half of all Americans drink bottled water, and about a third of the public consumes it regularly. Many different types of bottled water are now available including purified water, spring water, mineral water, and vitamin-enriched bottled water.

The water we drink from a home faucet is referred to as *tap water*. It comes from a city source and must be chlorinated for sanitation purposes. Tap water that has been processed by distillation or deionization is referred to as *purified water*. Some brands of purified water add minerals to mimic the taste of drinking water. Water obtained from natural springs that lie deep within the Earth is referred to as *spring water*. Although spring water does not require purification, the transportation of spring water can jeopardize purity. This fact has led some spring water suppliers to bottle the water directly at the source of the natural spring.

Consumers have been led to believe that bottled water is superior to tap water in terms of purity and health benefits. Surprisingly, in most cases bottled water adheres to purity standards that are less strict than tap water. While the Environmental Protection Agency (EPA) regulates the quality of public water supplies, the EPA has no authority over bottled water. Bottled water that crosses state lines is considered a food product and is overseen by the Food and Drug Administration (FDA), which does require that the water be bottled in sanitary conditions.

In a study by the Natural Resources Defense Council that tested 38 brands of California bottled water, eight samples were found to have measurable levels of contamination. In a study at Case Western Reserve University and Ohio State University, 57 samples of bottled water were compared to Cleveland's tap water. While 39 of the bottled water samples were purer than the tap water, 15 of the bottles had significantly higher bacteria levels. However, the scientists stated that all of the water was safe to drink.

An example of a sparkling mineral water is Perrier®, which comes from a source in the South of France. A distinguishing characteristic of sparkling mineral water is its natural carbonation. Carbonation is the result of volcanic activity within the Earth. Heat from a volcano produces steam that passes through limestone ($CaCO_3$), which in turn releases carbon dioxide gas into the mineral water.

A: The difference between bottled water and tap water depends on the source. Some bottled water originates from natural springs; whereas, other bottled water is distilled or deionized tap water, sometimes with minerals added.

Summary

Section 13.1 The liquid state has a variable shape but a fixed volume. Liquids usually flow readily but do not compress or expand as do gases. The densities of liquids vary but are approximately 1000 times greater than the densities of gases. Liquids mix and diffuse uniformly in a container, similar to gases.

Section 13.2 There are three basic types of intermolecular attraction: dispersion forces, dipole forces, and hydrogen bonds. A **dispersion force** has the weakest attraction and results from temporary dipoles in molecules. A **dipole force** has a stronger attraction and results from permanent dipoles in molecules. A **hydrogen bond** exerts a stronger attraction than dispersion and dipole forces, and occurs between molecules having H–O, H–N, or H–F bonds.

Section 13.3 When the **vapor pressure** of a liquid equals the atmospheric pressure, the liquid is at its **boiling point**. The normal boiling point of a liquid is the temperature at which the vapor pressure is 760 mm Hg. Liquids have a resistance to flow, and this property is called **viscosity**. The attraction between molecules at the surface of a liquid is called **surface tension**. Each of these properties is affected by the degree of attraction between molecules in the liquid. If the attraction is low, the vapor pressure is high. Conversely, if the attraction is high, the values for the boiling point, viscosity, and surface tension are also high.

Section 13.4 The solid state has a fixed shape and a fixed volume. A solid that has a highly defined structure is called a

crystalline solid. Solids do not compress or expand to any large degree. The density of a substance in the solid state is usually higher than that in the liquid state. Water and ammonia are two interesting exceptions to this rule. Solid ice and solid ammonia float on their respective liquids. Since the solid state contains particles in fixed positions, solids do not mix or diffuse.

Section 13.5 There are three basic types of crystalline solids. **Ionic solids** are made up of ions. The ions are attracted to each other and form repeating geometric patterns. **Molecular solids** form crystals made up of molecules. **Metallic solids** are made up of metal atoms arranged in a definite pattern. The properties of ionic, molecular, and metallic solids differ. Metallic solids are good conductors of electricity. Ionic solids conduct electricity only when they are melted or are in an aqueous solution. As a rule, molecular solids do not conduct electricity.

Section 13.6 Heat is necessary to change the physical state of a substance. The heat required to melt a substance is called the **heat of fusion**. The heat released when a substance freezes to a solid is called the *heat of solidification*. The heat of solidification of a substance has the same value as its heat of fusion. The heat required to convert a liquid to a gas at its boiling point is the **heat of vaporization**. When a vapor condenses to a liquid, it releases the same amount of heat energy that was necessary to vaporize the liquid. This energy is termed the *heat of condensation*.

Section 13.7 Water has unusual properties because of the two polar O–H bonds separated by a **bond angle** of 104.5°. The oxygen atom is the focus of a partial negative charge, and each hydrogen atom has a partial positive charge. The two polar bonds give rise to an overall **net dipole** for the water molecule.

Section 13.8 Hydrogen bonds between water molecules explain why water has extraordinarily high values for boiling point, viscosity, and surface tension. Hydrogen bonding also explains why the density of ice is less than that of water. For a small molecule, water has an unusually high melting point, boiling point, heat of fusion, and heat of vaporization.

Section 13.9 Water undergoes the following chemical reactions: (1) An electric current decomposes water into hydrogen and oxygen gases by **electrolysis**. (2) The active metals of Groups IA/1 and IIA/2 react with water to give a metal hydroxide and hydrogen gas. (3) A **metal oxide** reacts with water to give a metal hydroxide. (4) A **nonmetal oxide** combines with water to yield an acidic solution. (See Table 13.7).

Water is produced from (1) the reaction of hydrogen and oxygen gases, (2) burning hydrocarbons, (3) neutralizing an acid with a base, and (4) heating a **hydrate**. (See Table 13.8).

Section 13.10 The name of a hydrate consists of the name of the **anhydrous** compound followed by a Greek prefix and the word "hydrate"; for example, $BaCl_2 \cdot 2H_2O$ is named barium chloride dihydrate. Determining the chemical formula of a hydrate is similar to determining an empirical formula. From the percent water in a hydrate, we can calculate the moles of water and anhydrous compound. The ratio of moles of water to moles of the anhydrous compound gives the **water of hydration**, which is also termed *water of crystallization*.

TABLE 13.7 SUMMARY OF THE CHEMICAL REACTIONS OF WATER

water	$\xrightarrow{\text{electricity}}$	hydrogen + oxygen
active metal + water	$\longrightarrow$	metal hydroxide + hydrogen
metal oxide + water	$\longrightarrow$	basic solution
nonmetal oxide + water	$\longrightarrow$	acidic solution

TABLE 13.8 SUMMARY OF THE CHEMICAL REACTIONS PRODUCING WATER

hydrogen + oxygen	$\xrightarrow{\text{spark}}$	water
hydrocarbon + oxygen	$\xrightarrow{\text{spark}}$	carbon dioxide + water
acid + base	$\longrightarrow$	salt + water
hydrate	$\xrightarrow{\Delta}$	anhydrous compound + water

Key Terms *Answers to Key Terms are in Appendix H.*

Select the key term that corresponds to each of the following definitions.

_____ **1.** an intermolecular attraction based on temporary dipoles

_____ **2.** an intermolecular attraction based on permanent dipoles

_____ **3.** the attraction between a H atom bonded to O or N, and a nonbonding electron pair on an electronegative atom in another molecule

_____ **4.** the pressure exerted by gaseous molecules above a liquid in a closed container when the rates of evaporation and condensation are equal

(a) anhydrous *(Sec. 13.10)*
(b) boiling point (Bp) *(Sec. 13.3)*
(c) bond angle *(Sec. 13.7)*
(d) crystalline solid *(Sec. 13.4)*
(e) deionized water *(Sec. 13.7)*
(f) dipole force *(Sec. 13.2)*

_____ 5. the temperature at which the vapor pressure of a liquid is equal to the atmospheric pressure
_____ 6. the resistance of a liquid to flow
_____ 7. the tendency of a liquid to form spherical drops
_____ 8. a solid substance composed of ions or molecules that repeat in a regular geometric pattern
_____ 9. a crystalline solid composed of ions that repeat in a regular pattern
_____ 10. a crystalline solid composed of molecules that repeat in a regular pattern
_____ 11. a crystalline solid composed of metal atoms that repeat in a regular pattern
_____ 12. the heat required to convert a solid to a liquid at its melting point
_____ 13. the heat required to vaporize a liquid to a gas at its boiling point
_____ 14. the angle formed by two atoms bonded to a central atom in a molecule
_____ 15. the overall direction of partial negative charge in a molecule having two or more dipoles
_____ 16. the chemical reaction produced by the passage of electric current through an aqueous solution
_____ 17. a compound that reacts with water to form a basic solution; also termed a basic oxide
_____ 18. a compound that reacts with water to form an acidic solution; also termed an acidic oxide
_____ 19. a substance that contains a specific number of water molecules attached to a formula unit in a crystalline compound
_____ 20. refers to a compound that does not contain water
_____ 21. the number of water molecules bound to a formula unit in a hydrate; also termed the water of crystallization
_____ 22. water containing a variety of cations and anions such as Ca^{2+}, Mg^{2+}, Fe^{3+}, $CO_3{}^{2-}$, $SO_4{}^{2-}$, and $PO_4{}^{3-}$
_____ 23. water containing sodium ions and a variety of anions
_____ 24. water purified by removing ions using an ion exchange method; also termed demineralized water
_____ 25. water purified by boiling and collecting the condensed vapor

(g) dispersion force *(Sec. 13.2)*
(h) distilled water *(Sec. 13.7)*
(i) electrolysis *(Sec. 13.9)*
(j) hard water *(Sec. 13.7)*
(k) heat of fusion *(Sec. 13.6)*
(l) heat of vaporization *(Sec. 13.6)*
(m) hydrate *(Sec. 13.9)*
(n) hydrogen bond *(Sec. 13.2)*
(o) ionic solid *(Sec. 13.5)*
(p) metal oxide *(Sec. 13.9)*
(q) metallic solid *(Sec. 13.5)*
(r) molecular solid *(Sec. 13.5)*
(s) net dipole *(Sec. 13.7)*
(t) nonmetal oxide *(Sec. 13.9)*
(u) soft water *(Sec. 13.7)*
(v) surface tension *(Sec. 13.3)*
(w) vapor pressure *(Sec. 13.3)*
(x) viscosity *(Sec. 13.3)*
(y) water of hydration *(Sec. 13.10)*

Exercises *Answers to odd-numbered Exercises are in Appendix I.*

Properties of Liquids (Sec. 13.1)

1. Which of the following are observed general properties of liquids?
(a) variable shape **(b)** flow readily
(c) compress significantly **(d)** more dense than gases

2. Which of the following are observed general properties of liquids?
(a) fixed volume **(b)** mix homogeneously
(c) expand significantly **(d)** less dense than gases

3. Indicate the physical state (solid, liquid, gas) for each of the following noble gases at the designated temperature:
(a) Ne at –225 °C **(b)** Ne at –255 °C
(c) Ar at –175 °C **(d)** Ar at –200 °C
(e) Kr at –155 °C **(f)** Kr at –150 °C

The melting points and boiling points for neon, argon, and krypton are as follows:

Element	Melting Point	Boiling Point
neon, Ne	–249 °C	–246 °C
argon, Ar	–189 °C	–186 °C
krypton, Kr	–157 °C	–152 °C

◀ **The Liquid State** Maple syrup can be poured and its rate of flow depends on its viscosity.

4. Indicate the physical state (solid, liquid, or gas) for each of the following at the designated temperature:
(a) H_2O at –10 °C **(b)** H_2O at 110 °C
(c) NH_3 at –100 °C **(d)** NH_3 at –50 °C
(e) $CHCl_3$ at 100 °C **(f)** $CHCl_3$ at 50 °C

The melting points and boiling points for water, ammonia, and chloroform are as follows:

Element	Melting Point	Boiling Point
water, H_2O	0 °C	100 °C
ammonia, NH_3	–78 °C	–33 °C
chloroform, $CHCl_3$	–64 °C	62 °C

The Intermolecular Bond Concept (Sec. 13.2)

5. What is the strongest type of intermolecular attraction (dispersion force, dipole force, or hydrogen bond) that exists in each of the following liquids?
(a) C_6H_{14} (b) $C_2H_5–F$
(c) $CH_3–OH$ (d) $CH_3–O–CH_3$
6. What is the strongest type of intermolecular attraction (dispersion force, dipole force, or hydrogen bond) that exists in each of the following liquids?
(a) C_8H_{18} (b) $C_2H_5–Cl$
(c) HCOOH (d) $C_2H_5–O–C_2H_5$
7. Predict which liquid in each pair has the higher vapor pressure:
(a) CH_3COOH or C_2H_5Cl (b) C_2H_5OH or CH_3OCH_3
8. Predict which liquid in each pair has the higher boiling point:
(a) CH_3COOH or C_2H_5Cl (b) C_2H_5OH or CH_3OCH_3
9. Predict which liquid in each pair has the higher viscosity:
(a) CH_3COOH or C_2H_5Cl (b) C_2H_5OH or CH_3OCH_3
10. Predict which liquid in each pair has the higher surface tension:
(a) CH_3COOH or C_2H_5Cl (b) C_2H_5OH or CH_3OCH_3

Vapor Pressure, Boiling Point, Viscosity, and Surface Tension (Sec. 13.3)

11. Describe and illustrate the concept of vapor pressure using water as an example.
12. Describe and illustrate the concept of boiling point using water as an example.
13. Describe and illustrate the concept of viscosity using water as an example.
14. Describe and illustrate the concept of surface tension using water as an example.
15. If the molecules in a liquid are weakly attracted, which of the following properties has a high value?
(a) vapor pressure (b) boiling point
(c) viscosity (d) surface tension
16. If the molecules in a liquid are strongly attracted, which of the following properties has a high value?
(a) vapor pressure (b) boiling point
(c) viscosity (d) surface tension
17. What is the general relationship between the vapor pressure of a liquid and its temperature?
18. What is the general relationship between the boiling point of a liquid and its vapor pressure?
19. Refer to Figure 13.5 and estimate the vapor pressure of ethanol at each of the following temperatures:
(a) 45 °C (b) 60 °C
20. Refer to Figure 13.5 and estimate the vapor pressure of ether at each of the following temperatures:
(a) 15 °C (b) 30 °C
21. The vapor pressure of acetone is 1 torr at –59 °C, 10 torr at –31 °C, 100 torr at 8 °C, 400 torr at 40 °C, and 760 torr at 56 °C. What is the normal boiling point of acetone?
22. The vapor pressure of methanol is 1 atm at 65 °C, 2 atm at 84 °C, 5 atm at 112 °C, 10 atm at 138 °C, and 20 atm at 168 °C. What is the normal boiling point of methanol?

Properties of Solids (Sec. 13.4)

23. List five general properties of the solid state.
24. Distinguish between a solid and a liquid at the molecular level.
25. Indicate the physical state (solid, liquid, or gas) for gallium given its melting point (30 °C) and boiling point (2403 °C) under the following conditions:
(a) Ga in ice water (b) Ga in boiling water
26. Indicate the physical state (solid, liquid, or gas) for mercury given its melting point (–40 °C) and boiling point (357 °C) under the following conditions:
(a) Hg in ice water (b) Hg in boiling water

Crystalline Solids (Sec. 13.5)

27. List three examples of crystalline solids.
28. List three examples of noncrystalline solids.
29. State the type of particles that compose each of the following:
(a) ionic solid (b) molecular solid
(c) metallic solid

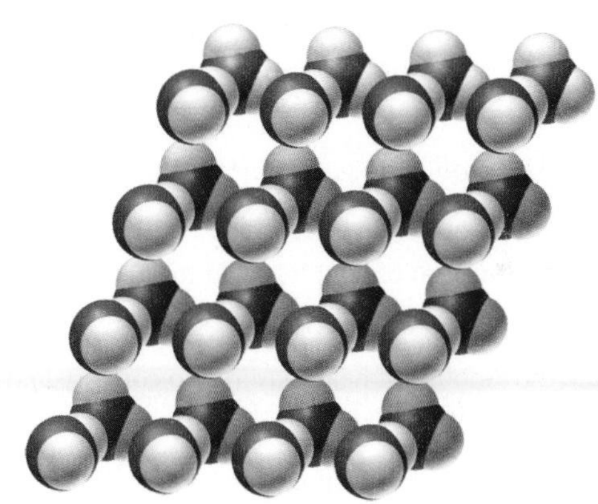

▲ A Molecular Solid, H_2O

30. State whether the following list of properties is most descriptive of an ionic, a molecular, or a metallic solid.
(a) malleable, ductile, insoluble in water
(b) high melting point, conductor of electricity when melted
(c) low melting point, nonconductor of electrical current
31. Classify each of the following crystalline solids as ionic, molecular, or metallic.
(a) zinc, Zn
(b) zinc oxide, ZnO
(c) phosphorus, P_4
(d) iodine monobromide, IBr

32. Classify each of the following crystalline solids as ionic, molecular, or metallic.
(a) sulfur, S_8
(b) sulfur dioxide, SO_2
(c) silver, Ag
(d) silver nitrate, $AgNO_3$

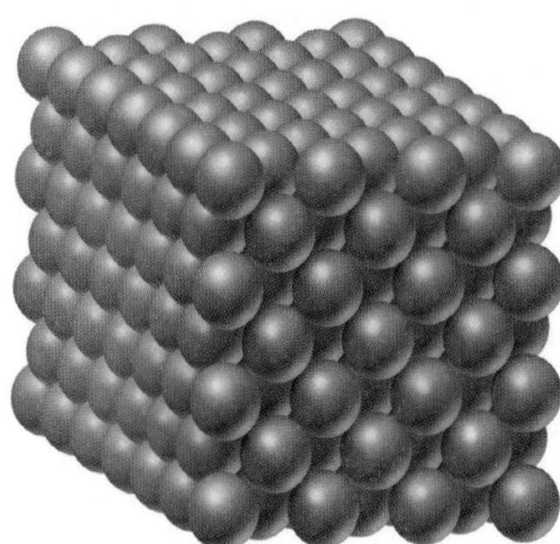

▲ A Metallic Solid, Cu

Changes of Physical State (Sec. 13.6)

33. Draw the general shape of the temperature–energy graph for the heating of ethanol from –120 °C to 120 °C (given that Mp = −117.3 °C; Bp = 78.5 °C).

34. Draw the general shape of the temperature–energy graph for the cooling of acetone from 100 °C to –100 °C (given that Mp = −95.4 °C; Bp = 56.2 °C).

35. Calculate the amount of heat required to change 50.0 g of ice to water at 0 °C.

36. Calculate the amount of heat released to change 75.0 g of water to ice at 0 °C.

37. Calculate the amount of heat required to convert 25.0 g of water at 25.0 °C to steam at 100.0 °C.

38. Calculate the amount of heat released when 65.5 g of water at 55.5 °C cools to ice at 0.0 °C.

39. Calculate the amount of heat required to convert 115 g of ice at 0.0 °C to steam at 100.0 °C.

40. Calculate the amount of heat released when 155 g of steam at 100.0 °C cools to ice at 0.0 °C.

41. Calculate the amount of heat required to convert 38.5 g of ice at –20.0 °C to steam at 100.0 °C.

42. Calculate the amount of heat released when 90.5 g of steam at 110.0 °C cools to ice at 0.0 °C.

43. Calculate the amount of heat required to convert 100.0 g of ice at –40.0 °C to steam at 125.0 °C.

44. Calculate the amount of heat released when 0.500 kg of steam at 150.0 °C cools to ice at –50.0 °C.

Structure of Water (Sec. 13.7)

45. How many bonding and nonbonding valence electron pairs are in a water molecule?

46. Draw the electron dot and structural formulas for a molecule of water.

47. What is the observed bond angle in a water molecule?

48. The center of a tetrahedron forms an angle of 109° with its corners. If an oxygen atom is at the center of the tetrahedron and a hydrogen atom is at two of the corners, explain why the observed bond angle is less than 109°.

49. Indicate the two dipoles in a water molecule using the delta convention.

50. Draw the net dipole in a molecule of water.

51. Draw two molecules of hydrogen fluoride, HF, and diagram a hydrogen bond.

52. Draw two molecules of ammonia, NH_3, and diagram a hydrogen bond.

Physical Properties of Water (Sec. 13.8)

53. If a water bottle is completely filled and capped, what will happen if the water in the bottle freezes to a solid?

54. Ammonia is less dense in the solid state than the liquid state. Does an "ammonia ice cube" float or sink in liquid ammonia?

55. Without referring to Table 13.6, predict which compound in each of the following pairs has the higher melting point:
(a) H_2O or H_2S
(b) H_2S or H_2Se

56. Without referring to Table 13.6, predict which compound in each of the following pairs has the higher boiling point:
(a) H_2O or H_2Se
(b) H_2S or H_2Te

57. Without referring to Table 13.6, predict which compound in each of the following pairs has the higher heat of fusion (cal/mol):
(a) H_2O or H_2S
(b) H_2S or H_2Se

58. Without referring to Table 13.6, predict which compound in each of the following pairs has the higher heat of vaporization (cal/mol):
(a) H_2O or H_2Se
(b) H_2S or H_2Te

59. In general, as the molar mass of Group VIA/16 hydrogen compounds increases, does each of the following increase or decrease?
(a) melting point
(b) boiling point
(c) heat of fusion
(d) heat of vaporization

60. Refer to the trends in Table 13.6 and estimate a value for the physical properties of radioactive H_2Po, that is, predict a value for the Mp, Bp, H_{fusion}, and H_{vapor}.

Chemical Properties of Water (Sec. 13.9)

61. Write a balanced equation for the reaction of hydrogen and oxygen gases.

62. Write a balanced equation for the decomposition of water by electrolysis.

63. Complete and balance the following equations:
(a) $Li(s) + H_2O(l) \longrightarrow$
(b) $Na_2O(s) + H_2O(l) \longrightarrow$
(c) $CO_2(g) + H_2O(l) \longrightarrow$

64. Complete and balance the following equations:
(a) $Rb(s) + H_2O(l) \longrightarrow$
(b) $Cs_2O(s) + H_2O(l) \longrightarrow$
(c) $P_2O_5(s) + H_2O(l) \longrightarrow$

65. Complete and balance the following equations:
(a) $Mg(s) + H_2O(l) \longrightarrow$
(b) $SrO(s) \longrightarrow H_2O(l) \longrightarrow$
(c) $N_2O_5(g) + H_2O(l) \longrightarrow$

66. Complete and balance the following equations:
(a) $Ba(s) + H_2O(l) \longrightarrow$
(b) $CaO(s) + H_2O(l) \longrightarrow$
(c) $N_2O_3(g) + H_2O(l) \longrightarrow$

67. Complete and balance the following equations:
(a) $C_3H_6(g) + O_2(g) \xrightarrow{\text{spark}}$
(b) $Na_2Cr_2O_7 \cdot 2H_2O(s) \xrightarrow{\Delta}$
(c) $HF(aq) + Ca(OH)_2(aq) \longrightarrow$

68. Complete and balance the following equations:
(a) $C_3H_7OH(l) + O_2(g) \xrightarrow{\text{spark}}$
(b) $Ca(NO_3)_2 \cdot 4H_2O(s) \xrightarrow{\Delta}$
(c) $H_2CO_3(aq) + KOH(aq) \longrightarrow$

69. Complete and balance the following equations:
(a) $C_4H_{10}(g) + O_2(g) \xrightarrow{\text{spark}}$
(b) $Co(C_2H_3O_2)_2 \cdot 4H_2O(s) \xrightarrow{\Delta}$
(c) $HNO_3(aq) + Ba(OH)_2(aq) \longrightarrow$

70. Complete and balance the following equations:
(a) $C_4H_9OH(l) + O_2(g) \xrightarrow{\text{spark}}$
(b) $KAl(SO_4)_2 \cdot 12H_2O(s) \xrightarrow{\Delta}$
(c) $H_3PO_4(aq) + NaOH(aq) \longrightarrow$

Hydrates (Sec. 13.10)

71. Supply a systematic name for each of the following hydrate compounds:
(a) $MgSO_4 \cdot 7H_2O$
(b) $Co(CN)_3 \cdot 3H_2O$
(c) $MnSO_4 \cdot H_2O$
(d) $Na_2Cr_2O_7 \cdot 2H_2O$

72. Supply a systematic name for each of the following hydrate compounds:
(a) $Sr(NO_3)_2 \cdot 6H_2O$
(b) $Co(C_2H_3O_2)_2 \cdot 4H_2O$
(c) $CuSO_4 \cdot 5H_2O$
(d) $Cr(NO_3)_3 \cdot 9H_2O$

73. Provide the chemical formula for each of the following hydrate compounds:
(a) sodium carbonate decahydrate
(b) nickel(II) nitrate hexahydrate
(c) cobalt(III) iodide octahydrate
(d) chromium(III) acetate monohydrate

74. Provide the chemical formula for each of the following hydrate compounds:
(a) sodium acetate trihydrate
(b) calcium sulfate dihydrate
(c) potassium chromate tetrahydrate
(d) zinc sulfate heptahydrate

75. Calculate the percentage of water in each of the following hydrates:
(a) $SrCl_2 \cdot 6H_2O$
(b) $K_2Cr_2O_7 \cdot 2H_2O$
(c) $Co(CN)_3 \cdot 3H_2O$
(d) $Na_2CrO_4 \cdot 4H_2O$

76. Calculate the percentage of water in each of the following hydrates:
(a) $MnSO_4 \cdot H_2O$
(b) $Sr(NO_3)_2 \cdot 6H_2O$
(c) $Co(C_2H_3O_2)_2 \cdot 4H_2O$
(d) $Cr(NO_3)_3 \cdot 9H_2O$

77. Determine the water of hydration for the following hydrates and write the chemical formula:
(a) $NiCl_2 \cdot XH_2O$ is found to contain 21.7% water.
(b) $Sr(NO_3)_2 \cdot XH_2O$ is found to contain 33.8% water.
(c) $CrI_3 \cdot XH_2O$ is found to contain 27.2% water.
(d) $Ca(NO_3)_2 \cdot XH_2O$ is found to contain 30.5% water.

78. Determine the water of hydration for the following hydrates and write the chemical formula:
(a) $SrCl_2 \cdot XH_2O$ is found to contain 18.5% water.
(b) $Ni(NO_3)_2 \cdot XH_2O$ is found to contain 37.2% water.
(c) $CoSO_4 \cdot XH_2O$ is found to contain 10.4% water.
(d) $Na_2B_4O_7 \cdot XH_2O$ is found to contain 30.9% water.

General Exercises

79. State the approximate percent of Earth's surface that is covered by water.

80. State the approximate percent of the human body that is water.

81. Refer to Figure 13.5 and determine the boiling point of water at an elevation where the atmospheric pressure is 650 torr.

82. Refer to Figure 13.5 and determine the boiling point of ethanol at an elevation where the atmospheric pressure is 0.5 atm.

83. The atmosphere on Venus has clouds of sulfuric acid. Explain why it is reasonable to predict that the "raindrops" on Venus are small spheres.

84. Explain why metallic solids are good conductors of electricity.

85. Ethylene glycol is a permanent antifreeze. Calculate the amount of heat released when 1250 g of liquid at 25.0 °C cools to a solid at its melting point. (For ethylene glycol specific heat = 0.561 cal/(g × °C), Mp = −11.5 °C, Bp = 197.6 °C, H_{fusion} = 43.3 cal/g, and H_{vapor} = 293 cal/g.)

86. Methanol is a temporary antifreeze because of its low boiling point. Calculate the amount of heat necessary to convert 1250 g of liquid at 25.0 °C to a vapor at its boiling point. (For methanol: specific heat = 0.610 cal/(g × °C), Mp = −97.9 °C, Bp = 65.2 °C, H_{fusion} = 23.7 cal/g, and H_{vapor} = 293 cal/g.)

13 Chapter Self-Test

Answers to Self-Test are in Appendix J.

1. Which of the following is an observed property of liquids? (Sec. 13.1)
(a) Liquids are more dense than gases.
(b) Liquids expand and compress significantly.
(c) Liquids have a fixed shape and variable volume.
(d) Liquids that are insoluble mix homogeneously.
(e) all of the above

2. What is the strongest intermolecular force in a liquid containing molecules with N–H bonds? (Sec. 13.2)
(a) covalent bonds
(b) dipole forces
(c) dispersion forces
(d) hydrogen bonds
(e) none of the above

3. If the molecules in a liquid have a strong attraction for each other, which of the following properties has a relatively high value? (Sec. 13.3)
(a) boiling point
(b) viscosity
(c) surface tension
(d) all of the above
(e) none of the above

4. Which of the following is an observed property of solids? (Sec. 13.4)
(a) Solids have a variable shape and fixed volume.
(b) Solids are usually less dense than liquids.
(c) Solids can have noncrystalline structures.
(d) Solids compress and expand significantly.
(e) none of the above

5. Which of the following is an example of a molecular crystalline solid? (Sec. 13.5)
(a) cuprite, Cu_2O
(b) dry ice, CO_2
(c) limestone, $CaCO_3$
(d) iron pyrite, FeS_2
(e) none of the above

6. Calculate the heat required to convert 10.0 g of water at 100 °C to steam at 100 °C. The specific heat of water is 1.00 cal/(g × °C); the heat of fusion is 80.0 cal/g; and the heat of vaporization is 540.0 cal/g. (Sec. 13.6)
(a) 8.0 cal
(b) 10.0 cal
(c) 54.0 cal
(d) 8.00×10^2 cal
(e) 5.40×10^3 cal

7. How many dipoles are in a water molecule? (Sec. 13.7)
(a) 0
(b) 1
(c) 2
(d) 3
(e) 4

8. Which of the following explains why ice floats on water? (Sec. 13.8)
(a) Ice has a greater density than water.
(b) Ice has a greater heat of fusion than water.
(c) Ice has a greater molar mass than water.
(d) Ice has a greater specific heat than water.
(e) Ice has a greater volume than an equal mass of water.

9. What is the product from the reaction of water and a nonmetal oxide? (Sec. 13.9)
(a) an acid
(b) a base
(c) a nonmetal hydroxide and hydrogen gas
(d) a nonmetal hydroxide and oxygen gas
(e) none of the above

10. An unknown hydrate of cobalt(II) cyanide, $Co(CN)_2 \cdot XH_2O$, is heated to give 32.7% water. What is the water of crystallization (X) for the hydrate?
(a) 2
(b) 3
(c) 4
(d) 6
(e) 10

Key Concepts

11. Which physical state can be described by the following analogy? A beehive is surrounded by flying honey bees. When approached with a lit torch, the honey bees move about much faster.

12. Why does hard boiling an egg take ten minutes at a mile-high elevation and only five minutes at sea level?

Critical Thinking

13. Observe the photograph closely and note water has a concave lens and mercury has a convex lens. Explain this observation in terms of the attractive forces between the glass cylinder and each liquid.

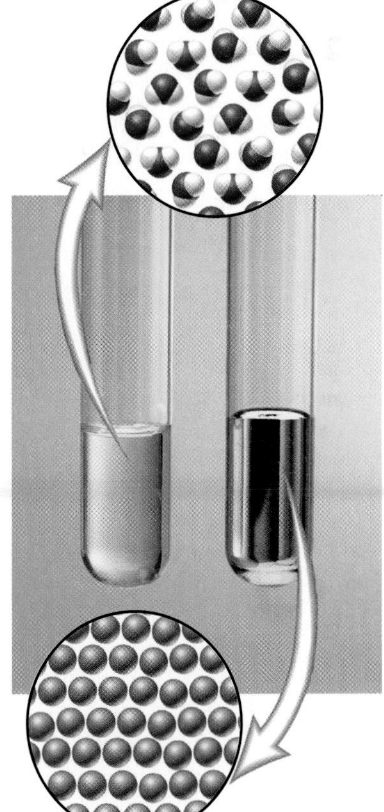

▶ **Water versus Mercury** The concave lens of water in a glass cylinder is shown on the left, and the convex lens of mercury is shown on the right.

CHAPTER

14 Solutions

As salt dissolves in the ocean, water molecules pull sodium ions and chloride ions into the solution. After dissolving, sodium and chloride ions are surrounded by water molecules in a solvent cage.

"Do not let what you cannot do interfere with what you can do."
John Wooden, UCLA Legendary Basketball Coach (1910–)

A solution is a homogeneous mixture. This means that a solution is the same throughout, and that every sample of it has the same properties. A solution consists of a solute dissolved in a solvent. For instance, the solute sugar dissolved in the solvent water gives an aqueous solution of sugar. Other examples of solutions are carbon dioxide dissolved in a soft drink and salt dissolved in the ocean. The proportion of solute and solvent can vary, but the solvent is always the greater quantity.

We usually think of solutions as liquids. However, solutes scattered throughout solvents exist in all three physical states; there are gaseous, liquid, and solid solutions. Table 14.1 lists some common examples of solutions whose physical state corresponds to that of the solvent.

TABLE 14.1 TYPES OF SOLUTIONS

SOLUTE	SOLVENT	SOLUTION
Gaseous Solutions —		
gas	gas	air (nitrogen, oxygen, argon gases)
liquid	gas	humid air (water vapor in air)
Liquid Solutions —		
gas	liquid	carbonated drinks (CO_2 in water)
liquid	liquid	vinegar ($HC_2H_3O_2$ in water)
solid	liquid	salt water (NaCl in water)
Solid Solutions —		
liquid	solid	amalgam dental fillings (Hg in Ag)
solid	solid	sterling silver (Cu in Ag)

14.1 GASES IN SOLUTION

OBJECTIVES

- To state the effect of temperature on the solubility of a gas in a liquid.
- To state the effect of pressure on the solubility of a gas in a liquid.

Muriatic acid is available from the supermarket and is used to increase the acidity of swimming pools. The systematic name for muriatic acid is hydrochloric acid. It is produced by dissolving hydrogen chloride gas in water. Like muriatic acid, household ammonia is another common example of a gas in solution. Ammonia solutions have ammonia gas dissolved in water. Champagne and soft drinks have dissolved carbon dioxide gas, and fish rely on dissolved oxygen gas in water.

Solubility and Temperature

Solutions of gases in liquids are greatly affected by changes in temperature. As the temperature increases, the kinetic energy of the solute gas becomes greater, and the gas molecules acquire more of a tendency to escape from the solvent. Thus, as the temperature goes up, the solubility of a gas in a liquid goes down.

A practical example of this principle is illustrated by a carbonated beverage. You have probably taken a soda from the refrigerator and let it warm to room temperature. When you opened the soda it foamed as gas escaped from the solution. Compare the foaming to what happens when you open a soda just after taking it from the refrigerator. At colder temperatures, carbon dioxide is more soluble and foaming is minimal. Table 14.2 indicates the effect of temperature on the solubility of carbon dioxide in water.

▲ **Deep-Sea Divers** As divers descend, each 33 feet of depth increases the total pressure by 1 atmosphere. Thus, at 33 feet the solubility of the air mixture in blood is two times that at sea level; at 66 feet the solubility of the air mixture is three times.

Solubility and Pressure

The solubility of a gas in a liquid is strongly influenced by pressure. In 1803, the English chemist William Henry conducted experiments on the solubility of gases in liquids. He found that the solubility of a gas was proportional to the partial pressure of the gas above the liquid. This statement is known as **Henry's law**. As an example, consider carbonated beverages. If we increase the partial pressure of CO_2, then we increase the solubility proportionally. If we decrease the partial pressure of CO_2, then we decrease the carbonation proportionally. Table 14.2 indicates the effect of pressure on the solubility of carbon dioxide in water.

If we examine the effect of pressure on the solubility of carbon dioxide as shown in Table 14.2, we see that the solubility increases when the pressure increases. Let's apply Henry's law to calculate the solubility of carbon dioxide when the pressure is 2.00 atm. The effect of increasing the pressure is to increase the solubility, which is given in the table as 0.348 g/100 mL water at 0 °C and 1.00 atm.

$$0.348\ \text{g}/100\ \text{mL} \times \frac{2.00\ \cancel{\text{atm}}}{1.00\ \cancel{\text{atm}}} = 0.696\ \text{g}/100\ \text{mL}$$

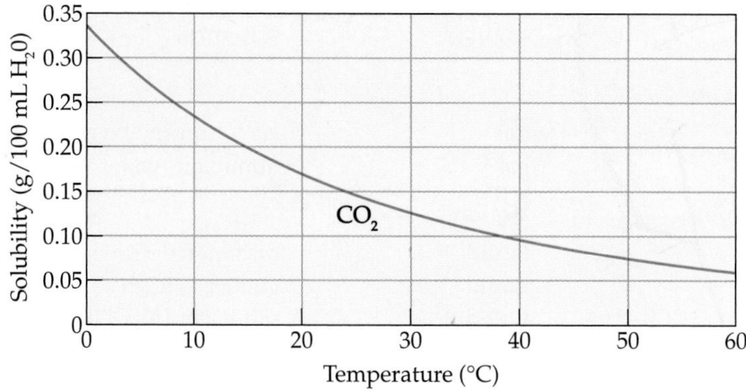

▶ **Solubility of a Gas in a Liquid** Notice that the solubility of carbon dioxide gas decreases as the temperature increases.

TABLE 14.2 SOLUBILITY OF CARBON DIOXIDE IN WATER

TEMPERATURE	PRESSURE	SOLUBILITY OF CO_2
Temperature Effect —		
0 °C	1.00 atm	0.348 g/100 mL water
20 °C	1.00 atm	0.176 g/100 mL water
40 °C	1.00 atm	0.097 g/100 mL water
60 °C	1.00 atm	0.058 g/100 mL water
Pressure Effect —		
0 °C	1.00 atm	0.348 g/100 mL water
0 °C	2.00 atm	0.696 g/100 mL water
0 °C	3.00 atm	1.044 g/100 mL water

Notice that higher temperatures decrease the solubility and that higher pressures increase the solubility.

Notice the solubility of the carbon dioxide in water given in Table 14.2 agrees with the calculated value, 0.696 g/100 mL. The following exercise further illustrates the application of Henry's law:

Example Exercise 14.1 Henry's Law

Calculate the solubility of carbon dioxide in water at 0 °C and a pressure of 3.00 atm. The solubility of carbon dioxide is 0.348 g/100 mL water at 0 °C and 1.00 atm.

Solution

The solubility of carbon dioxide gas is proportional to the pressure of the gas above the liquid.

$$\text{solubility} \times \text{pressure factor} = \text{solubility}$$

Since pressure increases, solubility increases, and the pressure factor is greater than one.

$$0.348 \text{ g/100 mL} \times \frac{3.00 \cancel{\text{atm}}}{1.00 \cancel{\text{atm}}} = 1.044 \text{ g/100 mL}$$

Practice Exercise

Oxygen is much less soluble in water than carbon dioxide, 0.00412 g/100 mL at 20 °C and 760 mm Hg. Calculate the solubility of oxygen gas in water at 20 °C and a pressure of 1150 mm Hg.

Answers: 0.00623 g/100 mL

Concept Exercise

The elevation of Lake Havasu is 400 feet, and Lake Tahoe is 6200 feet. If the water temperatures are the same, which lake has a lower concentration of oxygen gas?

Answer: See Appendix G.

14.2 LIQUIDS IN SOLUTION

OBJECTIVE

▶ To apply the *like dissolves like* rule to predict whether a liquid is soluble or insoluble in another liquid.

A **solution** is composed of a solute dissolved in a solvent. The **solute** is the lesser quantity, and the **solvent** is the greater quantity. Let's consider the factors that affect the solubility of a liquid solute in a liquid solvent. As we will see, water is the most common solvent on Earth and every living thing carries out reactions in aqueous solutions.

Dipoles

In Section 12.9, we learned that in water the more electronegative oxygen atom has a partial negative charge (δ^-). The two hydrogen atoms are less electronegative than the oxygen atom and have a partial positive charge (δ^+). A molecule having regions of a positive and a negative charge possesses a **dipole**. There are two dipoles in a water molecule. Together, the two dipoles produce a **net dipole** for the molecule. The symbol for a net dipole is an arrow pointing toward the negative end of the molecule, as shown in Figure 14.1.

Polar and Nonpolar Solvents

A liquid composed of polar molecules is called a **polar solvent**. Water molecules are polar, and water is the most common polar solvent. Although there are many exceptions, solvent molecules containing oxygen atoms, such as water, are usually polar. Methanol, CH_3OH, and ethanol, C_2H_5OH, are also examples of polar solvents.

A liquid composed of nonpolar molecules is a **nonpolar solvent**. Nonpolar solvents include hexane, C_6H_{14}; carbon tetrachloride, CCl_4; and ethyl ether, $C_4H_{10}O$. Table 14.3 lists some common polar and nonpolar solvents.

TABLE 14.3 SELECTED POLAR AND NONPOLAR SOLVENTS

POLAR SOLVENTS	NONPOLAR SOLVENTS
water, H_2O	hexane, C_6H_{14}
methanol, CH_3OH	heptane, C_7H_{16}
ethanol, C_2H_5OH	toluene, C_7H_8
acetone, C_3H_6O	carbon tetrachloride, CCl_4
methyl ethyl ketone, C_4H_8O	chloroform, $CHCl_3$
formic acid, $HCHO_2$	methylene chloride, CH_2Cl_2
acetic acid, $HC_2H_3O_2$	ethyl ether, $C_4H_{10}O$*

*The general rule that oxygen-containing solvents are polar has some exceptions. For example, ethyl ether contains oxygen and is a nonpolar solvent.

Like Dissolves Like Rule

Polar solvents such as water, H_2O, and ethanol, C_2H_5OH, dissolve in one another. Nonpolar solvents such as hexane, C_6H_{14}, and carbon tetrachloride, CCl_4, also dissolve in one another. If two solvents are both polar, or both nonpolar, they dissolve in one another. This principle is the ***like dissolves like*** **rule**.

Two solvents that dissolve completely in one another are said to be **miscible**. For example, water and ethanol are miscible. From the *like dissolves like* rule, it also follows

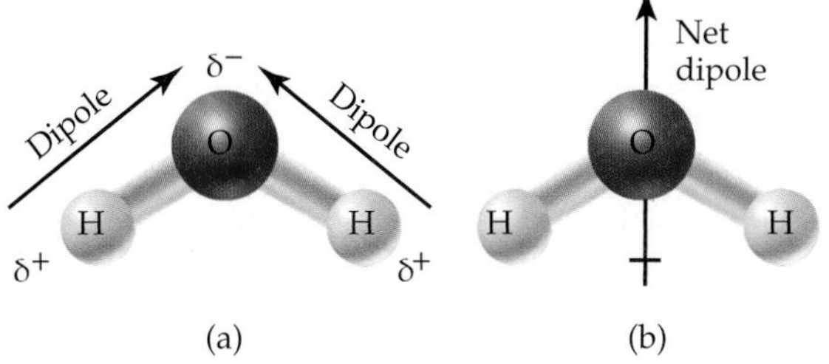

◀ **Figure 14.1 The Polar Water Molecule** The more electronegative oxygen atom polarizes the O–H bond, which in turn creates two dipoles in a water molecule. The two dipoles produce a net dipole for the entire water molecule.

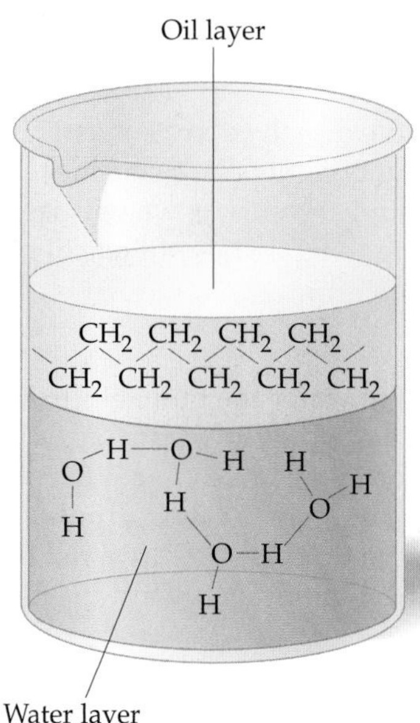

▶ **Immiscible Liquids** Oil and water are immiscible because oil molecules are nonpolar and water molecules are polar. A nonpolar solvent (oil) and a polar solvent (water) obey the *like dissolves like* rule.

▲ **Petroleum Oil Spill** Oil is less dense than water; thus, the "oil slick" floats on the ocean surface.

that a polar solvent and a nonpolar solvent are not miscible. These liquids are said to be **immiscible**. Experimentally, mixtures of immiscible liquids separate into layers. For example, oil and water are immiscible and if mixed will separate into two layers with the oil floating on the water. The *like dissolves like* rule is summarized in Table 14.4.

TABLE 14.4 SUMMARY OF THE *LIKE DISSOLVES LIKE* RULE FOR TWO LIQUIDS

LIQUID SOLUTE	POLAR SOLVENT	NONPOLAR SOLVENT
polar	miscible	immiscible
nonpolar	immiscible	miscible

The following example exercise illustrates the use of the *like dissolves like* rule to predict the miscibility of two liquids:

Example Exercise 14.2 Miscibility Predictions

Predict whether each of the following solvents is miscible or immiscible with water:

(a) methanol, CH_3OH
(b) toluene, C_7H_8

Solution

Let's use the simplifying assumption that most solvents containing oxygen are polar. Thus, methanol is polar and toluene is nonpolar. Applying the *like dissolves like* rule gives the following:

(a) CH_3OH is polar and therefore is *miscible* with H_2O.
(b) C_7H_8 is nonpolar and therefore is *immiscible* with H_2O.

Practice Exercise

Predict whether each of the following solvents is miscible or immiscible with water:

(a) methylene chloride, CH_2Cl_2 (b) glycerin, $C_3H_5(OH)_3$

Answers: (a) immiscible with H_2O; (b) miscible with H_2O

Concept Exercise

The *like dissolves like* rule states that two liquids are miscible if what property of the two liquids is alike?

Answer: See Appendix G.

Note In our discussion we have classified two solvents as either miscible or immiscible; that is, two solvents either dissolve or do not dissolve in each other. In practice, two solvents can dissolve partially in one another. For example, a nonpolar solvent can dissolve partially in a polar solvent. Although ethyl ether is generally considered a nonpolar solvent, it dissolves partially in water (~7% soluble).

14.3 SOLIDS IN SOLUTION

OBJECTIVE

▸ To apply the *like dissolves like* rule to predict whether a solid is soluble or insoluble in a liquid.

A solid substance such as sugar dissolves in a liquid, such as water, because the sugar and the water are attracted to each other. That is, the solute sugar particles are more strongly attracted to the solvent water molecules than they are to each other. The solute–solvent interaction is strongest when the polarities of the solute and the solvent are similar.

In general, a polar compound is likely to dissolve in a polar solvent such as water (Table 14.5). Sucrose, $C_{12}H_{22}O_{11}$, is ordinary table sugar and contains several oxygen atoms, and is therefore a polar compound. Applying the *like dissolves like* rule, we can correctly predict that sucrose, $C_{12}H_{22}O_{11}$, is soluble in water.

TABLE 14.5 SUMMARY OF THE *LIKE DISSOLVES LIKE* RULE FOR A SOLID IN A LIQUID

SOLID SOLUTE	POLAR SOLVENT	NONPOLAR SOLVENT
polar	soluble	insoluble
nonpolar	insoluble	soluble
ionic	soluble	insoluble

Similarly, molecules in a nonpolar compound are attracted by molecules in a nonpolar solvent. Thus, nonpolar compounds dissolve in nonpolar solvents. For example, grease dissolves in turpentine. This is because grease is a nonpolar compound and turpentine is a nonpolar solvent. Dissolving grease with turpentine is an illustration of the *like dissolves like* rule. Conversely, we cannot remove grease from our hands using water. Grease is a nonpolar compound and water is a polar solvent, so grease will not dissolve in water.

Since ionic compounds are composed of charged ions, they are similar to polar compounds in that they dissolve in polar solvents. That is, ionic compounds are more soluble in a polar solvent than in a nonpolar solvent. Consider ordinary table salt, NaCl, which dissolves readily in water. It does not dissolve in gasoline, which is a nonpolar solvent. This is because the solute (salt) is ionic and the solvent (gasoline) is nonpolar. It is important to note, however, that many ionic compounds are only slightly soluble in water (refer to the solubility rules in Section 8.9).

The following example exercise illustrates the *like dissolves like* rule for a solid compound in water:

Example Exercise 14.3 Solubility Predictions

Predict whether each of the following solid compounds is soluble or insoluble in water:

(a) fructose, $C_6H_{12}O_6$
(b) lithium carbonate, Li_2CO_3
(c) naphthalene, $C_{10}H_8$

Solution

Generally, we can apply the *like dissolves like* rule to determine if a compound is soluble. Since water is a polar solvent, we can predict that water dissolves polar compounds and many ionic compounds.

(a) Fructose has six oxygen atoms and is a polar compound. We can predict that $C_6H_{12}O_6$ is *soluble* in water.

(b) Lithium carbonate contains lithium and carbonate ions; it is therefore an ionic compound. We can predict that Li_2CO_3 is *soluble* in water.

(c) Naphthalene does not contain an oxygen atom and is a nonpolar compound. Thus, $C_{10}H_8$ is *insoluble* in water.

Practice Exercise

Predict whether each of the following solid compounds is soluble or insoluble in water:

(a) anthracene, $C_{14}H_{10}$ (b) cupric sulfate, $CuSO_4$

(c) lactic acid, $HC_3H_5O_3$

Answers: (a) insoluble; (b) soluble; (c) soluble

Concept Exercise

The *like dissolves like* rule states that a solid is soluble in a liquid if what property is alike?

Answer: See Appendix G.

Note Some compounds contain both polar and nonpolar bonding. Although cholesterol, $C_{27}H_{46}O$, contains an oxygen atom, it behaves as a nonpolar compound. It is nonpolar because it has such a large number of carbon and hydrogen atoms and only one oxygen atom. Using the *like dissolves like* rule, we can correctly predict that cholesterol is insoluble in water and soluble in a nonpolar substance. In the human body, cholesterol deposits in nonpolar fatty tissue.

CHEMISTRY CONNECTION Colloids

Q: Why is the beam of light not visible in the solution (left), but observed in the colloid (right)?

A colloid is similar to a solution in that both contain particles in a liquid. Familiar examples of colloids include milk, ink, and blood. The size of the dispersed particle distinguishes a colloid from a solution. A **colloid** contains large particles ranging in size from 1 to 100 nm, whereas a solution contains particles that are smaller than 1 nm in diameter. If we focus light on a colloid, the particles are large enough to show the beam of light. This phenomenon is called the **Tyndall effect**.

▲ **Tyndall Effect** The Tyndall effect is demonstrated as a beam of light passes through a red colloid.

Colloids and solutions are both stable, and their particles do not separate spontaneously from the liquid. However, we can separate colloid particles by centrifuging. For example, we can separate a colloid precipitate of AgCl by placing it in a test tube and spinning it in a centrifuge. Although colloid particles pass through ordinary filter paper, they are too large to pass through a natural cell membrane, or a synthetic membrane like cellophane.

Kidney dialysis is based on the separation of colloid particles using a synthetic membrane. During kidney dialysis, blood is circulated through an artificial kidney containing a synthetic membrane. Large protein molecules and red blood cells cannot pass through the membrane, but small toxin molecules do pass through the membrane and are removed. After 2–3 hours of dialysis, a patient has blood that is mostly free of small toxic molecules. We can summarize the distinctions between a solution and a colloid in the following ways:

CRITERION	SOLUTIONS	COLLOIDS
particle size	1 nm or less (small)	1–100 nm (large)
centrifuge separation	do not separate	separate
filter paper separation	do not separate	do not separate
membrane separation	do not separate	separate
Tyndall effect	light beam not visible	light beam visible

A: The beam of light is not observed as it passes through a true solution (left), but the Tyndall effect is visible as light passes through the colloid (right) with large solute particles.

14.4 THE DISSOLVING PROCESS

OBJECTIVE

- To illustrate how an ionic or a molecular compound dissolves in water.

Let's try to visualize the dissolving process. When a soluble crystal is dropped into water, the crystal begins to dissolve. This is because water molecules attack the crystal and begin pulling away part of it. Specifically, water molecules attack the edges and corners of the crystal.

As an example, suppose we drop a cube of table sugar into water. Water molecules are attracted to the polar sugar molecules, $C_{12}H_{22}O_{11}$, and pull molecules of sugar into the solution. Several water molecules surround each sugar molecule in the solution. The sugar molecules, which are held within a cluster of water molecules, are said to be in a **solvent cage**. The number of water molecules in the solvent cage varies, depending on the size and concentration of the solute (Figure 14.2).

As another example, consider what happens when crystals of table salt dissolve in water. According to the *like dissolves like* rule, polar water molecules are attracted to each ionic salt crystal. Once again, water molecules attack the edges of each crystal and begin pulling away part of it. For an ionic compound such as table salt, NaCl, the water molecules pull away positive and negative ions during the dissolving process. The negatively charged oxygen atom in a water molecule is attracted to the positively charged sodium ion, Na^+. The negatively charged chloride ion, Cl^-, is pulled into solution by the more positively charged hydrogen atoms in the water molecule. Figure 14.3 illustrates this process.

▲ **Figure 14.2 Table Sugar Dissolving in Water** Water molecules attack a sugar cube along the edges and at the corners. Each sugar molecule is pulled into the solution and surrounded by several water molecules.

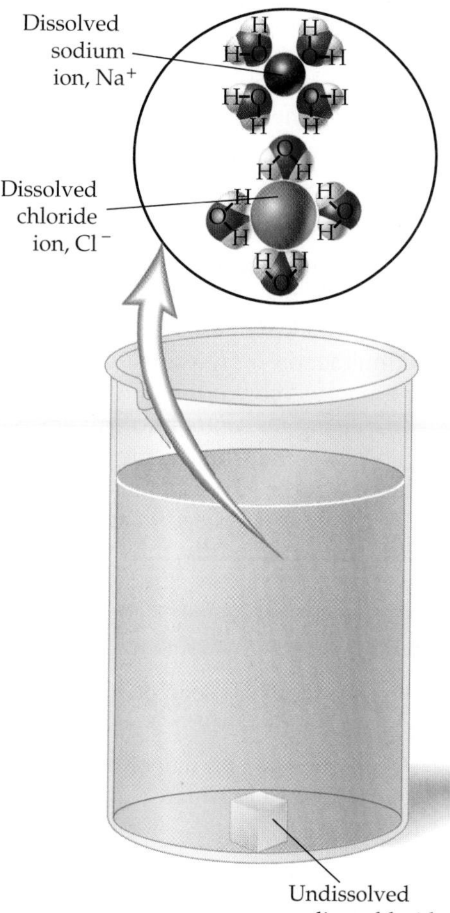

▲ **Figure 14.3 Table Salt Dissolving in Water** Note the orientation of the water molecules as they surround each ion. The Na^+ is attracted to the oxygen atom in a water molecule, whereas the Cl^- is attracted to the hydrogen atoms.

Example Exercise 14.4 The Dissolving Process

When potassium iodide, KI, dissolves in water, why does the oxygen atom of the water molecule attack the potassium ion, K^+?

Solution

Refer to Figure 14.1 and notice the oxygen atom bears a partial negative charge. Thus, the oxygen atom in a water molecule is attracted to the positive potassium ion.

Practice Exercise

When potassium iodide, KI, dissolves in water, why do the hydrogen atoms of the water molecule attack the iodide ion, I^-?

Answers: Refer to Figure 14.1 and notice the hydrogen atoms bear a partial positive charge. Thus, the hydrogen atoms in a water molecule are attracted to the negative iodide ion.

Concept Exercise

When an ionic compound dissolves in water, is the oxygen atom in a water molecule attracted to the cation or anion in the compound?

Answer: See Appendix G.

14.5 RATE OF DISSOLVING

OBJECTIVE

▸ To state the effect of temperature, stirring, and particle size on the rate at which a solid compound dissolves in water.

The rate at which a solid compound such as table sugar dissolves in a solution depends on three factors. We can increase the rate of dissolving by any one of the following:

1. heating the solution
2. stirring the solution
3. grinding the solid solute

By heating and stirring the solution or by grinding the solute, we increase the rate at which solvent molecules attack the solute. *Heating the solution* increases the kinetic energy of the solution, and the solvent molecules move faster. In an aqueous solution, water molecules attack the solute more frequently. Solute molecules are pulled into the solution faster, thus increasing the rate of dissolving. *Stirring the solution* increases the interaction between water molecules and the solute. Because the solute and solvent interact more often, the rate of dissolving is faster.

Grinding the solute into smaller particles creates more surface area. As the solute crystals become smaller, the total surface area increases. We know that solvent molecules attack the surface of crystals along the edges. When smaller crystals are created, more edges are exposed to attack by water molecules. The water molecules attack the solute more frequently, thus increasing the rate of dissolving (Figure 14.4).

Example Exercise 14.5 Rate of Dissolving

Why does heating a solution increase the rate of dissolving of a solid solute in water?

Solution

Heating increases the motion and energy of solvent molecules; thus, water molecules attack solid solute at a faster rate.

Practice Exercise

Why does stirring a solution increase the rate of dissolving of a solid solute in water?

Answers: Stirring increases the rate that solvent molecules come in contact with solute, and the rate that solvent cages are pulled from the solid solute.

Concept Exercise

Why does grinding solid crystals increase the rate of dissolving of a solid solute in water?

Answer: See Appendix G.

◀ **Figure 14.4 Rate of Dissolving Table Sugar** A cube of sugar suspended in water at 20 °C dissolves slowly. An equal amount of powdered sugar stirred in water at 100 °C dissolves immediately.

14.6 SOLUBILITY AND TEMPERATURE

OBJECTIVES

- To state the effect of temperature on the solubility of a solid compound in water.
- To interpret a graph that shows temperature versus solubility of a solid compound in water.

If we heat a solution, the interaction between the solute particles and the solvent increases. In general, the solubility of a solid compound becomes greater as the temperature increases. A few compounds are exceptions, however, and become less soluble as the temperature increases. We will define the **solubility** of a compound as the maximum amount of solute that can be dissolved in 100 g of water at a given temperature. Figure 14.5 graphs the solubility of various compounds in 100 grams of water at different temperatures.

Adding sugar to a hot beverage is an interesting illustration of the effect of temperature on solubility. Let's suppose you add sugar to a cup of hot tea and then allow

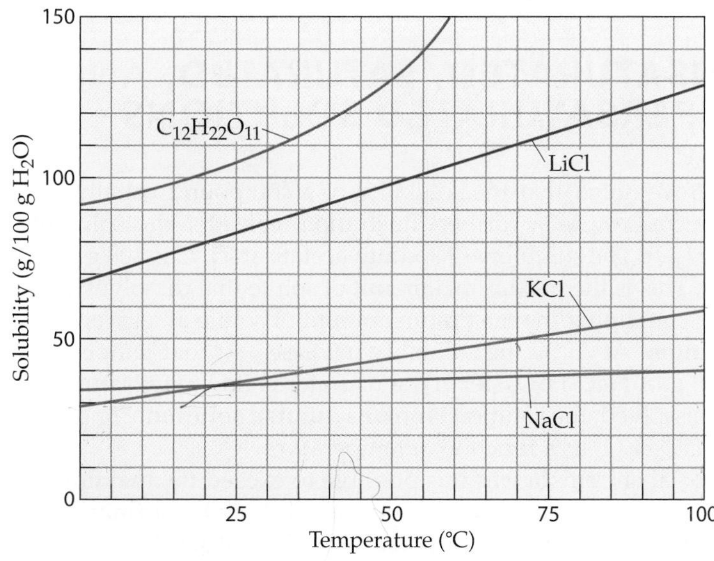

◀ **Figure 14.5 Solubility of Solid Compounds in Water** Although there are a few exceptions, solid compounds usually become more soluble as the temperature increases.

the tea to cool. The solubility of the sugar decreases as the tea cools. If you stir the cold tea, you may notice crystals of sugar at the bottom of the cup that have crystallized from solution at the cooler temperature.

Figure 14.5 illustrates that the effect of temperature on solubility varies with the compound. The solubility of table salt, NaCl, in water is only slightly affected by temperature. Its solubility is about 35 g of NaCl per 100 g of water at 20 °C and increases to 40 g/100 g water at 100 °C.

Interestingly, the solubility of table sugar, $C_{12}H_{22}O_{11}$, is greatly affected by an increase in temperature. At 20 °C the solubility of sugar is about 100 g/100 g water. At 55 °C the solubility is 140 g/100 g water. If you dissolve 140 g of table sugar in 100 g of hot water at 55 °C, and allow it to cool to 20 °C, you would see sugar crystals, just as in the previous cold tea example.

Example Exercise 14.6 Determining Solubility from a Graph

Determine the solubility of each of the following solid compounds at 50 °C as shown in Figure 14.5:

(a) NaCl
(b) KCl
(c) LiCl
(d) $C_{12}H_{22}O_{11}$

Solution

From Figure 14.5, let's find the point at which the solubility of the compound intersects 50 °C.

(a) The solubility of NaCl at 50 °C is about 38 g/100 g water.
(b) The solubility of KCl at 50 °C is about 45 g/100 g water.
(c) The solubility of LiCl at 50 °C is about 98 g/100 g water.
(d) The solubility of $C_{12}H_{22}O_{11}$ at 50 °C is about 130 g/100 g water.

Practice Exercise

Refer to the solubility behavior shown in Figure 14.5 and determine the minimum temperature required to obtain the following solutions:

(a) 35 g NaCl per 100 g of water
(b) 45 g KCl per 100 g of water
(c) 120 g LiCl per 100 g of water
(d) 109 g $C_{12}H_{22}O_{11}$ per 100 g of water

Answers: (a) 20 °C; (b) 55 °C; (c) 85 °C; (d) 30 °C

Concept Exercise

Given two cups of coffee, which can dissolve more sugar: hot coffee or coffee at room temperature?

Answer: See Appendix G.

14.7 UNSATURATED, SATURATED, AND SUPERSATURATED SOLUTIONS

OBJECTIVES

- To understand the distinction among solutions that are saturated, unsaturated, and supersaturated.
- To interpret a graph of temperature versus solubility and determine whether a solution is saturated, unsaturated, or supersaturated.

In Section 14.6 we noted that the solubility of a compound usually increases as the temperature increases. As a further illustration, consider the solubility of sodium acetate, $NaC_2H_3O_2$. The solubility of sodium acetate at 55 °C is 100 g of $NaC_2H_3O_2$ per 100 g of water. This is the maximum amount of solute that dissolves at 55 °C.

A solution containing the maximum amount of solute at a given temperature is a **saturated solution**. At 75 °C the solubility increases. At that temperature, a solution containing 100 g of $NaC_2H_3O_2$ per 100 g of water is no longer saturated. Since more solute can be dissolved, the solution is an **unsaturated solution**. Figure 14.6 graphs the solubility of $NaC_2H_3O_2$ as a function of temperature.

Under special circumstances it is possible to exceed the maximum solubility of a compound. A solution that contains more solute than ordinarily dissolves at a given temperature is said to be a **supersaturated solution**. For example, recall that

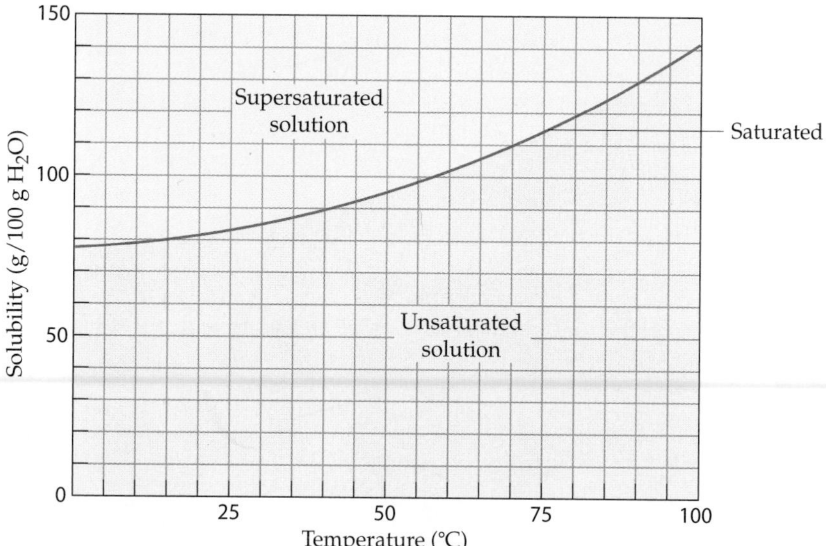

◀ **Figure 14.6 Solubility of Sodium Acetate in Water** The curve represents a saturated solution of sodium acetate, $NaC_2H_3O_2$, at various temperatures. The region below the curve represents unsaturated solutions, whereas that above represents supersaturated solutions.

100 g of $NaC_2H_3O_2$ can be dissolved in 100 g of water at 55 °C. This solution is saturated. If we allow the solution to cool, being careful not to disturb the apparatus, the excess solute will remain in solution. It is possible to cool the solution to 20 °C and still have 100 g of $NaC_2H_3O_2$ remain in the solution. At 20 °C the maximum solubility is only about 82 g/100 g water. However, since the excess solute remains in the solution, at 20 °C the solution is supersaturated.

Supersaturated solutions are unstable. In our example, one tiny crystal can cause massive crystallization of $NaC_2H_3O_2$ from solution. As another example, consider rock candy, which is made by suspending a small sugar crystal in a supersaturated sugar solution. The small crystal turns into a large rock of sugar as excess solute

◀ **Rock Candy** Rock candy is crystallized from a supersaturated sugar solution.

(a)

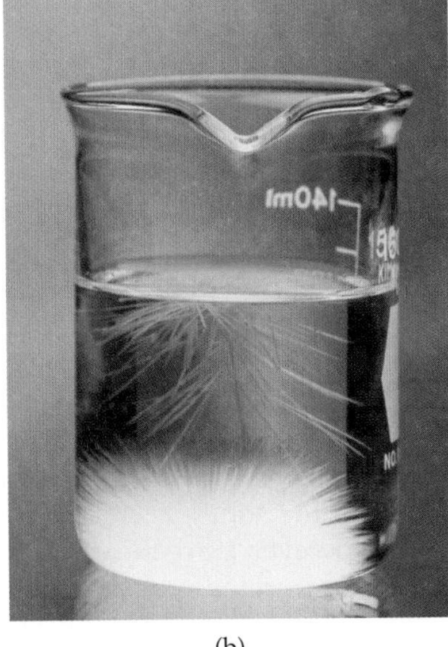
(b)

(c)

▲ **Illustration of Supersaturation** (a) A single crystal of sodium acetate, $NaC_2H_3O_2$, is dropped into a supersaturated solution. (b) The small crystal causes extensive crystallization, and eventually, (c) the solute forms a solid mass of $NaC_2H_3O_2$.

deposits from the supersaturated solution. The following example exercise provides practice in obtaining information from a solubility graph:

Example Exercise 14.7 Determining Saturation from a Graph

A sodium acetate solution contains 110 g of $NaC_2H_3O_2$ per 100 g of water. Refer to Figure 14.6 and determine whether the solution is unsaturated, saturated, or supersaturated at each of the following temperatures:

(a) 50 °C (b) 70 °C
(c) 90 °C

Solution

(a) At 50 °C the solubility of $NaC_2H_3O_2$ is about 97 g/100 g water. Since the solution contains more solute, 110 g/100 g water, the solution is *supersaturated.*
(b) At 70 °C the solubility is about 110 g/100 g water. Since the solution contains the same amount of solute, 110 g/100 g water, the solution is *saturated.*
(c) At 90 °C the solubility is about 130 g/100 g water. Since the solution has only 110 g/100 g water, the solution is *unsaturated.*

Practice Exercise

A sodium acetate solution contains 80 g of $NaC_2H_3O_2$ per 100 g water. Refer to Figure 14.6 and determine whether the solution is unsaturated, saturated, or supersaturated at each of the following temperatures:

(a) 0 °C (b) 15 °C
(c) 45 °C

Answers: (a) supersaturated; (b) saturated; (c) unsaturated

Concept Exercise

How is it possible to exceed the saturation of a solution and produce a supersaturated solution?

Answer: See Appendix G.

14.8 MASS/MASS PERCENT CONCENTRATION

OBJECTIVES

- To write three pairs of unit factors given the mass percent concentration of a solution.
- To perform calculations that involve the mass of solute, mass of solvent, and mass percent concentration of a solution.

The concentration of a solution tells us how much solute is dissolved in a given quantity of solution. We use the imprecise terms "dilute solution" and "concentrated solution" to describe the approximate concentration. A concentrated solution has a large amount of dissolved solute, whereas a dilute solution does not. Strong tea is a concentrated solution, and weak tea is a dilute solution.

We can also describe the precise concentration of a solution by comparing the mass of solute to the mass of solution. The **mass/mass percent** (symbol **m/m %**) concentration is the mass of solute dissolved in 100 g of solution. This concentration expression can be written as follows:

$$\frac{\text{mass of solute}}{\text{mass of solution}} \times 100\% = \text{m/m }\%$$

Since the total mass of the solution equals the mass of the solute plus solvent, we can also write the mass/mass percent concentration expression in the following form:

$$\frac{\text{g solute}}{\text{g solute} + \text{g solvent}} \times 100\% = \text{m/m }\%$$

If a student prepares a solution from 5.00 g of NaCl dissolved in 95.00 g of water, what is the mass/mass percent concentration? Since NaCl is the solute and water is the solvent, we can calculate the mass/mass percent concentration as follows:

$$\frac{5.00 \text{ g NaCl}}{5.00 \text{ g NaCl} + 95.00 \text{ g H}_2\text{O}} \times 100\% = \text{m/m }\%$$

$$\frac{5.00 \text{ g NaCl}}{100.00 \text{ g solution}} \times 100\% = 5.00\%$$

Writing Mass/Mass Percent Concentration Unit Factors

The unit analysis method of problem solving involves three steps: unknown units, relevant given value, and application of a unit factor. We can solve mass/mass percent calculations by using the solution concentration as a unit factor. For example, the aforementioned 5.00% solution contains 5.00 g of NaCl in each 100 g of solution. We can write this pair of unit factors as

$$\frac{5.00 \text{ g NaCl}}{100 \text{ g solution}} \quad \text{and} \quad \frac{100 \text{ g solution}}{5.00 \text{ g NaCl}}$$

Since the solution is composed of solute and solvent, we can write a second pair of unit factors. The solution contains 5.00 g of NaCl solute in 95.00 g of water solvent.

$$\frac{5.00 \text{ g NaCl}}{95.00 \text{ g water}} \quad \text{and} \quad \frac{95.00 \text{ g water}}{5.00 \text{ g NaCl}}$$

Furthermore, we can express the ratio of the mass of solvent to the mass of solution in a third pair of unit factors. The solution contains 95.00 g of solvent in 100 g of solution.

$$\frac{95.00 \text{ g water}}{100 \text{ g solution}} \quad \text{and} \quad \frac{100 \text{ g solution}}{95.00 \text{ g water}}$$

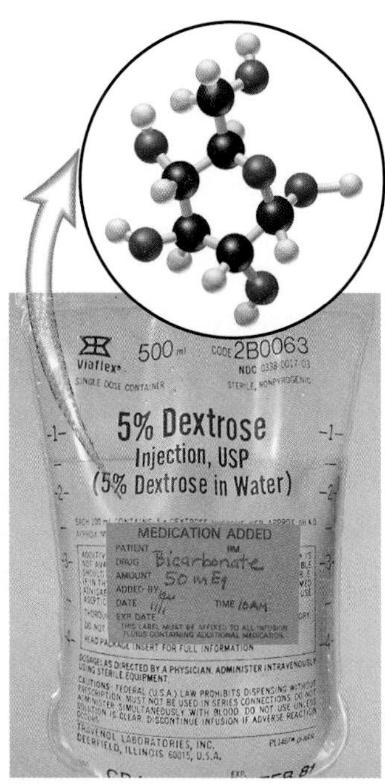

▲ **5% Dextrose** A 5% dextrose solution is normal 5.00% glucose solution. The sugar glucose, $C_6H_{12}O_6$, is also known by its common name, dextrose.

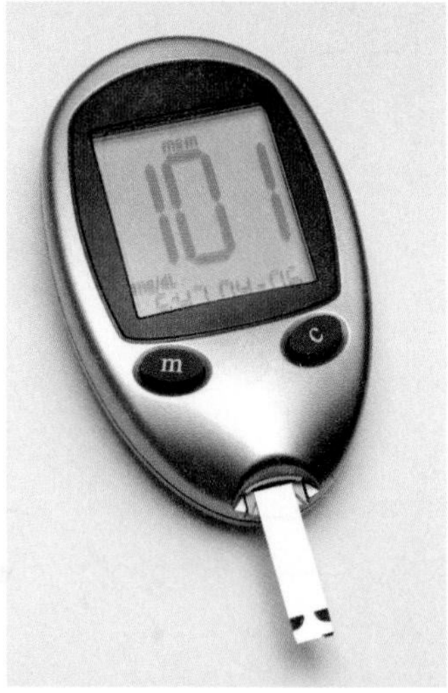

▲ **Blood Glucose Meter** Diabetes patients use a self-test meter for measuring glucose levels in blood. A normal blood glucose reading is 70–110 mg/dL.

Solving Mass/Mass Percent Concentration Problems

We can now apply unit factors derived from the mass/mass percent concentration to solution calculations. For example, intravenous (IV) injections of glucose are sometimes administered to patients with low blood sugar. If a normal IV glucose solution is 5.00%, what is the mass of solution that contains 1.25 g of glucose sugar?

Let's use the unit analysis method of problem solving. Step 1: The unknown quantity is g solution. Step 2: The relevant given value is 1.25 g glucose.

$$1.25 \text{ g glucose} \times \frac{\text{unit}}{\text{factor}} = \text{g solution}$$

The percent concentration provides the unit factor. Since the solution concentration is 5.00%, there is 5.00 g of solute in 100 g of solution. We can write the unit factors as

$$\frac{5.00 \text{ g glucose}}{100 \text{ g solution}} \quad \text{and} \quad \frac{100 \text{ g solution}}{5.00 \text{ g glucose}}$$

We will select the second unit factor so as to cancel units properly.

$$1.25 \text{ g glucose} \times \frac{100 \text{ g solution}}{5.00 \text{ g glucose}} = 25.0 \text{ g solution}$$

The following example exercises further illustrate calculations based on the mass/mass percent concentration of a solution:

Example Exercise 14.8 Mass/Mass Percent Concentration

Intravenous saline injections are given to restore the mineral balance in trauma patients. What is the mass of water required to dissolve 1.50 g of NaCl for a 0.90% normal IV saline solution?

Let's recall the unit analysis format of problem solving; that is,

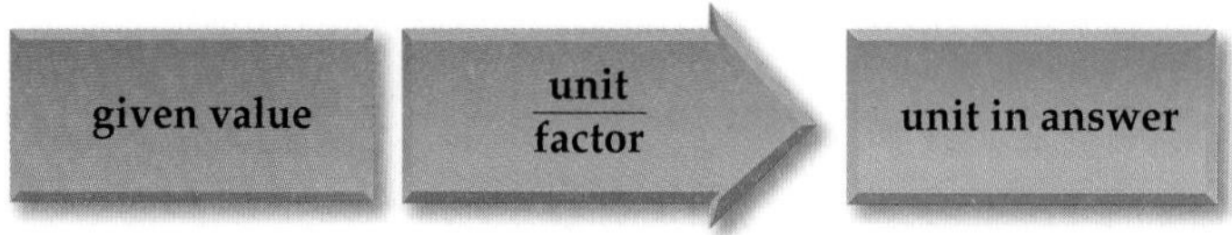

Solution

Step 1: The unit asked for in the answer is g water.

Step 2: The given value is 1.50 g NaCl.

Step 3: We apply a conversion factor to cancel units. Since the solution concentration is 0.90%, there is 0.90 g of solute in 100 g of solution. Therefore, there is 0.90 g NaCl for every 99.10 g water. We can write the unit factors

$$\frac{0.90 \text{ g NaCl}}{99.10 \text{ g water}} \quad \text{and} \quad \frac{99.10 \text{ g water}}{0.90 \text{ g NaCl}}$$

In this example, we select the second unit factor to cancel the given units.

$$1.50 \text{ g NaCl} \times \frac{99.10 \text{ g water}}{0.90 \text{ g NaCl}} = 170 \text{ g water}$$

We can summarize the conversion as follows:

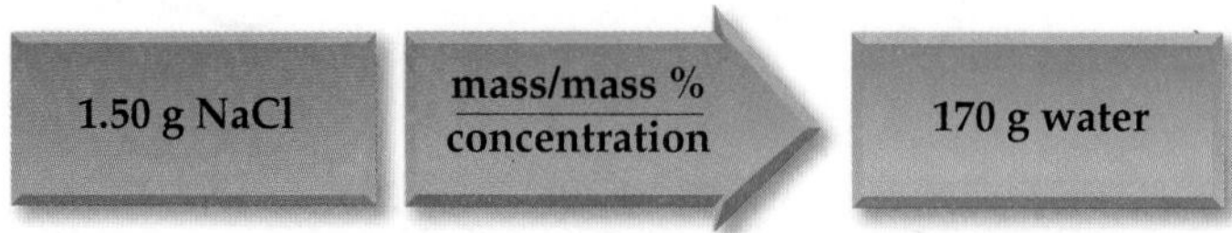

Practice Exercise

A 7.50% potassium chloride solution is prepared by dissolving enough of the salt to give 100.0 g of solution. What is the mass of water required?

Answer: 92.5 g water

Concept Exercise
Given the mass percent concentration of a solution, how many unit factors can we write?

Answer: See Appendix G.

Note The term "percent concentration" is ambiguous because it is used to express the mass of solute in 100 g of solution, the mass of solute in 100 mL of solution, and the volume of solute in 100 mL of solution. Thus, there are three ways to interpret percent concentration: mass/mass, mass/volume, and volume/volume. We usually assume that percent concentration refers to a mass/mass percent concentration.

14.9 MOLAR CONCENTRATION

OBJECTIVES

- To write a pair of unit factors given the molar concentration of a solution.
- To perform calculations that involve a mass of solute, volume of solution, and the molar concentration of a solution.

We can describe the precise concentration of medical saline and sugar solutions in terms of mass/mass percent concentration. However, most of the time we express the precise concentration of solutions in terms of molar concentration. The molar concentration, or **molarity** (symbol *M*), relates the amount of solute to a given volume of solution. Specifically, molarity is the number of moles of a solute dissolved in one liter of a solution. We can express the molarity as follows:

$$\frac{\text{moles solute}}{\text{liter solution}} = M$$

As an example, consider a household drain cleaner that is a solution of caustic sodium hydroxide, NaOH. If a manufacturer prepares a solution from 24.0 g of NaOH dissolved in 0.100 L of solution, what is the molarity? We can calculate the molarity of NaOH (40.00 g/mol) as follows:

$$\frac{24.0 \text{ g NaOH}}{0.100 \text{ L solution}} \times \frac{1 \text{ mol NaOH}}{40.00 \text{ g NaOH}} = \frac{\text{mol NaOH}}{\text{L solution}}$$
$$= 6.00\ M \text{ NaOH}$$

Notice that we started the calculation with a ratio of two units (g/L) to obtain an answer that is a ratio of two units (mol/L).

Writing Molar Concentration Unit Factors

The unit analysis method of problem solving involves the application of a unit factor. We can solve molarity calculations by using the solution concentration as a unit factor. For example, a 6.00 *M* solution of NaOH contains 6.00 mol of solute in each liter of solution. We can write this pair of unit factors as

$$\frac{6.00 \text{ mol NaOH}}{1 \text{ L solution}} \quad \text{and} \quad \frac{1 \text{ L solution}}{6.00 \text{ mol NaOH}}$$

In the laboratory, the volume of a solution is usually measured in milliliters. Since there are 1000 mL in a liter, we can substitute 1000 mL for 1 L in the unit factor.

$$\frac{6.00 \text{ mol NaOH}}{1000 \text{ mL solution}} \quad \text{and} \quad \frac{1000 \text{ mL solution}}{6.00 \text{ mol NaOH}}$$

Solving Molar Concentration Problems

We can now apply unit factors derived from the molar concentration to solution calculations. Consider a solution prepared by dissolving potassium dichromate in water. Let's calculate the mass of solute dissolved in 250.0 mL of 0.100 *M* $K_2Cr_2O_7$ solution.

(a)

(b)

(c)

▲ **Preparing 0.100 *M* Potassium Dichromate** (a) A 250 mL volumetric flask contains 7.36 g of solid orange $K_2Cr_2O_7$ in water. (b) The flask is swirled to dissolve the solid orange compound. (c) The volume of solution is adjusted to precisely 250 mL, giving a 0.100 *M* $K_2Cr_2O_7$ solution.

We begin with the unit analysis method of problem solving. Step 1: The unknown quantity is g $K_2Cr_2O_7$. Step 2: The relevant given value is 250.0 mL solution.

$$250.0 \text{ mL solution} \times \frac{\text{unit}}{\text{factor}} = \text{g } K_2Cr_2O_7$$

Since the solution concentration is 0.100 *M*, there is 0.100 mol of $K_2Cr_2O_7$ solute in each liter of solution. We can write two unit factors.

$$\frac{0.100 \text{ mol } K_2Cr_2O_7}{1000 \text{ mL solution}} \quad \text{and} \quad \frac{1000 \text{ mL solution}}{0.100 \text{ mol } K_2Cr_2O_7}$$

We choose the first unit factor so as to cancel units properly. In addition, we must use the molar mass of $K_2Cr_2O_7$ (294.20 g/mol) to convert from moles to grams.

$$250.0 \, \cancel{\text{mL solution}} \times \frac{0.100 \, \cancel{\text{mol } K_2Cr_2O_7}}{1000 \, \cancel{\text{mL solution}}} \times \frac{294.20 \text{ g } K_2Cr_2O_7}{1 \, \cancel{\text{mol } K_2Cr_2O_7}} = 7.36 \text{ g } K_2Cr_2O_7$$

The following example exercises further illustrate calculations based on the molar concentration of a solution:

Example Exercise 14.9 Molar Concentration

How many milliliters of 12.0 *M* hydrochloric acid contain 7.30 g of HCl solute (36.46 g/mol)?

In molarity calculations, we relate the given value to moles and then convert to the unit asked for in the answer.

Solution

Step 1: The unit asked for in the answer is mL acid.

Step 2: The relevant given value is 7.30 g HCl.

Step 3: The molar mass of HCl (36.46 g/mol) and the molarity of the solution (12.0 mol/1000 mL) provide the unit factors. We select unit factors so as to cancel units.

$$7.30\ \cancel{\text{g HCl}} \times \frac{1\ \cancel{\text{mol HCl}}}{36.46\ \cancel{\text{g HCl}}} \times \frac{1000\ \text{mL acid}}{12.0\ \cancel{\text{mol HCl}}} = 16.7\ \text{mL acid}$$

We can summarize the two conversions as follows:

Practice Exercise

What volume of 6.00 *M* hydrochloric acid contains 10.0 g of HCl solute (36.46 g/mol)?

Answer: 45.7 mL acid

Concept Exercise

Given the molar concentration of a solution, how many unit factors can we write?

Answer: See Appendix G.

CRITICAL THINKING Water Fluoridation

Q: In large cities, how expensive is fluoridating the public water supply to promote dental health?

Studies by the U.S. Public Health Service indicate that fluoride can help prevent dental cavities. The effect of fluoride is to make the enamel surface of teeth more resistant to decay.

Tooth enamel is mainly hydroxyapatite, $Ca_{10}(PO_4)_6(OH)_2$, the hardest substance in our bodies. However, bacteria in the mouth react with food to produce acids that dissolve tooth enamel. These acids can react with hydroxyapatite, causing pits in tooth enamel. Fluoride can prevent cavities by converting $Ca_{10}(PO_4)_6(OH)_2$ to $Ca_{10}(PO_4)_6F_2$, which is more resistant to acids in the mouth.

A U.S. Department of Health census reveals that every state has drinking water with natural or controlled fluoridation. In 2000, 42 of the 50 largest cities in the United States had fluoridated drinking water, and two-thirds of the American people drink water with fluoride. However, the percentage of people who drink fluoridated water varies widely from state to state. In Canada, a little less than half of the population drinks fluoridated water.

Fluoride has been proven to prevent tooth decay. In one study, children who grew up drinking fluoridated water had an average of three teeth with cavities. In the control group, children who grew up without fluoridation had an average of ten teeth with cavities. In other studies, it has been shown that fluoride is most effective when it is available while teeth are still developing.

Fluoridated water has a concentration of fluoride ion that is typically less than 1 ppm (1 mg/L). Despite proven benefits, not everyone agrees that fluoride should be added to the public water supply. One reason is that fluoride may have adverse effects. For example, fluoridation has been found to cause brown mottling of teeth in children. Accordingly, some cities have reduced the level of fluoride to 0.6 ppm.

The cost of fluoridating the public water supply ranges from about 30 cents per person per year in large population areas to about 2 dollars in small cities. In areas where the public water supply is not fluoridated, people can take inexpensive fluoride tablets. In addition, many toothpastes and mouthwashes are sources of fluoride. For example, sodium fluoride, NaF, and stannous fluoride, SnF_2, have been added to different toothpastes. Another alternative to fluoridated water is fluoride treatments in a dental office.

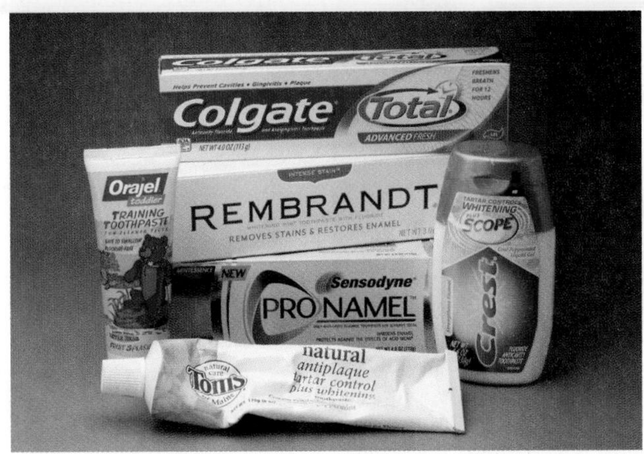

▲ **Fluoride Toothpaste** Many toothpastes contain a fluoride compound that strengthens the enamel surface of teeth and helps to prevent dental cavities.

A: In large cities, fluoridating the public water supply costs approximately 30 cents per person per year. In small cities, the expense can rise to about 2 dollars per person per year.

OBJECTIVES

- To understand the process of diluting the concentration of a solution.
- To perform calculations that involve a solution undergoing dilution.

14.10 DILUTION OF A SOLUTION

In your chemistry laboratory, you may have noticed bottles labeled 0.1 *M* HCl and 0.1 *M* NaOH. These solutions, whose concentrations are only approximate, are often prepared by diluting a more concentrated solution, called a stock solution. For example, a stock solution of 6 *M* NaOH may be available to prepare a 0.1 *M* NaOH solution for an experiment.

When calculating the concentration of a diluted solution, obey the following principle: *the amount of solute does not change during dilution.* Stated another way, the moles of solute in the concentrated solution is equal to the moles of solute in the diluted solution. Suppose we want to prepare 5.00 L of 0.1 *M* NaOH using a 6 *M* NaOH stock solution. First, we find the moles of NaOH in the diluted solution using 0.1 *M* NaOH as a unit factor.

$$5.00 \text{ L solution} \times \frac{0.1 \text{ mol NaOH}}{1 \text{ L solution}} = 0.5 \text{ mol NaOH}$$

▲ **Sodium Hydroxide, NaOH** Ordinary drain cleaner found in the home contains the same chemical as bottles of sodium hydroxide found in a chemistry laboratory.

Since we know the moles of solute remain constant during dilution, the amount of NaOH does not change. That is, we start with—and end up with—the same amount of NaOH. The moles of solute we start with in the concentrated solution is 0.5 mol NaOH. Using 6 *M* NaOH as a unit factor, we can calculate the volume of 6 *M* stock solution as follows:

$$0.5 \text{ mol NaOH} \times \frac{1 \text{ L solution}}{6 \text{ mol NaOH}} = 0.08 \text{ L NaOH} \qquad \textit{(stock solution)}$$

In this example, we demonstrated how to prepare 5.00 L of 0.1 *M* NaOH from a more concentrated stock solution. That is, we simply dilute 0.08 L (~80 mL) of 6 *M* NaOH stock solution to a total volume of 5.00 L with distilled water.

Alternatively, we can solve this problem using a simple algebraic equation. The product of molarity (mol/L) and volume (L) gives units of mole. Since *the moles of NaOH in the concentrated and the dilute solutions are equal,* their products are also equal.

$$M_1 \times V_1 = M_2 \times V_2$$

where M_1 and V_1 represent the molarity and the volume of the concentrated solution before dilution, and M_2 and V_2 are the molarity and volume after dilution. In the preceding problem, we found the volume of a 6 *M* NaOH stock solution that must be diluted to give 5.00 L of 0.1 *M* NaOH. We can also find the volume using the previous equation. After substituting the given values into the equation, we have

$$6\,M \times V_1 = 0.1\,M \times 5.00 \text{ L}$$

Solving for V_1,

$$\frac{0.1\,M \times 5.00\,L}{6\,M} = V_1$$

$$= 0.08 \text{ L NaOH}$$

▲ **Sodium Hydroxide, NaOH** Chemistry stockrooms prepare laboratory solutions of 6 *M* NaOH by weighing out sodium hydroxide and dissolving the pellets in a given volume of distilled water.

The following example exercise provides additional practice in solving dilution problems involving the concentration of a solution:

Example Exercise 14.10 Dilution of a Solution

Concentrated hydrochloric acid is available commercially as a 12 *M* solution. What is the molarity of an HCl solution prepared by diluting 50.0 mL of concentrated acid with distilled water to give a total volume of 2.50 L?

Conceptual Solution

Let's find the moles of concentrated HCl before dilution.

$$50.0 \text{ mL solution} \times \frac{12 \text{ mol HCl}}{1000 \text{ mL solution}} = 0.60 \text{ mol HCl}$$

We know that the moles of HCl do not change during dilution; the moles of diluted HCl solution must be the same as the moles of concentrated HCl. Since the volume of diluted HCl is 2.50 L, we can calculate the diluted concentration.

$$\frac{0.60 \text{ mol HCl}}{2.50 \text{ L solution}} = \frac{0.24 \text{ mol HCl}}{1 \text{ L solution}} = 0.24\ M \text{ HCl}$$

Algebraic Solution

Alternatively, we can solve this problem using the equation

$$M_1 \times V_1 = M_2 \times V_2$$

Substituting, we have 12 $M \times 50.0$ mL $= M_2 \times 2.50$ L

We must use the same units for volume; for example, we can convert 2.50 L to mL, which equals 2500 mL. Solving for M_2, we have

$$\frac{12\ M \times 50.0 \cancel{\text{mL}}}{2500 \cancel{\text{mL}}} = 0.24\ M \text{ HCl}$$

Practice Exercise

Battery acid is 18 M sulfuric acid. What volume of battery acid must be diluted with distilled water to prepare 1.00 L of 0.50 M H_2SO_4?

Answer: 28 mL H_2SO_4 (diluted to 1.00 L with distilled H_2O)

Concept Exercise

If equal volumes of 6 M sulfuric acid and distilled water are added together, what is the concentration of the diluted acid?

Answer: See Appendix G.

14.11 SOLUTION STOICHIOMETRY

OBJECTIVE

- To perform calculations that involve a balanced chemical equation and the molar concentration of a solution.

In Chapter 10, we performed mole calculations involving chemical equations. The relationship between quantities in a balanced equation is called stoichiometry. Now, let's apply stoichiometry to chemical reactions involving the molar concentration of a solution. Suppose we are given 4.24 g of sodium carbonate, Na_2CO_3, which we react with 0.150 M hydrochloric acid, HCl. The balanced equation for the reaction is

$$Na_2CO_3(s) + 2\ HCl(aq) \longrightarrow 2\ NaCl(aq) + H_2O(l) + CO_2(g)$$

Let's calculate the volume of 0.150 M HCl that reacts with 4.24 g of Na_2CO_3. We can solve this problem in three steps. In the first step, we calculate the moles of Na_2CO_3. The molar mass of Na_2CO_3 is 105.99 g/mol. Therefore,

$$4.24 \cancel{\text{g } Na_2CO_3} \times \frac{1 \text{ mol } Na_2CO_3}{105.99 \cancel{\text{g } Na_2CO_3}} = 0.0400 \text{ mol } Na_2CO_3$$

In the second step, we convert moles of Na_2CO_3 to moles of HCl by applying the coefficients from the balanced equation. The relationship between the two substances is 1 mol Na_2CO_3 = 2 mol HCl. The unit factor conversion is

$$0.0400 \cancel{\text{mol } Na_2CO_3} \times \frac{2 \text{ mol HCl}}{1 \cancel{\text{mol } Na_2CO_3}} = 0.0800 \text{ mol HCl}$$

In the third step, we use the molar concentration of the hydrochloric acid solution as a unit factor to obtain the volume of HCl that reacts with Na_2CO_3. In this example, 0.150 M HCl gives two unit factors: 0.150 mol HCl/1 L solution, and the reciprocal, 1 L solution/0.150 mol HCl.

$$0.0800 \cancel{\text{mol HCl}} \times \frac{1 \text{ L solution}}{0.150 \cancel{\text{mol HCl}}} = 0.533 \text{ L solution}$$

It is also possible to solve this stoichiometry problem in one continuous calculation. We can show the complete solution to the problem as

$$4.24 \text{ g } Na_2CO_3 \times \frac{1 \text{ mol } Na_2CO_3}{105.99 \text{ g } Na_2CO_3} \times \frac{2 \text{ mol HCl}}{1 \text{ mol } Na_2CO_3} \times \frac{1 \text{ L solution}}{0.150 \text{ mol HCl}}$$

$$= 0.533 \text{ L solution}$$

The following example exercise provides additional practice in solving calculations involving the concentration of a solution:

Example Exercise 14.11 Solution Stoichiometry Problem

Given that 37.5 mL of 0.100 *M* aluminum bromide solution reacts with a silver nitrate solution, what is the mass of AgBr (187.77 g/mol) produced? The balanced equation is

$$AlBr_3(aq) \quad + \quad 3\ AgNO_3(aq) \quad \rightarrow \quad 3\ AgBr(s) \quad + \quad Al(NO_3)_3(aq)$$

Solution

Step 1: The unit asked for in the answer is g AgBr.

Step 2: The given value is 37.5 mL solution.

Step 3: We apply unit conversion factors to cancel units. In this example, we can use molar concentration (0.100 mol $AlBr_3$/1000 mL solution) and molar mass (187.77 g/mol) as unit factors. From the balanced equation, we see that 1 mol $AlBr_3$ = 3 mol AgBr; therefore,

$$37.5 \text{ mL solution} \times \frac{0.100 \text{ mol } AlBr_3}{1000 \text{ mL solution}} \times \frac{3 \text{ mol AgBr}}{1 \text{ mol } AlBr_3} \times \frac{187.77 \text{ g AgBr}}{1 \text{ mol AgBr}} = 2.11 \text{ g AgBr}$$

We can summarize the three conversions as follows:

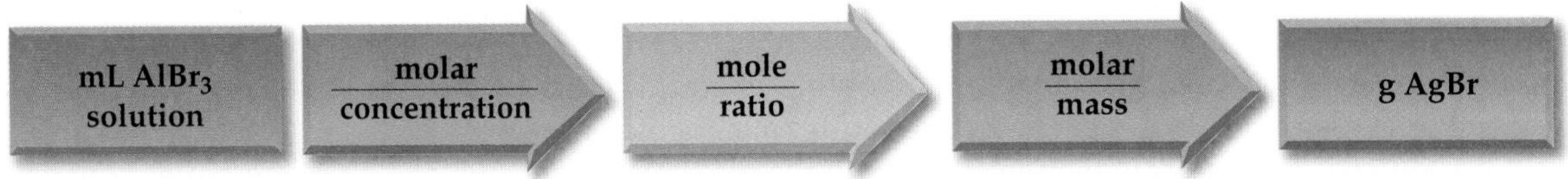

Practice Exercise

Given that 27.5 mL of 0.210 *M* lithium iodide solution reacts with 0.133 *M* lead(II) nitrate solution, what volume of $Pb(NO_3)_2$ solution is required for complete precipitation?

$$Pb(NO_3)_2(aq) \quad + \quad 2\ LiI(aq) \quad \rightarrow \quad PbI_2(s) \quad + \quad 2\ LiNO_3(aq)$$

Answers: 21.7 mL $Pb(NO_3)_2$

Concept Exercise

Before applying unit analysis and solving a solution stoichiometry problem, what must always be done first?

Answer: See Appendix G.

Summary

Section 14.1 The most common solutions are formed from a solid, a liquid, or a gas dissolved in water. If the solute is a gas, its solubility is affected by temperature and pressure. Raising the temperature of the solution decreases the solubility of the dissolved gas. On the other hand, raising the partial pressure of the gas above the solution increases the amount of dissolved gas in the solution. In fact, **Henry's law** states that the solubility of a gas in a liquid is proportional to the partial pressure of the gas above the liquid.

Section 14.2 A **solution** consists of a **solute** dissolved in a **solvent**. A solution results from the interaction of solute and solvent molecules. A polar molecule has a region of both partial positive and partial negative charge and is called a **dipole**. In water, there are two dipoles that combine to produce a single **net dipole** for the molecule. Water is composed of polar H_2O molecules, hence it is considered a **polar solvent**. Hexane is composed of nonpolar C_6H_{14} molecules, and so it is considered a **nonpolar solvent**. Although it is a simplification, polar solvents are often small molecules that contain one or more oxygen atoms, for example, H_2O, CH_3OH, and C_2H_5OH.

We can predict whether two liquids dissolve by applying the ***like dissolves like* rule**. This rule states that two liquids are **miscible** if they are both polar solvents or both nonpolar solvents. Furthermore, two liquids are **immiscible** if one is a polar solvent and the other a nonpolar solvent. Since CH_3OH is polar, it is miscible with water. Since C_6H_{14} is nonpolar, it is immiscible with water.

Section 14.3 We can also usually predict whether or not a solid compound is soluble in a given solvent. According to the *like dissolves like* rule, a polar solvent dissolves a polar compound, and a nonpolar solvent dissolves a nonpolar compound. Ionic compounds are not soluble in nonpolar solvents, and some ionic compounds are only slightly soluble in polar solvents.

Section 14.4 In dissolving a solid compound in water, solvent molecules attack solute particles and pull them into the solution. The solute particles are then surrounded by water molecules that form a **solvent cage**. Molecular compounds release molecules in the solution, whereas ionic compounds release ions in the solution. A **colloid** is similar to a solution except that the dissolved particles are larger. These larger colloid particles demonstrate an ability to scatter a beam of light; this phenomenon is called the **Tyndall effect**.

Section 14.5 The rate of dissolving a solid solute compound in a solution is increased by three factors: heating the solution, stirring the solution, and grinding the solute.

Section 14.6 Generally, raising the temperature of a solution increases the **solubility** of a solid compound. By referring to a graph of solubility, we can determine the amount of solute that can dissolve in 100 g of water at a given temperature.

Section 14.7 A solution containing the maximum amount of dissolved solute possible at a given temperature is a **saturated solution**. If the concentration of the solution is less than its maximum solubility, it is an **unsaturated solution**. Under special circumstances, it is possible to exceed the maximum solubility of a solution. Such a solution is unstable and it is said to be a **supersaturated solution**.

Section 14.8 Often, a solution concentration is expressed in terms of **mass/mass percent (m/m %)**. The mass/mass percent concentration is the mass in grams of dissolved solute in 100 g of solution.

Section 14.9 A solution concentration can be expressed as **molarity (*M*)**. The molar concentration, or molarity, is the number of moles of solute dissolved in one liter of solution (mol solute/L solution).

Section 14.10 Many dilute solutions in the chemistry laboratory are prepared from concentrated solutions. For example, a 6 *M* HCl solution can be diluted to 0.1 *M* HCl. When performing dilution calculations, it is important to realize that the amount of solute does not change when a solution is diluted.

Section 14.11 Solution stoichiometry calculations are similar to other stoichiometry calculations and require a balanced chemical equation. However, solution stoichiometry problems are different in that they involve the molar concentration of a solution.

PROBLEM–SOLVING ORGANIZER

TOPIC	PROCEDURE	EXAMPLE
Gases in Solution Sec. 14.1	Henry's law states that the solubility of a gas is directly proportional to the partial pressure of the gas above the liquid.	If the solubility of a gas is 0.100 g/100 mL at 1.00 atm, what is it at 3.50 atm? $0.100\text{ g}/100\text{ mL} \times \frac{3.50\ \cancel{\text{atm}}}{1.00\ \cancel{\text{atm}}} = 0.350\text{ g}/100\text{ mL}$
Mass/Mass Percent Concentration Sec. 14.8	1. Write down the unit asked for in the answer. 2. Write down the related given value. 3. Apply a unit factor to convert the given unit to the unit in the answer.	What mass of 10.0% table sugar solution contains 15.5 g of sugar? $15.5\ \cancel{\text{g sugar}} \times \frac{100\text{ g solution}}{10.0\ \cancel{\text{g sugar}}} = 155\text{ g solution}$
Molar Concentration Sec. 14.9	1. Write down the unit asked for in the answer. 2. Write down the related given value. 3. Apply unit factors to convert the given unit to the unit in the answer.	What is the mass of sucrose, $C_{12}H_{22}O_{11}$ (342.34 g/mol), in 0.500 L of a 0.100 *M* solution? $0.500\ \cancel{\text{L solution}} \times \frac{0.100\ \cancel{\text{mol sucrose}}}{1\ \cancel{\text{L solution}}} \times \frac{342.34\text{ g sucrose}}{1\ \cancel{\text{mol sucrose}}}$ $= 17.1\text{ g sucrose}$
Dilution of a Solution Sec. 14.10	Solve dilution problems by unit analysis or by the algebraic method: $M_1 \times V_1 = M_2 \times V_2$	If 50.0 mL of 6.0 *M* NaOH is diluted to a total volume of 250 mL, what is the molarity? $6.0\ M \times 50.0\text{ mL} = M_2 \times 250\text{ mL}$ Solving for M_2, $\frac{6.0\ M \times 50.0\ \cancel{\text{mL}}}{250\ \cancel{\text{mL}}} = 1.2\ M$

PROBLEM–SOLVING ORGANIZER *(Continued)*

TOPIC	PROCEDURE	EXAMPLE
Solution Stoichiometry Sec. 14.11	1. Write down the unit asked for in the answer. 2. Write down the related given value. 3. Apply unit factors to convert the given unit to the unit in the answer.	What is the mass of AgCl (143.32 g/mol) produced from 50.0 mL of 0.100 *M* KCl and excess $AgNO_3$? Given: $KCl(aq) + AgNO_3(aq) \longrightarrow AgCl(s) + KNO_3(aq)$ $50.0\ \cancel{mL\ KCl} \times \frac{0.100\ \cancel{mol\ KCl}}{1000\ \cancel{mL\ KCl}} \times \frac{1\ \cancel{mol\ AgCl}}{1\ \cancel{mol\ KCl}} \times \frac{143.32\ g\ AgCl}{1\ \cancel{mol\ AgCl}} = 0.717\ g\ AgCl$

Key Terms *Answers to Key Terms are in Appendix H.*

Select the key term that corresponds to each of the following definitions.

_____ **1.** the solubility of a gas in a liquid is proportional to the partial pressure of the gas
_____ **2.** the component of a solution that is present in the lesser quantity
_____ **3.** the component of a solution that is present in the greater quantity
_____ **4.** the general term for a solute dissolved in a solvent
_____ **5.** a region in a polar molecule with a partial negative and a partial positive charge
_____ **6.** the overall direction of partial negative charge in a molecule
_____ **7.** a dissolving liquid composed of polar molecules
_____ **8.** a dissolving liquid composed of nonpolar molecules
_____ **9.** the principle that solubility is greatest when the polarity of the solute is similar to that of the solvent
_____ **10.** refers to liquids that dissolve completely in one another
_____ **11.** refers to liquids that do not dissolve in one another and separate into two layers
_____ **12.** a cluster of solvent molecules surrounding a solute particle in solution
_____ **13.** a mixture in which the dispersed particles range from 1 to 100 nm
_____ **14.** the scattering of a beam of light by colloid particles
_____ **15.** the maximum amount of solute that can dissolve in a solvent at a given temperature
_____ **16.** a solution containing the maximum solute that can dissolve at a given temperature
_____ **17.** a solution containing less than the maximum solute that can dissolve at a given temperature
_____ **18.** a solution containing more than the maximum solute that can dissolve at a given temperature
_____ **19.** a solution concentration expression that relates the mass of solute in grams dissolved in each 100 g of solution
_____ **20.** a solution concentration expression that relates the moles of solute dissolved in each liter of solution

(a) colloid *(Sec. 14.3)*
(b) dipole *(Sec. 14.2)*
(c) Henry's law *(Sec. 14.1)*
(d) immiscible *(Sec. 14.2)*
(e) *like dissolves like* rule *(Sec. 14.2)*
(f) mass/mass percent (m/m %) *(Sec. 14.8)*
(g) miscible *(Sec. 14.2)*
(h) molarity (*M*) *(Sec. 14.9)*
(i) net dipole *(Sec. 14.2)*
(j) nonpolar solvent *(Sec. 14.2)*
(k) polar solvent *(Sec. 14.2)*
(l) saturated solution *(Sec. 14.7)*
(m) solubility *(Sec. 14.6)*
(n) solute *(Sec. 14.2)*
(o) solution *(Sec. 14.2)*
(p) solvent *(Sec. 14.2)*
(q) solvent cage *(Sec. 14.4)*
(r) supersaturated solution *(Sec. 14.7)*
(s) Tyndall effect *(Sec. 14.3)*
(t) unsaturated solution *(Sec. 14.7)*

Exercises *Answers to odd-numbered Exercises are in Appendix I.*

Gases in Solution (Sec. 14.1)

1. Indicate whether the solubility of oxygen gas in water increases or decreases for each of the following:
   (a) The temperature of the solution decreases from 20 °C to 10 °C.
   (b) The partial pressure of O_2 decreases from 50 mm Hg to 25 mm Hg.

2. Indicate whether the solubility of carbon dioxide gas in water increases or decreases for each of the following:
   (a) The temperature of the solution increases from 20 °C to 30 °C.
   (b) The partial pressure of CO_2 increases from 50 mm Hg to 75 mm Hg.

3. If the solubility of nitrogen gas is 1.90 mL/1 dL of blood at 1.00 atm, what is the solubility of nitrogen gas in a scuba diver's blood at a depth of 125 feet where the pressure is 4.50 atm?

◀ **Illustration of Henry's Law** After a bottle of soda is opened and poured into a glass, the soda fizzes as the carbonation escapes. The carbon dioxide gas is less soluble in the glass than when it was under pressure in the bottle.

4. If the solubility of nitrogen gas is 1.90 mL/1 dL of blood at 1.00 atm, what is the solubility of nitrogen gas in a scuba diver's blood at a depth of 225 feet where the pressure is 8.00 atm?
5. The solubility of nitrous oxide is 0.12 g/100 g water at 20 °C and 1.00 atm. What is the partial pressure required to dissolve 0.25 g of the gas in 100 g of water at 20 °C?
6. The solubility of nitrous oxide is 0.12 g/100 g water at 20 °C and 1.00 atm. What is the partial pressure required to dissolve 0.55 g of the gas in 100 g of water at 20 °C?

Liquids in Solution (Sec. 14.2)

7. State whether the solutes and solvents in each of the following combinations are miscible or immiscible:
(a) polar solute + polar solvent
(b) polar solute + nonpolar solvent
8. State whether the solutes and solvents in each of the following combinations are miscible or immiscible:
(a) nonpolar solute + polar solvent
(b) nonpolar solute + nonpolar solvent
9. Predict whether each of the following solvents is polar or nonpolar:
(a) water, H_2O (b) hexane, C_6H_{14}
(c) acetone, C_3H_6O (d) chloroform, $CHCl_3$
10. Predict whether each of the following solvents is polar or nonpolar:
(a) isopropyl alcohol, C_3H_7OH
(b) pentane, C_5H_{12}
(c) xylene, $C_6H_4(CH_3)_2$
(d) trichloroethane, $C_2H_3Cl_3$
11. Predict whether each of the following solvents is miscible or immiscible with water:
(a) ethanol, C_2H_5OH
(b) toluene, C_7H_8
(c) trichloroethylene, C_2HCl_3
(d) acetic acid, $HC_2H_3O_2$
12. Predict whether each of the following solvents is miscible or immiscible with hexane, C_6H_{14}:
(a) heptane, C_7H_{16}
(b) methanol, CH_3OH
(c) methyl ethyl ketone, C_4H_8O
(d) chloroform, $CHCl_3$
13. In the laboratory how could you quickly determine whether an unknown liquid is polar or nonpolar?
14. An oil slick forms when crude oil and water separate into two layers. Explain why the two liquids are immiscible using the *like dissolves like* rule.

▲ **Oil and Water** Salad oil is a less dense liquid than water; thus, salad oil floats on water.

Solids in Solution (Sec. 14.3)

15. State whether the following combinations of solute and solvent are generally soluble or insoluble:
(a) polar solute + polar solvent
(b) nonpolar solute + polar solvent
(c) ionic solute + polar solvent
16. State whether the following combinations of solute and solvent are generally soluble or insoluble:
(a) polar solute + nonpolar solvent
(b) nonpolar solute + nonpolar solvent
(c) ionic solute + nonpolar solvent
17. Predict whether each of the following compounds is soluble or insoluble in water:
(a) naphthalene, $C_{10}H_8$
(b) potassium hydroxide, KOH
(c) calcium acetate, $Ca(C_2H_3O_2)_2$
(d) trichlorotoluene, $C_7H_5Cl_3$
(e) glycine, $C_2H_5NO_2$
(f) lactic acid, $HC_3H_5O_3$

18. Predict whether each of the following compounds is soluble or insoluble in hexane, C_6H_{14}:
(a) trichloroethylene, C_2HCl_3
(b) iron(III) nitrate, $Fe(NO_3)_3$
(c) sulfuric acid, H_2SO_4
(d) dodecane, $C_{12}H_{26}$
(e) mothballs, $C_6H_4Cl_2$
(f) tartaric acid, $H_2C_4H_4O_6$

19. Predict whether each of the following compounds is water soluble or fat soluble:
(a) cholesterol, $C_{27}H_{46}O$
(b) citric acid, $C_6H_8O_7$
(c) fructose, $C_6H_{12}O_6$
(d) stearic acid, $CH_3(CH_2)_{16}COOH$
(e) lactic acid, $CH_3CH(OH)COOH$
(f) alanine, $CH_3CH(NH_2)COOH$

20. Predict whether each of the following vitamins is water soluble or fat soluble:
(a) vitamin B_1, $C_{12}H_{18}Cl_2N_4OS$
(b) vitamin B_3, $C_6H_6N_2O$
(c) vitamin B_6, $C_8H_{11}NO_3$
(d) vitamin C, $C_6H_8O_6$
(e) vitamin D, $C_{27}H_{44}O$
(f) vitamin K, $C_{31}H_{46}O_2$

The Dissolving Process (Sec. 14.4)

21. Diagram a molecule of fructose, $C_6H_{12}O_6$, dissolved in water.

22. Diagram a crystal of sucrose, $C_{12}H_{22}O_{11}$, dissolving in water.

23. Diagram a formula unit of each of the following substances dissolved in water:
(a) lithium bromide, LiBr (b) calcium chloride, $CaCl_2$

24. Diagram a formula unit of each of the following substances dissolved in water:
(a) cobalt(II) sulfate, $CoSO_4$
(b) nickel(II) nitrate, $Ni(NO_3)_2$

Rate of Dissolving (Sec. 14.5)

25. What three factors increase the rate of dissolving of a solid substance in solution?

26. Indicate whether each of the following increases, decreases, or has no effect on the rate at which 10.0 g of sugar dissolves in a liter of water:
(a) using water from the refrigerator
(b) shaking the sugar and water
(c) using powdered sugar rather than crystals
(d) using tap water rather than distilled water

Solubility and Temperature (Sec. 14.6)

27. How many grams of each of the following solutes can dissolve in 100 g of water at 20 °C? (Refer to Figure 14.5.)
(a) NaCl (b) KCl

28. How many grams of each of the following solutes can dissolve in 100 g of water at 30 °C? (Refer to Figure 14.5.)
(a) LiCl (b) $C_{12}H_{22}O_{11}$

29. Determine the maximum solubility of each of the following solid compounds at 40 °C. (Refer to Figure 14.5.)
(a) NaCl (b) KCl

30. Determine the maximum solubility of each of the following solid compounds at 50 °C. (Refer to Figure 14.5.)
(a) LiCl (b) $C_{12}H_{22}O_{11}$

31. What is the minimum temperature required to dissolve each of the following? (Refer to Figure 14.5.)
(a) 34 g NaCl in 100 g of water
(b) 50 g KCl in 100 g of water

32. What is the minimum temperature required to dissolve each of the following? (Refer to Figure 14.5.)
(a) 90 g LiCl in 100 g of water
(b) 120 g $C_{12}H_{22}O_{11}$ in 100 g of water

33. At what temperature is each of the following solutions saturated? (Refer to Figure 14.5.)
(a) 40 g NaCl/100 g water (b) 40 g KCl/100 g water

34. At what temperature is each of the following solutions saturated? (Refer to Figure 14.5.)
(a) 105 g LiCl/100 g water
(b) 140 g $C_{12}H_{22}O_{11}$/100 g water

Unsaturated, Saturated, and Supersaturated Solutions (Sec. 14.7)

35. State whether each of the following solutions is saturated, unsaturated, or supersaturated. (Refer to Figure 14.6.)
(a) 100 g $NaC_2H_3O_2$ in 100 g of water at 25 °C
(b) 100 g $NaC_2H_3O_2$ in 100 g of water at 55 °C
(c) 100 g $NaC_2H_3O_2$ in 100 g of water at 75 °C

36. State whether each of the following solutions is saturated, unsaturated, or supersaturated. (Refer to Figure 14.6.)
(a) 120 g $NaC_2H_3O_2$ in 100 g of water at 70 °C
(b) 120 g $NaC_2H_3O_2$ in 100 g of water at 80 °C
(c) 120 g $NaC_2H_3O_2$ in 100 g of water at 90 °C

37. State whether each of the following solutions is saturated, unsaturated, or supersaturated. (Refer to Figure 14.5.)
(a) 45 g KCl/100 g H_2O at 20 °C
(b) 45 g KCl/100 g H_2O at 50 °C
(c) 45 g KCl/100 g H_2O at 70 °C

38. State whether each of the following solutions is saturated, unsaturated, or supersaturated. (Refer to Figure 14.5.)
(a) 35 g NaCl/100 g H_2O at 0 °C
(b) 35 g NaCl/100 g H_2O at 25 °C
(c) 35 g NaCl/100 g H_2O at 100 °C

39. The solubility of rock salt at 30 °C is 40.0 g/100 g water. If a solution contains 10.0 g of rock salt in 25.0 g of water at 30 °C, is the solution saturated, unsaturated, or supersaturated?

40. The solubility of sugar at 50 °C is 130.0 g/100 g water. If a solution contains 95.0 g of sugar in 250 g of water at 50 °C, is the solution saturated, unsaturated, or supersaturated?

41. Assume that 100 g of LiCl is dissolved in 100 g of water at 100 °C and the solution is allowed to cool to 20 °C. (Refer to Figure 14.5.)
(a) How much solute remains in solution?
(b) How much solute crystallizes from the solution?

42. Assume that 120 g of $C_{12}H_{22}O_{11}$ is dissolved in 100 g of water at 100 °C and the solution is allowed to cool to 20 °C. (Refer to Figure 14.5.)
(a) How much solute remains in solution?
(b) How much solute crystallizes from the solution?

Mass/Mass Percent Concentration (Sec. 14.8)

43. Calculate the mass/mass percent concentration for each of the following solutions:
(a) 20.0 g KI in 100.0 g of water
(b) 2.50 g $AgC_2H_3O_2$ in 95.0 g of water
(c) 5.57 g $SrCl_2$ in 225.0 g of water
(d) 50.0 g $C_{12}H_{22}O_{11}$ in 200.0 g of water

44. Calculate the mass/mass percent concentration for each of the following solutions:
(a) 1.25 g NaCl in 100.0 g of solution
(b) 2.50 g $K_2Cr_2O_7$ in 95.0 g of solution
(c) 10.0 g $CaCl_2$ in 250.0 g of solution
(d) 65.0 g $C_6H_{12}O_6$ in 125.0 g of solution

45. Write three pairs of unit factors for each of the following aqueous solutions given the mass/mass percent concentration:
(a) 1.50% KBr (b) 2.50% $AlCl_3$
(c) 3.75% $AgNO_3$ (d) 4.25% Li_2SO_4

46. Write three pairs of unit factors for each of the following aqueous solutions given the mass/mass percent concentration:
(a) 3.35% $MgCl_2$ (b) 5.25% $Cd(NO_3)_2$
(c) 6.50% Na_2CrO_4 (d) 7.25% $ZnSO_4$

47. What mass of solution contains each of the following amounts of dissolved solute?
(a) 5.36 g of glucose in a 10.0% solution
(b) 25.0 g of sucrose in a 12.5% solution

48. What mass of solution contains each of the following amounts of dissolved solute?
(a) 35.0 g of sulfuric acid in a 5.00% solution
(b) 10.5 g of acetic acid in a 4.50% solution

49. How many grams of solute are dissolved in the following solutions?
(a) 85.0 g of 2.00% $FeBr_2$ solution
(b) 105.0 g of 5.00% Na_2CO_3 solution

50. How many grams of solute are dissolved in the following solutions?
(a) 10.0 g of 2.50% K_2CO_3 solution
(b) 50.0 g of 5.00% Li_2SO_4 solution

51. What mass of water is necessary to prepare each of the following solutions?
(a) 100.0 g of 5.00% nitric acid, HNO_3
(b) 250.0 g of 10.0% NaOH solution

52. What mass of water is necessary to prepare each of the following solutions?
(a) 250.0 g of 0.90% saline solution
(b) 100.0 g of 5.00% sugar solution

Molar Concentration (Sec. 14.9)

53. Calculate the molar concentration for each of the following solutions:
(a) 1.50 g NaCl in 100.0 mL of solution
(b) 1.50 g $K_2Cr_2O_7$ in 100.0 mL of solution
(c) 5.55 g $CaCl_2$ in 125 mL of solution
(d) 5.55 g Na_2SO_4 in 125 mL of solution

54. Calculate the molar concentration for each of the following solutions:
(a) 1.00 g KCl in 75.0 mL of solution
(b) 1.00 g Na_2CrO_4 in 75.0 mL of solution
(c) 20.0 g $MgBr_2$ in 250.0 mL of solution
(d) 20.0 g Li_2CO_3 in 250.0 mL of solution

55. Write two pairs of unit factors for each of the following aqueous solutions given the molar concentration:
(a) 0.100 *M* LiI (b) 0.100 *M* $NaNO_3$
(c) 0.500 *M* K_2CrO_4 (d) 0.500 *M* $ZnSO_4$

56. Write two pairs of unit factors for each of the following aqueous solutions given the molar concentration:
(a) 0.150 *M* KBr (b) 0.150 *M* $Ca(NO_3)_2$
(c) 0.33 *M* $Sr(C_2H_3O_2)_2$ (d) 0.333 *M* NH_4Cl

57. What volume of each of the following solutions contains the indicated amount of dissolved solute?
(a) 10.0 g solute in 0.275 *M* NaF
(b) 10.0 g solute in 0.275 *M* $CdCl_2$
(c) 10.0 g solute in 0.408 *M* K_2CO_3
(d) 10.0 g solute in 0.408 *M* $Fe(ClO_3)_3$

58. What volume of each of the following solutions contains the indicated amount of dissolved solute?
(a) 2.50 g solute in 0.325 *M* KNO_3
(b) 2.50 g solute in 0.325 *M* $AlBr_3$
(c) 2.50 g solute in 1.00 *M* $Co(C_2H_3O_2)_2$
(d) 2.50 g solute in 1.00 *M* $(NH_4)_3PO_4$

59. What is the mass of solute dissolved in the indicated volume of each of the following solutions?
(a) 2.25 L of 0.200 *M* $FeCl_3$
(b) 2.25 L of 0.200 *M* KIO_4
(c) 50.0 mL of 0.295 *M* $ZnSO_4$
(d) 50.0 mL of 0.295 *M* $Ni(NO_3)_2$

60. What is the mass of solute dissolved in the indicated volume of each of the following solutions?
(a) 1.00 L of 0.100 *M* NaOH
(b) 1.00 L of 0.100 *M* $LiHCO_3$
(c) 25.0 mL of 0.500 *M* $CuCl_2$
(d) 25.0 mL of 0.500 *M* $KMnO_4$

61. What is the molar concentration of a saturated solution of calcium sulfate that contains 0.209 g of solute in 100 mL of solution?

62. What is the molar concentration of a saturated solution of calcium hydroxide that contains 0.185 g of solute in 100.0 mL of solution?

63. A hospital glucose solution is analyzed to verify its concentration. A 10.0 mL sample with a mass of 10.483 g is evaporated to dryness. If the solid glucose residue has a mass of 0.524 g, what are (a) the mass/mass percent concentration, and the molar concentration of the glucose, $C_6H_{12}O_6$, solution?

64. A hospital saline solution is analyzed to confirm its concentration. A 50.0 mL sample with a mass of 50.320 g is evaporated to dryness. If the solid sodium chloride residue has a mass of 0.453 g, find (a) the mass/mass percent concentration and the molar concentration of the NaCl solution.

Dilution of a Solution (Sec. 14.10)

65. Concentrated hydrochloric acid is 12 *M*. What is the molarity of an HCl solution prepared by diluting 10.0 mL of concentrated acid to a total volume of 250.0 mL?

66. Concentrated acetic acid is 17 *M*. What is the molarity of an $HC_2H_3O_2$ solution prepared by diluting 50.0 mL of concentrated acid to a total volume of 500.0 mL?

67. Concentrated sulfuric acid is 18 *M*. What volume of H_2SO_4 must be diluted with distilled water to prepare 2.50 L of 0.10 *M* H_2SO_4?

68. Concentrated nitric acid is 16 *M*. What volume of HNO_3 must be diluted with distilled water to prepare 5.00 L of 0.10 *M* HNO_3?

Solution Stoichiometry (Sec. 14.11)

69. Given that 50.0 mL of 0.100 *M* magnesium bromide reacts with 13.9 mL of silver nitrate solution according to the unbalanced equation

$$MgBr_2(aq) + AgNO_3(aq) \longrightarrow AgBr(s) + Mg(NO_3)_2(aq)$$

(a) What is the molarity of the $AgNO_3$ solution?
(b) What is the mass of AgBr precipitate?

70. Given that 24.0 mL of 0.170 *M* sodium iodide reacts with 0.209 *M* mercury(II) nitrate solution according to the unbalanced equation

$$Hg(NO_3)_2(aq) + NaI(aq) \longrightarrow HgI_2(s) + NaNO_3(aq)$$

(a) What volume of $Hg(NO_3)_2$ is required for complete precipitation?
(b) What is the mass of HgI_2 precipitate?

General Exercises

71. In a concert auditorium, stage lights demonstrate the Tyndall effect. What can you conclude about the air in the auditorium?

72. In foggy weather, automobile headlights demonstrate the Tyndall effect. What is the approximate size of water droplets in the fog?

73. Explain why scuba divers use a special gas mixture of oxygen and helium, rather than compressed air, when they dive to depths greater than 125 feet.

74. Calculate the solubility of nitrogen in blood for a scuba diver at a depth of 125 feet and a pressure of 4.68 atm. The solubility of nitrogen gas is 0.0019 g/100 g blood at normal pressure.

75. Calculate the mass of sulfur dioxide gas in 1.00 L of saturated solution at 20 °C. The solubility of SO_2 at 20 °C is 22.8 g/100 mL.

76. Calculate the mass of chlorine gas in 500.0 mL of saturated solution at 20 °C. The solubility of Cl_2 at 20 °C is 0.63 g/100 mL.

77. Ethyl ether, $C_4H_{10}O$, is only 7% miscible in water. Explain why ether is only partially miscible even though ether contains a very electronegative oxygen atom.

78. Phenol, C_6H_5OH, is only 7% soluble in water. Explain why phenol is only partially soluble even though phenol contains a very electronegative oxygen atom.

79. Examine the list of solubilities of different alcohols in water. Explain the decrease in solubility.

Alcohol	Solubility
ethanol, C_2H_5OH	miscible
propanol, C_3H_7OH	97.2 g/100 mL H_2O
butanol, C_4H_9OH	7.9 g/100 mL H_2O
hexanol, $C_6H_{13}OH$	0.59 g/100 mL H_2O
decanol, $C_{10}H_{21}OH$	immiscible

80. Predict whether water or carbon tetrachloride, CCl_4, is a better solvent for each of the following household substances:
(a) grease **(b)** maple syrup
(c) food coloring **(d)** gasoline

81. Identify the solutes and solvents in the following solutions:
(a) 80-proof alcohol (40% ethanol in water)
(b) laboratory alcohol (95% ethanol in water)

82. Explain why grinding the solute increases the rate of dissolving of a solid substance in water.

83. If a household bleach solution contains 5.25% NaClO, what is the molarity of the sodium hypochlorite? (Assume that the density is 1.04 g/mL.)

84. If a vinegar solution contains 5.25% $HC_2H_3O_2$, what is the molarity of the acetic acid? (Assume that the density is 1.01 g/mL.)

14 Chapter Self-Test

Answers to Self-Test are in Appendix J.

1. Yeast and sugar are added to champagne to give the sparkle of carbonation. Under what conditions is carbon dioxide gas most soluble? (Sec. 14.1)
(a) low temperature, high pressure
(b) low temperature, low pressure
(c) high temperature, low pressure
(d) high temperature, high pressure
(e) none of the above

2. Apply the *like dissolves like* rule to predict which of the following liquids is *miscible* with water. (Sec. 14.2)
(a) benzene, C_6H_6 **(b)** chloroform, $CHCl_3$
(c) methanol, CH_3OH **(d)** all of the above
(e) none of the above

3. Apply the *like dissolves like* rule to predict which of the following is a water-soluble vitamin. (Sec. 14.3)
(a) vitamin A, $C_{20}H_{30}O$ **(b)** vitamin C, $C_6H_8O_6$
(c) vitamin D, $C_{27}H_{44}O$ **(d)** vitamin E, $C_{29}H_{50}O_2$
(e) none of the above

4. When glucose sugar, $C_6H_{12}O_6$, dissolves in water, which of the following is formed in the solution? (Sec. 14.4)
(a) hydrated clusters of H_2O
(b) hydrated clusters of CO_2
(c) hydrated clusters of H_2CO_3
(d) hydrated clusters of $C_6H_{12}O_6$
(e) none of the above

5. Which of the following decreases the rate of dissolving for a solid solute in a solvent? (Sec. 14.5)
   (a) grinding the solute **(b)** heating the solution
   (c) stirring the solution **(d)** all of the above
   (e) none of the above

6. What is the solubility of $C_{12}H_{22}O_{11}$ at 50 °C? (Refer to Figure 14.5.)
   (a) 50 g/100 g water **(b)** 55 g/100 g water
   (c) 130 g/100 g water **(d)** 140 g/100 g water
   (e) 150 g/100 g water

7. How concentrated is a solution containing 100 g of $NaC_2H_3O_2$ in 100 g water at 75 °C? (Refer to Figure 14.6.)
   (a) unsaturated **(b)** saturated
   (c) supersaturated **(d)** superunsaturated
   (d) none of the above

8. If 25.0 mL of seawater has a mass of 25.895 g and contains 1.295 g of solute, what is the mass/mass percent concentration of the sample? (Sec. 14.8)
   (a) 1.30% **(b)** 5.00%
   (c) 5.18% **(d)** 20.0%
   (e) 96.5%

9. What is the molarity of a saline solution that contains 0.900 g NaCl (58.44 g/mol) dissolved in 100.0 mL of solution? (Sec. 14.9)
   (a) 0.154 *M* **(b)** 0.527 *M*
   (c) 1.54 *M* **(d)** 1.90 *M*
   (e) 5.27 *M*

10. What is the molarity of a hydrochloric acid solution prepared by diluting 500.0 mL of 1.00 *M* HCl to a total volume of 2.50 L? (Sec. 14.10)
    (a) 0.100 *M* **(b)** 0.200 *M*
    (c) 0.500 *M* **(d)** 2.00 *M*
    (e) 5.00 *M*

11. What volume of 0.200 *M* hydrochloric acid neutralizes 0.400 g of baking soda, $NaHCO_3$ (84.01 g/mol)? (Sec. 14.11)

    $$NaHCO_3(s) + HCl(aq) \longrightarrow NaCl(aq) + H_2O(l) + CO_2(g)$$

    (a) 11.9 mL **(b)** 20.0 mL
    (c) 23.8 mL **(d)** 40.0 mL
    (e) 47.6 mL

Key Concepts

12. Explain why bubbles form on the inside surface of a pan of water when the water is heated.

13. Explain why peanut butter can remove grease from soiled clothing.

Critical Thinking

14. Lake Nuclear and Lake Green are adjoining at the same elevation. If water from Lake Nuclear cools a nuclear reactor power plant and Lake Green does not, which lake has more dissolved oxygen gas?

15. Examine the following illustrations of NaCl and AgCl dissolving in water. At the molecular level, explain why NaCl is more soluble than AgCl.

(a) NaCl very soluble

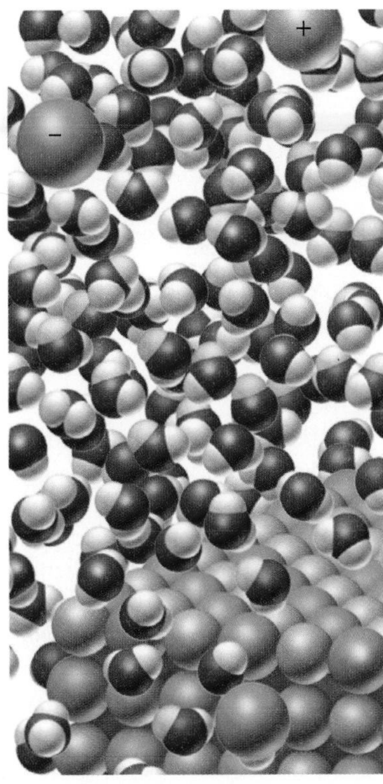

(b) AgCl slightly soluble

CHAPTER

15 Acids and Bases

Swimmers' hair often turns blond owing to sunlight and the chemicals in a swimming pool. Bleach is added to a swimming pool to destroy microorganisms; hydrochloric acid is added to increase acidity.

"Many of life's failures are people who did not realize how close they were to success when they gave up."
Thomas Alva Edison, Inventor and Scientist (1847–1931)

Acids and bases play an important role in our lives. The proper acidity of our blood and other body fluids is vital to our well-being and is carefully controlled by an elaborate buffering system. Many of the foods we eat are acidic. For instance, oranges and lemons contain citric acid, and vinegar is a solution of acetic acid. Vitamin C is ascorbic acid, and aspirin is acetylsalicylic acid. We add muriatic acid to our swimming pools and sulfuric acid to our car batteries.

Acids and bases are in a way opposites and neutralize each other. Baking soda and milk of magnesia are basic. Antacid tablets contain basic substances such as carbonates, bicarbonates, and hydroxides for neutralizing an acidic, upset stomach. We use a dilute solution of basic ammonia to clean floors and a concentrated solution of sodium hydroxide to clean drains and ovens.

15.1 PROPERTIES OF ACIDS AND BASES

OBJECTIVES

- To list the general properties of acids and bases.
- To classify a solution as strongly acidic, weakly acidic, neutral, weakly basic, or strongly basic, given its pH.

An *acid* is any substance that releases hydrogen ions, H^+, in water. This property of acids allows us to test for acidity. In the laboratory we can determine if a solution is acidic using a strip of litmus paper. Litmus paper turns red in the presence of hydrogen ions. Therefore, if we put a piece of litmus paper in a solution and the paper turns red, we know the solution is an acid. For example, litmus paper in lemon juice turns red because lemon juice is acidic.

Litmus paper is made from a plant pigment that is sensitive to changes in acid concentration. The colors of many plants are affected by the acidity of the soil in which they are grown. For example, the color of an orchid can vary from pale lavender to deep purple depending upon the acidity of the soil. The color of a rose fades from red to pink when the rose is placed in an acid solution.

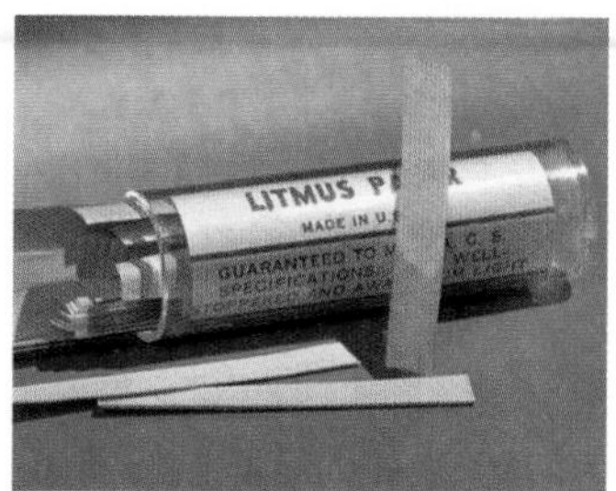

▲ **Litmus Paper** A strip of red litmus paper turns blue in a basic solution. A strip of blue litmus turns red in an acidic solution.

Acids have their own special properties. One property is that they have a sour taste. The tart taste of a lemon, an apple, or vinegar shows us that these are acidic foods. The taste buds that are sensitive to acids are located along the edge of our tongue, which is why we roll our tongue when we taste something that is sour.

A *base* is any substance that releases hydroxide ions, OH^-, in water. Bases also have special properties. For instance, they feel slippery or soapy to the touch and have a bitter taste. The bitter taste of milk of magnesia shows us that it is a basic substance. The taste buds that are sensitive to bases are located on the tongue toward the back of our throat, and we often react by sticking out our tongue when we taste something that is bitter.

In the laboratory we use litmus paper to determine if a solution is basic. Litmus paper turns blue in the presence of hydroxide ion, which is found in drain and oven cleaner. Another property of acids and bases is their ability to undergo a neutralization reaction, which we studied in Section 8.11. An acid and a base react to produce a salt and water. For example, hydrochloric acid neutralizes a potassium hydroxide solution to give potassium chloride and water.

$$HCl(aq) + KOH(aq) \longrightarrow KCl(aq) + H_2O(l)$$

The properties of acids and bases are summarized in Table 15.1.

TABLE 15.1 PROPERTIES OF ACIDS AND BASES

PROPERTY	ACIDIC SOLUTIONS	BASIC SOLUTIONS
taste	sour	bitter
feel	—	slippery, soapy
litmus paper	litmus turns red	litmus turns blue
pH value	less than 7	greater than 7
neutralization reaction	react with bases to give a salt and water	react with acids to give a salt and water

pH Scale

A pH value expresses the acidity or basicity of a solution. Most solutions have a pH between 0 and 14. For example, a 1 *M* HCl solution has a pH of 0 (strongly acidic), and a 1 *M* NaOH solution has a pH of 14 (strongly basic). A pH of 7 is considered neutral; pure distilled water has a pH of 7.

On the pH scale, acidic solutions have a pH less than 7. As the acidity increases, the pH value *decreases.* A solution having a pH of 3 is more acidic than a solution with a pH of 4. Basic solutions have a pH greater than 7. As the basicity increases, the pH value *increases.* A solution having a pH of 11 is more basic than a solution with a pH of 10.

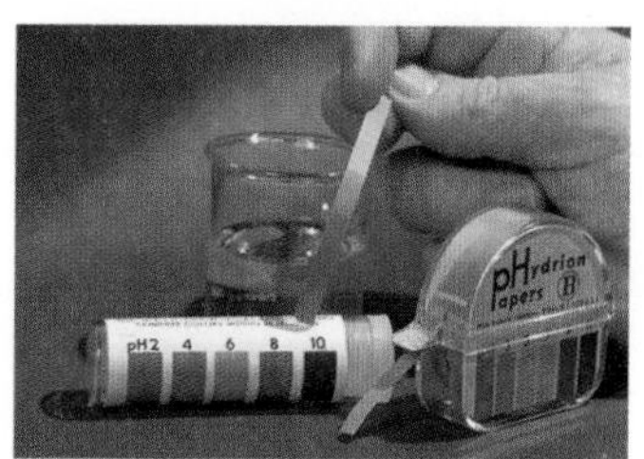

▲ **pHydrion Paper** A strip of pHydrion paper changes color when placed into a solution. The test strip is then compared to a color chart to estimate the pH of the solution.

A solution can be classified as strongly acidic, weakly acidic, neutral, weakly basic, or strongly basic on the basis of pH. A strongly acidic solution has a pH between 0 and 2. A weakly acidic solution has a pH between 2 and 7. A weakly basic solution

Acidic/Basic	pH	Example Solution
Strongly acidic	0	1 *M* HCl
	1	Stomach acid (1-3)
Weakly acidic	2	Lemon juice
	3	Vinegar, wine
	4	Grapes, orange juice
	5	Normal rain, coffee
	6	Milk, pH balanced shampoo
Neutral	7	Pure water
Weakly basic	8	Eggs, seawater
	9	Baking soda, antacids
	10	Milk of magnesia, soap
	11	Household ammonia
	12	Liquid bleach
Strongly basic	13	Drain cleaner
	14	1 *M* NaOH

▶ **Figure 15.1 The pH Scale** Note that solutions may be strongly acidic (pH ~1), weakly acidic (pH ~4), neutral (pH ~7), weakly basic (pH ~10), or strongly basic (pH ~14).

has a pH between 7 and 12. A strongly basic solution has a pH between 12 and 14. Figure 15.1 illustrates some common substances and their approximate pH.

The following example exercise illustrates the acid–base strength of a few common solutions:

Example Exercise 15.1 The pH Scale

Indicate whether each of the following solutions is considered strongly acidic, weakly acidic, neutral, weakly basic, or strongly basic:

(a) stomach acid, pH 1.5
(b) oven cleaner, pH 13.5
(c) orange juice, pH 4.5
(d) deionized water, pH 7.0
(e) eggs, pH 7.5
(f) carbonated soda, pH 4.0

Solution

Refer to the guidelines in Figure 15.1 to classify each solution.

(a) Gastric juice has a pH between 0 and 2. It is *strongly acidic.*
(b) Oven cleaner has a pH of 13.5. It is *strongly basic.*
(c) Orange juice has a pH between 2 and 7. It is *weakly acidic.*
(d) Deionized water has a pH of 7.0 and is *neutral.*
(e) Eggs have a pH of 7.5 and are *weakly basic.*
(f) Carbonated soda has a pH of 4.0 and is *weakly acidic.*

Practice Exercise

Indicate whether each of the following properties corresponds to an acid or to a base:

(a) sour taste
(b) slippery feel
(c) turns litmus paper red
(d) pH greater than 7

Answers: (a) acid; (b) base; (c) acid; (d) base

Concept Exercise

Which of the following has the highest acidity: pH 1, pH 5, or pH 9?

Answer: See Appendix G.

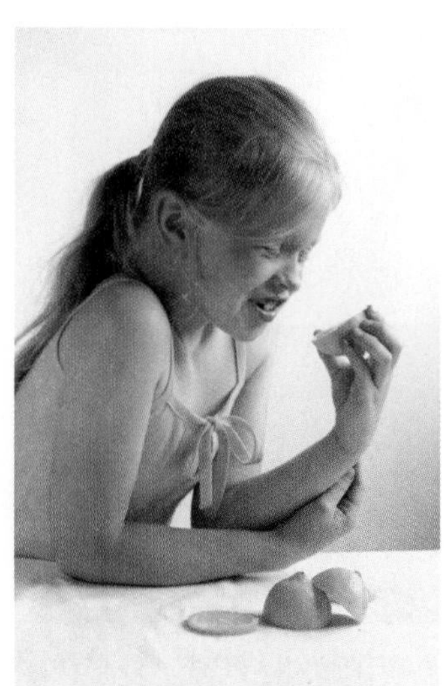

▲ **Lemons are Sour** Lemons, limes, oranges, and grapefruit contain citric acid, which is sour.

Buffers

Controlling acidity is important in the manufacture of food, paper, and chemicals. In agriculture, controlling the pH of the soil is necessary to obtain good crop yields. A **buffer** is a solution that resists changes and controls the pH when an acid or a base is added. A buffer solution functions by consuming excess hydrogen ions or hydroxide ions, and thereby maintains its pH.

A buffer solution is composed of an aqueous weak acid and one of its salts, for example, citric acid and sodium citrate. If we add acid to a solution, the citrate ion combines with excess hydrogen ions to form citric acid. Conversely, if we add base to a solution, citric acid neutralizes excess hydroxide ions to form a citrate salt. If the amount of acid or base added to a buffer solution is relatively small compared to the amount of buffer, there is very little change in pH.

Other examples of buffers include hydrogen carbonate ions and carbonate ions, which are present in seawater and help to preserve the pH near 8. A similar buffer system helps to control the pH of blood, which must be maintained in a very narrow range near a pH of 7.4.

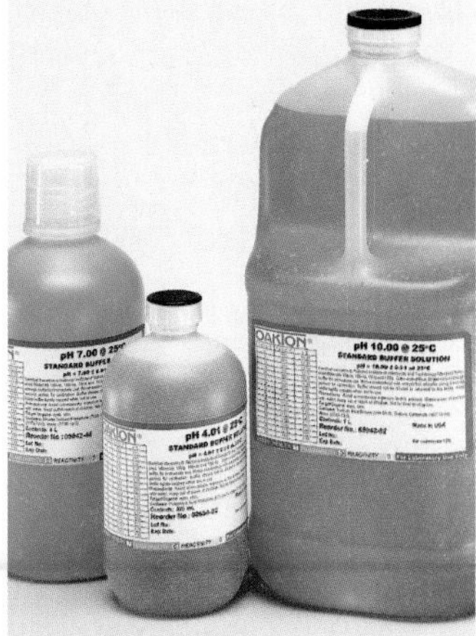

▲ **Buffer Solutions** A buffer solution resists a change in pH with the addition of an acid or a base.

15.2 ARRHENIUS ACIDS AND BASES

OBJECTIVES

- To understand the concept of an Arrhenius acid and base.
- To identify strong and weak acids and bases given the degree of ionization in aqueous solution.
- To state the acid and base that react to produce a given salt.

In 1884, the Swedish chemist Svante Arrhenius suggested the first definitions for an acid and a base. He proposed that an acid is a substance that ionizes in water to produce hydrogen ions. Conversely, he proposed that a base is a substance that dissociates in water to release hydroxide ions. These simple concepts define an **Arrhenius acid** and **Arrhenius base**.

Ionization versus Dissociation

Acids and bases have varying strengths. The strength of an Arrhenius acid is measured by the degree of ionization in solution. **Ionization** is the process by which molecules of a polar compound ionize to form cations and anions. Thus, a molecule of hydrochloric acid, HCl, ionizes into H^+ and Cl^-.

The strength of an Arrhenius base is measured by the degree of dissociation in the solution. **Dissociation** is the process whereby already existing ions of an ionic compound simply separate. Thus, a formula unit of sodium hydroxide, NaOH, dissociates into Na^+ and OH^-.

Strong and Weak Arrhenius Acids

An acid is considered either strong or weak depending on how much it ionizes in water. According to the Arrhenius definition, a strong acid ionizes extensively to release hydrogen ions in solution. Hydrochloric acid is considered a strong acid because it ionizes nearly 100%. Hydrofluoric acid is a weak acid because it is only slightly ionized, about 1%. Table 15.2 lists some common Arrhenius acids.

TABLE 15.2 COMMON ARRHENIUS ACIDS

AQUEOUS ACID	PERCENT IONIZATION	ACID STRENGTH
hydrochloric acid, $HCl(aq)$	~100%	strong
nitric acid, $HNO_3(aq)$	~100%	strong
sulfuric acid, $H_2SO_4(aq)$	~100%	strong
acetic acid, $HC_2H_3O_2(aq)$	slightly	weak
carbonic acid, $H_2CO_3(aq)$	slightly	weak
hydrofluoric acid, $HF(aq)$	slightly	weak
phosphoric acid, $H_3PO_4(aq)$	slightly	weak

▲ **Hydrofluoric Acid, HF** Although aqueous HF is a weak acid, it attacks glass. This design was created by coating glass with wax and then removing wax from the pattern areas. The glass was dipped into hydrofluoric acid, which etched the design.

All Arrhenius acids have a hydrogen atom attached to an acid molecule by a polar bond. Most acids are weak ternary oxyacids in which the H atom is attached to an O atom. When the acid molecule ionizes, it breaks this bond. In an aqueous solution, polar water molecules help the acid ionize by pulling the hydrogen ion away from the acid molecule. We can show the ionization for hydrochloric acid, HCl, and acetic acid, $HC_2H_3O_2$, as follows:

$$HCl(aq) + H_2O(l) \longrightarrow H_3O^+(aq) + Cl^-(aq) \qquad (\sim 100\%)$$

$$HC_2H_3O_2(aq) + H_2O(l) \longrightarrow H_3O^+(aq) + C_2H_3O_2{}^-(aq) \qquad (\sim 1\%)$$

Although chemists often refer to acids as solutions containing hydrogen ions, the actual species is H_3O^+. The **hydronium ion**, H_3O^+, results from a hydrogen ion attaching to a nonbonding electron pair on a water molecule. For simplicity, we will usually refer to a hydrogen ion in aqueous solution as $H^+(aq)$.

▲ **Ammonium Hydroxide, NH_4OH** Ammonia gas dissolved in water is commonly referred to as an ammonium hydroxide solution.

Strong and Weak Arrhenius Bases

Recall that a base is considered either strong or weak depending on how much of it dissociates. According to the Arrhenius definition, a strong base dissociates extensively to release hydroxide ions in solution. Sodium hydroxide is a strong base and dissociates nearly 100% in an aqueous solution. A weak base dissociates very little to release few hydroxide ions in solution. Ammonium hydroxide is a weak base because it provides relatively few hydroxide ions in solution. Table 15.3 lists some common Arrhenius bases.

TABLE 15.3 COMMON ARRHENIUS BASES

AQUEOUS BASE	PERCENT DISSOCIATION	BASE STRENGTH
barium hydroxide, $Ba(OH)_2(aq)$	~100%	strong
calcium hydroxide, $Ca(OH)_2(aq)$	~100%	strong
lithium hydroxide, $LiOH(aq)$	~100%	strong
potassium hydroxide, $KOH(aq)$	~100%	strong
sodium hydroxide, $NaOH(aq)$	~100%	strong
ammonium hydroxide, $NH_4OH(aq)$*	slightly	weak

*Ammonium hydroxide is prepared by dissolving ammonia gas in water. It is referred to as an aqueous ammonia solution, $NH_3(aq)$, or as ammonia water, $NH_3 \cdot H_2O$. The name "ammonium hydroxide" is somewhat misleading since there is no evidence for ammonium hydroxide molecules in a solution. For simplicity, however, we will refer to an ammonia solution as ammonium hydroxide, $NH_4OH(aq)$.

When we dissolve sodium hydroxide in water, we obtain aqueous NaOH. In the dissolving process, NaOH dissociates into aqueous sodium ions and aqueous hydroxide ions. Aqueous ammonium hydroxide, NH_4OH, is a weak base and provides relatively few hydroxide ions in solution. We can show this process for the two Arrhenius bases as follows:

$$NaOH(aq) \longrightarrow Na^+(aq) + OH^-(aq) \qquad (\sim 100\%)$$

$$NH_4OH(aq) \longrightarrow NH_4{}^+(aq) + OH^-(aq) \qquad (\sim 1\%)$$

The following example exercise illustrates the classification of acid–base strength based on the ability to donate ions in solution:

Example Exercise 15.2 Strong and Weak Acids

Classify each of the following solutions as a strong or a weak Arrhenius acid given the degree of ionization:

(a) perchloric acid, $HClO_4(aq)$ ~100%
(b) hypochlorous acid, $HClO(aq)$ ~1%

Solution

To classify an aqueous solution as a strong or a weak acid we must know the amount of ionization.

(a) Perchloric acid is extensively ionized and is considered a *strong acid.* An aqueous solution is primarily $H^+(aq)$ and $ClO_4^-(aq)$.

(b) Hypochlorous acid is slightly ionized and is considered a *weak acid.* An aqueous solution contains primarily molecules of HClO.

Practice Exercise

Classify each of the following solutions as a strong or a weak Arrhenius base given the degree of dissociation:

(a) magnesium hydroxide, $Mg(OH)_2(aq)$ ~1%

(b) strontium hydroxide, $Sr(OH)_2(aq)$ ~100%

Answers: (a) weak base; (b) strong base

Concept Exercise

Which of the following is a strong Arrhenius acid: $HNO_3(aq)$, $H_2CO_3(aq)$, $H_3PO_4(aq)$?

Answer: See Appendix G.

Neutralization Reactions

We learned in Section 8.11 that an acid neutralizes a base to give a **salt** and water. For example, hydrochloric acid reacts with sodium hydroxide to give the salt sodium chloride and water. The equation for the reaction is

$$HCl(aq) + NaOH(aq) \longrightarrow NaCl(aq) + H_2O(l)$$

This reaction produces the aqueous salt NaCl. Different acids give other salts. When we completely neutralize sulfuric acid using aqueous sodium hydroxide, for example, we obtain sodium sulfate. The equation is

$$H_2SO_4(aq) + 2\,NaOH(aq) \longrightarrow Na_2SO_4(aq) + 2\,H_2O(l)$$

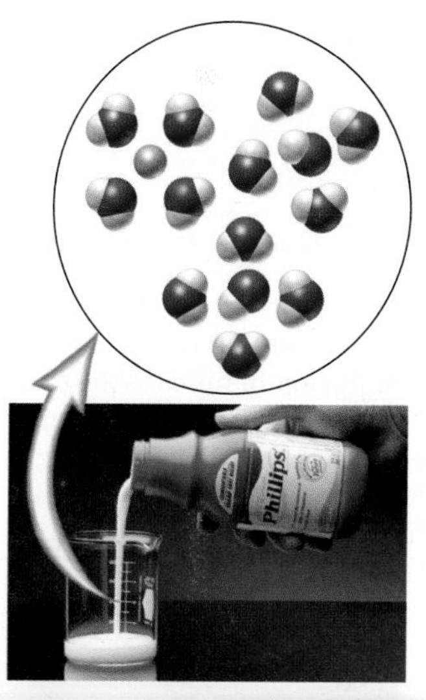

▲ **Stomach Antacid** Milk of Magnesia, $Mg(OH)_2$, is a weak base that can neutralize excess stomach acid.

The hydrogen ions from an acid are neutralized by the hydroxide ions from a base. The hydrogen ions and hydroxide ions combine to form water. Note from the preceding neutralization reaction that Na_2SO_4 is the salt produced.

We can identify the Arrhenius acid and base that produce any given salt because each salt is composed of the cation from the reacting base and the anion from the acid. As an example, we can predict the neutralization reaction that produces the salt potassium acetate, $KC_2H_3O_2$. Since the salt contains potassium, the neutralized base must be potassium hydroxide, KOH. The acetate ion came from acetic acid, $HC_2H_3O_2$. The equation for the reaction is

$$HC_2H_3O_2(aq) + KOH(aq) \longrightarrow KC_2H_3O_2(aq) + H_2O(l)$$

The following example exercise further illustrates the neutralization of acids and bases to give a salt and water:

Example Exercise 15.3 Neutralization Reactions

Determine the acid and base that produce each of the following salts. Write a balanced equation for the neutralization reaction.

(a) lithium fluoride, $LiF(aq)$ (b) calcium sulfate, $CaSO_4(aq)$

Solution

(a) The salt LiF is produced from the neutralization of the base LiOH and the acid HF. The equation for the reaction is

$$HF(aq) + LiOH(aq) \longrightarrow LiF(aq) + H_2O(l)$$

(b) The salt $CaSO_4$ is produced from the neutralization of the base $Ca(OH)_2$ and the acid H_2SO_4. The equation for the reaction is

$$H_2SO_4(aq) + Ca(OH)_2(aq) \longrightarrow CaSO_4(aq) + 2\,H_2O(l)$$

Practice Exercise

Determine the acid and base that produce each of the following salts. Write a balanced equation for the neutralization reaction.

(a) potassium iodide, $KI(aq)$ (b) barium nitrate, $Ba(NO_3)_2(aq)$

Answers:

(a) HI is the acid and KOH is the base:

$$HI(aq) + KOH(aq) \longrightarrow KI(aq) + H_2O(l)$$

(b) HNO_3 is the acid and $Ba(OH)_2$ is the base:

$$2\,HNO_3(aq) + Ba(OH)_2(aq) \longrightarrow Ba(NO_3)_2(aq) + 2\,H_2O(l)$$

Concept Exercise

Which acid and base undergo neutralization to give sodium acetate, $NaC_2H_3O_2$?

Answer: See Appendix G.

CHEMISTRY CONNECTION Svante Arrhenius

Q: What grade was Arrhenius awarded for his brilliant Ph.D. dissertation on the theory of ionic solutions?

Svante Arrhenius (1859–1927) was a child genius who taught himself to read by age three. After graduating from high school at the top of his class, he enrolled at the University of Uppsala in his native Sweden. He majored in chemistry and did research on the passage of electricity through aqueous solutions.

▲ **Svante August Arrhenius** (1859–1927)

At 22, Arrhenius began his doctoral study in Stockholm. For his graduate thesis, he continued to pursue his interest in the behavior of solutions. In particular, he found it puzzling that sodium chloride solutions conducted electricity, whereas sugar solutions did not. After carefully considering his observations, Arrhenius boldly proposed that a solution of sodium chloride conducts an electric current because it separates into charged particles in solution. Sugar does not conduct electricity because it does not form charged particles in solution.

Arrhenius also noted that the freezing point of water was lowered twice as much for a salt solution as it was for a sugar solution. He explained that when salt dissolves in solution, NaCl separates into sodium ions and chloride ions. When sugar dissolves in solution, it remains as molecules.

Arrhenius was aware, however, that there were difficulties with the concept of charged ions in solution. At the time, the scientific community considered atoms to be indivisible, electrically neutral particles.

In 1884, Arrhenius defended his Ph.D. dissertation on the theory of ionic solutions. After a grueling four-hour defense, he was awarded the lowest possible passing grade. The dissertation committee simply was not ready to accept the idea of ions in solution.

Arrhenius' career stagnated for almost a decade, and his ideas found little support. The ionic theory was eventually championed by Wilhelm Ostwald (1853–1932), a noted chemist, who invited Arrhenius to work with him. Gradually, evidence accumulated that supported the concept of ions. Most notably, the discovery of the electron in 1897 proved the existence of charged subatomic particles.

After the ionic theory gained credibility, Arrhenius was offered a professorship at the University of Stockholm. In 1899, he published the classic paper "On the Dissociation of Substances in Aqueous Solutions." In 1903, after struggling for 20 years to establish his ideas, Arrhenius was awarded the Nobel prize in chemistry for his ionic theory of solutions.

A: Arrhenius received the lowest possible passing grade; the examining committee simply did not believe in the ionic theory of solutions.

15.3 BRØNSTED–LOWRY ACIDS AND BASES

OBJECTIVE

- To identify the Brønsted–Lowry acid and base in a given neutralization reaction.

In 1923 the Danish chemist Johannes Brønsted (1879–1947) and the English chemist Thomas Lowry (1874–1936) independently proposed broader definitions than those of Arrhenius for an acid and a base. Whereas Arrhenius defined an acid as a substance that donates a hydrogen ion in water, Brønsted and Lowry defined an acid as a substance that donates a hydrogen ion to any other substance. Since a hydrogen ion is a single proton, a **Brønsted–Lowry acid** is also referred to as a **proton donor**.

Recall the Arrhenius definition of a base as a hydroxide ion donor in water. We know that hydroxide ions neutralize hydrogen ions to form water. There are, however, many substances that neutralize hydrogen ions in addition to hydroxide ions. Brønsted and Lowry proposed that a base is any substance that accepts a hydrogen ion. Thus, a **Brønsted–Lowry base** is a **proton acceptor**.

Although the Brønsted–Lowry definitions of an acid and a base are a bit different from the Arrhenius definitions, an acid and a base still neutralize each other. The following reactions illustrate the neutralization of a Brønsted–Lowry base by hydrochloric acid:

$$HCl(aq) + NaOH(aq) \longrightarrow NaCl(aq) + H_2O(l)$$

$$HCl(aq) + NH_3(aq) \longrightarrow NH_4Cl(aq)$$

$$HCl(aq) + H_2O(l) \longrightarrow H_3O^+(aq) + Cl^-(aq)$$

In the first equation, aqueous NaOH accepts a proton. Therefore, according to the Brønsted–Lowry definition, it is a base. In the second equation, aqueous NH_3 accepts a proton. It is also a base. As the third equation shows, even water can act as a Brønsted–Lowry base because it can accept a hydrogen ion.

Unlike Arrhenius acids and bases that are defined by hydrogen ions or hydroxide ions dissolved in water, Brønsted–Lowry acids and bases depend on a particular reaction. For example, aqueous $NaHCO_3$ can act as a *base* by accepting a proton. But in a different reaction, aqueous $NaHCO_3$ can act as an *acid* by donating a proton. The following examples illustrate $NaHCO_3$ acting as both an acid and a base:

$$HCl(aq) + NaHCO_3(aq) \longrightarrow NaCl(aq) + H_2CO_3(aq)$$

$$NaOH(aq) + NaHCO_3(aq) \longrightarrow Na_2CO_3(aq) + H_2O(l)$$

In the first reaction, $NaHCO_3$ accepts a proton from HCl and therefore acts as a Brønsted–Lowry base. In the second reaction, $NaHCO_3$ donates a proton to NaOH and therefore acts as a Brønsted–Lowry acid. A substance that is capable of both accepting and donating a proton is said to be **amphiprotic**. According to the Brønsted–Lowry theory, water is amphiprotic. That is, H_2O can act as an acid and donate a proton; H_2O can also act as a base and accept a proton to become the hydronium ion, H_3O^+. We can show the amphiprotic nature of water as follows:

$$\underset{\substack{\text{proton donor} \\ \textit{acid}}}{H_2O(l)} + \underset{\substack{\text{proton acceptor} \\ \textit{base}}}{H_2O(l)} \longrightarrow H_3O^+(aq) + OH^-(aq)$$

15.4 ACID–BASE INDICATORS

OBJECTIVES

- To understand the concept of an indicator to show the acidity of a solution.
- To state the color of the following indicators in a solution of a given pH: phenolphthalein, methyl red, and bromthymol blue.

We mentioned in Section 15.1 that litmus paper can be used to show whether a solution is acidic or basic. In addition, other paper test strips are available that indicate the pH of a solution. These test strips are permeated with substances that change color depending on the pH of the test solution. These strips contain substances extracted from plants. Some of the plant extracts come from grape juice, red cabbage, and flower petals such as those of roses, violets, and orchids.

A substance that is sensitive to a change in pH by exhibiting a different color is referred to as an **acid–base indicator**. Figure 15.2 illustrates the range in color of these indicators. Table 15.4 summarizes the pH range for each indicator.

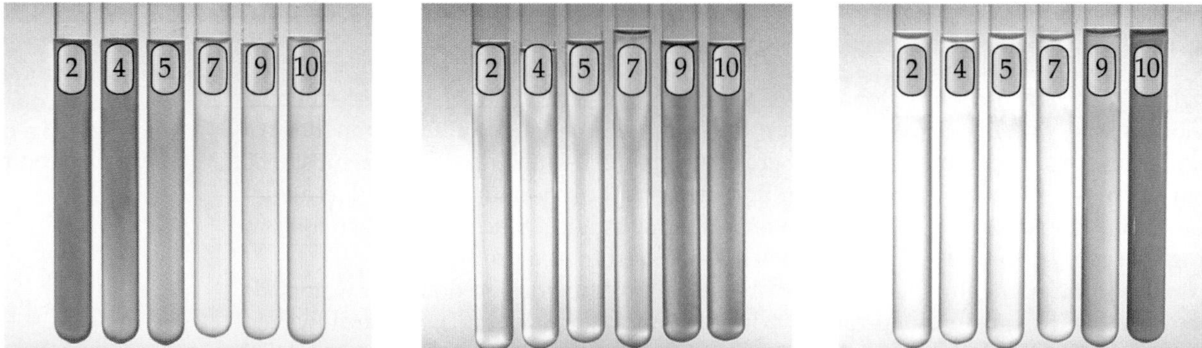

▲ **Figure 15.2 pH and Acid–Base Indicator Color** The color of three acid–base indicators at different pH values. Methyl red (left) changes from red to yellow at pH 5. Bromthymol blue (middle) changes from yellow to blue at pH 7. Phenolphthalein (right) changes from colorless to pink at pH 9.

TABLE 15.4 ACID–BASE INDICATORS

INDICATOR	COLOR CHANGE	COLOR
methyl red	pH ~5	below pH 5, red above pH 5, yellow
bromthymol blue	pH ~7	below pH 7, yellow above pH 7, blue
phenolphthalein	pH ~9	below pH 9, colorless above pH 9, pink

If a solution has a pH above 5, the indicator methyl red is yellow. If a solution has a pH below 5, the indicator is red. At a pH of 5, methyl red appears orange because it contains both the yellow and red forms of the indicator.

If a solution has a pH of 9 or above, the indicator phenolphthalein is pink. If a solution has a pH below 9, the indicator is colorless. Although there are many acid–base indicators available in the laboratory, phenolphthalein is the only common indicator that changes to colorless. This practical consideration and ease of observation makes phenolphthalein the indicator of choice for many applications.

The following example exercise illustrates the relationship of the color of an acid–base indicator and the pH of a given solution:

Example Exercise 15.4 Acid–Base Indicators

State the color of the acid–base indicator in each of the following solutions:

(a) A solution at pH 3 contains a drop of methyl red.
(b) A solution at pH 8 contains a drop of bromthymol blue.
(c) A solution at pH 11 contains a drop of phenolphthalein.

Solution

Refer to Table 15.4 to determine the color of each solution.

(a) A pH 3 solution containing methyl red indicator is *red*.
(b) A pH 8 solution containing bromthymol blue indicator is *blue*.
(c) A pH 11 solution containing phenolphthalein indicator is *pink*.

Practice Exercise

State the pH at which each of the following acid–base indicators changes color:

(a) methyl red (b) bromthymol blue
(c) phenolphthalein

Answers: (a) 5; (b) 7; (c) 9

Concept Exercise

Which common acid–base indicator is colorless in acidic and neutral solutions?

Answer: See Appendix G.

15.5 ACID–BASE TITRATIONS

OBJECTIVES

- To understand the concept of an acid–base titration.
- To perform calculations that involve acid–base titrations.

Vinegar tastes sour because it contains acetic acid. To find the molar concentration of acetic acid in vinegar we analyze the vinegar sample using a base solution of known concentration. In the laboratory, we can analyze acetic acid by a titration method. We use this **titration** process to deliver a measured volume of solution. To analyze the acetic acid in a vinegar sample we can titrate a sample by delivering a measured volume of sodium hydroxide solution. We can use phenolphthalein as an indicator to signal when we have neutralized all the acetic acid.

After the acetic acid is neutralized, an extra drop of NaOH increases the pH dramatically and the phenolphthalein indicator changes from colorless to pink. When the indicator changes color, we stop adding NaOH base and cease the titration. This turning point in the titration is the **endpoint**. Figure 15.3 illustrates the experimental procedure.

Consider the following titration of a vinegar sample for acetic acid. A 10.0 mL sample of vinegar containing acetic acid requires 37.55 mL of 0.223 *M* NaOH. The balanced equation for the reaction is

$$HC_2H_3O_2(aq) + NaOH(aq) \longrightarrow NaC_2H_3O_2(aq) + H_2O(l)$$

This is a solution stoichiometry problem (Section 14.11). To find the molarity of the acetic acid, we must first find the moles of NaOH. The volume of NaOH is 37.55 mL, and the molar concentration is 0.223 *M*. The molarity of the NaOH can be written as the unit factor 0.223 mol NaOH/1000 mL solution.

$$37.55 \text{ \sout{mL solution}} \times \frac{0.223 \text{ mol NaOH}}{1000 \text{ \sout{mL solution}}} = 0.00837 \text{ mol NaOH}$$

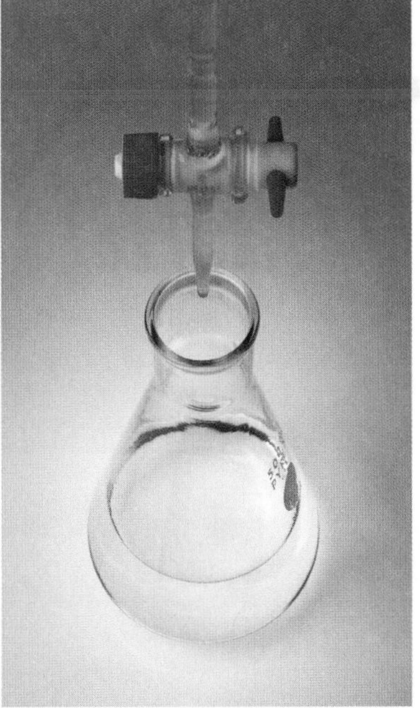

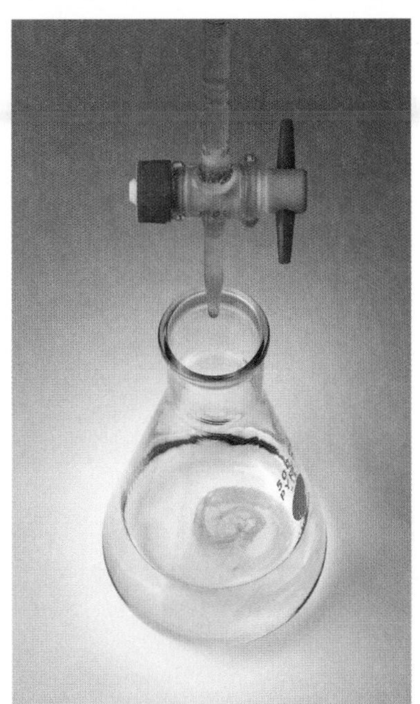

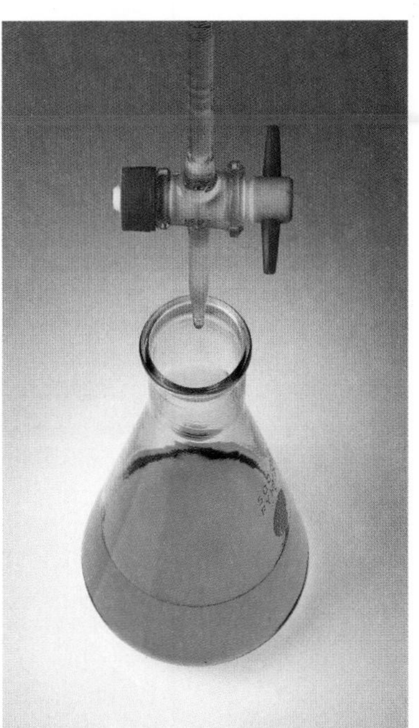

▲ **Figure 15.3 Titration of Acetic Acid with Sodium Hydroxide**
(a) The flask contains a sample of acetic acid and a drop of phenolphthalein indicator.
(b) The buret delivers a measured volume of sodium hydroxide solution into the flask.
(c) The titration is complete when the solution attains a permanent pink color.

Next, we find the moles of acetic acid in the acid solution. From the balanced equation we see that 1 mol of NaOH base neutralizes 1 mol of $HC_2H_3O_2$.

$$0.00837 \text{ mol NaOH} \times \frac{1 \text{ mol } HC_2H_3O_2}{1 \text{ mol NaOH}} = 0.00837 \text{ mol } HC_2H_3O_2$$

Finally, we can calculate the molarity of the acetic acid. The volume of the $HC_2H_3O_2$ solution is 10.0 mL.

$$\frac{0.00837 \text{ mol } HC_2H_3O_2}{10.0 \text{ mL solution}} \times \frac{1000 \text{ mL solution}}{1 \text{ L solution}} = \frac{0.837 \text{ mol } HC_2H_3O_2}{1 \text{ L solution}}$$

$$= 0.837\ M\ HC_2H_3O_2$$

Now let's express the molar concentration of the acetic acid as a mass percent (m/m %) concentration. Given that the density of the vinegar is 1.01 g/mL and that the molar mass of acetic acid is 60.06 g/mol, we can proceed as follows:

$$\frac{0.837 \text{ mol } HC_2H_3O_2}{1000 \text{ mL solution}} \times \frac{60.06 \text{ g } HC_2H_3O_2}{1 \text{ mol } HC_2H_3O_2} \times \frac{1 \text{ mL solution}}{1.01 \text{ g solution}} \times 100\%$$

$$= \frac{0.0498 \text{ g } HC_2H_3O_2}{1 \text{ g solution}} \times 100\% = 4.98\%\ HC_2H_3O_2$$

Therefore, the concentration of acetic acid in the vinegar solution is 0.837 *M*, which is equivalent to 4.98% $HC_2H_3O_2$. Typically, the label on a bottle of vinegar reads "diluted in water to 5% acidity."

Titrations can also be performed on basic solutions in the same way, but an acid is added from the buret instead of a base. We can calculate the concentration of ammonia in a household cleaning solution by titrating a sample. The following example exercise illustrates the analysis of an ammonia solution:

Example Exercise 15.5 Acid–Base Titration of Ammonia

If 25.30 mL of 0.277 *M* HCl is used to titrate 10.0 mL of aqueous ammonia to a methyl red endpoint, what is the molarity of the ammonia? When we write aqueous ammonia as NH_4OH, the balanced equation is

$$HCl(aq) + NH_4OH(aq) \longrightarrow NH_4Cl(aq) + H_2O(l)$$

Solution

To calculate the molarity of the NH_4OH, we must find the moles of NH_4OH. From the balanced equation, we see that 1 mol of HCl neutralizes 1 mol of NH_4OH. Since the acid solution is 0.277 *M* HCl, we have the unit factor 0.277 mol HCl/1000 mL solution.

$$25.30 \text{ mL solution} \times \frac{0.277 \text{ mol HCl}}{1000 \text{ mL solution}} \times \frac{1 \text{ mol } NH_4OH}{1 \text{ mol HCl}} = 0.00701 \text{ mol } NH_4OH$$

We can summarize the conversion as follows:

We obtain the molar concentration by dividing the moles of NH_4OH by the volume of aqueous ammonia titrated, 10.0 mL.

$$\frac{0.00701 \text{ mol } NH_4OH}{10.0 \text{ mL solution}} \times \frac{1000 \text{ mL solution}}{1 \text{ L solution}} = \frac{0.701 \text{ mol } NH_4OH}{1 \text{ L solution}}$$

$$= 0.701\ M\ NH_4OH$$

Practice Exercise

If 38.30 mL of 0.250 *M* NaOH is used to titrate 25.0 mL of phosphoric acid, what is the molarity of the acid? The balanced equation is

$$H_3PO_4(aq) + 3\ NaOH(aq) \longrightarrow Na_3PO_4(aq) + 3\ H_2O(l)$$

Answer: 0.128 *M*

Concept Exercise

Which of the following describes the titration of $H_3PO_4(aq)$ with NaOH(*aq*): strong acid with strong base, strong acid with weak base, or weak acid with strong base?

Answer: See Appendix G.

Let's try another type of problem and find the volume of base required to neutralize an acid. For example, we can find the volume of base needed to neutralize sulfuric acid in a battery acid sample. The following example exercise illustrates the analysis of battery acid by a titration method:

Example Exercise 15.6 Acid–Base Titration of Battery Acid

A 10.0 mL sample of 0.555 *M* battery acid, H_2SO_4, is titrated with 0.223 *M* NaOH. What volume of sodium hydroxide is required for the titration?

$$H_2SO_4(aq) + 2\ NaOH(aq) \longrightarrow Na_2SO_4(aq) + 2\ H_2O(l)$$

Solution

Step 1: The unit asked for in the answer is mL NaOH.

Step 2: The given value is 10.0 mL H_2SO_4.

Step 3: We apply unit conversion factors to cancel units. In this example, we use the molar concentration of an acid (0.555 mol H_2SO_4 per 1000 mL H_2SO_4) and the molar concentration of a base (0.223 mol NaOH per 1000 mL NaOH) as unit factors. From the balanced equation, we see that 1 mol H_2SO_4 = 2 mol NaOH.

$$10.0\ \text{mL solution} \times \frac{0.555\ \text{mol } H_2SO_4}{1000\ \text{mL solution}} \times \frac{2\ \text{mol NaOH}}{1\ \text{mol } H_2SO_4} \times \frac{1000\ \text{mL NaOH}}{0.223\ \text{mol NaOH}} = 49.8\ \text{mL NaOH}$$

We can summarize the overall solution as follows:

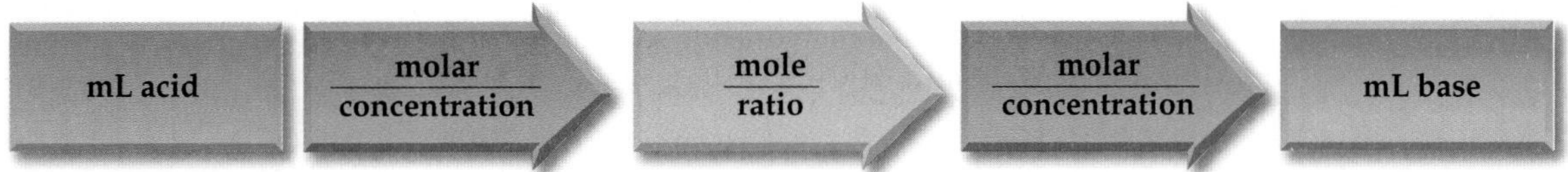

Practice Exercise

A 25.0 mL sample of hydrochloric acid is titrated with 0.125 *M* $Ba(OH)_2$. If 50.0 mL of barium hydroxide is required for the titration, what is the molar concentration of the acid? The balanced equation for the reaction is

$$2\ HCl(aq) + Ba(OH)_2(aq) \longrightarrow BaCl_2(aq) + 2\ H_2O(l)$$

Answer: 0.500 *M* HCl

Concept Exercise

Which of the following describes the titration of HCl(*aq*) with $Ba(OH)_2(aq)$: strong acid with strong base, strong acid with weak base, or weak acid with strong base?

Answer: See Appendix G.

OBJECTIVES

- To understand the concept of a standard acid or base solution.
- To perform calculations that involve a standard solution.

15.6 ACID–BASE STANDARDIZATION

A **standard solution** is a solution in which the concentration is known precisely, that is, three or four significant digits. For example, the concentration of NaOH used in the acetic acid titration was 0.223 *M*.

Chemists use standard solutions to repetitively analyze substances. For instance, a chemist may use a standard solution to analyze the neutralizing capacity of antacid tablets, or the tartness of soft drink flavors. Standard solutions are routinely used in the chemical industry to assure quality. Chemical samples are randomly selected and analyzed using a standard solution. This procedure is referred to as quality control (QC) or quality assurance (QA).

To standardize a solution of acid, chemists use a weighed quantity of a solid base. To standardize hydrochloric acid, for example, we can use solid sodium carbonate, Na_2CO_3. Let's find the molarity of hydrochloric acid if 25.50 mL of HCl solution is required to neutralize 0.375 g of Na_2CO_3. The balanced equation for the reaction of Na_2CO_3 dissolved in water is

$$2\ HCl(aq) + Na_2CO_3(aq) \longrightarrow 2\ NaCl(aq) + H_2O(l) + CO_2(g)$$

To calculate the molarity of the hydrochloric acid, we first find the number of moles of HCl. From the balanced equation we note that 2 mol of HCl react with 1 mol of Na_2CO_3 (105.99 g/mol).

$$0.375\ \cancel{g\ Na_2CO_3} \times \frac{1\ \cancel{mol\ Na_2CO_3}}{105.99\ \cancel{g\ Na_2CO_3}} \times \frac{2\ mol\ HCl}{1\ \cancel{mol\ Na_2CO_3}} = 0.00708\ mol\ HCl$$

To obtain the molarity of the acid, we divide the moles of HCl by the 25.50 mL of HCl required to neutralize the sodium carbonate.

$$\frac{0.00708\ mol\ HCl}{25.50\ \cancel{mL\ solution}} \times \frac{1000\ \cancel{mL\ solution}}{1\ L\ solution} = \frac{0.277\ mol\ HCl}{1\ L\ solution}$$

$$= 0.277\ M\ HCl$$

To standardize a solution of base, chemists can use a weighed quantity of a solid acid. For example, we can use crystals of oxalic acid, $H_2C_2O_4$, to standardize aqueous sodium hydroxide. We first dissolve a weighed sample of $H_2C_2O_4$ in water and then neutralize it with a measured volume of a base. The following example exercise illustrates how to determine the concentration of a standard NaOH solution:

Example Exercise 15.7 Acid–Base Standardization of a Base

What is the molarity of a sodium hydroxide solution if 32.15 mL of NaOH is required to neutralize 0.424 g of oxalic acid, $H_2C_2O_4$ (90.04 g/mol)?

$$H_2C_2O_4(aq) + 2\ NaOH(aq) \longrightarrow Na_2C_2O_4(aq) + 2\ H_2O(l)$$

Solution

To calculate the molarity of the NaOH, we find the number of moles of $H_2C_2O_4$. The mass of $H_2C_2O_4$ to be neutralized is 0.424 g, and the molar mass is 90.04 g/mol. From the balanced equation we see that 2 mol of NaOH reacts with 1 mol of $H_2C_2O_4$.

$$0.424\ \cancel{g\ H_2C_2O_4} \times \frac{1\ \cancel{mol\ H_2C_2O_4}}{90.04\ \cancel{g\ H_2C_2O_4}} \times \frac{2\ mol\ NaOH}{1\ \cancel{mol\ H_2C_2O_4}} = 0.00942\ mol\ NaOH$$

We can summarize the conversion as follows:

We can obtain the molar concentration of the base from the moles of NaOH divided by the volume of base required to neutralize the oxalic acid.

$$\frac{0.00942 \text{ mol NaOH}}{32.15 \text{ \sout{mL solution}}} \times \frac{1000 \text{ \sout{mL solution}}}{1 \text{ L solution}} = \frac{0.293 \text{ mol NaOH}}{1 \text{ L solution}} = 0.293\ M \text{ NaOH}$$

Practice Exercise

If an unknown sample of oxalic acid, $H_2C_2O_4$, is neutralized by 33.50 mL of 0.293 *M* NaOH, what is the mass of the sample? Refer to the previous balanced chemical equation for the neutralization of oxalic acid with sodium hydroxide.

Answer: 0.442 g $H_2C_2O_4$

Concept Exercise

What color is phenolphthalein indicator at the endpoint of the titration of oxalic acid with sodium hydroxide?

Answer: See Appendix G.

Molar Mass of a Solid Acid or Solid Base

One application of a standard solution is to identify a crystalline solid acid by determining its molar mass. For example, using a standard base solution, we could identify an unknown solid acid as citric acid by determining its molar mass. We begin by dissolving crystals of solid acid in water and then neutralizing the acid using a standard base solution. Given the mass of solid acid and the volume of standard base, we can calculate the molar mass of the acid.

The following example exercise illustrates the determination of molar mass for citric acid:

Example Exercise 15.8 Molar Mass of an Acid

Citric acid, abbreviated H_3Cit, gives a soft drink its tart taste. If 36.10 mL of 0.293 *M* NaOH neutralizes a 0.677 g sample of citric acid, what is the molar mass of citric acid? The balanced equation for the reaction is

$$H_3Cit(aq) + 3\ NaOH(aq) \longrightarrow Na_3Cit(aq) + 3\ H_2O(l)$$

Solution

The given volume of NaOH solution is 36.10 mL. Since the molar concentration is 0.293 *M*, we write the unit factor as 0.293 mol NaOH/1000 mL solution.

$$36.10 \text{ \sout{mL solution}} \times \frac{0.293 \text{ mol NaOH}}{1000 \text{ \sout{mL solution}}} = 0.0106 \text{ mol NaOH}$$

From the balanced chemical equation, we see that 3 mol of NaOH reacts with 1 mol of H_3Cit. Thus, the number of moles of citric acid is

$$0.0106 \text{ \sout{mol NaOH}} \times \frac{1 \text{ mol } H_3Cit}{3 \text{ \sout{mol NaOH}}} = 0.00353 \text{ mol } H_3Cit$$

We can summarize the conversion as follows:

Since the molar mass is defined as grams per mole, we have

$$\frac{0.677 \text{ g } H_3Cit}{0.00353 \text{ mol } H_3Cit} = 192 \text{ g/mol}$$

From titration data, we calculated the molar mass of citric acid as 192 g/mol. Given that the chemical formula of citric acid is $H_3C_6H_5O_7$, we can verify the molar mass by adding the atomic masses to obtain 192.14 g/mol.

Practice Exercise

Benzoic acid is a common food preservative, and its molecular formula is $HC_7H_5O_2$. If 22.55 mL of 0.293 *M* NaOH neutralizes a 0.806 g sample of benzoic acid, what is the molar mass of the acid? The balanced equation for the reaction is

$$HC_7H_5O_2(aq) + NaOH(aq) \longrightarrow NaC_7H_5O_2(aq) + H_2O(l)$$

Answer: 122 g/mol

Concept Exercise

What color is the phenolphthalein indicator at the endpoint of the titration of benzoic acid with sodium hydroxide?

Answer: See Appendix G.

15.7 IONIZATION OF WATER

OBJECTIVES

- To relate the ionization constant of water to the molar hydrogen ion and hydroxide ion concentrations.
- To calculate the molar hydroxide ion concentration given the molar hydrogen ion concentration.

We know that metals are good conductors of electricity. What would happen if we tested the electrical conductivity of an automobile key? We can predict that the key conducts electricity because it is made of metal. In Figure 15.4a, we see that the key completes the circuit in the conductivity apparatus and that the bulb lights up. The explanation is as follows. Electricity is defined as the flow of electrons. Recall from the electron sea model described in Section 13.5 that the valence electrons in metals are free to flow. An electric current can push electrons from one atom to another. Thus, electricity flows from one end of the metal key to the other as electrons move from atom to atom.

What happens when we test the electrical conductivity of pure water? In Figure 15.4b, we see that pure water is a very poor conductor of electricity. Evidence suggests that it is difficult to push electrons from one water molecule to another. How then do we explain the observation that some electricity flows in pure water? We explain this observation by suggesting that there are ions present in pure water.

Arrhenius found that salt solutions are good conductors of electricity. He correctly concluded that *ions* in the solution are responsible for electrical conductivity. Since water is a very poor conductor, we can conclude that only a few ions are present in pure water (Figure 15.5).

We know that water molecules in the liquid state are free to move about and collide with one another. Although it happens only rarely, on occasion two water

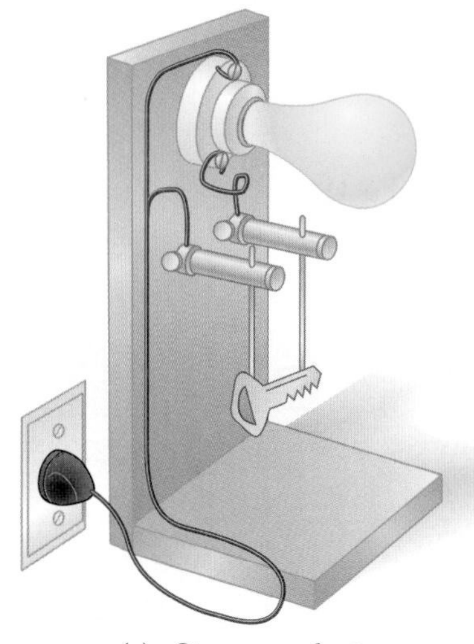

(a) Strong conductor

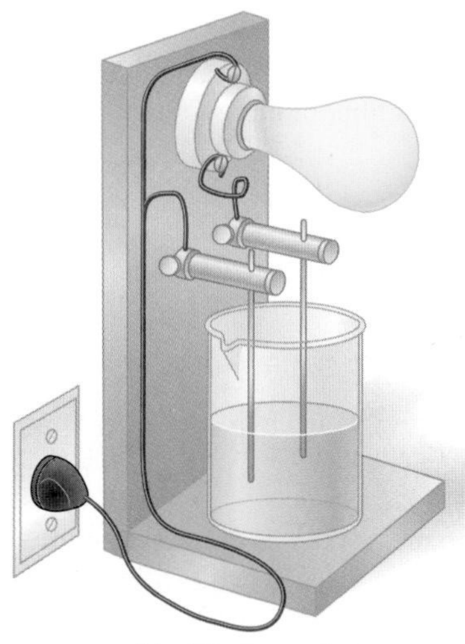

(b) Weak conductor

▶ **Figure 15.4 Conductivity Apparatus** (a) Metals are good conductors of electricity. A metal key completes the electrical circuit and the apparatus lights. (b) Pure water is a very poor conductor. It does, however, very weakly conduct an electric current.

molecules collide with sufficient energy for their bonds to break apart. This bond breaking produces a hydronium ion, H_3O^+, and a hydroxide ion, OH^-. Moreover, this occurrence is a dynamic process. While some molecules are breaking apart, other ions are combining to form water molecules. At any given moment, only 1 water molecule in 500 million molecules is ionized. Apparently, however, these few ions in water are sufficient to conduct a very weak electric current. The chemical equation for the collision reaction is

$$\underset{}{H_2O(l)} + H_2O(l) \longrightarrow \underset{\textit{hydronium ion}}{H_3O^+(aq)} + \underset{\textit{hydroxide ion}}{OH^-(aq)}$$

Alternatively, we can simplify the reaction by writing the ionization of water as

$$H_2O(l) \longrightarrow \underset{\textit{hydrogen ion}}{H^+(aq)} + \underset{\textit{hydroxide ion}}{OH^-(aq)}$$

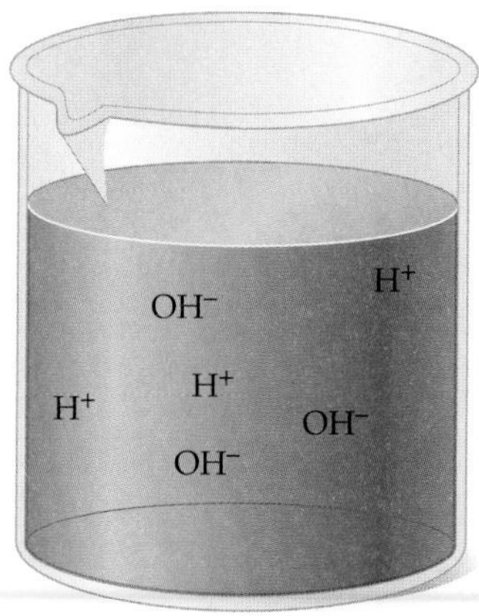

▲ **Figure 15.5 Ionization of Water** Pure water ionizes to give a few hydrogen ions and hydroxide ions. Water is a very weak conductor of electricity because only 1 molecule in about 500,000,000 forms ions.

The concentration of hydrogen ions in pure water is 1.0×10^{-7} mol/L at 25 °C. If we know the concentration of hydrogen ions, we also know the concentration of hydroxide ions. Since the ionization of water gives one H^+ and one OH^- for each molecule of water that ionizes, the two concentrations are equal. Therefore, the concentration of OH^- must also be 1.0×10^{-7} mol/L at 25 °C. Moreover, the molar concentration of H^+ multiplied by the molar concentration of OH^- equals a constant. This product is called the **ionization constant of water** (symbol K_w).

Let's calculate the value of K_w at 25 °C. For convenience, we use brackets to symbolize molar concentration. Hence, $[H^+]$ is the symbol for the molar concentration of hydrogen ions. We calculate a value for K_w as follows:

If $$[H^+] = 1.0 \times 10^{-7}$$

then $$[OH^-] = 1.0 \times 10^{-7}$$

and $$[H^+][OH^-] = (1.0 \times 10^{-7})(1.0 \times 10^{-7})$$

Since we add exponents when multiplying exponential numbers, we have

$$[H^+][OH^-] = 1.0 \times 10^{-14}$$

$$K_w = 1.0 \times 10^{-14} \text{ (at 25 °C)}$$

Notice that $[H^+]$ and $[OH^-]$ are inversely proportional. That is, if $[H^+]$ increases, $[OH^-]$ decreases, so that the constant K_w remains the same.

We should emphasize that water is neutral even though it contains small amounts of both H^+ and OH^-. Moreover, every aqueous solution has hydrogen ions and hydroxide ions. Even a hydrochloric acid solution contains H^+ and OH^-. Although an aqueous HCl solution has a high concentration of H^+, it has a few OH^- as well. In every aqueous solution, the product of $[H^+]$ and $[OH^-]$ is equal to the ionization constant of water.

$$[H^+][OH^-] = 1.0 \times 10^{-14}$$

If an aqueous HCl solution is 0.1 *M*, we can calculate its hydroxide ion concentration.

$$[0.1][OH^-] = 1.0 \times 10^{-14}$$

and $$[OH^-] = 1.0 \times 10^{-13}$$

Although $[OH^-]$ is only 0.000 000 000 0001 *M*, the hydrochloric acid solution contains a trace of hydroxide ion!

To summarize, in an acidic aqueous solution the $[H^+]$ is greater than $[OH^-]$. In a basic aqueous solution, $[H^+]$ is less than $[OH^-]$. In a neutral aqueous solution, $[H^+]$ is equal to $[OH^-]$. Example Exercise 15.9 illustrates the concept of hydrogen and hydroxide ions in aqueous solutions.

Example Exercise 15.9 The Ionization Constant, K_w

Given the following hydrogen ion concentrations in an aqueous solution at 25 °C, what is the molar concentration of hydroxide ion?

(a) $[H^+] = 1.4 \times 10^{-6}$ (b) $[H^+] = 5.2 \times 10^{-11}$

Solution

From the ionization constant of water, we know that $[H^+][OH^-] = 1.0 \times 10^{-14}$.

(a) If the $[H^+]$ is 1.4×10^{-6}, we can calculate $[OH^-]$ as follows:

$$K_w = [H^+][OH^-] = 1.0 \times 10^{-14}$$
$$1.4 \times 10^{-6}\,[OH^-] = 1.0 \times 10^{-14}$$

Dividing both sides of the equation by 1.4×10^{-6}, we have

$$[OH^-] = 7.1 \times 10^{-9}$$

(b) If the $[H^+]$ is 5.2×10^{-11}, we can calculate $[OH^-]$ as follows:

$$K_w = [H^+][OH^-] = 1.0 \times 10^{-14}$$
$$5.2 \times 10^{-11}\,[OH^-] = 1.0 \times 10^{-14}$$

Dividing both sides of the equation by 5.2×10^{-11}, we have

$$[OH^-] = 1.9 \times 10^{-4}$$

Practice Exercise

Given the following hydroxide ion concentrations in an aqueous solution, what is the molar concentration of hydrogen ion?

(a) $[OH^-] = 7.5 \times 10^{-4}$ (b) $[OH^-] = 2.1 \times 10^{-10}$

Answers: (a) $[H^+] = 1.3 \times 10^{-11}$; (b) $[H^+] = 4.8 \times 10^{-5}$

Concept Exercise

Explain why distilled water is neutral even though it contains hydrogen ions and hydroxide ions.

Answer: See Appendix G.

OBJECTIVES

- To explain the concept of pH for a solution.
- To relate the pH of a solution to molar hydrogen ion concentration.

15.8 THE pH CONCEPT

In Section 15.1, we introduced the term pH. Recall that pure water is neutral and has a pH of 7. We also know that an acid has a pH of less than 7. As the pH value *decreases*, the solution becomes more acidic. Thus, a pH of 3 is more acidic than a pH of 4. As the pH value *increases*, the solution becomes more basic. Thus, a pH of 11 is more basic than a pH of 10.

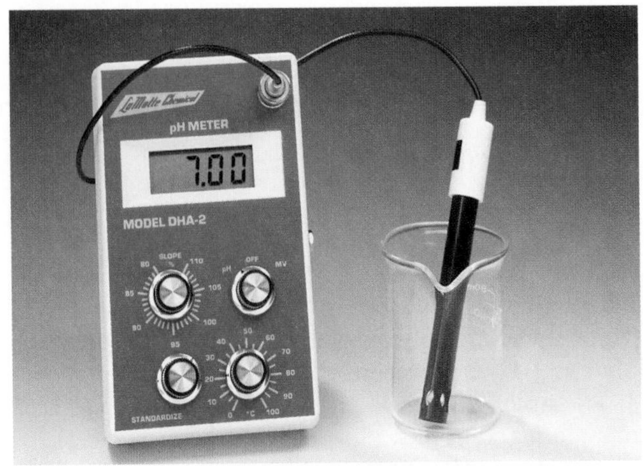

▶ **Digital pH Meter** The digital pH meter measures the hydrogen ion concentration in the beaker with the glass electrode (shown). Pure boiled water has a hydrogen ion concentration of 1.0×10^{-7} *M*, which corresponds to a pH of 7.00.

Converting from $[H^+]$ to pH

In an aqueous solution, the hydrogen ion concentration can vary from more than 1 *M* to less than 0.000 000 000 000 01 *M*. The pH scale is a convenient way to express this broad range of hydrogen ion concentrations. The pH scale expresses the molar hydrogen ion concentration, $[H^+]$, as a power of 10.

In other words, the pH scale uses powers of 10 to express the amount of acid in a solution. A solution with a pH of 3 is ten times more acidic than a solution with a pH of 4. Conversely, a solution with a pH of 11 is ten times less acidic than a solution with a pH of 10. In mathematical terms, the **pH** is the negative logarithm (log) of the molar hydrogen ion concentration.

$$pH = -\log [H^+]$$

For example, if the molar hydrogen ion concentration, $[H^+]$, is 0.1 *M*, then

$$pH = -\log 0.1$$

We do not need to understand logarithms to solve pH problems. All we need to know is that if we express the hydrogen ion concentration as a power of 10, the exponent equals the logarithm. Similarly, you may be familiar with the "Richter scale," which expresses the severity of an earthquake. The Richter scale is a logarithmic scale and an earthquake of magnitude 6 is ten times more severe than an earthquake of magnitude 5.

We can write 0.1 *M* hydrogen ion concentration as a power of 10 (that is, 10^{-1}) and find the pH as follows.

$$pH = -\log 10^{-1}$$
$$= -(-1) = 1$$

Thus, the pH of a solution with a hydrogen ion concentration of 0.1 *M* is 1.

The following example exercises provide practice in finding the pH of an aqueous solution given the molar hydrogen ion concentration:

Example Exercise 15.10 Converting $[H^+]$ to pH

Calculate the pH of the following solutions given the molar hydrogen ion concentration:
(a) vinegar, $[H^+] = 0.001\ M$ (b) antacid, $[H^+] = 0.000\ 000\ 001\ M$

Solution

(a) The pH of vinegar is equal to $-\log [H^+]$.

$$pH = -\log 0.001 = -\log 10^{-3}$$
$$= -(-3) \quad = 3$$

The sour taste of vinegar indicates that an acid is present. The acid in vinegar is acetic acid, $HC_2H_3O_2$, and the pH is approximately 3.

(b) The pH of the antacid is equal to $-\log [H^+]$.

$$pH = -\log 0.000\ 000\ 001 = -\log 10^{-9}$$
$$= \quad -(-9) \quad = 9$$

Practice Exercise
Calculate the pH for apple juice if the molar hydrogen ion concentration is 0.0001 *M*.

Answer: pH = 4

Concept Exercise
Classify apple juice as weakly acidic, neutral, or weakly basic given a pH of 4.

Answer: See Appendix G.

▲ **Vinegar** Acetic acid is found in vinegar, which gives it a sour taste.

Converting from pH to $[H^+]$

We can calculate the hydrogen ion concentration given the pH of a solution. Let's rearrange the definition of pH to express the molar hydrogen ion concentration. The $[H^+]$ is equal to 10 raised to a negative pH value. That is,

$$[H^+] = 10^{-pH}$$

If milk has a pH of 6, then the $[H^+]$ can be expressed as

$$[H^+] = 10^{-6} = 0.000\,001\ M$$

Example Exercise 15.11 further illustrates the conversion of a given pH to a molar hydrogen ion concentration.

Example Exercise 15.11 Converting pH to $[H^+]$

Calculate the molar hydrogen ion concentration given the pH of the following solutions:

(a) lemon juice, pH 2 (b) tomato juice, pH 5

Solution

(a) The hydrogen concentration of lemon juice is equal to 10 raised to the negative value of the pH. Since the pH is 2,

$$[H^+] = 10^{-pH}$$
$$= 10^{-2} = 0.01\ M$$

(b) If the pH of tomato juice is 5, $[H^+]$ is expressed as

$$[H^+] = 10^{-pH}$$
$$= 10^{-5} = 0.000\,01\ M$$

Notice that lemon juice and tomato juice differ by three pH units. Thus, lemon juice is 1000 times more acidic than tomato juice.

Practice Exercise

Calculate the hydrogen ion concentration for milk of magnesia if the pH is 10.

Answer: $[H^+] = 0.000\,000\,000\,1\ M$ ($1 \times 10^{-10}\ M$)

Concept Exercise

Classify milk of magnesia as weakly acidic, neutral, or weakly basic given a pH of 10.

Answer: See Appendix G.

▲ **Lemon Juice** Citric acid is found in lemon juice, which gives it a sour taste.

15.9 ADVANCED pH CALCULATIONS

OBJECTIVE

▸ To relate the pH of a solution to molar hydrogen ion concentration.

We defined pH as an exponential way of expressing the hydrogen ion concentration, but we considered only whole-number pH values. Many chemical reactions, however, must be carefully controlled to a fraction of a pH unit. Biochemical reactions, for example, are extremely sensitive to small pH changes. In fact, the pH of our blood must be maintained within the narrow range of 7.3–7.5. There is an elaborate buffering system that accomplishes this task, which involves carbonates, phosphates, and proteins.

Converting from $[H^+]$ to pH

Recall the mathematical definition of pH. That is, pH is the negative logarithm of the molar hydrogen ion concentration.

$$pH = -\log [H^+]$$

Suppose we wish to express the pH of a solution having a hydrogen ion concentration of 0.00015 *M*. The pH expression is

$$\text{pH} = -\log 0.00015$$
$$= -\log 1.5 \times 10^{-4}$$

We can easily obtain the logarithm for a number using a scientific calculator. If we enter 1.5×10^{-4} and touch the log key, the display shows –3.82. Since pH is the negative logarithm, we must change the sign to 3.82.

$$\text{pH} = -(-3.82)$$
$$= 3.82$$

Note that the number of significant digits in the given value equals the number of decimal places in the log expression. In this example, there are two significant digits and two decimal places in the log value. Example Exercise 15.12 provides practice in finding the pH of an aqueous solution given the molar hydrogen ion concentration.

Example Exercise 15.12 Converting [H⁺] to pH

Calculate the pH of the following solutions given the molar hydrogen ion concentration:
(a) stomach acid, $[\text{H}^+] = 0.020\ M$ (b) blood, $[\text{H}^+] = 0.000\ 000\ 048\ M$

Solution

(a) We can calculate the pH of stomach acid as follows:

$$\text{pH} = -\log 0.020$$
$$= -(-1.70)$$
$$= 1.70$$

The pH of stomach acid ranges from 1 to 3; hence, this sample is normal.

(b) We can calculate the pH of blood as follows:

$$\text{pH} = -\log 0.000\ 000\ 048$$
$$= -\log 4.8 \times 10^{-8}$$
$$= -(-7.32)$$
$$= 7.32$$

The pH of blood ranges from 7.3 to 7.5; hence, this sample is normal.

Practice Exercise

Calculate the pH of grape juice if the hydrogen ion concentration is 0.000 089 *M*.

Answer: pH = 4.05

Concept Exercise

Classify grape juice as weakly acidic, neutral, or weakly basic given a pH of 4.05.

Answer: See Appendix G.

▲ **Grape Juice** Tartaric acid is found in grape juice, which gives it a mild sour taste.

Converting from pH to [H⁺]

Previously, we expressed [H⁺] by raising 10 to the negative pH value.

$$[\text{H}^+] = 10^{-\text{pH}}$$

If orange juice has a pH of 2.75, then [H⁺] can be expressed as

$$[\text{H}^+] = 10^{-2.75}$$

CRITICAL THINKING Acid Rain

Q: What acids are found in "acid rain?"

Compare a 1935 photograph of a statue of George Washington with a current photograph. The striking difference is attributed to acid rain, which attacks limestone by reacting with the calcium carbonate in the statue.

The English chemist Robert Smith coined the term "acid rain" after studying the rainfall in London. He found that the air, heavily polluted from coal burning, produced rain that was abnormally acidic. The term "acid rain" has persisted and today usually refers to rain having a pH 5 or below.

The gases that most contribute to acid rain are oxides of sulfur and nitrogen. Sulfur dioxide and sulfur trioxide in the atmosphere are released mainly by industrial steel plants and electric power plants. These plants burn low-grade coal, which has a high sulfur content. Most of the oxides of nitrogen are emitted from automobiles. When oxides of sulfur and nitrogen are released into the air, they dissolve in atmospheric water vapor. Subsequently, the rain that falls is composed of sulfuric acid and nitric acid raindrops.

Normal rain and acid rain are both acidic, but their pH values are different. Normal rain has a pH of about 5.5. It is slightly acidic because carbon dioxide in the atmosphere dissolves in raindrops to form a dilute solution of carbonic acid. Severe acid rain can be a hundred times more acidic than normal rain, and an unhealthy pH of 2.8 has been recorded. In the northeastern United States, rainfall with a pH of 4 has been blamed on sulfur oxides released from the burning of coal.

Acid rain is a global problem, and countries all over the world are attempting to reduce it. One way is to reduce the emission of environmental pollutants. Canada has begun a program to reduce its sulfur dioxide emissions. The United States, Japan, and Germany are using high-grade coal to reduce their emissions. Ultimately, the problem of acid rain may be minimized by alternative energy sources. For instance, hydroelectric power and solar power are potential sources of energy that do not pollute the atmosphere.

◀ **Acid Rain** The photograph on the left shows a limestone statue of George Washington in 1935. The recent photograph on the right shows the effect of acid rain, which attacks the statue by dissolving away calcium carbonate in the marble.

A: Acid rain contains carbonic acid from burning fossil fuels, nitric acid from automobile emissions, and sulfuric acid from burning low-grade coal.

To find the number that corresponds to the fractional exponent, we obtain the inverse logarithm using a scientific calculator. If we enter –2.75 and touch the inverse log key, the display shows 0.0018, or 1.8×10^{-3}.

$$[H^+] = 10^{-2.75} = 0.0018\ M$$

The following example exercise provides further practice in converting a given pH to a molar hydrogen ion concentration:

Example Exercise 15.13 Converting pH to $[H^+]$

Calculate the molar hydrogen ion concentration given the pH of the following solutions:

(a) acid rain, pH 3.68 (b) seawater, pH 7.85

Solution

(a) If acid rain has a pH of 3.68, we can find the $[H^+]$ as follows:

$$[H^+] = 10^{-3.68}$$

Using the inverse log key on a scientific calculator, we have

$$[H^+] = 2.1 \times 10^{-4}$$
$$= 0.000\ 21\ M$$

(b) If seawater has a pH of 7.85, we can find the $[H^+]$ as follows:

$$[H^+] = 10^{-7.85}$$
$$= 1.4 \times 10^{-8}$$
$$= 0.000\ 000\ 014\ M$$

Practice Exercise

Calculate the hydrogen ion concentration in wine if the pH is 3.25.

Answer: $[H^+] = 0.000\ 56\ M\ (5.6 \times 10^{-4}\ M)$.

Concept Exercise

Classify wine as weakly acidic, neutral, or weakly basic given a pH of 3.25.

Answer: See Appendix G.

15.10 STRONG AND WEAK ELECTROLYTES

OBJECTIVE

▸ To determine the electrical conductivity of a strong electrolyte and a weak electrolyte in an aqueous solution.

Water is a weak conductor of electricity because it is very slightly ionized. In the laboratory, we can use a conductivity apparatus to determine whether a substance in an aqueous solution is a strong or a weak conductor of electricity. If an aqueous solution is a good conductor, it is called a **strong electrolyte**. If an aqueous solution is a poor conductor, it is called a **weak electrolyte**.

We can test the electrical conductivity of hydrochloric acid and acetic acid using the apparatus in Figure 15.6. From the observations, we can conclude that hydrochloric acid is highly ionized because it is a strong electrolyte. Acetic acid is only slightly ionized because it is a weak electrolyte. Similar experiments using the conductivity apparatus demonstrate the electrolyte behavior of aqueous solutions.

Degree of Ionization in an Aqueous Solution

We can measure the degree of ionization by testing the conductivity of aqueous solutions. Strong electrolytes are highly ionized. Strong electrolytes include strong acids, strong bases, and soluble ionic compounds. Soluble ionic compounds dissolve by dissociating into ions.

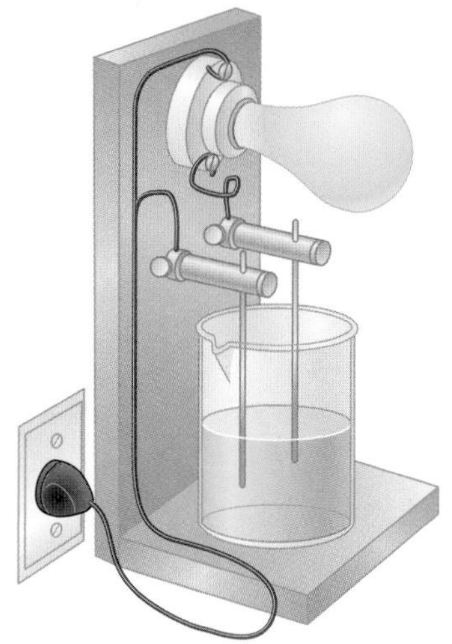

(a) Strong electrolyte

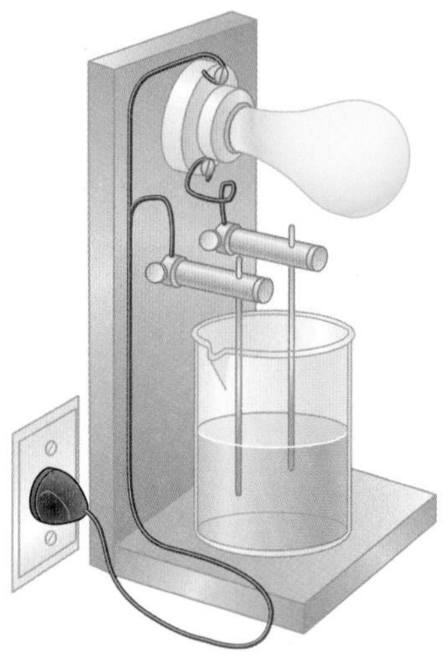

(b) Weak electrolyte

▲ **Figure 15.6 Conductivity of Aqueous Solutions** (a) An aqueous solution of hydrochloric acid is a strong electrolyte, and the bulb glows brightly. (b) An aqueous solution of acetic acid is a weak electrolyte, and the bulb glows dimly.

Weak electrolytes are slightly ionized. They include weak acids, weak bases, and slightly soluble ionic compounds. Although an ionic compound may be classified as insoluble, it is actually very slightly soluble. A sufficient amount of the insoluble ionic compound dissolves in a solution to act as a weak electrolyte. Table 15.5 lists examples of strong and weak electrolytes.

TABLE 15.5 STRONG ELECTROLYTES AND WEAK ELECTROLYTES

STRONG ELECTROLYTES	WEAK ELECTROLYTES
Strong Acids —	*Weak Acids —*
hydrochloric acid, $HCl(aq)$	hydrofluoric acid, $HF(aq)$
nitric acid, $HNO_3(aq)$	nitrous acid, $HNO_2(aq)$
sulfuric acid, $H_2SO_4(aq)$	sulfurous acid, $H_2SO_3(aq)$
perchloric acid, $HClO_4(aq)$	acetic acid, $HC_2H_3O_2(aq)$
	carbonic acid, $H_2CO_3(aq)$
	phosphoric acid, $H_3PO_4(aq)$
	most other acids
Strong Bases —	*Weak Bases —*
sodium hydroxide, $NaOH(aq)$	ammonium hydroxide, $NH_4OH(aq)$
potassium hydroxide, $KOH(aq)$	insoluble hydroxides such as
lithium hydroxide, $LiOH(aq)$	$Mg(OH)_2(s)$, $Al(OH)_3(s)$, $Fe(OH)_3(s)$
calcium hydroxide, $Ca(OH)_2(aq)$	most other bases
strontium hydroxide, $Sr(OH)_2(aq)$	
barium hydroxide, $Ba(OH)_2(aq)$	
Soluble Salts —	*Insoluble Salts* —*
sodium chloride, $NaCl(aq)$	silver chloride, $AgCl(s)$
potassium carbonate, $K_2CO_3(aq)$	calcium carbonate, $CaCO_3(s)$
zinc sulfate, $ZnSO_4(aq)$	barium sulfate, $BaSO_4(s)$

* Insoluble salts are very slightly soluble in water and dissociate into a sufficient number of ions to give a very weak electrolyte solution.

We can determine the degree of ionization in a solution from conductivity testing experiments. That is, we can distinguish between a highly ionized strong electrolyte and a slightly ionized weak electrolyte. Strong electrolytes ionize nearly completely in an aqueous solution. For example, sodium chloride and calcium chloride dissociate into ions as follows:

$$NaCl(aq) \longrightarrow Na^+(aq) + Cl^-(aq)$$

$$CaCl_2(aq) \longrightarrow Ca^{2+}(aq) + 2\,Cl^-(aq)$$

We can estimate the amount of ionization in an aqueous solution given the electrolyte strength. That is, the number of ions in solution is proportional to the conductivity. Strong electrolytes are highly ionized, and so we write their formulas in the ionized form. Weak electrolytes are only slightly ionized, and so we write their formulas in the nonionized form. The nonionized formula is sometimes referred to as the molecular formula. This is not quite accurate since ionic compounds exist as formula units, not molecules. The following example exercise illustrates how to write formulas for electrolytes in aqueous solution.

Example Exercise 15.14 Strong and Weak Electrolytes

Write the ionized or nonionized formula for each of the following aqueous solutions given the electrolyte strength:

(a) $HNO_3(aq)$, strong
(b) $NH_4OH(aq)$, weak
(c) $K_2CO_3(aq)$, strong

Solution

(a) A nitric acid solution is a strong electrolyte and therefore is highly ionized. It is written in the ionized form as $H^+(aq)$ and $NO_3^-(aq)$.
(b) An ammonium hydroxide solution is a weak electrolyte and therefore is slightly ionized. It is written in the nonionized form as $NH_4OH(aq)$.
(c) A potassium carbonate solution is a strong electrolyte and therefore is highly ionized. It is written as $2\ K^+(aq)$ and $CO_3^{2-}(aq)$.

Practice Exercise

Write the ionized or nonionized formula for each of the following aqueous solutions given the electrolyte strength:

(a) $HF(aq)$, weak
(b) $Ba(OH)_2(aq)$, strong
(c) $CaCO_3(s)$, weak

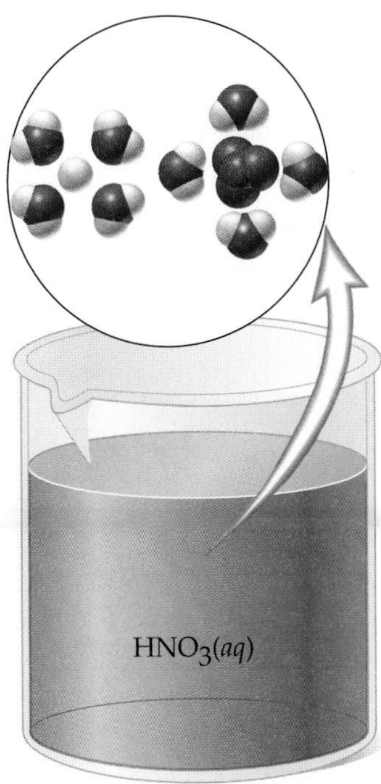

▲ **Nitric Acid, HNO_3** Nitric acid is a strong electrolyte, which is nearly 100% ionized in an aqueous solution.

▲ **Hydrofluoric Acid, HF** Hydrofluoric acid is a weak electrolyte, which is less than 1% ionized in an aqueous solution.

Answers: (a) $HF(aq)$; (b)$Ba^{2+}(aq)$ and $2\ OH^-(aq)$; (c) $CaCO_3(s)$

Concept Exercise

When the weak acid $HF(aq)$ neutralizes the weak base $NH_4OH(aq)$, will a conductivity apparatus glow brightly or dimly?

Answer: See Appendix G.

Note In the examples, we considered aqueous solutions to be either strong or weak electrolytes. Some substances, however, are nonelectrolytes. That is, they do not conduct electricity at all. Examples include organic liquids such as alcohols. Although deionized water may not give an observable conductivity test with the apparatus, deionized water is considered a very weak electrolyte.

15.11 NET IONIC EQUATIONS

OBJECTIVES

- To write the total ionic equation for a given chemical reaction.
- To simplify a total ionic equation to a net ionic equation.

In Section 15.10 we learned how to write ionized formulas for strong electrolytes and nonionized formulas for weak electrolytes. The concept of ionization allows us to portray ionic solutions more accurately. That is, we can now write chemical equations by showing strong electrolytes in the ionized form.

Let's consider the neutralization reaction of hydrochloric acid and sodium hydroxide. The nonionized equation for the reaction is

$$HCl(aq) + NaOH(aq) \longrightarrow NaCl(aq) + H_2O(l)$$

By writing the ionized formula for the strong acid, strong base, and soluble salt, we can describe the reaction more accurately. Each substance in the **total ionic equation** is written as it primarily exists in the solution. Table 15.5 shows highly ionized substances. Strong electrolytes include HCl, NaOH, and NaCl. Thus, the equation is

$$H^+(aq) + Cl^-(aq) + Na^+(aq) + OH^-(aq) \longrightarrow Na^+(aq) + Cl^-(aq) + H_2O(l)$$

Note that $Na^+(aq)$ and $Cl^-(aq)$ appear on both sides of the equation. These ions are called **spectator ions**. They are in the solution but do not participate in the reaction. We can simplify the total ionic equation by eliminating spectator ions. The resulting

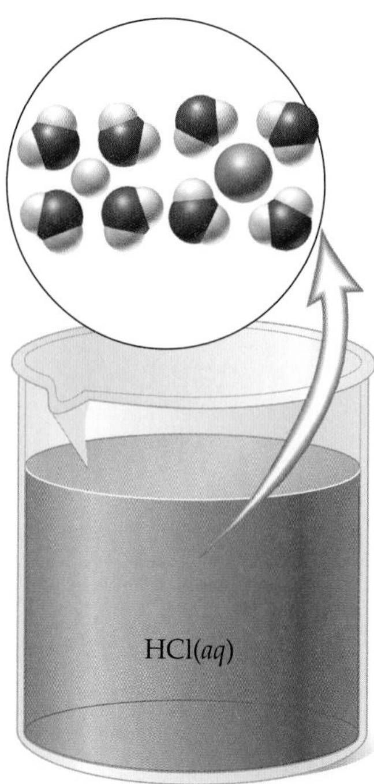

▲ **Hydrochloric Acid, HCl**
Hydrochloric acid is a strong acid, which is nearly 100% ionized in an aqueous solution.

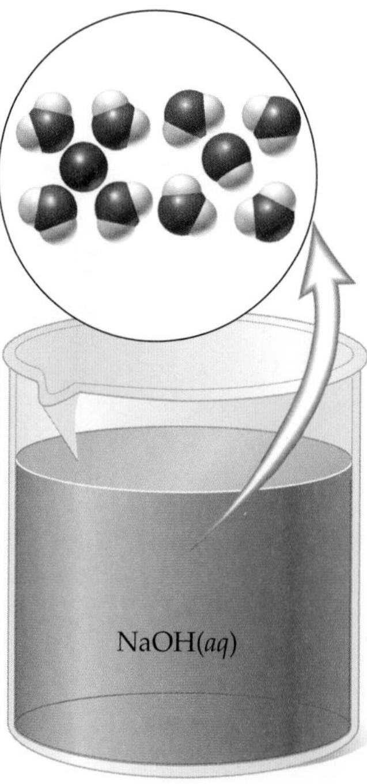

▲ **Sodium Hydroxide, NaOH**
Sodium hydroxide is a strong base, which is nearly 100% dissociated in an aqueous solution.

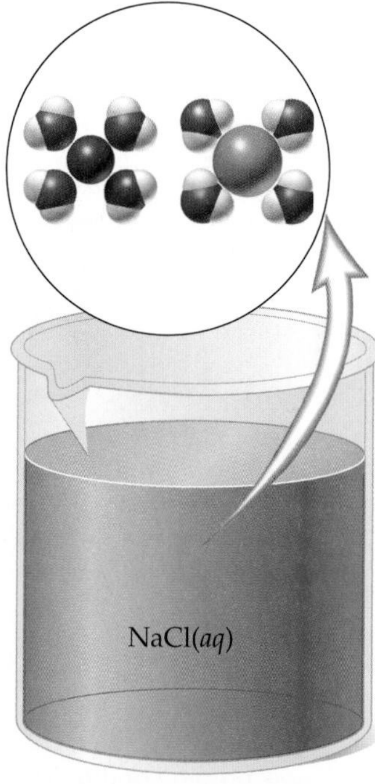

▲ **Sodium Chloride, NaCl**
Sodium chloride is a soluble salt, which is nearly 100% dissociated in an aqueous solution.

net ionic equation shows only the substances undergoing reaction. The net ionic equation for the preceding reaction is

$$H^+(aq) + OH^-(aq) \longrightarrow H_2O(l)$$

Note that we do not show the actual acid and base that were neutralized. This net ionic equation indicates that a strong acid and a strong base reacted to give water. In fact, the net ionic equation is identical for all strong acid and strong base reactions that yield a soluble salt.

Let's state a general procedure to be used when we write net ionic equations. Keep in mind that the net ionic equation gives us a good picture of substances undergoing reaction. That is, the net ionic equation helps us focus on only those substances that are undergoing reaction.

General Guidelines for Writing Net Ionic Equations

1. Complete and balance the nonionized chemical equation.
2. Convert the nonionized equation to a total ionic equation. Write strong electrolytes in the ionized form and weak electrolytes in the nonionized form. Write water and dissolved gases in the nonionized form. Refer to Table 8.2 to determine if a compound is soluble or insoluble.
3. Cancel spectator ions to obtain the net ionic equation.
   (a) If canceling spectator ions eliminates all species, there is no reaction.
   (b) If the coefficients can be simplified, do so to have the simplest whole number relationship.
4. Check (√) each ion or atom on both sides of the equation. The total charge (positive or negative) on the reactants side of the equation must equal the total charge on the products side of the equation.

The following example exercises illustrate the procedure for writing balanced net ionic equations:

Example Exercise 15.15 Net Ionic Equations for Neutralization Reactions

Write a net ionic equation for the reaction between nitric acid and aqueous potassium hydrogen carbonate.

$$HNO_3(aq) + KHCO_3(aq) \longrightarrow KNO_3(aq) + H_2O(l) + CO_2(g)$$

Solution

Step 1: We verify that the chemical equation is balanced. In this case, the coefficients are all 1 and the equation is balanced.

Step 2: We determine which species are highly ionized. From Table 15.5, we find that HNO_3 is a strong electrolyte. From Table 8.2, we find that $KHCO_3$ and KNO_3 are soluble strong electrolytes; H_2O and CO_2 are weak electrolytes. The total ionic equation is

$$H^+(aq) + NO_3^-(aq) + K^+(aq) + HCO_3^-(aq) \longrightarrow K^+(aq) + NO_3^-(aq) + H_2O(l) + CO_2(g)$$

Step 3: We obtain the net ionic equation by eliminating the spectator ions. In this example, K^+ and NO_3^- are in the aqueous solution, but do not participate in the reaction. The net ionic equation is

$$H^+(aq) + HCO_3^-(aq) \longrightarrow H_2O(l) + CO_2(g)$$

Step 4: We check to verify that the net ionic equation is balanced. The equation is balanced because (a) the number of atoms of each element is the same on each side and (b) the total charge on both sides is identical. In this case the net charge is zero.

Practice Exercise

Write a net ionic equation for the reaction between aqueous solutions of nitric acid and barium hydroxide.

$$2\ HNO_3(aq) + Ba(OH)_2(aq) \longrightarrow Ba(NO_3)_2(aq) + 2\ H_2O(l)$$

Answer: In this example, nitrate and barium are spectator ions and cancel.

$$H^+(aq) + OH^-(aq) \longrightarrow H_2O(l)$$

Concept Exercise

When the strong acid $HNO_3(aq)$ neutralizes the strong base $Ba(OH)_2(aq)$, will a conductivity apparatus glow brightly or dimly?

Answer: See Appendix G.

Example Exercise 15.16 Net Ionic Equations for Precipitation Reactions

Write a net ionic equation for the reaction between aqueous solutions of silver nitrate and aluminum chloride.

$$AgNO_3(aq) + AlCl_3(aq) \longrightarrow AgCl(s) + Al(NO_3)_3(aq)$$

Solution

Step 1: We balance the chemical equation as follows:

$$3\ AgNO_3(aq) + AlCl_3(aq) \longrightarrow 3\ AgCl(s) + Al(NO_3)_3(aq)$$

Step 2: We determine which of the species are strong electrolytes. From Table 8.2, we find that all the compounds are soluble except AgCl. Thus, AgCl is weakly ionized, and the total ionic equation is written as

$$3\ Ag^+(aq) + 3\ NO_3^-(aq) + Al^{3+}(aq) + 3\ Cl^-(aq) \longrightarrow 3\ AgCl(s) + Al^{3+}(aq) + 3\ NO_3^-(aq)$$

Step 3: We cancel spectator ions. In this example, Al^{3+} and NO_3^- do not participate in the reaction.

$$3\ Ag^+(aq) + 3\ Cl^-(aq) \longrightarrow 3\ AgCl(s)$$

Step 4: We verify that the equation is balanced. Note that each species has a coefficient of 3 in this equation. Thus, after simplifying, the net ionic equation is written as

$$Ag^+(aq) + Cl^-(aq) \longrightarrow AgCl(s)$$

Practice Exercise

Write a net ionic equation for the reaction between aqueous solutions of lithium sulfate and magnesium nitrate.

$$Li_2SO_4(aq) + Mg(NO_3)_2(aq) \longrightarrow MgSO_4(aq) + 2\ LiNO_3(aq)$$

Answer: $Li_2SO_4(aq) + Mg(NO_3)_2(aq) \longrightarrow$ no reaction

Concept Exercise

When the two strong electrolytes, $AgNO_3(aq)$ and $AlCl_3(aq)$, react to give insoluble $AgCl(s)$, will a conductivity apparatus glow brightly or dimly?

Answer: See Appendix G.

▲ **Silver Nitrate, $AgNO_3$** Silver nitrate is a soluble salt, which is nearly 100% dissociated in an aqueous solution.

▲ **Aluminum Chloride, $AlCl_3$** Aluminum chloride is a soluble salt, which is nearly 100% dissociated in an aqueous solution.

▲ **Silver Chloride, AgCl** Silver chloride is an insoluble salt, which is less than 1% dissociated in an aqueous solution.

Summary

Section 15.1 The properties of an acid include having a sour taste, turning litmus paper red, having a pH less than 7, and the ability to neutralize a base. The properties of a base include having a bitter taste, being slippery to the touch, turning litmus paper blue, and having a pH greater than 7. A solution that resists changes in pH when an acid or a base is added is referred to as a **buffer**.

Section 5.2 An **Arrhenius acid** is a substance that undergoes **ionization** in water to give hydrogen ions. Hydrochloric acid is a strong acid because it ionizes extensively. Acetic acid is a weak acid because it does not ionize extensively. A hydrogen ion attaches to a water molecule in solution to form a **hydronium ion**. An **Arrhenius base** is a substance that undergoes **dissociation** in water to give hydroxide ions. A **salt** is obtained from the reaction of an acid and a base. Barium nitrate, $Ba(NO_3)_2$, is obtained from the neutralization of barium hydroxide and nitric acid.

Section 15.3 A **Brønsted–Lowry acid** is a **proton donor** and a **Brønsted–Lowry base** is a **proton acceptor**. A substance capable of either donating or accepting a proton is said to be **amphiprotic**.

Section 15.4 We can use litmus paper to indicate if a solution is acidic or basic. In the laboratory, there are solutions that can act as an **acid–base indicator**. Three of the most common indicators are methyl red, bromthymol blue, and phenolphthalein. Methyl red changes color at a pH of 5, bromthymol blue changes at a pH of 7, and phenolphthalein at a pH of 9.

Section 15.5 A **titration** is a laboratory procedure for analyzing the amount of acid or base present in a solution. A measured volume of acid or base solution is delivered with a buret into a flask containing a known amount of sample. An indicator is used to signal the neutralization point in the titration. When the **endpoint** is reached, the indicator changes color and the titration is stopped. The titration data allows the chemist to calculate the amount of acid or base.

Section 15.6 The concentration of a **standard solution** is known accurately to at least three significant digits. Standard solutions are used to analyze the amount of acid or base in a given sample. For example, a chemist can determine the concentration of acetic acid in vinegar or the amount of baking soda in an antacid tablet.

Section 15.7 On the basis of conductivity experiments, pure water is found to be a very weak electrolyte. Water ionizes to give only a few hydrogen and hydroxide ions. $[H^+]$ and $[OH^-]$ are equal, and both have a value of 1×10^{-7}. The **ionization constant of water,** K_w, is 1×10^{-14} at 25 °C. In every aqueous solution, there is always a small number of both H^+ and OH^-.

Section 15.8 The **pH** of a solution expresses $[H^+]$ on an exponential scale. That is, a solution of pH 1 is 10 times more acidic than a solution of pH 2. To calculate the pH of a solution, we express $[H^+]$ as a power of 10 and change the sign of the exponent. For example, if $[H^+]$ is 10^{-2} *M*, the pH is 2. Conversely, to find [H^+] given the pH, we raise 10 to the negative pH. For example, if the pH is 3, [H^+] is 10^{-3} *M*, that is, 0.001 *M*.

Section 15.9 To calculate the pH of solutions whose hydrogen ion concentrations are not whole-number powers of 10, we find the negative logarithm values using a scientific calculator. That is, the pH is equal to −log [H^+]. Conversely, to calculate the hydrogen ion concentration that corresponds to a given pH, we find the inverse logarithm of −pH using a calculator.

Section 15.10 A strong acid, a strong base, and a soluble ionic compound are each examples of a **strong electrolyte**. Therefore, these solutions are highly ionized. Hydrochloric acid is a strong acid and is better represented in an aqueous solution as $H^+(aq)$ and $Cl^-(aq)$. Sodium chloride is soluble in water and is best represented as $Na^+(aq)$ and $Cl^-(aq)$. A weak acid, a weak base, and a slightly soluble ionic compound are each examples of a **weak electrolyte**. Therefore, these solutions are slightly ionized. Acetic acid is a weak acid and in an aqueous solution should be written as $HC_2H_3O_2(aq)$. The slightly soluble ionic compound silver chloride is written as $AgCl(s)$.

Section 15.11 A **net ionic equation** shows a solution reaction more accurately. The first step is to balance the chemical equation. Second, write a **total ionic equation** for the reaction. Strong electrolytes are written in the ionic form and weak electrolytes in the nonionized form. Third, cancel **spectator ions** that are identical on both sides of the equation. Finally, verify that the net ionic equation is balanced and simplify the coefficients when possible.

PROBLEM–SOLVING ORGANIZER

TOPIC	PROCEDURE	EXAMPLE
Acid–Base Titrations Sec. 15.5	1. Write down the unit asked for in the answer. 2. Write down the related given value. 3. Apply unit factors to convert the given unit to the unit in the answer.	If 10.00 mL of 0.100 *M* sulfuric acid neutralizes 15.50 mL of KOH, what is the molarity of the base? Given: $H_2SO_4(aq) + 2\ KOH(aq) \longrightarrow K_2SO_4(aq) + 2\ H_2O(l)$ $10.00\ \text{mL solution} \times \frac{0.100\ \text{mol}\ H_2SO_4}{1000\ \text{mL solution}} \times \frac{2\ \text{mol KOH}}{1\ \text{mol}\ H_2SO_4}$ $= 0.00200\ \text{mol KOH}$ $\frac{0.00200\ \text{mol KOH}}{15.50\ \text{mL solution}} \times \frac{1000\ \text{mL solution}}{1\ \text{L solution}} = 0.129\ \text{M KOH}$
Acid–Base Standardization Sec. 15.6	1. Write down the unit asked for in the answer. 2. Write down the related given value. 3. Apply unit factors to convert the given unit to the unit in the answer.	If 15.50 mL of KOH neutralizes 0.995 g of KHP (204.23 g/mol), $KHC_8H_4O_4$, what is the molarity of the base? Given: $KHP(aq) + KOH(aq) \longrightarrow K_2P(aq) + H_2O(l)$ $0.995\ \text{g KHP} \times \frac{1\ \text{mol KHP}}{204.23\ \text{g KHP}} \times \frac{1\ \text{mol KOH}}{1\ \text{mol KHP}}$ $= 0.00487\ \text{mol KOH}$ $\frac{0.0487\ \text{mol KOH}}{15.50\ \text{mL solution}} \times \frac{1000\ \text{mL solution}}{1\ \text{L solution}} = 0.314\ \text{M KOH}$
Ionization of Water Sec. 15.7	Solve for [H^+] or [OH^-] in an aqueous solution using the ionization of water equation. $K_w = [H^+][OH^-]$	If [H^+] is 0.0015 M in an aqueous solution, what is the hydroxide ion concentration? $[H^+][OH^-] = 1.0 \times 10^{-14}$ $[0.0015][OH^-] = 1.0 \times 10^{-14}$ $[OH^-] = 6.7 \times 10^{-12}$
The pH Concept Sec. 15.8	Convert from [H^+] to pH using the following equation: $pH = -\log[H^+]$	If [H^+] in an acid is 0.001 M, what is the pH? $pH = -\log[10^{-3}] = -(-3)$ $= 3$
Advanced pH Calculations Sec. 15.9	Convert from pH to [H^+] using the following equation: $[H^+] = 10^{-pH}$	If the pH of an acid is 3.20, M what is [H^+]? $[H^+] = 10^{-3.20} = 0.0063$

Key Terms *Answers to Key Terms are in Appendix H.*

Select the key term that corresponds to each of the following definitions.

_____ **1.** a solution that resists changes in pH when an acid or a base is added
_____ **2.** a substance that releases hydrogen ions when dissolved in water
_____ **3.** a substance that releases hydroxide ions when dissolved in water
_____ **4.** the process of a polar compound dissolving in water and forming positive and negative ions
_____ **5.** the process of an ionic compound dissolving in water and separating into positive and negative ions
_____ **6.** the ion that best represents the hydrogen ion in an aqueous solution
_____ **7.** a product obtained from a neutralization reaction in addition to water
_____ **8.** a substance that donates a proton in an acid–base reaction
_____ **9.** a substance that accepts a proton in an acid–base reaction
_____ **10.** a synonym for hydrogen ion donor
_____ **11.** a synonym for hydrogen ion acceptor
_____ **12.** a substance capable of either accepting or donating a proton in an acid–base reaction
_____ **13.** a chemical substance that changes color according to the pH of the solution
_____ **14.** a procedure for delivering a measured volume of solution through a buret
_____ **15.** the stage in a titration when the indicator changes color
_____ **16.** a solution whose concentration has been established precisely
_____ **17.** a constant that equals the product of the molar hydrogen ion concentration and the molar hydroxide ion concentration
_____ **18.** the molar hydrogen ion concentration expressed on an exponential scale
_____ **19.** an aqueous solution that is a good conductor of electricity
_____ **20.** an aqueous solution that is a poor conductor of electricity
_____ **21.** a chemical equation that represents highly ionized substances in the ionic form and slightly ionized substances in the nonionized form
_____ **22.** ions that are in an aqueous solution, but do not participate in a reaction and do not appear in the net ionic equation
_____ **23.** a chemical equation that represents an ionic reaction after spectator ions have been canceled

(a) acid–base indicator *(Sec. 15.4)*
(b) amphiprotic *(Sec. 15.3)*
(c) Arrhenius acid *(Sec. 15.2)*
(d) Arrhenius base *(Sec. 15.2)*
(e) Brønsted–Lowry acid *(Sec. 15.3)*
(f) Brønsted–Lowry base *(Sec. 15.3)*
(g) buffer *(Sec. 15.1)*
(h) dissociation *(Sec. 15.2)*
(i) endpoint *(Sec. 15.5)*
(j) hydronium ion (H_3O^+) *(Sec. 15.2)*
(k) ionization *(Sec. 15.2)*
(l) ionization constant of water (K_w) *(Sec. 15.7)*
(m) net ionic equation *(Sec. 15.11)*
(n) pH *(Sec. 15.8)*
(o) proton acceptor *(Sec. 15.3)*
(p) proton donor *(Sec. 15.3)*
(q) salt *(Sec. 15.2)*
(r) spectator ions *(Sec. 15.11)*
(s) standard solution *(Sec. 15.6)*
(t) strong electrolyte *(Sec. 15.10)*
(u) titration *(Sec. 15.5)*
(v) total ionic equation *(Sec. 15.11)*
(w) weak electrolyte *(Sec. 15.10)*

Exercises *Answers to odd-numbered Exercises are in Appendix I.*

Properties of Acids and Bases (Sec. 15.1)

1. Which of the following is a general property of acidic solutions?
   (a) taste bitter
   (b) turn litmus blue
   (c) pH greater than 7
   (d) neutralize bases

2. Which of the following is a general property of basic solutions?
   (a) taste sour
   (b) turn litmus red
   (c) pH less than 7
   (d) feel slippery

3. Classify the following aqueous solutions as strongly acidic, weakly acidic, neutral, weakly basic, or strongly basic:
   (a) lime juice, pH 2.7
   (b) gastric juice, pH 0.9
   (c) distilled water, pH 7.0
   (d) drain cleaner, pH 13.5

4. Classify the following aqueous solutions as strongly acidic, weakly acidic, neutral, weakly basic, or strongly basic:
   (a) vinegar, pH 3.2
   (b) battery acid, pH 0.5
   (c) shampoo, pH 7.0
   (d) ammonia, pH 11.1

▲ **Common Acids** An acid–base indicator is added to beakers of lime juice, vinegar, and club soda. The color of the indicator shows varying acidity in each beaker.

Arrhenius Acids and Bases (Sec. 15.2)

5. Classify each of the following Arrhenius acids as strong or weak given the degree of ionization:
   (a) chloric acid, $HClO_3(aq)$, ~100%
   (b) hypoiodous acid, $HIO(aq)$, ~1%
   (c) hydrobromic acid, $HBr(aq)$, ~100%
   (d) benzoic acid, $HC_7H_5O_2(aq)$, ~1%
6. Classify each of the following Arrhenius bases as strong or weak given the degree of ionization:
   (a) zinc hydroxide, $Zn(OH)_2(s)$, ~1%
   (b) lithium hydroxide, $LiOH(aq)$, ~100%
   (c) iron(II) hydroxide, $Fe(OH)_2(s)$, ~1%
   (d) barium hydroxide, $Ba(OH)_2(aq)$, ~100%
7. Classify each of the following as an Arrhenius acid, an Arrhenius base, or a salt:
   (a) $HNO_3(aq)$
   (b) $Mg(NO_3)_2(aq)$
   (c) $Ca(OH)_2(aq)$
   (d) $H_2SO_3(aq)$
8. Classify each of the following as an Arrhenius acid, an Arrhenius base, or a salt:
   (a) $HClO(aq)$
   (b) $KOH(aq)$
   (c) $K_2SO_4(aq)$
   (d) $Sr(OH)_2(aq)$
9. Identify the Arrhenius acid and Arrhenius base in each of the following neutralization reactions:
   (a) $HI(aq) + NaOH(aq) \longrightarrow NaI(aq) + H_2O(l)$
   (b) $HC_2H_3O_2(aq) + LiOH(aq) \longrightarrow LiC_2H_3O_2(aq) + H_2O(l)$
10. Identify the Arrhenius acid and Arrhenius base in each of the following neutralization reactions:
   (a) $2\,HClO_3(aq) + Ba(OH)_2(aq) \longrightarrow Ba(ClO_3)_2(aq) + 2\,H_2O(l)$
   (b) $H_2SO_4(aq) + 2\,KOH(aq) \longrightarrow K_2SO_4(aq) + 2\,H_2O(l)$
11. Determine the acid and base that were neutralized to produce each of the following salts:
   (a) sodium fluoride, $NaF(aq)$
   (b) magnesium iodide, $MgI_2(aq)$
   (c) calcium nitrate, $Ca(NO_3)_2(aq)$
   (d) lithium carbonate, $Li_2CO_3(aq)$
12. Determine the acid and base that were neutralized to produce each of the following salts:
   (a) potassium bromide, $KBr(aq)$
   (b) barium chloride, $BaCl_2(aq)$
   (c) cobalt(II) sulfate, $CoSO_4(aq)$
   (d) sodium phosphate, $Na_3PO_4(aq)$
13. Complete and balance the following neutralization reactions:
   (a) $HNO_3(aq) + Ca(OH)_2(aq) \longrightarrow$
   (b) $H_2CO_3(aq) + Ba(OH)_2(aq) \longrightarrow$
14. Complete and balance the following neutralization reactions:
   (a) $HC_2H_3O_2(aq) + Ca(OH)_2(aq) \longrightarrow$
   (b) $H_2SO_4(aq) + NH_4OH(aq) \longrightarrow$

Brønsted–Lowry Acids and Bases (Sec. 15.3)

15. Identify the Brønsted–Lowry acid and base in each of the following neutralization reactions:
   (a) $HC_2H_3O_2(aq) + LiOH(aq) \longrightarrow LiC_2H_3O_2(aq) + H_2O(l)$
   (b) $2\,NH_3(aq) + H_2SO_4(aq) \longrightarrow (NH_4)_2SO_4(aq)$
16. Identify the Brønsted–Lowry acid and base in each of the following neutralization reactions:
   (a) $2\,HClO_4(aq) + K_2CO_3(aq) \longrightarrow 2\,KClO_4(aq) + H_2O(l) + CO_2(g)$
   (b) $NaCN(aq) + HBr(aq) \longrightarrow NaBr(aq) + HCN(aq)$
17. Identify the Brønsted–Lowry acid and base in each of the following neutralization reactions:
   (a) $HI(aq) + H_2O(l) \longrightarrow H_3O^+(aq) + I^-(aq)$
   (b) $HC_2H_3O_2(aq) + HS^-(aq) \longrightarrow H_2S(aq) + C_2H_3O_2^-(aq)$
18. Identify the Brønsted–Lowry acid and base in each of the following neutralization reactions:
   (a) $HCO_3^-(aq) + OH^-(aq) \longrightarrow CO_3^{2-}(aq) + H_2O(l)$
   (b) $NO_2^-(aq) + HClO_4(aq) \longrightarrow HNO_2(aq) + ClO_4^-(aq)$
19. Complete and balance the following Brønsted–Lowry neutralization reactions:
   (a) $HF(aq) + NaHS(aq) \longrightarrow$
   (b) $HNO_2(aq) + NaC_2H_3O_2(aq) \longrightarrow$
20. Complete and balance the following Brønsted–Lowry neutralization reactions:
   (a) $H_3O^+(aq) + SO_4^{2-}(aq) \longrightarrow$
   (b) $H_2PO_4^-(aq) + NH_3(aq) \longrightarrow$

Acid–Base Indicators (Sec. 15.4)

21. Given the following pH values for solutions containing a drop of methyl red indicator, state the color of each solution:
   (a) pH 4
   (b) pH 8
22. What is the color of methyl red indicator in a solution of pH 5?
23. Given the following pH values for solutions containing a drop of bromthymol blue indicator, state the color of each solution:
   (a) pH 5 **(b)** pH 9
24. What is the color of bromthymol blue indicator in a solution of pH 7?
25. Given the following pH values for solutions containing a drop of phenolphthalein indicator, state the color of each solution:
   (a) pH 6 **(b)** pH 10

26. What is the color of phenolphthalein indicator in deionized water?

◀ **Natural Indicator** A rose contains a natural acid–base indicator. After the indicator is extracted from the dark-red rose (left beaker), the rose petals are a pale pink (right beaker).

Acid–Base Titrations (Sec. 15.5)

27. If the titration of a 10.0 mL sample of sulfuric acid requires 28.15 mL of 0.100 *M* sodium hydroxide, what is the molarity of the acid?

$$H_2SO_4(aq) + 2\,NaOH(aq) \longrightarrow Na_2SO_4(aq) + 2\,H_2O(l)$$

28. If the titration of a 25.0 mL sample of acetic acid requires 34.45 mL of 0.100 *M* calcium hydroxide, what is the molarity of the acid?

$$2\,HC_2H_3O_2(aq) + Ca(OH)_2(aq) \longrightarrow Ca(C_2H_3O_2)_2(aq) + 2H_2O(l)$$

29. If 34.45 mL of 0.210 *M* NaOH neutralizes 50.0 mL of phosphoric acid, what is the molarity of the acid?

$$H_3PO_4(aq) + 3\,NaOH(aq) \longrightarrow Na_3PO_4(aq) + 3\,H_2O(l)$$

30. If 29.50 mL of 0.175 *M* nitric acid neutralizes 50.0 mL of ammonium hydroxide, what is the molarity of the base?

$$HNO_3(aq) + NH_4OH(aq) \longrightarrow NH_4NO_3(aq) + H_2O(l)$$

31. How many milliliters of 0.100 *M* potassium hydroxide are required to completely neutralize 10.0 mL of 0.165 *M* H_2SO_3?

$$H_2SO_3(aq) + 2\,KOH(aq) \longrightarrow K_2SO_3(aq) + 2\,H_2O(l)$$

32. How many milliliters of 0.100 *M* barium hydroxide are required to completely neutralize 10.0 mL of 0.225 *M* HNO_2?

$$2\,HNO_2(aq) + Ba(OH)_2(aq) \longrightarrow Ba(NO_2)_2(aq) + 2\,H_2O(l)$$

33. Given the molarity and density for each of the following acidic solutions, calculate the mass/mass percent concentration:
(a) 6.00 *M* HCl (d = 1.10 g/mL)
(b) 1.00 *M* $HC_2H_3O_2$ (d = 1.01 g/mL)
(c) 0.500 *M* HNO_3 (d = 1.01 g/mL)
(d) 3.00 *M* H_2SO_4 (d = 1.18 g/mL)

34. Given the molarity and density for each of the following basic solutions, calculate the mass/mass percent concentration:
(a) 3.00 *M* NaOH (d = 1.12 g/mL)
(b) 0.500 *M* KOH (d = 1.02 g/mL)
(c) 6.00 *M* NH_3 (d = 0.954 g/mL)
(d) 1.00 *M* Na_2CO_3 (d = 1.10 g/mL)

Acid–Base Standardization (Sec. 15.6)

35. What is the molarity of nitric acid if 33.25 mL of HNO_3 is required to neutralize 0.500 g of sodium carbonate?

$$2\,HNO_3(aq) + Na_2CO_3(aq) \longrightarrow NaNO_3(aq) + H_2O(l) + CO_2(g)$$

36. What is the molarity of sulfuric acid if 27.75 mL of H_2SO_4 is required to neutralize 1.000 g of sodium hydrogen carbonate?

$$H_2SO_4(aq) + 2NaHCO_3(aq) \longrightarrow Na_2SO_4(aq) + 2H_2O(l) + 2CO_2(g)$$

37. What is the molarity of hydrochloric acid if 20.95 mL of HCl is required to neutralize 1.550 g of sodium oxalate, $Na_2C_2O_4$?

$$2\,HCl(aq) + Na_2C_2O_4(aq) \longrightarrow H_2C_2O_4(aq) + 2\,NaCl(aq)$$

38. What is the molarity of sodium hydroxide if 28.85 mL of NaOH is required to neutralize 0.506 g of oxalic acid, $H_2C_2O_4$?

$$H_2C_2O_4(aq) + NaOH(aq) \longrightarrow NaHC_2O_4(aq) + H_2O(l)$$

39. What volume of 0.479 *M* lithium hydroxide is required to neutralize 0.627 g of oxalic acid, $H_2C_2O_4$?

$$H_2C_2O_4(aq) + 2\,LiOH(aq) \longrightarrow Li_2C_2O_4(aq) + 2\,H_2O(l)$$

40. What volume of 0.167 *M* barium hydroxide is required to neutralize 1.655 g of potassium hydrogen phthalate, $KHC_8H_4O_4$ (204.23 g/mol)?

$$2\,KHC_8H_4O_4(aq) + Ba(OH)_2(aq) \longrightarrow BaK_2(C_8H_4O_4)_2(aq) + 2\,H_2O(l)$$

41. Ascorbic acid, HAsc, is the chemical name for vitamin C. If 30.95 mL of 0.176 *M* NaOH neutralizes 0.959 g of acid, what is the molar mass of vitamin C?

$$HAsc(aq) + NaOH(aq) \longrightarrow NaAsc(aq) + H_2O(l)$$

42. Tartaric acid, H_2Tart, gives grapes a sour taste. If 28.15 mL of 0.295 *M* NaOH neutralizes 0.623 g of acid, what is the molar mass of tartaric acid?

$$H_2Tart(aq) + 2\,NaOH(aq) \longrightarrow Na_2Tart(aq) + 2\,H_2O(l)$$

43. Alanine is an amino acid. If 21.05 mL of 0.145 *M* NaOH is required to neutralize 0.272 g of alanine, what is the molar mass of the amino acid? (*Note:* One mole of alanine neutralizes one mole of base.)

44. THAM is a base used to standardize acid solutions. If 21.35 mL of 0.115 *M* HCl is required to neutralize 0.297 g of THAM, what is the molar mass of the base? (*Note:* One mole of THAM neutralizes one mole of acid.)

Ionization of Water (Sec. 15.7)

45. Indicate each of the following for the ionization of pure water:
(a) the simplified ionization equation
(b) the ionization constant equation, K_w
(c) the ionization constant at 25 °C

46. Indicate each of the following for the ionization of pure water:
(a) the molecular collision equation
(b) the molar hydrogen ion concentration at 25 °C
(c) the molar hydroxide ion concentration at 25 °C

47. Given the molar concentration of hydrogen ion, calculate the concentration of hydroxide ion.
(a) $[H^+]$ = 0.025 **(b)** $[H^+]$ = 0.000 017

48. Given the molar concentration of hydrogen ion, calculate the concentration of hydroxide ion.
(a) $[H^+] = 6.2 \times 10^{-7}$
(b) $[H^+] = 4.6 \times 10^{-12}$

49. Given the molar concentration of hydroxide ion, calculate the concentration of hydrogen ion.
(a) $[OH^-] = 0.0016$
(b) $[OH^-] = 0.000\ 29$

50. Given the molar concentration of hydroxide ion, calculate the concentration of hydrogen ion.
(a) $[OH^-] = 8.8 \times 10^{-8}$
(b) $[OH^-] = 4.6 \times 10^{-13}$

The pH Concept (Sec. 15.8)

51. Calculate the pH of each of the following given the molar hydrogen ion concentration:
(a) soft drink, $[H^+] = 0.001\ M$
(b) coffee, $[H^+] = 0.000\ 01\ M$

52. Calculate the pH of each of the following given the molar hydrogen ion concentration:
(a) egg white, $[H^+] = 0.000\ 000\ 01\ M$
(b) sour milk, $[H^+] = 0.000\ 001\ M$

53. Calculate the molar hydrogen ion concentration of each of the following given the pH:
(a) shampoo, pH 6
(b) pH-balanced shampoo, pH 8

54. Calculate the molar hydrogen ion concentration of each of the following given the pH:
(a) phosphate detergent, pH 9
(b) nonphosphate detergent, pH 11

Advanced pH Calculations (Sec. 15.9)

55. Calculate the pH of each of the following given the molar hydrogen ion concentration:
(a) milk, $[H^+] = 0.000\ 000\ 30\ M$
(b) eggs, $[H^+] = 0.000\ 000\ 016\ M$

56. Calculate the pH of each of the following given the molar hydrogen ion concentration:
(a) carrots, $[H^+] = 0.000\ 007\ 9\ M$
(b) peas, $[H^+] = 0.000\ 000\ 39\ M$

57. Calculate the molar hydrogen ion concentration of each of the following biological solutions given the pH:
(a) gastric juice, pH = 1.80
(b) urine, pH = 4.75

58. Calculate the molar hydrogen ion concentration of each of the following biological solutions given the pH:
(a) saliva, pH = 6.55 **(b)** blood, pH = 7.50

59. Calculate the pH of each of the following given the molar hydroxide ion concentration:
(a) $[OH^-] = 0.11\ M$ **(b)** $[OH^-] = 0.000\ 55\ M$

60. Calculate the pH of each of the following given the molar hydroxide ion concentration:
(a) $[OH^-] = 0.000\ 031\ M$
(b) $[OH^-] = 0.000\ 000\ 000\ 66\ M$

61. Calculate the molar hydroxide ion concentration of each of the following solutions given the pH:
(a) pH = 0.90
(b) pH = 1.62

62. Calculate the molar hydroxide ion concentration of each of the following solutions given the pH:
(a) pH = 4.55
(b) pH = 5.20

Strong and Weak Electrolytes (Sec. 15.10)

63. State whether each of the following substances in an aqueous solution is highly ionized or slightly ionized:
(a) strong acids
(b) strong bases
(c) soluble ionic compounds

64. State whether each of the following substances in an aqueous solution is highly ionized or slightly ionized:
(a) weak acids
(b) weak bases
(c) slightly soluble ionic compounds

65. Classify each of the following bases as a strong or weak electrolyte:
(a) $KOH(aq)$
(b) $NH_4OH(aq)$
(c) $Ca(OH)_2(aq)$
(d) $Mg(OH)_2(s)$

66. Classify each of the following acids as a strong or weak electrolyte:
(a) $H_2CO_3(aq)$
(b) $H_2SO_4(aq)$
(c) $HI(aq)$
(d) $HNO_2(aq)$

67. Classify each of the following aqueous solutions as a strong or weak electrolyte:
(a) $ZnCO_3(s)$
(b) $Sr(NO_3)_2(aq)$
(c) $K_2SO_4(aq)$
(d) $PbI_2(s)$

68. Classify each of the following aqueous solutions as a strong or weak electrolyte:
(a) $Fe(C_2H_3O_2)_3(aq)$
(b) $AlPO_4(s)$
(c) $Ag_2CrO_4(s)$
(d) $CdSO_4(aq)$

69. Write the following acids in either the ionized or the nonionized form to best represent an aqueous solution:
(a) $HF(aq)$
(b) $HBr(aq)$
(c) $HNO_3(aq)$
(d) $HNO_2(aq)$

70. Write the following bases in either the ionized or the nonionized form to best represent an aqueous solution:
(a) $NaOH(aq)$
(b) $NH_4OH(aq)$
(c) $Ba(OH)_2(aq)$
(d) $Al(OH)_3(s)$

71. Write the following salts in either the ionized or the nonionized form to best represent an aqueous solution:
(a) $AgF(aq)$
(b) $AgI(s)$
(c) $Hg_2Cl_2(s)$
(d) $NiCl_2(aq)$

72. Write the following salts in either the ionized or the nonionized form to best represent an aqueous solution:
(a) $AlPO_4(s)$ (b) $Co(C_2H_3O_2)_3(aq)$
(c) $MnSO_4(aq)$ (d) $PbSO_4(s)$

Net Ionic Equations (Sec. 15.11)

73. List the four steps for writing a balanced net ionic equation.

74. What is the term for ions in the total ionic equation that are not present in the net ionic equation?

75. Write a balanced net ionic equation for each of the following acid–base reactions. Refer to Table 15.5 and Appendix D for electrolyte information.
(a) $HCl(aq) + KOH(aq) \longrightarrow KCl(aq) + H_2O(l)$
(b) $HC_2H_3O_2(aq) + Ca(OH)_2(aq) \longrightarrow Ca(C_2H_3O_2)_2(aq) + H_2O(l)$

76. Write a balanced net ionic equation for each of the following acid–base reactions. Refer to Table 15.5 and Appendix D for electrolyte information.
(a) $HF(aq) + Li_2CO_3(aq) \longrightarrow LiF(aq) + H_2O(l) + CO_2(g)$
(b) $H_2SO_4(aq) + Ba(OH)_2(aq) \longrightarrow BaSO_4(s) + H_2O(l)$

77. Write a balanced net ionic equation for each of the following solution reactions. Refer to Table 15.5 and Appendix D for electrolyte information.
(a) $AgNO_3(aq) + KI(aq) \longrightarrow AgI(s) + KNO_3(aq)$
(b) $BaCl_2(aq) + K_2CrO_4(aq) \longrightarrow BaCrO_4(s) + KCl(aq)$

78. Write a balanced net ionic equation for each of the following solution reactions. Refer to Table 15.5 and Appendix D for electrolyte information.
(a) $Zn(NO_3)_2(aq) + NaOH(aq) \longrightarrow Zn(OH)_2(s) + NaNO_3(aq)$
(b) $MgSO_4(aq) + NH_4OH(aq) \longrightarrow Mg(OH)_2(s) + (NH_4)_2SO_4(aq)$

General Exercises

79. A drop of methyl red indicator and a drop of phenolphthalein are both added to a beaker of distilled water. What is the resulting color of the water?

80. A drop of bromthymol blue indicator and a drop of phenolphthalein are both added to a beaker of distilled water. What is the resulting color of the water?

81. The acid–base indicator methyl orange changes color from pH 3.2 to pH 4.4. It appears red in strongly acidic solutions and yellow in basic solutions. Predict the color of the indicator in a solution having a pH of 3.8.

82. The acid–base indicator bromcresol green changes color from pH 3.8 to pH 5.4. It appears yellow in strongly acidic solutions and blue in basic solutions. Predict the color of the indicator in a solution having a pH of 4.6.

83. Identify an amphiprotic substance given the following reactions:

$$HCl(aq) + NaH_2PO_4(aq) \longrightarrow H_3PO_4(aq) + NaCl(aq)$$

$$NaOH(aq) + NaH_2PO_4(aq) \longrightarrow Na_2HPO_4(aq) + H_2O(l)$$

84. Identify an amphiprotic substance given the following reactions:

$$HCl(aq) + H_2O(l) \longrightarrow H_3O^+(aq) + Cl^-(aq)$$

$$NH_2^-(aq) + H_2O(l) \longrightarrow NH_3(aq) + OH^-(aq)$$

85. Cream of tartar is found in baking powder. If 42.10 mL of 0.100 *M* $Ba(OH)_2$ is required to neutralize 0.791 g of cream of tartar, what is the molar mass? (*Note:* One mole of cream of tartar neutralizes one mole of base.)

86. In 1 mL of water there are 3×10^{22} molecules of H_2O. How many hydrogen ions are in 1 mL of water? (*Hint:* In 1 billion water molecules, only 2 are ionized.)

87. What is the pH of a 0.50 *M* HCl solution?

88. What is the pH of a 0.25 *M* NaOH solution?

89. Write the net ionic equation for the reaction of the weak electrolytes acetic acid and ammonium hydroxide.

90. Write the net ionic equation for the reaction of the strong electrolytes nitric acid and potassium hydroxide.

15 Chapter Self-Test

Answers to Self-Test are in Appendix J.

1. Which of the following is a general property of a basic solution? (Sec. 15.1)
(a) tastes bitter (b) feels slippery
(c) turns litmus paper blue (d) pH greater than 7
(e) all of the above

2. Which of the following is a weak Arrhenius acid? (Sec. 15.2)
(a) HCl(aq) (~ 100% ionized)
(b) $HNO_3(aq)$ (~ 100% ionized)
(c) $HC_2H_3O_2(aq)$ (~ 1% ionized)
(d) all of the above
(e) none of the above

3. In the following reaction, which reactant is a Brønsted–Lowry base? (Sec. 15.3)

$$HCl(aq) + KHS(aq) \longrightarrow KCl(aq) + H_2S(aq)$$

(a) HCl (b) KHS
(c) KCl (d) H_2S
(e) none of the above

4. What is the color of phenolphthalein indicator at pH 7? (Sec. 15.4)
(a) red (b) blue
(c) pink (d) colorless
(e) none of the above

5. If 25.0 mL of 0.100 M HCl is titrated with 0.150 M $Ba(OH)_2$, what volume of barium hydroxide is required to neutralize the acid? (Sec. 15.5)

$$2\,HCl(aq) + Ba(OH)_2(aq) \longrightarrow BaCl_2(aq) + 2\,H_2O(l)$$

(a) 8.33 mL
(b) 16.7 mL
(c) 18.8 mL
(d) 33.3 mL
(e) 37.5 mL

6. What is the molarity of a hydrochloric acid solution if 20.00 mL of HCl is required to neutralize 0.424 g of sodium carbonate (105.99 g/mol)? (Sec. 15.6)

$$2\,HCl(aq) + Na_2CO_3(aq) \longrightarrow 2\,NaCl(aq) + H_2O(l) + CO_2(g)$$

(a) 0.100 M
(b) 0.200 M
(c) 0.300 M
(d) 0.400 M
(e) 0.500 M

7. Given an aqueous solution in which the $[H^+] = 2.5 \times 10^{-7}$, what is the molar hydroxide ion concentration? (Sec. 15.7)
(a) $[OH^-] = 2.5 \times 10^{-7}\,M$
(b) $[OH^-] = 2.5 \times 10^{-8}\,M$
(c) $[OH^-] = 4.0 \times 10^{-6}\,M$
(d) $[OH^-] = 4.0 \times 10^{-7}\,M$
(e) $[OH^-] = 4.0 \times 10^{-8}\,M$

8. What is the hydrogen ion concentration in a urine sample that registers a pH of 4 on a strip of pH paper? (Sec. 15.8)
(a) 0.001 M **(b)** 0.004 M
(c) 0.0001 M **(d)** 0.0004 M
(e) 0.00001 M

9. What is the hydrogen ion concentration in a blood sample that registers a pH of 7.30 using a pH meter? (Sec. 15.9)
(a) 0.000 000 020 M **(b)** 0.000 000 050 M
(c) 0.000 000 20 M **(d)** 0.000 000 50 M
(e) 0.000 073 M

10. If a drain cleaner solution is a strong electrolyte, which of the following must be true? (Sec. 15.10)
(a) Drain cleaner is highly reactive.
(b) Drain cleaner is slightly reactive.
(c) Drain cleaner is highly ionized.
(d) Drain cleaner is slightly ionized.
(e) none of the above

11. Nitric acid is a strong acid, sodium hydroxide is a strong base, and sodium nitrate is a soluble salt. Which of the following is the net ionic equation for the reaction? (Sec. 15.11)

$$HNO_3(aq) + NaOH(aq) \longrightarrow NaNO_3(aq) + H_2O(l)$$

(a) $H^+(aq) + OH^-(aq) \longrightarrow H_2O(l)$
(b) $NO_3^-(aq) + Na^+(aq) \longrightarrow NaNO_3(s)$
(c) $H^+(aq) + NaOH(aq) \longrightarrow Na^+(aq) + H_2O(l)$
(d) $HNO_3(aq) + OH^-(aq) \longrightarrow NO_3^-(aq) + H_2O(l)$
(e) none of the above

Key Concepts

12. What must be done first before applying unit analysis to solve an acid–base titration problem?

13. Beakers X and Y contain either nitric acid or nitrous acid. Based on the conductivity test shown, which beaker contains HNO_3? Which contains HNO_2?

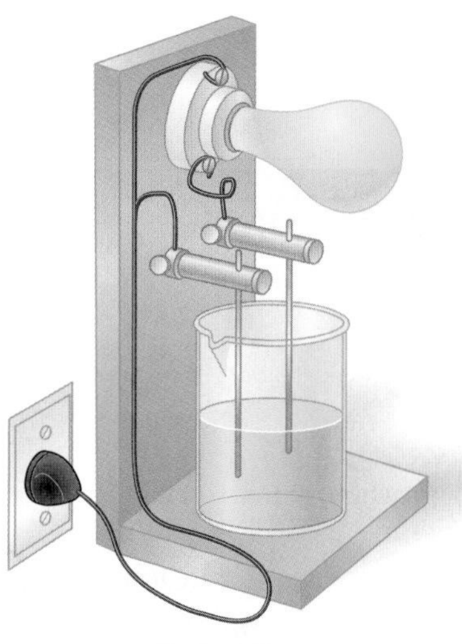

Beaker X

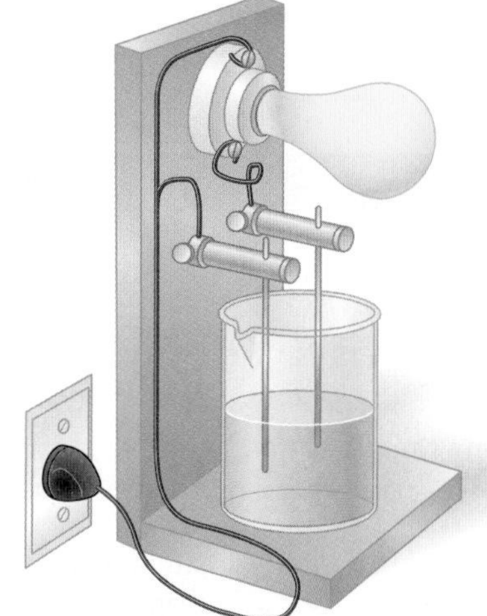

Beaker Y

Critical Thinking

14. The smell of fish is due to a basic organic compound related to ammonia. What common kitchen chemical can be used to neutralize an unpleasant fish odor?

15. What color is the solution in a test tube that contains 10 mL of acid, 1 drop of phenolphthalein, and 1 drop of bromthymol blue indicator?

16. Examine the following illustrations of HCl and HF ionizing in solution. At the molecular level, explain why HCl is more highly ionized than HF.

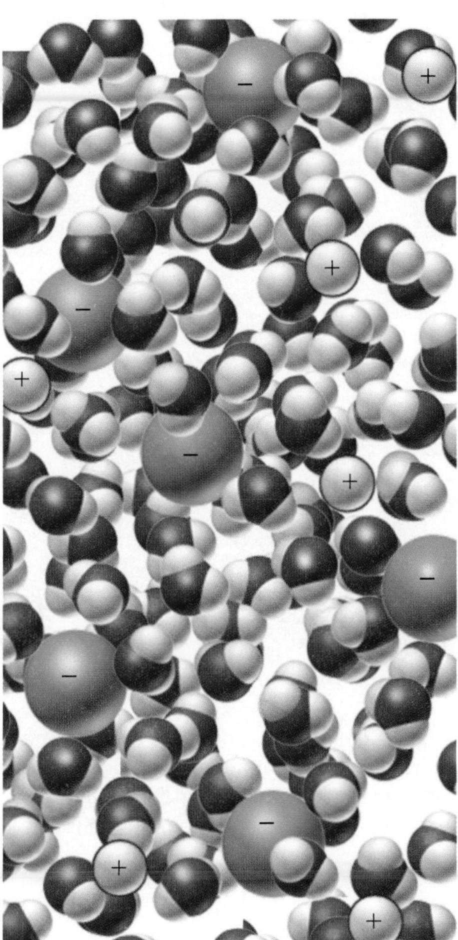

(a) HCl highly ionized

(b) HF slightly ionized

CHAPTER

16 Chemical Equilibrium

If cars go back and forth between a large city and a small city at the same rate, the number of cars in each city are not equal, but remain constant and traffic flow is at equilibrium. Similarly, if reactants and products in a chemical reaction go back and forth at the same rate, the amounts of reactant and product remain constant, and the reaction is at equilibrium.

"There is a single light of science, and to brighten it anywhere is to brighten it everywhere."
Isaac Asimov, Russian/American Science Writer (1920–1992)

We usually assume that a chemical reaction continues until the reactants are used up and the reaction stops. In reality, most reactions continue to occur as an ongoing, reversible process. That is, a reaction occurs in the forward direction to give products, and simultaneously a reverse reaction occurs in the opposite direction to give the original reactants.

As an example of a reversible reaction, let's consider a chemical change that gives a precipitate (Figure 16.1). Initially, insoluble particles form in the aqueous solution. Then a very small number of these particles dissolve and go back into the solution. Since some particles are precipitating while others are dissolving, the ongoing process is said to be dynamic. We can indicate this reversible chemical reaction as follows:

$$\text{soluble substance} \underset{\textit{dissolving}}{\overset{\textit{precipitating}}{\rightleftarrows}} \text{insoluble particles}$$

Most chemical reactions involve a dynamic, reversible change. As reactants undergo a reaction to form products, the products react with each other to give the original reactants. Initially, the rate of the forward reaction is rapid. However, as the reaction proceeds, the concentrations of the reactants decrease and the concentrations of the products increase. With this change in concentrations, the rate of the forward reaction slows down, and the rate of the reverse reaction speeds up.

When the forward reaction and the reverse reaction are taking place at the same rate, the reaction is at equilibrium. We can show these dynamic, reversible processes as follows:

$$\text{reactants} \underset{\textit{reverse reaction}}{\overset{\textit{forward reaction}}{\rightleftarrows}} \text{products}$$

It is important to keep in mind that a chemical change is a dynamic, reversible process. That is, in most chemical changes both the forward and reverse reactions continue to take place at the same time.

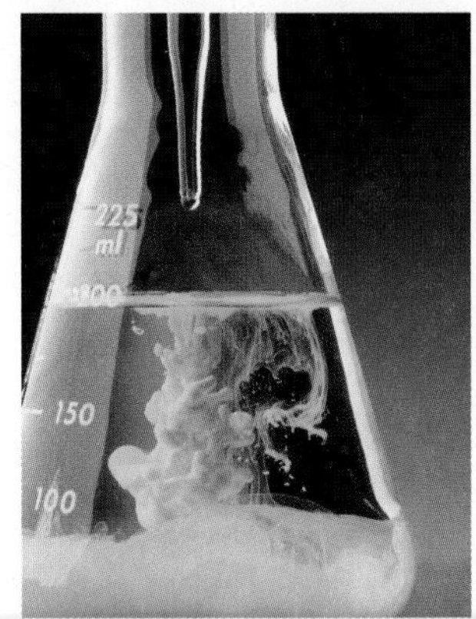

▲ **Figure 16.1 A Reversible Reaction** The reaction of lead(II) nitrate and potassium iodide produces insoluble particles of yellow lead(II) iodide precipitate. Simultaneously, slightly soluble particles of lead(II) iodide dissolve in the solution, while other yellow particles precipitate from the solution.

16.1 COLLISION THEORY

OBJECTIVES

- To state the effect of collision frequency, collision energy, and orientation of molecules on the rate of a chemical reaction.
- To state the effect of concentration, temperature, and a catalyst on the rate of a chemical reaction.

Chemists propose a theoretical model for reactions whereby molecules randomly collide and react. If a collision is successful, molecules of reactants are changed into products. In a successful collision, existing bonds in a molecule are broken and new bonds are formed. This model is referred to as the **collision theory** of reactions. In a successful collision, molecules go through a transition state in which bonds are rearranged. Figure 16.2 illustrates the collision theory by showing molecules of nitrogen and oxygen colliding to form two molecules of nitrogen monoxide.

Collision theory proposes that three factors affect the rate of a chemical reaction. That is, three factors affect the rate of effective collisions: (1) collision frequency,

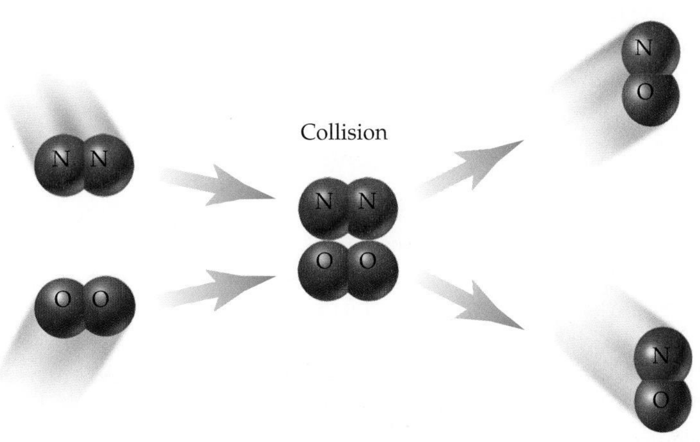

◀ **Figure 16.2 Collision Theory** Nitrogen and oxygen gas molecules collide to form a temporary transition state. If the collision is successful, a molecule of N_2 and a molecule of O_2 produce two molecules of NO.

(2) collision energy, and (3) orientation of the molecules. Let's discuss each factor in more detail.

1. **Collision Frequency.** When we increase the frequency of collision between molecules, we increase the rate of the reaction. One way to increase the collision frequency is to increase the concentrations of molecules. A second way is to increase the temperature. As the temperature increases, the molecules move at a faster velocity, and therefore collide more often.
2. **Collision Energy.** Even though molecules collide, they may not react. For a reaction to occur, molecules must collide with sufficient energy to break bonds. When we increase the temperature, the molecules move at a faster velocity, and the collisions between molecules are more energetic. Thus, the rate of a reaction increases as the energy of collision increases.
3. **Collision Orientation.** Even if molecules collide with a lot of energy, this does not assure that a reaction will occur. For a reaction to take place, the molecules must have a favorable orientation to have an effective collision. Otherwise, the molecules bounce off one another without reacting. Figure 16.3 illustrates (a) an effective collision, and (b) an ineffective collision.

Effects of Concentration, Temperature, and Catalyst

So far we have described the factors that influence the rate of reaction on the conceptual level. Experimentally, three factors have been identified that affect the speed at which a chemical reaction proceeds. These factors are the concentration of reactants, the temperature of the reaction, and the presence of a catalyst. Let's consider each of these factors.

1. **Reactant Concentration.** When we increase the concentration of a reactant, the molecules are closer together and collide more frequently. It therefore follows that a chemical reaction proceeds faster as the concentrations of the reactants increase.
2. **Reaction Temperature.** If we increase the temperature of the reaction, we increase the energy of the molecules. Increasing the temperature affects the rate of reaction in two ways. First, the collision frequency increases, and the reaction speeds up. Second, the collision energy is greater, and there are more molecules with enough energy to break bonds and form new molecules. Thus, as the temperature increases, the rate of reaction increases because of increased collision frequency and greater collision energy.
3. **Catalyst.** If we add a catalyst, we increase the rate of a chemical reaction. A catalyst speeds up a reaction by increasing the number of effective collisions. For example, in reactions involving hydrogen gas the metals Pt, Ni, and Zn can be used as catalysts. The rate of reaction increases because H_2 molecules are attracted to the surface of the metal catalyst, creating a more stable and favorable *collision orientation*.

▶ **Figure 16.3 Effective versus Ineffective Collisions** (a) An effective collision of gaseous NO and O_3 produces NO_2 and O_2. (b) An ineffective collision of NO and O_3 molecules does not give products. If molecules do not have the correct orientation, they simply bounce off each other without forming new molecules.

In the following example exercise, the variables that affect the rate of a chemical reaction are summarized:

Example Exercise 16.1 Rate of Reaction

According to collision theory, which of the following factors influences the rate of a chemical reaction?

(a) frequency of molecular collisions
(b) energy of molecular collisions
(c) orientation of molecular collisions

Solution

Each of the factors affects the rate of reaction as follows:

(a) Increasing collision frequency increases the rate of reaction.
(b) Increasing collision energy increases the rate of reaction.
(c) A favorable molecular orientation increases the rate of reaction.

Practice Exercise

State three experimental variables that can be changed to increase the rate of a chemical reaction.

Answer: The rate of reaction can be increased by (1) increasing the concentration of a reactant, (2) raising the reaction temperature, or (3) introducing a catalyst.

Concept Exercise

When temperature increases, the rate of effective collision increases for two reasons. Explain the two reasons.

Answer: See Appendix G.

16.2 ENERGY PROFILES OF CHEMICAL REACTIONS

OBJECTIVES

- To sketch the general energy profile for an endothermic and an exothermic reaction.
- To label the transition state, energy of activation, and heat of reaction on a given energy profile.

In air, nitrogen and oxygen gases do not react. In a hot automobile engine, however, the two gases react to produce nitrogen monoxide. To react, molecules of N_2 and O_2 must collide with sufficient energy to achieve the transition state before forming products. Since the transition state is a high-energy point, there is an energy barrier between the reactants and products. Figure 16.4 shows an analogy illustrating the energy changes that occur during a chemical reaction.

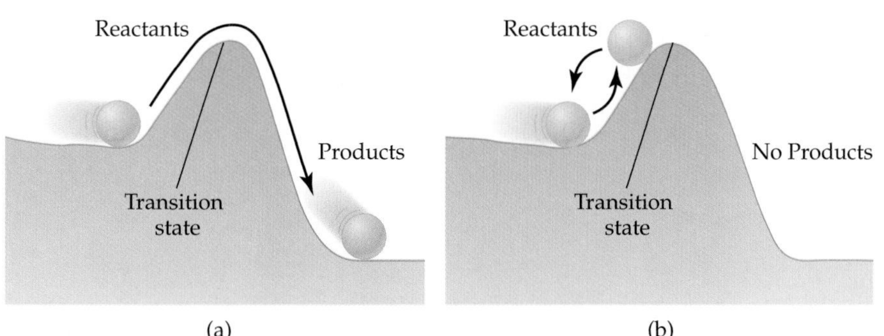

▲ **Figure 16.4 Energy Barrier Analogy for a Chemical Reaction** (a) The ball symbolizes molecules in a reaction that have sufficient energy to give products. If the ball has enough energy, it reaches the transition point at the top of the hill and rolls down the other side. (b) The ball symbolizes a reaction in which the molecules do not have sufficient energy to give products. If the ball does not have enough energy, it cannot reach the transition point and roll down the other side of the hill.

Endothermic Reaction Profiles

An **endothermic reaction** absorbs heat and proceeds by consuming heat energy. As an example, consider the formation of nitrogen monoxide from nitrogen and oxygen gases. The reaction is

$$N_2(g) + O_2(g) + \textit{heat} \rightleftharpoons 2\,NO(g)$$

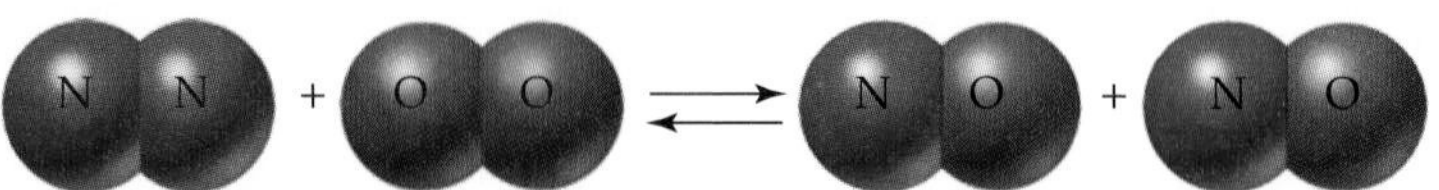

We can follow the progress of a chemical reaction by drawing an energy profile of reactants and products. A **reaction profile** shows the energy of reactants and products during the course of a reaction. The highest point on the reaction profile is the **transition state**, where bonds are breaking, and simultaneously other bonds are forming.

A reaction profile for the reaction of nitrogen and oxygen is shown in Figure 16.5. The energy required for the reactants to achieve the transition state is called the **activation energy** (symbol E_{act}). The energy difference between reactants and products is termed the **heat of reaction** (symbol ΔH). For endothermic reactions energy is absorbed, and the products end up at a higher potential energy level than the reactants.

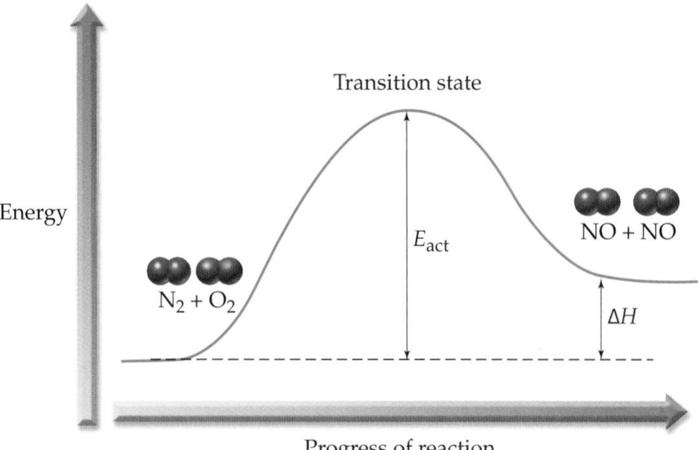

▶ **Figure 16.5 Reaction Profile for an Endothermic Reaction** Notice that the product of the reaction, NO, is at a higher energy level than the reactants, N_2 and O_2.

Exothermic Reaction Profiles

An **exothermic reaction** proceeds by releasing heat energy. As an example of a reaction that proceeds by liberating heat energy, let's again consider that automobiles produce nitrogen monoxide, NO. Also consider that air contains ozone, O_3. The ozone in air converts the NO to NO_2, a brown gaseous component of smog.

$$NO(g) + O_3(g) \rightleftharpoons NO_2(g) + O_2(g) + \textit{heat}$$

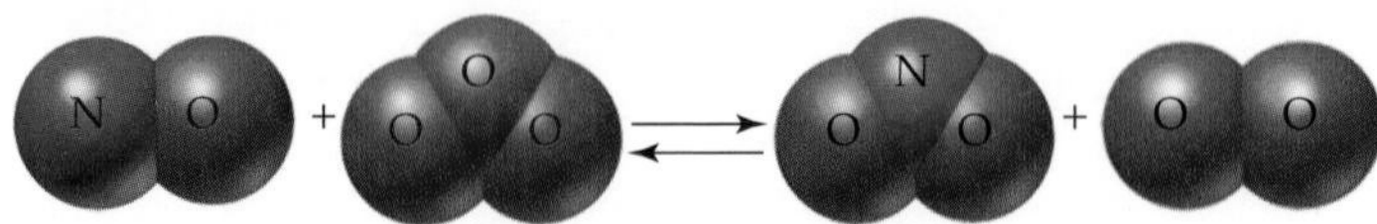

As you can see, the formation of nitrogen dioxide releases heat. Chemical processes that release heat energy are called exothermic reactions. As before, we can construct a reaction profile to follow the changes in energy for this reaction. Figure 16.6 illustrates the energy difference between reactants and products for an exothermic reaction. For exothermic reactions energy is released and the products end up at a lower energy level than the reactants.

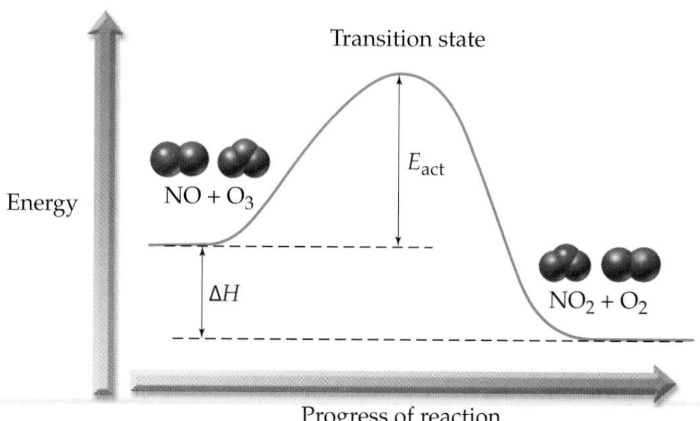

◀ **Figure 16.6 Reaction Profile for an Exothermic Reaction** Notice that the products of the reaction, NO_2 and O_2, are at a lower energy level than the reactants, NO and O_3.

For reactants to form products, they must have sufficient energy to overcome the activation energy barrier. The higher the activation energy, the more slowly the reaction proceeds. The reaction is slower because there are fewer molecules with enough energy to achieve the transition state.

To help a reaction proceed at a faster rate, we can use a catalyst. Recall that we previously defined a catalyst as a substance that speeds up the rate of a reaction without being consumed. Now, we can define a **catalyst** as a substance that allows a reaction to proceed faster by lowering the energy of activation. As an example, consider how water is produced from the exothermic reaction of hydrogen and oxygen gases. The equation for the reaction is

$$2\,H_2(g) + O_2(g) \rightleftharpoons 2\,H_2O(g) + \textit{heat}$$

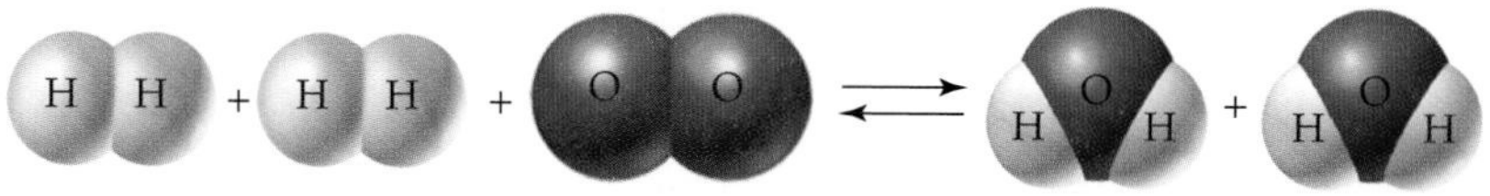

Although the reaction produces heat energy, it is very slow at normal temperatures. When a mixture of hydrogen and oxygen gases is exposed to a flame or spark, the reaction is instantaneous. In fact, the instantaneous release of a large amount of energy creates an explosion. At room temperature, a powdered metal such as zinc dust acts as a catalyst and causes the reaction to occur rapidly without explosion. Figure 16.7 illustrates the reaction profile for hydrogen and oxygen gases with and without a catalyst.

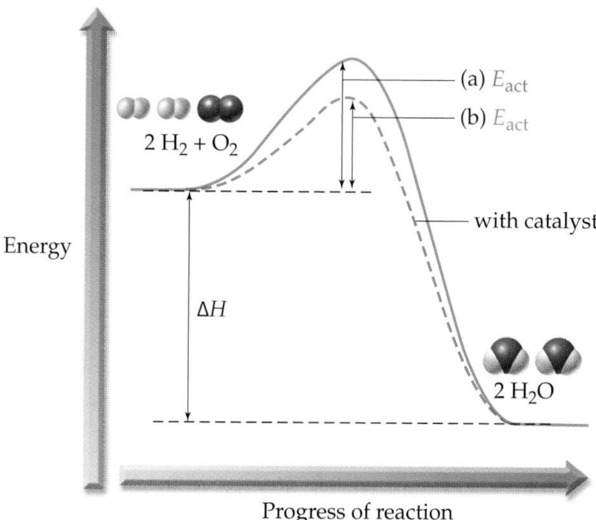

◀ **Figure 16.7 Effect of a Catalyst on Activation Energy** (a) Reaction profile for the reaction of H_2 and O_2 without a catalyst (solid line). (b) Reaction profile for the reaction of H_2 and O_2 with a zinc metal catalyst (dashed line). Notice the heat of reaction, ΔH, is not affected by the catalyst.

Note that the heat of reaction is the same with and without a catalyst and that ΔH does not depend on the catalyst. Thus, the rate of a reaction is not related to the heat of reaction. Table 16.1 lists the rate of reaction and the heat of reaction for the formation of water from hydrogen and oxygen gases in the presence of various catalysts.

TABLE 16.1 REACTION OF HYDROGEN AND OXYGEN GASES

CATALYST	RATE OF REACTION	HEAT OF REACTION*
none	very slow	57.8 kcal (242 kJ)
spark	explosive	57.8 kcal (242 kJ)
zinc dust	rapid	57.8 kcal (242 kJ)

*Energy released per mole of water produced at 25 °C

The following example exercise further illustrates the relationship between reactants and products and the reaction profile:

Example Exercise 16.2 Energy Profiles of Chemical Reactions

Ultraviolet light from the Sun converts oxygen molecules in the upper atmosphere to ozone molecules. Draw the reaction profile for the reaction:

$$3\,O_2(g) + heat \xrightleftharpoons{UV} 2\,O_3(g)$$

Solution

Since the reaction is endothermic, the reaction profile is

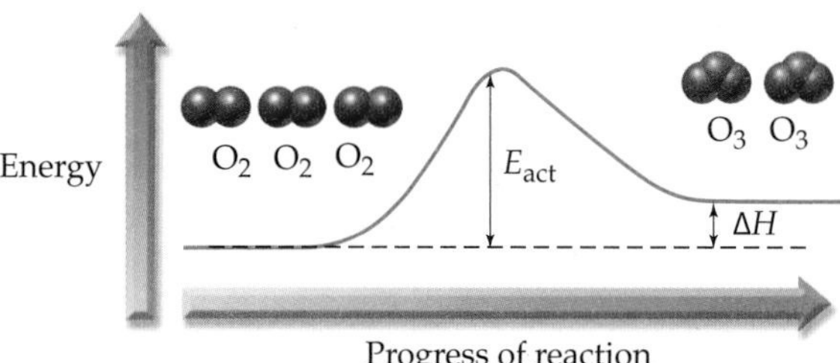

Practice Exercise

Nitrosyl bromide, NOBr, decomposes to give nitric oxide and bromine gas. Draw the reaction profile for the reaction:

$$2\,NOBr(g) \rightleftharpoons 2\,NO(g) + Br_2(g) + heat$$

Answer: Since the reaction is exothermic, the reaction profile is

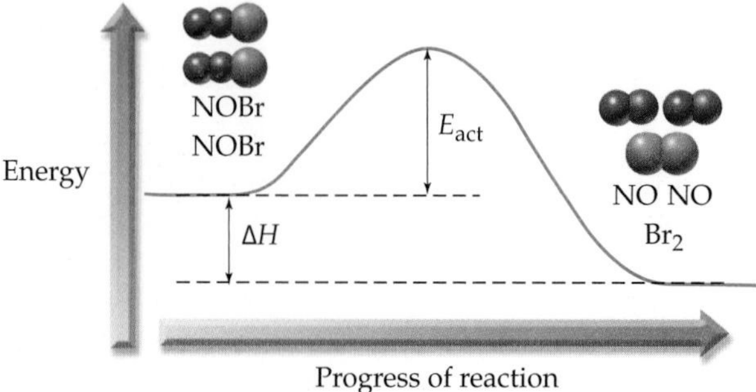

Concept Exercise

Which of the following reaction profiles illustrates an endothermic reaction?

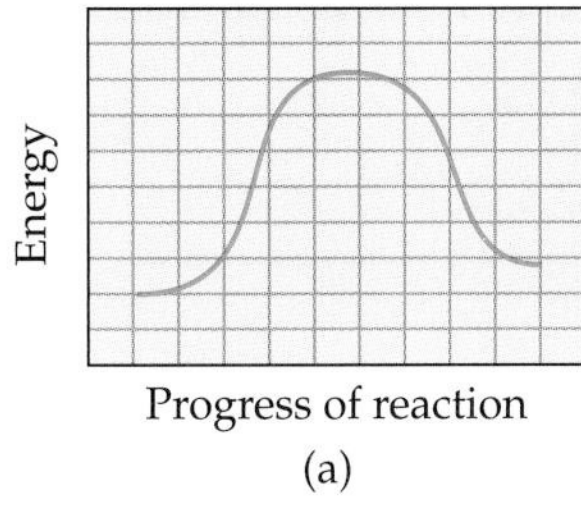

(a)

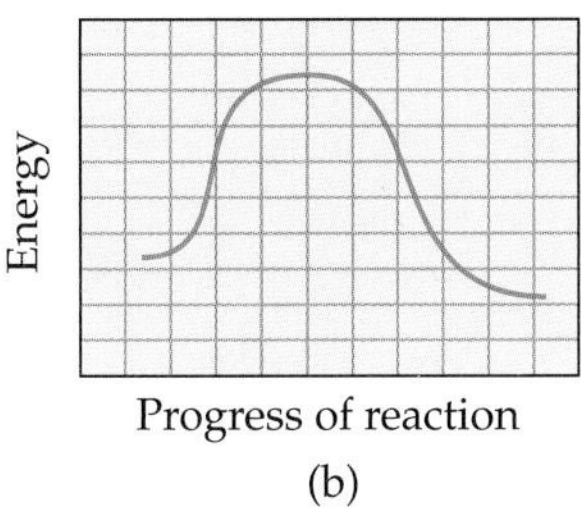

(b)

Answer: See Appendix G.

16.3 THE CHEMICAL EQUILIBRIUM CONCEPT

OBJECTIVES

- To describe the equilibrium concept for a reversible reaction.
- To express the law of chemical equilibrium as an equation.

We have said that most chemical reactions are reversible processes. That is, a chemical change is a **reversible reaction** that can proceed simultaneously both in the forward direction toward products and in the reverse direction toward reactants. When the rate of the forward reaction is equal to the rate of the reverse reaction, a reaction is said to be in a state of **chemical equilibrium**.

It is important to understand that when a reaction is at equilibrium, it does not mean that the reaction has stopped. It also does not mean that the amounts of reactants and products are equal. Rather, a chemical reaction at equilibrium implies a *dynamic, reversible process*. That is, an ongoing forward reaction ($\longrightarrow$) and an ongoing reverse reaction ($\longleftarrow$) are taking place at the same rate.

As an example, consider the conversion of oxygen molecules, O_2, to ozone molecules, O_3, in the upper atmosphere. Ultraviolet light initiates the reaction. The equation for the reversible reaction is

$$\underset{\textit{oxygen}}{3\,O_2(g)} \rightleftarrows \underset{\textit{ozone}}{2\,O_3(g)}$$

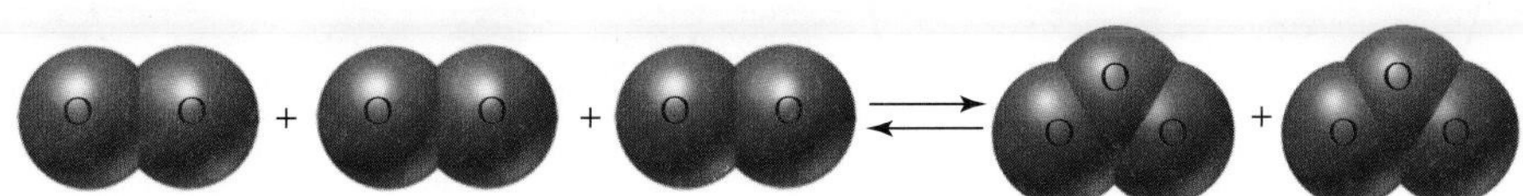

We can experimentally simulate this reversible reaction in a glass sphere. We begin by pumping oxygen gas into the glass sphere. Next, we radiate ultraviolet light on the oxygen gas in the sphere. Initially, the rate of the forward reaction (rate_f) is quite rapid and the rate of the reverse reaction (rate_r) is slow. As the reaction proceeds, the forward reaction slows down and the reverse reaction speeds up. When the gaseous reaction reaches chemical equilibrium, we have

$$\text{rate}_f\ (O_2 \text{ reaction}) = \text{rate}_r\ (O_3 \text{ reaction})$$

We can define the **rate of reaction** as the rate at which the concentrations of reactants decrease per unit time. Alternatively, we can define the rate of reaction as the rate at which the concentrations of products increase per unit time. Figure 16.8 shows the changes in the rates of the forward and reverse reactions as the reversible oxygen-ozone reaction progresses.

Figure 16.9 provides a model to help us understand the equilibrium process. Oxygen molecules, O_2, react in a closed container to give ozone molecules, O_3.

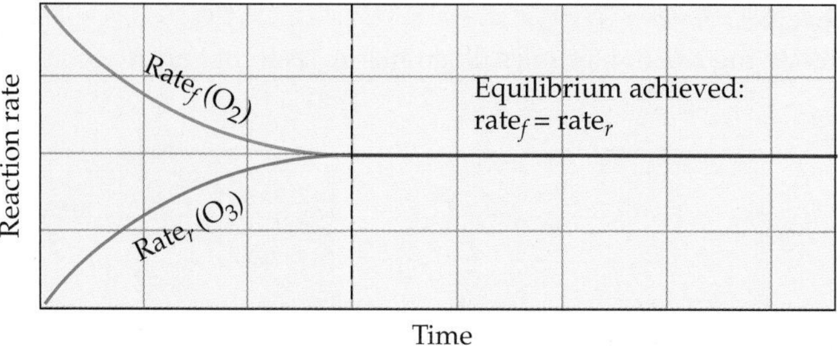

▶ **Figure 16.8 Chemical Equilibrium and Rates of Reaction** The rate of reaction for O_2 decreases with time. As the reaction proceeds, the rate of reaction for O_3 increases. At equilibrium, the rates of the forward and reverse reactions are the same.

Simultaneously, some of the O_3 molecules decompose to give O_2 molecules. Initially, the O_2 concentration is high and the forward reaction is rapid. The O_3 concentration is low and the reverse reaction is slow. As equilibrium is approached, the forward reaction slows down and the reverse reaction speeds up. When the rates of the forward and reverse reactions are equal, the reaction is at equilibrium.

Law of Chemical Equilibrium

In 1864, the Norwegian chemists Cato Guldberg and Peter Waage observed that a change in the amount of substance participating in a reversible reaction produced a shift in the equilibrium. When they added more of a reactant substance, the reaction shifted to form more product. Conversely, when they added a product substance, the reaction shifted to form more reactant. We now know that a shift in equilibrium can be produced for any reversible reaction.

$$a\,\mathrm{A} + b\,\mathrm{B} \rightleftharpoons c\,\mathrm{C} + d\,\mathrm{D}$$

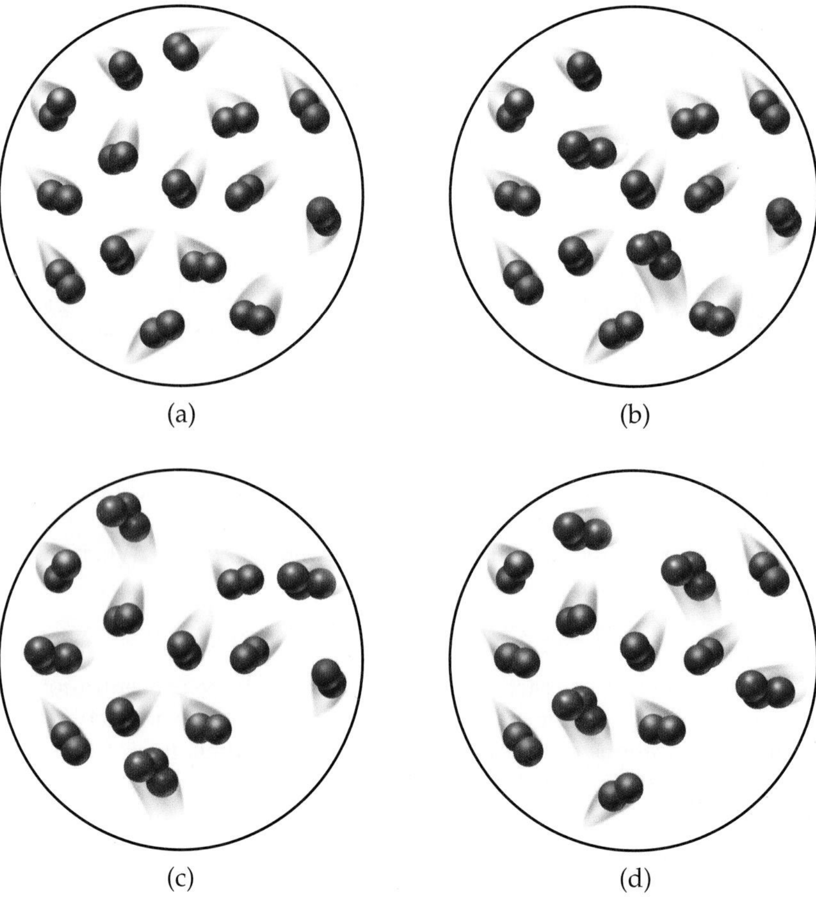

▶ **Figure 16.9 A Dynamic Equilibrium Model** (a) The container has only O_2 gas molecules. (b) As the reaction proceeds, O_3 is formed. (c) At equilibrium, the forward and reverse reactions occur at the same rate. (d) Later, the amounts of O_2 and O_3 are constant even though molecules of O_2 and O_3 continue to react.

where A and B represent reactants, C and D represent products, and *a*, *b*, *c*, and *d* represent the coefficients for a general reversible reaction.

Moreover, at equilibrium we would observe that the ratio of product concentrations to reactant concentrations remains constant at a given temperature. Guldberg and Waage found that the molar concentrations of the products (raised to the powers *c* and *d*), divided by the molar concentrations of reactants (raised to the powers *a* and *b*), always gave the same value. Regardless of the initial amounts of reactants, at equilibrium the molar concentration ratio is a constant. This relationship, which applies to every reversible reaction, is known as the **law of chemical equilibrium**. Mathematically, we can express the law of chemical equilibrium as follows:

$$K_{eq} = \frac{[C]^c\,[D]^d}{[A]^a\,[B]^b}$$

where [A], [B], [C], and [D] represent the molar concentrations of reactants and products, and *a*, *b*, *c*, and *d* correspond to the coefficients in the balanced chemical equation. The ratio is equal to the **general equilibrium constant** (symbol K_{eq}), which expresses the molar concentration of each substance participating in the reaction at a given temperature.

CHEMISTRY CONNECTION The Ozone Hole

Q: What animal population has been threatened as a result of the ozone hole over Antarctica?

All living species are protected from the Sun's harmful radiation by a layer of ozone molecules in the upper atmosphere. However, this ozone layer that shields Earth's fragile life forms is being depleted. For some time we have known about the ozone depletion over Antarctica referred to as an "ozone hole." Recently, scientists have found other areas of depletion in the ozone over heavily populated areas. Ozone depletion has been detected over Russia, Europe, Canada, and the United States.

In the stratosphere, about 50 miles above Earth, ozone molecules absorb high-energy ultraviolet radiation from the Sun. A hole in the ozone layer is of great concern because the ozone layer protects us from the Sun's powerful rays. It is well known that ultraviolet radiation can cause skin cancer and cataracts. The Sun's harmful rays can also affect agriculture and cause lower crop yields.

Scientists at the University of California, Irvine, were the first to propose that chlorofluorocarbons (CFCs) deplete the ozone layer. CFCs were used as propellants in aerosol cans and as coolants in air-conditioning units. These CFC compounds, such as Freon–12 (CF_2Cl_2), are quite inert. However, when released into the environment, they diffuse up into the atmosphere and destroy the ozone layer.

CFCs react with ozone molecules in steps. First, ultraviolet radiation from the Sun strips a chlorine atom away from a CFC molecule. The resulting chlorine atom has an unpaired electron (Cl•) and is referred to as a *free radical.*

Step 1: $\text{CFC} \xrightarrow{\text{UV}} \text{Cl}\bullet$

A chlorine free radical is very reactive and can act as a catalyst for the further decomposition of ozone molecules, O_3, into oxygen.

Step 2: $2\,O_3 \xrightarrow{\text{Cl}\bullet} 3\,O_2$

Since the chlorine free radical is not consumed in the reaction, this process occurs repeatedly, and a single chlorine free radical can destroy thousands of ozone molecules.

Ozone depletion is a worldwide problem. Many nations have halted the production and use of CFCs. The United States has resolved to phase out CFCs, but it is expensive to retrofit existing refrigeration units in automobiles and homes. However, in new homes and automobiles air-conditioning units are being equipped with non-CFC gases.

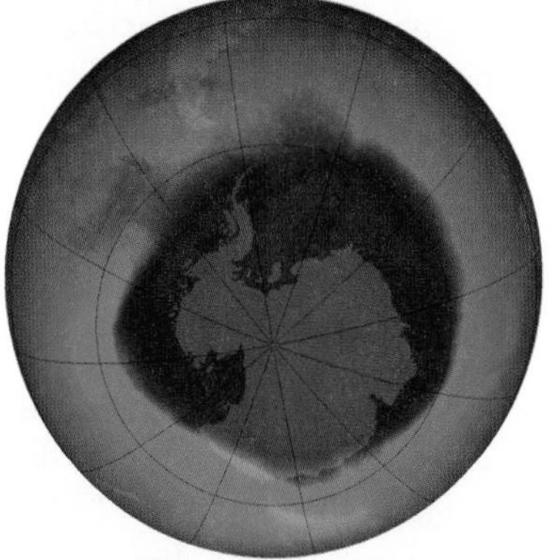

▲ The ozone levels are indicated for the stratosphere over Antarctica. The ozone concentration is low in the blue region and lowest in the violet area.

A: The penguin population has been threatened by the hole in the ozone layer over Antarctica.

OBJECTIVES

- To write the equilibrium constant expression for a reversible reaction.
- To calculate an equilibrium constant, K_{eq}, from experimental data.

16.4 GENERAL EQUILIBRIUM CONSTANT, K_{eq}

If K_{eq} is greater than 1, a reaction occurs that favors the product. Conversely, if K_{eq} is less than 1, a reaction favors the reactant.

Let's construct the equilibrium constant expression for the following reversible reaction:

$$2\,A \rightleftarrows B$$

To write the equilibrium expression, K_{eq}, we place the concentration of the product in the numerator and the concentration of the reactant in the denominator.

$$K_{eq} = \frac{[B]}{[A]}$$

The coefficients in the balanced equation are written as exponents in the K_{eq} expression. The coefficient of A is 2, and the coefficient of B is understood to be 1. These coefficients are written as the power of the corresponding concentration.

$$K_{eq} = \frac{[B]}{[A]^2}$$

The following example exercise provides additional practice in writing K_{eq} expressions for general cases of reversible reactions:

Example Exercise 16.3 General Equilibrium Constant, K_{eq}

Write the equilibrium constant expression for the following reversible reaction:

$$A + 2\,B \rightleftarrows 3\,C + D$$

Solution

Let's proceed in two steps. First, let's substitute into the K_{eq} expression. The ratio of product concentration to reactant concentration is

$$K_{eq} = \frac{[C]^c\,[D]^d}{[A]^a\,[B]^b}$$

Second, let's account for the coefficients of the balanced equation. The coefficients are 1, 2, 3, and 1, respectively.

$$K_{eq} = \frac{[C]^3\,[D]}{[A]\,[B]^2}$$

Practice Exercise

Write the equilibrium constant expression for the following reversible reaction:

$$2\,A + 3\,B \rightleftarrows C + 4\,D$$

Answer:

$$K_{eq} = \frac{[C]\,[D]^4}{[A]^2\,[B]^3}$$

Concept Exercise

Describe the dynamic reactant and product concentrations for a reversible reaction *after* the reaction reaches equilibrium.

Answer: See Appendix G.

Chemical Equilibria in the Same Physical State

In most reversible reactions, all the reactants and products are in the same physical state. In the ozone reaction, both the reactants and products are gases. A reversible reaction at equilibrium in which all the substances are in the same state is referred to as a **homogeneous equilibrium**. For example, low-grade coal has a high sulfur content, and burning it produces SO_2. In the atmosphere, oxygen gas converts SO_2 gas to SO_3 gas according to the following reversible reaction:

$$2\,SO_2(g) + O_2(g) \rightleftharpoons 2\,SO_3(g)$$

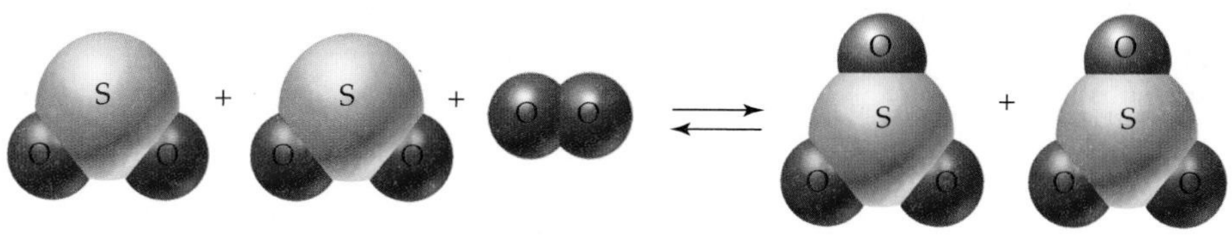

If we substitute the concentrations of reactants and products into the general equilibrium expression, we have

$$K_{eq} = \frac{[SO_3]}{[SO_2]\,[O_2]}$$

The respective coefficients in the balanced equation are 2, 1, and 2. Thus, the equilibrium expression for the reaction is

$$K_{eq} = \frac{[SO_3]^2}{[SO_2]^2\,[O_2]}$$

Chemical Equilibria in a Different Physical State

In the previous examples of gaseous equilibria, liquids and solids were not involved and did not appear in the K_{eq} expression. On occasion, a reactant or product may be in the solid or the liquid state. A reaction at equilibrium in which one of the substances is in a different physical state is referred to as a **heterogeneous equilibrium**.

As an example, consider hydrogen gas, which can be manufactured by passing steam (a gas) over hot charcoal (a solid). The equation for the reaction is as follows:

$$C(s) + H_2O(g) \rightleftharpoons CO(g) + H_2(g)$$

We can write the equilibrium constant expression as

$$K_{eq} = \frac{[CO][H_2]}{[C][H_2O]}$$

Charcoal is a solid but all the other substances in the reaction are gases. When the reaction is studied experimentally, it is found that the amount of charcoal has no effect on the equilibrium constant. Therefore, the concentration of charcoal, [C], is set equal to unity (1) and is omitted from the equilibrium expression, giving us

$$K_{eq} = \frac{[CO][H_2]}{[H_2O]}$$

In addition, it is found that liquids and solids have no effect on any gaseous equilibrium. Therefore, liquids and solids do not appear in the equilibrium expression. The following example exercise provides additional practice in writing K_{eq} expressions for homogeneous and heterogeneous equilibria:

Example Exercise 16.4 Writing K_{eq} Expressions

Write the equilibrium constant expression for each of the following reversible reactions:

(a) $2\,NO(g) + 2\,H_2(g) \rightleftarrows N_2(g) + 2\,H_2O(g)$

(b) $NH_4NO_3(s) \rightleftarrows N_2O(g) + 2\,H_2O(g)$

Solution

We can substitute into the K_{eq} expression as follows:

(a) Since the equation is balanced, we raise the concentration of each species to the power corresponding to the coefficient in the balanced equation.

$$K_{eq} = \frac{[N_2]\,[H_2O]^2}{[NO]^2\,[H_2]^2}$$

(b) In this reaction NH_4NO_3 is a solid. Since NH_4NO_3 is not a gas, it does not appear in the equilibrium expression; thus, the equilibrium expression is

$$K_{eq} = [N_2O]\,[H_2O]^2$$

Practice Exercise

Write the equilibrium constant expression for each of the following reversible reactions:

(a) $4\,NH_3(g) + 5\,O_2(g) \rightleftarrows 4\,NO(g) + 6\,H_2O(g)$

(b) $CaCO_3(s) \rightleftarrows CaO(s) + CO_2(g)$

Answers:

(a) $K_{eq} = \dfrac{[NO]^4\,[H_2O]^6}{[NH_3]^4\,[O_2]^5}$

(b) $K_{eq} = [CO_2]$

Concept Exercise

Describe the dynamic reactant and product concentrations for a reversible reaction *before* the reaction reaches equilibrium.

Answer: See Appendix G.

Determining K_{eq} by Experiment

One of the most thoroughly investigated equilibrium reactions is involved in the formation of hydrogen iodide from hydrogen gas and iodine vapor.

$$\underset{\text{colorless gas}}{H_2(g)} + \underset{\text{purple vapor}}{I_2(g)} \rightleftarrows \underset{\text{colorless gas}}{2\,HI(g)}$$

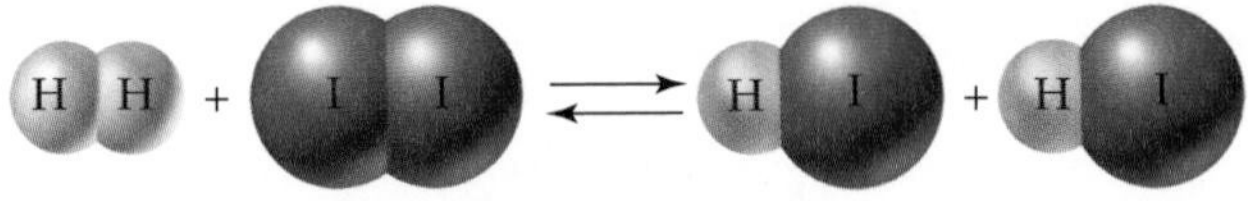

Experimentally, we can measure the molar concentration of each gas at equilibrium. If we start with 1.000 mol of H_2 and I_2 in a liter container, we find at equilibrium that $H_2 = 0.212$, $I_2 = 0.212$, and $HI = 1.576$. To calculate an experimental value for the equilibrium constant, K_{eq}, let's write the equilibrium expression.

$$K_{eq} = \frac{[HI]^2}{[H_2]\,[I_2]}$$

Substituting, we have

$$K_{eq} = \frac{[1.576]^2}{[0.212][0.212]} = 55.3$$

Interestingly, if we start with 1.000 mol of HI and approach equilibrium from the opposite direction, we obtain the same value for K_{eq}. In fact, the K_{eq} value for the reaction is 55.3 regardless of the initial concentrations of each gas. Table 16.2 presents data for four experiments involving the HI equilibrium.

TABLE 16.2 EXPERIMENTAL DETERMINATION OF K_{eq}

EXPERIMENT	INITIAL CONCENTRATION			EQUILIBRIUM CONCENTRATION			CALCULATED K_{eq}* AT 425 °C
	$[H_2]$	$[I_2]$	[HI]	$[H_2]$	$[I_2]$	[HI]	$\frac{[HI]^2}{[H_2][I_2]}$
1	1.000	1.000	0	0.212	0.212	1.576	55.3
2	0.500	0.500	0	0.106	0.106	0.788	55.3
3	0	0	1.000	0.106	0.106	0.788	55.3
4	0	0	2.000	0.212	0.212	1.576	55.3

*The units for an equilibrium constant are usually omitted. In this example, the units in the numerator and denominator of the K_{eq} expression cancel.

Conceptually, we can visualize a closed vessel containing the three gases H_2, I_2, and HI in equilibrium. When we change the amount of any gas, the reaction shifts forward or backward in such a way that the ratio among the concentrations of reactants and products remains constant. The following example exercise further illustrates the calculation of an equilibrium constant from experimental data:

Example Exercise 16.5 Experimental Equilibrium Constant, K_{eq}

In the upper atmosphere, nitric oxide and nitrogen dioxide participate in the following equilibrium:

$$2\,NO(g) + O_2(g) \rightleftharpoons 2\,NO_2(g)$$

Given the equilibrium concentrations for each gas at 25 °C, what is K_{eq}?

$$[NO] = 1.5 \times 10^{-11}$$

$$[O_2] = 8.9 \times 10^{-3}$$

$$[NO_2] = 2.2 \times 10^{-6}$$

Solution

We first write the equilibrium expression for the reaction.

$$K_{eq} = \frac{[NO_2]^2}{[NO]^2\,[O_2]}$$

Second, we substitute the concentration values into the expression.

$$K_{eq} = \frac{[2.2 \times 10^{-6}]^2}{[1.5 \times 10^{-11}]^2\,[8.9 \times 10^{-3}]}$$

$$= 2.4 \times 10^{12}$$

The large K_{eq} for this reaction indicates that the equilibrium overwhelmingly favors the formation of the product, NO_2. However, K_{eq} does not indicate the rate of reaction. In the atmosphere, the conversion of NO to NO_2 takes place slowly.

Practice Exercise

Given the equilibrium concentrations for the gas mixture at 100 °C, calculate the value of K_{eq} for the following reaction:

$$N_2O_4(g) \rightleftharpoons 2\,NO_2(g)$$

$$0.00140\,M \qquad 0.0172\,M$$

Answer: $K_{eq} = 0.211$

Concept Exercise

Write the equilibrium constant expression for the following reversible reaction:

$$2\,X_2(g) + 3\,Y_2(g) \rightleftharpoons 2\,X_2Y_3(g) + \text{heat}$$

Answer: See Appendix G.

Note In Section 11.3, we learned that gas pressure is proportional to the concentration of molecules. Therefore, we can write an equilibrium expression for a reversible gaseous reaction on the basis of partial gas pressures. The equilibrium expression for a reversible gaseous reaction can be written as

$$K_p = \frac{P_C{}^c \times P_D{}^d}{P_A{}^a \times P_B{}^b}$$

where K_p is the equilibrium constant, and P_A, P_B, P_C, and P_D are the partial pressures of the gases participating in the equilibrium reaction.

OBJECTIVE

- To apply Le Chatelier's principle to reversible reactions in the gaseous state.

16.5 EQUILIBRIA SHIFTS FOR GASES

Henri Louis Le Chatelier (1850–1936) was trained in France as a mining engineer. When he graduated from college, he became a professor of chemistry at a mining school. His first interest was in the study of flames in the hope of preventing mine explosions. Eventually, he studied the effect of heat on reversible reactions and in 1888 stated the principle for which he is famous: "*Every change of one of the factors of an equilibrium brings about a rearrangement of the system in such a direction as to minimize the original change.*"

The factors affecting a chemical equilibrium system are concentration, temperature, and pressure. A change in any of these factors shifts the equilibrium. For chemical reactions, **Le Chatelier's principle** can be stated as follows: when a reversible reaction at equilibrium is disturbed by a change in concentration, temperature, or pressure, the equilibrium shifts to relieve the stress.

Effect of Concentration

For any equilibrium system, we can experimentally shift the reaction forward or backward by changing the concentration of a reactant or product. Let's consider the following equilibrium:

$$\underset{\textit{colorless}}{N_2O_4(g)} \rightleftharpoons \underset{\textit{brown}}{2\,NO_2(g)}$$

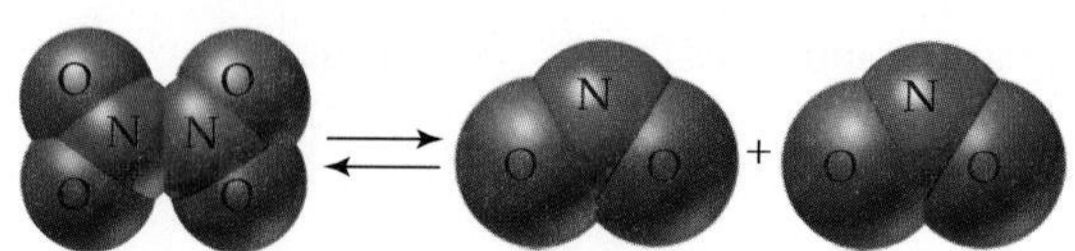

If we increase the N_2O_4 reactant concentration, the equilibrium shifts to the right and the NO_2 product concentration increases. If we decrease the N_2O_4 concentration, the equilibrium shifts to the left and the NO_2 concentration decreases.

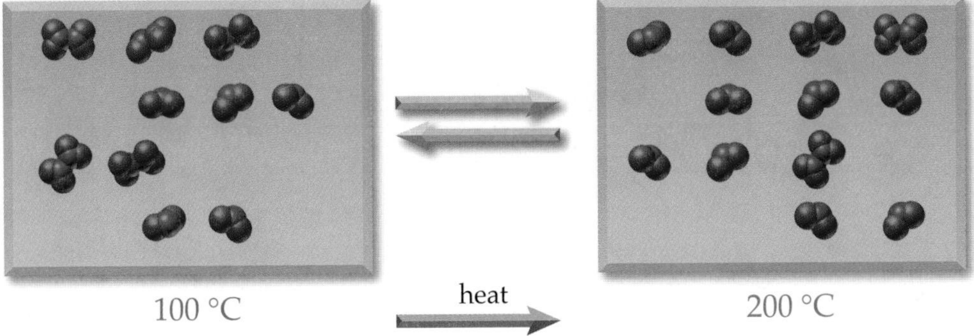

▲ **Figure 16.10 Effect of Temperature on Equilibrium** (a) Initially, there are five molecules of N_2O_4 and five molecules of NO_2 in the equilibrium mixture. (b) If the temperature increases, the equilibrium shifts to the right to relieve the stress. As a result, the number of N_2O_4 molecules decreases and the number of NO_2 molecules increases.

Effect of Temperature

Now, let's consider the effect of temperature on the following equilibrium:

$$\underset{\textit{colorless}}{N_2O_4(g)} + \textit{heat} \rightleftharpoons \underset{\textit{brown}}{2\,NO_2(g)}$$

When we change the temperature for a reversible reaction, we stress the equilibrium system, and the equilibrium shifts to relieve the stress. In this example, if we heat this endothermic reaction, the equilibrium shifts to the *right;* the N_2O_4 concentration decreases and the NO_2 concentration increases. If we cool the reaction, the equilibrium shifts to the *left.* As the equilibrium shifts to the left, the NO_2 concentration decreases and the N_2O_4 concentration increases.

In an endothermic reaction, heat causes the equilibrium to shift to the right toward the products to relieve the stress on the reactants side of the equation. In an exothermic reaction, heat causes the equilibrium to shift to the left toward the reactants to relieve the stress on the products side of the equation. Figure 16.10 illustrates the shift in equilibrium that results from heating the N_2O_4 reaction.

Note that N_2O_4 is a colorless gas and that NO_2 is a dark-brown gas. Since each gas has a distinctive color, we can observe an equilibrium shift by noting the color of the gaseous mixture. Figure 16.11 illustrates the shift in equilibrium resulting from heating the colorless N_2O_4 gas to give dark-brown NO_2 gas.

(a) (b) (c)

▲ **Figure 16.11 Heating the N_2O_4–NO_2 Equilibrium** (a) At low temperatures, the N_2O_4–NO_2 equilibrium favors N_2O_4, which is a colorless gas. (b) If we heat the reaction, the equilibrium shifts to produce more NO_2, which is a brown gas. (c) Further heating produces more NO_2, and the gas mixture becomes a darker brown.

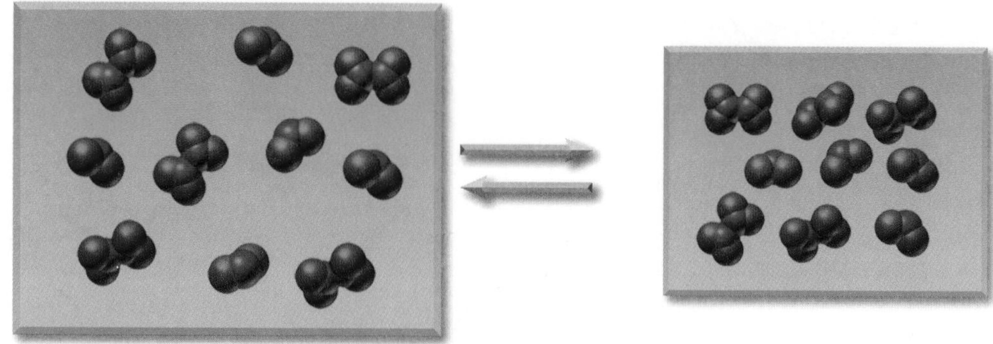

▶ **Figure 16.12 Effect of Pressure on Equilibrium**
(a) Initially, there are 10 molecules in the N_2O_4–NO_2 equilibrium. (b) When the volume decreases and the pressure increases, the equilibrium shifts to relieve the stress. As a result, NO_2 is converted to N_2O_4, and the total number of molecules decreases.

Effect of Pressure

For many gaseous equilibrium systems we can experimentally shift the reaction forward or backward by changing the volume, and therefore the pressures of the gases in the system. If we decrease the volume to increase the pressure, the equilibrium will shift to reduce the pressure change by producing fewer molecules. Once again, let's consider the following equilibrium:

$$\underset{\textit{colorless}}{N_2O_4(g)} \rightleftharpoons \underset{\textit{brown}}{2\,NO_2(g)}$$

Notice that 1 molecule of N_2O_4 produces 2 molecules of NO_2. If we decrease the volume of the container, the gas pressure increases. To relieve the stress, the equilibrium shifts toward fewer molecules. That is, the equilibrium shifts to the left because there is 1 molecule of N_2O_4 for every 2 molecules of NO_2. Figure 16.12 illustrates the equilibrium shift for a decrease in volume.

If we increase the volume of the container, the gas pressure decreases. To relieve the stress, the equilibrium shifts toward more molecules. That is, the equilibrium shifts to the right because there are 2 molecules of NO_2 for 1 molecule of N_2O_4.

Effect of an Inert Gas

Let's consider what happens to the N_2O_4–NO_2 equilibrium if we add an inert gas such as helium. Since helium does not participate in the N_2O_4–NO_2 equilibrium, the partial gas pressures of N_2O_4 and NO_2 do not change. Since the pressures of N_2O_4 and NO_2 remain constant, there is no shift in equilibrium. The following example exercise further illustrates Le Chatelier's principle as applied to gaseous systems:

Example Exercise 16.6 Gaseous State Equilibria Shifts

Charcoal reacts with steam according to the equation.

$$C(s) + H_2O(g) + \textit{heat} \rightleftharpoons CO(g) + H_2(g)$$

Predict the direction of equilibrium shift for each of the following stresses:

(a) increase $[H_2O]$
(b) decrease $[H_2O]$
(c) increase $[CO]$
(d) decrease $[H_2]$
(e) increase temperature
(f) decrease volume
(g) add a catalyst
(h) add charcoal

Solution

Let's apply Le Chatelier's principle to each of the stresses.

(a) If $[H_2O]$ increases, the equilibrium shifts to the *right*.

(b) If $[H_2O]$ decreases, the equilibrium shifts to the *left*.

(c) If [CO] increases, the equilibrium shifts to the *left*.
(d) If $[H_2]$ decreases, the equilibrium shifts to the *right*.
(e) If the temperature increases, the equilibrium shifts to the *right*.
(f) If the volume decreases, the equilibrium shifts toward fewer gas molecules. The equilibrium shifts to the *left*.
(g) Adding a catalyst increases the rate of reaction but has no affect on the amount of reactant or product. Thus, there is *no shift* in the equilibrium.
(h) Adding carbon has no effect on the gaseous equilibrium because carbon is a solid. Thus, there is *no shift* in equilibrium.

Practice Exercise

Methane, CH_4, reacts with limited oxygen according to the equation.

$$2\,CH_4(g) + O_2(g) \rightleftharpoons 2\,CO(g) + 4\,H_2(g) + heat$$

Predict the direction of equilibrium shift for each of the following stresses:

(a) increase $[CH_4]$
(b) increase [CO]
(c) decrease $[O_2]$
(d) decrease $[H_2]$
(e) decrease temperature
(f) decrease volume
(g) add a catalyst
(h) add gaseous neon

Answers: (a) shifts right; (b) shifts left; (c) shifts left; (d) shifts right; (e) shifts right; (f) shifts left; (g) no shift; (h) no shift

Concept Exercise

What is the effect of increasing (a) temperature and (b) pressure for the following equilibrium?

$$SO_3(g) + NO(g) + heat \rightleftharpoons SO_2(g) + NO_2(g)$$

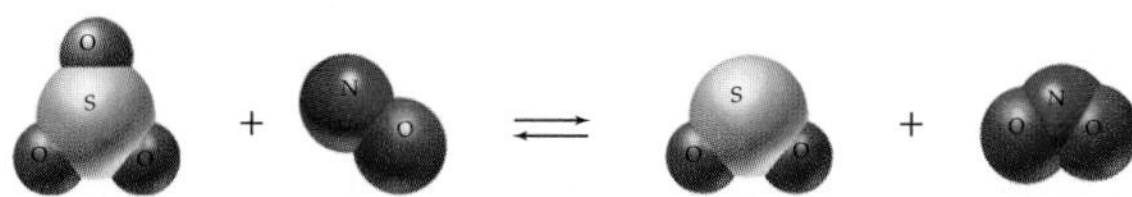

Answer: See Appendix G.

16.6 IONIZATION EQUILIBRIUM CONSTANT, K_i

OBJECTIVES

- To write the equilibrium constant expression for a weak acid or a weak base.
- To calculate an ionization constant, K_i, from experimental data.

In Section 16.3, we developed the concept of a general equilibrium constant expression, K_{eq}. Here we will develop a special equilibrium constant expression for a weak acid or weak base. This type of equilibrium is characterized by a weak acid or base in equilibrium with its ions.

Ionization of a Weak Acid

Let's begin with an aqueous solution of a weak acid. For example, acetic acid in an aqueous solution yields the following:

$$HC_2H_3O_2(aq) + H_2O(l) \rightleftharpoons H_3O^+(aq) + C_2H_3O_2^-(aq)$$

In aqueous solutions the concentration of water is constant. We can therefore simplify the reversible reaction as follows:

$$HC_2H_3O_2(aq) \rightleftharpoons H^+(aq) + C_2H_3O_2^-(aq)$$

In general, the equilibrium for weak acids lies overwhelmingly to the left. Typically, less than 1% of the parent molecules ionize. In the preceding reaction, very little ionization of acetic acid occurs. Although about 1% of the molecules form ions, 99% do not. Thus, the equilibrium strongly favors the reverse reaction.

▲ **Acetic Acid, $HC_2H_3O_2$** Acetic acid dissolved in water is a weak acid, which is less than 1% ionized.

When we substitute the acetic acid equilibrium into the general equilibrium constant expression, we obtain

$$K_i = \frac{[H^+]\,[C_2H_3O_2^-]}{[HC_2H_3O_2]}$$

where K_i is the **ionization equilibrium constant** for the weak acid. Since a weak acid ionizes only slightly, K_i is a very small number. The following example exercise provides practice in writing equilibrium constant expressions:

Example Exercise 16.7 Writing K_i Expressions

Write the equilibrium constant expression for the following weak acid:

$$HNO_2(aq) \rightleftharpoons H^+(aq) + NO_2^-(aq)$$

Solution

We substitute the product concentrations into the numerator and the reactant concentration into the denominator; thus,

$$K_i = \frac{[H^+]\,[NO_2^-]}{[HNO_2]}$$

Practice Exercise

Write the equilibrium constant expression for the following weak acid:

$$H_2CO_3(aq) \rightleftharpoons H^+(aq) + HCO_3^-(aq)$$

Answer: Carbonic acid, H_2CO_3, ionizes to give H^+ and HCO_3^-. Substituting into the ionization equilibrium expression, we have

$$K_i = \frac{[H^+]\,[HCO_3^-]}{[H_2CO_3]}$$

Concept Exercise

Write the equilibrium constant expression for the following weak acid:

$$H_3PO_4(aq) \rightleftharpoons H^+(aq) + H_2PO_4^-(aq)$$

Answer: See Appendix G.

▲ **Aqueous Ammonia, NH_4OH** Ammonia gas dissolved in water is a weak base, which is less than 1% ionized.

Ionization of a Weak Base

An aqueous solution of a weak base behaves similarly to an aqueous solution of a weak acid. That is, a weak base ionizes slightly and is in equilibrium with its ions. For example, aqueous ammonium hydroxide gives the following equilibrium:

$$NH_4OH(aq) \rightleftharpoons NH_4^+(aq) + OH^-(aq)$$

Typically, the equilibrium for weak bases lies overwhelmingly to the left. In ammonium hydroxide, less than 1% ionization occurs. Thus, the equilibrium strongly favors the reverse reaction. Substituting into the general equilibrium constant expression, we have

$$K_i = \frac{[NH_4^+]\,[OH^-]}{[NH_4OH]}$$

Determining K_i by Experiment

In the laboratory, we can determine the experimental equilibrium constant for a weak acid or weak base from pH measurements. For example, we can use a pH meter to measure $[H^+]$ in a solution. K_i is then calculated from the hydrogen ion concentration.

Let's calculate the ionization constant, K_i, for acetic acid. In an experiment, a 0.100 *M* solution of acetic acid is found to have a hydrogen ion concentration of 0.00134 *M*. We begin by analyzing the equilibrium concentrations. For each acetic acid molecule that ionizes, we obtain one H^+ and one $C_2H_3O_2^-$. Therefore, $[H^+]$ must equal $[C_2H_3O_2^-]$, that is, 0.00134 *M*. The equilibrium reaction is

$$\underset{0.100\ M}{HC_2H_3O_2(aq)} \rightleftharpoons \underset{0.00134\ M}{H^+(aq)} + \underset{0.00134\ M}{C_2H_3O_2^-(aq)}$$

Next, we write the equilibrium constant expression by placing the ion concentrations in the numerator and the molar concentration of the weak acid in the denominator. That gives

$$K_i = \frac{[H^+][C_2H_3O_2^-]}{[HC_2H_3O_2]}$$

Substituting into the equilibrium expression for each concentration, we have

$$K_i = \frac{[0.00134][0.00134]}{[0.100]}$$

$$= 1.80 \times 10^{-5}$$

The ionization constant for acetic acid is a small value, 1.80×10^{-5}, because acetic acid is a weak acid and ionizes only slightly in an aqueous solution.

16.7 EQUILIBRIA SHIFTS FOR WEAK ACIDS AND BASES

OBJECTIVE

▸ To apply Le Chatelier's principle to solutions of weak acids and weak bases.

In Section 16.5, we discussed Le Chatelier's principle applied to equilibria in the gaseous state. This principle also applies to equilibria in aqueous solutions. That is, a change in the concentration of any species in an aqueous solution disturbs the equilibrium and causes a shift between reactants and products.

Let's reconsider acetic acid, which ionizes very little in an aqueous solution. We can indicate the weak acid equilibrium as follows:

$$HC_2H_3O_2(aq) \rightleftharpoons H^+(aq) + C_2H_3O_2^-(aq)$$

If we increase the molar concentration of $HC_2H_3O_2$, we stress the left side of the equilibrium. The reversible reaction responds to the stress by shifting to the right. The final result is that more $HC_2H_3O_2$ molecules ionize and the molar concentrations of H^+ and $C_2H_3O_2^-$ increase.

If we increase the molar concentration of H^+ by adding a strong acid, we stress the right side of the equilibrium. The equilibrium responds to the stress by shifting to the left. The final result is that the molar concentration of $HC_2H_3O_2$ increases, whereas the molar concentration of $C_2H_3O_2^-$ decreases. Since additional acid was added, the molar concentration of H^+ also increases.

We can also indicate a change in $[H^+]$ that corresponds to a change in pH. Since $[H^+]$ and pH are inversely related, $[H^+]$ decreases as pH increases. Therefore, if the pH of an acetic acid solution increases, $[H^+]$ decreases. A decrease in the $[H^+]$ causes the equilibrium to shift to the right. Thus, $[HC_2H_3O_2]$ decreases, whereas $[C_2H_3O_2^-]$ increases.

If we increase the molar concentration of $C_2H_3O_2^-$ by adding solid $NaC_2H_3O_2$, we stress the right side of the equilibrium. As the $NaC_2H_3O_2$ dissolves, the equilibrium responds to the stress by shifting to the left. The final result is that the $[HC_2H_3O_2]$ increases and $[H^+]$ decreases. Since $NaC_2H_3O_2$ was added, $[C_2H_3O_2^-]$ also increases.

If we add solid NaOH, we neutralize H^+ and stress the right side of the equilibrium. The NaOH lowers the H^+ concentration, which shifts the equilibrium to the right. After the equilibrium shift, the $HC_2H_3O_2$ concentration has decreased, whereas the $C_2H_3O_2^-$ concentration has increased.

Finally, consider what happens if we add solid $NaNO_3$ to aqueous acetic acid. Since sodium nitrate is a soluble salt, it dissolves giving Na^+ and NO_3^- in solution. Neither of these ions participates in the acetic acid equilibrium. Thus, adding $NaNO_3$ has no effect, and there is no shift in the equilibrium. The following example exercise further illustrates Le Chatelier's principle:

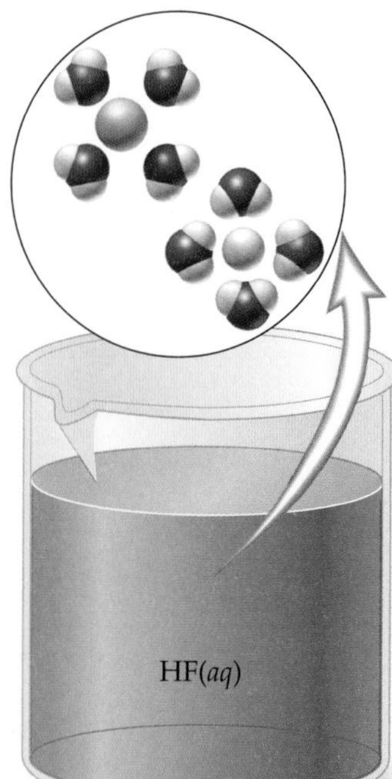

▲ **Hydrofluoric Acid, HF**
Hydrofluoric acid is a weak acid that is less than 1% ionized in an aqueous solution.

Example Exercise 16.8 Weak Acid Equilibria Shifts

Hydrofluoric acid, HF, ionizes according to the following equation:

$$HF(aq) \rightleftharpoons H^+(aq) + F^-(aq)$$

Predict the direction of equilibrium shift for each of the following stresses:

(a) increase [HF]
(b) increase $[H^+]$
(c) decrease $[F^-]$
(d) add gaseous HCl
(e) add solid NaF
(f) add solid NaCl
(g) add solid NaOH
(h) decrease pH

Solution

Let's apply Le Chatelier's principle to each of the stresses.

(a) If [HF] increases, the equilibrium shifts to the *right.*
(b) If $[H^+]$ increases, the equilibrium shifts to the *left.*
(c) If $[F^-]$ decreases, the equilibrium shifts to the *right.*
(d) If HCl is added, $[H^+]$ increases and the equilibrium shifts to the *left.*
(e) If NaF is added, $[F^-]$ increases and the equilibrium shifts to the *left.*
(f) If NaCl is added, $[Cl^-]$ increases, which has no effect on the equilibrium. Thus, there is *no shift.*
(g) If NaOH is added, the OH^- neutralizes H^+ and lowers its concentration. Thus, the equilibrium shifts to the *right.*
(h) If the pH decreases, $[H^+]$ increases, and the equilibrium shifts to the *left.*

Practice Exercise

Hydrocyanic acid, HCN, ionizes according to the following equation:

$$HCN(aq) \rightleftharpoons H^+(aq) + CN^-(aq)$$

Predict the direction of equilibrium shift for each of the following stresses:

(a) increase [HCN]
(b) decrease $[H^+]$
(c) increase $[CN^-]$
(d) add aqueous HNO_3
(e) add solid KCN
(f) add solid KCl
(g) add solid KOH
(h) increase pH

Answers: (a) shifts right; (b) shifts right; (c) shifts left; (d) shifts left; (e) shifts left; (f) no shift; (g) shifts right; (h) shifts right

Concept Exercise

What is the effect on the following equilibrium of adding (a) NH_4Cl, and (b) NaOH?

$$NH_4OH(aq) \rightleftharpoons NH_4^+(aq) + OH^-(aq)$$

Answer: See Appendix G.

16.8 SOLUBILITY PRODUCT EQUILIBRIUM CONSTANT, K_{sp}

OBJECTIVES

- To write the equilibrium constant expression for a slightly soluble ionic compound.
- To calculate a solubility product constant, K_{sp}, from experimental data.

Guldberg and Waage laid the foundation for the law of chemical equilibrium based on their study of insoluble barium carbonate in an aqueous solution. After adding soluble potassium sulfate, they found that the insoluble precipitate contained both barium carbonate and barium sulfate. The Norwegian chemists proposed the following reversible reaction in an aqueous solution:

$$BaCO_3(s) + K_2SO_4(aq) \rightleftharpoons BaSO_4(s) + K_2CO_3(aq)$$

Notice that $BaCO_3$ and $BaSO_4$ are both insoluble precipitates; however, each substance is very slightly soluble. In an aqueous solution, $BaCO_3$ dissociates to give

$$BaCO_3(s) \rightleftharpoons Ba^{2+}(aq) + CO_3^{2-}(aq)$$

When $BaCO_3$ dissociates into ions, Ba^{2+} can recombine with SO_4^{2-} to give an insoluble $BaSO_4$ precipitate. Therefore, they concluded that seemingly insoluble precipitates in an aqueous solution are actually *very slightly soluble*. Moreover, these insoluble ionic compounds are in dynamic equilibrium with their constituent ions.

Now, let's write the equilibrium reaction for very slightly soluble Ag_2SO_4 in an aqueous solution. Ag_2SO_4 precipitate dissociates to give

$$Ag_2SO_4(s) \rightleftharpoons 2\,Ag^+(aq) + SO_4^{2-}(aq)$$

Notice that Ag_2SO_4 dissociates into 2 Ag^+ and 1 SO_4^{2-}. We can substitute this equilibrium into the general K_{eq} expression. This gives

$$K_{eq} = \frac{[Ag^+]^2\,[SO_4^{2-}]}{[Ag_2SO_4]}$$

If an aqueous solution is saturated with insoluble Ag_2SO_4, the amount of solid precipitate has no effect on the equilibrium. As long as there is some undissolved precipitate, the concentrations of Ag^+ and SO_4^{2-} do not change. The amount of solid has no effect on the equilibrium and we can write a special equilibrium expression as

$$K_{sp} = [Ag^+]^2\,[SO_4^{2-}]$$

where K_{sp} is the **solubility product equilibrium constant**. We can define K_{sp} as a constant that expresses the molar equilibrium concentrations of ions in an aqueous solution for a slightly dissociated ionic compound. For an insoluble compound in an aqueous solution, only the ions appear in the equilibrium expression; the precipitate

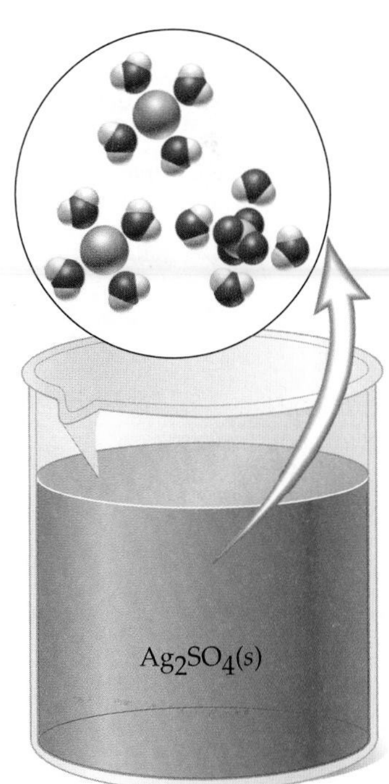

▲ **Silver Sulfate, Ag_2SO_4** Silver sulfate is an insoluble compound that is less than 1% dissociated in an aqueous solution.

never appears in the K_{sp} expression. The following example exercise provides further practice in writing solubility product expressions:

Example Exercise 16.9 Writing K_{sp} Expressions

Write the solubility product expression for slightly soluble calcium phosphate in an aqueous solution.

$$Ca_3(PO_4)_2(s) \rightleftharpoons 3\,Ca^{2+}(aq) + 2\,PO_4^{3-}(aq)$$

Solution

Calcium phosphate dissociates into 3 Ca^{2+} and 2 PO_4^{3-}. Thus, the Ca^{2+} concentration is raised to the third power and the PO_4^{3-} concentration to the second power.

$$K_{sp} = [Ca^{2+}]^3\,[PO_4^{3-}]^2$$

Practice Exercise

Write the solubility product expression for slightly soluble aluminum carbonate in an aqueous solution.

$$Al_2(CO_3)_3(s) \rightleftharpoons 2\,Al^{3+}(aq) + 3\,CO_3^{2-}(aq)$$

Answers: $K_{sp} = [Al^{3+}]^2\,[CO_3^{2-}]^3$

Concept Exercise

Write the equilibrium constant expression for the following slightly soluble ionic compound:

$$Al(OH)_3(s) \rightleftharpoons Al^{3+}(aq) + 3\,OH^-(aq)$$

Answer: See Appendix G.

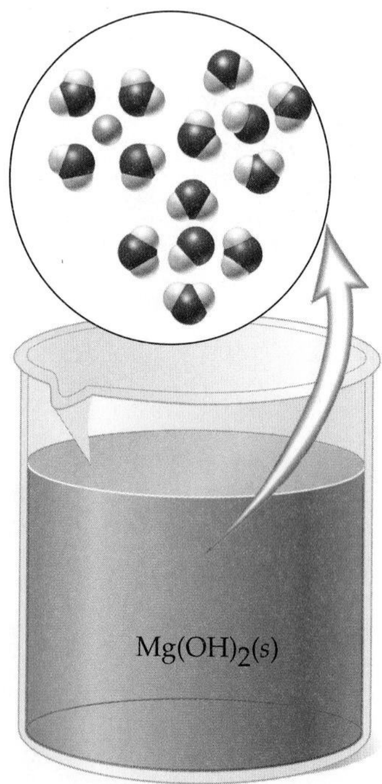

▲ **Magnesium Hydroxide, $Mg(OH)_2$** Magnesium hydroxide is an insoluble compound that is less than 1% dissociated in an aqueous solution.

Determining K_{sp} by Experiment

We can use several methods to find K_{sp} for a slightly soluble ionic compound. One method is to determine experimentally the ion concentrations in an aqueous solution and calculate a value for the solubility product constant. For example, in a saturated solution of milk of magnesia, $Mg(OH)_2$, the hydroxide ion concentration is found to be 0.00032 *M*. To calculate the solubility product constant, let's first analyze the equilibrium.

$$Mg(OH)_2(s) \rightleftharpoons Mg^{2+}(aq) + 2\,OH^-(aq)$$

Notice that $Mg(OH)_2$ dissociates into 1 Mg^{2+} and 2 OH^-. The equilibrium constant expression for the dissociation is

$$K_{sp} = [Mg^{2+}]\,[OH^-]^2$$

Since there is 1 Mg^{2+} for every 2 OH^-, $[Mg^{2+}]$ equals half of $[OH^-]$. The value of $[OH^-]$ is given as 0.00032 *M*, and so the $[Mg^{2+}]$ is 0.00016 *M*. Substituting into the K_{sp} expression, we have

$$K_{sp} = [1.6 \times 10^{-4}]\,[3.2 \times 10^{-4}]^2$$

Simplifying,

$$K_{sp} = 1.6 \times 10^{-11}$$

The solubility product constant for magnesium hydroxide is 1.6×10^{-11}.

16.9 EQUILIBRIA SHIFTS FOR SLIGHTLY SOLUBLE COMPOUNDS

OBJECTIVE

▶ To apply Le Chatelier's principle to a saturated solution of a slightly soluble ionic compound.

In Section 16.7, we discussed Le Chatelier's principle as it applies to ionization equilibria. This principle also applies to dissociation equilibria. That is, if we change the concentration of any ions participating in a dissociation equilibrium, there will be a shift to relieve the stress.

Let's consider the dissociation of an antacid tablet in an aqueous solution. The tablet contains aluminum hydroxide, $Al(OH)_3$, which is only slightly soluble. The dissociation equilibrium is

$$Al(OH)_3(s) \rightleftharpoons Al^{3+}(aq) + 3\,OH^-(aq)$$

If the solution is saturated, Al^{3+} and OH^- are at maximum concentration. Therefore, if we add more aluminum hydroxide, $Al(OH)_3$, to the solution, it has no effect. Hence, there is no shift in equilibrium.

Suppose we add solid $AlCl_3$ to the solution to increase the Al^{3+} concentration. The solid $AlCl_3$ dissolves to give more Al^{3+}. The aluminum ion stresses the right side of the reaction, and the equilibrium shifts to the left. Furthermore, more hydroxide ions precipitate and the concentration of OH^- decreases.

Suppose we add solid NaCl to the aqueous solution. Sodium chloride dissolves to give sodium ions and chloride ions. Since neither Na^+ nor Cl^- participate in the equilibrium, NaCl has no effect. Thus, there is no shift in equilibrium, and the concentrations of Al^{3+} and OH^- remain constant.

Let's consider a less obvious stress on the equilibrium. Let's add aqueous HCl and increase the H^+ concentration. Hydrochloric acid is a strong acid and ionizes completely to give H^+ and Cl^-. Although neither ion is shown in the equilibrium, H^+ neutralizes OH^-. Thus, the H^+ decreases the OH^- concentration.

As the concentration of OH^- decreases, we stress the right side of the equation and shift the equilibrium to the right. As the equilibrium shifts, more $Al(OH)_3$ dissolves and the Al^{3+} concentration increases. The following example exercise further illustrates Le Chatelier's principle:

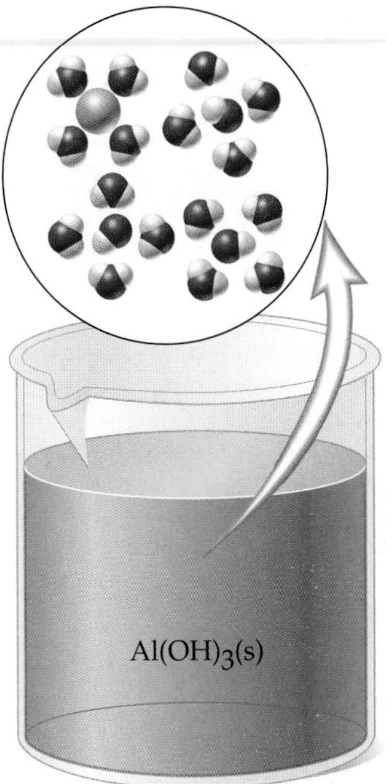

▲ **Aluminum Hydroxide, $Al(OH)_3$** Aluminum hydroxide is an insoluble compound that is less than 1% dissociated in an aqueous solution.

Example Exercise 16.10 Solubility Equilibria Shifts

A saturated solution of silver chloride dissociates according to the following equation:

$$AgCl(s) \rightleftharpoons Ag^+(aq) + Cl^-(aq)$$

Predict the direction of equilibrium shift for each of the following stresses:

(a) increase $[Ag^+]$
(b) decrease $[Ag^+]$
(c) increase $[Cl^-]$
(d) decrease $[Cl^-]$
(e) add solid $AgNO_3$
(f) add solid $NaNO_3$
(g) add solid AgCl
(h) decrease pH

Solution

Let's apply Le Chatelier's principle to each of the stresses.

(a) If $[Ag^+]$ increases, the equilibrium shifts to the *left*.
(b) If $[Ag^+]$ decreases, the equilibrium shifts to the *right*.
(c) If $[Cl^-]$ increases, the equilibrium shifts to the *left*.
(d) If $[Cl^-]$ decreases, the equilibrium shifts to the *right*.
(e) If $AgNO_3$ is added, $[Ag^+]$ increases and the equilibrium shifts to the *left*.
(f) If $NaNO_3$ is added, $[Na^+]$ and $[NO_3^-]$ increase, which has no effect on the equilibrium. Thus, there is *no shift*.
(g) If AgCl is added, there is no effect on the equilibrium because the solution is saturated with AgCl. Thus, there is *no shift*.
(h) If the pH decreases, $[H^+]$ increases. Since the hydrogen ion has no effect on the equilibrium, there is *no shift*.

▲ **Saturated Solution of AgCl** By adding various substances into the solution, we stress the solubility equilibrium. A stress may cause a Le Chatelier shift to the left (more precipitate) or to the right (less precipitate).

Practice Exercise

A saturated solution of magnesium hydroxide dissociates according to the following equation:

$$Mg(OH)_2(s) \rightleftharpoons Mg^{2+}(aq) + 2\,OH^-(aq)$$

Predict the direction of equilibrium shift for each of the following stresses:

(a) increase $[Mg^{2+}]$
(b) decrease $[Mg^{2+}]$
(c) increase $[OH^-]$
(d) decrease $[OH^-]$
(e) add solid KOH
(f) add solid KNO_3
(g) add solid $Mg(OH)_2$
(h) decrease pH

Answers: (a) shifts left; (b) shifts right; (c) shifts left; (d) shifts right; (e) shifts left; (f) no shift; (g) no shift; (h) shifts right (neutralizes OH^-)

Concept Exercise

What is the effect on the following equilibrium of adding (a) $Ca(NO_3)_2$, and (b) Na_2CO_3?

$$CaCO_3(s) \rightleftharpoons Ca^{2+}(aq) + CO_3^{\,2-}(aq)$$

Answer: See Appendix G.

Summary

Section 16.1 According to the **collision theory** of reaction rates, the rate of a reaction is regulated by the collision frequency, collision energy, and orientation of the molecules. We can speed up a reaction by (1) increasing the concentration of reactants, (2) raising the temperature, or (3) adding a catalyst.

Section 16.2 An **endothermic reaction** absorbs heat energy, and an **exothermic reaction** releases heat energy. A **reaction profile** graphs the change in energy as reactants are converted to products. For molecules to react, they must have enough energy to achieve the **transition state**. The energy necessary to reach the transition state is called the **activation energy,** E_{act}. A **catalyst** can be defined as any substance that speeds up a reaction by lowering the activation energy. The difference between the energy of the reactants and the products is called the **heat of reaction,** ΔH.

Section 16.3 A **reversible reaction** takes place simultaneously in both the forward and reverse directions. The **rate of reaction** is the rate at which the concentration of the reactants decreases, or the concentration of the products increases, per unit time. When the rates of the forward and reverse reactions are equal, the reaction is at **chemical equilibrium**. The **law of chemical equilibrium** states that the molar concentrations of products divided by reactants (each raised to a power corresponding to a coefficient in the balanced chemical equation) equals a **general equilibrium constant,** K_{eq}. The law of chemical equilibrium applies to every reversible reaction and is illustrated in Table 16.3.

TABLE 16.3 SELECTED TYPES OF CHEMICAL EQUILIBRIA

TYPE OF EQUILIBRIUM	EQUILIBRIUM CONSTANT EXPRESSION
General Equilibrium	
$a\,A + b\,B \rightleftharpoons c\,C + d\,D$	$K_{eq} = \dfrac{[C]^c\,[D]^d}{[A]^a\,[B]^b}$
Ionization Equilibrium	
$HC_2H_3O_2(aq) \rightleftharpoons H^+(aq) + C_2H_3O_2^-(aq)$	$K_i = \dfrac{[H^+]\,[C_2H_3O_2^-]}{[HC_2H_3O_2]}$
Dissociation Equilibrium	
$Al(OH)_3(s) \rightleftharpoons Al^{3+}(aq) + 3\,OH^-(aq)$	$K_{sp} = [Al^{3+}]\,[OH^-]^3$

Section 16.4 In a **homogeneous equilibrium**, all the participating substances are in the same physical state, for example, an equilibrium in which all the reactants and products are in the gaseous state. In a **heterogeneous equilibrium**, one of the participating substances is in a different physical state, for example, an equilibrium in which a solid substance decomposes to give a mixture of gases.

Section 16.5 A change in concentration, temperature, or pressure can cause a stress to a reversible reaction. According to **Le Chatelier's principle**, a chemical reaction at equilibrium shifts in order to relieve a stress. The shifts in equilibrium for various stresses are summarized in Table 16.4.

Section 16.6 The law of chemical equilibrium applies to aqueous solutions of weak acids and weak bases. The **ionization equilibrium constant**, K_i, is equal to the molar concentrations of the ions divided by the molar concentration of the weak acid or base.

Section 16.7 Le Chatelier's principle applies to aqueous solutions of weak acids and weak bases. If we change the concentration of one of the ions in an aqueous solution, the equilibrium will shift to relieve the stress. If we add a substance that does not effect the equilibrium, there will be no shift.

Section 16.8 The law of chemical equilibrium applies to saturated solutions of slightly soluble ionic compounds. The **solubility product equilibrium constant**, K_{sp}, is equal to the product of the molar concentrations of the ions in solution. The insoluble compound is not written in the K_{sp} expression because it has no effect on the equilibrium in a saturated solution.

Section 16.9 Le Chatelier's principle applies to saturated solutions of slightly soluble ionic compounds. If we change the concentration of one of the ions in an aqueous solution, the equilibrium will shift to relieve the stress. If we add a substance that does not participate in the equilibrium, there will be no shift.

TABLE 16.4 SUMMARY OF CHEMICAL EQUILIBRIA SHIFTS

STRESS ON EQUILIBRIUM	EFFECT ON EQUILIBRIUM
Concentration Increases (all systems)	
for a reactant	*shifts to the right*
for a product	*shifts to the left*
Temperature Increases (gaseous systems)	
for an endothermic reaction	*shifts to the right*
for an exothermic reaction	*shifts to the left*
Pressure Increases (gaseous systems)	
more reactant molecules	*shifts to the right*
more product molecules	*shifts to the left*
molecules of reactants and products are equal	*no shift*

Key Terms *Answers to Key Terms are in Appendix H.*

Select the key term that corresponds to each of the following definitions.

_____ **1.** the principle that the rate of a chemical reaction is controlled by the frequency and energy of molecules striking each other
_____ **2.** a chemical reaction that consumes heat energy
_____ **3.** a chemical reaction that liberates heat energy
_____ **4.** a graph of the energy of reactants and products as a reaction occurs
_____ **5.** the highest point on the reaction profile
_____ **6.** the energy required for reactants to reach the transition state
_____ **7.** the difference in energy between the reactants and the products
_____ **8.** a substance that speeds up a reaction by lowering the energy of activation
_____ **9.** a reaction that proceeds toward reactants and products simultaneously
_____ **10.** the rate at which the concentrations of reactants decrease per unit time
_____ **11.** a dynamic state for a reversible reaction in which the rates of the forward and reverse reactions are the same
_____ **12.** the principle that the molar concentrations of the products in a reversible reaction divided by the molar concentrations of the reactants (each raised to a power corresponding to a coefficient in the balanced equation) is equal to a constant

(a) activation energy (E_{act}) *(Sec. 16.2)*
(b) catalyst *(Sec. 16.2)*
(c) chemical equilibrium *(Sec. 16.3)*
(d) collision theory *(Sec. 16.1)*
(e) endothermic reaction *(Sec. 16.2)*
(f) exothermic reaction *(Sec. 16.2)*
(g) general equilibrium constant (K_{eq}) *(Sec. 16.3)*
(h) heat of reaction (ΔH) *(Sec. 16.2)*
(i) heterogeneous equilibrium *(Sec. 16.4)*
(j) homogeneous equilibrium *(Sec. 16.4)*

_____ 13. the constant that expresses the molar equilibrium concentration of each substance participating in a reversible reaction at a given temperature

_____ 14. a type of equilibrium in which all participating species are in the same physical state; e.g., gaseous state or aqueous solution

_____ 15. a type of equilibrium in which all participating species are not in the same physical state

_____ 16. the statement that a chemical equilibrium stressed by a change in concentration, temperature, or pressure, shifts to relieve the stress

_____ 17. the constant that expresses the molar equilibrium concentrations of ions in an aqueous solution for a slightly ionized acid or base

_____ 18. the constant that expresses the molar equilibrium concentrations of ions in an aqueous solution for a slightly dissociated ionic compound

(k) ionization equilibrium constant (K_i) *(Sec. 16.6)*
(l) law of chemical equilibrium *(Sec. 16.3)*
(m) Le Chatelier's principle *(Sec. 16.5)*
(n) rate of reaction *(Sec. 16.3)*
(o) reaction profile *(Sec. 16.2)*
(p) reversible reaction *(Sec. 16.3)*
(q) solubility product equilibrium constant (K_{sp}) *(Sec. 16.8)*
(r) transition state *(Sec. 16.2)*

Exercises *Answers to odd-numbered Exercises are in Appendix I.*

Collision Theory (Sec. 16.1)

1. Which of the following experimental factors influence the rate of a reaction?
   (a) concentration **(b)** temperature
   (c) catalyst **(d)** ultraviolet light
2. Which of the following theoretical factors influence the rate of a reaction?
   (a) collision frequency **(b)** collision energy
   (c) collision orientation **(d)** platinum container
3. Draw a diagram showing an effective collision orientation between a hydrogen molecule and a bromine molecule.
4. Draw a diagram showing an ineffective collision orientation between a hydrogen molecule and a bromine molecule.
5. State the effect on the rate of reaction for each of the following:
   (a) increase the concentration of a reactant
   (b) increase the temperature of the reaction
   (c) add a metal catalyst
6. State the effect on the rate of reaction for each of the following:
   (a) decrease the concentration of a reactant
   (b) decrease the temperature of the reaction
   (c) add an acid catalyst
7. The Haber process uses a metal oxide catalyst to produce ammonia gas. Does the catalyst increase the amount of ammonia gas? Explain.
8. The Contact process uses a metal catalyst to produce sulfur trioxide gas for the manufacture of sulfuric acid. Does the catalyst increase the amount of sulfur trioxide gas? Explain.

Energy Profiles of Chemical Reactions (Sec. 16.2)

9. Phosphorus pentachloride is used in the electronics industry to manufacture computer chips. Draw the energy profile for the following reaction:
   $$PCl_5(g) + heat \rightleftharpoons PCl_3(g) + Cl_2(g)$$
10. Boron trifluoride is used in the electronics industry to manufacture computer chips. Draw the energy profile for the following reaction:
   $$2\,B(s) + 3\,F_2(g) \rightleftharpoons 2\,BF_3(g) + heat$$
11. Draw the energy profile for the following endothermic reaction:
   $$H_2(g) + I_2(g) \rightleftharpoons 2\,HI(g)$$
   Label the axes "Energy" and "Progress of reaction" and indicate the reactants, products, transition state, activation energy, and energy of reaction.
12. Draw the energy profile for the following exothermic reaction:
   $$H_2(g) + Cl_2(g) \rightleftharpoons 2\,HCl(g)$$
   Label the axes "Energy" and "Progress of reaction" and indicate the reactants, products, transition state, activation energy, and energy of reaction.
13. State the effect of a catalyst on the energy of activation, E_{act}.
14. State the effect of a catalyst on the heat of reaction, ΔH.
15. Consider the energy profile for a reversible exothermic reaction. Is E_{act} greater for the forward or the reverse reaction?
16. Consider the energy profile for a reversible endothermic reaction. Is E_{act} greater for the forward or the reverse reaction?

The Chemical Equilibrium Concept (Sec. 16.3)

17. Define the rate of a forward reaction in terms of (a) the reactant concentration and (b) the product concentration.
18. Define chemical equilibrium using the symbols $rate_f$ and $rate_r$.
19. Which of the following statements is true regarding the general equilibrium expression?
   (a) K_{eq} for gaseous equilibria does not depend on temperature.
   (b) K_{eq} for gaseous equilibria does not include solid substances.
20. Which of the following statements is true regarding the general equilibrium expression?
   (a) K_{eq} can be determined experimentally.
   (b) K_{eq} can be determined theoretically.

General Equilibrium Constant, K_{eq} (Sec. 16.4)

21. Write the general equilibrium constant expression for each of the following:
   (a) $2\,A \rightleftharpoons C$ **(b)** $A + 2\,B \rightleftharpoons 3\,C$
   (c) $2\,A + 3\,B \rightleftharpoons 4\,C + D$

22. Write the general equilibrium constant expression for each of the following:
(a) $3\,A \rightleftarrows 2\,C$
(b) $A + B \rightleftarrows 2\,C$
(c) $3\,A + 5\,B \rightleftarrows C + 4\,D$

23. Does a substance in the solid state appear in the equilibrium expression for a gaseous state reaction?

24. Does a substance in the liquid state appear in the equilibrium expression for a gaseous state reaction?

25. Write the equilibrium constant expression for each of the following reversible reactions:
(a) $H_2(g) + F_2(g) \rightleftarrows 2\,HF(g)$
(b) $4\,NH_3(g) + 7\,O_2(g) \rightleftarrows 4\,NO_2(g) + 6\,H_2O(g)$
(c) $ZnCO_3(s) \rightleftarrows ZnO(s) + CO_2(g)$

26. Write the equilibrium constant expression for each of the following reversible reactions:
(a) $H_2(g) + Br_2(g) \rightleftarrows 2\,HBr(g)$
(b) $4\,HCl(g) + O_2(g) \rightleftarrows 2\,Cl_2(g) + 2\,H_2O(g)$
(c) $CO(g) + 2\,H_2(g) \rightleftarrows CH_3OH(l)$

27. Given the equilibrium concentrations for each gas at 850 °C, calculate the value of K_{eq} for the manufacture of sulfur trioxide.

$$2\,SO_2(g) + O_2(g) \rightleftarrows 2\,SO_3(g)$$
$$1.75\,M \quad 1.50\,M \quad 2.25\,M$$

28. Given the equilibrium concentrations for each gas at 500 °C, calculate the value of K_{eq} for the manufacture of ammonia.

$$N_2(g) + 3\,H_2(g) \rightleftarrows 2\,NH_3(g)$$
$$0.400\,M \quad 1.20\,M \quad 0.195\,M$$

Equilibria Shifts for Gases (Sec. 16.5)

29. Weather conditions affect the smog equilibrium in the atmosphere. What happens to the nitrogen dioxide concentration on (a) hot, sunny days and (b) cool, overcast days?

$$N_2O_4(g) + heat \rightleftarrows 2\,NO_2(g)$$

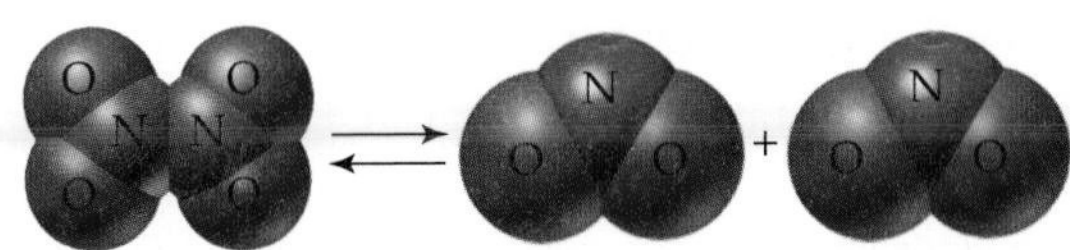

30. The conditions for producing ammonia industrially are 500 °C and 300 atm. What happens to the ammonia concentration if (a) the temperature increases and (b) the pressure increases?

$$N_2(g) + 3\,H_2(g) \rightleftarrows 2\,NH_3(g) + heat$$

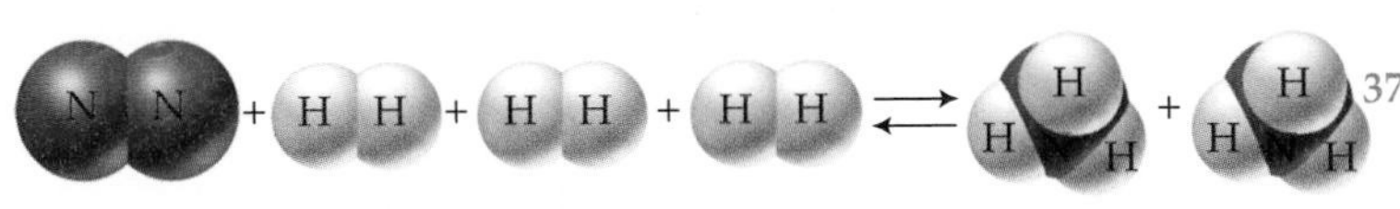

31. The industrial process for producing carbon monoxide gas involves passing carbon dioxide over hot charcoal.

$$C(s) + CO_2(g) + heat \rightleftarrows 2\,CO(g)$$

Predict the direction of equilibrium shift for each of the following stresses:
(a) increase $[CO_2]$ **(b)** decrease $[CO_2]$
(c) increase [CO] **(d)** decrease [CO]
(e) increase temperature **(f)** decrease temperature
(g) increase volume **(h)** decrease volume
(i) add C powder **(j)** add Kr inert gas

32. The industrial process for producing hydrogen gas involves reacting methane and steam at a high temperature.

$$CH_4(g) + H_2O(g) + heat \rightleftarrows CO(g) + 3\,H_2(g)$$

Predict the direction of equilibrium shift for each of the following stresses:
(a) increase $[CH_4]$ **(b)** decrease $[H_2O]$
(c) increase [CO] **(d)** decrease $[H_2]$
(e) increase temperature **(f)** decrease temperature
(g) increase volume **(h)** decrease volume
(i) add Zn dust **(j)** add Xe inert gas

33. Coal-burning power plants release sulfur dioxide into the atmosphere. The SO_2 is converted to SO_3 by nitrogen dioxide as follows:

$$SO_2(g) + NO_2(g) \rightleftarrows SO_3(g) + NO(g) + heat$$

Predict the direction of equilibrium shift for each of the following stresses:
(a) increase $[SO_2]$ **(b)** decrease $[NO_2]$
(c) increase $[SO_3]$ **(d)** decrease [NO]
(e) increase temperature **(f)** decrease temperature
(g) increase volume **(h)** decrease volume
(i) add He inert gas **(j)** ultraviolet light

34. Smog contains formaldehyde that is responsible for an eye-burning sensation. Formaldehyde, CH_2O, is produced from the reaction of ozone and atmospheric ethylene, C_2H_4, as follows:

$$2\,C_2H_4(g) + 2\,O_3(g) \rightleftarrows 4\,CH_2O(g) + O_2(g) + heat$$

Predict the direction of equilibrium shift for each of the following stresses:
(a) increase $[C_2H_4]$ **(b)** decrease $[O_3]$
(c) increase $[CH_2O]$ **(d)** decrease $[O_2]$
(e) increase temperature **(f)** decrease temperature
(g) increase volume **(h)** decrease volume
(i) add Ne inert gas **(j)** ultraviolet light

Ionization Equilibrium Constant, K_i (Sec. 16.6)

35. Write the equilibrium constant expression for each of the following weak acids:
(a) $HCHO_2(aq) \rightleftarrows H^+(aq) + CHO_2^-(aq)$
(b) $H_2C_2O_4(aq) \rightleftarrows H^+(aq) + HC_2O_4^-(aq)$
(c) $H_3C_6H_5O_7(aq) \rightleftarrows H^+(aq) + H_2C_6H_5O_7^-(aq)$

36. Write the equilibrium constant expression for each of the following weak bases:
(a) $NH_2OH(aq) + H_2O(l) \rightleftarrows NH_3OH^+(aq) + OH^-(aq)$
(b) $C_6H_5NH_2(aq) + H_2O(l) \rightleftarrows C_6H_5NH_3^+(aq) + OH^-(aq)$
(c) $(CH_3)_2NH(aq) + H_2O(l) \rightleftarrows (CH_3)_2NH_2^+(aq) + OH^-(aq)$

37. Nitrous acid, HNO_2, is used in the synthesis of organic dye compounds. If the hydrogen ion concentration of a 0.125 M solution is $7.5 \times 10^{-3}\,M$, what is the ionization constant for the acid?

38. Aqueous ammonium hydroxide, NH_4OH, is used as a household cleaning solution. If the hydroxide ion concentration of a 0.245 M solution is $2.1 \times 10^{-3}\,M$, what is the ionization constant for the base?

39. Hydrofluoric acid, HF, is used in the manufacture of silicon computer chips. If the pH of a 0.139 M solution is 2.00, what is the ionization constant of the acid?

40. Hydrazine, N_2H_4, is a weak base and is used as fuel in the space shuttle. If the pH of a 0.139 *M* solution is 11.00, what is the ionization constant of the base?

$$N_2H_4(aq) + H_2O(l) \rightleftharpoons N_2H_5^+(aq) + OH^-(aq)$$

Equilibria Shifts for Weak Acids and Bases (Sec. 16.7)

41. Given the chemical equation for the ionization of hydrofluoric acid

$$HF(aq) \rightleftharpoons H^+(aq) + F^-(aq)$$

Predict the direction of equilibrium shift for each of the following stresses:
(a) increase [HF]
(b) increase $[H^+]$
(c) decrease [HF]
(d) decrease $[F^-]$
(e) add solid NaF
(f) add gaseous HCl
(g) add solid NaOH
(h) increase pH

42. Given the chemical equation for the ionization of nitrous acid

$$HNO_2(aq) \rightleftharpoons H^+(aq) + NO_2^-(aq)$$

Predict the direction of equilibrium shift for each of the following stresses:
(a) decrease $[HNO_2]$
(b) decrease $[H^+]$
(c) increase $[HNO_2]$
(d) increase $[NO_2^-]$
(e) add solid KNO_2
(f) add solid KCl
(g) add solid KOH
(h) increase pH

43. Given the chemical equation for the ionization of acetic acid

$$HC_2H_3O_2(aq) \rightleftharpoons H^+(aq) + C_2H_3O_2^-(aq)$$

Predict the direction of equilibrium shift for each of the following stresses:
(a) increase $[HC_2H_3O_2]$
(b) increase $[H^+]$
(c) decrease $[HC_2H_3O_2]$
(d) decrease $[C_2H_3O_2^-]$
(e) add solid $NaC_2H_3O_2$
(f) add solid NaCl
(g) add solid NaOH
(h) increase pH

44. Given the chemical equation for the ionization of ammonium hydroxide

$$NH_4OH(aq) \rightleftharpoons NH_4^+(aq) + OH^-(aq)$$

Predict the direction of equilibrium shift for each of the following stresses:
(a) increase $[NH_4^+]$
(b) decrease $[OH^-]$
(c) increase $[NH_4OH]$
(d) decrease pH
(e) add gaseous NH_3
(f) add solid KCl
(g) add solid KOH
(h) add solid NH_4Cl

Solubility Product Equilibrium Constant, K_{sp} (Sec. 16.8)

45. Write the solubility product expression for each of the following slightly soluble ionic compounds in a saturated aqueous solution:
(a) $AgI(s) \rightleftharpoons Ag^+(aq) + I^-(aq)$
(b) $Ag_2CrO_4(s) \rightleftharpoons 2\,Ag^+(aq) + CrO_4^{2-}(aq)$
(c) $Ag_3PO_4(s) \rightleftharpoons 3\,Ag^+(aq) + PO_4^{3-}(aq)$

46. Write the solubility product expression for each of the following slightly soluble ionic compounds in a saturated aqueous solution:
(a) $Cu_2CO_3(s) \rightleftharpoons 2\,Cu^+(aq) + CO_3^{2-}(aq)$
(b) $ZnCO_3(s) \rightleftharpoons Zn^{2+}(aq) + CO_3^{2-}(aq)$
(c) $Al_2(CO_3)_3(s) \rightleftharpoons 2\,Al^{3+}(aq) + 3\,CO_3^{2-}(aq)$

47. The cobalt(II) ion concentration in a saturated solution of cobalt(II) sulfide, CoS, is 7.7×10^{-11} *M*. Calculate the value for the equilibrium constant.

48. The fluoride ion concentration in a saturated solution of magnesium fluoride, MgF_2, is 2.3×10^{-3} *M*. Calculate the value for the equilibrium constant.

49. The zinc ion concentration in a saturated solution of zinc phosphate, $Zn_3(PO_4)_2$, is 1.5×10^{-7} *M*. Calculate the value for the equilibrium constant.

50. The hydroxide ion concentration in a saturated solution of iron(III) hydroxide, $Fe(OH)_3$, is 2.2×10^{-10} *M*. Calculate the value for the equilibrium constant.

51. The K_{sp} values for $CaCO_3$ and CaC_2O_4 are 3.8×10^{-9} and 2.3×10^{-9}, respectively. In saturated solutions of $CaCO_3$ and CaC_2O_4, which has the higher calcium ion concentration?

52. The K_{sp} values for $MnCO_3$ and $Mn(OH)_2$ are 1.8×10^{-11} and 4.6×10^{-14}, respectively. In saturated solutions of $MnCO_3$ and $Mn(OH)_2$, which has the higher manganese(II) ion concentration?

Equilibria Shifts for Slightly Soluble Compounds (Sec. 16.9)

53. Teeth and bones are composed mainly of calcium phosphate, which dissociates slightly in an aqueous solution as follows:

$$Ca_3(PO_4)_2(s) \rightleftharpoons 3\,Ca^{2+}(aq) + 2\,PO_4^{3-}(aq)$$

Predict the direction of equilibrium shift for each of the following stresses:
(a) increase $[Ca^{2+}]$
(b) increase $[PO_4^{3-}]$
(c) decrease $[Ca^{2+}]$
(d) decrease $[PO_4^{3-}]$
(e) add solid $Ca(NO_3)_2$
(f) add solid KNO_3
(g) add solid $Ca_3(PO_4)_2$
(h) add H^+

54. Cadmium sulfide dissociates slightly in an aqueous solution as follows:

$$CdS(s) \rightleftharpoons Cd^{2+}(aq) + S^{2-}(aq)$$

Predict the direction of equilibrium shift for each of the following stresses:
(a) increase $[Cd^{2+}]$
(b) increase $[S^{2-}]$
(c) decrease $[Cd^{2+}]$
(d) decrease $[S^{2-}]$

(e) add solid $Cd(NO_3)_2$
(f) add solid $NaNO_3$
(g) add solid CdS
(h) add H^+

55. Strontium carbonate dissociates slightly in an aqueous solution as follows:

$$SrCO_3(s) \rightleftharpoons Sr^{2+}(aq) + CO_3^{2-}(aq)$$

Predict the direction of equilibrium shift for each of the following stresses:
(a) increase $[Sr^{2+}]$
(b) increase $[CO_3^{2-}]$
(c) decrease $[Sr^{2+}]$
(d) decrease $[CO_3^{2-}]$
(e) add solid $SrCO_3$
(f) add solid $Sr(NO_3)_2$
(g) add solid KNO_3
(h) decrease pH

56. Cupric hydroxide dissociates slightly in an aqueous solution as follows:

$$Cu(OH)_2(s) \rightleftharpoons Cu^{2+}(aq) + 2\,OH^-(aq)$$

Predict the direction of equilibrium shift for each of the following stresses:
(a) increase $[Cu^{2+}]$
(b) increase $[OH^-]$
(c) decrease $[Cu^{2+}]$
(d) decrease $[OH^-]$
(e) add solid $Cu(OH)_2$
(f) add solid NaOH
(f) add solid NaCl
(h) decrease pH

General Exercises

57. Although a clock with a swinging pendulum is not a chemical equilibrium system, explain how it represents a dynamic, reversible process.

58. Although a calculator charging and discharging is not a chemical equilibrium system, explain how it represents a dynamic, reversible process.

59. With regard to reaction rate, what is characteristic of a reversible reaction at equilibrium?

60. With regard to concentration, what is characteristic of a reversible reaction at equilibrium?

61. The N_2O_4–NO_2 reversible reaction is found to have the following equilibrium partial pressures at 100 °C. Calculate K_{eq} for the reaction.

$$N_2O_4(g) \rightleftharpoons 2\,NO_2(g)$$
$$4.5 \times 10^{-5}\,M \quad 3.0 \times 10^{-3}\,M$$

62. The N_2O_4–NO_2 reversible reaction is found to have the following equilibrium partial pressures at 100 °C. Calculate K_p for the reaction.

$$N_2O_4(g) \rightleftharpoons 2\,NO_2(g)$$
$$0.0014\text{ atm} \quad 0.092\text{ atm}$$

63. Given the chemical equation for the ionization of water

$$H_2O(l) \rightleftharpoons H^+(aq) + OH^-(aq)$$

Predict the direction of equilibrium shift for each of the following stresses:
(a) increase $[H^+]$
(b) decrease $[OH^-]$
(c) increase pH
(d) decrease pH

64. Given the chemical equation for the ionization of water

$$H_2O(l) \rightleftharpoons H^+(aq) + OH^-(aq)$$

Predict the direction of equilibrium shift for each of the following stresses:
(a) add gaseous HCl
(b) add solid NaOH
(c) add liquid H_2SO_4
(d) add solid NaF

65. Explain why ferric hydroxide, $Fe(OH)_3$, is more soluble in 0.1 *M* HCl than in water.

66. Explain why silver iodide, AgI, is more soluble in 0.1 *M* NaCl than in water.

67. A saturated solution of calcium hydroxide, $Ca(OH)_2$, has a pH of 12.35. Find the hydroxide ion concentration and calculate the equilibrium constant.

68. A saturated solution of zinc hydroxide, $Zn(OH)_2$, has a pH of 8.44. Find the hydroxide ion concentration and calculate the equilibrium constant.

16 Chapter Self-Test

Answers to Self-Test are in Appendix J.

1. Which of the following factors influences the rate of a chemical reaction? (Sec. 16.1)
(a) concentration
(b) temperature
(c) catalyst
(d) all of the above
(e) none of the above

2. If the heat of reaction is exothermic, which of the following is true? (Sec. 16.2)
(a) the reaction rate is fast
(b) the reaction rate is slow
(c) the energy of the reactants is greater than the products
(d) the energy of the reactants is less than the products
(e) none of the above

3. Which of the following is true after a reaction reaches equilibrium? (Sec. 16.3)
   (a) The amount of reactants and products are equal.
   (b) The amount of reactants and products are constant.
   (c) The amount of reactants is increasing.
   (d) The amount of products is increasing.
   (e) none of the above
4. Dinitrogen tetraoxide decomposes to produce nitrogen dioxide. Calculate the equilibrium constant for the reaction given the equilibrium concentrations at 100 °C: $[N_2O_4] = 0.800$ and $[NO_2] = 0.400$. (Sec. 16.4)

$$N_2O_4(g) \rightleftarrows 2\,NO_2(g)$$

   (a) $K_{eq} = 0.200$
   (b) $K_{eq} = 0.500$
   (c) $K_{eq} = 0.625$
   (d) $K_{eq} = 2.00$
   (e) $K_{eq} = 5.00$
5. Which of the changes listed has *no effect* on the equilibrium for the following reversible reaction? (Sec. 16.5)

$$CO(g) + H_2O(g) \rightleftarrows CO_2(g) + H_2(g) + \text{heat}$$

   (a) increase volume
   (b) decrease volume
   (c) add a catalyst
   (d) add neon gas
   (e) all of the above
6. If the hydrogen ion concentration of a 0.100 *M* $HC_3H_5O_3$ solution is 0.0038 *M*, what is the ionization constant for lactic acid? (Sec. 16.6)
   (a) $K_i = 1.4 \times 10^{-3}$
   (b) $K_i = 1.4 \times 10^{-4}$
   (c) $K_i = 1.4 \times 10^{-5}$
   (d) $K_i = 3.8 \times 10^{-3}$
   (e) $K_i = 3.8 \times 10^{-4}$
7. Which of the changes listed has *no effect* on the equilibrium for the following reversible reaction in an aqueous solution? (Sec. 16.7)

$$HC_2H_3O_2(aq) \rightleftarrows H^+(aq) + C_2H_3O_2^-(aq)$$

   (a) increase $[HC_2H_3O_2]$
   (b) increase $[H^+]$
   (c) add solid $NaC_2H_3O_2$
   (d) add solid $NaNO_3$
   (e) add solid NaOH
8. What is the K_{sp} for silver carbonate, Ag_2CO_3, if the silver ion concentration in a saturated solution is 0.00026 *M*? (Sec. 16.8)
   (a) $K_{sp} = 3.4 \times 10^{-8}$
   (b) $K_{sp} = 6.8 \times 10^{-8}$
   (c) $K_{sp} = 3.6 \times 10^{-11}$
   (d) $K_{sp} = 4.4 \times 10^{-12}$
   (e) $K_{sp} = 8.8 \times 10^{-12}$
9. Which of the changes listed has *no effect* on the equilibrium for the following reversible reaction? (Sec. 16.9)

$$Mg(OH)_2(s) \rightleftarrows Mg^{2+}(aq) + 2\,OH^-(aq)$$

   (a) increase $[OH^-]$
   (b) increase pH
   (c) add sulfuric acid
   (d) add solid $Mg(OH)_2$
   (e) add solid $Mg(NO_3)_2$

Key Concepts

10. Does the following reaction profile represent an exothermic or endothermic reaction?

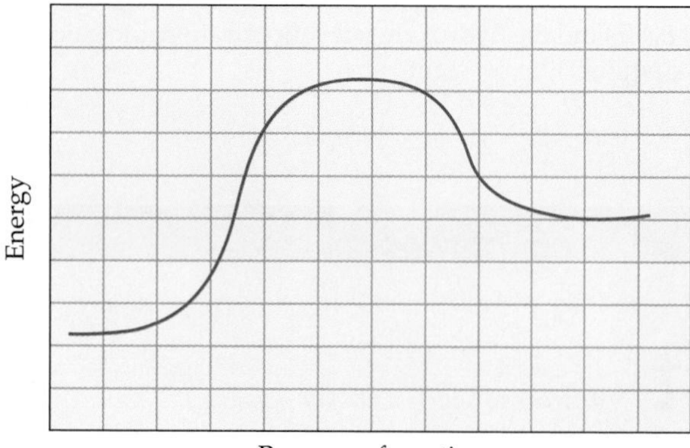

11. What is the equilibrium constant expression for the oxygen/ozone reaction?

$$3\,O_2(g) \rightleftarrows 2\,O_3(g)$$

12. What is the direction of equilibrium shift after heating the reaction?

$$3\,O_2(g) + \text{heat} \rightleftarrows 2\,O_3(g)$$

13. What is the direction of equilibrium shift after reducing the volume?

$$3\,O_2(g) \rightleftarrows 2\,O_3(g)$$

14. What is the direction of equilibrium shift after adding a catalyst?

$$3\,O_2(g) + \text{heat} \rightleftarrows 2\,O_3(g)$$

Critical Thinking

15. How can a spark ignite an explosion in a coal mine, but not in a charcoal barbecue?
16. A liquid in a sealed beaker begins to evaporate as shown in (a). When the volume above the liquid becomes saturated with vapor, the vapor begins to condense to a liquid as shown in (b). What is true of the rates of evaporation and condensation when the system achieves equilibrium (c)?

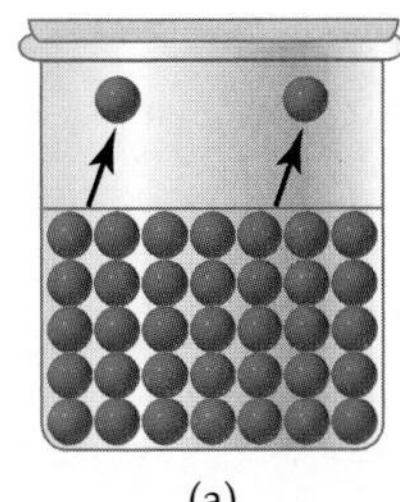
(a)

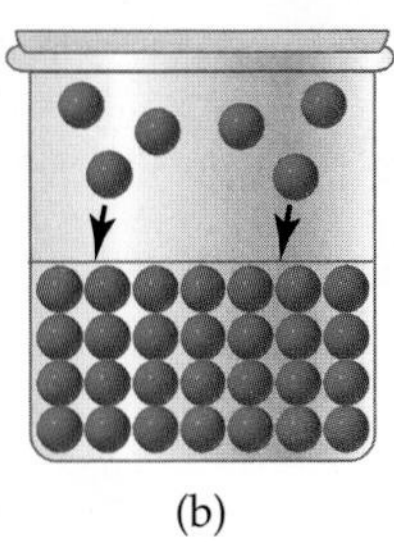
(b)

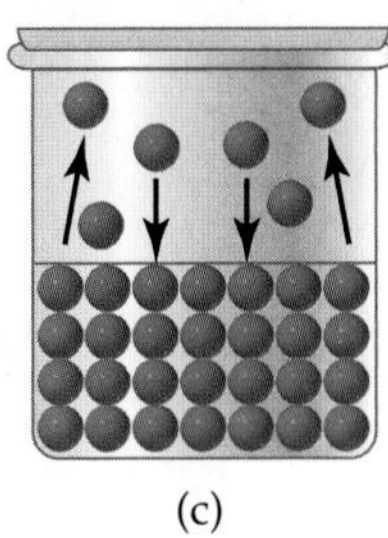
(c)

CHAPTER

17 Oxidation and Reduction

The space shuttle *Discovery* achieves liftoff using two long, white booster rockets. Liquid hydrogen and oxygen are stored in the large, brown tank. Fuel cells in the shuttle use hydrogen and oxygen to produce energy during a space mission.

"There are three things extremely hard: steel, a diamond, and to know one's self."
Benjamin Franklin, Scientist/Politician (1706–1790)

When electrons are transferred from one substance to another, one substance is said to be *oxidized* and the other is said to be *reduced*. In this chapter we will study reactions involving the transfer of electrons from one substance to another, that is, *oxidation–reduction* reactions.

Although we have not previously used the term "oxidation–reduction," many of the reactions we have already studied involve the transfer of electrons. In fact, whenever a metal reacts with a nonmetal, electrons are transferred. For instance, iron metal rusts when exposed to oxygen in the atmosphere (Figure 17.1). That is, iron reacts with oxygen to give iron(III) oxide. The equation for the reaction is

$$4\,Fe(s) + 3\,O_2(g) \longrightarrow 2\,Fe_2O_3(s)$$

(a) (b)

▲ **Figure 17.1 Oxidation of Iron** Examples of oxidation include (a) burning, which is the rapid combination of oxygen and steel wool, and (b) rusting, which is the slow combination of oxygen from the air and metal.

Iron(III) oxide is an ionic compound, so Fe_2O_3 exists as the ions Fe^{3+} and O^{2-}. In the process of becoming an ion, an iron atom loses three electrons. We say that the iron metal is *oxidized* because it loses electrons to the oxygen gas.

$$Fe \longrightarrow Fe^{3+} + 3\,e^-$$

As the reaction proceeds, we say that the oxygen is *reduced* because it gains electrons from the iron metal. In this example, each oxygen molecule reacts by gaining four electrons, two for each oxygen atom, and becoming two oxide ions.

$$O_2 + 4\,e^- \longrightarrow 2\,O^{2-}$$

Let's consider another example. When silver metal is exposed to traces of hydrogen sulfide in the air, the silver tarnishes. The equation for the reaction is

$$2\,Ag(s) + H_2S(g) \longrightarrow Ag_2S(s) + H_2(g)$$

Because silver sulfide is an ionic compound, we can write the ions as Ag^+ and S^{2-}. In the process of becoming an ion, a silver atom loses an electron. We say that the silver metal is *oxidized* by hydrogen sulfide gas.

$$Ag \longrightarrow Ag^+ + e^-$$

Simultaneously, hydrogen sulfide is *reduced* as silver is oxidized. Hydrogen sulfide reacts by gaining two electrons and yielding a hydrogen molecule.

$$H_2S + 2\,e^- \longrightarrow H_2 + S^{2-}$$

This overall process of one substance being oxidized while another is being reduced is called an oxidation–reduction reaction. Chemists refer to this process as simply a *redox reaction*.

OBJECTIVES

- To explain the concept of an oxidation number.
- To assign an oxidation number for an element in each of the following:
  - (a) metals and nonmetals
  - (b) monoatomic and polyatomic ions
  - (c) ionic and molecular compounds

17.1 OXIDATION NUMBERS

The redox reactions involving silver and iron show that an atom can lose or gain different numbers of electrons. In order to describe the number of electrons lost or gained by an atom, an **oxidation number** is assigned according to a set of rules (Table 17.1). All uncombined elements are said to be in the free state, and are electrically neutral. Their oxidation numbers have a value of zero. For example, Ag, Cu, Fe, S, and P have an oxidation number of zero. Even nonmetals that exist as molecules, such as O_2 and Cl_2, have an oxidation number of zero.

The oxidation number of a monoatomic ion is equal to its ionic charge. For example, the oxidation number for Ca^{2+} is positive two (+2), and for Cl^- negative one (–1). Note the distinction between the way ionic charges and oxidation numbers are written. Ionic charges are indicated by the number followed by the sign, and oxidation numbers are indicated by the sign followed by the number. For example, the oxidation numbers for Fe^{2+} and Fe^{3+} are +2 and +3, respectively.

Let's see how to assign oxidation numbers to elements in a compound. Recall that compounds are overall electrically neutral, and so the sum of the oxidation numbers of the individual atoms in a compound is 0. In binary ionic compounds, compounds that contain one metal and one nonmetal, the oxidation numbers for the metal and the nonmetal correspond to their ionic charges. For example, in NaCl the oxidation number of Na^+ is +1, and that of Cl^- is –1. In AlF_3, the oxidation number of Al^{3+} is +3, and that of F^- is –1. In ternary ionic compounds, compounds that contain three elements including at least one metal, the situation is more complex, but the sum of the oxidation numbers is still equal to 0.

In binary molecular compounds, compounds that contain two nonmetals, the more electronegative element is assigned an oxidation number equal to the ionic charge of its free ion.

Oxygen usually has an oxidation number of –2, and hydrogen usually has an oxidation number of +1. After assigning oxidation numbers to oxygen and hydrogen, we can determine the value for the other nonmetal in a binary molecular compound. For

▶ **Metals in the Free State** The oxidation number for any uncombined metal in the free state is 0. The metals shown are magnesium, copper and gold (clockwise).

◀ **Nonmetals in the Free State** The oxidation number for any uncombined nonmetal in the free state is 0. The nonmetals shown are yellow sulfur, violet iodine, and red phosphorus (clockwise).

example, in NO the oxidation number of nitrogen is +2; in NO_2 the oxidation number of nitrogen is +4. We can state general rules for assigning oxidation numbers as listed in Table 17.1.

TABLE 17.1 RULES FOR ASSIGNING OXIDATION NUMBERS

1. A metal or a nonmetal in the free state has an oxidation number of 0.
2. A monoatomic ion has an oxidation number equal to its ionic charge.
3. A hydrogen atom is usually assigned an oxidation number of +1.
4. An oxygen atom is usually assigned an oxidation number of –2.
5. For a molecular compound, the more electronegative element is assigned a negative oxidation number equal to its charge as an anion.
6. For a compound, the sum of the oxidation numbers for each of the atoms in the compound is equal to 0.
7. For a polyatomic ion, the sum of the oxidation numbers for each of the atoms in the compound is equal to the ionic charge on the polyatomic ion.

The rules for assigning oxidation numbers can be illustrated by listing some examples. Table 17.2 provides examples of oxidation numbers for elements in the free state, monoatomic ions, and polyatomic ions. Note that each example is referenced to the appropriate rule in Table 17.1.

Determining Oxidation Numbers in Compounds

Now, let's determine the oxidation number for an element in a compound. Oxalic acid, $H_2C_2O_4$, is used to clean jewelry. To find the oxidation number of carbon in $H_2C_2O_4$, we proceed as follows. First, we assign hydrogen an oxidation number of +1. Second, we assign oxygen, the most electronegative element in the compound, an oxidation number of –2. Because compounds are electrically neutral, the sum of the oxidation numbers (ox no) is equal to 0.

$$2\,(\text{ox no H}) + 2\,(\text{ox no C}) + 4\,(\text{ox no O}) = 0$$

TABLE 17.2 ASSIGNING OXIDATION NUMBERS

EXAMPLE	OXIDATION NUMBER	
magnesium metal, Mg	Mg = 0	(Rule 1)
bromine liquid, Br_2	Br = 0	(Rule 1)
potassium ion, K^+	K = +1	(Rule 2)
sulfide ion, S^{2-}	S = –2	(Rule 2)
water, H_2O	H = +1	(Rule 3)
	O = –2	(Rule 4)
carbon tetrachloride, CCl_4	Cl = –1	(Rule 5)
	C = +4	(Rule 6)
barium chloride, $BaCl_2$	Ba = +2	(Rule 6)
	Cl = –1	(Rule 6)
nitrate ion, NO_3^-	O = –2	(Rule 4)
	N = +5	(Rule 7)

▲ Diamond and coal are both examples of the element carbon.

Substituting the oxidation numbers of H and O, we have

$$2(+1) + 2(\text{ox no C}) + 4(-2) = 0$$

Simplifying and solving for the oxidation number of carbon, we have

$$+2 + 2(\text{ox no C}) + (-8) = 0$$
$$2(\text{ox no C}) = +6$$
$$\text{ox no C} = +3$$

We find that the oxidation number of carbon in $H_2C_2O_4$ is +3. The following example exercise provides additional illustrations of how to determine oxidation numbers for an element in a compound:

Example Exercise 17.1 Calculating Oxidation Numbers for Carbon

Calculate the oxidation number for carbon in each of the following compounds:
(a) diamond, C
(b) dry ice, CO_2
(c) marble, $CaCO_3$
(d) baking soda, $NaHCO_3$

Solution

Let's begin by recalling that uncombined elements, as well as compounds, are electrically neutral. Thus, free elements and compounds have no charge.

(a) In diamond, the oxidation number of carbon is zero.

(b) In dry ice, we assign oxygen an oxidation number of –2. We can determine the oxidation number of carbon in CO_2 as follows:

$$\text{ox no C} + 2(\text{ox no O}) = 0$$
$$\text{ox no C} + 2(-2) = 0$$
$$\text{ox no C} = +4$$

(c) In marble, we assign calcium ion an oxidation number of +2, and oxygen a value of –2. We can determine the value of carbon in $CaCO_3$ as follows:

$$\text{ox no Ca} + \text{ox no C} + 3(\text{ox no O}) = 0$$
$$+2 + \text{ox no C} + 3(-2) = 0$$
$$\text{ox no C} = +4$$

(d) In baking soda, we assign sodium ion an oxidation number of +1, hydrogen a value of +1, and oxygen a value of –2. We can determine the oxidation number of carbon in $NaHCO_3$ as follows:

$$\text{ox no Na} + \text{ox no H} + \text{ox no C} + 3\,(\text{ox no O}) = 0$$
$$+1 + +1 + \text{ox no C} + 3\,(-2) = 0$$
$$+1 + +1 + \text{ox no C} + -6 = 0$$
$$\text{ox no C} = +4$$

Practice Exercise

Calculate the oxidation number for iodine in each of the following compounds:

(a) iodine, I_2
(b) potassium iodide, KI
(c) silver periodate, $AgIO_4$
(d) zinc iodate, $Zn(IO_3)_2$

Answers: (a) 0; (b) –1; (c) +7; (d) +5

Concept Exercise

Calculate the oxidation number for nonmetal X in each of the following compounds:

(a) X_2
(b) X_2O
(c) CaX_2
(d) HXO_4

Answer: See Appendix G.

Determining Oxidation Numbers in Polyatomic Ions

Since many compounds contain polyatomic ions, let's determine the oxidation number of an element in a polyatomic ion. Consider the dichromate ion, $Cr_2O_7^{2-}$, which can be used as a fungicide in a fish aquarium. Unlike compounds, polyatomic ions are not neutral. The dichromate ion has a charge of 2–. To find the oxidation number of chromium in $Cr_2O_7^{2-}$, we assign oxygen an oxidation number of –2. Because $Cr_2O_7^{2-}$ has an overall charge of 2–, we can write the equation

$$2\,(\text{ox no Cr}) + 7\,(\text{ox no O}) = -2$$

Substituting –2 for the oxidation number of O, we have

$$2\,(\text{ox no Cr}) + 7\,(-2) = -2$$

Simplifying and solving for the oxidation number of Cr, we have

$$2\,(\text{ox no Cr}) = +12$$
$$\text{ox no Cr} = +6$$

◀ **Chromium Compounds** The oxidation number is +6 for Cr in yellow K_2CrO_4, as well as Cr in orange $K_2Cr_2O_7$.

The following example exercise provides additional practice in calculating the oxidation number for an element in a polyatomic ion:

Example Exercise 17.2 Calculating Oxidation Numbers for Sulfur

Calculate the oxidation number for sulfur in each of the following ions.
(a) sulfide ion, S^{2-}
(b) sulfite ion, $SO_3{}^{2-}$
(c) sulfate ion, $SO_4{}^{2-}$
(d) thiosulfate ion, $S_2O_3{}^{2-}$

Solution
We can begin by recalling that the charge on an ion corresponds to the sum of the oxidation numbers.
(a) In S^{2-}, the oxidation number of sulfur is –2.
(b) In $SO_3{}^{2-}$, the polyatomic anion has a charge of 2–. We assign oxygen an oxidation number of –2 and write the equation

$$\begin{aligned} \text{ox no S} + 3\,(\text{ox no O}) &= -2 \\ \text{ox no S} + 3\,(-2) &= -2 \\ \text{ox no S} &= +4 \end{aligned}$$

(c) In $SO_4{}^{2-}$, the polyatomic anion has a charge of 2–. We assign oxygen an oxidation number of –2 and write the equation

$$\begin{aligned} \text{ox no S} + 4\,(\text{ox no O}) &= -2 \\ \text{ox no S} + 4\,(-2) &= -2 \\ \text{ox no S} &= +6 \end{aligned}$$

(d) In $S_2O_3{}^{2-}$, the polyatomic anion has a charge of 2–. We assign oxygen an oxidation number of –2 and write the equation

$$\begin{aligned} 2\,(\text{ox no S}) + 3\,(\text{ox no O}) &= -2 \\ 2\,(\text{ox no S}) + 3\,(-2) &= -2 \\ 2\,(\text{ox no S}) &= +4 \\ \text{ox no S} &= +2 \end{aligned}$$

Practice Exercise
Calculate the oxidation number for chlorine in each of the following ions:
(a) hypochlorite ion, ClO^-
(b) chlorite ion, $ClO_2{}^-$
(c) chlorate ion, $ClO_3{}^-$
(d) perchlorate ion, $ClO_4{}^-$

Answers: (a) +1; (b) +3; (c) +5; (d) +7

Concept Exercise
Calculate the oxidation number for nonmetal X in each of the following ions:
(a) X^-
(b) XO^-
(c) $XO_2{}^-$
(d) $XO_3{}^-$

Answer: See Appendix G.

17.2 OXIDATION–REDUCTION REACTIONS

OBJECTIVES

- To identify the oxidized and reduced substances in a given redox reaction.
- To identify the oxidizing and reducing agents in a given redox reaction.

What do burning charcoal and burning magnesium have in common? The answer is that both reactions involve oxygen gas and the transfer of electrons. Charcoal is oxidized to carbon dioxide, and magnesium metal is oxidized to magnesium oxide. As we have said, a chemical change that involves the transfer of electrons is an oxidation–reduction reaction, or simply a **redox reaction**.

As an example, we can heat iron metal with sulfur powder to give the following redox reaction:

$$Fe(s) + S(s) \xrightarrow{\Delta} FeS(s)$$

The oxidation number of S changes from 0 to –2 (see Table 17.2). Simultaneously, Fe changes from 0 to +2. Therefore, each iron atom is oxidized and loses two electrons, and each sulfur atom is reduced and gains two electrons. We can diagram the process as follows:

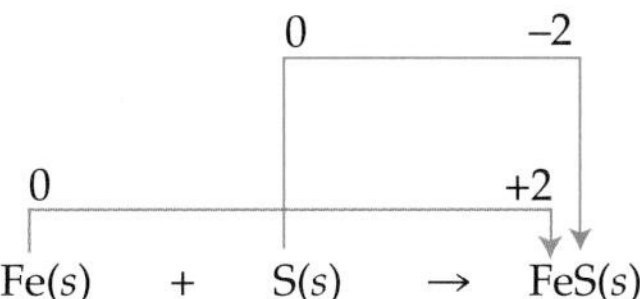

Because electrons are transferred during the chemical change, this is an example of a redox reaction. Notice that both iron and sulfur undergo a change in oxidation number.

Oxidizing and Reducing Agents

By definition, the process of **oxidation** is characterized by the loss of electrons. Conversely, the process of **reduction** is characterized by the gain of electrons. In the preceding example, iron undergoes oxidation because it loses two electrons; sulfur undergoes reduction because it gains two electrons.

An **oxidizing agent** is a substance that causes the oxidation in a redox reaction by accepting the transferred electrons. Thus, an oxidizing agent gains electrons. When sulfur reacts with iron, S causes Fe to be oxidized from 0 to +2. Hence, S is the oxidizing agent and gains electrons. A **reducing agent** is any substance that causes reduction by providing the transferred electrons. Thus, a reducing agent loses electrons. When iron reacts with sulfur, Fe causes S to be reduced from 0 to –2. Hence, Fe is the reducing agent. We can illustrate the redox process as follows:

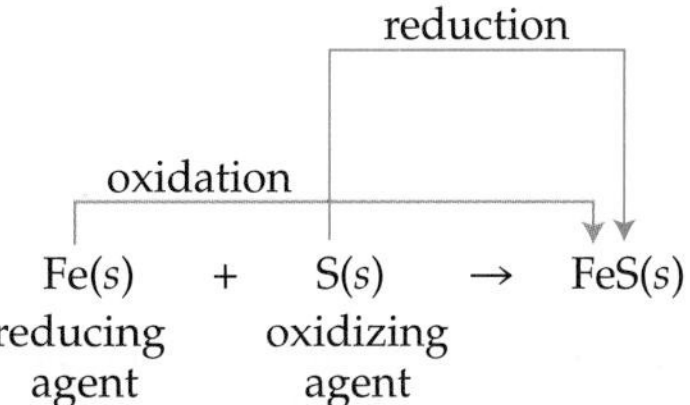

In a redox reaction one substance increases its oxidation number, while another substance decreases its oxidation number. In most instances, the maximum positive oxidation numbers range from +7 to –4. Figure 17.2 illustrates the relationship of oxidation numbers in a redox reaction.

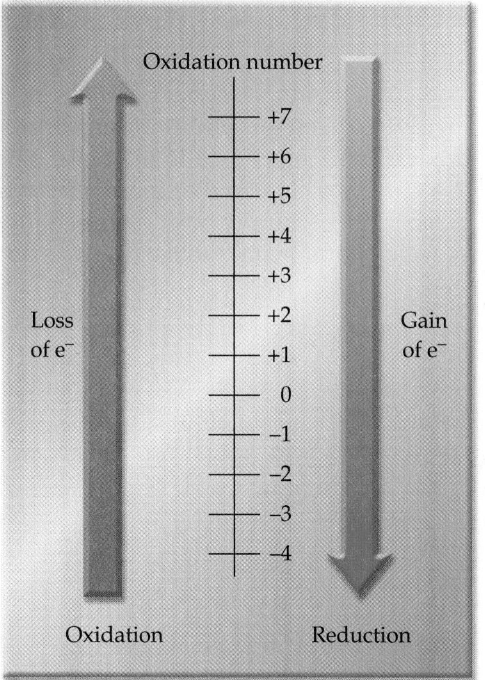

▶ **Figure 17.2 Oxidation and Reduction** The process of oxidation involves the loss of electrons (the oxidation number increases). The process of reduction involves the gain of electrons (the oxidation number decreases).

The following example exercise further illustrates redox reactions and the designation of oxidizing agents and reducing agents:

Example Exercise 17.3 Identifying Oxidizing and Reducing Agents

An oxidation–reduction reaction occurs when a stream of hydrogen gas is passed over hot copper(II) sulfide.

$$CuS(s) + H_2(g) \xrightarrow{\Delta} Cu(s) + H_2S(g)$$

Indicate each of the following for the above redox reaction:

(a) substance oxidized (b) substance reduced
(c) oxidizing agent (d) reducing agent

Solution

By definition, the substance oxidized loses electrons, and its oxidation number increases. The substance reduced gains electrons, and its oxidation number decreases. After assigning oxidation numbers to each atom, we have the following:

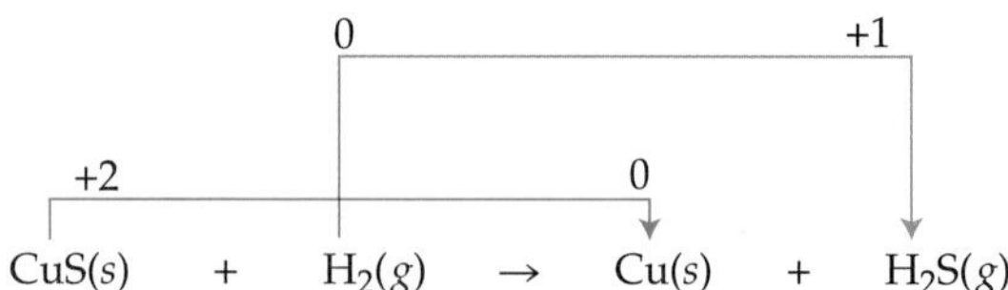

The oxidation number of hydrogen increases from 0 to +1. Thus, H_2 is oxidized. The oxidation number of copper decreases from +2 to 0. Thus, the Cu in CuS is reduced. Note that the oxidation number of S remains constant (–2).

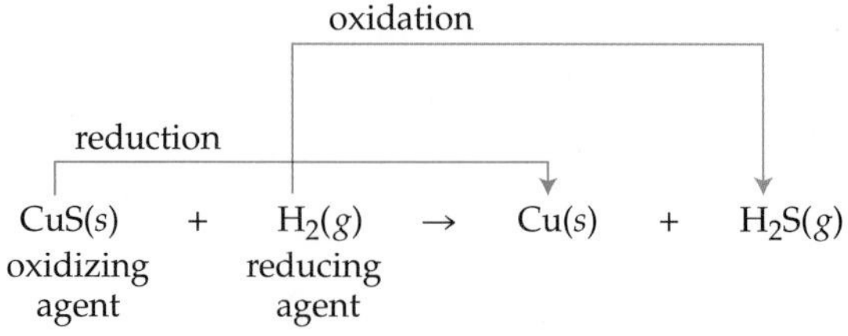

The oxidizing agent is CuS because it causes hydrogen to be oxidized from 0 to +1. The reducing agent is H_2 because it causes copper to be reduced from +2 to 0.

Practice Exercise

A redox reaction occurs when molten aluminum reacts with iron(III) oxide.

$$Fe_2O_3(l) + 2\ Al(l) \xrightarrow{\Delta} 2\ Fe(l) + Al_2O_3(l)$$

Indicate each of the following for the preceding redox reaction:

(a) substance oxidized (b) substance reduced

(c) oxidizing agent (d) reducing agent

Answers: (a) Al; (b) Fe_2O_3; (c) Fe_2O_3; (d) Al

Concept Exercise

Identify the oxidizing agent and reducing agent in the following redox reaction:

$$M(s) + 2\ HX(aq) \longrightarrow MX_2(aq) + H_2(g)$$

Answer: See Appendix G.

Ionic Equations in an Aqueous Solution

Recall that we learned to write ionic equations in Section 15.11. Redox reactions in an aqueous solution are most often shown in the ionic form. That is because ionic equations readily show us the change in oxidation number. As an example, the amount of iron in a sample is routinely analyzed by a redox method. A typical reaction uses a potassium permanganate solution as follows:

$$5\ Fe^{2+}(aq) + MnO_4^-(aq) + 8\ H^+(aq) \longrightarrow 5\ Fe^{3+}(aq) + Mn^{2+}(aq) + 4\ H_2O(l)$$

Although the reaction appears complex, we can systematically study the equation for the oxidized and reduced substances. First, notice from the ionic charges that the oxidation number of iron changes from +2 to +3. Because Fe^{2+} loses an electron, it is oxidized. Because Fe^{2+} is oxidized, it is the reducing agent in this reaction.

Next, let's identify the substance being reduced. When we calculate the oxidation number of manganese in permanganate ion, MnO_4^-, we find that the value is +7. On the product side of the equation, we find that the oxidation number of Mn^{2+} is +2. Because permanganate ion gains electrons, it is the substance reduced. Because MnO_4^- is reduced, it is the oxidizing agent in this reaction.

We can illustrate the oxidation and reduction processes for the redox reaction as follows:

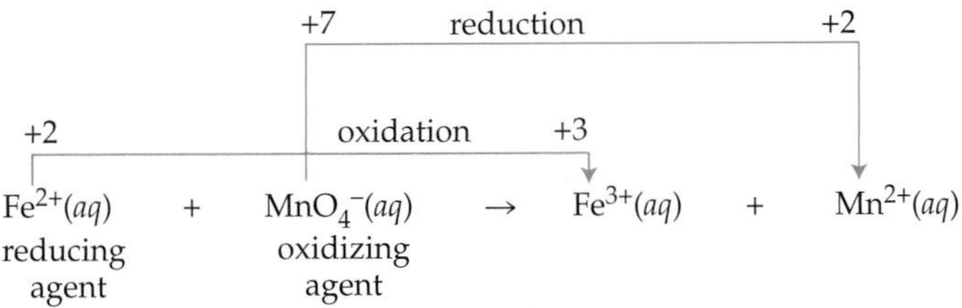

The following example exercise further illustrates a determination of an oxidizing agent and a reducing agent in a redox reaction:

Example Exercise 17.4 Identifying Oxidizing and Reducing Agents

The amount of iodine in a solution can be determined by a redox method using a sulfite solution:

$$I_2(s) + SO_3^{2-}(aq) + H_2O(l) \longrightarrow 2\ I^-(aq) + SO_4^{2-}(aq) + 2\ H^+(aq)$$

Indicate each of the following for the preceding reaction:

(a) substance oxidized (b) substance reduced
(c) oxidizing agent (d) reducing agent

Solution

Notice that iodine solid is converted to iodide ion in an aqueous solution. Since I_2 gains electrons, it is *reduced* and I_2 is the *oxidizing agent.*

The reducing agent is not as obvious. If we calculate the oxidation number for sulfur in SO_3^{2-} and SO_4^{2-}, we find sulfur changes from +4 to +6 and loses electrons. Thus, the sulfur in SO_3^{2-} is *oxidized*, and SO_3^{2-} is the *reducing agent.* We can illustrate the oxidation and reduction processes for the redox reaction as follows:

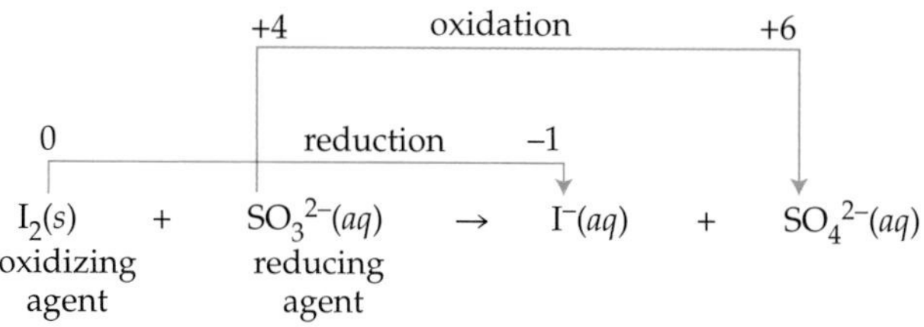

Practice Exercise

A redox reaction occurs when the tin(II) ion reacts with the iodate ion as follows:

$$6\,H^+(aq) + 3\,Sn^{2+}(aq) + IO_3^-(aq) \longrightarrow 3\,Sn^{4+}(aq) + I^-(aq) + 3\,H_2O(l)$$

Indicate each of the following for the preceding redox reaction:

(a) substance oxidized (b) substance reduced
(c) oxidizing agent (d) reducing agent

Answers: (a) Sn^{2+}; (b) IO_3^-; (c) IO_3^-; (d) Sn^{2+}

Concept Exercise

Identify the oxidizing agent and reducing agent in the following redox reaction:

$$Y_2(g) + 2\,X^-(aq) \longrightarrow 2\,Y^-(aq) + X_2(g)$$

Answer: See Appendix G.

17.3 BALANCING REDOX EQUATIONS: OXIDATION NUMBER METHOD

OBJECTIVE

▸ To write a balanced chemical equation for a redox reaction using the oxidation number method.

In Section 8.3, we learned to balance chemical equations by inspection. We can also balance simple redox reactions by inspection. It is often difficult, however, to balance complex redox reactions by this method. Therefore, we will use the following general guidelines for balancing redox equations with an *oxidation number method:*

Guidelines for Balancing Redox Equations: Oxidation Number Method

1. Inspect the reactants and products to determine the substances undergoing a change in oxidation number. In some cases, it is not obvious and you must calculate the oxidation number.
   (a) Write the oxidation number above the element that is oxidized and above the element that is reduced.
   (b) Diagram the number of electrons lost by the oxidized substance and gained by the reduced substance.

2. Balance each element in the equation using a coefficient. Keep in mind that the total electron loss by oxidation must equal the total electron gain by reduction.
   (a) In front of the oxidized substance, place a coefficient equal to the number of electrons gained by the reduced substance.*
   (b) In front of the reduced substance, place a coefficient equal to the number of electrons lost by the oxidized substance.*
   (c) Balance the remaining elements by inspection.
3. After balancing the equation, verify that the coefficients are correct.
   (a) Place a check mark (√) above the symbol for each element to indicate that the number of atoms is the same for reactants and products.
   (b) For ionic equations, verify that the total charge on the left side of the equation equals the total charge on the right side.

*If the number of electrons gained equals the number of electrons lost, a coefficient is not required.

Balancing Redox Equations by Oxidation Number

Let's apply the guidelines for balancing a redox reaction using the oxidation number method. Specifically, consider the reaction of a metal in an aqueous solution. For example, copper metal reacts with aqueous silver nitrate as follows:

$$Cu(s) + AgNO_3(aq) \longrightarrow Cu(NO_3)_2(aq) + Ag(s)$$

In this reaction, the oxidation number of copper increases from 0 to +2. Simultaneously, the oxidation number of silver decreases from +1 to 0. Thus, copper loses two electrons and silver gains one electron. We can diagram the loss and gain of electrons as follows:

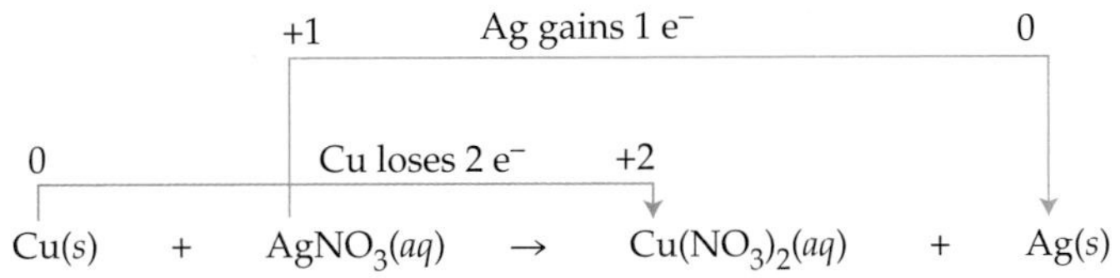

+1 Ag gains 1 e⁻ 0

0 Cu loses 2 e⁻ +2

$$\overset{0}{Cu}(s) + \overset{+1}{Ag}NO_3(aq) \rightarrow \overset{+2}{Cu}(NO_3)_2(aq) + \overset{0}{Ag}(s)$$

Because copper loses two electrons and silver gains only one electron, two silver atoms are required to balance the transfer of electrons. Therefore, we place the coefficient 2 in front of each silver substance. The balanced equation for the reaction is

$$Cu(s) + 2\ AgNO_3(aq) \longrightarrow Cu(NO_3)_2(aq) + 2\ Ag(s)$$

After placing the coefficients, we verify that the equation is balanced by checking (√) each element in the reaction. Because the nitrate ion did not change, we can check off the polyatomic ion as a single unit.

$$\overset{\surd}{Cu}(s) + 2\ \overset{\surd}{Ag}\overset{\surd}{NO_3}(aq) \longrightarrow \overset{\surd}{Cu}(\overset{\surd}{NO_3})_2(aq) + 2\ \overset{\surd}{Ag}(s)$$

Because all the elements are balanced, we have verified that this is a balanced equation. The following example exercise provides additional practice in balancing redox equations:

Example Exercise 17.5 Balancing Redox Equations by Oxidation Number

An industrial blast furnace reduces iron ore, Fe_2O_3, to molten iron. Balance the following redox equation using the oxidation number method:

$$Fe_2O_3(l) + CO(g) \longrightarrow Fe(l) + CO_2(g)$$

Solution

In this reaction, the oxidation number of iron decreases from +3 in Fe_2O_3 to 0 in Fe. Simultaneously, the oxidation number of carbon increases from +2 to +4. Thus, each Fe gains three electrons, while each C loses two electrons. We can diagram the redox process as follows:

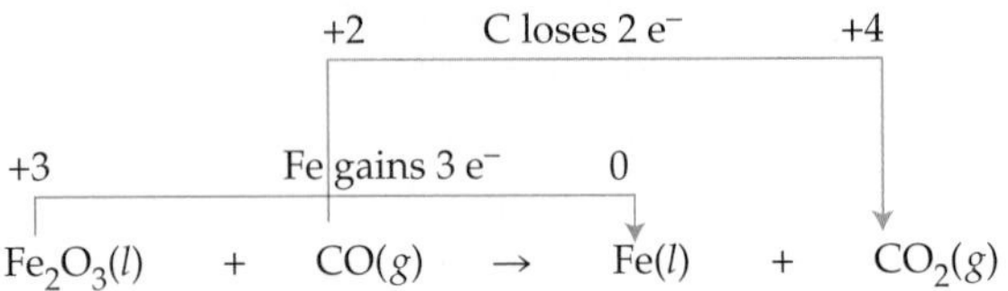

Because the number of electrons gained and lost must be equal, we find the lowest common multiple. In this case, it is 6. Each Fe gains three electrons, so we place the coefficient 3 in front of CO and CO_2.

$$Fe_2O_3(l) + 3\,CO(g) \longrightarrow Fe(l) + 3CO_2(g)$$

Each carbon atom loses two electrons, so we place the coefficient 2 in front of each iron atom. Because Fe_2O_3 has two iron atoms, it does not require a coefficient.

$$Fe_2O_3(l) + 3\,CO(g) \longrightarrow 2\,Fe(l) + 3CO_2(g)$$

Finally, we verify that the equation is balanced. We check (√) each element in the equation:

$$\overset{\surd\ \surd}{Fe_2O_3}(l) + 3\,\overset{\surd\ \surd}{CO}(g) \longrightarrow 2\,\overset{\surd}{Fe}(l) + 3\,\overset{\surd\ \surd}{CO_2}(g)$$

Because all the elements are balanced, we have a balanced redox equation.

Practice Exercise

Balance the following redox equation by the oxidation number method:

$$Cl_2O_5(g) + CO(g) \longrightarrow Cl_2(g) + CO_2(g)$$

Answer: $Cl_2O_5(g) + 5\,CO(g) \longrightarrow Cl_2(g) + 5CO_2(g)$

Concept Exercise

Balance the following redox reaction using the oxidation number method:

$$M_2O_3(s) + X_2(g) \longrightarrow MX_3(aq) + O_2(g)$$

Answer: See Appendix G.

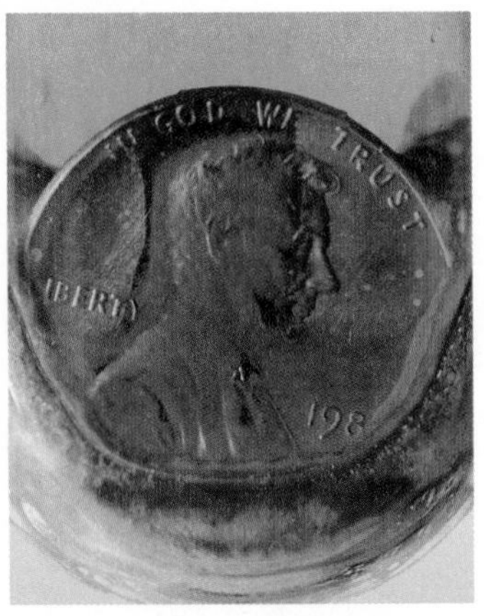

▲ **A Penny in Nitric Acid**
A copper penny reacts with concentrated nitric acid, HNO_3, to give a green Cu^{2+} solution and brown NO_2 gas.

Balancing Ionic Redox Equations by Oxidation Number

Now, let's apply the general guidelines for balancing a redox reaction to an ionic equation. As an example, we know that a copper penny reacts with nitric acid according to the following equation:

$$Cu(s) + H^+(aq) + NO_3^-(aq) \longrightarrow Cu^{2+}(aq) + NO_2(g) + H_2O(l)$$

In this ionic equation, we see that the oxidation number of copper increases from 0 to +2. Although it is not as evident, the oxidation number of nitrogen decreases from +5 to +4. Thus, Cu loses two electrons while N gains one electron. We can diagram the loss and gain of electrons as follows:

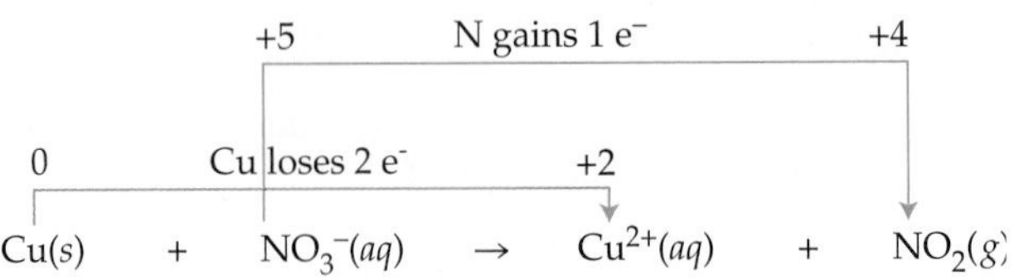

We can balance the loss and gain of electrons by adding the coefficient 2 in front of each nitrogen substance. It is understood that the coefficient of each copper is 1.

$$Cu(s) + H^+(aq) + 2\,NO_3{}^-(aq) \longrightarrow Cu^{2+}(aq) + 2\,NO_2(g) + H_2O(l)$$

In this example, H^+ and H_2O take part in the reaction. Thus, they also must be balanced. We complete balancing the equation by inspection. Because there is a total of six oxygen atoms as reactants, there must be six oxygen atoms in the products. Therefore, we need the coefficient 2 in front of H_2O. After balancing the remaining hydrogen atoms in the reaction, we have

$$Cu(s) + 4\,H^+(aq) + 2\,NO_3{}^-(aq) \longrightarrow Cu^{2+}(aq) + 2\,NO_2(g) + 2H_2O(l)$$

Now, we verify that the equation is balanced by checking (√) each element in the redox reaction.

$$\overset{\surd}{Cu}_{(s)} + 4\,\overset{\surd}{H}{}^+(aq) + 2\,\overset{\surd}{N}\overset{\surd}{O}{}_3^-(aq) \longrightarrow \overset{\surd}{Cu}{}^{2+}(aq) + 2\,\overset{\surd}{N}\overset{\surd}{O}{}_2(g) + 2\,\overset{\surd}{H}{}_2\overset{\surd}{O}(l)$$

Finally, we check to verify that the total charge on the reactants equals the total charge on the products. On the left side of the equation we have $+4 - 2 = +2$. On the right side of the equation we have +2. Because the ionic charge on each side is +2, the equation is balanced. The following example exercise provides additional practice in balancing ionic redox equations:

Example Exercise 17.6 Balancing Redox Equations by Oxidation Number

Aqueous sodium iodide reacts with a potassium dichromate solution. Write a balanced equation for the following redox reaction:

$$H^+(aq) + I^-(aq) + Cr_2O_7{}^{2-}(aq) \longrightarrow I_2(s) + Cr^{3+}(aq) + H_2O(l)$$

Solution

In this special example, $Cr_2O_7{}^{2-}$ and I_2 each contain a subscript that affects electron transfer. Note that there are two atoms of chromium in the reactant and two atoms of iodine in the product. Let's balance the chromium and iodine atoms first.

$$H^+(aq) + 2\,I^-(aq) + Cr_2O_7{}^{2-}(aq) \longrightarrow I_2(s) + 2\,Cr^{3+}(aq) + H_2O(l)$$

In this reaction, the oxidation number of each iodine atom increases from –1 to 0, and the oxidation number of each chromium atom decreases from +6 to +3. We can show the loss and gain of electrons as

+6 Cr gains 3 e^- +3

–1 I loses 1 e^- 0

$$I^-(aq) \quad + \quad Cr_2O_7{}^{2-}(aq) \quad \rightarrow \quad I_2(s) \quad + \quad Cr^{3+}(aq$$

There are two chromium atoms, and so the total electron gain is six electrons. Thus, the total electron loss must also be six electrons. Because there are two iodine atoms, and only one iodide ion, we place the coefficients as follows:

$$H^+(aq) + 6\,I^-(aq) + Cr_2O_7{}^{2-}(aq) \longrightarrow 3\,I_2(s) + 2\,Cr^{3+}(aq) + H_2O(l)$$

Next, we balance the oxygen and hydrogen atoms. Because there are 7 oxygen atoms as reactants, we place the coefficient 7 in front of H_2O. This gives 14 hydrogen atoms, and so we place the coefficient 14 in front of H^+.

$$14\,H^+(aq) + 6\,I^-(aq) + Cr_2O_7{}^{2-}(aq) \longrightarrow 3\,I_2(s) + 2\,Cr^{3+}(aq) + 7\,H_2O(l)$$

We can verify that the equation is balanced by checking (√) each element.

$$14\,\overset{\surd}{H}{}^+(aq) + 6\,\overset{\surd}{I}{}^-(aq) + \overset{\surd}{Cr}{}_2\overset{\surd}{O}{}_7{}^{2-}(aq) \longrightarrow 3\overset{\surd}{I}{}_2(s) + 2\,\overset{\surd}{Cr}{}^{3+}(aq) + 7\,\overset{\surd}{H}{}_2\overset{\surd}{O}(l)$$

Last, we verify that the ionic charges are balanced. On the left side of the equation, we have $+14 - 6 - 2 = +6$. On the right side of the equation, we have +6. Because the ionic charge on each side is +6, the equation is balanced.

Practice Exercise

Write a balanced equation for the following redox reaction:

$$H^+(aq) + MnO_4^-(aq) + NO_2^-(aq) \longrightarrow Mn^{2+}(aq) + NO_3^-(aq) + H_2O(l)$$

Answer:

$6\,H^+(aq) + 2\,MnO_4^-(aq) + 5\,NO_2^-(aq) \longrightarrow 2\,Mn^{2+}(aq) + 5\,NO_3^-(aq) + 3\,H_2O(l)$

Concept Exercise

Balance the following redox reaction in an acidic solution using the oxidation number method:

$$MnO_4^-(aq) + SO_3^{2-}(aq) \longrightarrow Mn^{2+}(aq) + SO_4^{2-}(aq)$$

Answer: See Appendix G.

17.4 BALANCING REDOX EQUATIONS: HALF-REACTION METHOD

OBJECTIVES

- To understand the concept of a half-reaction.
- To write a balanced chemical equation for a redox reaction using the half-reaction method
  - (a) in an acidic solution, and
  - (b) in a basic solution.

In the preceding section, we learned how to balance redox equations using oxidation numbers. In this section, we are going to balance redox equations with a different procedure. This alternative method is referred to as the half-reaction method.

A **half-reaction** shows the oxidation process and the reduction process of the redox reaction separately. The half-reaction method systematically balances the oxidation half-reaction and then balances the reduction half-reaction. We will use the following general guidelines for balancing redox equations with the half-reaction method:

Guidelines for Balancing Redox Equations: Half-Reaction Method

1. Write the half-reaction for both oxidation and reduction.
   (a) Identify the reactant that is oxidized and its product.
   (b) Identify the reactant that is reduced and its product.
2. Balance the atoms in each half-reaction using coefficients.
   (a) Balance all elements except oxygen and hydrogen.
   (b) Balance oxygen using H_2O.
   (c) Balance hydrogen using H^+.
   (*Note:* For reactions in a basic solution, add OH^- to neutralize H^+. For example, $2\,OH^-$ neutralizes $2\,H^+$ to give $2\,H_2O$.)
   (d) Balance the ionic charges using e^-.
3. Multiply each half-reaction by a whole number so the number of electrons lost by oxidation equals the number of electrons gained by reduction.
4. Add the two half-reactions together and cancel identical species, including electrons, on each side of the equation.
5. After balancing the equation, verify that the coefficients are correct.
   (a) Place a check mark (√) above the symbol of each element to indicate that the number of atoms is the same for reactants and products.
   (b) For ionic equations, verify that the total charge on the left side of the equation equals the total charge on the right side.

Balancing Redox Equations in an Acidic Solution

Let's apply the general guidelines for balancing a redox reaction in an acidic aqueous solution. As an example, zinc reacts in nitric acid to give nitrogen monoxide gas. The ionic equation for the redox reaction is

$$Zn(s) + NO_3^-(aq) \longrightarrow Zn^{2+}(aq) + NO(g)$$

Step 1: Write the oxidation and reduction half-reactions. Because the oxidation number of zinc increases, Zn is oxidized. Because the oxidation number of nitrogen decreases, NO_3^- is reduced.

$$\text{Oxidation:} \quad Zn \longrightarrow Zn^{2+}$$
$$\text{Reduction:} \quad NO_3^- \longrightarrow NO$$

Step 2: Balance the atoms and the charges for each half-reaction. The balanced half-reactions are

$$Zn \longrightarrow Zn^{2+} + 2\,e^-$$
$$3\,e^- + 4\,H^+ + NO_3^- \longrightarrow NO + 2\,H_2O$$

Step 3: Note that Zn loses two electrons and NO_3^- gains three electrons. So that the electron loss and gain will be equal, we multiply the Zn half-reaction by 3 and the NO_3^- half-reaction by 2.

$$3\,Zn \longrightarrow 3\,Zn^{2+} + 6\,e^-$$
$$6\,e^- + 8\,H^+ + 2\,NO_3^- \longrightarrow 2\,NO + 4\,H_2O$$

Step 4: Now, we add the two half-reactions together and cancel the 6 electrons.

$$3\,Zn \longrightarrow 3\,Zn^{2+} + \cancel{6\,e^-}$$
$$\cancel{6\,e^-} + 8\,H^+ + 2\,NO_3^- \longrightarrow 2\,NO + 4\,H_2O$$
$$3\,Zn + 8\,H^+ + 2\,NO_3^- \longrightarrow 3\,Zn^{2+} + 2\,NO + 4\,H_2O$$

Step 5: Finally, let's check the atoms and ionic charges to verify that the equation is balanced. We have

$$\text{Atoms: } 3\,Zn, 8\,H, 2\,N, 6\,O = 3\,Zn, 2\,N, 6\,O, 8\,H$$
$$\text{Charges: } 8(+1) + 2(-1) = 3(+2)$$
$$+6 = +6$$

Because the atoms and ionic charges are equal, the redox equation is balanced.

Balancing Redox Equations in a Basic Solution

Now, let's balance a redox reaction that takes place in a basic aqueous solution. For example, sodium iodide and potassium permanganate react in a basic solution to produce iodine and manganese(IV) oxide. The ionic equation is

$$I^-(aq) + MnO_4^-(aq) \longrightarrow I_2(s) + MnO_2(s)$$

The procedure for balancing a redox equation in a base is similar to the procedure given in the general guidelines for balancing a redox reaction in an acid.

Step 1: First, we write the oxidation and reduction half-reactions. We note that iodine is oxidized from –1 to 0. Simultaneously, manganese is reduced from +7 to +4.

$$\text{Oxidation:} \quad I^- \longrightarrow I_2$$
$$\text{Reduction:} \quad MnO_4^- \longrightarrow MnO_2$$

Step 2: Second, we balance each half-reaction for both atoms and charges. In the oxidation half-reaction, we place the coefficient 2 in front of I^- and add two electrons to the product side. We balance the reduction half-reaction as follows: add H_2O to balance O atoms, add H^+ to balance H atoms, and add $3\ e^-$ to balance charge. The balanced half-reactions are

$$2\,I^- \longrightarrow I_2 + 2\,e^-$$
$$3\,e^- + 4\,H^+ + MnO_4^- \longrightarrow MnO_2 + 2\,H_2O$$

Note that this redox reaction takes place in a basic solution, not in an acidic solution. Therefore, we must follow guideline 2(c), which states that the H^+ should be neutralized with OH^-. In this example we add $4\ OH^-$ to each side and simplify the equation.

$$3\,e^- + 4\,H^+ + 4\,OH^- + MnO_4^- \longrightarrow MnO_2 + 2\,H_2O + 4\,OH^-$$
$$3\,e^- + 4\,H_2O + MnO_4^- \longrightarrow MnO_2 + 2\,H_2O + 4\,OH^-$$
$$3\,e^- + 2\,H_2O + MnO_4^- \longrightarrow MnO_2 + 4\,OH^-$$

Step 3: Note that $2\ I^-$ lose two electrons and MnO_4^- gains three electrons. To equalize electron loss and gain, we multiply the I^- oxidation half-reaction by 3, and the MnO_4^- reduction half-reaction by 2.

$$6\,I^- \longrightarrow 3\,I_2 + 6\,e^-$$
$$6\,e^- + 4\,H_2O + 2\,MnO_4^- \longrightarrow 2\,MnO_2 + 8\,OH^-$$

Step 4: We add the two half-reactions together and cancel the six electrons.

$$6\,I^- \longrightarrow 3I_2 + \cancel{6e^-}$$
$$\cancel{6e^-} + 4H_2O + 2\,MnO_4^- \longrightarrow 2\,MnO_2 + 8\,OH^-$$
$$6\,I^- + 4\,H_2O + 2\,MnO_4^- \longrightarrow 3\,I_2 + 2\,MnO_2 + 8\,OH^-$$

Step 5: Finally, we check to verify that the equation is balanced.

Atoms: 6 I, 8 H, 12 O, 2 Mn = 6 I, 2 Mn, 12 O, 8 H

Charges: $6\,(-1) + 2\,(-1) = 8\,(-1)$

$-8 = -8$

Because the atoms and ionic charges are equal for the reactants and products, this redox equation is balanced. Example Exercise 17.7 illustrates balancing redox reactions by the half-reaction method.

Example Exercise 17.7 Balancing Redox Equations by Half-Reaction

Write a balanced ionic equation for the reaction of iron(II) sulfate and potassium permanganate in acidic solution. The ionic equation is

$$MnO_4^-(aq) + Fe^{2+}(aq) \longrightarrow Fe^{3+}(aq) + Mn^{2+}(aq)$$

Solution

We can balance the redox reaction by the half-reaction method as follows:

Step 1: Because Fe^{2+} is oxidized from +2 to +3, MnO_4^- must be reduced. The two half-reactions are as follows:

Oxidation: $Fe^{2+} \longrightarrow Fe^{3+}$

Reduction: $MnO_4^- \longrightarrow Mn^{2+}$

Step 2: We can balance each half-reaction as follows:

$$Fe^{2+} \longrightarrow Fe^{3+} + e^-$$
$$5\,e^- + 8\,H^+ + MnO_4^- \longrightarrow Mn^{2+} + 4\,H_2O$$

Step 3: Because Fe^{2+} loses 1 e^- and MnO_4^- gains 5 e^-, we multiply the Fe^{2+} half-reaction by 5.

$$5\,Fe^{2+} \longrightarrow 5\,Fe^{3+} + 5\,e^-$$

$$5\,e^- + 8\,H^+ + MnO_4^- \longrightarrow Mn^{2+} + 4\,H_2O$$

Step 4: Then, we add the two half-reactions together and cancel the 5 e^-.

$$\begin{array}{rcl} 5\,Fe^{2+} & \longrightarrow & 5Fe^{3+} + \cancel{5e^-} \\ \cancel{5e^-} + 8\,H^+ + MnO_4^- & \longrightarrow & Mn^{2+} + 4\,H_2O \\ \hline 5\,Fe^{2+} + 8\,H^+ + MnO_4^- & \longrightarrow & 5\,Fe^{3+} + Mn^{2+} + 4\,H_2O \end{array}$$

Step 5: Finally, let's check the atoms and ionic charges to verify that the equation is balanced. We have

Atoms: 5 Fe, 8 H, 1 Mn, 4 O = 5 Fe, 1 Mn, 8 H, 4 O

Charges: $5\,(+2) + 8\,(+1) + (-1) = 5\,(+3) + (+2)$

$$+17 = +17$$

Because the atoms and ionic charges are equal for reactants and products, the redox equation is balanced.

Practice Exercise

Write a balanced ionic equation for the reaction of sodium nitrite and potassium permanganate in an acidic solution. The ionic equation is

$$MnO_4^-(aq) + NO_2^-(aq) \longrightarrow Mn^{2+}(aq) + NO_3^-(aq)$$

Answer: $6\,H^+ + 2\,MnO_4^- + 5\,NO_2^- \longrightarrow 2\,Mn^{2+} + 5\,NO_3^- + 3\,H_2O$

Concept Exercise

Balance the following redox reaction in a basic solution using the half-reaction method:

$$MnO_4^-(aq) + SO_3^{2-}(aq) \longrightarrow MnO_2(s) + SO_4^{2-}(aq)$$

Answer: See Appendix G.

17.5 PREDICTING SPONTANEOUS REDOX REACTIONS

OBJECTIVES

- To predict the stronger oxidizing agent and reducing agent given a list of reduction potentials.
- To predict whether a redox reaction is spontaneous or nonspontaneous given a list of reduction potentials.

A redox reaction that occurs without any input of electrical energy is referred to as a *spontaneous reaction*. Let's consider the spontaneous reaction of a metal reacting in an aqueous solution. For example, consider the reaction of zinc metal in aqueous copper(II) sulfate. The net ionic equation for the reaction is

$$Zn(s) + Cu^{2+}(aq) \longrightarrow Zn^{2+}(aq) + Cu(s)$$

It is experimentally observed that the reaction of zinc metal in aqueous copper(II) sulfate proceeds spontaneously. Because copper and zinc ions can both be reduced, let's compare the two reduction half-reactions of copper and zinc:

$$Cu^{2+}(aq) + 2\,e^- \longrightarrow Cu(s)$$

$$Zn^{2+}(aq) + 2\,e^- \longrightarrow Zn(s)$$

According to the overall redox reaction of zinc in copper(II) sulfate solution, Zn(*s*) is oxidized and $Cu^{2+}(aq)$ is reduced. Therefore, we can conclude that the reduction of $Cu^{2+}(aq)$ has a greater tendency to occur than the reduction of $Zn^{2+}(aq)$.

Stated differently, Cu^{2+} has a greater tendency to gain electrons than Zn^{2+}. If we compare other combinations of metals and aqueous solutions, we can arrange a series of elements based on their ability to gain electrons. The strongest oxidizing agent is the substance that is most easily reduced, that is, the substance that has the greater tendency to gain electrons. Conversely, the strongest reducing agent is the substance that is most easily oxidized and has the greater tendency to lose electrons.

The tendency for a substance to gain electrons and undergo reduction is called its **reduction potential**. The reduction potential for a substance is its relative strength as an oxidizing agent, that is, its relative strength to pull electrons to itself. Table 17.3 lists the reduction potentials for several substances in order of their ability to undergo reduction. A substance that is high in the table has a strong reduction potential and is easily reduced because it powerfully attracts electrons. A substance low in the table has a weak reduction potential and is not easily reduced because it does not powerfully attract electrons.

TABLE 17.3 REDUCTION POTENTIALS OF SELECTED OXIDIZING AND REDUCING AGENTS

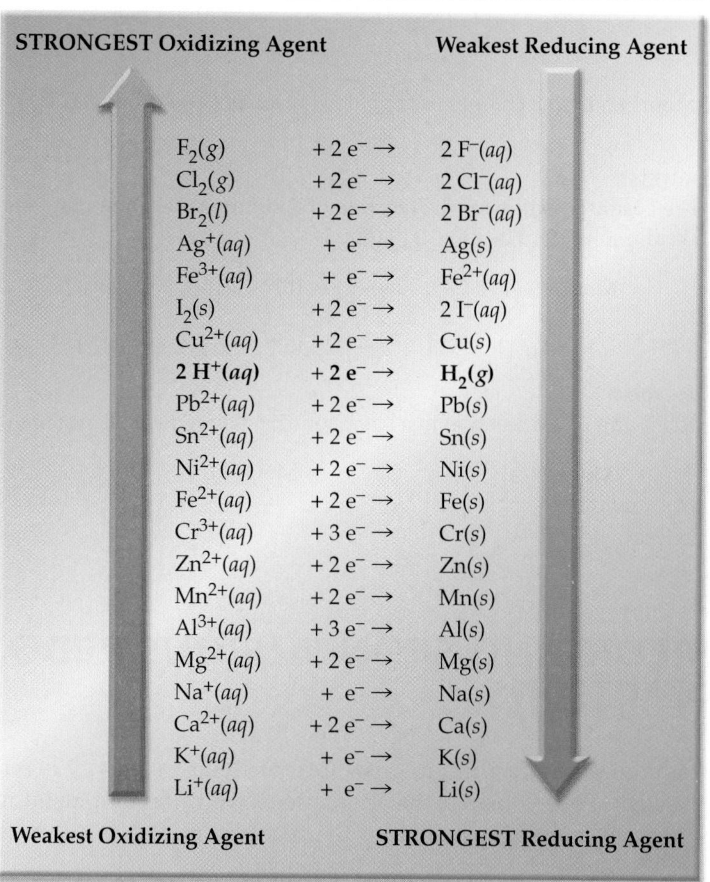

STRONGEST Oxidizing Agent		Weakest Reducing Agent
$F_2(g)$	$+ 2e^- \rightarrow$	$2F^-(aq)$
$Cl_2(g)$	$+ 2e^- \rightarrow$	$2Cl^-(aq)$
$Br_2(l)$	$+ 2e^- \rightarrow$	$2Br^-(aq)$
$Ag^+(aq)$	$+ e^- \rightarrow$	$Ag(s)$
$Fe^{3+}(aq)$	$+ e^- \rightarrow$	$Fe^{2+}(aq)$
$I_2(s)$	$+ 2e^- \rightarrow$	$2I^-(aq)$
$Cu^{2+}(aq)$	$+ 2e^- \rightarrow$	$Cu(s)$
$2H^+(aq)$	**$+ 2e^- \rightarrow$**	**$H_2(g)$**
$Pb^{2+}(aq)$	$+ 2e^- \rightarrow$	$Pb(s)$
$Sn^{2+}(aq)$	$+ 2e^- \rightarrow$	$Sn(s)$
$Ni^{2+}(aq)$	$+ 2e^- \rightarrow$	$Ni(s)$
$Fe^{2+}(aq)$	$+ 2e^- \rightarrow$	$Fe(s)$
$Cr^{3+}(aq)$	$+ 3e^- \rightarrow$	$Cr(s)$
$Zn^{2+}(aq)$	$+ 2e^- \rightarrow$	$Zn(s)$
$Mn^{2+}(aq)$	$+ 2e^- \rightarrow$	$Mn(s)$
$Al^{3+}(aq)$	$+ 3e^- \rightarrow$	$Al(s)$
$Mg^{2+}(aq)$	$+ 2e^- \rightarrow$	$Mg(s)$
$Na^+(aq)$	$+ e^- \rightarrow$	$Na(s)$
$Ca^{2+}(aq)$	$+ 2e^- \rightarrow$	$Ca(s)$
$K^+(aq)$	$+ e^- \rightarrow$	$K(s)$
$Li^+(aq)$	$+ e^- \rightarrow$	$Li(s)$
Weakest Oxidizing Agent		STRONGEST Reducing Agent

Spontaneous Reactions

We can interpret the information given in Table 17.3 as follows. The ability of a substance to be reduced, that is to attract electrons, decreases from top to bottom in the table. The strongest oxidizing agents are in the upper left portion of the table. Thus, F_2 has the highest potential for reduction and is the strongest oxidizing agent. The strongest reducing agents are in the lower right portion of the table. Thus, Li has the highest potential for oxidation and is the strongest reducing agent.

A metal ion can be reduced, that is gain electrons, by any reducing agent listed lower in the table. For example, Sn^{2+} can be reduced by Ni or Fe, but not by Pb. Any metal below H_2 in the table can react with an acid and be oxidized. A metal above H^+ in the table cannot react with acid. For example, Pb, Sn, and Ni react with acid and undergo oxidation, whereas Cu and Ag do not react with acid.

A nonmetal can oxidize any reducing agent listed lower in the table. For example, Cl_2 can oxidize Br^- or Ag, but not F^-. In Table 17.3, we see that F_2 is the strongest oxidizing agent and that Li is the strongest reducing agent. However, H_2 is the strongest *nonmetal* reducing agent listed in the table.

In Table 17.3 we note that $Cu^{2+}(aq)$ is a stronger oxidizing agent than $Zn^{2+}(aq)$. We also note that Zn metal is a stronger reducing agent than Cu metal. Therefore, we can explain the reaction as follows:

$Cu^{2+}(aq)$	+	$Zn(s)$	$\longrightarrow$	$Cu(s)$	+	$Zn^{2+}(aq)$
stronger oxidizing agent		stronger reducing agent		weaker reducing agent		weaker oxidizing agent

Moreover, we see that the reverse process has very little tendency to occur and is nonspontaneous. We can therefore conclude that there is no reaction between aqueous zinc ion and copper metal.

$Zn^{2+}(aq)$	+	$Cu(s)$	$\longrightarrow$	$Zn(s)$	+	$Cu^{2+}(aq)$
weaker oxidizing agent		weaker reducing agent		stronger reducing agent		stronger oxidizing agent

Now, let's consider a nonmetal reaction in an aqueous solution. For example, consider what happens if we bubble chlorine gas through an aqueous sodium bromide solution. The net ionic equation for the possible reaction is

$$Cl_2(g) + 2\,Br^-(aq) \longrightarrow 2\,Cl^-(aq) + Br_2(l)$$

In Table 17.3 we see that $Cl_2(g)$ is a stronger oxidizing agent than $Br_2(l)$. We also see that $Br^-(aq)$ is a stronger reducing agent than $Cl^-(aq)$. Therefore, we can explain the reaction as follows:

$Cl_2(g)$	+	$2\,Br^-(aq)$	$\longrightarrow$	$2\,Cl^-(aq)$	+	$Br_2(l)$
stronger oxidizing agent		stronger reducing agent		weaker reducing agent		weaker oxidizing agent

Because the reactants contain the stronger pair of oxidizing and reducing agents, the reaction takes place spontaneously. Is the reverse reaction spontaneous? Because the chloride ion and bromine liquid are the weaker pair of oxidizing and reducing agents, the reverse reaction is nonspontaneous. If a redox reaction is not spontaneous, we usually say that there is no reaction.

$2\,Cl^-(aq)$	+	$Br_2(l)$	$\longrightarrow$	$Cl_2(g)$	+	$2\,Br^-(aq)$
weaker oxidizing agent		weaker reducing agent		stronger reducing agent		stronger oxidizing agent

The following example exercise provides additional practice in predicting whether or not a redox reaction occurs spontaneously:

Example Exercise 17.8 Predicting Spontaneous Redox Reactions

Predict whether the following reaction is spontaneous or nonspontaneous:

$$Ni^{2+}(aq) + Sn(s) \longrightarrow Ni(s) + Sn^{2+}(aq)$$

Solution

Let's refer to the table of reduction potentials to predict whether or not the reaction is spontaneous. Table 17.3 lists $Ni^{2+}(aq)$ as a weaker oxidizing agent than $Sn^{2+}(aq)$. Moreover, $Sn(s)$ is a weaker reducing agent than $Ni(s)$.

$Ni^{2+}(aq)$	+	$Sn(s)$	$\longrightarrow$	$Ni(s)$	+	$Sn^{2+}(aq)$
weaker oxidizing agent		weaker reducing agent		stronger reducing agent		stronger oxidizing agent

Because the reactants are the weaker pair of oxidizing and reducing agents, the reaction is *nonspontaneous*. Conversely, the reverse reaction is spontaneous because the products are the stronger oxidizing and reducing agents.

Practice Exercise

Predict whether the following reaction is spontaneous or nonspontaneous:

$$Ni^{2+}(aq) + Al(s) \longrightarrow Ni(s) + Al^{3+}(aq)$$

Answer: spontaneous

Concept Exercise

Refer to Table 17.3 and predict which of the following metals reacts spontaneously in an aqueous $FeSO_4$ solution:

(a) Ag (b) Al
(c) Cr (d) Ni

Answer: See Appendix G.

OBJECTIVES

- To indicate the anode and cathode in a given voltaic cell.
- To indicate the oxidation and reduction half-reactions in a given spontaneous electrochemical cell.

17.6 VOLTAIC CELLS

Batteries play an important role in our daily lives. They enable us to start our automobile and use cordless electrical appliances. Although different devices may use different types of batteries, all batteries involve a spontaneous redox reaction that produces electrical energy, and a flow of electrons.

Redox Reactions in Voltaic Cells

The conversion of chemical energy to electrical energy, and vice versa, in a redox reaction is called **electrochemistry**. As an example, consider the reaction of zinc metal in a copper(II) sulfate solution.

$$Zn(s) + CuSO_4(aq) \longrightarrow Cu(s) + ZnSO_4(aq)$$

Experimentally, we can physically separate the oxidation half-reaction from the reduction half-reaction. We do this by placing aqueous solutions of zinc sulfate and copper(II) sulfate in separate containers or compartments. Then, we place a Zn metal electrode in the first compartment containing an aqueous solution of $ZnSO_4$, and a Cu metal electrode in the second compartment containing an aqueous solution of $CuSO_4$. The two electrodes are connected by a wire, so electrons are free to travel between the two compartments. Each compartment is called a **half-cell**, and together the apparatus is called an **electrochemical cell** (Figure 17.3).

As the reaction proceeds, the concentration of Zn^{2+} in the left compartment increases. The concentration of Cu^{2+} in the right compartment decreases. The

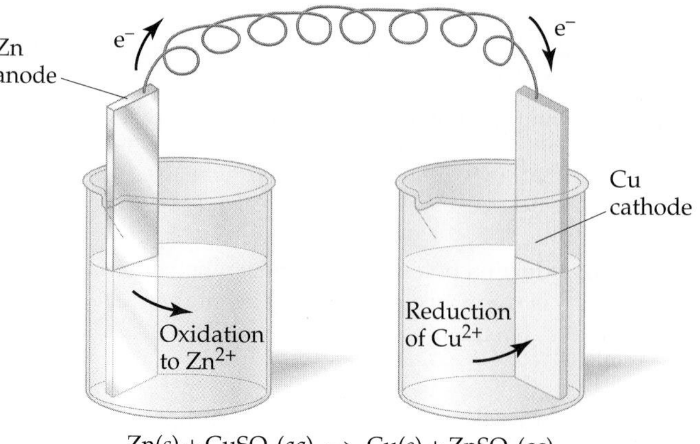

▶ **Figure 17.3 Redox Reaction in Half-Cells** In the compartment on the left, zinc metal is being oxidized. In the compartment on the right, copper ion is being reduced. This redox process allows electrons to flow from the zinc electrode to the copper electrode. As the reaction proceeds, the number of zinc ions in the left half-cell increases, and the number of copper(II) ions in the right half-cell decreases.

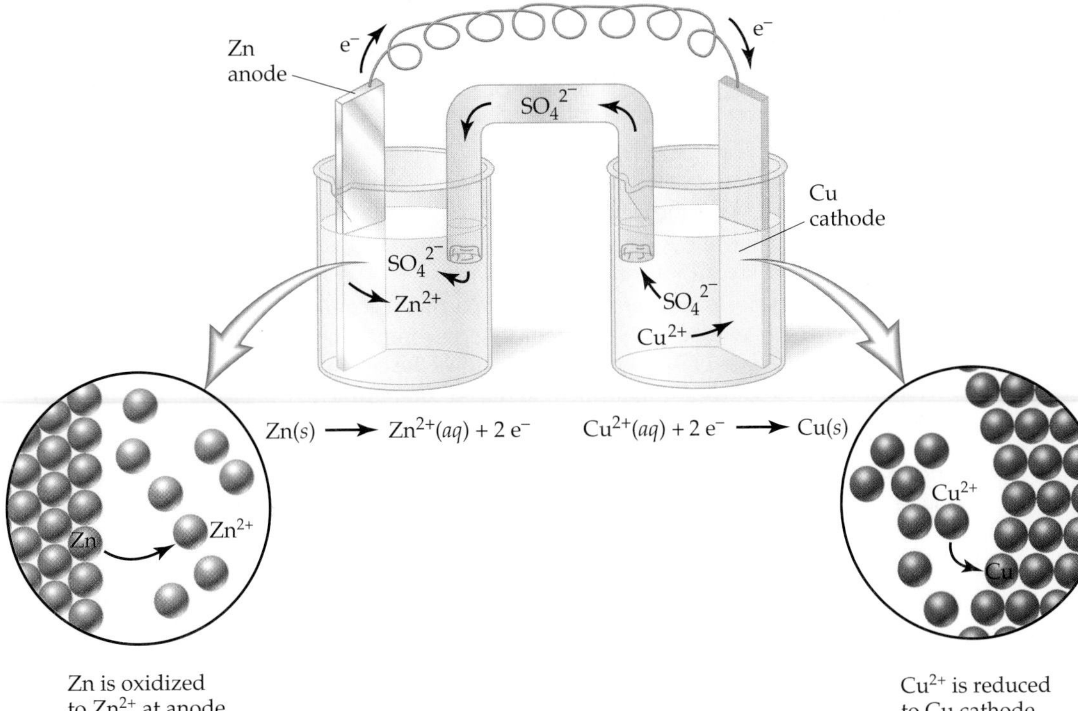

▲ **Figure 17.4 Voltaic Cell** The two half-cells are connected by a salt bridge, and the negative sulfate ions travel from the right half-cell to the left half-cell. The salt bridge reduces the positive charge buildup in the left half-cell and the negative charge buildup in the right half-cell. The cell continues to operate spontaneously as electrons flow from the zinc anode to the copper cathode.

concentration of SO_4^{2-} remains constant in both compartments. As a result, the left half-cell develops a net positive charge, as the right half-cell develops a net negative charge.

As the reaction continues, a charge develops in each compartment, which could halt the redox process. However, we can eliminate the charge buildup by introducing a **salt bridge**. A salt bridge placed between the compartments allows ions to travel between the two half-cells. The excess negative sulfate ions in the copper solution can move to the zinc solution, which has excess Zn^{2+}. Because a salt bridge eliminates charge buildup, the reaction can continue spontaneously. For spontaneous reactions that produce electrical energy, the cell is termed a **voltaic cell** (or galvanic cell). Figure 17.4 shows a voltaic cell based on the reaction of Zn in aqueous $CuSO_4$.

In an electrochemical cell, the electrode at which oxidation occurs is called the **anode**. The electrode at which reduction occurs is called the **cathode**. For the voltaic cell in Figure 17.4, the zinc electrode is the anode, and Zn is being oxidized. The copper electrode is the cathode, and Cu^{2+} is being reduced. The following example exercise further illustrates the processes of oxidation and reduction that take place in a spontaneous electrochemical cell:

Example Exercise 17.9 Voltaic Cells—Spontaneous Processes

Nickel can react with an aqueous silver nitrate solution according to the following ionic equation:

$$Ni(s) + 2\,Ag^{+}(aq) \longrightarrow 2Ag(s) + Ni^{2+}(aq)$$

Assume the half-reactions are separated into two compartments. A Ni electrode is placed in a compartment with 1.00 *M* $Ni(NO_3)_2$, and a Ag electrode is placed in a compartment with 1.00 *M* $AgNO_3$. Indicate each of the following:

(a) oxidation half-cell reaction
(b) reduction half-cell reaction
(c) anode and cathode
(d) direction of electron flow
(e) direction of NO_3^- in the salt bridge

Solution

Referring to Table 17.3, we see that Ag^+ has a higher reduction potential than Ni^{2+}. Therefore, the process is spontaneous and Ni is oxidized as Ag^+ is reduced. The two half-cell processes are

(a) Oxidation: $Ni \longrightarrow Ni^{2+} + 2\,e^-$

(b) Reduction: $Ag^+ + e^- \longrightarrow Ag$

(c) The anode is where oxidation occurs; thus, Ni is the anode. The cathode is where reduction occurs; thus, Ag is the cathode.

(d) Ni loses electrons while Ag^+ gains electrons, so the direction of electron flow is from Ni to Ag. (*Electrons flow from the anode to the cathode.*)

(e) As the Ni anode loses electrons, the Ni compartment acquires a net positive charge due to Ni^{2+}. As the Ag cathode gains electrons, the Ag compartment acquires a net negative charge due to excess NO_3^-. A salt bridge allows the cell to operate continuously as NO_3^- ions travel from the Ag compartment to the Ni compartment. (*Anions flow from the cathode to the anode.*)

Practice Exercise

Iron can react with an aqueous tin(II) nitrate solution according to the following ionic equation:

$$Fe(s) + Sn^{2+}(aq) \longrightarrow Sn(s) + Fe^{2+}(aq)$$

An Fe electrode is placed in a compartment with 1.00 *M* $Fe(NO_3)_2$, and a Sn electrode is placed in another compartment with 1.00 *M* $Sn(NO_3)_2$. Indicate each of the following:

(a) oxidation half-cell reaction
(b) reduction half-cell reaction
(c) anode and cathode
(d) direction of electron flow
(e) direction of NO_3^- in the salt bridge

Answers:

(a) Oxidation: $Fe \longrightarrow Fe^{2+} + 2\,e^-$

(b) Reduction: $Sn^{2+} + 2\,e^- \longrightarrow Sn$

(c) Fe is the anode; Sn is the cathode.

(d) Electrons flow from the Fe anode to the Sn cathode.

(e) A salt bridge allows NO_3^- ions to travel from the Sn cathode compartment to the Fe anode compartment.

Concept Exercise

Draw and illustrate the voltaic cell described in the practice exercise. Refer to Figure 17.4 and label the anode and cathode; show the direction of electron and anion flow.

Answer: See Appendix G.

By convention, the anode in an electrochemical cell is usually shown as the left half-cell. If the anode is found on the left, electrons flow from the left half-cell to the right half-cell. To avoid a buildup of negative charge in the right half-cell, negative anions exit the right half-cell, travel through the salt bridge, and enter the left half-cell.

We should note that cations are also present in the salt bridge. Conversely, cations exit the left half-cell, travel through the salt bridge, and enter the right half-cell. In summary, anions travel toward the half-cell compartment where electrons are exiting (anode), and cations travel toward the half-cell compartment where electrons are entering (cathode).

Batteries

A **battery** is a general term for any electrochemical cell that spontaneously produces electrical energy. A battery has stored chemical energy, which can produce electrical energy measured in volts. A battery is made up of one or more voltaic cells. For example, the ordinary lead storage battery in an automobile is a series of six cells, each of which produces 2 volts. Operating together, the six cells generate 12 volts.

Unlike an ordinary voltaic cell, the batteries we use in flashlights, electronic toys, and portable radios do not have electrodes immersed in a conducting solution. Rather, their electrodes are connected by a solid paste that can conduct electricity. A battery that does not use an electrolyte solution is called a **dry cell**. The common alkaline dry cell battery was invented more than 100 years ago by the French chemist Georges Leclanché.

An ordinary dry cell battery is not rechargeable, and the reactants stored in the battery are eventually depleted. However, a NiCad battery, which is a special type of dry cell, can be recharged almost indefinitely. That is, electricity can be used to reverse the spontaneous reaction, and the original reactants and products are reformed. A NiCad battery is so named because the redox reactions involve nickel and cadmium. NiCad batteries are used in rechargeable battery packs for calculators and portable electronic appliances. A collection of common batteries is shown in Figure 17.5.

▲ **Figure 17.5 Common Batteries** Batteries spontaneously supply electrical energy from a redox reaction.

17.7 ELECTROLYTIC CELLS

OBJECTIVES

- To indicate the anode and cathode in a given electrolytic cell.
- To indicate the oxidation and reduction half-reactions in a given nonspontaneous electrochemical cell.

So far we have discussed electrochemical cells that operate spontaneously. Equally as important are electrochemical cells that operate nonspontaneously. Although a redox reaction may be nonspontaneous, we can supply electricity and force the reaction to occur. This process is referred to as **electrolysis**. When a cell phone battery runs down, it is recharged by using electricity to force nonspontaneous reactions. Thus, the application of electricity creates reactants that will react spontaneously to provide the energy to operate your cell phone.

Chromium metal is commonly plated onto steel (iron) by an electrolytic method. Although the reaction of iron in an aqueous chromium solution is not spontaneous, we can force the reaction to occur. In this process, an iron or steel object is placed in an aqueous chromium solution, and a direct electric current is applied to overcome the nonspontaneous reaction.

Many important elements, including sodium and chlorine, are produced from their compounds by electrolysis. In the preparation of sodium metal and chlorine gas, first salt is obtained by the evaporation of seawater. Next, the crystalline NaCl is placed in a crucible and heated until it melts. Then, two inert platinum metal electrodes are dipped into the molten NaCl and electricity is applied. The electrodes, molten sodium chloride, and a source of electricity form an **electrolytic cell**.

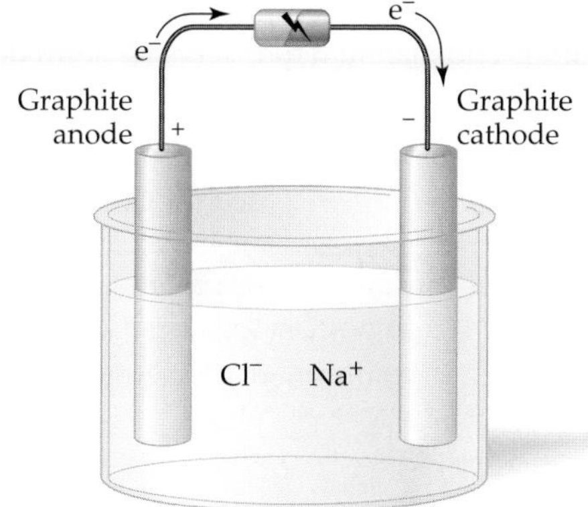

◀ **Figure 17.6 Electrolytic Cell** An electrolytic cell can produce sodium metal and chlorine gas from molten NaCl. A source of direct current forces electrons to travel toward the cathode for the reduction of sodium ions, as electrons are released from the anode from the oxidation of chloride ions.

An electrolytic cell is a system in which a nonspontaneous reaction occurs as a result of the input of direct electric current. The electric current forces the nonspontaneous reaction, and in this example, NaCl decomposes into its elements.

Cathode reaction (reduction):

$$2\,Na^{+}(l) + 2\,e^{-} \longrightarrow 2\,Na(l)$$

Anode reaction (oxidation):

$$2\,Cl^{-}(l) \longrightarrow Cl_2(g) + 2\,e^{-}$$

We can write the overall equation by adding the two half-reactions together. After canceling two electrons from each side of the equation, we have the equation for the nonspontaneous redox reaction.

$$2\,Na^+(l) + 2\,Cl^-(l) \longrightarrow 2\,Na(l) + Cl_2(g)$$

Example Exercise 17.10 illustrates additional nonspontaneous redox reactions that occur in an electrolytic cell.

Example Exercise 17.10 Electrolytic Cells—Nonspontaneous Processes

Aluminum metal is produced by passing an electric current through bauxite, which contains Al_2O_3 dissolved in the molten mineral cryolite. Graphite rods serve as electrodes, and we can write the redox equation as follows:

$$3\,C(s) + 2\,Al_2O_3(l) \xrightarrow{\text{electricity}} 4\,Al(l) + 3\,CO_2(g)$$

Indicate each of the following for this nonspontaneous process:

(a) oxidation half-cell reaction
(b) reduction half-cell reaction
(c) anode and cathode
(d) direction of electron flow

Solution

In the equation, we see that C is oxidized to CO_2 and that Al_2O_3 is reduced to Al metal. The two half-cell processes are

(a) Oxidation: $C + 2\,O^{2-} \longrightarrow CO_2 + 4\,e^-$
(b) Reduction: $Al^{3+} + 3\,e^- \longrightarrow Al$
(c) Oxidation occurs at the graphite anode, where CO_2 is released. Reduction occurs at the graphite cathode, where Al metal is produced.
(d) The electrons flow from the anode, where CO_2 gas is released, to the cathode, where Al metal is produced.

Practice Exercise

Magnesium metal is produced by passing an electric current through molten $MgCl_2$ obtained from evaporated seawater. Carbon and platinum rods serve as electrodes, and we can write the redox equation as follows:

$$C(s) + 2\,MgCl_2(l) \xrightarrow{\text{electricity}} 2\,Mg(l) + CCl_4(g)$$

Indicate each of the following for this nonspontaneous process:

(a) oxidation half-cell reaction
(b) reduction half-cell reaction
(c) anode and cathode
(d) direction of electron flow

Answers:

(a) Oxidation: $C + 4\,Cl^- \longrightarrow CCl_4 + 4\,e^-$
(b) Reduction: $Mg^{2+} + 2\,e^- \longrightarrow Mg$
(c) The C electrode is the anode; the Pt electrode is the cathode.
(d) The electrons flow from the C anode to the Pt cathode.

Concept Exercise

Draw and illustrate the electrolytic cell described in the practice exercise. Refer to Figure 17.6 and label the anode and cathode; show the direction of electron flow.

Answer: See Appendix G.

CHEMISTRY CONNECTION Hybrid Vehicles

Q: What is the estimated gas mileage of the General Motors Volt, which is a hybrid of gasoline engine and lithium ion batteries?

In an effort to obtain greater automobile gas mileage, companies are exploring a number of technologies. Japanese manufacturers currently offer hybrid vehicles that utilize a combination of traditional gasoline power and lightweight voltaic cells that produce electrical energy while the gasoline engine is operating.

In 2007, General Motors unveiled the Chevrolet Volt car, a hybrid vehicle. The General Motors Volt is touted to deliver 100+ miles per gallon, and can travel up to 640 miles without a fuel fill-up or a battery recharge.

Hybrids are designed to generate energy when braking or coasting, and store it as electricity in a battery pack. When accelerating, the vehicle uses the energy to power an electric motor, and reduce fuel consumption. While other hybrids use a battery-powered electric motor to supplement a gasoline engine, the Volt runs on electric power until the batteries run down. Then the gas engine turns on to regenerate and produce electricity for the batteries. Alternatively, the batteries can be recharged by plugging into an ordinary electrical outlet.

The General Motors Volt has a front-mounted electric motor that generates 160 horsepower. The batteries are placed beneath the floor, and a 12-gallon fuel tank is found in the rear of the car. The Volt can travel about 40 miles on pure electric power. The Volt accelerates to 60 mph in less than 9 seconds, and has a top speed of 120 mph. The Volt saves a typical motorist about 500 gallons of fuel a year. With regard to the environment, the Volt emits considerably less carbon dioxide gas into the air than a typical gasoline-powered automobile.

◀ **The General Motors Hybrid Gas–Electric Volt**
Under normal driving conditions, the General Motors Volt is estimated to get 100+ miles per gallon of fuel.

A: Under normal driving conditions, the General Motors Volt is estimated to get 100+ miles per gallon.

Summary

Section 17.1 An oxidation–reduction reaction is characterized by the transfer of electrons. When a substance undergoes oxidation, it loses electrons, and its **oxidation number** increases. Conversely, when a substance undergoes reduction, it gains electrons, and its oxidation number decreases. When Cu is oxidized to Cu^{2+}, its oxidation number changes from 0 to +2; when Cu^{2+} is reduced to Cu, its oxidation number changes from +2 to 0. Transition metals, as well as nonmetals, demonstrate a variety of oxidation numbers. For example, Mn has oxidation numbers of +7, +6, +4, +3, +2, and 0. The oxidation numbers of Cl include +7, +5, +3, +1, 0, and −1.

Section 17.2 In an oxidation–reduction reaction, or **redox reaction**, oxidation and reduction occur at the same time. In the process of **oxidation**, a substance loses electrons. In the process of **reduction**, a substance gains electrons. The substance that loses electrons causes reduction and is referred to as a **reducing agent**. On the other hand, the substance that gains electrons causes oxidation and is referred to as an **oxidizing agent**. We can illustrate the overall redox process as

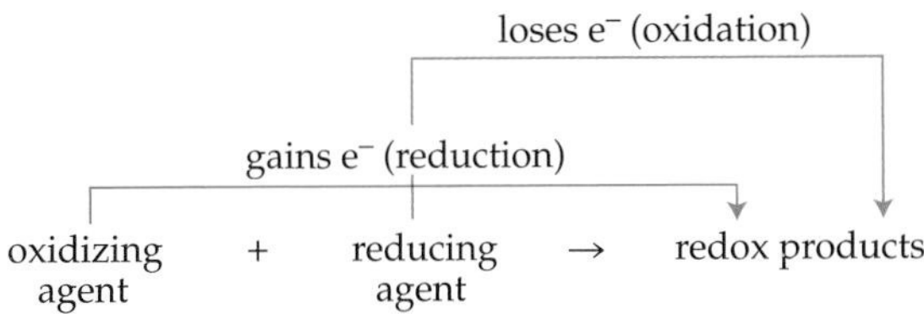

Section 17.3 To balance a redox equation, we keep track of the changes in oxidation number. This method allows us to balance a redox equation by determining changes in oxidation number for the oxidizing agent and reducing agent. If the oxidizing agent gains two electrons and the reducing agent loses one electron, the reducing agent must be multiplied by two to balance the electrons. The remaining elements in the equation are then balanced by inspection.

Section 17.4 Redox reactions can also be balanced by the **half-reaction** method. First, a balanced half-reaction is written for the oxidation process. Second, a balanced half-reaction is written for the reduction process. Each half-reaction is then multiplied by the number of electrons lost or gained in the opposite half-reaction. Thus, the number of electrons lost in oxidation equals the number of electrons gained in reduction. Finally, the two half-reactions are added together to give a balanced equation for the overall redox reaction.

Section 17.5 Half-reactions can be arranged according to their ability to undergo reduction. This tendency, called the **reduction potential**, is listed in Table 17.3 for selected oxidizing and reducing agents. A substance that is easily reduced is a strong oxidizing agent, and a substance that is easily oxidized is a strong reducing agent. For a redox reaction to be spontaneous, the oxidizing agent must have a stronger attraction for electrons than the substance being oxidized.

Section 17.6 **Electrochemistry** is the study of the interconversion of chemical and electrical energy. In an **electrochemical cell**, a chemical reaction produces electrical energy. An electrochemical cell is composed of two **half-cell** compartments. Oxidation occurs in one half-cell and reduction in the other. If the redox reaction occurs spontaneously, the cell is referred to as a **voltaic cell**. The **anode** is the electrode where oxidation occurs, and the **cathode** is the electrode where reduction occurs. In the operation of an electrochemical cell, the electrodes are connected by a wire and the two half-cells are joined by a **salt bridge** that allows ions to travel from one compartment to the other. A **battery** is any device that supplies electricity and is composed of one or more voltaic cells. A **dry cell** is a special type of battery in which there is no conducting electrolyte solution.

Section 17.7 Whereas the redox reaction is spontaneous in a voltaic cell, the redox reaction is not spontaneous in an **electrolytic cell**. In an electrolytic cell, it is necessary to supply electrical energy to produce a chemical reaction. For example, the **electrolysis** of water produces hydrogen and oxygen gases. Active metals, such as Na, K, and Ca, are obtained from electrolytic cells. In addition, chlorine gas and bromine liquid are produced commercially by passing a direct electric current through an electrolytic cell.

Key Terms *Answers to Key Terms are in Appendix H.*

Select the key term that corresponds to each of the following definitions.

_____ **1.** a value assigned to an atom in a substance that indicates whether the atom is electron-poor or electron-rich compared to a free atom
_____ **2.** a chemical reaction in which an electron transfer takes place
_____ **3.** a chemical process in which a substance loses electrons
_____ **4.** a chemical process in which a substance gains electrons
_____ **5.** the substance undergoing reduction in a redox reaction
_____ **6.** the substance undergoing oxidation in a redox reaction
_____ **7.** a reaction that represents either an oxidation or a reduction process
_____ **8.** the relative ability of a substance to undergo reduction
_____ **9.** the study of the interconversion of chemical and electrical energy
_____ **10.** a general term for an apparatus containing two electrodes in separate compartments connected by a wire
_____ **11.** a portion of an electrochemical cell having a single electrode where either oxidation or reduction occurs
_____ **12.** a porous device that allows ions to travel between two half-cells to maintain an ionic charge balance in each compartment
_____ **13.** an electrochemical cell in which a spontaneous redox reaction occurs and generates electrical energy
_____ **14.** the providing of electricity to a nonspontaneous redox process to cause a reaction
_____ **15.** an electrochemical cell in which a nonspontaneous redox reaction occurs as a result of the input of direct electric current
_____ **16.** the electrode in an electrochemical cell at which oxidation occurs
_____ **17.** the electrode in an electrochemical cell at which reduction occurs
_____ **18.** a general term for any electrochemical cell that produces electrical energy
_____ **19.** an electrochemical cell in which the anode and cathode reactions do not take place in an aqueous solution

(a) anode *(Sec. 17.6)*
(b) battery *(Sec. 17.6)*
(c) cathode *(Sec. 17.6)*
(d) dry cell *(Sec. 17.6)*
(e) electrochemical cell *(Sec. 17.6)*
(f) electrochemistry *(Sec. 17.6)*
(g) electrolysis *(Sec. 17.7)*
(h) electrolytic cell *(Sec. 17.7)*
(i) half-cell *(Sec. 17.6)*
(j) half-reaction *(Sec. 17.4)*
(k) oxidation *(Sec. 17.2)*
(l) oxidation number *(Sec. 17.1)*
(m) oxidizing agent *(Sec. 17.2)*
(n) redox reaction *(Sec. 17.2)*
(o) reducing agent *(Sec. 17.2)*
(p) reduction *(Sec. 17.2)*
(q) reduction potential *(Sec. 17.5)*
(r) salt bridge *(Sec. 17.6)*
(s) voltaic cell *(Sec. 17.6)*

Exercises *Answers to odd-numbered Exercises are in Appendix I.*

Oxidation Numbers (Sec. 17.1)

1. State the oxidation number for each of the following elements in the free state:
   (a) Ag **(b)** Mg
   (c) S **(d)** F_2
2. State the oxidation number for each of the following elements in the free state:
   (a) Au **(b)** Mn
   (c) P **(d)** O_2
3. State the oxidation number for each of the following monoatomic ions:
   (a) Ag^+ **(b)** Mg^{2+}
   (c) S^{2-} **(d)** F^-
4. State the oxidation number for each of the following monoatomic ions:
   (a) Au^{3+}
   (b) Mn^{2+}
   (c) P^{3-}
   (d) O^{2-}

5. Calculate the oxidation number for nitrogen in the following compounds:
(a) NH_3
(b) N_2O_4
(c) Li_3N
(d) KNO_3

6. Calculate the oxidation number for silicon in the following compounds:
(a) Si_2H_6
(b) SiO_2
(c) Si_3N_4
(d) $CaSiO_3$

7. Calculate the oxidation number for sulfur in the following polyatomic ions:
(a) SO_3^{2-}
(b) HSO_4^-
(c) HS^-
(d) $S_2O_8^{2-}$

8. Calculate the oxidation number for carbon in the following polyatomic ions:
(a) CO_3^{2-}
(b) HCO_3^-
(c) CN^-
(d) CNO^-

Oxidation–Reduction Reactions (Sec. 17.2)

9. Supply the term that corresponds to each of the following:
(a) a redox process characterized by electron loss
(b) a substance undergoing oxidation in a redox reaction

10. Supply the term that corresponds to each of the following:
(a) a redox process characterized by electron gain
(b) a substance undergoing reduction in a redox reaction

11. State the substances undergoing oxidation and reduction in the following redox reactions:
(a) $Co(s) + S(s) \longrightarrow CoS(s)$
(b) $Cd(s) + Cl_2(g) \longrightarrow CdCl_2(s)$

12. State the substances undergoing oxidation and reduction in the following redox reactions:
(a) $Mg(s) + S(s) \longrightarrow MgS(s)$
(b) $Mn(s) + Br_2(l) \longrightarrow MnBr_2(s)$

13. Indicate the oxidizing and reducing agents in the following redox reactions:
(a) $CuO(s) + H_2(g) \longrightarrow Cu(s) + H_2O(l)$
(b) $PbO(s) + CO(g) \longrightarrow Pb(s) + CO_2(g)$

14. Indicate the oxidizing and reducing agents in the following redox reactions:
(a) $Ca(s) + 2H_2O(l) \longrightarrow Ca(OH)_2(aq) + H_2(g)$
(b) $Mg(s) + 2\,HCl(aq) \longrightarrow MgCl_2(aq) + H_2(g)$

15. Indicate the substances undergoing oxidation and reduction in the following ionic redox reactions:
(a) $Al(s) + Cr^{3+}(aq) \longrightarrow Al^{3+}(aq) + Cr(s)$
(b) $F_2(g) + 2\,Cl^-(aq) \longrightarrow 2\,F^-(aq) + Cl_2(g)$

16. Indicate the substances undergoing oxidation and reduction in the following ionic redox reactions:
(a) $Fe^{3+}(aq) + SO_3^{2-}(aq) \longrightarrow Fe^{2+}(aq) + SO_4^{2-}(aq)$
(b) $Cl_2(g) + 2\,I^-(aq) \longrightarrow 2\,Cl^-(aq) + I_2(s)$

17. Indicate the oxidizing and reducing agents in the following redox reactions:
(a) $Cr^{2+}(aq) + AgI(s) \longrightarrow Cr^{3+}(aq) + Ag(s) + I^-(aq)$
(b) $Sn^{2+}(aq) + Hg^{2+}(aq) \longrightarrow Sn^{4+}(aq) + Hg_2^{2+}(aq)$

18. Indicate the oxidizing and reducing agents in the following redox reactions:
(a) $H_2O_2(aq) + SO_3^{2-}(aq) \longrightarrow H_2O(l) + SO_4^{2-}(aq)$
(b) $IO_3^-(aq) + Cu^+(aq) \longrightarrow Cu^{2+}(aq) + I_2(s)$

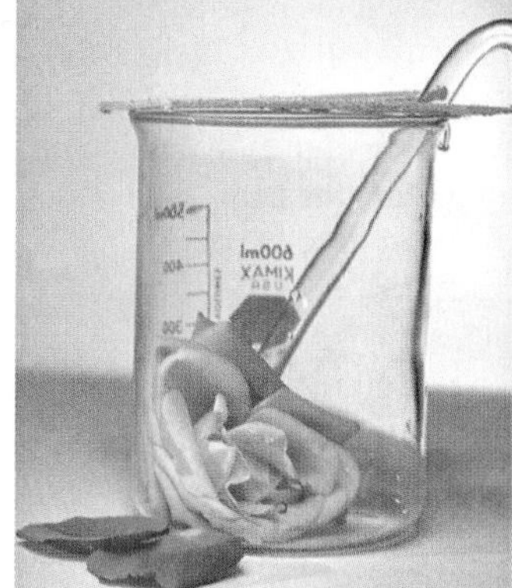

◀ Reduction of a Rose A red rose (top) undergoes a redox reaction with colorless SO_2 gas. After a few minutes, the color of the rose changes from red to pink (bottom).

Balancing Redox Equations: Oxidation Number Method (Sec. 17.3)

19. Is it always true that the electron loss by oxidation equals the electron gain by reduction in a balanced redox equation?

20. Is it always true that the total ionic charge on the reactants equals the total ionic charge on the products in a balanced redox equation?

21. Write a balanced equation for each of the following redox reactions using the oxidation number method:
(a) $Br_2(l) + NaI(aq) \longrightarrow I_2(s) + NaBr(aq)$
(b) $PbS(s) + O_2(g) \longrightarrow PbO(aq) + SO_2(g)$

22. Write a balanced equation for each of the following redox reactions using the oxidation number method:
(a) $Cl_2(g) + KI(aq) \longrightarrow I_2(s) + KCl(aq)$
(b) $Fe_2O_3(s) + CO(g) \longrightarrow Fe(s) + CO_2(g)$

23. Write a balanced equation for each of the following redox reactions using the oxidation number method:
(a) $MnO_4^-(aq) + I^-(aq) + H^+(aq) \longrightarrow Mn^{2+}(aq) + I_2(s) + H_2O(l)$
(b) $Cu(s) + H^+(aq) + SO_4^{2-}(aq) \longrightarrow Cu^{2+}(aq) + SO_2(g) + H_2O(l)$

24. Write a balanced equation for each of the following redox reactions using the oxidation number method:
(a) $Fe^{2+}(aq) + H_2O_2(aq) + H^+(aq) \longrightarrow Fe^{3+}(aq) + H_2O(l)$
(b) $Cr_2O_7^{2-}(aq) + Br^-(aq) + H^+(aq) \longrightarrow Cr^{3+}(aq) + Br_2(l) + H_2O(l)$

Balancing Redox Equations: Half-Reaction Method (Sec. 17.4)

25. Write a balanced half-reaction for each of the following in an *acidic* solution:
(a) $SO_2(g) \longrightarrow SO_4^{2-}(aq)$
(b) $BrO_3^-(aq) \longrightarrow Br_2(l)$

26. Write a balanced half-reaction for each of the following in an *acidic* solution:
(a) $H_2O_2(aq) \longrightarrow H_2O(l)$
(b) $AsO_3^{3-}(aq) \longrightarrow AsO_3^-(aq)$

27. Write a balanced half-reaction for each of the following in a *basic* solution:
(a) $ClO^-(aq) \longrightarrow Cl^-(aq)$
(b) $MnO_4^-(aq) \longrightarrow MnO_2(s)$

28. Write a balanced half-reaction for each of the following in a *basic* solution:
(a) $Ni(OH)_2(s) \longrightarrow NiO_2(s)$
(b) $NO_2^-(aq) \longrightarrow N_2O(g)$

29. Write a balanced equation for each of the following redox reactions in an *acidic* solution using the half-reaction method:
(a) $Zn(s) + NO_3^-(aq) \longrightarrow Zn^{2+}(aq) + NO(g)$
(b) $Mn^{2+}(aq) + BiO_3^-(aq) \longrightarrow MnO_4^-(aq) + Bi^{3+}(aq)$

30. Write a balanced equation for each of the following redox reactions in an *acidic* solution using the half-reaction method:
(a) $Sn^{2+}(aq) + IO_3^-(aq) \longrightarrow Sn^{4+}(aq) + I_2(s)$
(b) $AsO_3^{3-}(aq) + Br_2(l) \longrightarrow AsO_4(aq) + Br^-(aq)$

31. Write a balanced equation for each of the following redox reactions in a *basic* solution using the half-reaction method:
(a) $MnO_4^-(aq) + S^2(aq) \longrightarrow MnO_2(s) + S(s)$
(b) $Cu(s) + ClO^-(aq) \longrightarrow Cu^{2+}(aq) + Cl^-(aq)$

32. Write a balanced equation for each of the following redox reactions in a *basic* solution using the half-reaction method:
(a) $Cl_2(g) + BrO_2^-(aq) \longrightarrow Cl^-(aq) + BrO_3^-(aq)$
(b) $MnO_2(s) + O_2(g) \longrightarrow MnO_4^-(aq) + H_2O(l)$

33. Chlorine can undergo a redox reaction in which it is simultaneously oxidized and reduced. Write a balanced equation for the following in an *acidic* solution:

$$Cl_2(aq) \longrightarrow Cl^-(aq) + HOCl(aq)$$

34. Nitrous acid, HNO_2, can undergo a redox reaction in which it is simultaneously oxidized and reduced. Write a balanced equation for the following in an *acidic* solution:

$$HNO_2(aq) \longrightarrow NO_3^-(aq) + NO(g)$$

35. Chlorine can undergo a redox reaction in which it is simultaneously oxidized and reduced. Write a balanced equation for the following in a *basic* solution:

$$Cl_2(aq) \longrightarrow ClO_2^-(aq) + Cl^-(aq)$$

36. Sulfur can undergo a redox reaction in which it is simultaneously oxidized and reduced. Write a balanced equation for the following in a *basic* solution:

$$S(s) \longrightarrow SO_3^{2-}(aq) + S^{2-}(aq)$$

Predicting Spontaneous Redox Reactions (Sec. 17.5)

37. Refer to Table 17.3 and indicate which substance in each of the following pairs has the greater tendency to be reduced:
(a) $Pb^{2+}(aq)$ or $Zn^{2+}(aq)$ (b) $Fe^{3+}(aq)$ or $Al^{3+}(aq)$
(c) $Ag^+(aq)$ or $I_2(s)$ (d) $Cu^{2+}(aq)$ or $Br_2(l)$

38. Refer to Table 17.3 and indicate which substance in each of the following pairs has the greater tendency to be oxidized:
(a) $Li(s)$ or $K(s)$ (b) $Al(s)$ or $Mg(s)$
(c) $Fe^{2+}(aq)$ or $I^-(aq)$ (d) $Br^-(aq)$ or $Cl^-(aq)$

39. Refer to Table 17.3 and indicate which substance in each of the following pairs is the stronger oxidizing agent:
(a) $F_2(g)$ or $Cl_2(g)$
(b) $Ag^+(aq)$ or $Br_2(l)$
(c) $Cu^{2+}(aq)$ or $H^+(aq)$
(d) $Mg^{2+}(aq)$ or $Mn^{2+}(aq)$

40. Refer to Table 17.3 and indicate which substance in each of the following pairs is the stronger reducing agent:
(a) $Cu(s)$ or $Cr(s)$
(b) $H_2(g)$ or $Cu(s)$
(c) $Cu(s)$ or $I^-(aq)$
(d) $Cl^-(aq)$ or $H_2(g)$

41. Refer to Table 17.3 and state whether each of the following reactions is spontaneous or nonspontaneous:
(a) $Br_2(l) + LiF(aq) \longrightarrow F_2(s) + LiBr(aq)$
(b) $Al(NO_3)_3(aq) + Mn(s) \longrightarrow Mn(NO_3)_2(aq) + Al(s)$

42. Refer to Table 17.3 and state whether each of the following reactions is spontaneous or nonspontaneous:
(a) $Cr(s) + HCl(aq) \longrightarrow CrCl_3(aq) + H_2(g)$
(b) $FeCl_3(aq) + NaI(aq) \longrightarrow FeCl_2(aq) + NaCl(aq) + I_2(s)$

43. Refer to Table 17.3 and state whether each of the following ionic redox reactions is spontaneous or nonspontaneous:
(a) $Mg(s) + Sn^{2+}(aq) \longrightarrow Mg^{2+}(aq) + Sn(s)$
(b) $H^+(aq) + Ni(s) \longrightarrow H_2(g) + Ni^{2+}(aq)$

44. Refer to Table 17.3 and state whether each of the following ionic redox reactions is spontaneous or nonspontaneous:
(a) $H^+(aq) + I^-(aq) \longrightarrow H_2(g) + I_2(s)$
(b) $Fe^{3+}(aq) + I^-(aq) \longrightarrow Fe^{2+}(aq) + I_2(s)$

45. Given the following redox reactions, list A, B, and C according to which is *higher* in the table of reduction potentials:

$$A(s) + B^+(aq) \longrightarrow B(s) + A^+(aq)$$
$$B(s) + C^+(aq) \longrightarrow C(s) + B^+(aq)$$

46. Given the following redox reactions, list X, Y, and Z according to which is *higher* in the table of reduction potentials:

$$X_2(aq) + 2\,Y^-(aq) \longrightarrow Y_2(aq) + 2\,X^-(aq)$$
$$Y_2(aq) + 2\,Z^-(aq) \longrightarrow Z_2(aq) + 2\,Y^-(aq)$$

Voltaic Cells (Sec. 17.6)

47. Sketch the following voltaic cell showing the two compartments, the anode and cathode, the wire connecting the two electrodes, and a salt bridge:

$$Ni(s) + 2\,AgNO_3(aq) \longrightarrow 2Ag(s) + Ni(NO_3)_2(aq)$$

48. Sketch the following voltaic cell showing the two compartments, the anode and cathode, the wire connecting the two electrodes, and a salt bridge:

$$Mn(s) + Cd(NO_3)_2(aq) \longrightarrow Cd(s) + Mn(NO_3)_2(aq)$$

49. Diagram the direction of electron flow and the movement of nitrate ions in the salt bridge that connects the two half-cells in Exercise 47.

50. Diagram the direction of electron flow and the movement of nitrate ions in the salt bridge that connects the two half-cells in Exercise 48.

51. The spontaneous redox reaction of manganese and aqueous iron(II) sulfate occurs according to the following ionic equation:

$$Mn(s) + Fe^{2+}(aq) \longrightarrow Mn^{2+}(aq) + Fe(aq)$$

The half-reactions are separated into two compartments. A Mn electrode is placed in 1.00 *M* $MnSO_4$, and an Fe electrode in 1.00 *M* $FeSO_4$. Indicate each of the following for the voltaic cell:
(a) oxidation half-cell reaction
(b) reduction half-cell reaction
(c) anode and cathode
(d) direction of electron flow
(e) direction of SO_4^{2-} in the salt bridge

52. The spontaneous redox reaction of tin and aqueous copper(II) sulfate occurs according to the following ionic equation:

$$Sn(s) + Cu^{2+}(aq) \longrightarrow Cu(s) + Sn^{2+}(aq)$$

The half-reactions are separated into two compartments. A Sn electrode is placed in 1.00 *M* $SnSO_4$, and a Cu electrode in 1.00 *M* $CuSO_4$. Indicate each of the following for the voltaic cell:
(a) oxidation half-cell reaction
(b) reduction half-cell reaction
(c) anode and cathode
(d) direction of electron flow
(e) direction of SO_4^{2-} in the salt bridge

53. The spontaneous redox reaction of magnesium and aqueous lead(II) nitrate occurs according to the following equation:

$$Mg(s) + Pb(NO_3)_2(aq) \longrightarrow Pb(s) + Mg(NO_3)_2(aq)$$

The half-reactions are separated into two compartments. A Mg electrode is placed in 1.00 *M* $Mg(NO_3)_2$, and a Pb electrode in 1.00 *M* $Pb(NO_3)_2$. Indicate each of the following for the voltaic cell:
(a) oxidation half-cell reaction
(b) reduction half-cell reaction
(c) anode and cathode
(d) direction of electron flow
(e) direction of NO_3^- in the salt bridge

54. The spontaneous redox reaction of cobalt and aqueous manganese(II) nitrate occurs according to the following equation:

$$Mn(s) + Co(NO_3)_2(aq) \longrightarrow Co(s) + Mn(NO_3)_2(aq)$$

The half-reactions are separated into two compartments. A Mn electrode is placed in 1.000 *M* $Mn(NO_3)_2$, and a Co electrode in 1.00 *M* $Co(NO_3)_2$. Indicate each of the following for the voltaic cell:
(a) oxidation half-cell reaction
(b) reduction half-cell reaction
(c) anode and cathode
(d) direction of electron flow
(e) direction of NO_3^- in the salt bridge

Electrolytic Cells (Sec. 17.7)

55. The nonspontaneous redox reaction of nickel and aqueous iron(II) sulfate occurs electrolytically according to the following equation:

$$Ni(s) + FeSO_4(aq) \longrightarrow Fe(s) + NiSO_4(aq)$$

The half-reactions are separated into two compartments. A Ni electrode is placed in 1.00 *M* $NiSO_4$, and an Fe electrode in 1.00 *M* $FeSO_4$. Indicate each of the following for the electrolytic cell:
(a) oxidation half-cell reaction
(b) reduction half-cell reaction
(c) anode and cathode
(d) direction of electron flow
(e) direction of SO_4^{2-} in the salt bridge

56. The nonspontaneous redox reaction of tin and aqueous cobalt(II) sulfate occurs electrolytically according to the following equation:

$$Sn(s) + CoSO_4(aq) \longrightarrow Co(s) + SnSO_4(aq)$$

The half-reactions are separated into two compartments. A Sn electrode is placed in 1.00 *M* $SnSO_4$, and a Co electrode in 1.00 *M* $CoSO_4$. Indicate each of the following for the electrolytic cell:
(a) oxidation half-cell reaction
(b) reduction half-cell reaction
(c) anode and cathode
(d) direction of electron flow
(e) direction of SO_4^{2-} in the salt bridge

57. The nonspontaneous redox reaction of lead metal and aqueous zinc nitrate occurs electrolytically according to the following ionic equation:

$$Pb(s) + Zn^{2+}(aq) \longrightarrow Zn(s) + Pb^{2+}(aq)$$

The half-reactions are separated into two compartments. A Pb electrode is placed in 1.00 *M* $Pb(NO_3)_2$, and a Zn electrode in 1.00 *M* $Zn(NO_3)_2$. Indicate each of the following for the electrolytic cell:
(a) oxidation half-cell reaction
(b) reduction half-cell reaction
(c) anode and cathode
(d) direction of electron flow
(e) direction of NO_3^- in the salt bridge

58. The nonspontaneous redox reaction of chromium metal and aqueous aluminum nitrate occurs electrolytically according to the following ionic equation:

$$Cr(s) + Al^{3+}(aq) \longrightarrow Al(s) + Cr^{3+}(aq)$$

The half-reactions are separated into two compartments. A Cr electrode is placed in 1.00 *M* $Cr(NO_3)_3$, and an Al electrode in 1.00 *M* $Al(NO_3)_3$. Indicate each of the following for the electrolytic cell:
(a) oxidation half-cell reaction
(b) reduction half-cell reaction
(c) anode and cathode
(d) direction of electron flow
(e) direction of NO_3^- in the salt bridge

59. The nonspontaneous redox reaction of chlorine gas and aqueous sodium fluoride occurs electrolytically according to the following equation:

$$Cl_2(g) + 2\,NaF(aq) \longrightarrow 2\,NaCl(aq) + F_2(g)$$

The half-reactions are separated into two compartments. A Pt electrode is placed in 1.00 *M* NaCl, and a second Pt electrode in 1.00 *M* NaF. Indicate each of the following for the electrolytic cell:
(a) oxidation half-cell reaction
(b) reduction half-cell reaction
(c) anode and cathode
(d) direction of electron flow
(e) direction of Na^+ in the salt bridge

60. The nonspontaneous redox reaction of bromine and aqueous sodium chloride occurs electrolytically according to the following equation:

$$Br_2(l) + 2\ NaCl(aq) \longrightarrow 2\ NaBr(aq) + Cl_2(g)$$

The half-reactions are separated into two compartments. A Pt electrode is placed in 1.00 *M* NaBr, and a second Pt electrode in 1.00 *M* NaCl. Indicate each of the following for the electrolytic cell:

(a) oxidation half-cell reaction
(b) reduction half-cell reaction
(c) anode and cathode
(d) direction of electron flow
(e) direction of Na^+ in the salt bridge

General Exercises

61. Calculate the oxidation number for sulfur in calcium thiosulfate, CaS_2O_3.

62. Calculate the oxidation number for sulfur in sodium tetrathionate, $Na_2S_4O_6$.

63. The redox reaction between cobalt metal and aqueous mercury(II) nitrate produces cobalt(II) nitrate and droplets of liquid mercury metal. Write a net ionic equation for the reaction.

64. The redox reaction between liquid mercury and aqueous gold(III) acetate produces mercury(I) acetate and gold metal. Write a net ionic equation for the reaction.

65. The redox reaction between zinc metal and sulfuric acid produces zinc sulfate and hydrogen gas. Write a net ionic equation for the reaction.

66. The redox reaction between aluminum metal and hydrochloric acid produces aluminum chloride and hydrogen gas. Write a net ionic equation for the reaction.

67. The redox reaction between lithium metal and water produces lithium hydroxide and hydrogen gas. Write a net ionic equation for the reaction.

68. The redox reaction between calcium metal and water produces calcium hydroxide and hydrogen gas. Write a net ionic equation for the reaction.

69. Write a balanced equation for the reaction of a copper penny in a dilute nitric acid solution.

$$Cu(s) + NO_3^-(aq) \longrightarrow Cu^{2+}(aq) + NO(g)$$

70. Write a balanced equation for the reaction of ethyl alcohol, C_2H_5OH, and an acidic potassium dichromate solution.

$$C_2H_5OH(aq) + Cr_2O_7^{2-}(aq) \longrightarrow HC_2H_3O_2(aq) + Cr^{3+}(aq) + H_2O(l)$$

17 Chapter Self-Test

Answers to Self-Test are in Appendix J.

1. What is the oxidation number of solid iodine crystals? (Sec. 17.1)

(a) 0 **(b)** –1
(c) –2 **(d)** –3
(e) none of the above

2. What substance is reduced in the following redox reaction? (Sec. 17.2)

$$Zn(s) + Cu^{2+}(aq) \longrightarrow Zn^{2+}(aq) + Cu(s)$$

(a) Zn **(b)** Cu^{2+}
(c) Zn^{2+} **(d)** Cu
(e) none of the above

3. What is the coefficient of CO after balancing the following redox reaction using the oxidation number method? (Sec. 17.3)

$$Co_2O_3(s) + CO(g) \longrightarrow Co(s) + CO_2(g)$$

(a) 1 **(b)** 2
(c) 3 **(d)** 6
(e) none of the above

4. What is the coefficient of H_2O after balancing the following redox reaction in *acidic* solution using the half-reaction method? (Sec. 17.4)

$$Cu(s) + NO_3^-(aq) \longrightarrow Cu^{2+}(aq) + NO(aq)$$

(a) 1 **(b)** 2
(c) 3 **(d)** 4
(e) none of the above

5. Which substance listed is the *strongest* oxidizing agent given the following *spontaneous* redox reaction? (Sec. 17.5)

$$Mg(s) + Pb^{2+}(aq) \longrightarrow Mg^{2+}(aq) + Pb(s)$$

(a) Mg **(b)** Pb^{2+}
(c) Mg^{2+} **(d)** Pb
(e) none of the above

6. Which of the statements listed is true regarding the following redox reaction occurring in a *spontaneous* electrochemical cell? (Sec. 17.6)

$$Mg(s) + Cu^{2+}(aq) \longrightarrow Cu(s) + Mg^{2+}(aq)$$

(a) Mg is oxidized at the anode.
(b) Cu^{2+} is reduced at the cathode.
(c) Electrons flow from the Mg electrode to the Cu electrode.
(d) Anions in the salt bridge flow from the Cu half-cell to the Mg half-cell.
(e) all of the above

7. Which of the statements listed is true regarding the following redox reaction occurring in a *nonspontaneous* electrochemical cell? (Sec. 17.7)

$$3\,C(s) + 2\,Co_2O_3(l) \xrightarrow{\text{electricity}} 4\,Co(l) + 3\,CO_2(g)$$

(a) oxidation half-reaction: $C + 2\,O^{2-} \longrightarrow CO_2 + 4\,e^-$
(b) reduction half-reaction: $Co^{3+} + 3\,e^- \longrightarrow Co$
(c) CO_2 gas is produced at the anode
(d) Co metal is produced at the cathode
(e) all of the above

Key Concepts

8. Refer to the redox reaction and identify the following items:

$$H_2O_2(aq) + NO_2^-(aq) \longrightarrow H_2O(l) + NO_3^-(aq)$$

(a) oxidation half-reaction
(b) reduction half-reaction
(c) substance oxidized
(d) substance reduced
(e) oxidizing agent
(f) reducing agent

9. Refer to the voltaic cell illustration and identify the following items:
(a) oxidation half-cell reaction
(b) reduction half-cell reaction
(c) anode and cathode
(d) direction of electron flow
(e) direction of SO_4^{2-} ion flow

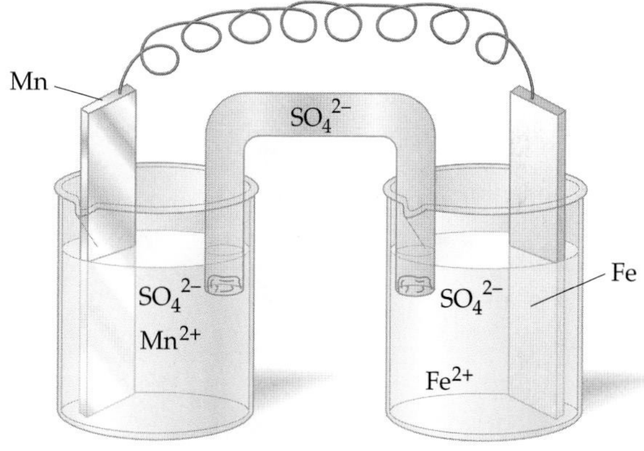

Critical Thinking

10. Nickel-cadmium batteries are used in rechargeable electronic calculators. Given the reaction for a *recharging* NiCad battery, what substance is being *oxidized*?

$$Cd(OH)_2(s) + Ni(OH)_2(s) \longrightarrow Cd(s) + NiO_2(s) + 2\,H_2O(l)$$

11. Nickel-cadmium batteries are used in rechargeable electronic calculators. Given the reaction for a *discharging* NiCad battery, what substance is being *oxidized*?

$$Cd(s) + NiO_2(s) + 2\,H_2O(l) \longrightarrow Cd(OH)_2(s) + Ni(OH)_2(s)$$

12. Silver-zinc "button cells" are used as watch batteries. Given the following reaction, indicate the substance undergoing *oxidation*, and the substance undergoing *reduction.*

$$Zn(s) + Ag_2O(s) \longrightarrow ZnO(s) + 2\,Ag(s)$$

◀ **Silver-Zinc Batteries**

CHAPTER

18 Nuclear Chemistry

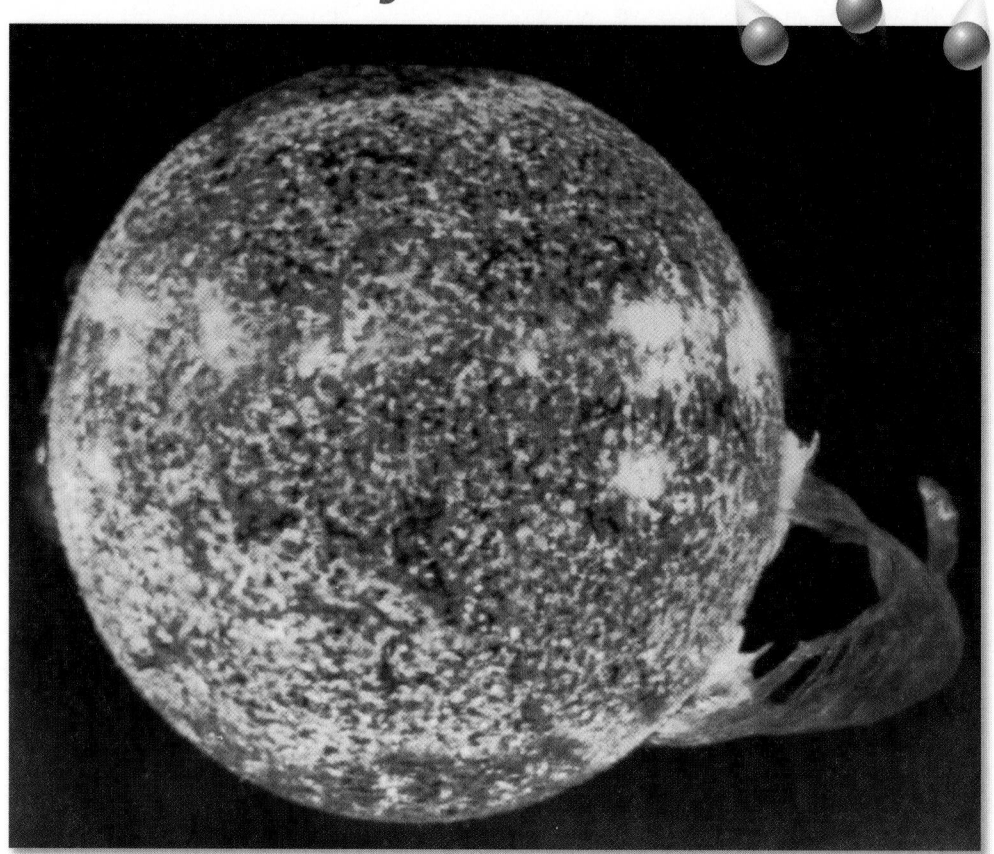

The Sun is an immense nuclear reactor that combines two hydrogen nuclei into a helium nucleus by a process termed nuclear fusion. In fact, the name helium comes from the Greek word for sun, that is, *helios.*

"I am among those who think that science has great beauty."
Marie Curie, Polish/French Physicist (1867–1934)

The energy requirements of industrial nations and the rapid consumption of nonrenewable fossil fuels have created a need to develop additional energy sources. To compensate for declining oil and natural gas resources, there is great interest in developing alternative energy sources. Nuclear energy is one possibility. The energy released from the splitting of atomic nuclei, most notably uranium nuclei, is called nuclear fission energy, or simply nuclear energy. Although the United States has enough uranium to meet its nuclear energy needs through the middle of this century, the threat of nuclear accident and the disposal of radioactive waste are problems of great public concern.

Approximately 15% of all the world's electrical power comes from nuclear power plants. Interestingly, the United States generates the most nuclear energy, and about 20% of its electricity comes from nuclear power plants. In England, Germany, and Japan, the percentage is somewhat less than the United States. In China the percentage of electricity from nuclear power is only 1%, but in France the percentage is almost 80%, and in regions of Eastern Europe it is even higher.

18.1 NATURAL RADIOACTIVITY

OBJECTIVE

- To state the properties of alpha, beta, and gamma radiation.

In 1896, the French physicist Henri Becquerel (1852–1908) noticed a fogged image on an unexposed photographic plate he had just removed from its heavy black protective paper. Moreover, the image was that of a uranium-containing crystal that had been resting on the paper wrapping around the photographic plate. Becquerel concluded that the uranium crystal had given off high-energy radiation to produce the fogged image. After further experiments, he discovered that the uranium crystal emitted small, charged particles in addition to high-energy radiation.

▲ **Marie Curie** In 1903, Marie Curie shared the Nobel prize in physics with her husband and Henri Becquerel. In 1911, she won an unshared Nobel prize in chemistry.

Radioactive Decay of Unstable Atoms

The rays Becquerel discovered were of immediate interest to the French physicist Pierre Curie (1859–1906) and his soon-to-be wife, Marie Sklodowska (1867–1934). In fact, it was Marie who coined the name **radioactivity** for Becquerel's rays. A radioactive element decays by emitting particles and/or energy from an unstable atomic nucleus. In 1898, the Curies discovered two other radioactive elements, polonium and radium. However, it took 4 years of meticulous labor to isolate a single gram of radium from 8 tons of uranium ore. In 1903, Pierre and Marie Curie shared the Nobel prize in physics with Becquerel for their work with radioactivity.

By 1903, the British physicist Ernest Rutherford had identified and named three different types of radioactivity. One type, an alpha ray, consists of a stream of positive particles. A second type, a beta ray, consists of negative particles. The third type, a gamma ray, does not consist of particles, but rather is high-energy light similar to X rays. Rutherford was able to separate alpha, beta, and gamma rays experimentally by passing the radiation from a radioactive sample through an electric field. The apparatus he used is illustrated in Figure 18.1.

After studying the radiation from unstable atoms, Rutherford eventually identified alpha particles as helium nuclei. That is, an **alpha particle** (**α**) is identical to a helium nucleus containing two protons and two neutrons. He found that a **beta particle** (**β**) is identical to an electron. A **gamma ray** (**γ**) is a stream of very high-energy photons. Table 18.1 lists the properties of alpha, beta, and gamma radiation.

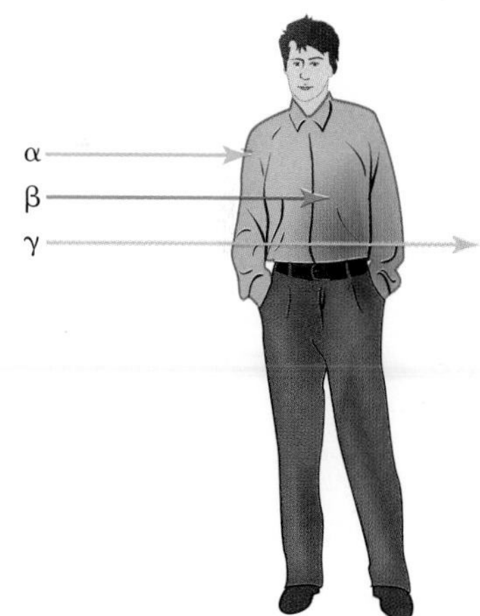

▲ **Radiation's Penetrating Power** Alpha particles are stopped by the skin, while beta particles can penetrate about a centimeter. Gamma rays easily pass through the human body.

TABLE 18.1 PROPERTIES OF NUCLEAR RADIATION

RADIATION	IDENTITY	APPROXIMATE VELOCITY*	SHIELDING REQUIRED	PENETRATING POWER
alpha, α	helium nucleus	≤10% c	paper, clothing	low, stopped by the skin
beta, β	electron	≤90% c	30 cm wood, aluminum foil	medium, ~1 cm of flesh
gamma, γ	high-energy radiation	100% c	10 cm lead, 30 cm concrete	high, passes through body

*The letter **c** is the symbol for the velocity of light, 3×10^8 m/s.

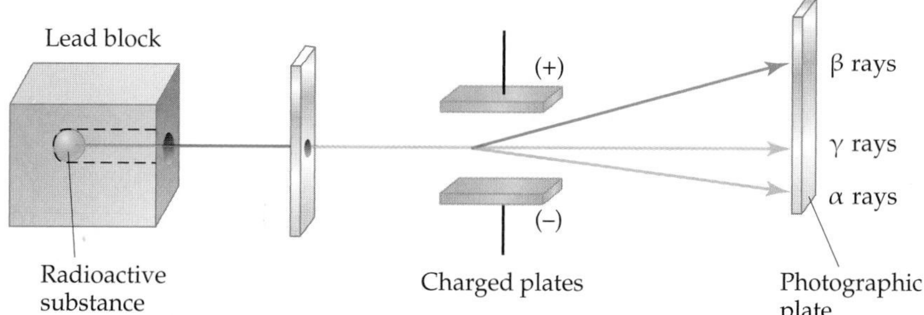

▲ **Figure 18.1 Separation of Alpha, Beta, and Gamma Rays** Notice that the alpha ray is deflected toward the negative electrode, and the beta ray toward the positive electrode. This indicates that the alpha ray is positively charged and the beta ray is negatively charged. The gamma ray is not affected because it has no charge.

OBJECTIVES

- To illustrate the following types of radiation using atomic notation: alpha, beta, gamma, positron, neutron, and proton.
- To write balanced nuclear equations involving natural radioactivity.

18.2 NUCLEAR EQUATIONS

In Section 5.4, we introduced the atomic notation used to indicate the composition of an atomic nucleus. In atomic notation, the symbol for the element is preceded by a subscript and superscript. The subscript indicates the number of protons (atomic number). The superscript indicates the number of protons and neutrons (mass number). In the following example, the **atomic number** (symbol *Z*) appears as a subscript, and the **mass number** (symbol *A*) appears as a superscript.

mass number (p^+ and n^0) → $^{A}_{Z}\mathrm{Sy}$ — symbol of the element

atomic number (p^+) → $^{A}_{Z}\mathrm{Sy}$

In this chapter, we will use the term **nuclide** when referring to the nucleus of a specific atom. For example, the nucleus of strontium-90 is a radioactive nuclide that contains a total of 90 protons and neutrons. From the periodic table, we find that strontium has an atomic number of 38. Thus, the atomic notation for strontium-90 is

$$^{90}_{38}\mathrm{Sr}$$

If we subtract the atomic number from the mass number ($90 - 38 = 52$), we find that strontium-90 has 52 neutrons in addition to 38 protons. Although strontium-90 is radioactive, other strontium nuclides are not. For example, Sr-86, Sr-87, and Sr-88 are stable. Thus, we conclude that the number of neutrons can affect the stability of a Sr nucleus and perhaps the nuclides of other elements. Nuclear equations represent changes in nuclides due to radioactivity, and just like chemical equations they need to be balanced. Unlike chemical changes that involve valence electrons, nuclear reactions involve atomic nuclei.

Balancing Nuclear Equations

A **nuclear reaction** involves a high-energy change in an atomic nucleus. For example, a uranium-238 nucleus changes into a thorium-231 nucleus by releasing a helium-4 particle, and a great deal of energy.

$$^{235}_{92}\mathrm{U} \longrightarrow {}^{231}_{90}\mathrm{Th} + {}^{4}_{2}\mathrm{He}$$

To write a balanced **nuclear equation**, we must account for all atomic numbers and mass numbers. That is, the sums of the atomic numbers for reactants and products must be equal. Similarly, the sums of the mass numbers for reactants and products must be equal. The guidelines for balancing a nuclear equation are as follows:

General Guidelines for Balancing a Nuclear Equation

1. The sum of the atomic numbers (subscripts) on the left side of the equation must equal the sum of the atomic numbers on the right side.

sum of atomic number for reactants	=	sum of atomic number for products

2. The sum of the mass numbers (superscripts) on the left side of the equation must equal the sum of the mass numbers on the right side.

sum of mass number for reactants	=	sum of mass number for products

3. After completing the equation by writing all the nuclear particles in atomic notation, a coefficient may be necessary to balance the following:
   (a) atomic numbers of reactants and products
   (b) mass numbers of reactants and products

To balance the atomic numbers and mass numbers in a nuclear equation, we must be able to write the atomic notation for each particle in the reaction. Table 18.2 lists some common particles involved in nuclear reactions.

TABLE 18.2 PARTICLES IN NUCLEAR REACTIONS

PARTICLE	NOTATION	MASS	CHARGE
alpha, α	${}^{4}_{2}\mathrm{He}$	~4 amu	+2
beta, β^-	${}^{0}_{-1}\mathrm{e}$	~0 amu	–1
gamma*, γ	${}^{0}_{0}\gamma$	0 amu	0
positron, β^+	${}^{0}_{+1}\mathrm{e}$	~0 amu	+1
neutron, n^o	${}^{1}_{0}\mathrm{n}$	~1 amu	0
proton, p^+	${}^{1}_{1}\mathrm{H}$	~1 amu	+1

*Unlike particle radiation, a gamma ray is light radiation and has no mass.

Alpha Particle Emission

Let's write a balanced nuclear equation for the radioactive decay of radium-226 by alpha emission. We can write the decay process using symbols as follows:

$$ {}^{226}_{88}\mathrm{Ra} \longrightarrow {}^{A}_{Z}\mathrm{X} + {}^{4}_{2}\mathrm{He}\ (\alpha \text{ particle}) $$

To balance the number of protons, the atomic number (represented by Z) of the unknown element X must be 86 (88 = 86 + 2). Therefore, we can write

$$ {}^{226}_{88}\mathrm{Ra} \longrightarrow {}^{A}_{86}\mathrm{X} + {}^{4}_{2}\mathrm{He} $$

When we refer to the periodic table, we find that element 86 is radon, Rn. Thus,

$$ {}^{226}_{88}\mathrm{Ra} \longrightarrow {}^{A}_{86}\mathrm{Rn} + {}^{4}_{2}\mathrm{He} $$

To balance the number of protons and neutrons, the mass number of X (represented by *A*) must be 222 (226 = 222 + 4). The balanced equation is

$$ {}^{226}_{88}\mathrm{Ra} \longrightarrow {}^{222}_{86}\mathrm{Rn} + {}^{4}_{2}\mathrm{He} $$

▲ **Natural Radioactivity** Before the hazards of radiation exposure were fully understood, radioactive radium was used on the face of a watch. Radiation from the radium causes the watch to glow in the dark.

Beta Particle Emission

Unlike radium-226 that decays by alpha emission, radium-228 decays by beta emission. In beta emission, a neutron essentially decays into a proton and releases an electron. We can write this decay process in symbols as follows:

$$ {}^{228}_{88}\mathrm{Ra} \longrightarrow {}^{A}_{Z}\mathrm{X} + {}^{0}_{-1}\mathrm{e}\ (\beta^- \text{ particle}) $$

To balance the number of protons, the atomic number of X must be 89 (88 = 89 − 1). Since nuclide X has an atomic number of 89, the periodic table shows us that the element is actinium, Ac. The equation is

$$ {}^{228}_{88}\mathrm{Ra} \longrightarrow {}^{A}_{89}\mathrm{Ac} + {}^{0}_{-1}\mathrm{e} $$

To balance the mass numbers for reactants and products, the *A* value for the Ac nuclide must be 228 since a beta particle has a mass number of zero (228 = 228 + 0). The balanced equation for the reaction is

$$^{228}_{88}\text{Ra} \longrightarrow {}^{228}_{89}\text{Ac} + {}^{0}_{-1}\text{e}$$

Note that beta emission increases the atomic number by 1, thus producing the next higher atomic number element in the periodic table. The net result is that one neutron in the radium-228 nucleus decays into a proton as it releases an electron.

Gamma Ray Emission

Now let's try a more complex example. Uranium-233 decays by releasing both alpha particles and gamma rays. We can write the decay process as follows:

$$^{233}_{92}\text{U} \longrightarrow {}^{A}_{Z}\text{X} + {}^{4}_{2}\text{He} + {}^{0}_{0}\gamma\ (\gamma\ \text{ray})$$

To balance subscripts, the atomic number of X must be 90 (92 = 90 + 2 + 0). Since nuclide X has an atomic number of 90, we see from the periodic table that the element is thorium, Th. The equation is

$$^{233}_{92}\text{U} \longrightarrow {}^{A}_{90}\text{Th} + {}^{4}_{2}\text{He} + {}^{0}_{0}\gamma$$

To balance the mass numbers for reactants and products, the *A* value for the Th nuclide must be 229 (233 = 229 + 4 + 0). The balanced equation for the reaction is

$$^{233}_{92}\text{U} \longrightarrow {}^{229}_{90}\text{Th} + {}^{4}_{2}\text{He} + {}^{0}_{0}\gamma$$

Note that the atomic number and mass number of a gamma ray are zero and have no effect on the equation. The gamma ray is included, however, to indicate that high-energy radiation is emitted during the decay process.

Positron Particle Emission

In 1932, a new particle was discovered. Its mass is identical to that of an electron, but its charge is +1 rather than –1. The newly discovered particle was given the name **positron** ($\boldsymbol{\beta}^{+}$), for positive electron. When positrons strike electrons, they immediately annihilate each other. A collision between a positron and an electron causes mutual destruction and the release of two gamma rays.

Scientists have studied and cataloged the types of emission from radioactive nuclides. Sodium-22 emits positrons and is used in medicine to diagnose small lesions in the brain. The nuclear equation for the decay of Na-22 is as follows:

$$^{22}_{11}\text{Na} \longrightarrow {}^{A}_{Z}\text{X} + {}^{0}_{+1}\text{e (positron)}$$

To balance the atomic numbers of reactants and products in this reaction, the atomic number of X must be 10 (11 = 10 + 1). Referring to the periodic table, we see that the atomic number 10 corresponds to neon, Ne. The equation is

$$^{22}_{11}\text{Na} \longrightarrow {}^{A}_{10}\text{Ne} + {}^{0}_{+1}\text{e}$$

To balance the mass numbers of reactants and products, the mass number of Ne must be 22 since a positron has a mass number of zero (22 = 22 + 0). The balanced equation for the reaction is

$$^{22}_{11}\text{Na} \longrightarrow {}^{22}_{10}\text{Ne} + {}^{0}_{+1}\text{e}$$

Note that positron emission decreases the atomic number by 1, thus producing the next lower element in the periodic table. The net result is that one proton in the Na-22 nucleus decays into a neutron while releasing a positron.

Electron Capture

A few large, unstable nuclides decay by a process called **electron capture (EC)**. In this process, a heavy, positively charged nucleus strongly attracts an electron that lies close to the nucleus. If the attraction is sufficient, the negatively charged electron is captured by the nucleus and converts a positive proton to a neutral neutron. Since the number of protons is less, the atomic number decreases by one and a different element is formed. Consider a lead-205 nuclide that captures an electron as follows:

$$^{205}_{82}\text{Pb} + {}^{0}_{-1}\text{e} \longrightarrow {}^{A}_{Z}\text{X}$$

To balance the atomic numbers, the atomic number of X must be 81 (82 − 1 = 81). From the periodic table we see that element 81 is thallium, Tl. Since the mass number of an electron is 0, the mass number does not change (205 + 0 = 205). The balanced equation for the reaction is

$$^{205}_{82}\text{Pb} + {}^{0}_{-1}\text{e} \longrightarrow {}^{205}_{81}\text{Tl}$$

The following example exercise further illustrates how to write balanced nuclear equations for radioactively decaying nuclides:

Example Exercise 18.1 Writing Nuclear Equations for Natural Radiation

Write a balanced nuclear equation for each of the following radioactive decay reactions:

(a) Radon-222 decays by alpha and gamma emission.
(b) Barium-133 decays by electron capture.

Solution

To write a balanced nuclear equation we must have

1. equal sums of atomic numbers for reactants and products, and
2. equal sums of mass numbers for reactants and products.

(a) We can write the equation for the decay of radon-222 by alpha and gamma emission as follows:

$$^{222}_{86}\text{Rn} \longrightarrow {}^{218}_{84}\text{Po} + {}^{4}_{2}\text{He} + {}^{0}_{0}\gamma$$

Since the sums of the atomic numbers (86) and mass numbers (222) are the same for reactants and products, the equation is balanced.

(b) We can write the balanced nuclear equation for the decay of barium-133 by electron capture as follows:

$$^{133}_{56}\text{Ba} + {}^{0}_{-1}\text{e} \longrightarrow {}^{133}_{55}\text{Cs}$$

Since the sums of the atomic numbers (55) and mass numbers (133) are the same for reactants and products, the equation is balanced.

Practice Exercise

Identify nuclide X in the following nuclear reactions that represents the radioactive decay product from (a) beta emission and (b) positron emission:

(a) $^{99}_{43}\text{Tc} \longrightarrow {}^{A}_{Z}\text{X} + {}^{0}_{-1}\text{e}$ (b) $^{30}_{15}\text{P} \longrightarrow {}^{A}_{Z}\text{X} + {}^{0}_{+1}\text{e}$

Answers: (a) $^{99}_{44}\text{Ru}$; (b) $^{30}_{14}\text{Si}$

Concept Exercise

Which type of natural radiation corresponds to each of the following descriptions?

(a) stopped by human skin and requires heavy cloth shielding
(b) penetrates human skin and requires aluminum shielding
(c) passes through the body and requires lead shielding

Answer: See Appendix G

OBJECTIVES

- To understand the concept of a radioactive decay series.
- To identify products in the 14-step decay series for uranium-238.
- To identify products in the 11-step decay series for uranium-235.
- To identify products in the 10-step decay series for thorium-232.

18.3 RADIOACTIVE DECAY SERIES

In Section 18.2, we discussed the radioactive decay of unstable nuclides. During this process, unstable nuclides stabilize by emitting radiation. Many nuclides become stable by emitting radiation in a single step. Heavy nuclides, such as uranium–238, must go through a series of decay steps to reach a nuclide that is stable. This stepwise disintegration of a radioactive nuclide until a stable nucleus is reached is called a **radioactive decay series.**

Natural Decay Series

The decay series for three naturally occurring radioactive nuclides have been studied extensively by physicists and chemists. They are uranium-238, uranium-235, and thorium-232. The following diagram suggests that these radioactive nuclides decay to lead in a series of steps:

$$\text{U-238} \longrightarrow \quad \longrightarrow \quad \longrightarrow \text{Pb-206}$$
$$\text{U-235} \longrightarrow \quad \longrightarrow \quad \longrightarrow \text{Pb-207}$$
$$\text{Th-232} \longrightarrow \quad \longrightarrow \quad \longrightarrow \text{Pb-208}$$

We use the term **parent–daughter nuclides** to describe a decaying parent nucleus and the resulting daughter nucleus. Although we have not shown the number of steps necessary to reach a stable nuclide, the actual sequence has been determined for each nuclide. In the case of uranium-238, 14 decay steps are required to become stable. Uranium-235 requires 11 steps, and thorium-232 requires 10 steps.

The first step in the radioactive decay of uranium-238 is the emission of an alpha particle to give thorium-234.

$$\underset{\textit{parent nuclide}}{^{238}_{92}\text{U}} \longrightarrow \underset{\textit{daughter product}}{^{234}_{90}\text{Th}} + \underset{\textit{alpha particle}}{^{4}_{2}\text{He}}$$

The second step is the radioactive decay of Th-234 into Pa-234 by the emission of a beta particle.

$$^{234}_{90}\text{Th} \longrightarrow {}^{234}_{91}\text{Pa} + {}^{0}_{-1}\text{e}$$

The third step in the series is the decay of Pa-234 by beta emission. Several more steps in the decay series are required before a stable nuclide, Pb-206, is reached. Figure 18.2

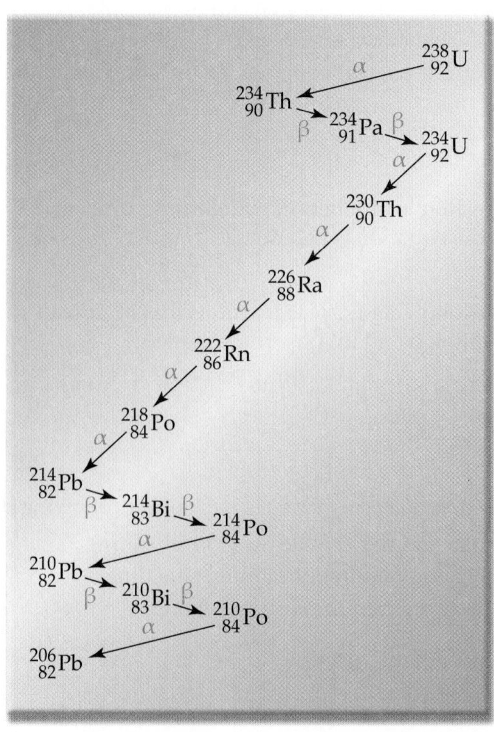

▶ **Figure 18.2 Radioactive Decay Series for Uranium-238** Uranium-238 requires 14 decay steps before reaching a stable nuclide. Notice that several radioactive nuclides of Po, Pb, and Bi are unstable and decay. Only Pb-206, which ends the series, is stable.

illustrates the entire decay series for uranium-238. Notice that the atomic number *decreases* by 2 after each α emission ($^{4}_{2}$He), whereas the atomic number *increases* by 1 for each β emission ($^{0}_{-1}$e).

Note When we examine the periodic table, we see that a mass number in parentheses, rather than an atomic mass, is listed for all the elements following bismuth. This indicates that there are no stable nuclides of any elements higher in atomic number than bismuth. All the elements beyond atomic number 83 are unstable and therefore undergo spontaneous radioactive decay.

18.4 RADIOACTIVE HALF-LIFE

OBJECTIVES

- To understand the concept of half-life for a radioactive sample.
- To relate the amount of radioactive sample, and its radioactivity, to a given half-life.

In 1908, Hans Geiger, while working for Rutherford, developed an instrument for measuring radiation. Geiger filled a tube with a gas that ionizes when struck by radiation. The ionized gas particles allow passage of a pulse of electricity that can be converted to an audio signal heard as a click. Each click of a *Geiger counter* indicates an emission from the radioactive sample (Figure 18.3). A digital counter registers the number of clicks, providing a measure of the number of atomic nuclei undergoing radioactive decay. The number of nuclei that disintegrate in a given period of time is called the **activity** of the sample.

▲ **Figure 18.3 Geiger Counter** A Geiger counter gives a "click" each time a nucleus decays and releases a radioactive particle.

Half-life of a Radioactive Sample

Scientists have observed that the level of radioactivity for all radioactive samples decreases with time. If we begin, for example, with a sample that has an activity of 1000 disintegrations per minute (dpm), after a given period of time the radioactivity will drop to 500 dpm, which is half the original level. Moreover, after the same period of time, the activity will drop from 500 dpm to 250 dpm.

The amount of time required for the activity of a sample to decrease by half is called the **half-life** (symbol $t_{1/2}$). If the half-life of a sample is 4 hours, it requires 4 hours for the activity to drop by half. That is, it takes 4 hours for each of the following half-life changes: 1000 to 500 dpm, 500 to 250 dpm, and 250 to 125 dpm. Since three half-lives are required for the activity to drop from 1000 to 125 dpm, the total time required is

$$3\,t_{1/2} \times \frac{4 \text{ hours}}{1\,t_{1/2}} = 12 \text{ hours}$$

The decrease in activity can be illustrated by the decay curve in Figure 18.4. The curve shows that each time the activity drops by 50%, one half-life has elapsed.

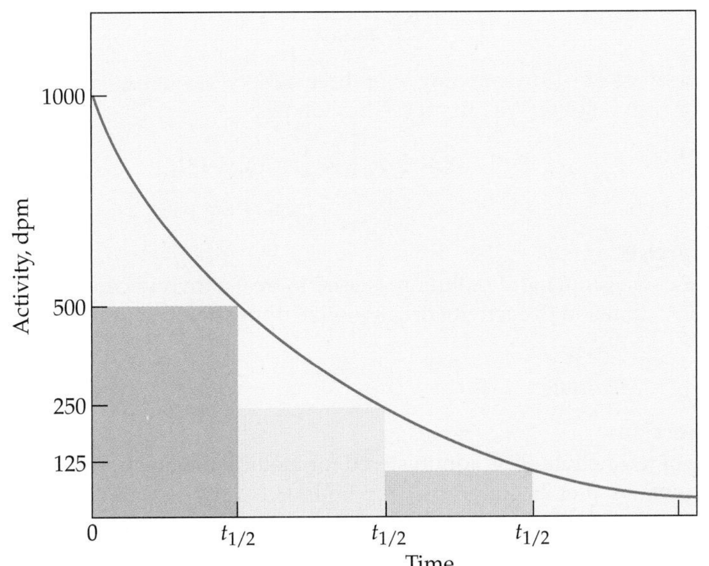

◀ **Figure 18.4 Decay Curve and Half-Life** The time required for the activity of a radioactive sample to decrease from 1000 to 500 dpm is one half-life. Notice that it requires the same amount of time, one half-life, for the activity to drop from 500 to 250 dpm.

Radioactive Waste

Disposing of radioactive waste material is a major concern because dangerous levels of radioactivity can persist for thousands of years. For example, the radioactive fuel from a plutonium-239 nuclear reactor emits high levels of radioactivity long after it is too weak to run a nuclear reactor. Let's find out how long it takes for the radioactivity of plutonium-239 waste to decrease from 20,000 to 625 dpm.

The half-life of Pu-239 is 24,000 years. By definition, we know that the activity of a radioactive sample drops by one-half after each half-life. We can list the activity of the sample corresponding to the elapsed time.

ACTIVITY	ELAPSED TIME	ELAPSED TIME
20,000 dpm	—	0
10,000 dpm	1 $t_{1/2}$	24,000 y
5,000 dpm	2 $t_{1/2}$	48,000 y
2,500 dpm	3 $t_{1/2}$	72,000 y
1,250 dpm	4 $t_{1/2}$	96,000 y
625 dpm	5 $t_{1/2}$	120,000 y

We see that it takes five half-lives for the radioactivity of plutonium to drop from 20,000 to 625 dpm. Given that the half-life is 24,000 years, we can calculate the amount of time required for the radiation to drop to 625 dpm. The equation is

$$5\ t_{1/2} \times \frac{24\ 000\ \text{y}}{1\ t_{1/2}} = 120{,}000\ \text{y}$$

The activity of a sample is proportional to the number of radioactive nuclei, and thus to the mass of the sample. The time required for radioactive substances to decay to a safe level varies according to the nuclide. When debating the use of nuclear energy, we must consider the problem of radioactive waste disposal and its potential effect on the environment for thousands of years.

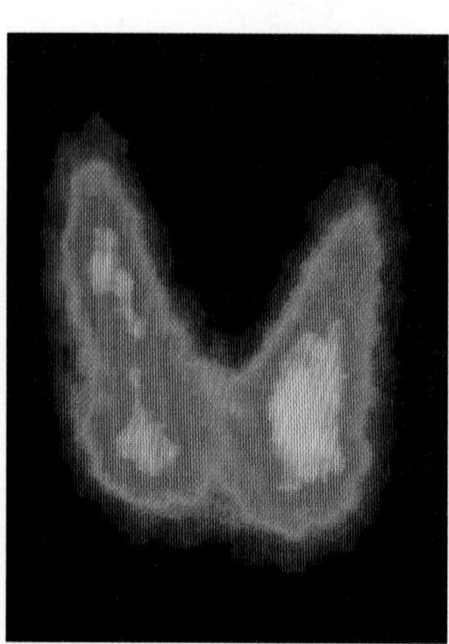

▲ **Thyroid Scan** Physicians use radioactive iodine-131 as a diagnostic tool to assess activity of the thyroid gland, which regulates metabolism. A doctor can obtain a PET scan image of the thyroid gland, which has a shape that resembles a butterfly.

Example Exercise 18.2 Half-LIfe of a Radioactive Nuclide

Iodine-131, in a sodium iodide tablet, is used to measure the activity of the thyroid gland. If a tablet initially contains 88 mg of I-131, what is the mass of the nuclide that remains after 24 days? The half-life of I-131 is 8 days.

Solution

This problem asks us to find the mass of I-131 remaining after 24 days. Let's begin by determining the number of half-lives that elapse. One half-life equals 8 days. Thus,

$$24\ \cancel{\text{days}} \times \frac{1\ t_{1/2}}{8\ \cancel{\text{days}}} = 3\ t_{1/2}$$

To calculate the mass of I-131 remaining after three half-lives, we decrease the mass of I-131 by one-half for each half-life. We can proceed as follows:

$$88\ \text{mg I-131} \times \frac{1}{2} \times \frac{1}{2} \times \frac{1}{2} = \text{mg I-131}$$
$$= 11\ \text{mg I-131}$$

Practice Exercise

Iodine-128, in a potassium iodide solution, is used to treat a thyroid condition. If the initial activity is 10,000 dpm and the activity drops to 1250 dpm after 75 minutes, what is the half-life of I-128?

Answer: $t_{1/2}$ = 25 minutes

Concept Exercise

When 500 mg of technetium-99 is administered for medical diagnosis, how much of the radionuclide remains after 24 hours? ($t_{1/2}$ = 6 hours)

Answer: See Appendix G.

18.5 APPLICATIONS OF RADIONUCLIDES

OBJECTIVES

- To describe the radiocarbon dating method.
- To describe the uranium–lead dating method.
- To state applications of selected radionuclides.

A nuclide that is unstable is called a **radionuclide**. Different radionuclides decay by emitting different types of radiation. These different types of radioactive emission have many applications in agriculture, industry, medicine, and research.

Radiocarbon Dating

Carbon occurs naturally as two stable nuclides, carbon-12 and carbon-13. It also occurs as an unstable nuclide, carbon-14, which decays by beta emission. The fact that carbon-14 is unstable allows its half-life to be used to estimate the age of any substance containing carbon. Scientists date objects by measuring the level of carbon-14 radioactivity in a sample. This process is called *carbon-14 dating*, or *radiocarbon dating*.

Among other things, radiocarbon dating has been used to establish the age of human remains that may be thousands of years old, the age of artifacts such as ancient wooden bowls, and the age of carbon dioxide gas from the atmosphere that has been trapped for thousands of years in glacial ice. This technique is considered reliable up to 50,000 years. Older objects usually emit too little radiation for accurate measurement.

Carbon-14 is continuously produced in Earth's upper atmosphere. The process begins when cosmic rays strike molecules and scatter high-energy neutrons. These neutrons in turn collide with nitrogen in the atmosphere. These collisions produce carbon-14 and protons. The nuclear equation is as follows:

$$^{14}_{7}\text{N} + ^{1}_{0}\text{n} \longrightarrow ^{14}_{6}\text{C} + ^{1}_{1}\text{H}$$

Carbon-14 atoms in the atmosphere combine with oxygen molecules to give radioactive CO_2. This carbon dioxide is then incorporated into plant life through the process of photosynthesis. The food chain begins with plants and continues through animals and eventually humans. A substance that contains carbon-14 emits beta radiation according to the following nuclear equation:

$$^{14}_{6}\text{C} \longrightarrow ^{14}_{7}\text{N} + ^{\ 0}_{-1}\text{e}$$

In a living plant or animal, the radiation from radiocarbon is 15.3 dpm per gram of carbon. When a plant or animal dies, it ceases to intake carbon-14, and the radiation level begins to decrease. After one half-life has elapsed, the activity declines to about 7.7 dpm. The half-life of carbon-14 is 5730 years. Therefore, if a radioactive sample has an activity of 7.7 dpm, its estimated age is 5730 years.

When a piece of wood was found on Mt. Ararat in eastern Turkey it was proposed that the wood might be from Noah's Ark. The wood sample underwent carbon-14 dating and gave an activity of 12.5 dpm per gram of carbon. This activity corresponds to that of a sample about 1000 years old. Thus, using radiocarbon dating, scientists concluded that the wood was not part of Noah's Ark.

When the Dead Sea Scrolls were found, radiocarbon dating was used to verify their authenticity. Actually, it was the linen wrapping around the Old Testament writings that gave a dating value of 2000 years. This value corresponds to a time that is consistent with biblical texts. It was this evidence from radiocarbon dating that led scholars to believe that the Dead Sea Scrolls are authentic writings.

In a similar fashion, Stonehenge in England was dated using charcoal from various campsites. The charcoal samples gave an activity of 9.5 dpm per gram of carbon, which corresponds to a sample about 3800 years old. Thus, the Stonehenge site is believed to have been active in about 1800 B.C. Figure 18.5 illustrates radiocarbon dating.

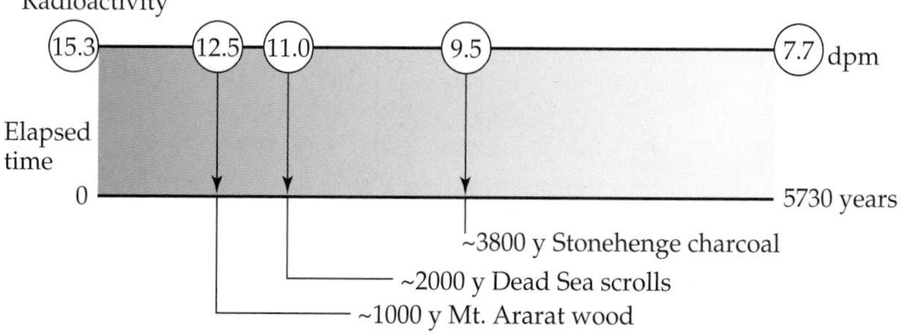

◀ **Figure 18.5 Radiocarbon Dating** The activity of carbon-14 in living tissue is about 15.3 dpm. When tissue expires, it continues to decay radioactively, but the tissue no longer incorporates carbon-14. After 5730 years, half of the carbon-14 nuclides in the sample have decayed and the activity is about 7.7 dpm.

The accuracy of radiocarbon dating depends on a fundamental assumption: *the amount of carbon-14 in the atmosphere has remained constant for the last 50,000 years.* If this assumption is not true, radiocarbon dating is not valid.

CRITICAL THINKING Nuclear Medicine

Q: Which radionuclide emits gamma rays that are used for radiation therapy on tumors in cancer patients?

The term *nuclear medicine* refers to the use of radionuclides for medical purposes. For example, iodine-131 is used to measure the activity of the thyroid gland, which requires iodine to regulate metabolism. Patients diagnosed as having a thyroid condition are given an iodine-131 tablet, and after 24 hours a radiation scan is performed. The amount of radiation detected at the base of the throat, where the thyroid is located, is an indication of thyroid activity.

Xenon-133 is used to diagnose respiratory problems. A patient inhales air containing the radionuclide. Both oxygen and xenon-133 are taken up by healthy lung tissue. As the chest is scanned for radiation, healthy areas of the lungs are revealed. An area where respiration is impaired is indicated by less radiation.

Iron-59 is used to diagnose anemia, a condition marked by a low red blood cell count. Red blood cells contain hemoglobin, which in turn contains iron atoms. A patient is given iron-59, and the radionuclide bonds to the hemoglobin. When the patient is checked for radiation, a scan reveals radioactive hemoglobin, providing an indirect count of red blood cells.

Brain tumors can be diagnosed using technetium-99, which concentrates in brain tissue. When a patient is given technetium-99, rapidly dividing cancer cells incorporate more of the radionuclide than normal cells. A scan shows radiation hot spots that can help locate the tumor. After locating the tumor, a cobalt-60 source of gamma radiation is focused on the tumor, which destroys the cancer cells and a small portion of the surrounding healthy tissue.

Breast cancer can be treated with iridium-192. A hypodermic needle is first injected into the tumor, followed by insertion of an iridium-192 bead through the needle. The needle is then withdrawn, and gamma rays from the radionuclide destroy the tumor. However, similar to radiation therapy, the gamma rays destroy some healthy tissue at the periphery of the tumor.

Although there is public concern about the use of radioactive substances, nuclear medicine is considered safe. In fact, diagnoses and treatments involving medical radionuclides save or prolong thousands of lives each year.

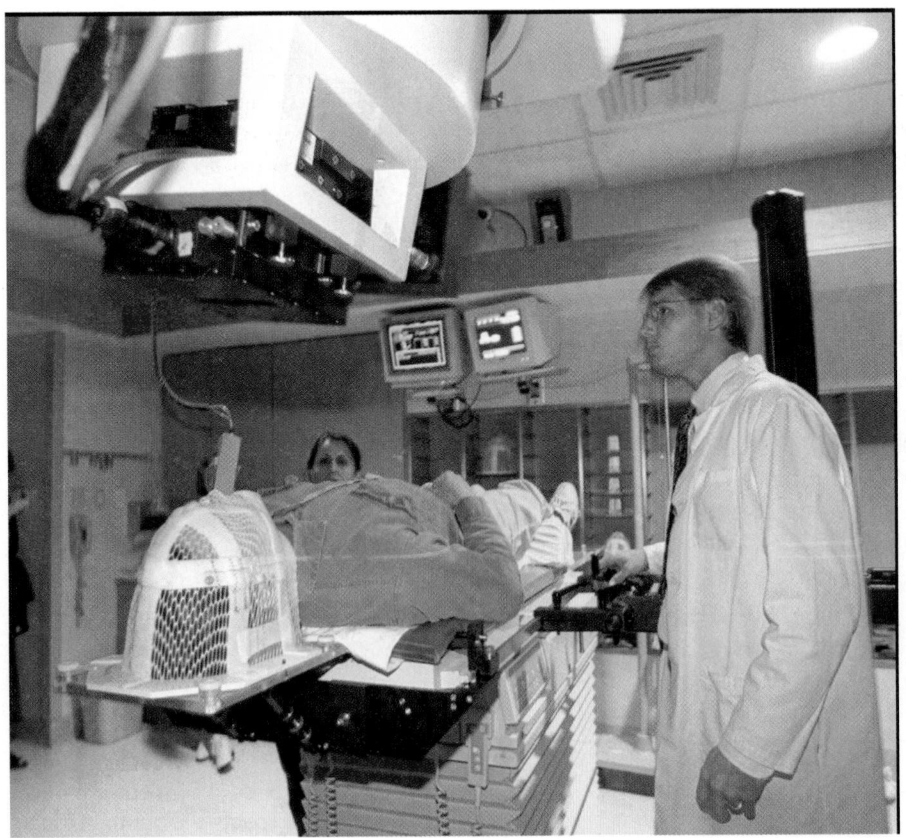

◀ **Radiation Therapy** A cancer patient receiving radiation therapy for a brain tumor.

A: Cobalt-60 is radioactive and emits gamma rays for radiation therapy treatment that destroys tumors in cancer patients.

Uranium–Lead Dating

Naturally occurring uranium-238 radioactively decays to lead-206 in a series of 14 steps. The half-life for the overall process is 4.5 billion years. We can indicate the net reaction according to the following equation:

$$^{238}_{92}\text{U} \longrightarrow {}^{206}_{82}\text{Pb} + 8\,^{4}_{2}\text{He} + 6\,^{0}_{-1}\text{e}$$

The ratio of uranium-238 to lead-206 in uranium samples provides an estimate of the age of very old geological events. For example, a meteorite that fell in Mexico is found to have a uranium-238/lead-206 ratio of 1:1. A ratio of 1:1 corresponds to a single half-life. Thus, the meteorite is about 4.5 billion years old.

The uranium–lead dating technique has also been used to estimate the age of Earth and of lunar rock samples collected during the *Apollo* missions to the Moon. Uranium–lead dating indicates that Earth and the Moon are about the same age. Thus, scientists have evidence for the theory that the Moon may have originally been part of Earth and broke away while Earth was in an early molten stage.

Agricultural Applications of Radionuclides

Pesticides like DDT can effectively control insects, but they often persist in the environment for a long time. Some of these pesticides are toxic to animals and humans. As an alternative, insect populations can be controlled using radiation. Gamma rays from a radionuclide such as cobalt-60 have been employed to sterilize male insects. When large numbers of sterilized males are released in an infested area, they mate with the females, but fertilization does not occur. The insect population is thus reduced and effectively controlled. A decrease in the insect population results in an increase in agricultural harvests.

Gamma irradiation of processed food destroys microorganisms. For example, cobalt-60 irradiation of pork destroys the parasite that causes trichinosis; irradiation of chicken destroys the bacteria that cause salmonella. Irradiation of other foods can extend shelf life without the use of preservatives.

▲ Irradiation of Food Gamma irradiation destroys microorganisms in food and extends freshness. All the strawberries shown were picked at the same time; however, the strawberries on the left were irradiated.

Medical Applications

Today, more than 100 radionuclides are available for medical diagnoses and treatments. An interesting example is plutonium-238, which decays by gamma emission. The gamma rays are used to power pacemakers in patients with irregular heartbeats. The radionuclide is sealed in a stainless steel case and implanted in the chest of the patient. The radiation from a plutonium-238 source can power a pacemaker for up to 10 years before it needs to be replaced.

18.6 INDUCED RADIOACTIVITY

OBJECTIVE

▸ To write balanced nuclear equations involving induced radioactivity.

When a nuclide is bombarded with an atomic particle, the nuclide is converted to a different element. That is, a nuclear reaction occurs. Causing a nuclear reaction by particle bombardment is termed induced radioactivity. The conversion of one element to another by an induced nuclear reaction is called **transmutation**.

In 1919, Ernest Rutherford discovered the first transmutation reaction. When he bombarded nitrogen-14 with alpha particles, he detected the release of a proton, $^{1}_{1}\text{H}$. After careful analysis, he concluded that oxygen had also been produced. The equation for the transmutation is

$$^{14}_{7}\text{N} + {}^{4}_{2}\text{He} \longrightarrow {}^{17}_{8}\text{O} + {}^{1}_{1}\text{H}$$

In 1932, James Chadwick, a student of Rutherford's, fired alpha particles at a sheet of beryllium metal. He found that some of the beryllium was changed to carbon and that a neutral particle was released. Chadwick called the neutral particle a neutron.

Interestingly, Rutherford had predicted a neutral particle two decades earlier; however, at that time the neutron was too elusive to be confirmed experimentally. The equation for the transmutation is

$$^{9}_{4}\text{Be} + ^{4}_{2}\text{He} \longrightarrow ^{12}_{6}\text{C} + ^{1}_{0}\text{n}$$

Synthesis of New Elements

▲ **Cyclotron** The cyclotron was invented at the University of California, Berkeley, and was responsible for synthesizing new elements.

After the discovery of the neutron, it was proposed that a neutral particle could easily invade another nucleus. In 1940, this proposal proved correct, as uranium was transmuted into a new element. The new element was named neptunium for the planet Neptune. (Neptune lies just beyond Uranus, the planet for which uranium is named.) The equation for the synthesis of neptunium is

$$^{238}_{92}\text{U} + ^{1}_{0}\text{n} \longrightarrow ^{239}_{93}\text{Np} + ^{0}_{-1}\text{e}$$

The elements beyond uranium in the periodic table do not occur naturally. Rather, they have been created in the laboratory. These transuranium elements are synthesized by accelerating a light projectile particle to a high velocity and firing it at a heavy nuclide that serves as a target nucleus. One instrument used to accelerate light particles to sufficient energy to overcome the repulsion of the target nucleus is a *cyclotron.* Most of the elements beyond uranium have been synthesized using the cyclotron at the University of California at Berkeley. In fact, element 103 is named lawrencium (Lr) for the Berkeley physicist who invented the cyclotron, Ernest O. Lawrence.

Another instrument for accelerating particles is the *linear accelerator.* The linear accelerator at Stanford University was completed in 1966 and is about two miles long. It is capable of accelerating positive or negative projectile particles to a velocity approaching the speed of light. Linear accelerators and cyclotrons are sometimes referred to as "atom smashers" because they send high-velocity particles smashing into target nuclides (Figure 18.6).

In 1964 Soviet physicists claimed a successful synthesis of element 104. They named the element kurchatovium (Ku) in honor of their team leader, Igor Kurchatov. They used a linear accelerator to send a neon-22 projectile smashing into a plutonium target nucleus. Although the half-life of the new element was only 0.3 s, the Russians proposed the following reaction:

$$^{242}_{94}\text{Pu} + ^{22}_{10}\text{Ne} \longrightarrow ^{260}_{104}\text{Ku} + ^{1}_{0}\text{n}$$

Notice that the mass numbers in the equation are not balanced. The total of the reactant superscripts is 264 (242 + 22), but the total of the product superscripts is only 261 (260 + 1). To balance the equation, we must use 4 neutrons. The balanced nuclear equation is

$$^{242}_{94}\text{Pu} + ^{22}_{10}\text{Ne} \longrightarrow ^{260}_{104}\text{Ku} + 4\,^{1}_{0}\text{n}$$

▶ **Figure 18.6 Fermi National Accelerator** The Fermi accelerator, outside Chicago, is one of the most powerful in the world. It accelerates particles to a high velocity around an oval path that is about 4 miles in diameter.

When the Americans tried to confirm the Russian experiment, they were unsuccessful, which cast doubt on the credibility of element 104. In 1969, however, scientists at the University of California at Berkeley reported a different nuclide of element 104 that has a half-life of 5 s. The Berkeley group named the element rutherfordium (Rf) in honor of Ernest Rutherford. They induced the reaction by firing a carbon-12 projectile at a californium target nucleus.

$$^{249}_{98}\text{Cf} + ^{12}_{6}\text{C} \longrightarrow ^{257}_{104}\text{Rf} + ^{1}_{0}\text{n}$$

When we check the equation, we find that the atomic numbers are balanced but the mass numbers are not. The superscript total is 261 on the left side, but only 258 on the right side. To balance the equation, we place the coefficient 4 before the neutrons. The balanced nuclear equation is

$$^{249}_{98}\text{Cf} + ^{12}_{6}\text{C} \longrightarrow ^{257}_{104}\text{Rf} + 4\,^{1}_{0}\text{n}$$

The following example exercise provides additional practice in writing balanced nuclear equations for the synthesis of new elements:

Example Exercise 18.3 Writing Nuclear Equations for Induced Radiation

Write a balanced nuclear equation for the following nuclear reaction induced by an iron nuclide:

Bi-209 + Fe-? ⟶ Mt-266 + 1 neutron

Solution

To write a nuclear equation, we must balance atomic numbers and mass numbers. Let's begin by writing the equation using atomic notation. We obtain the atomic number of each nuclide from the periodic table.

$$^{209}_{83}\text{Bi} + ^{?}_{26}\text{Fe} \longrightarrow ^{266}_{109}\text{Mt} + ^{1}_{0}\text{n}$$

We see that the sum of the atomic numbers is 109 on both sides of the equation. Since the sum of the mass numbers on the right side is 267, the sum on the left side must be 267. Thus, the mass number for Fe must be 58 (209 + 58 = 267). The balanced nuclear equation is

$$^{209}_{83}\text{Bi} + ^{58}_{26}\text{Fe} \longrightarrow ^{266}_{109}\text{Mt} + ^{1}_{0}\text{n}$$

Practice Exercise

Write a balanced nuclear equation for the following nuclear synthesis reaction:

Cf-249 + O-18 ⟶ Sg-? + 4 neutrons

Answer:

$$^{249}_{98}\text{Cf} + ^{18}_{8}\text{O} \longrightarrow ^{263}_{106}\text{Sg} + 4\,^{1}_{0}\text{n}$$

Concept Exercise

In 1994, physicists at Darmstadt, Germany synthesized element 110. If Ds-269 and a neutron were produced by firing a projectile into Pb-208, what was the projectile nuclide?

Answer: See Appendix G

18.7 NUCLEAR FISSION

OBJECTIVES

- To describe the process in a nuclear fission reaction.
- To write a nuclear equation for a given fission reaction.

Spontaneous **nuclear fission** is the process whereby a heavy nucleus, such as uranium-235, splits into lighter nuclei while releasing energy. Many heavy nuclei are unstable and decay by spontaneous nuclear fission. For example, californium-252 can split apart spontaneously into lighter nuclei as follows:

$$^{252}_{98}\text{Cf} \longrightarrow ^{142}_{56}\text{Ba} + ^{106}_{42}\text{Mo} + 4\,^{1}_{0}\text{n} + \text{energy}$$

There are only a few nuclides that can be made to undergo fission, for example, uranium-235, uranium-233, and plutonium-239. The fission process is induced by a slow-moving neutron. When the neutron strikes a heavy nucleus and is captured, the resulting nucleus becomes unstable. The unstable nucleus then decays by splitting into lighter nuclei and releasing a few neutrons. The fission of uranium-235 can occur as follows:

$$^{1}_{0}\text{n} + ^{235}_{92}\text{U} \longrightarrow ^{137}_{52}\text{Te} + ^{97}_{40}\text{Zr} + 2\,^{1}_{0}\text{n} + \text{energy}$$

After uranium-235 is struck by a neutron, the resulting unstable nucleus can undergo fission in a number of ways, one of which is depicted in Figure 18.7. As an analogy, a glass jar can split apart in various ways after being shot by a pellet. For example, uranium-235 can give the following nuclei as products:

$$^{1}_{0}\text{n} + ^{235}_{92}\text{U} \longrightarrow ^{142}_{56}\text{Ba} + ^{91}_{36}\text{Kr} + 3\,^{1}_{0}\text{n} + \text{energy}$$

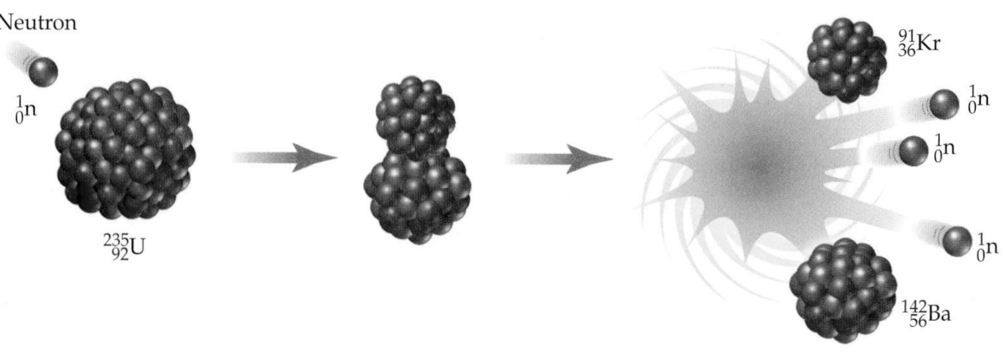

▲ **Figure 18.7 Nuclear Fission** A slow-moving neutron collides with a uranium-235 nucleus to form uranium-236. The resulting nucleus is unstable and splits into two lighter nuclei, releasing a few neutrons. In addition to particles, the fission reaction releases a huge burst of energy.

Nuclear Chain Reaction

The process of nuclear fission is analogous to a single domino knocking over three dominos, which in turn can ultimately topple an entire row of dominos. For each nucleus that undergoes fission, two or three neutrons are usually released. The emitted neutrons can cause additional fission reactions. In turn, the neutrons released from these fission reactions can initiate more fission reactions. If the fission process continues to repeat itself and is self-sustaining, it is called a **chain reaction**. We can indicate the growth in the number of neutrons in a chain reaction of uranium-235 as follows:

$$1\,^{1}_{0}\text{n} \longrightarrow 3\,^{1}_{0}\text{n} \longrightarrow 9\,^{1}_{0}\text{n} \longrightarrow 27\,^{1}_{0}\text{n}$$

We can illustrate a chain reaction as shown in Figure 18.8. Starting with a single uranium-235 nucleus and 1 neutron, the chain reaction quickly escalates to give fission products and 9 neutrons.

For a chain reaction to occur, there must be a fairly large mass of fissionable nuclide. Recall that atoms are mostly empty space. If the mass is too small, fission neutrons simply pass through the fissionable substance without striking a nucleus and causing it to fission. The minimum mass of fissionable nuclide required to sustain a chain reaction is called the **critical mass**. The critical mass depends on the purity of the sample and other factors, but for uranium-235 the critical mass is about 50 kg.

In 1939, nuclear fission was discovered in Germany. The possibility for a sustained chain reaction was experimentally confirmed three years later. In 1942, the Italian physicist Enrico Fermi directed an international group of scientists who carried out the first self-sustaining nuclear chain reaction. This landmark event took place on a converted squash court beneath the University of Chicago and is commemorated by a plaque at the university (Figure 18.9).

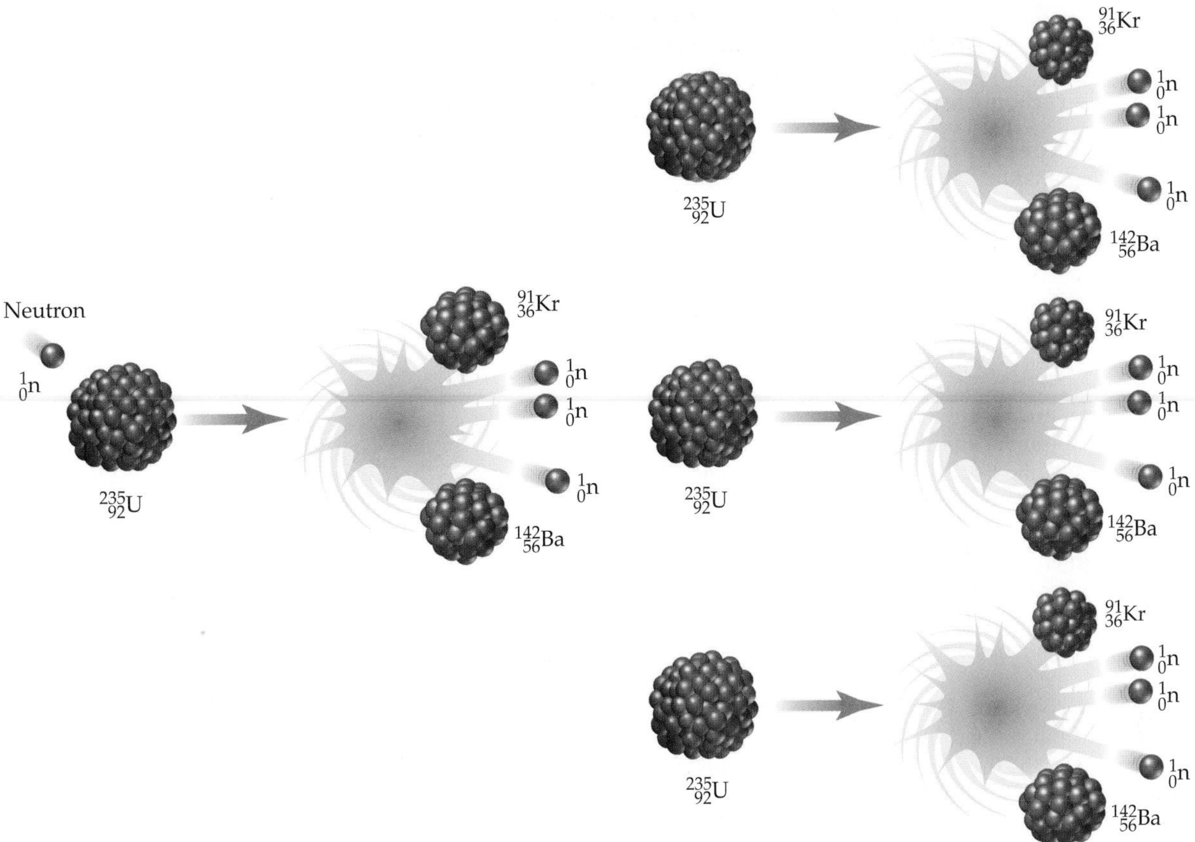

▲ **Figure 18.8 Nuclear Chain Reaction** A single neutron can cause the fission of a uranium-235 nucleus that releases three neutrons. In turn, the three emitted neutrons can cause three additional nuclei to fission with the release of three neutrons each (nine total), and so on.

Uranium-235 Enrichment

Naturally occurring uranium is only 0.7% fissionable uranium-235. Most of the uranium, 99.3%, is nonfissionable uranium-238. To obtain a chain reaction, it is necessary to enrich the natural mixture with the fissionable nuclide. That is, the mixture must be enriched from 0.7% to about 3% uranium-235. The enrichment process is started by reacting uranium with fluorine gas to give gaseous uranium hexafluoride.

$$^{235}U(s) + 3\ F_2(g) \longrightarrow {}^{235}UF_6(g)$$

$$^{238}U(s) + 3\ F_2(g) \longrightarrow {}^{238}UF_6(g)$$

The nuclides $^{235}UF_6$ and $^{238}UF_6$ are too similar to separate chemically. However, the two gaseous nuclides can be separated physically. The lighter nuclide, $^{235}UF_6$, diffuses slightly faster than the heavier nuclide, $^{238}UF_6$. When the first fraction that diffuses through the tube is collected, it is richer in the faster-diffusing nuclide, $^{235}UF_6$. The collected uranium fraction is then reduced to the element, which is enriched in the nuclide uranium-235.

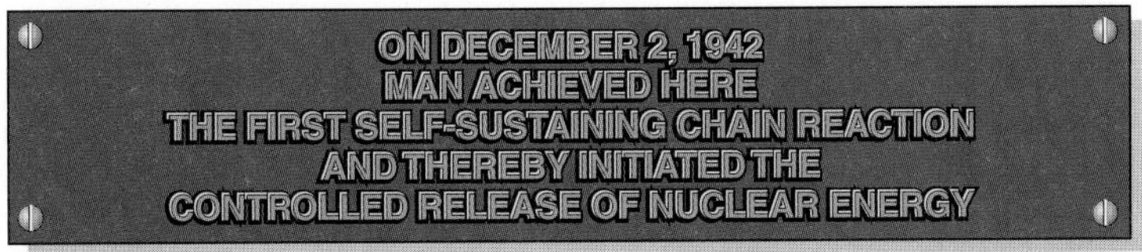

▲ **Figure 18.9 Commemoration Plaque** This large plaque at the University of Chicago signifies the ushering in of the nuclear age.

CHEMISTRY CONNECTION Nuclear Power Plant

Q: What fissionable radionuclides can be used in the fuel rods of a nuclear reactor?

Nuclear energy is an attractive source of energy because of its enormous potential compared to that of energy from ordinary reactions. For example, the fission of 1 g of uranium-235 produces about 12 million times more energy than the combustion of 1 g of gasoline!

A nuclear power plant generates electrical energy by a nuclear chain reaction. Nuclear fission occurs in the *reactor core*, which contains cylindrical metal *fuel rods*. Each fuel rod is filled with a fissionable nuclide, such as uranium-235, plutonium-239, or uranium-233. The fuel rods are separated by *control rods* made of boron or cadmium. These rods absorb neutrons and thus regulate the rate of fission. When the control rods are raised, more of the fuel rods are exposed, and the fission rate increases. When the control rods are lowered, more neutrons are absorbed and the rate slows, thus preventing the reactor core from overheating.

A liquid *coolant* such as water circulates through the reactor core to absorb heat. If the coolant is D_2O, the reactor is called a "heavy water" reactor. If the coolant is ordinary H_2O, the reactor is called a "light water" reactor. The coolant also serves to slow neutrons released from nuclear fission. Slower neutrons are more easily captured by the fissionable nuclides in the fuel rods. Since the coolant affects the rate of capture, it acts as a *moderator* for the fission reaction.

In a nuclear power plant, the nuclear reactor is housed in a concrete building to lessen the chance of an accidental release of radiation into the environment. When the fuel rods are exhausted, they are still highly radioactive and must be stored safely. Many nuclear power plants have on-site underwater vaults for temporary storage. Later, the fuel rods are recycled and the nuclear fuel recovered.

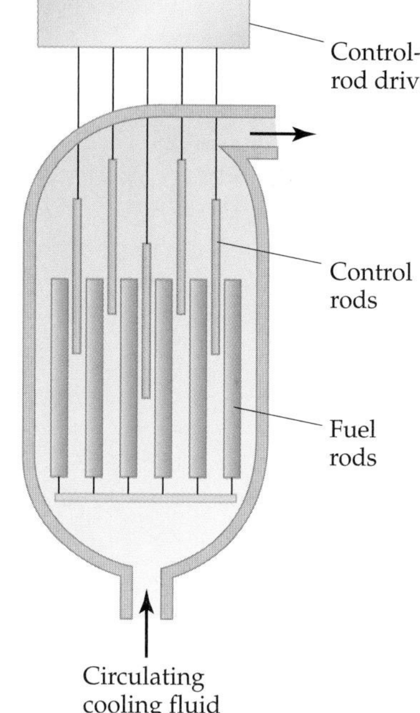

▶ **Nuclear Reactor Core** Notice the control rods inserted between the fuel rods. Lowering the control rods slows the chain reaction by absorbing neutrons released by the fuel rods.

◀ **Nuclear Power Plant** A nuclear power plant in southern California that generates electrical energy.

A: The fissionable radionuclides that are used for fuel in nuclear reactors are U-235, Pu-239, and U-233.

OBJECTIVES

- To describe the process of nuclear fusion using deuterium and tritium.
- To write a nuclear equation for a given fusion reaction.

18.8 NUCLEAR FUSION

Nuclear fusion is the process of combining two light nuclei into a heavier nucleus. It is more difficult to initiate a fusion reaction than a fission reaction, but a fusion reaction releases more energy than a comparable fission reaction. In addition, nuclear fusion is a cleaner process than fission because it produces very little radioactive waste.

The Sun is a huge nuclear fusion reactor operating at millions of degrees. The Sun is made up of 73% hydrogen, 26% helium, and only 1% of all other elements. Three of the fusion reactions thought to occur on the Sun are

$$^{1}_{1}\text{H} + ^{1}_{1}\text{H} \longrightarrow ^{2}_{1}\text{H} + ^{\ 0}_{+1}\text{e} + \text{energy}$$

$$^{2}_{1}\text{H} + ^{1}_{1}\text{H} \longrightarrow ^{3}_{2}\text{He} + \text{energy}$$

$$^{3}_{2}\text{He} + ^{1}_{1}\text{H} \longrightarrow ^{4}_{2}\text{He} + ^{\ 0}_{+1}\text{e} + \text{energy}$$

Extreme temperatures, greater than 10,000,000 °C, are needed to overcome the strong repulsive forces of the positively charged nuclei. Experimentally, a powerful laser beam is used to generate the necessary heat to initiate nuclear fusion. A major practical problem is how to confine a reaction that takes place at millions of degrees. Since ordinary materials vaporize at this temperature, a strong magnetic field is used to confine the fusion reaction. Thus, nuclear fusion is said to occur in a "magnetic bottle."

Nuclear fusion has been achieved in research fusion reactors on a limited scale (Figure 18.10). To date, continuous nuclear fusion has not been achieved, and no commercial fusion reactors are in operation. If the technology obstacles can be overcome, the fusion of two deuterium nuclei could be used to supply the world with electrical energy.

▲ **Figure 18.10 Fusion Reactor** A nuclear fusion reactor is essentially a large "magnetic bottle" that contains light nuclei heated to millions of degrees by a powerful pulse laser.

Deuterium, $^{2}_{1}\text{H}$, is the nuclide of hydrogen that contains one proton and one neutron. Deuterium is available from the sea in the form of heavy water molecules, D_2O (Section 5.5). It is estimated that more energy is potentially available from the deuterium in one cubic mile of seawater than from all the current petroleum reserves in the world. A typical fusion reaction of deuterium is as follows:

$$^{2}_{1}\text{H} + ^{2}_{1}\text{H} \longrightarrow ^{4}_{2}\text{He} + \text{energy}$$

Another promising nuclear fusion reaction involves deuterium and tritium. **Tritium**, $^{3}_{1}\text{H}$, is the nuclide of hydrogen that contains one proton and two neutrons. Here again, the fusion of nuclei requires temperatures above 10,000,000 °C and is initiated by a powerful laser. A typical fusion reaction of deuterium and tritium produces an alpha particle and a neutron, while releasing a huge amount of energy.

$$^{3}_{1}\text{H} + ^{2}_{1}\text{H} \longrightarrow ^{4}_{2}\text{He} + ^{1}_{0}\text{n} + \text{energy}$$

Summary

Section 18.1 An unstable atomic nucleus disintegrates naturally by emitting **radioactivity** in the form of alpha, beta, or gamma radiation. An **alpha particle** (α) has a positive charge and is identical to a helium-4 nucleus. A **beta particle** (β) has a negative charge and is identical to an electron. A **gamma ray** (γ) is a form of high-energy light that has neither mass nor charge.

Section 18.2 The composition of a nucleus is indicated using atomic notation, where the **atomic number** of the element is indicated by a subscript and the **mass number** by a superscript. In this chapter, we have used the term **nuclide** to refer to the nucleus of an atom with a specific composition of protons and neutrons.

A **nuclear reaction** involves a change in a nucleus, for example, **electron capture** or **positron** emission. A **nuclear equation** represents the nuclear change using atomic notation for each particle. To have a balanced nuclear equation, two criteria are necessary: (1) the sums of the atomic numbers must be the same on each side of the equation, and (2) the sums of the mass numbers must be equal for reactants and products.

Section 18.3 If a nuclide disintegrates through the emission of radiation in more than one step, the overall process is called a **radioactive decay series**. The decaying nucleus is referred to as the parent nuclide, and the resulting nucleus is the daughter nuclide. Together, they are referred to as **parent–daughter nuclides**. In nature, uranium-235, uranium-238, and thorium-232 decay naturally in a series of steps. Each of these nuclides continues to decay in a series of steps until a stable nuclide of lead is produced.

Section 18.4 We can keep track of the number of decaying nuclei in a sample using a Geiger counter. The **activity** of the sample refers to the number of unstable nuclei that decay in a given period of time. The activity of a sample decreases with time because there are fewer nuclei to decay. However, the time required for 50% of the radioactive nuclei to decay is constant and is called the **half-life**. After each half-life, only 50% of the radioactive nuclei remain.

Section 18.5 Scientists have found important applications for a broad selection of radioactive nuclides. **Radionuclides** are available for dating objects, for improving agricultural harvests, and for providing energy. In addition, there is a long and growing list of radionuclides that are used for medical diagnosis and treatment.

Section 18.6 Scientists can induce radioactivity by firing light, accelerated nuclides at heavy nuclei. In the event that the atomic number changes and the new nucleus corresponds to a different element, the change is called a **transmutation** reaction. With particle accelerators, it is possible to cause hundreds of nuclear reactions. A particle accelerator, such as a *cyclotron* or a *linear accelerator*, accelerates a projectile nuclide and sends it crashing into a target nucleus. This technique is called "atom smashing" and has created hundreds of synthetic nuclides.

Section 18.7 A neutron striking an unstable nucleus can induce **nuclear fission**; that is, the nucleus can split into two or more particles while releasing energy. The three most important nuclides that undergo fission are uranium-235, plutonium-239, and uranium-233. When a nucleus fissions, two or three neutrons are usually released. These neutrons can in turn cause a second and a third nuclear fission. If a single neutron initiates continuous fission, the process is called a **chain reaction**. For a chain reaction to occur, there must be a sufficient amount of fissionable nuclide. The minimum size of sample that can support a chain reaction is called the **critical mass**.

Section 18.8 The process of combining two light nuclei into a single nucleus is called **nuclear fusion**. In the interior of the Sun at temperatures above 10,000,000 °C, there is continuous nuclear fusion. In the laboratory, scientists have had limited success and to date have not been able to create continuous nuclear fusion. One promising fusion reaction is that involving two **deuterium**, ^{2}H, nuclei. Another possibility is the fusion of deuterium and **tritium**, ^{3}H, nuclei. The technological problems are considerable because the fusion process requires temperatures of millions of degrees. However, nuclear fusion does not produce radioactive waste and is therefore an attractive possibility as a future energy source.

Key Terms *Answers to Key Terms are in Appendix H.*

Select the key term that corresponds to each of the following definitions.

____ 1. the emission of particles or energy from an unstable nucleus
____ 2. nuclear radiation identical to a helium-4 nucleus
____ 3. nuclear radiation identical to an electron
____ 4. nuclear radiation identical to high-energy light
____ 5. the value that indicates the number of protons in a nucleus
____ 6. the value that indicates the number of protons and neutrons in a nucleus
____ 7. an atom with a specific number of protons and neutrons
____ 8. a high-energy change involving the atomic nucleus
____ 9. a shorthand representation using atomic notation to describe a nuclear reaction
____ 10. nuclear radiation identical in mass but opposite in charge to that of an electron
____ 11. a nuclear reaction in which a heavy nuclide attracts one of its inner core electrons into the nucleus

(a) activity *(Sec. 18.4)*
(b) alpha particle (α) *(Sec. 18.1)*
(c) atomic number (Z) *(Sec. 18.2)*
(d) beta particle (β^-) *(Sec. 18.1)*
(e) chain reaction *(Sec. 18.7)*
(f) critical mass *(Sec. 18.7)*
(g) deuterium *(Sec. 18.8)*
(h) electron capture (EC) *(Sec. 18.2)*
(i) gamma ray (γ) *(Sec. 18.1)*
(j) half-life ($t_{1/2}$) *(Sec. 18.4)*

_____ 12. the stepwise disintegration of a radioactive nucleus until a stable nuclide is reached
_____ 13. the relationship of a decaying nuclide and the resulting nuclide that is produced
_____ 14. the number of radioactive nuclei in a sample that disintegrate in a given period of time, for example, 500 dpm
_____ 15. the amount of time required for 50% of the radioactive nuclei in a given sample to decay
_____ 16. an atom whose nucleus is unstable and decays radioactively
_____ 17. the conversion of one element to another by a nuclear reaction
_____ 18. a nuclear reaction in which a nucleus splits into two or more lighter nuclei
_____ 19. a fission reaction in which the neutrons released initiate a second reaction, which in turn initiate a third reaction, and so on
_____ 20. the minimum mass of a fissionable nuclide necessary to sustain a chain reaction
_____ 21. a nuclear reaction in which two light nuclei combine into a single nucleus
_____ 22. the nuclide of hydrogen having one neutron
_____ 23. the nuclide of hydrogen having two neutrons

(k) mass number (*A*) *(Sec. 18.2)*
(l) nuclear equation *(Sec. 18.2)*
(m) nuclear fission *(Sec. 18.7)*
(n) nuclear fusion *(Sec. 18.8)*
(o) nuclear reaction *(Sec. 18.2)*
(p) nuclide *(Sec. 18.2)*
(q) parent–daughter nuclides *(Sec. 18.3)*
(r) positron (β^+) *(Sec. 18.2)*
(s) radioactive decay series *(Sec. 18.3)*
(t) radioactivity *(Sec. 18.1)*
(u) radionuclide *(Sec. 18.5)*
(v) transmutation *(Sec. 18.6)*
(w) tritium *(Sec. 18.8)*

Exercises *Answers to odd-numbered Exercises are in Appendix I.*

Natural Radioactivity (Sec. 18.1)

1. Which type of radiation is a helium nucleus and is deflected toward the negative electrode as it passes between electrically charged plates?
2. Which type of radiation cannot penetrate human skin and requires heavy paper or clothing as minimum protective shielding?
3. Which type of nuclear radiation is identical to an electron and is deflected toward the positive electrode as it passes between electrically charged plates?
4. Which type of radiation can penetrate about 1 cm of human flesh and requires wood or aluminum as minimum protective shielding?
5. Which type of nuclear radiation is powerful light energy that is *not* deflected as it passes between electrically charged plates?
6. Which type of radiation can pass through the human body and requires thick lead or concrete as minimum protective shielding?

◀ **Radon Test Kit** The earth releases radioactive radon gas that may collect beneath a home. Although the gas is only present in trace amounts, it may accumulate in basements and areas with minimal ventilation and pose a health hazard.

Nuclear Equations (Sec. 18.2)

7. What is the approximate mass for each of the following?
   (a) alpha particle **(b)** beta particle
   (c) gamma ray **(d)** positron
   (e) neutron **(f)** proton
8. What is the relative charge for each of the following?
   (a) α **(b)** β^-
   (c) γ **(d)** β^+
   (e) n^o **(f)** p^+
9. Write an equation for each of the following natural radioactive decay processes:
   (a) W-160 decays by alpha emission
   (b) P-32 decays by beta emission
   (c) Co-55 decays by positron emission
   (d) Ti-44 decays by electron capture
10. Write an equation for each of the following natural radioactive decay processes:
    (a) U-238 decays by alpha emission
    (b) Al-28 decays by beta emission
    (c) O-15 decays by positron emission
    (d) Fe-55 decays by electron capture
11. Identify X, an unknown radioactive nuclide, given the following information:
    (a) Nuclide X decays by alpha emission to give ^{217}Rn.
    (b) Nuclide X decays by beta emission to give ^{43}Ca.
    (c) Nuclide X decays by positron emission to give ^{73}Br.
    (d) Nuclide X decays by electron capture to give ^{133}Cs.
12. Identify X, an unknown radioactive nuclide, given the following information:
    (a) Nuclide X decays by alpha emission to give ^{218}Ra.
    (b) Nuclide X decays by beta emission to give ^{56}Fe.
    (c) Nuclide X decays by positron emission to give ^{19}F.
    (d) Nuclide X decays by electron capture to give ^{37}Cl.

Radioactive Decay Series (Sec. 18.3)

13. The uranium-238 decay series begins with the emission of an alpha particle. The daughter product emits a beta particle to give which nuclide?
14. The uranium-235 decay series begins with the emission of an alpha particle. The daughter product emits a beta particle to give which nuclide?
15. In the final step of the uranium-238 disintegration series, the parent nuclide decays into lead-206 and an alpha particle. Identify the parent nuclide.
16. In the final step of the uranium-235 disintegration series, the parent nuclide decays into lead-207 and a beta particle. Identify the parent nuclide.
17. Supply each of the following emission particles in the 10-step decay series for radioactive thorium-232:

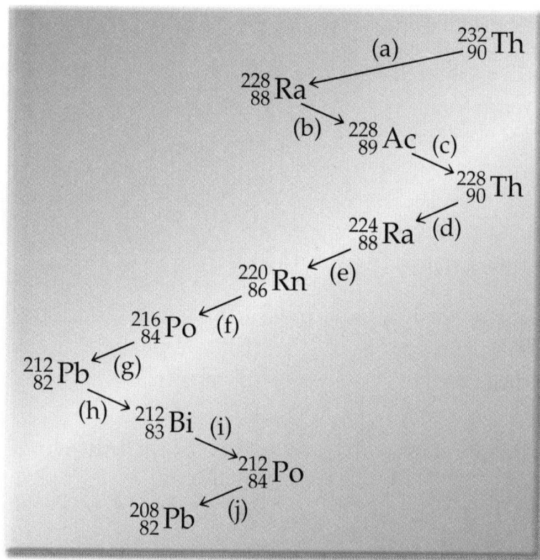

18. Supply each of the following decaying nuclides in the 11-step decay series for radioactive neptunium-237:

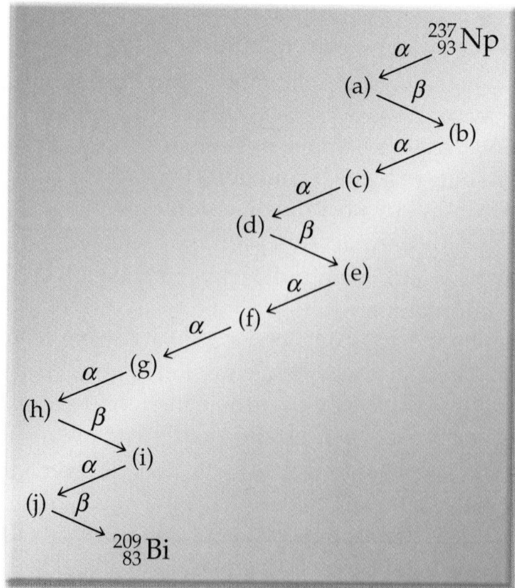

19. Plutonium-241 decays in a series of 13 steps, beginning with the emission of a beta particle, followed by an alpha particle, followed by another alpha particle. Identify the radionuclide produced in Step 3 of the Pu-241 series.
20. Plutonium-241 decays in a series of 13 steps. In Step 11, bismuth-213 emits a beta particle, followed by an alpha particle, and ending with a beta particle. Identify the radionuclide produced in Step 13 of the Pu-241 series.

Radioactive Half-Life (Sec. 18.4)

21. What percentage of U-238 radionuclides in a sample remain after two half-lives?
22. What percentage of U-238 radionuclides in a sample remain after three half-lives?
23. A nuclear reactor produces spent Pu-239 fuel rods that are highly radioactive after they no longer support fission. If 15 half-lives are required for the nuclide to reach a safe level, how long must the rods be stored ($t_{1/2}$ = 24,400 years)?
24. A nuclear fission reactor produces Sr-90 radioactive waste. If 20 half-lives are required for the nuclide to reach a safe level, how long must the nuclear waste be stored ($t_{1/2}$ = 28.8 years)?
25. If the carbon-14 radioactivity of an ancient wooden artifact is 6.25% that of a reference sample, what is the estimated age of the artifact ($t_{1/2}$ = 5730 years)?
26. If the carbon-14 reading of a fossil bone is 60 dpm, and a recent bone is 240 dpm, what is the estimated age of the fossil ($t_{1/2}$ = 5730 years)?
27. Sodium-24 in the form of NaCl is given as an injection to measure the sodium electrolyte balance. If 80 mg of the medical radionuclide is injected, how much Na-24 remains after 60 hours ($t_{1/2}$ = 15 hours)?
28. If 160 mg of technetium-99 is administered for a medical diagnosis, how much of the nuclide remains after 24 hours ($t_{1/2}$ = 6 hours)?
29. Small beads of iridium-192 are sealed in a plastic tube and inserted through a needle into breast tumors. If an Ir-192 sample has an initial activity of 560 dpm, how much time is required for the activity to drop to 35 dpm ($t_{1/2}$= 74 days)?
30. If an iron-59 sample has an initial activity of 200 dpm, how much time is required for the activity to drop to 25 dpm ($t_{1/2}$= 45 days)?
31. The initial radioactivity of a cobalt-60 sample was 1200 dpm and after 21.2 years the activity dropped to 75 dpm. What is the half-life of the radionuclide?
32. If 2400 μg of hydrogen-3 decay to 600 μg after 24.8 years, what is the half-life of this radionuclide that is used as a chemical tracer?

Applications of Radionuclides (Sec. 18.5)

33. What radionuclide technique can be used to estimate the age of geological events up to a few billion years?
34. What radionuclide technique can be used to estimate the age of fossils up to 50,000 years old?
35. What radionuclide can be used to sterilize male insects and control pests?
36. What radionuclide can be used to irradiate food and destroy microorganisms?
37. What γ-emitting radionuclide is used to power pacemakers for heart patients?
38. What γ-emitting radionuclide can be used to diagnose and locate inactive lung tissue in the respiratory system?
39. What β-emitting radionuclide can be used to measure the activity of the thyroid gland?

40. What β-emitting radionuclide can be used to diagnose anemia by attaching to a hemoglobin molecule?
41. What γ-emitting radionuclide can be used to diagnose and locate brain tumors?
42. What γ-emitting radionuclide can be used to treat breast cancer?

Induced Radioactivity (Sec. 18.6)

43. Bombarding Br-81 with gamma rays gives a radioactive nuclide and a neutron. Identify the radionuclide.
44. Bombarding Na-23 with a proton gives a radioactive nuclide and a neutron. Identify the radionuclide.
45. Bombarding U-238 with a hydrogen-2 nucleus produces a radioactive nuclide, a beta particle, and two neutrons. Identify the radionuclide.
46. Bombarding Li-6 with a neutron produces a radioactive nuclide and an alpha particle. Identify the radionuclide.
47. Firing a hydrogen-2 nucleus at a target nucleus gives I-131 and a neutron. What is the target nuclide?
48. Firing a neutron at a target nucleus gives Mn-56 and an alpha particle. What is the target nuclide?
49. Firing an accelerated particle at a boron-10 target nucleus produces N-14 and a gamma ray. What is the projectile particle?
50. Firing an accelerated particle at a Mg-26 target nucleus produces Mg-27 and a proton. What is the projectile particle?
51. In 1967, a Russian team smashed a neon nucleus into an americium target and claimed it had synthesized a new element, X. Given the equation for the reaction, what nuclide did the team create?

$$^{243}_{95}\text{Am} + ^{22}_{10}\text{Ne} \longrightarrow ^{A}_{Z}\text{X} + 5\,^{1}_{0}\text{n}$$

52. In 1970, a group at the University of California at Berkeley fired a nitrogen nucleus into a californium target and claimed it had synthesized a new element, X. Given the equation for the reaction, what nuclide did the group create?

$$^{249}_{98}\text{Cf} + ^{15}_{7}\text{N} \longrightarrow ^{A}_{Z}\text{X} + 4\,^{1}_{0}\text{n}$$

53. In 1974, a team of Russian physicists smashed a chromium-54 nucleus into a lead-207 target. If a new element, X, and one neutron were produced from the collision, what nuclide did the team create?
54. In 1976, a team of German physicists smashed a chromium-54 nucleus into a bismuth-209 target. If a new element, X, and one neutron were produced from the collision, what nuclide did the team create?

Nuclear Fission (Sec. 18.7)

55. If a neutron causes a fission reaction that releases two neutrons, how many neutrons have been produced from the third fission step? (Assume each step in the fission process releases two neutrons.)
56. If a neutron causes a fission reaction that releases three neutrons, how many neutrons have been produced from the third fission step? (Assume each step in the fission process releases three neutrons.)
57. The fission of uranium-235 produces 2.4 neutrons per nucleus. Why is there a fractional value for the number of neutrons released?
58. Why must uranium ore be enriched in U-235 before it can be used in the fuel rods of a nuclear reactor?
59. How many neutrons are produced from the following fission reaction?

$$^{235}_{92}\text{U} + ^{1}_{0}\text{n} \longrightarrow ^{142}_{56}\text{Ba} + ^{91}_{36}\text{Kr} + ^{1}_{0}\text{n}$$

60. How many neutrons are produced from the following fission reaction?

$$^{235}_{92}\text{U} + ^{1}_{0}\text{n} \longrightarrow ^{144}_{54}\text{Xe} + ^{90}_{38}\text{Sr} + ^{1}_{0}\text{n}$$

61. How many neutrons are produced from the following fission reaction?

$$^{239}_{94}\text{Pu} + ^{1}_{0}\text{n} \longrightarrow ^{137}_{55}\text{Cs} + ^{100}_{39}\text{Y} + ^{1}_{0}\text{n}$$

62. How many neutrons are produced from the following fission reaction?

$$^{239}_{94}\text{Pu} + ^{1}_{0}\text{n} \longrightarrow ^{142}_{57}\text{La} + ^{94}_{37}\text{Rb} + ^{1}_{0}\text{n}$$

63. What nuclide undergoes nuclear fission to give barium-143, krypton-88, and three neutrons?
64. What nuclide undergoes nuclear fission to give xenon-142, strontium-90, and two neutrons?

▲ **Nuclear Fission** In a uranium-235 atomic bomb explosion, the large uranium nucleus splits (fissions) into two smaller fragments, and releases neutrons and energy.

Nuclear Fusion (Sec. 18.8)

65. The nuclear fusion of two deuterium nuclei gives a tritium nucleus, a positron, and particle X. Identify particle X.
66. The nuclear fusion of two helium-3 nuclei gives two protons and particle X. Identify particle X.
67. The nuclear fusion of a helium-3 nucleus and particle X releases an alpha particle and a positron. Identify particle X.
68. The nuclear fusion of a lithium-7 nucleus and particle X releases two alpha particles and a neutron. Identify particle X.
69. The nuclear fusion of two identical nuclides produces an alpha particle and a gamma ray. Identify the fusion nuclide.
70. The nuclear fusion of two identical nuclides produces a deuterium nucleus and a positron. Identify the fusion nuclide.

General Exercises

71. Plutonium-241 undergoes a 13-step decay series. If an alpha particle is emitted in the first step, what is the daughter product?
72. Plutonium-241 undergoes a 13-step decay series. If an alpha particle is emitted in the first step followed by a beta particle, what is the granddaughter product?
73. Nuclide X has a half-life of 23 s and decays by beta emission to produce Xe-136 and a neutron. Write a nuclear equation for the reaction.
74. Nuclide X has a half-life of 2.6 years and decays by electron capture to produce Mn-55 and a gamma ray. Write a nuclear equation for the reaction.
75. Crater Lake in Oregon was formed by volcanic explosion. If a tree charred by the explosion gave an activity of ~7.7 dpm and the half-life of C-14 is 5730 years, what is the approximate age of Crater Lake?
76. Uranium–lead dating of lunar rock samples gave a U-238 to Pb-206 ratio of ~1.0. If the half-life of U-238 is 4.5 billion years, what is the approximate age of the lunar rocks?
77. A radionuclide ionizes the air in a smoke detector, which in turn allows the flow of electricity. When smoke particles enter the detector, they disrupt the electric current and set off an alarm. What α-emitting radionuclide is used to ionize the air in a smoke detector? (*Hint:* The radionuclide decays into Np-237.)
78. A PET scan is a medical imaging technique for the circulatory system. First, a radionuclide is injected that decays by positron emission. When the positron is emitted, it strikes an electron, causing mutual annihilation and the release of two gamma rays. Write an equation for the annihilation reaction.
79. Element 99 was first synthesized by bombarding uranium-238 with neutrons. The reaction yielded Es-253 and seven beta particles. How many neutrons are necessary to balance the equation?
80. Element 100 was first synthesized by bombarding uranium-238 with neutrons. The reaction yielded Fm-255 and eight beta particles. How many neutrons are necessary to balance the equation?
81. In 1955, a single atom of a new element was synthesized using the cyclotron at the University of California at Berkeley. If Es-253 reacted with an alpha particle to release a neutron, what new nuclide was produced?
82. In the Middle Ages, alchemists attempted to transmute base metals into gold. Recently, platinum has been transmuted into gold by bombarding Pt-198 with a deuterium nucleus to give a platinum nuclide and a proton. If the platinum nuclide decays by beta emission, what nuclide results?
83. The nuclear fusion of deuterium and tritium in a hydrogen bomb releases an alpha particle, energy, and nuclide X. Identify nuclide X.
84. The nuclear fission of lithium-6 in an atomic bomb releases an alpha particle, energy, and nuclide X. Identify nuclide X.

18 Chapter Self-Test

Answers to Self-Test are in Appendix J.

1. Which type of nuclear radiation is identical to an electron and is deflected toward the positive electrode as it passes between electrically charged plates? (Sec. 18.1)
   (a) alpha (b) beta
   (c) gamma (d) all of the above
   (e) none of the above
2. What is the approximate mass and relative charge of gamma radiation? (Sec. 18.2)
   (a) 4 amu and +2 (b) 1 amu and +1
   (c) 1 amu and 0 (d) 0 amu and 0
   (e) 0 amu and –1
3. What is the name of the particle having the following notation: ${}^{0}_{+1}e$? (Sec. 18.2)
   (a) electron (b) neutron
   (c) positron (d) proton
   (e) none of the above
4. What particle is emitted when a Ra-221 nucleus decays to Rn-217? (Sec. 18.2)
   (a) alpha (b) beta
   (c) neutron (d) positron
   (e) proton
5. In the final step of the uranium-235 decay series, the parent nuclide decays to a lead-207 nuclide and a beta particle. What is the parent nuclide? (Sec. 18.3)
   (a) ${}^{206}_{82}Pb$
   (b) ${}^{208}_{82}Pb$
   (c) ${}^{207}_{83}Bi$
   (d) ${}^{207}_{81}Tl$
   (e) none of the above
6. If an iron-59 sample has an initial activity of 160 dpm, how much time is required for the activity to drop to 20 dpm ($t_{1/2}$ = 45 days)? (Sec. 18.4)
   (a) 45 days
   (b) 90 days
   (c) 135 days
   (d) 180 days
   (e) 210 days
7. What is the principle assumption for radiocarbon dating to be reliable? (Sec. 18.5)
   (a) The amount of nitrogen-14 in the atmosphere has remained constant.
   (b) The amount of carbon-14 in the atmosphere has remained constant.

(c) The number of neutrons in the atmosphere has remained constant.
(d) The number of protons in the atmosphere has remained constant.
(e) The temperature of the atmosphere has remained constant.

8. In 1934 Irene Curie (daughter of Marie and Pierre) and her husband Frederic Joliot synthesized the first radionuclide. They bombarded aluminum-27 with an alpha particle and obtained a neutron and what other radionuclide? (Sec. 18.6)
(a) phosphorus-30 (b) phosphorus-31
(c) silicon-30 (d) silicon-31
(e) none of the above

9. How many neutrons are produced from the following fission reaction? (Sec. 18.7)

$$^{235}_{92}\text{U} + ^{1}_{0}\text{n} \longrightarrow ^{141}_{56}\text{Ba} + ^{91}_{36}\text{Kr} + ^{1}_{0}\text{n}$$

(a) 1 (b) 2
(c) 3 (d) 4
(e) none of the above

10. The nuclear fusion of a helium-3 nucleus and nuclide X releases an alpha particle and a positron. What is nuclide X? (Sec. 18.8)
(a) deuterium (b) tritium
(c) neutron (d) positron
(e) proton

Key Concepts

11. Write a nuclear equation for each of the following natural decay reactions:
(a) U-238 decays by alpha emission
(b) U-239 decays by beta emission
(c) K-40 decays by positron emission
(d) K-40 decays by electron capture

12. A single neutron causes uranium-235 to undergo nuclear fission and release three neutrons. Assuming each of the neutrons released causes a fission that releases three more neutrons, how many neutrons are released in the second step?

13. The nuclear fusion of two protons releases a deuterium nucleus and nuclide X. Identify nuclide X.

Critical Thinking

14. Write a nuclear equation corresponding to the following illustration of radioactive decay:

82 p^+ 128 n^0 → 83 p^+ 127 n^0 + ○

15. In 2003 an international team of scientists at the Lawrence Berkeley Laboratories claimed the synthesis of a new element by firing a krypton-86 nucleus into a lead-208 target. If one neutron was released from the collision, identify the new element by its nuclide.

CHAPTER

19 Organic Chemistry

As the world's petroleum reserves decrease, society must make difficult choices over energy needs versus environmental concerns. The oil drilling platform shown is in the Santa Barbara channel off the coast of a beautiful California beach area.

"Life exists in the universe only because the carbon atom possesses certain exceptional properties."
Sir James Jeans, English Physicist (1877–1946)

In the early 1800s, chemists believed that inorganic compounds were different from organic compounds. Inorganic compounds originate from rocks and minerals. Organic compounds, on the other hand, originate from plants and animals. Since organic compounds are derived from living organisms, they were thought to contain a life force. This popular idea was referred to as the *vital force theory*.

In 1828, the German chemist Friedrich Wöhler (1800–1882) heated ammonium cyanate and obtained urea, a compound present in urine. The result was surprising because ammonium cyanate was considered an *inorganic* compound, whereas urea from animals is an *organic* compound. Wöhler's experiment helped disprove the vital force theory, although reluctant skeptics insisted that the vital force from Wöhler's hands had contaminated the result.

From simple and somewhat controversial beginnings, organic chemistry has grown to the point that many of our modern day materials and medicines are the result of research in organic chemistry. Plastics; fibers for fabrics and carpets; medical products ranging from artificial joints to pharmaceuticals; and parts for appliances, cars, and airplanes all derive from discoveries in organic chemistry. In addition, the molecules associated with animal and plant life are organic molecules.

19.1 HYDROCARBONS

OBJECTIVES

- To note that hydrocarbons can have single, double, or triple bonds.
- To classify a hydrocarbon as saturated, unsaturated, or aromatic.
- To classify a hydrocarbon as an alkane, alkene, alkyne, or arene.

The definition of **organic chemistry** is the study of compounds that contain carbon. Currently, about 10 million organic compounds account for about 90% of all known substances. In addition, over 50,000 new organic compounds are synthesized each year. One reason that there are so many organic compounds is that carbon atoms have the ability to link together and form long chains. The chief sources of carbon are petroleum, natural gas, and coal, all of which are referred to as *fossil fuels.*

An organic compound composed of hydrogen and carbon is called a **hydrocarbon**. Hydrocarbons can be classified as saturated or unsaturated. A **saturated hydrocarbon** has a single bond between each of its carbon atoms. Since a carbon atom has four valence electrons, it can bond to four other atoms. That is, a carbon atom can form four single covalent bonds. If a hydrocarbon compound has all single bonds, it belongs to the *alkane* family.

An **unsaturated hydrocarbon** has either a double bond or a triple bond between two carbon atoms. If a compound has a double bond, it is a member of the *alkene* family. If it has a triple bond, it belongs to the *alkyne* family. An **aromatic hydrocarbon** is usually characterized by having a benzene ring. If a compound contains a benzene ring, it is a member of the *arene* family. Figure 19.1 shows the classification of hydrocarbons.

We can construct molecular models of hydrocarbons to gain an appreciation of their three-dimensional structure. Ball-and-stick models are commonly used to show the arrangement of atoms in a molecule. Figure 19.2 illustrates the four classes of compounds with selected molecular models.

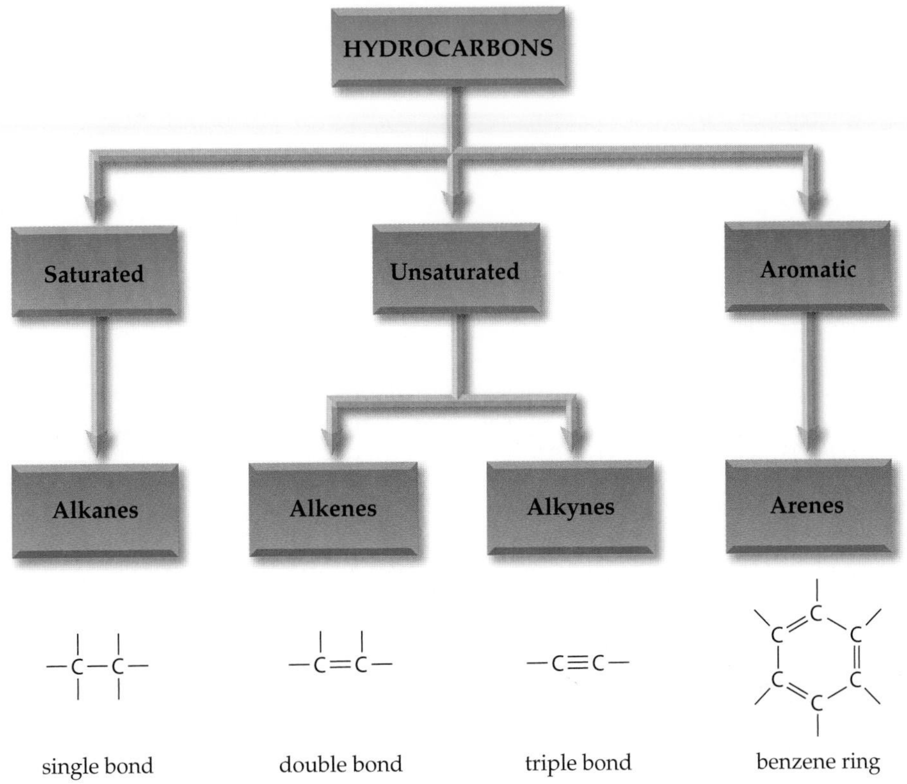

◀ **Figure 19.1 Classification of Hydrocarbons** A hydrocarbon can be classified as (a) an alkane, (b) an alkene, (c) an alkyne, or (d) an arene.

(a)

(b)

(c)

(d)

▲ **Figure 19.2 Molecular Models of Hydrocarbons** The photographs show molecular models of (a) an alkane (saturated), (b) an alkene (unsaturated), (c) an alkyne (unsaturated), or (d) an arene (aromatic hydrocarbon).

OBJECTIVES

- To write names and formulas for simple alkanes.
- To write combustion reactions for alkanes.

19.2 ALKANES

Alkanes are saturated hydrocarbons whose names end in the suffix *-ane*. The general molecular formula for the alkanes is C_nH_{2n+2}, where n is the number of carbon atoms. The alkane in Figure 19.2(a) has 2 carbon atoms ($n = 2$), 6 hydrogen atoms ($2n + 2 = 6$), and the molecular formula is C_2H_6. The sixth member of the alkane family has 6 carbon atoms ($n = 6$), 14 hydrogen atoms ($2n + 2 = 14$), and the molecular formula is C_6H_{14}.

Alkane Family of Compounds

The smaller members of the alkane family, ranging from 1 to 10 carbons, are used as gaseous and liquid fuels. The larger members, ranging from 20 to 40 carbons, are waxy solids. These solid alkanes are referred to as paraffin and are used to make waxes and candles. Table 19.1 lists the first 10 members of the alkane family.

TABLE 19.1 ALKANES

IUPAC NAME*	MOLECULAR FORMULA	STRUCTURAL FORMULA	BOILING POINT
methane	CH_4	CH_4	–161 °C
ethane	C_2H_6	$CH_3{-}CH_3$	–89 °C
propane	C_3H_8	$CH_3{-}CH_2{-}CH_3$	–44 °C
butane	C_4H_{10}	$CH_3{-}CH_2{-}CH_2{-}CH_3$	–1 °C
pentane	C_5H_{12}	$CH_3{-}CH_2{-}CH_2{-}CH_2{-}CH_3$	36 °C
hexane	C_6H_{14}	$CH_3{-}CH_2{-}CH_2{-}CH_2{-}CH_2{-}CH_3$	68 °C
heptane	C_7H_{16}	$CH_3{-}CH_2{-}CH_2{-}CH_2{-}CH_2{-}CH_2{-}CH_3$	98 °C
octane	C_8H_{18}	$CH_3{-}CH_2{-}CH_2{-}CH_2{-}CH_2{-}CH_2{-}CH_2{-}CH_3$	125 °C
nonane	C_9H_{20}	$CH_3{-}CH_2{-}CH_2{-}CH_2{-}CH_2{-}CH_2{-}CH_2{-}CH_2{-}CH_3$	151 °C
decane	$C_{10}H_{22}$	$CH_3{-}CH_2{-}CH_2{-}CH_2{-}CH_2{-}CH_2{-}CH_2{-}CH_2{-}CH_2{-}CH_3$	174 °C

*Note that the Greek prefixes we learned previously (*penta-, hexa-, hepta-, octa-, nona-, deca-*) are used to form the root names of alkanes C_5H_{12} through $C_{10}H_{22}$. The first four members of the alkanes are gases at normal temperature (25 °C); pentane through decane are liquids.

The molecular and structural formulas of alkanes are shown in Table 19.1. For example, the molecular formula of propane is C_3H_8. The condensed **structural formula** is $CH_3{-}CH_2{-}CH_3$. To see the structure even more clearly, we can write an expanded structural formula. For simplicity, we can omit the hydrogen atoms and write a skeletal structural formula. Each of the following formulas can be used to represent a molecule of propane, C_3H_8:

```
                          H  H  H
                          |  |  |
CH3—CH2—CH3           H—C—C—C—H          —C—C—C—
                          |  |  |            |  |  |
                          H  H  H
```

condensed structural formula | expanded structural formula 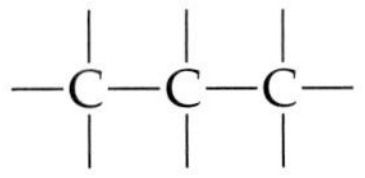 | skeletal structural formula

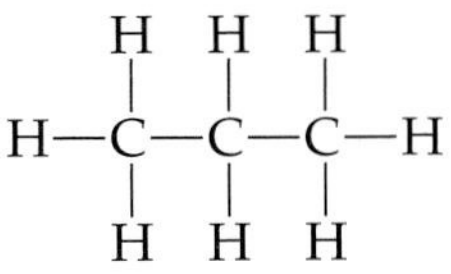

Branched Hydrocarbon Chains

Each structural formula in Table 19.1 represents a molecule that is a straight chain and does not have branches. However, most carbon chains do have branches. When branching occurs, a different compound results. Two compounds that have the same molecular formula, but a different structural formula are called structural **isomers**.

This text places names such as "isobutane" in quotes to indicate it is a common name. The systematic name for "isobutane" is 2–methylpropane.

Butane has the molecular formula C_4H_{10}. However, we can draw the structural formula in two ways: $CH_3CH_2CH_2CH_3$ and $CH_3CH(CH_3)CH_3$. These formulas represent the two isomers—butane and "isobutane." In addition to different structural formulas, isomers have different physical properties such as melting points and boiling points.

```
                                   CH3
                                   |
CH3—CH2—CH2—CH3              CH3—CH—CH3
```

butane
Mp = –138 °C, Bp = –0.5 °C

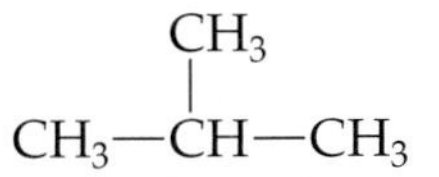

"isobutane"
Mp = –159 °C, Bp = –12 °C

As the number of carbons in the molecule increases, the number of possible isomers increases. For example, C_5H_{12} has 3 isomers, C_6H_{14} has 5 isomers, and $C_{10}H_{22}$ has 75 isomers! To draw the structural isomers corresponding to a molecular formula, we use the following guidelines:

Guidelines for Drawing Hydrocarbon Isomers

1. Draw a continuous chain of carbon atoms. For C_5H_{12}, the chain is five carbon atoms long. To simplify, we omit hydrogen atoms and show the skeletal formula.

   C – C – C – C – C

2. To construct a different isomer, draw a shorter continuous chain; in this case, the chain is now four carbon atoms long. Attach the fifth carbon to either carbon atom in the middle of the chain.

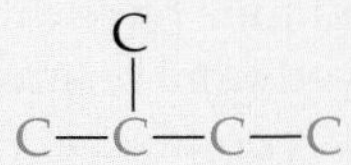

```
   C
   |
C—C—C—C
```

   It makes no difference which middle carbon atom we attach the fifth carbon to since we can flip the molecule over without changing the structure. The branch is still on the second carbon from the end. The following structure is identical to the previous one:

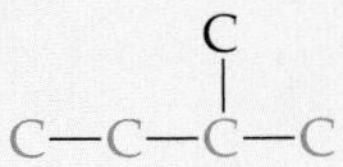

```
      C
      |
C—C—C—C
```

3. To construct another isomer, shorten the chain to three carbon atoms. Attach the fourth and fifth carbon atoms to the central carbon atom.

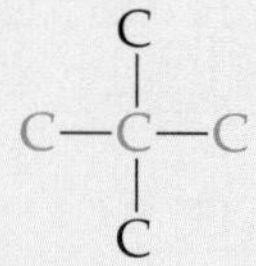

```
     C
     |
 C — C — C
     |
     C
```

4. For higher members of the alkane series that contain more than five carbons, apply a similar procedure. Although the process can become complex, this systematic procedure is very effective.

Alkyl and Aryl Groups

When a hydrogen atom is removed from an alkane, an **alkyl group** (symbol **R**–) results. Because an alkyl group is missing a hydrogen atom, it can bond to another atom or molecule. The name of the alkyl group is formed by changing the suffix *-ane* of the alkane to the suffix *-yl*. For example, methane becomes a methyl group.

$$\underset{\text{meth}\mathbf{ane}}{CH_4} \quad - H \quad = \quad \underset{\text{meth}\mathbf{yl}}{CH_3-}$$

Similarly, removing a hydrogen atom from ethane gives an ethyl group.

$$\underset{\text{eth}\mathbf{ane}}{CH_3-CH_3} \quad - H \quad = \quad \underset{\text{eth}\mathbf{yl}}{CH_3-CH_2-}$$

Removal of a hydrogen atom from propane can occur in two ways—from the end of the molecule or from the middle carbon atom. The structural formula of propane, $CH_3-CH_2-CH_3$, shows that there are six hydrogen atoms on the end carbons and two hydrogen atoms on the central carbon. Since the end carbon atoms are identical, we can remove any one of the six hydrogen atoms to produce a propyl group.

$$\underset{\text{prop}\mathbf{ane}}{CH_3-CH_2-CH_3} \quad - H \quad = \quad \underset{\text{prop}\mathbf{yl}}{CH_3-CH_2-CH_2-}$$

Since the middle two hydrogen atoms are equivalent, we can remove either of these atoms to form an isopropyl group.

$$\underset{\text{propane}}{CH_3-CH_2-CH_3} \; - H \quad = \quad \underset{\text{isopropyl}}{CH_3-\overset{|}{C}H-CH_3}$$

It is easy to see that we can substitute a chlorine atom onto a propane molecule at two different positions. If the chlorine is attached to an end carbon, the result is "propyl chloride," $CH_3-CH_2-CH_2-Cl$. If the chlorine atom is attached to the central carbon, the result is "isopropyl chloride," $CH_3-CH(Cl)-CH_3$. "Propyl chloride" and "isopropyl chloride" are isomers, and thus have different properties.

When a hydrogen atom is removed from an aromatic hydrocarbon, the result is an **aryl group** (symbol **Ar—**). The most common aryl group results from removal of a hydrogen atom from a benzene ring. The resulting aryl group is called a **phenyl group**.

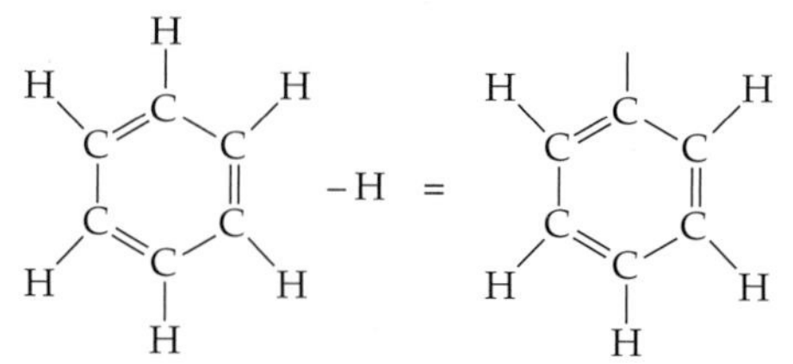

```
        H                          
        |                          |
  H     C     H              H     C     H
   \  //  \  /                \  //  \  /
    C      C                   C      C
    |      ||    - H  =        |      ||
    C      C                   C      C
   /  \\  /  \                /  \\  /  \
  H     C     H              H     C     H
        |                          |
        H                          H
```

benzene, C_6H_6 phenyl, $C_6H_5—$

Now, let's summarize the structures of alkyl groups. The names and formulas of selected alkyl and aryl groups are shown in Table 19.2.

TABLE 19.2 ALKYL AND ARYL GROUPS

NAME	MOLECULAR FORMULA	STRUCTURAL FORMULA	EXAMPLE
methyl	$CH_3—$	$CH_3—$	$CH_3—Cl$ "methyl chloride"
ethyl	$C_2H_5—$	$CH_3—CH_2—$	$CH_3—CH_2—Cl$ "ethyl chloride"
propyl	$C_3H_7—$	$CH_3—CH_2—CH_2—$	$CH_3—CH_2—CH_2—Cl$ "propyl chloride"
isopropyl	$C_3H_7—$	$CH_3—CH—$ $\|$ CH_3	$CH_3—CH—Cl$ $\|$ CH_3 "isopropyl chloride"
phenyl	$C_6H_5—$	H H C=C H—C C— C—C H H	H H C=C H—C C—Cl C—C H H "phenyl chloride"

Nomenclature of Alkanes

In 1892, the International Union of Chemistry met in Geneva, Switzerland, and recommended a systematic set of rules for naming organic compounds. Although these rules are revised periodically by the International Union of Pure and Applied Chemistry (IUPAC), the basic guidelines for naming organic compounds are still referred to as the Geneva convention. In naming an alkane, we will apply the following nomenclature guidelines:

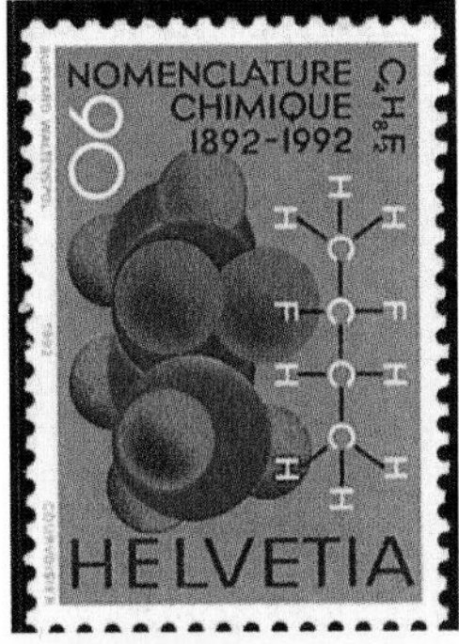

▲ **Organic Nomenclature**
This Swiss stamp commemorates the 100th anniversary of the systematic nomenclature rules for organic compounds that were first proposed at an IUPAC meeting in Geneva, Switzerland.

Guidelines for Naming Alkanes

1. Name an alkane for its longest continuous carbon chain, which is referred to as the "parent" chain. If the longest chain is five carbon atoms, the compound is a *pentane*. Regardless of the branches on the carbon chain, the alkane is named a pentane.

$$\begin{array}{c} \qquad\qquad\qquad\qquad\qquad CH_3 \\ \qquad\qquad\qquad\qquad\qquad | \\ CH_3—CH_2—CH_2—CH—CH_3 \end{array}$$

2. Number the longest continuous chain of carbon atoms starting from the end closest to the first branch on the chain. In the preceding example, we number from right to left.

$$\begin{array}{ccccc} & & & CH_3 & \\ & & & | & \\ CH_3— & CH_2— & CH_2— & CH— & CH_3 \\ 5 & 4 & 3 & 2 & 1 \end{array}$$

3. Indicate the position of the alkyl group(s) by name and number. In the preceding example, a methyl group is on the second carbon atom. The IUPAC name is thus *2-methylpentane*.

4. If there are two or more of the same group attached to the chain, use the prefix *di-* for two, *tri-* for three, or *tetra-* for four.

$$\begin{array}{ccccccccc} & & & & CH_3 & & CH_3 & & \\ & & & & | & & | & & \\ CH_3 & - & CH_2 & - & CH & - & CH & - & CH_3 \\ 5 & & 4 & & 3 & & 2 & & 1 \end{array}$$

Since there are two methyl groups on the second and third carbon atoms, the name of the hydrocarbon is *2,3-dimethylpentane*.

Example Exercise 19.1 Nomenclature of Alkanes

Give the IUPAC name for the following structure:

$$\begin{array}{ccccccccccc} & & & & & & CH_3 & & & & \\ & & & & & & | & & & & \\ CH_3 & - & CH & - & CH_2 & - & C & - & CH_2 & - & CH_3 \\ & & | & & & & | & & & & \\ & & CH_3 & & & & CH_3 & & & & \end{array}$$

Solution

Let's apply the systematic rules for naming alkanes.

1. The longest continuous carbon chain has six carbon atoms, and so the compound is a *hexane*.
2. Since the first branch is closest to the left end of the chain, we number from left to right.

$$\begin{array}{ccccccccccc} & & & & & & CH_3 & & & & \\ & & & & & & | & & & & \\ CH_3 & - & CH & - & CH_2 & - & C & - & CH_2 & - & CH_3 \\ & & | & & & & | & & & & \\ & & CH_3 & & & & CH_3 & & & & \\ 1 & & 2 & & 3 & & 4 & & 5 & & 6 \end{array}$$

3. Each of the three attached groups is a methyl. Thus, the compound is a *methylhexane*.
4. Since there are three methyl groups, we use the prefix *tri-* and indicate the position of each methyl. The IUPAC name for this hydrocarbon is *2,4,4-trimethylhexane*.

Practice Exercise

Draw the condensed structural formula for 2,3-dimethyldecane.

Answer:

$$\begin{array}{ccccccccccccccccccc} CH_3 & - & CH & - & CH & - & CH_2 & - & CH_2 & - & CH_2 & - & CH_2 & - & CH_2 & - & CH_2 & - & CH_3 \\ & & | & & | & & & & & & & & & & & & & & \\ & & CH_3 & & CH_3 & & & & & & & & & & & & & & \end{array}$$

Concept Exercise

Which of the following compounds is an example of an alkane: decane ($C_{10}H_{22}$), decene ($C_{10}H_{20}$), decyne ($C_{10}H_{18}$), or naphthalene ($C_{10}H_8$)?

Answer: See Appendix G

Reactions of Alkanes

Since alkanes have strong C–C and C–H bonds, they are unreactive at room temperature. At high temperatures, however, alkanes burn rapidly with oxygen in the air. The reaction is called a **combustion reaction**, and burning alkanes supply much of the world's energy.

The products from the complete combustion of an alkane are carbon dioxide and water. If the amount of oxygen is limited, carbon monoxide may be produced. For example, the equation for the complete combustion of methane gas is

$$CH_4(g) \;+\; 2\,O_2(g) \xrightarrow{\text{spark}} CO_2(g) \;+\; 2\,H_2O(g)$$

19.3 ALKENES AND ALKYNES

OBJECTIVES

- To write names and formulas for simple alkenes.
- To write names and formulas for simple alkynes.
- To write addition reactions for alkenes and alkynes.

Alkenes are unsaturated hydrocarbons that contain a carbon–carbon double bond. The general molecular formula for alkenes with one double bond is C_nH_{2n}. The alkene in Figure 19.2(b) has 2 carbon atoms ($n = 2$), 4 hydrogen atoms ($2n = 4$), and the molecular formula is C_2H_4. The sixth member of the alkene family has 6 carbon atoms ($n = 6$), 12 hydrogen atoms ($2n = 12$), and the molecular formula is C_6H_{12}.

Alkynes are unsaturated hydrocarbons that contain a carbon–carbon triple bond. The general molecular formula for alkynes with one triple bond is C_nH_{2n-2}. The alkyne in Figure 19.2(c) has 2 carbon atoms ($n = 2$), 2 hydrogen atoms ($2n - 2 = 2$), and the molecular formula is C_2H_2. The sixth member of the alkyne family has 6 carbon atoms ($n = 6$), 10 hydrogen atoms ($2n - 2 = 10$), and the molecular formula is C_6H_{10}.

Alkene Family of Compounds

The first, smallest member of the alkene family is $CH_2{=}CH_2$, as shown in Figure 19.2(b). Its IUPAC name is ethene, but it is commonly called "ethylene." The second member of the alkene family is $CH_2{=}CH{-}CH_3$. Its IUPAC name is propene, but it is commonly referred to as "propylene." "Ethylene" and "propylene" are used to manufacture the plastics polyethylene and polypropylene.

The next member of the alkene family, C_4H_8, is butene. Butene has two structural isomers because there are two possible positions for the double bond. The double bond can be between the first and second carbon atoms, $CH_2{=}CH{-}CH_2{-}CH_3$, or between the second and third, $CH_3{-}CH{=}CH{-}CH_3$. The name of the first isomer is 1–butene and the name of the second is 2-butene. Table 19.3 lists the first few members of the alkene family.

TABLE 19.3 ALKENES

IUPAC NAME	COMMON NAME	MOLECULAR FORMULA	STRUCTURAL FORMULA	BOILING POINT
ethene	"ethylene"	C_2H_4	$CH_2{=}CH_2$	–104 °C
propene	"propylene"	C_3H_6	$CH_2{=}CH{-}CH_3$	–47 °C
1–butene	"butylene"	C_4H_8	$CH_2{=}CH{-}CH_2{-}CH_3$	–6 °C

Nomenclature of Alkenes

An alkene is named after the corresponding alkane. That is, the name of an alkene is formed by changing the suffix *-ane* to the suffix *-ene*. For example, ethane is changed to ethene. The rules for naming alkenes are similar to the rules for naming alkanes. We will use the following nomenclature guidelines:

Guidelines for Naming Alkenes

1. Name an alkene for the longest continuous carbon chain that contains the double bond. Consider the following alkene:

$$\begin{array}{l} \quad\;\; CH_2{-}CH_3 \\ \quad\;\; | \\ CH_2{=}C{-}CH_2{-}CH_2{-}CH_3 \end{array}$$

Although the longest chain is six carbon atoms, only five carbon atoms are in the chain with the double bond. Thus, the compound is a *pentene*.

2. Number the longest continuous chain starting from the end closest to the double bond. Thus, we number from left to right to give *1-pentene*.

$$\begin{array}{l} \quad\;\; CH_2{-}CH_3 \\ \quad\;\; | \\ CH_2{=}C{-}CH_2{-}CH_2{-}CH_3 \\ \;\;1 \quad\;\; 2 \quad\; 3 \quad\;\;\; 4 \quad\;\;\; 5 \end{array}$$

3. Indicate the position of each attached alkyl group(s) by name and number. Since there is an ethyl group, CH_3-CH_2-, on the second carbon, the compound is named *2-ethyl-1-pentene.*
4. If there are two or more of the same group on the chain, use the prefix *di-* for two, *tri-* for three, or *tetra-* for four. In this case there is one group, so the name remains *2-ethyl-1-pentene.*

Example Exercise 19.2 Nomenclature of Alkenes

Give the IUPAC name for the following structure:

$$\begin{array}{ccccccc} & CH_3 & & & & & \\ & | & & & & & \\ CH_3- & C & -CH_2- & C & =CH- & CH_3 \\ & | & & | & & \\ & CH_3 & & CH_3 & & \end{array}$$

Solution

Let's apply the systematic rules for naming alkenes.

1. The longest continuous carbon chain containing the double bond has six carbons, and so the compound is a *hexene.*
2. To assign the double bond the lowest value, we start numbering from the right. This compound is a *2-hexene.*

$$\begin{array}{cccccc} & CH_3 & & & & \\ & | & & & & \\ CH_3- & C & -CH_2- & C & =CH- & CH_3 \\ & | & & | & & \\ & CH_3 & & CH_3 & & \\ 6 & 5 & 4 & 3 & 2 & 1 \end{array}$$

3. Each of the three branches is a methyl group. Thus, the compound is a *methyl-2-hexene.*
4. Since there are three methyl groups, we use the prefix *tri-* and indicate the position of each group. The IUPAC name for this alkene is *3,5,5-trimethyl-2-hexene.*

Practice Exercise

Draw the condensed structural formula for 2-methyl-2-pentene.

Answer:

$$\begin{array}{ccc} CH_3- & C & =CH-CH_2-CH_3 \\ & | & \\ & CH_3 & \end{array}$$

Concept Exercise

Which of the following compounds is an example of an alkene: decane ($C_{10}H_{22}$), decene ($C_{10}H_{20}$), decyne ($C_{10}H_{18}$), or naphthalene ($C_{10}H_8$)?

Answer: See Appendix G

Alkyne Family of Compounds

The first, smallest member of the alkyne family is $CH\equiv CH$, see Figure 19.2(c). Its IUPAC name is ethyne, but it is commonly called "acetylene." Oxygen and acetylene gases are used for oxyacetylene welding. When oxygen and acetylene gases are ignited, the temperature reaches 3000 °C. The second member of the alkyne family is $CH\equiv C-CH_3$. Its IUPAC name is propyne, but it is commonly referred to as "methyl acetylene."

The next member of the alkyne family, C_4H_6, is butyne. Butyne has two structural isomers because there are two possible positions for the triple bond. The triple bond can be between the first and second carbon atoms, $CH\equiv C-CH_2-CH_3$, or between the second and third, $CH_3{-}C \equiv C{-}CH_3$. The name of the first isomer is 1-butyne and the name of the second is 2-butyne. Table 19.4 lists the first few members of the alkyne family.

TABLE 19.4 ALKYNES

IUPAC NAME	COMMON NAME	MOLECULAR FORMULA	STRUCTURAL FORMULA	BOILING POINT
ethyne	"acetylene"	C_2H_2	$CH{\equiv}CH$	–84 °C
propyne	"methyl acetylene"	C_3H_4	$CH_3{-}C{\equiv}CH$	–23 °C
1–butyne	"ethyl acetylene"	C_4H_6	$CH_3{-}CH_2{-}C{\equiv}CH$	8 °C
2–butyne	"dimethyl acetylene"	C_4H_6	$CH_3{-}C{\equiv}C{-}CH_3$	27 °C

Nomenclature of Alkynes

Like alkenes, alkynes are named after the corresponding alkane. That is, the name of an alkyne is formed by changing the suffix *-ane* to the suffix *-yne.* For example, ethane is changed to ethyne. The rules for naming complex alkynes are similar to the general rules for naming alkenes. The following example exercise illustrates these nomenclature rules:

Example Exercise 19.3 Nomenclature of Alkynes

Write the IUPAC name for the following structure:

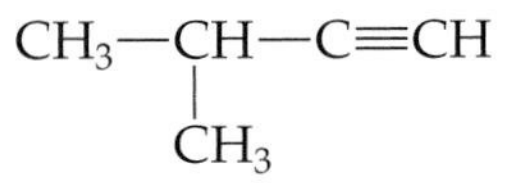

$$\begin{array}{l} CH_3{-}\underset{\displaystyle CH_3}{\underset{|}{CH}}{-}C{\equiv}CH \end{array}$$

Solution

We can follow the systematic rules for naming alkenes, but we substitute the suffix *-yne* for alkynes.

1. The longest continuous carbon chain containing the triple bond has four carbons, and so the compound is a *butyne.*
2. To assign the triple bond the lowest value, we start numbering from the right. This compound is *1-butyne.*

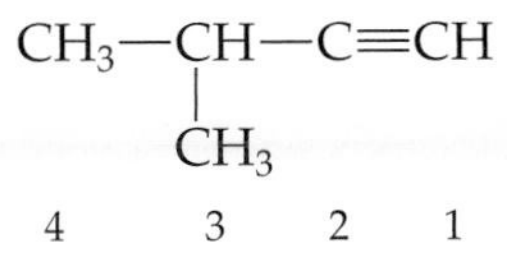

$$\begin{array}{l} CH_3{-}\underset{\displaystyle CH_3}{\underset{|}{CH}}{-}C{\equiv}CH \\ \;\;4 \qquad\quad 3 \qquad 2 \quad\; 1 \end{array}$$

3. Since there is a methyl group on the third carbon, the complete name of the compound is *3-methyl-1-butyne.*

Practice Exercise

Draw the condensed structural formula for 1-pentyne.

Answer: $CH \equiv C{-}CH_2{-}CH_2{-}CH_3$

Concept Exercise

Which of the following compounds is an example of an alkyne: decane ($C_{10}H_{22}$), decene ($C_{10}H_{20}$), decyne ($C_{10}H_{18}$), or naphthalene ($C_{10}H_8$)?

Answer: See Appendix G

Reactions of Alkenes and Alkynes

Like alkanes, alkenes and alkynes also burn with oxygen to provide energy. For example, acetylene, C_2H_2, and oxygen gases give the following combustion reaction in oxyacetylene welding:

$$2\,C_2H_2(g) + 5\,O_2(g) \xrightarrow{\text{spark}} 4\,CO_2(g) + 2\,H_2O(g)$$

Alkenes and alkynes are more reactive than alkanes because of their double and triple bonds. Unlike an alkane, an alkene or alkyne has unsaturated bonds that can readily undergo an **addition reaction**. That is, an alkene or alkyne can add molecules to its double bond or triple bond. When hydrogen molecules are added, the reaction is called a *hydrogenation reaction.* Hydrogenation reactions occur readily at room temperature and atmospheric pressure when a metal catalyst, such as Ni or Pt, is introduced.

Consider the following stepwise addition of hydrogen gas to acetylene. First, ethyne is converted to ethene; then, ethene is converted to ethane.

$$\underset{\text{ethyne}}{CH \equiv CH(g)} + H_2(g) \xrightarrow{Ni} \underset{\text{ethene}}{CH_2 = CH_2(g)}$$

$$\underset{\text{ethene}}{CH_2 = CH_2(g)} + H_2(g) \xrightarrow{Ni} \underset{\text{ethane}}{CH_3{-}CH_3}$$

In addition to adding hydrogen, alkenes and alkynes can react by adding halogen molecules. Such a reaction is called a *halogenation reaction.* For example, alkenes and alkynes can add bromine, and a catalyst is generally not necessary. Consider the following stepwise addition of bromine to ethyne:

$$\underset{\text{ethyne}}{CH \equiv CH(g)} + Br_2(l) \longrightarrow \underset{\text{1,2-dibromoethene}}{CHBr = CHBr(l)}$$

$$\underset{\text{1,2-dibromoethene}}{CHBr = CHBr(l)} + Br_2(l) \longrightarrow \underset{\text{1,1,2,2-tetrabromoethane}}{CHBr_2{-}CHBr_2(l)}$$

Polymerization Reactions of Alkenes

A **polymer** is a giant molecule made up of many small molecules, each of which is called a **monomer**. A polymer can be formed when an alkene molecule reacts with a second molecule, which in turn reacts with a third molecule, and so on. For instance, ethylene, $CH_2 = CH_2$, can react to give a giant molecule called *polyethylene.* We can

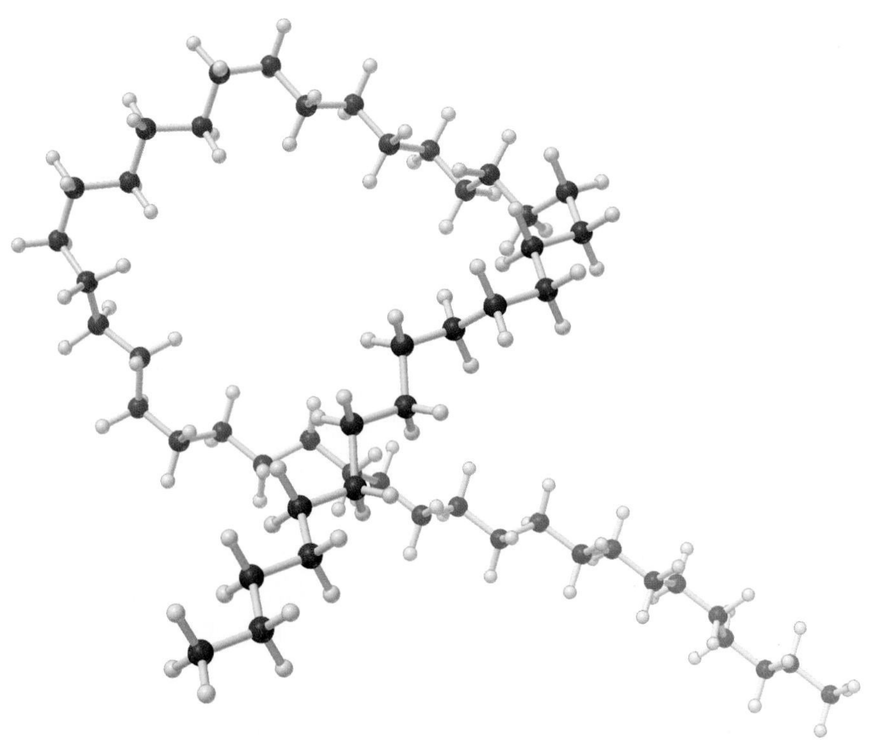

▶ **Polyethylene** A polyethylene chain can contain over a million ethylene molecules.

represent the polymer as $[-CH_2-CH_2-]_n$, where n indicates the number of repeating ethylene monomers. During the polymerization reaction, the double bond in each ethylene molecule "springs open" and joins to another ethylene molecule as follows:

monomer	+	*monomer*	+ .... ⟶	*polymer*
$CH_2{=}CH_2$	+	$CH_2{=}CH_2$	+ .... ⟶	$[-CH_2-CH_2-]_n$
ethylene	+	ethylene	+ .... ⟶	polyethylene

19.4 ARENES

OBJECTIVE

▸ To write names and formulas for simple arenes.

In 1825, the English scientist Michael Faraday (1791–1867) isolated a substance from the fuel oil used in lamps. After analysis, the formula of the compound was determined to be C_6H_6. A few years later, the same compound was isolated from the natural product benzoin, a topical antiseptic. This compound, C_6H_6, was originally called benzin, but the name evolved into benzine, and eventually benzene.

Arene Family of Compounds

As other organic compounds were isolated and identified, chemists observed that those containing a benzene structure often had a fragrant odor. Thus, these compounds were termed aromatic because of their pleasant aroma. Further analysis revealed that the benzene structure and the fragrant odor were not related. The term persisted, however, and today an aromatic hydrocarbon is one that contains the benzene structure. **Arenes** are a family of compounds that are aromatic hydrocarbons and contain a benzene ring.

In 1865, the German chemist August Kekulé (1829–1896) proposed that benzene, C_6H_6, had a cyclic structure with three double bonds. He suggested that six carbon atoms formed a ring with alternating single and double bonds. The structure is called the Kekulé structure of benzene. Since the double bonds are not in fixed positions and shift between carbon atoms, the electrons are said to be delocalized. Because of this phenomenon, and for simplicity, the benzene ring is usually shown by a hexagon enclosing a circle to represent C_6H_6.

▲ **August Kekulé** The stamp commemorates Kekulé's discovery of the cyclic nature of benzene in 1865.

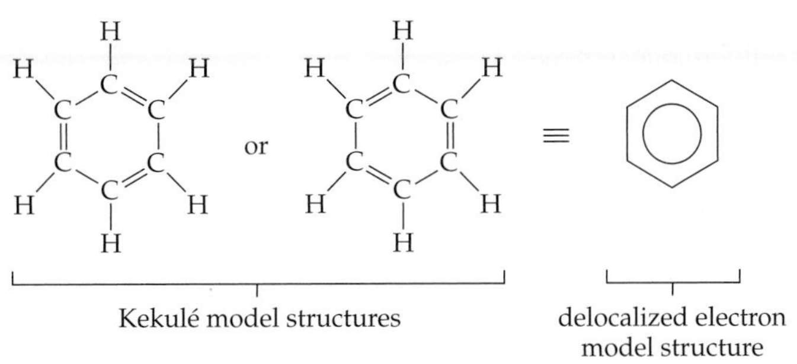

Isomers of Benzene

Using the delocalized electron model of benzene, we can represent the structure of dichlorobenzene as follows:

Cl
Cl

Notice that the two chlorine atoms are next to each other on the benzene ring. When two groups are adjacent on a benzene ring, their positions are indicated by the prefix *ortho*-. If the groups are separated by a carbon, their positions are indicated by the prefix *meta*-. If the groups are opposite each other on the benzene ring, their positions are indicated by the prefix *para*-.

Since *ortho*-, *meta*-, and *para*- compounds have the same molecular formula but are different, they are structural isomers. Recall that isomers have different physical properties. We can illustrate these isomers as follows:

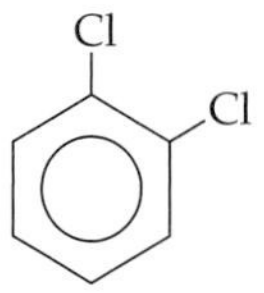

ortho-dichlorobenzene
Mp = –17 °C, Bp = 181 °C

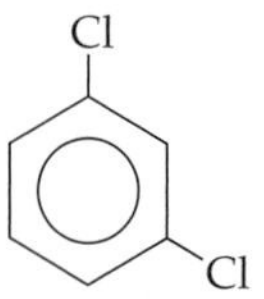

meta-dichlorobenzene
Mp = –25 °C, Bp = 174 °C

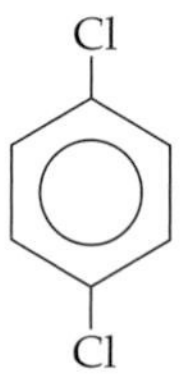

para-dichlorobenzene
Mp = 53 °C, Bp = 175 °C

Example Exercise 19.4 Benzene Isomers

Draw the structural formula for *meta*-dimethylbenzene.

Solution

The compound contains a benzene ring. It also has two $-CH_3$ groups on the benzene ring in the *meta* position.

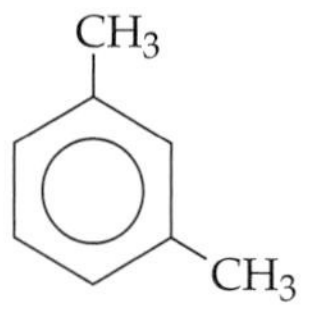

Practice Exercise

Give the systematic IUPAC name for the following:

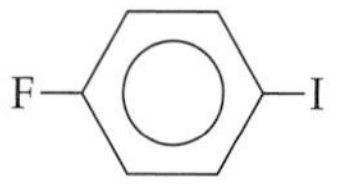

Answer: *para*-fluoroiodobenzene

Concept Exercise

Which of the following compounds is an example of an arene: decane ($C_{10}H_{22}$), decene ($C_{10}H_{20}$), decyne ($C_{10}H_{18}$), or naphthalene ($C_{10}H_8$)?

Answer: See Appendix G

19.5 HYDROCARBON DERIVATIVES

OBJECTIVES

- To understand the concept of an organic functional group.
- To identify the functional group in each of the following: organic halide, alcohol, phenol, ether, and amine.
- To identify the functional group in each of the following: aldehyde, ketone, carboxylic acid, ester, and amide.

In the 1800s, chemists began to realize that there was a huge number of organic compounds. As more compounds were identified and studied, the subject became overwhelming. According to the German chemist Friedrich Wöhler, organic chemistry seemed to be a *"dreadful endless jungle."* Gradually, chemists began to realize that organic compounds could be classified into only a few categories. Each of these categories is referred to as a **class of compounds** and all the compounds in a given class have similar names and properties.

Hydrocarbon Derivatives

To comprehend the vast subject of organic chemistry, chemists have divided a massive amount of information into divisions of organic compounds. The first division separates hydrocarbons and their derivatives. A hydrocarbon contains only hydrogen and carbon. A **hydrocarbon derivative** is derived from a hydrocarbon, but contains other elements, such as oxygen, nitrogen, or a halogen (F, Cl, Br, I).

We will study 10 classes of hydrocarbon derivatives. Our task is simplified by the fact that each member in a class has similar names and properties. The classes are organic halides (R–X), alcohols (R–OH), phenols (Ar–OH), ethers (R–O–R′), amines ($R-NH_2$), aldehydes (R–CHO), ketones (R–COR′), carboxylic acids (R–COOH), esters (R–COOR′), and amides ($R-CONH_2$). The use of R and R′ indicates that the groups may or may not be the same.

The symbol **R** represents an alkyl group and **Ar** represents an aryl group. The designations R and R′ indicate different alkyl groups.

The structural feature in a molecule that characterizes a class of compounds is called the **functional group**. For example, in an alcohol (R–OH), the functional group is –OH attached to a carbon atom. For an ether (R–O–R′), the functional group is the oxygen atom between two carbon atoms.

We can further divide these 10 hydrocarbon derivatives into five classes that have a carbon atom double bonded to an oxygen atom (C=O), and five classes that do not (Figures 19.3 and 19.4). A carbon atom joined to an oxygen atom by a double bond is called a **carbonyl group**. The presence of a carbonyl group gives distinct properties to a given class of compounds, such as amides that have a carbonyl group and amines that do not.

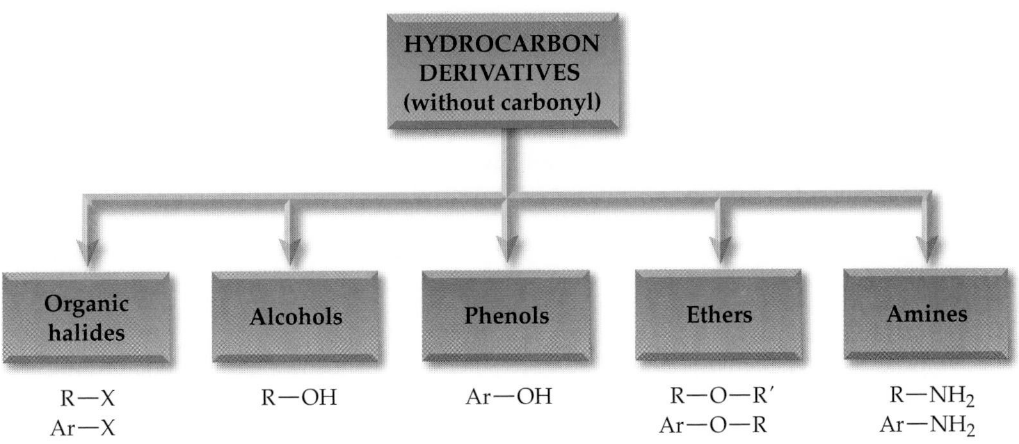

▲ **Figure 19.3 Hydrocarbon Derivatives (without Carbonyl)** For organic halides, X can be F, Cl, Br, or I. For alcohols, R can be any alkyl group that is not an aryl (Ar); for phenols, the –OH group must be attached directly to a benzene ring.

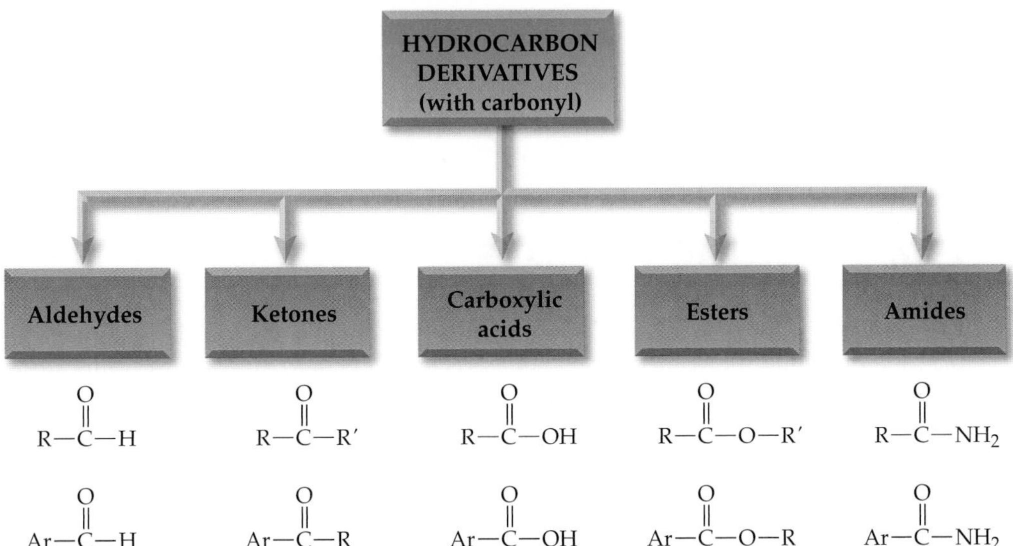

▲ **Figure 19.4 Hydrocarbon Derivatives (with Carbonyl)** Aldehydes, ketones, carboxylic acids, esters, and amides all contain a carbonyl group with different attached groups. The symbols R and Ar can be any alkyl or aryl group. Except in ketones, R can also be a hydrogen atom.

Example Exercise 19.5 Classifying Hydrocarbon Derivatives

Indicate the class of hydrocarbon derivative for each of the following compounds:

(a) CH_3-Br

(b) $CH_3-CH_2-NH_2$

(c) 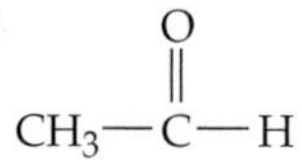 $CH_3-C(=O)-H$

(d) 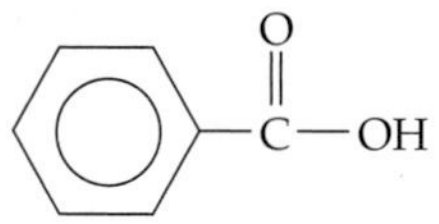 $C_6H_5-C(=O)-OH$

Solution

Let's first identify the functional group in each compound. We can then refer to the general formulas to determine the class of compound.

(a) This hydrocarbon derivative does not have a carbonyl group. Since the functional group is a halogen (–Br),

CH_3-Br is an *organic halide.*

(b) This hydrocarbon derivative does not have a carbonyl group. Since the functional group is an amine ($-NH_2$),

$CH_3-CH_2-NH_2$ is an *amine.*

(c) This hydrocarbon derivative has a carbonyl group. Since the functional group has a carbonyl attached to a H atom,

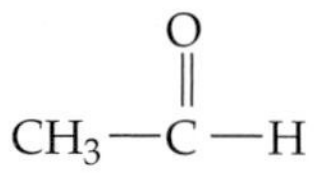

$CH_3-C(=O)-H$ is an *aldehyde.*

(d) This hydrocarbon derivative has a carbonyl group attached to a benzene ring. Since the functional group has a carbonyl attached to an OH,

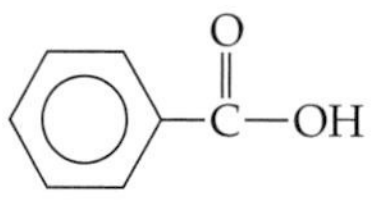

$C_6H_5-C(=O)-OH$ is a *carboxylic acid.*

Practice Exercise

Indicate the class of hydrocarbon derivative for each of the following compounds:

(a) $(CH_3)_2CH-OH$

(b) 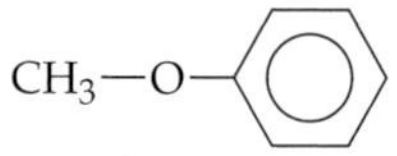 $CH_3-O-C_6H_5$

(c)

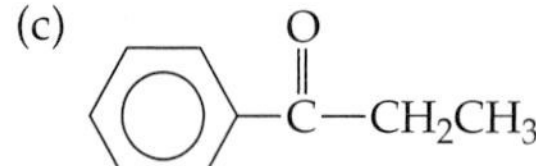

(d) 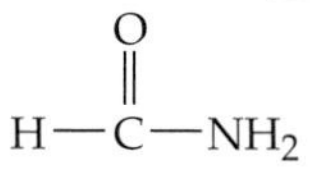 $H-C(=O)-NH_2$

Answers: (a) alcohol; (b) ether; (c) ketone; (d) amide

Concept Exercise

Which classes of hydrocarbon derivatives contain a carbonyl group? Which classes of hydrocarbon derivatives do not contain a carbonyl group?

Answer: See Appendix G

OBJECTIVE

▶ To write names and formulas for simple organic halides.

19.6 ORGANIC HALIDES

If a halogen (F, Cl, Br, I) atom replaces a hydrogen atom in a hydrocarbon, it gives rise to a class of compounds called *organic halides* (Figure 19.5). In this class of compounds, the hydrocarbon group can be alkyl, R–X, or aryl, Ar–X. For example,

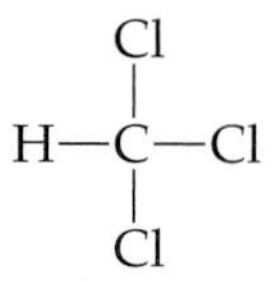

(a) "chloroform" $CHCl_3$

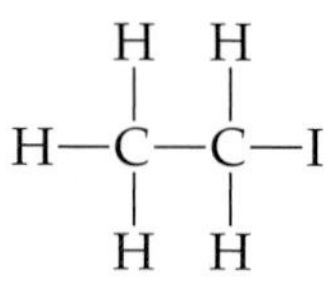

(b) "ethyl iodide" CH_3-CH_2-I

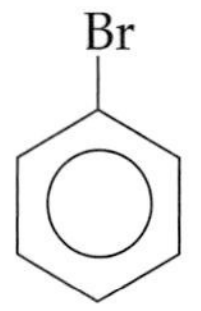

(c) "phenyl bromide" C_6H_5-Br

(a)

(b)

(c)

▲ **Figure 19.5 Molecular Models of Organic Halides** Model structures of (a) "chloroform," $CHCl_3$; (b) "ethyl iodide," C_2H_5I; and (c) "phenyl bromide," C_6H_5Br.

Organic halides are used primarily as industrial and household solvents. Although "carbon tetrachloride," CCl_4, was once widely used for dry cleaning and spot removing, it has been replaced because of its toxicity. Now less toxic solvents, such as "trichloroethane," $C_2H_3Cl_3$, are used in dry cleaning.

"Chloroform," $CHCl_3$, is a common solvent originally used for general anesthesia. When its vapor was found to be harmful to the respiratory system, it was replaced by safer anesthetics such as halothane, $CF_3CHClBr$. "Ethyl chloride,"CH_3CH_2Cl, is used as a mild topical anesthetic. After it is sprayed on the skin, ethyl chloride rapidly evaporates and produces a cold sensation that numbs the skin.

Chlorinated fluorocarbons (CFCs) are organic halides composed of chlorine, fluorine, and carbon. The best known example is Freon–12, CF_2Cl_2, which has been used as an aerosol propellant and a refrigerant gas. Carbon dioxide has replaced Freon-12 in most aerosol cans. Currently, CFCs are being replaced by hydrofluorocarbons (HFCs) in refrigeration and air-conditioning units. This change was mandated after CFCs in the upper atmosphere were found to be responsible for depletion of the ozone layer.

Organic halides are found in many pesticides, including Aldrin, Chlordane, and DDT. In 1972, the Environmental Protection Agency banned DDT because of its threat to wildlife. Before the ban, DDT was responsible for boosting agricultural harvests and reducing insect populations that carried malaria and yellow fever.

According to IUPAC nomenclature, organic halides are named for the halogen attached to the parent alkane. For example, CH_3CH_2Br is named bromoethane. Alternatively, we can designate an acceptable common name for the compound by stating the alkyl group attached to the bromine atom; that is, we can also name CH_3CH_2Br "ethyl bromide."

Most organic halides are nonpolar, although some are slightly polar. In general, organic halides have physical properties similar to those of alkanes. That is, they have low boiling points, are soluble in hydrocarbon solvents, and are insoluble in water. Table 19.5 briefly describes a few simple organic halides.

TABLE 19.5 ORGANIC HALIDES

IUPAC NAME	COMMON NAME	STRUCTURAL FORMULA	BOILING POINT*	SOLUBILITY IN WATER
chloromethane	"methyl chloride"	$CH_3{-}Cl$	–24 °C	insoluble
chloroethane	"ethyl chloride"	$CH_3{-}CH_2{-}Cl$	12 °C	insoluble
1–chloropropane	"propyl chloride"	$CH_3{-}CH_2{-}CH_2{-}Cl$	47 °C	insoluble
2–chloropropane	"isopropyl chloride"	$CH_3{-}CH(Cl){-}CH_3$	36 °C	insoluble
chlorobenzene	"phenyl chloride"	$C_6H_5{-}Cl$	132 °C	insoluble

*Most organic halides with low molecular mass are liquids at room temperature.

OBJECTIVES

- To write names and formulas for simple alcohols.
- To write names and formulas for simple phenols.
- To write names and formulas for simple ethers.

19.7 ALCOHOLS, PHENOLS, AND ETHERS

Alcohols, phenols, and ethers all contain oxygen in addition to hydrogen and carbon. Alcohols and phenols contain an –OH group and, in general, are polar. Ethers, however, contain a C–O–C functional group and are mostly nonpolar.

Alcohol and Phenol Families of Compounds

When an –OH group replaces a hydrogen atom on an alkane, the result is an *alcohol*, R–OH. When an –OH group replaces a hydrogen atom on a benzene ring, the result is a *phenol*, Ar–OH. The –OH group on an alcohol or a phenol is called a **hydroxyl group** (Figure 19.6). For example,

(a) "methyl alcohol" (b) "ethyl alcohol" (c) phenol

Alcohols are everyday substances with many practical uses. The simplest family member, "methyl alcohol," CH_3OH, is often used as an antifreeze, solvent, and in the production of biodiesel fuel. It is referred to as "wood alcohol" because it can be produced from the distillation of wood. "Wood alcohol" has a infamous history because it is found in "moonshine" whiskey. Unlike grain alcohol, small amounts of wood alcohol can cause blindness, and even death, if taken internally.

"Ethyl alcohol," CH_3CH_2OH, is blended with gasoline to give gasohol. In fact, the popular E85 fuel at gas stations is 85% ethanol. It is sometimes called grain alcohol because it is produced by the fermentation of sugars found in various grains such as corn, rye, and barley. "Isopropyl alcohol," $CH_3CH(OH)CH_3$, is common in most homes and is referred to as rubbing alcohol.

Alcohols may have two or more hydroxyl groups in the same molecule. Ethylene glycol, $CH_2(OH)CH_2(OH)$, and glycerol, $CH_2(OH)CH(OH)CH_2(OH)$, are important examples. Ethylene glycol is the main ingredient in permanent automobile antifreeze. Glycerol, also called glycerin, is a sweet syrupy liquid used as a moisturizer in cosmetics and candy.

According to IUPAC nomenclature, the name of an alcohol is derived from the name of the parent alkane. The name is formed by dropping the final *-e* and adding the suffix *-ol.* For example, CH_3OH has an –OH group on a methane molecule; the IUPAC name is methanol, and the common name is "methyl alcohol." In CH_3CH_2OH, the –OH is on an ethane molecule; the systematic name is ethanol, and the common name is "ethyl alcohol."

If a hydroxyl group is attached to a benzene ring, the resulting alcohol belongs to a special class of compounds called *phenols*, Ar–OH. The simplest member of this

(a) (b) (c)

▲ **Figure 19.6 Molecular Models of Alcohols and Phenol** Model structures of (a) "methyl alcohol," CH_3OH; (b) "ethyl alcohol," C_2H_5OH; and (c) phenol, C_6H_5OH.

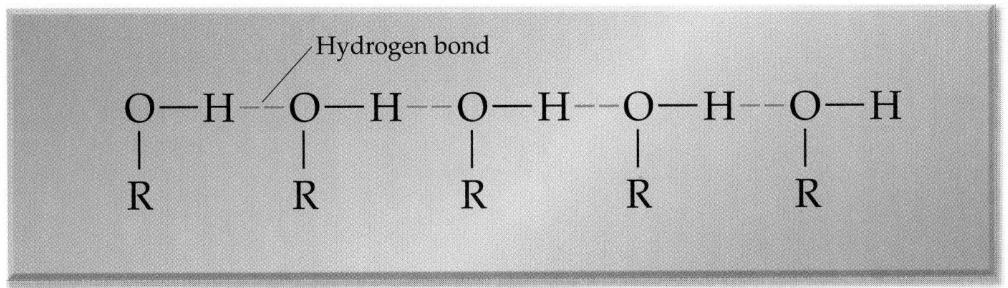

◀ **Figure 19.7 Hydrogen Bonding in Alcohols** Notice that hydrogen bonds connect alcohol molecules in a long chain.

class, phenol, C_6H_5OH, is an important industrial chemical. Phenol is used in the manufacture of plastics and for the preparation of dyes. It is somewhat acidic and is referred to as "carbolic acid." Phenol derivatives have antimicrobial properties and are used in medical and dental offices, and in cosmetics as preservatives.

Most alcohols are quite polar because of the –OH group. Alcohols, like water, can form hydrogen bonds between an oxygen atom and a hydrogen atom in separate molecules. Figure 19.7 depicts hydrogen bonding between alcohol molecules.

As a result of hydrogen bonding, the physical properties of alcohols are quite different from those of the parent alkane. Alcohols have higher boiling points than alkanes, and are soluble, or partially soluble, in polar solvents such as water. Table 19.6 lists a few simple alcohols as well as phenol.

TABLE 19.6 ALCOHOLS AND PHENOL

IUPAC NAME	COMMON NAME	STRUCTURAL FORMULA	BOILING POINT*	SOLUBILITY IN WATER
methanol	"methyl alcohol"	$CH_3{-}OH$	65 °C	soluble
ethanol	"ethyl alcohol"	$CH_3{-}CH_2{-}OH$	78 °C	soluble
1-propanol	"propyl alcohol"	$CH_3{-}CH_2{-}CH_2{-}OH$	97 °C	soluble
2-propanol	"isopropyl alcohol"	$CH_3{-}CH(OH){-}CH_3$	82 °C	soluble
phenol	"carbolic acid"	$C_6H_5{-}OH$	182 °C	~7%

*Most alcohols with low molecular mass are liquids at room temperature; phenol is a solid.

Ether Family of Compounds

If two alkyl groups are attached to an oxygen atom, we have a class of compounds called *ethers*, R–O–R′. Commercially, the most significant member of this class is diethyl ether, $CH_3CH_2{-}O{-}CH_2CH_3$. It is usually referred to as "ethyl ether," or simply "ether."

"Ethyl ether" is a common laboratory solvent for organic reactions. It was once used as a general surgical anesthetic, but has been replaced because it is highly flammable and causes patients to become nauseous. "Ethyl ether" is synthesized by dehydrating "ethyl alcohol." In the reaction sulfuric acid is used to split out a molecule of water from two molecules of alcohol as follows:

$$\underset{\text{"ethyl alcohol"}}{CH_3CH_2{-}\mathbf{OH}} + \underset{\text{"ethyl alcohol"}}{\mathbf{H}{-}OCH_2CH_3} \xrightarrow{H_2SO_4} \underset{\text{"diethyl ether"}}{CH_3CH_2{-}O{-}CH_2CH_3} + \mathbf{H_2O}$$

In the ether class of compounds, the attached groups may be alkyl or aryl. That is, an ether may be R–O–Ar, or Ar–O–Ar′, as well as R–O–R′ (Figure 19.8). The IUPAC names of ethers are not often used, and we will refer to an ether by indicating the two attached groups. For example, the name of $CH_3{-}O{-}CH_2CH_3$ is "methyl ethyl ether." Similarly, the name of $CH_3{-}O{-}C_6H_5$ is "methyl phenyl ether." Figure 19.8 illustrates the model structure of selected ethers.

In general, the physical properties of ethers lie between those of nonpolar hydrocarbons and those of polar alcohols of comparable molecular mass. Ethers are relatively nonpolar and do not form hydrogen bonds between molecules. Table 19.7 briefly describes a few simple ethers.

(a)

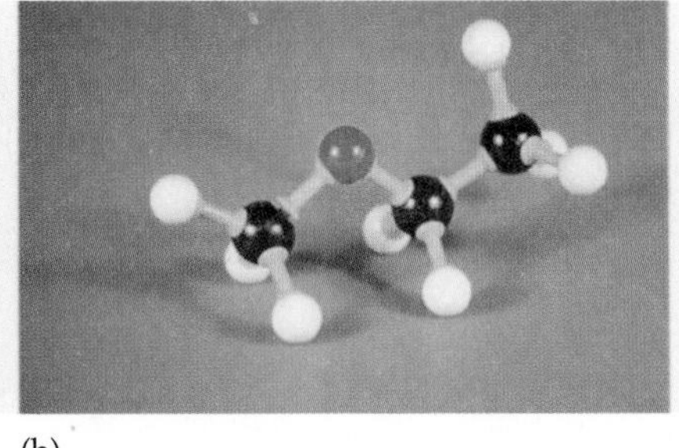

(b)

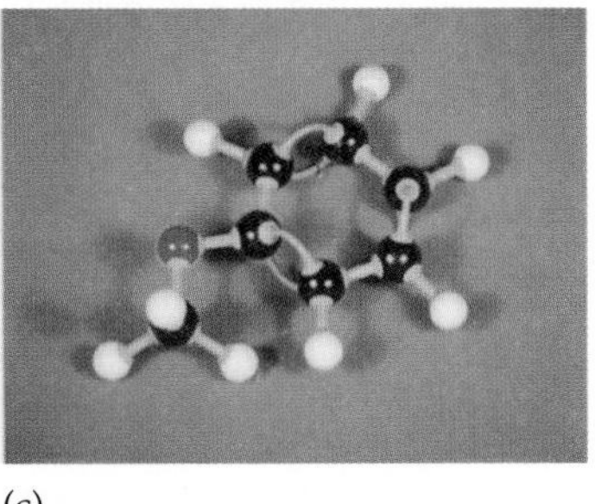

(c)

▲ **Figure 19.8 Molecular Models of Ethers** Model structures of (a) "diethyl ether," $C_2H_5{-}O{-}C_2H_5$; (b) "methyl ethyl ether," $CH_3{-}O{-}C_2H_5$; and (c) "methyl phenyl ether," $CH_3{-}O{-}C_6H_5$.

TABLE 19.7 ETHERS

IUPAC NAME	COMMON NAME	STRUCTURAL FORMULA	BOILING POINT*	SOLUBILITY IN WATER
—	"dimethyl ether"	$CH_3{-}O{-}CH_3$	–25 °C	soluble
—	"methyl ethyl ether"	$CH_3{-}O{-}CH_2CH_3$	8 °C	soluble
—	"diethyl ether"	$CH_3CH_2{-}O{-}CH_2CH_3$	35 °C	~7%
—	"diphenyl ether"	$C_6H_5{-}O{-}C_6H_5$	259 °C	insoluble

*Most ethers with low molecular mass are liquids at room temperature; "diphenyl ether" is a solid.

OBJECTIVE

▸ To write names and formulas for simple amines.

19.8 AMINES

The amines often have an unpleasant odor. The simplest amine, CH_3NH_2, smells like ammonia. As the carbon chain increases, the odor becomes progressively unpleasant, and $CH_3CH_2CH_2CH_2NH_2$ has an offensive smell. The amines putrescine and cadaverine have a revolting smell, and are found in decaying flesh. Other examples of amines include the dental anesthetic novocaine, the hormone epinephrine, nicotine, amphetamine, and cocaine.

Amine Family of Compounds

If an alkyl group (R–) replaces a hydrogen in ammonia, NH_3, the resulting $R{-}NH_2$ is called an *amine* (Figure 19.9). Further substitution may occur to give amines with two or three attached alkyl groups, that is, $R_2{-}NH$ or $R_3{-}N$. In this class of compounds, the hydrocarbon group can be alkyl or aryl. For example,

$CH_3{-}CH_2{-}NH_2$

(a) "ethylamine"

$CH_3{-}NH{-}CH_3$

(b) "dimethylamine"

$C_6H_5{-}NH_2$

(c) "phenylamine"

The IUPAC names of amines are often not used; thus, we will refer to the amines by their common name. For example, the common name of CH_3NH_2 is "methylamine" and the name of $CH_3CH_2NH_2$ is "ethylamine." Similarly, we can refer to $C_6H_5NH_2$ as "phenylamine."

Most amines are polar because of the $-NH_2$ group. In fact, amines are similar to alcohols in that they can form intermolecular hydrogen bonds. As we might expect, the physical properties of amines and alcohols are comparable. That is, they both have high boiling points and are soluble in water. Table 19.8 briefly describes a few simple amines.

(a)

(b)

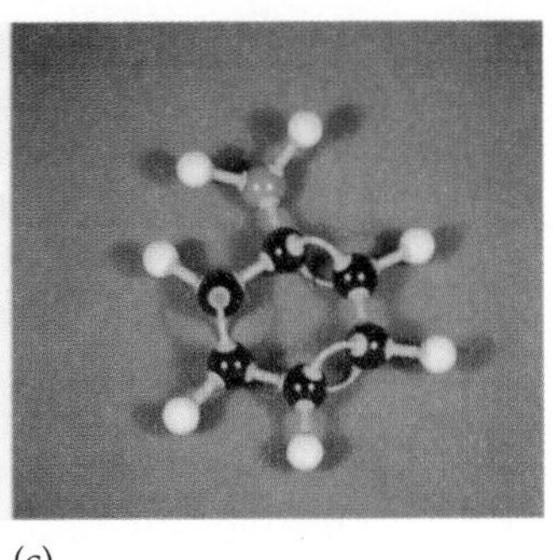
(c)

▲ **Figure 19.9 Molecular Models of Amines** Model structures of (a) "ethylamine," $C_2H_5NH_2$; (b) "dimethylamine," $(CH_3)_2NH$; and (c) "phenylamine," $C_6H_5NH_2$.

TABLE 19.8 AMINES

IUPAC NAME	COMMON NAME	STRUCTURAL FORMULA	BOILING POINT*	SOLUBILITY IN WATER
—	"methylamine"	$CH_3{-}NH_2$	–7 °C	soluble
—	"ethylamine"	$CH_3{-}CH_2{-}NH_2$	17 °C	soluble
—	"propylamine"	$CH_3{-}CH_2{-}CH_2{-}NH_2$	48 °C	soluble
—	"isopropylamine"	$CH_3{-}CH(NH_2){-}CH_3$	33 °C	soluble
—	"phenylamine"	$C_6H_5{-}NH_2$	185 °C	~5%

*Most amines with low molecular mass are liquids at room temperature.

19.9 ALDEHYDES AND KETONES

OBJECTIVES

- To write names and formulas for simple aldehydes.
- To write names and formulas for simple ketones.

Many aldehydes and ketones have an appealing taste and a fragrant odor. For this reason they are frequently used as flavorings in food and candy and as fragrance in inhalants and perfumes. They are also used as solvents and for the manufacture of other organic compounds.

Aldehyde Family of Compounds

Aldehydes contain a carbonyl group (C=O) and, in general, are polar. In *aldehydes*, RCHO, the carbonyl group is attached to a hydrogen atom and either an alkyl or an aryl group (Figure 19.10). For example,

$$\mathrm{H{-}\overset{\overset{\displaystyle O}{\|}}{C}{-}H}$$

(a) "formaldehyde"

$$\mathrm{CH_3{-}\overset{\overset{\displaystyle O}{\|}}{C}{-}H}$$

(b) "acetaldehyde"

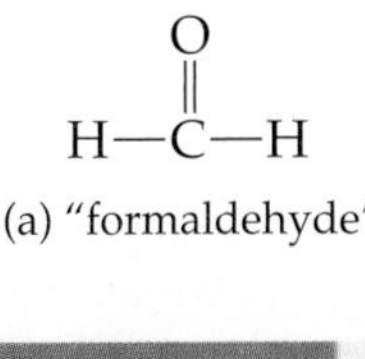

(c) benzaldehyde

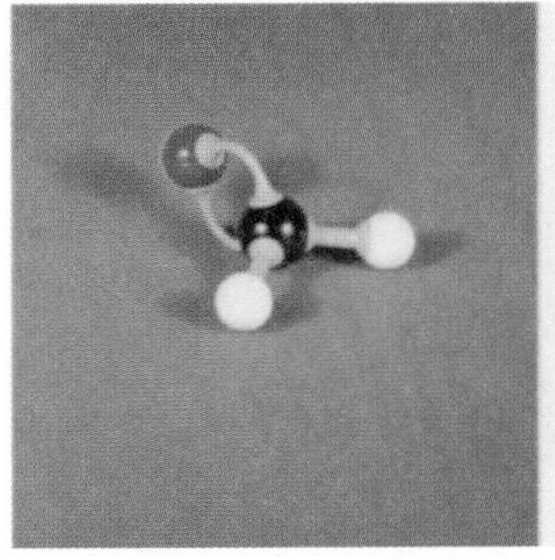
(a)

(b)

(c)

◄ **Figure 19.10 Molecular Models of Aldehydes** Model structures of (a) "formaldehyde," HCHO; (b) "acetaldehyde," CH_3CHO; and (c) benzaldehyde, C_6H_5CHO.

The simplest aldehyde, "formaldehyde," HCHO, is one of the top 50 industrial chemicals. Formaldehyde is perhaps best known as the solution used to preserve specimens in biology classes. The preservative is actually formalin, an aqueous solution of formaldehyde. Formaldehyde is manufactured by the oxidation of methanol using a variety of metal catalysts.

According to IUPAC nomenclature, the name of an aldehyde is derived from the name of the parent alkane. The name is formed by dropping the final *-e* and adding the suffix *-al*. For example, the IUPAC name for HCHO is methanal, and the name of CH_3CHO is ethanal. We can also refer to CH_3CHO by its common name "acetaldehyde."

Aldehydes with low molecular mass are slightly polar because of the carbonyl group. They are not as polar as alcohols and do not form hydrogen bonds. Accordingly, the physical properties of aldehydes fall between those of polar alcohols and nonpolar hydrocarbons. In general, aldehydes are less soluble in water and have lower boiling points than similar alcohols. Table 19.9 lists a few simple aldehydes.

TABLE 19.9 ALDEHYDES

IUPAC NAME	COMMON NAME	STRUCTURAL FORMULA	BOILING POINT*	SOLUBILITY IN WATER
methanal	"formaldehyde"	H–CHO	–21 °C	soluble
ethanal	"acetaldehyde"	CH_3-CHO	21 °C	soluble
propanal	"propionaldehyde"	CH_3-CH_2-CHO	49 °C	soluble
benzaldehyde	"benzoic aldehyde"	C_6H_5-CHO	179 °C	~1%

*Most aldehydes with low molecular mass are liquids at room temperature.

Ketone Family of Compounds

Ketones contain a carbonyl group (C=O) and, in general, are polar. In *ketones*, RCOR′, either an alkyl group or an aryl group is attached on each side of the carbonyl group (Figure 19.11). For example,

$$CH_3-\overset{\overset{\displaystyle O}{\|}}{C}-CH_3 \qquad CH_3-\overset{\overset{\displaystyle O}{\|}}{C}-CH_2-CH_3 \qquad CH_3-\overset{\overset{\displaystyle O}{\|}}{C}-C_6H_5$$

(a) "dimethyl ketone" or "acetone"

(b) "methyl ethyl ketone" or butanone

(c) "methyl phenyl ketone" or acetophenone

"Acetone," CH_3COCH_3, is the simplest ketone and is one of the top 50 industrial chemicals. It evaporates rapidly, and its vapor is quite flammable. Although acetone is soluble in water, it can dissolve a wide range of polar and nonpolar compounds. "Methyl ethyl ketone," $CH_3COCH_2CH_3$, is a solvent in the paint industry. Ketones are produced by the oxidation of alcohols. For example, large quantities of acetone are obtained from the oxidation of isopropyl alcohol.

(a) (b) (c)

▲ **Figure 19.11 Molecular Models of Ketones** Model structures of (a) "acetone," CH_3COCH_3; (b) butanone, $CH_3COC_2H_5$; and acetophenone, $CH_3COC_6H_5$.

According to IUPAC nomenclature, the name of a ketone is derived from the name of the parent alkane. The name is formed by dropping the final *-e* and adding the suffix *-one*. For example, the IUPAC name for CH_3COCH_3 is propanone, but propanone is also referred to by its common names, "dimethyl ketone" or "acetone." The IUPAC name for $CH_3COCH_2CH_3$ is butanone, and it is referred to by its common name, "methyl ethyl ketone." The next member of the series is pentanone. However, the carbonyl group may be on the second carbon, $CH_3COCH_2CH_2CH_3$, or the third carbon, $CH_3CH_2COCH_2CH_3$, of the parent chain. Thus, there are two isomers that are named 2–pentanone and 3–pentanone, respectively.

Ketones with low molecular mass are slightly polar because of the carbonyl group. They are not as polar as alcohols and do not form hydrogen bonds. In general, ketones are less soluble in water and have lower boiling points than similar alcohols. Table 19.10 lists three of the most common ketones.

TABLE 19.10 KETONES

IUPAC NAME	COMMON NAME	STRUCTURAL FORMULA	BOILING POINT*	SOLUBILITY IN WATER
propanone	"acetone" or "dimethyl ketone"	$CH_3-CO-CH_3$	56 °C	soluble
butanone	"methyl ethyl ketone"	$CH_3-CO-CH_2CH_3$	80 °C	soluble
acetophenone	"methyl phenyl ketone"	$CH_3-CO-C_6H_5$	202 °C	~1%

*Most ketones with low molecular mass are liquids at room temperature; acetophenone melts at 20 °C.

19.10 CARBOXYLIC ACIDS, ESTERS, AND AMIDES

OBJECTIVES

- To write names and formulas for simple carboxylic acids.
- To write names and formulas for simple esters.
- To identify the parent alcohol and acid in an ester.
- To write names and formulas for simple amides.

Carboxylic acids, esters, and amides are similar in that members of each class of compound contain a carbonyl group (C=O) and, in general, are polar. Carboxylic acids undergo many reactions, some of which produce esters and amides. Thus, esters and amides are derived from carboxylic acids.

Carboxylic Acid Family of Compounds

In a *carboxylic acid*, RCOOH, we find a carbonyl group and a hydroxyl group bonded together to form a **carboxyl group**, –COOH (Figure 19.12). In this class of compounds, the hydrocarbon group can be an alkyl or an aryl. For example,

$$\mathrm{H{-}\overset{\overset{\large O}{\|}}{C}{-}OH} \qquad \mathrm{CH_3{-}\overset{\overset{\large O}{\|}}{C}{-}OH} \qquad \mathrm{C_6H_5{-}\overset{\overset{\large O}{\|}}{C}{-}OH}$$

(a) "formic acid" (b) "acetic acid" (c) benzoic acid

The simplest carboxylic acid is "formic acid," HCOOH. Formic acid was first extracted from ants. Its name, in fact, is derived from the Latin *formica*, meaning "ant." Formic acid is responsible for the stinging sensation of a red ant sting. It is used industrially to tan leather and to coagulate rubber latex.

(a)

(b)

(c)

▲ **Figure 19.12 Molecular Models of Carboxylic Acids** Model structures of (a) "formic acid," HCOOH; (b) "acetic acid," CH_3COOH; and (c) benzoic acid, C_6H_5COOH.

"Acetic acid" was first isolated from vinegar, and its name is derived from the Latin *acetum,* meaning vinegar. Acetic acid gives vinegar its sour taste and is the most common carboxylic acid. It is used for manufacturing other chemicals and for preparing foods such as pickles and salad dressing.

According to IUPAC nomenclature, the name of a carboxylic acid is derived from the name of the parent alkane. The name is formed by dropping the final *-e* and adding the suffix *-oic acid.* For example, the IUPAC name for HCOOH is methanoic acid, and for CH_3COOH is ethanoic acid. We can also refer to CH_3COOH by its common name, "acetic acid."

Inasmuch as carboxylic acids have a polar –COOH group, they are capable of forming intermolecular hydrogen bonds. The properties of carboxylic acids are similar to those of alcohols, and they have high boiling points. Since the carboxyl group is quite polar, we would predict that carboxylic acids are usually soluble in water. When we refer to Table 19.11, we see that our predictions are correct for low molecular mass carboxylic acids.

TABLE 19.11 CARBOXYLIC ACIDS

IUPAC NAME	COMMON NAME	STRUCTURAL FORMULA	BOILING POINT*	SOLUBILITY IN WATER
methanoic acid	"formic acid"	H–COOH	101 °C	soluble
ethanoic acid	"acetic acid"	CH_3-COOH	118 °C	soluble
propanoic acid	"propionic acid"	CH_3-CH_2-COOH	141 °C	soluble
benzoic acid	—	C_6H_5-COOH	249 °C	~1%

*Most carboxylic acids with low molecular mass are liquids at room temperature; benzoic acid is a solid.

Ester Family of Compounds

In *esters,* RCOOR′, the carbonyl is attached to an –R and to an –OR′ group (Figure 19.13). In this class of compounds, the hydrocarbon group can be an alkyl or an aryl. For example,

$$H-\overset{O}{\overset{\|}{C}}-O-CH_2CH_3 \qquad CH_3-\overset{O}{\overset{\|}{C}}-O-C_6H_5 \qquad C_6H_5-\overset{O}{\overset{\|}{C}}-O-CH_3$$

(a) "ethyl formate" (b) "phenyl acetate" (c) methyl benzoate

Esters typically have a pleasant fruity odor. For example, "butyl acetate" has the aroma of banana, "ethyl butyrate" has the smell of pineapple, and "ethyl formate" has the taste of rum.

An ester is produced by the reaction of a carboxylic acid and an alcohol. In the presence of sulfuric acid catalyst, the reaction takes place by splitting out a molecule of water. For example, "acetic acid" and "ethyl alcohol" react to give "ethyl acetate" and water.

$$CH_3-\overset{O}{\overset{\|}{C}}-\mathbf{OH} + \mathbf{H}O-CH_2CH_3 \xrightarrow{H_2SO_4} CH_3-\overset{O}{\overset{\|}{C}}-O-CH_2CH_3 + \mathbf{H_2O}$$

ethanoic acid "acetic acid" — ethanol "ethyl alcohol" — ethyl ethanoate "ethyl acetate"

▲ **Ester Fragrance** The aroma of banana is "butyl acetate," that of apple is "methyl butyrate," and that of orange is "octyl acetate."

According to IUPAC nomenclature, the name of an ester is formed from the names of the parent alcohol and carboxylic acid and the suffix *-oate.* For example, the IUPAC name for $HCOOCH_2CH_3$ is formed from the parent alcohol (ethanol) and the parent acid (methanoic acid). After adding the suffix *-oate,* the resulting name is ethyl methanoate. The common name for the ester $HCOOCH_2CH_3$ is "ethyl formate."

Esters are slightly polar because of the carbonyl group. They are not as polar as alcohols and do not form hydrogen bonds. Consequently, their physical properties fall

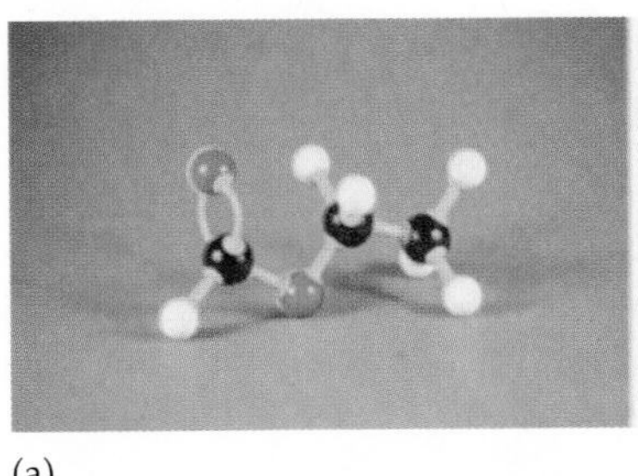
(a)

(b)

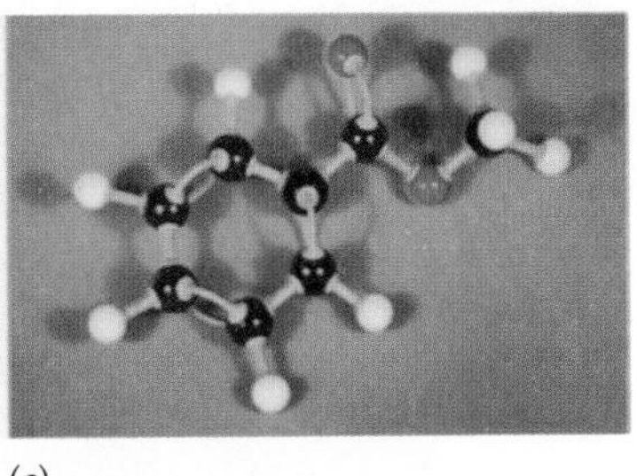
(c)

▲ **Figure 19.13 Molecular Models of Esters** Model structures of (a) "ethyl formate," $HCOOC_2H_5$; (b) "phenyl acetate," $CH_3COOC_6H_5$; and (c) methyl benzoate, $C_6H_5COOCH_3$.

between those of polar alcohols and those of nonpolar hydrocarbons. In general, esters are not soluble in water and have lower boiling points than similar alcohols. Table 19.12 describes a few common esters.

TABLE 19.12 ESTERS

IUPAC NAME	COMMON NAME	STRUCTURAL FORMULA	BOILING POINT	SOLUBILITY IN WATER
methyl methanoate	"methyl formate"	$H{-}CO{-}OCH_3$	32 °C	soluble
methyl ethanoate	"methyl acetate"	$CH_3{-}CO{-}OCH_3$	57 °C	soluble
ethyl ethanoate	"ethyl acetate"	$CH_3{-}CO{-}OCH_2CH_3$	77 °C	soluble
methyl benzoate	—	$C_6H_5{-}CO{-}OCH_3$	200 °C	~1%

* Most esters with low molecular mass are liquids at room temperature.

Amide Family of Compounds

In *amides*, $RCONH_2$, the carbonyl is attached to an –R group and to an $-NH_2$ group (Figure 19.14). In this class of compounds, the hydrocarbon group can be an alkyl or an aryl. For example,

$$\underset{\text{(a) "formamide"}}{H{-}\overset{\overset{\displaystyle O}{\|}}{C}{-}NH_2} \qquad \underset{\text{(b) "acetamide"}}{CH_3{-}\overset{\overset{\displaystyle O}{\|}}{C}{-}NH_2} \qquad \underset{\text{(c) benzamide}}{C_6H_5{-}\overset{\overset{\displaystyle O}{\|}}{C}{-}NH_2}$$

An amide is synthesized from a carboxylic acid and ammonia, NH_3. In the presence of heat, a molecule of water is lost as the amide is formed. For example, "acetic acid" and ammonia react to split out water and give "acetamide."

$$\underset{\substack{\text{ethanoic acid}\\ \text{"acetic acid"}}}{CH_3{-}\overset{\overset{\displaystyle O}{\|}}{C}{-}\mathbf{OH}} + \underset{\text{ammonia}}{\mathbf{H}{-}NH_2} \xrightarrow{\Delta} \underset{\substack{\text{ethanamide}\\ \text{"acetamide"}}}{CH_3{-}\overset{\overset{\displaystyle O}{\|}}{C}{-}NH_2} + \mathbf{H_2O}$$

According to IUPAC nomenclature, the name of an amide compound ends in the suffix *-amide.* The name of an amide is formed by dropping *-oic acid* from the name of the parent acid and adding the suffix *-amide.* For example, the amide derivative of ethanoic acid is CH3CONH_2; it is named ethanamide. The amide derivative of propanoic acid is $CH_3CH_2CONH_2$; it is named propanamide.

Inasmuch as amides have a polar $-CONH_2$ group, they are capable of forming intermolecular hydrogen bonds. The properties of amides are similar to those of carboxylic acids, and amides are usually soluble in water. If we refer to Table 19.13, we see that amides are solids at room temperature.

(a)

(b)

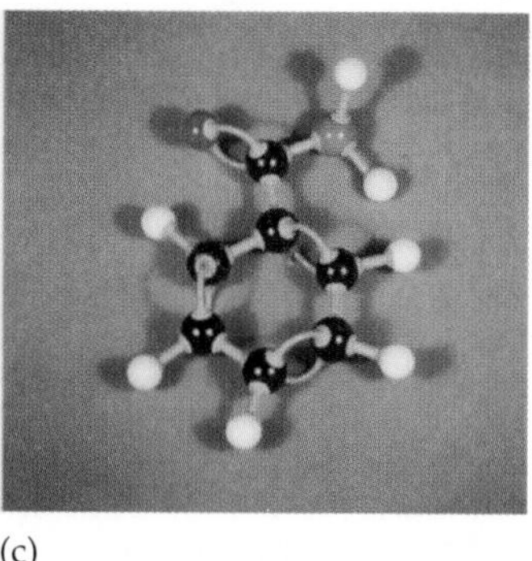
(c)

▲ **Figure 19.14 Molecular Models of Amides** Model structures of (a) "formamide," $HCONH_2$; (b) "acetamide," CH_3CONH_2; and (c) benzamide, $C_6H_5CONH_2$.

TABLE 19.13 AMIDES

IUPAC NAME	COMMON NAME	STRUCTURAL FORMULA	MELTING POINT*	SOLUBILITY IN WATER
methanamide	"formamide"	$H-CONH_2$	3 °C	soluble
ethanamide	"acetamide"	CH_3-CONH_2	82 °C	soluble
propanamide	"propionamide"	$CH_3CH_2-CONH_2$	79 °C	soluble
benzamide	"benzoic acid amide"	$C_6H_5-CONH_2$	132 °C	~1%

*Most amides are solids at room temperature; methanamide is a liquid.

CHEMISTRY CONNECTION Forensic Chemistry

Q: What are the duties of a forensic chemist?

The *CSI: Crime Scene Investigation* program depicts forensic chemistry and is one of the most popular shows on television. Interestingly, the show offers realistic insight into the work of forensic chemists who help to solve crimes.

A forensic chemist, trained in chemistry and biology, performs analyses to identify materials associated with crime scene evidence. A forensic chemist can determine the composition of the material evidence, and identify its source by matching samples. The evidence may include hair samples, paint chips, glass fragments, or blood evidence. The reliance on DNA analysis is making a knowledge of genetics increasingly important to a forensic chemist. Modern "crime-solvers" employ a wide range of techniques using both traditional "wet chemistry" procedures, and high-tech instrumental methods of analysis.

In the forensic laboratory, chemists measure blood alcohol content. They also microscopically examine and identify blood or tissue matter. If police investigators suspect drugs are involved, forensic chemists examine the materials taken from a crime scene to confirm the presence of illegal narcotics. In a drug raid where a controlled substance is confiscated, the substance must be analyzed and identified by a chemist using known reference samples. If fiber evidence is collected, a forensic chemist attempts to identify the source by comparing the fibers against a catalog of known fibers.

Forensic chemists may be called upon to give courtroom testimony and impartially explain the nature of the physical evidence and how it was analyzed. This is a challenging task as the scientific methods are often sophisticated and the forensic chemist must convey the analyses in layman's terms.

Most forensic chemists work in government jobs associated with a federal, state, or local police department, or a medical examiner's office. However, there are some private labs that carry out forensic analyses. Finding a job in the field of forensics is promising and two-year, four-year, and private schools are currently offering programs leading to a degree in the field.

▼ A forensic chemist obtains the DNA sequence (...GTAGAAGGTTCA...) for a sample.

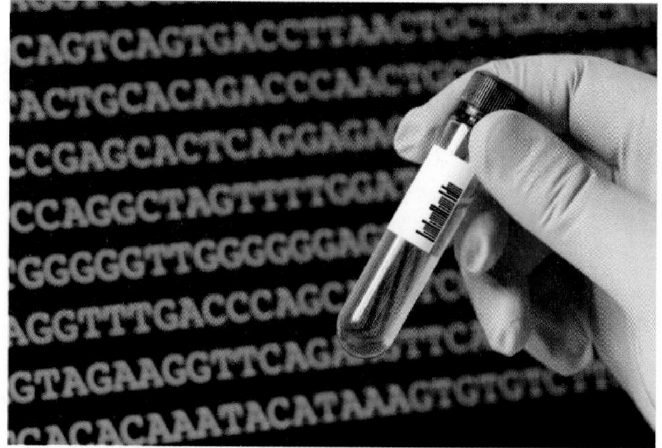

A: A forensic chemist analyzes crime scene evidence using chemical and biological techniques, and often presents their analyses at court trials.

Summary

Section 19.1 **Organic chemistry** is the study of carbon compounds. A **hydrocarbon** is a carbon and hydrogen compound. **Saturated hydrocarbons** have single bonds and compose a class of compounds called alkanes. **Unsaturated hydrocarbons** have either a double bond and compose the family alkenes, or a triple bond and compose the family alkynes. **Aromatic hydrocarbons** typically contain a benzene ring and compose the family arenes.

Section 19.2 **Alkanes** have single bonds and the general molecular formula C_nH_{2n+2}. The **structural formula** shows the arrangement of atoms in a molecule. Molecules having the same molecular formula, but a different structural formula, are called **isomers**. The name of each member of the alkane family ends in the suffix *-ane*. If a H atom is removed from an alkane, an **alkyl group** results. If a H atom is removed from an aromatic hydrocarbon, an **aryl group** results. If a H atom is removed from a benzene molecule, a **phenyl group** results. When an alkane burns with oxygen in air, it is said to undergo a **combustion reaction**.

Section 19.3 **Alkenes** have at least one carbon–carbon double bond and the general molecular formula C_nH_{2n}. **Alkynes** have at least one carbon–carbon triple bond and the general molecular formula C_nH_{2n-2}. The name of each member of the alkene family ends in the suffix *-ene*. The names of alkynes end in the suffix *-yne*. Alkenes and alkynes can undergo an **addition reaction**. That is, hydrogen and the halogens react with alkenes and alkynes by adding a molecule of H_2, Cl_2, Br_2, or I_2 to an unsaturated bond. An alkene can react with itself to form a long chain molecule, which is called a **polymer**, and each repeating molecular unit is called a **monomer**.

Section 19.4 **Arenes** contain a benzene ring. If there are two groups attached to the ring, there are three possible isomers: *ortho-, meta-, and para-*. For example, if two bromine atoms are in adjacent positions on the ring, the isomer is called *ortho*-dibromobenzene. If the bromine atoms are separated by a carbon on the ring, the isomer is called *meta*-dibromobenzene. If the bromine atoms are on opposite sides of the ring, the isomer is called *para*-dibromobenzene.

Section 19.5 A **hydrocarbon derivative** contains carbon, hydrogen, and an additional element such as O, N, or a halogen. The hydrocarbon derivatives can be divided into families, each of which is called a **class of compounds**. There are 10 classes of hydrocarbon derivatives, 5 of which have a **carbonyl group** (C=O) and 5 of which do not. The atom or group of atoms that characterizes a class of compounds is referred to as a **functional group**.

Section 19.6 *Organic halides* are compounds containing hydrogen, carbon, and a halogen. They are generally nonpolar and insoluble in water. The IUPAC systematic name for an organic halide is derived from the names of the halogen and the parent alkane. For example, CH_3Br is named bromomethane, and its common name is "methyl bromide."

Section 19.7 *Alcohols* and *phenols* are generally polar and are soluble in water. Alcohols have high boiling points because the **hydroxyl group** (–OH) can attract other molecules to form a hydrogen bond. The systematic name for an alcohol is formed by applying the suffix *-ol* to the name of the parent alkane. For example, CH_3OH is named methanol, and CH_3CH_2OH is named ethanol. The most important member of the phenol class is C_6H_5OH; its systematic name is phenol. *Ethers* are generally nonpolar and only slightly soluble in water. Ethers are usually referred to by their common names; for example, $CH_3CH_2{-}O{-}CH_2CH_3$ is referred to as "diethyl ether."

Section 19.8 *Amines* are generally polar and are soluble in water. The amines have unusually high melting and boiling points because they can hydrogen bond. The common name for an amine is formed by adding "amine" to the name of the parent alkyl group. For example, the name of CH_3NH_2 is "methylamine"; $CH_3CH_2NH_2$ is named "ethylamine." The name of the aromatic amine $C_6H_5NH_2$ is "phenylamine."

Section 19.9 *Aldehydes* and *ketones* are slightly polar, and those with low molecular mass are soluble in water. The systematic name for an aldehyde is formed by applying the suffix *-al* to the name of the parent alkane. For example, HCHO is named methanal, but is usually referred to by its common name "formaldehyde." The systematic name for a ketone is formed by applying the suffix *-one* to the name of the parent alkane. For example, CH_3COCH_3 is named propanone, but it is referred to as "acetone" or "dimethyl ketone."

Section 19.10 *Carboxylic acids* are soluble in water and have high boiling points because the **carboxyl group** (–COOH) can hydrogen bond. The systematic name for an acid is formed by applying the suffix *-oic acid* to the name of the parent alkane. For example, CH_3COOH is named ethanoic acid, but it is usually referred to as "acetic acid." *Esters* are made from the reaction of a carboxylic acid and an alcohol. The systematic name for an ester is formed by applying the suffix *-oate* to the name of the parent carboxylic acid. For example, CH_3COOCH_3 is named methyl ethanoate. *Amides* are made from the reaction of a carboxylic acid and ammonia. The systematic name for an amide is formed by applying the suffix *–amide* to the name of the parent carboxylic acid. For example, CH_3CONH_2 is synthesized from ethanoic acid and is named ethanamide.

TABLE 19.14 SUMMARY OF HYDROCARBON DERIVATIVES

FAMILY NAME	GENERAL FORMULA	FUNCTIONAL GROUP	EXAMPLE
Organic halide	$R-X$	$-X$	CH_3-CH_2-Cl "ethyl chloride"
Alcohol	$R-OH$	$-OH$	CH_3-CH_2-OH "ethyl alcohol"
Phenol	$Ar-OH$	$-OH$	⌬—OH phenol
Ether	$R-O-R'$	$-O-$	CH_3-O-CH_3 "dimethyl ether"
Amine	$R-NH_2$	$-NH_2$	$CH_3-CH_2-NH_2$ "ethyl amine"
Aldehyde	$R-\overset{O}{\overset{\|}{C}}-H$	$-\overset{O}{\overset{\|}{C}}-H$	$CH_3-\overset{O}{\overset{\|}{C}}-H$ "acetaldehyde"
Ketone	$R-\overset{O}{\overset{\|}{C}}-R'$	$-\overset{O}{\overset{\|}{C}}-$	$CH_3-\overset{O}{\overset{\|}{C}}-CH_3$ "acetone"
Carboxylic acid	$R-\overset{O}{\overset{\|}{C}}-OH$	$-\overset{O}{\overset{\|}{C}}-OH$	$CH_3-\overset{O}{\overset{\|}{C}}-OH$ "acetic acid"
Ester	$R-\overset{O}{\overset{\|}{C}}-O-R'$	$-\overset{O}{\overset{\|}{C}}-O-$	$CH_3-\overset{O}{\overset{\|}{C}}-O-CH_3$ "methyl acetate"
Amide	$R-\overset{O}{\overset{\|}{C}}-NH_2$	$-\overset{O}{\overset{\|}{C}}-NH_2$	$CH_3-\overset{O}{\overset{\|}{C}}-NH_2$ "acetamide"

Key Terms *Answers to Key Terms are in Appendix H.*

Select the key term that corresponds to each of the following definitions.

______ **1.** the study of carbon compounds

______ **2.** a compound containing only hydrogen and carbon

______ **3.** a hydrocarbon compound containing only single bonds

______ **4.** a hydrocarbon compound containing a double bond or triple bond

______ **5.** a hydrocarbon compound typically containing a benzene ring

______ **6.** a family of saturated hydrocarbon compounds

______ **7.** a family of unsaturated hydrocarbon compounds having a double bond

______ **8.** a family of unsaturated hydrocarbon compounds having a triple bond

______ **9.** a giant molecule composed of small molecules bonded together repeatedly in a long, continuous chain

______ **10.** a small molecule that bonds to itself repeatedly to form a long, continuous chain

______ **11.** a family of aromatic hydrocarbon compounds

______ **12.** a chemical formula that shows the arrangement of atoms in a molecule

______ **13.** compounds with the same molecular formula, but different structures

______ **14.** the part that remains after a hydrogen atom is removed from an alkane

______ **15.** the part that remains after a hydrogen atom is removed from an arene

(a) addition reaction *(Sec. 19.3)*
(b) alkanes *(Sec. 19.2)*
(c) alkenes *(Sec. 19.3)*
(d) alkyl group (R–) *(Sec. 19.2)*
(e) alkynes *(Sec. 19.3)*
(f) arenes *(Sec. 19.4)*
(g) aromatic hydrocarbon *(Sec. 19.1)*
(h) aryl group (Ar–) *(Sec. 19.2)*
(i) carbonyl group *(Sec. 19.5)*
(j) carboxyl group *(Sec. 19.10)*
(k) class of compounds *(Sec. 19.5)*
(l) combustion reaction *(Sec. 19.2)*
(m) functional group *(Sec. 19.5)*
(n) hydrocarbon *(Sec. 19.1)*

_____ 16. the part that remains after a hydrogen atom is removed from benzene
_____ 17. a chemical reaction in which a hydrocarbon reacts rapidly with oxygen
_____ 18. a chemical reaction in which an unsaturated hydrocarbon adds hydrogen or a halogen
_____ 19. a family of compounds in which all members have similar names and properties
_____ 20. a compound containing carbon, hydrogen, and another element
_____ 21. an atom or group of atoms that characterizes a class of compounds
_____ 22. the C=O group
_____ 23. the –OH group
_____ 24. the –COOH group

(o) hydrocarbon derivative *(Sec. 19.5)*
(p) hydroxyl group *(Sec. 19.7)*
(q) isomers *(Sec. 19.2)*
(r) monomer *(Sec. 19.3)*
(s) organic chemistry *(Sec. 19.1)*
(t) phenyl group *(Sec. 19.2)*
(u) polymer *(Sec. 19.3)*
(v) saturated hydrocarbon *(Sec. 19.1)*
(w) structural formula *(Sec. 19.2)*
(x) unsaturated hydrocarbon *(Sec. 19.1)*

Exercises *Answers to odd-numbered Exercises are in Appendix I.*

Hydrocarbons (Sec. 19.1)

1. What are the three types of fossil fuels?
2. What is the primary source of hydrocarbons?
3. Which class of hydrocarbon is saturated?
4. Which class of hydrocarbon has only single bonds?
5. Which classes of hydrocarbon are unsaturated?
6. Which class of hydrocarbon has a triple bond?

Alkanes (Sec. 19.2)

7. Which of the following molecular formulas represents an alkane?
   (a) $C_{10}H_{24}$ (b) $C_{10}H_{22}$
8. What is the molecular formula for each of the following alkanes?
   (a) $C_{11}H_?$ (b) $C_{12}H_?$
9. State the name for each of the following alkanes:
   (a) $CH_3-CH_2-CH_3$
   (b) $CH_3-CH_2-CH_2-CH_2-CH_3$
   (c) $CH_3-CH_2-CH_2-CH_2-CH_2-CH_2-CH_3$
   (d) $CH_3-CH_2-CH_2-CH_2-CH_2-CH_2-CH_2-CH_2-CH_3$
10. Draw the condensed structural formula for each of the following alkanes:
    (a) ethane (b) butane
    (c) hexane (d) octane
11. Draw the condensed structural formula for the five isomers of hexane, C_6H_{14}.
12. Draw the condensed structural formula for the three isomers of pentane, C_5H_{12}.
13. Draw the two isomers of chloropropane, C_3H_7Cl.
14. Draw the four isomers of dichloropropane, $C_3H_6Br_2$.
15. Draw each of the following alkyl groups:
    (a) methyl (b) ethyl
16. Draw the structure for each of the following alkyl groups:
    (a) propyl (b) isopropyl
17. Give the IUPAC names for each of the following structural formulas:

(a)
```
                     CH3
                     |
CH3—CH2—CH2—CH—CH3
```

(b)
```
                         CH2—CH3
                         |
CH3—CH—CH2—CH—CH2—CH3
        |
        CH3
```

(c)
```
                         CH3
                         |
CH3—CH—CH2—C—CH2—CH2—CH3
        |               |
        CH3             CH3
```

(d)
```
                                     CH3
                                     |
CH3—CH2—CH2—CH—CH—CH—CH2—CH3
                     |         |
                     CH3       CH3
```

18. Draw the condensed structural formula for each of the following compounds:
    (a) 2-methylpropane (b) 2,2-dimethylbutane
    (c) 3-ethylheptane (d) 3,3-diethylpentane
19. Write a balanced equation for the complete combustion for each of the following:
    (a) methane (b) propane
    (c) hexane (d) octane

◀ **Combustion Reaction** Charcoal reacts with oxygen in air to give carbon dioxide, water, and heat.

20. Write a balanced equation for the complete combustion for each of the following:
    (a) ethane (b) butane
    (c) pentane (d) heptane

Alkenes and Alkynes (Sec. 19.3)

21. Which of the following molecular formulas represents an alkene?
(a) $C_{10}H_{22}$ (b) $C_{10}H_{20}$

22. What is the molecular formula for each of the following alkenes?
(a) $C_{11}H_?$ (b) $C_{12}H_?$

23. State the name for each of the following alkenes:
(a) $CH_2{=}CH{-}CH_2{-}CH_3$
(b) $CH_2{=}CH{-}CH_2{-}CH_2{-}CH_3$
(c) $CH_3 - CH{=}CH{-}CH_2{-}CH_2{-}CH_3$
(d) $CH_3{-}CH_2{-}CH_2{-}CH{=}CH{-}CH_2{-}CH_2{-}CH_3$

24. Draw the condensed structural formula for each of the following alkenes:
(a) 2-pentene (b) 3-heptene
(c) 1-propene (d) 4-nonene

25. Which of the following molecular formulas represents an alkyne?
(a) $C_{10}H_{20}$ (b) $C_{10}H_{18}$

26. What is the molecular formula for each of the following alkynes?
(a) $C_{11}H_?$ (b) $C_{12}H_?$

27. State the name for each of the following alkynes:
(a) $CH_3{-}C{\equiv}C{-}CH_2{-}CH_3$
(b) $CH{\equiv}C{-}CH_2{-}CH_2{-}CH_3$
(c) $CH_3{-}CH_2{-}C{\equiv}C{-}CH_2{-}CH_3$
(d) $CH_3{-}CH_2{-}CH_2{-}CH_2{-}C{\equiv}C{-}CH_3$

28. Draw the condensed structural formula for each of the following alkynes:
(a) 1-propyne (b) 4-octyne
(c) 2-butyne (d) 3-decyne

29. Draw the two structural isomers of straight-chain pentene, C_5H_{10}.

30. Draw the two structural isomers of straight-chain pentyne, C_5H_8.

31. Give the IUPAC name for each of the following structural formulas:
(a) $CH_3{-}CH{=}CH{-}CH(CH_3){-}CH_3$
(b) $CH_3{-}CH(CH_3){-}CH_2{-}C(CH_3){=}CH{-}CH_3$

32. Draw the condensed structural formula for each of the following compounds:
(a) 2-methyl-1-propene (b) 3,3-dimethyl-1-butyne

33. Give the IUPAC names for each of the following structural formulas:
(a) $CH_3{-}C{\equiv}C{-}C(CH_3)_2{-}CH_2{-}CH_2{-}CH_3$
(b) $CH_3{-}CH_2{-}CH_2{-}CH(CH_3){-}CH(CH_3){-}CH(CH_3){-}C{\equiv}CH$

34. Draw the condensed structural formula for each of the following compounds:
(a) 3-ethyl-2-heptene
(b) 4,4-dimethyl-2-pentyne

35. Write a balanced equation for each of the following reactions:
(a) $CH_2{=}CH_2 + O_2 \longrightarrow$
(b) $CH_3{-}CH{=}CH_2 + H_2 \longrightarrow$
(c) $CH_3{-}CH{=}CH{-}CH_3 + Br_2 \longrightarrow$

36. Write a balanced equation for each of the following reactions:
(a) $CH{\equiv}C{-}CH_3 + O_2 \longrightarrow$
(b) $CH_3{-}C{\equiv}CH + 2\,H_2 \longrightarrow$
(c) $CH_3{-}C{\equiv}C{-}CH_3 + 2\,Br_2 \longrightarrow$

Arenes (Sec. 19.4)

37. Draw and name the three isomers of the solvent xylene, $C_6H_4(CH_3)_2$.

38. Draw and name the three isomers of dinitrobenzene, $C_6H_4(NO_2)_2$.

39. What are the two products from the complete combustion of benzene?

40. Write a balanced chemical equation for the complete combustion of benzene.

Hydrocarbon Derivatives (Sec. 19.5)

41. Identify the class of compound for each of the following hydrocarbon derivatives:
(a) R—O—R′ (b) R—X
(c) Ar—OH (d) $R{-}NH_2$

42. Identify the class of compound for each of the following hydrocarbon derivatives:
(a) $R{-}C({=}O){-}OH$ (b) $R{-}C({=}O){-}H$
(c) $R{-}C({=}O){-}O{-}R'$ (d) $R{-}C({=}O){-}NH_2$

43. Identify the class of compound for each of the following hydrocarbon derivatives:
(a) $CH_3{-}NH_2$ (b) $CH_3{-}CH_2{-}F$
(c) $CH_3{-}C({=}O){-}O{-}CH_3$ (d) $H{-}C({=}O){-}OH$
(e) $C_6H_5{-}C({=}O){-}H$ (f) $CH_3{-}CH_2{-}O{-}C_6H_5$
(g) $C_6H_5{-}C({=}O){-}CH_3$ (h) $C_6H_5{-}C({=}O){-}NH_2$

44. Identify the class of compound for each of the following hydrocarbon derivatives:
(a) $CH_3{-}OH$ (b) $CH_3{-}CH_2{-}O{-}CH_2{-}CH_3$
(c) $CH_3{-}C({=}O){-}NH_2$ (d) $CH_3{-}C({=}O){-}CH_2{-}CH_3$
(e) $CH_3{-}C({=}O){-}O{-}C_6H_5$ (f) $CH_3{-}C_6H_4{-}OH$
(g) $C_6H_5{-}Cl$ (h) $C_6H_5{-}CH_2{-}OH$

Organic Halides (Sec. 19.6)

45. Give the common name for each of the following organic halides:
(a) $CH_3{-}I$ (b) $CH_3{-}CH_2{-}Br$
(c) $CH_3{-}CH_2{-}CH_2{-}F$ (d) $(CH_3)_2{-}CH{-}Cl$

46. Draw the structure for each of the following organic halides:
(a) dichloromethane (b) iodoethane
(c) 2-fluoropropane (d) 1,2-dibromopropane

47. Draw the structure for "trichloroethylene," a common solvent used for degreasing electronic and automobile parts.

48. Draw the structure for 1,1,1-trichloroethane, a solvent formerly used for dry cleaning clothes.

49. Indicate whether organic halides are generally soluble or insoluble in each of the following solvents:
(a) water (b) hydrocarbons

50. Indicate whether organic halides have boiling points that are generally higher, lower, or similar to those of the following. (Assume that the molecular mass is comparable.)
(a) alcohols (b) hydrocarbons

Alcohols, Phenols, and Ethers (Sec. 19.7)

51. Give the systematic IUPAC name for each of the following alcohols:
(a) $CH_3-CH_2-CH_2-CH_2-OH$
(b) $CH_3-CH_2-CH(OH)-CH_3$
(c) $CH_2(OH)-CH_2-CH_2-CH_3$
(d) $CH_3-CH(OH)-CH_2-CH_3$

52. Draw the structure for each of the following alcohols:
(a) "methyl alcohol" (b) "ethyl alcohol"
(c) "propyl alcohol" (d) "isopropyl alcohol"

53. Give an acceptable name for each of the following phenols:
(a)
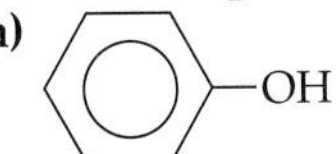

(b)

54. Draw the structure for each of the following phenols:
(a) *ortho*-chlorophenol (b) *meta*-ethylphenol

55. Give the common name for each of the following ethers:
(a) CH_3-O-CH_3
(b) $CH_3-O-CH_2CH_3$
(c) $CH_3CH_2CH_2-O-CH_2CH_2CH_3$
(d)
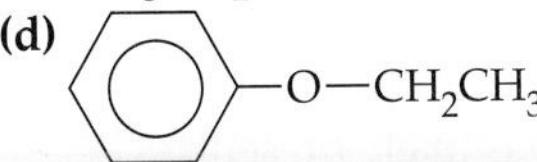

56. Draw the structure for each of the following ethers:
(a) "diisopropyl ether"
(b) "diphenyl ether"
(c) "ethyl propyl ether"
(d) "methyl phenyl ether"

57. "Ethyl alcohol" and "dimethyl ether" have the same molecular mass. Explain why the boiling point of "ethyl alcohol" (78 °C) is much higher than the boiling point of "dimethyl ether" (–25 °C).

58. Methanol, ethanol, and propanol are soluble in water. Explain why butanol is only slightly soluble.

59. There are two compounds that have the molecular formula C_2H_6O. Draw the structure for each isomer and classify the hydrocarbon derivative.

60. There are three compounds that have the molecular formula C_3H_8O. Draw the structure for each isomer and classify the hydrocarbon derivative.

Amines (Sec. 19.8)

61. Give the common name for each of the following amines:
(a) $CH_3CH_2NH_2$ (b) $CH_3CH_2CH_2NH_2$

62. Draw the structure for each of the following amines:
(a) "methylamine" (b) "isopropylamine"

63. "Ethylamine" and propane have about the same molecular mass. Explain why the boiling point of "ethylamine" (17 °C) is much higher than that of propane (–44 °C).

64. "Ethylamine" and "diethylamine" are soluble in water. Explain why "triethylamine" is only slightly soluble in water.

Aldehydes and Ketones (Sec. 19.9)

65. Give the systematic IUPAC name for each of the following aldehydes:
(a) $H-C(=O)-H$
(b)
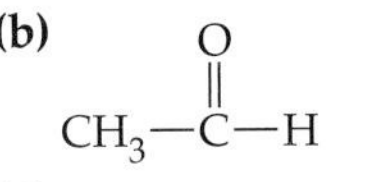

(c) $CH_3-CH_2-C(=O)-H$
(d)
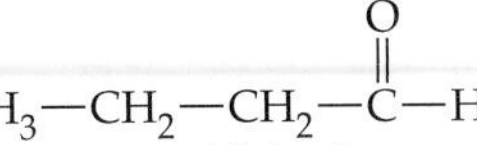

66. Draw the structure for each of the following aldehydes:
(a) "formaldehyde" (b) "acetaldehyde"
(c) "propionaldehyde" (d) "benzaldehyde"

67. Give the systematic IUPAC name for each of the following ketones:
(a) $CH_3-C(=O)-CH_3$
(b)
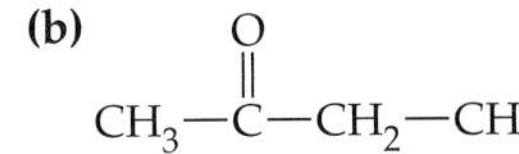

(c) $CH_3-CH_2-CH_2-C(=O)-CH_3$
(d) $CH_3-CH_2-C(=O)-CH_2-CH_3$

68. Draw the structure for each of the following ketones:
(a) "methyl ethyl ketone" (b) "diethyl ketone"
(c) "methyl phenyl ketone" (d) "diphenyl ketone"

69. Indicate whether the following compounds have boiling points that are generally higher or lower than that of a hydrocarbon of comparable molecular mass.
(a) aldehydes (b) ketones

70. Indicate whether the following compounds are generally soluble or insoluble in water.
(a) aldehydes (b) ketones

71. An aldehyde and a ketone each have the molecular formula C_3H_6O. Draw the structure for each isomer.

72. An aromatic aldehyde and an aromatic ketone each have the molecular formula C_8H_8O. Draw the structure for each isomer.

Carboxylic Acids, Esters, and Amides (Sec. 19.10)

73. Give the systematic IUPAC name for each of the following carboxylic acids:
(a) $H-C(=O)-OH$
(b) $CH_3-C(=O)-OH$
(c) $CH_3-CH_2-C(=O)-OH$
(d)
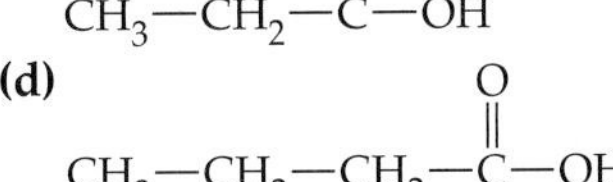

74. Draw the structure for each of the following carboxylic acids:
(a) "formic acid" (b) "acetic acid"
(c) "propionic acid" (d) "benzoic acid"

75. Give the systematic IUPAC name for each of the following esters:

(a) $H-\overset{O}{\overset{\|}{C}}-O-CH_2-CH_3$ (b) $CH_3-\overset{O}{\overset{\|}{C}}-O-CH_3$

(c) $CH_3-CH_2-\overset{O}{\overset{\|}{C}}-O-CH_2-CH_3$

(d) $H-\overset{O}{\overset{\|}{C}}-O-C_6H_5$ (benzene ring)

76. Draw the structure for each of the following esters:
(a) "propyl formate" (b) "ethyl acetate"
(c) "phenyl propionate" (d) "ethyl benzoate"

77. Give an acceptable name for each of the following amides:

(a) $H-\overset{O}{\overset{\|}{C}}-NH_2$ (b) $C_6H_5-\overset{O}{\overset{\|}{C}}-NH_2$

78. Draw the structure for each of the following amides:
(a) "acetamide" (b) "propionamide"

79. Which of the following classes of compounds can form hydrogen bonds?
(a) carboxylic acids (b) esters
(c) amides

80. Which of the following classes of compounds is generally soluble in water?
(a) carboxylic acids (b) esters
(c) amides

81. "Propionic acid" and "methyl acetate" have the same molecular formula, $C_3H_6O_2$. Predict which isomer has the higher boiling point.

82. "Butyric acid" and "ethyl acetate" have the same molecular formula, $C_4H_8O_2$. Predict which isomer has the higher boiling point.

83. Write a balanced equation for each of the following reactions:

(a) $CH_3-\overset{O}{\overset{\|}{C}}-OH + CH_3-OH \xrightarrow{H_2SO_4}$

(b) $C_6H_5-\overset{O}{\overset{\|}{C}}-OH + NH_3 \xrightarrow{\Delta}$ (benzene ring)

84. Write a balanced equation for each of the following reactions:

(a) $C_6H_5-\overset{O}{\overset{\|}{C}}-OH + CH_3OH \xrightarrow{H_2SO_4}$ (benzene ring)

(b) $H-\overset{O}{\overset{\|}{C}}-OH + NH_3 \xrightarrow{\Delta}$

85. What ester is produced from the reaction of phenol and ethanoic acid?

86. What carboxylic acid and alcohol react to give ethyl methanoate?

87. What amide is produced from the reaction of ammonia and methanoic acid?

88. What carboxylic acid reacts with ammonia to give propanamide?

General Exercises

89. Predict whether each of the following compounds is saturated or unsaturated as indicated by the name:
(a) eicosane (b) dodecene

90. Predict whether each of the following compounds is saturated or unsaturated as indicated by the name:
(a) cyclohexene (b) cyclopropane

91. The following pair of compounds have about the same molecular mass. Predict which compound in each pair has the higher boiling point.
(a) CH_3OCH_3 or CH_3CH_2OH
(b) $CH_3CH_2CH_2NH_2$ or $CH_3CH_2CH_2F$

92. The following pairs of compounds have about the same molecular mass. Predict which compound in each pair has the higher boiling point.
(a) CH_3COOH or $HCOOCH_3$
(b) $CH_3CH_2CONH_2$ or CH_3COOCH_3

93. Acetylsalicylic acid is the active ingredient in aspirin. Based on its name and suffix, which functional group is likely present?

94. Sulfanilamide is an antibacterial. Based on its name and suffix, which functional group is likely present?

95. Androsterone is a male hormone. Based on its name and suffix, which functional group is likely present?

96. Progesterone is a female hormone. Based on its name and suffix, which functional group is likely present?

97. Chloral hydrate is a veterinary anesthetic. Based on its name, which two functional groups are likely present?

98. Ethynyl estradiol is an oral contraceptive. Based on its name, which two functional groups are likely present?

19 Chapter Self-Test

Answers to Self-Test are in Appendix J.

1. Which of the following hydrocarbons is saturated? (Sec. 19.1)
(a) alkanes (b) alkenes
(c) alkynes (d) all of the above
(e) none of the above

2. What is the IUPAC name for $CH_3{-}CH_2{-}CH_2{-}CH_2{-}CH_2{-}CH_3$? (Sec. 19.2)
(a) decane (b) heptane
(c) hexane (d) octane
(e) none of the above

3. What is the product from the reaction of one mole of acetylene and one mole of hydrogen gas using platinum catalyst? (Sec. 19.3)
(a) ethane (b) ethene
(c) propane (d) propene
(e) none of the above

4. How many representations of benzene, C_6H_6, exist according to the delocalized electron model? (Sec. 19.4)
(a) one (b) two
(c) three (d) four
(e) none of the above

5. Which of the following classes of compounds has a carbonyl group? (Sec. 19.5)
(a) alcohol (b) amine
(c) ether (d) phenol
(e) none of the above

6. What is the condensed structural formula for "propyl bromide"? (Sec. 19.6)
(a) $CH_3{-}Br$ (b) $CH_3{-}CH_2{-}Br$
(c) $CH_3{-}CH_2{-}CH_2{-}Br$ (d) $CH_3{-}CHBr{-}CH_3$
(e) none of the above

7. What is the condensed structural formula for 2–propanol? (Sec. 19.7)
(a) $CH_3{-}CH_2{-}CH_2{-}OH$ (b) $CH_3{-}CH(OH){-}CH_3$
(c) $CH_3{-}CH_2{-}O{-}CH_3$ (d) $CH_3{-}O{-}CH_2{-}CH_3$
(e) none of the above

8. Which of the following compounds can form hydrogen bonds? (Sec. 19.8)
(a) $CH_3{-}CH_2{-}CH_2{-}CH_2{-}Cl$
(b) $CH_3{-}CH_2{-}CH_2{-}CH_2{-}CH_3$
(c) $CH_3{-}CH_2{-}CH_2{-}CH_2{=}CH_3$
(d) $CH_3{-}CH_2{-}CH_2{-}CH_2{-}NH_2$
(e) none of the above

9. What is the condensed structural formula for 2–pentanone? (Sec. 19.9)
(a) $CH_3{-}CH_2{-}CH_2{-}CH_2{-}CH_3$
(b) $CH_3{-}CH{=}CH{-}CH_2{-}CH_3$
(c) $CH_3{-}CH_2{-}CO{-}CH_2{-}CH_3$
(d) $CH_3{-}CH_2{-}CH_2{-}CH_2{-}COOH$
(e) none of the above

10. What are the products from the reaction of ethanoic acid and methanol with sulfuric acid catalyst? (Sec. 19.10)
(a) $CH_3COOH + CH_3OH \xrightarrow{H_2SO_4} CH_3COCH_3 + H_2O$
(b) $CH_3COOH + CH_3OH \xrightarrow{H_2SO_4} CH_3CH_2CHO + H_2O$
(c) $CH_3COOH + CH_3OH \xrightarrow{H_2SO_4} CH_3CH_2COOH + H_2O$
(d) $CH_3COOH + CH_3OH \xrightarrow{H_2SO_4} CH_3COOCH_3 + H_2O$
(e) $CH_3COOH + CH_3OH \xrightarrow{H_2SO_4}$ no reaction

Key Concepts

11. Draw the alcohol and ether isomers with the molecular formula C_2H_6O.

12. Draw the aldehyde and ketone isomers with the molecular formula C_3H_6O.

13. Draw the carboxylic acid and ester isomers with the molecular formula $C_2H_4O_2$.

14. Identify five functional groups in thyroxine (thyroid hormone).

I I O
HO– –O– –CH_2–CH–C–OH
NH_2
I I

Critical Thinking

15. Draw the condensed structural formula for each of the following models:
(a)

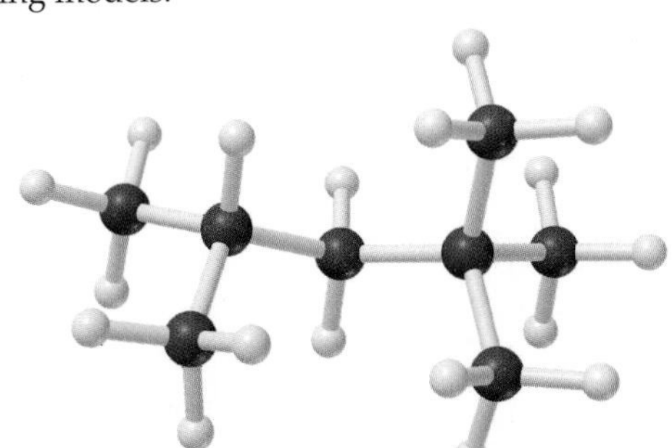

(b)

16. The organic compound caffeine is a stimulant found in coffee. Based on its suffix, in which class of compound does caffeine most likely belong?

17. The organic compound cholesterol has been linked to arterial plaque. Based on its suffix, in which class of compound does cholesterol most likely belong?

18. The organic compound cortisone is an anti-inflammatory drug. Based on its suffix, in which class of compound does cortisone most likely belong?

CHAPTER

20 Biochemistry

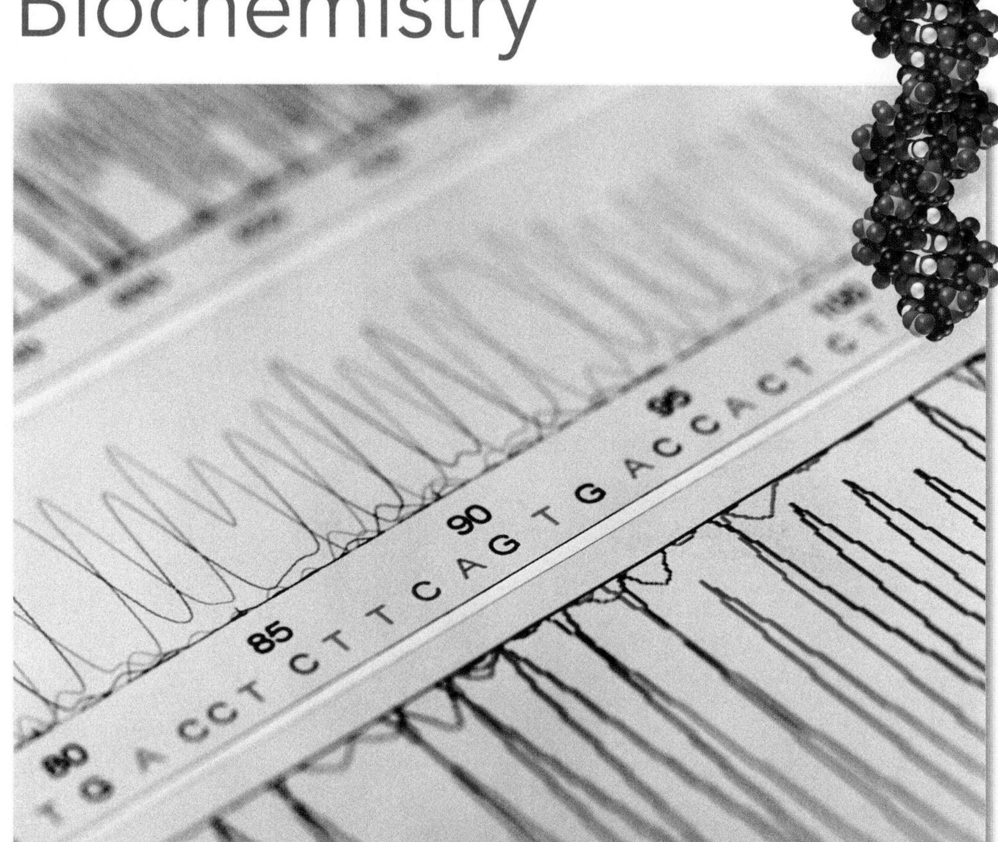

Every living system, plant and animal, has a unique DNA profile indicated by its sequence of adenine (A), thymine (T), cytosine (C), and guanine (G).

"Art is I; science is we."
Claude Bernard, French Physiologist (1813–1878)

There are about 30 chemical elements essential for life. The pictorial periodic table of elements in Figure 4.7 indicates the *essential elements* that are necessary to support life. The elements found in significant amounts in humans are oxygen, carbon, hydrogen, nitrogen, phosphorus, chlorine, sulfur, calcium, magnesium, sodium, and potassium. In addition, there are several *trace elements* that are present in only small amounts, but nonetheless are necessary to sustain life. These trace elements include iron, zinc, cobalt, iodine, and a few others.

The study of the chemistry of living things is referred to as *biochemistry*. Many biochemists are employed in the pharmaceutical industry, which manufactures drugs that affect chemical reactions in plants and animals. Some biochemists are referred to as molecular biologists because they study the relationships among different types of biological molecules and their role in life processes. As we proceed into the twenty-first century, applications of biotechnology and genetic engineering offer the potential for transforming our lives in ways that previously were unimaginable.

20.1 BIOLOGICAL COMPOUNDS

OBJECTIVES

- To recognize a protein given its structural formula.
- To recognize a carbohydrate given its structural formula.
- To recognize a lipid given its structural formula.
- To recognize a nucleic acid given its structural formula.

Biochemistry can be defined as the study of biological compounds and their chemical reactions. Although biological compounds are often large and complex, they have the same types of functional groups that we encountered in simple organic compounds. For instance, biochemical compounds frequently contain an alcohol, an amine, an aldehyde, a ketone, a carboxylic acid, an ester, or an amide functional group.

Biological compounds are often large molecules with molar masses greater than 1,000,000 g/mol. These large molecules are composed of small molecules linked together in a long continuous chain. A giant molecule containing many repeating smaller molecules is called a *polymer.* Proteins, starch, and nucleic acids are examples of naturally occurring polymers found in plants and animals.

In this chapter, we will study the following types of biochemical compounds: *proteins, carbohydrates, lipids,* and *nucleic acids.* These four types of compounds are classified in Figure 20.1.

Before discussing biological compounds in detail, let's briefly describe the general features of proteins, carbohydrates, lipids, and nucleic acids. A **protein** is a naturally occurring polymer composed of many amino acids. An amino acid has

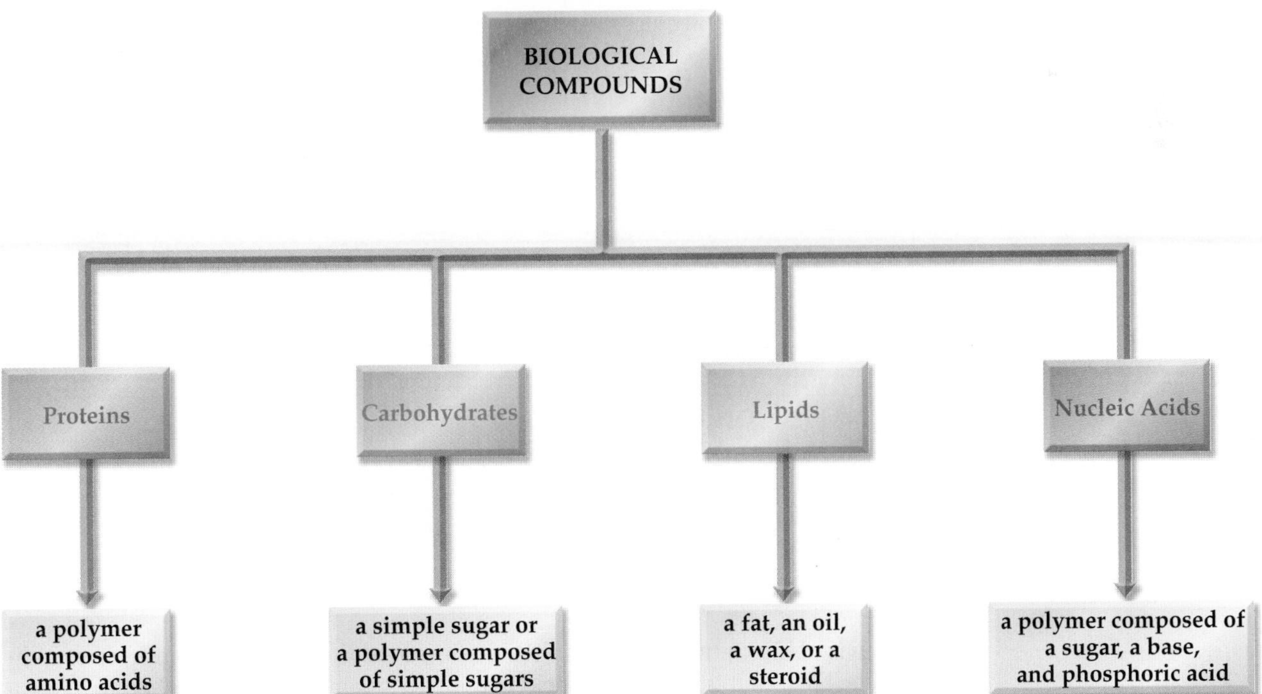

▲ **Figure 20.1 Classification of Biological Compounds** Biochemical compounds can be classified as one of the following: protein, carbohydrate, lipid, or nucleic acid. In this chapter, protein structures will be shown in red, carbohydrates in green, lipids in blue, and nucleic acids in pink.

both an amine and a carboxylic acid functional group. In a protein, amino acid molecules are linked together by amide bonds. Each of these amide bonds is called a *peptide linkage*, which we can show as follows:

$$\text{--- Amino acid} - \overset{\overset{\displaystyle O}{\|}}{C} - NH - \text{Amino acid ---}$$

↑ Peptide linkage

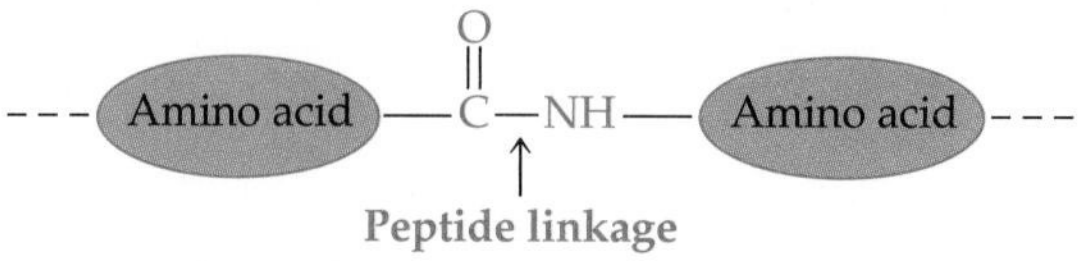

A **carbohydrate** is a compound that can be either a simple sugar or a polymer composed of many simple sugar molecules. A carbohydrate usually contains either an aldehyde or a ketone functional group. Furthermore, a typical carbohydrate is characterized by having several alcohol groups: glucose is a simple sugar that contains an aldehyde and five alcohol groups; fructose is a simple sugar that contains a ketone and five alcohol groups.

Starch is a carbohydrate polymer composed of glucose units linked together in a large molecule. In carbohydrates containing more than one sugar unit, the sugar molecules are joined together by an —O— linkage. The —O— bond between sugar molecules in a carbohydrate is called a *glycoside linkage*, which we can show as follows:

Sugar —O— Sugar

↑ Glycoside linkage

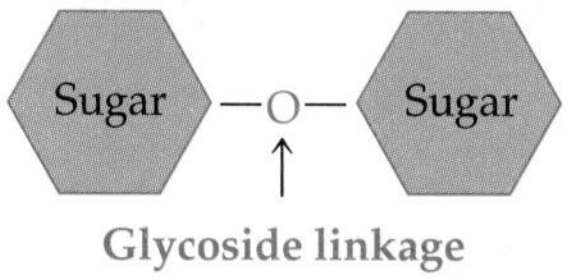

A **lipid** is a water-insoluble compound usually composed of an alcohol and one or more carboxylic acid molecules. The lipids include animal fats and vegetable oils. Fats and oils are esters of glycerol, an alcohol that has three –OH groups. As a result, each molecule of a fat or an oil contains three ester groups resulting from three carboxylic acid molecules joined to one glycerol molecule. We can show the *ester linkages* in a fat or oil molecule as follows:

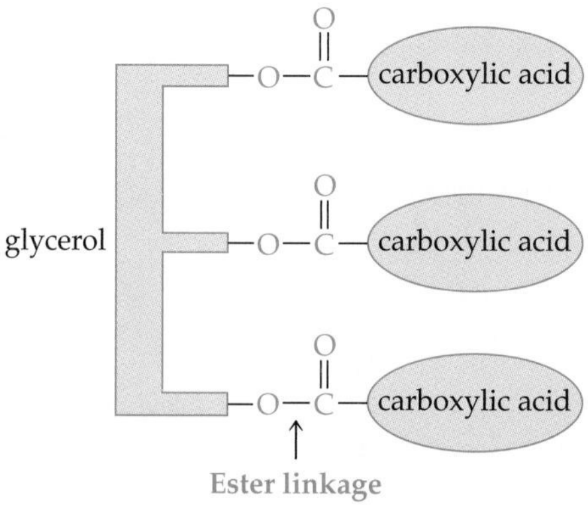

glycerol
—O—C(=O)— carboxylic acid
—O—C(=O)— carboxylic acid
—O—C(=O)— carboxylic acid

↑ Ester linkage

A **nucleic acid** is a biochemical polymer composed of a very large number of individual units. Each unit in the nucleic acid contains a sugar molecule attached to a nitrogen base and a phosphate group. The nitrogen base has an amine functional group and is simply referred to as a *base*. The backbone of the nucleic acid is formed by sugar molecules joined together in a long polymer chain. In a nucleic acid molecule, there may be millions of sugar molecules, each with an attached base, linked by phosphate ester bonds. We can show the *phosphate linkage* in a typical nucleic acid as follows:

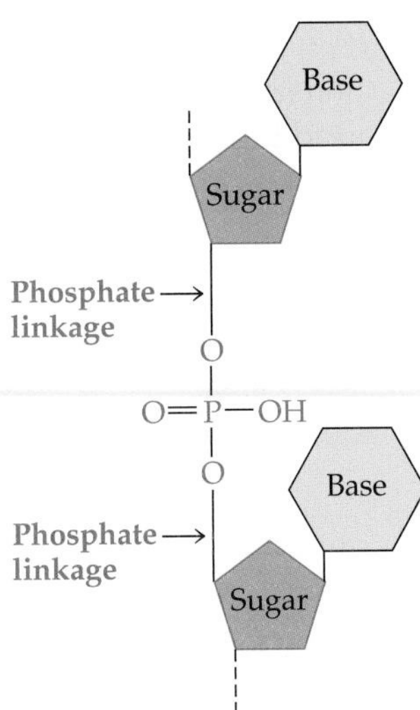

We will now examine the composition of these four types of biological compounds in more detail. That is, we will discuss the structure and function of proteins, carbohydrates, lipids, and nucleic acids.

20.2 PROTEINS

OBJECTIVES

- To describe the primary structure of a protein.
- To describe the secondary structure of a protein.
- To describe the tertiary structure of a protein.

Proteins play a crucial role in virtually all biological processes. In fact, the word "protein" is derived from the Greek word *proteios,* meaning "of first importance." The shapes of protein molecules have been observed to be either long fibers or compact globules. Moreover, biochemists have found that the molecular shape of a protein is related to its biological function.

Long protein fibers are responsible for the structure, support, and motion of an organism. Accordingly, these protein fibers are found in muscle tissue and cartilage. In addition, protein fibers compose the hair, skin, and nails. On the other hand, compact protein globules are responsible for metabolic processes. For example, many hormones are globular proteins such as those secreted by the pituitary gland and the pancreas. Globular proteins are involved in controlling growth, transporting small molecules and ions, and providing immune protection from foreign substances.

◀ **Protein Foods** Meat, fish, poultry, cheese, and milk products are all food sources rich in protein.

Human protein is composed of 20 different amino acids. Of these 20 amino acids, about half are essential and must be included in the diet. The other half are nonessential, and the human body can synthesize them from one of the essential amino acids. An **amino acid** has an amine group, $-NH_2$, attached to a carbon that is also bonded to a carboxyl group, –COOH. The carbon atom in the molecule that is attached to the carboxyl group is called the alpha carbon (α-carbon). Since the amine group is attached to the α-carbon, an amino acid is referred to as an α-amino acid. The α-amino acids have the following general structural formula:

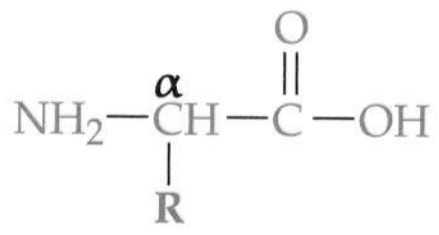

$$NH_2-\overset{\alpha}{\underset{\text{R}}{\text{CH}}}-\overset{\text{O}}{\overset{\|}{\text{C}}}-OH$$

Alpha (α) amino acid

Amino acids differ from one another according to the R– group attached to the α-carbon. This R– group is referred to as the *side chain*. In some amino acids, the R– group is neutral, and in other amino acids the side chain has either acidic or basic properties. The structures of the 20 α-amino acids found in human protein are shown in Figure 20.2. For convenience, chemists abbreviate the names of amino acids by a three-letter symbol; for example, the name of the amino acid glycine is abbreviated Gly.

We mentioned previously that proteins consist of amino acids linked together by amide bonds. In a protein, this amide bond is called a **peptide linkage** or a peptide bond. A peptide linkage has the following general structural features:

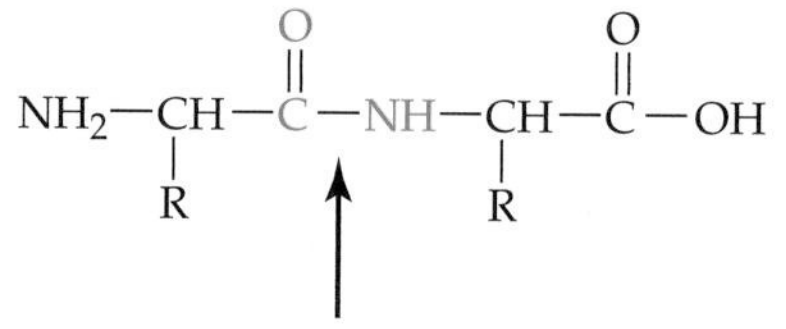

$$\text{(amine end)}\quad NH_2-\underset{\text{R}}{\text{CH}}-\overset{\text{O}}{\overset{\|}{\text{C}}}-NH-\underset{\text{R}}{\text{CH}}-\overset{\text{O}}{\overset{\|}{\text{C}}}-OH\quad\text{(carboxyl end)}$$

Peptide linkage

In a **dipeptide**, 2 amino acid units are joined by an amide bond. In a tripeptide, there are 3 amino acids, and in a **polypeptide**, there are up to 50 amino acids. Human insulin, for example, is a polypeptide composed of 48 amino acid units. We can illustrate a specific dipeptide by joining the 2 amino acids glycine and alanine.

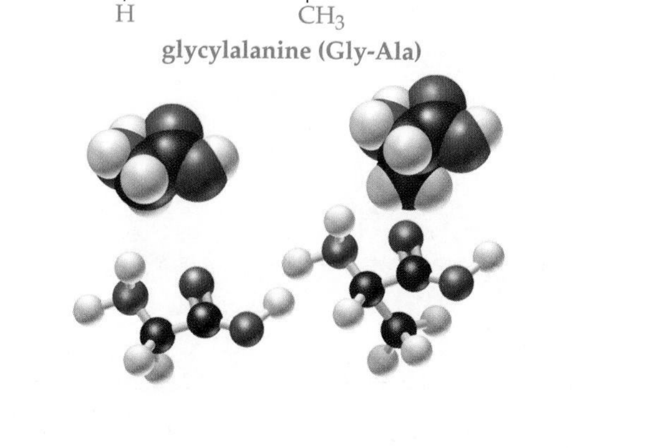

$$H_2N-\underset{\text{H}}{\text{CH}}-\overset{\text{O}}{\overset{\|}{\text{C}}}-\textbf{OH} \;+\; \textbf{H}-\underset{\text{H}}{\text{N}}-\underset{CH_3}{\text{CH}}-\overset{\text{O}}{\overset{\|}{\text{C}}}-OH \longrightarrow$$

glycine (Gly) alanine (Ala)

$$H_2N-\underset{\text{H}}{\text{CH}}-\overset{\text{O}}{\overset{\|}{\text{C}}}-NH-\underset{CH_3}{\text{CH}}-\overset{\text{O}}{\overset{\|}{\text{C}}}-OH + H_2O$$

glycylalanine (Gly-Ala)

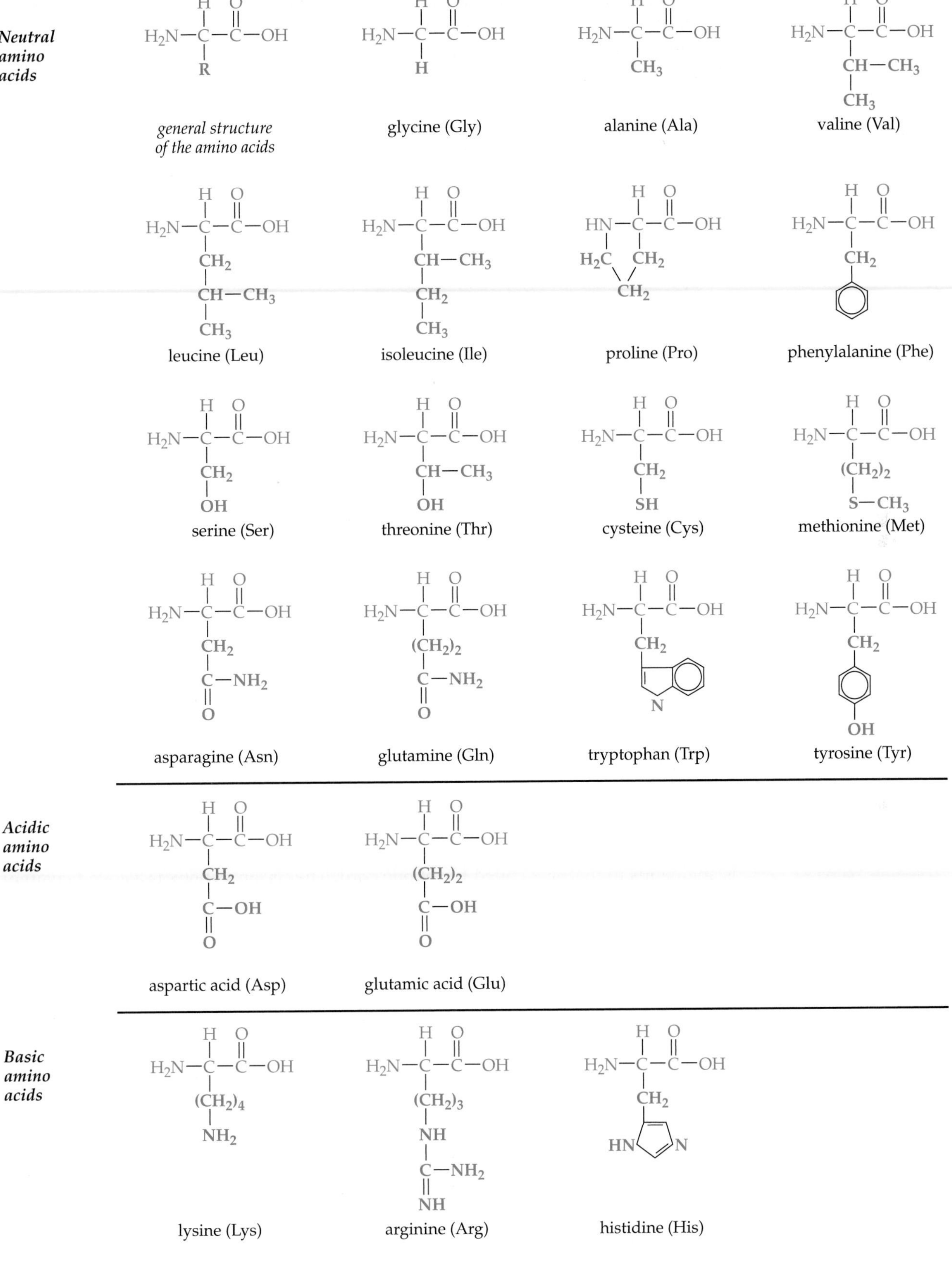

▲ **Figure 20.2 Common Amino Acids** Notice that the amino acids are similar except for the R– group. The side chain is responsible for the amino acid having neutral, acidic, or basic properties.

The dipeptide of glycine and alanine is named glycylalanine and is abbreviated Gly–Ala. The sequence of amino acids in a peptide or protein is always listed beginning with the amine end of the chain on the left. The other amino acids in the chain are listed in order, and the amino acid with the free carboxyl group is mentioned last. In the previously mentioned dipeptide example, the amine end of the peptide chain is glycine (Gly) and the carboxyl end is alanine (Ala).

Primary Structure of Proteins

Proteins are polypeptides that may contain hundreds or thousands of amino acid units linked together in a long-chain molecule. Egg whites, for example, contain a protein chain with nearly 400 amino acid molecules linked by peptide linkages.

One of the most amazing scientific achievements of the twentieth century was the unraveling of the mystery of protein synthesis in living organisms. In the 1960s, biochemists discovered how living organisms synthesize huge protein molecules in which thousands of amino acids are always in exactly the same sequence. When an organism synthesizes a protein, the sequence of amino acids is determined by a genetic code embodied in the structure of a nucleic acid.

Using currently available instruments, biochemists can determine the exact sequence of amino acids in small proteins. The sequence of amino acids in a peptide chain is referred to as the *primary structure* of the peptide. Figure 20.3 shows the primary structure of a polypeptide with ten amino acids.

The replacement of a single amino acid by another amino acid in the primary structure of a protein can alter its biological activity. Even a small change in the primary sequence of amino acids can drastically alter the biochemistry of a protein. For example, the disease sickle-cell anemia results from an amino acid substitution in hemoglobin. Hemoglobin is the protein responsible for oxygen transport through the circulatory system. Sickle-cell anemia, an extremely serious, often fatal condition, results when a single amino acid, glutamic acid, is replaced by the amino acid valine.

Secondary Structure of Proteins

When a peptide chain twists and bends, the protein acquires a *secondary structure.* One type of secondary structure exhibited by proteins is referred to as an α–helix. This is the structure of protein molecules found in hair and wool fibers. An α–helix structure is analogous to that of a coiled telephone cord. That is, an α–helix is a coiled chain of amino acids. Figure 20.4 illustrates the α–helix secondary structure of a hypothetical polypeptide composed of ten amino acids.

Another type of secondary structure exhibited by proteins is referred to as a pleated sheet. In this arrangement, protein chains are antiparallel; that is, the carboxyl ends of two adjacent protein chains run in opposite directions. This is the structure of protein molecules found in silk fibers. A pleated sheet structure is analogous to a sheet of paper folded into an accordion shape. Figure 20.5 illustrates the pleated sheet secondary structure for a pair of hypothetical polypeptides, such as silk and collagen.

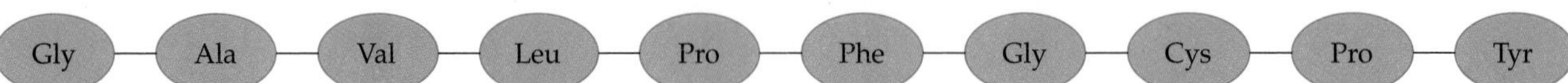

▲ **Figure 20.3 Primary Structure of a Protein** The primary structure of a protein refers to the sequence of amino acids present in the polymer chain. (Each ellipse represents an amino acid.)

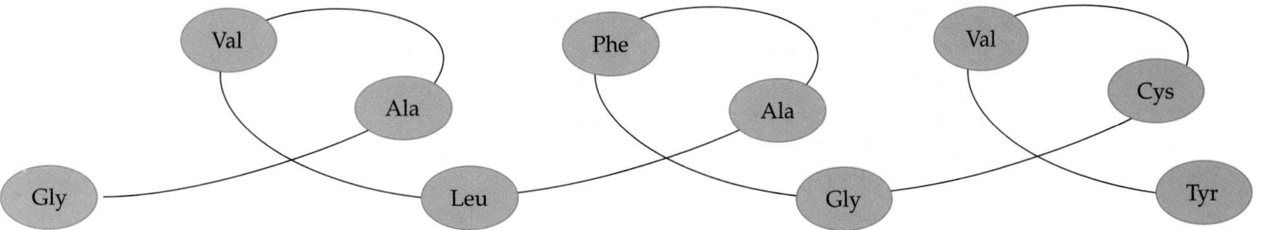

▲ **Figure 20.4 Secondary Structure of a Protein: α-Helix** The secondary structure of a protein may be in the shape of a coiled chain of amino acids referred to as an α-helix. (Each ellipse represents an amino acid.)

What maintains the secondary structure of a protein? The answer is hydrogen bonding (Section 13.2). Hydrogen bonds between amino acids are responsible for the secondary structure of a protein. In an α-helix, the hydrogen bonds are between nearby amino acid units in the same coiled protein chain. In a pleated sheet, the hydrogen bonds are between amino acids in neighboring protein chains.

Tertiary Structure of Proteins

The overall three-dimensional structure of a protein is referred to as its *tertiary structure.* The tertiary structure of a protein may be long and extended or compact and folded. As an analogy, the tertiary structure of a protein may be compared to a telephone cord that is straight and extended or twisted about itself. The twisting of an α-helix illustrates a tertiary structure. The formation of a twisted helix results in a protein molecule with the appearance of a globule.

The tertiary structure of a protein is held together by intermolecular forces, primarily hydrogen bonds. Figure 20.6 illustrates the tertiary structure of myoglobin. Myoglobin is a globular protein that stores oxygen in animal tissue until it is required for a metabolic activity.

Hydrogen bonds are weaker than covalent bonds, and the tertiary structure of a protein can collapse with a change in pH or temperature. The process of breaking hydrogen bonds in a protein molecule is called *denaturation*. A familiar example of denaturing a protein is boiling an egg in hot water.

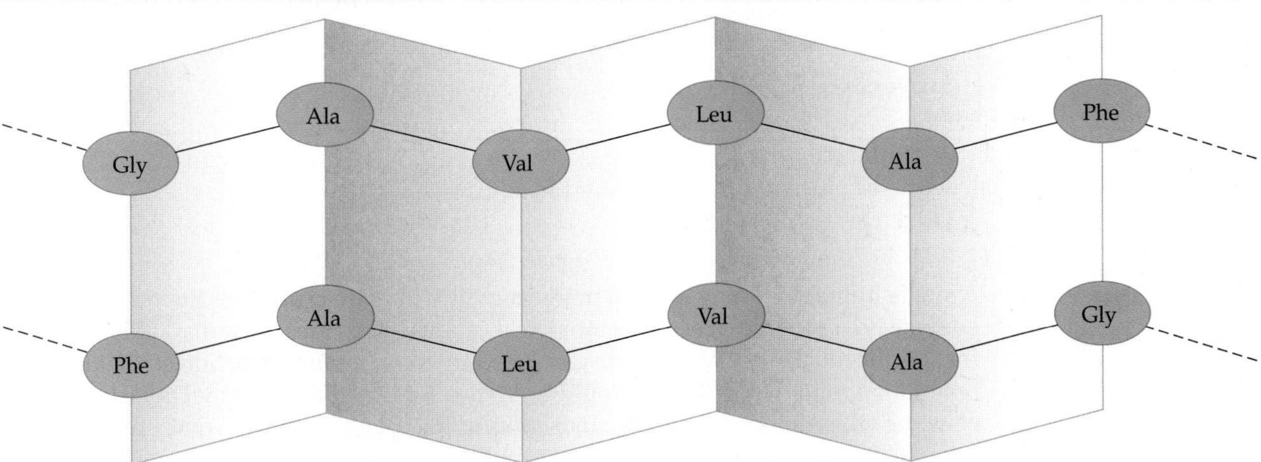

▲ **Figure 20.5 Secondary Structure of a Protein: Pleated Sheet** The secondary structure of a protein may be in the shape of folded chains of amino acids referred to as a pleated sheet. Notice that the protein chains are parallel, but aligned in opposite directions. (Each ellipse represents an amino acid.)

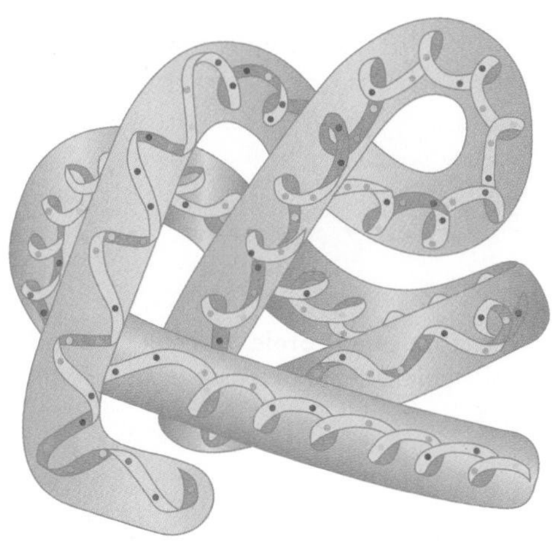

▶ **Figure 20.6 Tertiary Structure of a Protein** A twisted α–helix can create the tertiary structure of a protein. This model structure shows the coiled secondary structure of myoglobin twisted into a tertiary structure.

Example Exercise 20.1 Proteins

Identify the amino acids in the following dipeptide:

$$H_2N-\underset{\displaystyle H}{\underset{|}{CH}}-\overset{\displaystyle O}{\overset{\|}{C}}-NH-\underset{\displaystyle CH_2-OH}{\underset{|}{CH}}-\overset{\displaystyle O}{\overset{\|}{C}}-OH$$

Solution

A dipeptide has two amino acids joined by a peptide linkage. If we refer to Figure 20.2, we can identify the two amino acids in the dipeptide as glycine (left) and serine (right).

Practice Exercise

Classify glycine and serine as neutral, acidic, or basic amino acids.

Answer: If we refer to Figure 20.2, we find that glycine and serine are both examples of neutral amino acids. Note that neither amino acid has a carboxyl group or an amine group on the R— portion of the amino acid structure.

Concept Exercise

The structure of a protein is described as a helix of amino acids in a spiral protein chain. Does this description correspond to the primary, secondary, or tertiary structure of a protein molecule?

Answer: See Appendix G.

OBJECTIVES

- To explain the role of an enzyme as a catalyst.
- To explain the lock-and-key model for an enzyme.

20.3 ENZYMES

A single human cell may contain many enzymes that are crucial for life processes. An **enzyme** is a protein that acts as a biological catalyst. Enzymes are remarkably selective for specific molecules. An enzyme can speed up a biochemical reaction so that the rate is a million times faster than it would be in the absence of the enzyme. In fact, without enzymes, these biochemical reactions would take place so slowly that life would be impossible. Furthermore, only a very small amount of enzyme catalyst is necessary because a biochemical reaction can occur in microseconds, and the enzyme is then free to repeat its role as catalyst.

Biochemical reactions require sensitive conditions and cannot withstand the conditions of extreme pH or high temperature often associated with inorganic and organic chemical reactions. If the pH or temperature is too high or too low, an enzyme becomes inactive and a biochemical reaction cannot take place. A chemical that renders an enzyme inactive is called an *inhibitor*. Sarin and phosgene are inhibitors that have been used as chemical weapons. They act as nerve agents, which operate by blocking an enzyme responsible for breaking down lethal toxins that accumulate naturally in the body.

Lock-and-Key Model

The details of an enzyme's function are complex, but we can use a **lock-and-key model** to describe the general mechanism. The lock-and-key theory proposes that an enzyme molecule has a given structural feature complementary in shape to the shape of the reacting molecule. This location on the enzyme where a reaction can occur is called the *active site* and the reacting molecule is referred to as the *substrate*. Figure 20.7 illustrates the lock-and-key model using a lock and key as an analogy.

An enzyme catalyzed reaction takes place in two steps. First, the substrate (S) binds to the active site on the enzyme (E). Second, the substrate molecule undergoes a reaction and the enzyme releases two or more products (P_1 and P_2).

$$\text{Step 1:} \quad E + S \rightarrow ES$$

$$\text{Step 2:} \quad ES \rightarrow E + P_1 + P_2$$

Figure 20.8 illustrates this two-step process for an enzyme catalyzed reaction of a substrate molecule to give two products.

After the enzyme releases the products, the enzyme is then free to bind to another substrate molecule and repeat the process all over again. Enzyme catalysis is quite rapid, and a trace amount of enzyme can yield a large amount of product in a short period of time.

You can demonstrate enzyme catalysis with an ordinary saltine cracker. Although the cracker is not sweet, it contains starch, which is a polymer of glucose sugar. If you place the cracker in your mouth, in a short time you will notice a sweet taste. The explanation is that saliva contains the enzyme ptyalin, which begins the digestion process by breaking down starch molecules into sugar units. If the sweet taste takes a minute to develop with the enzyme, it would take about a million minutes for the taste to develop without the enzyme; that is almost two years!

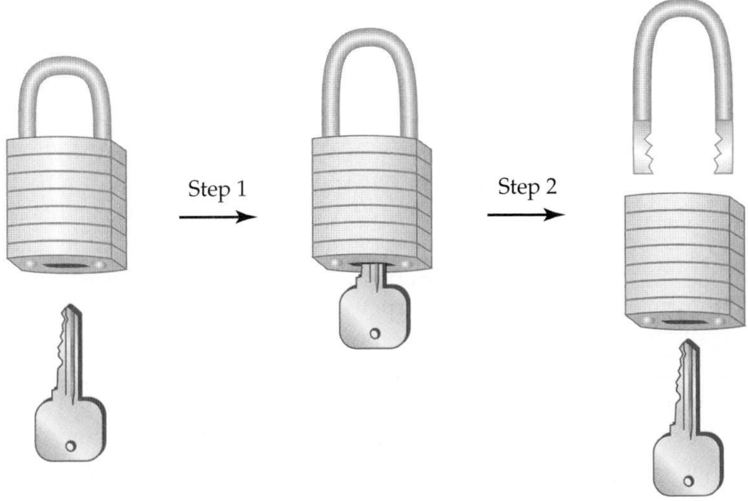

◀ **Figure 20.7 Analogy for Enzyme Catalysis** In Step 1, the key (enzyme) is inserted into the lock (substrate). In Step 2, the key opens the lock and releases the two parts (products).

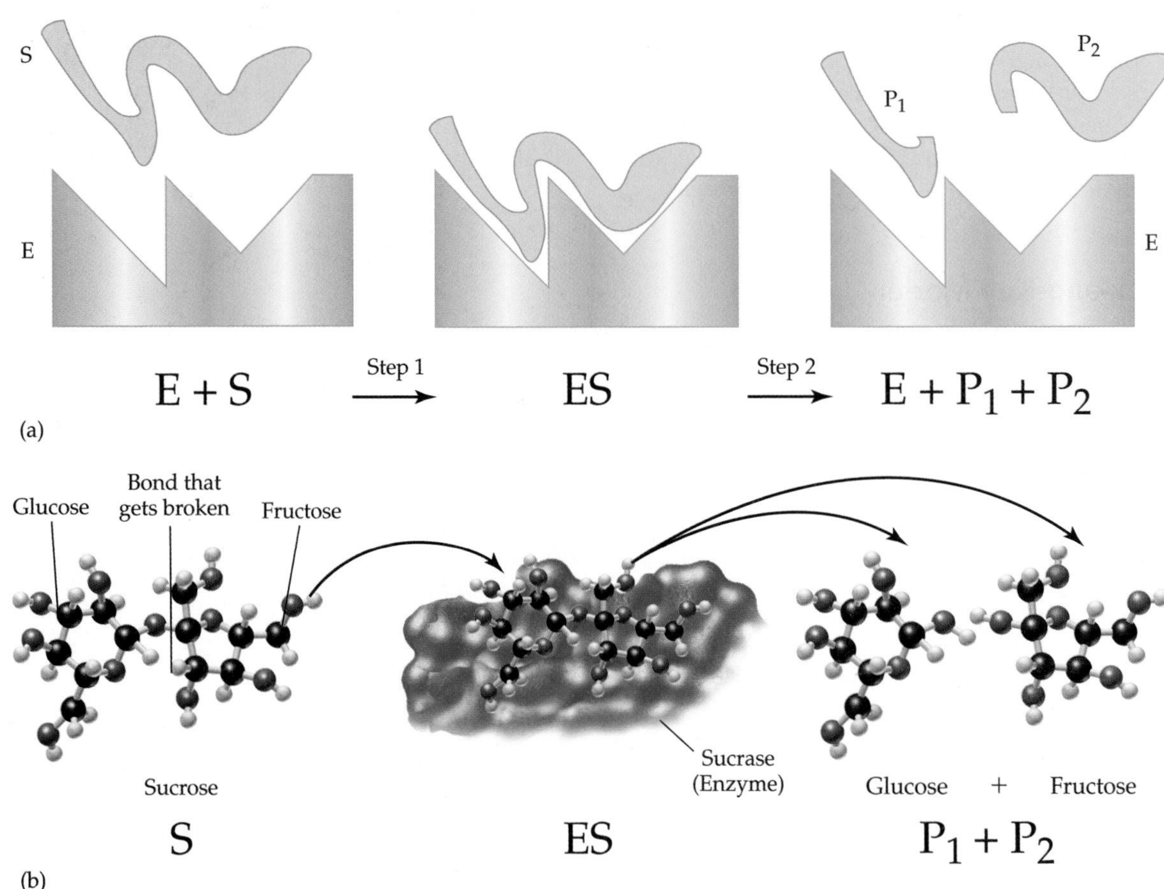

▲ **Figure 20.8 Enzyme Catalysis** In Step 1, the substrate molecule (S) binds to the active site on the enzyme (E). In Step 2, the enzyme cleaves the substrate and releases the two products, P_1 and P_2.

20.4 CARBOHYDRATES

OBJECTIVES

- To recognize the structure of mono-, di-, and polysaccharides.
- To identify a glycoside linkage between monosaccharides.

Carbohydrates are an important food source for animals and provide the structure of plants. The word "carbohydrate" literally means "hydrates of carbon" ($C \cdot H_2O$), and indeed many carbohydrates have the empirical formula CH_2O. The names of carbohydrates usually end in the suffix *-ose,* for example, glucose, fructose, and sucrose. A carbohydrate is characterized by having an aldehyde or ketone functional group and, in addition, several hydroxyl (–OH) groups.

A simple sugar molecule typically has three to six carbons, an aldehyde or ketone group, and a few hydroxyl groups. A simple sugar molecule is referred to as a **monosaccharide**. If a monosaccharide has an aldehyde group, it is referred to as an *aldose;* if it has a ketone group, it is called a *ketose.* If the sugar molecule contains five carbons, it is referred to as a *pentose;* if it contains six carbons, it is called a *hexose.*

▲ Bread, fruits, and vegetables are all food sources rich in carbohydrates.

Glucose, $C_6H_{12}O_6$, and fructose, $C_6H_{12}O_6$, are both hexoses. However, glucose has an aldehyde group and is an aldose; fructose has a ketone group and is a ketose. Ribose, $C_5H_{10}O_5$, is an important pentose found in nucleic acids. Figure 20.9 illustrates the structures of these three simple sugars.

Glucose occurs in grapes and blood and, accordingly, is referred to as grape sugar or blood sugar. Fructose occurs naturally in fruit and honey and is commonly referred to as fruit sugar. Although they have the same molecular formula, fructose is twice as sweet as glucose.

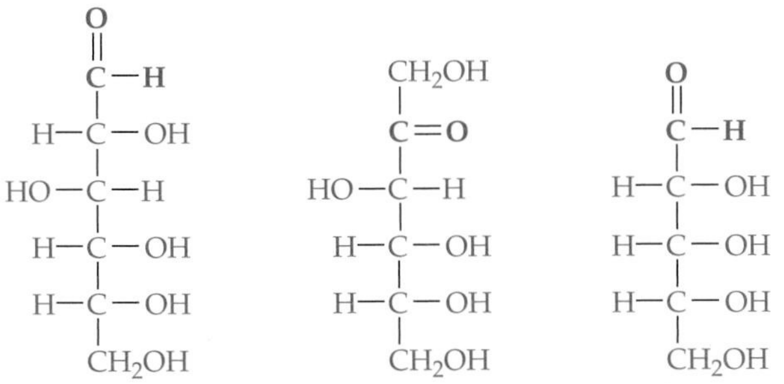

Glucose	Fructose	Ribose
O=C—H	CH_2OH	O=C—H
H—C—OH	C=O	H—C—OH
HO—C—H	HO—C—H	H—C—OH
H—C—OH	H—C—OH	H—C—OH
H—C—OH	H—C—OH	CH_2OH
CH_2OH	CH_2OH	
Glucose	**Fructose**	**Ribose**
(an aldose)	(a ketose)	(an aldose)
(a hexose)	(a hexose)	(a pentose)

◀ **Figure 20.9 Common Monosaccharides** Glucose and fructose are both hexoses, but notice that glucose is an aldose and fructose is a ketose. Glucose and ribose are both aldoses, but notice that glucose is a hexose and ribose is a pentose.

A **disaccharide** is composed of two monosaccharides bonded together. For instance, maltose is malt sugar, which is composed of two glucose molecules. Lactose is milk sugar, which is comprised of glucose and galactose. Sucrose is ordinary table sugar derived from sugarcane and sugar beets. Sucrose is a disaccharide comprised of glucose and fructose. In the presence of an enzyme or aqueous acid, a disaccharide breaks down into its two component monosaccharides. The equation for the acid *hydrolysis* of maltose, lactose, and sucrose is

$$C_{12}H_{22}O_{11} + H_2O \xrightarrow{H^+} C_6H_{12}O_6 + C_6H_{12}O_6$$

maltose	+	water	→	glucose	+	glucose
lactose	+	water	→	galactose	+	glucose
sucrose	+	water	→	glucose	+	fructose

Structures of Sugars in an Aqueous Solution

In an aqueous solution, sugar molecules usually exist as ringlike structures. In water, glucose forms a cyclic structure by joining a hydroxyl group to the aldehyde group. The resulting ring structure contains five carbon atoms and one oxygen atom. Figure 20.10 illustrates the formation of a glucose ring in the solution.

In the formation of a disaccharide, two simple monosaccharides split out water and are joined together by a special —O— bond called a **glycoside linkage**. For instance, a molecule of galactose and glucose can split out water and form a molecule of lactose.

In the process of digestion, the enzyme lactase breaks down lactose to yield galactose and glucose. It is interesting to note that adults have a less active form of this enzyme than infants. As a result, some adults cannot break down lactose and consequently may suffer indigestion. This condition, known as lactose intolerance, is treated by avoiding milk products or by taking a lactase supplement in the form of a liquid or tablet. Figure 20.11 illustrates the formation and breakage of a glycoside linkage in lactose.

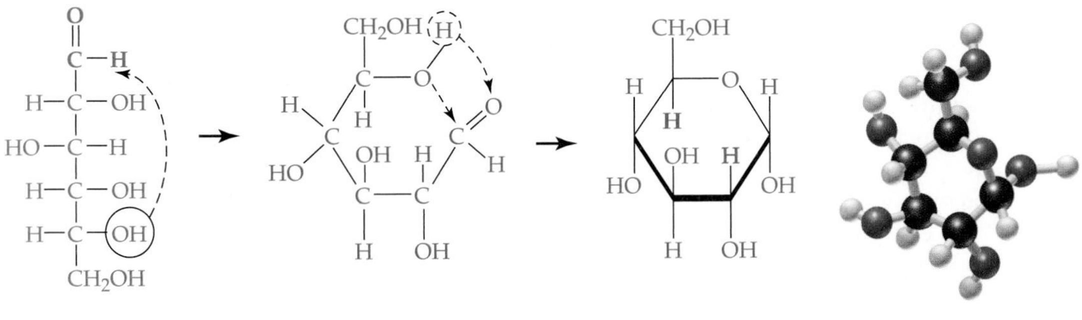

▲ **Figure 20.10 Glucose in an Aqueous Solution** In an aqueous solution, a glucose molecule forms a ring structure containing five carbon atoms and one oxygen atom.

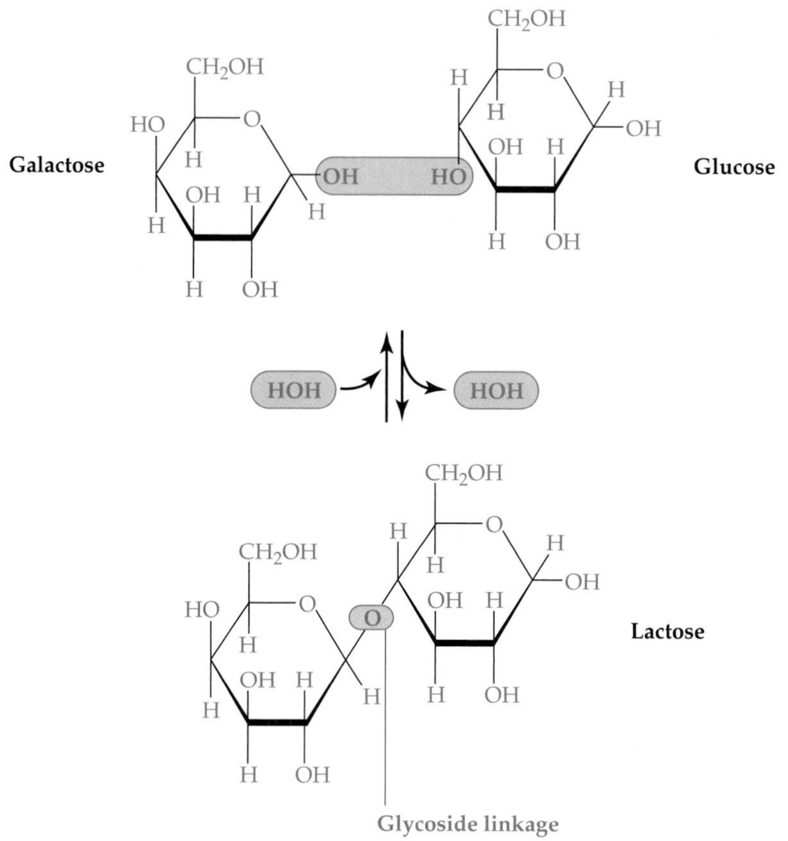

▶ **Figure 20.11 Formation and Breakage of a Glycoside Linkage** Lactose is composed of two simple sugars, glucose and galactose. In the formation of a glycoside linkage, a molecule of galactose and glucose split out a water molecule. In the reverse reaction, the glycoside linkage is cleaved in the presence of an enzyme or acid.

Polysaccharides

A **polysaccharide** is a polymer composed of many, often thousands, of monosaccharide units joined by glycoside linkages. Starch and cellulose are important examples of polysaccharides. Starch is a polysaccharide composed of the same repeating unit. When starch undergoes hydrolysis and is broken down with dilute acid, or an enzyme, the only product obtained is glucose. Figure 20.12 illustrates a segment of a starch polymer showing a simplified representation of glucose molecules.

In the human body, glucose monomers can bond to one another to form *glycogen*, a small polysaccharide stored in the muscles and liver as an energy resource. Excess glucose, however, is converted to lipids and stored in the body as fat. When energy is expended for physical activity, glycogen is released. The glycogen in turn is broken down to release glucose molecules. Glucose molecules are further broken down to produce carbon dioxide, water, and energy for the body.

Cellulose is a polysaccharide found in plants and trees. When cellulose is broken down with dilute acid, the only product obtained is glucose. Thus, starch and cellulose are both polysaccharides composed exclusively of glucose units. Although the distinction is subtle, the glucose units in starch and cellulose are not linked together in exactly the same way. In cellulose every other glucose unit is flipped upside down. Figure 20.13 illustrates a segment of a cellulose polymer showing a simplified version of glucose molecules.

The subtle structural difference between starch and cellulose reveals that nature is highly selective. That is, most animals possess an enzyme that digests starch, but not cellulose. However, bacteria in the digestive tract of grazing animals and termites have enzymes that can cleave the inverted glycoside linkages in cellulose.

▶ **Figure 20.12 A Segment of Starch** Starch is a polymer of repeating glucose units. In addition to the straight chain molecule shown, a starch molecule has numerous branches that are also composed of glucose units.

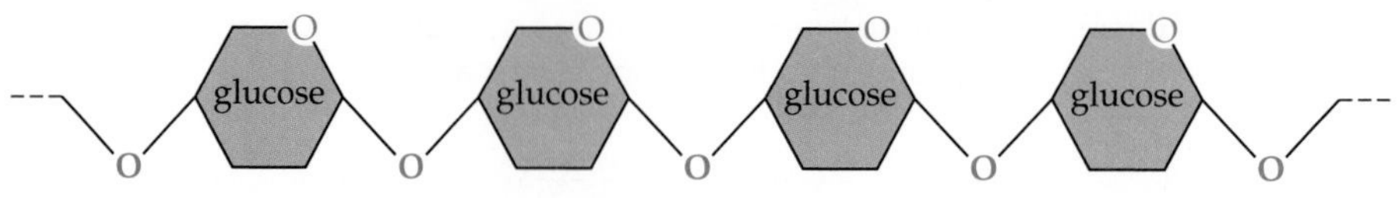

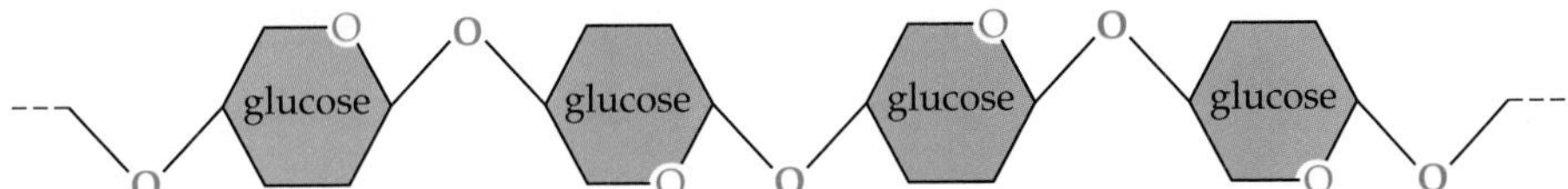

▲ **Figure 20.13 A Segment of Cellulose** Cellulose is a polymer of repeating glucose units. Note the position of the "O" in each glucose ring, which indicates that every other glucose unit is flipped upside down.

Thus, cows can digest the cellulose in grass, and termites can digest wood. Humans, however, cannot digest either grass or wood because we lack the enzyme to cleave the —O— bonds in cellulose.

20.5 LIPIDS

OBJECTIVE

- To recognize the structure of a triglyceride.
- To recognize the structure of a phospholipid.
- To recognize the structure of a lipid wax.
- To recognize the ring structure of a steroid.

Lipids are biochemical compounds that are not soluble in water. This property sets them apart from proteins, carbohydrates, and nucleic acids that are soluble in water. Lipids include the familiar compounds we refer to as fats, oils, and waxes. Lipids also include steroids, such as cholesterol, testosterone, and progesterone.

Triglycerides (Triacylglycerols)

A **triglyceride** is a lipid formed from glycerol, a trihydroxy alcohol, and three long-chain carboxylic acids. A carboxylic acid with a long hydrocarbon chain is referred to as a **fatty acid**. When three fatty acids (RCOOH, R'COOH, R"COOH) react with glycerol, $C_3H_5(OH)_3$, a triglyceride is produced that has three ester linkages. We can represent the general reaction for the formation of a triglyceride as follows:

Although the term *triglyceride* is common, the term triacylglycerol is currently more acceptable.

$$
\begin{array}{lll}
CH_2\text{—OH} & + \; \text{HO—}\overset{\overset{O}{\|}}{C}\text{—R} & \\
| & & \\
CH\text{—OH} & + \; \text{HO—}\overset{\overset{O}{\|}}{C}\text{—R}' & \\
| & & \\
CH_2\text{—OH} & + \; \text{HO—}\overset{\overset{O}{\|}}{C}\text{—R}'' & \\
\text{glycerol} & \text{fatty acids} &
\end{array}
\longrightarrow
\begin{array}{l}
CH_2\text{—O—}\overset{\overset{O}{\|}}{C}\text{—R} \\
| \\
CH\text{—O—}\overset{\overset{O}{\|}}{C}\text{—R}' \; + \; 3\;\text{HOH} \\
| \\
CH_2\text{—O—}\overset{\overset{O}{\|}}{C}\text{—R}'' \\
\text{triglyceride (triacylglycerol)}
\end{array}
$$

The notations R, R', and R" represent long hydrocarbon chains on the three fatty acids. If the triglyceride is obtained from an animal source, the fatty acid chains are mostly saturated and have few double bonds. A semisolid lipid obtained from an animal source is called a **fat**. Table 20.1 lists some common saturated fatty acids present in animal fat.

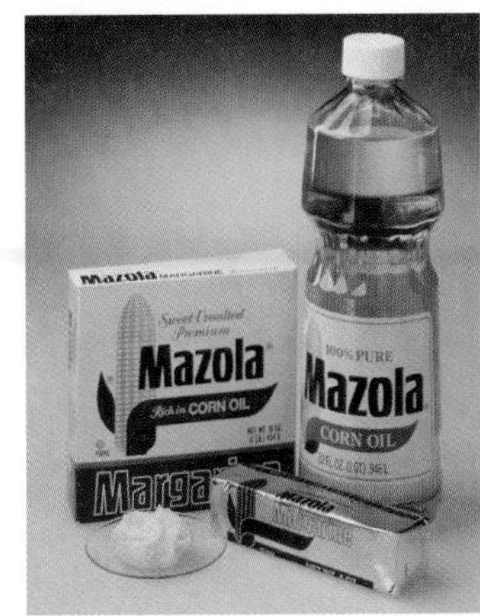

▲ **Lipid Foods** Vegetable oils, margarine, and "rich, creamy" foods are all sources of lipids.

TABLE 20.1 COMMON SATURATED FATTY ACIDS

NAME	STRUCTURE
lauric acid	$CH_3–CH_2–CH_2–CH_2–CH_2–CH_2–CH_2–CH_2–CH_2–CH_2–CH_2–COOH$ $CH_3–(CH_2)_{10}–COOH$
myristic acid	$CH_3–(CH_2)_{12}–COOH$
palmitic acid	$CH_3–(CH_2)_{14}–COOH$
stearic acid	$CH_3–(CH_2)_{16}–COOH$

The following structure illustrates an example of a saturated triglyceride found in animal fat:

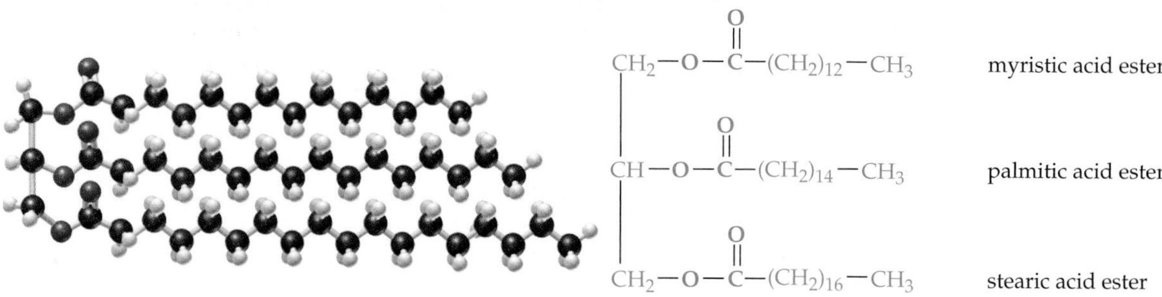

A liquid lipid obtained from a plant source is called an **oil**. If a triglyceride is obtained from a plant source, the fatty acid chains are mostly unsaturated and have one or more double bonds. Table 20.2 lists some common unsaturated fatty acids that occur in vegetable oils.

TABLE 20.2 COMMON UNSATURATED FATTY ACIDS

NAME	STRUCTURE
oleic acid	$CH_3{-}CH_2{-}CH_2{-}CH_2{-}CH_2{-}CH_2{-}CH_2{-}CH_2{-}CH{=}CH{-}(CH_2)_7{-}COOH$
linoleic acid	$CH_3{-}CH_2{-}CH_2{-}CH_2{-}CH_2{-}CH{=}CH{-}CH_2{-}CH{=}CH{-}(CH_2)_7{-}COOH$
linolenic acid	$CH_3{-}CH_2{-}CH{=}CH{-}CH_2{-}CH{=}CH{-}CH_2{-}CH{=}CH{-}(CH_2)_7{-}COOH$

The following structure illustrates an example of an unsaturated triglyceride found in vegetable oils. Note that the fatty acids have one, two, and three double bonds, respectively.

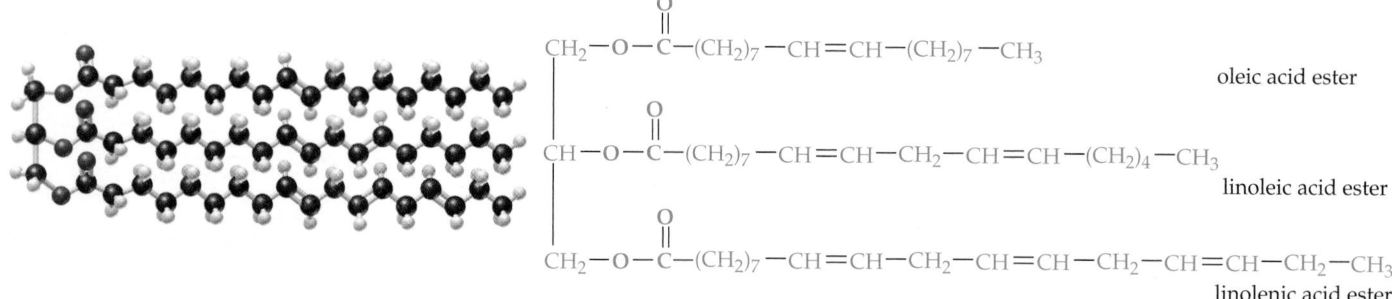

The distinction between a fat and an oil, based on the saturation of the fatty acids, is a generalization with numerous exceptions. That is, even fats have a small percentage of unsaturated fatty acids, and oils contain a few saturated fatty acids. Moreover, a natural source of a fat or oil is a complex mixture of triglycerides.

A detailed analysis of a fat or oil indicates the average percentage of each fatty acid. Corn oil, for example, averages about 2% myristic acid, 8% palmitic acid, 5% stearic acid, 35% oleic acid, and 50% linoleic acid. Thus, corn oil averages about 15% saturated fatty acids and 85% unsaturated fatty acids.

Saponification of Fats and Oils

The ester linkage in a triglyceride can be broken by treatment with aqueous sodium hydroxide. The products of the reaction are glycerol and three sodium salts of fatty acids. Sodium salts of fatty acids constitute what we refer to as soap. This reaction is called **saponification**, which literally means "soap making." We can represent the general reaction for the saponification of a triglyceride as follows:

$$\begin{array}{l} CH_2-O-\overset{O}{\overset{\|}{C}}-R \\ | \\ CH-O-\overset{O}{\overset{\|}{C}}-R' \\ | \\ CH_2-O-\overset{O}{\overset{\|}{C}}-R'' \\ \text{triglyceride} \end{array} + 3\,NaOH \longrightarrow \begin{array}{l} CH_2-OH \\ | \\ CH-OH \\ | \\ CH_2-OH \\ \text{glycerol} \end{array} \begin{array}{l} + \quad R\mathbf{COO^-Na^+} \\ \\ + \quad R'\mathbf{COO^-Na^+} \\ \\ + \quad R''\mathbf{COO^-Na^+} \\ + \quad \text{soap} \end{array}$$

Soap has a polar ionic "head" (COO^-Na^+) and a long nonpolar "tail" (R, R', or R"). Although oil and water do not mix, we can use soap to remove an oily residue from our hands. Soap orients its fatty acid nonpolar "tail" toward the oily layer on our hands, and its polar "head" toward the water. Soap thereby provides a link between nonpolar oil molecules and polar water molecules (Figure 20.14). Thus, washing our hands with soap and water can remove an oily residue even though oil and water are not miscible.

Phospholipids

A **phospholipid** is a glyceride found in living cells and cell membranes. Unlike fats and oils, which are triglycerides, phospholipids have *two* fatty acids esterified to glycerol. The third –OH in glycerol is linked to phosphoric acid. Thus, the properties

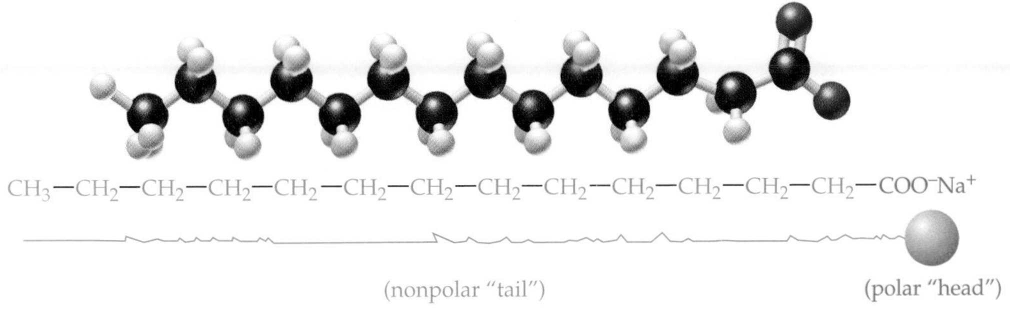

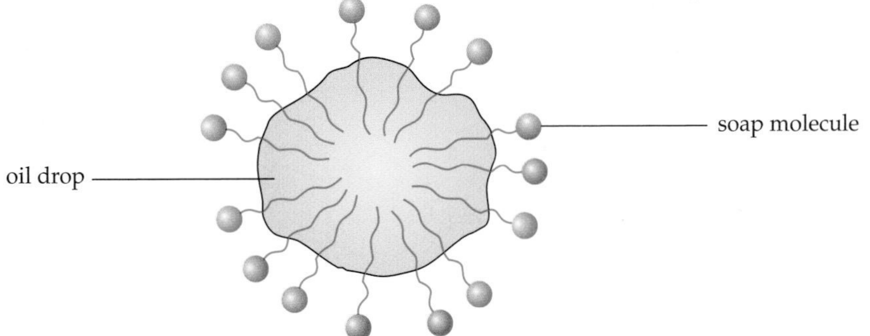

◀ **Figure 20.14 The Action of Soap** A soap molecule ($R-COO^-Na^+$) has a long nonpolar "tail" and a polar "head" ($-COO^-Na^+$). Soap is effective in removing oil because soap has a "tail" that is soluble in oil, and a "head" that is soluble in water.

of phospholipids are different from those of fats and oils. With the addition of a phosphate group, a phospholipid has an ionic "head" that is soluble in water, as well as a nonpolar "tail" that is insoluble. In fact, the phospholipid lecithin is used to emulsify margarine and chocolate, which are otherwise insoluble in water. The following example structure illustrates a typical composition of lecithin:

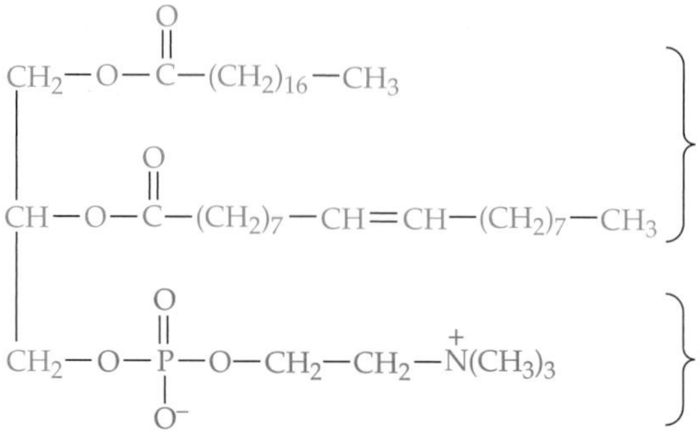

$$\begin{array}{l} CH_2-O-\overset{O}{\overset{\|}{C}}-(CH_2)_{16}-CH_3 \\ | \\ CH-O-\overset{O}{\overset{\|}{C}}-(CH_2)_7-CH=CH-(CH_2)_7-CH_3 \\ | \\ CH_2-O-\underset{O^-}{\underset{|}{\overset{O}{\overset{\|}{P}}}}-O-CH_2-CH_2-\overset{+}{N}(CH_3)_3 \end{array}$$

nonpolar "tail" (first two chains); ionic "head" (phosphate–choline group)

(choline)

The phosphate ester in lecithin is bonded to choline, $HO{-}CH_2CH_2N^+(CH_3)_3$, as well as glycerol. Choline is a B-complex vitamin found in egg yolks and is essential to liver function.

Waxes

A lipid **wax** is a mixture of naturally occurring esters of fatty acids and long-chain alcohols. An example is beeswax, which is found in honeycombs and is used to make candles and shoe polish. Another example is carnauba wax, an ingredient in car polish and furniture polish. A lipid wax is water insoluble and occurs naturally as the slippery coating on fruits, vegetables, and plant leaves. Since waxes are insoluble in water, they can be used to waterproof fabrics and wood. The following example structure illustrates one of the typical esters found in beeswax:

$$CH_3-(CH_2)_{34}-\overset{O}{\overset{\|}{C}}-O-(CH_2)_{35}-CH_3 \qquad \text{Beeswax}$$

Waxes are esters found in nature and, like triglycerides, demonstrate variety in their fatty acid composition. A typical analysis of carnauba wax reveals a 12-carbon unsaturated fatty acid and a broad mixture of long-chain alcohols.

Steroids

A **steroid** belongs to a special class of lipids that has a structure composed of four rings of carbon atoms fused together. Cholesterol is an example of an important steroid and is the precursor of many other steroid molecules essential to human life. On the other hand, high levels of cholesterol have been linked to the formation of plaque in arteries, which can contribute to heart disease. We can diagram the ring structure of cholesterol as follows:

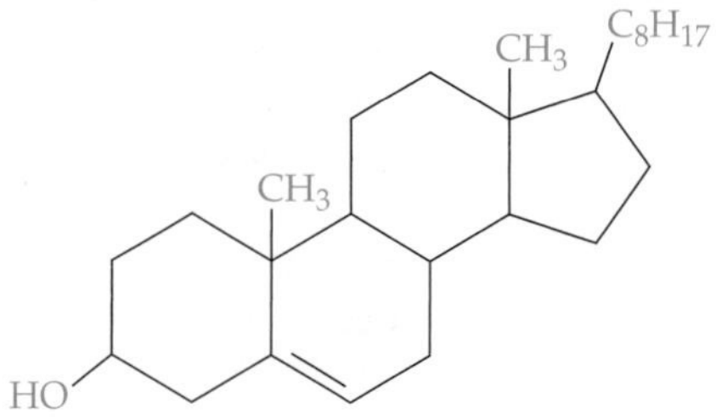

CH_3, C_8H_{17}, CH_3, HO

Cholesterol

As mentioned previously, sex hormones are steroids. This includes the hormones testosterone and progesterone secreted by the testes and ovaries, respectively. Testosterone is responsible for the sex characteristics of males, while progesterone helps regulate ovulation in females. It is interesting to note that the two hormones are quite similar and differ only slightly in molecular structure.

Testosterone

Progesterone

CRITICAL THINKING Vitamins

Q: Which vitamins are fat soluble and should not be taken in excessive amounts?

Vitamins are biochemical compounds required for normal metabolism. Vitamins are not synthesized in the human body, and therefore must be included in the diet. If a vitamin is not present in the diet, its absence can lead to a vitamin-deficiency disease.

The U.S. Food and Nutrition Board has established minimal vitamin requirements. The board periodically publishes a new listing of the Recommended Dietary Allowance (RDA) for vitamins based on current research. These RDA values are intended as a guideline for estimating the needs of adults and children and may vary depending on individual circumstances. Ingesting polluted air or water can increase vitamin requirements. Some infectious diseases, accompanied by a high fever, may also result in increased vitamin requirements.

In most instances, a diet containing fresh fruits, vegetables, dairy products, and meat ensures an adequate intake of necessary vitamins. Vitamin supplements, however, may be recommended by a physician during periods of stress, growth, or pregnancy. There is a growing body of evidence that vitamins A, C, and E may lessen the risk of cancer. These vitamins are antioxidants, and the so-called "free-radical theory" proposes that oxidizing molecular radicals build up in tissues and can alter the structure of DNA. Vitamins C and E, in conjunction with the mineral selenium, help to clear the tissues of toxic free radicals.

The consumption of megadoses of vitamin supplements has recently raised controversy. Excess amounts of water-soluble vitamins (B_1, B_2, B_3, B_6, B_{12}, and C) are rapidly excreted in the urine and are rarely toxic. Although vitamin C is relatively nontoxic, daily doses in excess of a gram may lead to skin rashes and diarrhea in some individuals. A deficiency of vitamin C, ascorbic acid, can lead to the disease scurvy, which is characterized by loose teeth and bleeding gums. Citrus fruits prevent and cure this disease. In the early 1800s, the British navy began issuing rations of lemons and limes to sailors who acquired the nickname "limeys."

Excess amounts of fat-soluble vitamins (A, D, E, and K) can accumulate in body fat tissue. Excess vitamin A, retinol, can lead to symptoms of toxin poisoning. The symptoms include irritability, vomiting, loss of appetite, headache, and dry skin. On the other hand, a deficiency of vitamin A has long been known to cause night blindness.

Vitamin D is synthesized by the body when skin is exposed to sunlight. An excess of vitamin D can cause weakness, fatigue, headache, nausea, and vomiting. Vitamin D deficiencies are rare in tropical climates and are more common in northern regions. A vitamin D deficiency results in failure to absorb calcium and phosphorus, causing malformation of the bones. In children, the disease is known as rickets and is characterized by a deformed rib cage and bowed legs.

A supplement of vitamin E, alpha tocopherol, has been advocated for many disorders, especially coronary artery disease. In moderate excess, it is considered relatively nontoxic. Vitamin K is available in milk products and is essential for the clotting of blood. In addition, a wide variety of foods such as vegetables, eggs, and fish contain vitamin K, so a deficiency of this vitamin is rare.

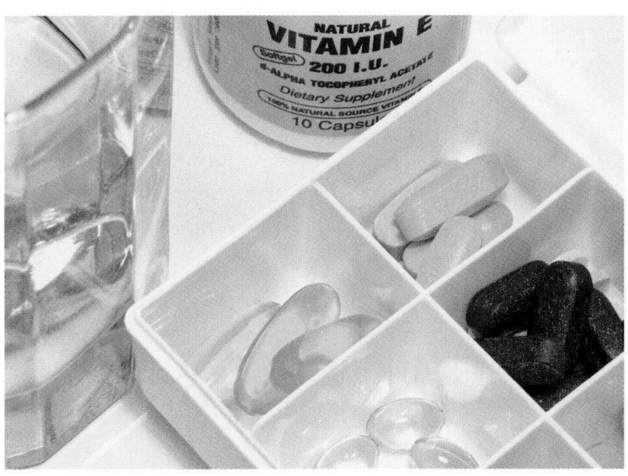

▲ **Vitamins** A sampling of various vitamins that are available as tablets and gelatin capsules.

A: Vitamins A, D, E, and K are fat-soluble and should not be ingested in excessive amounts.

OBJECTIVES

- To state the three components of a nucleotide.
- To distinguish between DNA and RNA nucleotides.
- To explain the process of protein synthesis.

20.6 NUCLEIC ACIDS

Nucleic acids are biochemical compounds found in every living cell. Nucleic acids contain the genetic information responsible for the reproduction of a species. There are two types of nucleic acids: deoxyribonucleic acid (DNA) and ribonucleic acid (RNA).

A nucleic acid is a polymer composed of many repeating units, each of which is called a **nucleotide**. All nucleotides are similar and consist of a sugar, a base, and phosphoric acid. These three components form the general structure of a nucleotide shown in the Figure 20.15.

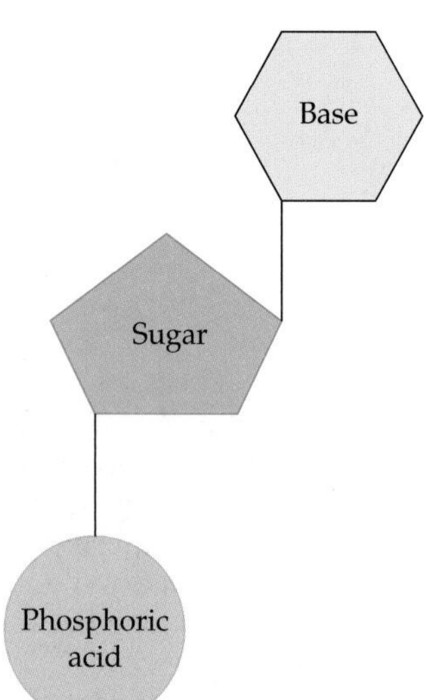

▲ **Figure 20.15 Nucleic Acid Nucleotide** A nucleic acid is a polymer of nucleotides. A nucleotide is composed of ribose or deoxyribose sugar and an attached nitrogen base. Nucleotides are joined together in the nucleic acid polymer by phosphoric acid linkages.

The sugars found in DNA and RNA differ slightly. That is, a DNA nucleotide contains the sugar deoxyribose, and RNA contains the sugar ribose. These sugars are identical, except for a hydroxyl group (–OH).

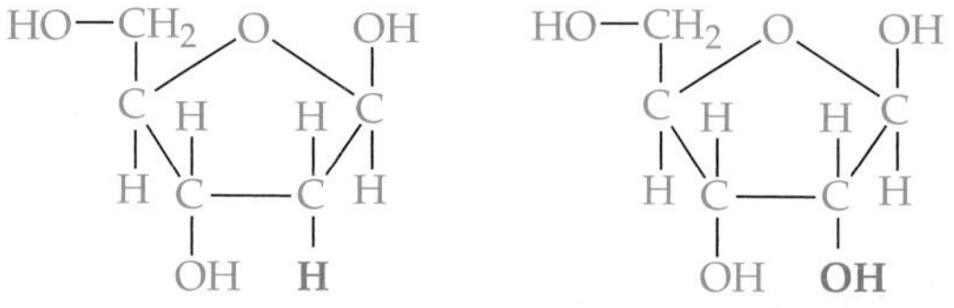

Deoxyribose (DNA) | Ribose (RNA)

▲ Notice that deoxyribose, the sugar in DNA, and ribose, the sugar in RNA, differ by a single hydroxyl group (–OH).

With one exception, the bases found in DNA and RNA are the same. DNA and RNA each contain adenine (A), cytosine (C), and guanine (G). The difference is that DNA contains the base thymine (T), while RNA contains the base uracil (U). Thymine and uracil are identical, except for a methyl group. Figure 20.16 shows the structures for the bases in DNA and/or RNA.

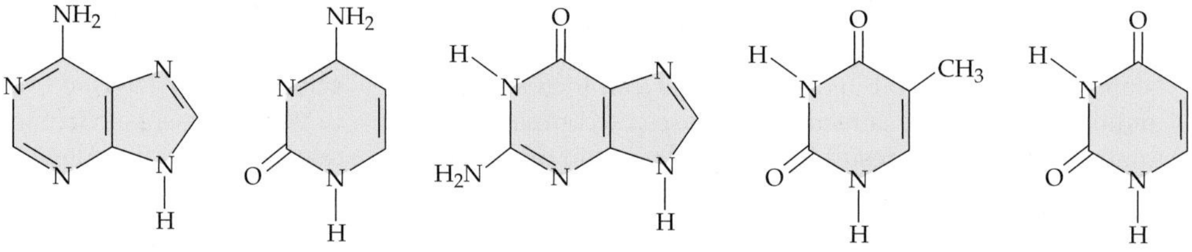

Adenine (A)	Cytosine (C)	Guanine (G)	Thymine (T)	Uracil (U)
[DNA]	[DNA]	[DNA]	[DNA]	
[RNA]	[RNA]	[RNA]		[RNA]

▲ **Figure 20.16 Nucleic Acid Bases** Note that DNA contains the four bases adenine (A), cytosine (C), guanine (G), and thymine (T). RNA contains the four bases A, C, G, and uracil (U).

Now that we are familiar with the three components of a nucleotide, let's see how the sugar, base, and phosphoric acid groups are arranged. Figure 20.17 illustrates a DNA nucleotide and an RNA nucleotide.

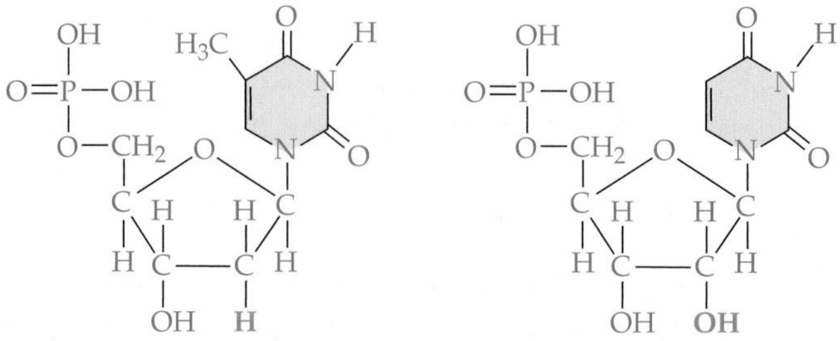

DNA nucleotide | RNA nucleotide

▶ **Figure 20.17 DNA and RNA Nucleotides** Note that the DNA nucleotide contains deoxyribose sugar and that the RNA nucleotide contains ribose sugar. In this example, deoxyribose in the DNA nucleotide has an attached thymine base, and ribose has an attached uracil base.

We can draw a short segment of a nucleic acid. That is, we can diagram a few nucleotides joined together in a chain. For simplicity, we will not specify the bases. Figure 20.18 illustrates a small chain of nucleotides found in a segment of a nucleic acid.

```
       OH
       |
     O=P—OH
       |                   Base
       O—CH2  O
         |         \
         C  H    H  C
         | \|    |/ |
         H  C————C  H
            |    |
            O    H
            |
  [       O=P—OH                  ]
            |            Base
            O—CH2  O
              |         \
              C  H    H  C
              | \|    |/ |
              H  C————C  H
                 |    |
                 O    H
                 |                  n
               O=P—OH
                 |            Base
                 O—CH2  O
                   |         \
                   C  H    H  C
                   | \|    |/ |
                   H  C————C  H
                      |    |
                      OH   H
```

◀ **Figure 20.18 A Chain of Nucleotides** A nucleic acid is composed of many nucleotide molecules joined by phosphate linkages. Here the central nucleotide can vary, but is repeated a number of times (*n*) in the long chain.

The DNA Double Helix

A landmark scientific achievement was the determination of the structure of DNA by the English chemist Francis Crick (1916–2004) and the American biochemist James Watson (1928–). They determined that a DNA molecule is composed of two long chains of DNA nucleotides. These two strands of DNA are wound together in a spiral structure that is referred to as a "double helix."

◀ **DNA Spiral Helix**

The two strands of the DNA helix are held together by hydrogen bonds between bases on the two separate chains. Specifically, adenine (A) on one strand *always* hydrogen bonds to thymine (T) on the other strand. In Figure 20.19, note that there are two hydrogen bonds between each pair of these bases (A=T or T=A). Moreover, cytosine (C) on one strand *always* hydrogen bonds to guanine (G) on the other strand. Note that there are three hydrogen bonds between each pair of these bases (C≡G or G≡C).

The term **replication** describes the process by which a single molecule of DNA synthesizes two molecules of DNA. First, the DNA molecule unwinds by breaking hydrogen

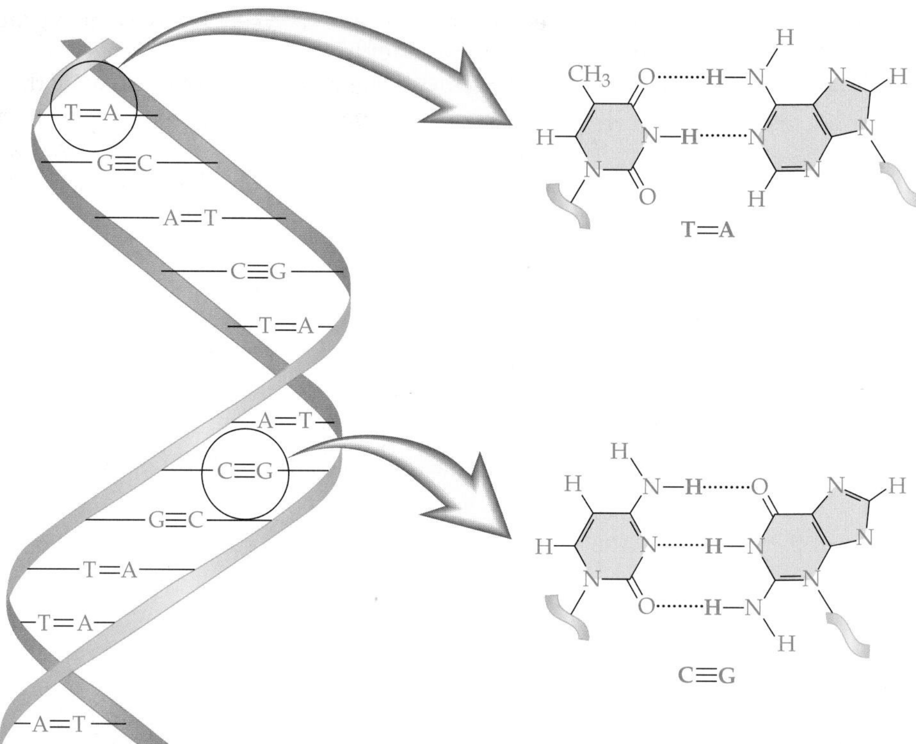

▶ **Figure 20.19 DNA Double Helix** The two DNA strands of nucleotides are held together by hydrogen bonds. The hydrogen bonds are always between the same two bases. Notice that there are two hydrogen bonds between thymine (T) and adenine (A) and that there are three hydrogen bonds between cytosine (C) and guanine (G).

bonds between bases on the two strands of the double helix. Then, each single strand of DNA acts as a template to synthesize a new strand of DNA, one nucleotide at a time.

Since the bases always pair in the same way, A=T, T=A, C≡G, or G≡C, each base in the DNA template strand always codes for its complementary base on the growing strand of DNA. That is, adenine (A) on the template strand codes for thymine (T) on the growing strand, thymine (T) codes for adenine (A), cytosine (C) codes for guanine (G), and guanine (G) codes for cytosine (C).

When replication is complete, each strand in the original molecule of DNA has produced a complementary strand of DNA. Each template strand is hydrogen bonded to its complementary strand in a new DNA double helix. The final result is two molecules of DNA, each of which is identical to the original molecule of DNA.

Protein Synthesis

DNA is the parent molecule responsible for transmitting genetic information. DNA can transfer the genetic code to RNA, which contains the instructions for protein synthesis. First, a molecule of DNA in the cell nucleus synthesizes a molecule of RNA. That is, a DNA molecule transcribes its genetic code into an RNA molecule. Next, the RNA molecule moves out of the cell nucleus into the surrounding cytoplasm, where protein synthesis takes place.

Unlike DNA, RNA is a single stranded polymer of nucleotides. The term **transcription** describes the process by which a DNA double helix unwinds and synthesizes two complementary strands of RNA. That is, a string of nucleotides in each DNA strand codes for a complementary strand of nucleotides in a growing RNA chain. Specifically, thymine (T) on the DNA template strand codes for adenine (A) on the growing RNA chain, cytosine (C) codes for guanine (G), and guanine (G) codes for cytosine (C). But unlike replication, adenine (A) codes for uracil (U), not thymine (T), on a growing RNA chain.

The genetic code involves three consecutive nucleotides in an RNA molecule, and each of these trinucleotides is referred to as a *codon*. Each codon specifies a particular amino acid to be used in building a protein. For example, the GCU codon represents three consecutive nucleotides having the bases guanine, cytosine, and uracil, respectively.

The GCU codon is specific for the amino acid alanine, and is responsible for adding alanine to a growing protein chain. Other examples include the CGU codon, which specifies arginine; and the UUU codon, which specifies phenylalanine. Thus, each RNA codon designates a specific amino acid, in a given sequence, in a growing protein chain. This process is termed *translation*.

There are 64 codons corresponding to all the possible combinations of the four bases (4^3). Since there are only 20 amino acids, more than one codon may specify a given amino acid; for example, GCU, GCC, GCA, and GCG all code for the amino acid alanine. There are three so-called "stop codons" (UAA, UAG, UGA) and their function is to terminate a growing protein chain.

CRITICAL THINKING Left- and Right-Handed Molecules

Q: Which of the cups shown below has a nonidentical mirror image?

One of the reasons that there are so many different organic compounds is due to the phenomenon of isomers. In Chapter 19, we learned about structural isomers such as butane and isobutane, and functional group isomers such as alcohols and ethers having the same molecular formula.

In addition to structural isomers, organic compounds can exist as "stereoisomers" that differ in their spatial arrangement of atoms about a central carbon atom. That is, if a carbon atom has four different attached groups (for example, F, Cl, Br, and I), there are two possible spatial orientations and two molecules exist that are nonidentical mirror images. Human hands are an example of nonidentical mirror images. Although our left and right hands are similar, they are not superimposable. If you hold a mirror in front of your left hand, the reflected image in the mirror is not identical to your left hand, but is similar to your right hand. Carefully observe the figure (right) and you will see that the two organic molecules are nonidentical mirror images.

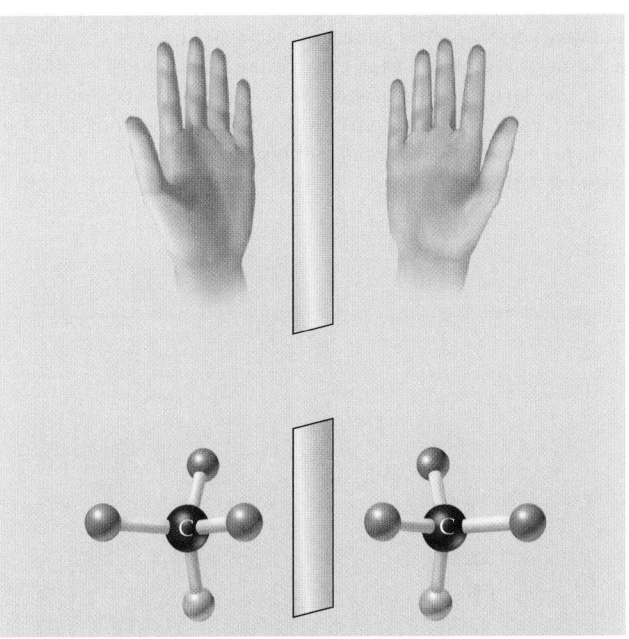

In this chapter we learned that amino acids have an amine group on the alpha (α) carbon. Examine the figure (below) and note that there are four nonidentical groups attached to the alpha carbon atom:

$$\mathrm{NH_2{-}\underset{\displaystyle R}{\underset{|}{\overset{\displaystyle H}{\overset{|}{C}}}}H{-}\overset{\displaystyle O}{\overset{\|}{C}}{-}OH}$$

Alpha (α) amino acid

Since amino acids have four nonidentical groups on the alpha carbon, we know that stereoisomers are possible. When chemists synthesize amino acids in the laboratory, they obtain both the left- and right-handed stereoisomers, unless a special left- or right-handed reagent is used in the synthesis. However, only one of the stereoisomers is found in nature. Nature produces only the left-handed amino acid and not the right-handed molecule. Similarly, sugar molecules like glucose have four nonidentical groups attached to carbon atoms, which gives rise to stereoisomers. Interestingly, nature produces only the right-handed sugar and not the left-handed molecule.

When we studied isomers in Chapter 19, we learned that isomers had different physical and chemical properties. Stereoisomers have identical properties with one exception. A beam of polarized light passed through a solution containing a stereoisomer rotates the light either to the left or to the right. The nonidentical mirror image rotates the polarized light the same number of degrees in the opposite direction. A solution containing equal concentrations of each stereoisomer exhibits zero rotation of polarized light.

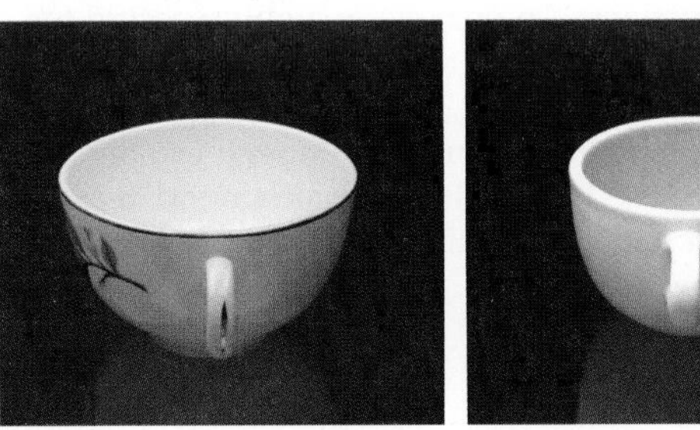

▲ Unsymmetrical and Symmetrical Cups.

A: The cup on the left has a design on one side; the cup on the right does not. Thus, the cup on the left has a nonidentical mirror image; the cup on the right has an identical mirror image.

Summary

Section 20.1 **Biochemistry** is the study of biological compounds and their chemical reactions. A **protein** is a polymer composed of amino acids. A **carbohydrate** is a compound consisting of one or more simple sugar molecules. A **lipid** is a water-insoluble biological compound such as a fat, oil, wax, or steroid. A **nucleic acid** is a polymer composed of a sugar molecule, a nitrogen base, and phosphoric acid. Although biological compounds can be large and complex, they have familiar functional groups such as amines, alcohols, and esters.

Section 20.2 A carboxylic acid having an amine group is called an **amino acid**. A **dipeptide** is two amino acids joined by an amide bond called a **peptide linkage**. A **polypeptide** has several amino acids joined by peptide linkages. A protein is a large polypeptide that can contain thousands of amino acids. The structure of a protein is related to its biological function. Extended protein molecules are responsible for shape and structure, whereas globular proteins often regulate hormonal activity.

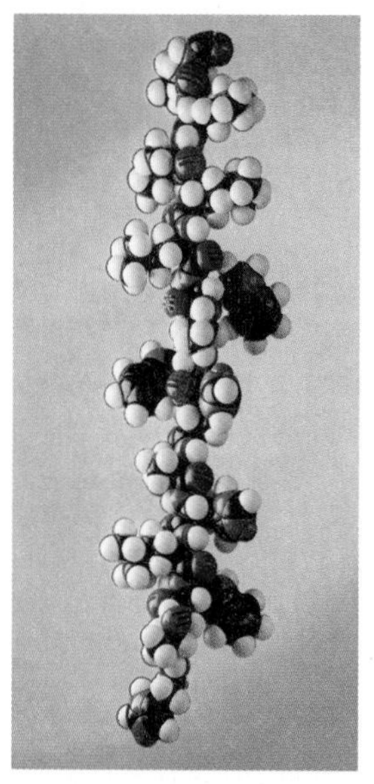

▶ α-**Helix Model** The molecular model shows a portion of a protein chain whose secondary structure is twisted into an alpha helix.

Section 20.3 An **enzyme** is a protein that acts as a catalyst for biochemical reactions. An enzyme is highly specific for a given biological molecule and is quite sensitive to changes in pH and temperature. Although the mechanism of enzyme activity is complex, the **lock-and-key model** explains the selectivity of enzymes. To speed up a biochemical reaction, the shape of the substrate molecule must fit the contour of the active site on an enzyme.

Section 20.4 A carbohydrate is a biological compound that contains one or more sugar molecules. Glucose is a **monosaccharide** corresponding to one simple sugar molecule. Sucrose is a **disaccharide** and contains molecules of glucose and fructose joined by a **glycoside linkage**. Starch is a **polysaccharide** and contains many molecules of glucose joined by glycoside linkages.

Section 20.5 Many lipids contain glycerol, an alcohol with three hydroxyl groups. A **fatty acid** is a carboxylic acid bearing a long hydrocarbon chain. A **triglyceride** is an ester of glycerol and three fatty acids. An animal **fat** is a triglyceride with mostly saturated fatty acids. A vegetable **oil** is a triglyceride with mostly unsaturated fatty acids. Triglycerides react with sodium hydroxide to produce soap and glycerol in a chemical reaction called **saponification**.

A **phospholipid** contains two fatty acids and phosphoric acid joined to glycerol by ester linkages. A lipid **wax** is a simple ester of a fatty acid and a long-chain alcohol. A **steroid** is a lipid with a characteristic structure of four rings of carbon atoms. Steroids include many hormones and fat-soluble vitamins.

Section 20.6 A nucleic acid is a biological compound that carries genetic information. Nucleic acids are polymers, which have repeating units called nucleotides. A **nucleotide** is composed of a sugar, a base, and phosphoric acid. Deoxyribonucleic acid (DNA) and ribonucleic acid (RNA) are two types of nucleic acids. DNA and RNA have slightly different nucleotides whose composition is summarized in Table 20.3.

When a single molecule of DNA synthesizes an identical molecule of DNA, the process is termed **replication**. First, the DNA molecule unwinds by breaking hydrogen bonds between bases in the double helix. Then, each strand of DNA synthesizes a complementary strand of DNA, one nucleotide at a time. Similarly, when a molecule of DNA unwinds and acts as a template for the synthesis of a complementary strand of RNA, the process is termed **transcription**. The addition of an amino acid to a growing protein chain during protein synthesis is specified by an RNA trinucleotide, which is called a *codon*, in a process termed *translation*.

TABLE 20.3 DNA VERSUS RNA NUCLEOTIDES

COMPONENT	DNA	RNA
Sugar	deoxyribose	ribose
Base	adenine (A)	adenine (A)
	cytosine (C)	cytosine (C)
	guanine (G)	guanine (G)
	thymine (T)	uracil (U)
Inorganic Acid	phosphoric acid	phosphoric acid

Key Terms *Answers to Key Terms are in Appendix H.*

Select the key term that corresponds to each of the following definitions.

_____ **1.** the study of biological compounds
_____ **2.** a biological compound that is a polymer of amino acids
_____ **3.** a biological compound composed of one or more sugar molecules
_____ **4.** a biological compound that is insoluble in water, such as a fat, oil, or wax
_____ **5.** a biological polymer compound that can transmit genetic information
_____ **6.** a carboxylic acid with an amine group on the alpha carbon
_____ **7.** an amide bond that joins two amino acids
_____ **8.** two amino acids joined by a peptide linkage
_____ **9.** 10–50 amino acids joined by peptide linkages
_____ **10.** a protein molecule that catalyzes a biochemical reaction
_____ **11.** a theory that explains enzyme catalysis
_____ **12.** a carbohydrate composed of a simple sugar molecule
_____ **13.** a carbohydrate composed of two simple sugar molecules
_____ **14.** a carbohydrate composed of many simple sugar molecules
_____ **15.** an –O– bond that joins two simple sugars
_____ **16.** a carboxylic acid with a long hydrocarbon chain
_____ **17.** a lipid composed of glycerol and three fatty acids
_____ **18.** a triglyceride from an animal source that has mostly saturated fatty acids
_____ **19.** a triglyceride from a plant source that has mostly unsaturated fatty acids
_____ **20.** a chemical reaction of a fat or oil with sodium hydroxide to produce soap
_____ **21.** a lipid composed of glycerol, two fatty acids, and phosphoric acid
_____ **22.** a lipid composed of a fatty acid and a long-chain alcohol
_____ **23.** a lipid hormone composed of four rings of carbon atoms
_____ **24.** a repeating unit in a nucleic acid composed of a sugar, a base, and phosphoric acid
_____ **25.** the process by which a DNA molecule synthesizes an identical molecule of DNA
_____ **26.** the process by which a DNA molecule synthesizes a complementary single strand of RNA

(a) amino acid *(Sec. 20.2)*
(b) biochemistry *(Sec. 20.1)*
(c) carbohydrate *(Sec. 20.1)*
(d) dipeptide *(Sec. 20.2)*
(e) disaccharide *(Sec. 20.4)*
(f) enzyme *(Sec. 20.3)*
(g) fat *(Sec. 20.5)*
(h) fatty acid *(Sec. 20.5)*
(i) glycoside linkage *(Sec. 20.4)*
(j) lipid *(Sec. 20.1)*
(k) lock-and-key model *(Sec. 20.3)*
(l) monosaccharide *(Sec. 20.4)*
(m) nucleic acid *(Sec. 20.1)*
(n) nucleotide *(Sec. 20.6)*
(o) oil *(Sec. 20.5)*
(p) peptide linkage *(Sec. 20.2)*
(q) phospholipid *(Sec. 20.5)*
(r) polypeptide *(Sec. 20.2)*
(s) polysaccharide *(Sec. 20.4)*
(t) protein *(Sec. 20.1)*
(u) replication *(Sec. 20.6)*
(v) saponification *(Sec. 20.5)*
(w) steroid *(Sec. 20.5)*
(x) transcription *(Sec. 20.6)*
(y) triglyceride *(Sec. 20.5)*
(z) wax *(Sec. 20.5)*

Exercises *Answers to odd-numbered Exercises are in Appendix I.*

Biological Compounds (Sec. 20.1)

1. Identify the type of biological compound having each of the following structures:

(a) --- Amino acid —C(=O)—NH— Amino acid ---

(b) --- Sugar(Base) —O—P(=O)(OH)—O— Sugar(Base) ---

2. Identify the type of biological compound having each of the following structures:

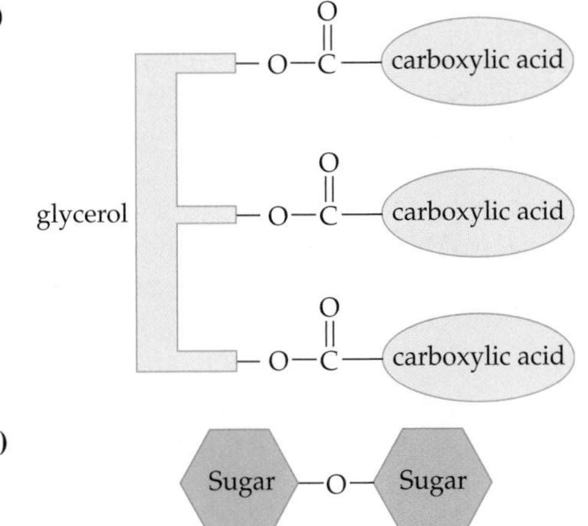

3. Identify the type of linkage that joins repeating units in each of the following:
   (a) protein (b) nucleic acid
4. Identify the type of linkage that joins each of the following:
   (a) two sugars in a carbohydrate
   (b) glycerol and a carboxylic acid in a lipid
5. Identify one or more organic functional groups that characterize each of the following biological compounds:
   (a) protein (b) carbohydrate
   (c) lipid (d) nucleic acid
6. Identify one or more organic functional groups present in each of the following biological compounds:
   (a) amino acid (b) ribose sugar
   (c) fatty acid (d) nitrogen base

Proteins (Sec. 20.2)

7. What type of bonds are responsible for the primary structure of a protein?
8. What type of bonds are responsible for the secondary structure of a protein?
9. Explain the difference between the primary and secondary structure of a protein.
10. Explain the difference between the secondary and tertiary structure of a protein.
11. Draw the structure of the dipeptide formed by two molecules of the amino acid alanine. Circle the peptide linkage.
12. Draw the structure of the dipeptide formed by two molecules of the amino acid cysteine. Circle the peptide linkage.
13. Draw the structure of the dipeptide seryltyrosine (Ser–Tyr).
14. Draw the structure of the dipeptide tyrosylserine (Tyr–Ser).
15. Indicate all the possible sequences for a tripeptide containing glycine, histidine, and tryptophan; use the standard abbreviation for each amino acid.
16. Indicate all the possible sequences for a tripeptide containing alanine, glutamine, and methionine; use the standard abbreviation for each amino acid.
17. A segment of a protein chain contains the following tripeptide. Identify the three amino acids, given the structure for the tripeptide.

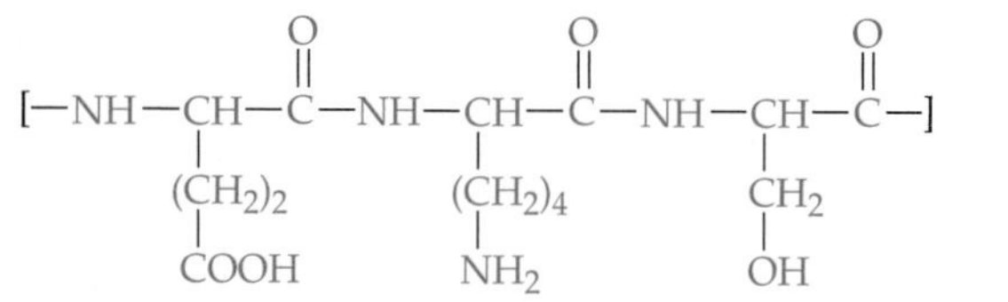

[—NH—CH(—(CH$_2$)$_2$—COOH)—C(=O)—NH—CH(—(CH$_2$)$_4$—NH$_2$)—C(=O)—NH—CH(—CH$_2$—OH)—C(=O)—]

18. The artificial sweetener NutraSweet is the methyl ester of a dipeptide. Identify the two amino acids, given the structure for NutraSweet.

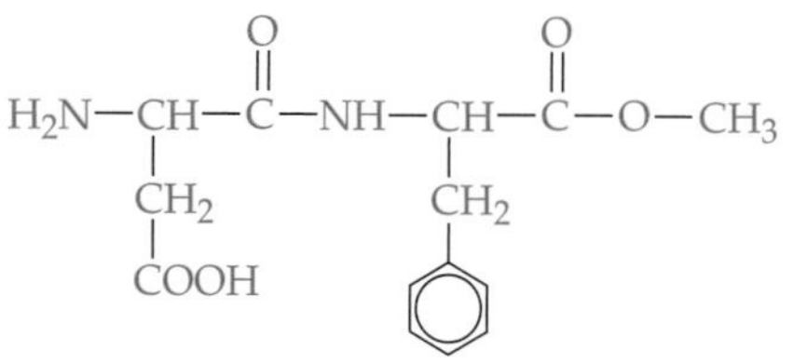

H$_2$N—CH(—CH$_2$—COOH)—C(=O)—NH—CH(—CH$_2$—C$_6$H$_5$)—C(=O)—O—CH$_3$

19. The two hormones vasopressin and oxytocin are polypeptides with a similar amino acid composition. Given the primary sequence for each hormone, identify the amino acids that differ in each polypeptide.

vasopressin: Cys–Tyr–Phe–Gln–Asn–Cys–Pro–Arg–Gly
oxytocin: Cys–Tyr–Ile–Gln–Asn–Cys–Pro–Leu–Gly

20. Vasopressin raises blood pressure in humans, and oxytocin induces lactation in mammals. Suggest a reason why the biological activity of the polypeptides differs.

Enzymes (Sec. 20.3)

21. In the lock-and-key model of enzyme catalysis, what does the lock represent?
22. In the lock-and-key model of enzyme catalysis, what does the key represent?
23. What is the term for a location on an enzyme that conforms to the shape of the substrate?
24. What do the "teeth" on the key represent in the lock-and-key model of enzyme catalysis?
25. Describe Step 1 in the enzyme catalysis for the reaction of a substrate.
26. Describe Step 2 in the enzyme catalysis for the reaction of a substrate.
27. What are the experimental conditions that make an enzyme inactive?
28. What is the term for a molecule that blocks an enzyme and makes it inactive?

Carbohydrates (Sec. 20.4)

29. What are the functional groups in an aldose sugar?
30. What are the functional groups in a ketose sugar?
31. How many carbon atoms are in a molecule of a hexose sugar?
32. How many carbon atoms are in a molecule of a pentose sugar?
33. What is the distinction between a monosaccharide and a disaccharide?
34. What is the distinction between a disaccharide and a polysaccharide?
35. Draw the open-chain structure for a fructose molecule.
36. Draw the open-chain structure for a ribose molecule.
37. Draw the cyclic structure for a galactose molecule in an aqueous solution.
38. Draw the cyclic structure for a glucose molecule in an aqueous solution.
39. Complete the following reaction for the acid hydrolysis of maltose and label the two products:

$$\underset{\text{maltose}}{C_{12}H_{22}O_{11}} + \underset{\text{water}}{H_2O} \xrightarrow{H^+}$$

40. Complete the following reaction for the acid hydrolysis of lactose and label the two products:

$$\underset{\text{lactose}}{C_{12}H_{22}O_{11}} + \underset{\text{water}}{H_2O} \xrightarrow{H^+}$$

41. What is the monomer monosaccharide in glycogen?

42. What is the biological function of glycogen in the body?

Lipids (Sec. 20.5)

43. State whether each of the following constituents typifies a fat, an oil, or both:
(a) saturated fatty acids
(b) unsaturated fatty acids
(c) glycerol
(d) ester linkages

44. State whether each of the following characteristics indicates a fat, an oil, or both:
(a) animal source
(b) plant source
(c) semisolid
(d) insoluble in water

45. Draw the structural formula for tristearin, the triglyceride of stearic acid.

46. Draw the structural formula for triolein, the triglyceride of oleic acid.

47. Identify the fatty acids in the following triglyceride:

$$\begin{array}{l} CH_2-O-\overset{O}{\overset{\|}{C}}-(CH_2)_{14}-CH_3 \\ | \\ CH-O-\overset{O}{\overset{\|}{C}}-(CH_2)_{10}-CH_3 \\ | \\ CH_2-O-\overset{O}{\overset{\|}{C}}-(CH_2)_{12}-CH_3 \end{array}$$

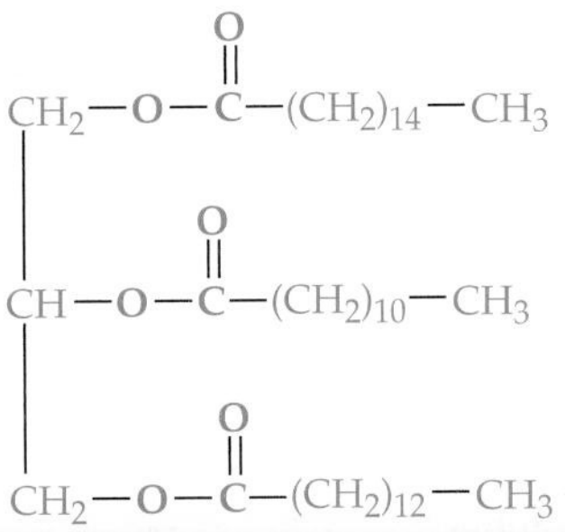

48. Identify the fatty acids in the following triglyceride:

$$\begin{array}{l} CH_2-O-\overset{O}{\overset{\|}{C}}-(CH_2)_{16}-CH_3 \\ | \\ CH-O-\overset{O}{\overset{\|}{C}}-(CH_2)_7-CH=CH-CH_2-CH=CH-(CH_2)_4-CH_3 \\ | \\ CH_2-O-\overset{O}{\overset{\|}{C}}-(CH_2)_7-CH=CH-CH_2-CH=CH-CH_2-CH=CH-CH_2-CH_3 \end{array}$$

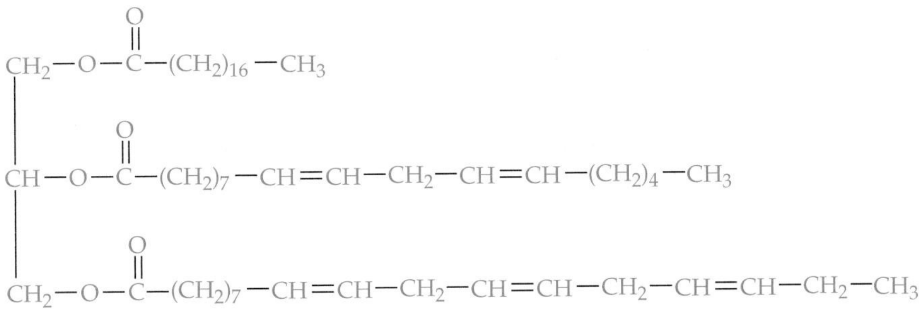

49. Carnauba wax is a lipid obtained from the leaves of palm trees and used in polishes for metal and wood surfaces. If the given structure is treated with aqueous NaOH, what are the formulas of the two products?

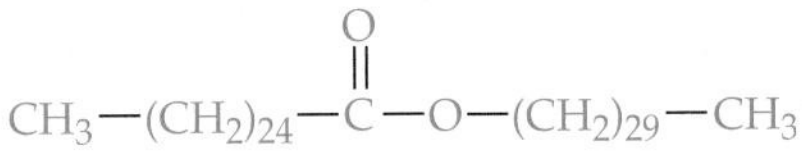

$$CH_3-(CH_2)_{24}-\overset{O}{\overset{\|}{C}}-O-(CH_2)_{29}-CH_3$$

Carnauba wax

50. Lanolin is a lipid wax obtained from sheep's wool and is used in ointments and cosmetic lotions. If lanolin is treated with aqueous NaOH, what two general types of organic structures are obtained?

51. Which vitamins belong to the lipid class of biological compounds?

52. Draw the unique ring structure that is typical of steroid compounds.

Nucleic Acids (Sec. 20.6)

53. What are the three general components of a DNA nucleotide?

54. What are the three general components of an RNA nucleotide?

55. What are the four bases in DNA?

56. What are the four bases in RNA?

57. Structurally, what is the difference between a molecule of deoxyribose (in DNA) and a molecule of ribose (in RNA)?

58. Structurally, what is the difference between a molecule of thymine (in DNA) and a molecule of uracil (in RNA)?

59. How many strands of nucleotides are in a DNA molecule?

60. How many strands of nucleotides are in an RNA molecule?

61. How many hydrogen bonds are there between adenine and thymine nucleotides in the DNA double helix?

62. How many hydrogen bonds are there between cytosine and guanine nucleotides in the DNA double helix?

63. During DNA replication, an adenine base (A) in the template strand codes for which base in the complementary strand?

64. During DNA replication, a cytosine base (C) in the template strand codes for which base in the complementary strand?

65. During RNA transcription, an adenine base (A) in the template strand codes for which base on the growing strand?

66. During RNA transcription, a cytosine base (C) in the template strand codes for which base on the growing strand?

General Exercises

67. What is the overall shape of a protein that provides strength, such as a protein in muscle tissue?

68. What is the overall shape of a protein that serves a metabolic role, such as a protein for oxygen transport?

69. What happens to a polysaccharide when it undergoes hydrolysis in an aqueous acid?

70. What happens to a protein when it undergoes denaturation in an aqueous acid?

71. Predict the type of biological compound that is the main component of insect skeletons, which is a polyhydroxy aldehyde polymer.

72. Predict the type of biological compound that is the main component of animal horns and hooves, which is an amino acid polymer.

73. How many codons are possible for an RNA trinucleotide having an adenine base, a cytosine base, and guanine base?

74. How many codons are possible for an RNA trinucleotide having an adenine base, and two guanine bases?

75. Explain the process of replication and how one molecule of DNA synthesizes two identical molecules of DNA.

76. Explain the process of transcription and how a molecule of DNA synthesizes a molecule of RNA.

77. Explain the process of protein synthesis and how an RNA codon specifies a given amino acid on a growing protein chain.

20 Chapter Self-Test

Answers to Self-Test are in Appendix J.

1. Which of the following types of linkages is found in a protein? (Sec. 20.1)
(a) peptide linkage
(b) glycoside linkage
(c) ester linkage
(d) phosphate linkage
(e) none of the above

2. What type of bonds shape the secondary structure of a protein? (Sec. 20.2)
(a) amide bonds
(b) ester bonds
(c) hydrogen bonds
(d) ionic bonds
(e) none of the above

3. How many possible tripeptides contain glycine (Gly), valine (Val), and proline (Pro) amino acids? (Sec. 20.3)
(a) 1
(b) 3
(c) 6
(d) 9
(e) none of the above

4. An enzyme is an example of which type of biological compound? (Sec. 20.3)
(a) protein
(b) carbohydrate
(c) lipid
(d) nucleic acid
(e) none of the above

5. Which of the following functional groups are found in a ketose sugar? (Sec. 20.4)
(a) alcohol and aldehyde
(b) alcohol and ketone
(c) aldehyde and ketone
(d) aldehyde and phenol
(e) none of the above

6. Which monosaccharides are produced from the hydrolysis of sucrose? (Sec. 20.4)
(a) glucose and fructose
(b) glucose and galactose
(c) glucose and ribose
(d) glucose and threose
(e) none of the above

7. Which of the following is characteristic of a lipid fat? (Sec. 20.5)
(a) animal source
(b) saturated fatty acids
(c) semisolid
(d) water insoluble
(e) all of the above

8. Which of the following is *not* found in a lipid wax? (Sec. 20.5)
(a) ester linkage
(b) glycerol
(c) long–chain alcohol
(d) saturated fatty acid
(e) unsaturated fatty acid

9. Which of the following is found in a DNA nucleotide? (Sec. 20.6)
(a) deoxyribose sugar
(b) nitrogen base
(c) phosphoric acid
(d) all of the above
(e) none of the above

10. Which of the following bases is *not* found in a DNA nucleotide? (Sec. 20.6)
(a) adenine
(b) cytosine
(c) guanine
(d) thymine
(e) uracil

Key Concepts

11. What type of biological compound is a water-insoluble fat, oil, or wax?
12. What type of biological compound is a polymer of a sugar attached to a base, joined together by phosphoric acid?
13. Draw the structure for the dipeptide glycylvaline (Gly–Val).
14. Complete the following reaction for the acid hydrolysis of sucrose and label the two products:

$$\underset{\text{sucrose}}{C_{12}H_{22}O_{11}} + \underset{\text{water}}{H_2O} \xrightarrow{H^+}$$

15. Complete the following reaction for the saponification of an animal fat. Label the two products and identify the fatty acids in the animal fat.

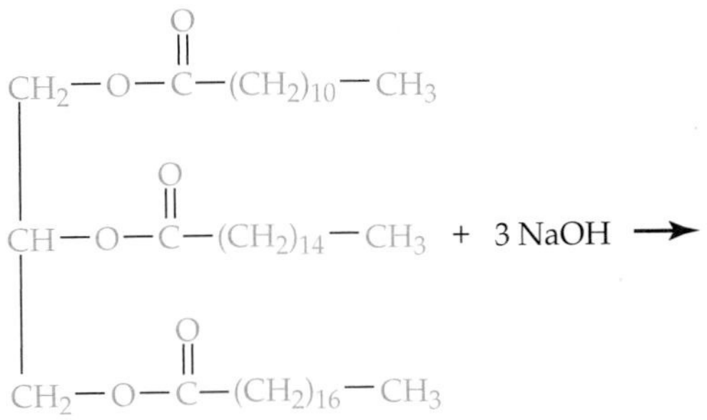

$$\begin{array}{l} CH_2{-}O{-}\overset{O}{\overset{\|}{C}}{-}(CH_2)_{10}{-}CH_3 \\ | \\ CH{-}O{-}\overset{O}{\overset{\|}{C}}{-}(CH_2)_{14}{-}CH_3 \quad + \ 3\ NaOH \longrightarrow \\ | \\ CH_2{-}O{-}\overset{O}{\overset{\|}{C}}{-}(CH_2)_{16}{-}CH_3 \end{array}$$

16. One of the components of beeswax is an ester composed of the following fatty acid and alcohol. Draw the structure of the beeswax ester.

Fatty acid: $CH_3{-}(CH_2)_{34}{-}COOH$ Alcohol: $CH_3{-}(CH_2)_{35}{-}OH$

17. During DNA replication, an adenine base on the template strand will code for which base on the complementary strand?
18. During DNA transcription, an adenine base on the template strand will code for which base on the growing RNA strand?

Critical Thinking

19. What is the difference between the shape of a protein that forms muscle tissue versus the shape of a protein that serves a metabolic role, such as transporting oxygen from the lungs to muscle tissue?
20. Identify the sugar and base in each of the following:
(a) nucleotide A **(b)** nucleotide B

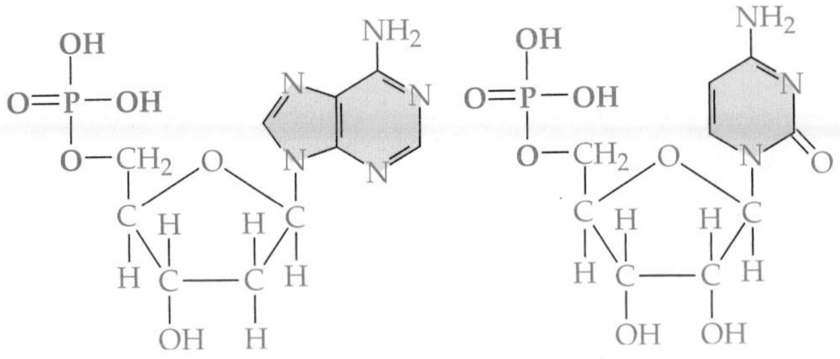

Weights and Measures

METRIC SYSTEM EXACT EQUIVALENTS

Length:	1 meter (m)	=	100 centimeters (cm)
	1 meter (m)	=	1000 millimeters (mm)
	1 kilometer (km)	=	1000 meters (m)
Mass:	1 gram (g)	=	1000 milligrams (mg)
	1 kilogram (kg)	=	1000 grams (g)
	1 metric ton	=	1000 kilograms (kg)
Volume:	1 liter (L)	=	1000 milliliters (mL)
	1 liter (L)	=	10 deciliters (dL)
	1 milliliter (mL)	=	1 cubic centimeter (cm^3)

ENGLISH SYSTEM EXACT EQUIVALENTS

Length:	1 foot (ft)	=	12 inches (in.)
	1 mile (mi)	=	1760 yards (yd)
	1 mile (mi)	=	5280 feet (ft)
Mass:	1 pound (lb)	=	16 ounces (oz)
	1 troy pound (t lb)	=	12 troy ounces (t oz)
	1 ton	=	2000 pounds (lb)
Volume:	1 quart (qt)	=	32 fluid ounces (fl oz)
	1 quart (qt)	=	2 pints (pt)
	1 gallon (gal)	=	4 quarts (qt)

ENGLISH–METRIC APPROXIMATE EQUIVALENTS

Length:	1 inch (in.)	=	2.54 centimeters (cm)
Mass:	1 pound (lb)	=	454 grams (g)
Volume:	1 quart (qt)	=	946 milliliters (mL)
Time:	1 second (sec)	=	1.00 second (s)

TEMPERATURE EQUIVALENTS

$$\frac{9}{5}\,°C + 32 = °F$$

$$\frac{5}{9}\,(°F - 32) = °C$$

$$°C + 273 = K$$

ENERGY EQUIVALENTS

1 calorie (cal)	=	4.184 joules (J)
1 kilocalorie (kcal)	=	1000 calories (cal)
1 kilocalorie (kcal)	=	4.184 kilojoules (kJ)

APPENDIX

B Physical Constants

Avogadro's number	=	6.02×10^{23}
Absolute zero		
Kelvin scale	=	0 K
Celsius scale	=	−273.15 °C
Standard temperature and pressure (STP)		
standard temperature	=	0 °C (273 K)
standard atmospheric pressure	=	1 atm
	=	760 mm Hg (760 torr)
	=	76 cm Hg
	=	29.9 in. Hg
	=	14.7 psi
	=	101 kPa
Molar volume of a gas (STP)	=	22.4 L/mol
Ideal gas constant, R	=	0.0821 L · atm/mol · K
Ionization constant of water, K_w	=	1.00×10^{-14}
Mass of proton	=	1.0073 amu
Mass of neutron	=	1.0087 amu
Mass of electron	=	0.00055 amu
Velocity of light	=	3.00×10^{8} m/s

APPENDIX

Activity Series of Metals

Most active:	$Li \rightarrow Li^{+} + e^{-}$
	$K \rightarrow K^{+} + e^{-}$
	$Ba \rightarrow Ba^{2+} + 2\,e^{-}$
	$Sr \rightarrow Sr^{2+} + 2\,e^{-}$
	$Ca \rightarrow Ca^{2+} + 2\,e^{-}$
	$Na \rightarrow Na^{+} + e^{-}$
	$Mg \rightarrow Mg^{2+} + 2\,e^{-}$
	$Al \rightarrow Al^{3+} + 3\,e^{-}$
	$Mn \rightarrow Mn^{2+} + 2\,e^{-}$
	$Zn \rightarrow Zn^{2+} + 2\,e^{-}$
	$Cr \rightarrow Cr^{3+} + 3\,e^{-}$
	$Fe \rightarrow Fe^{2+} + 2\,e^{-}$
	$Cd \rightarrow Cd^{2+} + 2\,e^{-}$
	$Co \rightarrow Co^{2+} + 2\,e^{-}$
	$Ni \rightarrow Ni^{2+} + 2\,e^{-}$
	$Sn \rightarrow Sn^{2+} + 2\,e^{-}$
	$Pb \rightarrow Pb^{2+} + 2\,e^{-}$
	$H_2 \rightarrow 2\,H^{+} + 2\,e^{-}$
	$Cu \rightarrow Cu^{2+} + 2\,e^{-}$
	$Ag \rightarrow Ag^{+} + e^{-}$
	$Hg \rightarrow Hg^{2+} + 2\,e^{-}$
Least active:	$Au \rightarrow Au^{3+} + 3\,e^{-}$

APPENDIX

Solubility Rules for Ionic Compounds

Compounds containing the following ions are generally *soluble* in water:

1. alkali metal ions and the ammonium ion, Li^+, Na^+, K^+, NH_4^+
2. acetate ion, $C_2H_3O_2^-$
3. nitrate ion, NO_3^-
4. halide ions ($X = Cl^-, Br^-, I^-$)
(AgX, Hg_2X_2, and PbX_2 are exceptions and are *insoluble*)
5. sulfate ion, SO_4^{2-}
($SrSO_4$, $BaSO_4$, and $PbSO_4$ are exceptions and are *insoluble*)

Compounds containing the following ions are generally *insoluble* in water:*

6. carbonate ion, CO_3^{2-}
(see Rule 1 exceptions, which are *soluble*)
7. chromate ion, CrO_4^{2-}
(see Rule 1 exceptions, which are *soluble*)
8. phosphate ion, PO_4^{3-}
(see Rule 1 exceptions, which are *soluble*)
9. sulfide ion, S^{2-}
(CaS, SrS, BaS, and Rule 1 exceptions are *soluble*)
10. hydroxide ion, OH^-
[$Ca(OH)_2$, $Sr(OH)_2$, $Ba(OH)_2$, and Rule 1 exceptions are *soluble*]

* These compounds are actually slightly soluble, or very slightly soluble in water.

Vapor Pressure of Water

TEMPERATURE (°C)	VAPOR PRESSURE (mm Hg)	TEMPERATURE (°C)	VAPOR PRESSURE (mm Hg)	TEMPERATURE (°C)	VAPOR PRESSURE (mm Hg)
0	4.6	21	18.7	35	41.2
5	6.5	22	19.8	40	55.3
10	9.2	23	21.1	45	71.9
12	10.5	24	22.4	50	92.5
14	12.0	25	23.8	55	118.0
16	13.6	26	25.2	60	149.4
17	14.5	27	26.7	70	233.7
18	15.5	28	28.4	80	355.1
19	16.5	29	30.0	90	525.8
20	17.5	30	31.8	100	760.0

APPENDIX

F Properties of Water

Density of H_2O:	0.99987 g/mL at 0 °C
	1.00000 g/mL at 4 °C
	0.99707 g/mL at 25 °C

Heat of fusion at 0 °C:	80.0 cal/g (335 J/g)
Heat of solidification at 0 °C:	80.0 cal/g (335 J/g)

Heat of vaporization at 100 °C:	540 cal/g (2260 J/g)
Heat of condensation at 100 °C:	540 cal/g (2260 J/g)

Specific heat of water (solid, liquid, and gas)

ice:	0.50 cal/(g × °C) [2.1 J/(g × °C)]
water:	1.00 cal/(g × °C) [4.18 J/(g × °C)]
steam:	0.48 cal/(g × °C) [2.0 J/(g × °C)]

Ionization constant, K_w:	1.00×10^{-14} at 25 °C

Answers to Concept Exercises

Chapter 1

1. Although the perception is that line segment AB is shorter than BC, the two lines are of equal length. Measure each line segment to confirm this statement.

Chapter 2

1. No instrument is capable of an exact measurement due to uncertainty.
2. No measurement is exact due to uncertainty.
3. No measurement has infinite significant digits because no measurement is exact. Significant digits are limited by the uncertainty of the measurement.
4. The rules for significant digits do not apply to exact numbers; the rules only apply to measurements.
5. When *adding* or *subtracting* measurements, the answer is limited by the decimal place with the most uncertainty. That is, the measurement that is least certain in a set of data limits the answer.
6. When *multiplying* or *dividing* measurements, the answer is limited by the number of significant digits in the data. That is, the measurement with the least number of significant digits limits the number of digits in the answer.
7. The length 1×10^{-3} cm is less; note the negative power of ten.
8. The mass 0.000 001 g is less; note the extra placeholder zero.
9. The mass 0.000 01 g is greater; that is, 1×10^{-5} g is greater than 1×10^{-6} g.
10. There are infinite significant digits in the unit equation since 1 hour exactly equals 60 minutes, and 1 minute exactly equals 60 seconds by definition.
11. There are infinite significant digits in the metric unit equation since 1 L exactly equals 1000 mL by definition.
12. There are two significant digits in the unit equation since 1 km approximately equals 0.62 mi; mile and kilometer are from different systems of measurement.
13. The percent of gold in the alloy is 75%; that is, pure gold is 100% minus 20% silver and 5% copper.
14. The unit factors are $\frac{11.2 \text{ g hydrogen}}{100 \text{ g water}}$ and $\frac{100 \text{ g water}}{11.2 \text{ g hydrogen}}$.

Chapter 3

1. The basic units of the metric system are meter, gram, and liter.
2. Since 1 m equals 1000 mm exactly, there is an infinite number of significant digits in the unit equation.
3. Since 1 L = 1000 mL exactly, there is an infinite number of significant digits in the unit factor.
4. The volume of a cube 1 cm on a side (1 cm^3) equals 1 mL.
5. Since 1 g = 1000 mg exactly, there is an infinite number of significant digits in the unit factor.
6. Since 1 qt = 946 mL approximately, there are only three significant digits in the unit factor.
7. Since 1 kg = 2.20 lb approximately, there are only three significant digits in the unit factor.
8. Since 1 mi is a greater distance than 1 km, 65 m/h is a faster speed.
9. The volume of a cube 10 cm on a side (1000 cm^3) equals 1 L.
10. The thicknesses are all equal; that is, 1 mm = 0.1 cm = 0.001 m.
11. The thicknesses are all equal; that is, 1 L = 1000 mL = 1000 cm^3.
12. The volumes are equal; that is, 1 mL = 1 cm^3.
13. Since ice floats in water, water has the greater density.
14. Since 1 kg = 1000 g and 1 L = 1000 mL, the two densities are equal.
15. The density of a human body is about the same as the density of water, 1 g/mL.
16. The Celsius and centigrade temperature scales are equivalent; that is, 20° Celsius and 20° centigrade are the same temperature.
17. The Kelvin temperature scale is assigned a value of zero for the coldest possible temperature; thus, −100 K cannot exist.
18. If the can feels cold, heat is flowing away from your hand into the can.
19. If a pie crust cooks faster than the apple filling, the filling has a higher specific heat. Food cooks faster at a higher temperature and the apple filling requires more heat to raise its temperature.

Chapter 4

1. (a) solid, (b) liquid, (c) gas
2. (a) element, (b) element, (c) compound, (d) mixture
3. Calcium is a solid metal under normal conditions.
4. Bromine is a liquid nonmetal under normal conditions.
5. The molecular formula for vitamin C is $C_6H_8O_6$.
6. Nitroglycerin has a total of 20 atoms in one molecule.
7. Copper turning black when heated is an example of a chemical change.
8. An Alka-Seltzer® tablet dissolving in water and producing gas bubbles is an example of a chemical change.
9. Applying the conservation of mass law, the mass of sulfur is found by difference to be 0.252 g (1.252 g − 1.000 g).

10. As the temperature increases, the kinetic energy of air molecules increases. As kinetic energy increases, the velocity of air molecules increases.

11. Chemical, electrical, heat, light, mechanical, and nuclear are all basic forms of energy; solar energy is an example of light energy.

Chapter 5

1. Atoms of different elements have a different number of protons, and *cannot have* the same atomic number.

2. Atoms of different elements *can have* isotopes with the same mass number.

3. (a) Simple Average Mass: $\dfrac{2.00\text{ g} + 1.00\text{ g}}{2} = 1.50\text{ g}$

(b) Weighted Average Mass:

$$2.00\text{ g}\left(\frac{75\text{ large marbles}}{100\text{ total marbles}}\right) + 1.00\text{ g}\left(\frac{25\text{ small marbles}}{100\text{ total marbles}}\right)$$

$$= 1.50\text{ g} + 0.25\text{ g} = 1.75\text{ g}$$

4. Atomic number and mass number are always whole numbers; the atomic mass is never a whole number because it is the result of a mass measurement.

5. The energy *decreases* as the wavelength increases; but the energy *increases* as frequency increases.

6. A water wave is to a drop of water, as a light wave is to a *photon*.

7. All three statements are true.

8. The theoretical number of sublevels corresponds to the energy level. Thus, the tenth energy level has 10 sublevels.

9. The sublevels are closer together and the energy difference between sublevels *decreases* moving away from the nucleus. (Refer to Figure 5.13.)

10. The atomic number of Mn is one more than Cr, so Mn has one more *d* electron in the outermost sublevel.

11. All three statements are true.

Chapter 6

1. Note that Ar precedes K in the periodic table, even though the atomic mass of Ar (39.95) is greater than K (39.10).

2. The group number for the each family of elements is as follows: (a) IA/1 (b) IIA/2 (c) VIIA/17 (d) VIIIA/18.

3. Period 1 at the top of the periodic table represents the smallest atomic radius. Group IA/1 at the far left of the periodic table represents the most metallic character.

4. Since Li is directly above Na in the periodic table, we can apply periodic trends to estimate values for Li. The atomic radius is less (~0.15 nm), the density is less (~0.50 g/mL), and melting point is greater (~150 °C). Refer to Table 6.2 and note that the estimated values are reasonable.

5. Since Zn and Cd are in the same group, they have very similar chemical properties.

6. Since Ag is in Period 5 and Group 1B/11, the highest energy sublevel is 4*d*.

7. Since Pt is the 8th element in the 5*d* sublevel, we predict the number of 5*d* electrons is 8.

8. Representative elements have valence electrons in the outermost *s* and *p* sublevels. Transition elements have electrons in the outermost *d* and *f* sublevels, which complicates making reliable predictions.

9. Since Mg has two valence electrons in a 3 sublevel, we can draw the two electrons next to each other, or clockwise. Either method is acceptable; however, remember that valence electrons are not stationary particles, and that the electron dot formula is only a tool for keeping track of valence electrons.

Mg: which may be preferable to ·Mg·

10. The noble gases have the highest ionization energies since they have filled valence shells.

11. Group IA/1 elements have a charge of 1+; Group IIA/2 elements have a charge of 2+; and Group VIIA/17 elements have a charge of 1−.

12. The next ion in the isoelectronic series Se^{2-}, Br^-, Kr, Rb^+, Sr^{2+} is Y^{3+}.

13. The electron configuration is simply [Rn].

Chapter 7

1. A binary acid is in an aqueous solution and is indicated by (*aq*), e.g., HCl(*aq*).

2. Hg^{2+} is a monoatomic cation; whereas, Hg_2^{2+} is a polyatomic cation.

3. The Stock system specifies the charge on a cation using a Roman numeral in parentheses.

4. Iodide is a monoatomic anion because it has an −*ide* suffix.

5. A monoatomic anion has an −*ide* suffix; a polyatomic anion has an −*ate* or −*ite* suffix.

6. The formula for the metal oxide is M_2O_3.

7. The formula for the metal carbonate is $M_2(CO_3)_3$.

8. The charge on an oxide ion is 2−, and there are two oxide ions. The total negative charge is 4−. Thus, the positive charge on metal M is 4+.

9. Sodium chloride is a binary ionic compound as indicated by the −*ide* suffix.

10. LiCl is a binary ionic compound because it has two elements.

11. Since potassium and sodium are in the same family, and fluorine and iodine are in the same family, the formula for potassium iodide is KI.

12. The charge on a carbonate ion is 2−. Thus, the positive charge on metal M is 2+.

13. Potassium nitrate and potassium nitrite are both ternary ionic compounds as indicated by the −*ate* and −*ite* suffixes.

14. LiClO and $LiClO_2$ are both ternary ionic compounds because they have three elements.

15. Since calcium and strontium are in the same family, the formula for strontium sulfate is $SrSO_4$.

16. P_2O_3 is a binary molecular compound and requires Greek prefixes to specify the nonmetal ratio, that is, diphosphorus trioxide. Fe_2O_3 is an ionic compound and does not use Greek prefixes, that is, iron(III) oxide.

17. Nitrogen oxide is a binary molecular compound because it contains a nonmetal and oxygen.

18. HBr(*aq*) is a binary acid named hydrobromic acid.

19. $HBrO_3$(*aq*) is a ternary oxyacid named bromic acid, similar to chloric acid, $HClO_3$(*aq*). $HBrO_2$(*aq*) is a ternary oxyacid named bromous acid, similar to chorus acid, $HClO_2$(*aq*).

Chapter 8

1. Evidence for a chemical reaction include the following: (1) a gas is produced; (2) an insoluble solid is produced in solution; (3) a permanent color change is observed; (4) an energy change is observed.

2. The nonmetals that occur naturally as diatomic molecules are H_2, N_2, O_2, F_2, Cl_2, Br_2, and I_2.

3. No, changing a subscript gives an incorrect chemical formula.

4. Writing water as HOH often makes a neutralization equation easier to balance, as it is a special type of a double replacement reaction.

5. Decomposition reactions are easily recognized as they only have one reactant, unlike other types of reactions.

6. Heating a metal and oxygen produces a metal oxide.

7. Heating a nonmetal and oxygen produces a nonmetal oxide.

8. Heating a metal and nonmetal produces an ionic compound.

9. Heating a metal hydrogen carbonate produces a metal carbonate and carbon dioxide gas.

10. Heating a metal carbonate produces a metal oxide and carbon dioxide gas.

11. Heating an oxygen-containing compound often releases oxygen gas.

12. The metals that react with aqueous $AgNO_3$ (above Ag in the activity series) include Al and Cu, but not Au.

13. The metals that react with aqueous acid (above H in the activity series) include Zn, but not Cu and Au.

14. $Pb(NO_3)_2$ is soluble in water; $PbCl_2$ and $PbCO_3$ are insoluble.

15. $AgCl(s)$ is the insoluble product.

16. $NaCl(aq)$ is the salt product.

Chapter 9

1. A platinum atom has a mass of 196.97 amu. Avogadro's number of platinum atoms has a mass of 196.97 g.

2. The number of molecules in 1.00 mol of chlorine gas is 6.02×10^{23}.

3. The number of molecules in 1.00 mol of iodine crystals is 6.02×10^{23}.

4. The mass of Avogadro's number of water molecules is 18.02 g.

5. The mass of one mole of S_8 crystals is 256.56 g (8×32.07 g).

6. The mass of Avogadro's number of ozone molecules is 48.00 g.

7. The mass of a carbon monoxide molecule is 28.01 amu.

8. The He balloon rises because the density of helium gas is much less than air. The N_2 balloon does not rise because the density of nitrogen gas is about the same as air, which is composed of 78% N_2 and 21% O_2.

9. The volume of 1.00 mol of any gas at STP is 22.4 L.

10. The number of molecules in 1.00 mol of ozone gas is 6.02×10^{23}.

11. The mass of 1.00 mol of hydrogen gas is 2.02 g.

12. The percent oxygen is 53.3% (100.0% − 40.0% − 6.7%).

13. The empirical formula is SO_2.

14. The empirical formula is SO_3.

15. The empirical formula is HO.

Chapter 10

1. The coefficients of a balanced equation indicate the ratio of moles and volumes of gases, but *not* the masses of gas, which vary with the molar mass of each gas.

2. One unit factor is required to convert the moles given to moles in the answer.

3. The three types of stoichiometry problem are mass–mass, mass–volume, and volume–volume.

4. The three steps in the unit analysis method of problem solving are as follows: (1) write down the unit asked for in the answer; (2) write down the given value that is related to the answer; (3) apply unit factor(s) to convert the given value to the units in the answer.

5. Three unit factors are usually required to convert the mass given to the mass in the answer.

6. The three steps in unit analysis are as follows: (1) write the unit in the answer; (2) write the given value; (3) apply unit factor(s) to convert the given value to the answer.

7. Three unit factors are usually required to convert the mass given to the volume in the answer.

8. One unit factor is usually required to convert the volume given to the volume in the answer.

9. Each bicycle frame requires two tires for a total of six tires. Thus, only three bicycles can be assembled.

10. Zn metal is the limiting reactant.

11. S powder is the limiting reactant.

12. A percent yield greater than 100% is possible as some experimental errors lead to high results, and others to low results.

Chapter 11

1. An atmosphere is the largest unit, and the greatest gas pressure is 1 atm.

2. Increasing the volume produces a *decrease* in gas pressure because molecules are further apart and collide less frequently.

3. The gas pressure *increases* because the volume decreases, and there are more molecular collisions.

4. The volume *increases* because the temperature increases, gas molecules move faster, and the balloon expands to maintain constant pressure.

5. The pressure *increases* because the temperature increases; the air molecules move faster and have more energetic collisions.

6. The pressure doubles when the temperature doubles, but the pressure halves when the volume doubles. Thus, the volume does not change.

7. The partial pressure of a NO gas is found by subtracting the partial pressure of N_2 and O_2 from 2.50 atm. Thus, the partial pressure of NO is 0.50 atm.

8. No real gases are ideal; however, He behaves most like an ideal gas.

9. The proportionality constant (R) in the ideal gas law equals PV/nT.

Chapter 12

1. The attraction between a metal cation and a nonmetal anion gives an *ionic bond*; the attraction between two nonmetal atoms produces a *covalent bond*.

2. Xenon is isoelectronic with both Ba^{2+} and I^-.

3. All three statements are true.

4. All three statements are true.

5. There are two pairs of nonbonding electrons in H_2S.

6. There are six pairs of nonbonding electrons in SO_2.

7. There are 10 pairs of nonbonding electrons in SO_3^{2-}.

8. The general trends for increasing electronegativity in the periodic table is from left to right in a period, and from bottom to top in a group.

9. HF has the greatest difference in electronegativity $(4.0 - 2.1 = 1.9)$; thus, the bond in H−F is the most polar.

10. The bond in C−I is the most nonpolar because there is no difference in the electronegativity of carbon and iodine $(2.5 - 2.5 = 0)$.

11. The structural formula for the molecule is H−Cl−O. Since the oxygen atom can be removed with its six valence electrons, the Cl−O bond is coordinate covalent.

Chapter 13

1. Since CH_3-CH_2-OH can hydrogen bond, it has a stronger intermolecular attraction than CH_3-O-CH_3.

2. Since CH_3-CH_2-OH can hydrogen bond, it has a higher boiling point than CH_3-O-CH_3.

3. A paper clip can float on water due to the surface tension between water molecules. If you disturb the paper clip, it will sink because it is more dense.

4. An *ionic crystalline solid* has a high melting point, and is a conductor of electricity only when melted. A *molecular crystalline solid* has a low melting point, and is a nonconductor of electricity.

5. Since the heat of vaporization is greater than the heat of fusion, it requires more heat to vaporize 1 g of water (540 cal) than to melt 1 g of water (80 cal).

6. The systematic names for the hydrates are calcium chloride *monohydrate*, calcium chloride *dihydrate*, and calcium chloride *hexahydrate*, respectively.

7. The monohydrate, $CaCl_2 \cdot H_2O$, has the lowest percentage of water.

8. The hexahydrate, $CaCl_2 \cdot 6H_2O$, has the highest water of hydration (6).

Chapter 14

1. Since atmospheric pressure is less at higher elevations, Lake Tahoe has less dissolved oxygen gas than Lake Havasu.

2. According to the *like dissolves like* rule, two liquids are miscible when both liquids are polar, or both liquids are nonpolar.

3. According to the *like dissolves like* rule, a solid is soluble in a liquid when the solid and liquid are both polar, or both nonpolar. An ionic compound is more soluble in a polar liquid than a nonpolar liquid.

4. Since the oxygen atom in a water molecule is partially negative, it is attracted to the positive cation in the solid solute.

5. Grinding the solute gives smaller particles with more surface area. Since solvent molecules attack the particle surface, the solute dissolves faster.

6. More sugar dissolves in hot coffee than coffee at a lower temperature.

7. By saturating a solution at a higher temperature, and allowing the solution to cool without disturbing it, the solute remains in the solution at a lower temperature and is said to be supersaturated.

8. There are six mass percent concentration unit factors, that is, three unit factors and their reciprocals.

9. There are four molar concentration unit factors, that is, two unit factors and their reciprocals.

10. Since equal volumes of 6 *M* sulfuric acid and distilled water are added together, the concentration is halved, and the diluted acid is 3 *M*.

11. Before solving a stoichiometry problem, first write a balanced chemical equation for the reaction.

Chapter 15

1. The solution with the lowest pH (1) has the highest acidity.

2. Aqueous HNO_3 is a strong Arrhenius acid. (Refer to Table 15.2.)

3. Aqueous $HC_2H_3O_2$ and NaOH react to give sodium acetate, $NaC_2H_3O_2$.

4. Phenolphthalein is colorless in acidic and neutral solutions. (Refer to Table 15.4.)

5. The titration of H_3PO_4 with NaOH is the titration of a weak acid with a strong base.

6. The titration of HCl with $Ba(OH)_2$ is the tritration of a strong acid with a strong base.

7. Phenolphthalein is pink at the endpoint of the titration of oxalic acid.

8. Phenolphthalein is pink at the endpoint of the titration of benzoic acid.

9. Distilled water is neutral because the concentration of hydrogen ions and hydroxide ions is equal.

10. Apple juice at pH 4 is weakly acidic.

11. Milk of magnesia at pH 10 is weakly basic.

12. Grape juice at pH 4.05 is weakly acidic.

13. Wine at pH 3.25 is weakly acidic.

14. The reaction of HF(*aq*) and NH_4OH(*aq*) produces NH_4F and H_2O; NH_4F is a soluble salt, so the conductivity apparatus will glow brightly.

15. The reaction of HNO_3(*aq*) and $Ba(OH)_2$(*aq*) produces $Ba(NO_3)_2$ and H_2O; $Ba(NO_3)_2$ is a soluble salt, so the apparatus will glow brightly.

16. The reaction of $AgNO_3$(*aq*) and $AlCl_3$(*aq*) gives AgCl and $Al(NO_3)_3$. Although AgCl is insoluble, $Al(NO_3)_3$ is soluble, so the apparatus will glow brightly.

Chapter 16

1. (a) At higher temperatures molecules have a greater velocity, and the collision frequency increases. As the number of collisions increases, there are more effective collisions.

(b) At higher temperatures molecules have more kinetic energy, and collision energy increases. If the collisions are more energetic, more molecules have the necessary energy to react.

2. Notice that the progress of reaction goes from lower to higher energy, which characterizes an endothermic reaction.

3. After a reaction reaches chemical equilibrium, the concentrations of reactants and products are constant.

4. Before a reaction reaches chemical equilibrium, the concentration of reactants is decreasing, and the concentration of products is increasing.

5. The equilibrium constant expression is:
$K_{eq} = [X_2Y_3]^2 / [X_2]^2 [Y_2]^3$

6. (a) Increasing temperature for an endothermic reaction shifts the equilibrium to the right, toward the products.

(b) Increasing pressure for this reaction has no effect on the equilibrium because there are the same number of reactant and product molecules.

7. The ionization constant expression is $K_i = [H^+] [H_2PO_4^-]/[H_3PO_4]$.

8. (a) Adding NH_4Cl increases the NH_4^+ concentration, thus, shifting the equilibrium to the left toward the reactants.

(b) Adding NaOH increases the OH^- concentration, thus, shifting the equilibrium to the left toward the reactants.

9. The solubility product constant expression is $K_{sp} = [Al^{3+}] [OH^-]^3$.

10. (a) Adding $Ca(NO_3)_2$ increases the Ca^{2+} concentration, thus, shifting the equilibrium to the left toward the reactants.

(b) Adding Na_2CO_3 increases the CO_3^{2-} concentration, thus, shifting the equilibrium to the left toward the reactants.

Chapter 17

1. (a) 0; (b) +1; (c) −1; (d) +7
2. (a) −1; (b) +1; (c) +3; (d) +5
3. oxidizing agent: HX; reducing agent: M
4. oxidizing agent: Y_2; reducing agent: X^-
5. $2\,M_2O_3(s) + 6\,X_2(g) \rightarrow 4\,MX_3(aq) + 3\,O_2(g)$
6. $2\,MnO_4^-(aq) + 5\,SO_3^{2-}(aq) + 6\,H^+(aq) \rightarrow$
$2\,Mn^{2+}(aq) + 5\,SO_4^{2-}(aq) + 3\,H_2O(l)$
7. $2\,MnO_4^-(aq) + 3\,SO_3^{2-}(aq) + H_2O(l) \rightarrow$
$2\,MnO_2(s) + 3\,SO_4^{2-}(aq) + 2\,OH^-(aq)$
8. Table 17.3 lists Cr and Al as stronger reducing agents than $Fe^{2+}(aq)$. Thus, Cr and Al react spontaneously with $FeSO_4(aq)$; Ag and Ni are weaker and do not react.
9. (voltaic cell) $Fe(s) + Sn^{2+}(aq) \rightarrow Sn(s) + Fe^{2+}(aq)$

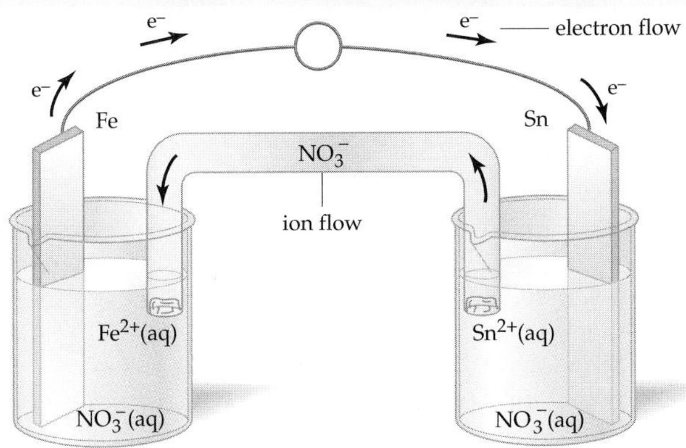

10. (electrolytic cell) $C(s) + 2\,MgCl_2(l) \rightarrow 2\,Mg(l) + CCl_4(g)$

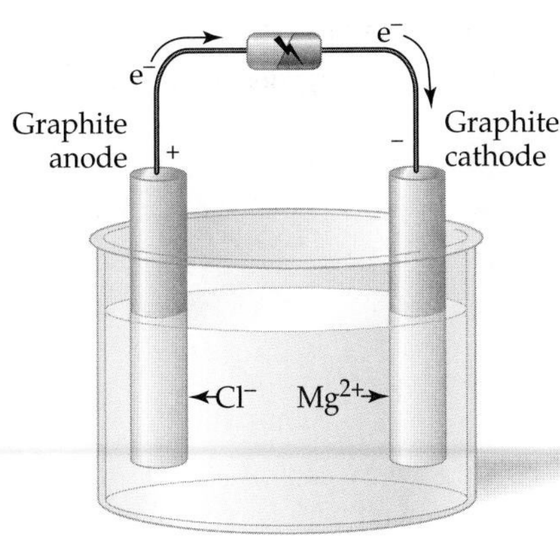

Chapter 18

1. (a) alpha ray, (b) beta ray, (c) gamma ray
2. 31 mg $^{208}_{82}Pb + ^{62}_{28}Ni \rightarrow ^{269}_{110}Ds + ^{1}_{0}n$

Chapter 19

1. Note that decane ($C_{10}H_{22}$) has an *−ane* suffix, and fits the general molecular formula for an alkane (C_nH_{2n+2}).
2. Note that decene ($C_{10}H_{20}$) has an *−ene* suffix, and fits the general molecular formula for an alkene (C_nH_{2n}).
3. Note that decyne ($C_{10}H_{18}$) has a *−yne* suffix, and fits the general molecular formula for an alkyne (C_nH_{2n-2}).
4. Note that decene ($C_{10}H_{20}$) and naphthalene ($C_{10}H_8$) both have an *−ene* suffix, but decene fits the general molecular formula for an alkene (C_nH_{2n}).
5. Aldehydes, ketones, carboxylic acids, esters, and amides contain a carbonyl group. Organic halides, alcohols, phenols, ethers, and amines do not.

Chapter 20

1. A spiral helix describes the *secondary* structure of a protein. The sequence of amino acids in the protein corresponds to the primary structure, and a twisted helix describes the tertiary structure.

APPENDIX

Answers to Key Term Exercises

Chapter 1

1. i, **2.** d, **3.** j, **4.** e, **5.** k, **6.** g, **7.** a, **8.** c, **9.** f, **10.** h, **11.** b

Chapter 2

1. g, **2.** a, **3.** d, **4.** h, **5.** e, **6.** p, **7.** f, **8.** t, **9.** o, **10.** i, **11.** m, **12.** c, **13.** k, **14.** n, **15.** r, **16.** s, **17.** l, **18.** b, **19.** q, **20.** j

Chapter 3

1. e, **2.** o, **3.** n, **4.** h, **5.** m, **6.** q, **7.** j, **8.** v, **9.** f, **10.** w, **11.** p, **12.** u, **13.** c, **14.** x, **15.** d, **16.** r, **17.** i, **18.** t, **19.** g, **20.** b, **21.** l, **22.** s, **23.** a, **24.** k

Chapter 4

1. z, **2.** cc, **3.** i, **4.** l, **5.** m, **6.** dd, **7.** a, **8.** h, **9.** k, **10.** b, **11.** g, **12.** t, **13.** s, **14.** j, **15.** v, **16.** bb, **17.** c, **18.** w, **19.** r, **20.** u, **21.** e, **22.** y, **23.** f, **24.** d, **25.** x, **26.** aa, **27.** n, **28.** p, **29.** o, **30.** q

Chapter 5

1. h, **2.** t, **3.** q, **4.** d, **5.** e, **6.** p, **7.** c, **8.** n, **9.** b, **10.** a, **11.** y, **12.** m, **13.** o, **14.** x, **15.** v, **16.** g, **17.** s, **18.** f, **19.** k, **20.** j, **21.** l, **22.** i, **23.** w, **24.** u, **25.** r

Chapter 6

1. r, **2.** h, **3.** q, **4.** b, **5.** c, **6.** i, **7.** p, **8.** t, **9.** u, **10.** o, **11.** a, **12.** j, **13.** s, **14.** v, **15.** f, **16.** d, **17.** w, **18.** g, **19.** k, **20.** m, **21.** l, **22.** n, **23.** e

Chapter 7

1. i, **2.** h, **3.** d, **4.** p, **5.** e, **6.** b, **7.** c, **8.** q, **9.** f, **10.** a, **11.** l, **12.** n, **13.** o, **14.** j, **15.** m, **16.** g, **17.** k

Chapter 8

1. h, **2.** o, **3.** n, **4.** d, **5.** q, **6.** g, **7.** s, **8.** r, **9.** f, **10.** l, **11.** i, **12.** v, **13.** j, **14.** k, **15.** u, **16.** m, **17.** p, **18.** c, **19.** b, **20.** a, **21.** e, **22.** t

Chapter 9

1. a, **2.** g, **3.** e, **4.** f, **5.** j, **6.** b, **7.** d, **8.** i, **9.** c, **10.** h

Chapter 10

1. j, **2.** b, **3.** d, **4.** h, **5.** m, **6.** f, **7.** g, **8.** o, **9.** i, **10.** l, **11.** c, **12.** e, **13.** a, **14.** n, **15.** k

Chapter 11

1. k, **2.** v, **3.** b, **4.** d, **5.** u, **6.** c, **7.** p, **8.** i, **9.** e, **10.** f, **11.** l, **12.** g, **13.** t, **14.** w, **15.** h, **16.** r, **17.** x, **18.** q, **19.** m, **20.** s, **21.** j, **22.** a, **23.** o, **24.** n

Chapter 12

1. e, **2.** bb, **3.** v, **4.** p, **5.** g, **6.** n, **7.** r, **8.** c, **9.** b, **10.** d, **11.** t, **12.** k, **13.** z, **14.** y, **15.** j, **16.** aa, **17.** s, **18.** x, **19.** m, **20.** h, **21.** w, **22.** u, **23.** i, **24.** f, **25.** o, **26.** cc, **27.** l, **28.** q, **29.** a

Chapter 13

1. g, **2.** f, **3.** n, **4.** w, **5.** b, **6.** x, **7.** v, **8.** d, **9.** o, **10.** r, **11.** q, **12.** k, **13.** l, **14.** c, **15.** s, **16.** i, **17.** p, **18.** t, **19.** m, **20.** a, **21.** y, **22.** j, **23.** u, **24.** e, **25.** h

Chapter 14

1. c, **2.** n, **3.** p, **4.** o, **5.** b, **6.** i, **7.** k, **8.** j, **9.** e, **10.** g, **11.** d, **12.** q, **13.** a, **14.** s, **15.** m, **16.** l, **17.** t, **18.** r, **19.** f, **20.** h

Chapter 15

1. g, **2.** c, **3.** d, **4.** k, **5.** h, **6.** j, **7.** q, **8.** e, **9.** f, **10.** p, **11.** o, **12.** b, **13.** a, **14.** u, **15.** i, **16.** s, **17.** l, **18.** n, **19.** t, **20.** w, **21.** v, **22.** r, **23.** m

Chapter 16

1. d, **2.** e, **3.** f, **4.** o, **5.** r, **6.** a, **7.** h, **8.** b, **9.** p, **10.** n, **11.** c, **12.** l, **13.** g, **14.** j, **15.** i, **16.** m, **17.** k, **18.** q

Chapter 17

1. l, **2.** n, **3.** k, **4.** p, **5.** m, **6.** o, **7.** j, **8.** q, **9.** f, **10.** e, **11.** i, **12.** r, **13.** s, **14.** g, **15.** h, **16.** a, **17.** c, **18.** b, **19.** d

Chapter 18

1. t, **2.** b, **3.** d, **4.** i, **5.** c, **6.** k, **7.** p, **8.** o, **9.** l, **10.** r, **11.** h, **12.** s, **13.** q, **14.** a, **15.** j, **16.** u, **17.** v, **18.** m, **19.** e, **20.** f, **21.** n, **22.** g, **23.** w

Chapter 19

1. s, **2.** n, **3.** v, **4.** x, **5.** g, **6.** b, **7.** c, **8.** e, **9.** u, **10.** r, **11.** f, **12.** w, **13.** q, **14.** d, **15.** h, **16.** t, **17.** l, **18.** a, **19.** k, **20.** o, **21.** m, **22.** i, **23.** p, **24.** j

Chapter 20

1. b, **2.** t, **3.** c, **4.** j, **5.** m, **6.** a, **7.** p, **8.** d, **9.** r, **10.** f, **11.** k, **12.** l, **13.** e, **14.** s, **15.** i, **16.** h, **17.** y, **18.** g, **19.** o, **20.** v, **21.** q, **22.** z, **23.** w, **24.** n, **25.** u, **26.** x

APPENDIX

I Answers to Odd-Numbered Exercises

Chapter 1

1. yin and yang
3. air, earth, fire, and water
5. plan an experiment, make observations, and record data
7. conduct additional experiments to test the hypothesis
9. A scientific *theory* explains the behavior of nature using a model or concept, whereas, a *natural law* states a relationship that is measurable under various experimental conditions.
11. Statement (a) is a *natural law* because mass can be measured.

Statement (b) is a *theory* because gas molecule collisions cannot be measured.

Statement (c) is a *natural law* because gas volumes can be measured.

Statement (d) is a *theory* because it is a model that explains the nucleus.

13. Chemistry is considered the central science because a knowledge of chemistry is essential to a full understanding of biology, physics, geology, and medicine.
15. Scientists and engineers must have formal training in chemistry, as well as doctors, nurses, dentists, physical therapists, chiropractors, dietitians, and veterinarians.
17. A solution to the nine-dot problem using three straight lines is shown below; the assumption regards the angle of the lines and the size of the dots.

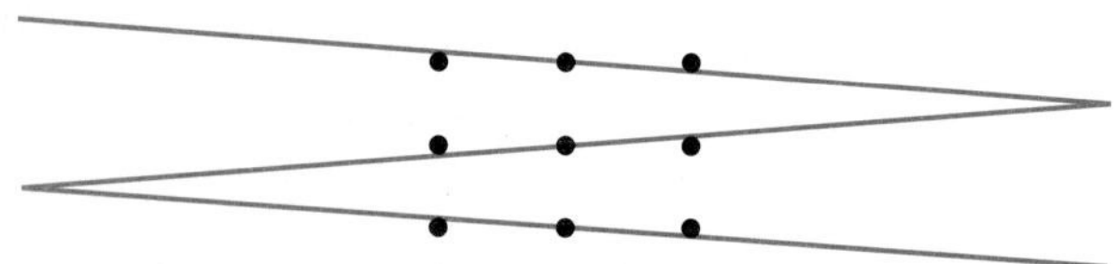

19. By staring at the point where the blocks intersect, we can "flip" the image and view the blocks as stacking upward, or hanging downward.

Chapter 2

1. (a) length; (b) mass; (c) volume; (d) time
3. (b) 10.0 cm; (c) 10.5 cm
5. (c) 75.22 g
7. (b) 25.0 mL; (c) 25.5 mL
9. (a) 1; (b) 2; (c) 3; (d) 4
11. (a) 1; (b) 2; (c) 3; (d) 1
13. (a) 4; (b) 1; (c) 3; (d) 2
15. (a) 10.3; (b) 10.2; (c) 0.0103; (d) 10,200
17. (a) 1.45×10^{1}; (b) 1.46×10^{2}; (c) 1.51×10^{-3}; (d) 1.50×10^{-4}
19. (a) 55.0 cm; (b) 22.3 cm
21. (a) 67 g; (b) 13.01 g
23. (a) 7.67 cm^2; (b) 10.1 cm^2; (c) 28 cm^2; (d) 400 cm^2
25. (a) 2.57 cm; (b) 66 cm; (c) 6.846 cm^2; (d) 6.86 cm^2
27. (a) 10^4; (b) $(1/10)^3$ or 10^{-3}
29. (a) 2^3; (b) $(1/2)^3$ or 2^{-3}
31. (a) 1000; (b) 0.000 000 1
33. (a) 1×10^9; (b) 1×10^{-8}
35. (a) 10; (b) 0.1
37. (a) 1.01×10^{15}; (b) 4.56×10^{-13}; (c) 9.45×10^{16}; (d) 1.950×10^{-17}
39. 2.69×10^{22} helium atoms
41. 3.35×10^{-23} g/neon atom
43. (a) 1 mile = 5280 feet; (b) 1 ton = 2000 pounds
45. (a) $\frac{1 \text{ mile}}{5280 \text{ feet}}$ and $\frac{5280 \text{ feet}}{1 \text{ mile}}$; (b) $\frac{1 \text{ ton}}{2000 \text{ pounds}}$ and $\frac{2000 \text{ pounds}}{1 \text{ ton}}$
47. (a) 1 mile = 5280 feet is an exact equivalent.
49. (a) 1 liter = 1000 milliliters is an exact equivalent.
51. (a) infinite; (b) 3
53. (a) infinite; (b) 3
55. 91.4 centimeters
57. 0.150 grams
59. 5.0 cups
61. 66 months
63. 2.29×10^{10} troy pounds
65. 1.08×10^{12} meters
67. 4.91 g copper
69. 40.5%
71. 16.9 g water
73. 1.17×10^{25} g oxygen; 6.09×10^{24} g silicon; 1.78×10^{24} g aluminum
75. 94.99% copper
77. 10.00 mL (± 0.01 mL)
79. 186,000 miles per second
81. 258.1 g
83. (a) 3.52×10^6; (b) 1.91×10^{-6}

85. 3.3475×10^{-24} g
87. 2.00×10^2 lb
89. 3.5×10^{13} hours
91. 34.0% A+ blood type
93. 454 grams of feathers
95. A pound of feathers is heavier.

Chapter 3

1. (b) gram and (c) liter
3. (a) km; (b) Gg; (c) dL; (d) ps
5. (a) length, (b) mass, (c) volume, and (d) time
7. (a) length, (b) mass, (c) volume, and (d) time
9. (a) tera (T); (b) giga (G); (c) milli (m); (d) micro (μ)
11. (a) terameter; (b) gigagram; (c) milliliter; (d) microsecond
13. (a) 1×10^{12} m = 1 Tm; (b) 1×10^9 g = 1 Gg; (c) 1 L = 1×10^3 mL; (d) 1 s = 1×10^6 μs
15. (a) $\frac{1 \times 10^{12}\text{ m}}{1\text{ Tm}}$ and $\frac{1\text{ Tm}}{1 \times 10^{12}\text{ m}}$; (b) $\frac{1 \times 10^{9}\text{ g}}{1\text{ Gg}}$ and $\frac{1\text{ Gg}}{1 \times 10^{9}}$; (c) $\frac{1\text{ L}}{1 \times 10^{3}\text{ mL}}$ and $\frac{1 \times 10^{3}\text{ mL}}{1\text{ L}}$; (d) $\frac{1\text{ s}}{1 \times 10^{6}\text{ μs}}$ and $\frac{1 \times 10^{6}\text{ μs}}{1\text{ s}}$
17. (a) 1.00×10^{-3} km; (b) 1×10^4 cg; (c) 1.00 dL; (d) 1.00×10^4 ns
19. (a) 6.50×10^6 Mm; (b) 6.50×10^8 kg; (c) 6.50×10^{-2} dL; (d) 0.650 ps
21. (a) 2.54 cm = 1 in.; (b) 454 g = 1 lb; (c) 946 mL = 1 qt; (d) 1.00 s = 1 sec
23. (a) 34 in.; (b) 0.079 lb; (c) 473 mL; (d) 8.00×10^2 s
25. (a) 1.8 m; (b) 79.5 kg; (c) 0.473 L; (d) 6.30×10^{-6} ks
27. 31 mi/gal
29. 62,500 mm^3 (6.25×10^4 mm^3)
31. 1.26 cm
33. (a) 1000 mL; (b) 1000 cm^3
35. 372 $in.^3$
37. 3.0 mL
39. 130 mL
41. (a) sink; (b) float
43. (a) rise; (b) drop
45. (a) 35.0 mL; (b) 18.9 mL
47. (a) 16.4 g; (b) 780 g (7.80×10^2 g)
49. (a) 0.788 g/mL; (b) 2.8 g/mL
51. (a) 32 °F; (b) 273 K; (c) 0 °C
53. (a) 38 °C; (b) −137 °C
55. (a) 768 K; (b) 88 K
57. Temperature is a measure of the *average kinetic energy* of a molecule, while heat is the *total kinetic energy* of the molecules.
59. 1.9×10^4 cal
61. 67.5 cal
63. 0.0317 cal/(g × °C)
65. 1.5×10^6 MB
67. (a) infinite; (b) 3; (c) infinite; (d) 3
69. 3.5×10^4 μs
71. 5.87×10^{12} miles/year
73. The 1500-meter race (0.93 mile) is shorter.
75. 7690 tablets
77. 435 m^2
79. 5.0 g/mL
81. 0.801
83. 62.4 lb/ft^3
85. −252 °C, 21 K
87. 1.0×10^2 kcal
89. 56.5 g
91. 59.2 °C
93. 52.9 °C
95. 1.36×10^4 kg/m^3
97. 3.90 cm

Chapter 4

1. liquid state
3. liquid state
5. liquid state
7. (a) condensing; (b) freezing
9. (a) cooling; (b) heating
11. heating
13. (a) element; (b) element; (c) compound; (d) mixture
15. (a) element; (b) heterogeneous mixture; (c) compound; (d) homogeneous mixture
17. (a) Li; (b) Ar; (c) Mg; (d) Mn; (e) Be; (f) Si; (g) Hg; (h) Ti
19. (a) chlorine; (b) neon; (c) cadmium; (d) germanium; (e) cobalt; (f) radium; (g) chromium; (h) tellurium
21. (a) 1; (b) 5; (c) 13; (d) 22; (e) 33; (f) 38; (g) 50; (h) 83
23. (a) metal; (b) nonmetal; (c) metal; (d) nonmetal
25. (a) metal; (b) nonmetal; (c) nonmetal; (d) metal
27. (a) metal; (b) semimetal; (c) metal; (d) nonmetal
29. (a) metal; (b) semimetal; (c) nonmetal; (d) nonmetal
31. (a) solid; (b) gas; (c) solid; (d) gas
33. (a) gas; (b) gas; (c) solid; (d) liquid
35. (a) 12 carbon atoms; 18 hydrogen atoms; 2 chlorine atoms; 8 nitrogen atoms; 1 oxygen atom; 1 sulfur atom (b) 17 carbon atoms; 20 hydrogen atoms; 4 nitrogen atoms; 6 oxygen atoms
37. (a) $C_{20}H_{30}O$; (b) $C_{31}H_{46}O_2$
39. (a) 56 atoms; (b) 167 atoms
41. The mass ratio of the elements in water is constant: 1:8.
43. (a) physical; (b) physical; (c) physical; (d) chemical
45. (a) physical; (b) chemical; (c) chemical; (d) physical
47. (a) chemical; (b) physical; (c) physical; (d) physical
49. (a) chemical; (b) physical; (c) physical; (d) chemical
51. (a) chemical; (b) physical; (c) physical; (d) chemical
53. (a) physical; (b) chemical; (c) physical; (d) chemical

55. (a) physical; (b) chemical; (c) physical; (d) chemical
57. (a) physical; (b) chemical; (c) chemical; (d) physical
59. 7.77 g iron phosphide
61. 0.055 g oxygen gas
63. gaseous state
65. increase
67. The motion of helium atoms will *increase.*
69. 1500 kcal
71. 807 cal
73. 48.0 kcal
75. (a) nuclear → heat; (b) heat → mechanical; (c) mechanical → mechanical; (d) mechanical → electrical
77. (a) heat → mechanical; (b) mechanical → mechanical; (c) mechanical → mechanical; (d) mechanical → electrical
79. A pre-1982 penny is an alloy and a homogeneous mixture.
81. homogeneous mixture
83. germanium
85. oxygen, silicon, and aluminum
87. (a) 47; (b) 82; (c) 50; (d) 79
89. physical change
91. 3250 cal
93. The mass is slightly more.
95. When a playground swing is at its lowest point, the potential energy is *minimum,* and kinetic energy is *maximum.*
97. E = energy, m = mass, c = speed of light

Chapter 5

1. (c) Atoms are indivisible.
3. (1) Robert Boyle; (2) Antoine Lavoisier; (3) Joseph Proust
5. electron (e^-)
7. electron (−1)
9. electrons
11. 1×10^{-13} cm
13. −1 and +1
15. Electrons are found outside the nucleus.
17. (a) 12 n°; (b) 14 n°; (c) 35 n°; (d) 60 n°
19. (a) 1 n°; (b) 8 n°; (c) 33 n°; (d) 78 n°
21.

Atomic Notation	Atomic Number	Mass Number	Number of Protons	Number of Neutrons	Number of Electrons
$^{11}_{5}B$	5	11	5	6	5
$^{15}_{7}N$	7	15	7	8	7
$^{40}_{20}Ca$	20	40	20	20	20
$^{200}_{80}Hg$	80	200	80	120	80

23. (a) $4n^0$ $3p^+$ $3e^-$

(b)
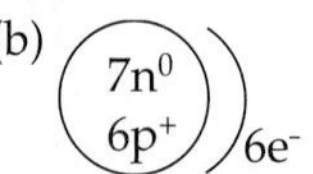

(c)
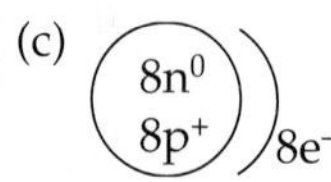

(d)
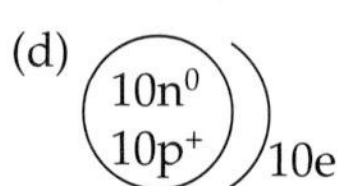

25. Atomic mass is the weighted average of the naturally occurring isotopic masses for an element.
27. The current reference isotope is carbon-12.
29. 22.99 amu
31. 19.00 amu
33. 6.941 amu
35. 55.85 amu
37. Br-81
39. violet light
41. violet light
43. violet light
45. 550 nm
47. photon
49. (a) continuous; (b) quantized
51. (a) continuous; (b) quantized
53.

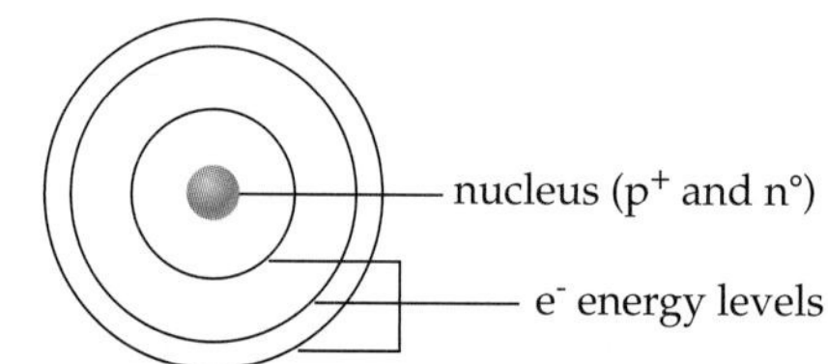

55. 5 → 2
57. blue-green
59. ultraviolet
61. red line
63. (a) 1; (b) 1
65. (a) ultraviolet; (b) red; (c) infrared
67. spectral emission lines
69. (a) 1 sublevel; (b) 2 sublevels; (c) 3 sublevels; (d) 4 sublevels
71. (a) 2 e^-; (b) 6 e^-; (c) 10 e^-; (d) 14 e^-
73. 8 e^-
75. 3*d* sublevel
77. (a) $1s^2$;
(b) $1s^2\ 2s^2$;
(c) $1s^2\ 2s^2\ 2p^6\ 3s^2\ 3p^6\ 4s^2\ 3d^7$;
(d) $1s^2\ 2s^2\ 2p^6\ 3s^2\ 3p^6\ 4s^2\ 3d^{10}\ 4p^6\ 5s^2\ 4d^{10}$
79. (a) Li; (b) Si; (c) Ti; (d) Sr
81. An orbit is the path traveled by an electron of given energy about the nucleus of an atom; an orbital is a region about the nucleus in which there is a high probability of finding an electron of given energy.

83. (a)

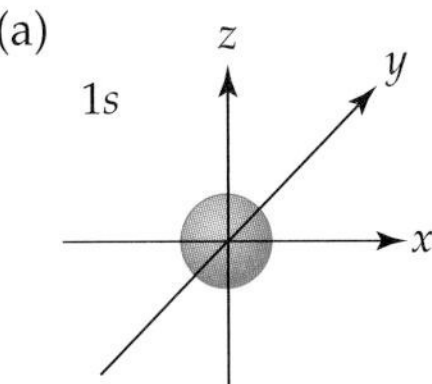

(b)

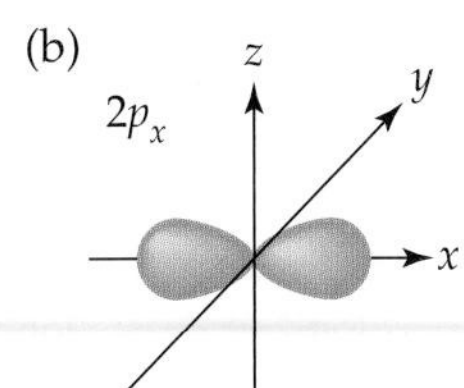

(c)

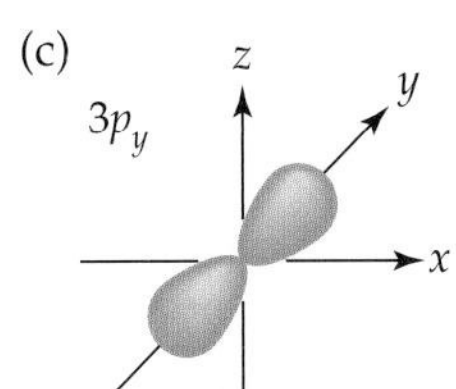

(d)

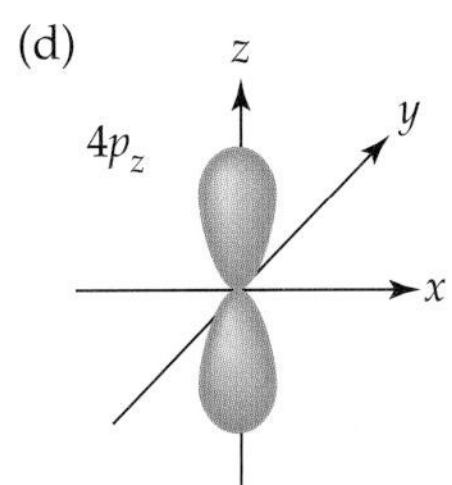

85. (a) $3s$; (b) $3p_x$; (c) equal; (d) equal

87. (a) $5s$; (b) $4p$

89. (a) 2 e^-; (b) 2 e^-; (c) 2 e^-; (d) 2 e^-

91. 9.09×10^{-28} g

93. 106.9 amu

95. no stable isotope

97. (a) ultraviolet; (b) visible; (c) infrared

99. The $3d$ sublevel is filled and is therefore more stable.

Chapter 6

1. The noble gases were not discovered until 1894–1898.

3. Group IV corresponds to the oxide formula GeO_2, and germanium oxide is an example.

5. increasing atomic number

7. Co and Ni

9. periods or series

11. representative elements

13. the lanthanides

15. inner transition elements

17. metals

19. B, Si, Ge, As, Sb, and Te

21. (a) IA/1; (b) IIA/2; (c) VIIA/17; (d) VIIIA/18

23. (a) 1; (b) 13; (c) 15; (d) 17; (e) 11; (f) 3; (g) 5; (h) 7

25. (a) Ge; (b) Na; (c) I; (d) Pm

27. (a) Mg; (b) Br; (c) Lu; (d) Th

29. decreases

31. decreases

33. (a) Na; (b) P; (c) Ca; (d) Kr

35. (a) Al; (b) K; (c) Ba; (d) Fe

37. 0.230 nm; 1.38 g/mL; 14.5 °C

39. 0.097 nm; 3.27 g/mL; 152.2 °C

41. (a) KF; (b) CaF_2; (c) BBr_3; (d) GeI_4

43. (a) CaF_2; (b) $CaCl_2$; (c) $CaBr_2$; (d) CaI_2

45. (a) As_2O_3 and As_2O_5; (b) Sb_2O_3 and Sb_2O_5

47. *s* sublevels

49. *d* sublevels

51. $4f$ sublevel

53. (a) $1s$; (b) $3s$; (c) $4f$; (d) $4p$; (e) $5s$; (f) $2p$; (g) $5p$; (h) $6s$

55. (a) $1s^2\ 2s^1$
(b) $1s^2\ 2s^2\ 2p^5$
(c) $1s^2\ 2s^2\ 2p^6\ 3s^2$
(d) $1s^2\ 2s^2\ 2p^6\ 3s^2\ 3p^3$
(e) $1s^2\ 2s^2\ 2p^6\ 3s^2\ 3p^6\ 4s^2$
(f) $1s^2\ 2s^2\ 2p^6\ 3s^2\ 3p^6\ 4s^2\ 3d^5$
(g) $1s^2\ 2s^2\ 2p^6\ 3s^2\ 3p^6\ 4s^2\ 3d^{10}\ 4p^1$
(h) $1s^2\ 2s^2\ 2p^6\ 3s^2\ 3p^6\ 4s^2\ 3d^{10}\ 4p^6\ 5s^1$

57. (a) 1; (b) 3; (c) 5; (d) 7

59. (a) 1; (b) 3; (c) 5; (d) 7; (e) 2; (f) 4; (g) 6; (h) 8

61. H· ;
·B· ;
·N: ;
:F: ;
Ca· ;
·Si· ;
·O: ;
:Ar:

63. increases

65. Group VIIIA/18

67. (a) Mg; (b) S; (c) Sn; (d) N

69. (a) Cs; (b) Ar; (c) Al; (d) I

71. (a) 1+; (b) 2+; (c) 3+; (d) 4+

73. (a) 1+; (b) 3+; (c) 2−; (d) 1−

75. (b) Ca^{2+}; (c) S^{2-}

77. (a) [Ne]
(b) [Ar]
(c) [Ar] $3d^6$
(d) [Kr] $4d^2$

79. (a) [He] $2s^2\ 2p^6$, or [Ne]
(b) [Ne] $3s^2\ 3p^6$, or [Ar]
(c) [He] $2s^2\ 2p^6$, or [Ne]
(d) [Kr] $5s^2\ 4d^{10}\ 5p^6$, or [Xe]

81. scandium (Sc)

83. (a) Group IA; (b) Group IB; (c) Group IIIB; (d) Group IIIA

85. 0. 284 nm; 2.21 g/mL; 17.9 °C

87. (a) [Kr] $5s^2$; (b) [Kr] $5s^2\ 4d^6$; (c) [Kr] $5s^2\ 4d^{10}\ 5p^3$; (d) [Xe] $6s^1$

89. When an alkali metal atom loses one electron, it assumes a noble gas electron configuration, but an alkaline earth metal does not.

91. Although hydrogen and the Group IA/1 metals each have one valence electron in an *s* sublevel, in hydrogen the negatively charged electron occupies the 1*s* sublevel, which is closest to the positively charged nucleus.

Chapter 7

1. (a) binary molecular; (b) ternary ionic; (c) binary ionic; (d) ternary oxyacid

3. (a) polyatomic cation; (b) monoatomic cation; (c) monoatomic anion; (d) polyatomic anion

5. (a) potassium ion; (b) barium ion; (c) iron(II) ion; (d) cobalt(III) ion

7. (a) Li^+ ; (b) Al^{3+} ; (c) Pb^{2+}; (d) Ni^{2+}

9. (a) cuprous ion; (b) ferric ion; (c) stannic ion; (d) plumbous ion

11. (a) Co^{2+} ion; (b) Co^{3+} ion

13. (a) fluoride ion; (b) iodide ion; (c) oxide ion; (d) phosphide ion

15. (a) chlorate ion; (b) nitrite ion; (c) carbonate ion; (d) hydrogen sulfate ion

17. (a) $Cr_2O_7^{2-}$; (b) PO_4^{3-}; (c) OH^-; (d) ClO^-

19. (a) LiCl; (b) Ag_2O; (c) CrI_3; (d) Sn_3N_2

21. (a) KNO_3; (b) $(NH_4)_2Cr_2O_7$; (c) $Al_2(SO_3)_3$; (d) $Cr(ClO)_3$

23. (a) $Sr(NO_2)_2$; (b) $Zn(MnO_4)_2$; (c) $CaCrO_4$; (d) $Cr(ClO_4)_3$

25. (a) magnesium oxide; (b) zinc oxide;
(c) cadmium oxide; (d) barium oxide

27. (a) lithium bromide; (b) strontium iodide;
(c) sodium nitride; (d) aluminum fluoride

29. (a) Cu_2O; (b) FeN; (c) $HgCl_2$; (d) PbS_2

31. (a) RbCl; (b) NaBr

33. (a) GaN; (b) AlAs

35. (a) lithium permanganate; (b) strontium chlorite;
(c) barium carbonate; (d) ammonium dichromate

37. (a) copper(II) sulfate; (b) iron(II) chromate;
(c) mercury(II) nitrite; (d) lead(II) acetate

39. (a) $Mn(C_2H_3O_2)_2$; (b) $Cu(ClO_2)_2$;
(c) $Sn_3(PO_4)_2$; (d) $Fe(ClO)_3$

41. (a) Fr_2SO_4; (b) Na_2SO_3

43. (a) $Ra(ClO_3)_2$; (b) $Ba(BrO_3)_2$

45. (a) bromine monofluoride; (b) carbon tetrafluoride;
(c) diiodine tetraoxide; (d) dichlorine trioxide

47. (a) N_2O_5; (b) CCl_4; (c) IBr; (d) H_2S

49. (a) hydrobromic acid; (b) hydroiodic acid

51. (a) chlorous acid; (b) phosphoric acid

53. (a) $HC_2H_3O_2(aq)$; (b) $HNO_3(aq)$

55. (a) HIO(*aq*); (b) HBrO(*aq*)

57. (a) 0; (b) 2+; (c) 3+; (d) 0

59. SiO_3^{2-}

61.

Ions	fluoride Ion	oxide Ion	nitride Ion
copper(II) ion	CuF_2 copper(II) fluoride	CuO copper(II) oxide	Cu_3N_2 copper(II) nitride
cobalt(II) ion	CoF_2 cobalt(II) fluoride	CoO cobalt(II) oxide	Co_3N_2 cobalt(II) nitride
lead(II) ion	PbF_2 lead(II) fluoride	PbO lead(II) oxide	Pb_3N_2 lead(II) nitride

63.

Ions	hydroxide Ion	sulfate Ion	phosphate Ion
mercurous ion	$Hg_2(OH)_2$ mercurous hydroxide	Hg_2SO_4 mercurous sulfate	$(Hg_2)_3(PO_4)_2$ mercurous phosphate
ferrous ion	$Fe(OH)_2$ ferrous hydroxide	$FeSO_4$ ferrous sulfate	$Fe_3(PO_4)_2$ ferrous phosphate
stannous ion	$Sn(OH)_2$ stannous hydroxide	$SnSO_4$ stannous sulfate	$Sn_3(PO_4)_2$ stannous phosphate

65. (a) *-ide*; (b) *-ic* acid

67. (a) *-ite*; (b) *-ous* acid

69. (a) *-ate*; (b) *-ic* acid

71. (a) H_2O; (b) NaClO; (c) NaOH; (d) $NaHCO_3$

73. (a) boron trifluoride; (b) silicon tetrachloride; (c) diarsenic pentaoxide; (d) diantimony trioxide

75. TiO_2

77. $AcCl_3$

Chapter 8

1. (a) chemical reaction; (b) chemical reaction

3. (a) chemical reaction; (b) physical change

5. (a) chemical reaction; (b) chemical reaction

7. $Mg(s) + Br_2(l) \rightarrow MgBr_2(s)$

9. $Zn(HCO_3)_2(s) \rightarrow ZnCO_3(s) + CO_2(g) + H_2O(g)$

11. $Cd(s) + Co(NO_3)_2(aq) \rightarrow Cd(NO_3)_2(aq) + Co(s)$

13. $LiI(aq) + AgNO_3(aq) \rightarrow AgI(s) + LiNO_3(aq)$

15. $HNO_3(aq) + NaOH(aq) \rightarrow NaNO_3(aq) + HOH(l)$

17. (a) $3\,H_2(g) + N_2(g) \rightarrow 2\,NH_3(g)$
(b) $Al_2(CO_3)_3(s) \rightarrow Al_2O_3(s) + 3\,CO_2(g)$
(c) $Sr(s) + 2\,H_2O(l) \rightarrow Sr(OH)_2(aq) + H_2(g)$
(d) $K_2SO_4(aq) + Ba(OH)_2(aq) \rightarrow BaSO_4(s) + 2\,KOH(aq)$
(e)
$2\,H_3PO_4(aq) + 3\,Mn(OH)_2(s) \rightarrow Mn_3(PO_4)_2(s) + 6\,HOH(l)$

19. (a)
$2\,H_3PO_4(aq) + 3\,Mn(OH)_2(s) \rightarrow Mn_3(PO_4)_2(s) + 6\,HOH(l)$
(b)
$Hg_2(NO_3)_2(aq) + 2\,NaBr(aq) \rightarrow Hg_2Br_2(s) + 2\,NaNO_3(aq)$
(c) $Mg(s) + 2\,HC_2H_3O_2(aq) \rightarrow Mg(C_2H_3O_2)_2(aq) + H_2(g)$
(d) $2\,LiNO_3(s) \rightarrow 2\,LiNO_2(s) + O_2(g)$
(e) $2\,Pb(s) + O_2(g) \rightarrow 2\,PbO(s)$

21. (a) combination; (b) decomposition; (c) single replacement; (d) double replacement; (e) neutralization

23. (a) neutralization; (b) double replacement; (c) single replacement; (d) decomposition; (e) combination

25. metal + oxygen gas $\rightarrow$ metal oxide
(a) $2\,Ni(s) + O_2(g) \rightarrow 2\,NiO(s)$
(b) $4\,Fe(s) + 3\,O_2(g) \rightarrow 2\,Fe_2O_3(s)$

27. (a) $2\,C(s) + O_2(g) \rightarrow 2\,CO(g)$
(b) $4\,P(s) + 5\,O_2(g) \rightarrow 2\,P_2O_5(s)$

29. (a) $2\,Cu(s) + Cl_2(g) \rightarrow 2\,CuCl(s)$
(b) $Co(s) + S(s) \rightarrow CoS(s)$

31. (a) $4\,Cr(s) + 3\,O_2(g) \rightarrow 2\,Cr_2O_3(s)$
(b) $2\,Cr(s) + N_2(g) \rightarrow 2\,CrN(s)$

33. (a) $4\,Li + O_2 \rightarrow 2\,Li_2O$
(b) $2\,Ca + O_2 \rightarrow 2\,CaO$

35. (a) $2\,Na + I_2 \rightarrow 2\,NaI$
(b) $3\,Ba + N_2 \rightarrow Ba_3N_2$

37. (a) $2\,KHCO_3(s) \rightarrow K_2CO_3(s) + H_2O(g) + CO_2(g)$
(b) $Sr(HCO_3)_2(s) \rightarrow SrCO_3(s) + H_2O(g) + CO_2(g)$

39. (a) $Co_2(CO_3)_3(s) \rightarrow Co_2O_3(s) + 3\,CO_2(g)$
(b) $Sn(CO_3)_2(s) \rightarrow SnO_2(s) + 2\,CO_2(g)$

41. (a) $Ca(NO_3)_2(s) \rightarrow Ca(NO_2)_2(s) + O_2(g)$
(b) $2\,Ag_2SO_4(s) \rightarrow 2\,Ag_2SO_3(s) + O_2(g)$

43. (a) $2\,AgHCO_3(s) \rightarrow Ag_2CO_3(s) + H_2O(g) + CO_2(g)$
(b) $Zn(HCO_3)_2(s) \rightarrow ZnCO_3(s) + H_2O(g) + CO_2(g)$

45. (a) $2\,NaClO_3(s) \rightarrow 2\,NaCl(s) + 3\,O_2(g)$
(b) $Ca(NO_3)_2(s) \rightarrow Ca(NO_2)_2(s) + O_2(g)$

47. (c) Co; (d) Mn

49. (a) Ni; (b) Zn; (d) Al

51. (a) Li; (c) Ca

53. (a) $Cu(s) + Al(NO_3)_3(aq) \rightarrow$ NR
(b)
$2\,Al(s) + 3\,Cu(NO_3)_2(aq) \rightarrow 3\,Cu(s) + 2\,Al(NO_3)_3(aq)$

55. (a) $Ni(s) + Pb(C_2H_3O_2)_2(aq) \rightarrow Pb(s) + Ni(C_2H_3O_2)_2(aq)$
(b) $Pb(s) + Ni(C_2H_3O_2)_2(aq) \rightarrow$ NR

57. (a) $Mg(s) + 2\,HNO_3(aq) \rightarrow Mn(NO_3)_2(aq) + H_2(g)$
(b) $Mn(s) + 2\,HCl(aq) \rightarrow MgCl_2(s) + H_2(g)$

59. (a) $2\,Li(s) + 2\,H_2O(l) \rightarrow 2\,LiOH(aq) + H_2(g)$
(b) $Ba(s) + 2\,H_2O(l) \rightarrow Ba(OH)_2(aq) + H_2(g)$

61. (a) $Zn(s) + Pb(NO_3)_2(aq) \rightarrow Pb(s) + Zn(NO_3)_2(aq)$
(b) $Cd(s) + Fe(NO_3)_2(aq) \rightarrow$ NR

63. (a) $Zn(s) + 2\,HNO_3(aq) \rightarrow Zn(NO_3)_2(aq) + H_2(g)$
(b) $Cd(s) + 2\,HNO_3(aq) \rightarrow Cd(NO_3)_2(aq) + H_2(g)$

65. (a) $2\,K(s) + 2\,H_2O(l) \rightarrow 2\,KOH(aq) + H_2(g)$
(b) $Ba(s) + 2\,H_2O(l) \rightarrow Ba(OH)_2(aq) + H_2(g)$

67. (a) $Al(NO_3)_3$; (b) Na_2SO_4

69. (a) insoluble; (c) insoluble; (d) insoluble

71. (a) $CrI_3(aq) + 3\,NaOH(aq) \rightarrow Cr(OH)_3(s) + 3\,NaI(aq)$
(b)
$NiSO_4(aq) + Hg_2(NO_3)_2(aq) \rightarrow Hg_2SO_4(s) + Ni(NO_3)_2(aq)$

73. (a) $2\,AgC_2H_3O_2(aq) + SrI_2(aq)$
$\rightarrow 2\,AgI(s) + Sr(C_2H_3O_2)_2(aq)$
(b) $FeSO_4(aq) + Ca(OH)_2(aq) \rightarrow Fe(OH)_2(s) + CaSO_4(aq)$

75. (a) $H_2CO_3(aq) + 2\,LiOH(aq) \rightarrow Li_2CO_3(aq) + 2\,HOH(l)$
(b) $2\,HNO_3(aq) + Ca(OH)_2(aq)$
$\rightarrow Ca(NO_3)_2(aq) + 2\,HOH(l)$

77. (a) $HNO_3(aq) + Ba(OH)_2(aq)$
$\rightarrow Ba(NO_3)_2(aq) + 2\,HOH(l)$
(b) $H_2SO_4(aq) + 2\,NaOH(aq) \rightarrow Na_2SO_4(aq) + 2\,HOH(l)$

79. If a chemical equation is impossible to balance, it is most likely that the equation has a chemical formula with an incorrect subscript.

81. (a) $3\,Fe(s) + 4\,H_2O(g) \rightarrow Fe_3O_4(s) + 4\,H_2(g)$
(b) $4\,FeS(s) + 7\,O_2(g) \rightarrow 2\,Fe_2O_3(s) + 4\,SO_2(g)$

83. (a) $F_2(g) + 2\,NaBr(aq) \rightarrow Br_2(l) + 2\,NaF(aq)$
(b) $Sb_2S_3(s) + 6\,HCl(aq) \rightarrow 2\,SbCl_3(aq) + 3\,H_2S(g)$

85. (a) $CH_4(g) + 2\,O_2(g) \rightarrow CO_2(g) + 2\,H_2O(g)$
(b) $C_3H_8(g) + 5\,O_2(g) \rightarrow 3\,CO_2(g) + 4\,H_2O(g)$

87. $Fe_2O_3(l) + 3\,CO(g) \rightarrow 2\,Fe(l) + 3\,CO_2(g)$

89. (1) $4\,NH_3(g) + 5\,O_2(g) \rightarrow 4\,NO(g) + 6\,H_2O(g)$
(2) $2\,NO(g) + O_2(g) \rightarrow 2\,NO_2(g)$
(3) $3\,NO_2(g) + H_2O(l) \rightarrow 2\,HNO_3(aq) + NO(g)$

Chapter 9

1. (a) 9.01 amu; (b) 79.90 amu; (c) 44.96 amu; (d) 78.96 amu

3. (a) 9.01 g; (b) 79.90 g; (c) 44.96 g; (d) 78.96 g

5. (a) 6.02×10^{23} atoms Li; (b) 6.02×10^{23} molecules F_2

7. (a) 1 mol NaCl formula units; (b) 1 mol NH_3 molecules

9. (a) 7.52×10^{22} atoms Zn; (b) 1.51×10^{23} molecules Cl_2; (c) 4.06×10^{23} formula units $ZnCl_2$

11. (a) 0.0208 mol Mn; (b) 0.415 mol SO_3; (c) 11.2 mol $MnSO_4$

13. (a) 26.98 g/mol; (b) 28.09 g/mol; (c) 32.00 g/mol; (d) 256.56 g/mol

15. (a) 72.15 g/mol; (b) 136.15 g/mol; (c) 173.95 g/mol; (d) 357.49 g/mol

17. (a) 98.3 g Hg; (b) 0.540 g N_2; (c) 1.75 g $BaCl_2$

19. (a) 2.31×10^{22} atoms K; (b) 8.84×10^{21} molecules O_2; (c) 1.75×10^{21} formula units $AgClO_3$

21. (a) 1.50×10^{-23} g/atom; (b) 3.82×10^{-23} g/atom; (c) 9.79×10^{-23} g/atom; (d) 1.24×10^{-22} g/atom

23. 0 °C and 1 atm

25. (a) 0.178 g/L; (b) 3.17 g/L; (c) 0.717 g/L; (d) 1.34 g/L

27. (a) 32.0 g/mol; (b) 34.0 g/mol; (c) 44.1 g/mol; (d) 121 g/mol

29.

Gas	Molecules	Mass	Volume at STP
chlorine, Cl_2	6.02×10^{23}	70.90 g	22.4 L
hydrogen chloride, HCl	6.02×10^{23}	36.45 g	22.4 L
sulfur tetrachloride, SCl_4	6.02×10^{23}	173.87 g	22.4 L

31. (a) 1.40 L He; (b) 4.04 L N_2

33. (a) 1.60 g H_2S; (b) 18.1 g N_2O_3

35. (a) 2.69×10^{21} molecules H_2; (b) 1.89×10^{21} molecules NH_3

37.

Gas	Molecules	Mass	Volume at STP
hydrogen, H_2	**1.50×10^{23}**	0.503 g	5.58 L
ammonia, NH_3	1.50×10^{23}	**4.25 g**	5.58 L
methane, CH_4	1.50×10^{23}	4.00 g	**5.58 L**

39. The law of definite composition states that a compound always contains the same elements in the same proportion by mass. Thus, the percent hydrogen by mass in a kilogram of water is 11%.

41. 42.10% C; 6.49% H; 51.42% O

43. 40.24% C; 7.444% H; 9.388% N; 21.49% S; 21.44% O

45. 26.722% Hg; 6.125% Na; 31.999% C; 1.08% H; 21.288% Br; 12.79% O

47. SnO_2

49. CuO

51. Co_2S_3

53. CuCl

55. $SnBr_2$

57. C_2HCl_3

59. $C_9H_8O_4$

61. $C_6H_{16}N_2$

63. CH_3O, $C_2H_6O_2$

65. CHCl, $C_6H_6Cl_6$

67. C_5H_7N, $C_{10}H_{14}N_2$

69. 6.02×10^{23} photons

71. 5 g nickel coin

73. Earth

75. Ga_2O_3

77. 2.99×10^{-23} cm^3 H_2O

79. 2.11×10^{22} atoms C

81. 11 atoms C

83. 6.04×10^{23} atoms Cu/mol

Chapter 10

1. (a) 3 mol C; (b) 2 L B

3. (a) 2.50 g C; (b) 1.75 g B

5. (a) 88.04 g → 88.04 g; (b) 92.02 g → 92.02 g

7. 2.50 mol O_2

9. 0.150 mol Cl_2 react; 0.100 mol $FeCl_3$

11. 0.500 mol C_3H_8 react; 1.50 mol CO_2

13. mass–volume problem

15. volume–volume problem

17. mass–mass problem

19. 3.39 g Ag

21. 1.24 g ZnO

23. 5.21 g $BiCl_3$

25. 6.38 g Hg

27. 1.68 g $Ca_3(PO_4)_2$

29. 262 mL CO_2

31. 1620 mL O_2

33. 0.244 g Mg

35. 0.154 g Na

37. 125 mL I_2

39. 22.5 mL N_2

41. 5.00 L O_2

43. 1750 mL Cl_2

45. 50.0 L SO_3

47. 500.0 cm^3 N_2

49. The limiting reactant is NO, which produces 1.00 mol of NO_2.

51. Neither N_2 or O_2 is a limiting reactant because each produces the same amount of product (10.0 mol NO).

53. The limiting reactant is H_2, which produces 5.00 mol of H_2O.

55. The limiting reactant is C_2H_6, which produces 3.00 mol of H_2O.

57. (a) Neither Co or S is a limiting reactant; the reaction produces 1.00 mol CoS. After the reaction Co = 0.00 mol; S = 0.00 mol; and CoS = 1.00 mol.

(b) The limiting reactant is Co, which produces 2.00 mol of CoS. After the reaction Co = 0.00 mol; S = 1.00 mol; and CoS = 2.00 mol.

59. 23.0 g Fe

61. 77.6 g Fe

63. 0.742 g $MgSO_4$

65. 0.693 g H_2O

67. 90.0 mL NO_2

69. 30.0 mL N_2O_3

71. 2.50 L SO_3

73. 20.0 L Cl_2

75. 89.9%

77. 106%

79. g/mol

81. 0.929 g Cr_2O_3

83. 655 g Al

85. 0.593 L H_2S

87. 124 L H_2

89. 16.3 g H_2O

91. The limiting reactant is gasoline; the excess reactant is oxygen.

Chapter 11

1. (a), (b), and (d)

3. (a) 760 mm Hg; (b) 76 cm Hg

5. (a) 73.5 psi; (b) 3.80×10^2 cm Hg

7. (a) 104 kPa; (b) 783 torr

9. (a) pressure decreases; (b) pressure increases; (c) pressure increases

11. (a) pressure decreases; (b) pressure increases; (c) pressure decreases

13. Increasing the volume of a gas causes gas molecules to be further apart; thus they have fewer collisions and the pressure decreases.

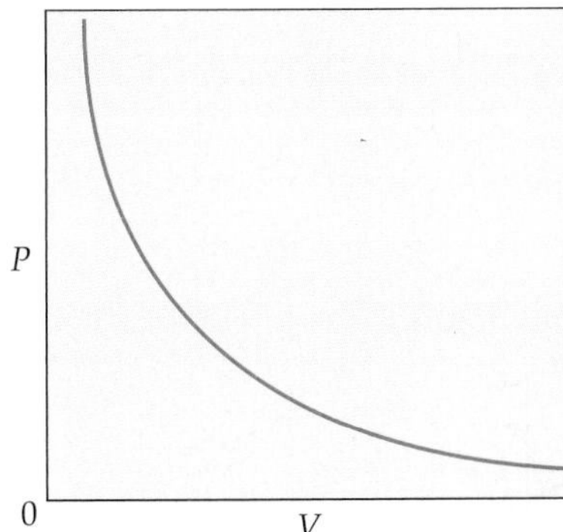

17. 3.50 atm

19. 1.19 L

21.

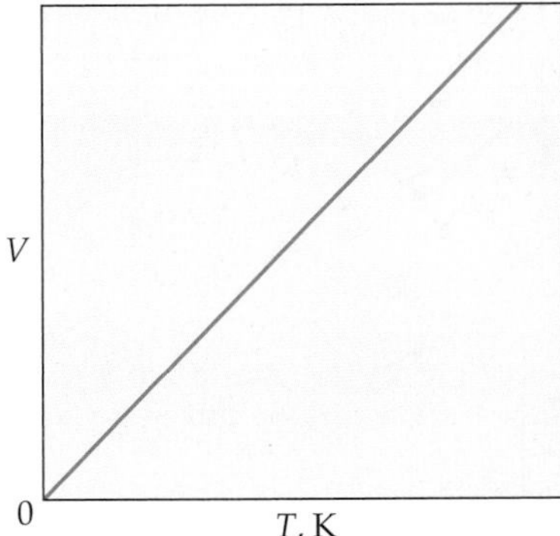

23. 602 mL

25. 144 °C

23.

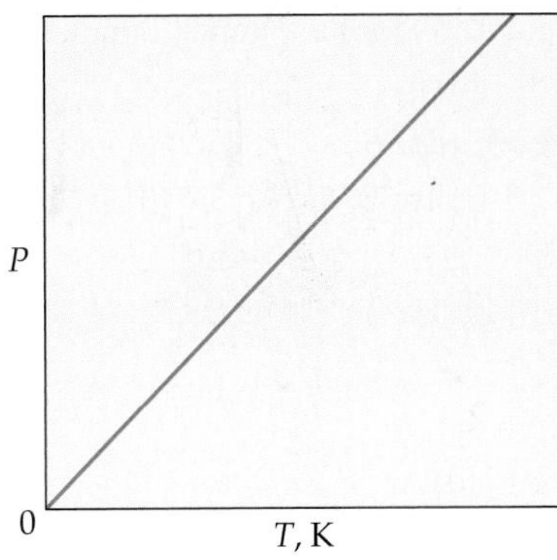

29. 1.31 atm

31. 215 °C

33. 8.83 L

35. 1.33 L

37. 8.19 atm

39. 5190 °C

41. 756 mm Hg

43. The vapor pressure of a liquid increases as the temperature increases.

45. (a) 23.8 mm Hg; (b) 92.5 mm Hg

47. 760 mm Hg

49. 135 mm Hg

51. A gas may be collected in a flask that contains water, and the gas will displace an equal volume of the liquid.

53. 749 mm Hg

55. Refer to Section 11.1 in the text.

57. high temperature, low pressure

59. (a) all gases are equal; (b) all gases are equal; (c) He atoms; (d) Ar atoms

61. 0 mm Hg

63. 2.45 atm

65. 2.78 mol

67. 37,500 lb

69. 1030 cm H_2O

71. 2350 psi

73. 38.5 mL

75. 1.23 atm

77. CH_4 molecules

79. O_3 (48.0 g/mol)

81. 12.5 g/mol

83. 3.92 g Cl_2

85. 62.4 L · torr/mol · K

87. Since helium gas is much lighter than nitrogen and oxygen gases, the helium atoms move faster over the vocal cords and produce a high-pitched voice.

Chapter 12

1. K (1 valence e^-); I (7 valence e^-)

3. K^+ (0 valence e^-); I^- (8 valence e^-)

5. (a) ionic; (b) covalent; (c) covalent; (d) ionic

7. (a) molecule; (b) molecule; (c) formula unit; (d) formula unit

9. (a) molecule; (b) atom; (c) formula unit; (d) molecule

11. (a) 1+; (b) 2+; (c) 3+; (d) 4+

13. (a) 1−; (b) 1−; (c) 2−; (d) 3−

15. (a) $1s^2$ (b) $1s^2\,2s^2\,2p^6$ (c) $1s^2\,2s^2\,2p^6\,3s^2\,3p^6$ (d) $1s^2\,2s^2\,2p^6$

17. (a) $1s^2\,2s^2\,2p^6\,3s^2\,3p^6$

(b) $1s^2\,2s^2\,2p^6\,3s^2\,3p^6\,4s^2\,3d^{10}\,4p^6\,5s^2\,4d^{10}\,5p^6$

(c) $1s^2\,2s^2\,2p^6\,3s^2\,3p^6$

(d) $1s^2\,2s^2\,2p^6\,3s^2\,3p^6$

19. (a) Ar; (b) Ar; (c) Ar; (d) Ar

21. (a) He; (b) Ne; (c) Ar; (d) Ne

23. (a) Ar; (b) Xe; (c) Ar; (d) Ar

25. (a) Li atom; (b) Mg atom; (c) F ion; (d) O ion

27. The false statements are (b) and (c).

29. (a) The bond length in H−Cl is less. (b) The bond length in O=O is less.

31. The false statements are (a) and (d).

33. (a) H:H H−H

(b)  :F:F: :F−F:

(c)  H:Br: H−Br:

(d) 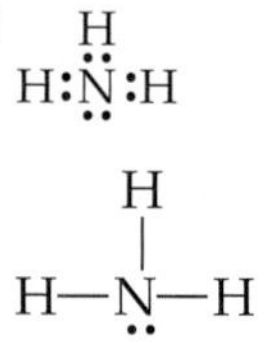 H:N:H (with H above N) H−N−H (with H bonded above N)

35. (a)  H:O:N::O H−O−N=O:

(b) 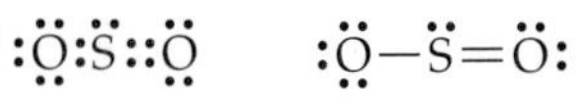 :O:S::O :O−S=O:

(c) 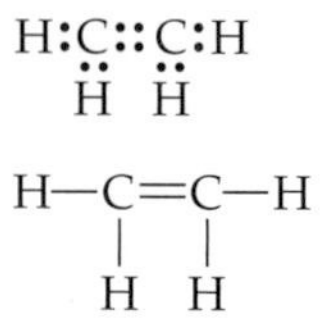 H:C::C:H (each C with H below) H−C=C−H (each C bonded to H below)

(d) H:C:::C:H H−C≡C−H

37. (a) 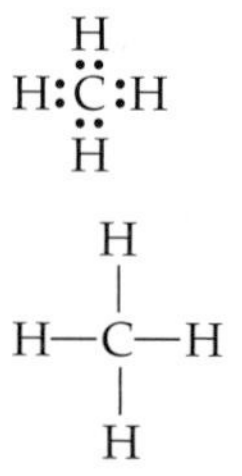 H:C:H (with H above and below C) H−C−H (with H bonded above and below C)

(b) :F:O:F: :F−O−F:

(c) H:O:O:H H−O−O−H

(d) 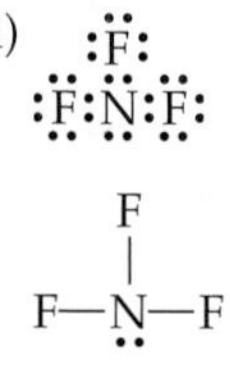 :F:N:F: (with :F: above N) F−N−F (with F bonded above N)

39. (a) 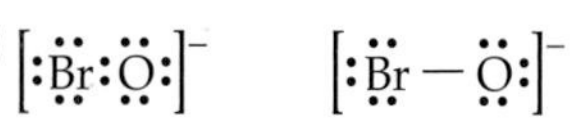 [:Br:O:]⁻ [:Br−O:]⁻

(b) 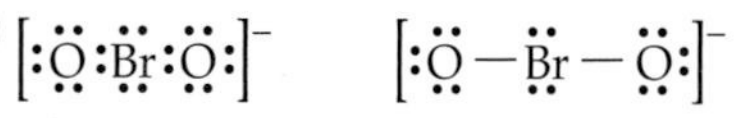 [:O:Br:O:]⁻ [:O−Br−O:]⁻

(c) 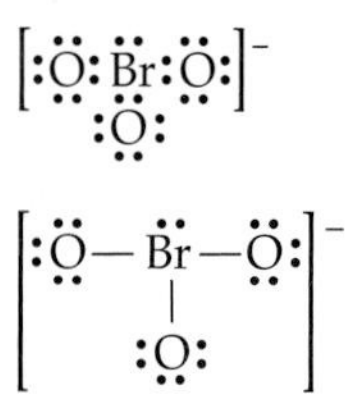 [:O:Br:O:]⁻ (with :O: below Br) [:O−Br−O:]⁻ (with :O: bonded below Br)

(d) 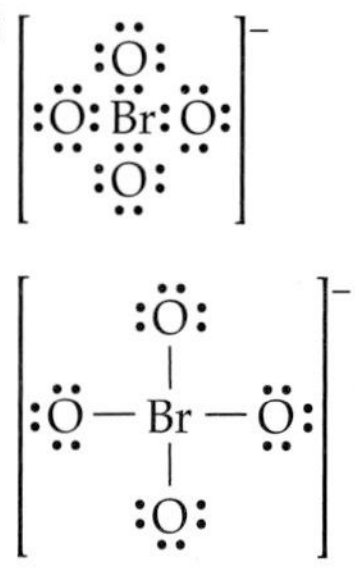 [:O:Br:O:]⁻ (with :O: above and below Br) [:O−Br−O:]⁻ (with :O: bonded above and below Br)

41. (a) 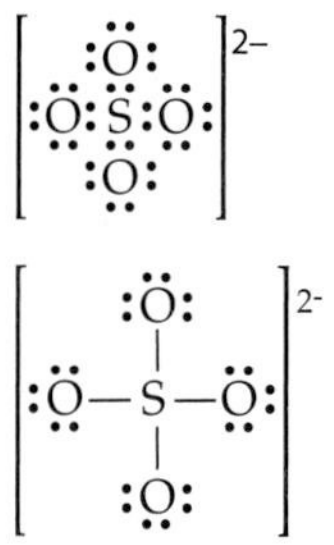 [:O:S:O:]²⁻ (with :O: above and below S) [:O−S−O:]²⁻ (with :O: bonded above and below S)

(b) 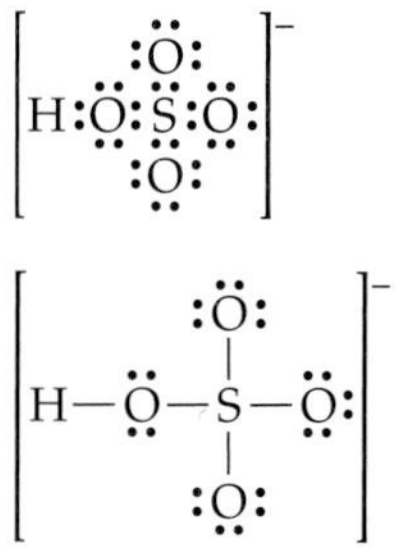 [H:O:S:O:]⁻ (with :O: above and below S) [H−O−S−O:]⁻ (with :O: bonded above and below S)

(c) 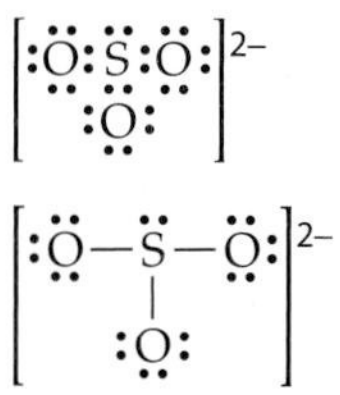 [:O:S:O:]²⁻ (with :O: below S) [:O−S−O:]²⁻ (with :O: bonded below S)

(d) 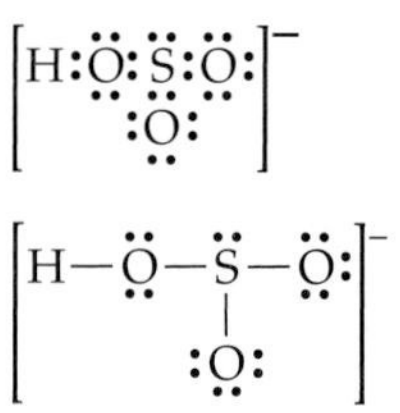 [H:O:S:O:]⁻ (with :O: below S) [H−O−S−O:]⁻ (with :O: bonded below S)

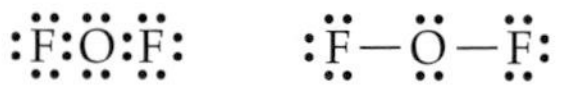

43. (a) $[H:\ddot{O}:H]^+$ with H below O; $[H—\ddot{O}—H]^+$ with H bonded below O

(b) $[:\ddot{O}:H]^-$ $[:\ddot{O}—H]^-$

(c) $[H:\ddot{S}:]^-$ $[H—\ddot{S}:]^-$

(d) $[:C:::N:]^-$ $[:C \equiv N:]^-$

45. Electronegativity increases from bottom to top in the periodic table.

47. nonmetals

49. (a) Cl; (b) O; (c) Se; (d) F

51. (a) 0.2; (b) 1.2; (c) 0.5; (d) 0.3

53. (a) δ^+ H—S δ^-; (b) δ^- O—S δ^+; (c) δ^+ N—F δ^-; (d) δ^+ S—Cl δ^-

55. (a), (b), and (d)

57. H_2, N_2, Cl_2, and Br_2

59. $H:\ddot{Br}:\ddot{O}:$

$H:\ddot{Br}—\ddot{O}:$

61. $H:\ddot{Br}:\ddot{O}:$ with $:\ddot{O}:$ below Br

$H:\ddot{Br}—\ddot{O}:$ with $:\ddot{O}:$ bonded below Br

63. $[H:\ddot{N}:H]^+$ with H above and H below N

$[H:\ddot{N}:H]^+$ with H above and H bonded below N

65. $[:\ddot{O}:\ddot{N}::\ddot{O}:]^-$ with $:\ddot{O}:$ above N

$[:\ddot{O}:N::\ddot{O}:]^-$ with $:\ddot{O}:$ bonded above N

67. H—N: ········· H—N: (each N also bonded to H above and H below); arrow labeled "H bond"

69. The bond energy is less in a hydrogen bond than in a polar covalent bond.

71. tetrahedral, tetrahedral, 109.5°

73. tetrahedral, trigonal pyramidal, 107°

75. tetrahedral, bent, 104.5°

77. Each C—F bond is polar, but the symmetrical tetrahedral arrangement of the four bonds produces a nonpolar molecule.

79. (a) atom; (b) molecule; (c) formula unit; (d) molecule

81. (a) Sr_3As_2; (b) RaO; (c) $Al_2(CO_3)_3$; (d) $Cd(OH)_2$

83. The sodium ion has one less energy level (3(*s*)).

85. 1.0 (polar)

87. 0 (nonpolar)

89. δ^+ Ge—Cl δ^-

91. $H:\ddot{Si}:H$ with H below Si; H—Si—H with H bonded above and below Si

93. $[:\ddot{O}:\ddot{As}:\ddot{O}:]^{3-}$ with $:\ddot{O}:$ below As; $[:\ddot{O}—\ddot{As}—\ddot{O}:]^{3-}$ with $:\ddot{O}:$ bonded below As

95. linear, linear, 180°

97. trigonal planar, trigonal planar, 120°

99. $:\ddot{O}:\ddot{Xe}:\ddot{O}:$ $:\ddot{O}—\ddot{Xe}—\ddot{O}:$

Chapter 13

1. (a) true; (b) true; (c) false; (d) true

3. (a) gas; (b) solid; (c) gas; (d) solid; (e) liquid; (f) gas

5. (a) dispersion force; (b) dipole force; (c) hydrogen bond; (d) dipole force

7. (a) C_2H_5Cl; (b) CH_3OCH_3

9. (a) CH_3COOH; (b) C_2H_5OH

11. The vapor pressure of water is the pressure exerted by water vapor molecules when the rate of evaporation is equal to the rate of condensation.

13. The intermolecular attraction in water is strong; thus the viscosity of water is higher than other liquids with molecules the same size.

15. (a) high; (b) low; (c) low; (d) low

17. As the temperature of a liquid increases, the vapor pressure increases.

19. (a) ~180 mm Hg; (b) ~330 mm Hg

21. 56 °C

23. Refer to Section 13.4 in the text.

25. (a) solid; (b) liquid

27. NaCl, Cl_2, and Na

29. (a) ions; (b) molecules; (c) atoms

31. (a) metallic; (b) ionic; (c) molecular; (d) molecular

33.

35. 4.00×10^4 cal (40.0 kcal)

37. 15,400 cal (15.4 kcal)

39. 82,800 cal (82.8 kcal)

41. 28,100 cal (28.1 kcal)

43. 75,200 cal (75.2 kcal)

45. two bonding electron pairs, two nonbonding electron pairs

47. 104.5°

49.

δ−
O
H H
δ+ δ+

51. H—F··········H—F
↑
H bond

53. The cap will pop off, or the bottle will break, as the volume expands.

55. (a) H_2O; (b) H_2Se

57. (a) H_2O; (b) H_2Se

59. (a) increases; (b) increases; (c) increases; (d) increases

61. $2\,H_2(g) + O_2(g) \rightarrow 2\,H_2O(l)$

63. (a) $2\,Li(s) + 2\,H_2O(l) \rightarrow 2\,LiOH(aq) + H_2(g)$
(b) $Na_2O(s) + H_2O(l) \rightarrow 2\,NaOH(aq)$
(c) $CO_2(g) + H_2O(l) \rightarrow H_2CO_3(aq)$

65. (a) $Mg(s) + H_2O(l) \rightarrow NR$
(b) $SrO(s) + H_2O(l) \rightarrow Sr(OH)_2(aq)$
(c) $N_2O_5(s) + H_2O(l) \rightarrow 2\,HNO_3(aq)$

67. (a) $2\,C_3H_6(g) + 9\,O_2(g) \rightarrow 6\,CO_2(g) + 6\,H_2O(g)$
(b) $Na_2Cr_2O_7{\cdot}2\,H_2O(s) \rightarrow Na_2Cr_2O_7(s) + 2\,H_2O(g)$
(c) $2\,HF(aq) + Ca(OH)_2(aq) \rightarrow CaF_2(aq) + 2\,HOH(l)$

69. (a) $2\,C_4H_{10}(g) + 13\,O_2(g) \rightarrow 8\,CO_2(g) + 10\,H_2O(g)$
(b) $Co(C_2H_3O_2)_2{\cdot}4\,H_2O(s) \rightarrow Co(C_2H_3O_2)_2(s) + 4\,H_2O(g)$
(c) $2\,HNO_3(aq) + Ba(OH)_2(aq)$
$\rightarrow Ba(NO_3)_2(aq) + 2\,HOH(l)$

71. (a) magnesium sulfate heptahydrate; (b) cobalt(III) cyanide trihydrate; (c) manganese(II) sulfate monohydrate; (d) sodium dichromate dihydrate

73. (a) $Na_2CO_3 \cdot 10H_2O$; (b) $Ni(NO_3)_2 \cdot 6H_2O$; (c) $CoI_3 \cdot 8H_2O$; (d) $Cr(C_2H_3O_2)_3 \cdot H_2O$

75. (a) 40.55% H_2O; (b) 10.91% H_2O; (c) 28.28% H_2O; (d) 30.80% H_2O

77. (a) $NiCl_2 \cdot 2\,H_2O$; (b) $Sr(NO_3)_2 \cdot 6\,H_2O$; (c) $CrI_3 \cdot 9\,H_2O$; (d) $Ca(NO_3)_2 \cdot 4\,H_2O$

79. ~75%

81. ~95 °C

83. spherical "raindrops"

85. 79,700 cal (79.7 kcal)

Chapter 14

1. (a) increases; (b) decreases

3. 8.55 mL N_2/1 dL blood

5. 2.1 atm

7. (a) miscible; (b) immiscible

9. (a) polar; (b) nonpolar; (c) polar; (d) nonpolar

11. (a) miscible; (b) immiscible; (c) immiscible; (d) miscible

13. Add several drops of the unknown to water. If the unknown liquid is polar, it will be miscible; if it is nonpolar, it will be immiscible.

15. (a) soluble; (b) insoluble; (c) soluble

17. (a) insoluble; (b) soluble; (c) soluble; (d) insoluble; (e) soluble; (f) soluble

19. (a) fat soluble; (b) water soluble; (c) water soluble; (d) fat soluble; (e) water soluble; (f) water soluble

21.

H_2O
H_2O---$C_6H_{12}O_6$---H_2O
H_2O

23. (a)

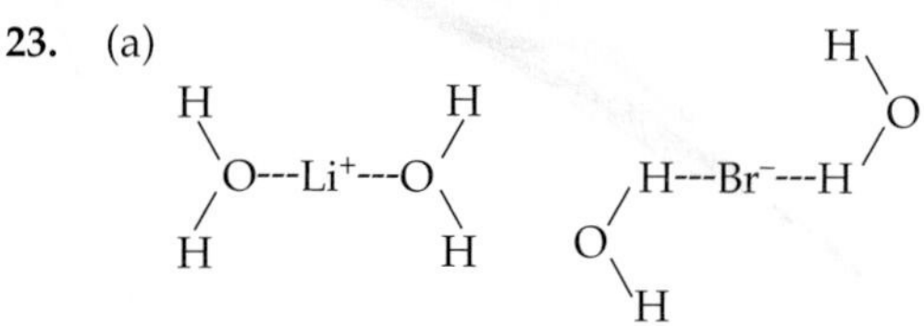

(b)

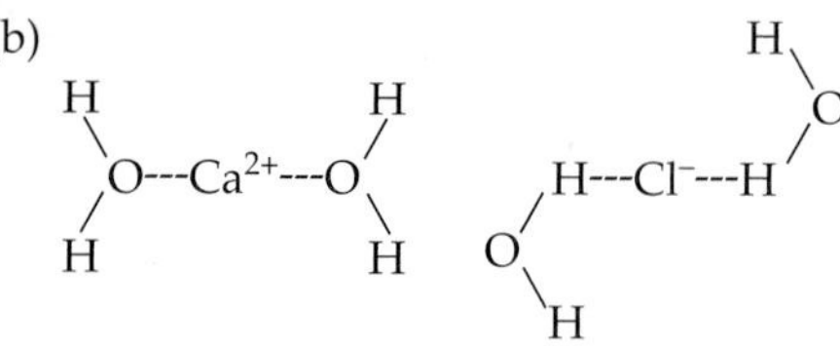

25. (1) heating the solution; (2) stirring the solution; (3) grinding the solute

27. (a) ~35 g NaCl; (b) ~35 g KCl

29. (a) ~36 g NaCl/100 g H_2O; (b) ~41 g KCl/100 g H_2O

31. (a) ~0 °C; (b) ~70 °C

33. (a) ~100 °C; (b) ~35 °C

35. (a) supersaturated; (b) saturated; (c) unsaturated

37. (a) supersaturated; (b) saturated; (c) unsaturated

39. saturated

41. (a) ~80 g solute; (b) ~20 g solute

43. (a) 16.7%; (b) 2.56%; (c) 2.42%; (d) 20.0%

45. (a) $\frac{1.50\text{ g KBr}}{100\text{ g solution}}$ and $\frac{100\text{ g solution}}{1.50\text{ g KBr}}$

$\frac{98.50\text{ g }H_2O}{100\text{ g solution}}$ and $\frac{100\text{ g solution}}{98.50\text{ g }H_2O}$

$\frac{1.50\text{ g KBr}}{98.50\text{ g }H_2O}$ and $\frac{98.50\text{ g }H_2O}{1.50\text{ g KBr}}$

(b) $\frac{2.50\text{ g }AlCl_3}{100\text{ g solution}}$ and $\frac{100\text{ g solution}}{2.50\text{ g }AlCl_3}$

$\frac{97.50\text{ g }H_2O}{100\text{ g solution}}$ and $\frac{100\text{ g solution}}{97.50\text{ g }H_2O}$

$\frac{2.50\text{ g }AlCl_3}{97.50\text{ g }H_2O}$ and $\frac{97.50\text{ g }H_2O}{2.50\text{ g }AlCl_3}$

(c) $\frac{3.75\text{ g }AgNO_3}{100\text{ g solution}}$ and $\frac{100\text{ g solution}}{3.75\text{ g }AgNO_3}$

$\frac{96.25\text{ g }H_2O}{100\text{ g solution}}$ and $\frac{100\text{ g solution}}{96.25\text{ g }H_2O}$

$\frac{3.75\text{ g }AgNO_3}{96.25\text{ g }H_2O}$ and $\frac{96.25\text{ g }H_2O}{3.75\text{ g }AgNO_3}$

(d) $\frac{4.25\text{ g }Li_2SO_4}{100\text{ g solution}}$ and $\frac{100\text{ g solution}}{4.25\text{ g }Li_2SO_4}$

$\frac{95.75 \text{ g } H_2O}{100 \text{ g solution}}$ and $\frac{100 \text{ g solution}}{95.75 \text{ g } H_2O}$

$\frac{4.25 \text{ g } Li_2SO_4}{95.75 \text{ g } H_2O}$ and $\frac{95.75 \text{ g } H_2O}{4.25 \text{ g } Li_2SO_4}$

47. (a) 53.6 g solution; (b) 2.00×10^2 g solution

49. (a) 1.70 g $FeBr_2$; (b) 5.25 g Na_2CO_3

51. (a) 95.0 g H_2O; (b) 225 g H_2O

53. (a) 0.257 *M* NaCl; (b) 0.0510 *M* $K_2Cr_2O_7$; (c) 0.400 *M* $CaCl_2$; (d) 0.313 *M* Na_2SO_4

55. (a) $\frac{0.100 \text{ mol LiI}}{1 \text{ L solution}}$ and $\frac{1 \text{ L solution}}{0.100 \text{ mol LiI}}$

$\frac{0.100 \text{ mol LiI}}{1000 \text{ mL solution}}$ and $\frac{1000 \text{ mL solution}}{0.100 \text{ mol LiI}}$

(b) $\frac{0.100 \text{ mol } NaNO_3}{1 \text{ L solution}}$ and $\frac{1 \text{ L solution}}{0.100 \text{ mol } NaNO_3}$

$\frac{0.100 \text{ mol } NaNO_3}{1000 \text{ mL solution}}$ and $\frac{1000 \text{ mL solution}}{0.100 \text{ mol } NaNO_3}$

(c) $\frac{0.500 \text{ mol } K_2CrO_4}{1 \text{ L solution}}$ and $\frac{1 \text{ L solution}}{0.500 \text{ mol } K_2CrO_4}$

$\frac{0.500 \text{ mol } K_2CrO_4}{1000 \text{ mL solution}}$ and $\frac{1000 \text{ mL solution}}{0.500 \text{ mol } K_2CrO_4}$

(d) $\frac{0.500 \text{ mol } ZnSO_4}{1 \text{ L solution}}$ and $\frac{1 \text{ L solution}}{0.500 \text{ mol } ZnSO_4}$

$\frac{0.500 \text{ mol } ZnSO_4}{1000 \text{ mL solution}}$ and $\frac{1000 \text{ mL solution}}{0.500 \text{ mol } ZnSO_4}$

57. (a) 0.866 L solution; (b) 0.198 L solution; (c) 0.177 L solution; (d) 0.0800 L solution

59. (a) 73.0 g $FeCl_3$; (b) 104 g KIO_4; (c) 2.38 g $ZnSO_4$; (d) 2.69 g $Ni(NO_3)_2$

61. 0.0154 *M* $CaSO_4$

63. (a) 5.00%; (b) 0.291 *M* $C_6H_{12}O_6$

65. 0.48 *M* HCl

67. 0.014 L (14 mL)

69. 0.719 *M* $AgNO_3$; 1.88 g AgBr

71. The auditorium air contains colloidal-size particles of smoke and dust.

73. Helium gas is less soluble in blood and allows for faster decompression.

75. 228 g SO_2

77. Ethyl ether is not miscible in polar water because it is a nonpolar molecule owing to the nonpolar bonds between carbon and hydrogen atoms.

79. The polar –OH on an alcohol can hydrogen bond with water, causing it to be soluble. If the nonpolar C_xH_y– portion of the molecule is large, the molecule is less polar and is immiscible in water.

81. (a) ethyl alcohol (solute), water (solvent)
(b) ethyl alcohol (solvent), water (solute)

83. 0.733 *M* NaClO

Chapter 15

1. (d) acidic solutions neutralize bases

3. (a) weakly acidic; (b) strongly acidic; (c) neutral; (d) strongly basic

5. (a) strong acid; (b) weak acid; (c) strong acid; (d) weak acid

7. (a) Arrhenius acid; (b) salt; (c) Arrhenius base; (d) Arrhenius acid

9. (a) $HI(aq)$, $NaOH(aq)$; (b) $HC_2H_3O_2(aq)$, $LiOH(aq)$

11. (a) $HF(aq)$, $NaOH(aq)$; (b) $HI(aq)$, $Mg(OH)_2(aq)$; (c) $HNO_3(aq)$, $Ca(OH)_2(aq)$; (d) $H_2CO_3(aq)$, $LiOH(aq)$

13. (a)
$2\,HNO_3(aq) + Ca(OH)_2(aq) \rightarrow Ca(NO_3)_2(aq) + 2\,H_2O(l)$
(b) $H_2CO_3(aq) + Ba(OH)_2(aq) \rightarrow BaCO_3(aq) + 2\,H_2O(l)$

15. (a) $HC_2H_3O_2(aq)$, $LiOH(aq)$; (b) $H_2SO_4(aq)$, $NH_3(aq)$

17. (a) $HI(aq)$, $H_2O(l)$; (b) $HC_2H_3O_2(aq)$, $HS^-(aq)$

19. (a) $HF(aq) + NaHS(aq) \rightarrow H_2S(aq) + NaF(aq)$
(b) $HNO_2(aq) + NaC_2H_3O_2(aq) \rightarrow NaNO_2(aq) + HC_2H_3O_2(aq)$

21. (a) red; (b) yellow

23. (a) yellow; (b) blue

25. (a) colorless; (b) pink

27. 0.141 *M* H_2SO_4

29. 0.0482 *M* H_3PO_4

31. 33.0 mL KOH

33. (a) 19.9% HCl;
(b) 5.95% $HC_2H_3O_2$;
(c) 3.12% HNO_3;
(d) 24.9% H_2SO_4

35. 0.284 *M* HNO_3

37. 1.104 *M* HCl

39. 29.0 mL LiOH

41. 176 g/mol

43. 89.2 g/mol

45. (a) $H_2O(l) \rightarrow H^+(aq) + OH^-(aq)$;
(b) $K_w = [H^+][OH^-]$;
(c) $K_w = 1.0 \times 10^{-14}$

47. (a) $[OH^-] = 4.0 \times 10^{-13}\,M$; (b) $[OH^-] = 5.9 \times 10^{-10}\,M$

49. (a) $[H^+] = 6.3 \times 10^{-12}\,M$; (b) $[H^+] = 3.4 \times 10^{-11}\,M$

51. (a) pH = 3; (b) pH = 5

53. (a) $[H^+] = 0.000\,001\,M$; (b) $[H^+] = 0.000\,000\,01\,M$

55. (a) pH = 6.52; (b) pH = 7.80

57. (a) $[H^+] = 0.016\,M$; (b) $[H^+] = 0.000\,018\,M$

59. (a) pH = 13.04; (b) pH = 10.74

61. (a) $[OH^-] = 7.7 \times 10^{-14}\,M$; (b) $[OH^-] = 4.2 \times 10^{-13}\,M$

63. (a) highly ionized; (b) highly ionized; (c) highly ionized

65. (a) strong; (b) weak; (c) strong; (d) weak

67. (a) weak; (b) strong; (c) strong; (d) weak

69. (a) $HF(aq)$; (b) $H^+(aq)$ and $Br^-(aq)$; (c) $H^+(aq)$ and $NO_3^-(aq)$; (d) $HNO_2(aq)$

71. (a) $Ag^+(aq)$ and $F^-(aq)$; (b) $AgI(s)$; (c) $Hg_2Cl_2(s)$; (d) $Ni^{2+}(aq)$ and $2\,Cl^-(aq)$

73. (1) Complete and balance the nonionized equation.
(2) Convert the nonionized equation to the total ionic equation.
(3) Cancel spectator ions to obtain the net ionic equation.
(4) Check (√) each ion or atom on both sides of the equation.

75. (a) $H^+(aq) + OH^-(aq) \rightarrow H_2O(l)$
(b) $HC_2H_3O_2(aq) + OH^-(aq) \rightarrow C_2H_3O_2^-(aq) + H_2O(l)$

77. (a) $Ag^+(aq) + I^-(aq) \rightarrow AgI(s)$
(b) $Ba^{2+}(aq) + CrO_4^{2-}(aq) \rightarrow BaCrO_4(s)$

79. yellow

81. orange (red + yellow)

83. NaH_2PO_4

85. 188 g/mol

87. pH = 0.30

89. $NH_4OH(aq) + HC_2H_3O_2(aq)$
$\rightarrow NH_4^+(aq) + C_2H_3O_2^-(aq) + H_2O(l)$

Chapter 16

1. (a), (b), (c), and (d) for selected reactions.

3.

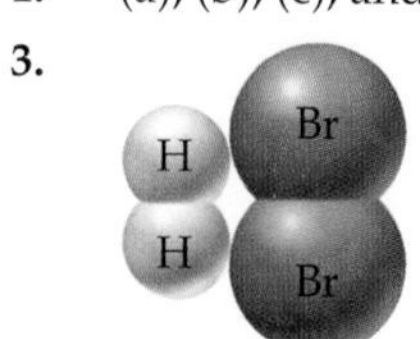

5. (a) increases the rate; (b) increases the rate; (c) increases the rate

7. No, a catalyst speeds up the rate of reaction, but does not effect the amount of product.

9.

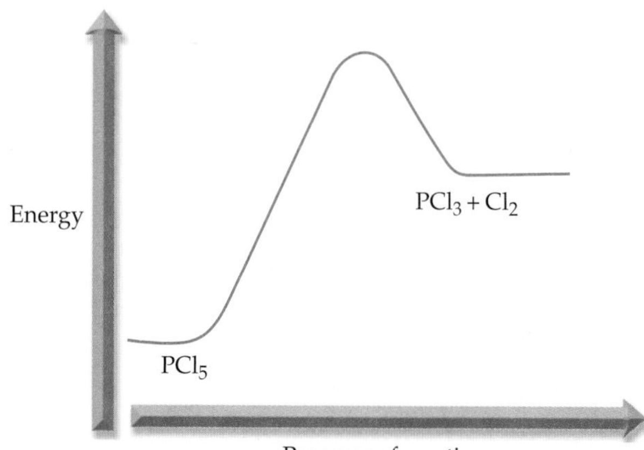

11.

Transition state
Energy
E_{act}
2 HI
DH
$H_2 + I_2$
Progress of reaction

13. lowers E_{act}

15. reverse direction

17. (a) The rate of the forward reaction corresponds to the rate of *decrease* in the concentration of reactants.
(b) The rate of the forward reaction corresponds to the rate of *increase* in the concentration of products.

19. (b) The K_{eq} for a gaseous equilibrium does not include solid substances; thus, solids are omitted from the equilibrium expression.

21. (a) $K_{eq} = \dfrac{[C]}{[A]^2}$
(b) $K_{eq} = \dfrac{[C]^3}{[A][B]^2}$
(c) $K_{eq} = \dfrac{[C]^4[D]}{[A]^2[B]^3}$

23. no

25. (a) $K_{eq} = \dfrac{[HF]^2}{[H_2][F_2]}$
(b) $K_{eq} = \dfrac{[NO_2]^4[H_2O]^6}{[NH_3]^4[O_2]^7}$
(c) $K_{eq} = [CO_2]$

27. $K_{eq} = 1.10$

29. (a) increases; (b) decreases

31. (a) right; (b) left; (c) left; (d) right; (e) right; (f) left; (g) right; (h) left; (i) no shift; (j) no shift

33. (a) right; (b) left; (c) left; (d) right; (e) left; (f) right; (g) no shift; (h) no shift; (i) no shift; (j) no shift

35. (a) $K_i = \dfrac{[H^+][CHO_2^-]}{[HCHO_2]}$
(b) $K_i = \dfrac{[H^+][HC_2O_4^-]}{[H_2C_2O_4]}$
(c) $K_i = \dfrac{[H^+][H_2C_6H_5O_7^-]}{[H_3C_6H_5O_7]}$

37. $K_i = 4.5 \times 10^{-4}$

39. $K_i = 7.2 \times 10^{-4}$

41. (a) right; (b) left; (c) left; (d) right; (e) left; (f) left; (g) right; (h) right

43. (a) right; (b) left; (c) left; (d) right; (e) left; (f) no shift; (g) right; (h) right

45. (a) $K_{sp} = [Ag^+][I^-]$
(b) $K_{sp} = [Ag^+]^2[CrO_4^{2-}]$
(c) $K_{sp} = [Ag^+]^3[PO_4^{3-}]$

47. $K_{sp} = 5.9 \times 10^{-21}$

49. $K_{sp} = 3.4 \times 10^{-35}$

51. The calcium ion concentration in a saturated solution of $CaCO_3$ is slightly greater than in a saturated solution of CaC_2O_4.

53. (a) left; (b) left; (c) right; (d) right; (e) left; (f) no shift; (g) no shift; (h) right

55. (a) left; (b) left; (c) right; (d) right; (e) no shift; (f) left; (g) no shift; (h) right

57. A swinging pendulum represents a *dynamic, reversible process* as it is in constant motion and swings back and forth in opposite directions.

59. rate of forward reaction = rate of reverse reaction

61. $K_{eq} = 0.20$

63. (a) left; (b) right; (c) right; (d) left

65. $Fe(OH)_3$ is more soluble in aqueous HCl than in water because of the following reaction:
$Fe(OH)_3(s) + 3\,H^+(aq) \rightleftharpoons Fe^{3+}(aq) + 3\,H_2O(l)$.

67. $K_{sp} = 5.3 \times 10^{-6}$

Chapter 17

1. (a) 0; (b) 0; (c) 0; (d) 0
3. (a) +1; (b) +2; (c) −2; (d) −1
5. (a) −3; (b) +4; (c) −3; (d) +5
7. (a) +4; (b) +6; (c) −2; (d) +7
9. (a) oxidation; (b) reducing agent
11. (a) oxidized—Co, reduced—S;
(b) oxidized—Cd, reduced—Cl_2
13. (a) oxidizing agent—CuO; reducing agent—H_2;
(b) oxidizing agent—PbO; reducing agent—CO
15. (a) oxidized—Al, reduced—Cr^{3+};
(b) oxidized—Cl^-, reduced—F_2
17. (a) oxidizing agent—AgI; reducing agent—Cr^{2+};
(b) oxidizing agent—Hg^{2+}; reducing agent—Sn^{2+}
19. yes
21. (a) $Br_2(l) + 2\,NaI(aq) \rightarrow 2\,NaBr(aq) + I_2(s)$
(b) $2\,PbS(s) + 3\,O_2(g) \rightarrow 2\,PbO(s) + 2\,SO_2(g)$
23. (a) $2\,MnO_4^-(aq) + 10\,I^-(aq) + 16\,H^+(aq) \rightarrow 2\,Mn^{2+}(aq) + 5\,I_2(s) + 8\,H_2O(l)$
(b) $Cu(s) + SO_4^{2-}(aq) + 4\,H^+(aq) \rightarrow Cu^{2+}(aq) + SO_2(g) + 2\,H_2O(l)$
25. (a) $SO_2(g) + 2\,H_2O(l) \rightarrow SO_4^{2-}(aq) + 4\,H^+(aq) + 2\,e^-$
(b) $2\,BrO_3^-(aq) + 12\,H^+(aq) + 10\,e^- \rightarrow Br_2(l) + 6\,H_2O(l)$
27. (a) $ClO^-(aq) + H_2O(l) + 2\,e^- \rightarrow Cl^-(aq) + 2\,OH^-(aq)$
(b) $MnO_4^-(aq) + 2\,H_2O(l) + 3\,e^- \rightarrow MnO_2(s) + 4\,OH^-(aq)$
29. (a) $3\,Zn(s) + 2\,NO_3^-(aq) + 8\,H^+(aq) \rightarrow 3\,Zn^{2+}(aq) + 2\,NO(g) + 4\,H_2O(l)$
(b) $2\,Mn^{2+}(aq) + 5\,BiO_3^-(aq) + 14\,H^+(aq) \rightarrow 2\,MnO_4^-(aq) + 5\,Bi^{3+}(aq) + 7\,H_2O(l)$
31. (a) $3\,S^{2-}(aq) + 2\,MnO_4^-(aq) + 4\,H_2O(l) \rightarrow 3\,S(s) + 2\,MnO_2(s) + 8\,OH^-(aq)$
(b) $Cu(s) + ClO^-(aq) + H_2O(l) \rightarrow Cu^{2+}(aq) + Cl^-(aq) + 2\,OH^-(aq)$
33. $Cl_2(g) + H_2O(l) \rightarrow Cl^-(aq) + HOCl(aq) + H^+(aq)$
35. $4\,Cl_2(g) + 8\,OH^-(aq) \rightarrow 2\,ClO_2^-(aq) + 6\,Cl^-(aq) + 4\,H_2O(l)$
37. (a) $Pb^{2+}(aq)$; (b) $Fe^{3+}(aq)$; (c) $Ag^+(aq)$; (d) $Br_2(l)$
39. (a) $F_2(g)$; (b) $Br_2(l)$; (c) $Cu^{2+}(aq)$; (d) $Mn^{2+}(aq)$
41. (a) nonspontaneous; (b) spontaneous
43. (a) spontaneous; (b) spontaneous
45. C > B > A
47.

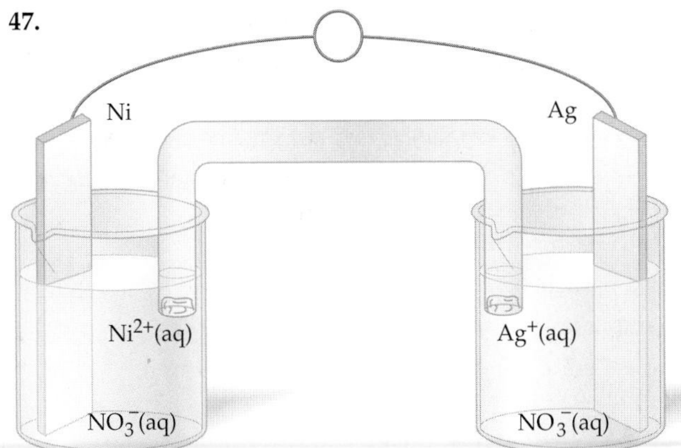

49.

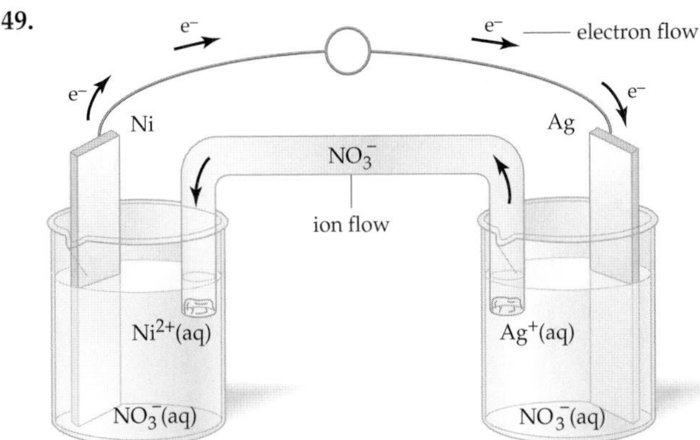

51. (a) $Mn(s) \rightarrow Mn^{2+}(aq) + 2\,e^-$
(b) $Fe^{2+}(aq) + 2\,e^- \rightarrow Fe(s)$
(c) Mn anode, Cu cathode
(d) Mn anode → Fe cathode
(e) Fe half − cell → Mn half − cell
53. (a) $Mg(s) \rightarrow Mg^{2+}(aq) + 2\,e^-$
(b) $Pb^{2+}(aq) + 2\,e^- \rightarrow Pb(s)$
(c) Mg anode, Pb cathode
(d) Mg anode → Pb cathode
(e) Pb half − cell → Mg half − cell
55. (a) $Ni(s) \rightarrow Ni^{2+}(aq) + 2\,e^-$
(b) $Fe^{2+}(aq) + 2\,e^- \rightarrow Fe(s)$
(c) Ni anode, Fe cathode
(d) Ni anode → Fe cathode
(e) Fe half − cell → Ni half − cell
57. (a) $Pb(s) \rightarrow Pb^{2+}(aq) + 2\,e^-$
(b) $Zn^{2+}(aq) + 2\,e^- \rightarrow Zn(s)$
(c) Pb anode, Zn cathode
(d) Pb anode → Zn cathode
(e) Fe half − cell → Pb half − cell
59. (a) $2\,F^-(aq) \rightarrow F_2(g) + 2\,e^-$
(b) $Cl_2(g) + 2\,e^- \rightarrow 2\,Cl^-(aq)$
(c) $Pt(F_2)$ anode, $Pt(Cl_2)$ cathode
(d) $Pt(F_2)$ anode → $Pt(Cl_2)$ cathode
(e) $Pt(F_2)$ half − cell → $Pt(Cl_2)$ half − cell
61. +2
63. $Co(s) + Hg^{2+}(aq) \rightarrow Co^{2+}(aq) + Hg(l)$

65. $Zn(s) + 2\,H^+(aq) \rightarrow Zn^{2+}(aq) + H_2(g)$

67. $2\,Li(s) + 2\,H_2O(l) \rightarrow 2\,Li^+(aq) + 2\,OH^-(aq) + H_2(g)$

69. $3\,Cu(s) + 8\,H^+(aq) + 2\,NO_3^-(aq)$
$\rightarrow 3\,Cu^{2+}(aq) + 2\,NO(g) + 4\,H_2O(l)$

Chapter 18

1. alpha particle (α)
3. beta particle (β)
5. gamma ray (γ)
7. (a) ~4 amu; (b) ~0 amu; (c) 0 amu; (d) ~0 amu; (e) ~1 amu; (f) ~1 amu
9. (a) ${}^{160}_{74}W \rightarrow {}^{156}_{72}Hf + {}^{4}_{2}He$
   (b) ${}^{32}_{15}P \rightarrow {}^{32}_{16}S + {}^{0}_{-1}e$
   (c) ${}^{55}_{27}Co \rightarrow {}^{55}_{26}Fe + {}^{0}_{+1}e$
   (d) ${}^{44}_{22}Ti + {}^{0}_{-1}e \rightarrow {}^{44}_{21}Sc$
11. (a) ${}^{221}_{88}Ra$; (b) ${}^{43}_{19}K$; (c) ${}^{73}_{36}Kr$; (d) ${}^{133}_{56}Ba$
13. ${}^{210}_{84}Po$
15. ${}^{207}_{81}Tl$
17. (a) α; (b) β; (c) β; (d) α; (e) α; (f) α; (g) α; (h) β; (i) β; (j) α
19. ${}^{233}_{91}Pa$
21. 25%
23. 366,000 years
25. 22,900 years old
27. 5 mg
29. 296 days
31. 5.30 years/half-life
33. uranium–lead dating
35. cobalt-60
37. plutonium-238
39. iodine-131
41. technetium-99
43. ${}^{80}_{35}Br$
45. ${}^{238}_{94}Pu$
47. ${}^{130}_{52}Te$
49. ${}^{4}_{2}He$
51. ${}^{260}_{105}X$
53. ${}^{260}_{106}X$
55. 8 neutrons
57. The number of neutrons varies, but the *average* number is 2.4.
59. 3 neutrons
61. 3 neutrons
63. ${}^{233}_{92}U$
65. ${}^{1}_{0}n$
67. ${}^{1}_{1}H$
69. ${}^{2}_{1}H$
71. ${}^{237}_{92}U$
73. ${}^{137}_{53}I$
75. ~6000 years
77. ${}^{241}_{95}Am$
79. 15 neutrons
81. ${}^{256}_{101}Md$
83. ${}^{1}_{0}n$

Chapter 19

1. petroleum, coal, and natural gas
3. alkane class of compounds
5. alkene and alkyne classes of compounds
7. (b) $C_{10}H_{22}$
9. (a) propane; (b) pentane; (c) heptane; (d) nonane
11. $CH_3{-}CH_2{-}CH_2{-}CH_2{-}CH_2{-}CH_3$

```
        CH3
        |
CH3—CH—CH2—CH2—CH3

            CH3
            |
CH3—CH2—CH—CH2—CH3

    CH3 CH3
    |   |
CH3—CH—CH—CH3

    CH3
    |
CH3—C—CH2—CH3
    |
    CH3
```

13. $CH_3{-}CH_2{-}CH_2{-}Cl$ and $CH_3{-}CH(Cl){-}CH_3$
15. (a) $CH_3{-}$; (b) $CH_3CH_2{-}$
17. (a) 2-methylpentane;
    (b) 4-ethyl-2-methylhexane;
    (c) 2,4,4-trimethylheptane;
    (d) 3,4,5-trimethyloctane
19. (a) $CH_4 + 2\,O_2 \rightarrow CO_2 + 2\,H_2O$
    (b) $C_3H_8 + 5\,O_2 \rightarrow 3\,CO_2 + 4\,H_2O$
    (c) $2\,C_6H_{14} + 19\,O_2 \rightarrow 12\,CO_2 + 14\,H_2O$
    (d) $2\,C_8H_{18} + 25\,O_2 \rightarrow 16\,CO_2 + 18\,H_2O$
21. (b) $C_{10}H_{20}$
23. (a) 1-butene; (b) 1-pentene; (c) 2-hexene; (d) 4-octene
25. (a) $C_{10}H_{18}$
27. (a) 2-pentyne; (b) 1-pentyne; (c) 3-hexyne; (d) 2-heptyne
29. $CH_2 = CH{-}CH_2{-}CH_2{-}CH_3$ and $CH_3{-}CH = CH{-}CH_2{-}CH_3$
31. (a) 4-methyl-2-pentene; (b) 3,5-dimethyl-2-hexene
33. (a) 4,4-dimethyl-2-heptyne; (b) 3,4,5-trimethyl-1-octyne
35. (a) $CH_2 = CH_2 + 3\,O_2 \rightarrow 2\,CO_2 + 2\,H_2O$
    (b) $CH_3{-}CH = CH_2 + H_2 \rightarrow CH_3{-}CH_2{-}CH_3$
    (c)
    $CH_3{-}CH = CH{-}CH_3 + Br_2 \rightarrow CH_3{-}CHBr{-}CHBr{-}CH_3$
37. CH_3, CH_3 (*ortho*); CH_3, CH_3 (*meta*); CH_3, CH_3 (*para*)

ortho *meta* *para*

39. CO_2 and H_2O
41. (a) ether; (b) organic halide; (c) phenol; (d) amine
43. (a) amine; (b) organic halide; (c) ester; (d) carboxylic acid; (e) aldehyde; (f) ether; (g) ketone; (h) amide
45. (a) "methyl iodide"; (b) "ethyl bromide"; (c) "propyl fluoride"; (d) "isopropyl chloride"
47. $CCl_2 = CHCl$
49. (a) insoluble; (b) soluble
51. (a) 1-butanol; (b) 2-butanol; (c) 1-butanol; (d) 2-butanol
53. (a) phenol; (b) para-methylphenol
55. (a) "dimethyl ether"; (b) "methyl ethyl ether"; (c) "dipropyl ether"; (d) "phenyl ethyl ether"
57. "Ethyl alcohol" molecules can hydrogen bond.
59. CH_3-CH_2-OH and CH_3-O-CH_3
61. (a) "ethylamine"; (b) "propylamine"
63. "Ethylamine" molecules can hydrogen bond.
65. (a) methanal; (b) ethanal; (c) propanal; (d) butanal
67. (a) 2-propanone; (b) 2-butanone; (c) 2-pentanone; (d) 3-pentanone
69. (a) higher boiling point; (b) higher boiling point
71. $CH_3CH_2-\overset{O}{\overset{\|}{C}}-H$ and $CH_3-\overset{O}{\overset{\|}{C}}-CH_3$
73. (a) methanoic acid; (b) ethanoic acid; (c) propanoic acid; (d) butanoic acid
75. (a) ethyl methanoate; (b) methyl ethanoate; (c) ethyl propanoate; (d) phenyl methanoate
77. (a) methanamide ("formamide"); (b) benzamide
79. (a) and (c)
81. propionic acid
83. (a)

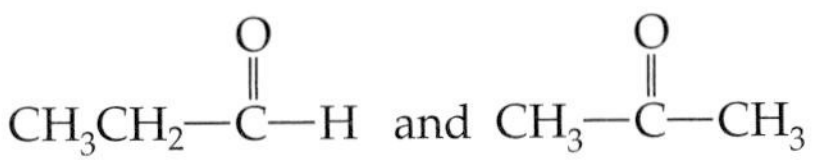

$$CH_3-\overset{O}{\overset{\|}{C}}-OH + CH_3-OH \xrightarrow{H_2SO_4} CH_3-\overset{O}{\overset{\|}{C}}-O-CH_3 + H_2O$$

(b)

$$C_6H_5-\overset{O}{\overset{\|}{C}}-OH + NH_3 \xrightarrow{\Delta} C_6H_5-\overset{O}{\overset{\|}{C}}-NH_2 + H_2O$$

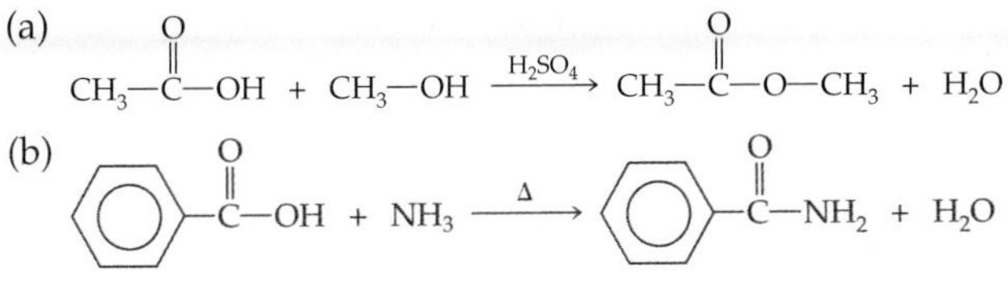

85. phenyl ethanoate ("phenyl acetate")
87. methanamide ("formamide")
89. (a) saturated; (b) unsaturated
91. (a) CH_3CH_2-OH; (b) $CH_3CH_2CH_2-NH_2$
93. carboxylic acid
95. ketone
97. organic halide, aldehyde

Chapter 20

1. (a) protein; (b) nucleic acid
3. (a) peptide; (b) phosphate
5. (a) carboxylic acid, amine, amide
   (b) aldehyde or ketone, alcohol
   (c) carboxylic acid, alcohol, ester
   (d) aldehyde, alcohol, amine
7. covalent amide bond
9. The primary structure of a protein refers to the amino acid sequence. The secondary structure of a protein refers to the shape of the polymer chain.
11. $H_2N-\underset{CH_3}{\underset{|}{CH}}-\overset{O}{\overset{\|}{C}}-NH-\underset{CH_3}{\underset{|}{CH}}-\overset{O}{\overset{\|}{C}}-OH$ (amide bond $-\overset{O}{\overset{\|}{C}}-NH-$ circled)
13. $H_2N-\underset{CH_2-OH}{\underset{|}{CH}}-\overset{O}{\overset{\|}{C}}-NH-\underset{CH_2-C_6H_4-OH}{\underset{|}{CH}}-\overset{O}{\overset{\|}{C}}-OH$
15. (1) Gly–His–Trp; (2) Gly–Trp–His; (3) His–Gly–Trp; (4) His–Trp–Gly; (5) Trp–Gly–His; (6) Trp–His–Gly
17. glutamic acid, lysine, and serine
19. In oxytocin, *isoleucine* replaces phenyl alanine and *leucine* replaces arginine.
21. The lock represents the *substrate* (S) molecule.
23. The enzyme location that conforms to the shape of the substrate molecule is called an *active site.*
25. In Step 1, the substrate molecule binds to the active site on the enzyme.
27. An enzyme is inactive under conditions where the pH or temperature is too high or too low.
29. aldehyde and alcohol groups
31. six carbon atoms
33. A disaccharide has two monosaccharide molecules joined by a glycoside linkage.
35. See Figure 20.9.
37. See Figure 20.11.
39. $\underset{\text{maltose}}{C_{12}H_{22}O_{11}} + \underset{\text{water}}{H_2O} \xrightarrow{H^+} \underset{\text{glucose}}{C_6H_{12}O_6} + \underset{\text{glucose}}{C_6H_{12}O_6}$
41. a glucose molecule
43. (a) fat; (b) oil; (c) fat and oil; (d) fat and oil
45.

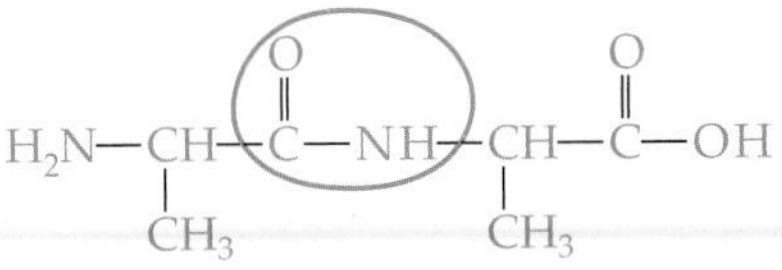

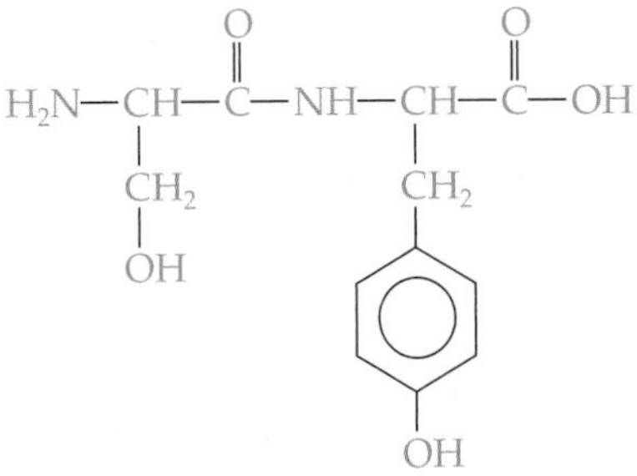

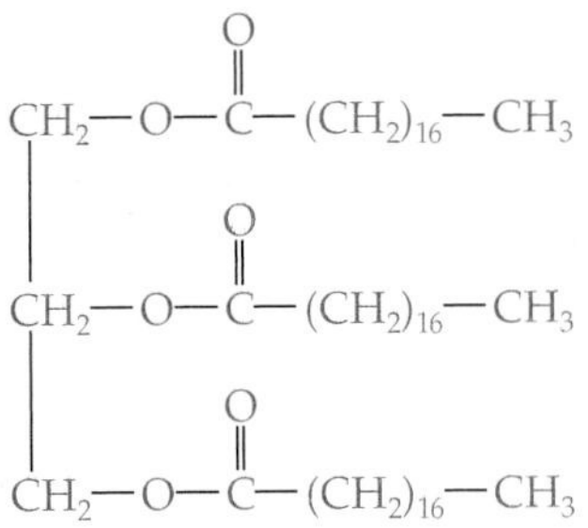

$$\begin{array}{l} CH_2-O-\overset{O}{\overset{\|}{C}}-(CH_2)_{16}-CH_3 \\ | \\ CH-O-\overset{O}{\overset{\|}{C}}-(CH_2)_{16}-CH_3 \\ | \\ CH_2-O-\overset{O}{\overset{\|}{C}}-(CH_2)_{16}-CH_3 \end{array}$$

47. (a) palmitic acid, $CH_3-(CH_2)_{14}-COOH$;
(b) lauric acid, $CH_3-(CH_2)_{10}-COOH$;
(c) myristic acid, $CH_3-(CH_2)_{12}-COOH$

49. $CH_3-(CH_2)_{24}-COO^-Na^+$ and $CH_3-(CH_2)_{29}-OH$

51. A, D, E, and K

53. deoxyribose sugar, a base, and phosphoric acid

55. adenine (A), cytosine (C), guanine (g), and thymine (T)

57. The structures are identical except that deoxyribose is missing an $-OH$ group.

59. 2 strands of nucleotides

61. 2 hydrogen bonds

63. thymine (T)

65. uracil (U)

67. long and extended

69. When a polysaccharide undergoes hydrolysis in an aqueous acid, glycoside linkages are cleaved, releasing the monosaccharide sugars in the polymer.

71. Since the compound is a polyhydroxy aldehyde polymer, the main component in an insect skeleton is carbohydrate.

73. There are six possible codons: ACG, AGC, CAG, CGA, GAC, and GCA.

75. The first step in *replication* is for a DNA molecule to unwind by breaking the hydrogen bonds (A = T and C ≡ G) holding together the double helix. Next, each strand of DNA acts as a template to synthesize a complementary strand. For example, adenine (A) on the template strand codes for thymine (T) on the growing strand; cytosine (C) codes for guanine (G); guanine (G) codes for cytosine (C); and thymine (T) codes for adenine (A).

At conclusion, each pair of template strands and complementary strands are joined by hydrogen bonds (A = T and C ≡ G), and the two resulting structures are exact duplicates of the original DNA molecule. Thus, two identical molecules of DNA are produced from one molecule of DNA.

77. An RNA molecule carries instructions for protein synthesis. The process involves trinucleotides in an RNA molecule, each of which is referred to as a *codon*. For example, the CAU codon represents three consecutive nucleotides that contain the bases cytosine, adenine, and uracil, respectively. The CAU codon is specific for the amino acid histidine and is responsible for adding histidine to a growing protein chain. Other examples include the GAG codon, which specifies glutamine; and UUU, which specifies phenylalanine.

APPENDIX

Answers to Chapter Self-Tests

Chapter 1—Self-Test

1. (d); 2. (e); 3. (e)

Key Concepts

4. The image of a cube creates a dual perception. By mentally pulling the red dot, the front face of the cube appears to be down and to the right. By mentally pushing the red dot, the front face of the cube appears to be up and to the left.

Critical Thinking

5. Biological evolution is a *scientific theory* because it is not measurable, and it cannot be stated as a mathematical equation.

6. Chemical evolution is a *scientific theory* because it is not measurable. Although scientists can determine the number of protons in an atom, the relationship between elements is not expressed as a mathematical equation.

Chapter 2—Self-Test

1. (a); 2. (c); 3. (e); 4. (b); 5. (c); 6. (a); 7. (c); 8. (b); 9. (c); 10. (e)

Key Concepts

11. 2 cm (a little less than an inch)

12. 0.1 cm (about the thickness of a thumbnail)

13. 5 g (although larger than 5¢, the mass is about the same)

14. Write down the unit asked for in the answer.

15. Write down the given value related to the answer.

16. Select the unit factor that cancels units in the given value.

17. Round off the answer in the display after the final calculation.

Critical Thinking

18. Note the exponent for the mass of an alpha particle (10^{-24} g) and the mass of a beta particle (10^{-28} g). Since 10^{-24} is four powers of ten greater than 10^{-28}, the alpha particle is much heavier.

19. In the English system of measurement, precious metals such as silver and gold are weighed in troy ounces, while copper and zinc are weighed in avoirdupois ounces. In the metric system, the unit of gram is constant. Thus, an ounce of silver (31.1 g) weighs more than an ounce of copper (28.4 g).

20. The percent of gold is zero (27% + 63% + 10% = 100%). Although jewelry may appear yellow in color, the amount of gold may be zero.

Chapter 3—Self-Test

1. (a); 2. (d); 3. (b); 4. (c); 5. (e); 6. (a); 7. (b); 8. (d); 9. (b)

Key Concepts

10. Fool's gold (d = 5.00 g/mL) sinks in water, but rests on liquid mercury. The gold nugget (d = 18.3 g/mL) sinks in both water and liquid mercury (d = 13.6 g/mL), and comes to rest at the bottom of the glass cylinder.

11. Since the absolute scale begins at zero Kelvin, the coldest possible temperature is 0 K; for comparison, 0 °F is 523 K, and 0 °C is 273 K.

Critical Thinking

12. Add a few drops of each liquid to water and observe if they sink or float. Hexane (d = 0.70 g/mL) is *less dense* than water (d = 1.00 g/mL), so it floats. Chloroform (d = 1.5 g/mL) is *more dense* than water, so it sinks.

13. A 10¢ silver dime weighs more than a 1¢ zinc penny because the silver metal in a dime coin is more dense than the zinc metal in a penny coin.

Chapter 4—Self-Test

1. (d); 2. (d); 3. (b); 4. (e); 5. (c); 6. (a); 7. (c); 8. (a); 9. (b); 10. (e)

Key Concepts

11. (a) liquid; (b) gas; (c) solid

12. (a) heterogeneous mixture; (b) compound; (c) element; (d) homogeneous mixture

13. The photo shows a mineral and illustrates a heterogeneous mixture. The model illustrates a molecule of S_8, which is an element.

14. The molecular formula for aspirin is $C_9H_8O_4$.

15. The *potential energy* of the cannonball is maximum at the top of its flight.

16. From the conservation of mass law, we can predict that the mass of a gallon of gasoline is less than the mass of the gaseous products, that is CO_2 and H_2O. The products weigh more because oxygen gas has combined with the gasoline.

Critical Thinking

17. Ice floating in water is an example of a heterogeneous mixture because ice and water have different densities, and can be separated by physical methods.

18. The properties of an element (Fe) and its compounds (FeO, Fe_2O_3, $Fe(NO_3)_3$) are not related, and their physical and chemical properties are different.

Chapter 5—Self-Test

1. (d); 2. (a); 3. (e); 4. (b); 5. (d); 6. (e); 7. (b); 8. (a); 9. (c); 10. (b); 11. (c)

Key Concepts

12. In the analogy "like a marble in the Astrodome" the marble represents the atomic nucleus, and the Astrodome represents the entire atom.

13. In the analogy "like a missile shot through our solar system" the missile represents an alpha particle shot through a metal foil with atomic nuclei that represent planets in our solar system.

14. An *electron* is to electricity, as a *photon* is to light.

15. The evidence for electrons occupying fixed energy states comes from the "atomic fingerprint" shown in the emission line spectrum of an element.

16. The Bohr model and quantum mechanical model of the atom both view an electron in a fixed energy state. However, the quantum mechanical atom addresses the wave-like properties of an electron, and assigns an orbital boundary for the electron rather than a simple orbit described by the Bohr model of the atom.

Critical Thinking

17. If the two isotopes, ^{63}Cu and ^{65}Cu, were each 50% abundant, the atomic mass would be approximately 64 amu. Note the atomic mass of copper is 63.55 amu; thus, the lighter isotope, ^{63}Cu, must be more abundant.

18. No, atoms of different elements have different atomic numbers, and cannot have the same number of protons. Yes, atoms of different elements can have the same mass number, and the same number of neutrons.

Chapter 6—Self-Test

1. (c); 2. (a); 3. (b); 4. (e); 5. (c); 6. (a); 7. (d); 8. (d); 9. (a); 10. (b)

Key Concepts

11. The atomic radius *decreases* up a group of elements, and generally *decreases* from left to right in a period.

12. The metallic character *decreases* up a group of elements, and generally *decreases* from left to right in a period.

13. The ionization energy *increases* up a group of elements, and generally *increases* from left to right in a period.

14. Elements in the same group react to form compounds with similar chemical formulas; for example, P and As are in Group VA/15 and react with oxygen to form P_2O_5 and As_2O_5.

Critical Thinking

15. *No*, ions of different elements cannot have the same number of protons. *Yes*, ions of different elements can have the same number of electrons. For example, Na^+ and Mg^{2+} each have 10 electrons, but 9 and 10 protons, respectively.

16. The yellow substance appears to be common sulfur powder. The gray substance appears metallic, so it must be the semimetal antimony. By elimination, the orange substance must be antimony sulfide, Sb_2S_3.

Chapter 7—Self-Test

1. (c); 2. (b); 3. (a); 4. (c); 5. (a); 6. (d); 7. (a); 8. (b); 9. (d)

Key Concepts

10. NaCl is binary ionic; $NaClO_3$ is ternary ionic; HCl is binary molecular; HCl(*aq*) is a binary acid; $HClO_3$(*aq*) is a ternary oxyacid

11. NaCl is named using an *-ide* suffix, that is, sodium chlor**ide**.

12. $NaClO_3$ is named using an *-ate* suffix, that is, sodium chlor**ate**.

13. $NaClO_2$ is named using an *-ite* suffix, that is, sodium chlor**ite**.

14. HCl(*aq*) is named as a *hydro* + *nonmetal stem* + *ic* acid, that is, hydrochloric acid.

15. $HClO_3$(*aq*) is named *nonmetal stem* + *ic* acid, that is, chloric acid.

16. $HClO_2$(*aq*) is named *nonmetal stem* + *ous* acid, that is, chlorous acid.

Critical Thinking

17. Copper predictably loses two 4*s* electrons giving Cu^{2+}. Copper can also lose one 4*s* electron giving Cu^+. This ion is especially stable as the remaining 4*s* electron moves into the 3*d* sublevel, thus filling the third energy level.

18. Iron predictably loses two 4*s* electrons giving Fe^{2+}. Iron can also lose one 3*d* electron giving Fe^{3+}. This ion is especially stable as five electrons remain in the 3*d* sublevel; this half-filled 3*d* sublevel has one electron in each 3*d* orbital.

Chapter 8—Self-Test

1. (d); 2. (e); 3. (c); 4. (a); 5. (a); 6. (d); 7. (c); 8. (c); 9. (b); 10. (e); 11. (d)

Key Concepts

12. $4\ Fe(s) + 3\ O_2(g) \rightarrow 2\ Fe_2O_3(s)$

13. $Ca(HCO_3)_2(s) \rightarrow CaCO_3(s) + H_2O(g) + CO_2(g)$

14. $Zn(s) + Ni(NO_3)_2(aq) \rightarrow Zn(NO_3)_2(aq) + Ni(s)$

15. $HNO_3(aq) + NH_4OH(aq) \rightarrow NH_4NO_3(aq) + HOH(l)$

16. $C_2H_5OH(l) + 3\ O_2(g) \xrightarrow{\text{spark}} 2\ CO_2(g) + 3\ H_2O(g)$

Critical Thinking

17. $HCl(aq) + NaHCO_3(aq) \rightarrow NaCl(aq) + H_2O(l) + CO_2(g)$

18. $Pb(NO_3)_2(aq) + 2\ KI(aq) \rightarrow PbI_2(s) + 2\ KNO_3(aq)$

Chapter 9—Self-Test

1. (a); 2. (c); 3. (b); 4. (c); 5. (e); 6. (b); 7. (c); 8. (b); 9. (d)

Key Concepts

10. The mass is found in the periodic table. One atom of F has a mass of 19.00 amu, and Avogadro's number of atoms has a mass of 19.00 g.

11. One mole of fluorine gas, F_2, has a mass of 38.00 g, a volume of 22.4 L, and contains 6.02×10^{23} molecules.

12. Avogadro's number of steel balls has an estimated mass equal to Earth.

13. Avogadro's number of small marbles has a total volume equal to the Moon.

Critical Thinking

14. No. Ten trillion (1×10^{13}) copper atoms weigh 1×10^{-9} g, which is one thousandth of a microgram (1×10^{-6} g).

15. A googol (1×10^{100}) of atoms is many times more than 1×10^{22}, the number of carbon atoms in a 1-carat diamond.

Chapter 10—Self-Test

1. (a); 2. (d); 3. (b); 4. (c); 5. (b); 6. (a); 7. (b); 8. (b); 9. (c)

Key Concepts

10. We must always start with a balanced chemical equation before solving a stoichiometry problem.

11. Typically, *three* unit factors are required to solve a mass–mass problem.

12. Typically, *three* unit factors are required to solve a mass–volume problem.

13. Typically, only *one* unit factor is required to solve a volume–volume problem because the coefficients of the balanced equation are in the same ratio as the volumes of gaseous reactants and products.

14. A gas must be at STP because the molar volume is 22.4 L/mol at STP.

15. It is *not* necessary for reactants and products to be at STP because Avogadro's theory states equal volumes of gases contain equal numbers of molecules at the same temperature and pressure.

Critical Thinking

16. (a) 2 Ag_2S, 2 S (Ag is limiting reactant)
 (b) 2 Ag_2S, 2 Ag (S is limiting reactant)

Chapter 11—Self-Test

1. (c); 2. (e); 3. (a); 4. (b); 5. (b); 6. (d); 7. (c); 8. (a); 9. (b); 10. (e); 11. (a)

Key Concepts

12. As the steam cools inside the capped can, it condenses to a liquid, thus reducing gaseous molecules and the pressure inside the can. The atmospheric pressure outside the can is greater than inside, and crushes the can.

13. If you check tire pressure after rolling several miles, the pressure is greater because road friction has heated the tire and the air molecules inside are moving faster and colliding more frequently.

14. You would be repulsed first because H_2S (34 amu) has less molecular mass than $C_3H_6O_2$ (74 amu), and the molecules of H_2S move faster.

Critical Thinking

15. The balloon rises because the density of air inside the balloon is less than outside. Heating air in a hot-air balloon causes air molecules to move faster and escape so as to equalize pressure inside and outside the balloon.

16. If you release a helium balloon into the air, the volume of the balloon increases as it ascends because atmospheric pressure outside the balloon is less at higher elevations.

17. On a hot day the water bottle bulges because the vapor pressure of water inside the bottle increases, while the outside atmospheric pressure remains constant.

18. On a cold night the water bottle crushes because the vapor pressure of water inside the bottle decreases, while the outside atmospheric pressure remains constant.

Chapter 12—Self-Test

1. (c); 2. (d); 3. (e); 4. (d); 5. (b); 6. (e); 7. (c); 8. (b); 9. (c); 10. (a)

Key Concepts

11. The three diagrams represent (a) atoms, (b) molecules, and (c) formula units, for example, (a) Na atoms, (b) H_2O molecules, and (c) NaCl formula units.

12. The HF molecule is most polar because F is the most electronegative element in Group VIIA/17.

13. A water molecule has two pairs of nonbonding electrons on the oxygen atom, and two hydrogen atoms. Thus, *four* H_2O molecules can attach directly to a single H_2O molecule by hydrogen bonds. (See Figure 13.12.)

14. The electron pair geometry of $CHCl_3$ is *tetrahedral*, and the molecular shape is *tetrahedral*.

Critical Thinking

15. The two resonance structures for sulfur dioxide are O = **S**−O and O−**S** = O.

Due to resonance, the two bonds in an SO_2 molecule are equivalent and have the same bond energy and bond length.

16. The electron pair geometry in both H_2O and H_3O^+ is tetrahedral. However, in H_3O^+ the molecular shape is *trigonal pyramidal*, and in H_2O the molecular shape is *bent*.

17. The electron pair geometry in both NH_3 and NH_4^+ is tetrahedral. However, in NH_3 the molecular shape is *trigonal pyramidal* and in NH_4^+ the molecular shape is *tetrahedral*.

Chapter 13—Self-Test

1. (a); 2. (d); 3. (d); 4. (c); 5. (b); 6. (e); 7. (c); 8. (e); 9. (a); 10. (b)

Key Concepts

11. A beehive surrounded by flying honey bees is analogous to the gaseous physical state. The honey bees move about faster when heated, similar to the movement of gas molecules.

12. At sea level, water boils at 100 °C. At a mile-high elevation, water boils at 90 °C due to lower atmospheric pressure. Food cooks faster at higher temperatures, and eggs hard-boil faster at 100 °C than at 90 °C.

Critical Thinking

13. Water has a concave lens because there is intermolecular *attraction* between polar water molecules and glass. Mercury has a convex lens because there is *repulsion* between nonpolar mercury atoms and glass.

Chapter 14—Self-Test

1. (a); 2. (c); 3. (b); 4. (d); 5. (e); 6. (c); 7. (a); 8. (b); 9. (a); 10. (b); 11. (c)

Key Concepts

12. Bubbles form on the inside of a pan when the water is heated because water contains air and the solubility of a gas in water is less as the temperature increases.

13. Peanut butter can dissolve grease in soiled clothing because peanut butter and grease are both nonpolar.

Critical Thinking

14. Lake Green has more dissolved oxygen gas. Lake Nuclear cools a nuclear reactor and the heat raises the temperature of the water. At a higher temperature (in Lake Nuclear), the solubility of oxygen gas is less.

15. At the conceptual level, we can explain the greater solubility of NaCl versus AgCl as follows. With NaCl there is less attraction between Na^+ and Cl^- ions than the ion attraction for solvent water molecules. Conversely, with AgCl there is more attraction between Ag^+ and Cl^- ions than the ion attraction for solvent water molecules.

Chapter 15—Self-Test

1. (e); 2. (c); 3. (b); 4. (d); 5. (a); 6. (d); 7. (e); 8. (c); 9. (b); 10. (c); 11. (a)

Key Concepts

12. We must always start with a balanced equation before applying unit analysis to solve an acid–base titration problem.

13. The bright light indicates Beaker X contains a strong electrolyte such as HNO_3, which is highly ionized. The dim light indicates Beaker Y has a weak electrolyte such as HNO_2, which is slightly ionized.

Critical Thinking

14. A "fishy smell" due to a basic organic compound can be neutralized using an acid. Household vinegar contains acetic acid, which we can use to neutralize the fish odor.

15. Since phenolphthalein is colorless in acid, and bromthymol blue is yellow in acid, the color of the solution in the test tube will be yellow.

16. At the conceptual level, we can explain the higher ionization of HCl versus HF as follows. It requires less energy to break H–Cl bonds than to break H–F bonds. In addition, more energy is released by the solvent water molecules surrounding H^+ and Cl^- than H^+ and F^-.

Chapter 16—Self-Test

1. (d); 2. (c); 3. (b); 4. (a); 5. (e); 6. (b); 7. (d); 8. (e); 9. (d)

Key Concepts

10. The reaction profile shows products at a higher energy level than reactants; thus, the profile represents an endothermic reaction.

11. $K_{eq} = [O_3]^2/[O_2]^3$

12. After heating an endothermic reaction, the equilibrium shifts to the right.

13. After reducing the volume of a reaction in which products have fewer molecules than reactants, the equilibrium shifts to the right.

14. After adding a catalyst, the rate increases but there is no shift in equilibrium.

Critical Thinking

15. In a coal mine, small particles of coal dust in the air provide more surface area for reaction. In the barbecue, large briquettes have less surface area to contact oxygen in air.

16. At equilibrium, the rate of evaporation of the liquid equals the rate of condensation of the vapor.

Chapter 17—Self-Test

1. (a); 2. (b); 3. (c); 4. (d); 5. (b); 6. (e); 7. (e)

Key Concepts

8. In the redox reaction shown
   (a) $H_2O(l) + NO_2^-(aq) \rightarrow NO_3^-(aq) + 2\,H^+(aq) + 2\,e^-$
   (b) $2\,e^- + 2\,H^+(aq) + H_2O_2(aq) \rightarrow 2\,H_2O(l)$
   (c) $NO_2^-(aq)$
   (d) $H_2O_2(aq)$
   (e) $H_2O_2(aq)$
   (f) $NO_2^-(aq)$

9. In the voltaic cell shown
   (a) $Mn(s) \rightarrow Mn^{2+}(aq) + 2\,e^-$
   (b) $Fe^{2+}(aq) + 2\,e^- \rightarrow Fe(s)$
   (c) anode: Mn; cathode: Fe
   (d) Electrons flow from Mn anode to Fe cathode.
   (e) SO_4^{2-} ions flow from Fe compartment to Mn compartment.

Critical Thinking

10. As a NiCad battery is *recharging*, $Ni(OH)_2$ is being *oxidized*.

11. As a NiCad battery is *discharging*, Cd is being *oxidized*.

12. Zinc metal is being *oxidized*, while silver oxide is *reduced*.

Chapter 18—Self-Test

1. (b); 2. (d); 3. (c); 4. (a); 5. (d); 6. (c); 7. (b); 8. (a); 9. (d); 10. (e)

Key Concepts

11. (a) ${}^{238}_{92}U \rightarrow {}^{234}_{92}U + {}^{4}_{2}He$
    (b) ${}^{239}_{92}U \rightarrow {}^{239}_{93}Np + {}^{0}_{-1}e$
    (c) ${}^{40}_{19}K \rightarrow {}^{40}_{18}Ar + {}^{0}_{+1}e$
    (d) ${}^{40}_{19}K + {}^{0}_{-1}e \rightarrow {}^{40}_{18}Ar$

12. $1\,n^0 \rightarrow 3\,n^0 \rightarrow 9\,n^0$

13. positron, ${}^{0}_{+1}e$

Critical Thinking

14. ${}^{210}_{82}Pb \rightarrow {}^{210}_{83}Bi + {}^{0}_{-1}e$

15. ${}^{86}_{36}Kr + {}^{208}_{82}Pb \rightarrow {}^{293}_{118}X + {}^{1}_{0}n$

Chapter 19—Self-Test

1. (a); 2. (c); 3. (b); 4. (a); 5. (e); 6. (c); 7. (b); 8. (d); 9. (e); 10. (d)

Key Concepts

11. alcohol: CH_3-CH_2-OH; ether: CH_3-O-CH_3

12. aldehyde: CH_3-CH_2-CHO; ketone: $CH_3-CO-CH_3$

13. carboxylic acid: $CH_3-CO-OH$; ester: $H-CO-OCH_3$

14. Thyroxine contains the following functional groups (left to right): phenol, organic halide, ether, amine, carboxylic acid.

Critical Thinking

15. (a) $CH_3-CH(CH_3)-CH_2-C(CH_3)_2-CH_3$
    (b) $CH_3-CH(OH)-CH_2-CH_3$

16. Caffeine is an *amine*; an *-ine* suffix usually denotes an amine.

17. Cholesterol is an *alcohol*; an *-ol* suffix usually denotes an alcohol.

18. Cortisone is a *ketone*; an *-one* suffix usually denotes a ketone.

Chapter 20—Self-Test

1. (a); 2. (c); 3. (c); 4. (a); 5. (b); 6. (a); 7. (e); 8. (b); 9. (d); 10. (e)

Key Concepts

11. A *lipid* is a water-insoluble fat, oil, or wax.

12. A *nucleic acid* is a polymer of a sugar, a base, and phosphoric acid.

13.

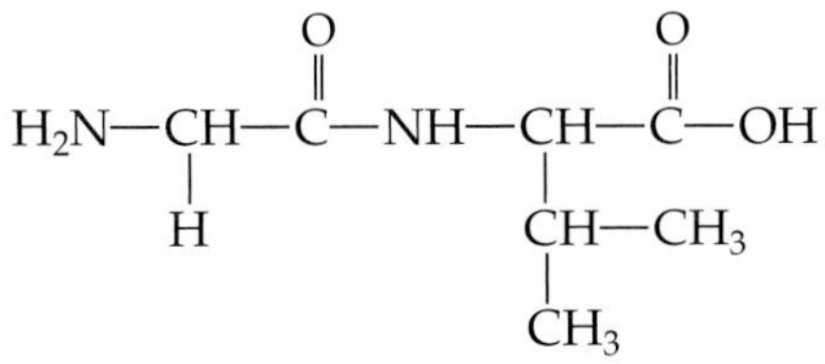

$$H_2N-\underset{\displaystyle H}{\underset{|}{CH}}-\overset{\displaystyle O}{\overset{\|}{C}}-NH-\underset{\underset{\displaystyle CH_3}{|}}{\underset{\displaystyle CH-CH_3}{\underset{|}{CH}}}-\overset{\displaystyle O}{\overset{\|}{C}}-OH$$

14.

$$\underset{\text{sucrose}}{C_{12}H_{22}O_{11}} + \underset{\text{water}}{H_2O} \xrightarrow{H^+} \underset{\text{glucose}}{C_6H_{12}O_6} + \underset{\text{fructose}}{C_6H_{12}O_6}$$

15.

$$\text{fat} + 3\,\text{Na OH} \rightarrow \begin{array}{l} CH_2-OH + CH_3-(CH_2)_{10}-COO^-Na^+ \\ | \\ CH-OH \;+ CH_3-(CH_2)_{14}-COO^-Na^+ \\ | \\ CH_2-OH + CH_3-(CH_2)_{16}-COO^-Na^+ \end{array}$$

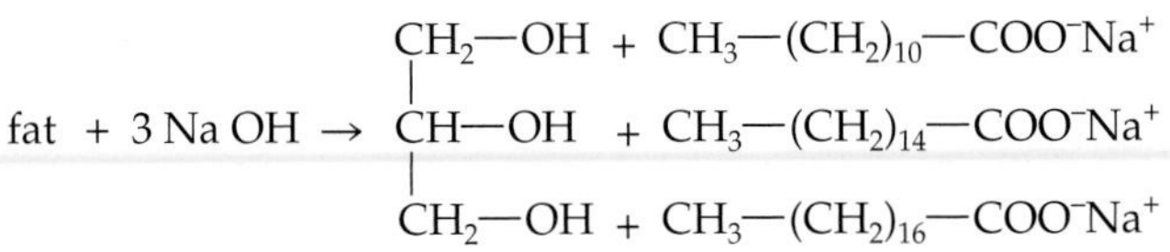

The products are glycerol and soap. The soap is composed of the sodium salts of lauric acid, palmitic acid, and stearic acid.

16.

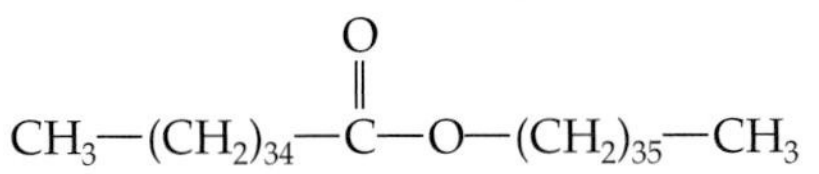

$$CH_3-(CH_2)_{34}-\overset{\displaystyle O}{\overset{\|}{C}}-O-(CH_2)_{35}-CH_3$$

17. During DNA replication, an adenine base will code for *thymine*.

18. During DNA transcription, an adenine base will code for *uracil*.

Critical Thinking

19. The shape of a protein that forms muscle tissue is long and extended. The shape of a protein that has a metabolic role is compact and folded.

20. (a) Nucleotide A shows the sugar deoxyribose and the base adenine; thus, nucleotide A is found in DNA.

(b) Nucleotide B shows the sugar ribose and the base cytosine; thus, nucleotide B is found in RNA.

Glossary

A

absolute zero The theoretical temperature at which the kinetic energy of a gas is zero. *(Sec. 11.10)*

acid A substance that releases hydrogen ions (H^+) when dissolved in water. *(Sec. 8.11)*

acid–base indicator A chemical substance that changes color according to the pH of the solution. *(Sec. 15.4)*

acidic oxide See *nonmetal oxide.*

actinide series The elements with atomic numbers 90–103. *(Sec. 6.3)*

activation energy (E_{act}) The energy necessary for reactants to achieve the transition state and form products. *(Sec. 16.2)*

active metal A metal that is sufficiently active to react with water at room temperature; the active metals include Li, Na, K, Ca, Sr, and Ba. *(Sec. 8.7)*

activity The number of radioactive nuclei in a given sample that decay per unit time, for example, 500 disintegrations per minute (dpm). *(Sec. 18.4)*

activity series A relative order of metals arranged according to their ability to undergo a reaction; also called an *electromotive series.* *(Sec. 8.7)*

actual yield The amount of product that is experimentally obtained from given amounts of reactants. *(Sec. 10.9)*

addition reaction A chemical reaction in which an unsaturated compound adds hydrogen or a halogen. *(Sec. 19.3)*

alchemy A pseudoscience that attempted to convert a base metal, such as lead, into gold; a medieval science that sought to discover a universal cure for disease and a magic potion for immortality. *(Sec. 1.2)*

alkali metals The Group IA/1 elements, excluding hydrogen. *(Sec. 6.3)*

alkaline earth metals The Group IIA/2 elements. *(Sec. 6.3)*

alkanes A family of saturated hydrocarbon compounds. *(Sec. 19.2)*

alkenes A family of unsaturated hydrocarbon compounds having a double bond. *(Sec. 19.3)*

alkyl group (R–) The part that remains after a hydrogen atom is removed from an alkane. *(Sec. 19.2)*

alkynes A family of unsaturated hydrocarbon compounds having a triple bond. *(Sec. 19.3)*

alloy A homogeneous mixture of two or more metals. *(Sec. 4.2)*

alpha particle (α) A particle in an alpha ray that is identical to a helium-4 nucleus. *(Sec. 18.1)*

amino acid A carboxylic acid with an amine group on the alpha carbon; also referred to as an α-amino acid. *(Sec. 20.2)*

amphiprotic A substance that is capable of either accepting a proton or donating a proton in acid–base reactions. *(Sec. 15.3)*

anhydrous A term for a compound that does not contain water. *(Sec. 13.10)*

anion A negatively charged ion. *(Sec. 7.1)*

anode The electrode in an electrochemical cell at which oxidation occurs. *(Sec. 17.6)*

aqueous solution (*aq*) A homogeneous mixture of a substance dissolved in water. *(Secs. 7.1 and 8.1)*

arenes A family of aromatic hydrocarbon compounds. *(Sec. 19.4)*

aromatic hydrocarbon A hydrocarbon compound typically containing a benzene ring. *(Sec. 19.1)*

Arrhenius acid A substance that releases hydrogen ions when dissolved in water. *(Sec. 15.2)*

Arrhenius base A substance that releases hydroxide ions when dissolved in water. *(Sec. 15.2)*

aryl group (Ar–) The part that remains after a hydrogen atom is removed from an arene. *(Sec. 19.2)*

atmospheric pressure The pressure exerted by the air molecules in Earth's atmosphere. The atmospheric pressure at sea level is 1 atm or 760 mm of Hg. *(Sec. 11.2)*

atom The smallest particle that represents an element. *(Sec. 4.3)*

atomic mass The weighted average mass of all the naturally occurring isotopes of an element. *(Sec. 5.5)*

atomic mass unit (amu) A unit of mass exactly equal to 1/12 the mass of a carbon-12 atom. *(Sec. 5.5)*

atomic notation A symbolic method for expressing the composition of an atomic nucleus; the mass number and atomic number are indicated to the left of the chemical symbol for the element. *(Sec. 5.4)*

atomic nucleus A region of very high density in the center of the atom that contains protons and neutrons. *(Sec. 5.3)*

atomic number (*Z*) A number that identifies a specific element. *(Sec. 4.4)* A value that indicates the number of protons in the nucleus of an atom. *(Secs. 5.4 and 18.2)*

Avogadro's number (*N*) The value that corresponds to the number of carbon atoms in 12.01 g of carbon; 6.02×10^{23} particles. *(Sec. 9.1)*

Avogadro's theory The principle that equal volumes of gases, at the same temperature and pressure, contain equal numbers of molecules; also referred to as *Avogadro's law* or *Avogadro's hypothesis.* *(Secs. 9.5, 10.1, and 11.3)*

B

barometer An instrument for measuring atmospheric pressure. *(Sec. 11.2)*

base A substance that releases hydroxide ions (OH^-) when dissolved in water. *(Sec. 8.11)*

basic oxide See *metal oxide.*

battery A general term for any electrochemical cell that produces electrical energy spontaneously. *(Sec. 17.6)*

beta particle (β) A particle in a beta ray that is identical to an electron. *(Sec. 18.1)*

binary acid A compound that contains hydrogen and a nonmetal dissolved in water. *(Sec. 7.1)*

binary ionic compound A compound that contains one metal and one nonmetal. *(Sec. 7.1)*

binary molecular compound A compound that contains two nonmetals. *(Sec. 7.1)*

biochemistry The study of chemical substances derived from plants and animals. *(Sec. 1.2)* The study of biological compounds. *(Sec. 20.1)*

Bohr atom A model of the atom that describes an electron circling the atomic nucleus in an orbit of specific energy. *(Sec. 5.8)*

boiling point (Bp) The temperature at which the vapor pressure of a liquid is equal to the atmospheric pressure. *(Sec. 13.3)*

bond angle The angle formed by two atoms bonded to the central atom in a molecule. *(Secs. 12.10 and 13.7)*

bond energy The amount of energy required to break a covalent bond between two atoms. *(Sec. 12.3)*

bond length The distance between the nuclei of two atoms that are joined by a covalent bond. *(Sec. 12.3)*

bonding electrons The valence electrons in a molecule that are shared between two atoms. *(Sec. 12.4)*

Boyle's law The statement that the pressure and the volume of a gas are inversely proportional at constant temperature, that is, $P_1V_1 = P_2V_2$. *(Sec. 11.4)*

Brønsted–Lowry acid A substance that donates a proton in an acid–base reaction. *(Sec. 15.3)*

Brønsted–Lowry base A substance that accepts a proton in an acid–base reaction. *(Sec. 15.3)*

buffer A solution that resists changes in pH when an acid or a base is added. *(Sec. 15.1)*

C

calorie (cal) The amount of heat required to raise the temperature of 1 g of water 1 °C. *(Sec. 3.9)*

carbohydrate A biological compound that contains one or more sugar molecules; a sugar is a polyhydroxy aldehyde or ketone. *(Sec. 20.1)*

carbonyl group (−C=O) The C=O group, which is present in aldehydes, ketones, carboxylic acids, esters, and amides. *(Sec. 19.5)*

carboxyl group (−COOH) The −COOH group, which is present in carboxylic acids. *(Sec. 19.10)*

catalyst A substance that speeds up a reaction without being permanently changed. *(Sec. 8.2)* A substance that allows a reaction to proceed faster by lowering the energy of activation. *(Sec. 16.2)*

cathode The electrode in an electrochemical cell at which reduction occurs. *(Sec. 17.6)*

cation A positively charged ion. *(Sec. 7.1)*

Celsius degree (°C) The basic unit of temperature in the metric system. *(Sec. 3.8)*

centimeter (cm) A common unit of length in the metric system that is equal to one-hundredth of a meter. *(Sec. 2.1)*

chain reaction A fission reaction in which the neutrons released initiate a second reaction, which in turn initiates a third reaction, and so on. *(Sec. 18.7)*

Charles's law The statement that the volume and the Kelvin temperature of a gas are directly proportional at constant pressure, that is, $V_1/T_1 = V_2/T_2$. *(Sec. 11.5)*

chemical bond The general term for the attraction between two ions or the affinity between two atoms. *(Sec. 12.1)*

chemical change A modification of a substance that alters the chemical composition. *(Sec. 4.7)*

chemical equation A shorthand representation using formulas and symbols to describe a chemical reaction. *(Sec. 8.2)*

chemical equilibrium A dynamic state for a reversible reaction in which the rates of the forward and reverse reactions are equal and the amounts of reactants and products are constant. *(Sec. 16.3)*

chemical formula An abbreviation for the name of a chemical compound that indicates the number of atoms of each element; for example, H_2O is the formula for water. *(Sec. 4.5)*

chemical property A characteristic of a substance that cannot be observed without changing the chemical formula of the substance. *(Sec. 4.6)*

chemical reaction The process of undergoing a chemical change. *(Sec. 8.1)*

chemical symbol An abbreviation for the name of a chemical element; for example, Cu is the symbol for copper. *(Sec. 4.3)*

chemistry The branch of science that studies the composition and properties of matter. *(Sec. 1.2)*

class of compounds A family of compounds in which all the members have the same structural feature (that is, an atom or group of atoms) and similar chemical properties. *(Sec. 19.5)*

coefficient A digit in front of a chemical formula that helps to balance an equation; a digit that helps to balance the number of atoms of each element on the left side of an equation with those on the right side of the equation. *(Sec. 8.3)*

collision theory The principle that the rate of a chemical reaction is regulated by the collision frequency, collision energy, and the orientation of molecules striking each other. *(Sec. 16.1)*

colloid A homogeneous mixture in which the diameters of the dispersed particles range from 1 to 100 nm. *(Sec. 14.3)*

combination reaction A type of reaction in which two substances produce a single compound. *(Sec. 8.4)*

combined gas law The statement that the pressure exerted by a gas is inversely proportional to its volume and directly proportional to its Kelvin temperature; that is, $P_1V_1/T_1 = P_2V_2/T_2$. *(Sec. 11.7)*

combustion reaction A chemical reaction in which a hydrocarbon reacts rapidly with oxygen (burns) to produce carbon dioxide, water, and heat. *(Sec. 19.2)*

compound A pure substance that can be broken down into two or more simpler substances by a chemical reaction. *(Sec. 4.2)*

continuous spectrum A single, broad band of radiant energy. *(Sec. 5.6)*

coordinate covalent bond A bond in which an electron pair is shared, but both electrons have been donated by a single atom. *(Sec. 12.8)*

core The portion of the atom that includes the nucleus and inner electrons that are not available for bonding; also termed the *kernel* of the atom. *(Sec. 6.8)*

core notation A method of writing the electron configuration in which the inner core electrons are represented by a noble gas symbol in brackets followed by valence electrons, for example, [Ne] $3s^2$. *(Sec. 6.10)*

covalent bond A chemical bond characterized by the sharing of one or more pairs of valence electrons. *(Sec. 12.1)*

critical mass The minimum mass of a fissionable nuclide necessary to sustain a continuous chain reaction. *(Sec. 18.7)*

crystalline solid A solid substance composed of ions or molecules that repeat in a regular geometric pattern. *(Sec. 13.4)*

cubic centimeter (cm^3) A unit of volume occupied by a cube 1 cm on a side; 1 cm^3 is exactly equal to 1 mL. *(Sec. 3.5)*

D

Dalton's law of partial pressures The statement that the pressure exerted by a mixture of gases is equal to the sum of the individual pressures exerted by each gas, that is, $P_1 + P_2 + P_3 + ... = P_{total}$. *(Sec. 11.9)*

decay series See *radioactive decay series.*

decomposition reaction A type of reaction in which a single compound produces two or more substances. *(Sec. 8.4)*

degree Celsius See *Celsius degree.*

degree Fahrenheit See *Fahrenheit degree.*

deionized water Water purified by removing ions using an ion exchange method; also termed *demineralized water*. *(Sec. 13.7)*

delta (δ) notation A method of indicating the partial positive charge (δ^+) and the partial negative charge (δ^-) in a polar covalent bond. *(Sec. 12.6)*

density (*d*) The amount of mass in one unit volume of matter. *(Sec. 3.7)*

deposition A direct change of state from a gas to a solid without forming a liquid. *(Sec. 4.1)*

deuterium The nuclide of hydrogen with one neutron in the nucleus. *(Sec. 18.8)*

diatomic molecule A particle composed of two nonmetal atoms. *(Sec. 8.2)* A molecule composed of two nonmetal atoms held together by a covalent bond. *(Sec. 12.7)*

dipeptide A molecule containing two amino acids joined by an amide bond. *(Sec. 20.2)*

dipole A region in a molecule having partial negative charge and partial positive charge resulting from a polar bond. *(Sec. 14.2)*

dipole force A type of attraction between two molecules that have permanent dipoles. *(Sec. 13.2)*

directly proportional The association between two variables that have a direct relationship; for example, if one variable doubles the other doubles, or if one variable is halved the other is halved. *(Sec. 11.5)*

disaccharide A carbohydrate composed of two simple sugar molecules joined by a glycoside linkage. *(Sec. 20.4)*

dispersion force A type of attraction between two molecules that have temporary dipoles. A temporary dipole arises from the unequal distribution of charge as electrons move around a molecule; also called a *London dispersion force*. *(Sec. 13.2)*

dissociation The process of an ionic compound dissolving in water and separating into positive and negative ions; for example, NaOH dissolves in water to give sodium ions and hydroxide ions. *(Sec. 15.2)*

distilled water Water purified by boiling and collecting the condensed vapor. *(Sec. 13.7)*

double bond A bond between two atoms composed of two shared electron pairs; a double bond is represented as two dashes between the symbols of two atoms. *(Sec. 12.4)*

double-replacement reaction A type of reaction in which two cations in different compounds exchange anions. *(Sec. 8.4)*

dry cell An electrochemical cell in which the anode and cathode reactions do not take place in an aqueous solution. *(Sec. 17.6)*

ductile The property of a metal that allows it to be drawn into a wire. *(Sec. 4.4)*

E

elastic collision An impact between gas molecules in which the total energy remains constant. *(Sec. 11.10)*

electrochemical cell A general term for an apparatus containing two solutions with electrodes in separate compartments that are connected by a conducting wire and salt bridge. *(Sec. 17.6)*

electrochemistry The study of the interconversion of chemical and electrical energy from redox reactions. *(Sec. 17.6)*

electrolysis The chemical reaction produced by passing electricity through an aqueous solution. *(Sec. 13.9)* The providing of electricity to a nonspontaneous redox process to cause a reaction. *(Sec. 17.7)*

electrolytic cell An electrochemical cell in which a nonspontaneous redox reaction occurs as a result of the input of direct electric current. *(Sec. 17.7)*

electromotive series See *activity series.*

electron (e^-) A negatively charged subatomic particle having a tiny mass. *(Sec. 5.2)*

electron capture (EC) A nuclear reaction in which a heavy nuclide attracts one of its inner core electrons into the nucleus. *(Sec. 18.2)*

electron configuration A shorthand description of the arrangement of electrons by sublevels according to increasing energy. *(Secs. 5.10 and 6.6)*

electron dot formula A representation for an atom and its valence electrons that shows the chemical symbol surrounded by a dot for each valence electron. *(Sec. 6.8)* A diagram of a molecule or polyatomic that shows the chemical symbol of each atom and a dot representing each bonding and nonbonding electron. An electron dot formula is also called a *Lewis structure*. *(Secs. 6.8 and 12.4)*

electron pair geometry The geometric shape formed by bonding and nonbonding electron pairs surrounding the central atom in a molecule. *(Sec. 12.9)*

electronegativity The ability of an atom to attract a pair of electrons in a chemical bond. *(Sec. 12.6)*

element A pure substance that cannot be broken down by ordinary chemical reaction. *(Sec. 4.2)*

emission line spectrum Several narrow bands of radiant energy that result from excited atoms releasing energy. *(Sec. 5.8)*

empirical formula A chemical formula that expresses the simplest whole-number ratio of atoms of each element in a molecule or of ions in a formula unit. *(Sec. 9.8)*

endothermic reaction A chemical reaction that absorbs heat energy. *(Secs. 8.1 and 16.2)*

endpoint The stage in a titration when the indicator changes color permanently. *(Sec. 15.5)*

energy level An orbit of specific energy that electrons occupy at a fixed distance from the nucleus; also referred to as a *main energy level*; designated as 1, 2, 3, 4 ... *(Sec. 5.8)*

energy sublevel An electron energy level that results from splitting a main energy level; designated as *s, p, d, f* ... *(Sec. 5.9)*

English system A nondecimal system of measurement without any basic unit for length, mass, or volume. *(Sec. 3.1)*

enzyme A protein molecule that catalyzes a specific biochemical reaction. *(Sec. 20.3)*

equilibrium constant See *general equilibrium constant* (K_{eq}), *ionization equilibrium constant* (K_i), or *solubility product equilibrium constant* (K_{sp}).

evaporation See *vaporization.*

exact equivalent A statement that relates two values that are exactly equal, for example, 1 yd = 36 in. and 1 m = 100 cm. *(Secs. 2.8 and 3.2)*

exothermic reaction A chemical reaction that releases heat energy. *(Secs. 8.1 and 16.2)*

experiment A scientific procedure for collecting data and recording observations under controlled conditions. *(Sec. 1.1)*

exponent A number written as a superscript indicating that a value is multiplied by itself, for example, $10^4 = 10 \times 10 \times 10 \times 10$ and $cm^3 = cm \times cm \times cm$. *(Sec. 2.6)*

F

Fahrenheit degree (°F) A basic unit of temperature in the English system. *(Sec. 3.8)*

fat A triglyceride from an animal source that contains mostly saturated fatty acids. *(Sec. 20.5)*

fatty acid A carboxylic acid with a long hydrocarbon chain. *(Sec. 20.5)*

formula unit The simplest representative particle in a compound composed of ions. *(Sec. 7.4)* The simplest representative particle in a substance held together by ionic bonds. *(Sec. 12.1)*

frequency The number of times a light wave travels a complete cycle in one second. *(Sec. 5.6)*

functional group An atom or group of atoms that characterizes a class of compounds and contributes to their similar chemical properties. *(Sec. 19.5)*

G

galvanic cell See *voltaic cell.*

gamma ray (γ) A powerful type of radiant energy with a very short wavelength. *(Sec. 18.1)*

gas density The ratio of mass per unit volume for a gas, usually expressed in grams per liter. *(Sec. 9.5)*

gas pressure A measure of the frequency and energy of gas molecules colliding against the walls of their container. *(Sec. 11.2)*

Gay-Lussac's law The statement that the pressure and the Kelvin temperature of a gas are directly proportional at constant volume, that is, $P_1/T_1 = P_2/T_2$. *(Sec. 11.6)*

Gay-Lussac's law of combining volumes See *law of combining volumes.*

general equilibrium constant (K_{eq}) A constant that expresses the molar equilibrium concentration of each substance participating in a reversible reaction at a given temperature. *(Sec. 16.3)*

glycoside linkage An –O– bond that joins two simple sugar molecules. *(Sec. 20.4)*

gram (g) The basic unit of mass in the metric system. *(Secs. 2.1 and 3.1)*

group A vertical column in the periodic table; a family of elements with similar properties. *(Sec. 6.3)*

H

half-cell A portion of an electrochemical cell having a single electrode where either oxidation or reduction occurs. *(Sec. 17.6)*

half-life ($t_{1/2}$) The amount of time required for 50% of the radioactive nuclei in a given sample to decay. *(Sec. 18.4)*

half-reaction A reaction that represents either an oxidation or reduction process separately and indicates the number of electrons lost or gained. *(Sec. 17.4)*

halogens The Group VIIA/17 elements. *(Sec. 6.3)*

hard water Water containing a variety of cations and anions such as Ca^{2+}, Mg^{2+}, Fe^{3+}, CO_3^{2-}, SO_4^{2-}, and PO_4^{3-}. *(Sec. 13.7)*

heat The flow of energy from an object at a higher temperature to an object at a lower temperature. *(Sec. 3.9)*

heat of fusion (H_{fusion}) The heat required to convert a solid to a liquid at its melting point. Conversely, the heat released when a liquid changes to a solid is called the *heat of solidification* (H_{solid}). *(Sec. 13.6)*

heat of reaction (Δ*H*) The difference in heat energy between the reactants and the products for a given chemical reaction. *(Sec. 16.2)*

heat of vaporization (H_{vapor}) The heat required to convert a liquid to a gas at its boiling point. Conversely, the heat released when a gas changes to a liquid is called the *heat of condensation* (H_{cond}). *(Sec. 13.6)*

Henry's law The principle that the solubility of a gas in a liquid is proportional to the partial pressure of the gas above the liquid. *(Sec. 14.1)*

heterogeneous equilibrium A type of equilibrium in which all of the participating species are not in the same physical state, for example, an equilibrium between an insoluble precipitate and its ions in an aqueous solution. *(Sec. 16.4)*

heterogeneous mixture Matter having an indefinite composition and variable properties. *(Sec. 4.2)*

homogeneous equilibrium A type of equilibrium in which all of the participating species are in the same physical state, for example, an equilibrium between reactants and products in the gaseous state. *(Sec. 16.4)*

homogeneous mixture Matter having a definite composition, but variable properties from sample to sample; examples include alloys and solutions. *(Sec. 4.2)*

hydrate A substance that contains a specific number of water molecules attached to a formula unit in a crystalline compound. *(Sec. 13.9)*

hydrocarbon A compound containing only hydrogen and carbon. *(Sec. 19.1)*

hydrocarbon derivative A compound containing carbon, hydrogen, and another element such as oxygen, nitrogen, or a halogen. *(Sec. 19.5)*

hydrogen bond The intermolecular attraction between a H atom bonded to O or N, and a nonbonding electron pair on an electronegative atom in another molecule. *(Secs. 12.9 and 13.2)*

hydronium ion (H_3O^+) The ion that results when a hydrogen ion attaches to a water molecule by a coordinate covalent bond. The hydronium ion is the predominant form of the hydrogen ion in an aqueous acid solution. *(Sec. 15.2)*

hydroxyl group The –OH group, which is present in alcohols and phenols. *(Sec. 19.7)*

hypothesis An initial, tentative proposal of a scientific principle that attempts to explain the meaning of a set of data collected in an experiment. *(Sec. 1.1)*

I

ideal gas A theoretical gas that obeys the kinetic theory under all conditions of temperature and pressure. *(Sec. 11.10)*

ideal gas constant The proportionality constant R in the equation $PV = nRT$. *(Sec. 11.11)*

ideal gas law The principle stated by the relationship $PV = nRT$; also called the *ideal gas equation*. *(Sec. 11.11)*

immiscible A term that refers to liquids that do not dissolve in one another and separate into two layers. *(Sec. 14.2)*

inner transition elements The elements in the lanthanide and actinide series. *(Sec. 6.3)*

inorganic chemistry The study of chemical substances that do not contain the element carbon. *(Sec. 1.2)*

inorganic compound A compound that does not contain carbon. *(Sec. 7.1)*

instrument A device for recording a measurement such as length, mass, volume, time, or temperature. *(Sec. 2.1)*

International System (SI) A sophisticated system of measurement that is more comprehensive than the metric system and has seven base units. *(Sec. 3.1)*

inversely proportional The association between two variables that have an inverse relationship; for example, if one variable doubles, the other variable is halved. *(Sec. 11.4)*

ion An atom (or group of atoms) that has a negative or positive charge resulting from the gain or loss of valence electrons. *(Sec. 6.9)*

ionic bond A chemical bond characterized by the attraction between a cation and an anion. *(Sec. 12.1)*

ionic charge A term for the positive charge on a metal atom that has lost electrons, or the negative charge on a nonmetal atom that has gained electrons. *(Sec. 6.10)*

ionic solid A crystalline solid composed of ions that repeat in a regular pattern. *(Sec. 13.5)*

ionization The process of a polar compound dissolving in water and forming positive and negative ions; for example, HCl dissolves in water to give hydrogen ions and chloride ions. *(Sec. 15.2)*

ionization constant of water (K_w) A constant that equals the product of the molar hydrogen ion concentration and the molar hydroxide ion concentration in water, $K_w = 1.0 \times 10^{-14}$ at 25 °C. *(Sec. 15.7)*

ionization energy The amount of energy necessary to remove an electron from a neutral atom in the gaseous state. *(Sec. 6.9)*

ionization equilibrium constant (K_i) A constant that expresses the molar equilibrium concentrations of ions in an aqueous solution for a slightly ionized acid or base. *(Sec. 16.6)*

isoelectronic A term for two or more ions (or ions and an atom) having the same electron configuration; for example, Na^+ and F^- each have 10 electrons, and their electron configurations are identical to the noble gas neon. *(Sec. 6.10)*

isomers Compounds with the same molecular formula, but with different structural formulas. Isomers have different physical and chemical properties. *(Sec. 19.2)*

isotopes Atoms having the same atomic number, but a different mass number; atoms of the same element that differ by the number of neutrons in the nucleus. *(Sec. 5.4)*

IUPAC nomenclature The system of rules for naming chemical compounds set forth by the International Union of Pure and Applied Chemistry. *(Sec. 7.1)*

J

joule (J) A unit of energy in the SI system, 1 cal = 4.184 J. *(Sec. 3.9)*

K

Kelvin unit (K) The basic unit of temperature in the SI system. *(Sec. 3.8)*

kernel See *core*.

kinetic energy (KE) The energy associated with the motion of particles; the energy associated with the mass and velocity of a particle. *(Sec. 4.9)*

kinetic theory A theoretical description of gas molecules demonstrating ideal behavior; also referred to as the *kinetic molecular theory*. *(Sec. 11.10)*

L

lanthanide series The elements with atomic numbers 58–71. *(Sec. 6.3)*

Latin system A naming system that designates the variable charge on a metal cation with an *-ic* or *-ous* suffix attached to the stem of the Latin name. *(Sec. 7.2)*

law See *scientific law*.

law of chemical equilibrium The principle that the molar concentrations of the products in a reversible reaction divided by the molar concentrations of the reactants (each raised to a power corresponding to a coefficient in the balanced equation) is equal to a constant. The law of chemical equilibrium can be written mathematically as $K_{eq} = [C]^a [D]^d / [A]^a [B]^b$. *(Sec. 16.3)*

law of combining volumes The principle that volumes of gases that combine in a chemical reaction, at the same temperature and pressure, are in the ratio of small whole numbers; also called *Gay-Lussac's law of combining volumes*. *(Sec. 10.6)*

law of conservation of energy The principle that states energy can neither be created nor destroyed. *(Sec. 4.10)*

law of conservation of mass The principle that states mass can neither be created nor destroyed. *(Sec. 4.8)* The statement that the total mass of the reactants in a chemical reaction is equal to the total mass of the products. *(Sec. 10.1)*

law of conservation of mass and energy The principle that states the total mass and energy in the universe is constant. The statement that the total mass and energy, before and after a chemical change, is constant. *(Sec. 4.10)*

law of definite composition The principle that states a compound always contains the same elements in the same proportion by mass. *(Sec. 4.5)*

Le Chatelier's principle The statement that any reversible reaction at equilibrium when stressed by a change in concentration, temperature, or pressure, shifts to relieve the stress. *(Sec. 16.5)*

Lewis structure See *electron dot formula*.

light A specific term that refers to visible radiant energy, that is, violet to red; a general term that refers to both visible and invisible radiant energy. *(Sec. 5.6)*

***like dissolves like* rule** The general principle that solute–solvent interaction is greatest when the polarity of the solute is similar to that of the solvent. *(Sec. 14.2)*

limiting reactant The substance in a chemical reaction that controls or limits the maximum amount of product formed. *(Sec. 10.7)*

line spectrum See *emission line spectrum.*

lipid A biological compound such as a fat, oil, wax, or steroid that is insoluble in water. *(Sec. 20.1)*

liter (L) The basic unit of volume in the metric system equal to the volume of a cube 10 cm on a side. *(Sec. 3.1)*

lock-and-key model A theory that explains enzyme catalysis; the theory states that the shape of the substrate molecule must fit the contour of the active site on an enzyme. *(Sec. 20.3)*

M

main-group elements See *representative elements.*

malleable The property of a metal that allows it to be hammered or machined into a foil. *(Sec. 4.4)*

mass The quantity of matter in an object that is measured by a balance. *(Sec. 2.1)*

mass/mass percent (m/m %) A solution concentration expression that relates the mass of solute in grams dissolved in each 100 g of solution. *(Sec. 14.8)*

mass–mass problem A type of stoichiometry calculation that relates the masses of two substances according to a balanced equation. *(Sec. 10.3)*

mass number (*A*) A value that indicates the number of protons and neutrons in the nucleus of a given atom. *(Secs. 5.4 and 18.2)*

mass–volume problem A type of stoichiometry calculation that relates the mass of a substance to the volume of a gas according to a balanced equation. *(Sec. 10.3)*

measurement A numerical value with an attached unit that expresses a physical quantity such as length, mass, volume, time, or temperature. *(Sec. 2.1)*

metal An element that is generally shiny in appearance, has a high density, high melting point, and is a good conductor of heat and electricity. *(Sec. 4.4)*

metal oxide A compound that reacts with water to form an alkaline solution; also termed a *basic oxide. (Sec. 13.9)*

metallic solid A crystalline solid composed of metal atoms that repeat in a regular pattern. *(Sec. 13.5)*

metalloid See *semimetal.*

meter (m) The basic unit of length in the metric system of measurement. *(Sec. 3.1)*

metric system A decimal system of measurement using prefixes and a basic unit to express physical quantities such as length, mass, and volume. *(Sec. 3.1)*

milliliter (mL) A common unit of volume in the metric system of measurement that is equal to one-thousandth of a liter. *(Sec. 2.1)*

miscible A term that refers to liquids that dissolve completely in one another. *(Sec. 14.2)*

molar mass (*MM*) The mass of 1 mol of substance expressed in grams; the mass of Avogadro's number of atoms, molecules, or formula units. *(Secs. 9.3 and 10.1)*

molar volume The volume occupied by 1 mol of gas at STP; at 0 °C and 1.00 atm the volume of 1 mol of gas is 22.4 L. *(Secs. 9.5 and 10.5)*

molarity (*M*) A solution concentration expression that relates the moles of solute dissolved in each liter of solution. *(Sec. 14.9)*

mole (mol) The amount of substance that contains Avogadro's number of particles, that is, an amount of substance that contains 6.02×10^{23} particles. *(Sec. 9.2)*

mole ratio The ratio of moles of reactants and products according to the coefficients in the balanced chemical equation. *(Sec. 10.1)*

molecular formula A chemical formula that expresses the actual number of atoms of each element in a molecule. *(Sec. 9.9)*

molecular shape The geometric shape formed by the atoms bonded to the central atom in a molecule; also called *molecular geometry. (Sec. 12.10)*

molecular solid A crystalline solid composed of molecules that repeat in a regular pattern. *(Sec. 13.5)*

molecule A single particle composed of nonmetal atoms. *(Sec. 4.5)* The simplest representative particle in a compound composed of nonmetals. *(Sec. 7.7)* The simplest particle in a substance held together by covalent bonds. *(Sec. 12.1)*

monoatomic ion A single atom that bears a negative or a positive charge as the result of gaining or losing valence electrons. *(Secs. 7.1 and 12.5)*

monomer A small molecule that bonds to itself repeatedly to form a long, continuous chain. *(Sec. 19.3)*

monosaccharide A carbohydrate composed of a simple sugar molecule. A simple sugar is characterized by an aldehyde or ketone functional group and usually several hydroxyl groups. *(Sec. 20.4)*

N

natural law An extensively tested proposal of a scientific principle that states a measurable relationship under different experimental conditions. A natural law is often expressed as an equation, for example, $P_1V_1 = P_2V_2$. *(Sec. 1.1)*

net dipole The overall direction of partial negative charge in a molecule having two or more dipoles. *(Secs. 13.7 and 14.2)*

net ionic equation A chemical equation that portrays an ionic reaction after spectator ions have been canceled from the total ionic equation. *(Sec. 15.11)*

neutralization reaction A type of reaction in which an acid and a base produce a salt and water. *(Sec. 8.4)*

neutron (n^0) A neutral subatomic particle having a mass of 1 amu, which is approximately equal to the mass of a proton. *(Sec. 5.3)*

noble gases The relatively unreactive Group VIIIA/18 elements. *(Sec. 6.3)*

nonbonding electrons The valence electrons in a molecule that are not shared. *(Sec. 12.4)*

nonmetal An element that is generally dull in appearance, has a low density, low melting point, and is not a good conductor of heat and electricity. *(Sec. 4.4)*

nonmetal oxide A compound that reacts with water to form an acidic solution; also termed an *acidic oxide. (Sec. 13.9)*

nonpolar bond A covalent bond in which one or more pairs of electrons are shared equally. *(Sec. 12.7)*

nonpolar solvent A dissolving liquid composed of nonpolar molecules. *(Sec. 14.2)*

nonsignificant digits The digits in a measurement that exceed the certainty of the instrument. *(Sec. 2.3)*

nuclear equation A shorthand representation using atomic notation to describe a nuclear reaction. *(Sec. 18.2)*

nuclear fission A nuclear reaction in which a heavy nucleus splits into two or more lighter nuclei and releases energy. *(Sec. 18.7)*

nuclear fusion A nuclear reaction in which two light nuclei combine into a single heavy nucleus and releases energy. *(Sec. 18.8)*

nuclear reaction A high-energy change involving the atomic nucleus. *(Sec. 18.2)*

nucleic acid A biological polymer compound that carries genetic information. *(Sec. 20.1)*

nucleotide A repeating unit in a nucleic acid composed of a sugar, a base, and phosphoric acid. *(Sec. 20.6)*

nucleus See *atomic nucleus.*

nuclide A specific atom with a designated number of protons and neutrons in its nucleus. *(Sec. 18.2)*

O

octet rule The statement that an atom tends to bond in such a way so as to acquire eight valence electrons. A hydrogen atom is an exception to the rule and acquires only two valence electrons. *(Sec. 12.1)*

oil A triglyceride from a plant source that contains mostly unsaturated fatty acids. *(Sec. 20.5)*

orbital A region in space surrounding the nucleus of an atom in which there is a high probability (~95%) of finding an electron with a given energy. *(Sec. 5.11)*

organic chemistry The study of chemical substances that contain the element carbon. *(Sec. 1.2)* The study of carbon compounds. *(Sec. 19.1)*

oxidation A process in which a substance undergoes an increase in oxidation number; a process characterized by losing electrons. *(Sec. 17.2)*

oxidation number A positive or negative value assigned to an atom in a substance according to a set of rules; a value that indicates whether an atom is electron-poor or electron-rich compared to a free atom. *(Sec. 17.1)*

oxidation–reduction reaction See *redox reaction.*

oxidizing agent A substance that causes the oxidation of another substance in a redox reaction; the substance that is reduced in a redox reaction. *(Sec. 17.2)*

oxyanion A polyatomic anion that contains one or more elements combined with oxygen; for example, NO_3^- and SO_4^{2-}. *(Sec. 7.3)*

P

parent–daughter nuclides A term for a decaying nuclide and the resulting nuclide that is produced. *(Sec. 18.3)*

partial pressure The pressure exerted by an individual gas in a mixture of two or more gases. *(Sec. 11.9)*

peptide linkage An amide bond that joins two amino acid molecules in a small peptide, polypeptide, or protein. *(Sec. 20.2)*

percent (%) An expression for the amount of a single quantity compared to an entire sample; an expression of parts per hundred parts. *(Sec. 2.10)*

percent composition A list of the mass percent of each element in a compound. *(Sec. 9.7)*

percent yield An expression for the amount of the actual yield compared to the theoretical yield. *(Sec. 10.9)*

period A horizontal row in the periodic table; a series of elements with properties that vary from metallic to nonmetallic. *(Sec. 6.3)*

periodic law The properties of the elements recur in a repeating pattern when arranged according to increasing atomic number. *(Sec. 6.2)*

periodic table A chart that arranges elements according to their properties and includes metals, nonmetals, and semimetals. *(Sec. 4.4)*

pH The molar concentration of hydrogen ions expressed on an exponential scale; the negative logarithm of the molar hydrogen ion concentration. *(Sec. 15.8)*

phenyl group The part that remains after a hydrogen atom is removed from a benzene molecule. *(Sec. 19.2)*

phospholipid A lipid that has two fatty acids and phosphoric acid joined to glycerol by ester bonds. *(Sec. 20.5)*

photon A particle of radiant energy. *(Sec. 5.7)*

physical change A modification of a substance that does not alter the chemical composition, for example, a change in physical state. *(Sec. 4.7)*

physical property A characteristic that can be observed without changing the chemical formula of a substance. *(Sec. 4.6)*

physical state The condition of matter existing as a solid, liquid, or gas. *(Sec. 4.1)*

polar bond A covalent bond in which one or more pairs of electrons are shared unequally. *(Sec. 12.6)*

polar solvent A dissolving liquid composed of polar molecules. *(Sec. 14.2)*

polyatomic ion A group of atoms bound together and having an overall positive or negative charge. *(Sec. 7.1)* A group of atoms held together by covalent bonds and having an overall positive or negative charge. *(Sec. 12.5)*

polymer A giant molecule composed of small molecules bonded together repeatedly in a long, continuous chain. *(Sec. 19.3)*

polypeptide A large molecule composed of 10–50 amino acids joined by amide bonds. *(Sec. 20.2)*

polysaccharide A carbohydrate polymer composed of simple sugar molecules joined by glycoside linkages. *(Sec. 20.4)*

positron (β^+ or ${}^{0}_{+1}e$) A nuclear radiation identical in mass, but opposite in charge, to that of an electron. The term positron is derived from "positive electron." *(Sec. 18.2)*

potential energy (PE) The stored energy that matter possesses owing to its position or chemical composition. *(Sec. 4.9)*

power of 10 A positive or negative exponent of 10. *(Sec. 2.6)*

precipitate An insoluble solid substance produced by a chemical reaction in an aqueous solution. *(Sec. 8.1)*

product A substance resulting from a chemical reaction. *(Sec. 8.2)*

protein A biological compound that is a polymer composed of amino acids joined by peptide linkages. *(Sec. 20.1)*

proton (p^+) A positively charged subatomic particle having a mass of 1 amu, which is approximately 1836 times the mass of an electron. *(Sec. 5.2)*

proton acceptor A term for a Brønsted–Lowry base, which is a synonym for hydrogen ion acceptor. *(Sec. 15.3)*

proton donor A term for a Brønsted–Lowry acid, which is a synonym for hydrogen ion donor. *(Sec. 15.3)*

pure substance See *substance.*

Q

quantum level See *energy level.*

quantum mechanical atom A model of the atom that describes an electron in terms of its probability of being found in a particular location around the nucleus. *(Sec. 5.11)*

R

radiant energy spectrum A range of light energy extending from short-wavelength gamma rays to long-wavelength microwaves. *(Sec. 5.6)*

radioactive decay series The stepwise disintegration of an unstable nuclide until a stable nucleus is reached; also called *decay series. (Sec. 18.3)*

radioactivity The emission of particles or energy from an unstable atomic nucleus. *(Sec. 18.1)*

radionuclide An atom whose nucleus is unstable and decays radioactively. *(Sec. 18.5)*

rare earth elements The elements with atomic numbers 21, 39, 57, and 58–71. *(Sec. 6.3)*

rate of reaction The rate at which the concentrations of reactants decrease, or the concentrations of products increase, per unit time. *(Sec. 16.3)*

reactant A substance undergoing a chemical reaction. *(Sec. 8.2)*

reaction profile A graph of the energy of reactants and products as a reaction progresses. *(Sec. 16.2)*

real gas An actual gas that deviates from ideal behavior at a low temperature and high pressure. *(Sec. 11.10)*

reciprocal The relationship between a fraction and its inverse, for example, 1 yard/3 feet and 3 feet/1 yard. *(Sec. 2.8 and 3.2)*

redox reaction A chemical reaction that involves electron transfer between two reacting substances. *(Sec. 17.2)*

reducing agent A substance that causes the reduction of another substance in a redox reaction; the substance that is oxidized in a redox reaction. *(Sec. 17.2)*

reduction A process in which a substance undergoes a decrease in oxidation number; a process characterized by gaining electrons. *(Sec. 17.2)*

reduction potential The relative ability of a substance to undergo reduction; the relative strength of an oxidizing agent. *(Sec. 17.5)*

replication The process by which a DNA molecule synthesizes an identical molecule of DNA. *(Sec. 20.6)*

representative elements The Group A (1, 2, and 13–18) elements in the periodic table; also termed *main-group elements. (Sec. 6.3)*

reversible reaction A reaction that proceeds simultaneously in both the forward direction toward products and the opposite direction toward reactants. *(Sec. 16.3)*

rounding off The process of eliminating digits that are not significant. *(Sec. 2.3)*

S

salt An ionic compound produced by an acid–base reaction; the product of a neutralization reaction in addition to water. *(Secs. 8.11 and 15.2)*

salt bridge A porous device that allows ions to travel between two half-cells to maintain an ionic charge balance in each compartment. *(Sec. 17.6)*

saponification A chemical reaction of a fat or oil with sodium hydroxide to produce soap and glycerol. *(Sec. 20.5)*

saturated hydrocarbon A hydrocarbon compound containing only single bonds. *(Sec. 19.1)*

saturated solution A solution containing the maximum solute that can dissolve at a given temperature. *(Sec. 14.7)*

science The methodical exploration of nature and the logical explanation of the observations. *(Sec. 1.1)*

scientific method A systematic investigation that involves performing an experiment, proposing a hypothesis, testing the hypothesis, and finally, stating a theory or law that explains a scientific principle. *(Sec. 1.1)*

scientific notation A method for expressing numbers by moving the decimal place after the first significant digit and indicating the number of decimal moves by a power of 10. *(Sec. 2.7)*

second (s) The basic unit of time in the metric system. *(Sec. 3.1)*

semimetal An element that is generally metallic in appearance and has properties midway between those of a metal and a nonmetal; also called a *metalloid. (Sec. 4.4)*

SI See *International System.*

significant digits The digits in a measurement known with certainty plus one digit that is estimated; also referred to as *significant figures. (Sec. 2.2)*

single bond A bond between two atoms composed of one shared electron pair. A single bond is represented as a dash between the symbols of two atoms. *(Sec. 12.4)*

single-replacement reaction A type of reaction in which a more active element displaces a less active element from a solution or compound. *(Sec. 8.4)*

soft water Water containing sodium cations, Na^+, and a variety of anions such as CO_3^{2-}, SO_4^{2-}, and PO_4^{3-}. *(Sec. 13.7)*

solubility The maximum amount of solute that can dissolve in a solvent at a given temperature; usually expressed in grams of solute per 100 g of solvent. *(Sec. 14.6)*

solubility product equilibrium constant (K_{sp}) A constant that expresses the molar equilibrium concentrations of ions in an aqueous solution for a slightly dissociated ionic compound. *(Sec. 16.8)*

solute The component of a solution that is the lesser quantity. *(Sec. 14.2)*

solution The general term for a solute dissolved in a solvent. A solution is an example of a homogeneous mixture. *(Sec. 14.2)*

solvent The component of a solution that is the greater quantity. *(Sec. 14.2)*

solvent cage A cluster of solvent molecules surrounding a solute particle in a solution. *(Sec. 14.4)*

specific gravity (sp gr) The ratio of the density of a liquid compared to the density of water at 4 °C; a unitless expression. *(Sec. 3.7)*

specific heat The amount of heat required to raise the temperature of 1 g of any substance 1 °C; the specific heat of water is 1.00 cal/(g × °C). *(Sec. 3.9)*

spectator ions Those ions that are in an aqueous solution, but do not participate in a reaction and do not appear as reactants or products in the net ionic equation. *(Sec. 15.11)*

standard conditions See *standard temperature and pressure (STP).*

standard solution A solution whose concentration has been established accurately (usually by titration to 3 or 4 significant digits). *(Sec. 15.6)*

standard temperature and pressure (STP) A temperature of 0 °C and a pressure of 1 atm. *(Secs. 9.5 and 10.5)* A temperature of 273 K and a pressure of 760 mm Hg for a gas. *(Sec. 11.7)*

steroid A lipid hormone composed of four rings of carbon atoms fused into a single molecular structure. *(Sec. 20.5)*

Stock system A naming system that designates the variable charge on a metal cation with Roman numerals in parentheses. *(Sec. 7.2)*

stoichiometry The relationship between quantities (mass of substance or volume of gas) in a chemical reaction according to the balanced chemical equation. *(Sec. 10.3)*

strong electrolyte An aqueous solution that is a good conductor of electricity, for example, strong acids, strong bases, and soluble salts. *(Sec. 15.10)*

structural formula A diagram of a molecule or polyatomic ion that shows the chemical symbol of each atom and a dash representing each pair of bonding electrons. *(Sec. 12.4)* A chemical formula that shows the arrangement of atoms in a molecule. *(Sec. 19.2)*

sublimation A direct change of state from a solid to a gas without forming a liquid. *(Sec. 4.1)*

subscript A digit in a chemical formula that represents the number of atoms or ions occurring in the substance. *(Sec. 8.3)*

substance Matter having a definite composition and constant properties; examples include elements and compounds. It is also referred to as *pure substance. (Sec. 4.2)*

supersaturated solution A solution containing more than the maximum solute that can dissolve at a given temperature. *(Sec. 14.7)*

surface tension The resistance of a liquid to spread out, and its tendency to form spherical drops with minimum surface area. *(Sec. 13.3)*

T

temperature The average energy of molecules in motion. *(Sec. 3.8)*

ternary ionic compound A compound that contains three elements, including at least one metal. *(Sec. 7.1)*

ternary oxyacid A compound that contains hydrogen, a nonmetal, and oxygen dissolved in water. *(Sec. 7.1)*

theoretical yield The amount of product that is calculated to be obtained from given amounts of reactants. *(Sec. 10.9)*

theory An extensively tested proposal of a scientific principle that explains the behavior of nature. A theory offers a model, for example the atomic theory, to describe nature. *(Sec. 1.1)*

titration A laboratory procedure for delivering a measured volume of solution through a buret. *(Sec. 15.5)*

torr A unit of gas pressure equal to 1 mm Hg. *(Sec. 11.2)*

total ionic equation A chemical equation that portrays highly ionized substances in the ionic form and slightly ionized substances in the nonionized form. *(Sec. 15.11)*

transcription The process by which a DNA molecule synthesizes a complementary single strand of RNA. *(Sec. 20.6)*

transition elements The Group B (3–12) elements in the periodic table. *(Sec. 6.3)*

transition state The highest point on the reaction profile at which there is the greatest potential energy. *(Sec. 16.2)*

transmutation The conversion of one element to another by inducing a nuclear reaction using particle bombardment. *(Sec. 18.6)*

transuranium elements The elements with atomic numbers beyond 92. All of the elements following uranium are synthetic. *(Sec. 6.3)*

triglyceride A lipid that has three fatty acids joined to glycerol by ester bonds; also called a triacylglycerol. *(Sec. 20.5)*

triple bond A bond between two atoms composed of three shared electron pairs. A triple bond is represented as three dashes between the symbols of two atoms. *(Sec. 12.4)*

tritium The nuclide of hydrogen with two neutrons in the nucleus. *(Sec. 18.8)*

Tyndall effect The phenomenon of scattering a beam of light by colloid particles. *(Sec. 14.3)*

U

uncertainty The degree of inexactness in a measurement obtained from an instrument. *(Sec. 2.1)*

uncertainty principle The statement that it is impossible to precisely measure both the location and energy of a particle at the same time. *(Sec. 5.11)*

unit analysis method A systematic procedure for solving problems that converts the units in a given value to the units in the answer; also referred to as *dimensional analysis* or *factor label method. (Secs. 2.9 and 3.3)*

unit equation A statement that relates two values that are equivalent, for example, 1 ft = 12 in. and 1 in. = 2.54 cm. *(Secs. 2.8 and 3.2)*

unit factor A ratio of two quantities that are equivalent and can be applied to convert from one unit to another, for example, 1 m/100 cm. *(Secs. 2.8 and 3.2)*

unsaturated hydrocarbon A hydrocarbon compound containing a double bond or triple bond. *(Sec. 19.1)*

unsaturated solution A solution containing less than the maximum solute that can dissolve at a given temperature. *(Sec. 14.7)*

V

vacuum A volume of space that does not contain gas molecules. *(Sec. 11.2)*

valence electrons The electrons that occupy the outermost *s* and *p* sublevels of an atom. *(Sec. 6.7)* The electrons in the highest *s* and *p* sublevels in an atom that undergo reaction and form chemical bonds. *(Sec. 12.1)*

valence shell electron pair repulsion See *VSEPR theory.*

vapor pressure The pressure exerted by gaseous vapor above a liquid in a sealed container when the rates of vaporization and condensation are equal. *(Secs. 11.8 and 13.3)*

viscosity The resistance of a liquid to flow. *(Sec. 13.3)*

visible spectrum A range of light energy observed as violet, blue, green, yellow, orange, or red; the wavelengths of light from 400–700 nm. *(Sec. 5.6)*

voltaic cell An electrochemical cell in which a spontaneous redox reaction occurs and generates electrical energy; also called a *galvanic cell. (Sec. 17.6)*

volume by displacement A technique for determining the volume of a solid or a gas by measuring the volume of water it displaces. *(Sec. 3.6)* A technique for determining the volume of a gas by measuring the volume of water it displaces. *(Sec. 11.9)*

volume–volume problem A type of stoichiometry calculation that relates the volumes of two gases (at the same temperature and pressure) according to a balanced equation. *(Sec. 10.3)*

VSEPR theory A model that explains the shapes of molecules as a result of bonding and nonbonding electron pairs around the central atom repelling each other. *(Sec. 12.10)*

W

water of hydration A term for water molecules bound to a formula unit in a hydrate; also called the *water of crystallization. (Sec. 13.10)*

wavelength The distance a light wave travels to complete one cycle. *(Sec. 5.6)*

wax A lipid that contains a fatty acid joined to an alcohol by an ester linkage. *(Sec. 20.5)*

weak electrolyte An aqueous solution that is a poor conductor of electricity, for example, weak acids, weak bases, and slightly soluble salts. *(Sec. 15.10)*

weight The force exerted by gravity on an object. The *mass* of an object is fixed, whereas the *weight* of an object varies with elevation and the distance from the center of Earth. *(Sec. 2.1)*

Photo Credits

Chapter 1: Page 2 Luca Galuzzi **Page 3** iStock **Page 4** C. Marvin Lang **Page 5** Getty Images **Page 6** PAL **Page 7** C. Marvin Lang

Chapter 2: Page 12 National Geographic Images **Page 13** Public Domain **Page 15** Ohaus Corporation **Page 15** Ohaus Corporation **Page 15** Mettler Toledo corporation **Page 18** iStock **Page 19** Texas Instruments Incorporated **Page 23** NASA Headquarters **Page 23** Custom Medical Stock Photo, Inc. **Page 24** Alcoa Technical Center **Page 26** Pearson Education/ PH College **Page 28** Author Supplied **Page 29** AP Wide World Photos **Page 30** Getty Images **Page 31** Shutterstock **Page 32** Corbis RF **Page 33** United States Mint Headquarters **Page 33** United States Mint Headquarters **Page 33** Shutterstock **Page 33** Shutterstock **Page 33** United States Mint Headquarters **Page 33** United States Mint Headquarters **Page 37** Jupiterimages Unlimited **Page 38** Jupiterimages Unlimited **Page 38** NASA **Page 39** iStock

Chapter 3: Page 40 Fundamental Photographs **Page 42** Bureau International des Poids et Mesures **Page 43** US Department of Energy **Page 46** Shutterstock **Page 49** Pearson Owned **Page 50** Author Supplied **Page 51** Shutterstock **Page 52** Getty Images **Page 54** Dorling Kindersley Media Library **Page 54** Fundamental Photographs, NYC **Page 55** Redux Pictures **Page 59** Dorling Kindersley Media Library **Page 61** Bloomberg News **Page 63** National Standards Commission **Page 64** Fundamental Photographs, NYC **Page 65** Shutterstock **Page 68** Shutterstock **Page 72** iStock **Page 72** Getty Images, Inc.- Photodisc./Royalty Free **Page 73** Shutterstock

Chapter 4: Page 74 Getty Images Inc.—Stone Allstock **Page 75** Shutterstock **Page 75** Shutterstock **Page 75** Shutterstock **Page 76** Fundamental Photographs, NYC **Page 78** Fundamental Photographs, NYC **Page 78** Pearson Education/PH College **Page 79** iStock **Page 79** iStock **Page 79** iStock **Page 79** iStock **Page 82** iStock **Page 86** CERN **Page 89** Shutterstock **Page 89** Getty Images, Inc.—PhotoDisc **Page 89** iStock **Page 90** Shutterstock **Page 92** Fundamental Photographs, NYC **Page 92** PAL **Page 92** Fundamental Photographs, NYC **Page 92** Visuals Unlimited **Page 94** Fundamental Photographs, NYC **Page 94** Fundamental Photographs, NYC **Page 94** Fundamental Photographs, NYC **Page 95** Pearson Education/PH College **Page 95** iStock **Page 98** Corbis RF **Page 101** Pearson Education/PH College **Page 102** C. Marvin Lang **Page 104** Shutterstock **Page 105** Fundamental Photographs, NYC **Page 108** iStock **Page 109** Pearson Education/PH College

Chapter 5: Page 110 SPMage07 **Page 111** CORBIS-NY **Page 117** Fundamental Photographs **Page 118** Reuters Limited **Page 125** C. Marvin Lang **Page 127** Author Supplied **Page 128** Phillip Slawinski **Page 128** Phillip Slawinski **Page 128** Phillip Slawinski **Page 128** Phillip Slawinski **Page 128** Phillip Slawinski **Page 133** American Institute of Physics/Emilio Segre Visual Archives **Page 138** Author Supplied **Page 141** Getty Images **Page 141** Author Supplied

Chapter 6: Page 142 Arena Reed **Page 144** C. Marvin Lang **Page 151** PAL **Page 162** NASA **Page 168** Spanish Postal Service **Page 169** Pearson Education/ PH College

Chapter 7: Page 170 Photos.com **Page 172** iStock **Page 173** Fundamental Photographs, NYC **Page 174** PAL **Page 179** Shutterstock **Page 180** Shutterstock **Page 181** Fundamental Photographs, NYC **Page 182** Raimond Spekking **Page 183** Photo Researchers **Page 183** Chip Clark **Page 185** Shutterstock **Page 186** Photo Researchers **Page 187** iStock **Page 188** Fundamental Photographs, NYC **Page 189** Charles H. Corwin **Page 189** Pearson Owned **Page 190** Fundamental Photographs, NYC **Page 193** Pearson Education/PH College **Page 194** Photolibrary **Page 196** Phototake NYC **Page 197** Fundamental Photographs, NYC

Chapter 8: Page 198 Shutterstock **Page 199** Shutterstock **Page 199** H. Eugene LeMay, Jr. **Page 200** PAL **Page 200** Fundamental Photographs **Page 200** Fundamental Photographs **Page 205** Fundamental Photographs **Page 206** Photo Researchers **Page 209** Pearson Education/ PH College **Page 209** Pearson Education/PH College **Page 210** Pearson Education/PH College **Page 213** Fundamental Photographs, NYC **Page 215** Pearson Education/PH College **Page 216** Pearson Education/PH College **Page 217** Pearson Education/PH College **Page 218** Fundamental Photographs, NYC **Page 218** Fundamental Photographs **Page 219** Pearson Education/PH College **Page 220** Pearson Education/ PH College **Page 222** Fundamental Photographs, NYC **Page 223** PAL **Page 226** Fundamental Photographs, NYC **Page 227** Pearson Education/PH College **Page 228** Pearson Education/PH College **Page 231** Fundamental Photographs, NYC **Page 231** Fundamental Photographs, NYC **Page 231** Fundamental Photographs, NYC **Page 231** Pearson Education/PH College **Page 231** Pearson Education/ PH College **Page 231** Pearson Education/PH College

Chapter 9: Page 232 Photo Researchers, Inc. **Page 236** Fundamental Photographs, NYC **Page 238** Pearson Education/PH College **Page 242** Shutterstock **Page 242** Dr. Charles Corwin **Page 243** Fundamental Photographs, NYC **Page 253** Fundamental Photographs, NYC

Chapter 10: Page 260 Shutterstock **Page 261** Getty Images—Iconica **Page 265** Shutterstock **Page 266** Fundamental Photographs, NYC **Page 267** Pearson Education/PH College **Page 268** Pearson Education/ PH College **Page 269** Getty Images Inc.—Stone Allstock **Page 271** Pearson Owned **Page 273** Fundamental Photographs, NYC **Page 274** Photo Researchers **Page 275** Getty Images **Page 275** Pearson Education/PH College **Page 277** Fundamental Photographs, NYC **Page 280** Photo Researchers, Inc. **Page 281** Pearson Education/PH College **Page 282** Pearson Education/PH College **Page 285** Fundamental Photographs, NYC **Page 286** PAL

Chapter 11: Page 290 NOAA **Page 291** Shutterstock **Page 294** iStock **Page 299** The Granger Collection **Page 300** Fundamental Photographs, NYC **Page 300** Fundamental Photographs, NYC **Page 308** CORBIS- NY **Page 317** iStock **Page 317** iStock **Page 319** iStock

Chapter 12: Page 322 Shutterstock **Page 324** iStock **Page 324** Shutterstock **Page 340** AP Wide World Photos **Page 347** General Electric Corporate Research & Development Center

Chapter 13: Page 354 Shutterstock **Page 355** Pearson Education Corporate Digital Archive **Page 360** PAL **Page 360** Photo Researchers, Inc. **Page 361** PAL **Page 363** Richard M. Busch **Page 363** Fundamental Photographs, NYC **Page 363** PAL **Page 364** PAL **Page 364** Fundamental Photographs, NYC **Page 370** Photoshot Holdings Ltd./Bruce Coleman **Page 371** Fundamental Photographs **Page 371** Fundamental Photographs **Page 371** Fundamental Photographs **Page 372** Phototake NYC **Page 373** Fundamental Photographs, NYC **Page 376** Charles H. Corwin **Page 378** iStock **Page 383** PAL

Chapter 14: Page 384 Shutterstock **Page 385** Shutterstock **Page 388** AP **Page 390** PAL **Page 393** Fundamental Photographs, NYC **Page 395** Pearson Owned **Page 396** Fundamental Photographs, NYC **Page 396** Fundamental Photographs, NYC **Page 396** Fundamental Photographs, NYC **Page 398** Pearson Education/PH College **Page 398** iStock **Page 400** Pearson Education/PH College **Page 400** Pearson Education/PH College **Page 400**

Pearson Education/ PH College **Page 401** PAL **Page 402** Pearson Owned **Page 402** Pearson Education/PH College **Page 407** Pearson Owned **Page 407** Pearson Owned

Chapter 15: Page 412 Reuters **Page 413** Pearson Education/PH College **Page 413** Pearson Education/ PH College **Page 414** iStock **Page 415** Cole-Parmer Instrument Company **Page 415** Shutterstock **Page 416** Fundamental Photographs **Page 417** Pearson Owned **Page 418** PAL **Page 420** Pearson Education/PH College **Page 420** Pearson Education/PH College **Page 420** Pearson Education/PH College **Page 421** Fundamental Photographs, NYC **Page 421** Fundamental Photographs, NYC **Page 421** Fundamental Photographs, NYC **Page 428** Fundamental Photographs, NYC **Page 429** Pearson Owned **Page 430** Pearson Owned **Page 431** Pearson Owned **Page 432** Fundamental Photographs, NYC **Page 432** Fundamental Photographs, NYC **Page 432** nadp.sws.uiuc.edu **Page 442** Fundamental Photographs, NYC **Page 443** Pearson Education/PH College

Chapter 16: Page 448 iStock **Page 449** Fundamental Photographs **Page 457** NASA **Page 463** Fundamental Photographs, NYC **Page 463** Fundamental Photographs, NYC **Page 463** Fundamental Photographs, NYC **Page 472** Fundamental Photographs, NYC

Chapter 17: Page 480 NASA/John F. Kennedy Space Center **Page 481** Fundamental Photographs **Page 481** iStock **Page 482** PAL **Page 483** Fundamental Photographs **Page 484** Getty Images **Page 485** Fundamental Photographs, NYC **Page 492** Fundamental Photographs, NYC **Page 503** Pearson Owned **Page 505** Getty Images **Page 507** Pearson Education/PH College **Page 507** Pearson Education/ PH College **Page 511** Pearson Owned

Chapter 18: Page 512 NASA Headquarters **Page 513** PAL **Page 515** Fundamental Photographs, NYC **Page 519** Shutterstock **Page 520** Photo Researchers, Inc. **Page 522** AP **Page 523** Photo Researchers **Page 524** Lawrence Berkeley National Laboratory **Page 524** Fermilab Visual Media Services **Page 528** Comstock Images **Page 529** Science Faction **Page 531** Fundamental Photographs, NYC **Page 533** CORBIS-NY

Chapter 19: Page 536 Alamy Images **Page 538** Charles H. Corwin **Page 541** Government of Switzerland **Page 547** C. Marvin Lang **Page 551** Charles H. Corwin **Page 552** Charles H. Corwin **Page 554** Charles H. Corwin **Page 555** Charles H. Corwin **Page 555** Charles H. Corwin **Page 556** Charles H. Corwin **Page 557** Charles H. Corwin **Page 558** Getty Images, Inc. - PhotoDisc **Page 559** Charles H. Corwin **Page 560** Charles H. Corwin **Page 560** iStock **Page 563** iStock

Chapter 20: Page 568 Photo Researchers **Page 571** Alamy Images Royalty Free **Page 578** Shutterstock **Page 581** Fundamental Photographs, NYC **Page 585** Shutterstock **Page 587** iStock **Page 589** Pearson Owned **Page 589** Pearson Owned **Page 590** Fundamental Photographs, NYC

Index

Note: Page numbers in italics denote graphic material. Page numbers followed by an italicized *t* denote a table. Parenthetic entries following a heading or a subheading denote an abbreviation, symbol, chemical symbol, or a chemical formula, following the practice used in text.

A

B

D

E

F

G

H

I

Q

R

Common Monoatomic Cations

—Stock System

Cation	Name	Cation	Name
Al^{3+}	aluminum ion	Li^{+}	lithium ion
Ba^{2+}	barium ion	Mg^{2+}	magnesium ion
Cd^{2+}	cadmium ion	Mn^{2+}	manganese(II) ion
Ca^{2+}	calcium ion	Hg_2^{2+}	mercury(I) ion
Co^{2+}	cobalt(II) ion	Hg^{2+}	mercury(II) ion
Co^{3+}	cobalt(III) ion	Ni^{2+}	nickel(II) ion
Cu^{+}	copper(I) ion	K^{+}	potassium ion
Cu^{2+}	copper(II) ion	Ag^{+}	silver ion
Cr^{3+}	chromium(III) ion	Na^{+}	sodium ion
H^{+}	hydrogen ion	Sr^{2+}	strontium ion
Fe^{2+}	iron(II) ion	Sn^{2+}	tin(II) ion
Fe^{3+}	iron(III) ion	Sn^{4+}	tin(IV) ion
Pb^{2+}	lead(II) ion	Zn^{2+}	zinc ion
Pb^{4+}	lead(IV) ion		

—Latin System

Cation	Name	Cation	Name
Co^{2+}	cobaltous ion	Pb^{2+}	plumbous ion
Co^{3+}	cobaltic ion	Pb^{4+}	plumbic ion
Cu^{+}	cuprous ion	Hg_2^{2+}	mercurous ion
Cu^{2+}	cupric ion	Hg^{2+}	mercuric ion
Fe^{2+}	ferrous ion	Sn^{2+}	stannous ion
Fe^{3+}	ferric ion	Sn^{4+}	stannic ion